T0258943

BEHAVIOUR OF STEEL STRUCTURES IN SEISMIC AREAS

PROCEEDINGS OF THE THIRD INTERNATIONAL CONFERENCE STESSA 2000
MONTREAL / CANADA / 21 - 24 AUGUST 2000

Behaviour of Steel Structures in Seismic Areas

Edited by

Federico M. Mazzolani
University 'Federico II', Naples, Italy

Robert Tremblay
Ecole Polytechnique, Montreal, Canada

A.A.BALKEMA / ROTTERDAM / BROOKFIELD / 2000

The texts of the various papers in this volume were set individually by typists under the supervision of either each of the authors concerned or the editor.

Authorization to photocopy items for internal or personal use, or the internal or personal use of specific clients, is granted by A.A.Balkema, Rotterdam, provided that the base fee of US$1.50 per copy, plus US$0.10 per page is paid directly to Copyright Clearance Center, 222 Rosewood Drive, Danvers, MA 01923, USA. For those organizations that have been granted a photocopy license by CCC, a separate system of payment has been aranged. The fee code for users of the Transactional Reporting Service is: 90 5809 130 9/00 US$1.50 + US$0.10

Published by
A.A.Balkema, P.O.Box 1675, 3000 BR Rotterdam, Netherlands
fax: +31.10.413.5947; E-mail: balkema@balkema.nl
internetsite: www.balkema.nl
A.A.Balkema Publishers, Old Post Road, Brookfield, VT 0536-9704, USA
fax: 802.276.3837; E-mail: info@ashgate.com

ISBN 90 5809 130 9
© 2000 A.A.Balkema, Rotterdam
Printed in the Netherlands

Behaviour of Steel Structures in Seismic Areas, Mazzolani & Tremblay (eds) © 2000 Balkema, Rotterdam, ISBN 90 5809 130 9

Table of contents

Connections

Codification, design and application

Base isolation and energy dissipation

Global behaviour

Performance based design and moment resisting frames

SAC steel project

Behaviour of Steel Structures in Seismic Areas, Mazzolani & Tremblay (eds) © 2000 Balkema, Rotterdam, ISBN 90 5809 130 9

Preface

STESSA is the first International Speciality Conference on the behaviour and design of steel structures in seismic areas. The first edition of STESSA was held in 1994 in Timisoara, Romania, under the auspices of the European Convention for Constructional Steelwork (E.C.C.S.). The Conference included a workshop as well as a Seminar and nearly one hundred experts from all over the world (17 countries) contributed in presenting 80 papers which have been published in the Proceedings of the Conference.

This success encouraged the organisers in pursuing this effort and it was decided to hold the Conference every three years in three different geographical areas concerned with seismic hazard: Europe, Japan, and North America. In 1997, more than 150 participants coming from 18 countries attended the second STESSA Conference in Kyoto, Japan. A total of 100 papers were presented at the Conference and collected in a second volume of Proceedings.

Since then, significant progress has been made in the understanding of the seismic response of steel structures and new analytical, design, and construction methods have been developed which can now be implemented in practice. In particular, the vigorous research efforts that have been initiated following the 1994 Northridge and 1995 Kobe earthquakes have permitted to improve our knowledge on the behaviour of moment resisting frames and resulted in enhanced beam-to-column connections for ductile response. Major developments also took place in several other areas such as the response to pulse-type ground motions, low-cycle fatigue, welding techniques, overall inelastic dynamic response including stability and torsion effects, performance criteria, and testing procedures. Innovative structural and energy dissipation systems have also been proposed to minimise the impact of earthquakes on steel structures.

The Third International STESSA Conference was organised to bring together the people involved in these changes and share and discuss this new information in view of further improving seismic design and construction practices for steel structures. Montreal, Canada, was selected for the venue of this event as it is located in a seismic active region of eastern North America. The STESSA 2000 Conference also hosted the first meeting of the International Seismic Steel Joint Action Group, a new organisation aiming at the construction of better earthquake-resistant steel structures worldwide through active collaboration and exchange among the international engineering community.

A total of 96 papers prepared by experts from 18 countries have been collected in this volume of Proceedings. They are subdivided according to the following chapters, which correspond to the working Sessions of the Conference:

Material and member behaviour

Material properties, special steel grades, strain rate effects, localised yielding, brittle fracture, rotation capacity, local buckling, overall buckling, classification of sections, strength and stiffness deterioration, composite members.

Connections

Cyclic behaviour of joints including beam-to-column (bare steel and composite) connections and column bases, analytical models, test results, welded and bolted connections, testing procedure, dynamic response, semi-rigid behaviour, quality assurance.

Codification, design and application

Seismic design, safety issues and principles, up-dating of national codes, national practices, case studies, analysis of damage, seismic retrofitting.

Base isolation and energy dissipation

Behaviour of isolated structures, bridge bearings, special energy dissipation devices, criteria for detailing.

Global behaviour

Inelastic dynamic response of structures, P-Δ effect, collapse mechanisms, evaluation of force reduction factors, bracing systems, influence of non-structural elements, modelling of deterioration, damage assessment.

Performance-based design and moment resisting frames

Engineering description of performance levels, conceptual design for multiple performance objectives, methods for analytical prediction of performance, influence of fully- and partially-restrained connections, seismic demand and capacity of frames with welded and bolted connections, examination of damage, retrofitting of existing structures.

SAC Steel Project

Papers related to the SAC Steel Project initiated in the US following the Northridge earthquake to develop guidelines for the seismic design, construction, and/or retrofit of steel moment resisting frames, including the assessment of the seismic demand, global response, beam-to-column connections, material and acceptance test procedures.

Federico M. Mazzolani and Robert Tremblay
Chairmen of the International Scientific Committee

Behaviour of Steel Structures in Seismic Areas, Mazzolani & Tremblay (eds) © 2000 Balkema, Rotterdam, ISBN 90 5809 130 9

Organization

HONORARY CHAIRMAN

D.J. Laurie Kennedy, University of Alberta, Canada

CHAIRMAN

Federico M. Mazzolani, University of Naples, Italy

CO-CHAIRMAN

Robert Tremblay, Ecole Polytechnique, Canada

ORGANIZING COMMITTEE

Robert Tremblay, Ecole Polytechnique, Canada
Michel Bruneau, University of New York, Buffalo, USA
Michael Gilmor, CISC, Canada
Helmut Prion, University of British Columbia, Canada

SCIENTIFIC SECRETARIAT

Bruno Calderoni, University of Naples, Italy

INTERNATIONAL SCIENTIFIC COMMITTEE

Hiroshi Akiyama, University of Tokyo, Japan
Jean Marie Aribert, INSA Rennes, France
Vitelmo Bertero, University of California, Berkeley, USA
Michel Bruneau, University of New York, Buffalo, USA
Luis Calado, Technical University of Lisbon, Portugal
Antonella De Luca, University of Naples, Italy
André Filiatrault, University of California, S.Diego, USA
Victor Gioncu, Technical University of Timisoara, Romania
Subhash Goel, University of Michigan, USA
Aurelio Ghersi, University of Catania, Italy
Le-Wu Lu, Lehigh University, Bethlehem, USA
Raffaele Landolfo, University of Chieti, Pescara, Italy
Stephen Mahin, University of California, Berkeley, USA
André Plumier, University of Liège, Belgium
Koichi Takanashi, Chiba University, Japan
Tsutomu Usami, Nagoya University, Japan
Ioannis Vayas, Nat. Techn. University of Athens, Greece
Carlos Ventura, University of British Columbia, Canada

Material and member behaviour

Behaviour of Steel Structures in Seismic Areas, Mazzolani & Tremblay (eds) © 2000 Balkema, Rotterdam, ISBN 90 5809 130 9

New trends in the evaluation of available ductility of steel members

A. Anastasiadis & V. Gioncu
Department of Architecture, University 'Politehnica' Timisoara, Romania

F. M. Mazzolani
Department of Structural Analysis and Design, University of Naples, Italy

ABSTRACT: The ductility of steel I-section members is determined for monotonic and seismic loads using the collapse plastic mechanisms. In addition, fracture ductility is defined considering the rotation corresponding the cracking of buckled flanges. The influence of different factors is considered: gravitational loads, earthquake type, strain rate and cyclic loads. Simplified relationships for practical design are given.

1 INTRODUCTION

In modern earthquake design of steel structures, inelastic excursions of members and redistribution of internal actions are permitted when the structure undergoes severe seismic actions. In this context, the condition for a structure to avoid collapse is that the members sustaining plastic deformation should possess available ductility greater than ductility requirements. So, the available ductility of structural components is a very important parameter in seismic design, but despite of many years of research it is not completely quantified. In fact no clear definition of the available ductility is now available in the current aseismic codes which consider the cross-section ductility only, giving width-to-thickness flange and web limitations, but neglecting the member ductility, which is of the primary importance. Concerning this latter issue, lessons learned from recent earthquakes (Northridge, Kobe) have demonstrated that both joint and member ductility must be evaluated. But the modern codes impose that plastic deformations occur only at the beam ends and at the column bases without considering the joints. However, in the paper only member ductility was investigated. In order to evaluate the aforementioned problems a great research effort was made (Gioncu & Petcu 1997, Mazzolani & Piluso 1992, 1993, 1996, Anastasiadis, 1999, Anastasiadis & Gioncu 1999). A significant contribution on the prediction of available ductility has to be mentioned in the frame of activity of the INCO-COPERNICUS European research program (Gioncu et al. 1999, Mazzolani 1999).

For the first time the concept of ductility was used in the static plastic design in order to assure the inelastic redistribution of internal actions. So, a monotonic ductility is intensively studied for cross-sections and EC 3 gives a classification of cross-sections in four ductility classes. For seismic design the interest in ductility concept is directed to the dissipation of the input seismic energy. The seismic actions are characterized by cyclic loading and velocity conditions. Thus, the reference of seismic code EC 8 to the monotonic ductility given in EC 3 is a very doubtful provision.

In the paper the factors affecting member ductility has been investigated, by using the local plastic mechanism method and DUCTROT M computer program (Petcu & Gioncu 1999). The obtained results show that the available ductility may be now quantified and it is the right moment to work out an annex of EC 8 code which is referred to structure ductility checks in addition to the strength and rigidity ones:

$$\gamma_r D_{req} \le \frac{D_{av}}{\gamma_a} \qquad (1)$$

where D_{req} and D_{av} are the required and available ductilities, respectively, and γ_r, γ_a the partial safety factors allowing to account for some uncertainties in their evaluation.

2 AVAILABLE MONOTONIC DUCTILITY

2.1 Member plastic rotation capacity

An accurate method to evaluate the member plastic capacity is based on the actual behaviour M-θ curve of a member, which is restrained against flexural-torsional buckling (Figure 1). For instance, girders belonging to a framed structure are restained by floor slab and secondary beams.

A quantitative evaluation of the local member ductility can be done by means of the available plastic rotation capacity, $R_{av.p}$, which defined in the lowering postbuckling curve at the intersection with the theoretical full plastic moment (Figure1):

$$R_{av.p} = \frac{\theta_u}{\theta_p} - 1 \qquad (2)$$

where θ_u is the ultimate rotation, while θ_p is the plastic rotation corresponding to the M_p. By using such a definition, it is considered that after local buckling the member strength is not drastically reduced, what has been observed from the experimental evidence, leading to a better exploitation of the plastic capacity reserves. For the evaluation of the rotation capacity a variety of methods has been proposed, such theoretical, approximate and empirical ones. Among these methods, the plastic collapse mechanism approach seems to be the best due to its simplicity and quite good correlation with the experimental results (Gioncu & Petcu 1997, Anastasiadis & Gioncu 1999). A model of local mechanism composed by plastic zones and yield lines is presented in Figure 2. Some results obtained using the local plastic mechanism methodology are presented in Figure 3. Two different classifications are plotted: one after EC 3 *(cross-section classes)*, and the other proposed by Mazzolani and Piluso (1993) *(member classes)*. The great difference between the two classifications is evident. It is clear that code provisions have to consider the member classes, instead of cross-section classification, being more adequate for checking the structure ductility.

For practical design, the rotation capacity of beams and beam-columns, can be evaluated by means of some simplified design relationships, which has been proposed on the based of experimental and numerical tests:

(i) For beams (Mazzolani & Gioncu 1997, Anastasiadis & Gioncu 1999):

$$R_{av.mon} = 3x10^4 c_r \frac{t_f}{bL_{sb}} \varepsilon \left(0.8 + 0.2 \frac{f_{yw}}{f_{yf}} \right) \qquad (3)$$

where $\varepsilon = 235/ f_{yw}$, c_r is a coefficient taking into account the influence of flange-web junction:

$$c_r = \left(\frac{b}{b - 0.5t_w - 0.8r} \right) \qquad (4)$$

t_w, t_f being the web and flange thickness, respectively, b half width of flange, L_{sb} span of the standard beam, f_{yw} and f_{yf} the nominal yield stresses for web and flange, respectively.

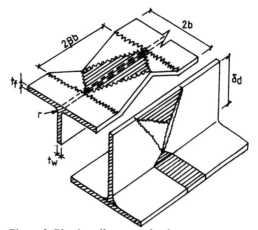

Figure 2. Plastic collapse mechanism

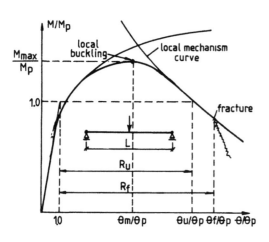

Figure 1. Definition of rotation capacity

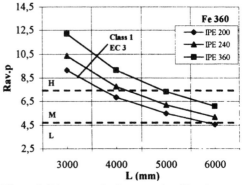

Figure 3. Discrepancies between ductility classes

(ii) For beam-columns (Anastasiadis 1999, Anastasiadis & Gioncu 1999):

$$R_{av.mom} = \eta c_r \left(\bar{\lambda} \frac{b}{t_f} \sqrt{f_y} \right)^{r_2} ; n_p = 0.10...0.40 \quad (5a)$$

where:

$$r_1 = 275(1+44.2n_p) ; 0.1 \leq n_p \leq 0.4 \quad (5b)$$
$$r_2 = -1.25(1+0.72\ n_p) ; 0.1 \leq n_p \leq 0.4 \quad (5c)$$

while b/t_f is the flange slenderness, f_y is the yield stress and $\bar{\lambda}$, the nondimensional slenderness:

$$\bar{\lambda} = \left(\frac{N}{N_{cr}} \right)^{1/2} \quad (6)$$

with:

$$N_{cr} = \frac{\pi^2 EI_e}{(L_0)^2} \quad (7)$$

where L_0 is the buckling length, being equal with μL_{sb} (Anastasiadis & Gioncu 1999). The results obtained with the equations (5a,b) cover the practical domain of double T shapes from HE 100 to HE 600A and B, with a moment ratio $M_{sup} / M_{inf} = 0...1$, as well as considering the influence of the axial force level $n_p = N/N_p$. For $n_p < 0.1$, an interpolation between the values obtained for $n_p=0$ and $n_p=0.1$ it is possible.

2.2 Influence of gravitational loads

Generally, the member rotation capacity is determined by using the three-point beam, which introduces the influence of lateral forces, so the member can be considered as isolated. But an actual structure is subjected simultaneously to both lateral and gravitational loads the latter influencing the zero point of the bending moment diagram and consequently the beam deformation capacity. Furthermore, the beam belongs to a framed structure and should be treated as a component of this frame. In order to take into account all these factors the concept of standard beam has been introduced. For the beam negative moment, where the variation is strong, the three-point beam, (SB1), with concentrated load should be used, while for the positive one the SB2 beam with uniformly distributed load should be used (Figure 4).

By using the standard beam concept and DUCTROT M computer program, the effect of gravitational loads has been investigated. In Figure 5 the influence of gravitational loads on member ductility is plotted. One can observe an asymmetry on the rotation capacity of the two plastic hinges at the end of the beam due to shifting of the moment zero point. This asymmetry is stronger in case of high gravitational loads ($M_p / qL^2 < 0.35$), while in

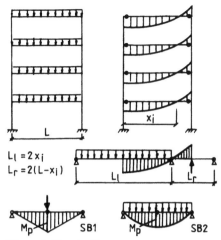

Figure 4. Standard beam concept

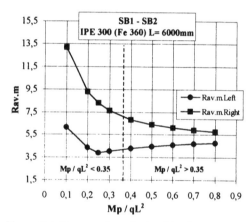

Figure 5. Influence of gravitational loads

case of smaller gravitational loads ($M_p / qL^2 > 0.35$) it is reduced, but it is still not negligible. So, also the gravitational loads must be considered in seismic analyses in order to calculate a more realistic value of the available member ductility.

2.3 Fracture rotation capacity

During experimental tests some cracks along the yield lines in the buckled flanges sometimes arise due to the attainment of the ultimate strain, what has been also confirmed during severe earthquakes. The fracture could be developed after or before the attainment of θ_u in the unstable part of the M-θ curve, (Figure1), in the latter case leading to a premature failure and incomplete formation of the available plastic ductility.

According to the capacity-demand concept, the

5

evaluation of section fracture rotation, θ_f, requires the definition of the rotation of the flange, α, as a function of rotation θ_u, as well as the fracture rotation of the flange, α_f. Examining the collapse mechanism of Figure 6 the following relation can be obtained from its geometry:

$$\alpha = 2\frac{\Delta}{\beta b} = 2\frac{(\beta b\Delta)^{1/2}}{\beta b} = 2\left(\frac{\delta d\theta}{\beta b}\right)^{1/2} = 2\left(\frac{\theta}{\chi}\right)^{1/2} \quad (8)$$

where $\chi = \beta b / \delta d$ is a nondimensional parameter of the buckled shape. The δ factor has a value between 0.81 and 0.820 for IPE hot rolled profiles while for HE-A and HE-B is equal to 1.0 (Anastasiadis 1999). The β coefficient can be calculated, after Gioncu & Petcu (1997), from the following relation:

$$\beta \approx 0.6\left(\frac{t_f}{t_w}\right)^{3/4}\left(\frac{d}{b}\right)^{1/4} \quad (9)$$

where b, d, t_f, t_w, are the cross-section geometrical dimensions (Figure1).

The rupture rotation, can be calculated considering the buckled wave length and the length of the plastic zone, resulting the following relation (Figure 7):

$$\alpha_f = 2\left(\frac{1}{\rho_y} - 1\right)\frac{\ell}{t}\varepsilon_u \quad (10)$$

where $\rho_y = f_y / f_u$ is the yield ratio, ε_u is the nominal ultimate strain of the material, and ℓ is the distance between the inflection points of the buckled flange, which is equal with βb.

The fracture ductility strongly depends on the yield ratio, buckled length, ultimate strain of the material as well as the cross-section geometry.

Equating the relation (8) and (10) we obtain the fracture rotation of the section, which can be considered as a measure of the total inelastic deformation:

$$\theta_f = \chi\left(\frac{1}{\rho_y} - 1\right)^2\left(\frac{\ell}{t}\right)^2(\varepsilon_u)^2 \quad (11)$$

Using the relation (11) the fracture rotation could be determined:

$$R_f = \frac{\theta_f}{\theta_p} - 1 \quad (12)$$

For a suitable member behaviour the plastic hinges must be assure the above two conditions simultaneously:

$$R_{av.mon} \le R_{req} \quad (13a)$$

$$R_{av.mon} \le R_f \quad (13b)$$

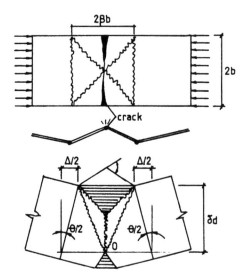

Figure 6. Determination of flange rotation

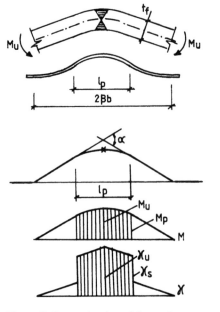

Figure 7. Determination of flange fracture rotation.

In Figure 8 the fracture condition for IPE beams as a function of steel grade and cross-section geometry is plotted. One can see that profiles with medium geometrical dimensions, IPE 300 to IPE 500, with low steel grade (Fe 360), have the best behaviour against fracture. The influence of the increased yield strength due to random material variability (Mazzolani et al. 1991, 1992, 1994) is illustrated in Figure 9. A gradually reduction of fracture ductility with the increasing of yielding strength can be observed. As a consequence, it is very important to consider the maximum yield strength in the evaluation of the member ductility, in order to achieve a better local ductility control.

3 AVAILABLE SEISMIC MEMBER DUCTILITY

3.1 Influence of earthquake type

For the previous section the rotation capacity has been studied just under static loading conditions. But during the seismic action the members are affected by some physical phenomena, such as fracture due

to the effect of strain-rate, or cumulative damages due to cyclic loading. Therefore, the performance determined under monotonic conditions can not be extrapolated to seismic condition.

The use of equation (1) for checking the ductility under of seismic loading can be acceptable provided that some modifications for the required and available ductilities are introduced. For the required ductility an amplification of static determined values is necessary, while for static available ductility an erosion should be determined in function of the ground motion characteristics. In the last years some very important new information has been obtained, thanks to the development of a large network of instrumentation all over the World. This new situation offers the possibility to introduce new concepts in structural design: the influence of the type of ground motion and the different behaviour of structures in near-source and far-source fields (Gioncu 1999), which is mainly influenced by the velocity pulse and the cyclic loading, respectively (Figure 10). For near-source actions the fracture takes place at the first cycle and the strain-rate has a very important role, while for the far-source ones, the fracture occurs after some cycles due to the accumulation of deformations. Therefore the performance of members will be very different in the two cases.

Another very important factor influencing the available ductility is the presence, in case of near-source earthquakes, of superior vibration modes. Due to the change of the inflexion points position, a decreasing of the available ductility may occurs.

3.2 Strain-rate influence

In case of near-source earthquakes, high velocity values occurs giving rise to very important strain-rates in structure.

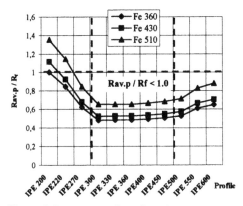

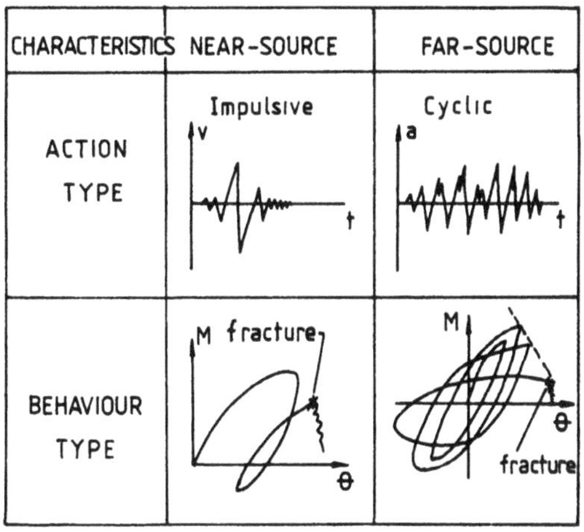

Figure 8. Fracture rotation of IPE beams

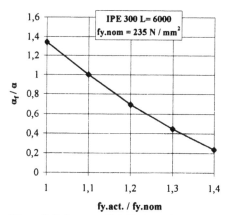

Figure 9. Influence of yield strength

CHARACTERISTICS	NEAR-SOURCE	FAR-SOURCE
ACTION TYPE	Impulsive	Cyclic
BEHAVIOUR TYPE	M fracture	M fracture

Figure 10. Near-source and far-source earthquakes.

7

It is well known that the strain-rate has the main influence on the increasing of the yield stress, especially for values greater than 10^{-1} /s (Gioncu 2000). At the same time the increasing of the ultimate strength is moderate. Consequently, the yield ratio increases with the increasing of strain-rate (Figure 11). Using the Sorousian and Choi (1987) relation, for Fe 360 steel grade results:

$$\rho_{ysr} = \frac{1.46 + 0.0925 \cdot \log \dot{\varepsilon}}{1.15 + 0.0496 \cdot \log \dot{\varepsilon}} \rho_y \qquad (14)$$

where $\rho_{y.sr}$, ρ_y are the yield ratios for strain-rate and static loading, respectively, ε being the strain-rate. Taking into account that the randomness of the yield ratio for usual steel is about 0.60 to 0.73, and that the strain-rate in the field of near-source strong earthquakes varies between $10^{-1}...10^{1}$, results an increasing of the yield ratio of about $0.75...0.95$ can be foreseen. Using these values of ρ_{ysr}, the fracture rotation may be determined from the equation (10). Figure 12 shows the ratio between fracture rotation and ultimate monotonic rotation. One can see that in case of near source earthquakes, a local fracture of section produced at the first cycle, what reduces the plastic ductility. The sections with thick flanges are more dangerous for fracture than the ones with thinner flanges. Thus, the use of equation (13b) for near-field earthquakes must be done using an increased yield ratio, in function of the predicted strain-rate.

3.3 Influence of superior vibration modes

In case of near-source earthquakes the superior vibration modes interact with the first mode and the moment diagram shows some irregularities in its distribution, especially at the middle of the frame height (Gioncu 1999) (Figure 13a). The rotation capacity for a column in a frame with different moment distributions is depicted in Figure 13b. One can see that for one curvature moment variation, M_{sup} / $M_p = -0.5$ to 0, the rotation capacity

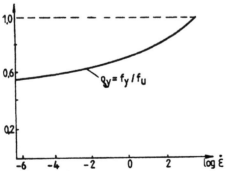

Figure 11. Influence of strain-rate on yield ratio.

decreases. As a result, the superior vibration modes may produce a dramatic reduction of the available beam-column ductility at the stories situated at the middle height of structures.

3.4 Cyclic loading influence

The behaviour of I-sections has a particular feature under cyclic loading. The first semi-cycle produces buckling of the compression flange and the section rotates around a point located in or near the opposite flange. As a result, the tensile forces are very small. During the reversal semi-cycle, the compressed flange buckles also. The most important observation is that the opposite flange remains unchanged because of the small tension force, which is not able to straighten the buckled flange (Figure 14a).

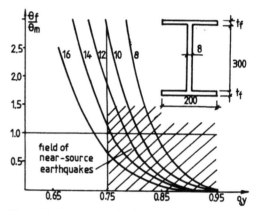

Figure 12. Fracture to plastic rotation ratio.

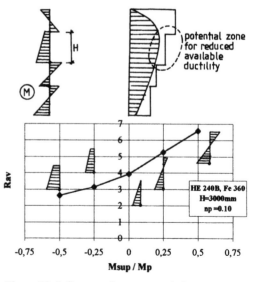

Figure 13. Influence of moment variation.

8

Therefore, during the next cycle, the section works with an initial geometrical imperfection resulted from the previous cycle. In this way after each cycle an additional rotation is superposed on the previous one, according to the plastic mechanism which is assumed in the web and in the flanges.

Experimental investigations show that the plastic mechanism in the web is not completely developed, a part of yield lines remaining in the elastic range (Figure 14b). As the beam ductility is mainly based on the web plastic mechanism, the incomplete plastification produces an important reduction in member ductility (Figure 15). So, as the influence of the accumulation of residual displacements in flanges and incomplete formation of yield lines in web increases gradually, the moment-rotation curve in the softening branch shows a degradation with respect to the monotonic case (Figure 16). The reduction of rotation capacity is about:

$$\theta_{av.cyc.} \approx 0.60\theta_{av.mon} \qquad (15)$$

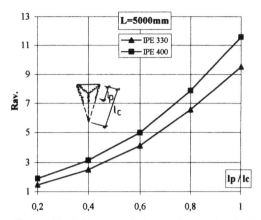

Figure 15. Influence of incomplete web plastic mechanism on rotation capacity.

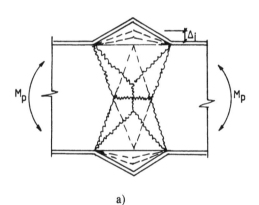

a)

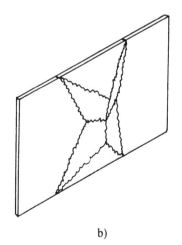

b)

Figure 14. Plastic mechanism for cyclic loading.

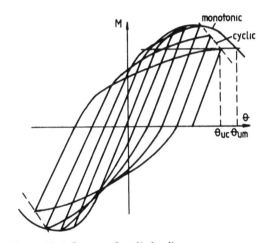

Figure 16. Influence of cyclic loading.

4 CONCLUSIONS

In order to clarify the physical meaning of available ductility of a steel member the monotonic and seismic ductilities are defined. For monotonic ductility the rotation capacity is evaluated using the collapse plastic mechanisms. The dimensions of standard beams are determined in function of gravitational loads. The flange fracture may limit, in some cases, the member rotation capacity. For seismic ductility, the influence of earthquake type must be considered. For near-source earthquakes, the effects of strain-rate and superior vibration modes must be introduced, while for far-source earthquakes the cyclic loading condition has to be taken into account.

Due to the quantitative results obtained in the last period, it is possible now to elaborate some code provisions for checking the structural ductility of steel members in a more modern way, by using also the results presented in this paper.

REFERENCES

Anastasiadis, A. 1999. *Ductility problems of steel moment resisting frames*. Ph.D thesis. University Politehnica Timisoara.

Anastasiadis, A. & V. Gioncu 1999. Ductility of IPE and HEA beams and beam-columns. In D. Dubina & M.Ivanyi, *Stability and Ductility of Steel Structures*: 249-258. London. Elsevier.

Gioncu, V. & Petcu. D. 1997. Available rotation capacity of wide-flange beams and beam-columns. Part1: Theoretical Approaches. Part 2. Experimental and numerical tests. *Journal. of Constructional Steel Research*. 43: 1-3, 161-217, 219-244.

Gioncu, V. & F.M. Mazzolani 1997. Simplified approach for evaluating the rotation capacity of double T steel sections. In F.M. Mazzolani & H. Akiyama, *Behaviour of Steel Structures in Seismic Areas, STESSA' 97 (Kyoto)*. Ed. 10/17 Salerno, 303-310.

Gioncu, V. 1999. Framed structures. Ductility and seismic response. General report. In D. Dubina & M.Ivanyi, *Stability and Ductility of Steel Structures*.

Gioncu, V., G. Mateescu, D. Petcu, A. Anastasiadis. 2000a. Prediction of available ductility by means of local plastic mechanism method: DuctRot computer program. In "Moment Resistant Connections of Steel Building Frames in Seismic Areas". Ed. F.M. Mazzolani. London: E &FN Spon.

Gioncu, V. 2000. Influence of strain-rate. In F.M. Mazzolani & R.Tremblay, *Behaviour of Steel Structures in Seismic Areas, STESSA 2000, Montreal*. Rotterdam: Balkema.

Mazzolani, F.M., E. Mele & V. Piluso 1991. Random variability effects in seismic resistant steel structures. *Journal of Singapore Structural Steel Society*. Vol. 2, no. 1.

Mazzolani, F.M., E. Mele & V. Piluso 1992. The seismic behaviour of steel frames with random material variability. *Proc. 10th World Conference on Earthquake Engineering, Madrid*. Rotterdam: Balkema

Mazzolani, F.M. & V. Piluso 1992. Evaluation of the rotation capacity of beams and beam-columns. Proceedings of 1st State-of-the-Art Workshop COST 1. Strasburg. 517-529.

Mazzolani, F.M. & V. Piluso 1993. Member behavioural classes of steel beams and beam columns. *XIV Congresso CTA*, Viareggio, Italy, Ricerca Teoretica e sperimentale, 405-416.

Mazzolani, F.M., B. Calderoni & V. Piluso 1992. Quality control of material properties for seismic purposes. In F.M. Mazzolani and V. Gioncu. *Behaviour of Steel Structures in Seismic Areas, STESSA' 94 (Timisoara)*. London: Champan & Hall.

Mazzolani, F.M. &V. Piluso 1996. *Theory and Design of Seismic Resistant Steel Frames*. London: Champan & Hall.

Mazzolani, F.M. 1999. Reliability of moment resistant connections of steel frames in seismic areas: the first year of activity of the Recos project. *Proceedings of the Conference EUROSTEEL' 99*, Prague.

Petcu, D. &V. Gioncu 1999. DUCTROT M. Plastic rotation capacity of steel members. Guide for users. INCERC Timisoara.

Soroushian, P. & K.B. Choi 1987. Steel mechanical properties at different strain rates. *J. Struct. Eng.* 113: 863-872.

Behaviour of Steel Structures in Seismic Areas, Mazzolani & Tremblay (eds) © 2000 Balkema, Rotterdam, ISBN 90 5809 130 9

ASTM A913 grades 50 and 65: Steels for seismic applications

Serge Bouchard
Trade ARBED, New York, USA

Georges Axmann
ProfilARBED, Luxembourg

ABSTRACT: This paper presents the physical and mechanical properties of ASTM A913 steels, which are produced by an advanced thermo-mechanical controlled process, namely Quenching and Self-Tempering (QST). This process allows even for very thick rolled shape products to combine three important but formerly incompatible properties: high strength, good toughness and easy weldability.

As such, a detailed review of the weldability, the toughness, the ductility and the through-thickness properties of A913 steels is given. New requirements have been introduced in A913 since its original publication in 1993. Grade 50 is available with a yield strength limited to 65 ksi and with a maximum yield to tensile ratio of 0.85. Very heavy beams and columns, called "Jumbo" shapes, are available with a guaranteed toughness in the core area of the material, as required by the SAC Seismic Design Guidelines.

Finally, this paper shows that the use of A913 Grade 65 in columns and Grade 50 in the beams is an economical way to achieve the "strong column – weak beam" concept. The combination of this solution and the advantages of the dog bone connection, also known as Reduced Beam Section (RBS), originally patented by ARBED and released in 1995, is also included.

1 INTRODUCTION

For designing welded steel moment resisting frames (SMRF) in seismic zones, the economical strong column - weak beam concept has been extensively used for many years.

After the Northridge earthquake (1994), although no steel building collapsed, some connections showed cracks raising questions on the connection design, fabrication and the steel quality. In order to investigate these topical areas, a joint venture of all concerned parties had been created.

After a first phase of investigations, surveys and researches, the following has been concluded:

➤ The reliability of the connection performance is lower than assumed.
➤ The fabrication had in many cases not been respecting the code provisions.
➤ Applicable material standards lack specific requirements for earthquake like loading.

Thus, for achieving a better behavior of welded steel moment resisting frames, supplementary efforts are needed to improve design and fabrication procedures and material characteristics.

Concerning the conclusion on material properties, a statistical survey of the mechanical properties of steel shapes showed that the actual yield strength of material as delivered could be well in excess of the minimum values specified by the standard.

In fact, the average yield strength of steel A36 and A572 Grade 50 were 49 ksi respectively 57 ksi. This difference of the average yield strength between the two grades would not allow to take for granted that the yield strength of the columns is always superior to that of the beams. So the basic conditions for the strong column - weak beam concept may not always be satisfied.

In order to solve this problem, it has been recognized that in addition to Grade 50 for beams, it is useful to have a higher strength grade for columns, Grade 65 for instance.

It is to be noted that ASTM A913 covers besides Grade 50 also a Grade 65 up to the shape size group 5. In recent years, these two grades have been introduced in the following design codes:

- AISC/ ASD&LRFD and Seismic Provisions for Structural Steel Buildings,
- Uniform Building Code (UBC),
- International Building Code (IBC),
- SAC Seismic Design Guidelines for SMFS (Steel Moment Frame Structures).

The inclusion of A913 Grades in those codes enables the building designers to economically use the strong column - weak beam concept throughout the future.

2 ASTM A913

A913 is the Standard Specification for High-Strength Low-Alloy Steel Shapes of Structural Quality, Produced by the Quenching and Self-Tempering Process (QST).

The QST is an improvement of "Thermo-Mechanical Controlled Processes" (TMCP) that have been known since the seventies. The QST-process has been developed in the eighties in order to satisfy the market demands for steel shapes exhibiting a combination of three important properties previously incompatible for heavy shapes:

- high strength,
- good toughness at low temperatures and
- excellent weldability.

A913 covers Grades 50 [345 MPa], 60 [415 MPa], 65 [450 MPa] and 70 [485 MPa]. Currently Grades 50 and 65 are available in the U.S. *Tables 1 and 2*

Table1. Chemical composition

	Maximum content in %	
	Grade 50	Grade 65
Carbon	0.12	0.16
Manganese	1.60	1.60
Phosphorus	0.040	0.030
Sulfur	0.030	0.030
Silicon	0.40	0.40
Copper	0.45	0.35
Nickel	0.25	0.25
Chromium	0.25	0.25
Molybdenum	0.07	0.07
Columbium	0.05	0.05
Vanadium	0.06	0.06
Carbon Equivalent	0.38	0.43

show the chemical composition and the tensile properties of Grades 50 and 65 in accordance with ASTM A913/913M-97 included in ASTM Volume 01.04.

2.1 Main advantages of A913 steels vs. hot rolled steels to A36, A572 and A992

• *Toughness*

Since the Northridge earthquake, it has been generally recognized that a sufficient level of toughness is required in order to avoid the initiation and propagation of cracks in the brittle fracture mode. The SAC guidelines mention (in chapter B2.4): "A Charpy V - notch (CVN) value of 20 ft.lbs [27 J] at 70°F [21°C] should be specified when toughness is deemed necessary for an application".

According to A673/A673M the specimen is located at 1/4th of the flange thickness at 1/6th of the total flange width.

The main text of A913 specifies the following:
"§ 6.2: Charpy V-notch tests shall be made in accordance with spec. A673/A673M, frequency H."
"§ 6.2.2: The test results of full-size specimens shall meet an average of 40 ft.lbs [54 J] at 70°F [21°C]."

Thereby all A913 grades meet and exceed the above toughness requirement, whereas no requirement for Charpy V-notch test exists in the main text of ASTM A36, A 572 and A992.

• *Weldability*

The weldability of steel is commonly characterised by the carbon equivalent (CE). Its calculation is shown in *figure 3*. A913 grades respect restrictive CE max. values of 0.38 % for Grade 50 resp. 0.43 % for Grade 65. Due to the low CE values, these steels are easily weldable, as they can be welded without preheating (at temperatures over 32°F).

It is to be noted that the main texts of ASTM A36, A572 do not limit the CE. A992 has a specified max. CE of 0.47% for shape groups 4 and 5.

• *Supplementary advantages of A913 Grades for seismic applications*

On top of the general advantages of A913 steels, described in the previous chapter, A913 has been modified by the addition of "Additional Supplementary Requirements" that provide engineers with all the necessary guarantees for a safe seismic design. These features are the following:

> The yield point of A913 Grade 50 can be restricted to 65 ksi maximum.
> The yield to tensile strength ratio of A913 Grade 50 can be limited to 0.85 maximum.

The upper limit of the yield for the Grade 50 will help the designer to control the formation of the plastic hinge in the beams under earthquake loads. Particularly when used in combination with columns in Grade 65 it also guarantees him that the actual yield strength in the column will be always higher than the actual yield strength in the beams, and thus the strong column - weak beam concept is always satisfied.

> The groups 4 and 5 structural shapes can be supplied to the so-called AISC Supplement 2. This requirement specifies a Charpy V-notch test min. value of 20 ft.lbs [27 J] at 70°F [21°C] on a specimen taken from the core of the web-flange intersection.

This requirement is to ensure that heavy shapes have the necessary toughness in the web-flange intersection when submitted to a high level of stresses, particularly for welded members in tension. SAC has extended this requirement to group 3 shapes with flanges 3 ½ in. or thicker.

3 PRODUCTION OF A913 STEELS

3.1 Description of the Quenching and Self-Tempering (QST) process

The superior properties of A913 steels are obtained by the QST treatment, which leads to high yield and ultimate tensile strength with remarkable low carbon content thus reduced CE values leading to excellent weldability and superior toughness. In the QST process, directly after the last rolling pass, an intense water-cooling is applied to the whole surface of the beam so that the skin is quenched. Cooling is interrupted before the core of the material is affected and the outer layers are tempered by the flow of heat from the core to the surface.

Figure 1 schematically illustrates the QST treatment. At the exit of the finishing stand, directly at the entry of the cooling bank, temperatures are typically 1600°F [850°C]. After the short cooling phase, the self-tempering temperature is 1100°F [600°C].

3.2 Typical actual properties of A913 steels

• *Toughness*

Table 3 shows the statistical distribution of the toughness for QST steels in Grades 50 and 65 at

Table 2. Tensile properties

Grade	Yield Strength Min.		Tensile Strength Min.		Elongation (%), Min.	
					200 mm 8 in.	50 mm 2 in.
	Ksi	[MPa]	ksi	[MPa]		
50 [345]	50	[345]	65	[450]	18	21
65 [450]	65	[450]	80	[550]	15	17

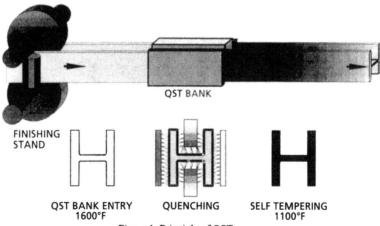

Figure 1. Principle of QST

Table 3. Charpy V-notch impact energies

CVN test at 70°F [21°C]	Grade 50 ft. lbs [J]		Grade 65 ft. lbs [J]	
Average	176	[238]	167	[226]
Standard deviation	22	[30]	25	[33]
Min.	101	[136]	91	[124]
Max.	225	[304]	231	[312]

70°F [21°C]. These excellent values based on normal production and not on special requirements (such as offshore applications) show that A913 grades have an outstanding toughness.

• *Weldability*

The American Welding Institute (AWI) assessed the weldability of "Jumbo" shapes in A913 Grade 65 welded without preheating (AWI: Report N° 91-002, December 1992). This study confirms the outstanding weldability of A913 beams. The ATLSS Center at Lehigh University (Prof. Fisher) also evaluated the mechanical properties of full section weld splices fabricated from Jumbo sections (Report N° 92-06, September 1992). The following conclusions demonstrate the excellent material performance:

➢ Heavy A913 Grade 65 beams can be welded successfully using no preheat and according to AWS D1.1 Structural Welding Code.
➢ Filler metal is commercially available for Shielded Metal Arc Welding (SMAW), Flux-Cored Arc Welding (FCAW), and Submerged Arc Welding (SAW) for welding A913 Grade 65 beams.
➢ As-welded toughness is substantially better than code requirements.

➢ The weld joint tensile strength exceeds the minimum tensile strength requirements for ASTM A913 Grade 65 in all the tested joint types (SMAW, FCAW and SAW processes at several heat inputs).
➢ The toughness of the weld metal in all tested joint types is well in excess of the specification requirements.
➢ Charpy V-Notch tests performed on base material at the position specified in the AISC requirements for heavy shapes near the web-flange intersection (AISC Suppl. 2) and at other locations across the flange cross sections show very high levels of fracture toughness well in excess of the specification requirements.

Figure 2 shows the welding of two pieces of "Jumbo" shapes W14x730 in A913 Grade 65 welded by AWI without preheating. In this case, the welder needed 140 passes to weld the two beams together. Thanks to the low CE of the A913 steel, he did not need to preheat the beams and saved about 4 hours that would have been necessary for preheating. As the need for preheating decreases for smaller thicknesses, the demonstration that the heaviest size can be welded without preheating was the evidence that all sizes needed no preheating.

Figure 3 shows the calculation formula of the carbon equivalent (CE) and compares the CE values of conventional steels (hot rolled) to A913 steels. It exhibits that all A913 Grades can be welded without preheating thanks to their low CE.

• *Through thickness strength*

One of the major concerns of designers in steel construction is the through thickness behavior under

Figure 2. CJP welding of Jumbo shapes

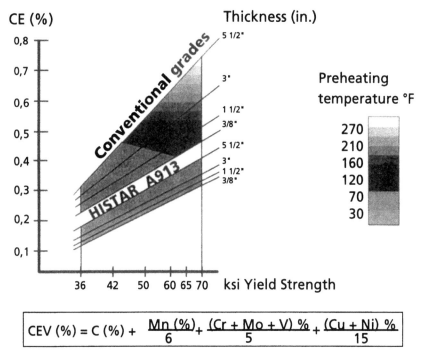

CE (%)

0,8
0,7
0,6
0,5
0,4
0,3
0,2
0,1

Conventional grades

HISTAR A913

Thickness (in.)

5 1/2"
3"
1 1/2"
3/8"
5 1/2"
3"
1 1/2"
3/8"

Preheating
temperature °F

270
210
160
120
70
30

36   42   50   60 65 70   ksi Yield Strength

$$CEV\ (\%) = C\ (\%) + \frac{Mn\ (\%)}{6} + \frac{(Cr + Mo + V)\ \%}{5} + \frac{(Cu + Ni)\ \%}{15}$$

Figure 3. Weldability of A913 Grades

Figure 4. Basic tee-joint pull-plate specimen

Figure 5. Fracture in pull-plate

seismic loading of welded moment frame connections using the strong column - weak beam concept.

In parallel with the SAC testing program, covering 33 tests on Grade 50 column sections including A913 Grade 50, eight tee-joint pull-plate tests were performed on A913 Grade 65 heavy column sections.

The tee-joint specimens were loaded in tension through high-strength 100 ksi [690MPa] MSYS

pull-plates. The welds were made at two heat-input levels: 15 and 35 KJ/cm. The higher heat-input level was intended to have the worst possible Heat Affected Zone (HAZ) in the column flange.

Figures 4 and 6 show the 2 types of tee-joint specimens that have been tested. These are:

- basic tee-joint pull-plate specimen with the complete column section *(fig. 4)* and
- tee-joint specimen with flange only *(fig. 6).*

Figure 6. Tee-joint specimen with flange only, showing broken pull-plate on the bottom after test

The type of tested specimens and the results are given in *Table 3*. All specimens broke in either the weld or the pull plates at nominal pull-plate stress levels in excess of 100 ksi [690 MPa].

Figure 5 shows a typical fracture in the pull-plate after substantial necking. This fracture in the pull plate at higher stresses than the strength of the column flange can be explained by the existence of tri-axial constraint of the column flange material, which creates hydrostatic tension stresses, raising the apparent through-thickness strength.

None of these joints failed because of inadequate strength or ductility of the column section. Lamellar tearing did not occur in any of these joints.

In summary, through-thickness failure of A913 Grade 65 column sections could not be induced in the tee-joint test, despite applying stress greater than 100 ksi [690 MPa], i.e. well above the strength of any structural steel.

Since the through-thickness failure of column sections is very unlikely in moment connections or other types of tee-joints, it is recommended that the through-thickness strength does not need to be explicitly checked in the design of welded beam-to-column connections.

4 A913 AND THE CODES

Since 1993, QST-steels in accordance with ASTM A913/A913M have been offered and successfully used worldwide in numerous projects. The SAC guidelines mentioned that designers might wish to begin incorporating ASTM A913 Grade 65 steel in order to continue to design with strong column - weak beam concept. Since 1995, ASTM A913 is included in AWS D1.1 American Welding Code. Moreover, ASTM A913 is approved by AISC and A913 grades 50 and 65 are recognized and included the design codes ASD, LRFD, IBC 2000 and UBC. The SAC Seismic Design Criteria and Recommended Specifications for Moment-Resisting Steel Frames for Seismic Applications recognize A913 Grades 50 and 65 as prequalified structural steels.

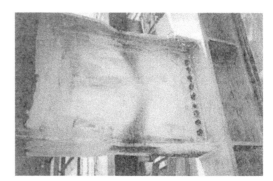

Figure 7: Reduced Beam Section (RBS) after test

Table 3. "Tee-joint" test results on A913 Grade 65

Test #	Column Size	Peak load (MN)	Pull-plate area (m^2)	Pull-plate Stress (MPa)	Weld area** (m^2)	Weld stress** (MPa)	Failure Mode	Type of "Tee-joint" test
34	W14x257	2.098	0.0026	813	0.003	699	Pull plate	Basic specimen
35	W14x605	2.070	0.0026	796	0.003	690	Pull plate	Low heat input weld
36	W14x257	2.066	0.0026	801	0.003	689	Pull plate	Flange only
37	W14x605	2065	0.0026	800	0.003	688	Pull plate	Low heat input weld
38	W14x257	2.017	0.0026	782	0.003	672	Weld	Basic specimen
39	W14x605	2.077	0.0026	805	0.003	692	Pull plate	High heat input weld
40	W14x257	1.903	0.0026	737	0.003	634	Weld	Flange only
41	W14x605	1.989	0.0026	771	0.003	663	Weld	High heat input weld

**This weld area is measured on a plane 1.6 mm above the column surface, and weld stress is the peak load divided by this weld area.

5 REDUCED BEAM SECTION (RBS): THE "DOG BONE" CONNECTION

The Luxembourg-based steel manufacturing company, ARBED held a 1992 US patent on the RBS. Following the Northridge earthquake, ARBED waived in 1995 all patent and claim rights associated with the RBS for the benefit of the profession. This gracious gesture allowed further development of the concept for use in post-Northridge SMF buildings.

The RBS is one type of connection that has been developed to force the plastic hinge away from the beam-column interface. *Figure 7* shows a RBS connection after being tested. This connection relies on the selective removal of beam flange material adjacent to the beam-to-column connection to reduce the cross sectional area of the beam. The flange cut-out can easily be done in the fabrication shop as a constant, tapered or radius cut, the latter turns out to be the most economical today [1].

Current RBS design procedures assume the minimum specified yield stress of the beam to be 50 ksi or less, and that the minimum specified yield stress of the column is 50 ksi or greater, i.e. Grade 50 or Grade 65 columns.

The main advantages of the RBS are:

➢ The smaller moment generated at the column face limits the demand and thereby reduces the possibility of fracture.
➢ Helps satisfying "strong column - weak beam" requirements.
➢ Minimizes column doubler-plate requirements.
➢ No local undermatching of column when actual properties of beam in Grade 50 are higher than in the column.
➢ Avoids uneconomical addition of strengthening plates and special weldments.
➢ Cost effective, consistently performing connection.
➢ Established performance based on extensive testing.

6 CONCLUSIONS

A913 Grades 50 and 65 eliminate concerns that engineers may have had about rolled sections used in seismic applications. These steels are easily weldable due to their low carbon equivalent, have an outstanding toughness and excellent ductile behavior. For plastic design purpose, Grade 50 beams can be supplied with limits for maximum yield strength and yield strength to tensile strength ratio.

Heavy columns in Grade 65 can be supplied with a guaranteed toughness in the web-flange intersection. Thus A913 Grades 50 and 65 are tailored to allow a safe and economic design of steel moment resisting frames using the strong column - weak beam concept.

REFERENCES

[1] Moore K.S., Malley J.O., Engelhardt M.D. 1999. Steel Tips-Design of Reduced Beam Section Moment Frame Connections. Structural Steel Educational Council (ed.)

Behaviour of Steel Structures in Seismic Areas, Mazzolani & Tremblay (eds) © 2000 Balkema, Rotterdam, ISBN 90 5809 130 9

Influence of strain-rate on the behaviour of steel members

V.Gioncu
Department of Architecture, University 'Politehnica' Timisoara, Romania

ABSTRACT: The paper reviews the current status of research works concerning the influence of strain-rate on the behaviour of steel members. The strain-rate is very important in the case of near-source earthquakes, where very high horizontal and vertical velocities are recorded. The range of strain-rate for these cases is 10^{-1} to 10^{1} sec^{-1}, higher than the values currently considered in design. The main effect of strain rate for monotonic or cyclic loading is to transform a ductile plastic collapse into a brittle fracture.

1 INTRODUCTION

Seismic-resistant steel structures are usually designed relying on their ability to sustain high plastic deformations. The earthquake input energy is dissipated through the hysteretic behaviour of the plastic hinges. For the moment resisting frames (MRFs) the formation of these plastic hinges is designed in localized positions, so, that structure develop a global mechanism. The plastic rotation must show a stable hysteretic behaviour with a sufficient ductility to allow for an inelastic dissipation of input energy. Accordingly with this design philosophy, the structure may be designed for lower forces than it have to resist.

But the 1994 Northridge and 1995 Kobe earthquakes, both being near-source events, produced widespread and unexpected damage to moment-resisting frames. While modern steel special MRFs were designed to dissipate energy through the hysteretic behaviour at beams and joints, this was not the case during these events. Instead of plastic rotation of predestine zones, brittle fractures were relevant, the most common occurrence being the crack propagation in weldments, heat affected zones and webs. In some cases no signs of plastic deformations were recorded. In other cases, beams exhibited little evidence of plastification and local buckling, indicating that they dissipated some energy before fracture. But the most stupefying establishments were that, in spite that there exist some differences in design assumption, constructional details, failure modes, etc. between Northridge and Kobe joints, there are many similitude showing a common cause for both events. In these circumstances, questions have been raised if

the design philosophy, based on the dissipation of a great amount of seismic input energy by means of plastic rotations, is effective in the case of near-source earthquakes. Several causes were assumed to be the reason of this bad behaviour: shortcomings in welding procedures and constructional details or the presence of important vertical components or the effect of high velocities, producing in structure important strain-rate.

The community of specialists is divided in considering the causes which produced these unexpected fractures, and especially on the measures which are necessary to eliminate the possibility of similar damage occurrence during the future strong earthquakes. Some of them consider that the actual design philosophy is proper and only some improvement of constructional details and welding technologies should enough. But in the same time, there are other specialists who consider that these damage is mainly produced due to the characteristics of near-source earthquakes, combined with influence of welding and inadequate details.

The paper presents a state-of-the art on the strain rate influence on the behaviour of steel members. The main effect is the possibility to transform the ductile plastic deformation into a brittle fracture due to the important increasing of yield stress.

2. LEVEL OF STRAIN-RATE DURING THE STRONG EARTHQUAKES

The main characteristic of the dynamic loading at the level of ground is the velocity, while at the level of structure is the strain-rate. A relationship between these values must be established.

(i) *Recorded velocities.* Many strong motion records has been obtained, since strong motion observation was started since 1932. Owing the recent efforts for dense installation of strong motion accelerographs in seismic areas, the number of strong motion records observed in epicentral areas is increased. The 1994 Northiridge and 1995 Kobe earthquakes produced the largest addition to the collection of ground motion records. The maximum horizontal velocity during the California event was about 177cm.sec[-1] (Rinaldi station) and during the Japanese event 176cm.sec[-1] (Takatori station) (Gioncu et all 2000a, Hall et all., 1995, Midorikawa, 1995). These values are about 5 times of that for the famous 1940 El Centro record, which was for long time, the reference earthquake. The velocities of vertical ground motions were more reduced than the horizontal ones: 72 cm.sec[-1] (Northridge, Tarzana) and 62 cm.sec[-1] (Kobe, Port Island). The velocity response spectra for Northridge (Tarzana station) and Kobe (Fukiai JMA and Fukiai stations), where the peak ground velocities were about 100cm/sec, are shown in Figure 1, (Akiyama, 1996, Kurobane et al., 1997), for a damping coefficient of 2%. For the Northridge earthquake an important amplification of velocity occurrs for period range shorter than 1 sec, while for Kobe earthquake, the amplification is very important for periods till 2.0 sec. Spectra velocities of about 400 cm.sec[-1] can be reached. These results show that many structures in Northridge and Kobe sustained very great energy inputs. These velocity values are about ten times than the velocities recorded in far-source areas, which for the surface earthquakes frame in the range of 30-50 cm.sec[-1].

(ii) *Strain-rate.* In order to establish the relationship between velocity and strain-rate, the behaviour of SDOF system is studied (Figure 2a). The

displacements in the plastic range are (Gioncu 1995):

$$z_p = v_g \left(t - \frac{\sin \dfrac{2\pi t}{T_g}}{\dfrac{2\pi}{T_g}} \right) + z_0 \qquad (1)$$

where v_g and T_g are the velocity and natural vibration period of ground motion, respectively, t is the time measured from the moment of formation of the plastic hinge and z_0 is the displacement of system at t=0. In (1) the damping effect is neglected due to its very weak influence on motion in plastic range. One can see that the displacement is composed by a continuous increasing one (corresponding to the displacement and velocity reached at the t=0) and an oscillatory one due to the ground motion (mainly depending on the natural vibration of the ground). The velocity of rotation of the plastic mechanism results from (1):

$$\theta_p = \frac{v_g}{H} \left(1 - \cos \frac{2\pi t}{T_g} \right) \qquad (2)$$

and the strain-rate of the buckled flange of a I-profile (Figure 2b), considering the length of a plastic hinge as equal with the width of flanges:

$$\varepsilon = v_g \frac{b}{h} \frac{1}{H} \left(1 - \cos \frac{2\pi t}{T_g} \right) \qquad (3)$$

The maximum value is obtained for $t = T_g$:

$$\varepsilon = 2 v_g \frac{b}{h} \frac{1}{H} \qquad (4)$$

So, the strain-rate depends directly on the ground velocity, sectional dimensions and mass distribution. The strain-rate for a SDOF system with b≈h and H=400cm results as $\varepsilon = v_g /200$ (sec[-1]). For range of velocity between 40–400 cm.sec[-1], the strain-rate occur in the range of 0.25... 2.5 sec[-1].

In the case of MDOF systems with equal masses at all stories, the strain-rate can be determined with the relationship (4), considering the H as the distance from base to the mass centroid (global mechanism), or the first level height (storey mechanism) (Figure 3). One can see that the strain-rate is higher in the case of storey mechanism than in the case of global mechanism. This is an additional argument to avoid the concentration of plastic deformations on a single level of structure. In the case of more general considerations, the range of strain-rate is between

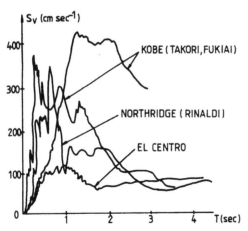

Figure 1 Velocity spectra for Northridge and Kobe earthquakes

$10^{-1}...10^1$ sec^{-1}, the limits referring to the moderate and high velocities. This means that the upper limits is larger than it is usually considered in some research works (10^{-1} sec^{-1}) (Uang & Bondad, 1996, Nakashima et al., 1998).

3 INFLUENCE OF STRAIN-RATE ON STEEL MECHANICAL PROPERTIES

The first research work concerning the influence of strain-rate on the behaviour of metals was performed by Manjoin (1944). Tests were conducted at room temperature for the strain-rate from 9.5×10^{-7} sec^{-1} till 3×10^2 sec^{-1}, with testing durations between 24h till a fraction of second (Rostrepo-Posada et al., 1994).

The results, reproduced in Figure 4a, indicate a very important increasing of yield stress with an increase of strain-rate, especially for strain-rate greater than 10^{-1}sec^{-1}, which corresponds with the range of strong earthquakes values. The increase of ultimate tensile strength is moderate, the influence of strain-rate being less important. Consequently, the yield ratio, defined by:

$$\rho_y = \frac{f_y}{f_u} \quad (5)$$

where f_y and f_u are the yield stress and tensile strength, respectively, increase as far as the strain-rate increases, with the tendency to reach the value 1 (Figure 4b). So, a reduction of material ductility occurs, especially for strain-rate greater than 10^{-1} sec^{-1}. More recent results (Wright &Hall, 1964, Kaneta et al., 1986, Soroushian &Choi, 1987, Fujimoto et al., 1988, Kassar &Yu, 1992, Obata et al., 1996) have experimental confirmed the previous results of Manjoine.

Different constitutive laws modelling the influence of strain-rate $\dot{\varepsilon}$ in some range of validity are proposed:

-Wright & Hall (1964), $10^{-6} < \dot{\varepsilon} < 10^3$:

$$\frac{f_{ys}}{f_y} = 1 + 2.77 \exp[0.162\left(\log \varepsilon - 3.74\right)] \quad (6a)$$

-Rao et al., (1966), $0 < \dot{\varepsilon} < 1.4 \times 10^0$:

$$\frac{f_{ys}}{f_y} = 1 + 0.021(\varepsilon)^{0.26} ; \text{ (A36)} \quad (6b)$$

-Soroushian & Choi (1987): $10^{-4} < \dot{\varepsilon} < 10^1$

$$\frac{f_{ys}}{f_y} = 1.46 - 4.51 \times 10^{-7} f_y + \left(0.0927 - 9.20 \times 10^{-7} f_y\right)\log\varepsilon$$

$$(6c)$$

$$\frac{f_{us}}{f_u} = 1.15 - 7.71 \times 10^{-7} f_y + \left(0.0497 - 2.44 \times 10^{-7} f_y\right)\log\varepsilon$$

-Wallace &Krawinkler (1989), $10^{-5} < \dot{\varepsilon} < 10^{-1}$:

$$\frac{f_{ys}}{f_y} = 0.973 + 0.45(\varepsilon)^{0.53} \quad (6d)$$

-Kassar & Yu (1992), $10^{-4} < \dot{\varepsilon} < 10^0$:

$$\frac{f_{ys}}{f_y} = 1.289 + 0.109\log\varepsilon + 0.009(\log\varepsilon)^2 \quad f_y = 320 \text{N/mm}^2;$$

$$\frac{f_{ys}}{f_y} = 1.104 + 0.3021\log\varepsilon + 0.0021(\log\varepsilon)^2 \quad f_y = 495 \text{ N/mm}^2$$

$$(6e)$$

-Kaneko et al. (1996): $10^{-4} < \dot{\varepsilon} < 10^1$

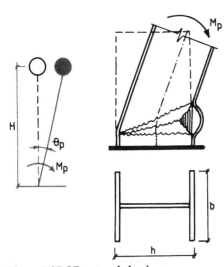

Figure 2 SDOF system behaviour.

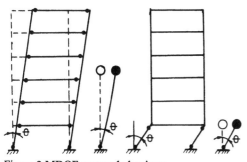

Figure 3 MDOF system behaviour.

$$\frac{f_{ys}}{f_y} = 1 + \frac{21}{f_y} \log \frac{\dot\varepsilon}{\varepsilon_0} \qquad \dot\varepsilon_0 = 10^{-4}/\sec \quad \text{(MPa)}$$

$$\frac{f_{us}}{f_u} = 1 + \frac{7.4}{f_y} \log \frac{\dot\varepsilon}{\varepsilon_0} \qquad \dot\varepsilon_0 = 10^{-4}/\sec \quad \text{(MPa)} \tag{6f}$$

In these equations the f_{ys}, f_{us}, represent the values of yield stress and tensile strength, respectively, considering the effect of strain-rate. The comparison of these different laws is presented in Figure 5. One can see that only (6a) and (6c) and (6f) cover the field of near-source earthquakes.

The influence of welding on the material behaviour was one of the main cause of brittle fracture of connections during the last important seismic events. As a consequence, many research works have been devoted to this purpose. The influence of strain-rate for unwelded and welded specimens are tested by Kaneta et al. (1986) and Kohzu &Suita (1996). Under the static loading the influence of welding is not particularly significant. On the contrary it is very important in the case of strain-rate. A remarkable increasing of yield stress of butt welded joints of about 10-20 percent is observed in comparison with

the parent steel. At the same time, the ultimate strength is not sensitive to strain-rate. So, the welding influence magnifies the sensitivity to the strain-rate effect.

The temperature has also a great influence on the effect of the strain-rate, the decreasing of temperature increasing the yield stress. The nil ductility temperature (NDT) is defined as temperature for which the ductility is zero (Akiyama, 1999). The fracture-transition plastic temperature (FTP) is the temperature limit to eliminate brittle fracture in the plastic range. The influence of strain-rate is to increase the FTP temperature, (Figure 6), and and the danger of biitle fracture (Kurobane et al., 1997). The increasing of yield stress and tensile strengths for temperature less than room temperature ($+20^0$) is studied by Beg et al. (2000).

The influence of strain-rate on the yield ratio, given by the relation (5), may be determined from the following proposals:

-Souroushian & Choi (1987):

$$\frac{\rho_{ys}}{\rho_y} = \frac{1.46 + 0.0925 \log \dot\varepsilon}{1.15 + 0.00496 \log \dot\varepsilon} \quad \text{(Fe 360)} \tag{7a}$$

-Kaneko et al. (1986):

$$\frac{\rho_{ys}}{\rho_y} = \frac{1 + \dfrac{21}{f_y} \log \dfrac{\dot\varepsilon}{\dot\varepsilon_0}}{1 + \dfrac{7.4}{f_y} \log \dfrac{\dot\varepsilon}{\dot\varepsilon_0}} \quad \text{(MPa)} \tag{7b}$$

A comparison of these two relationships is presented in Figure 7. For the case of welded section and in the

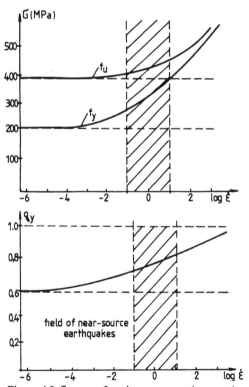

Figure 4 Influence of strain-rate on steel properties.

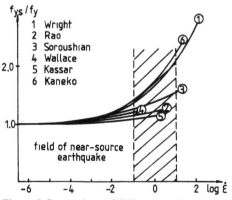

Figure 5 Comparison of different strain-rate laws.

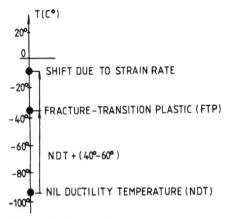

Figure 6 Influence of low temperature.

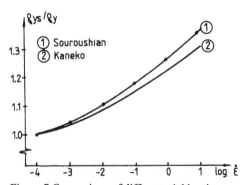

Figure 7 Comparison of different yield ratios.

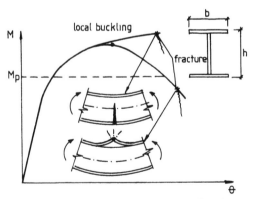

Figure 8 Crack failure due to the effect of strain-rate.

condition of low temperature, the values determined from equations (7) must be corrected:

$$\rho_{ys}^r = c_w c_T \rho_{ys} \qquad (8)$$

where c_w and c_T considers the effect of welding and temperature, respectively, by increasing the yield stress ($c_w \approx 1.1...1.2$, $c_T \approx 1.1...1.2$).

4 EFFECT OF STRAIN-RATE FOR THE MONOTONIC LOADS

During experimental test under impulsive loading some cracks may be noticed in tensile zones or along the boundary lines of buckled areas (Figure 8), due to the reaching of ultimate strains. The high strain-rates of actions accentuate this behaviour, in many cases the ductile collapse being replaced by brittle fracture.

Considering the standard beam presented in Figure 9, the length of the plastic zone can be determined from:

$$L_p = \left(\frac{M_u}{M_p} - 1 \right) L \approx \left(\frac{1}{\rho_y} - 1 \right) L \qquad (9)$$

where M_u is the ultimate moment determined from a simplified relation $M_u \approx (f_u / f_y) M_p$, M_p being the plastic moment. The curvature of plastic zone is given by:

$$\chi = \frac{\theta}{L_p} = \frac{1}{1/\rho_y - 1} \frac{\theta}{L} \qquad (10)$$

and taking into account that:

$$\varepsilon = \frac{h}{2} \chi = \frac{1}{2} \frac{1}{1/\rho_y - 1} \frac{h}{L} \theta \qquad (11)$$

results for the ultimate rotation which produced the tension fracture:

$$\theta_f = 2 \left(\frac{1}{\rho_y} - 1 \right) \frac{L}{h} \varepsilon_u \qquad (12)$$

ε_u being the ultimate strain.

In the case of crack of buckled flange (Figure 10), the fracture rotation of buckled flange is:

$$\alpha_f = 2 \left(\frac{1}{1/\rho_u - 1} \right) \frac{L_f}{t_f} \varepsilon_u \qquad (13)$$

where L_f is (Figure 10):

$$L_f = \left(\frac{1}{1/\rho_y - 1} \right) b \qquad (14)$$

23

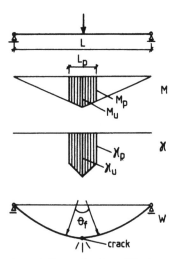

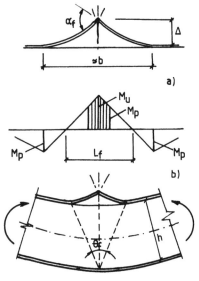

Figure 9 Determination of fracture rotation.

Figure 10 Effect of compression flange crack.

considering that the length of plastic hinge is equal with the flange depth. The fracture rotation of beam (Figure 10c) results, after some geometrical considerations:

$$\theta_f = \frac{L_f^3}{t_f^2 h}\left(\frac{1}{\rho_y}-1\right)^2 \varepsilon_u^2 \qquad (15)$$

One can see that the fracture rotation depends on cross-section characteristics, yield ratio and steel quality. Figure 11a shows the influence of yield ratio

on member rotation capacity. For a determined yield ratio, the ductile rotation capacity is changed in a fracture collapse. Figure 11b shows the influence of strain-rate on the rotation capacity. One can see that in the field of strong earthquakes high strain-rate it is possible the occurring of fracture collapse. So a nilductility can occur due to high velocity.

5 HIGH VELOCITY CYCLE LOADING

During both strong earthquake in near-source areas and cyclic experimental tests of members or joints performed with high velocity (Suita et al., 1996) the same cracks as the ones from Figure 8 occurs after some reduced number of cycles.

Some research works have classified these failures as the effects of low-cycle fatigue. But, due to reduced number of cycles producing high plastic deformations, the classifying these fractures, as belonging to the category of fatigue-failure is questionable. The cycles with high plastic deformations cause an accumulation of these deformations along the yield lines, inducing cracks or rupture in the buckled plates, rather than a reduction of the material strength.

The behaviour of I-section has a particular feature during the cyclic loading (Figure 12). The first semi-cycle, producing buckling of compression flange, rotates the section around a point located in, or near,

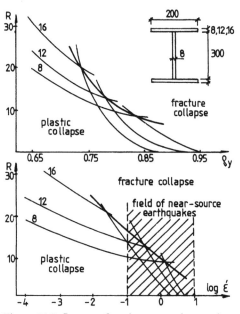

Figure 11 Influence of strain-rate on the member rotation.

the opposite flange. As a result, the tension forces are very small. For the reversal semi-cycle, the compression flange buckles also. But due to the reduced tension forces, this opposite rotation is not able to straighten the buckled flange. Therefore, during the next cycle, the section works as having initial geometrical imperfections resulted from the previous cycle. In this manner, after each cycle a new deformation is superimposed over the previous one. For the web, a continuous reducing in the length of yield lines occurs, due to the presence of

the deformed shape produced by previous cycle. So, after 3-4 complete cycles, the plastic mechanism is composed by two shapes, each one for the corresponding flange (Gioncu et al., 2000b). Considering the value of rotation which produces the flange fracture, it is possible to determine the number of cycles producing the flange rupture.

Figure 13 shows the reduction of this number when the strain-rate increases. In the case of high velocity just the first cycle may produce the fracture collapse. A reduction of detrimental effects of strain-rate is due to the increasing of local temperature of plastic hinges during the cycle rotations (Nakashima et al, 1998). The increase of yield ratio given in previous sections is determined for monotonic loads. So, in the case of cyclic loads it may be expected a reduced rise of yield ratio, corresponding to an increasing of temperature with 20^0–25^0 C.

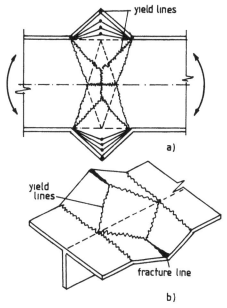

Figure 12 Plastic mechanism under cycling loading.

6 CONCLUSIONS

During the ground motions recorded in the near-source areas very high values of horizontal and vertical velocities were noticed. Due to this fact, very important strain-rates are induced in structures, exceeding the usual values considered as normal limit for strain-rate. The main effect of strain-rate is the increasing the values of yield ratio. So, for some values which correspond very well with the ones of field of near-source earthquakes, the ductile plastic deformations can be shifted in a brittle fracture. In the same time, in the case of high velocity cycling loading, the number of cycles, which produces the fracture collapse is substantially reduced. So, the effect of strain-rate must be considered in the design of steel structures in the near-source areas.

7 REFERENCES

Akiyama, H. 1996. Damage of structures in the Hyogoken-Nanbu earthquake. In " *La citta sicurre: Terremoti, eruzioni e protezioni civile:, Napoli, 10-13 Febbraio, 1996.*

Akiyama, H. 1999. Evaluation of fractural mode of failure in steel structures following Kobe lessons. Key Lecture. In *Stability and Ductility of Steel Structures,* Eds. D.Dubina & M. Ivanyi, *Timisoara, 9-11 Sptember 1999.*

Beg, D., C Remec, & A. Plumier 2000. Influence of strain-rate. In Moment Resistant Connections of Steel Building Frames in Seismic Areas, Ed. F.M. Mazzolani. London: E&FN Spon.

Fujimoto, M., T. Naruba, & S. Sasaki 1988. Strength and deformation capacity of steel brace under high-speed loading. In *9th World Conference on Earthquake Engineering, Tokyo-Kyoto, 2-9 August 1988*, Vol. IV, 139-144.

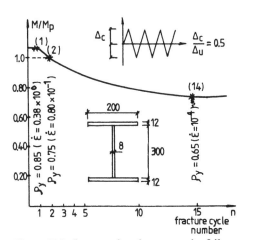

Figure 13 Influence of strain-rate on the failure cycle number.

Gioncu, V. 1995. Local and global ductility interaction in seismic design of MR frames. In *Steel Structures*, EUROSTEEL'95, Ed. A Kounadis, *Athens, 18-20 May 1995*, Rotterdam: Balkema.

Gioncu, V., G. Mateescu, L.Tirca, A. Anastasiadis. 2000a. Influence of the type of seismic ground motions. In "Moment Resistant Connections of Steel Building Frames in Seismic Areas". Ed. F.M. Mazzolani. London: E &FN Spon.

Gioncu, V., G. Mateescu, D. Petcu, A. Anastasiadis. 2000b. Prediction of available ductility by means of local plastic mechanism method: DUCTROT computer program. In "Moment Resistant Connections of Steel Building Frames in Seismic Areas". Ed. F.M. Mazzolani. London: E &FN Spon.

Hall, J.F. T.H. Heaton, & M. W. Halling, J.D. Wald 1995. Near-source ground motion and its effect on flexible buildings. *Earthquake Spectra*, Vol. 11, No. 4, 569-605.

Kaneko, H. 1997. Influnce of strain-rate on yield ratio. In *Kobe Earthquake Damage to Steel Moment Connections and Suggested Improvement*. Ed. B.Kato, JSSC Technical Report No 39/96.

Kaneta, K., I. Kohzu, I. & K. Fujimura 1986. On the strength and ductility of steel structural joints subjected to high speed monotonic tensile loading. In *8th European Conference on Earthquake Engineering*, Lisbon, 12-17 September, 1986.

Kassar, M. & W.W. Yu 1992. Effect of strain-rate on material properties on sheet steels. *Journal of Structural Engineering*, Vol. 118, Np. 11, 3136-3150.

Kohzu I., & Suita K 1996. Single or few excursions failure of steel structural joints due to impulsive shocks. In the 1995 Hyogoken-Nanbu earthquake. In 11th *World Conference on Earthquake Engineering*. Acapulco, 26-28 June 1996 CD-ROM. Paper 412.

Kurobane, Y., K. Azuma, & K. Ogawa 1997. Brittle fracture in steel building frames. Comparative study of Northridge and Kobe earthquake damage. International Institute of Welding , Annual Assembly, *San Francisco, 13-18 July*, 1997, 1-30.

Manjoine M.J. 1944. Influence of strain-rate and temperature on yield stress of mild steel. *Journal of Applied Mechanics*, Vol. 66, No. 11, 211-218.

Midorikawa, S. 1995. Ground motion intensity in epicentral area. In *New Direction in Seismic Areas.Tokyo, 9-10 October 1995*, 247-250.

Nakashima, M., K. Suita, & K. Morisako, Y. Marioka. 1998. Tests of welded beam-column subassemblies I. Global behaviour. II Detailed behaviour. *Journal of Structural Engineering*, Vol. 124, No. 11, 1236-1244, 1245-1252.

Obata, M., Y. Goto, & S. Matsura, H. Fujiwara. 1996. Ultimate behaviour of tie plates at high-speed tension. *Journal of Structural Engineering*, Vol. 122, No. 4, 416-422.

Rao, N.R.N., M. Lohramann, & L. Tall. 1966. Effect of strain-rate on the yield stress of structural steels. *ASTM Journal of Materials*, Vol. 1, No.1.

Restrepo-Posada J.I., L.L. Dodd, & N.Cooke 1994. Variable affecting cyclic behaviour of reinforcing steel. *Journal of Structural Engineering*, Vol. 120, No. 11, 3178-3196.

Soroushian, P. & K.B. Choi 1987. Steel mechanical properties at different strain rates. *Journal of Structural Engineering* Vol. 113, No. 4, 863-872.

Suita, K., I. Kohzu, & Yasutomi. 1996. The effect of strain-rate in restoring force characteristics of steel braced frames under high-speed cycling loadings. In 11th *World Conference on Earthquake Engineering*. Acapulco, 26-28 June 1996, CD-ROM. Paper 1220.

Uang C.M, & D.M. Bordad. 1996. Dynamic testing of full-scale steel moment connections. In 11th *World Conference on Earthquake Engineering*. Acapulco, 26-28 June 1996, CD-ROM. Paper 407.

Wallace, B.J., & H. Krawinkler. 1989. Small-scale model tests of structural steel assemblies. *Journal of Structural Engineering*, Vol. 115, No. 8, 1999-2015.

Wright, R. N., & W.J. Hall. 1964. Loading rate effects in structural steel design. Journal of Structural Division, Vol. 90, No. 55, 11-37.

Behaviour of Steel Structures in Seismic Areas, Mazzolani & Tremblay (eds) © 2000 Balkema, Rotterdam, ISBN 90 5809 130 9

Modeling of cracking and local buckling in steel frames by using plastic hinges with damage

P. Inglessis, S. Medina, A. López & J. Flórez-López
Facultad de Ingeniería, Universidad de Los Andes, Mérida, Venezuela

ABSTRACT: A model of the process of cracking and local buckling in steel structural elements is presented. The model is based on the concept of plastic hinge and the methods of continuum damage mechanics. It is assumed that all these degrading phenomena can be lumped at the hinges. This new kind of hinge is characterized by two state variables: the plastic rotation and the damage. Based on this damage model, a new finite element is formulated. An elastic beam-column and two inelastic hinges at the ends constitute the element. The stiffness matrix, the yield functions and the damage laws of both hinges define the new finite element. In order to verify the model, several small-scale frames were tested in the laboratory under monotonic loading. A lateral load at the top was applied in a stroke-controlled mode until the ultimate capacity of the frame was reached. These tests were simulated with the damage model and the comparison between model and tests is presented and discussed.

1 INTRODUCTION.

Conventionally, inelastic analysis of steel frames is done on the basis of plasticity theory. However, in the last decades a new and powerful tool for the study of solids and structures has been developed: continuum damage mechanics (CDM). This theory is based on the introduction of a new internal variable: the damage. This variable measures the density of microcracks and microvoids in a continuum. Isotropic damage can be characterized by a scalar quantity that takes values between zero and one. Zero corresponds to an intact or non-damaged material; one represents a totally damaged volume element or perhaps a macrocrack initiation. A general presentation of CDM can be seen in (Lemaitre 1992).

In some previously published papers (Flórez-López, 1995 Cipollina et al. 1995, Inglessis et al. 1999), a theory that combines CDM with the concept of plastic hinge has been proposed. This theory has been called Lumped Damage Mechanics (LDM) in order to underline the difference with conventional damage mechanics.

The main purpose of this paper is the proposition of a simplified damage model for planar steel frames. The new model was developed within the framework of LDM. This model was included in the library of finite elements of a commercial structural analysis program and allows the numerical simulation of the process of damage and, possible collapse of steel frames.

2 EXPERIMENTAL RESULTS OBTAINED IN STEEL MEMBERS SUBJECTED TO BENDING

An experimental program, in which some specimens representing steel frame members were subjected to bending, was carried out at the Materials and Test Laboratory of the Los Andes University (Gómez 1996). Specimens of different length, area and cross-section were supported by an enlarged end block simulating a rigid column and loaded at the tip as cantilever beams. The loading of the test included elastic unloadings and reloadings as in conventional damage identification tests of CDM, such as are represented en Figure 1.

Figure 2 shows the results obtained in a test with a tube of circular section (38 mm exterior radius, 3 mm thick and 665 mm long).

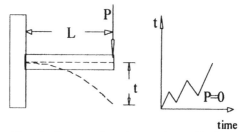

Figure 1. Test on steel members: specimen and loading.

The behavior of the tube can be separated in three stages: a zone of quasilinear response followed by a phase of plastic hardening that seems to stabilize and then a softening period, that presents a behavior that could be represented by a straight line

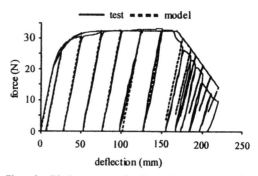

Figure 2. Displacement at the tip vs force and numerical simulation, tube of circular section.

3 SIMPLIFIED MODELS OF DAMAGE FOR STEEL FRAME MEMBERS.

The purpose of this section is to extend the conventional lumped plasticity models for frames to include concepts of damage mechanics.

3.1 Generalized stresses and deformations for a frame member

A constitutive model relates strains or deformations and stresses. In the case of a frame member, generalized deformations $\{\Phi\}^t = (\Phi_i, \Phi_j, \delta)$ and stresses $\{M\}^t = (M_i, M_j, N)$ can be defined as indicated in Figure 3. The former variable is then the equivalent of the strain tensor in continuum mechanics and the latter would correspond to the Cauchy stress tensor for the case of frame members.

In an elastic member, the constitutive model can be written in matrix from as follows:

$$\{M\} = [S^0]\{\Phi\} \text{ or } \{\Phi\} = [F^0]\{M\} \tag{1}$$

where $[S^0]$ and $[F^0]$ are, respectively, the stiffness and flexibility matrix of the member.

In order to include inelastic effects, the member is assumed to be the assemblage of an elastic beam-column and two inelastic hinges as indicated in Figure 4. This is a conventional hypothesis in the plastic analysis of frames. Generalized deformations can now be split into two terms, beam or elastic deformations and hinge deformations $\{\Phi^h\}$:

$$\{\Phi\} = [F^0]\{M\} + \{\Phi^h\} \tag{2}$$

We assume that hinge deformations result from plastic deformations $\{\Phi^p\} = (\Phi^p_i, \Phi^p_j, \delta^p)$, as defined in conventional plastic theories for frames and an additional damage related term $\{\Phi^d\}$:

$$\{\Phi^h\} = \{\Phi^p\} + \{\Phi^d\} \tag{3}$$

where, the latter variable will be defined by the use of CDM concepts as described in the next section.

3.2 Elasticity law of a frame member with plastic-damage hinges

In presence of flexural effects we postulate the existence of a set $\{D\} = (d_i, d_j, d_n)$, taking values between zero and one, so that the damaged induced deforma-

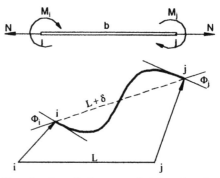

Figure 3. Generalized stress and deformations in a frame member.

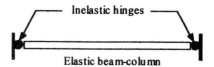

Figure 4. Lumped inelasticity model of a frame member.

tions of the hinges can be defined (Flórez-López 1995) as:

$$\{\Phi^d\} = [C_{(D)}]\{M\} \tag{4}$$

where $[C_{(D)}]$ is a diagonal matrix, whose non-negative terms are: $C_{11} = d_i/[(1 - d_i)S^0_{11}]$; $C_{22} = d_j/[(1 - d_j)S^0_{22}]$ and $C_{33} = d_n/[(1 - d_n)S^0_{33}]$

It must be underlined that these variables are a global measure of damage without any distinction as to their physical causes.

Taking into account the equations 2, 3 and 4, we obtain the state law for a damaged elastoplastic frame member:

$$\{\Phi - \Phi^p\} = [F_{(D)}]\{M\} \text{ or } \{M\} = [S_{(D)}]\{\Phi - \Phi^p\} \qquad (5)$$

where $[F_{(D)}] = [F^0] + [C_{(D)}]$ and $[S_{(D)}] = [F_{(D)}]^{-1}$. Matrices $[F_{(D)}]$ and $[S_{(D)}]$ are respectively the flexibility and stiffness matrices of a damaged member.

The elasticity law does not describe the relationship between generalized stresses and deformations since it includes two internal variables: plastic deformations and damage. Two additional equations are then needed; they will be called Internal Variable Evolution Laws.

3.3 Time-independent evolution laws

Let us consider again the test of Figure 2. The lumped inelasticity model of this test is presented in Figure 5.

Taking into account the following conditions (t is the deflection at the free end of the member, t^p is the permanent deflection and P the force):

$$M_i = PL \ ; \ M_j = 0 \ ; \ \Phi_i = t/L \ ; \ \Phi_i^p = t^p/L$$
$$d_i = d \ ; \ d_j = 0 \qquad (6)$$

and the elasticity law (5), we can write the relationship between the force and the displacement as follows:

$$P = Z_{(d)}(t - t^p) \ ; \ Z_{(d)}(1-d)Z^0 \ ; \ Z^0 = 3EI/L^3 \qquad (7)$$

The term Z(d) can be interpreted as the slope of the elastic unloading in the tests of figure 2, the plastic deformation and the damage of the hinge can be expressed as:

$$\Phi^p = t^p/L \ ; \ d = 1 - Z/Z^0 \qquad (8)$$

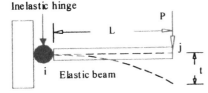

Figure 5. Lumped inelasticity model of the test.

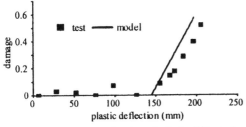

Figure 6. Damage as a function of plastic deflection in a hinge

It is now possible to construct the damage curve as a function of plastic deflections of Figure 6. It can be noticed that a straight line can represent damage evolution in the hinge, as indicated in the same figure.

At the beginning of damage evolution, the slope of the force-deflection graph of figure 2 becomes negative and the softening phase of the behavior starts. Physically this point coincides with the beginning of the local buckling in the member.

We then propose the following constitutive equations for a steel frame member with plastic-damage hinges.

Elasticity law:

$$\{M\} = [S_{(D)}]\{\Phi - \Phi^p\} \qquad (9)$$

Yield functions:

$$f_i = |M_i/(1-d_i) - X_i| - M_e$$
$$f_j = |M_j/(1-d_j) - X_j| - M_e \qquad (10)$$

This expression of the yield function can be justified on the basis of LDM concepts (see Inglessis et al. 1999)

Plastic deformation evolution laws:

$$d\Phi_i^p = 0 \text{ if } f_i < 0 \text{ or } df_i < 0$$
$$d\Phi_i^p \neq 0 \text{ if } f_i = 0 \text{ or } df_i = 0$$
$$d\Phi_j^p = 0 \text{ if } f_j < 0 \text{ or } df_j < 0 \qquad (11)$$
$$d\Phi_j^p \neq 0 \text{ if } f_j = 0 \text{ or } df_j = 0$$

Damage evolution laws:

$$d_i = c\langle p_i - p_{cr}\rangle \ ; \ d_j = c\langle p_j - p_{cr}\rangle \qquad (12)$$

where p_{cr} and c are member dependent constants, and p is the cumulated generalized plastic deformations:

$$dp_i = |d\Phi_i^p| \ ; \ dp_j = |d\Phi_j^p| \qquad (13)$$

X represents the kinematic hardening term. A nonlinear kinematic hardening is assumed expressed by the following evolution law:

$$dX_i = \alpha(X_\infty d\Phi_i^p - Xdp_i); X_i = 0 \text{ for } p_i = 0$$
$$dX_j = \alpha(X_\infty d\Phi_j^p - Xdp_j); X_j = 0 \text{ for } p_j = 0 \qquad (14)$$

Constants M_e, X_∞ and α are member dependent parameters. M_e is the moment needed to initiate plasticity in the cross-section.

During a monotonic loading, the equation (14) characterizes an exponential evolution of X as a function of the cumulated plastic deformation, which has the following expression:

$$X = X_\infty \left(1 - e^{-\alpha p}\right) \qquad (15)$$

that is, X tends to stabilize to the limit value X_∞ when the cumulated plastic deformation tends to infinite. Therefore, there is the following relationship between the yield moment M_y (moment needed for the entire plastification of the cross-section) and the constants of the model:

$$M_y = M_e + X_\infty \qquad (16)$$

A simulation of the test with the model is superimposed on the same Figure 2, with the following parameters c = 6.9, p_{cr} = 0.22, EI = 1.363×10^7 mm^2-N, M_e = 13300 mm-N, X_∞ = 7700 mm-N, α = 0.00135.

4 FORMULATION OF A FINITE ELEMENT

Figure 7 describes the generalized displacements {q} and internal forces {Q} in a frame member. The relationship between generalized deformations and member displacement can be obtained by simple geometrical considerations. In the general case, including nonlinear geometrical effects, (see figure 8) we have:

$$\Phi_i = q_3 - (\alpha_0 - \alpha_{(q)}) \; ; \; \Phi_j = q_6 - (\alpha_0 - \alpha_{(q)})$$
$$\delta = L_{(q)} - L_0 \qquad (17)$$

where

$$\alpha_{(q)} = \tan^{-1}\left((\Delta Y_0 + q_5 - q_2)/(\Delta X_0 + q_4 - q_1)\right)$$
$$L_{(q)} = \sqrt{(\Delta Y_0 + q_5 - q_2)^2 + (\Delta X_0 + q_4 - q_1)^2} \qquad (18)$$

the terms with the index 0 represent quantities in the original configuration.

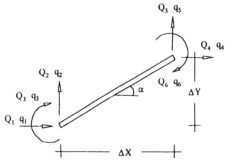

Figure 7. Generalized displacements and forces

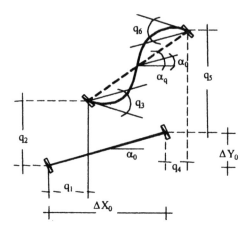

Figure 8. Physical configurations of member

The relationship between generalized stresses and internal forces can be obtained by considering the equilibrium of the member.

In the general case, we obtain:

$$Q_1 = (M_i + M_j)\left(\sin \alpha_{(q)}/L_{(q)}\right) - N \cos \alpha_{(q)}$$

$$Q_2 = -(M_i + M_j)\left(\cos \alpha_{(q)}/L_{(q)}\right) - N \sin \alpha_{(q)}$$

$$Q_3 = M_i \qquad (19)$$

$$Q_4 = -(M_i + M_j)\left(\sin \alpha_{(q)}/L_{(q)}\right) + N \cos \alpha_{(q)}$$

$$Q_5 = (M_i + M_j)\left(\cos \alpha_{(q)}/L_{(q)}\right) + N \sin \alpha_{(q)}$$

$$Q_6 = M_j$$

Equations (18, 19) and the model, represented by (9, 10, 11, 12, 13, 14, 15, 16), constitute a set of equations that define a conventional finite element. This finite element has been implemented in a commercial FE program that allows nonlinear analysis (ABAQUS 1998).

5. VERIFICATION OF THE MODEL

5.1 Experimental Model of Planar Frame

The frame tested for model verification is presented in figure 9. The specimen consists of a steel frame of two levels and two spans. The elements have rectangular hollow cross section and are

Table 1. Nominal characterics of mem.tested frame			
	H (cm)	B (cm)	e (mm)
frame	4.10	2.40	2.5
1	sect. (cm^2)	Ix (cm^4)	Iy (cm^4)
	3.00	6.3970	2.6655

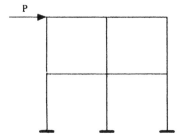

P

Figure 9.Tested frame

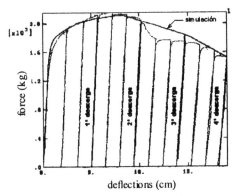

Figure 10. Test and simulation on a planar frame.

Table 3. Nominal charact. of steel members				
element	H (cm)	B (cm)	e (mm)	sect. (cm^2)
beam 1	4.10	2.40	2.50	3.00
beam 2	4.10	2.40	2.50	3.00
colum 1	4.10	2.40	2.50	3.00
colum 2	4.10	2.40	2.50	3.00
element	Ix (cm^4)	Iy (cm^4)	long (cm)	
beam 1	6.3970	2.6655	30.00	
beam 2	6.3970	2.6655	30.00	
colum 1	6.3970	2.6655	20.00	
colum 2	6.3970	2.6655	20.00	

Table 4. Parameters model			
test	M_u (kg-cm)	c	p_{cr}
beam 1	6350.00	1.30	0.210
beam 2	7300.00	1.60	0.170
colum 1	5950.00	1.30	0.200
colum 2	7800.00	1.60	0.190
beams	6825.00	1.45	0.190
columns	6875.00	1.45	0.195
test	X_- (kg-cm)	a	
beam 1	1400.00	18.00	
beam 2	2300.00	21.00	
colum 1	1250.00	18.50	
colum 2	2000.00	20.00	
beams	1850.00	19.50	
columns	1625.00	19.25	

Table 2. Frame dimensions		
	L_1 (cm)	L_2 (cm)
frame 1	30.00	30.00
	L_3 (cm)	L_4 (cm)
	20.00	20.00

welded at the joints. Nominal characteristics and dimensions of the frame and its members are shown in tables 1 and 2 respectively. The experimental results of one of the monotonic tests are shown in figure 10

5.2 Identification of model's parameters

Nominal characteristics of steel members (see Table 3) and the parameters of the model were identified by testing a single element of the frame (see Figure 1). The parameters are presented in Table 4.

5.3 Numerical simulation

Figure 10 shows the comparison between model and experimental results. For the sake of clarity, only four of the elastic unloadings are represented in the simulation. The two curves show an excellent fit.. Changes in the tangent slopes that are the conse-

quence of damage appearing in the inelastic hinges can be seen in the graph of the simulation.

Figure 11 indicates the state of damage at four different phases of the simulation. The figures beside the hinges represent the damage values at the end of the four elastic unloading indicated in Figure 10. The first distribution presents 6 plastic hinges with no damage. The remaining figures show the increase in the values of damage accompanying plastic deformation. The model is very powerful because not only global behavior but also local buckling is represented. The maps of damage distribution can be used to obtain the residual capacity for seismic analysis.

6 CONCLUSIONS AND FINAL REMARKS.

A model for steel frame members that generalizes the concept of the plastic hinge has been proposed. The model includes concepts of CDM and allows the representation of the behavior of members under monotonic loadings. Specifically, the model takes into account damage-softening behavior and the loss of stiffness and strength due to damage in the inelastic hinges.

This model permitted the development of a finite element for steel frame members that was included in a nonlinear finite element program. This element allows the description of the general behavior of a frame under severe overloads.

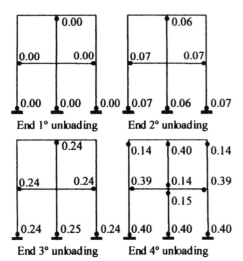

End 1° unloading   End 2° unloading

End 3° unloading   End 4° unloading

Figure 11. Damage distribution.

An effort was made to keep the models as simple as possible but still correctly represent the behavior of the structure. We believe that only simple but effective models can gain some acceptance among practitioners.

This model can be extended to represent other effects like hysteretic loading, axial force and biaxial flexure.

REFERENCES

Abaqus User's manual, 1998 Hibbit, Karlsson & Sorensen, Inc.

Cipollina A, López-Inojosa A, & Flórez-lópez J. A 1995 simplified damage mechanics approach to nonlinear analysis of frames. Comput Struct;54(6):1113-26

Flórez-López J. 1995 Unilateral model of damage for RC frames. J Struc Eng ASCE; 121(12):1765-72

Gómez G. 1996 Un modelo simplificado de daño para estructuras de acero [in Spanish]. Thesis for partial fulfilment of the requirements for the degree of Master Scientie in Structural Engineering. Mérida, Venezuela: University of Los Andes

Inglessis P, Gómez G, Quintero G & Flórez-lópez J. 1999 Model of damage for steel frame members. Engineering Structures 21 954-964

Lemaitre J. 1992 A course on damage mechanics. Berlin: Springer-Verlag,

Behaviour of Steel Structures in Seismic Areas, Mazzolani & Tremblay (eds) © 2000 Balkema, Rotterdam, ISBN 90 5809 130 9

Buckling-restrained braces as hysteretic dampers

Mamoru Iwata & Takashi Kato
Department of Architecture and Building Engineering, Kanagawa University, Yokohama, Japan

Akira Wada
Structural Engineering Research Center, Tokyo Institute of Technology, Yokohama, Japan

ABSTRACT: Braces are effective as seismic members of building structures and are widely used in low- to high-rise buildings. However, conventional braces sustain buckling under a compressive force. To solve these shortcomings surrounding conventional braced structures, a number of studies have been reported on the use of buckling-restrained braces. In this study, first the said four types are designed on the condition that braces will not buckle at least by the frame deformation of 1/100 story deformation angle. Then, the hysteresis characteristics, final fracture characteristics and cumulative absorbed energy are investigated.

1 INTRODUCTION

The buckling-restrained brace is a brace whose core plate is covered with a restraining part to prevent buckling. An unbonded material or a clearance is provided between the core plate and restraining part so that the axial force borne by the core plate is not transmitted to the restraining part (Figure 1). Theoretically, a buckling-restrained brace does not buckle if the maximum compressive force arising in the core plate remains smaller than the Euler's buckling load of the restraining part. The buckling-restrained brace produces equal yield strengths against both tensile and compressive forces and ensures a stable hysteresis.

A design method that is becoming increasingly popular in Japan keeps columns and beams of buildings in elastic regions as much as possible by using buckling-restrained braces as hysteretic dampers and allowing the hysteretic absorbed energy to produce a damping effect. According to this design method, buckling-restrained braces plasticize to provide hysteresis damping when medium earthquake occurs, thereby keeping columns and beams in elastic regions when large earthquake occurs.

This significantly reduces the plastic strain of connections of columns and beams, and keeps them undamaged, even when shaken by incredible large earthquakes. This design method has several economic advantages over conventional methods which include weight saving throughout the whole structure of a building, confining the need for after-earthquake inspection, repair and replacement to buckling-restrained braces, and permitting continued use of whole building (Wada 1998).

2 PURPOSE OF THE STUDY

The ductility factor and cumulative ductility factor of buckling-restrained braces used as hysteretic dampers reach 10 to 20 and 40 to 100 in large earthquakes. The average strain rate becomes two to three times greater than that of conventional structural members. Thus they receive fairly rigorous input energies. In incredible large earthquakes, the above values could be more than doubled.

The study aims at evaluating the critical-state performance of buckling-restrained braces used as hysteretic dampers. Various methods to impose buckling restraint on braces have been conceived and tested. However, such methods do not allow direct comparison because sizes and types of specimens and testing methods employed are not the same. In this study, four commercially available buckling-restrained braces are tested and evaluated. To make equal the yield strength of braces, the sectional area and length of core plates are equalized. Restraining parts are designed to have an equal geometrical moment of inertia that is used as a variable in imposing

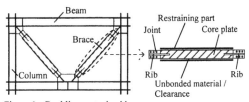

Figure 1. Buckling-restrained brace

buckling restraint. The four different buckling-restrained braces are prepared under the same conditions so that they cause no buckling and exhibit sufficient hysteresis even when they are subjected to a large earthquake in which the maximum ground motion of 50 cm/sec with a story deformation angle of 1/100. The hysteresis characteristics and final fracture characteristics of the individual buckling-restrained braces are determined by experiments and the cumulative absorbed energy by them is calculated.

3 EXPERIMENT PROGRAM

3.1 Test specimens

Four types of specimens, Types 1, 2, 3 and 4 are prepared. Types 1, 2 and 3 have a core plate PL-16 x 176, whereas Type 4 has a core plate BH-136 x 136 x 9 x 6. The core plates are designed to have substantially the same sectional area. The core plates are made of the SN400B steel for building structures. This steel has good weldability and impact properties, with the upper and lower limits of yield point defined. The yield stresses measured in the test are 262.6 N/mm² for Types 1, 2 and 3 and 289.1 N/mm² for Type 4. The restraining parts are made of the STKR400 and SS400 steels, with their geometrical moment of inertia being designed to become substantially equal. Both ends of the core plates protruding from the restraining parts are cruciform-shaped to prevent the occurrence of local buckling.

Tables 1, 2 and 3 show a list of specimens used, their theoretical yield strengths and their material properties.

3.1.1 Specimen Type 1
This specimen comprises a core plate covered with a rectangular hollow section that serves as a restraining part, with mortar filled in between the core plate and hollow section. An unbonded soft-rubber sheet is provided between the core plate and mortar (Figure 2(a)). The author designed and prepared this specimen based on the description in the reference (Fujimoto 1988).

3.1.2 Specimen Type 2
This specimen comprises a core plate that is covered with a rectangular hollow section alone. No other restraining part and material is used (Figure 2(b)). The author designed this specimen based on the description in the reference (Kamiya 1997) and prepared under the same conditions that were employed in the preparation of Specimen Type 1.

3.1.3 Specimen Type 3
This specimen comprises a core plate that is covered with a restraining part that is formed by joining to-

Table 1. Test specimens

Test Specimen	Core plate	
	Size (mm)	Sectional area (mm²)
Type 1 Type 2 Type 3	PL-16×176	2816
Type 4	BH-136×136×9×6	2748

Test Specimen	Restraining part		
	Size (mm)	Sectional area (mm²)	Geometrical moment of inertia ×10⁴(mm⁴)
Type 1	□-210×150×3.2	3366	864 Mortar(250)
Type 2	□-150×150×6	2263	1150
Type 3	2[-180×75×7×10.5 +2PL-160×6	7360	1171
Type 4	□-150×150×6	3456	1196

Table 2. Calculated strength

Test Specimen	Core plate		Restraining part	P_E / P_y
	Yield load P_y (kN)	Yield strain ε_y (%)	Buckling load P_E (kN)	
Type1	741.8	0.128	8929.8	12.0
Type2	741.8	0.128	9218.9	12.4
Type3	741.8	0.128	8841.6	11.9
Type4	791.7	0.140	9587.3	12.1

Table 3. Material properties & Chemical components of core plates

Test Specimen	Material properties			
	Yield stress (N/mm²)	Tensile strength (N/mm²)	Yield ratio (%)	Elongation (%)
Type 1 Type 2 Type 3	262.6	432.5	61	32
Type 4	289.1	451.3	64	29

Test Specimen	Chemical components (%)				
	C	Si	Mn	P	S
Type 1 Type 2 Type 3	0.15	0.09	0.90	0.011	0.002
Type 4	0.20	0.10	0.60	0.016	0.007

gether the channel and flat steels with high-strength bolts. An unbonded soft-rubber sheet is provided between the core plate and restraining part (Figure 2(c)). The author designed this specimen based on the description in the reference (Fukuda 1999) and prepared under the same conditions that were employed in the preparation of Specimen Type 1.

3.1.4 Specimen Type 4

This specimen comprises a core plate consisting, unlike the core plate of other specimens, of a built-up wide-flange beam that is covered with a rectangular hollow section serving as a restraining part. A clearance is left between the core plate and rectangular hollow section, with no unbonded material filled (Figure 2(d)). The author designed this specimen based on the description in the reference (Suzuki 1994) and prepared under the same conditions that were employed in the preparation of Specimen Type 1.

3.2. Loading Method

A loading test machine (an electrically powered hydraulic actuator) is used for the application of load. Figure 3 shows the test equipment used in the experiment. The test jig and specimen are carefully aligned so that the force applied from the actuator is in the axial direction of the brace without eccentricity. The lower part of the H-400 x 400 x 13 x 21 jig is pin-supported.

Now that the specimen is placed in the direction of 45 degrees, the axial force developing in the specimen is square root of 2 times greater than the horizontal force from the actuator.

The expansion and contraction of the specimen is one over square root of 2 greater than its horizontal deformation. The length of the specimen is square root of 2 times greater than the story height. Therefore, the axial strain arising in the specimen is one half the story deformation angle. However, the plastic strain occurring in the specimen becomes substantially equal to the story deformation angle. Because, the length of a portion of the specimen that undergoes plastic deformation is one half its overall length.

The strain of the specimen corresponding to a story deformation angle 1/200 in a medium earthquake (in which the maximum ground motion of 25 cm/sec) is 0.5 percent. The strain of the specimen corresponding to a story deformation angle 1/100 in a large earthquake (in which the maximum ground motion of 50 cm/sec) is 1.0 percent. It is considered that buckling-restrained braces must have a capacity to maintain a strain of up to 1.0 percent. And, then, tests are conducted with strains between 1.0 percent and 3.0 percent, which is equivalent to a story deformation angle 0.03 radian to determine critical-state performance.

Tests are conducted by applying increasing loads that are applied alternately in positive and negative directions (tension and compression). Control is exercised by varying load until the elastic region is reached and, then, by changing the axial deformation of the core plate beyond that region. One each load equal to 1/3 and 2/3 of the yield strain is applied before yielding. After yielding, tests are made with a 0.25 percent strain (once), 0.5 and 0.75 percent

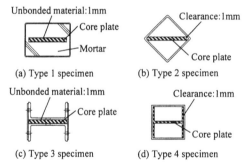

Unbonded material:1mm
Core plate
Mortar

(a) Type 1 specimen

Clearance:1mm
Core plate

(b) Type 2 specimen

Unbonded material:1mm
Core plate

(c) Type 3 specimen

Clearance:1mm
Core plate

(d) Type 4 specimen

Figure 2. Section of specimens

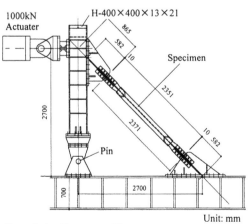

1000kN Actuater
H-400×400×13×21
865
582
10
Specimen
2351
2371
10 582
2700
Pin
700
2700
Unit: mm

Figure 3. Loading test machine

Table 4. Loading plan

Number of times	State	Strain ε (%)		Load P (kN)	Displacement δ (mm)
1	$\varepsilon_y/3$	0.043	(Type 1~3)	249.2	0.64
		0.047	(Type 4)	265.8	0.75
1	$2\varepsilon_y/3$	0.085	(Type 1~3)	492.6	1.36
		0.093	(Type 4)	526.0	1.48
1		0.25	(Type 1~3)		3.58
			(Type 4)		3.61
2	1/200	0.5	(Type 1~3)		6.71
			(Type 4)		6.74
2		0.75	(Type 1~3)		9.73
			(Type 4)		9.86
5	1/100	1.0	(Type 1~3)		12.96
			(Type 4)		12.99
2		1.5	(Type 1~3)		19.22
			(Type 4)		19.25
2		2.0	(Type 1~3)		25.47
			(Type 4)		25.50
2		2.5	(Type 1~3)		31.73
			(Type 4)		31.76
		3.0	(Type 1~3)		37.98
			(Type 4)		38.01

strains (twice each), 1.0 percent strain (five times), and 1.5, 2.0 and 2.5 percent strains (twice each).

Then, a 3.0 percent strain is applied until strength dropped or the specimen fractured. Each load after yielding is applied twice (except 0.25 and 1.0 percent) to determine the stability of the loop. Table 4 shows a list of the loads applied.

4 EXPERIMENTAL RESULTS

4.1 Specimen Type 1

Loading was continued until the 14th application of a 3.0 percent strain (Figure 4(a)).

On the 14th compression with a 3.0 percent strain, the upper mortar collapsed under pressure and local buckling occurred in the core plate in the vicinity of the rib end. While the local buckling proceeded until the mortar ultimately collapsed under pressure, stable hysteresis was observed in both tension and compression. Maximum strength was 1,155 kN in tension and 1,296 kN in compression.

When the rectangular hollow section and mortar were removed, the exposed core plate exhibited a sign of high degree buckling mode (Figure 5(a)).

4.2 Specimen Type 2

Loading was continued until the first application of a 2.5 percent strain (Figure 4(b)).

On the first application of a 0.5 percent strain, the load dropped slightly and displacement in the direction of the weak axis of the lower portion of the brace increased. On the second compression with a 1.0 percent strain, local buckling occurred in an area slightly above the middle of the core plate. On the fifth compression with a 1.0 percent strain, the local buckling in the area slightly above the middle of the core plate proceeded and the rectangular hollow section was deformed. On the first tension with a 2.5 percent strain, the core plate near the upper rib end fractured. The maximum strength was 850 kN in tension and 806 kN in compression.

The core plate exhibited a sign of light buckling throughout, in addition to the area where local buckling proceeded (Figure 5(b)). The rectangular hollow section used as a restraining part underwent a substantial cross-sectional deformation near the point of local buckling in the core plate. Cracks were observed in the same region.

4.3 Specimen Type 3

Loading was continued until the first application of a 3.0 percent strain (Figure 4(c)).

Sufficiently stable hysteresis was observed up to the application of a 1.0 percent strain. Local buckling started near the rib end on the first compression with

a 2.5 percent strain and the clearance between the restraining part and core plate increased.

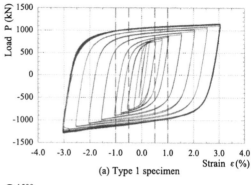

(a) Type 1 specimen

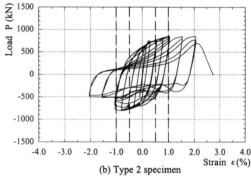

(b) Type 2 specimen

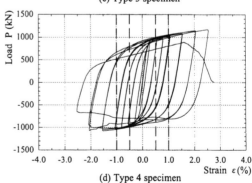

(c) Type 3 specimen

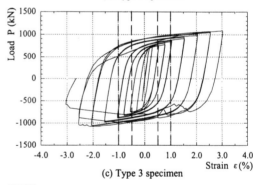

(d) Type 4 specimen

Figure 4. Load-Strain relations

36

Even then, hysteresis was substantially stable. On the first compression with a 3.0 percent strain, a shear fracture occurred in the high-strength bolts fastening the restraining part near the rib end. The maximum strength was 1,084 kN in tension and 1,078 kN in compression.

The core plate exhibited a sign of high degree buckling mode, in addition to the area where severe local buckling occurred (Figure 5(c)).

4.4 Specimen Type 4

Loading was continued until the second application of a 2.5 percent strain (Figure 4(d)).

Hysteresis was sufficiently stable up to the application of a 1.0 percent strain, and remained substantially stable up to a 2.0 percent strain. On the second compression with a 2.0 percent strain, local buckling occurred in a region slightly below the middle of the core plate leading to the development of cracks in the rectangular hollow section. Then, local buckling in the flange of the core plate resulted in the bulging of the rectangular hollow section. On the first compression with a 2.5 percent strain, the local buckling slightly below the middle of the core plate proceeded and the cracks in the rectangular hollow section expanded and propagated to the opposite side of the section. Then, the web of the core plate bulged greatly downward along the weak axis as a result of local buckling. On the second tension with a 2.5 percent strain, the core plate fractured at a point slightly below the middle. The maximum strength was 1,169 kN in tension and 1,076 kN in compression.

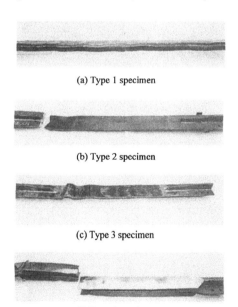

(a) Type 1 specimen

(b) Type 2 specimen

(c) Type 3 specimen

(d) Type 4 specimen

Figure 5. Critical state of core plates

The core plate exhibited a sign of high degree buckling mode in the flange, whereas no sign of deformation was observed in the web except at the fractured point (Figure 5(d)). Some cracks were observed in the flange of the core plate.

5 DISCUSSIONS

5.1 Hysteresis characteristics

Substantially stable hysteresis was observed up to a 1.0 percent strain in all specimens but Type 2 in which a slight drop in strength began.

Under high strains, all specimens but Type 1 exhibited a trough of load in compression. This is due to the local buckling of the core plate that proceeds to the extent allowed by the clearance between the core plate and the restraining part, thus causing a temporary drop in load. When the core plate comes in touch with the restraining part and deformation stops, load resumes its former course of rising. The degree of buckling mode rises by repeating this cycle. Thus, the core plates tested exhibited a sign of high degree buckling mode. In all specimens, a drop in strength started in a loop following the appearance of the trough of load. Therefore it can be said the deformation of the core plate becomes larger with an increase in the number of the trough of load, which, in turn, accelerates decline in strength.

5.2 Final fracture characteristics

Local buckling or tensile fracture occurred near the rib end in all specimens but Type 4. Specimen Type 4 also fractured, but, unlike other specimens, between the rib end and the center of the core plate. In Specimen Type 1, the mortar in the upper portion collapsed under pressure and local buckling occurred in the core plate near the rib end. It was only Specimens Type 2 and 4 that fractured in the test. The tensile fracture of Specimens Type 2 and 4 is considered to be ascribable to the absence of the unbonded material that prevented sufficient restraining of local buckling and allowed continued progress of plastic deformation. In Specimen Type 3, the high-strength bolts fastening together the channel and flat steels constituting a restraining part fractured by the action of shear force resulting from the progress of local buckling in the core plate.

5.3 Cumulative absorbed energy

The cumulative energies absorbed by the individual specimens throughout the test are as follows:
  (1) Specimen Type 1 : 2,632 kN·m
  (2) Specimen Type 2 :   292 kN·m
  (3) Specimen Type 3 :   752 kN·m
  (4) Specimen Type 4 :   617 kN·m

Specimen Type 1 that allowed applying a 3.0 percent strain fourteen times absorbed a much greater amount of energy than the other specimens. Specimens Type 2 and 4 fractured in the first and second tension with a 2.5 percent strain. Although the difference was only one cycle, Specimen Type 4 absorbed more than twice as much energy than Specimen Type 2. This result demonstrates that Specimen Type 2 did not exhibit good hysteresis under high strain.

6 CONCLUSIONS

1) The author designed specimens demonstrating four types of commercially available buckling-restrained braces, which are considered to resist buckling even at a story deformation angle of 1/100, based on the description in their references. The specimens were prepared under the same conditions and put to a test.

2) As was aimed at by the design, all specimens were empirically found to have sufficient hysteresis to withstand strains up to 1.0 percent.

3) Under high strains exceeding 1.0 percent, the four specimens exhibited significantly different performances due to their buckling-restraining methods. While the difference in hysteresis is as pointed out in Discussions. Type 1, 3, 4 and 2 cumulatively absorbed greater amounts of energy in the listed order.

4) Specimens Type 2 and 4 having no unbonded material finally fractured as a result of tensile fracture caused by a rapid progress in local buckling. The unbonded material prevents the transmission of the axial force of the core plate to the restraining part. Furthermore, it is considered to keep a uniform clearance between the core plate and restraining part and prevent the occurrence of abrupt local buckling.

ACKNOWLEDGMENT

The authors wish to thank Dr. H. Ogawa, Mr. M. Murai, Mr. S. Kusunoki and Mr. D. Ichikawa of Kanagawa University for their cooperation in the performance of tests and compilation of experimental data.

REFERENCES

Fujimoto, M. & A.Wada, et al. 1988.
    Unbonded brace encased in buckling-restraining concrete and tube, Structural engineering, Vol. 34B, AIJ
Fukuda, K. & T.Ishibashi, et al. 1999.
    Seismic retrofit of over-track buildings using brace-type hysteretic dampers, Annual meeting of the AIJ
Iwata, M. 1994. Damage tolerant structures,
    Journal of architecture and building science, Vol.109, No.1352, AIJ
Kamiya, M. & H.Shimokawa, et al. 1997.
    Elasto-plastic behavior of flat-bar brace stiffened by square steel tube, Annual meeting of the AIJ
Suzuki, N. & R. Kono, et al. 1994.
    H-section steel brace encased in RC or steel tube, Annual meeting of the AIJ
Wada, A. & M.Iwata, et al. 1998.
    Damage-controlled design for buildings, Maruzen, Tokyo

Behaviour of Steel Structures in Seismic Areas, Mazzolani & Tremblay (eds) © 2000 Balkema, Rotterdam, ISBN 90 5809 130 9

Seismic performance of brace fuse elements for concentrically steel braced frames

M. Rezai & H.G.L. Prion
University of British Columbia, Vancouver, Canada

R. Tremblay & N. Bouatay
École Polytechnique, Montreal, Quebec, Canada

P. Timler
Canadian Institute of Steel Construction, Surrey, B.C., Canada

ABSTRACT: This paper presents the results of a pilot experimental study performed to examine the possibility of using ductile fuse elements to dissipate seismic input energy and reduce the force demand on brace connections in concentrically braced steel frames. These fuse elements are introduced in the bracing members and are designed to yield prior to buckling or yielding of the braces. A total of 21 tests were performed on different fuse elements detailed for two types of bracing members: tapered fuse plates for HSS and double angle bracing members, localized reduction in the cross section of HSS bracing members, and angle splices, also for HSS bracing members. All specimens were subjected to a cyclic quasi-static loading protocol with stepwise deformation increments except two specimens for which a loading history matching more closely the inelastic response of a braced frame structure was used. Some of the specimens experienced local and/or global buckling and exhibited a rather poor performance. Fuse elements for which adequate support had been provided exhibited a stable hysteretic response with high energy dissipation characteristics, large inelastic deformation capacity and beneficial strain hardening behaviour.

1 INTRODUCTION

Over the last decade or so, Canada's National Standard for Limit States Design of Steel Structures, CAN/CSA-S16.1 (S16.1) has evolved from a fairly simplistic approach for concentric brace and connection design to one that reflects greater attention required for their selection and detailing with respect to seismic resistance. S16.1's code clauses for concentric brace application in seismic zones has been founded, since 1989, on a deterministic philosophy known as the capacity design procedure. The focus of this philosophy is the control of locations for energy absorption in the lateral load resisting system rather than merely member strength considerations. In the latest edition of S16.1 (CSA, 1994), changes to the design force expectations for brace end connections have had a major impact on the practical design and fabrication of steel braced structures, in particular, within the low-rise industry where hollow structural sections (HSS) are the members of choice. Currently, end connections must be designed for the full tensile capacity of the selected brace element, i.e., A_gF_y. As

concentric braces are selected based on their compression capacity, large disparities between design force and tensile capacity are common. While previous editions of the code allowed for a reasonable connection force margin over the design level, today's code force level for connections can be many multiples of its predecessor, and in some instances, an order of magnitude greater. This has significant cost implications in the development of projects as not only do material and labour increase for the structural steel system, but foundation sizes increase as well.

Spawned by these realities in new design application, the consulting profession in British Columbia, one of Canada's highest seismic regions, and local industry, launched a pilot experimental program to examine the potential of reducing the disparity between the tension and compression capacities of typical brace components. The objective of the program was to establish if merit existed in the development of controlled fuses within HSS braces intended for the majority of the building market. The Canadian Institute of Steel Construction (CISC), in collaboration with the University of British Columbia and École Polytechnique, conducted this investigation thorough the sponsorship of numerous professional and industry associations and donations by local fabricator members of the CISC. A synopsis of the findings from this work is presented herein.

2 SCOPE OF STUDY

The experimental study on brace fuses was done in three phases. During the first phase three fuse details were tested in full scale cross-braced frames. This study served to do the planning of the second phase, which consisted of cyclic tension-compression tests on fuse details only. The third phase consisted of full scale tests on chevron type braced frames with e best fuse details from phase 2 selected and improved upon. The guidelines outlined in ATC-24 (1992) for cyclic testing of structural components were followed for the cyclic tests. This paper will give an overview of the test series and discuss details of only a few selected tests. More detailed information on phases I, II and III testing can be found in Tremblay and Bouatay, 1999 & Rezai et al., 1999 & Tremblay and Bouatay, 1999, respectively. A full summary of all the tests with brief results are listed in Table 1.

Table 1. Scope of Testing Program

Spec. No.	Description of fuse	Fuse detail photo	Ductility and failure mode	Load-Displacement Loops	Remarks	Damaged fuse detail photo
Phase 1						
B1	HSS 102×102×4.8, 25 mm long steel plate 6.4×120 mm		$5\Delta_y$ degradation after $3\Delta_y$, lateral buckling and tearing of fuse zone		Out of plane support needs to be improved	
B2	HSS 102×102×4.8, Tapered 260 mm long, 10×90 mm, laterally braced with angles		$3\Delta_y$, fracture of fuse		High energy dissipation Sudden fracture, not desirable	
B3	HSS 102×102×4.8, Similar to B2, Corners of taper rounded, fuse length 220 mm		$3\Delta_y$, fracture of fuse		High energy dissipation Sudden fracture, not desirable	
Phase 2						
1A, 1B	HSS 102×102×4.8 Oval cut-out, four sides, 125 mm and 175 mm long		1A: $3\Delta_y$, 2.5 mm fracture across fuse zone, 1B not tested		Limited ductility before plastic buckling of fuse, therefore longer fuse detail not tested	
2A, 2B	HSS 102×102×4.8 Oval cut-out, 125 mm and 175 mm long, four sides sleeved		2A not tested, 2B: $6\Delta_y$, 3 mm, binding during compression cycles		Longer fuse detail more promising, lateral support adequate, but higher compression loads due to binding	
3	HSS 102×102×4.8 Oval cut-out, 175 mm long, two sides, slotted end plate		Approx. $3\Delta_y$, unsymmetrical loading protocol, overall buckling		Too high capacity, needs additional lateral support, such as a sleeve	

Table 1. Scope of Testing Program

4A, 4B	HSS 102×102×4.8 Staggered holes, 30 mm dia., four sides, specimen 4B with sleeve		4A: 3Δ_y, 3.5 mm, net section fracture 4B: 3Δ_y, 2 mm, binding increased compression load, causing overall member buckling	Promising detail with good energy dissipation When using sleeve, binding causes problems as sleeve participates as load carrying member
5	HSS 102×102×4.8 300W plate 125 × 6 mm similar to B1 with additional lateral support		1Δ_y, 1 mm unsymmetrical, overall buckling of the member	Needs more lateral support as yielded fuse zone cannot withstand any buckling load
6	HSS 102×102×4.8 300W dogbone plate 9 × 90 mm, 220 mm long, angle lateral support (simplified version of B2 and B3)		No ductility, premature elastic buckling of fuse and support angles	Lateral support needs to be strengthened
7A, 7B, 7C	HSS 102×102×4.8, 25×25×5 angles on four corners, specimen A and B: 50 and 75 mm long, specimen C: 75mm with batten straps		5Δ_y, 2.5 mm all specimens, plastic buckling of corner angles followed by tension fracture	Good hysteretic behaviour, need to increase yield displacements
Phase 3				
B4	HSS 102×102×4.8 Oval cut-out, four sides, 175 mm long (similar to spec. 2, phase 2), inside-outside sleeve		1Δ_y, 4 mm local buckling inside sleeve	Inadequate ductility, binding action inside sleeves initiated local straining
B5	HSS 102×102×4.8 Oval cut-out, four sides, 350 mm long (similar to spec. B4), inside-outside sleeve		3Δ_y, 13 mm more overall yielding in fuse area, reduced ductility (2Δ_y) for second test with near-fault protocol	Better energy dissipation for regular testing protocol, early fracture during initial thrust of near fault earthquake protocol reduced the ductility
B6	HSS 102×102×4.8 25×25×5 angles on four corners, 325 mm long, 3 batten straps, not attached, interior sleeve		2Δ_y, 10 mm, plastic angle buckling and tension fracture	Local buckling of angles occurred due to shifting of batten plates. These need to be attached or a continuous batten plate
B7	2-L127×89×13 angle brace 300W dogbone fuse plate 13 × 100 mm, 400 mm long		3Δ_y, 12 mm, net section fracture of fuse plate	Good energy dissipation, adequate lateral support

3 PHASE 1 TESTS

Tests in Phase I were performed in the full scale test frame shown in Figure 1. In test B1, the splice plates (6.4 mm thick × 120 mm wide) that connected each half-segment of the two HSS 102×102×4.8 members were used as fuse elements. Both splice plates were connected to each other at their intersection point by means of a single bolt. A 25 mm × 100 mm × 200 mm reinforcing plate was welded in the central portion of both fuse plates to improve their buckling strength. The splice plates then had only a 25 mm long unsupported length between each HSS and the reinforcing plate. Despite of this detailing, out-of-plane buckling of the plates developed early in the test, in the first cycle of amplitude $1.0\Delta_y$, and accentuated in the subsequent cycles. Tearing of the plates initiated during the second cycle at $2\Delta_y$ and fracture eventually took place when reaching $5\Delta_y$. This specimen exhibited a severely pinched hysteretic response mainly due to the poor capacity of the fuse plates in carrying compression loads.

Specimen B2 was similar to B1 except that the fuse plates had a dog-bone shape with a 260 mm long × 90 mm wide reduced section. They were also supported over their full length by a set of cover plates. These cover plates were attached to the HSS bracing members by means of angles spanning between the two segments of each brace. Slotted holes were used in the connections of the angles to the HSS members to ensure that all the brace loads would flow in the fuse plates. Specimen B3 was identical to B2 except that large radius transitions were specified at the changes in width along the fuse plates. The observed performance of both specimens was nearly the same: stable hysteretic response up to a ductility of 3.0, at which point fracture of the fuse plates occurred at the width transition. The addition of smoother radii in specimen B3 only permitted to undergo one more loading cycle when compared to specimen B2. During the test, significant out-of-plane movement of the bracing assembly was also observed due to the flexibility of the various parts and connectors attaching the HSS segments, the cover plates, and the angles in the middle portion of the braces. It was suggested to eliminate this behaviour by moving the fuse elements towards the ends of the braces and to change the bracing geometry for a chevron (inverted-V) bracing. These recommendations were applied in the subsequent phases of the project.

4 PHASE 2 TESTS

Ten single HSS brace elements with various fuse detail configurations were tested cyclically in the Structural Engineering Laboratory at the University of British Columbia (Rezai *et al.*, 1999). All test specimens were made of HSS 102 × 102 × 4.8 G40.21-350W class C

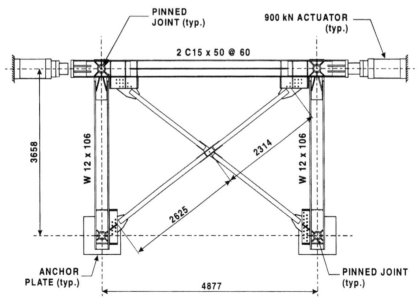

Figure 1. Test frame used in Phases I and III.

steel material. A short description of each test was presented in Table 1. In the preceding, the cyclic behaviour of fuse details 2B, 6 and 7 that were subsequently used in the third phase of testing is discussed.

For specimen 2B, an elliptical shape 55 mm wide and 175 mm long was removed from all four sides of the HSS brace element. Both inside and outside supports were provided along the fuse detail to prevent premature buckling. Specimen 6 was a tapered fuse detail with a long (220 mm) fuse length. Specimens 7A, 7B and 7C had similar fuse detail configurations comprised of four L25 × 25 × 5 angle sections welded on all four corners of the HSS brace element. The length of the fuse detail was selected as 50 mm and 75 mm for specimens 7A and 7B, respectively. For specimen 7C steel straps were provided across the mid-length of the fuse which had a fuse length of 75 mm. The result of axial load versus averaged displacement measured by two LVDTs mounted on two opposite sides of the HSS members for specimens 2B, 6 and 7B is shown in Figure 2.

The behaviour of specimen 2B under the first nine cycles of loading prior to yielding was nearly perfectly linear. The specimen, however, was slightly stiffer in compression than in tension. The yield strength in tension was achieved at about 310 kN as opposed to 375 kN in compression. The reason for this behaviour was attributed to the participation of the steel jacket around the fuse detail in resisting the axial load through a jamming effect. The hysteresis loops are the fullest indicating a very robust and stable energy dissipation mechanism throughout the cyclic loading. Strength degradation and stiffness decay started during the tensile axial displacement loading of six times the yield displacement. This was mainly attributed to the reduction in the cross-sectional area of the fuse detail as a result of tearing across the HSS corner legs.

The sudden out-of-plane buckling of the fuse detail plate in specimen 6 terminated the test. This was mainly attributed to the excess length of the fuse and inadequate out-of-plane support provided by the angles. The available gap between the fuse plate and the angles supporting the fuse could have caused the premature buckling. As the brace underwent compression, buckling of the fuse plate developed in the form of a small out-of-plane movement thereby closing the gap. Then, with a further increase in load, the plate pushed outward against the adjacent angle. Due to inherent eccentricity in the system and weak

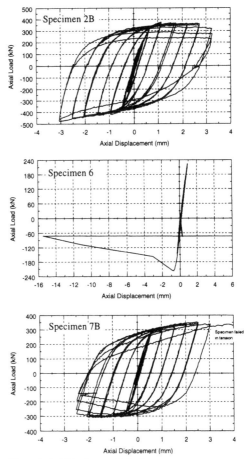

Figure 2. Axial load vs. axial displacement response of specimens 2B, 6 and 7B.

buckling resistance of the angle, it bent in the out-of-plane direction and thereby caused a sudden elastic buckling failure of the specimen. The global out-of-plane buckling of the specimen is shown in Table 1. It is noted that this shortcoming of the specimen 6 was addressed during the third phase of testing as explained in the Phase 3 Testing.

In general, the cyclic behaviour of specimen 7B was comparable to the behaviour observed for specimen 7A. The yield strength of the specimen was achieved at an axial load of about 270 kN with the corresponding yield displacement of about 0.55 mm. The cyclic loops were the fullest for displacement ductility ratios of up to five times the measured yield displacement. However, considerable strength degradation during compression loading was observed at post-yield displacement of about five times the yield displacement. The specimen eventually failed in tension during the second loading cycle of six times yield displacement. It is noted that increasing the length of the fuse detail for specimen 7B improved the ductile behaviour of the specimen to a displacement ductility factor of six as opposed to a displacement ductility factor of five measured for specimen 7A.

5 PHASE III

In this phase, it was decided to explore further the potential of the elliptical cut-out scheme and the splice angle system for HSS bracing members. The first detail was used in specimens B4 and B5. Only the length of the elliptical cuts was different between these two specimens: 175 mm for **B4** and 350 mm for B5. The latter is shown in Figure 3. In specimen B6, four angles 25 mm × 25 mm × 5 mm were used to span a 175 mm long gap between two bracing member segments. Fuse specimens B4, B5, and B6 were all introduced in HSS 102×102×4.8 bracing members arranged in a chevron configuration in the test frame used in Phase I. Local buckling of the HSS corners in fuses B4 and B5 was prevented by an HSS 89×89 interior sleeve and an HSS 127×127 exterior sleeve. The former one was provided over the full length of the brace to avoid any kinks of the bracing members upon yielding of the fuses in compression.

The exterior sleeve was slightly longer than the oval cuts. In specimen B6, an HSS 89×89 interior sleeve was also used but it was only 375 mm long. Buckling of the angles towards the exterior was prevented by 4-25 mm wide steel straps uniformly distributed over the length of the fuse (175 mm). These straps were not welded to the angles in order to minimize stress concentrations in the yielding portion of the angles. Specimen B7 represented a third fuse design for the third phase of testing. It included a dog-bone fuse plate similar to the one used in specimens B2 and B3 in Phase I. This time, the narrower portion was made longer (400 mm vs. 260 mm) to improve the low-cycle fatigue life of the detail and the bracing member was a double angle 127×89×13. One end of the fuse was connected to the columns of the test frame while the other one was bolted to the brace. As shown in Figure 3, the angles extended over the full length of the fuse to provide support against buckling.

In other to examine the influence of the loading history on the fracture life of the fuses, replica of specimens B5 and B7 (B5-B and B7-B) were fabricated and subjected to a loading history obtained from nonlinear dynamic analyses of a typical braced frame building. Figure 4 shows both loading histories as well as the fracture life for all Phase III specimens. Figure 5 shows axial load vs. axial displacement response of specimens B5-A, B6 and B7-A. The overall performance of specimens B4, B5-A, and B7-A was very much the same as they all developed good hysteretic behaviour with strain hardening response and no strength degradation until fracture

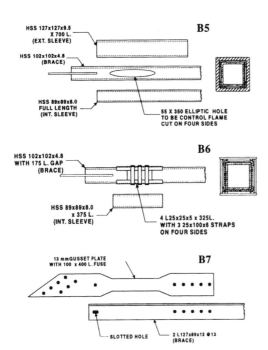

Figure 3. Specimens B5, B6 and B7.

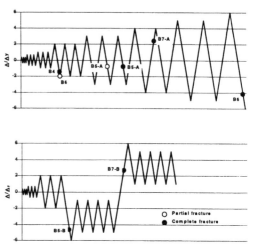

Figure 4. Loading history and fracture life of Phase III specim

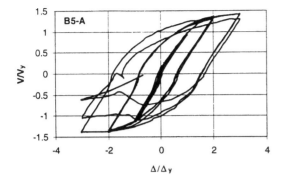

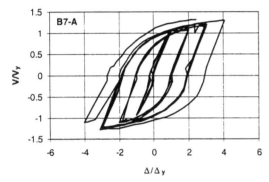

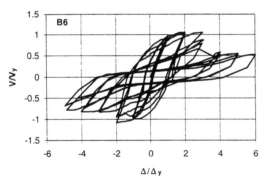

Figure 5. Axial load vs. axial displacement response of specimens B5-A, B6 and B7-A.

occurred. The fracture life (and maximum ductility) increased with the length of the fuse: 1st cycle at $2\Delta_y$ (B4), 3rd cycle at $3\Delta_y$ (B5-A), and 2nd cycle at $4\Delta_y$ (B7-A). The same trend was observed or specimens B5-B and B7-B which were subjected to the seismic type of loading. Fracture in specimen B6 developed progressively during the test as the connecting angles started to break in sequence at a ductility of 2.0. Such premature failure was attributed to the high strain demand that developed near the ends of the angles. Complete fracture of all four angles on one brace occurred in the first cycle of $6\Delta_y$. During the test, the

steel straps also started to slide along the angles at a ductility of 2.0 and local buckling of the angles eventually occurred as a result of their longer unsupported length. In addition, a kink developed at the location of the fuse in the bracing members of specimen B6. This can be explained by the fact that the interior sleeve was too short to maintain the brace in a straight position upon yielding and local buckling of the fuse angles in compression. Local buckling and kinking of the braces also contributed to the high strain demand in the angles. Premature fracture of the angles and instability resulted in a pinched hysteresis loop with strength degradation. However, in view of the very large ductility demonstrated by some of the angles, it is believed that this detail could perform very well if instability was mitigated.

6 CONCLUSIONS

This pilot study clearly demonstrated the potential and the feasibility of introducing brace fuse elements in concentrically braced frames for steel buildings. The systems examined in this study exhibited a stable hysteretic response under cyclic loading when local and global buckling response could be properly prevented. The fracture life and ductility levels of the fuse elements both increased with the length of the fuse, i.e., when the inelastic demand experienced by the fuse for a given storey drift was reduced. All three systems examined in Phase III represent good candidates for seismic applications. The fracture life of the cut-out system would likely be improved by changing the shape of the cut from an ellipse to a rectangle with round corners to increase the length of the yielding region. The energy dissipation of the fuse with splice angles should be enhanced by having a longer interior sleeve, by providing a continuous exterior support to the angles, and by reducing slightly the cross section of the angles away from the welded connections to the HSS members. These recommendations should be verified through further testing. The study also showed that the fracture life of fuse elements strongly depends upon the loading history. It is therefore recommended that a loading protocol representative of the response of steel braced frames be developed and used in future test programs. Methods for predicting the low-cycle fatigue life of fuse elements should also be examined in future research.

7 ACKNOWLEDGEMENTS

Financial assistance of this project was provided by

the National Research Council of Canada and from the following organizations: Department of Civil Engineering at the University of British Columbia, Association of Professional Engineers and Geoscientists of British Columbia, Division of Structural Engineers, Structural Engineering Consultants of British Columbia, Steel Structures Education Foundation, Vancouver Structural Engineers Group, Vancouver Island Chapter of the Canadian Society of Civil Engineers, Canadian Welding Bureau and Herold Engineering.

The contribution of BC steel fabricators: Canron, George Third and Sons, Agra Coast Steel, Empire Iron Works, Les Constructions Beauce-Atlas, Solid Rock Steel and X.L. Iron Works for the fabrication of the specimens is greatly appreciated.

The helpful suggestions and recommendations of the numerous west-coast engineers and steel fabricators in selecting the appropriate fuse details are acknowledged.

The authors' sincere appreciation goes to the staff of the Structural Engineering Laboratory of the University of British Columbia and École Polytechnique in Montreal for their assistance in conducting the tests.

8 REFERENCES

Applied Technology Council 1992. Guidelines for Cyclic Seismic Testing of Components of Steel Structures. Redwood City: ATC-24.

CSA. 1994. CAN/CSA-S16.1-94, Limit States Design of Steel Structures. Canadian Standard Association, Rexdale: Ont.

Tremblay, R. & Bouatay, N. 1999a. Pilot Testing on Ductile Yield Plate Fuses for HSS Braces Intended for Low-rise Buildings - Phase I Full-scale Testing of Prototype Brace Fuse Detail. Report CDT/ST99-05, Dept. of Civil Eng., Struct. Div. École Polytechnique, Montreal: Canada.

Tremblay, R. and Bouatay, N. 1999b. Pilot Testing on Ductile Yield Plate Fuses for HSS Braces Intended for Low-rise Buildings - Phase III Full-scale Testing of Prototype Brace Fuse Detail. Report CDT/ST99-16, Dept. of Civil Eng., Struct. Div. École Polytechnique, Montreal: Canada.

Rezai, M. Prion H.G.L. & Timler P. 1999. Pilot Testing of Fuse Details for HSS Bracing Members (Phase II), Dept. of Civil Engineering, University of British Columbia, Vancouver: Canada.

Behaviour of Steel Structures in Seismic Areas, Mazzolani & Tremblay (eds) © 2000 Balkema, Rotterdam, ISBN 90 5809 130 9

Cyclic elasto-plastic analysis of steel members

Iraj H. P. Mamaghani
Department of Civil Engineering, Kanazawa University, Japan

ABSTRACT: This paper is concerned with the cyclic elasto-plastic analysis of steel members subjected to the combined axial force and bending moment by employing the two-surface plasticity model in force space (2SM-FS). First the basic concepts of the 2SM-FS and procedures for determining its parameters are presented. Then the accuracy of the 2SM-FS is verified by comparing cyclic nonlinear sectional behavior of steel members obtained from the 2SM-FS with those of the experiments and analysis using the two-surface model in stress space (2SM-SS). Finally by employing the 2SM-FS for sectional nonlinear behavior in a beam-column element formulation, cyclic elasto-plastic analysis of steel members, such as fixed-ended beam-columns and cantilever columns, are carried out. The cyclic plasticity performance of the 2SM-FS is discussed and is verified against the experiments and numerical results using 2SM-SS. It is shown that both the 2SM-SS and 2SM-FS accurately predict cyclic behavior of steel members while the 2SM-FS has the advantage of larger computational speed than that of the 2SM-SS.

1 INTRODUCTION

During severe earthquakes, structural steel members such as bridge piers, columns in buildings and offshore structures are subjected to cyclic lateral loads in addition to the constant axial load of the supported structures. This leads to their damage and consequent collapse of the structures. That is, from the viewpoint of limit state design of steel structures, investigating and clear understanding of the cyclic behavior of steel members and structures are important to prevent their collapse under severe earthquakes.

Recently, cyclic behavior of steel members and structures has extensively been studied by taking into consideration material nonlinearity as well as geometrical nonlinearity. As for material nonlinearity, the two-surface plasticity model in stress space (2SM-SS) for structural steels (Mamaghani et al. 1995, Shen et al. 1995), among many constitutive models in the stress space, has been developed and applied to the cyclic nonlinear analyses of steel structures. The 2SM-SS has shown an excellent capability to predict the cyclic behavior of steel structures (Mamaghani 1996, Mamaghani et al. 1996). However, the finite element structural analysis with constitutive model in the stress space requires generally much longer computational time than that with

the constitutive model in the force space. Therefore, application of constitutive model in the stress space may be limited to the analyses of simple structures or members.

The present paper is concerned with the analysis of steel members under cyclic loading using the two-surface plasticity model in force space (2SM-FS), recently developed by the author and his coworkers; Suzuki et al. (1995), Mizuno et al. (1996), Mamaghani & Kajikawa (1997). First the fundamental concepts of the 2SM-FS, which has been developed with extension of concepts used in the 2SM-SS, are briefly presented and discusses. The 2SM-FS considers with the force space plasticity modeling of the cyclic cross-sectional behavior of steel beam-columns subjected to the combined axial force and bending moment. It takes accurately into account cyclic elastoplastic behavior of structural steels such as the yield plateau, Bauschinger effect and strain hardening. Then, the procedure for determining of the 2SM-FS parameters is presented and the values of the model parameters for circular, I and box sections are given. Later, the accuracy of the 2SM-FS is examined and is verified by comparing the cyclic nonlinear sectional behavior of steel members obtained from the 2SM-FS with those of the experiments and analysis using the 2SM-SS. Finally by employing the 2SM-FS for sectional nonlinear behavior in a beam-column element formulation, cyclic inelastic analysis of steel members such as fixed-ended beam-columns and cantilever columns are carried out. The cyclic plasticity performance of the 2SM-FS is discussed and verified against the experiments and analytical results using 2SM-SS. It is shown that the 2SM-FS provides reasonably accurate results while it has the advantage of larger computational speed than that of the 2SM-SS.

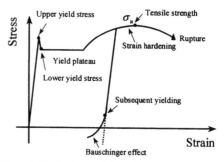

Figure 1. Cyclic characteristics of structural steels.

2 BASIC CONCEPTS OF THE TWO-SURFACE PLASTICITY MODEL IN FORCE SPACE

Figure 1 shows the key behavioral characteristics of the structural steel response under cyclic loading, such as existence of the yield plateau, the Bauschinger effect and strain hardening (Mamaghani 1996). The concepts of the 2SM-FS are based on the experimental evidences that these important characteristics of the material response under cyclic loading exhibited at the stress level by the steel members, thus propagates to the force (stress-resultant) level provided that no local buckling occurs (Mamaghani 1996). Therefore, the 2SM-FS has been developed with extension of concepts used in the 2SM-SS. The major assumptions taken in developing the 2SM-FS are:

1. The cross-section remains in plane after deformation;
2. There is no distortion of the cross-section;
3. Only normal stress acting on the cross-section;
4. No local buckling occurs.

The 2SM-FS, similar to the 2SM-SS, utilizes the concept of bounding surface formulation in force space. The 2SM-FS formulation uses two nested curves: an inner loading curve, and an outer bounding curve, as schematically shown in Figure 2 in two-dimensional normalized force space; axial force $n = N / N_y$ and bending moment $m = M/M_y$. N_y and M_y denote the yield axial load and yield bending moment of the cross-section, respectively. The inner loading curve represents the locus of loads and moments that causes the initiation of yielding at some point on the cross-section. The outer curve represents the sate of load at which a limiting stiffness of the cross-section is achieved. As shown in Figure 2a, at the initial unloaded state the loading and bounding curves coincide with the initial yield curve and yield plateau curve, respectively. The loading curve takes the same shape as the yield plateau curve and bounding curve, respectively, when it is in contact with these curves, Figure 2.

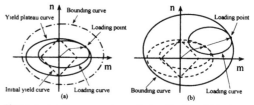

Figure 2. Loading curve in contact with: (a) yield plateau curve; (b) bounding curve.

Once the loading point has contacted the loading curve, the response is governed by a number of hardening rules, which determine subsequent elastoplastic behavior. As the cross-section is loaded inelastically, both the loading and bounding curves may translate (kinematic hardening), contract or expand (isotropic hardening), to model phenomena such as strength degradation, the Bauschinger effect and cyclic strain hardening. The degree of plasticity at the cross-section is a function of the distance between the two curves. In the following, some important features of the 2SM-FS will be presented and discussed.

2.1 Plastic modulus

The plastic modulus E^p associated with loading curve is used to prescribe the plastic flow under the assumption of the associated flow rule. In 2SM-FS, the same equation for E^p in the 2SM-SS is used. That is, the value of E^p is a function of the distance δ (Fig. 3a) between the two loading and bounding curves and is taken as:

$$E^p = E_0^p + h(\delta)\frac{\delta}{\delta_{in} - \delta} \qquad (1)$$

where E_0^p, which is a function of plastic deformation, and $h(\delta)$ are the current plastic modulus of bounding curve and the shape parameter as in the

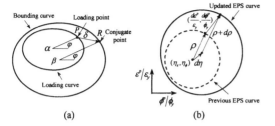

(a)   (b)

Figure 3. (a) Definition of δ; (b) effective plastic strain (EPS) curve.

2SM-SS (Shen et al. 1995), respectively. δ is the distance from loading point P to conjugate point R on the bounding curve and is measured in Euclidean norm, Figure 3a. δ_{in} is the value of δ at first contact with the loading curve.

As shown in Figure 3a, the conjugate point R on the bounding curve is defined to have the same direction from the center of bounding curve as the direction of the loading point P from the center of the loading curve. Note that the value of δ, measured in the dimensionless force space n versus m, is used in Equation 1 after multiplying by the yield stress σ_y.

2.2 Effective plastic strain curve

Based on the definition of the effective plastic strain surface in plastic strain space for multiaxial 2SM-SS, the effective plastic strain (EPS) curve is defined for the 2SM-FS in the nondimensional plastic strain space; $\varepsilon^p / \varepsilon_y$ versus ϕ^p / ϕ_y (Fig. 3b) as follows:

$$\Phi(\varepsilon^p / \varepsilon_y, \phi^p / \phi_y) = (\frac{\varepsilon^p}{\varepsilon_y} - \eta_\varepsilon)^2$$
$$- (\frac{\phi^p}{\phi_y} - \eta_\phi)^2 - \rho^2 = 0 \qquad (2)$$

In which $(\eta_\varepsilon, \eta_\phi)$ and ρ are the center and radius of the curve, respectively. ε_y and ϕ_y denote the yield values of the axial strain and curvature of cross-section respectively. The EPS curve, which represents a memory of maximum plastic deformation that the cross-section has ever experienced through the loading history, expands and translates if $\Phi\{(\varepsilon^p + d\varepsilon^p)/\varepsilon_y, (\phi^p + d\phi^p)/\phi_y\} > 0$. The evolution of loading and bounding curves in 2SM-FS is related to the size ρ of EPS curve.

2.3 Hardening rule

The hardening rule adopted in the 2SM-FS defines evolution of loading and bounding curves in a way to ensure that the two curves be tangential to each other when they contact. The loading curve at the initial unloaded state coincides with initial yield curve F_0 (Fig. 2a) defined by:

$$F_0(m,n) = |m| + |n| - 1 = 0 \qquad (3)$$

When the cross-section is loaded it undergoes elastic deformation until the loading point reaches the initial yield curve. At this state plastic flow starts and subsequent loading curve evolves and begins moving towards the yield plateau curve F_y (Fig. 4a), which is defined by:

$$F_y(m,n) = \left(\frac{m}{f_y}\right)^{c_1} + n^{c_2} - 1 = 0 \qquad (4)$$

where, c_1 and c_2 are constant values related to the type of cross-section and material; and f_y is a shape parameters. As shown in Figure 4a, the yield plateau curve, which represents the locus of loads causing elastic-perfectly plastic load-deformation behavior, governs the evolution of the loading curve before it ceases to exist as it does in the 2SM-SS. Subsequent loading curve before the yield plateau disappears is given by:

$$f(m,n,\alpha_m,\alpha_n,r) = \theta_1 \left[\left|\frac{m - \alpha_m}{r}\right| + \left|\frac{n - \alpha_n}{r}\right|\right]$$
$$+ (1 - \theta_1)\left[\left|\frac{m - \alpha_m}{rf_y}\right|^{c_1} + \left|\frac{n - \alpha_n}{r}\right|^{c_2}\right] - 1 = 0 \qquad (5)$$

where (α_m, α_n) is the center of loading curve; $r = \kappa / \kappa_0$ is the size of loading curve. κ is the radius of loading surface and $\kappa_0 = \sigma_y$ is the radius of initial yield surface in 2SM-SS. The same expression for κ as in 2SM-SS is used just by replacing the radius ρ of the effective plastic strain surface in 2SM-SS with products of the radius of EPS curve ρ and yield strain ε_y; $\rho\varepsilon_y$, that is:

$$\kappa = \kappa_0\{\alpha - a\exp(-200b\rho\varepsilon_y) - (\alpha - a - 1)\exp(-200c\rho\varepsilon_y)\} \qquad (6)$$

where the material parameters a, b, c and α in Equation 6 are inherited from the 2SM-SS (Shen et al. 1995). Similar to the 2SM-SS, the loading curve

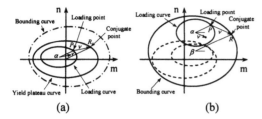

Figure 4. Evolution of the loading and bounding curves: (a) before yield plateau disappears; (b) after yield plateau disappears.

in the 2SM-FS softens isotropically, as a function of the size of ρ, to provide experimentally observed decreasing zone of elastic behavior of the steel and its effects at the force level (Mamaghani 1996).

In Equation 5, θ_1 whose value changes from 1 at the initial unloaded state to zero when the loading curve contacts the yield plateau curve, is the evolution tracer of the loading curve before the yield plateau disappears and is defined by: $\theta_1 = \min(\delta^{yp} / \delta^{yp}_{in})$, in which δ^{yp} is the value of δ measured to the yield plateau curve; and δ^{yp}_{in} is the δ^{yp} value of the current loading path at first contact with the loading curve. The sign "min" indicates that θ_1 assumes the minimum value through the whole loading history to guarantee transformation of loading curve is irreversible. The loading curve progressively changes in shape and assumes the same shape with that of the yield plateau curve when they tangentially contact at the loading point, as shown in Figure 1a. After the yield plateau curve disappears as a function of the size ρ of EPS curve and the cumulative plastic work as in the 2SM-SS, the loading curve progressively changes in shape and moves towards the bounding curve F_b (Fig. 4b) defined by:

$$F_b(m,n,\beta_m,\beta_n,r_b) = \left(\frac{m-\beta_m}{r_b f_b}\right)^{c_3} + $$

$$\left(\frac{n-\beta_n}{r_b}\right)^{c_4} - 1 = 0 \qquad (7)$$

where c_3 and c_4 are constant values related to the type of cross section and material; f_b is a shape parameters; $r_b = \overline{\kappa} / \kappa_0$ is the size of bounding curve; $\overline{\kappa}$ = the radius of bounding surface in 2SM-SS and is defined in a manner similar to κ described above. It is given as:

$$\overline{\kappa} = \overline{\kappa}_\infty + \left(\overline{\kappa}_0 - \overline{\kappa}_\infty\right)\exp[-\zeta(\rho\varepsilon_y)^2] \qquad (8)$$

where the parameters κ_∞ and ζ are the limit size of the bounding surface in 2SM-SS and material constant, respectively. They are inherited from the 2SM-SS. In Equation 7, (β_m,β_n) is the coordinate of the center of bounding curve. As shown in Figure 4b, the bounding curve governs the evolution of the subsequent loading curve, which is expressed by:

$$f(m,n,\alpha_m,\alpha_n,r) = \theta_2\left[\left|\frac{m-\alpha_m}{rf_y}\right|^{c_1} + \left|\frac{n-\alpha_n}{r}\right|^{c_2}\right]$$

$$+ (1-\theta_2)\left[\left|\frac{m-\alpha_m}{rf_b}\right|^{c_3} + \left|\frac{n-\alpha_n}{r}\right|^{c_4}\right] - 1 = 0 \qquad (9)$$

in which θ_2, whose value changes from 1 when the yield plateau disappears to zero when the loading curve hits the bounding curve, is another evolution tracer of the loading curve and is defined by: $\theta_2 = \min(\delta^{bs} / \delta^{bs}_{in})$, where δ^{bs} is the δ value measured to the bounding curve; δ^{bs}_{in} is the δ^{bs} value of the current loading path at first contact with the loading curve. θ_2 assumes the minimum value through the whole loading history and plays the same role as θ_1 does before the yield plateau disappears. The loading curve tangentially contacts the bounding curve at the loading point and assumes the same shape with that of the bounding curve as shown in Figure 1b. The two curves remain tangent on further loading until unloading occurs.

The instantaneous translation of the loading curve associated with the load increment (dm, dn) occurs along PR following the Mroz type of hardening rule given by $(\Delta\alpha_m, \Delta\alpha_n) = C_\alpha(v_m, v_n)$, in which (v_m, v_n) is the unit vector in the direction of PR as shown in Figure 4. C_α is the step size of translation and is determined through the consistency condition of $df = 0$ (Shen et al. 1995).

To consider random cycling and identify smaller plastic excursions relative to previous larger excursions, the concepts of memory curve and virtual bounding curve are used in the 2SM-FS in the same way they are used in 2SM-SS. Similar to 2SM-SS, the termination of the yield plateau is judged by the same expression as in the 2SM-SS through the size of EPS curve ρ in Equation 2 and the plastic work.

2.4 Constitutive equations

From the incremental theory of plasticity and adopting the associated flow rule, the incremental force-deformation relationship for 2SM-FS can be expressed as:

$$\begin{Bmatrix} dn \\ dm \end{Bmatrix} = \begin{bmatrix} 1.0 - \dfrac{\dfrac{i_n^2}{N_y \varepsilon_y}}{\lambda^p} & -\dfrac{\dfrac{i_n i_m}{N_y \varepsilon_y}}{\lambda^p} \\ -\dfrac{\dfrac{i_n i_m}{M_y \phi_y}}{\lambda^p} & 1.0 - \dfrac{\dfrac{i_m^2}{M_y \phi_y}}{\lambda^p} \end{bmatrix} \begin{Bmatrix} \dfrac{d\varepsilon}{\varepsilon_y} \\ \dfrac{d\phi}{\phi_y} \end{Bmatrix}$$

$$\lambda^p = K^p + \frac{i_n^2}{N_y \varepsilon_y} + \frac{i_m^2}{M_y \phi_y} \qquad (10)$$

in which (i_m, i_n) is the unit normal vector on the loading curve at the current loading point. K^p is a scalar related to the plastic hardening modulus E^p in Equation 1. For example, under uniaxial tension and compression loading $(i_n = 1, i_m = 0)$, K^p can be derived from the incremental plastic deformation vector conjugate to the incremental force vector as:

$$K^p = \frac{E^p A}{N_y^2} \qquad (11)$$

3 PARAMETERS OF 2SM-FS

The material properties for SS400 (ASTM A36) and the corresponding 2SM-FS parameters related to the 2SM-SS are given in Shen et al. (1995). In this section, the direct integration method (Minagawa et al. 1988) is used to determine the 2SM-FS parameters related to the strength curves (loading, yield plateau and bounding curves) discussed above. In this approach, the section analyzed is divided into elemental areas, as shown in Figure 5 for a hollow box section. The incremental stress-strain relation for each elemental area is described by the uniaxial 2SM-SS. The normalized stress resultants of axial force n and bending moment m are calculated simply by summing the contribution of each elemental area over the cross-section.

The normalized axial strain $\varepsilon / \varepsilon_y$ and curvature ϕ / ϕ_y, with a prescribed ratio as shown in Figure 6, are increased incrementally and the axial force and bending moment are calculated. From the $n - \varepsilon^p$ and $m - \phi^p$ curves the values of n_i and m_i corresponding to the yield plateau and bounding curves are determined for a specific i-th loading path, cross-section and material, example of which is shown in Figure 6 for a box section. Here, ε^p and ϕ^p denote the plastic strain and plastic curvature respectively.

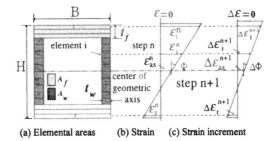

(a) Elemental areas (b) Strain (c) Strain increment

Figure 5. Subdivision of cross-section and strain distribution for a box section.

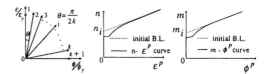

Figure 6. Loading paths and definition of initial bounding line.

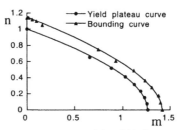

Figure 7. Definition of the yield plateau and initial bounding curves.

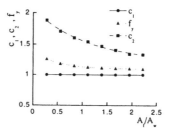

(a) Yield plateau curve

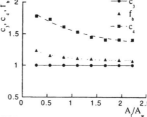

(b) Bounding curve

Figure 8. 2SM-FS parameters for a box section.

51

The results for different loading paths are plotted in the n versus m coordinate system and the values of c_1 to c_4, f_y and f_b are determined by fitting the yield plateau and bounding curves using the least square method, as shown in Figure 7 for a box section with steel SS400. The values of model parameters are examined for different sectional parameters; ratio of flange area to web area A_f / A_w for box and I sections, and ratio of diameter to thickness D / t for a circular section. The results for a typical example with a box section are shown in Figure 8.

The model parameters determined for the circular, I and box sections corresponding to steel SS400 are given in Table 1.

Table 1. 2SM-FS strength curves parameters (SS400).

Parameter	I or box section	Circular section
c_1	1	1
c_2	$1.23+0.83\exp\{-0.92(A_f/A_w)\}$	1.73
f_y	$1.1+0.31\exp\{-2.29(A_f/A_w)\}$	$1.23+0.83\exp\{-0.92(A_f/A_w)\}$
c_3	1	1
c_4	$1.29+0.66\exp\{-0.93(A_f/A_w)\}$	1.67
f_b	$1.08+0.26\exp\{-1.92(A_f/A_w)\}$	$1.36-1.19\text{X}10^{-3}\exp(D/t)$

4 VERIFICATION OF 2SM-FS

The cyclic sectional behavior of steel members are analyzed using the 2SM-FS and the results are compared with those of the direct integration method (DI) using 2SM-SS and experiments by Minagawa et al. (1988). The results for two typical examples will be presented in this section.

4.1 *Example 1*

The first example is a hollow box section subjected to the combined proportional axial load and bending moment. The box section has a size of $B = H = 125mm$; flange thickness of $t_f = 8.7mm$; web thickness of $t_w = 6.1mm$. The assumed material is steel SS400. Figure 9 compares the normalized axial strain $\varepsilon / \varepsilon_y$ versus axial load n and curvature ϕ / ϕ_y versus bending moment m for the 2SM-FS and direct integration method using 2SM-SS. As shown in this figure, a good correlation between the two analytical models is achieved indicating the accuracy of the 2SM-FS.

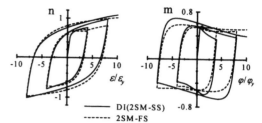

Figure 9. 2SM-FS versus direct integration method using 2SM-SS (box section).

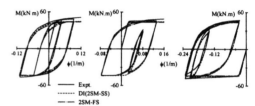

Figure 10. Comparison with experiments, SS400.

4.2 *Example 2*

The second example is the H-shaped section of $H125\times125\times6.5\times9$ with steel grade SS400 that was tested by Minagawa et al. (1988). Figure 10 illustrates comparison between the two analytical models and experiments for the normalized load-deflection curves associated with three different loading histories. As shown in this figure, both the 2SM-FS and direct integration method using 2SM-SS provide an excellent prediction of the cyclic elasto-plastic behavior for the entire hysteresis curves under random cyclic loading histories.

5 APPLICATION

The elasto-plastic behavior of steel members of beam-column type subjected to cyclic loading is predicted by the general purpose finite element analysis program FEAP (Zienkiewicz 1977) incorporating the 2SM-FS for sectional elasto-plastic behavior through Bernolli-Euler beam element. The modified approximate updated Lagrangian formulation is adopted for geometrical nonlinearity in the beam-column element formulation. The effects of local buckling, initial geometrical imperfection and residual stress are not considered in the analysis. The details of the numerical analysis can be found in the author work (Mamaghani 1996). The results by the 2SM-FS are compared with the experimental data and the numerical results from the 2SM-SS. In this section the results for two typical examples will be presented and discussed.

The first example is a fixed-ended tubular beam-column, which is subjected to constant lateral load Q at midspan and cyclic axial displacements u, as shown in Figure 11a. The geometrical parameters of the tubular beam-column are: outer diameter $D = 114mm$; wall-thickness $t = 2.3mm$; length $L = 5720mm$; and the effective slenderness ratio $KL / r = 72$. Major material parameters are: the yield stress $\sigma_y = 289MPa$; and the Young's modulus $E = 200GPa$. The parameters for strength curves in 2SM-FS are: $c_1 = 1.0$; $c_2 = 1.73$; $f_y = 1.3$; $c_3 = 1.0$; $c_4 = 1.67$; and $f_b = 1.3$. The beam-column is divided into ten elements along its length.

After applying the lateral load $Q = 0.4Q_y$ (where Q_y is the value of lateral load Q that causes the first yielding in the member in the absence of axial load) a compressive axial load (assumed as positive) is exerted first. Figures 11b and c show the normalized axial load-axial shortening and axial load-midspan deflection, respectively. The results are compared with the test data reported by Sherman (1980).

The 2SM-FS and 2SM-SS simulate the experiments quite well in the pre-and post-buckling stages of axial deformation, Figure 11b. Upon reversal of the axial deformation in tension and reloading in compression, both of the models provide a relatively

close fit to the experimental data, owing to the reasons that they: (a) take accurately into account the Bauschinger effect, which has the effect of softening and reduction in stiffness on the hysteresis curve; (b) correctly treat the yield plateau and cyclic strain hardening of the material.

As shown in Figure 11c, both the 2SM-FS and 2SM-SS lead to a valuable prediction of the axial load-midspan deflection for the beam-column, namely, the residual deflection of the beam-column at the end of the previous tensioning which has a large effect on the buckling load capacity and subsequent cyclic behavior. It is worth noting that the 2SM-FS has the advantages of its simplicity and calculation speed, which is about ten times faster as compared with the 2SM-SS.

5.2 Cantilever column

The second example analyzed is a cantilever column tested by Sueda et al. (1997); test C1. The tested column, shown in Figure 12a, had a slenderness ratio parameter of $\bar{\lambda} = 0.38$, diameter-to-thickness ratio of $D / t = 30$ $(D = 2R)$. The parameter $\bar{\lambda}$ controls the global instability and is defined as: $\bar{\lambda} = 2h / r\pi\sqrt{\sigma_y / E}$, in which, $E =$ the Young's modulus; $r =$ the radius of gyration; $h = 3073mm =$ the height of the column; $R = 300mm =$ the outer radius of the tube; and $t = 20mm =$ thickness of the tube. The material used was SS400 with the measured yield stress of $\sigma_y = 285MPa$.

The tested column was subjected to a constant axial load of $P / P_y = 0.086$ ($P_y =$ squash load) and cyclic lateral displacement of increasing amplitude at the tip. The D / t ratio inhibits local buckling of the tube. Owing to the smaller value of D / t ratio, local buckling did not occur until lateral displacement reaches at $\delta / \delta_{y0} = 6$. During the test a slight local buckling appeared in the compression side at loading point $\delta / \delta_{y0} = 6$, however, the strength did not deteriorate until lateral displacement of $\delta / \delta_{y0} = 9$, as shown in Figure 12b.

Figure 12b shows the normalized lateral load H / H_{y0} versus lateral displacement δ / δ_{y0} responses from the experiment and the 2SM-FS. The notations $H_{y0} = M_y / h$ and $\delta_{y0} = H_{y0}h^3 / 3EI$ indicate, respectively, the yield load and yield displacement (neglecting shear deformation) corresponding to zero axial load. Here, $I =$ moment of inertia. Figure 12b shows that the overall shape of the hysteresis loops from the 2SM-FS are significant close to

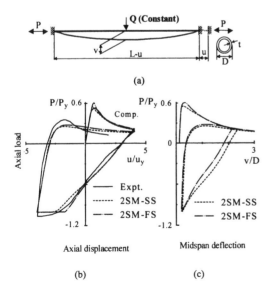

(a)

(b)　　　　(c)

Figure 11. Experiment versus analytical models. (a) fixed-ended tubular beam-column; (b) axial load P-axial displacement u; (c) axial load P-midspan deflection v.

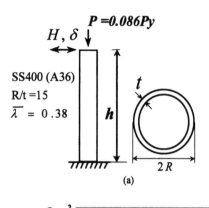

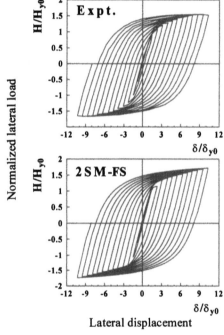

Figure 12. (a) Steel tubular column; (b) comparison of the normalized lateral load H / H_{y0} versus lateral displacement δ / δ_{y0}.

the experimental results. These results indicate that the developed formulation is accurate enough. An observed small discrepancy between experimental and analytical 2SM-FS hysteresis loops is that the predicted cyclic load capacities (Fig. 12b), in both directions of loading, are slightly larger than those of the experiment. The two possible reasons are: (1) the initial residual and geometrical imperfections are not considered in the analysis; (2) the effect of local buckling is ignored in the analysis.

6 CONCLUSIONS

This paper was concerned with the cyclic elasto-plastic analysis of steel members subjected to the combined axial force and bending moment by employing the two-surface plasticity model in force space (2SM-FS). The basic concepts and formulation of the 2SM-FS, which has been developed with extension of concepts used in the two-surface plasticity model in stress space (2SM-SS), were presented and discusses. The 2SM-FS accounts for the cyclic elasto-plastic behavior of structural steels such as the yield plateau, Bauschinger effect and strain hardening through a number of hardening rules. Cyclic plasticity performance of the 2SM-FS was verified and discussed, both at cross-sectional and structural levels, by comparing the results obtained from the 2SM-FS with those of the experiments and analyses using 2SM-SS. It has been concluded that the 2SM-FS can be used as an efficient and accurate alternative model to predict cyclic behavior of steel members. For further study local buckling effect in thin-walled members needs to be included in this model.

REFERENCES

Mamaghani, I.H.P. 1996. Cyclic elastoplastic behavior of steel structures: theory and experiments. *Doctoral Dissertation,* Department of Civil Engineering, Nagoya University, Nagoya, Japan.
Mamaghani, I.H.P. & Kajikawa Y. 1997. Cyclic inelastic sectional behavior of steel members. *Proc. of the 52th Annual Meeting, JSCE, Japan,* Vol. I-A78: 156-157.
Mamaghani, I.H.P., Shen, C., Mizuno & E., Usami, T. 1995. Cyclic behavior of structural steels. I: Experiments. *J. Engrg. Mech., ASCE,* Vol. 121(11): 1158-1164
Mamaghani, I.H.P., Usami, T. & Mizuno, E. 1996. Inelastic large deflection analysis of structural steel members under cyclic loading. *Engineering Structures,* UK, Elsevier Science, Vol. 18(9): 659-668.
Minagawa, M. Nishiwaki, T. & Masuda, N. 1988. Prediction of hysteresis moment-curvature relations of steel beams. *J. of Structural Engineering, JSCE,* Vol. 34A: 111-120.
Mizuno, E., Mamaghani, I.H.P. & Usami, T. 1996. Cyclic large displacement analysis of steel structures with two-surface model in force space. *Proc. of Int. Conf. on Advances in Steel Structures,* Pergamon, Vol. 1: 183-188.
Shen, C., Mamaghani, I. H. P., Mizuno, E. & Usami, T. 1995. Cyclic Behavior of Structural Steels. II: Theory. *J. Engrg. Mech., ASCE,* Vol. 121:11, 1165-1172.
Sueda, A., Yasunami, H., Mizutani, S., Kobayashi, Y. & Nakagawa, T. 1997. A study on ductility of steel pier under cyclic loading. In Usami, T. (eds), *Proc. 5th Int. Coll. stability and ductility of steel structures,* Nagoya, Japan, 237-244.
Suzuki, T., Mamaghani, I.H.P., Mizuno, E. & Usami, T. 1995. Finite displacement analysis of steel structures with two-surface plasticity model in stress resultant space. *Proc. Of the 50th Annual Meeting, JSCE, Japan,* Vol. 1(A): 110-111.
Zienkiewicz, O.C. 1977. *The finite element method.* 3rd Ed., McGraw-Hill, New York.

Behaviour of Steel Structures in Seismic Areas, Mazzolani & Tremblay (eds) © 2000 Balkema, Rotterdam, ISBN 90 5809 130 9

Member response to strong pulse seismic loading

G. Mateescu
Building Research Institute INCERC Timisoara, Romania

V. Gioncu
Department of Architecture, University 'Politehnica' Timisoara, Romania

ABSTRACT: The member response under strong pulse seismic loading is studied both theoretical and experimental. This loading type is typically for near-source earthquakes, where the horizontal components are accompanied with important vertical components. The pulse loading induces in the structural members a fracture collapse during the first cycles, impending a good dissipation of the input seismic energy.

1 INTRODUCTION

In the evaluation of seismic design philosophy the most effective teacher is the examination of the impact of an earthquake on full-scale buildings. No theory or design method can be accepted unless it correctly explains what happens during these natural tests. The process of learning from earthquakes typically involves three major steps: (i) detection of new aspects that are not considered in the existing codes; (ii) theoretical evaluation of these new revelations; (iii) experimental verification of the theoretical developments. Once the three-step process is completed and a consensus among the earthquake designers is established, the lessons have been properly learned, so the implementation in the form of a code provision may be performed (Mc Clure, 1989). In this way, after the Northridge and Kobe earthquakes, there are many discussions about the adequacy of the existing codes for these events. Because both earthquakes occurred beneath a heavily urbanized and very well instrumented area, they provided much new information about the behaviour of structures located near to sources. The conclusions of these debates are that for short distances from the epicenter the present code provisions cannot predict in proper manner the recorded loading and the behaviour of structures. Only for far from the source areas these provisions give a reasonable estimation.

This paper deals with the member response to strong seismic loading, characteristic for near-source earthquakes. Both theoretical studies and experimental tests were performed in order to find out the main characteristics of the member collapse subjected to this loading type.

2 STRONG PULSE SEISMIC LOADING

The near-source area can be defined as the region within a few kilometers of either the surface rupture or the projection on the ground surface of the fault rupture zone. This region is also referred as the near-field region (Iwan, 1995, Gioncu, 1999). The far-source regions are situated at some hundred kilometers from the source.

Unfortunately, the design methods adopted in the majority of codes are mainly based on records obtained from far-source fields. Only the last UBC 97 has introduced some supplementary provisions concerning the near-source earthquakes, considering the lessons learned from the last great events.

The main characteristics of near-source earthquakes (Gioncu et al, 2000) are: (i) directivity of the actions along the direction of fault rupture; (ii) very strong pulse loading; (iii) important vertical components, in some cases higher than the horizontal ones; (iv) high velocity of the ground motion. Due to these unusual actions, numerous steel structural buildings from Northridge and Kobe were severely damaged.

It has been observed that the flanges of the welded zones at the connections were fractured in brittle manner, without evidences of plastic deformations. From the damage analyse, correlated with the above mentioned characteristics of the ground motion, it has been recognised that the members and joints may suffer high pulse forces; because of the simultaneous action of both horizontal and vertical ground motions, the failure could result after a single or few failure plastic excursions (Kohzu and Suita, 1996). Therefore, instead of a cyclic behaviour of the structure, able to dissipate an important amount of seismic energy, a premature failure occurs and consequently, the dissipated energy is under the code estimation.

The pulse loading may arise from acceleration, velocity or displacement ground motion pulse. Examining a great number of earthquakes recorded in epicentral areas (Gioncu et al, 2000, Hall et al, 1995) the conclusions were that the pulse characteristic of the ground motion is related to velocity (Fig.1). The pulses can have an asymmetry related to the maximum value. One till three adjacent pulses, or some distinguished pulses could compose the ground motion. A detailed examination of the pulse periods for maximum intensities have shown that the range for horizontal components is within 0.25... 1.0 sec, while for vertical components, within 0.15...0.5 sec. Generally, in case of intraplate earthquakes, the pulse periods are shorter than the ones of interplate earthquakes.

3 SPECTRA FOR PULSE LOADING

The seismic response of a structure subjected to a pulse loading can be correctly determined only through direct non-linear dynamic time-history analysis. As a result of the extensive monitoring of the areas with potential sources it was possible to collect a large number of records. But all these records have very high random characteristics due to the particularities of source, travel path, soil conditions, magnitude, influence of neighbouring buildings, etc. There exist a large variability in the accelerogram characteristics able to influence the seismic response of structures, causing a very high scattering of the response values. Therefore, it is difficult to select an accelerogram to have similar characteristics with those expected at the structure site.

So, as an alternative to earthquake recordings, the use of artificial generated accelerograms has been allowed by some recent codes. The control parameters of these accelerograms should be chosen with regard to the characteristics of the expected earthquakes at the site, depending on the generic mechanisms and the position of the site in relation to the focal area. Fig.2 shows a Northridge recorded accelerogram (Elysian Park, Hall, 1995) and the artificial generated one. One can see that is more adequate to use an artificial accelerogram for pulse

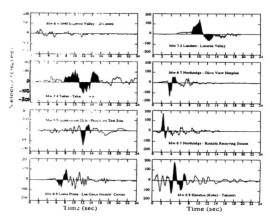

Figure 1. Velocity pulses (after Hall et al)

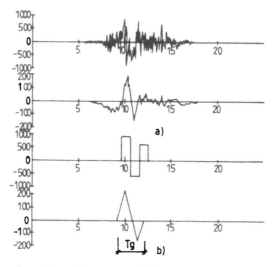

Figure 2. Recorded vs. generated seismogram

loading than for cyclic ones, because the number of velocity peaks is reduced and the earthquake duration is shorter.

The main simulated characteristic of the ground motion is the pattern of velocity pulse, defined by the velocity peak, pulse period and number of pulses. The pulse periods are very short ($T_g < 0.5$ sec) for intraplate earthquakes and long ($T_g > 1.0$ sec) for interplate earthquakes. So, the following artificial ground motions were generated for:
(i) Direction of the actions:
- horizontal;
- vertical.
(ii) Ratio of positive and negative velocity peaks:
- $\upsilon = 0.6$; 1.0 (symmetry); 1.6.

(iii) Duration of the pulse:
- horizontal components: $T_g = 0.1; 0.2; 0.3; 0.4;$ 0.5 and 1.0 sec;
- vertical components: $T_g = 0.05; 0.1; 0.2; 0.3; 0.4;$ and 0.5 sec.

(iv) Number of pulses:
- one single pulse;
- two adjacent pulses;
- two distinguished pulses.

For a single pulse the relation between velocities $v_{max, min}$ and the acceleration peaks are:

$$a_{max1} = \frac{4v_{max}}{T_g} = \alpha g \; ; \qquad 0 < t < \frac{T_g}{4} \quad (1)$$

$$a_{min} = \frac{2(v_{max} + |v_{min}|)}{T_g} = \frac{1+v}{2}\alpha g \; ; \quad \frac{T_g}{4} < t < \frac{3T_g}{4} \quad (2)$$

$$a_{max2} = \frac{4|v_{min}|}{T_g} = v\alpha g \; ; \qquad \frac{3T_g}{4} < t < T_g \quad (3)$$

where

$$v = \frac{v_{max}}{|v_{min}|} \; ; \qquad \alpha = \frac{a_g}{g} \qquad (4a,b)$$

v being a characteristic of the velocity pulse asymmetry and α, a measure of the acceleration.

Figure 3 shows the elastic spectra for horizontal and vertical components of ground motions for different pulse periods and patterns. The comparison with the EC8 spectrum shows in the range of reduced periods, a higher amplification than the one considered in the code (Fig.3a,b). In exchange, for the horizontal component periods of steel structural significance, the code design spectra appear to be very conservative. In case of vertical components, the high amplification field corresponds with the vertical structure periods (Gioncu et al, 2000). Parametric studies have shown that the natural period and the asymmetry of the pulse are the main parameters of artificial accelerograms. It is important to notice that the pulse with highest value of first peak gives the maximum amplification. In case of multiple pulses, the amplification is maximum for adjacent pulses (Fig.3c,d).

The comparison of spectra obtained from artificial accelerograms and the ones from recorded ground motions (Fig.4) shows a very good correspondence, especially for the cases of intraplate earthquakes.

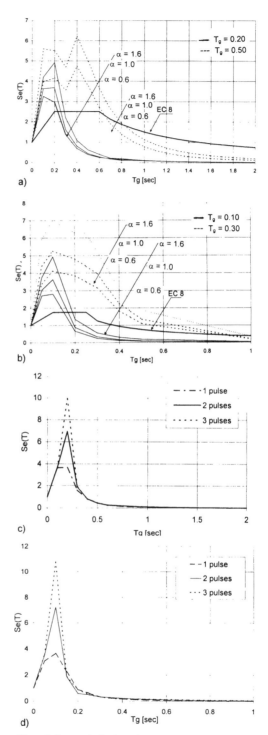

Figure 3. Spectra for horizontal and vertical pulse earthquakes

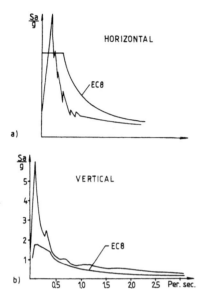

Figure 4. Spectra for recorded intraplate earthquakes

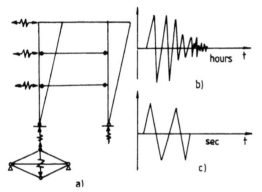

Figure 5. Member response into a structure

4 STANDARD BEAM RESPONSE FOR PULSE LOADING

The member behaviour in a structure is studied using the standard beam (three-point beam), determined in function of the plastic hinges positions and the inflection points. Neglecting the influence of gravitational loads, in this study, the standard beam span is determined by the span of the structure beam (Fig.5a). The beam moments increase if an asychronism of vertical components occurs. The loading modeling depends on the adopted design method. If a static method is used, the loading system must be static alternate but, opposed to ECCS rules, decreasing values must be adopted, the first cycle being the strongest (Fig.5b). In this case,

the main problems of the design are the determination of accumulated plastic deformations and the fracture rotation (Anastasiadis et al, 2000). If a dynamic method is considered, the loading must be of pulse type. In this case the influence of the strain-

Table 1. Dimension of beam specimens

Specimen	B (mm)	d (mm)	t_f (mm)	t_w (mm)	L (mm)
A0	150	168	6.4; 6.6	6.4	1900
A1			6.9	6.4	
A2			6.4	6.4	
A3			6.1; 6.4	6.9	
A4			6.5; 6.6	6.6	
A5			6.5; 6.6	6.7	
A6			6.6; 6.7	6.6	
A7			6.2; 6.4	6.4	
A8			6.4; 6.9	7.0	
A9			6.9	6.9	
A10			6.4	6.4	

Table2. Mechanical properties

Specimen	f_y (N/mm^2)	f_u (N/mm^2)	Elong. %	YR (=f_y/f_u)
A0 – f*	322	410	55.7	0.785
A0 – f	322	393	66.7	0.819
A0 – w**	412	532	49.7	0.774
A0 – w	397	532	45	0.746
A3 – f	358	500	66.7	0.716
A3 – f	344	431	59.3	0.798
A3 – w	364	507	53.3	0.718
A3 – w	364	513	48.0	0.710
A4 – f	386	530	45.0	0.728
A4 – f	384	527	48.0	0.729
A4 – w	412	532	55.7	0.774
A4 – w	397	532	49.7	0.746

* - flange; ** - web

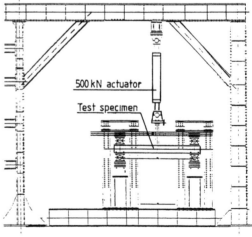

Figure 6. Testing set up

rate and the number of pulses till the fracture are the most important characteristics of the beam behaviour (Fig.5.c) (Gioncu, 2000).

5 EXPERIMENTAL RESULTS

In order to determine the response of beams subjected to different loading types, a number of 11 beams were tested. The specimens are welded 150x180x6 I-shaped (Fe 360) beams. The plate dimensions of the profile were proportioned to admit buckling in the post-elastic range only ($b/2t_f$=12.5, d/t_w=28). The measured cross-sectional dimensions for the two flanges, web and beam span are listed in Table1. Table 2 shows the mechanical properties of some specimen material derived from tension tests.

The beams were simply supported and central loaded by a 500 kN actuator, as shown in Fig.6. The loading types were monotonic, cyclic-alternate with increasing or decreasing forces (Fig.7a,b) and displacement pulse with one, two or three cycles (Fig.7c,d,e).

Beams A2 and A3 were used only to calibrate the pulse tests, in order to establish the maximum possible velocity, without loosing the registration computer program. During the monotonic and cyclic alternate tests, deformations (deflection on the mid span and support rotations) were adequately measured and automatically registered. By the pulse tests, only the loading-displacement curve of the actuator could be measured.

Fig.8 presents the moment-rotation curve for the beam A0. The experimental values of the beam characteristics are: M_p = 69.348 kNm, θ_p = 0.0094 and θ_u = 0.093 rad, compared to the theoretical values from Table 3 show a very good correspondence. The failure mode (Fig.11) shows the local buckling of the compressed flange.

The next beams were loaded cyclic alternate, with increasing branch (A1) and decreasing (A4), respectively. The failure mode of beam A1 was symmetric, with flange local buckling (Fig.9a and

Table 3. Theoretical rotation parameters

Loading Type	Specimen	M_p [kNm]	θ_p [rad]	θ_u [rad]
Monotonic	A0	69.32	0.009	0.0834
cyclic	A1	72.83	0.009	0.0834
alternate	A4	84.17	0.0096	0.0834
1 pulse	A5	84.44	0.0108	0.0834
	A6	85.22	0.0108	0.0834
2 adjacent	A7	81.00	0.0109	0.0834
pulses	A9	88.67	0.0109	0.0837
3 adjacent	A8	86.27	0.0109	0.0834
pulses	A10	82.06	0.0108	0.0834

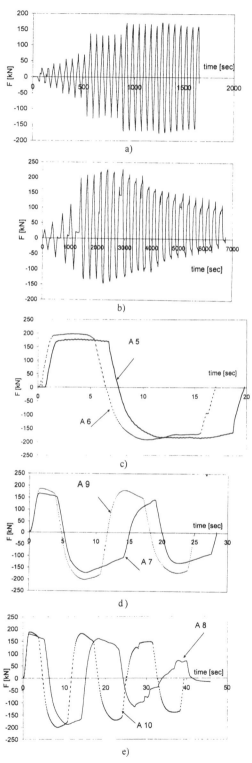

Figure 7. Loading types

59

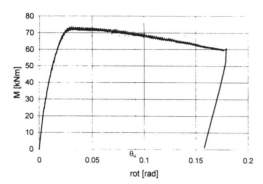

Figure 8. Moment-rotation curve for beam A1

Fig.10a,b shows the load displacement curves for the beams A5 and A6, with strain-rates of 0.065 and 0.0417 sec^{-1}, respectively. Both beams revealed a symmetric mode shape with flange local buckling.

The two displacement pulses involved a more severe failure with fractures of the flanges during the second pulse at the beam A7 (Fig.10c) and cracks along the flange-web connection, close to the

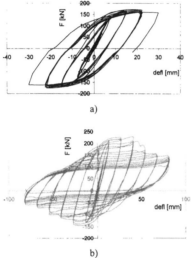

Figure 9. Force-displacement curves

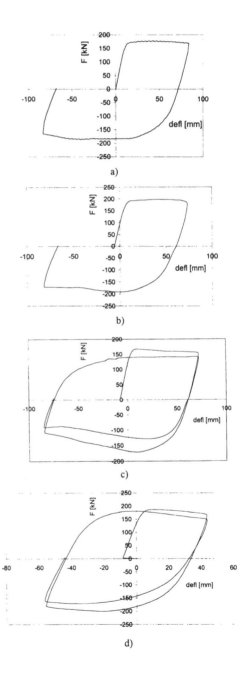

Fig.11). The failure mode of A4-beam (Fig.9b and Fig.11) was a highly unsymmetric out of plane mode shape, with flange fractures in one half-span only. Cracks occurred in the flange-to-web welds at maximum displacement values, followed by flange fractures during the decreasing cycles.

If the cyclic behaviour of beams in bending is already known from a great number of tests, the behaviour under displacement pulses represented a quite different problem to be solved. Taking into account of the maximum deflection of the actuator, the displacement amplitude was chosen equal to 167.6 mm (± 83.8 mm for each semi-cycle) for beams A5, A7 and A8. For the beams A6, A9 and A10, the loading were realised with a smaller amplitude, of 112,4 mm (± 56.2 mm), in order to have a comparison term for the behaviour under different velocities and strain-rates.

60

e)

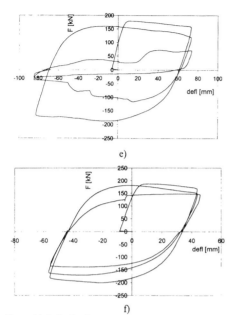

f)

Figure 10. Pulse loading

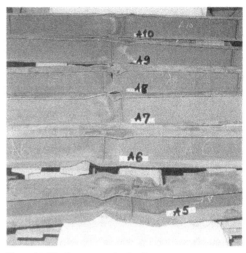

Figure 12. Displacement pulse loading

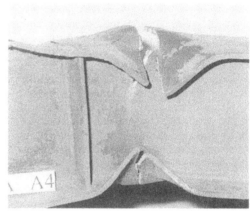

Figure 13. Details of beam A4

Figure 11. Static monotonic and alternate loading

stiffener in one half-span, at A9 (Fig.10d). The strain-rate was of about 0.085 sec^{-1} for both beams.

Finally, the three pulses test conducted to the total fracture (flanges and web) of the beam A8 (Fig.10e) during the third cycle, in one half-span, as could be seen also from Fig.14. At beam A10 (Fig.10f), with smaller amplitude of the displacement, there were observed only cracks in the flanges. The strain-rate was equal to 0.095 sec^{-1}.

Figure 14. Details of beam A8

6 CONCLUSIONS

The examination of near-source earthquakes has shown that the main characteristic of the ground motion is the velocity or displacement pulse.

The artificial spectra for horizontal and vertical pulse points out in evidence the high amplification of accelerations in the field of short structure periods. Due to this characteristic, in many cases the common ductile cyclic behaviour is replaced by a brittle fracture after the first or second cycle. The experimental tests performed for monotonic loading, cyclic with increasing or decreasing loading steps, or for one, two or three pulses with significant strain rate have shown very different behaviour types.

Welded symmetric I-beams were tested on bending under pulse loading, and the effects of the pulse amplitude and velocity were evaluated. It was found that a drastic failure occurs in case of three adjacent pulses with fractures, which affect the flanges and web. While one pulse action involves only cracks in the flange-to-web weld zone, two displacement pulses conduct to fractures of the flanges, beginning from the middle axis directed to the borders of the plate; three pulses produce the fracture of all the cross-section. Generally, the failure mode developed in one half-span only, with a highly unsymmetric buckling region.

So, the use of static performance of the members cannot be applied to the structural analysis for seismic loading, and the use of the EC3 provisions for the seismic checking, as it is provided in EC8 is a very doubtful methodology.

REFERENCES

Anastasiadis, A., V.Gioncu & F.M.Mazzolani 2000. New trends in the evaluation of available ductility of steel members. In F.M.Mazzolani & R.Tremblay (eds) *Behaviour of Steel Structures in Seismic Areas, STESSA 2000, Montreal, 21-24 August, 2000.*

McClure, F.E. 1989. Lessons learned from recent moderate earthquakes. In K.H Jacob & C.J Turkstra (eds) *Earthquake Hazards and the Design of Constructed Facilities in the Eastern United States. Annales of the New York Academy of Sciences.* 558: 251-258.

ECCS – CECM 1986. Recommended Testing procedure for Assessing the Behaviour of Structural Steel Elements under Cyclic Loading. *Doc.ECCS TWS 1.3 N.45.*

Gioncu, V. 1999. Design criteria for seismic resistant steel structures. *Seismic Resistant Steel Structures: Progress and Challenge, CIM Courses, Udine, 18-22 October, 1999.*

Gioncu, V. 2000. Influence of strain-rate in the behaviour of steel members. In F.M.Mazzolani & R.Tremblay (eds) *Behaviour of Steel Structures in Seismic Areas, STESSA 2000, Montreal, 21-24 August, 2000.*

Gioncu, V., G.Mateescu, L.Tirca & A.Anastasiadis 2000. Influence of the type of seismic ground motions. In F.M.Mazzolani (ed) *Moment Resistant Connections of Steel Building Frames in Seismic Areas.* London, E&FN Spon.

Hall, J.F, T.H.Heaton, M.W. Halling & D.J.Wald 1995. Near source ground motion and its effects on flexible buildings. *Earthquake Spectra,* 11 (4): 569-605.

Hall. J.F. 1995. Parameter study of the response of moment resisting steel frame buildings to near-source ground motions. *Technical Report SAC* 95-05: 1.1 – 1.83.

Iwan, W.D. 1995. Drift demand spectra for selected Northridge sites. *Technical Report SAC.* 95-05: 2.1-2.40.

Kohzdu, I. & K.Suita 1996. Single or few excursions failure of steel structural joints due to impulsive shocks. *11th World Conference on Earthquake Engineering, Acapulco 26-28 June, 1996.* CD-ROM Paper 412.

Behaviour of Steel Structures in Seismic Areas, Mazzolani & Tremblay (eds) © 2000 Balkema, Rotterdam, ISBN 90 5809 130 9

Calibration of thin-walled members ductility

A. Moldovan
Building Research Institute, Timisoara, Romania

D. Petcu
West University, Timisoara, Romania

V. Gioncu
Polytechnic University, Timisoara, Romania

ABSTRACT: The aim of this paper is to provide data on the use of thin-walled members in seismic resistant structures. The ductility of thin-walled sections in the plastic area is studied by evaluating the plastic moment, the ultimate moment, the post-critical behaviour and the rotation capacity of thin-walled members. The evaluation of the rigid-plastic curve shape of the plate is an important step for estimating the ductility of thin-walled members. This curve was theoretically determined and validated by experimental results. For a rapid determination of the ultimate rotation and the ductility of thin-walled members, a computer program named DUCTROT TWM has been elaborated. In the paper the behaviour factor q is determined for built-up beams formed by two channels or lipped channels.

1. INTRODUCTION

It is well known that thin-walled sections have a reduced ductility because their bending behaviour is strongly affected by the elastic local buckling of the compressed walls. Due to this sensitivity at instability phenomena, thin-walled structures are included in practice of seismic design in the category of non-dissipative systems. But the examination of the complete load-deformation history of thin-walled members pointed-out that their exclusion from the category of structures able to dissipate a certain amount of seismic energy must be revised.

Even if elastic buckling of the compression walls occurs, an important amount of plastic deformations is concentrated in the proximity of the corners of the profiles in the so-called "effective width". So, for structures with one or two levels and for low or medium earthquakes with a reduced number of incursions in the plastic range, thin-walled structures are able to dissipate a reduced seismic energy, which can be considered in the seismic design.

2. ANALYSIS METHODS FOR THIN-WALLED MEMBERS DUCTILITY

To calculate the ductility of a member, the rigid-plastic curve shape is important to be drawn. It depends on the type of plastic mechanism of the element, which is essentially governed by the local plastic mechanism of buckled plates of the section.

Thin walled steel plates can develop three types of local plastic mechanism, namely the pyramid-shaped, the roof-shaped and the flip-disc mechanism. For a simply supported plate they are observed in Figure 1 (a- pyramid-shaped, b- roof-shaped, c- flip-disc).

The rigid-plastic curve can be determined knowing the average axial stress σ, whose characteristic equation for each type of mechanism was presented in a previous paper (Moldovan et. al. 1999).

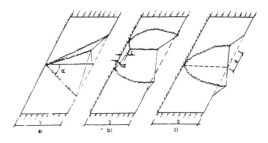

Figure 1. Types of plastic mechanisms for simply supported thin-walled plates.

For the idealised roof mechanism the stress σ is given in function of the deflection of the flange □, the yield stress f_y and the basic geometric parameters α and c, which were assumed to be 30^0 and 0,2b respectively, to obtain the rigid-plastic curve (Mahendran & Murray 1991).

For the idealised flip-disc mechanism of parabolic shape, the average axial stress σ is obtained considering the ratio a/b=0.1 (Mahendran & Murray 1991). For the pyramid mechanism α is taken 45^0 (Feldman 1994).

It should be noted that the equations for the average axial stress were determined for doubly supported plates, but they can be also used for simply supported plates if half of the mechanism shape is used (Gioncu & Mazzolani 1999). So these mechanism types can describe the collapse shapes of beam flanges.

In evaluating the ductility of thin-walled beams, two-types of section commonly used for beams and columns were considered: I section formed of two channel profiles and I section formed of two lipped channel profiles (Fig. 2). Since the webs of these sections have the thickness of 2t, the local buckling will only affect the compressed flanges. The three types of plastic mechanisms discussed above were considered.

The local buckling of the compressed flange causes the decrease of its capacity, which is expressed by:

$$C(\theta) = C_p \frac{\sigma}{f_y} \qquad (1)$$

where C_p is the load at which the flange is fully plastified and σ/f_y depends on the type of the mechanism. Local buckling causes the reduction in the width of the compressed flange and neutral axes to shift from the midpoint of the web. Distance z_0 from the compressed flange to the neutral axes is determined from internal forces equilibrium (fig.2). The rotation θ and the plastic moment M should be calculated. Plotting the plastic moment M versus rotation θ can draw the rigid-plastic curve.

The intersection of the elastic and rigid-plastic curves can be used to estimate the ultimate load and the framework around the actual behaviour of thin-walled members can be drawn.

For the I section formed by two lipped channels it is considered that stiffeners of the compressed flange are buckled, so their capacity is not taken into account, which is conservative for the analysis.

The ultimate rotation of thin-walled beams can be established by intersection of the rigid-plastic curve and the reduced plastic moment curve. The reduced

plastic moment is the plastic bending moment for the effective section of the member affected by the safety factor γ_M.

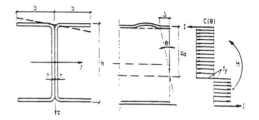

a. I section formed by channels

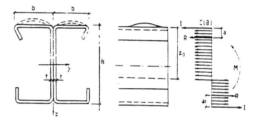

b. I section formed by lipped channels

Figure 2. Internal forces equilibrium for I section formed of channels.

The above equations are determined for first-order rigid-plastic theory in which the influence of deformations is neglected. This is an appropriate hypothesis for the profiles that belong to class 1 and 2, questionable for class 3 and wrong for class 4. The thin-walled cross sections are affected by geometrical imperfections. Their influence on local collapse mechanism is increasing during the post-critical behaviour due to the second-order effects, producing a reduction of beam capacity and rigidity. For this reason, the theoretical values of plastic moments obtained for flip-disc mechanism must be rectified:

$$M_c(\theta) = f(i)M(\theta) \qquad (2)$$

where $M_c(\theta)$ is the reduced moment and f(i) a function considering the influence of initial geometrical imperfections and the second-order effects. Because this function is very difficult to be determined theoretically, a calibration must consider the post-critical values experimentally obtained.

3. COMPARISON OF THEORETICAL RESULTS WITH EXPERIMENTS

3.1. Improvement of theoretical rigid-plastic curves

We have the results obtained on 5 beams with built-up section formed by two channels and 5 beams with built-up section formed of two lipped channels. Two equal forces with a distance of 1.0 m between them were applied symmetrically on a beam of 3.0 m span. Their post-critical behaviours were followed and the complete moment-rotation (or load-defection) curves have been plotted (De Martino et al. 1992).

The load-deflection and moment-rotation curves put in evidence the different behaviour of stiffened versus unstiffened sections. For sections formed of channels, flange buckling is more progressive, with a slower reduction of strength in the softening branch. On the contrary, the presence of edge stiffeners increases the load carrying capacity, but their buckling causes an immediate loss of load-carrying capacity (Fig. 3).

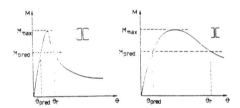

Figure 3. Features of moment-rotation curves experimentally determined

The theoretical rigid-plastic curves were plotted together with experimental curve for each specimen. It should be noted that pyramid-shaped mechanism gives a very close curve with roof-shaped mechanism, so only roof mechanism and flip-disc mechanism curves were represented. Flip-disc mechanism curve gives smaller moments for the same plastic rotation and is closer to the experimental curves.

The imperfection function is proposed to have the expression:

$$f(i) = \frac{1}{\xi(\theta)} \qquad (3)$$

where $\xi(\theta)$ is obtained similar to the imperfection factor of the Ayrton-Perry relation. For the section formed of channel profiles:

$$\xi(\theta) = \frac{\dfrac{y_{0\,max}}{t}}{1 - \dfrac{C(\theta)}{I}} = \frac{\dfrac{y_{0\,max}}{t}}{\eta\,\dfrac{C(\theta)}{I}} \qquad (4)$$

where y_{0max} is the maximum imperfection of the flange in accordance with EC3 and I is the tension capacity of the flange. To be in accordance with experimental results, for this section $y_{0max} = t$ and $\eta = 1,8$. Hence, the plastic moment becomes:

$$M_c(\theta) = 1,8\,\frac{C(\theta)}{I}\,M(\theta) \qquad (5)$$

For the section formed by lipped channel profiles the authors propose the following expression:

$$\xi(\theta) = \frac{\left(\dfrac{y_{0\,max}}{t}\right)^2}{1,8\,\dfrac{C(\theta)}{I}} \qquad (6)$$

and the rectified plastic moment is:

$$M_c(\theta) = 1,8\,\frac{C(\theta)}{I}\left(\frac{t}{y_{0\,max}}\right)^{\frac{1}{2}}M(\theta) \qquad (7)$$

In Fig.4 (for sections formed of lipped channels) and Fig.5 (for sections formed of channels) theoretical and experimental curves for tested specimens are plotted. The rectified curve is very close to the experimental one. Between curves (3) and (4) an area of influence of the imperfections is created, which is greater for sections formed by lipped channels. In some cases (P7, P8), actual imperfections were smaller than the considered ones.

3.2. Ductility of thin-walled beams

Having considered the rectified curves, the rotation θ_{pred} corresponding to the first yielding and the ultimate rotation θ_r of thin-walled beams were determined (Fig.3). The rotation capacity R (which expresses the ductility) has the expression:

$$R = \frac{\theta_r}{\theta_{pred}} - 1 \qquad (8)$$

For these 10 tested specimens, the main results are presented in Table 1.

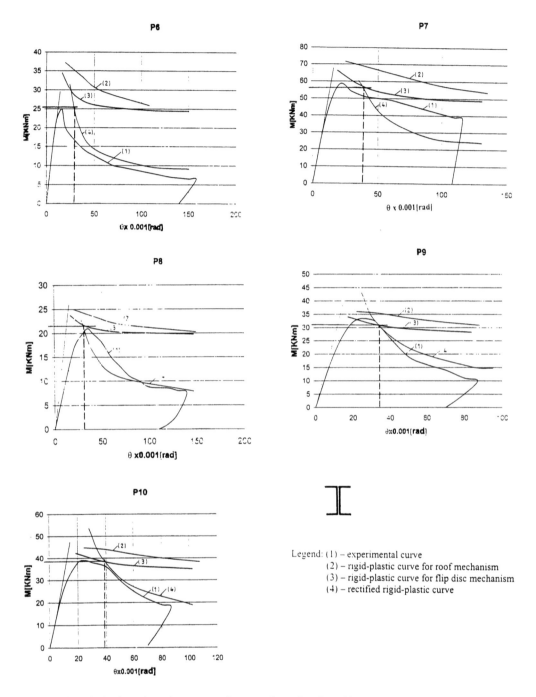

Figure 4. Theoretical and experimental moment-rotation curves for sections formed by two channels.

Legend: (1) – experimental curve
(2) – rigid-plastic curve for roof mechanism
(3) – rigid-plastic curve for flip disc mechanism
(4) – rectified rigid-plastic curve

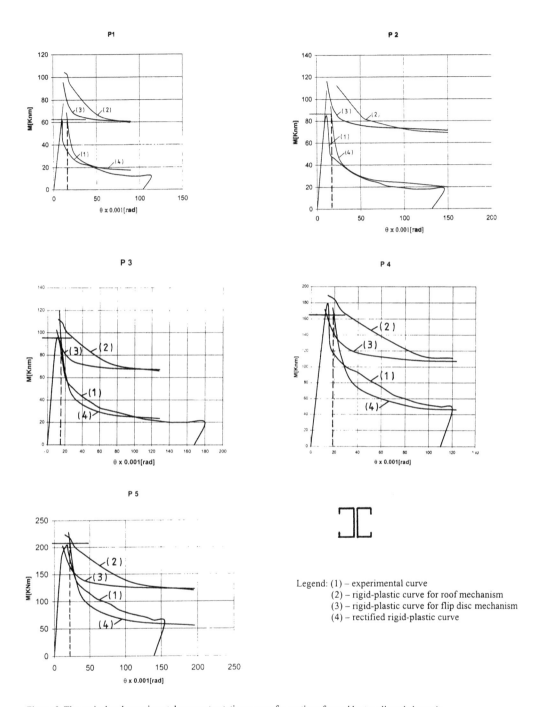

Legend: (1) – experimental curve
(2) – rigid-plastic curve for roof mechanism
(3) – rigid-plastic curve for flip disc mechanism
(4) – rectified rigid-plastic curve

Figure 5. Theoretical and experimental moment-rotation curves for sections formed by two lipped channels.

Table 1. The ultimate rotation and the rotation capacity of tested beams.

Test	Dimensions [mm]	θ_r rad	R
P1	300x200x50x2	0,0164	0.56
P2	300x200x60x2.5	0.0169	0.64
P3	300x170x70x3	0.0149	0.77
P4	300x200x80x4	0.0194	1.04
P5	300x200x90x5	0.0222	1.44
P6	200x100x2.5	0.0297	1.12
P7	200x100x5	0.0385	1.83
P8	200x40x3	0.0306	1.20
P9	200x50x4	0.0343	1.58
P10	200x50x5	0.0404	1.99

4. DUCTROT TWM COMPUTER PROGRAM

Since the method presented above for evaluating the plastic rotation capacity is complicated enough, the authors of the present paper have elaborated a computer program named DUCTROT TWM (DUCTility ROTation of Thin-Walled Members). The program consists of 8 sheets.

Some of the most important of them are presented in Fig.6. To obtain the thin-walled members ductility the following initial data are to be introduced: type of considered section,

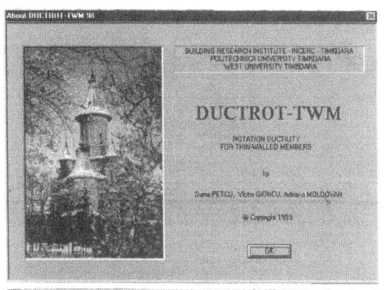

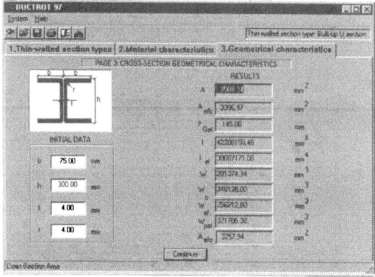

Figure 6. Sheets of DUCTROT TWM computer program.

material and geometrical characteristics and loading system. The program gives: plastic moment and reduced plastic moment, plastic and reduced plastic axial forces, slenderness of the element, plastic rotation, ultimate rotation and rotation capacity. The moment-rotation curve is also plotted.

DUCTROT TWM program can figure a parametric variation of ductility by representing the ultimate rotation and the rotation capacity in function of the following parameters: width of the flange, depth of the web, thickness of the section, axial load, yield limit and length of the beam.

5. PARAMETRICAL STUDIES

The obtained ductility values for experimental tests, as well as parametrical studies performed with DUCTROT TWM computer program have permitted to establish the parameters with a great influence on the ductility of thin-walled beams.

These are: the thickness of the sheet and the span of the beam and moreover for sections composed of lipped channels, the depth of the web.

The spans of the beams in correlation with the thickness of the sheet for which the q factor can be considered equal to 2 are presented in fig.7. In

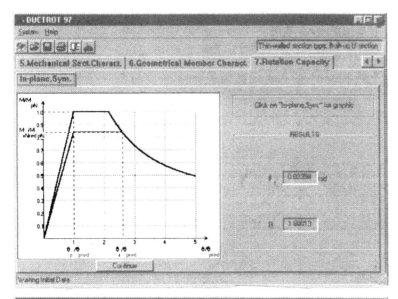

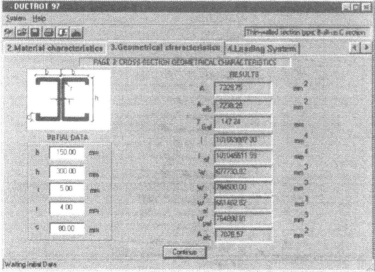

Figure 6 (continued).

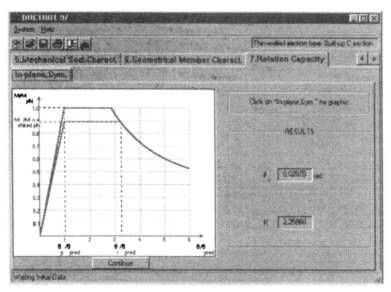

Figure 6 (continued).

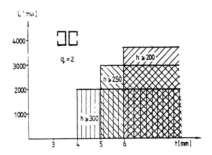

a. Sections formed of two channels.

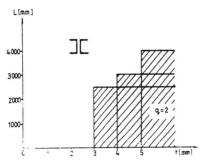

b. Sections formed of two lipped channels.

Figure 7. The range for which q = 2.

fig.7b the depth of the web is also taken into account.

6. CONCLUSIONS

The ductility of thin-walled beams with built-up sections is evaluated with theoretical methods in order to provide useful data on their use in seismic resistant structures. The comparison between theoretically possible rigid-plastic curves for the post-critical range of thin-walled beams and the experimental curves has established that flip-disc mechanism is characteristic for the types of sections which were considered, deciding their plastic behaviour.

The influence of imperfections was taken into account by rectifying the theoretical curves established with the mechanism equations in concordance with experimental results. More experimental tests have to be carried on to emphasise this conclusion and to calibrate more accurately the imperfection functions.

The obtained values of the ductility of the considered sections put in evidence that the adding of the stiffeners to a section increases its load carrying capacity, but on the contrary, the ductility of the section decreases. As a consequence, sections composed of lipped channels have a much lower ductility than sections formed of channel profiles.

The behaviour of other types of sections used at thin-walled members (for example box sections) and the influence of the axial load in the post-critical range must be analysed for the future.

REFERENCES

De Martino A., Ghersi A., Landolfo R., Mazzolani F.M. 1992: Calibration of a bending model for cold-formed sections. *XIth International Speciality Conference on Cold-Formed Steel Structures*: 503-511. St.Louis.

Feldman M. 1994. Zur Rotazionskapazität von I – Profilen statisch und dynamisch belasteter Träger. *Ph.Doctoral Thesis*. Aachen Universität.

Gioncu V., Mazzolani F.M.1999. Ductility of Seismic Resistant Steel Structures. *E&FN Spon* (manuscript).

Mahendran M., Murray N.W. 1991. Effect of initial imperfections on local plastic mechanisms in thin steel plates with in-plane compression. *International Conference on Steel and Aluminium Structures*: 491-500. Singapore.

Moldovan A., Petcu D., Gioncu V., 1999. Ductility of thin-walled members. *The 6th International Colloquium on Stability & Ductility of Steel Structures*: 299-307. Elsevier.

Behaviour of Steel Structures in Seismic Areas, Mazzolani & Tremblay (eds) © 2000 Balkema, Rotterdam, ISBN 90 5809 130 9

Fatigue damage modelling of steel structures

R. Perera, S. Gómez & E. Alarcón
Department of Structural Mechanics, Technical University of Madrid (UPM), Spain

ABSTRACT: In the present work a constitutive model is developed which permits the simulation of the low cycle fatigue behavior in steel framed structures. In the elaboration of this model, the concepts of the mechanics of continuum medium are applied on lumped dissipative models. In this type of formulation an explicit coupling between the damage and the structural mechanical behavior is employed, allowing the possibility of considering as a whole different coupled phenomena. A damage index is defined in order to model elastoplasticity coupled with damage and fatigue damage

1 INTRODUCTION

The modern approach to the seismic design of structures accounts for dissipation of the seismic energy input through plastic deformations. A ductile response is characterized by the structure's ability to undergo large inelastic displacements without loss in the load carrying capacity. The evaluation of the structural performance requires the definition of parameters to characterize the structural damage. Traditionally, ductility has been employed as the principal criterion for design. However ductility does not account for the duration of ground shaking which is very important in inelastic design since the combined effects of ductility and energy absorption may lead to failure even at modest ductility demands. To suitably account for the performance of the structures in the design procedure, it is necessary to assess accurately the damage accumulation which progressively reduces the mechanical properties of the structural components subjected to plastic strains under earthquakes.

As an alternative to ductility based design, energy may be used as the basis for design. The simplest way to evaluate the cumulative damage using an energy approach consists on summing the inelastic deformations. However, this approach does not take into account the fact that the damage due to a large number of small plastic deformations may be less than one due to a smaller number of large plastic deformations. To overcome this problem, another way of thinking about energy is to use the concept of low cycle fatigue. Since the deformation histories are composed of random cycles, the structural

damage is governed both by the maximum plastic displacement and by the dissipated energy. Then the low cycle fatigue approach appears to be a very interesting approach. It is possible to express the duration effects of an earthquake in terms of a effective number of cycles of loading and, in a similar way, to consider the energy absorption capacity in terms of a number of displacement cycles.

On the other hand, in the last years the fatigue study has been reorientated through its incorporation in the Continuum Damage Mechanics (CDM) (Lemaitre and Chaboche 1985; Lemaitre 1993). The same concepts used in CDM to model ductile failure can be extended to the low cycle fatigue damage processes, where plasticity is the key mechanism for crack initiation. Damage Mechanics deals with damage as a continuum variable and, because of it, CDM models including plasticity and damage can predict ductile crack initiation. An extension of themselves including the number of cycles could be suitable to simulate the low cycle fatigue damage. According to it, Chaboche (1985) developed a formulation for damaged materials where the fatigue phenomenon was incorporated in the CDM. However, only harmonic loads were considered being the hypotheses of fatigue cumulative damage suitable.

In this paper, a non linear damage model extended to consider fatigue effects is proposed. The model is formulated according to the concepts and theories of the CDM and is particularly well suited in the case of steel structures where damage accumulation due to local buckling and low cycle fatigue reduces the mechanical behavior of the steel structural components.

2 ELASTOPLASTIC DAMAGE MODEL

2.1. *Constitutive Equations*

Damage in Continuum Damage Mechanics takes into account the degradation of materials resulting in a stiffness reduction.

According to the Strain Equivalence Principle proposed by Lemaitre (1971) and using the Kachanov's definition of effective stress, the stiffness of a damaged material can be obtained as E(1-d) being E the initial Young's modulus and d a scalar representing the isotropic damage. Assuming a damaged elastic material, the strain due to damage can be obtained as (Ortiz 1985; Ju 1989):

$$\varepsilon^d = \frac{\sigma d}{E(1-d)} \tag{1}$$

which is consistent with the response of steel under uniaxial monotonic loading.

Equation 1 can be applied to a member of constant area A sujected to an axial load:

$$\delta^d = \frac{NL}{EA} \frac{d_a}{1-d_a} \tag{2}$$

where N is the axial force and δ^d the elongation due to the axial damage d_a.

Equation 2 can be generalized in order to take into account the flexural damage effects in a frame member. Thus , we consider an element where the stress distribution is described by a three component vector, $q=[M_i,M_j,N]^T$, collecting the bending moments at the two ends and the axial force (Fig. 1), which is associated to the corresponding kinematic variables $u=[\theta_i,\theta_j,\delta]^T$. The constitutive equations expressing the relations between the flexural moments and the corresponding rotations due to da-

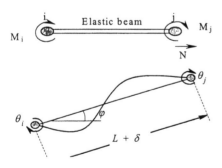

Figure 1. Generalized stresses and strains for the model

mage, $u^d = \left[\theta_i^d, \theta_j^d, \delta^d\right]$, are obtained as:

$$\theta_i^d = \frac{d_i}{1-d_i} \frac{L}{4EI} M_i \tag{3}$$

$$\theta_i^d = \frac{d_i}{1-d_i} \frac{L}{4EI} M_i \tag{4}$$

being di and dj the damage variables due to flexural effects at both ends of the member. Therefore, the damage vector for each member will be defined as $D^t = (d_i \quad d_j \quad d_a)$.

More details about the formulation of the constitutive equations for this model can be found in (Florez-Lopez 1995; Perera *et al* 1998).

2.2. *Plastic Dissipative Potential*

In order to completely define the model, the evolution laws for plastic strains and damage must be specified. In order to obtain the plastic evolution law it is necessary to define a plastic dissipative potential. For it, by analogy with the effective stress concept proposed by Rabotnov in 1968, the three component stress vector q proposed in the last paragraph can be redefined as an effective generalized stress vector using the following expression:

$$\bar{q} = \left(\frac{M_i}{1-d_i} \quad \frac{M_j}{1-d_j} \quad \frac{N}{1-d_a} \right)^t \tag{5}$$

According to the strain equivalence principle, any constitutive equation for a damaged material may be derived in the same way as for a virgin material replacing the usual stress by the effective stress (Lemaitre 1971).

Therefore the plastic dissipation potential for each plastic hinge of the member may be expressed using the same expression employed for undamaged materials replacing the moment by the corresponding effective moment.

Then, when damage occurs, if we do not consider the effect of the axial plastic strains, the plastic function can be written at each end as:

$$f_i = \left| \frac{M_i}{1-d_i} - X_i \right| - M_y \tag{6}$$

where X_i is the kinematic hardening term and M_y is the yield moment.

To define the evolution of X, the following expression is proposed:

$$X = X_\infty (1 - e^{-\alpha\theta^p}) \tag{7}$$

being X_∞ and α parameters to be identified for each material and geometry; from the expresion it can be observed that X tends to saturate to some value X_∞ with a velocity controlled by the value of α.

Being defined the plastic potential, the Principle of Maximum Plastic Dissipation implies the normality of the plastic flow rule in the generalized stress space:

$$du^p = d\lambda^p \frac{\partial f}{\partial q} \qquad (8)$$

where $d\lambda^p$ is a plastic parameter which can be obtained enforcing the plastic consistency condition.

3 CUMULATIVE DAMAGE LAW

To completely define the model, the damage evolution law has to be specified. In cyclically loaded materials it is convenient to use cumulative damage models. Since seismic loads induce severe inelastic cycles at relatively large ductilities, the concept of using low-cycle fatigue theories to model damage is logical.

Assuming linear damage accumulation, the total cyclic fatigue damage may be obtained using the principle formulated by Palmgren (1924) and Miner (1945). Damage functions due to each individual cycle are summed until fracture occurs. Failure is assumed to occur when these damage functions sum up to or exceed unity:

$$D = \sum \frac{n_i}{N_f} \geq 1 \qquad (9)$$

where n_i is the number of cycles for the current amplitude and N_f is the number of cycles to failure for this amplitude.

The quantification of the number of cycles to failure N_f is usually performed through the Manson-Coffin relationship (1953):

$$N_f = C(\Delta\delta^p)^K \qquad (10)$$

where $\Delta\delta^p$ is the plastic strain amplitude of the hysteretic cycles (Fig. 2) and C and K are parameters depending on the materials. Some authors (Kunnath *et al* 1997; Koh and Stephen 1991) suggested that total strain amplitude could be used instead of plastic strain.

Damage values evaluated according to Equation 9 through the loading history are introduced in the elastoplastic damage model pre-sented in Section 2. For it, first of all, it is necessary to evaluate the plas

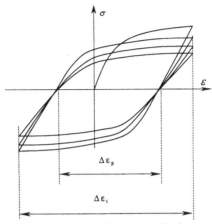

Figure 2. Total and plastic strain amplitude

tic strain increment at every load step using the flow rule of Equation 8 and enforcing the plastic consistency condition. With the plastic strain amplitude obtained and using Equations 9 and 10 the progressive damage increments are obtained.

Therefore, we need to determine the Manson-Coffin law in order to complete the model. Then, the keypoint of the cumulative damage law is related to the identification of the structural damage parameters C and K appearing in Equation 10.

Usually, through experimental tests performed on beams made of different profiles some results are obtained to calibrate the Manson-Coffin relationship (Ballio & Castiglioni 1994). The specimens are subjected to displacement cycles of constant amplitude up to collapse. The results obtained allow the definition of a relationship between the amplitude of the displacement cycles imposed and the number of cycles performed to reach the failure (N_f). Performing the tests for different amplitudes a linear relationship amplitude-N_f is obtained on a log-log scale which allows the determination of the parameters C and K.

These results, combined with the Miner law, may be useful as a criterion to predict the failure of structural elements. However, in the model proposed in Section 2, the damage is defined as an internal variable affecting the mechanical behavior and, basically, incorporatig the gradual loss of stiffness. Therefore, the limiting value d=1.0 of the damage variable may be identified with complete loss of stiffness. Due to it, in the definition of the parameters appearing in the Manson-Coffin law and, therefore, in the damage evaluation would be more convenient to keep the consistency with the definition of the damage index in the model as a variable measuring the progressive loss of stiffness.

Krawinkler & Zohrei (1983) performed several

experimental tests of constant amplitude cyclic loading on steel structures in order to characterize the cumulative damage. In the experimental work developed, they consider damage associated to several different phenomena such as strength deterioration, energy dissipation and, as in Continuum Damage Mechanics, stiffness deterioration. The constant amplitude tests of several wide flange shapes (W 6x9) of ASTM A36 steel provided the relationship between damage increment per reversal (in terms of stiffness deterioration), and plastic rotation range. This relation is assumed to be constant within a certain range of the number of reversals. For it, three deterioration ranges were identified according to the deterioration rate. In the first and third ranges, deterioration grows rapidly while in the second range deterioration proceeds at a slow and almost constant rate. More details about it can be found in Krawinkler & Zohrei (1983).

For each range, the rate of stiffness deterioration per reversal, Δd_k, for constant amplitude cycling is expressed by a function of the form:

$$\Delta d_k = A(\Delta\theta_p)^a \qquad (11)$$

where A and a are determined through experimental tests and $\Delta\theta_p$ is the plastic rotation range. From Equation 11, assuming linear damage accumulation for reversals with variable amplitude, the accumulated damage can be expressed as:

$$d = \sum_{i=1}^{n}(\Delta d_k)_i = A\sum_{i=1}^{n}(\Delta\theta_p)^a_i \qquad (12)$$

where n is the number of reversals.

Denoting as K_o and K the undamaged and damaged stiffnesses, respectively, the rate Δd_k represented by Krawinkler & Zohrei (1983) corresponds to the relation $(K_o-K)/K_o$. In order to employ Equation 11 in the model presented in Section 2 a suitable relationship between the rate Δd_k defined in this equation and the rate Δd corresponding to the model has to be deduced. After some calculations, the following expression for the damage in the model is deduced:

$$d = \frac{4(1-K/K_o)}{4-K/K_o} \qquad (13)$$

from which the following relationship is obtained

$$\Delta d = \frac{(4-d)^2}{12}\Delta d_k \qquad (14)$$

or, applying Equation 11:

$$\Delta d = \frac{(4-d)^2}{12}A(\Delta\theta^p)^a$$

Therefore Equation 15 will be employed in our model to evaluate the damage rate per reversal.

On next section, to check the efficiency of the model proposed in this paper, numerical results are compared with the experimental results presented in Krawinkler & Zohrei (1983).

4 NUMERICAL SIMULATION

Figure 3 shows the results of one experimental test performed on a beam with W6x9 section subjected to a cyclic loading of constant amplitude equal to 1.7 in.

From the test, in deterioration range II the following rate of deterioration per reversal has been obtained:

$$\Delta d_k^{II} = 0.446[\Delta\theta^p]^{1.415} \qquad (16)$$

where range II includes from cycle 10 to cycle 40. This rate has been employed in Equation 14 to evaluate the damage increment in the numerical simulation.

In the same way, the following values have been employed in the plastic dissipative potential: M_y=23 kN m; X_∞ =20 kN m; α=150.

Figure 4 shows the results obtained in the numerical simulation. As it has been commented before, only the function corresponding to the wider range of number of reversals has been used, which implies a certain deviation from the experimental results for the first and the last cycles.

In Figure 5, the damage evolution through the number of cycles obtained numerically is represented. The last numerical value (d=0.6) can be compared with the last experimental value (d=0.65) which has been obtained through Equation 13 measuring the relation K/K_o in the last cycle of range II. As it is logical, the numerical value is a little smaller than the experimental value since in the numerical results the range I, for which the deterioration proceeds at high rate, has not been considered.

5 CONCLUSIONS

The results obtained are very hopeful. The model performs very well under cyclic loading of constant amplitude. The model appears to be very interesting since it applies the concepts of the CDM in a simplified way to simulate the cumulative damage.

The approach presented is amenable of further generalizations and it would be convenient to obtain experimental results for more complex loading histo

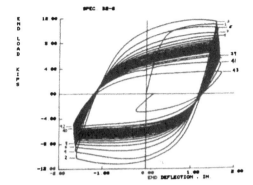

Figure 3. Experimental results

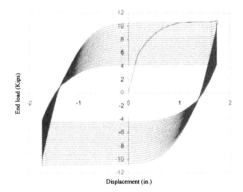

Figure 4. Numerical results

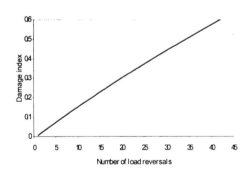

Figure 5. Damage evolution

ries (cyclic loading with variable deflection amplitudes, seismic loading) in order to check the efficiency of the model in more realistic loading cases.

In the model proposed, damage is related to the stiffness degradation. A very interesting possibility would be to try performing through some experimental tests a calibration of this damage index in order to formulate a failure criterion of the structure.

REFERENCES

Ballio, G. & Castiglioni, C., 1994. "An approach to the seismic design of steel structures based on cumulative damage criteria", *Earthq. Eng. And Struct. Dyn.*, Vol 23, 969-986

Chaboche, J. 1985. " Continuum damage mechanics and its application to structural lifetime prediction", *Rech.Aerosp.*, 4, pp37-53

Florez-López, J. 1995.. "Simplified model of unilateral damage for RC frames." *J. Struc. Eng.* Vol. 121, No.12 ,pp1765-1772.

Ju, J. W. 1989 . "On energy-based coupled elastoplastic damage theories: constituve modeling and computational aspects". *Int. J. Solids Structures.* Vol. 25, No.7 , pp803-833.

Koh, S.K. & Stephen, R.I. 1991. "Mean stress effects on low cycle fatigue for high strength steel", *Fatigue Fract. Engng. Mater.*, Vol.14, pp413-428

Krawinkler, H. & Zohrei, H., 1983. "Cumulative damage in steel structures subjected to earthquake ground motions", Computers & Structures, Vol. 16, 531-541

Kunnath, S.K.; El-Bahy, A.; Taylor, A. & Stone, W. 1997., "Cumulative seismic damage of reinforced concrete bridge piers". Technical Report NCEER-97-0006. *National Center for Earthquake Engineeing Research.* State University of New York at Buffalo.

Lemaitre, J. & Chaboche, J., 1985. *Mecanique des materiaux solides.*,Dunod, Paris

Lemaitre, J. 1996. *A course on damage mechanics.*, Springer-Verlag, Berlin

Manson, S.S. 1953. "Behavior of materials under conditions of thermal stress", *Heat Transfer Symposium*, University of Michigan Engineering research Institute, Ann Harbor, Michigan, pp9-75.

Miner, M.A. 1945. "Cumulative damage in fatigue". *Jour. Appl. Mech.* Vol. 12: A-159

Ortiz, M. 1985. " A constitutive theory for the inelastic behavior of concrete", *Mech. Mater.*, Vol. 4, pp67-93

Palmgren, A. 1924. "Die lebensdauer von kugellagern", *Verfahrentechinik*, Berlin, 68, pp339-341

Perera, R. ,Carnicero,A., Alarcón, E. & Gómez, S. 1998. " A damage model for seismic retrofitting of structures", *Advances in Civil and structural Engineering*, CIVIL-COMP Press, pp309-315

Behaviour of Steel Structures in Seismic Areas, Mazzolani & Tremblay (eds) © 2000 Balkema, Rotterdam, ISBN 90 5809 130 9

Performance of high strength CFT columns under seismic loading

A. H. Varma, J. M. Ricles, R. Sause & L. W. Lu
Department of Civil and Environmental Engineering, Lehigh University, Bethlehem, Pa., USA

ABSTRACT: The behavior of high strength square CFT beam-columns was experimentally investigated. The CFT specimens were 305 mm square tubes made from conventional (A500 Grade-B) or high strength (A500 Grade-80) steel with nominal width-to-thickness (b/t) ratios of 32 or 48 and filled with high strength (110 MPa) concrete. The effect of the stress-strain characteristics and nominal b/t ratio of the steel tube on the stiffness, strength and ductility of the high strength CFT specimens was evaluated. Four stub columns, eight monotonic beam-columns, and three cyclic beam-column tests were conducted. The experimental results indicate that the concrete infill delays the local buckling of the steel tube and that the steel tube offers some confinement to the infill concrete thus increasing its compressive ductility. The monotonic curvature ductility of the CFT beam-column specimens decreases significantly with an increase in the axial load level and the b/t ratio. Cyclic loading does not have a significant influence on the monotonic flexural stiffness and moment capacity of the beam-column specimens. However, the post-peak moment resistance of the beam-column specimens decreases more rapidly under cyclic loading. This has a significant influence on the cyclic curvature ductility of the specimens. The axial load capacity of the stub column specimens was conservatively predicted by superposition of the yield strength of the steel tube with 85% of the compressive strength of the concrete infill. The moment capacity of axially loaded beam-column specimens was conservatively predicted by using the American Concrete Institute provisions for conventional strength CFT beam-columns.

1 INTRODUCTION

Composite building construction involves steel moment resisting frames (MRFs) or braced frame systems with composite columns or composite bracing members. Composite columns include structural shapes encased in concrete (SRCs) and concrete filled steel tubes (CFTs). In a CFT column, the steel tube acts as a formwork for placing the concrete, and provides some confinement to the cured concrete thus increasing its compressive ductility. The concrete infill delays the local buckling of the steel tube and increases the flexural stiffness of the column, thus enabling the drift requirements for MRFs to be met more economically. This composite interaction of both materials in the CFT cross-section provides a unique opportunity to effectively utilize both high strength steel and concrete in building construction. The current lack of experimental research on the seismic behavior of high strength CFT beam-columns has motivated the research program described herein.

This research program is currently in progress at Lehigh University, and is a part of the U.S.-Japan Cooperative Earthquake Research Program on Composite and Hybrid Structures. It involves experimental and analytical investigations of the effects of the stress-strain characteristics and steel tube b/t ratio on the flexural stiffness, strength and ductility of high strength CFT beam-columns. Results from completed tests of the research program are presented.

2 TEST PROGRAM

The CFT specimens were two-third scale models of a base column of a 12-story perimeter MRF in the U.S.-Japan CFT theme structure. The specimens were 305 mm square steel tubes filled with high strength (110 MPa) concrete. The square tubes were made from either conventional strength (A500 Grade-B, nominal yield stress = 317 MPa) or high strength (A500 Grade-80, nominal yield stress = 552

MPa) steel, with nominal b/t ratios of 32 and 48. The tubes from each grade of steel were made from the same heat. The tubes were filled with high strength concrete from the same batch-mix.

The experimental test matrix is shown in Table 1. The nomenclature used to identify each specimen consists of the specimen type (SC designates stub columns, BC designates monotonic beam-columns, and CBC designates cyclic beam-columns), nominal b/t ratio, nominal yield stress of the steel tube in ksi, and the nominal level of axial load as a percentage of P_o. In Table 1 σ_y is the steel yield stress, P is the applied axial load and P_o is the experimentally obtained axial load capacity of the stub column specimens.

The material properties of the steel were determined by conducting uniaxial tension tests in accordance with American Society of Testing Materials (ASTM E8, 1997) specifications on coupons that were removed and machined from the walls of the tubes. Typical stress-strain curves from the uniaxial tension tests are shown in Figure 1. A500 Grade-80 is a high strength low alloy steel with less inelastic strain deformation capacity and strain hardening than conventional strength A500 Grade-B steel.

The 28-day compressive strength of the concrete cylinders, tested in accordance with ASTM (ASTM C39, 1997) specifications, was 110 MPa. Concrete cylinders were cored from an untested CFT specimen and tested in accordance with ASTM specifications to determine the in-situ properties of the concrete. The cylinders were cored and tested during the time the stub column and beam-column specimens were tested and had an average compressive strength (f'_c) of 110 MPa. A typical stress-strain curve from the cored concrete cylinder tests is shown in Figure 2. High strength concrete has greater stiffness and strength than conventional strength concrete, and has a nearly linear stress-strain response up to failure. The peak stress (f'_c)

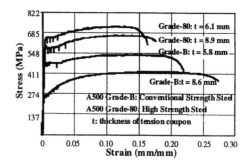

Figure 1. Steel stress-strain curves.

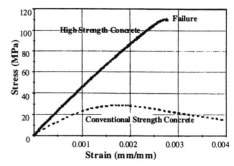

Figure 2. Concrete stress-strain curve.

was reached at an average compressive strain (ε_{uc}) of 0.0027 mm/mm. The failure of the concrete cylinders was brittle and explosive.

2.1 Stub column tests

Four stub column tests were conducted. The specimen parameters are summarized in Table 1. The tests were conducted under load control, at a rate of 225 kN/minute. Alignment of the axial load was ascertained by monitoring the longitudinal strains around the perimeter of the tube and at different locations along the height. Shortening of the stub columns (over gage lengths of 300 mm) and the overall machine head travel were measured by displacement transducers. Failure was defined as the abrupt loss in axial load resistance due to extensive local buckling and concrete crushing. After failure the specimen was subjected to increasing axial deformations at a rate of 1.25 mm/minute.

The experimental axial load-head travel response of Specimens SC-32-46 and SC-48-46 is shown in Figure 3. The experimental axial load-head travel response was nearly linear up to the peak load (P_o). Local buckling and concrete crushing occurred at P_o.

Table 1. Experimental test matrix.

Specimen	Measured		
	b/t ratio	σ_y MPa	Axial load level P/P_o
SC-32-46	35.29	250	---
SC-32-80	34.29	560	---
SC-48-46	52.17	453	---
SC-48-80	50.00	624	---
BC-32-46-20 ; BC-32-46-40	35.29	257	0.21; 0.43
BC-32-80-20 ; BC-32-80-40	34.29	560	0.20 ; 0.41
BC-48-46-20 ; BC-48-46-22	52.17	453	0.18 ; 0.22
BC-48-80-20 ; BC-48-80-40	50.00	624	0.19 ; 0.38
CBC-32-46-20;CBC-32-46-30	35.29	257	0.21 ; 0.30
CBC-48-46-20;CBC-48-46-30	52.17	453	0.18 ; 0.30
CBC-32-80-20;CBC-32-80-30	34.29	560	0.18 ; 0.30
CBC-48-80-20;CBC-48-80-30	50.00	624	0.19 ; 0.30

After P_o was reached the specimen axial load resistance gradually deteriorated before failure occurred at $0.97P_o$ and $0.99P_o$ for specimens SC-32-46 and SC-48-46, respectively, as shown in Figure 3. The stub columns maintained their axial load resistance after failure with increasing axial deformations.

Table 2 shows the experimental results obtained from the stub column tests and the comparison of the experimental results with predictions of strength and stiffness. In Table 2 P_o, K_{ax}, A_s, A_c, EA_t are the experimental axial load capacity, experimental section axial stiffness, area of steel in the cross-section, area of concrete in the cross-section and the section axial stiffness predicted by transformed section properties, respectively. The axial load capacity of the stub column specimens was conservatively predicted by superposition of the yield strength of the steel tube with the compressive strength of the concrete infill, while using $0.85f'_c$ for the concrete strength. The experimental axial load capacity of specimen SC-48-80 is less than the value predicted by superposition of strengths because the steel tube buckled elastically and did not develop full yielding at peak load (Varma et al., 1998). The section axial stiffness of the stub column specimens was predicted with reasonable accuracy by using transformed section properties.

2.2 Monotonic beam-column tests

Eight monotonic beam-column tests were conducted. The specimen parameters are summarized in Table 1. The CFT specimens were subjected to a constant axial load (P) and monotonically increasing flexural loading using the test-setup shown in Figure 4. Monotonically increasing flexural rotations were applied at the ends

Table 2. Stub column test results.

Specimen	P_o (kN)	K_{ax} x 10^6 (KN)	$\dfrac{P_o}{A_s\sigma_y+0.85f'_cA_c}$	$\dfrac{K_{ax}}{EA_t}$
SC-32-46	11390	5.203	1.10	1.00
SC-32-80	11568	4.864	1.02	1.01
SC-48-46	14116	5.474	1.04	1.05
SC-48-80	12307	4.949	0.96	1.02

of the test-length of the CFT specimen by the use of actuators as shown in Figure 4. These actuators were controlled to apply equal and opposite loads (Q). The test-length of the specimen was subjected to a constant axial load and constant primary bending moment. The specimen ends were attached to cylindrical bearings that were free to rotate in-plane thus simulating pin-pin end conditions. In-plane displacements of the cylindrical bearings and various locations along the test-length of the specimen were monitored by displacement transducers. These displacements were used to evaluate second-order moments. Rotations along the test-length of the specimen were monitored by rotation transducers; these rotations were used to evaluate the end rotations of the test zone and the curvature of different segments along the length. The longitudinal and transverse strains were monitored around the perimeter at different locations along the test-length of the specimen.

The experimental mid height moment - average end rotation (M-θ) response of Specimens BC-32-46-20 and BC-32-80-20 is shown in Figure 5. The specimens maintained their initial flexural stiffness till concrete tension cracking occurred. With increasing deformations, the flexural stiffness of the specimen gradually decreased due to concrete tension cracking and material nonlinearity. The extreme steel and concrete fibers at the mid height cross-section had reached ε_y (tensile yield strain) and ε_{uc}, respectively, before the peak moment (M_u) was reached. Local buckling of the steel tube compression flange occurred at M_u. The specimen moment capacity decreased with increasing flexural deformations due to extensive local buckling and concrete crushing. Local buckling of the webs of the steel tube occurred during the post-peak response. Photographs of the steel tube and concrete infill following the test of Specimen BC-32-80-20 are shown in Figure 6. Local buckling of the compression flange and webs can be seen in the removed portion of the tube. Extensive concrete crushing can be seen directly under the buckled regions of the tube.

The moment-curvature (M-ϕ) response of the failure segment of the specimen was used to estimate the flexural section stiffness and the curvature ductility. Table 3 gives a summary of the experimental results from the eight beam-column tests. In Table 3 K_{f-ini}, K_{f-sec}, M_u, and μ_ϕ are the

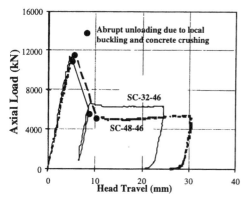

Figure 3. Axial load-head travel response of Specimens SC-32-46 and SC-48-46.

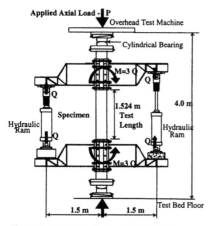

Figure 4. Monotonic beam-column test-setup.

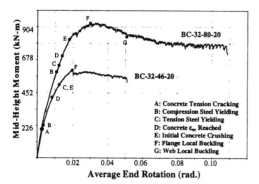

Figure 5. Moment – rotation response of Specimens BC-32-46-20 and BC-32-80-20.

Figure 6. Specimen BC-32-80-20 after testing.

measured initial section flexural stiffness, the measured section flexural stiffness corresponding to 60% of the peak moment, the peak moment, and the curvature ductility. The curvature ductility μ_ϕ is

Table 3. Monotonic beam-column test results.

Specimen	$K_{f-ini} \times 10^{10}$ (kN-mm^2)	$K_{f-sec} \times 10^{10}$ (kN-mm^2)	M_u (kN-m)	μ_ϕ
BC-32-46-20	5.202	4.345	606	15.3
BC-32-46-40	5.000	4.345	532	2.8
BC-32-80-20	5.570	4.201	933	15.5
BC-32-80-40	5.104	5.219	806	3.2
BC-48-46-20	5.049	3.575	597	3.2
BC-48-46-22	5.015	3.814	629	6.6
BC-48-80-20	4.975	3.742	700	5.9
BC-48-80-40	4.828	4.952	574	1.9

equal to ϕ_u/ϕ_y where ϕ_u is the curvature of the failure segment when the section moment is $0.9M_u$ in the post-peak response and ϕ_y is the curvature of the failure segment corresponding to $0.9M_u$ in the pre-peak response divided by the stiffness K_{f-sec} (i.e., ϕ_y = $0.9M_u/K_{f-sec}$).

The results summarized in Table 3 indicate that increasing the nominal yield stress of the steel tube or the level of axial load does not influence K_{f-ini}. Increasing the b/t ratio of the steel tube replaces some area of steel in the cross-section by concrete, which results in a slight reduction in K_{f-ini}. Except for Specimen BC-32-46-40, K_{f-sec} increased with higher levels of axial load because more area of the concrete is in compression and thereby actively participating in resisting axial and flexural loads at $0.60M_u$. The steel tube of Specimen BC-32-46-40 yielded in compression at $0.60M_u$. As a result, K_{f-sec} did not increase and was almost the same as the value for Specimen BC-32-46-20.

The moment capacity (M_u) of the specimens increases when the nominal yield stress of the steel tube was increased, while maintaining the b/t ratio and the axial load level. For specimens with a b/t ratio of 32 this increase is approximately 52%. For specimens with a b/t ratio of 48 this increase is only 17%, because the steel tube did not develop appreciable inelastic deformation before the peak moment was reached due to local buckling. Increasing the b/t ratio of the steel tube while maintaining the nominal yield stress of the steel tube and axial load level decreases M_u of the specimens. For the specimens made from A500 Grade-B steel this reduction is only 1.5%. For specimens made from A500 Grade-80 steel this reduction is almost 27%, because the thinner tube did not develop appreciable inelastic deformation before the peak moment was reached due to local buckling (Varma et al., 1998).

The results summarized in Table 3 indicate that increasing the nominal b/t ratio of the steel tube from 32 to 48, while maintaining the nominal yield stress and axial load level significantly decreases the curvature ductility. Increasing the level of axial load acting on the specimens, while maintaining the nominal b/t ratio and yield stress of the steel tube also significantly decreases the curvature ductility. The results also indicate that variability in μ_ϕ can occur, as observed for Specimens BC-48-46-20 and BC-48-46-22. These specimens had identical material and geometric parameters, and the axial load differed by 4%. With a nominal b/t ratio of 48, the onset of local buckling was sensitive to initial geometric imperfections. This may have resulted in significant variation in μ_ϕ for a small variation in axial load (Varma et al., 1998).

2.3 Cyclic beam-column tests

Three of the eight cyclic beam-column tests have been completed. These are Specimens CBC-32-46-20, CBC-48-46-20, and CBC-48-80-20, whose parameters are summarized in Table 1. The CFT specimens were subjected to a constant axial load with cyclically applied lateral loads by the test-setup shown in Figure 7.

Figure 7. Cyclic beam-column test-setup.

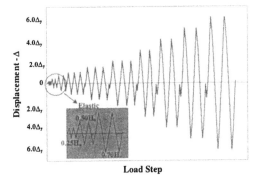

Figure 8. Cyclic loading history.

As shown in Figure 7 the specimens are fixed at the base and the axial load (P) is applied concentrically at the top of the specimen. The lateral load (H) is also applied at the top of the specimen. The test-length of the specimen (1.486 m) is subjected to a constant axial load and linearly varying primary bending moment, with maximum bending moment at the base.

The cyclic loading history is shown in Figure 8. The loading history consists of elastic and inelastic cycles. The monotonic moment capacity (M_u) corresponding to the applied axial load was used to estimate the lateral load capacity ($H_u = M_u$/test-length). The elastic cycles were conducted under load control at load levels of $0.25H_u$, $0.50H_u$, and $0.70H_u$. The secant flexural stiffness (K_{sec-70}) of the specimen was measured during the first cycle at $0.70H_u$ to estimate the yield displacement Δ_y (i.e., $\Delta_y = H_u/K_{sec-70}$). The inelastic cycles were conducted under displacement control at displacement levels of $1.0\Delta_y$, $1.5\Delta_y$, $2.0\Delta_y$, $3.0\Delta_y$, $4.0\Delta_y$, $5.0\Delta_y$, and $6.0\Delta_y$. The inelastic loading cycles were imposed until the lateral load capacity decreased to $0.60H_u$ or failure occurred due to tension fracture of the steel tube.

Lateral displacements were measured at various locations along the test-length of the specimen by displacement transducers to evaluate second-order moments. Rotations were also measured at various locations along the test-length of the specimens by rotation transducers to determine the curvature over the failure segment at the base of the specimen.

The cyclic lateral load-lateral displacement at the top of Specimen CBC-32-46-20 is shown in Figure 9. The moment capacity of the corresponding monotonic Specimen BC-32-46-20 was 606 kN-m. H_u and Δ_y were estimated as 408 kN and 14 mm, respectively. With increasing lateral displacements after Δ_y, the flexural stiffness of the specimen decreased due to concrete tension cracking and yielding at the base of the steel tube. The peak load was reached in the first $2\Delta_y$ cycle, and was accompanied by crushing of the concrete infill and local buckling of the compression flanges at the base of the steel tube. The lateral load resistance of the specimen deteriorated with the additional cycles due to concrete crushing and flange local buckling. Web local buckling at the base occurred during the first $4\Delta_y$ cycle. The specimen lateral load resistance deteriorated very rapidly after local web buckling had occurred. Buckling of the steel tube corners occurred during the first $5\Delta y$ cycle. The local buckles grew more rapidly with additional cycles, and the CFT cross section at the base became extremely distorted. The specimen lateral load resistance increased slightly during the $6\Delta_y$ cycles. This may have been caused by strain hardening of the A500 Grade-B steel with thickness t = 8.6 mm (see Figure 1) and extensive distortion of the CFT cross-section at the base. At the end of the loading history the specimen lateral load resistance was 62% of the peak load. Therefore, the specimen was subjected to additional loading cycles corresponding to displacement levels of 8Δ$_y$ and 6Δ$_y$, respectively. The specimen lateral load resistance increased slightly with the additional cycles. After the first half cycle corresponding to 6Δ$_y$ a tension fracture developed in a corner of the steel tube. This last 6Δ$_y$ cycle is shown by a dashed line in Figure 9.

The envelop of the cyclic moment-curvature response of the 300 mm failure segment at the base of the specimen is shown in Figure 10. The monotonic moment-curvature response of the failure segment of Specimen BC-32-46-20 is also shown in Figure 10. Cyclic loading causes a more rapid decrease of the post-peak moment resistance. This decrease is quite rapid after web local buckling occurs. A photograph of the steel tube and the concrete infill following the test of Specimen CBC-32-46-20 are shown in Figure 11, where a portion of the steel tube has been removed. Local buckling of the steel tube and extensive crushing of the concrete under the locally buckled regions of the tube can be seen.

The cyclic lateral load-lateral displacement response of Specimens CBC-48-46-20 and CBC-48-80-20 were similar to the behavior of Specimen CBC-32-46-20. The envelop of the cyclic moment-curvature response of the 300 mm failure segment at the base of these specimens is also shown in Figure 10, along with the monotonic moment-curvature response of Specimens BC-48-46-20 and BC-48-80-20. CBC-48-80-20 developed a tension fracture in a corner of the steel tube in the first half cycle corresponding to a displacement level of 5Δ$_y$. This specimen was made from an A500 Grade-80 steel tube with a nominal b/t ratio of 48 and t = 6.1 mm, which had the smallest inelastic deformation capacity as shown in Figure 1. The cyclic lateral load resistance of Specimen CBC-48-46-20 had reduced to 53% of the peak value at the end of the first 5Δ$_y$ cycle.

The envelop of the cyclic moment-curvature response was used to estimate the flexural section stiffness, the moment capacity and the curvature ductility. Table 4 gives a summary of the experimental results from the three cyclic beam-column tests. In Table 4 $K_{f\text{-ini-c}}$, $K_{f\text{-sec-c}}$, $M_{u\text{-c}}$, and $\mu_{\phi-c}$ are equal to the measured initial section flexural stiffness, the measured section flexural stiffness corresponding to 60% of the average peak moment, the average peak moment, and the curvature ductility. The curvature ductility $\mu_{\phi-c}$ is equal to the average value of ϕ_u/ϕ_y where ϕ_u is the curvature of the failure segment when the section moment is 0.9$M_{u\text{-c}}$ in the post-peak response and ϕ_y is the curvature of the failure segment corresponding to 0.9$M_{u\text{-c}}$ in the pre-peak response divided by the stiffness $K_{f\text{-sec-c}}$ (i.e., $\phi_y = 0.9M_{u\text{-c}}/K_{f\text{-sec-c}}$).

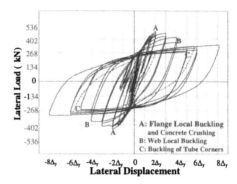

Figure 9. Lateral load-lateral displacement response of Specimen CBC-32-46-20.

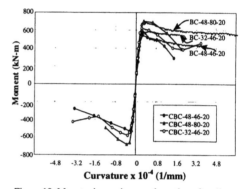

Figure 10. Monotonic specimen and envelop of cyclic specimen moment-curvature response.

Figure 11. Specimen CBC-32-46-20 after testing.

Table 4. Cyclic beam-column test results.

Specimen	$K_{f\text{-ini-c}}$ x 10^{10} (kN-mm^2)	$K_{f\text{-sec-c}}$ x 10^{10} (kN-mm^2)	$M_{u\text{-c}}$ (kN-m)	$\mu_{\phi-c}$
CBC-32-46-20	5.236	4.086	612	5.5
CBC-48-46-20	4.972	3.720	548	5.4
CBC-48-80-20	5.260	3.730	705	4.8

Increasing the nominal yield stress of the steel tube, while maintaining the nominal b/t ratio at 48 and the level of axial load, increases the moment capacity by 29% and reduces $\mu_{\phi-c}$ by 11%. Increasing the nominal b/t ratio, while maintaining the nominal yield stress (317 MPa) and the level of axial load, decreases the moment capacity by 10% but does not have a significant influence on $\mu_{\phi-c}$. In general there is not much variation in $\mu_{\phi-c}$ of the specimens because the post-peak moment resistance decreases rapidly under cyclic loading.

A comparison of the monotonic and cyclic beam-column test results summarized in Tables 3 and 4, respectively, and Figure 10 indicates that cyclic loading does not have a significant influence on the flexural section stiffness and moment capacity of the CFT specimens. However, cyclic loading has a significant influence on the curvature ductility of the CFT specimens, specially the specimen made from A500 Grade-B steel tube with a nominal b/t ratio of 32.

3 COMPARISON WITH CODE PROVISIONS

The current code provisions were developed based on the existing experimental database for CFT columns made from conventional strength materials. The following code provisions are relevant: American Institute of Steel Construction – Load and Resistance Factor Design (AISC-LRFD, 1994), American Concrete Institute (ACI, 1995), Architectural Institute of Japan (AIJ, 1987), and Eurocode 4 (EC4, 1996). Axial load-moment capacity (P-M) interaction curves for the beam-column specimens were developed based on the methods outlined in these code provisions (Varma et al., 2000). The P-M interaction curve for the specimen made from an A500 Grade-B steel tube with a nominal b/t ratio of 32 is shown in Figure 12. The stub column as well as the monotonic and cyclic beam-column test results are indicated on the P-M interaction curve. A comparison of the experimental moment capacity with the values predicted by the code provisions is given in Table 5.

The AISC-LRFD provisions significantly underestimate the moment capacity (M_{LRFD}) of the high strength CFT beam-column specimens. These provisions do not appropriately account for the contribution of the concrete infill to the moment capacity of a composite cross-section. The results indicate that the ACI predicted moment capacities (M_{ACI}) are accurate for specimens with b/t ratios of 48 because the extreme concrete compressive fiber strain at failure of these specimens was close to 0.003. The ACI predictions are conservative for the specimens with b/t ratios of 32 because the extreme concrete compressive fiber strain at failure of these specimens was much larger than 0.003, due to the confinement offered by the thicker steel tube (Varma et al., 1998).

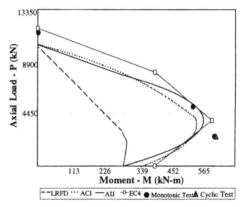

Figure 12. P-M interaction curves for a CFT specimen made from A500 Grade-B steel tube with a nominal b/t ratio of 32.

The experimental results indicate that the AIJ (M_{AIJ}) and EC4 (M_{EC4}) predictions are accurate (but sometimes unconservative) for specimens with a b/t ratio of 32 because the stress-state of the cross-section of these specimens at failure was similar to the AIJ and EC4 assumptions, respectively. The AIJ and EC4 predictions are generally unconservative for specimens made from A500 Grade-80 steel with a nominal b/t ratio of 48 because the steel tube did not develop its plastic moment capacity before reaching the peak moment (Varma et al., 1998).

Comparison of the experimental results with predictions based on the current code provisions indicates that the ACI provisions are appropriate for calculating a lower bound estimate of the moment capacity of high strength square CFT beam-columns until more rational strength prediction methods that account for the experimentally observed limit states become available.

Table 5. Experimental-to-Predicted Moment Capacities.

Specimen	M_u	M_{u-c}	$\dfrac{M_u}{M_{LRFD}}$	$\dfrac{M_u}{M_{ACI}}$	$\dfrac{M_u}{M_{AIJ}}$	$\dfrac{M_u}{M_{EC4}}$
	(kN-m)					
BC-32-46-20	606	612	2.05	1.14	1.13	1.14
BC-32-46-40	532	---	2.58	1.08	0.98	0.98
BC-32-80-20	933	---	1.42	1.13	1.02	1.01
BC-32-80-40	806	---	1.68	1.21	0.96	0.92
BC-48-46-20	597	548	1.56	1.00	0.98	0.99
BC-48-46-22	630	---	2.04	1.07	1.01	1.03
BC-48-80-20	700	705	1.27	1.00	0.87	0.87
BC-48-80-40	574	---	1.40	0.99	0.72	0.72

4 SUMMARY AND CONCLUSIONS

The behavior of CFT columns made from high strength materials and subjected to monotonic axial loading and combined axial and flexural loading was experimentally investigated. The CFT specimens were 305 mm square steel tubes filled with high strength (110 MPa) concrete. The steel tubes had nominal b/t ratios of 32 and 48 and were made from either conventional or high strength material. Fifteen CFT specimens, including four stub columns, eight monotonic beam-columns and three cyclic beam-columns were tested. The moment capacity of a CFT beam-column increases with an increase in the nominal yield stress of the steel tube. However, this increase is comparatively smaller for tubes with a b/t ratio of 48. Increasing the b/t ratio of the steel tube decreases the moment capacity of the CFT beam-column. This reduction in moment capacity is more significant for CFT beam-columns made from high strength steel tubes due to local buckling. The monotonic curvature ductility of the CFT beam-column specimens decreases significantly with an increase in the axial load level and the b/t ratio. However, it is not influenced by the nominal yield stress of the steel tube. Cyclic loading does not have a significant influence on the flexural stiffness and moment capacity of the beam-column specimens, however it has a significant influence on ductility. The post-peak moment resistance of the beam-column specimens decreases more rapidly under cyclic loading. The axial load capacity of the stub column specimens was conservatively predicted by superposition of the yield strength of the steel tube with 85% of the compressive strength of the concrete infill. The moment capacity of axially loaded beam-column specimens was conservatively predicted by using the American Concrete Institute provisions for conventional strength CFT beam-columns.

5 ACKNOWLEDGEMENTS

The research program reported herein was supported by the National Science Foundation (Grant No. CMS-9632911) under the U.S.-Japan Cooperative Earthquake Research Program: Phase V – Composite and Hybrid Structures (Dr. Shih-Chi Liu – cognizant NSF program official). The steel tubes were donated by Bull Moose Tube; Prairie Materials of Chicago donated the high strength concrete.

6 REFERENCES

American Concrete Institute (1995), *Building Code Requirements for Reinforced Concrete and Commentary*, ACI 318-95, Detroit, Michigan.

American Institute of Steel Construction (1994), *Manual of Steel Construction – Load and Resistance Factor Design*, Second Edition, Chicago, IL.

Architectural Institute of Japan (1987), *Structural Calculations of Steel Reinforced Concrete Structures*, Tokyo, Japan.

ASTM C39-89 (1997), "Standard Test Method for Compressive Strength Test of Cylindrical Concrete Specimens," American Society of Testing Materials.

ASTM E8 (1997) "Test Methods for Tension Testing of Metallic Materials," American Society of Testing Materials.

Eurocode 4 (1996), *Eurocode 4: Design of Steel and Concrete Structures, Part 1.1, General Rules and Rules for Buildings*, European Committee for Standardization, Brussels, Belgium.

Varma, A. H., Hull, B. K., Ricles, J.M., Sause, R., Lu, L.W. (1998), "Behavior of High Strength Square CFT Columns," *ATLSS Report No. 98-10*.

Varma, A.H., Ricles, J.M., Sause, R., Hull, B. K., Lu, L.W. (2000), "Seismic Behavior of High Strength Square CFT Columns," *ACI Special Publication on Composite Steel-Reinforced Concrete Structures - In honor of the late Dr. Walter P. Moore, Jr.*

Connections

Behaviour of Steel Structures in Seismic Areas, Mazzolani & Tremblay (eds) © 2000 Balkema, Rotterdam, ISBN 90 5809 130 9

Monotonic and cyclic analysis of bolted T-stubs

S. Ádány
Department of Structural Mechanics, Technical University of Budapest, Hungary

L. Dunai
Department of Steel Structures, Technical University of Budapest, Hungary

ABSTRACT: This paper deals with the numerical modelling and analysis of bolted T-stub components of steel-to-steel and steel-to-concrete end-plate type joints. The applied model is an advanced finite element model, considering both monotonic and cyclic loading. The axial stiffness of T-stubs is calculated and verified by test results. Monotonic loading ultimate analysis of base-plate T-stub is completed simulating experimental testing. Cyclic loading analyses are performed on T-stubs in a parametric study. The results are evaluated and the typical behaviour modes are presented. The conclusions on the T-stub monotonic and cyclic loading analyses are summarised.

1 INTRODUCTION

The importance of the joint behaviour is well-known in the static and seismic design of steel framed structures. In this respect there have been an intensive research on the monotonic and cyclic behaviour of steel-to-steel bolted end-plate and steel-to-concrete base-plate type joints in the last decade. The research work is started mainly by experimental studies. On this background empirical design formulation is derived and implemented in new design standards. The other direction of the research activities is on the application of numerical approaches to study the joint behaviour. These numerical models are generally result in a complicated computational solution technique what is the major obstacle of the practical application. In the recent years, however, the significant development in the computer technology provides with the opportunity to extend the application of the numerical models to perform parametric studies by virtual experiments.

The authors of this paper have been working on advanced numerical models of steel-to-steel and steel-to-concrete structural joints. The main idea of the model development is to analyse the joints' monotonic and cyclic behaviour modes and their interaction on the basis of the joints' geometry and cyclic material properties. In the first phase of the research the complex model and the computational method is developed and reported (Dunai 1992, Ádány & Dunai 1997a, Ádány & Dunai 1997b). The main features of the model are tested separately by demonstrative applications on cyclic plasticity of steel material, plate buckling under cyclic loading,

cyclic behaviour of supporting concrete and anchor bolt (Dunai & Ádány 1997).

This paper presents the results of the second phase of the research in which the developed model is applied on bolted components. The most typical bolted component that can represent the end-plate and base-plate type joint behaviour is the bolted T-stub. Two monotonic loading studies are completed on steel and steel-to-concrete T-stubs in parallel with experimental testing. The results are compared and discussed. The model is applied then for cyclic parametric study of T-stubs with different bolt – plate stiffness ratios. The typical results are presented and compared to each other.

2 NUMERICAL MODEL

The model and the solution method is presented in detail in other papers of the authors (Dunai et al. 1996, Ádány & Dunai 1997a, Dunai & Ádány 1997). Here only the main features are summarised.

The model is based on a layered degenerated shell finite element which can effectively model the steel plates. The support restraint and the bolts are modelled by an additional foundation layer that is linked to the shell elements. This extra layer works only as a Winkler-type foundation, which means that both the support restraint and the bolts are modelled by a system of independent springs. The spring characteristics can be chosen in order to cover a wide range support restraints: rigid foundation, elastic restraint, various kind of cyclic steel models, cyclic deterioration concrete model, different kinds of anchor bar models. The uni-lateral nature of the restraint and the bolts are taken into consideration by also the appropriate definition of the material model of foundation layer. Note that the assumption of the one-dimensional spring system excludes the consideration of shear forces.

In the numerical model the cyclic plasticity material model has main importance. In the steel material behaviour the combined isotropic and kinematic hardening should be modelled due to the continuous changes of the stiffness and strength according to the prior history of straining. In the current research a Mroz-type multi-surface model is applied in which finite number of yield surfaces are defined. The hardening is controlled between the adjacent surfaces by the actual plastic moduli of the material in function of the accumulated plastic strain; the sizes of each intermediate surface changes according to the prior straining history. The formulation of the above constitutive model is applied in the numerical model to follow the staining history in the layer mid-points of the Gauss-integration points of the elements.

Since the numerical model intends to simulate whole joints there is a need to handle the plate-buckling problem under cyclic loading, which can be a characterising phenomenon of the behaviour. For this reason the geometric non-linearity is introduced into the numerical model, by adopting the Total Lagrangian approach with the assumption of small strains and large deflections. However, in the analysis of T-stubs, which is the topic of this paper, the effect of geometric non-linearity is negligible.

In the non-linear solution method an incremental-iterative Newton-Raphson-type technique is applied. Some sub-types of this method is developed depending on the recalculation of the stiffness matrix. From practical point of view an important feature of the method that the whole analysis is completely automatic. The importance of this question comes from the time-consuming nature of the cyclic loading simulation. To make the procedure automatic several problems are to be solved: the load step generation, back-step option (to be applied if the iteration process fails), the stop condition of the iteration process (may need to be changed depending on the load level) etc.

During the model development several tests are done to verify the results on the separated behaviour components. In the following chapters the model is applied to analyse the behaviour of T-stubs under monotonic and cyclic loading.

3 AXIAL STIFFNESS OF T-STUBS

3.1 Experimental study

An experimental program is done at Salerno University on T-stub assemblies. The main goal of the experimental study is to determine the stiffness of different T-stubs with snug tightened and pre-tensioned bolts. Altogether 16 T-stub specimens are tested in a standardised arrangement. The specimens are obtained from hot-rolled profiles (HEA and HEB). Two bolt diameters are used: 20 and 12 mm, with different level of pre-tensioning.

On the basis of the test results the rotational stiffness of end-plate connections are derived. The results are detailed in a research report (Faella et al. 1996).

3.2 Numerical study

A numerical study is performed by the model detailed in the previous Chapter to predict the axial stiffness of the T-stubs in parallel with the test procedure. Six specimens are modelled, the main characteristics of which are summarised in Table 1. Note that β is a parameter expressing the ratio between the flange flexural stiffness and the bolt axial stiffness, as described in Faella et. al (1996).

The models for the above cases are developed and the analyses are done for snug-tightening cases. As an example, the applied finite element mesh for T-Stub 9 specimen is presented in Figure 1, together with the deformed shape of the model.

Table 1. Specimens main characteristics

	T-stub section	bolt diameter	β
T-stub 4	HEA160	20 mm	0.36
T-stub 5	HEA160	20 mm	0.36
T-stub 7	HEB200	12 mm	0.80
T-stub 8	HEB200	12 mm	0.80
T-stub 9	HEB200	12 mm	0.85
T-stub 10	HEB200	12 mm	0.85

Figure 1. Deformed and undeformed FE mesh for T-Stub 9

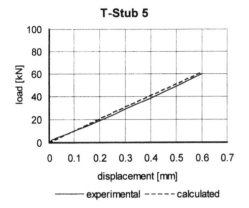

T-Stub 5

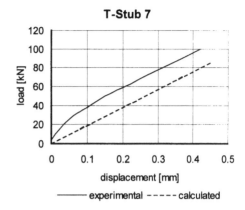

T-Stub 7

Figure 2. Experimental and calculated load - displacement curves

Table 2. Calculated and measured axial stiffness

	calculated secant stiffness	experimental secant stiffness
T-stub 4	103 kN/mm	103 kN/mm
T-stub 5	103 kN/mm	99 kN/mm
T-stub 7	189 kN/mm	237 kN/mm
T-stub 8	189 kN/mm	213 kN/mm
T-stub 9	212 kN/mm	214 kN/mm
T-stub 10	212 kN/mm	266 kN/mm

The axial load-displacement relationships are calculated for each case. Some examples are shown in Figure 2, together with the experimental curves. The stiffness values are calculated and compared to the experimental secant stiffness values in Table 2.

It can be concluded that the coincidence between the experimental and numerical stiffness values is excellent when the experimental load - displacement relationship is linear. If non-linearities are appeared in the experimental results (probably due to some non-adjusted bolt pre-tensioning), the difference is bigger between the secant stiffness values. It is important to observe, however, that the experimental load-displacement curve has always a linear part, the tangent stiffness of which is practically equal to the calculated one.

4 BASE-PLATE T-STUB IN TENSION

4.1 Benchmark tests on base-plate connections

Experimental study is completed at the Czech Technical University on basic components of column-base joints (Sokol & Wald, 1997). In this benchmark experimental program T-stubs in tension, T-stubs in compression and anchor bars are tested. The main purpose of the tests is to provide experimental data for the calibration of different column-base models. In a joint research the numerical simulation of the experiments are done (Sokol et al., 1999).

4.2 Numerical study

The 3D plate/shell model – detailed in Chapter 2 – is applied to simulate base-plate T-stub behaviour in tension. In this section the details of the numerical model and the results are demonstrated on a T-sub specimen W97-03 of (Sokol & Wald, 1997).

The specimen can be seen in Figure 3, while its finite element model is presented in Figure 4. Due to symmetry, only half of the specimen is modelled by the shell finite elements. The measured steel material properties are used in the pure kinematic hardening material model while the concrete properties are assumed by a bi-linear stress-strain relationship. The measured anchor bolt behaviour is approximated by also a bi-linear relationship in the model. The load is applied in about 300 small increments up to about 4 mm displacement of the vertical plate.

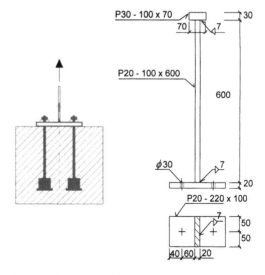

Figure 3. Specimen W97-03

91

Figure 4. Deformed and undeformed FE mesh for W97-03

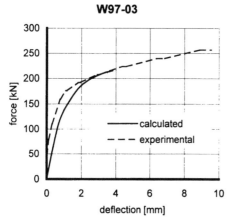

Figure 5. Load - displacement curves for W97-03

The calculated load-displacement relationship is presented in Figure 5, together with the one obtained from the benchmark test.

It can be seen that the measured and calculated curves are highly similar to each other, although a certain difference can be found in the initial part of the curves. The reason of this difference can be assumed as a reasonable pre-loading of the bolt. The phenomenon is very similar to that one mentioned in the previous Chapter when significant snug-tight is applied in the bolt.

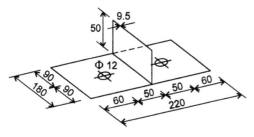

Figure 6. Main dimensions of specimen

5 T-SUBS UNDER CYCLIC LOADING

5.1 *Virtual experimental program*

A parametric study is performed on T-stubs by applying the above-described numerical model. The purpose of this study is the examination of the bolt - end-plate interaction, and its effect on the cyclic behaviour of T-stubs. Primarily the phenomena are analysed and discussed qualitatively rather then quantitatively. Another purpose is to test the model capabilities before applying it to more complex problems like the cyclic analysis of whole joints.

The main geometrical dimensions of the analysed T-stubs are shown in Figure 6. It is to be noted that a relatively thick web is chosen to avoid the effect of plate buckling which is a potential failure mode of the web.

To be able to change the relative load-bearing capacity of the bolt and the end-plate the following solution is applied: two parameters are changing during the analyses, these are the yield stress of the bolt material and the thickness of the end-plate, while the main geometrical dimensions of the specimens remain unchanged. In this way it is possible to use exactly the same finite element mesh in order to exclude the unwanted effects which may be come from the difference between the meshes.

The choice of the finite element mesh is the result of a compromise between accuracy and efficiency. On the one hand it is necessary to use a mesh dens enough to be able to follow the behaviour accurately. Moreover, also the number of the layers can have a considerable influence on it. On the other hand it is important to get the results within a reasonable time According to these considerations six layers were chosen (plus the additional foundation layer) with a mesh presented in Figures 10 and 12. Due to symmetry only one half of the T-stub assembly is modelled and analysed.

Simple kinematic hardening material models are applied both for the steel plates and the bolts. The yield stress of the plate is equal to 320 MPa, while the yield stress of the bolt material varies between

Table 3. Parameters of specimens for cyclic analysis

Specimen	Yield stress of bolt material	End-plate thickness
TS1 ("Strong bolt")	900 MPa	10 mm
TS2 ("Medium bolt")	720 MPa	10 mm
TS3 ("Medium bolt")	540 MPa	10 mm
TS4 ("Weak bolt")	900 MPa	16 mm

360 and 900 MPa. This wide range ensures to cover all the practical cases, as it will be shown in the followings. For both cases the tangent modulus is assumed to be 1/100 of the Young's modulus. The foundation layer is modelled by a linear elastic material model with a stiffness large enough to get a practically rigid foundation.

The loading history is in accordance with ECCS (1986), with the difference that the constant amplitude repetitions are neglected. It is governed by the displacement u/u_e ratio, where u is the actual value of displacement, while u_e is the displacement corresponding to the first yield. Ten full cycles of loading-unloading are simulated in case of each specimens, with a monotonously increasing amplitude of displacements, as it is shown in Figure 7.

In the frame of the parametric study several different specimens are analysed. In this paper four characterising cases are presented, the parameters of which are summarised in Table 3. These representative examples cover all the practical cases, even cases of extremely strong or weak bolts. It is to be noted that TS1 is chosen to show Mode 1 behaviour according to Eurocode 3 (1991) definitions, TS2 and TS3 correspond to Mode 2, while TS4 to Mode 3.

5.2 Analysis and results

During the analysis of each specimen the force-displacement relationship is established. Here, term "displacement" means the prescribed displacement of the web edge, while term "force" is the sum of the reaction forces measured at the same place. Figure 8 shows the diagrams for all the four specimens. The diagrams are plotted in a non-dimensionalised coordinate system, u_e and F_e being the displacement and the force belonging to the limit of elastic region,

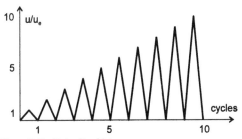

Figure 7. Cyclic loading history

as they are defined in the previous Section. To be able to easier observe the behaviour two cycles of each diagram are drawn with thicker line, these are cycles no. 3 and no 9, in each cases.

5.3 Evaluation of results

The first important general observation can be the strong difference between the various diagrams in the sense of rigidity, resistance, as well as the absorbed energy.

As far as the resistance is concerned the tendency is obvious: the weaker the bolt (comparing to the end-plate), the less the maximum resistance of the T-stub. The variation is considerable: in case of strong bolt (TS1) the maximum resistance F_{max} is almost three times greater then F_e, whilst in the other extreme case (TS4) F_{max} equals to $1.5 \cdot F_e$ only.

From rigidity point of view the differences are conspicuous, too. For TS1 the rigidity degradation during the cycles is almost negligible. However, in every case when considerable bolt yielding also occurs, (that is TS2, TS3, and TS4) the rigidity can be equal to zero, which is caused by the rigid-body displacement of the specimen, as being discussed later.

Concerning the absorbed energy the tendency is the same than that of the resistance: the weaker the bolt, the less the absorbed energy. Without detailed calculation it can be stated that the differences between the various cases are significant.

If the diagrams are studied in a more detailed way, the following observations can be done.

TS1 corresponds to the case when the bolt strong enough to remain elastic during the whole loading history. Thus, the force-displacement curve is determined basically by two phenomena: the elastic/plastic deformations of the end-plate, and the separation/re-contact between the end-plate and the foundation.

The effect of plastification is clearly can be seen, since the shape of the cycles is very similar to the cycles of a pure steel material test with a similar loading history. However, a certain unsymmetry of the F-u curve can also be observed, that is the tensile forces are always larger than the compressive ones at the end of the half-cycles. A simple explanation of this unsymmetry can be given with the help of Figure 9, where the most significant phases of the behaviour are illustrated on a simple 2D T-model. In case of loading plastic deformations of the end-plate take place in both the web region and bolt region. During the unloading, however, plastic deformations are concentrated mainly to the web region, while the bolt region remains elastic, which is due to the separation of the edge region from the foundation. Thus, the hardening in the tensile side can be more significant than that in the compressive one. The two characteristic states (at the end of half-cycles) are also presented in the real 3D model, in Figure 10.

93

TS4 corresponds to the other extreme case. Here, this is the end-plate which remains mostly elastic. The behaviour is governed by the elastic/plastic elongation of the bolt, and the separation/re-contact which can take place not only between the end-plate and the foundation, but between the bolt-head and the end-plate, too.

The behaviour again can be well illustrated in a 2D model, see Figure 11. The first part of the figure shows the loading process: due to the large deformation of the bolts, the basically elastic end-plate separates from the foundation. When the loads begin to decrease an elastic shortening of the bolts are take place together with the elastic straightening of the end-plate. At the zero force level, however, a gap remains between the end-plate and the foundation. Thus, in the next phase, the end-plate can move as a rigid body until it makes contact with the foundation

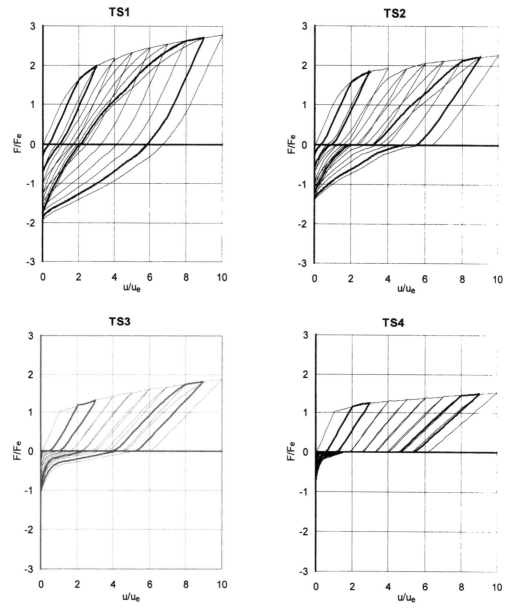

Figure 8. Load - displacement curves

again. In the final stage of the cycle a smaller compressive force is obtained, because of some previous end-plate plastification. Two characteristic states (at the end of half-cycles) are also presented in the real 3D model, in Figure 12.

Between the two extreme cases a successive transformation can be observed (Fig. 8, TS2 and TS3). The behaviour is a kind of mixture of the two extreme cases, depending on the parameters of the T-stub. However, two main groups can be defined.

In the first group the plastic deformations first appear in the end-plate (TS2). In this region the shape of the diagram is similar to that of TS1. After a certain number of cycles also the bolt material begins to flow due to the hardening of plate material, and the behaviour tends to that of TS4, as the rigid-body-type displacement appears.

In case of the second group this is the bolt which first yields (TS3). That is why the shapes of the first cycles are similar to those of TS4. However, after a certain number of cycles the shape of the hysteresis loops become "fatter", and this is caused by the plastification of the end-plate due to the hardening effect.

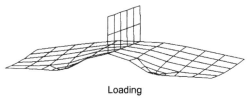
Loading

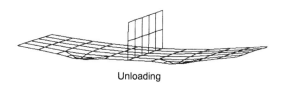
Unloading

Figure 9. Illustration of "strong bolt" behaviour

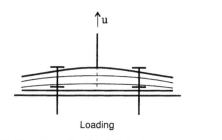
Loading

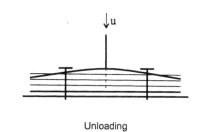
Unloading

Figure 10. Characteristic states of "strong bolt" behaviour

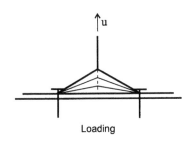
Loading

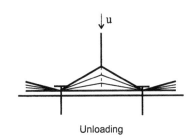
Unloading

Figure 11. Illustration of "weak bolt" behaviour

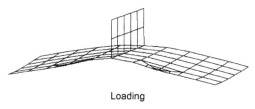
Loading

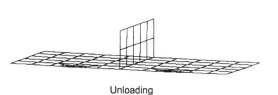
Unloading

Figure 12. Characteristic states of "weak bolt" behaviour

95

6 CONCLUDING REMARKS

In the paper a 3D plate/shell model is applied to study the monotonic and cyclic behaviour of steel-to-steel and steel-to-concrete bolted T-sub assemblies. The applications are related to experimental simulations and virtual experiments. On the basis of the results and model application experiences the following conclusions can be drawn.

The model predicts the initial tension stiffness of the T-stub components with a high accuracy if preloading is not applied in the bolts. The model should be refined and verified when different level of pretensioning is used in the bolts, which can be an important topic of further investigations.

It is promising to apply the model for the ultimate load analysis of end-plate joints where the steel plate behaviour is dominant. In this case a moderate size of numerical model within a reasonable time provides with very good accuracy in following the non-linear behaviour of the joint. However, the local phenomena of the bolts (e.g. bolt head/nut bending) can not be simulated accurately by the model.

The proposed model is able to follow the bolt – plate interaction of T-stubs for cyclic loading, too. According to the presented results the model can be applied even in extreme cases, that is in case of extremely strong or weak bolts. It is found that the parameters' ratio of the T-stub assembly (namely bolt-to-plate ratio) have significant effect on the cyclic behaviour. The rigidity, the resistance, as well as the energy absorption capacity of the T-stub are strongly influenced by the bolt-to-plate ratio. The virtual experiments on T-stub components with different bolt-to-plate ratios show the transfer between the different cyclic behaviour modes.

As far as the model applicability is concerned the following concluding remarks can be drawn.

- The above results can be achieved by relatively simple numerical model within a reasonable time by a completely automatic non-linear solution procedure.
- By applying the proposed model the monotonic and cyclic parametric virtual experiments on T-stub components can be done under a practically good accuracy – efficiency condition.
- The model application on joint components gives the opportunity to extend it for complex joint analysis.

ACKNOWLEDGEMENTS

The research work has been conducted under the financial support of OTKA T020738 project.

REFERENCES

Ádány, S. & Dunai, L., 1997a. "A Modified Multi-Surface Model for Structural Steel under Cyclic Loading," *Proceedings of the 5th International Colloquium on Stability and Ductility of Steel Structures* (ed. T. Usami), July 29-31, 1997, Nagoya, Japan, Vol. 2, pp. 841-846.

Ádány S. & Dunai L., 1997b. "Modelling of Steel-to-Concrete End-plate Connections under Monotonic and Cyclic Loading," *Periodica Polytechnica*, Technical University of Budapest, Vol. 41, No. 1, pp. 3-16.

Dunai, L., 1992. "Modelling of Cyclic Behaviour of Steel Semi-rigid Connections," *Proccedings of State-of-the-Art Workshop on Semi-rigid Behaviour of Civil Engineering Structural Connections: COST C1,(ed. A. Colson)*, pp. 394-405.

Dunai, L., Ádány, S., Wald, F. & Sokol, Z., 1996. "Numerical Modelling of Column-Base Connections," *Advances in Computational Techniques* (ed. B. H. V. Topping), Civil-Comp Press, Edinburgh, 171-177.

Dunai, L. & Ádány, S. 1997. "Cyclic Deterioration Model for Steel-to-Concrete Joints," Second International Conference on the Behaviour of Steel Structures in Seismic Areas (STESSA '97), August 3-8, 1997, Kyoto, Japan, *Proceedings, ed. F. M. Mazzolani, H. Akiyama)*, pp. 564-571.

ECCS 1986. "Recommended Testing Procedure for Assessing the Behaviour of Structural Steel Elements under Cyclic Loads", Technical Committee 1, TWG 1.3, No. 45.

Eurocode 3 1991. "Design Rules for Steel Structures, Part 1, General Rules and Rules for Buildings."

Faella, C., Piluso, V. & Rizzano, G., 1996. "Experimental analysis of T-stub Assemblies with Snug Tightened or Pretensioned Bolts," *Technical Report*, No 73, University of Salerno, Department of Civil Engineering, 1996.

Sokol Z. & Wald F., 1997. "Experiments with T-stubs in Tension and Compression," *Research Report*, ČVUT, Praque.

Sokol Z., Ádány S., Dunai L. & Wald F., 1999. "Column Base Finite Element Modelling," *Acta Polytechnica – Eurosteel '99*, Czech Technical University in Prague, Vol. 39, No.5/1999, pp 51-63.

Behaviour of Steel Structures in Seismic Areas, Mazzolani & Tremblay (eds) © 2000 Balkema, Rotterdam, ISBN 90 5809 130 9

Experimental studies on cyclic behaviour modes of base-plate connections

S. Ádány
Department of Structural Mechanics, Technical University of Budapest, Hungary

L. Calado
Department of Civil Engineering, Instituto Superior Técnico, Lisbon, Portugal

L. Dunai
Department of Steel Structures, Technical University of Budapest, Hungary

ABSTRACT: In this paper an experimental study is reported performed on end-plate type joints. The test arrangement represents a column-base joint of a steel frame. Altogether five specimens were tested, each of them subject to cyclic loading. The specimens were designed to present the typical behaviour types of end-plate joints. On the basis of the experimentally established moment-rotation relationship the cyclic characteristics of each specimen have been calculated and compared to one another. The results are evaluated, qualitative and quantitative conclusions are drawn.

1 INTRODUCTION

End-plate-type joints are widely used in steel frame structures, connecting either two steel elements (like beam-to-column, beam-to-beam or column-to-column joints) or a steel and a concrete/reinforced concrete element (like column-base joints or joints of a steel beam and a reinforced concrete column). Although these joints have some practical advantages, their application results in a more complicated structural behaviour and, consequently, requires more complex design.

In the recent decade lots of experimental investigations have been performed to analyse the behaviour of the various kinds of end-plate joints, focused mainly on the monotonic behaviour. To the cyclic behaviour, however, much less efforts have been devoted. Nevertheless, certain number of experimental programs have been performed (Dunai et al. 1996, Calado et al. 1999, Ballio et al. 1997, Calado & Lamas 1998, Calado et al. 1998, Bernuzzi et al. 1996).

The complete understanding of the cyclic behaviour of end-plate joints is essential, especially in the seismic design. The importance of the problem was clearly justified during the recent earthquake events, where significant structural damage of steel frames took place in the connection zones in several cases. Thus, it is important to understand and simply but reliably asses the behaviour of the joints in case of seismic actions, in order to satisfy the required resistance, rigidity, ductility and energy absorption demands.

In this paper an experimental program is reported, carried out in the Instituto Superior Técnico, Lisbon, Portugal. The test program is devoted to the cyclic behaviour of column-base type end-plate joints with the primary aim of providing information on the typical behaviour modes as illustrated in Figure 1, including

- behaviour governed by the *end-plate behaviour,*
- behaviour governed by the *bolt behaviour,*
- behaviour governed by the *column behaviour,*
- behaviour governed by the *interaction of them.*

It is important to underline that the concrete behaviour is out of the scope of this study.

In Section 2 the preliminary work is summarised, in Section 3 the results are presented, finally some conclusions are drawn.

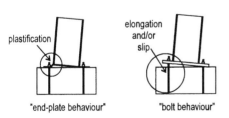
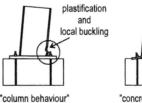

Figure 1. Behaviour components of column-base joints

2 DESIGN OF THE TESTS

2.1 *Test equipment*

The test equipment, on which the present experimental program is carried out, is basically developed to test beam-to-column joints of steel frames. The global arrangement is illustrated in Figure 2. In addition, there is a lateral frame to make possible the lateral support of the specimen, avoiding its lateral movement or twisting. The whole testing process is managed by a personal computer, by means of a data acquisition unit which commands the actuator and reads the data from the load cell and displacement transducers. Another important note that displacement control is used. More information about the test equipment can be found in Ferreira (1994) and Calado & Mele (1999).

In designing the test, the specimen characteristics are determined in accordance with the parameters of the existing test set-up, by considering the geometrical properties, the load capacity of the actuator and load cell, as well as the displacement capacity of the inductive displacement transducers. The main geometrical dimensions of the specimens are presented in Figure 2. The arrangement represents a column base joint, with an H-shaped column and a practically rigid base. The top part of the specimen has the role to ensure the restraint against lateral movement and twisting of the column.

2.2 *Preliminary calculations*

To be able to achieve the intended phenomena of the specimens preliminary calculations were done. The moment resistance of the joint can be determined as the minimum of resistances belonging to the possible failure modes.

Neglecting transverse effects, the resistance of the connection (end-plate + bolts) can be determined on a simple two-dimensional connection model. Four modes of failure can be defined as illustrated in Figure 3. Mode 1 represents the pure end-plate failure without failure of bolts. Mode 4 corresponds to the pure bolt failure, without any failure of the end-plate. Mode 2 and 3 are two cases of combined bolt and end-plate failure.

Moreover, the resistance of the column can be easily calculated according to Eurocode 3 (1991). The calculation depends on the classification of the cross-section, considering plastic reserve or local buckling of the section.

For the material properties, S235 steel and high-strength bolts of grade 8.8 are assumed. In principle, the characteristic values of yield strength, as defined in Eurocode 3 (1991), are adopted for the calculations.

The main geometrical dimensions are given in Figure 2. The only undefined parameter is the end-plate thickness, which is chosen to control the connection behaviour. The connection resistance was

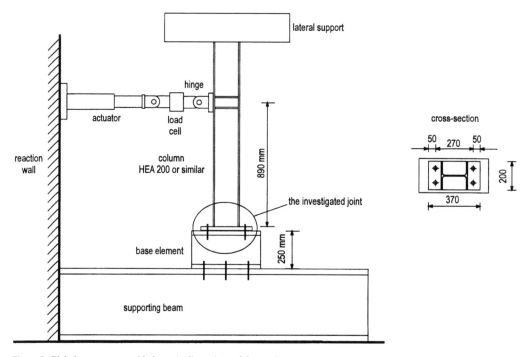

Figure 2. Global arrangement with the main dimensions of the specimens

Table 1. Specimens main characteristics

Specimen	Column section	End-plate thickness	Bolt tightening	Anticipated behaviour
CB1 (CB1R)	HEA200	25 mm	hand-tightened	Mode 3
CB2	HEA200	16 mm	hand-tightened	Mode 2
CB3	welded	25 mm	hand-tightened	local buckling
CB4	HEA200	25 mm	pre-tensioned	Mode 3
CB5	HEA200	12 mm	hand-tightened	Mode 1

Table 2. Loading history

Cycle nr.	1	2	3	4	5	6	7	8	9	10	11	
Displacement amplitude	$\frac{1}{4}e_y$	$\frac{1}{2}e_y$	$\frac{3}{4}e_y$	e_y	$2e_y$	$2e_y$	$3e_y$	$3e_y$	$4e_y$	$4e_y$	$5e_y$	etc.

calculated for various end-plate thickness values. It was found that pure bolt failure is not realistic to achieve since it occurs only in case of extremely thick end-plate. For that reason, pure bolt failure (Mode 4) was eliminated from the study and, finally, three pieces of end-plate thickness were chosen, according to Modes 1, 2 and 3.

Two types of column cross-section are designed. One is a HEA 200 hot-rolled profile, which stocky enough to avoid local buckling. At the same time, however, another section was designed to study the effect of local buckling, by applying a slender welded column profile with less resistance than that of the connection itself. It is to be noted that the welded section is designed so that its system lines (mid-lines of the plane elements) would be identical with those of HEA 200 section.

2.3 The specimens

Altogether five specimens were designed. The main characteristics of the specimens are summarised in Table 1.

Four characteristic behaviour types are anticipated, as well as the effect of bolt pre-tensioning is also studied.

An additional note that butt welds are applied between the column and the end-plate in order to minimise the risk of weld failure.

2.4 Displacement transducers

To measure the displacements inductive displacement transducers were used. The number and the position of the transducers were defined so as to get as

much as possible information on the behaviour, considering also the place required for each transducer, which gives a limitation of their maximal number. Finally, altogether 13 transducers were used. 12 pieces to record the displacement of the joint and one at the force application point to control the loading process.

2.5 Loading history

The loading is controlled by the yielding displacement, that is the displacement belonging to the limit of elasticity (e_y). The loading history is defined in accordance with the ECCS Recommendations (ECCS 1986), however, with two differences, see Table 2.

In this study two cycles are applied in the plastic range instead of three as proposed in the ECCS (1986), because it was observed in previous tests (Calado & Lamas 1998) that the third cycle does not give additional information relatively to the previous two cycles.

The other difference is as follows: after the limit of elasticity is reached, $2e_y$, $3e_y$, $4e_y$, $5e_y$, etc. displacement amplitude was applied instead of $2e_y$, $4e_y$, $6e_y$, etc. The reason is to have more cycles and more information in the plastic range even if the displacement capacity of the specimen is not very high.

The limit of elasticity is calculated by using advanced finite element models. Details about the model can be found in Dunai & Ádány, 1997. The initial part of the moment-rotation characteristic of the joint was established by FEM analysis, then the limit of elasticity is defined according to ECCS (1986).

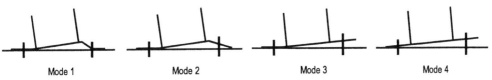

Mode 1   Mode 2   Mode 3   Mode 4

Figure 3. Failure modes

99

3 RESULTS

3.1 *Moment-rotation curves*

After the tests are performed, the moment-rotation relationship of the joint can be established on the basis of the measured forces and displacements. In order to be able to show the results in a unified way, a "joint reference section" is introduced, which is used to calculate the moments and rotations. This section is defined at a distance of twice the column section depth from the top surface of the base-plate in order to be not disturbed by the intensive deformation due

to local buckling. Since the depth is equal to 180 mm for all the specimens, the reference section is situated 360 mm from the base-plate.

The moment-rotation curves are presented in Figure 4. Six curves are plotted, because one of the tests was repeated. In case of the CB1, due to the deterioration of the nut thread, the nuts of the tensioned bolts were "jumped" from the bolt at the beginning of plastic cycles. Since the specimen has not significantly deteriorated it was decided to replace the bolts, applying more nuts (3), and repeating the test. The repeated test is referred as CB1R.

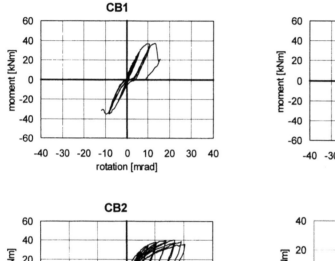

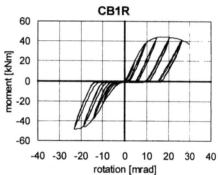

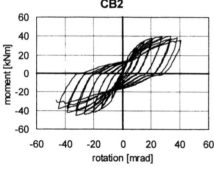

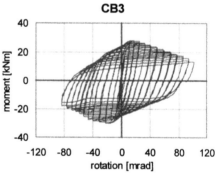

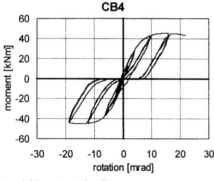

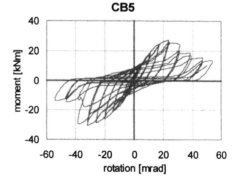

Figure 4. Moment-rotation diagrams

3.2 Slip between the base-plate and base element

Figure 5 shows the slip-force diagram for CB2 and CB3 specimens. (Here, the term "force" is the shear force, which is equal to the applied force.) It can be observed that the development of the slip begins at the zero-force level of each cycle, and very rapidly reaches its maximum value. This observation can be easily explained by taking into consideration that, at the zero-force level, the contact area between the base-plate and the base element is reduced due to the residual deformations of base-plate.

Similar diagrams can be obtained for all the cases where significant base-plate or bolt deformation is experienced (CB1, CB1R, CB2, CB4, and CB5). However, in case of CB3, the effect of slip is almost negligible, since the base-plate and the bolts are not subjected to plastic deformations.

3.3 Bolt elongation

Figure 6 presents the bolt elongation in case of CB2 and CB3 specimens.

The diagram for CB2 well demonstrates the yielding of the bolts, as well as the residual defor-

mations. Similar diagram can be obtained for CB1R and CB4, too, where significant bolt elongation occurred.

In case of CB3 and CB5, however, the bolts remain practically elastic, without large deformations. This is also well demonstrated in the diagram for CB3 (Fig. 6).

3.4 Cyclic parameters

On the basis of the moment-rotation curve some cyclic parameters are calculated, as it is proposed in ECCS (1986). Four parameters are calculated:
– the full ductility ratio,
– the resistance ratio,
– the rigidity ratio,
– the absorbed energy ratio.

In the followings, these ratios are plotted in function of the partial ductility. The calculated cyclic parameters are presented in Figure 7 for the positive hemicycles of the moment-rotation diagrams. Note that CB1 test is not evaluated here, since, due to the early failure of the bolt and nut thread, the loading history is too short.

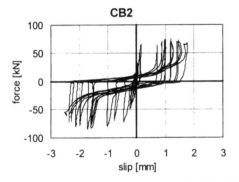 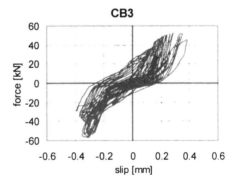

Figure 5. Slip between the base and base-plate for CB2 and CB3

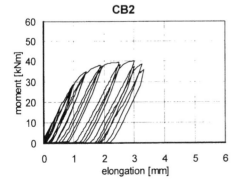 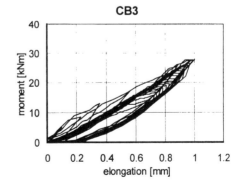

Figure 6. Bolt elongation for CB2 and CB3

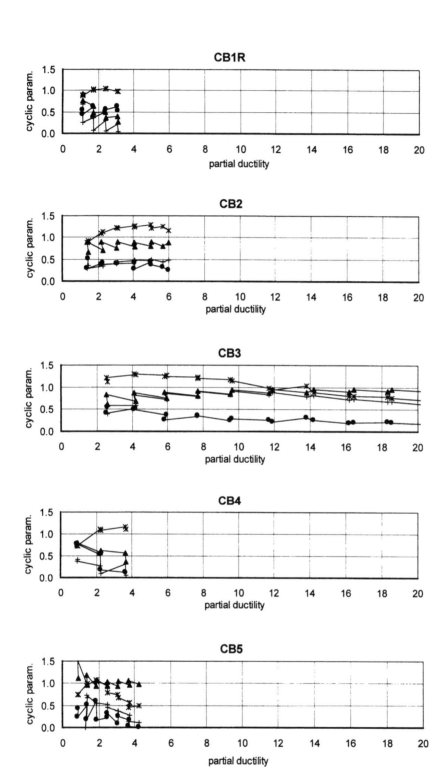

Figure 7. Cyclic parameters of the moment-rotation diagrams

4 RESULTS EVALUATION

4.1 *Effect of bolt pre-tensioning*

CB1R and CB4 specimens are identical. However, in case of CB1R the bolts are hand-tightened, while in case of CB4 pre-tensioning is applied. Thus, the effect of bolt pre-tensioning can be analysed by comparing the two cases.

The behaviour modes that are observed during the two tests are similar, as well as the calculated cyclic parameters have similar tendencies, according to Figure 5. Thus, it can be stated that the bolt pre-tensioning has no important effect on the cyclic behaviour in the analysed cases.

4.2 *Deformation capacity*

An important observation is that the deformation capacities of the various tested joints strongly differ from each other.

If the governing behaviour is the bolt behaviour (Mode 3, see CB4 or CB1R), the maximal value of the partial ductility is approx. equal to 4. In this case the failure of the specimen is caused by the failure of the bolts. Since the bolts are of high-strength steel, the deformation capacity of the bolts is limited, which results in a small deformation capacity of the joint.

In case of CB2 there is a strong interaction between the bolt and end-plate (Mode 2). The maximal value of the partial ductility is 6. In this case the failure is caused by the crack occurred at the flange to end-plate welds. It means, that although the behaviour itself is governed by the end-plate and bolt deformations, the failure is caused by the weld crack, which limits the joint deformation capacity.

CB 5 corresponds to Mode 1, when the behaviour is basically governed by the end-plate deformations. The maximal partial ductility is slightly more than 4. It is to be noted, however, that the failure is caused by the failure of the flange weld again, which reduced the deformation capacity.

If the governing behaviour is the column flange/web buckling (CB3), the maximal partial ductility is much more than any of the other cases (more than 20). Moreover, it should be mentioned that CB3 joint has not reached its deformation capacity during the test. The test was finished because of the measuring devices.

4.3 *Ductility*

In case of bolt behaviour (CB4) the full ductility ratio is rapidly decreasing during the consecutive cycles. This degradation is in connection with the rigid-body-type rotation of the joint, as it is described later. In the other cases the value of the full ductility ratio is near to 1, which corresponds to a ductile behaviour.

4.4 *Resistance*

Generally it can be stated that all the cases represent good cyclic behaviour from the viewpoint of resistance. The value of the resistance ratio is usually more than 1, without considerable degradation, which means that the moment resistance of the joint does not change significantly even after several plastic cycles of loading.

The only exception is the case of the end-plate behaviour (CB5). In this case the decreasing tendency is definitely caused by the flange weld crack, which resulted in a reduced end-plate cross-section at the welds, consequently, a reduced resistance of the end-plate and the joint.

4.5 *Rigidity*

In practically all the cases considerable rigidity degradation can be observed.

In case of governing bolt behaviour (CB4, CB1R) a rigid-body-type rotation clearly can be observed in the moment-rotation diagram. This kind of rigid-body rotation is designated by the horizontal part of the diagram at the zero force level.

When the bolt behaviour is combined with the end-plate behaviour the rigid-body-type rotation does not occur. However, the rigidity is strongly reduced.

From rigidity point of view CB3 shows the best behaviour. There is no rigid-body-type rotation at all, although the rigidity is continuously decreasing due to the deterioration of the whole column section.

4.6 *Absorbed energy*

In case of governing bolt behaviour (CB1R, CB4) the absorbed energy ratio rapidly decreases during the consecutive cycles. Another important observation that in the repeated cycles (with the same maximal displacement) there is a significant drop of the ratio, which indicates that there is almost no energy dissipation in these repeated cycles. The phenomenon can be drawn back to the rigid-body-type displacement, as discussed above.

In case of CB2 (combined end-plate and bolt behaviour) the value of the absorbed energy ratio is approx. equal to 0.5, constantly. The important thing is that there is no degradation, in this case.

The diagram for CB5 shows a decreasing tendency. This degradation is certainly caused by the weld cracks, which reduced the area of the end-plate, resulted in reduced moment resistance and energy dissipation capacity.

The most advantageous behaviour belongs to CB3, when the local buckling of column flanges/web determine the behaviour. Even after several plastic cycles the absorbed energy ratio is more than 0.5, although it has a decreasing tendency if the partial ductility is greater than 10-12.

5 CONCLUSION

In this paper an experimental program on steel bolted end-plate joints is presented with the primary aim of providing information on the behavioural components which determine the joint behaviour. Here, some general conclusions are drawn.

It can be stated that the experienced behaviour of each specimen is in accordance with the expected behaviour. Thus, the applied method for the preliminary calculation is justified. The five different specimens cover a wide range of behaviour, including governing bolt behaviour (CB1 and CB4), governing base-plate behaviour (CB5), column local buckling (CB3) and a combined base-plate/bolt behaviour (CB2).

The tests justified the existence of the three basic types of behaviour. (Note that the concrete behaviour was not investigated in the present study.) However, the important effect of weld cracks is also highlighted. Whenever there is intensive end-plate deformation the failure is caused by the cracks occurred at the flange to end-plate welds. The cracks also influence the cyclic characteristics causing significant degradation of the moment resistance and the energy absorption capacity. Thus, although the end-plate behaviour would have good cyclic characteristics (since it is determined by the steel material behaviour), the weld cracks can strongly modify the behaviour.

The most advantageous behaviour is experienced if the deformations are concentrated in the column section, forming a plastic hinge (CB3, governing column behaviour). In this case the behaviour is extremely ductile, with considerable energy absorption capacity. On the other hand, whenever there is significant bolt elongation, the rigidity and energy dissipation capacity of the joint considerably decrease, due to the rigid-body-type rotation of the joint. In this case also the deformation capacity is reduced, as a consequence of the limited elongation capacity of the bolts.

The obtained results are applicable for the verification and calibration of numerical models. Detailed experimental data are provided for various behaviour types corresponding to the same joint topology. It is important, however, to study the concrete behaviour, which can be the topic of further investigations.

ACKNOWLEDGEMENTS

The research work has been conducted under the financial support of the following projects:
- OTKA T020738 (Hungarian National Scientific Research Foundation),
- FCT - Fundação para a Ciência e a Tecnologica,
- TEMPUS JEP 11236/96.

REFERENCES

Ballio, G., Calado, L. & Castiglioni, C. A. 1997. "Low Cycle Fatigue Behaviour of Structural Steel Members and Connections", *Fatigue & Fracture of Engineering Materials & Structures*, Vol. 20, No. 8, pp. 1129-1146.

Bernuzzi, C., Zandonini, R.& Zanon, P. 1996. "Experimental Analysis and Modelling of Semi-Rigid Steel Joints under Cyclic Reversal Loading", *Journal of Constructional Steel Research*, Vol. 38, pp. 95-123, 1996.

Calado, L., Bernuzzi, C. & Castiglioni, C. A. 1998. "Structural Steel Components under Low-cycle Fatigue: Design Assisted by Testing", Structural Engineering World Congress, SEWC, San Francisco.

Calado, L. & Lamas, A. 1998. "Seismic Modelling and Behaviour of Steel Beam-to-Column Connections", 2nd World Conference on Steel Construction, San Sebastian, Spain.

Calado, L. & Mele, E. 1999. "Experimental Research Program on Steel Beam-to-Column Connections ", Report ICIST, DT no 1/99, ISSN:0871-7869

Calado, L., Mele, E. & De Luca, A. 1999. "Cyclic Behaviour of Steel Semirigid Beam-to-Column Connections", *to be published in ASCE*.

Dunai, L., Fukumoto, Y. & Ohtani, Y. 1996. "Behaviour of Steel-to-Concrete Connections under Combined Axial Force and Cyclic Bending", *Journal of Constructional Steel Research*, Vol. 36, No. 2, pp. 121-147, 1996.

Dunai, L. & Ádány, S. 1997. "Cyclic Deterioration Model for Steel-to-Concrete Joints," Second International Conference on the Behaviour of Steel Structures in Seismic Areas (STESSA '97), August 3-8, 1997, Kyoto, Japan, *Proceedings, ed. F. M. Mazzolani, H. Akiyama)*, pp. 564-571.

ECCS 1986. "Recommended Testing Procedure for Assessing the Behaviour of Structural Steel Elements under Cyclic Loads", Technical Committee 1, TWG 1.3, No. 45.

Eurocode 3, ENV 1993-1-1 1992. "Design Rules for Steel Structures, Part 1, General Rules and Rules for Buildings."

Ferreira, J. 1994. "Characterisation of the Behaviour of Semi-Rigid Steel Connections" MSc Thesis, Instituto Superior Técnico, Lisbon, Portugal. (*in portuguese*)

Behaviour of Steel Structures in Seismic Areas, Mazzolani & Tremblay (eds) © 2000 Balkema, Rotterdam, ISBN 90 5809 130 9

Cyclic behaviour of the shear connection component in composite joints

J. M. Aribert & A. Lachal
Laboratory of Structures, INSA-Rennes, France

ABSTRACT: Two series of experimental tests to study the cyclic behaviour of usual types of shear connectors in buildings and their influence on the global behaviour of beam-to-column composite joints are presented in this paper. The first series involves 30 push-pull tests to evaluate the shear resistance, the slip capacity and the low-cycle fatigue resistance of two types of shear connectors associated with solid or composite slabs. The second series of tests involves 11 full scale composite joints. Main results and their interpretation deal with moment resistance, rotational stiffness, absorbed energy and rotation capacity linked with risks of rupture by low cycle fatigue either in the bolted end plate steelwork part or in the shear connection of the beam adjacent to the joint. An equivalent static approach seems significant to evaluate the skeleton curve of the cyclic moment-rotation characteristics. From the so-interpreted results a preliminary proposal may be suggested for the seismic design of the shear connection of frame spans.

1 INTRODUCTION

Steel-concrete composite structures appear competitive today in comparison with other types of structure and their dynamic performances in seismic areas are generally recognized as efficient although not well formulated in seismic design codes. Their advantages are the clear increase in stiffness and resistance of the steel elements (beams and columns) due to contribution of the concrete, the capacity to dissipate energy and the reduction of risks of buckling especially when the steel elements are encased by concrete. Obviously, other well-known advantages are the improved fire resistance, also the ease and rapidity of erection leading to low construction costs.

But often designers are in front of many questions for which no convincing answers or minimal guidance exist. In Annex D of Eurocode 8 Part 1-3 (1995) where the same behaviour factors q are adopted for composite structures as for steel structures (apparently due to lack of knowledge), odd concepts put forward without real scientific background e.g. recommending the use of partial shear connection.

The overstrength of joints in dissipative zones of steel structures to satisfy the principle of capacity design under seismic actions is difficult to apply to composite building frames where composite joints are generally semi-rigid and partial strength and therefore contribute to dissipate energy.

Even in the case of quasi-static loads, a major problem persists with two aspects: on the one hand, many types of joint can be imagined; on the other hand, design codes do not give detailed rules at the present for composite joints. So, Eurocode 4-Part 1.1 (1992) is content with describing what is meant by a composite joint, leaving the designers to adapt the detailed rules given in Annex J of Eurocode 3, part 1-1 (1992) for the steel components. Nevertheless, the Technical Committee 11 of ECCS has published recently a detailed model code for the design of composite joints subject to quasi-static loads (ECCS-TC11 1999); parts of this paper may be a useful reference to simplify the interpretation of the cyclic behaviour of composite joints.

The main objective is to provide experimental information on the risk of degradation of the shear connection and on the consequent global behaviour (expressed by moment-rotation curves) of composite joints subject to cyclic repeated loads.

2 BEHAVIOUR CHARACTERIZATION OF ISOLATED CONNECTORS UNDER REPEATED CYCLIC SHEAR

2.1 *Experimental program*

30 tests of both Push-Out type and Push-Pull type were carried out at the Laboratory of Structures

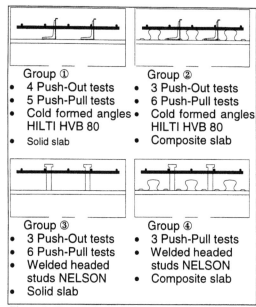

Figure 1. Types of tested shear connectors

Group ①
- 4 Push-Out tests
- 5 Push-Pull tests
- Cold formed angles HILTI HVB 80
- Solid slab

Group ②
- 3 Push-Out tests
- 6 Push-Pull tests
- Cold formed angles HILTI HVB 80
- Composite slab

Group ③
- 3 Push-Out tests
- 6 Push-Pull tests
- Welded headed studs NELSON
- Solid slab

Group ④
- 3 Push-Pull tests
- Welded headed studs NELSON
- Composite slab

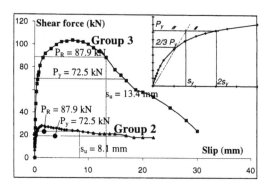

Figure 2. Monotonic curves

in INSA-Rennes (France), distinguishing four groups as presented in figure 1. All the specimens were fabricated according to the recommendations given in Eurocode 4-1-1 (Chapter 10 for Push-Out tests) consisting of one HEB 200 steel section and two concrete slabs each having *120 x 400 mm* cross-section. A single reinforcement layer with *4ϕ10 mm* longitudinal rebars of *400 mm* length and *5ϕ8 mm* transverse rebars of *350 mm* length was included in each slab, noting that a single reinforcement layer will also be used later in the slab of the tested composite joints.

2.2 *Monotonic behaviour of reference*

Examples of average curves deduced from the monotonic Push-Out tests for Groups ② and ③ are presented in figure 2.

In order to follow a specific procedure for cyclic tests (ECCS 1985) elastic limit shear forces P_y^- and P_y^+ as well as associated slip values s_y^- and s_y^+ have to be determined from the monotonic curves of Push-Out and Pull-Out tests respectively. Taking into account the symmetrical behaviour observed for Push and Pull actions, only one couple of elastic characteristics P_y and s_y can be given here. Moreover considering the practical difficulty to draw the initial tangent to the monotonic curve the authors propose to adopt a similar definition to the one given in figure 2.

Revised Annex J of Eurocode 3-1-1 (CEN, 1997), in which the elastic limit P_y corresponds to a slip twice the elastic one (figure 2) and the initial stiffness is defined as the slope of the secant line joining the origin and the point on the monotonic curve located at ordinate $2/3$ P_y (this geometrical construction requires a short iterative procedure).

For the characteristic resistance P_R which should be considered when designing a shear connection at ultimate limit state, as specified in Eurocode 4-1-1, the deviation from the mean value obtained from three identical specimens has led to adopt $P_R = 0.85$ P_{max} where P_{max} is the mean maximum shear resistance of the connectors.

Elastic limit characteristics P_y and s_y, as well as the shear resistance P_R and the slip capacity s_u defined as the slip for which the shear force falls below P_R on the monotonic decreasing branch (figure 2) are shown for all the groups in table 1.

The P_R values are not very different from the design values specified either by Eurocode 4-1-1 for welded headed studs (*82.5 kN* according to clause 6.3.2) or by the manufacturer for cold formed angles (namely *22 kN*). All the slip capacities s_u are higher than the required *6 mm* that permits these connector types to be considered as ductile (see clause 6.1.2 in EC4-1-1).

Table1: Characteristics of reference

GROUP	P_y (kN)	s_y (mm)	P_{max} (kN)	P_R (kN)	s_u (mm)
①	29.5	0.6	39	33.2	12.9
②	19	0.12	27.8	23.6	8.1
③	72.5	0.3	103.4	87.9	13.4
④	58	0.3	(85)	(72)	(>6)

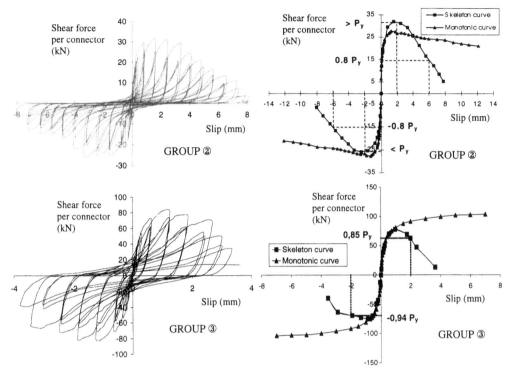

Figure 3: Cyclic test - ECCS type of procedure

Figure 4: Skeleton curves

2.3 Full reversed cyclic behaviour according to ECCS procedure

The ECCS procedure consists of generating 4 cycles successively for the ranges of slip displacement $\pm s_y/4$, $\pm s_y/2$, $\pm 3s_y/4$, $\pm s_y$ followed up to failure by series of 3 cycles each with a range $\pm 2n\,s_y$ where $n = 1,2,3,...$

Examples of test results are presented in figure 3 for Groups ② and ③. It appears that the responses of the specimens subject to full reversed displacements express a reduction both of maximum strength and ductility due to yielding and fatigue of the shear connectors and damage of the concrete. Different failure modes were observed: a ductile failure on the 30th cycle for Group ② and a rupture by shear of the stud cross-section located just above the weld on the 20th cycle for Group ③.

To propose a pragmatic interpretation of ECCS cyclic curves, it is possible to determine, from the peak strengths, the skeleton curve in each group of tests and to use it as an equivalent monotonic curve. So, for Groups ② and ③, the skeleton curves are plotted in figure 4 with the corresponding real monotonic curves (given above in figures 2). When the slip range exceeds a value of about ±1 mm or ±2 mm as the case may be, a rapid decrease of the

skeleton curve occurs with approximate linear variation up to failure. From the skeleton curves (figure 4), shear resistances expressed as a fraction of P_y have been determined for two slip amplitudes: ±2 mm and ±6 mm. These values are given in table 2.

Table 2: Skeleton curve interpretation

GROUP	Shear resistance measured for the slip displacement :		Shear resistance measured for the slip displacement:	
	+ 2 mm	- 2 mm	+ 6 mm	- 6 mm
②	$> P_y$	$< -P_y$	$0.8\,P_y$	$-0.8\,P_y$
③	$0.85\,P_y$	$-0.94\,P_y$	Failure before ± 6 mm	

2.4 Cycle fatigue behaviour

An example of test in low-cycle fatigue with a constant slip range is presented in figure 5 for one of the tests of Group ③. In all the cases, a reduction of the shear resistance appears and depends on both the slip range Δs and the number of cycles N.

According to Ballio and Castiglioni (1994), also to Bernuzzi, Calado et al (1997) for steel beam-to-column joints under cyclic reversal loading, the

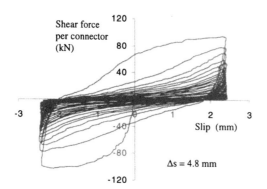

Figure 5 : Low-cycle fatigue test
(Welded headed stud with solid slab - Group ③)

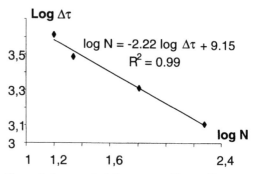

Figure 6: Low-cycle fatigue curve (Group ③)

following damage model which generalize the usual fatigue rule for design of steel structures has been tried in order to propose an analytical formulation for all the low-cycle fatigue results:

$$N[\Delta\tau]^m = K \quad \text{with } \Delta\tau = \frac{\Delta s}{s_y}\tau_y \quad (1)$$

where: τ_y is the shear stress of the connector corresponding to its elastic limit shear resistance P_y,

s_y is the corresponding limit slip of the shear connector,

Δs the constant slip range applied to the shear connector,

K a constant depending on the fatigue strength category of the considered detail,

m a constant exponent.

When cycles of different amplitudes Δs_i are applied to the specimen, the Miner's rule based on linear damage cumulation may be assumed introducing the following damage index:

$$D = \frac{1}{K}\sum_{i=1}^{\ell} N_i \left[\frac{\Delta s_i}{s_y}\tau_y\right]^m \quad (2)$$

where: N_i is the number of cycles with same range Δs_i and ℓ is the number of different ranges Δs_i to be considered.

The low-cycle fatigue results have been plotted in log-log co-ordinates, then a linear regression associated to a coefficient of determination R^2 has been determined for each group as shown for Group ③ in figure 6. For connectors of Groups ① and ③ with a solid slab, in comparison with current fatigue strength curves for steel structures (chapter 9 of Eurocode 3-1-1), a more severe damage due to concrete weakness appears (the slope constant m of the fatigue strength curves is about 2.2, therefore clearly lower than 5, and the detail category

corresponding to the constant $logK$, is about: $\Delta\tau_c = 20$ N/mm^2).

For connectors of Group ④ (welded headed stud with a composite slab in figure 4.3.10c), the slope constant is slightly greater (about 2.8) and the detail category is about *60 N/mm²*.

Results of Group ② (cold formed angle with composite slab) are too scattered to be fitted by a linear curve. In this case, the effects of the steel profiled sheeting deformation needs to be clarified.

2.5 Partial conclusion

At the present stage of investigation, the Push-Pull tests demonstrate that shear connectors defined as ductile (according to Eurocode 4-1-1) under monotonic actions should be considered as non ductile under cyclic repeated actions. Moreover, the requirement of ± 2 mm as minimum slip capacity sometimes considered for non ductile connectors may lead to a reduction of their static shear resistance.

Regarding the Push-Pull test results on the whole, a severe weakness under fatigue actions should be taken into account if large slip ranges beyond the elastic domain were accepted (for example, $\Delta s = 4.8$ mm in figure 6). Damage models similar to (1) and (2) seem appropriate to control the fatigue resistance of shear connectors and ought to be investigated more thoroughly.

However, it would be premature to specify practical provisions at the present stage without having tested composite beam-to-column joints to know the real cyclic behaviour of the zone of shear connection adjacent to the joint.

3 CYCLIC BEHAVIOUR OF COMPOSITE BEAM-TO-COLUMN BOLTED JOINTS

3.1 Experimental program

The experimental program on composite beam-to-column bolted joints has involved 5 cyclic tests completed by other monotonic tests of reference on bare steel joints and composite joints under sagging or hogging bending. Only the tests under cyclic loading are presented in this paper following the ECCS procedure based on two different elastic limit rotations of joint ϕ_y^+ and ϕ_y^-. Three groups may be distinguished in table 3: the first two groups deal with a cruciform beam-to-column type of joint with symmetrical loading so that no distortion effect of the steel column web panel occurs as a supplementary component in the joint rotation. On the contrary, the third group corresponds to a T arrangement. The main investigated parameters are: the degree of shear connection for group 1 (J4, J5, J6), the beam and column depths and the different type of slab for group 2 (J10) and the possible influence of the distortion of the column web panel on the partial shear connection for group 3 (J11).

3.2 Experimental results dealing with groups 1 and 2 (cruciform test without distortion effect)

The moment-rotation curves obtained for both groups are presented in figure 7 (test J5) and figure 8 (test J10). For all the tests, hysteresis loops exhibit an unstable behaviour with a pinching phenomenon associated with degradation of strength, stiffness and energy absorption capacity. This behaviour is not only due to concrete damage but also to permanent deformation in tension of the steelwork part in front of the bolt rows.

For all specimens failure occurred by rupture of bolts in the zone of the end plate close to the steel flange on the opposite side to the slab while observing a large inelastic deflection of the end plate.

Slip distributions measured along the beam at the hogging bending moment peaks for different cycles of rotation ranges ($\Phi_y^\pm, 2n\Phi_y^\pm, ...$) tend to demonstrate that outside a zone of 400 mm from the column axis, the slip becomes quasi-uniform

Table 3: Tested composite joints under cyclic ECCS procedure

Test reference	J4	J5	J6	J10	J11
Group		1		2	3
TYPE of specimen	Cruciform (symmetrical loading)				T arrangement
Column (S 235)	HEB 200			HEB 240	HEB 300
Steel beam	IPE 360 $(f_y=282 \ N/mm^2)$			IPE 450 $(f_y=347 \ N/mm^2)$	IPE 360 $(f_y=282 \ N/mm^2)$
Steelwork part: -end plate (S 235) -bolts (tightened for 1/3 nominal preload)	450×200×15 HR 10.9 - Φ 18 mm			580×240×20 HR 10.9 - Φ 22mm	450×200×20 HR 10.9 - Φ 22 mm
Shear connectors : -type -number per half span or cantilever span	Cold-formed angles (HILTI HVB 80 20 10 30			Welded headed stud (NELSON) Φ19 mm; h =100 mm 10	Welded headed studs (NELSON) Φ19 mm; h =100 mm 5
-degree of connection	Full (1.04)	Partial (0.53)	More than full (1.59)	Full (1.01)	Partial (0.80)
Type of slab: -concrete material -profiled sheeting	1000 × 120 (mm²) $(f_{cm}=24.3 \ N/mm^2)$ COFRASTRA 40			1000 × 120 (mm²) $(f_{cm}=27.0 \ N/mm^2)$ SOLID SLAB	1000 × 120 (mm²) $(f_{cm}=25.9 \ N/mm^2)$ COFRASTRA 40
Reinforcement: -Longitudinal -Transversal	10 Φ 10 mm rebars $(f_{ys}=568 \ N/mm^2)$ Φ8 each 10 cm			10 Φ14 mm rebars $(f_{ys}=568 \ N/mm^2)$ Φ8 each 10 cm	10 Φ 10 mm rebars $(f_{ys}=568 \ N/mm^2)$ 5Φ10 each 5.3 cm, then Φ10 each 10 cm

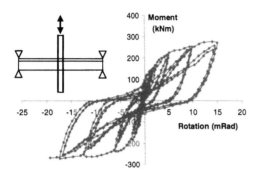

Figure 7: Moment-rotation curve in cyclic loading (test J5 – Group 1)

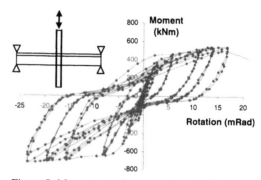

Figure 8: Moment-rotation curve in cyclic loading (test J10 – Group 2)

along the beam allowing to define one significant value of slip for each test. The maximum slip values so-obtained just before failure are given in table 4.

Table 4: Maximum slip values

TEST	J4	J5	J6	J10
Maximum slip (mm)	2	3.5	0.6	1.2

As it could be foreseen the higher value appears for test J5. Taking account of the Push-Pull test results, it is not excluded that for this test a beginning of failure, hidden by a premature failure of bolts in low-cycle fatigue, was set in motion in the shear connection. For the other tests with full or more than full shear connection, the risk of connector rupture is not probable.

3.3 Interpretation of tests of Groups 1 and 2 by means of non-dimensional ECCS parameters

A first interpretation of the cyclic results has been made using the non-dimensional ECCS parameters determined in accordance with the definitions proposed by Mazzolani and Piluso (1992)

Showing here only the variations of the absorbed energy ratio $\eta_i^{\pm}(\mu_i^{\pm})$ expressed as a function of the partial ductility $\mu_i^{\pm}$, it appears in figure 9 that the use of partial shear connection in composite joints under cyclic loading should not be advised; for test J5, the absorbed energy ratio is clearly 30% lower than for test J4 with a full shear connection. Moreover, the stiffness ratio curve $\xi_i^{\pm}(\mu_i^{\pm})$ has demonstrated a more pronounced decreasing for test J5 than for the other tests.

However a question remains, which concerns the possible unfavourable effect of the reinforcement position in the slab located approximately at the top of the angle connectors used in specimen J5; an insufficient force transfer between the reinforcement and the shear connection might speed up the joint degradation. On the other hand, the use of a group of shear connectors concentrated on a short width in a rib might lead to a loss of shear connection efficiency probably due to a greater damage of concrete in the ribs specially under cyclic loading. That might explain the very slight increasing of performance of test J6 in comparison with test J4.

The performance of test J10 with a solid slab fully connected by welded headed studs appears as good as the one of test J4; although the fatigue characteristics of the shear connectors are weak (see 2.4), the maximum slip range inside the joint does not exceed 2 mm.

3.4 Skeleton curves and monotonic curves for cruciform tests

A second mode of interpretation of experimental results concerns the skeleton curves obtained from the peak hogging and sagging bending moments. These curves have been drawn in figure 10 for tests J4, J5 and J6 (group 1) and compared to the monotonic M-Φ curve when the shear connection is full. The skeleton curves appear less favourable, especially under hogging bending and for partial shear connection (test J5). Another aspect of comparison concerns the limited rotation capacity of the skeleton curves, about 15 mrd in sagging bending, due to the rupture of the steelwork part leading the rotation in hogging bending to be truncated. On the contrary, it should be mentioned that the skeleton curve of Group 2 (not presented here) appears in good agreement with the corresponding monotonic M-Φ curve.

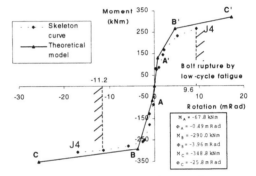

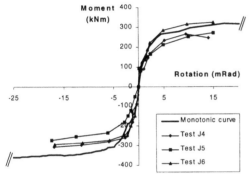

Figure 9: Absorbed ECCS energy ratio (Group 1)

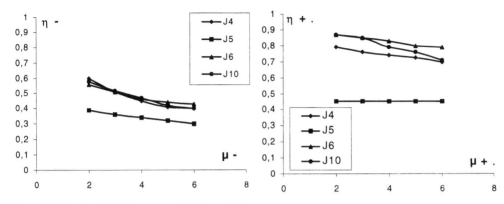

Figure 10: Skeleton curves (Group 1)

Figure 11: Simplified theoretical approach

3.5 *Simplified theoretical approach*

A simplified approach has been developed to evaluate suitably the skeleton curve of moment-rotation cycles when the following assumptions are satisfied:
- the shear connection is full (or more than full);
- the moment-rotation monotonic curves of the bare steel joint in sagging bending and hogging bending are known either by appropriate tests or by sophisticated numerical modelling (J.M. Aribert, A.Lachal and O.Dinga 1999);
- the resistance in low-cycle fatigue of details of the bare steel joint, specially the fatigue resistance of the tension bolts is well controlled.

Figure 11 illustrates the validity of this theoretical approach compared to the experimental curve of test J4. Detailed calculations may be found in reference (J.M.Aribert, A.Lachal 2000)

3.6 *Experimental results dealing with Group 3 (T-arrangement with distortion effect)*

The moment-rotation curve for test J11, here defined at the load-introduction cross-section (i.e. the interface between end plate and column flange), is presented in figure 12. Compared to cyclic tests of Group 1 and 2 (without distortion effect) and more particularly to test J5 with partial shear connection the so-defined cyclic curve exhibits a greater capacity of rotation which is explainable by the more appropriate design of the bolted end plate. In addition, the absorbed energy ratio $\eta_i^{\pm}(\mu_i^{\pm})$ is quasi-constant and greater than for test J5 as well as in sagging bending ($\eta_i^+ = 0.75$) than in hogging bending ($\eta_i^- = 0.5$). The failure of the joint occurred during the 19th cycle by rupture of the shear connection in low cycle fatigue. Figure 13 shows the evolution of the

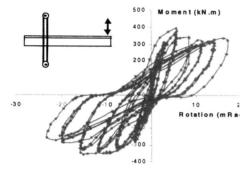

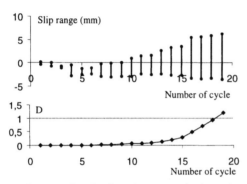

Figure 12: Moment-rotation curve in cyclic loading (test J11 – Group 3)

Figure 13: Cyclic damage of the shear connection (Group 3 - J11)

slip range (measured in the zone of quasi-uniform distribution) in the course of cycles; below, the corresponding values of damage index D are calculated according to relationship (2) with the fatigue characteristics given in 2.4 ($m=2.8$ and $\Delta\tau_c=60$ N/mm^2). The shear connection rupture is perfectly demonstrated by the damage model.

4 CONCLUSION

The present results on Push-Pull and beam-to-column composite joint tests lead to the conclusion that only full shear connection under seismic actions so that the maximum slip at ends of frame spans is reduced sufficiently (at least less than 2 mm) on account of the loss of connector ductility under cyclic actions. Moreover, care should be taken in reducing a little the static shear resistance P_{Rd} ; on the base of results obtained in 2.2; the

value 0.8 P_{Rd} might be proposed. It is also required to well control the steel details of composite joints with respect to low-cycle fatigue to ensure appropriate values of rotation capacity and absorbed energy ratio. The unsymmetrical character of loading does not seem to affect these values considered at the load-introduction cross-section; but this last conclusion needs other tests to be confirmed.

REFERENCES

Aribert J.M. 1996 Influence of slip of the shear connection on composite joint behaviour. *Connections in Steel Structures* III, 11-22. Pergamon, Trento.

Aribert J.M. 1999. Theoretical solutions relating to partial shear connection of steel-concrete composite beams and joints. *Steel and Composite Structures International Conf. – Session 3*, Delft, 7/1-7/16.

Aribert J.M., Lachal A. and Dinga O. 1999. Modélisation du comportement d'assemblages métalliques semi-rigides de type poutre-poteau boulonnés par platine d'extrémité. *Revue Construction Métallique* 1, 25-46.

Aribert J.M.and Lachal A. 2000. Chapter 4 of book "Moment Resistance Connection of Steel Building Frames in Seismic Areas" *Edited by F.M.Mazzolani* E & FN Spon.

Ballio G. and Castiglioni C.A. 1994. Seismic behaviour of steel sections. *Journal of Constructional Steel Research* 29, 21-24.

Bernuzzi C, Calado L. and Castiglioni C.A. 1997. Behaviour of steel beam-to-column joint under cyclic reversal loading: an experimental study. *5th Int. Col. On Stability and Ductility of Steel Structures*, Nagoya, Japan.

CEN 1992. Eurocode 3 (ENV 1993-1-1). *Design of composite steel structures – general rules and rules for buildings*, Brussels.

CEN 1992. Eurocode 4 (ENV 1994-1-1). *Design of composite steel and concrete structures – general rules and rules for buildings*, Brussels.

CEN 1995. Eurocode 8 (ENV 1998-1-3). *Design provisions for earthquake resistance of structures - Annexe D : specific rules for steel-concrete composite buildings*, Brussels.

ECCS 1985. *Recommended testing procedure for assessing the behaviour of structural steel element under cyclic loads*, Committee TWG 1.3.

ECCS Technical Committee 11 – 1999 Design of Composite Joints for Buildings – Publication N° 109.

Mazzolani F.M. and Piluso V. 1996. *Theory and Design of Seismic Resistant Steel Frames*, E & FN Spon.

Behaviour of Steel Structures in Seismic Areas, Mazzolani & Tremblay (eds) © 2000 Balkema, Rotterdam, ISBN 90 5809 130 9

The behaviour of compression welds and carrying capacity with earthquake effects

K. Badamchi

Department of Civil Engineering, University of Tabriz Tarbiat Moallem, Iran

ABSTRACT: In the current plastic and elastic standards, in relation to this subject, and in many other foreign standards, there are no specific rules concerning compression flange welds of beam-to-column connections. Calculations made with respect to such welds are worked out as if they are equivalent of tension flange welds. The results of an experimental study on 36 welded connections, reveal that a significant reduction in weld thickness is possible for compression flanges.

1 INTRODUCTION

In this paper, the results of an experimental research on the behaviour of some typical fully welded beam-to-column connections are given. The experiments were conducted in Istanbul Technical University, Civil Engineering Department Structural Laboratories.

In the plastic theory for plastic design it is assumed that, in such connections (figure 1), bending moment is carried by both flange welds and the shearing force by web welds only, but the axial force by is carried the whole weld area. d' being a reduced beam depth due to the presence of $a_2 l_{w2}$ welds. The welds calculations in plastic design have been obtained from formulaes 2 and 3 (Badamchi 1988).

$$d' = d - 2 \frac{2 a_2 l_{w2} t_f}{a_1 l_{w1} + 2 a_2 l_{w2}} \qquad (1)$$

For flange welds

$$\frac{1}{0.85} \left(\frac{N}{\Sigma a l_w} \pm \frac{M_x / d'}{a_1 l_{w1} + 2 a_2 l_{w2}} \right) \le \sigma_{yd} \qquad (2)$$

For web welds

$$\sqrt{\left(\frac{N}{0.85 \Sigma a l_w} \right)^2 + \left(\frac{Q_y}{0.75 \times 2 a_3 l_{w3}} \right)^2} \le \sigma_{yd} \qquad (3)$$

σ_{yd} = yield stress of the main steel members

One group of the results gathered from the experiments performed, throw light upon this subject. In the following, an evaluation of these experimental results have been made, and a proposal with regard to a possible reduction in the compression flange weld thickness of beam-to-column connection is put forward.

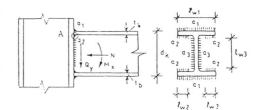

Figure 1. Moment transferring beam-to-column connection

2 THE EXPERIMENTS

The experimental research was conducted on a total of 18 beam-to-column connection specimens, of which the first 6 specimens are Just for getting some ideas. Since each specimen had two joints, a total of 36 beam-to-column connections were tested.

A welded beam-to-column connections sample is shown in Figure 2.

During experiments, the load was first increased, step by step, up to 80% of the calculated moment carrying capacity. Then, the load was removed from the specimen in order to establish its behaviour and its residual deformation up to the proportional limit. The specimens were subjected to load for a second time, up to the yielding limit, and then unloaded again. Finally, force was reapplied to the specimen for a third time, until the joint fails.

During the first two loadings, at eath step, deformation values were measured through the comparametres. For the main experimental series, also the values of the stresses in the welds were measured by means of strain gauges, attached very close to the connection fillet welds. In the third and final loadings, the comparametres measuring deformation were removed for instrumental safety, and only the strain gauges were used, until such time they stopped working.

The quality of the steel profiles used in the experiments was Steel 37 nominally, with a yield strength of 2400kg/cm^2. The cross sections of the beams were standard I profiles, and the cross sections of columns were wide flange IPB profiles. The dimensions of columns and beams were different in each specimen. An experimental sample is shown in Figure 4.

In the initial experimental series of six specimens, which include 12 welded beam-to-column joints, the thickness and areas of tension and compression flange welds were equal. For this test series, it was observed that while the tension flange welds and web welds ruptured, compression flange welds showed no sign of any cracking.

In the main experimental series of twelve specimens, comprising 24 joints, the compression flange welds were thinner. In some specimens, the weld thickness was only 3 mm, which is absolutly minimum. The type of electrode used in welding is AS 43.33 ruty 1. This type normally used in Turkey as standard one. The welding has been done by hand carefully.

Figure 3. Experiment sistem

Figure 4. An experimental sample

Two typical deformation curves obtained from the two strain gauges connected to the compression and tension flang welds in the specimen 8 shown in the figure 5.

figure 6 also shows the residual deformation in specimen 8.

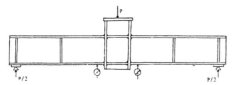

Figure 2. A welded beam-to-column connections sample

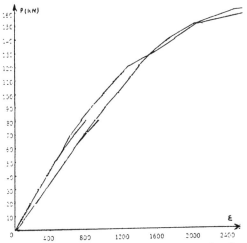

Figure 5a. The ε-p diagram in the compression flang welds.

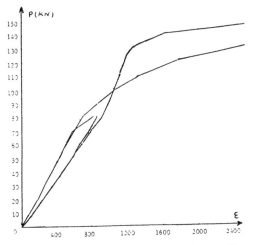

Figure 5b. The ε-p diagram in the tension flang welds.

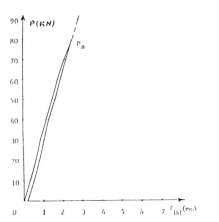

Figure 6. deformation-loading diagram in specimen 8.

3 THE ANALYSIS OF EXPERIMENTAL RESULTS

The experimental results relevant to this paper can be summarized as follows.

3.1 The Comparison of Theoretical Yielding Values and Those Observed in the Experiment

Calculations with respect to weld dimensions are carried out according to plastic design (Badamchi 1988). The level of the load at which yielding occurred in experimental conditions, and the level at which this yielding was expected according to theoretical calculations are compared in Table 1.

If Table 1 is examined, it is easily seen that the experimental load level is, with one minor

Tabel 1a. The beam and weld sizes in the specimens

Sample	beam size	column size	weld size (mm)	
			Tension	compression
7	I 240	IPB 120	7	5
8	I 240	IPB 120	5	3
9	I 180	IPB 120	5	3
10	I 240	IPB 280	9	5
11	I 240	IPB 280	9	3
12	I 240	IPB 280	7	5
13	I 240	IPB 280	7	3
14	I 180	IPB 280	7	3
15	I 180	IPB 280	7	3
16	I 180	IPB 220	7	3
17	I 180	IPB 220	7	3
18	I 180	IPB 220	7	3

Tabel 1b. Theoretical and experimental load levels at which joints yielded

Sample	$P_a \times 10^{-3}$					
	In Tension			In Compression		
	plasitc theory	elastic theory	Experimental	plasitc theory	elastic theory	Experimental
	kg	kg	kg	kg	kg	kg
7	10,90	11,48	11,06	5	6,42	9
8	7,80	8,42	9,23	4,79	5,71	8,26
9	4,63	4,96	6,135	2,86	3,44	5,12
10	13,40	14,56	14,29	7,80	9,35	8,60
11	13,40	13,66	13,23	4,80	6,13	10,64
12	10,68	11,82	13,28	7,80	9,13	9,76
13	10,67	11,16	11,85	4,79	5,93	9,69
14	6,30	6,56	6,90	2,80	5,53	4,92
15	6,30	6,56	7,56	2,80	5,53	-
16	6,30	6,56	7,08	2,80	5,53	4,55
17	6,30	6,56	5,59	2,80	5,53	5,57
18	6,30	6,56	7,20	2,80	5,53	6,97

exception, always superior or equal to the theoretical level. This shows that the design requirement by plastic theory allow for a sufficient margin of safety.

If calculations are made according to formula $\sigma_{yd} = (M/I).y \rightarrow p$ extrapolated from elastic theory, the theoretical results are a little bit higher than the experimental ones, but still very close, If these results are examined, it is concluded that theoretical values found according to plastic theory, are closer to the experimental values than values derived from elastic theory, and are the most suitable. The use of plastic theory is therefore recommended, as its results are more conservative. In the aforementioned formula, σ_{yd} represents yield stress, M is the bending moment, I the moment of inertia of the weld, and y the distance of the weld from the neutral axis. In the table, p_{yd} represents the value of the load at which yielding occurs and ε_d the unit deformation observed in the experiments.

3.2 Welds Connecting the Compression Flange of Beams to columns

Welds connecting the compression flange of a beam to the column have different effects on the load carrying capacity of a joint and on the rate of joint deformation under bending moment. With respect to load carrying capacity, the use of minimum weld thickness in compression flange does not significantly reduce this capacity. On the other hand, the extend of deformation is more greatly affected if minimum weld thickness is used.

Table 2. Rupture to yield loads ratio

Sample	Rupture load P_u	In Tension		In Compression	
		yield load P_a	P_u/P_a	yield load P_a	P_u/P_a
—	kg	kg	—	kg	—
7	$24,0\times10^3$	$11,06\times10^3$	2,17	$9,00\times10^3$	2,66
8	$18,5\times10^3$	$9,23\times10^3$	2,00	$8,26\times10^3$	2,24
9	$11,5\times10^3$	$6,13\times10^3$	1,87	$5,12\times10^3$	2,24
10	$30,0\times10^3$	$14,29\times10^3$	2,10	$8,60\times10^3$	3,49
11	$30,0\times10^3$	$13,23\times10^3$	2,27	$10,64\times10^3$	2,82
12	$25,0\times10^3$	$13,28\times10^3$	1,88	$9,76\times10^3$	2,56
13	$23,5\times10^3$	$14,31\times10^3$	1,64	$9,69\times103$	2,42
14	$13,5\times103$	$6,90\times10^3$	1,96	$4,92\times10^3$	2,74
15	$13,5\times10^3$	$7,56\times10^3$	1,78	—	—
16	$15,0\times10^3$	$7,08\times10^3$	2,12	$4,55\times10^3$	3,29
17	$14,0\times10^3$	$5,59\times10^3$	2,50	$5,57\times10^3$	2,51
18	$15,0\times10^3$	$7,20\times10^3$	2,08	$6,97\times10^3$	2,15
Avg.			2,03		2,65

This fact can be explained as follows : Even if a joint yields, the beam's compression flange will remain in direct contact with the column flange and can thereby transfer much of the load upon it. In this way, even after a joint yields, no cracking is caused in the compression flange weld. But, with earthquake effect from the deformation point of view, a weak compression flange weld, increases the speed and order of the deformation.

Consequently, with the exception of some structural members in which deformation must be avoided, such as cantilever beams, it is proposed that the thickness of compression flange welds can be reduced.

4 CARRING CAPACITY WELDED BEAM-TO-COLUMN CONNECTIONS

Rupture to yield load ratio in beam-to-column Connections is shown in table 2. The ratio approximately is around 2 which is quite high. The Table 2 clearly indicate that carring capacity up to rupture point is much greater than the yield strengths. This rupture point is reached after a large deformation. Although it is useless for ordinary service loads but it shows more safety in earthquakes.

5 CONCLUSION

Welds connecting a beam's tension and compression flanges to the flanges or web of a column, are considered equal in the current standards, and are calculated in the same manner. However, the compression flange welds have shown a different behaviour as compared to tension flange welds in our experiments. The load carrying capacity of the joint is not greatly affected if the compressive flange weld is less substantial than standards recommend, The only effect of such a reduction is an increase in the earthquake resistance and rate of deformation. For this reason, with the exception of joints very sensitive to deformation, such as those of cantilever beams, and also joints where the bending moment changes its sign such as in earthquake loads, a reduction in compression flange welds thickness is possible.

In suitable beam-to-column connections, the minimum weld thickness recommended by the standards as being a lower limit, it may be considered that the design of compression flange welds on the assumption that they bear about 50% of the transferred loads, is safe. Further margin of safety is provided through the 25% increment over the percentage stated in the experiment in connection with the axial compressive force transfer from a beam to column.

Carrying capacity reached by the test specimen at the instant of rupture is generally very high in comparison with the values. the weld yield strength. From the point of view of the usage limits, this limit reached only after quite a lot of deformation is rather useless for the ordinary service loads, but it has a definitive importance in extraordinary loadings as earthquakes since it represents an additional safety. The repetition of the same high values in all experiments proves the reliability of the results.

REFERENCES

Badamchi, K. 1988. Determination of the restraint degrees and the ultimate loads of some typical planar welded beam-column connections. Institute of I.T.U. Istanbul (in Turkish)

Eurocode No. 3, 1984. common unified rules for communities, report EUR 8849.

Eurocode 3, 1992. Design of steel structures part 1.1: general rules and rules for buildings. ENV 1993-1-1, Brussels.

Behaviour of Steel Structures in Seismic Areas, Mazzolani & Tremblay (eds) © 2000 Balkema, Rotterdam, ISBN 90 5809 130 9

Monotonic and cyclic behaviour of composite beam-column connections

J.G. Beutel & D.P. Thambiratnam
Department of Civil Engineering, Queensland University of Technology, Australia

N. Perera
Robert Bird and Partners

ABSTRACT: An investigation into the behaviour of composite column-to-beam connections using experimental and analytical techniques is currently being conducted, the main objective of which is the development of a deterministic procedure for the connections design. Seven large-scale connections have been tested, four under monotonic loading and three under cyclic loading. All connections consisted of a concrete-filled steel tubular column (circular), a compact universal beam (class 1 to Eurocode 3) and a shop fabricated connection stub. Four reinforcing bars were welded to the top and bottom flanges of the beam and embedded into the concrete core, and each specimen had different bars and embeddment details. It was found that connection strength increased as the capacity of the embedded bars increased, to a stage where no connection failure occurred, and the beam formed a plastic hinge outside the zone of influence of the bars. This mode of specimen response gave improved cyclic response characteristics and superior flexural ductility response limits compared to those which underwent yielding in the connection itself.

1 INTRODUCTION

The present trend in the construction of tall buildings clearly indicates the structural and economic superiority of composite steel/concrete construction. Concrete filled steel tubular columns (CFSTC's) are one of the most recent composite members becoming increasingly popular in tall building construction. This popularity is due to a variety of reasons, the most important of these being their inherent strength and ductility, and the advanced construction techniques that are used in their erection, resulting in lower column and overall structure cost. These advantages make them ideally suited to tall structures, particularly those located in regions of high seismic risk.

When CFSTC's are combined with compact structural steel beams a potentially excellent elasto-plastic energy dissipating system can be created. Because these members are usually mated within a structural typology incorporating full strength connections, it is connection performance that will ultimately control how effectively the structure performs under seismic loads. This connection must be capable of facilitating the construction and performance of each member and the structure as a whole,

without significant strength degradation or loss of load transfer capability under cyclic loads. A lack of design guidance and understanding into the behaviour of suitable connections has limited the use of CFSTC's in such applications.

Due to the increasing popularity of CFSTC's a wide variety of different beam to column connections have been investigated quite recently. External connection details which attach to the outside of the tube only, include: direct connection of the steel beam to the outside of the tube only (Shakir-Khalil, 1994), and using external stiffening rings at the level of the beam flanges (Kato et. al., 1992, Choi et.al., 1995). It has been found that many of these details impose a severe elastic demand on the tube wall resulting in excessive deformation and poor cyclic performance. Modifications which have attempted to alleviate these problems have included through bolting and endplates (Kanatani et.al. 1987), continuing the steel beam through the column (Aziznamini, 1995) and the use of reinforcing bars welded to the top and bottom flanges and embedded into the concrete core (Alostaz, 1996, Ashadi, 1995). While many of these connections were successful in reducing the demand on the tube wall, their complexity and inherent cost have severely limited their use. This research in most cases has only provided information on general performance, and very little actual design information still exists on such connections.

Because of this lack of design guidance, a collaborative research project has been undertaken by the Queensland University of Technology and Robert Bird & Partners Pty. Ltd. to develop a deterministic design procedure, using experimental and analytical techniques, for an economically buildable joint configuration. This paper briefly discusses some of the findings from the experimental testing.

2 SPECIMEN DETAILS AND EXPERIMENTAL PROCEDURE

Connections such as these when incorporated into a moment resisting frame (MRF) are typically designed for a plastic hinge to form adjacent to or at the beam-to-column connection. Therefore only connections that could generate the plastic bending strength of the beam, potentially provide good ductility and post-elastic response, and were relatively cheap and easy to construct where considered. After a comprehensive review of the previous research, and upon consultation with the industry partner the following detail was selected for this study (Figure 1). Relevant background information regarding this review and the selection of this detail can be found in an earlier paper (Beutel, 1998).

Table 1 gives a brief description of each of the specimens tested. Each was fabricated from a universal beam, 360 mm deep with a nominal yield of 300 MPa, and a 406mm diameter pipe with a 6 mm wall thickness and 350 MPa nominal yield. The beam section was connected directly to the steel column using flange connection plates, full strength butt welds, and a web cleat plate. Four reinforcing bars were welded to the top and bottom flanges and embedded into the concrete core. Each of the specimens had varying diameter bars during the monotonic stage of testing, and different shaped bars were also tested during the cyclic testing phase. All specimens were filled with 40MPa concrete. Table 1 also shows the relative strengths of the flange connections for each specimen, taking into account both the through shear strength of the flange plate to tube connection and the tensile strength of the bars (using nominal yield strength values). This has been quoted

as a percentage against the nominal yield strength of the flange. Table 2 shows the theoretical and actual strengths of each connection. The theoretical capacities given are based on the through shear strength of the flange to column connection and the yield strength of the bars (actual material strengths). Also shown are expected hinge locations based on whether the connection had the capacity to generate the plastic strength of the framing beam (M_p=290kNm) past the end of the bars.

Table 1 - Specimen Description.

Specimen	Bar Size[**]	Flange Conn. Capacity (%)[*]	Loading
SM12s	12 s	78	Mono.
SM16s	16 s	107	Mono.
SM20s	20 s	143	Mono.
SM24s	24 s	188	Mono.
SC20s	20 s	143	Cyclic
SC24s	24 s	188	Cyclic
SC24c	24 c	188	Cyclic

[*] All percentages are based on the nominal yield capacity of the flange.
[**] s-straight bar, c-cogged bar.

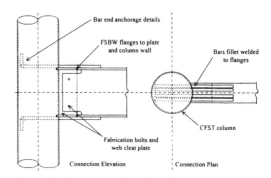

Figure 1 - Connection Configuration.

All specimens were tested in the vertical plane, in a specially designed test rig. The supports were constructed to simulate inflection points and were designed to rotate about the plane of bending of the specimen. The column axial load was applied through the top of the column via a 150 tonne hydraulic jack. This load was 1000kN being 12.5% of the squash load of the column, which is a fairly typical loading level for a corner column in a moment resisting frame, and was kept constant throughout the duration of each test. A 30 tonne hydraulic actuator was used to apply the beam tip load. This was applied as one constant push until specimen destruction for those specimens tested under monotonic loading. For those specimens tested under cyclic loading, a load history was applied to the beam tip, which was calculated according to ATC24 (1992), 'Guidelines for Cyclic Seismic Testing of Components of Steel Structures'. The first six cycles were elastic and conducted under load control. Two cycles each at 25, 50 and 75% of the yield load, determined during the monotonic loading stage of testing, were applied. These cycles were then followed by inelastic cycles, completed under deflection control and in stages, each stage made up of three cycles. The amplitude for each successive stage was increased by the yield deflection for the specimen.

3 TEST RESULTS

The load-tip deflection $(P\text{-}\delta)$, and moment-connection rotation $(M\text{-}\theta)$ response have been used to compare the performance of each of the connections tested. The moments are those at the face of the column, and are normalised with respect to M_p (being the plastic bending strength of the connection beam) calculated using actual material strength and cross sectional properties.

The Flexural Ductility Ratio (FDR) quoted for each cyclic specimen is the peak deflection, divided by its yield deflection. This peak deflection is the largest deflection achieved, provided that the specimen resistance has not fallen below a specified resistance during that stage of inelastic loading. This resistance will be taken as that suggested by Roeder and Fouch (1996) being 80% of the load to produce M_p at the face of the column.

3.1 Monotonic Tests

Figures 2 and 3 shows the $M\text{-}\theta$ response including connection classification limits according to Bjhovde et al (1990), and the beam tip response of the monotonic specimens. In general, as bar size (hence flange connection capacity) increased, so too did the ultimate strength of the connection, with significant improvements in initial rotational stiffness and the energy absorbed by the specimen. SM16s, SM20s and SM24s all had adequate rotational stiffness and strength to classify them as full strength rigid connections.

Specimens SM12s and SM16s were able to mobilise

approximately 70 and 90 % of M_p respectively and underwent connection yield. These capacities and the resulting hinge locations correlate well with those predicted. The strengths were achieved after considerable connection rotation producing connection yield and strain hardening ie. at approximately .015 radians.

Table 2 - Monotonic specimen results.

Specimen	Calculated Capacity	Actual Capacity	Expected Hinge Location
SM12s	161 kNm	174 kNm	Conn.
SM16s	220 kNm	236 kNm	Conn.
SM20s	294 kNm	304 kNm	Conn.
SM24s	390 kNm	Na[*]	Beam

[*] Connection capacity not reached or exceeded

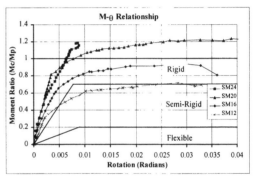

Figure 2 - $M\text{-}\theta$ response - Monotonic loading.

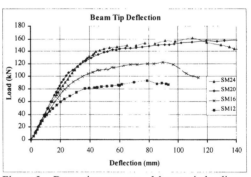

Figure 3 – Beam tip response – Monotonic loading.

SM12s and SM16s had similar failure modes, with the initial yield of the top flange bars (occurring at 0.0018 and 0.0023 radians connection rotation respectively) causing the onset of connection softening. This softening resulted in excessive connection rotation under increasing load, producing a crack in the steel tube at the corner of the beam flange to column connection, after which no further significant strength increase was achieved. The tube wall in the vicinity of the top bars under-went noticeable deformation (bulging) due to excessive load being transferred to the flange-to-tube weld. By the end of the test this weld had cracked along its full length. The bottom flange exhibited no signs of distress. The bottom bars reached yield at larger connection rotations than the top bars (at approximately 0.004 radians) for both specimens. No other signs of distress were evident at any other location.

Both SM20s and SM24s had the capacity to achieve M_p, with better than $1.2M_p$ being generated at the connection for each specimen. The plastic capacity of the beam was achieved at relatively low connection rotation levels (approximately 0.0077 and 0.0063 radians for SM20s and SM24s respectively). Both connections produced a degree of yielding in the connection beam away from the connection zone, in a region adjacent to the end of the bars. SM20s achieved full section yield at this location but only minor bottom flange buckling, after which yield zone stiffening transferred a larger moment to the connection causing failure in an identical fashion to specimens SM12s and SM16s. Top and bottom bar yield occurred at connection rotations of approximately 0.003 and 0.004 radians respectively. Connection capacity predictions for this specimen again seemed accurate.

SM24s had the capacity to form a full plastic hinge in the connection beam adjacent to the end of the bars, and because of this the connection's full plastic capacity was not achieved. This can be seen in Figure 2, which shows that the connection rotation halted and leveled out to a constant due to yield of the beam and Figure 3, which shows this specimens beam tip response was practically identical to SM20s. SM20s and SM24s therefore, absorbed a similar amount of energy, but SM24s did so without noticeable damage to the connection or column. Some minor cracks began to form in the corner of the tube-to-column weld but only in the surface at the root of the weld. These cracks had not propagated through the tube itself. Stress levels in the tube around the connection reached yield, as did both the top and bottom bars (at 0.0055 and 0.0045 radians connection rotation respectively), however connection integrity was not breached. It should be noted that for all specimens tested, there was no evidence of bar fracture, pullout or cracking of the fillet weld

Table 3 - Cyclic specimen results.

Specimen	FDR	Beam Hinge Rotation (rads.)[*]	No. Plastic Cycles
SC20s	1.93	NA	6
SC24s	2.01	NA	6
SC24c	4.14	0.033	11

[*] Maximum rotation under 3*beam tip yield deflection.

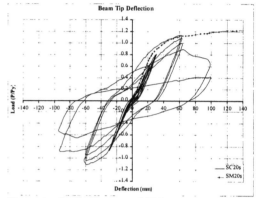

Figure 4 - Cyclic Response SC20s.

connecting the bar to the flange.

3.2 *Cyclic Tests*

Figures 4, 5, and 6 show the beam tip deflection responses from specimens SC20s, SC24s, and SC24c respectively, and Table 3 shows important performance parameters for each specimen. Each specimen was able to mobilise M_p at the face of the column, however it is obvious that the cyclic performance of both SC20s and SC24s was very poor due to a rapid deterioration in specimen resistance and stiffness during the third stage of inelastic loading. This resulted in these specimens only being able to achieve a FDR of approximately 2.

Specimen SC20s underwent a very similar failure mode to its monotonic counterpart, suffering from the formation and propagation of a crack in the tube wall at the corner of the beam flange to column connection in both the top and bottom flanges. SC24s however underwent an altered failure mode from that experienced under monotonic loading. This was due to an anchorage failure of the bars in both the top and bottom flanges. This caused the specimen to suffer from tube wall fracture in a similar fashion to SC20s.

Specimen SC24c was then tested with cogged bars in an effort to improve bar anchorage. This specimen was able to generate upwards of 140% of the beams plastic capacity at the face of the column, and form a plastic hinge in the beam adjacent to the end of the flange bars. It was able to achieve an FDR of 4.1 and under four times the yield deflection the specimen resistance gradually dropped below the yield load ie. reduced to 93% during the last cycle. The specimen exhibited no damage to the connection itself, only the yielding of the reinforcing bars and the formation of a surface crack similar to that found in the monotonic specimen.

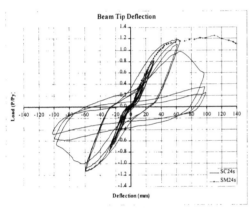

Figure 5 - Cyclic Response SC24s

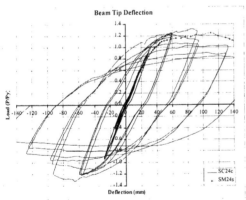

Figure 6 - Cyclic Response SC24c

4 DISCUSSION

Reinforcing bars welded off the top and bottom flanges and embedded into the concrete were effective in transferring both compressive and tensile loads directly into the concrete core under both monotonic and cyclic loads. All bars reached yield at relatively low levels of connection rotation. For specimens SM12 to SM20 the top bars reached yield before the bottom, with the rotation to cause first yield in the top bars increasing with increasing bar size. The compression bars for each specimen however, all reached yield at approximately the same rotation (at approximately 0.004 radians). This sequence altered for SM24, where the bottom bars reached yield before the top bars due to the increased connection rotation required to produce top bar yield and the position of the neutral axis being higher in the section.

Top flange and bar strains behaved in a linear fashion, until bar yield was reached. At this point flange strains began increasing at a faster rate and a tear in the tube began to form at the corner of the flange to tube connection at approximately 700 micro-strain. Flange strain reached a peak value of around 800 micro-strain, which corresponds to the maximum load that the flange to column weld is capable of transferring via its through shear strength (ie 214kN through shear strength using actual tube yield, cf. 265kN flange load at 800 micro-strain). After top bar yield occurred very little strength increase was achieved (typically in the order of 20%), although the connection still underwent significant rotation (typically two to three times top bar yield rotation).

Bottom flange strains behaved in a linear fashion until top bar yield. At this point they tended to level out, however bottom bar strains continue to increase in a linear manner. This was because very little extra tensile load was being produced in the top of the section, to be balanced by the compressive zone. The larger plateau strain for each specimen in the bottom flange relates to the higher load being supported by these specimens at the same connection rotation.

Under cyclic loads only specimen SC24c behaved in a satisfactory manner. Both SC20s and SC24s underwent very rapid strength deterioration reflected in the low FDR's achieved for these specimens. For each specimen the decrease in stiffness between each loading stage was due primarily to a decrease in the stiffness of the connection itself. SC24c also suffered from a strength reduction, but this was very gradual, and allowed a FDR of 4.14 to be achieved. This reduction was due to a decrease in the strength of the hinge that formed in the beam itself, rather than any failing of the connection. The deterioration of the hinge was due to buckling of both the bottom and top flanges of the beam.

The beam hinge still however, went through 0.0033 radians of rotation. Many researchers in this field consider 0.0015 radians of rotation to be the minimum required if it is to dissipate the energy required through plastic deformation needed to justify the use of $R_w=12$ (DeLuca and Mele, 1996).

According to Roeder and Fouch (1996) a FDR of a between 4 and 6 can be related to a R_w factor of 12, which is the required ductility demand form a moment resisting frame designed according to the uniform Building Code (1998). Whilst this should not be taken as the only measure of a connection's/specimen's likely seismic performance, the excellent form of the hysteretic behaviour would suggest that a system utilising this connection would behave very well during a seismic event.

5 CONCLUSIONS

The externally stiffened bar connection detailed above was tested both monotonically and cyclically. It was found that under monotonic loads as the bar size increased, so too did connection strength with improvements in rotational stiffness. These increases occurred up to a point were the connection was strong enough to form a plastic hinge in the beam, outside of the influence of the bars. The bars were very effective in transferring both tensile and compressive loads directly into the column core, and the capacity predictions used were accurate in assessing specimen strength and hinge location. It was found that such a connection can be designed with adequate strength and stiffness to classify it as a full strength rigid connection.

Under cyclic loads it was found that forcing the hinge to form in the beam rather than the connection resulted in significantly improved seismic performance properties. Adopting a minimum resistance level of 80% of the yield load, specimen SC24c achieved a FDR of 4.14 and beam hinge rotations in the order of 3.3%, suggesting that this connection is suitable for application into a special moment resisting frame.

This study will now lead into further tests on similar connections, using cyclic load application. Analytical testing will complement the experimental testing and the information gathered will lead to the development of a deterministic design procedure. Refer to Beutel *et al* (2000) for a full description of all of the experimental testing conducted in this study.

REFERENCES

ACT - 24 (1992) Guidelines for Cyclic Testing of Components of Steel Structures, California: Applied Technology Council.

Alostaz, Y. & Schneider, S. (1996) 'Connections to Concrete Filled Tubes', *11th World Conference on Earthquake Engineering*, Acapulco.

Ashadi, H.W. & Bouwkamp, J.G. (1995) 'Behaviour of hybrid composite structural earthquake resistant joints', *10th European Conference on Earthquake Engineering*, Rotterdam, pp.1619-1624.

Azizinamini, A. & Shekar, Y. (1995) 'Design of through beam connection detail for circular composite columns', *Engineering Structures*, Vol. 17, No. 3, pp. 209-213.

Beutel, J.G., Thambiratnam, D.P. & Perera,N.J. (2000) 'Experimental testing of steel beam to composite column connections under monotonic and cyclic loads', *Physical Infrastructure Centre Research Monograph 2000-2*, Queensland University of Technology, Brisbane, Australia.

Beutel, J. G., Thambiratnam, D.P., & Perera, N.J. (1998) 'On the behaviour and design of composite column to beam connections under seismic loads', *Tubular Structures VIII – Proceedings of the Eight International Symposium on Tubular Structures*, Singapore, pp.615-625.

Bjorhovde, R., Brozzetti, J., and Colson, A. (1990) 'Classification System for Beam-Column Connections', *Journal of Structural Engineering*, Vol. 116, No. 11, pp. 3059-3076.

Choi, S.M., Shin, I.B., Eom, C.H., Kim, D.K. & Kim, D.J. (1995) 'Elasto-Plastic Behaviour of the Beam to Concrete Filled Circular Steel Column Connections with External Stiffener Rings', *Building for the 21st Century - Proceedings of the Fifth East Asia-Pacific Conference on Structural Engineering and Construction*, Griffith University, pp. 451-456.

De Luca, A. & Mele, E. (1996) 'The recent Kobe earthquake: General data and lessons learned on steel structures', *Earthquake performance of civil engineering structures, COST C1 – Seismic working group report*, 1996, Brussels, pp.1-21.

Kanatani, H., Tabuchi, M., Kamba, T., Hsiaolien, J. & Ishikawa, M. (1987) 'A Study on Concrete Filled RHS Column to H-Beam Connections Fabricated with HT Bolts in Rigid Frames', *Composite Construction in Steel and Concrete - Proceedings of an Engineering Foundation Conference*, New Hampshire, pp. 614-635.

Kato, B., Kimura, M., Ohta, H. & Mizutani, N. (1992) 'Connection of Beam Flange to Concrete-Filled Tubular Column', *Composite Construction in Steel and Concrete II - Proceedings of an Engineering Foundation Conference*, Missouri, pp. 528-38.

Roeder, C.W, & Foutch, D.A (1996) 'Experimental Results for Seismic Resistant Steel Moment Frame Connections', *Journal of Structural Engineering*, vol.122, No. 6, pp.581 – 588.

Shakir-Khalil, H. (1994) 'Beam connection to concrete-filled tubes', *Tubular Structures VI - Proceedings sixth International Symposium on Tubular Structures*, Melbourne, pp. 357-364.

Uniform Building Code (1988). Int. Conf. Of Build. Officials, Whittier, Calif.

Behaviour of Steel Structures in Seismic Areas, Mazzolani & Tremblay (eds) © 2000 Balkema, Rotterdam, ISBN 90 5809 130 9

Effect of column straightening protocol on connection performance

R. Bjorhovde
The Bjorhovde Group, Tucson, Ariz., USA

L. J. Goland
Southwest Research Institute, San Antonio, Tex., USA

D. J. Benac
Bryant-Lee Associates, San Antonio, Tex., USA

ABSTRACT: The major findings of a recently completed investigation into the performance of several types of beam-to-column connections are presented. The primary aim of the project was to resolve questions about the performance of the steel in rotary straightened wide-flange shapes. Full-scale test specimens utilized common beam and column sizes as well as welded flanges and cover plates, and bolted webs; they were designed using the AISC seismic criteria. Certain specimens used rotary straightened columns; others had gag straightened columns or unstraightened columns. The loading was either quasi-static in accordance with ATC-24 or dynamic (1 Hz). A total of 17 connections were tested; the results show that the form of column straightening protocol has no effect on the performance of the connections. Quasi-static testing is adequate for seismic purposes, although dynamic testing is more severe and probably closer to seismic conditions.

INTRODUCTION

Steel has been the primary construction material for a very large number of buildings, bridges and other structures for more than 100 years. Its elastic and inelastic responses to loads and load effects make it a material with predictable and reliable behavior under a wide range of service conditions. The ease and speed of fabrication and erection provide significant economies of construction.

As with other types of construction, the most complex elements of steel structures are the connections. For buildings this is particularly true for the beam-to-column connections, where structural details and fabrication processes combine to produce three-dimensional conditions. To facilitate construction and allow for economical usage, over the years certain connection types became common. They proved their adequacy through service in many buildings, designers were confident about design methods and details, and fabricators produced high-quality structures. In particular, beam-to-column connections utilizing welded joints between the beam and column flanges and a bolted beam web were used extensively.

Tests and analyses showed that these connections were capable of developing appropriate moment and shear capacities, and their deformation characteristics were very good.

Use of steel structures in areas of high seismicity was considered especially advantageous, due to the inherent inelastic deformation capacity of the material. Design and fabrication practices proved their worth through a number of minor and major earthquakes. However, the understanding of seismic effects and structural behavior has advanced significantly over the past ten to twenty years, and tools such as computers have facilitated increasingly fine-tuned designs. Prompted by owners and architects, structural systems have also changed, to allow for differing working and living space arrangements. As a result, structures in some ways have become simpler, with fewer primary load-carrying elements, but at the cost of reduced redundancy. The effects of events such as earthquakes would therefore have to be accommodated by fewer structural members and especially by fewer connections.

The 1994 Northridge earthquake had a significant effect on state-of-the-art thinking in the United States about ductile structural response. A number of steel-framed structures were found to have cracks in their beam-to-column connections, and it was thought that the earthquake had caused these failures. It is now clear that many of the cracks had occurred before the earthquake, and that such cracking also has taken place in structures where no seismicity has been present. However, it was also evident that material and structural behavior and design and fabrication approaches needed careful re-examination.

As a specific example of the non-earthquake-related cracking incidences, the fabricator for a large project in California experienced cracking in the column of beam-to-column assemblies during the shop fabrication. The connections were of the welded flange, bolted web type, but they also utilized welded beam cover plates as well as continuity plates for the column. Details of the connection are shown in Fig. 1. Although the beam and column sizes shown in the figure are not identical to those of the project in question, they are representative of what was used.

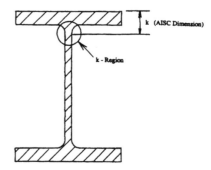

Figure 2. Wide-Flange Shape with k-Area Shown

Additional problems evolved as subsequent examinations found that the k-area of W-shapes was prone to exhibit high strength and hardness, but lower ductility and fracture toughness. These properties were related to the fact that many sizes of wide-flange shapes are rotary straightened in the steel mill in order to meet the straightness requirements of ASTM (1998). This is typical practice for mills all over the world, but is nevertheless a phenomenon that merited study.

These issues and events form the background for the study that is presented here. Complete data regarding the testing program is provided in the report by Bjorhovde et al. (1999).

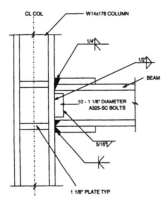

Figure 1. Original (As-Built) Connection

The cracks in the as-built connection were found in the web of the column, in the region of the cross section now commonly known as the "k-area". This is a small area of the web of the wide-flange shape surrounding the location where the transition fillet from the flange enters the web. The k-dimension measures the distance from the outside of the flange to the end of the transition fillet. Figure 2 illustrates these terms.

Originally the earthquake- and fabrication-related cracks were thought to have taken place in part as a result of inadequate material properties. Much discussion took place about yield stresses significantly higher than the specified minimum values, and some investigators tended to state that current steel mill practices therefore were faulty.

SCOPE OF TESTING PROGRAM

Table 1 gives the details of the research program. A total of 17 full-scale beam-to-column connections were fabricated and tested. All specimens used W14x176 columns and W21x122 beams in ASTM A572 Grade 50 (A992) steel.

Eight of the test connections were the same as those of the California structure where the original cracking had been detected. These used cover-plated, complete joint penetration (CJP) welds to attach the beam to the column, and CJP welds were used for the continuity plates that were placed in the web area of the column. The beam web connection was provided by ten 1-1/8 inch diameter ASTM A325 high strength bolts. Further, it was decided to examine the potential influence of the column straightening protocol that had been used by the steel producer to achieve members that would meet the steel standard's straightness criteria (ASTM, 1998). Thus, five of the eight specimens had rotary straightened columns, two used gag straightened columns, and one used an unstraightened member.

Table 1.   Key Features of Connection Testing Program

Number of Tests	Connection Type	Connection Description*	Straightening Protocol**	Loading Protocol***
8	As-Built (AB)	CJP welds for beam to column and continuity plates to column; 1-5/8" cover plates; 1-1/8" A325 bolts for beam web	5 R 2 G 1 U	4 Q 4 D
6	Revised Type 1 (RAB 1)	As AB, but with 1 inch cover plate; fillet welds for continuity plates	5 R 1 U	4 Q 2 D
1	Revised Type 2 (RAB 2)	As AB, plus 1/2 inch fillet weld transition from beam to column	1 R	1 D
2	Revised Type 3 (RAB 3)	As RAB 1, plus 1/2 inch fillet weld transition and repositioned continuity plates	2 R	2 D

*   CJP = complete joint penetration
**   Straightening protocols are designated as R = rotary straightened; G = gag straightened; U = unstraightened
***   Loading protocols are designated as Q = slow cyclic (quasi-static) testing; D = dynamic (1 Hz frequency) testing

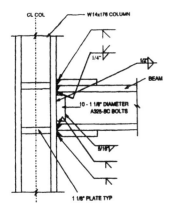

Figure 3.   Revised As-Built Connection Type 1

Six specimens utilized a revised (Revised Type 1) connection, with thinner (1 inch) cover plates, and fillet welds instead of the CJP welds for the continuity plates.   The use of fillet welds, in particular, was a major departure from current practice.   Five of the specimens used rotary straightened columns; one had an unstraightened member.   Figure 3 shows the details of the Revised Type 1 connection.

Of the remaining three specimens, one (Revised Type 2) was identical to the original specimens, with the only change being a 1/2 inch transition fillet weld between the cover plates and the column flange.   It was felt that this would provide a better

force and deformation transfer path for the connection in an area where cracks had been proven to initiate the eventual failures.   The column was rotary straightened.   Figure 4 shows the details of the Revised Type 2 connection.

The final two specimens were further revised (Revised Type 3) from the other designs.   It was identical to the Type 1 revised connections, but also used the 1/2 inch transition fillet weld that was utilized for Type 2.   In addition, the continuity plates were repositioned, to allow for an improved load and fracture path for the connection.   The columns were rotary straightened.   Figure 5 shows the details of the Revised Type 3 connection.

CONNECTION DESIGN

The connections were designed in accordance with the criteria of the AISC LRFD Specification (1994), including the 1997 and 1999 seismic design requirements.   To achieve optimal seismic performance of structures and their elements, current US principles utilize the "strong column, weak beam" concept, where plastic hinges will form in the beams at the ultimate limit state.   This provides for improved structural redundancy and ductile failure modes for a structure as a whole. However, for these tests it was decided to impose the most demanding conditions possible on the column material, primarily since the cracking that had been observed occurred in the columns.   It was

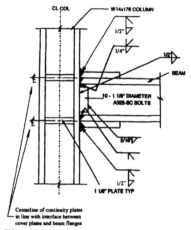

Figure 4. Revised As-Built Connection Type 2

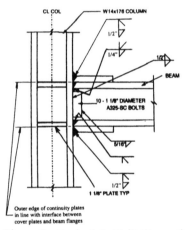

Figure 5. Revised As-Built Connection Type 3

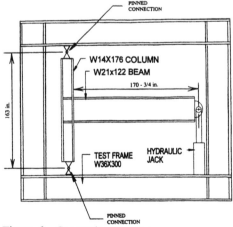

Figure 6. Connection Test Setup

felt that this would represent a worst case scenario. The test specimens therefore reflect assemblies with strong beams and weak columns, where plastic hinges will form in the column panel zones.

Figure 6 shows the test setup that was used. No axial load was applied to the column, and the beam tip load was applied a distance of 170-3/4 inches from the column face. The column was 163 inches long between the pinned ends.

LOADING PROTOCOLS

A number of beam-to-column connection tests have been conducted in past research projects. Many of these used slowly increasing or effectively static loads. Recognizing the importance of dynamic and especially seismic response characteristics, the US Applied Technology Council (ATC) has developed testing criteria that are based on cyclic loads (ATC, 1992). These are often referred to as quasi-static testing conditions, since it is not attempted to model earthquake loading input. Rather, using a displacement control approach, the cyclic load is applied in alternate directions, with increasing amplitudes of the load application point. This is the quasi-static loading protocol that was utilized for eight of the tests of the connection tests; it is identified by the letter Q in Table 1.

More recent studies by seismologists and structural engineers have emphasized the need to have the test loading simulate seismic conditions as closely as possible. This led to the development of criteria that focused on loads applied at certain loading or strain rates, to mimic the earthquake response of the structure. Although opinions still differ as to whether dynamic loads impose more realistic conditions than quasi-static loads, the former in all likelihood reflects a worst case scenario. It was therefore decided to test the second group of nine specimens dynamically, using a frequency of 1 Hz. The tests are identified by the letter D in Table 1.

Figure 7 illustrates the load-displacement profile, indicating that 3 cycles were used for each beam tip deflection. The maximum displacement was ± 10 inches.

COLUMN STRAIGHTENING PROTOCOLS

One of the issues that led to the decision to perform this program of connection tests was the performance of the steel itself in the web of the

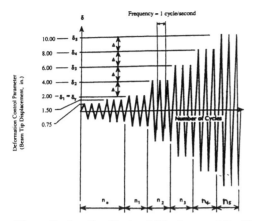

Figure 7. Loading Protocol Displacement Profile

column. Specifically, cracks had developed in the k-region of some columns during welding, and some had been found in structures after the Northridge earthquake. It was also noted that the k-area is deformed significantly because of the rotary straightening that is used in steel mills for small to medium size shapes to meet the straightness requirements of the material delivery standard (ASTM, 1998). As a result, the k-area steel of rotary straightened shapes tends to have higher strength and hardness than other areas of the cross section, but also lower ductility and toughness. Since this form of straightening is applied continuously, the localized areas of changed material characteristics appear along the full column length.

Another factor in the rotary straightening process is the initial out-of-straightness of the shape as it comes from the cooling bed in the steel mill. It is possible to have shapes that are very straight after the rolling; these require less straightening effort to meet the delivery criteria. On the other hand, some shape lengths may turn out to be fairly out-of-straight; these require higher straightening loads. For ease of production operations, current US steel mill practice dictates that all shapes be straightened.

Heavier shapes are straightened by the application of concentrated loads at discrete points; this is termed gag straightening. Gag straightening therefore does not introduce continuous areas of high strength and hardness in the manner of rotary straightening; it is a localized phenomenon.

To investigate the effect of straightening per se, two of the specimens were also fabricated with unstraightened columns. This was done for demonstration purposes only, and to establish the other extreme condition for straightening.

Unstraightened shapes are not commercially available products.

DISPLACEMENTS AND ROTATIONS

Measure of Performance

The key measure for the performance of a connection is its plastic rotation capacity or angle, θ_p. This gages the ability of a connection to sustain plastic deformations prior to failure, and is therefore regarded as a criterion by which the connection can be evaluated for seismic performance and suitability.

Key Assumptions

If the test frame is infinitely stiff, the true beam tip displacement would be defined as the vertical deflection of the beam end, measured relative to its undeformed original position. However, the frame is not rigid, and deformations of some magnitude will take place. If the specimen displacements are measured in relation to the test frame, the frame deformations would be added to the true specimen displacements in some fashion, resulting in incorrect beam tip and other deformations. To account for these effects during the tests, vertical and horizontal displacements were measured at the centers of the top and bottom column supports (pins), the center of the pin at the beam end, and at the center of the column panel zone. These displacements were used to calculate the actual beam tip deflections and to eliminate the effects of the frame flexibility.

In addition to the vertical translations of the column end supports, horizontal movements occur at both column ends as well as at the panel zone center. The sense of these displacements will indicate whether the entire test specimen rotates as a rigid body around a point on the column close to the bottom pin. Using an additional, "shake-down" specimen for verification of the testing system performance, it was found that the top column pin moved horizontally, in direct proportion to the beam end deflection, with magnitudes up to ± 0.3 inches. The lower column pin also moved horizontally, but in the opposite direction; the magnitudes of this translation were never larger than ± 0.04 inches. It was decided to treat this deformation as insignificant, leading to the conclusion that the entire specimen rotated as a rigid body about the bottom column pin support.

During an earthquake, the deformation demand is partly accommodated by the elastic displacements of the frame. Additional deformations have to be provided by the structure in the form of plastic hinge rotations in the beams and by plastic deformations in the column panel zones. The FEMA Interim Guidelines (FEMA, 1995) that have been developed over the past several years recommend that new steel-framed construction should be able to accommodate plastic rotations of at least 0.030 radians in the connection regions. Minimum rotation capacities of 0.025 radians are recommended for retrofitted structures.

For connection testing, either as proof of performance of existing construction or for the acceptance of new designs, the minimum plastic rotation capacities indicated above must be sustained for at least one full cycle of loading.

Cumulative and Normalized Cumulative Plastic Rotations

Most research on connections has limited the presentation of the results to the requisite plastic rotations and the accompanying number of cycles. Included are also hysteresis loops, observations of failure modes and whether the FEMA acceptance criterion were met. However, beyond the appearance of the hysteresis loops there is nothing provided to permit an analysis of the important measure of energy absorption capacity of a connection. For seismic performance this is a key measure of suitability.

For the past few years Japanese research reports have included data on cumulative plastic rotations and normalized cumulative plastic rotations (Nakashima et al., 1998). The former is simply the sum of the plastic rotations associated with each cycle of loading until failure occurs; the latter is a relative measure of the same. However, it is evident that the cumulative plastic rotation for a connection reflects its energy absorption capacity, and therefore provides key information on its performance ability. The survival of a connection for one cycle says little about its potential response under sustained seismic activity. It is understood that updated Japanese seismic criteria now utilize a cumulative plastic rotation capacity of 0.3 radians as the measure of acceptable performance of a connection (Malley, 1999).

The cumulative plastic rotation, $\Theta_p = \Sigma\theta_p$, is defined as the sum of the individual plastic rotations occurring during each complete half cycle of the test. The quantity also includes the excursion amount occurring at failure.

MATERIALS TESTING

Extensive testing was performed for the column materials in the connection specimens, since the cracking and eventual connection failure would occur within these members. The steel grade was ASTM A572 Grade 50 (A992), and the tensile property and chemical analysis tests showed that the steel in all of the specimens was satisfactory. The tension specimens were taken from flanges, web and k-area material. As expected, the tensile properties of the web and flange steel met and reasonably exceeded the minimum requirements. The k-area of the rotary straightened columns had higher yield and tensile strengths and lower ductility. The gag straightened and unstraightened members showed nearly uniform strength and ductility properties at all locations. These results were all as expected.

Charpy V-Notch (CVN) specimens for impact testing were taken from the flanges as well as the web, core and k-region of the columns. As expected, the flange and web materials in the rotary straightened columns exhibited excellent toughness; the core and especially the k-area steel was much less tough. Gag straightened and unstraightened columns did not display the k-region decreases in toughness. The minimum CVN value recorded for any of the tests was found in the k-area, as expected; this toughness was 5.0 ft-lbs.

Rockwell B hardness tests were conducted for the steel in flange-to-web T-intersection, mapping the hardness variability within this area of the column shapes. As expected, the k-area of rotary straightened shapes exhibited higher hardness in the areas where the CVN toughness was low. The data also delineated the size of the high hardness area, showing that it extended approximately 3/4 inch into the web beyond the end of the transition fillet.

PERFORMANCE OF TESTED CONNECTIONS

Figures 3 through 5 show the appearances of the connections, and Table 1 gives the details of the various types. The heavy cover plates and complete joint penetration welds of the original (as-built) connection were the same as those of the structure that prompted the research project. The continuity plates were located with mid-thickness at the level of the interface between the beam flange and cover plate. Type 2 is identical to the as-built connection, with the exception of the 1/2 inch transition fillet weld between the cover plate and the column flange.

The Revised Type 1 connection uses a thinner, 1 inch cover plate as well as fillet welded continuity plates. The placement of the continuity plates is the same as for the as-built and the Type 2 connection.

For the Type 3 connection, of particular interest are (1) the smaller thickness (1 inch) cover plate, which is the same as that of Type 1; (2) the use of the 1/2 inch transition fillet weld between the top and bottom cover plates and column flanges; (3) the fillet welded continuity plates; and (4) the repositioned continuity plates. It was decided that a repositioning of the continuity plates would allow for an improved fracture path following crack initiation, based on the knowledge gained during the testing of the other specimens in the research program. The continuity plates were placed with their outside edges in line with the interface between the beam cover plate and the beam flange.

Test Results

Figure 8 shows the moment vs. plastic rotation hysteresis loops for Specimen A4. This was an as-built connection with a rotary straightened column, and the testing protocol was dynamic. The somewhat "ragged" nature of the curves is a result of the dynamic testing, as well as the very high rate of data recording. Failure was initiated through a crack at the toe of the bottom cover plate to column weld. The full stiffness was maintained through one complete cycle of ± 0.028 radians plastic rotation.

Figure 9 shows the hysteresis loops for Test A7. This was an as-built connection with a gag straightened column, and the testing protocol was dynamic. Failure was initiated through a crack at the toe of the top cover plate to column weld. The full stiffness was maintained through one complete cycle of ± 0.028 radians plastic rotation.

Figure 10 shows the hysteresis loops for Test A8. This had an unstraightened column, and testing was quasi-static (this explains the smooth appearance of the curves). Failure was initiated through a crack at the toe of the top cover plate to column weld; a crack also developed at the edge of the cover plate, close to the edge of the column flange. The full stiffness was maintained through one complete cycle of ± 0.028 radians plastic rotation. It is important to note that the behavior of the specimen was identical to those of A4 and A7, demonstrating that

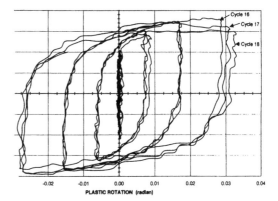

Figure 8. Column Face Moment vs. Plastic Rotation Hysteresis Curves for Test A4

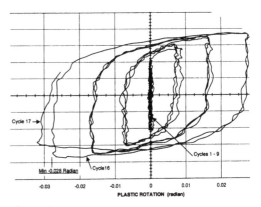

Figure 9. Hysteresis Loops for Test A7

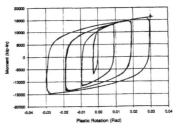

Figure 10. Hysteresis Loops for Test A8

the column straightening protocol has no influence on the performance of a connection.

Figure 11 shows the moment vs. plastic rotation hysteresis loops for Specimen R1-4. This was one of the Revised Type 1 connections with a rotary straightened column, and the testing protocol was

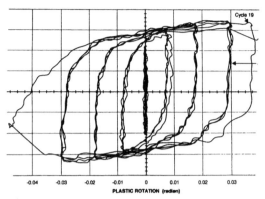

Figure 11. Hysteresis Loops for Test R1-4

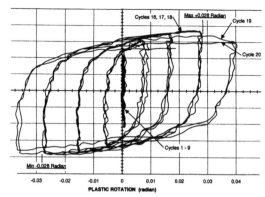

Figure 12. Hysteresis Loops for Test R2

dynamic. Failure was initiated through a crack at the toe of the top cover plate to column weld. The full stiffness was maintained through three complete cycles of ± 0.031 radians plastic rotation; failure took place during the second half of the first cycle of ± 0.038 radians plastic rotation.

Figure 12 shows the hysteresis loops for Specimen R2. This was the only Revised Type 2 that was tested; it had a rotary straightened column, and the testing protocol was dynamic. Failure was initiated through a crack at the toe of the bottom cover plate to column weld. The fracture extended across the k-area and then along the web, with no extension along the k-area. Small cracks were also found between the cover plate and the beam flange, as well as at the fillet weld "access hole" at the edge of the continuity plate. The continuity plate had been cropped at 1 inch to allow for the placement of the plate within the W-shape; this positioned the end of the fillet weld within the k-area of the W14x176. Despite this detail, no k-area cracking took place. The full stiffness was maintained through two complete cycles of ± 0.038 radians plastic rotation; failure took place during the second half of the first cycle of ± 0.038 radians plastic rotation.

Figure 13 shows hysteresis loops for Test R3-2. This was a Revised Type 3 connection; it had a rotary straightened column, and the testing protocol was dynamic.

As shown in Fig. 13, specimen R3-2 maintained full stiffness and integrity for all of the ± 0.030 radians cycles as well as the three first ± 0.038 radians cycles (nos. 19-21). The stiffness and beam tip load decreased slightly and uniformly for each cycle after no. 21, although the connection maintained its integrity, with no observed cracking.

The test of R3-2 continued with plastic rotations of ± 0.038 radians and with complete integrity

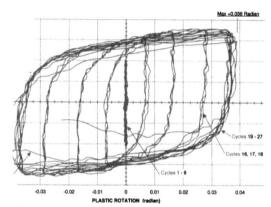

Figure 13. Hysteresis Loops for Test R3-2

during 26-1/2 cycles. It failed due to cracking in the top flange cover plate to column flange region. Local buckling did not occur.

By the end of the 21 cycles of the test of Specimen R3-1, plastic rotations of ± 0.038 radians were recorded. By the end of the 26 first cycles of test R3-2 the maximum and minimum plastic rotations were +0.038 and -0.040 radians; during the first half of the 27th cycle the plastic rotation was +0.039 radians. Comparing the results for connection R3-2 after 21 cycles with those of R3-1, R3-2 had undergone cumulative plastic rotations of +0.986 and -0.985 radians. These compare very closely with the cumulative plastic rotation of 1.004 radians for specimen R3-1 after 21 cycles.

Failure Characteristics for R3-1 and R3-2

The performance of the two Type 3 connections warrants additional comments. For R3-1, cracks initiated at the weld root between the top and bottom cover plate and the beam flange. The crack in the top connection extended about 1-1/2 inches through the column flange. Overall failure had not taken place at the end of 21 cycles. Fractographic evaluations showed that the crack arrested and never went completely through the flange, which explains why this connection did not fail. There were no welding anomalies.

For connection R3-2, a crack initiated at the weld root between the top cover plate and the beam flange. This crack eventually extended through the column flange and intersected with a secondary crack that had initiated at the weld access hole adjacent to the toe of the continuity plate weld. The crack at the access hole had the appearance of stable growth, before it intersected with the flange crack. Following intersection of the cracks, the fracture extended from the top continuity plate towards the bottom continuity plate.

It was found that the location of the secondary crack initiation for R3-2 was at the continuity plate weld that coincided with the k-area. The fracture markings indicated that the crack started at the intersection of the continuity plate fillet weld and the k-area. Additional fractographic evaluation shows that there were 7 crack arrest marks. This indicates that the crack had been propagating for at least 7 cycles, meaning that the initiation occurred before the 20th cycle. This is consistent with the results for other specimens. These results also indicate that the k-area cracking at the weld for the continuity plate is a secondary fracture, and therefore not a primary failure location. Finally, although the crack was propagating in the k-area, the arrest marks demonstrate that this was a slow and ductile cracking phenomenon.

ASSESSMENT OF PERFORMANCE

Overall, the connections performed well to very well. However, the responses of the Type 2 and especially the Type 3 connections were excellent. In particular, the performance of the two Type 3 connections point to a number of significant results. Primary among these is the fact that crack initiation did not occur in the k-area, but at the toe of the cover plate weld. A secondary crack did initiate in the k-area, but this propagated in a stable manner until it intersected with the crack from the toe of the cover plate weld. The stable crack growth in the k-area negates the perceived problem that the k-area and its high hardness and low toughness will always result in unstable brittle fracture. Further, some of

the cracking initiated in the near vicinity of the continuity plate access holes. It is therefore clear that enlarging these access holes to move the weld termination farther away from the k-region will offer additional performance benefits.

The pronounced effect of the relocation of the continuity plates further emphasizes the critical nature of the connection detailing. By providing for improved load paths in the connection, the rotation capacity and endurance of the assembly were increased very significantly. Finally, the close correlation between the two tests attests to the quality of the materials and the connection design and fabrication as a whole.

CONCLUSIONS AND RECOMMENDATIONS

17 full-scale beam-to-column connection tests were conducted. The effects of steel mill straightening practices were evaluated through the use of rotary straightened, gag straightened and unstraightened columns, and the results demonstrate conclusively that there are no significant performance differences between assemblies using rotary straightened, gag straightened and unstraightened columns.

The specimens included flange and cover-plate complete joint penetration (CJP) welded beam to column joints with bolted web connections, and CJP-welded continuity plates. One such connection also had a fillet welded transition between the cover plate and the column flange. Other specimens used thinner cover plates and fillet welded continuity plates. Finally, two specimens used the thinner cover plates and fillet welded continuity plates, in addition to the transition fillet weld, and the continuity plates were repositioned.

Eight of the 17 connection specimens in the program were tested under quasi-static, displacement-controlled loading conditions. The other nine specimens were tested under 1 Hz dynamic loading, also using displacement-controlled amplitudes. The results show that dynamic loading is a more severe condition, but the differences in performance between otherwise identical specimens are not significant.

Whereas the FEMA single cycle plastic rotation criterion is one measure of ductility and deformation capacity, it does not reflect the energy absorption characteristics of a connection. The cumulative and normalized cumulative plastic rotation capacities are significantly better and more realistic measures of performance. Using the updated Japanese acceptance criterion of a cumulative plastic rotation capacity of 0.3 radians, all of the connections but one were satisfactory. This was specimen A3, one of the as-built connections; it reached a cumulative plastic rotation of 0.219 radians. A detailed fractographic evaluation showed an imperfection in one of the CJP welds between a continuity plate and the column web.

The tests of the Type 3 connections demonstrated excellent plastic rotation and energy absorption capacities under the most demanding of testing conditions; their performance outpaced that of all other connections by significant margins. For these connections it was also found that although cracks developed and propagated through portions of the column material, the propagation was slow and stable, with numerous crack arrest events during the tests. This also occurred for the cracks that propagated into the k-area of the columns, demonstrating that a crack in this region will propagate in stable fashion, given appropriate connection details and fracture paths. Further, fabrication is much easier with the thinner cover plates and fillet welded and repositioned continuity plates. The cropping of the continuity plates is important, to the effect that the ends of the welds should be moved away from the k-area of the column. Fabrication and construction economies will be derived with the revised Type 3 connections.

REFERENCES

American Society for Testing and Materials (ASTM) (1998), *"Specification for General Requirements for Rolled Structural Steel Bars, Plates, Shapes and Sheet Piling"*, ASTM Standard A6. ASTM, Conshohocken, PA.

Bjorhovde, Reidar, Goland, L. J., and Benac, D. J. (1999), *"Tests of Full-Scale Beam-to-Column Connections"*, Technical Report, Nucor-Yamato Steel Company, Blytheville, AR, August.

American Institute of Steel Construction (AISC) (1994), *"Specification for the Load and Resistance Factor Design of Steel Building Structures"*, 2nd Edition. AISC, Chicago, IL.

American Institute of Steel Construction (AISC) (1997), *"Seismic Provisions for Structural Steel Buildings"*, and *"Supplement 1 to the 1997 Seismic Provisions"* (1999), AISC, Chicago, IL.

Applied Technology Council (ATC) (1992), *"Guidelines for Cyclic Seismic Testing of Components of Steel Structures"*, ATC Guideline No. ATC-24. ATC, Redwood City, CA.

Federal Emergency Management Agency (FEMA) (1995), *"Interim Guidelines: Evaluation, Repair, Modification and Design of Steel Moment Frames"*, FEMA Report No. FEMA-267. FEMA, Washington, D.C.

Nakashima, M.; Suita, K.; Morisako, K. and Maruoka, Y. (1998), "Tests of Welded Beam-Column Assemblies. I: Global Behavior", *Journal of Structural Engineering*, ASCE, Vol. 124, No. 11, November (pp. 1236-1244).

Malley, J. O. (1999), *"Private Communication"*, San Francisco, CA.

Cyclic testing of flush end-plate semi-rigid steel connections

B. M. Broderick & A. W. Thomson

Department of Civil, Structural and Environmental Engineering, University of Dublin, Trinity College, Ireland

ABSTRACT: There is increasing interest in developing and employing steel beam-to-column connections other than the full-strength heavily-welded type. One such alternative is the flush end-plate connection, for which considerable gravity-design guidance already exists, but is necessarily partial strength. This paper reports the initial results of a series of cyclic tests carried out on flush end-plate connections to determine their suitability for use in earthquake resistant steel moment frames. The test specimens tested consist of a standard size connection in which the thickness of the end-plate is varied. This allows the three failure modes defined in Eurocode 3: Annex J to be examined in relation to their seismic response. Parameters such as ductility, the resistance drop ratio and stiffness hardening characteristics are compared for the different end-plate thicknesses and conclusions are drawn as to the connections' ability to handle seismic response demands.

1 INTRODUCTION

In the design of steel frames for use in high seismic risk areas, it is common for designers to favour full-strength heavily-welded connections over less resistant typologies. However, following the inspection of such connections after recent earthquakes, a large number of brittle failures were observed. An alternative to the full-strength option is the end-plate connection that is advocated for use in wind-moment design in the United Kingdom, for which there is considerable design guidance available (BSCA, 1995). These connections are characterised as semi-rigid, partial strength joints that will yield in a controlled manner under the demands of seismic loading. In order to ensure a ductile response, careful selection of the design details, such as the end-plate thickness, bolt size and spacing among others, is necessary. Although the available guidance already considers these criteria as part of the ductility requirements, it is uncertain whether sufficient rotation capacity for seismic engineering applications is always provided.

The use of semi-rigid connections offers a number of advantages over full-strength typologies. The connections and the structures incorporating them usually have a lower initial stiffness, resulting in a longer natural period, which may in turn lead to lower seismic design forces. The fact that the connections are partial strength and should yield before any of their connected members allows the yield point of the structural system to be defined at the connection design stage, therefore allowing the ductility demands under severe loading conditions to be predicted. In many cases, these partial strength connections can also increase the energy dissipation characteristics of the frame, resulting in greater economy.

However, because many designers prefer to consider connections as either perfectly pinned or perfectly rigid, the advantages of semi-rigid construction are not widely utilised. The most obvious of these is that beam moments are reduced, leading to smaller and less expensive members. This can be best understood if a single span beam section is considered with connections at both ends. If simple connections are assumed, beam moments are critical at the mid-span of the section, while if perfectly rigid connections are assumed, end-moments are critical. However, with the use of semi-rigid connections, the moments are more evenly balanced between the mid-point and the ends of the span. In the seismic design of moment resisting frames, the degree to which this feature may be exploited depends upon the relative magnitude of the gravity and seismic design forces acting at each floor level. A further advantage results from these end-plate connections being simpler to fabricate, offering improved economy and reliability. Even should designers still wish to specify fully rigid connections for the frame, the use of the semi-rigid design techniques in the Eurocode 3: Annex J (CEN, 1998) will allow considerable savings. So long as the initial stiffness of the connection remains in the rigid zone (as defined in the Eurocode), unnecessary bolt-rows, haunches, and stiffeners may be removed without compromising the rigidity of the connection. On the down side, the flexibility of semi-rigid connections may pose difficulties at the serviceability limit state, which is often the critical feature in the seismic design of moment-resisting steel frames. Column design may also be penalised, both due to increased effective lengths and higher over-strength factors for capacity design.

The wind-moment design is a well-established method used for the design of unbraced steel frames. It relies on the rotational stiffness of the beam-column connection to resist horizontal loading, but assumes that these joints are pinned for the vertical loading. This 'simple' approach has been allowed by BS 5950, Eurocode 3: Annex J and the AISC codes of practice (AISC, 1993, BSI, 1990). As the wind-moment method is a type of semi-rigid design, it is possible avail of the above advantages offered by the semi-rigid connection. It is anticipated that the use of wind-moment connections may be extended for use in seismic engineering so long as certain design restrictions and recommendations are followed.

This paper describes a series of experimental tests carried out on flush end-plate connections. Beam-to-column subassemblies representing an external moment connection are subjected to cyclic displacements of increasing amplitude. This allows the rotational and moment capacities of the connections to be evaluated under conditions similar to those experienced during the response of a building to strong ground motion. The connections have all been designed to conform with the guidelines set out in Eurocode 3: Annex J and are based around a standard industry template (BSCA, 1995). However, in order to investigate the three failure modes that are defined in the codes, the thicknesses of the end-plates used are varied.

The objective of these tests is to investigate whether the predicted mode of failure are observed under cyclic response conditions, and to determine typical rotational ductility capacities for connections designed to respond in each of these modes. The results of the tests are presented and a number of key parameters such ductility and resistance drop ratio are discussed for each connection configuration. The initial stiffness and ultimate moment predicted by Eurocode 3: Annex J are also presented for each connection and the differences between these values and the experimental values are discussed.

2 EXPERIMENTAL PROCEDURES

2.1 Experimental Test Set-up

To facilitate placing and loading, the specimens were set-up with the column orientated horizontally in the loading rig as shown in Figure 1. The specimens are held in place laterally by means of an adjustable restraint. The servo-hydraulic actuator is computer-controlled and pinned at both ends to reduce bending effects. This testing system works on a feedback loop where the actuator command console outputs a voltage to move the actuator to a required position, whether that is load or displacement, and receives a return voltage which corresponds to the actual position. The console uses this return value to calculate the difference in the actual and required position and adjusts the output voltage as needed. This continues until the actuator is within 0.1mm of the correct load or displacement. It is capable of load or displacement control and can provide a maximum stroke of ±56mm or 100kN. For the purposes of the experimental work described here it is operated exclusively on displacement control.

A set of strain gauges and LVDTs are placed on the specimen to measure strains and displacements in those areas that were felt to be of prime importance. These concentrate around the end-plate, although measurements were also taken along the length of the column and beam members.

2.2 Experimental Procedures and Control

The experimental work carried out consists of a series of cyclic displacement waves of increasing amplitude. This series is based on the procedures published by the ECCS (ECCS, 1986) for the cyclic testing of structural steel elements. These procedures describe two methods, and the short testing method was employed. This procedure involves imposing at least four cycles of a displacement less than the yield displacement on the specimen. With the data from these cycles, the initial stiffness and hence the yield rotation (e_y) and moment (M_y) for the specimen may be calculated. Once the yield rotation of the specimen has been determined, it is recommended that the following cyclic series is followed:

– One cycle each in the $e_y/4$, $e_y/2$, $3e_y/4$ and e_y range;
– Three cycles in the $2e_y$ range;
– Three cycles in the $2(n+1)e_y$ range (for n = 1, 2, 3, etc.)

This series is continued until the actuator limit is reached, or until the specimen has failed. Should the specimen not fail before the actuator limit is obtained, a further six cycles at the actuator limit are completed before the end of the test.

The control system used for the series of test described below was developed in Trinity College Dublin and is based on the ECCS procedures described above. This system centres around a data acquisition card which is controlled using a program written in the graphical programming environment LabVIEW. The control system is capable of controlling two actuators simultaneously and is capable of scanning up to forty strain gauges as well as seven LVDTs or load cells. Options such as scan rate, output rate, and the number of scans to write to file at a time may all be set independently for each test, as the predicted response suggests. This system is accurate to within 0.5mm over the entire stroke distance of the actuator which corresponds to a maximum displacement error of 0.9% between the required displacement and the actual displacement.

3 EXPERIMENTAL TEST SPECIMENS

The specimens tested are shown in Figure 2. They are all variations on a standard wind-moment flush end-plate connection (BSCA, 1995). The specimens consist of a one metre length of 203 x 203 x 86 kg/m^3 UC section and a one metre length of 254 x 102 x 22 kg/m^3 UB section. The end-plates are 200 x 277 mm with the varying thickness, as discussed below. The bolt holes are located 45mm from the centre of the beam web and 60mm from the outer edge of the beam flanges. All of the steel is Grade 43 (S275) and all bolts are of size M20 and Grade 8.8 (tensile yield strength = 640 N/mm^2). The bolts were all preloaded to a torque of 300 Nm to ensure that all the tests were performed under identical conditions.

Three different end-plate thicknesses were selected to allow the failure modes that are detailed in Eurocode 3: Annex J to be evaluated under cyclic loading. The thicknesses were 8, 12 and 20mm; the corresponding specimens are designated as EP1, EP2 and EP3 respectively. The three failure modes defined in Eurocode 3: Annex J (revised) are shown in Figure 3. Specimen EP1 was designed to give a Mode 1 failure. This occurs when plastic hinges are formed at the bolt line and at the beam flange as seen in Figure 3(a). Mode 2 failure is defined as the formation of a plastic hinge at the beam flange followed by yielding of the bolts as shown in Figure 3(b). It was predicted that this would occur for specimen EP2. This end-plate was also used as a benchmark as it is the plate thickness detailed by the BCSA moment connections specifications (BSCA, 1995). The third failure mode occurs when the bolts yield while the end-plate remains elastic (Figure 3(c)), resulting in a brittle failure which is not recommended for design.

4 RESULTS OF EXPERIMENTAL WORK

Table 1 presents a summary of the results obtained from the series of cyclic tests. All three specimens behaved in the manner predicted by Eurocode 3: Annex J (CEN, 1998) as discussed above. However, there are significant differences between the experimental and predicted values for the initial stiffness and ultimate moment capacity. The initial stiffnesses for specimens EP1 and E2 are very similar in magnitude and are approximately three times that of specimen EP3. In comparison, the predicted stiffnesses of the three connections increase in stiffness from 7.38 kNm/mrad for specimen EP1 to 13.11 kNm/mrad for specimen EP3. The similarity in the initial stiffness for EP1 and EP2 may be attributed to the relative thicknesses of the end-plates combined with the initial preload on the bolts. This preload could conceivably increase the initial stiffness of EP1 to that of EP2, but should not have any influence on the yield or ultimate moment resistance of

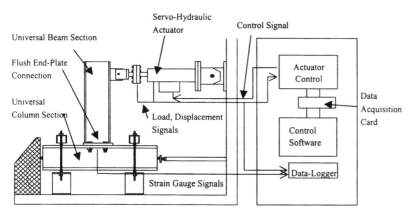

Figure 1. Experimental Test Specimen Set-up

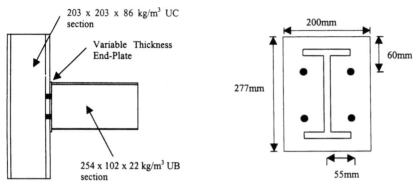

203 x 203 x 86 kg/m³ UC section

Variable Thickness End-Plate

254 x 102 x 22 kg/m³ UB section

200mm

60mm

277mm

55mm

Figure 2. Plan and elevation of test specimens

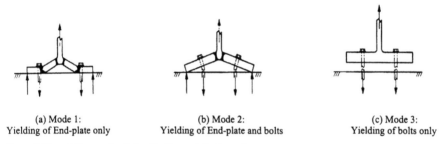

(a) Mode 1:
Yielding of End-plate only

(b) Mode 2:
Yielding of End-plate and bolts

(c) Mode 3:
Yielding of bolts only

Figure 3. Three failure modes defined in Eurocode 3: Annex J

Table 1. Experimental results

Specimens	Initial Stiffness	Experimental Yield Values		Experimental Ultimate Values		Predicted Initial Stiffness	EC3 Moment Capacity
		Rotation	Moment	Rotation	Moment		
	kNm/mrad	mrad	kNm	mrad	kNm	kNm/mrad	kNm
EP1	10.01	2.94	29.44	67.1*	54.7	7.38	24.61
EP2	10.51	4.49	47.2	35.1	72.7	11.12	47.33
EP3	3.58	18.15	65.1	56.2	79.4	13.11	55.53

* Equipment displacement limits were reached before specimen EP1 failed.

the connections. However, preloading of the bolts should not have caused such a significant difference in the initial stiffness of EP3 and further testing may be required to determine the reason for such a dramatic difference between the predicted and experimental values. Figure 4 shows the monotonic moment-rotation relationships derived from Eurocode 3: Annex J (revised). As can be seen, the yield rotations for specimens EP1 and EP2 are reasonably similar to those calculated by Eurocode 3, 2.2 mrad and 3.1 mrad respectively. However, it should be noted that the stiffness and moment-rotation relationships derived are only valid for bolts that are assumed to have no preloading. It can also be seen that the yield moment resistance measured in each test is reasonably close to the ultimate moment resistance predicted by Eurocode 3: Annex J (revised) and that

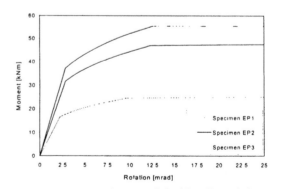

Figure 4. Moment-rotation curves derived from Eurocode 3: Annex J (revised)

138

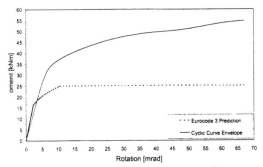

Figure 5. Comparison of Eurocode 3 values and cyclic curve envelope for specimen EP1

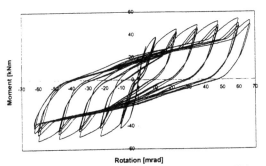

Figure 6. Moment-rotation hysteresis curve for specimen EP1

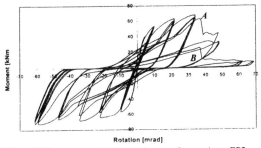

Figure 7. Moment-rotation hysteresis curve for specimen EP2

ultimate experimental values are much higher than expected, as shown in Figure 5.

Specimen EP1 behaved in a highly ductile manner, which is consistent with complete yielding of the end-plate (i.e. Mode 1 failure). The amplitude of the imposed displacement cycles were continually increased in accordance with the procedures described above until the equipment limits were reached. At this point, a rotation ductility of 22.5 had been achieved. The moment-rotation hysteresis curve obtained from the test is shown in Figure 6. It was anticipated that the ultimate moment of the connection would not increase with further testing and

that the end-plate would continue to display a ductile response at larger displacements.

Specimen EP2 did not respond in as a symmetrical manner as EP1, possibly due to material or fabrication imperfections. This is suggested by the manner in which the specimen failed. As the imposed displacement was increased to more than 30 mrad, the threads on one nut in the tension bolt-row stripped (shown by point A on Figure 7), resulting in a fall of the moment resistance of the connection from 60 kNm to 40 kNm. Even with the failure of one bolt, the connection was still capable of resisting a reasonably large moment for the remaining two cycles in the group. However, when the imposed displacement increased further in the next group of cycles, the second bolt in the same row failed in the same manner (point B) and the moment resistance of the connection decreased to nearly zero. Although the connection was still capable of resisting displacement in the negative direction, there was virtually no resistance to positive rotation. For all practical purposes, and for the analysis carried out in this experimental study, the connection is considered to have failed when the first of the bolts fails at point A. This failure mechanism is consistent with that predicted by Eurocode 3 with yielding of the end-plate being followed by yielding of the bolts. When failure of specimen EP2 occurred, a ductility ratio of 7.8 had been achieved with failure occurring at a rotation of 35.1 mrad. For most purposes, this is sufficient for plastic analysis of the connection and/or structure to be carried out. However, as the final failure of the bolts happened in a brittle manner, it may not be advisable to employ this connection detail as a dissipative element in an earthquake-resistant structure. Specimen EP3 also failed as predicted by Eurocode 3: Annex J (CEN, 1998) with complete yielding of the bolts occurring while the end-plate remained elastic (i.e. Mode 3). However, the yield rotation and initial stiffness for this specimen were radically different from the expected values, with a yield rotation of 18.15 mrad and an initial stiffness of 3.58 kNm/mrad. This contributed in part to the low observed ductility ratio of 3.1. This brittle failure mechanism does not provide adequate rotational capacity for energy dissipation purposes. It should, however, be noted that the ultimate moment capacity reported above in Table 1 is at the limit of the actuator load cell and may in fact be slightly higher.

A large degree of stiffness hardening was noted in the hysteretic responses, as can be seen in Figures 6 and 7. This hardening is attributed to the cyclic response of the slight extension in the end-plate beyond the beam flange. As the imposed displacement on the specimen is increased for each group of three cycles, the end-plate will deform slightly. As this displacement is removed, there will be both elastic rebound and plastic deformation in the plate. This elastic rebound will result in some stiffening on the next cycle when the specimen is returned to the same displacement, but due to its plastic deformation, this doesn't begin at the same rotation. The plastic deformation in the end-plate toe therefore causes a drop in the resistance of the specimen which is quantified as the resistance drop ratio. This parameter is measured over each group of three cycles of the same amplitude. It can be used as a measurement of the ability of the connection to resist repeated rotations. EP1 had an average resistance drop of 0.9 over the six groups of cycles imposed. EP2 had an average resistance drop of 0.95 up to failure of the joint, showing that it is more capable of resisting repeated loads at the expense of ductility. This is also observed with specimen EP3, which had a resistance drop ratio of 0.97, nearly no drop in load over a group of three cycles. Therefore, the specimen with the highest level of ductility was also the least capable of repeating moment.

5 CONCLUSIONS

Results from a series of cyclic experiments on semirigid partial-strength flush end-plate connections with varying end-plate thicknesses have been presented. These results show a significant difference between experimental data and the values predicted from Eurocode 3: Annex J (CEN, 1998). Some of the differences in initial stiffness may be due to preloading of the bolts in the experimental tests, but this cannot explain all of the inconsistencies, especially those concerning specimen EP3 where the initial stiffness was much lower than expected. However, the experimental study did show that the three failure modes predicted by Eurocode 3 remain valid under cyclic response conditions, as all of the specimens behaved accordingly. It is clear that Mode 3 failure is not suitable for dissipative connections in steel frames as the failure mechanism is too brittle and sudden, resulting in insufficient rotational capacity. Specimens that undergo Mode 2 failure appear to have the required level of rotation capacity, although the sudden failure of the bolts is a cause for concern. One possible solution to this may be to employ nuts of a higher grade. This approach may be advantageous as connections that undergo Mode 1 failure display much lower moment resistances and may therefore be incapable of resisting design forces for severe earthquake loading.

It was also seen that the connection resistance drop ratio is reduced when the thickness of the end-plate is increased. This in attributed to the relative proportion of elastic and plastic deformation in the end-plate. However, this reduction in resistance drop ratio is offset by a lowering of the ductility of the connection, It may be possible to determine an optimum thickness for both parameters so that the connection retains sufficient rotational capacity.

The test results have also shown that Eurocode 3: Annex J appears to give highly conservative predictions of the ultimate moment capacity when calculating the moment-rotation relationship. This is due to the fact that Eurocode 3 uses a conservative value for the yield strength of the steel when calculating the ultimate resistance of the connections. This has serious design implications for capacity design of columns. If the moment capacity of the connection is under-estimated, yielding is likely to occur in column members.

Presently, further experimental work is being planned to investigate the effects of the pre-loading of the bolts on the initial stiffness of the connections, as well as testing specimens with higher grade nuts with standard bolts. It is also intended to conduct further tests on specimens with Mode 3 failures to determine the cause for the unexpected stiffness result presented above. These further test series are designed to enable the development of a component-based model for predicting the behaviour of semirigid partial-strength flush end-plate connections in cyclic and dynamic loading conditions. It is anticipated that this model will follow the guidelines set out in Eurocode 3: Annex J (revised) and will be suitable for incorporation in a standard finite element or structural analysis packages.

REFERENCES

American Institute of Steel Construction (AISC). 1993. *Load and Resistance Factor Design Specification for Structural Steel Buildings.*

British Constructional Steelwork Association (BSCA). 1995. Wind-moment Connections. *Joints in Steel Construction: Moment Connections.* Steel Construction Institute: 50–54.

British Standards Institute (BSI). 1990. BS 5950: Structural use of Steelwork in building. London.

CEN. 1998. Annex J: Joints in Building Frames (revised). *Eurocode 3: Design of Steel Structures – Part 1.1: General Rules and rules for Buildings.* ENV 1993-1-1.

Davison, J.B., Kirby, P.A. & Nethercot, D.A. 1987. Rotational Stiffness Characteristics of Steel Beam-to-Column Connections. *Journal of Constructional Steel Research.* 8: 17–54

European Convention for Constructional Steelwork (ECCS). 1986. Recommended Testing Procedure for Assessing the Behaviour of Structural Steel Elements under Cyclic Loads. Number 45. 1st Edition.

Leon, R.T. & Deierlein, G.G. 1996. Considerations for the Use of Quasi-Static Testing. *Earthquake Spectra.* 12(1): 87–110.

Macken, C. 1997. Testing of Steel Moment Connections. *BAI Thesis, University of Dublin, Trinity College, Dublin.*

Thomson, A.W. & Broderick, B.M. 2000. Cyclic Testing and Analysis of Steel End-plate Connections. *Abnormal Loading of Structures: Experimental and Numerical Modelling; Proc. Intern. Conf., London, 17-19 April 2000.* In Press.

Weynand, K., Jaspart, J.P. & Steenhuis M. 1995. The Stiffness Model of revised Annex J of Eurocode 3. In R. Bjorhovde (ed.). *Connections in Steel Structures III: Behaviour, Strength and Design.*: 441–452.

Behaviour of Steel Structures in Seismic Areas, Mazzolani & Tremblay (eds) © 2000 Balkema, Rotterdam, ISBN 90 5809 130 9

Welded plate and T-stub tests and impact on structural behavior of moment frame connections

F.W. Brust, P. Dong & T. Kilinski
Battelle Memorial Institute, Columbus, Ohio, USA

ABSTRACT: Simple-welded plate specimens (SWPS) and T-Stub tension specimens were used to investigate the effects of weld strength mismatch and weld metal toughness on both mechanical and fracture behavior of welded moment connections. A large number of tests were performed under various base material (BM) and weld metal (WM) matching conditions. Selected finite element analyses were also performed to shed light on the structural behavior of these specimens for interpretation of the experimental results and of structural frame behavior. Implications on structural behavior in welded moment connections will be discussed in detail with a special emphasis on constraint and weld residual stress effects.

1 INTRODUCTION

It is well known that constraint can have a significant effect on failure ductility of metals. Figure 1 illustrates a typical failure locus of test data. Here we define constraint as σ_m/σ_{eq} with:

$$\sigma_m = (1/3)(\sigma_1 + \sigma_2 + \sigma_3)$$

$$\sigma_{eq} = \sqrt{(3/2)S_{ij}S_{ij}}$$

$$S_{ij} = \sigma_{ij} - \sigma_m.$$

As seen in Figure 1, the plastic strain at failure decreases (ductility decreases) as constraint increases. For base metal, the constraint varies throughout a structure and depends on a number of factors including geometry, loading, local stress state, material, amount of plasticity, etc.

Weld induced residual stresses and weld metal strength mismatch effects can significantly influence constraint conditions in a welded structure. This effect is investigated here by testing and analyzing small coupon (uniaxial), wide plate, and T-Stub specimens. Many researchers have studied the weld strength mismatch effect in recent years as discussed by Dong and Gordon (1992) and the references cited therein. The effect of weld induced residual stresses on constraint has only recently been considered: see for instance Zhang and Dong (1999) and a companion paper by Dong, Zhang, and Brust (2000) in this volume. Small tensile coupons, e.g., ASTM

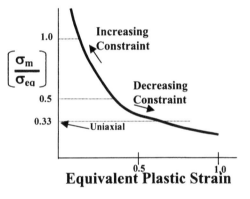

Figure 1. Typical failure locus for metals.

E8 type, are designed for uni-axial stress-strain curve characterization and provide negligible lateral constraint. The wide plate specimens (i.e., SWPS – Figure 2), with an adequate width (W), can subject the middle section of the specimen width into plane-strain (high constraint) deformation conditions, and therefore, provide insight on structural failure behavior at regions of high constraint. In such specimens, weld residual stress effects can also be retained as discussed by Brust, Dong, and Zhang (1997a and b). In addition, T-Stub specimens subjected to tension as shown in Fig. 1 provide a means for assessing load transfer (from the horizontal beam flange to the vertical column flange) effects on three dimensional stress-strain

(a) Uniaxial

(b) SWPS (Simple Welded Plate Specimen)

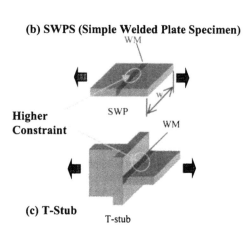

Higher Constraint

(c) T-Stub

Figure 2. Three weld test specimens considered.

Table 1. Strength mismatch conditions for tests.

Weld σ_y / σ_{ult} (ksi)	A36 42 71	A572 Gr. 50 54 81	A572 Gr. 60 60 87	HPS 70W 85 96
Plate Yield (σ_y) Ultimate(σ_{ult}) (ksi)				
NR203 NiC 65 / 79	+55% +11%	+20% -2%	+8% -9%	-24% -18%
NR311 Ni 81/93		+50% +15%	+35% +7%	-5% -3%

states at the joint. This paper investigates the effect of constraint on welded moment frame structures both experimentally and by performing a few selected finite element (FE) analyses.

2   EXPERIMENTAL RESULTS

2.1 Materials

The test matrix of materials is listed in Table 1. Four different base plate and weld filler materials were used. The strength mismatches, defined as the ratio of uni-axial weld metal to uni-axial base material yield (or ultimate) stress, is listed in Table 1. For practical purposes, the A572 Grade 60 plate with NR203 NiC weld filler metal, and HPS-70W plate and NR311 Ni weld filler are matched materials. We present some of these test results below. More details can be found in Dong et al (1999).

2.2 Test Procedure.

Testing on the SWPS plates was performed using a 4-post servo-hydraulic test frame with a rated capacity of 1,000,000 pounds. Load was measured using a load cell in series with the loading axis of the specimen. Displacement was measured using an LVDT internal to the hydraulic actuator. The T-stub

specimens utilized the same fixturing for the "beam" segment of the specimen (see Figure 2c). Instrumentation for the tests consisted of load (1,000,000 pound load cell), actuator displacement (± 2.5-inch LVDT), average strain across the test weld in two locations (a 1-inch gage/50% extensometer and a 2-inch gage/50% extensometer), and remote strain in the base metal (1 inch gage/15% extensometer). Data was collected using LabTech Notebook® software running on a desktop PC equipped with a 16 channel, 16-bit D/A board.

Two types of tests were performed. One at a slow rate, which amounted to an applied strain rate of 0.00005 in/in/sec. The other, termed a 'fast' rate, was 0.02 in/in/sec. It will be seen later that this 'fast' rate is actually not as fast as one could expect in a dynamic earthquake event. Data was collected at 1 Hz for the slow rate tests, and at 1kHz for the fast tests.

2.3 Significant Results.

A complete summary of all of the test results can be found in Dong, et al. (1999). In particular, a summary of the test failure results for all 36 tests performed both with and without pre-cracks in the welds, is listed in Table 5 of Dong et al. (1999). Here we elaborate on only a few results and provide summaries.

An example of the load versus load line displacement for the A572 Gr. 60/NR203 NiC slow and fast load cases is shown in Figure 3. From Table 1 it is seen that this case represents a nearly matched weld and base metal property case. Notice that the yield and ultimate load for the 'fast' load case is slightly higher than that for the 'slow' load case. This trend occurred for nearly all of the cases and the implications are discussed in the Summary section.

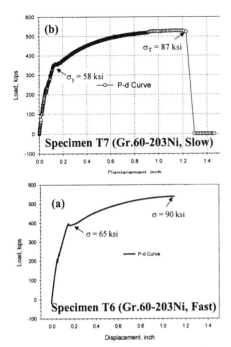

Figure 3. Measured load-displacement curves for T-stub specimens (a) Quasi-static loading; (b) Fast loading.

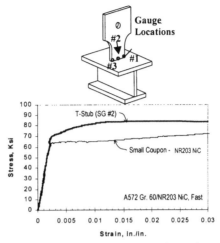

Figure 4: Local stress-strain development in T-Stub specimen (measured at mid-width within WM).

Figure 4 illustrates the applied (global) stress versus local strain (measured in the weld) for the A572 Gr. 60/NR203 NiC T-stub case. One of the SWPS and one T-stub specimen were instrumented with strain gages during testing to get a handle on the local strain and constraint effect. Figure 4 clearly illustrates that the ductility attained in the T-stub test is significantly less than that achievable in a uniaxial weld test (low constraint). For the same amount of load direction strain, the stress in the corresponding T-stub is much higher. In fact, the constraint in the T-stub case is larger than that attainable within the SWPS specimens. As such, one expects the failure of the T-stub specimens to be less ductile compared with the SWPS specimen, and much less ductile compared with a uniaxial weld specimen, as discussed next.

The ultimate weldment strengths, as observed from the different tests, were dominated by very complex interactions among WM, HAZ, and BM and final failure path. For the SWPS specimen, the matched and overmatched cases (i.e., A572 Gr. 60/NR203 NiC, HPS-70W/NR311 Ni, and A572 Gr. 60/NR311 Ni), the ultimate specimen strengths were largely dominated by final failure occurring in the base material. For T-Stub specimens, however, most of the final failures (five out of six specimens) occurred in the weld. Furthermore, the fracture surfaces clearly indicate brittle fracture features in the five specimens. The plastic deformations associated with the final fractures were significantly less than those in SWPS specimens with identical BM/WM combinations. To contrast the difference in fracture mode between SWPS and SWPS specimens, Fig. 5 summarizes the fracture surfaces for cases with final fracture in the weld. It can be seen that with identical BM/WM combinations and loading conditions, there is a change in fracture

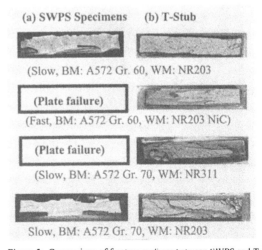

Figure 5: Comparison of fracture surfaces between SWPS and T-Stub specimens for failure in the weld.

mode from dominantly ductile (SWPS shown in Fig. 5a) to brittle[*] (T-Stubs in Fig. 5b). The top three cases in Figure 5 represent matched properties and the bottom case represents a weld under match case. This phenomenon can be attributed to the severe joint constraint conditions seen in T-Stub specimens, regardless the weld metal used. It is interesting to note that the brittle feature on the fracture surfaces in the T-Stub specimens did not result in any noticeable reduction in the peak measured loads for most cases.

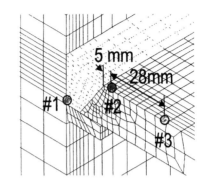

3 ANALYSIS RESULTS

The introduction section discussed the effects of constraint on crack nucleation and final structural failure. The greater the constraint, the lower the ductility expected at failure. From the experimental results it was observed that the failures for the most highly constrained cases (T-stub) resulted in a more brittle fracture feature. The source of the constraint comes from a number of sources. One source is the geometry of the T-stub specimen. The T-stub specimen is meant to simulate the conditions inherent in at a beam-column connection typically seen in steel fabrication. For the case considered here (see Figure 6), and typical in construction of many larger buildings, the column is thick and provides additional constraint to the beam flange compared with simply a plate (the beam thickness is also 1").

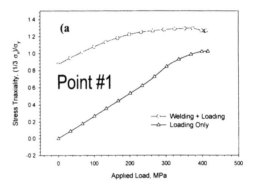

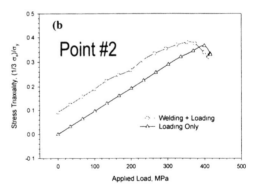

Figure 7: Stress tri-axiality development as a function of remote loading in T-Stub specimen: (a) Point #1 - weld root; (b) Point #2 - near HAZ

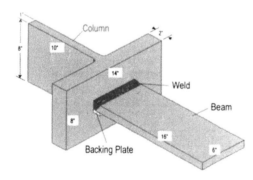

Figure 6. T-stub specimen dimensions.

Another additional source of constraint that is often neglected is the weld induced residual stresses that

develop in welded connections. The above observations can be further demonstrated using a 3D finite element model for T-Stub specimen as shown in Fig. 7, where weld residual stress effects were also considered by assuming a five-pass weld. The predicted residual stresses are summarized in Dong, Zhang, and Brust (2000). Near the weld root, the residual stresses in all three directions are of high magnitude (Dong, Zhang, and Brust (2000)). The

---

[*] The failures were not truly brittle, but were clearly associated with much less plasticity compared with the SWPS specimen.

yield strength of the weld metal was assumed to be the same as the column material (A572 Gr. 50). The T-Stub model was then used to simulate tension loading conditions used in the experiments.

Three locations were examined in detail, as shown in Fig. 7. The stress triaxiality normalized by yield strength $((\sigma_1+\sigma_2+\sigma_3)/3\sigma_y)$ as a function of remote nominal stress (F/A) are plotted in Fig. 7. As it can been seen, the weld residual stresses played a significant role in the high stress triaxiality buildup within the weld region, e.g., Points #1 (Fig. 7a) and Point #2 (Fig. 7b). For instance, the stress triaxiality at Point #1 is already near unity (see the curve with circle symbols) even before loading is applied if weld residual stresses are considered. Away from the weld, such as Point 3, where weld residual stresses become insignificant (actually slightly compressive), the stress triaxility increases linearly with the external loading due to the structural constraints along the width direction.

4 SUMMARY AND DISCUSSION

The following summarize the results presented in this paper.

- The structural constraint conditions in small tensile coupons, SWPS specimens, and T-Stub specimens are seen to be significantly different, as demonstrated by the finite element results. The T-Stub specimens possess the most severe structural constraints based on the stress triaxiality characterization. Although a direct stress triaxiality characterization for special moment frame connection joints was not performed here, it can be argued with confidence that the joint constraint conditions in T-Stub specimens resemble more closely the actual constraint conditions in moment frame conditions. As such, the T-Stub specimens should be more effective in mimicking the structural constraint conditions in actual moment frames than SWPS specimens.

- The ultimate weldment strengths, as observed from the tests, were dominated by very complex interactions among WM, HAZ, and BM and final failure path. For the matched and overmatched cases (i.e., A572 Gr. 60/NR203 NiC, HPS-70W/NR311 Ni, and A572 Gr. 60/NR311 Ni), the ultimate weld strengths were largely dominated by final failure occurring in the base material in SWPS specimens. For T-Stub specimens, however, most of the final failures (five out of six specimens) occurred in the weld. Furthermore, the fracture surfaces clearly indicate brittle fracture features in the five specimens.

4.1 Loading Rate Effects.

One purpose of this paper was to investigate the potential failure modes in dynamic 'earthquake' type loading conditions in moment frame structures. Loading rates effects were not significant within the "slow" (about 0.00001 in/in/sec) and "fast" (about 0.02/in/in/sec) loading rates tested. A consistent increase in both yield load and peak load before failure can be seen under the fast loading conditions. There is only one exception for the case of A572 Gr. 60/NR311 Ni, in which the unusual low peak load was resulted from the lack of penetration at the weld root (see Dong et al. (1999)). There is no clear trend identified in either failure location or fracture appearance between the two extreme loading rates.

It is well known that at high strain rates the yield strength of steel increases, and if the strain rates are high enough, fracture toughness can decrease. However, as discussed earlier, the current test results showed that the peak loads before final fracture tend to increase in a consistent manner under 'fast' load rate conditions. Charpy tests were also performed (see Dong et al. (1999)). The CVN toughness is considerably lower in the A572 Grade 50 and 60 steels compared to the other steels and weld metals, even at temperature considerable higher than room temperature. One possible explanation is that the strain rates in the "fast" loading rate tests are actually much lower than those strain rates experienced in the Charpy tests, resulting in the apparent increase in failure loads. It is important to note that the strain rates experienced in the Charpy tests conducted in this study were approximately estimated at about 5-10 in/in/sec near the notch (Brust, et al., 1993).

Some early work in studying nuclear piping systems under seismic loading conditions [e.g see Brust et al. (1993), Marschall et al. (1994), Scott et al. (1994), and Rudland and Brust (1997)] examined the effects of loading rates on the tensile and fracture properties of nuclear steels. The slow loading rates used in this Nuclear work were about the same level as used in the present study (strain rates were about .00005 in/in/sec). However, the 'fast' loading rates in the present study were only .02 in/in/sec while the strain rates considered in the nuclear work were up to 10 in/in/sec considering seismic loading. In Marschall (1994), it is observed that, for ductile stainless steels studied, loading rate has little effect on fracture loads. However, in the carbon steels studied, the fracture toughness of the steels was markedly decreased at the high loading rates (1 to 10 in/in/sec). These tests were conducted at light water reactor conditions of 288 C where the carbon steels were susceptible to dynamic strain aging. However, it may still be inferred that a marked decrease in fracture load (or fracture toughness) at a much higher high load rates (e.g., above 1 in/in/sec) may be expected for the SWPS and T-Stub specimens here. Additional work is required to further examine the effect of strain rate of damage development and fracture of structural steels.

5 REFERENCES

Brust, F. W., et al., NUREG/CR-6098, "Loading Rate Effects on Strength and Fracture Toughness of Pipe Steels Used in Task 1 of the IPIRG Program", Prepared by Battelle for the US NRC, October 1993.

Brust, F. W., Zhang, J., and Dong, P., 1997a. "Pipe and Pressure Vessel Cracking: The Role of Weld Induced Residual Stresses and Creep Damage during Repair", _Transactions of the 14th International Conference on Structural Mechanics in Reactor Technology (SMiRT 14)_, Lyon, France, Vol. 1, pp. 297-306.

Brust, F. W., Dong, P., Zhang, J., 1997b. "Crack Growth Behavior in Residual Stress Fields of a Core Shroud Girth Weld", _Fracture and Fatigue_, H. S. Mehta, Ed., PVP-Vol. 350, pp. 391-406.

Dong, P. and Gordon, J. R., "The Effect of Weld Metal Mismatch on the Fracture Behavior of Center-Cracked Panels", 1992 ASME Pressure Vessels & Piping Conference, June 21-25, 1992, New Orleans, Louisiana, PVP Vol. 241, _Fatigue, Fracture, and Risk_, pp. 59-67.

Dong, P., Kilinski, T., Zhang, J., and Brust, F.W., 1999. "Effects of Strength/Toughness Mismatch on Structural and Fracture Behaviors in Weldments (Sub-Tasks 5.2.1 And 5.2.2), Battelle Final Report to SAC Steel Project.

Dong, P., Zhang, J., and Brust, F. W., "Fracture Mechanics Analysis of Moment Frame Joints Incorporating Welding Effects", Proceedings of STESSA 2000.

Marschall, C. W., and others, "Effect of Dynamic Strain Aging on the Strength and Toughness of Nuclear Ferritic Piping at LWR Temperatures," NUREG/CR-6226, October 1994.

Rudland D.L, Brust, F., "The Effect of Cyclic Loading During Ductile Tearing on the Fracture Resistance of Nuclear Pipe Steels," Fatigue and Fracture Mechanics: 27th Volume, ASTM STP 1296, R.S. Piascik, J.C. Newman, and N.E. Dowling Eds., American Society of Testing and Materials, 1997, pp. 406-426.

Scott, P., Olson, R., and Wilkowski, G., "The IPIRG- 1 Pipe System Fracture Tests – Experimental Results' in Fatigue, Flaw Evaluation, and Leak-Before-Break Assessments, ASME PVP Vol. 280, pp. 135-151, June 1994.

Zhang, J. and Dong, P., "Residual Stresses in Welded Moment Frames and Implications on Structural Performance," in press, ASCE _Journal of Structural Engineering_, 1999.

Behaviour of Steel Structures in Seismic Areas, Mazzolani & Tremblay (eds) © 2000 Balkema, Rotterdam, ISBN 90 5809 130 9

Cyclic behaviour of steel beam-to-column joints with a concrete slab

C.A.Castiglioni & C.Bernuzzi
Structural Engineering Department, Politecnico di Milano, Italy

L.Calado
Civil Engineering Department, Instituto Superior Tecnico, Lisboa, Portugal

ABSTRACT: The paper presents the experimental results of constant amplitude cyclic tests carried out on 10 specimens of welded beam-to-column connections, in order to characterize their low-cycle fatigue behaviour. In 5 of the 10 specimens, a concrete slab 100 mm thick is present, cast in a corrugated sheet, in order to compare the different behaviors of the steel frame alone as well as the one of the real structure, in the presence of the deck.

1 INTRODUCTION

The Northridge and Kobe earthquakes, showed that Moment Resisting Steel Frames, usually conceived as having a ductile behaviour, might suffer of unexpected brittle fractures. As a large number of these fractures were detected in welded connections of beam-to-column joints, research programs were undertaken, all over the world, in order to:
- investigate the behaviour of such types of connections under seismic loading;
- identify new connection typologies, developing a ductile behaviour under an earthquake;
- set up safe design rules for new structures in seismic areas;
- identify suitable damage assessment methods and repair procedures to be adopted in engineering practice.

The authors gathered experience in more than fifteen years of researches on the cyclic and seismic behaviour of connections in steel structures, taking part in various research programs, aimed at answering to some of the unanswered questions raised after the aforementioned earthquakes (Ballio et al. 1997)

In particular, an experimental study is presently being carried out on 10 specimens of welded beam-to-column connections made with HEB180 profiles in FE510 (columns) and IPE160 shapes in Fe430 (beams), subjected to constant amplitude cyclic tests, in order to characterize their low-cycle fatigue behaviour. Although of small size, the adopted shapes are usual in the current design of low-rise buildings in southern European countries.

In 5 of the 10 specimens, a concrete slab 100 mm thick is present, cast in a corrugated sheet, in order to compare the different behaviors of the steel frame alone as well as the one of the real structure, in the presence of the deck. In fact, very few experimental results are presently available in the literature on the behaviour of connections in composite steel-concrete constructions (Bernuzzi et al. (1997)).

The research is part of a wider research program, sponsored by both the European Community and the Italian Ministry for University and Research. In fact, after characterization of the cyclic behaviour of the joints, a number of full-scale two floors, two bays frames have been designed and fabricated, and they will be tested on the shaking table facility of the Laboratory for Earthquake Engineering at the National Technical University of Athens.

The cyclic tests on the joints are presently completed at the Structural Engineering Department of Politecnico di Milano; other cyclic tests on different typologies of beam-to-column connections fabricated adopting the same steel shapes will be carried out in the near future at the Laboratory of the Instituto Superior Tecnico in Lisbon.

The shaking table tests are presently underway in Athens.

The paper presents the results of constant amplitude tests aimed at characterization of the low-cycle fatigue behaviour of the connections; the first shaking table test was carried out in Athens during January 2000, and the results are presently under re-elaboration, and will be presented in the next future.

2 EXPERIMENTAL PROGRAM

2.1 *Test Specimens*

The specimens consist of an IPE 160 beam attached to a HEB180 column by a welded connection. Two types of welds were considered: Specimens type C1 are representative of a typical site welding, with single bevel V groove welds and a backing bar; specimens type C4 are representative of a typical shop welding, with double bevel K groove welds (Fig. 1).

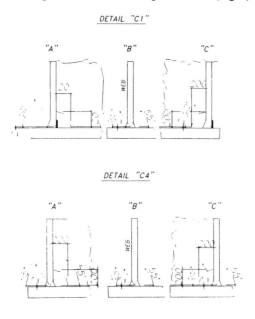

Figure 1. Welding details.

Two series of 6 specimens each were fabricated, representative of an external beam-to-column connection in a MR frame; hence the specimens were of the ⊥ type.

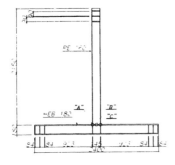

Figure 2. Steel beam-to-column connection specimen

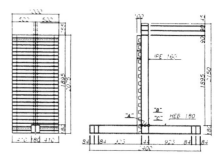

Figure 3. Specimen of steel beam-to-column connection with concrete slab.

The first series (identified as "ns") consists of the sole steel beam-to-column connection (Fig.2); the second one, (identified as "s") encompassing the presence of a 100 mm r.c. slab, cast in a corrugated steel sheeting 55 mm deep (Fig. 3). The slab was connected to the steel beam only by means of three studs positioned near the free end of the beam, only in order to prevent separation and detachment of the slab from the model. Hence, the specimens cannot be considered as acting as composite, but the presence of the slab is considered only in order to simulate real structural conditions and to prevent local buckling of the upper flange of the beam. Of the 6 specimens of each series, three were welded according to the C1 detail, the other three according to the C4 detail.

2.2 *Test set-up*

Specimens of beam-to-column connection were tested at the Laboratory for testing and Materials of the Structural Engineering Department of Politecnico di Milano, adopting an equipment (Fig. 4) designed by Ballio and Zandonini (1985).

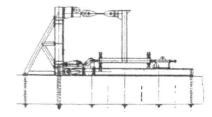

Figure 4. Testing set-up

During the test, the column is laying horizontallywhile the beam is standing in the vertical position, and subjected to an horizontal displacement imposed by a jack governed by an electric engine.

Both the imposed displacement and the corresponding horizontal load at the jack were measured, as well as, through LVDT's, the displacements at the plastic hinge location and in the panel zone.(Fig. 5)

Figure 5. Instrumentation of the specimens

Furthermore, in the case of specimens with the concrete slab, out of plane rotations of the slab were measured by means of six LVDT's positioned, three on each side, at the edges of the slab itself, spaced in the vertical direction.

2.3 Test Results

Only 5 specimens of each series have been tested at present; the sixth one was left intact, for eventual future experimental verifications. The following tables 1 and 2 summarize the experimental results for both the specimens with slab and without slab.

In the tables, Δv is the imposed cycle amplitude, v_y^s and F_y^s are respectively the yield displacement and the yield strength in the case of negative moment (concrete in tension), while v_y^c and F_y^c are respectively the yield displacement and the yield strength in the case of positive moment (concrete in compression).

Table 1. Test results for specimens with r.c. slab

Specimen	Δv [mm]	v_y^s [mm]	F_y^s [kN]	v_y^c [mm]	F_y^c [kN]	$N_{tot.}$
C4-s-3	350	38.8	19.1	35.6	34.1	4
C4-s-2	200	36.2	22.9	30.0	29.1	17
C1-s-1	160	37.6	22.2	31.3	29.9	37
C1-s-2	140	35.8	22.9	30.6	29.2	56
C1-s-3	120	39.1	20.7	32.4	33.9	57

The following Figure 6 shows a comparison between the hysteresis loops for the two specimens C4-S-2 and C1-ns-1, that were tested under cycles of the same amplitude (200 mm). It can be noticed

that, specimen c4s2, in the first cycles, when the concrete in compression is undamaged, shows a larger strength than specimen c1ns1 (made of steel only).

Table 2 Test results for specimens without r.c. slab

Specimen	Δv [mm]	v_y^s [mm]	F_y^s [kN]	$N_{tot.}$
C1-ns-2	300	41.4	18.6	5
C1-ns-3	240	43.3	18.8	22
C1-ns-1	200	43.0	19.1	25
C4-ns-1	160	39.9	18.4	51
C4-ns-2	120	41.6	18.8	73

There is also an evident difference between the stiffness of specimen c4-s-2 between the case of positive and negative bending. However, immediately after the first large cycle beyond the elastic limit, the concrete slab cracked in compression (Figure 7), and the specimen's strength shows an evident reduction becoming comparable with that of specimen c1-ns-1 (without the concrete slab).

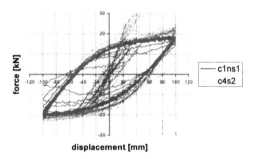

Figure 6. Comparison of hysteresis loops for specimens with and without concrete slab.

Figure 7. Damage in the concrete slab.

The following Figure 8 shows a comparison among the various specimens tested, in terms of dissipated energy per cycle (E) normalized over the value of the energy dissipated in the first stable cycle in the plastic range (E_0).

It can be noticed that, in non dimensional terms, the specimens with the concrete slab show a clear deterioration in the first three to five cycles, followed by a stabilization and a rapid final stage of deterioration, leading to collapse. On the contrary, the specimens without the concrete slab do not evidence the initial deterioration stage, but show a rather stable behaviour, until a final stage is reached when rapid deterioration takes place, leading to collapse. These considerations can be applied to all the specimens, except to those (c4-s-3 and c1-ns-2) tested under very large amplitudes, that do not present any stabilization stage, but collapse in a few cycles.

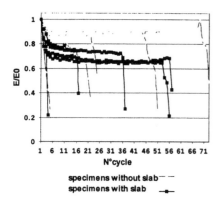

Figure 8. Comparison of energy dissipation capacity for specimens with and without concrete slab

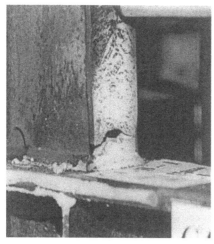

Figure 9. Typical failure mode, with weld cracking

From the same Figure 8, it is also evident that practically all the specimens tested collapsed in a "brittle" failure mode, that is without showing any steady deterioration but evidencing an abrupt deterioration only one or two cycles before complete

failure, after a relatively long period in which the capacity of the specimen to dissipate energy remains nearly constant.

Figure 9 shows a typical failure mode for all the specimens tested. Independently on the presence or absence of the concrete slab, failure was always attained by cracking at the welded joint, with limited buckling evidence in the beam flange. This confirms the already mentioned "brittleness" of the failure mode. In fact, the rather limited slenderness ratios of the web and flanges of the beam is such that the cycle amplitudes Δv imposed to the specimens result always lower than the threshold value Δv_{Th} associated with local buckling of the beam flanges and the formation of a plastic hinge. This fact causes overloading of the welded joint and leads to its ("brittle") failure after a few cycles, as already observed for steel joints by Calado et al. (1998)

3 RE-ANALYSIS OF EXPERIMENTAL DATA

As proposed by Ballio and Castiglioni (1995) the methodology used to assess the fatigue endurance of beam-to-column connections was the S-N line approach, (Equation. 1), assuming as strain range S a parameter related to the global behaviour of the specimen (such as displacement or a rotation):

$$NS^m = K \qquad (1)$$

Both parameters S and the number of cycles to failure (N) should be clearly defined, in order to apply Equation.1 in a consistent re-elaboration of test data, in accordance with the basic assumptions of the selected damage model. The number of cycles to failure N can be identified on the basis of the failure criterion, while S can be defined with reference to the proposals presented in the literature by various authors. The slope $(-1/m)$ of the S-N line (Figure 10) as well as the value of $Log\ K$, representing the

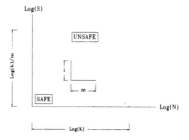

Fig. 10 – Fatigue resistance line in Log(S)-Log(N) space.

intersection of the line with the horizontal ($Log\ N$) axis can then be defined by fitting the test data plotted in a $Log\ S$-$Log\ N$ scale.

3.1. Definition of the Strain Range (S)

As discussed by Calado et al (1997), there are some relevant proposals available in literature for the definition of the strain range S. Although the various authors consider different definitions for the relevant parameter S, the same S-N line type of approach can interpret their proposals. Of course, due to these differences in the original formulations for parameter S, it is to be expected that the exponent m and the constant K of Equation.1 are different from one formulation to the other. In this research the Ballio & Castiglioni (1995) proposal was assumed. The fundamental hypothesis of this proposal is the validity of the following equation:

$$\frac{\Delta\varepsilon}{\varepsilon_y} = \frac{\Delta\delta}{\delta_y} \qquad (2)$$

where $\Delta\varepsilon$ represents the strain, $\Delta\delta$ a generalised displacement component (e.g. a displacement v, or a rotation θ), Δ is the range of variation in a cycle and the subscript y identifies yielding of the material $(\varepsilon_y = f_y/E)$ as well as conventional yielding with reference to the generalised displacement component (δ_y). According to the Ballio & Castiglioni (1995) proposal, parameter S can be identified as:

$$S = \Delta\sigma^* = \frac{\Delta\delta}{\delta_y} f_y \qquad (3)$$

$\Delta\sigma^*$ representing an effective stress range, associated with the real strain range in an ideal component made of an indefinitely linear elastic material. Hence, Equation. 1 can be re-written as:

$$N(\Delta\sigma^*)^m = K \qquad (4)$$

3.2. Definition of the Fatigue Endurance (N)

The failure criterion adopted is the *"Energy Reduction failure Criterion"*, proposed by Calado et al (1989, 1995) and is based on parameters (i.e. stiffness, strength or dissipated energy) associated with the response of the component. This criterion is characterised by a general validity for structural steel components under both constant and variable amplitude loading histories, and can be written as:

$$\frac{\eta_f}{\eta_0} = \alpha_f \qquad (5)$$

Term η_f represents the ratio of the energy absorbed by the considered component at the last cycle before collapse (E_f) to the energy that might be absorbed by the same component, in the same cycle, if the material has an elastic-perfectly plastic behaviour (E_{eppf}) while η_0 represents the same ratio but related

to the first cycle in the plastic range ($\eta_0 = E_0/E_{eppo}$). In case of variable amplitude loading histories the same criterion remains valid but should be applied with reference to the hemi-cycles, which can be defined in plastic range as the part of the hysteresis loop under positive or negative loads or as two subsequent load reversal points. In the case of constant amplitude loading, the ratio $\alpha_f = \eta_f/\eta_0 = E_f/E_0$, because $E_{eppf} = E_{eppo}$. The value of α_f which, in general should be determined by fitting the experimental result, depends on several factors (as the type of joint and the steel grade of the component). As it is particularly interesting to identify α_f a priori, in order to define a unified failure criterion for all types of steel components (members and joints), a value of $\alpha_f = 0.5$ is recommended by Calado and Castiglioni (1995) for a satisfactory and conservative appraisal of the fatigue life.

Later, Ballio, Calado and Castiglioni (1997) showed that the ratio η_f/η_0 at collapse depends on the cycle amplitude, although the available results did not allow a clear identification of the relationship.

During an extensive re-analysis of data of constant amplitude cyclic tests carried out on more than 150 specimens of both members and joints, three types of collapse were observed. Independently on the structural component (member or joint), such failure modes occurred for different values of the ductility range $\Delta v/v_y$, defined as the ratio between the imposed displacement range (Δv) and the conventional yield displacement (v_y). The observed collapse modes are:

brittle failure mode, which was usually achieved under cycles corresponding to small values of $\Delta v/v_y$. Specimen response was not affected by remarkable deterioration of strength, stiffness and/or energy absorption capability. Collapse was sudden, without warning signs, owing to a crack either at weld toes or in the base material;

ductile failure mode, generally associated with a large ductility range $\Delta v/v_y$. The component performance was significantly affected by a remarkable and progressive deterioration of the key behavioural parameters, with the formation of a plastic hinge associated with local buckling of the beam flanges. Collapse was caused by the gradual propagation of a crack initiating at surface striations forming, due to attainment of the maximum tensile strain of the material, at the buckles at the plastic hinge location.

mixed failure mode, which appears as a combination of the two previous failure modes. A progressive deterioration of the key behavioural parameters, such as stiffness, strength and dissipated energy, was observed. It was usually associated with both plastic hinge formation and local buckling phenomena. However collapse was generally due to a crack at the weld toes.

Re-analysis of constant amplitude test data showed that a suitable threshold displacement, Δv_{Th}, can be identified, separating the two different failure modes.

In absence of experimental data, an approach to estimate by means of a simple equation the value of Δv_{Th} has been developed and validated (Castiglioni et al 1997). In particular, it has been assumed that Δv_{Th} depends on the following parameters: the conventional elastic displacement, v_y; the beam web slenderness ratio $\lambda_w = d/t_w$, defined as the ratio between the depth of the profile, d, and the web thickness, t_w; the beam flange slenderness ratio $\lambda_f = c/t_f$, where c represents half width of the flange and t_f is its thickness; the weld quality and/or the severity of the detail, globally accounted for by the introduction of the numerical coefficient ξ. This term ranges from, $\xi=1.0$ for good quality (or no) welds to $\xi=0.5$ for poor quality welds.

The threshold value, Δv_{th}, has been hence defined as:

$$\Delta v_{th} = \frac{\gamma v_y}{\xi \lambda_f \lambda_w} \qquad (6)$$

As to the non-dimensional coefficient γ, on the basis of the available experimental data, a value of $2000 \pm 15\%$ (i.e., in the range 1700 - 2300) was suggested, independently on the considered component (Castiglioni et al., 1997).

As previously said, Calado and Castiglioni recommended $\alpha_f = 0.5$ for a satisfactory appraisal of the fatigue life. This value leads to a very good appraisal of the fatigue endurance. In fact, it allows a definition of the number of cycles to failure N_f in good agreement with the experimental evidence for the steel components collapsed in a "ductile" or "mixed" mode. However, such criterion is not applicable in the case of "brittle" failures, which usually occurred for values of α_f ranging between 1.0 and 0.8, but always greater than 0.5.

Hence, a new a-priori failure criterion was proposed, to extend the range of validity of the Calado and Castiglioni criterion also to "brittle" failure modes.

A conservative and satisfactory appraisal of the fatigue life can be obtained assuming (Castiglioni 1999) a value of α_f in the failure criterion determined by fitting the experimental data (Figure 11) so that all the performed tests plot below the $\alpha_f \div \Delta v/\Delta v_{th}$ line.

$$\alpha_f = 1-0.235*(\Delta v/\Delta v_{th}) \quad \text{if} \quad \Delta v/\Delta v_{th}< 0.85$$
$$\alpha_f = 1.65- (\Delta v/\Delta v_{th}) \quad \text{if} \quad 0.85<\Delta v/\Delta v_{th}< 1.15 \quad (7)$$
$$\alpha_f = 0.5 \quad \text{if} \quad \Delta v/\Delta v_{th} >1.15$$

It can be seen that, in the range $0.85< \Delta v/\Delta v_{th} <1.15$,

a linear variation of α_f in the range $0.5< \alpha_f <0.8$ is proposed; in this range of $\Delta v/\Delta v_{th}$, a mixed failure mode is to be expected.

For $\Delta v/\Delta v_{th}>1.15$, a ductile failure mode is to be expected, and a value of $\alpha_f = 0.5$ is assumed, in agreement with the Calado and Castiglioni proposal.

For $\Delta v/\Delta v_{th}<0.85$ a brittle failure mode is to be expected, and a linear variation of α_f in the range $0.8< \alpha_f <1$ is proposed.

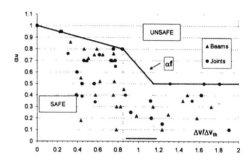

Figure 11. Definition of parameter α_f.

Of course, in the present case, the "Energy Reduction Failure Criterion" presents some uncertainty, as the value of Δv_{th} cannot be estimated experimentally because all the imposed cycles had amplitudes lower than Δv_{th} and because the approximated value given by eq. (6) applies only to steel joints, and its extension to the case of specimens with a concrete slab has to be investigated yet.

So, in order to assess the validity of such failure criterion, as well as in order to evaluate the Number of cycle to failure to adopt in a low-cycle fatigue re-analysis of the experimental data, in the following reference is also made to the "Relative Energy Drop Criterion" (Bernuzzi et al., 1997).

According to this criterion, failure occurs either when the relative energy drop, defined as:

$$\Delta W_r^{\,i} = \frac{E_0 - E_i}{E_0} \qquad (8)$$

(where E_i is the absorbed energy at i-th cycle) shows an evident increment relatively to $\Delta W_r^{\,i-1}$ or in correspondence of the last cycle of the loading history (if the energy drop is not remarkably evident).

3.3. Definition of m

Parameter m identifies the slope $(-1/m)$ of the line interpreting, in a Log-Log scale, the relationship between number of cycles to failure (N) and the stress (strain) range *(S)*. Using the Ballio & Castiglioni method it is suggested to adopt a value of *m=3*.

3.4. Low-cycle fatigue analysis

The following Tables 3, 4 and 5 summarise the tests results in terms of values for the number of cycles to failure as well as of parameter $\alpha_f = \eta_f / \eta_0 = E_f/E_0$ to be eventually assumed in the fatigue analysis for the various specimens, according to the previously described failure criteria.

N_{tot} is the total number of cycles experimentally imposed to the specimen N_f is the number of cycles to failure according to the *"Relative Energy Drop"* criterion, N_α is the number of cycles to failure according to the *"Energy Reduction Failure Criterion"*, α_f is the value assumed by parameter α at failure (experimentally defined) while α is the value for the ratio $\eta_f / \eta_0 = E_f/E_0$ estimated according to equation (7)

Table 3 Specimens without slab

Spec.	Δv[mm]	N_{tot}.	N_f	α_f
C1-ns-2	300	5	5	0.50
C1-ns-3	240	22	20	0.65
C1-ns-1	200	25	22	0.86
C4-ns-1	160	51	45	0.81
C4-ns-2	120	73	69	0.84

Table 4 Specimens without slab

Spec.	α_f	α	$N\alpha$	85%α	N85%α	N_{tot}.	N_f
C1-ns-2	0.50	0.89	1	0.75	3	5	5
C1-ns-3	0.65	0.91	4	0.77	14	22	20
C1-ns-1	0.86	0.92	4	0.78	24	25	22
C4-ns-1	0.81	0.94	3	0.79	47	51	45
C4-ns-2	0.84	0.95	1	0.81	71	73	69

Table 5 Specimens with slab

Spec.	Δv[mm]	N_{tot}.	N_f	α_f
C4-s-3	350	4	3	0.50
C4-s-2	200	17	15	0.65
C1-s-1	160	37	35	0.86
C1-s-2	140	56	53	0.81
C1-s-3	120	57	56	0.84

Examining the previous tables, it can be noticed that, except for the specimens tested under the largest cycles amplitudes (C1-ns-2 and C4-s-3), at failure was always $\alpha_f > 0.5$; this confirms that, in case of brittle failures, a larger value should be assumed for α_f. It is also confirmed that, the proposal for α_f given in Equation (7) gives always results on the safe side. In the present case, however, due to the particular behaviour of the specimens, that showed a long , nearly horizontal plateau in the diagrams relating the Energy Absorption Capacity to the number of cycles

(showed in Figure 8) as well as to the uncertainties related to the definition of Δv_{th} (that was estimated not experimentally but according to Equation (6)) the definition of the number of cycles to failure associated with parameter α_f as defined from Equation (7) (indicated as $N\alpha$ in the tables) seems to be very conservative. It should however be noticed that a small variation in the values of α_f leads to very good estimates of N_α .In fact, assumung a reduction of the 15% on parameter α_f the associated values for the number of cycles to failure range within the two values N_{tot} and N_f . This confirms again the validity of the proposed failure criterion.

In order to obtain low-cycle fatigue S-N lines, and to compare results for specimens with and without r.c. slab, as the proposed failure criterion has been validated only for steel specimens, it is assumed,, as parameter N, the value N_f associated with *the Relative Energy Drop* criterion.

Figure 12 and 13shows respectively the S-N lines obtained as best fits of the experimental data and the "design" S-N lines, obtained according to the procedures proposed in by IIW (1994). In order to obtain these figures, test data were re-analysed according to the definition of S given in Equation (3).

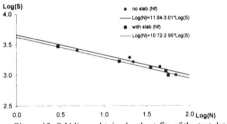

Figure 12. S-N lines obtained as best-fits of the test data

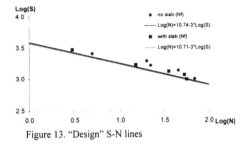

Figure 13. "Design" S-N lines

The definition of S given in Equation (3) was originally proposed for steel specimens. Hence, there is uncertainty in the values to be assumed for parameters v_y and f_y in Equation (3), in the case of specimens with the concrete slab. The data plotted in the previous Figures 12 and 13 were obtained making reference, in the case of specimens with the r.c.

slab, to the average values for v_y and f_y obtained for the specimens without slab.

In order to avoid any bias of the results, the S-N lines for the same specimens were also obtained assuming as parameter S the displacement range Δv. These lines are presented in the following Figures 14 and 15. It can be noticed that the two couple of Figures lead to similar conclusions: the low-cycle fatigue strength of the specimens with the r.c. slab is lower than the one of the specimens made of the steel profiles only.

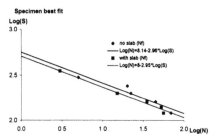

Figure 14. Best fit S-N lines assuming $S=\Delta v$.

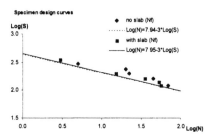

Figure 15. Design S-N lines assuming $S=\Delta v$.

This result seems to confirm and justify the unexpected brittle fractures reported in steel MR frames after the recent Northridge and Kobe earthquakes.

4 CONCLUSIONS

An experimental study was carried out on beam-to-column connections typical of MR steel frames. Some specimens consisted only of the steel profiles, others had also a r.c. slab. The specimens were subjected to cyclic quasi-static constant amplitude displacement histories. Failure was always attained in a brittle mode, by fracture of the welds between the beam and the column flange. In the presence of the slab, failure was always attained at the lower flange. Test data were re-analyzed in order to assess the low-cycle fatigue strength of these structural details, according to a proposal by Ballio and Castiglioni (1995), and adopting the failure criteria presented in this paper. It is shown that the failure criterion proposed by Castiglioni (1999) as well as the proposal

by Castiglioni et al (1997) give conservative results, that can lead to close estimates (a-priori) of the actual failure conditions, both in terms of number of cycles to failure and of energy dissipation capacity of the specimen.

Test results evidenced that the presence of the concrete slab reduces the low-cycle fatigue strength of the specimens. This fact is probably due to the increment of strains in the lower flange caused by the shift upward of the neutral axis in the presence of the concrete slab.

ACKNOWLEDGEMENT

Authors wishes to acknowledge the financial support to this research by the Italian Ministry of University and Research (MURST) Co-financed Research Program 1997, by the Italian Consiglio Nazionale delle Ricerche (CNR) and by the Portuguese Instituto por la Cooperacion Cientifica e Tecnologica Internacional (ICCTI)

REFERENCES

Ballio G., Castiglioni C.A., 1995, "A Unified Approach for the Design of Steel Structures under Low and High Cycle Fatigue", *Journal of Constructional Steel Research*, vol. 34, pp. 75-101.

Ballio G., Zandonini R., (1985) An experemental equipment to test steel structural members and subassemblages subjected to cyclic loads, Ingegneria Sismica, Vol 3, n.2

Ballio G., Calado L., Castiglioni C.A. (1997), Low cycle fatigue behaviour of structural steel members and connections, *Fatigue and Fracture of Engineering Materials and Structures*, vol. 20, n.8, pp. 1147-1157

Bernuzzi C., Calado L., Castiglioni C.A., (1997), Ductility and Load Carrying Capacity Prediction of Steel Beam-to-Column Connections under Cyclic Reversal Loading, Journal of Earthquake Engineering, vol. 1, n.2, pp.401-432

Calado L., Castiglioni C.A., 1995, "Low Cycle Fatigue Testing of Semi-Rigid Beam-to-Column Connections", Proc. of 3rd International Workshop on Connections in Steel Structures, Trento, 28-31 May, pp. 371-380

Calado L., Castiglioni C.A., Bernuzzi C., (1997) Cyclic behaviour of structural steel elements. Method for re-elaboration of test data, Proc. Of the 1st National Colloquium on Steel and composite Construction, Porto, Portugal, Nov., pp. 633-660

Calado L., Castiglioni C.A., Barbaglia P., Bernuzzi C., (1998) Procedures for the assessment of low-cycle fatigue resistance for steel connections", Proceedings of the COST Conference, Liege, September Castiglioni C.A., Bernuzzi C., Calado L., Agatino M.R., (1997), Experimental study on steel beam-to-column joints under cyclic reversal loading, Proc. Of the Northridge Earthquake Research Conference, Los Angeles, August, pag. 526-533.

Castiglioni C.A. (1999) Failure Criteria and Cumulative Damage Models for steel components under low-cycle fatigue, Proc. XVII C.T.A., Naples, October

International Welding Institute, IIW, JWG XIII-XV, 1994, "Fatigue Recommendations", Doc. XIII-1539-94/XV-845-94, September.

Zandonini R., Bernuzzi C., Bursi O., (1997), Steel and Steel-concrete composite joints subjected to seismic actions, Proc. of Behaviour of Steel Structures in Seismic Areas, Kyoto, August.

Behaviour of Steel Structures in Seismic Areas, Mazzolani & Tremblay (eds) © 2000 Balkema, Rotterdam, ISBN 90 5809 130 9

Characterization of components in steel and composite connections under cyclic loading

P.J.S.Cruz
Department of Civil Engineering, University of Minho, Guimarães, Portugal

L.Calado
Department of Civil Engineering, Instituto Superior Técnico, Lisboa, Portugal

L.Simões da Silva
Department of Civil Engineering, University of Coimbra, Portugal

ABSTRACT: An analytical description of the behaviour of a joint has to cover all sources of deformabilities, local yielding, plastic redistribution within the joint itself and local instabilities. Due to the multitude of influencing parameters, a macroscopic inspection of the complex joint - by subdividing it into components - has proved to be the most appropriate.

As the component tests are relatively cheap, a variety of influencing parameters can be covered. With the help of comprehensive parametric studies, using non-linear finite elements models, the sophisticated and relatively complex formulae describing stiffness, resistance and deformation ability of each component can be reduced to easy-to-handle formats.

Reliability techniques combined with non-linear analysis of structures will be applied to interpret the effects of parameter variability on the global structural behaviour and to quantify the risk of failure. Simplified safety rules are derived from this probabilistic high-level approach.

1. INTRODUCTION

The behaviour of the connections is extremely important for the structural design, influencing its price (Cruz 1999) and its response to both static and dynamic loads.

A considerable effort was undertaken over the past fifteen years to give consistent predictions of the behaviour of steel connections. However, until now, most research studies on the behaviour of semi-rigid joints were focused on determining resistance and stiffness characteristics, leading to the component method for the evaluation of strength and stiffness of steel and composite connections that were prepared for Eurocodes 3 and 4 (CEN 1996, 1997).

Current application of the component method is almost limited to the evaluation of the stiffness and resistance characteristics for each individual basic component. Widening the scope of this approach to the evaluation of the full force-displacement response of each component is the object of the present investigation, which is part of a wider research project on the dynamic behaviour of composite structures accounting for the real behaviour of the beam-to-column connections. This project, resulting from a partnership among the University of Coimbra, University of Minho, Instituto Superior Técnico of Lisbon, University of Beira Interior and Martifer comprises three major tasks, namely the characterisation of the dynamic behaviour at the structure level, the evaluation of the dynamic response of major and minor axis composite connections and the investigation of the behaviour of connection components. This last topic constitutes the main objective of this work.

A fundamental numerical study is as important as a detailed experimental study. A deep research of the mesh sensivity, the identification of the critical parameters and the evaluation of its variability, is crutial to evaluate the reliability of the numerical results.

The T-stub is one of the most significant components as it has been proved to model adequately parts of the connection under tension.

Comprehensive parametric studies of different T-stubs were done. Sophisticated and relatively complex formulae describing it response have been reduced to easy-to-handle formats.

2. METHOD OF UNCERTAINTY ANALYSIS

To simulate the behaviour of a system it is necessary to prescribe a set of parameters and then analyse the relations between this set and the results. The most obvious way to create such sets is the Monte-Carlo method. The disadvantage of this method is the number of computer runs needed for reaching reliable results. Since the non-linear model considered is quite complex this method would be extremely slow and would need huge computer capabilities. An alternative approach is based on a modified Monte-Carlo method. The Latin hypercube is a Monte-Carlo method type. However, by constraining the random results, it reduces the number of needed computer runs (Fig. 1).

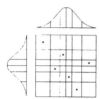

Figure 1. Latin hypercube.

In the Monte-Carlo method the samples are completely ramdom with the only restriction of having the same probabilities distribution of the simulated variables. In the Latin hypercube method the interval of possible values of each variable is divided into N disjunct intervals in order to have equal probabilities. Each interval can be represented by its centroid.

The N observations of each input variable are associated with a sequence of integers. Independent permutations of these integers are considered resulting in N input samples. For each one of these samples, simulation is carried out and N results for each output variable are obtained. Statistical parameters can be evaluated from these experiments.

The tables used in this method are purely random. However, this procedure can create statistical correlation between table columns, having a considerable influence in the final results of the simulation (Florian 1992).

In this work several sample sets were considered. The one with the smallest correlation matrix norm was chosen (Henriques et al. 1994).

3. STEEL AND COMPOSITE CONNECTIONS

3.1 Monotonic behaviour

The characteristics of the connections must be taken under consideration to accurately predict the behaviour of a structure. For assessing the behaviour of a connection three methods have been used in the past. The first and most obvious is by perform experimental tests. However, performing experimental tests is extremely expensive and time-consuming, therefore, such method is unsuitable for designing purposes.

To take advantage of all tests already done worldwide databases of experimental results were built like the one developed in Portugal at the University of Minho and the University of Coimbra under Cost Project (Cruz et al. 1998). The drawback with a database is that it needs constant management in order to be updated.

The third method is the advanced numerical modelling. However, this method may become very complex due to a multitude of phenomena, ranging from material non-linearity (plasticity, strain hardening), non-linear contact and slip, geometrical non-linearity (local instability) to residual stress conditions and complicated geometrical configurations.

Identification of the various components that constitute a connection (bolts, welds, stiffeners) gives a good picture of the complexity of its analysis. The so-called component method corresponds precisely to a simplified model composed of extensional springs and rigid links, whereby the springs (components) represent a specific part of a joint that, dependent on the type of loading, make an identified contribution to one or more of its structural properties. Recent extensive research has widened the scope of the component method from bare steel joints to steel-concrete composite joints (Anderson 1999).

Several components contribute to the overall response of a beam-to-column steel connection, namely: (i) column web in shear (ii) column web in compression, (iii) column web in tension, (iv) column flange in bending, (v) end-plate in bending, (vi) flange cleat in bending, (vii) beam flange in compression, (viii) beam web in tension or compression, (ix) plate in tension or compression, (x) bolts in tension, (xi) bolts in shear, (xii) bolts in bearing and (xiii) welds.

In a beam-to-column composite connection the layers of reinforcement are assumed to behave like bolt-rows in tension, but with different deformation characteristics. Is is assumed that full shear connection is provided and full interaction is achieved. Current pre-normative specifications in preparation (CEN 1997) already cover end-plate configurations under hogging moment subject to predominantly static loading where shear buckling of the column web is not a design criterion. However, until now, external nodes, asymmetric loading conditions where the moments can even have opposite signs from one side of the column to the other and joints with composite columns still remain quite unexplored in the literature.

Steel joints may present a variety of geometries, with different numbers of bolt rows and connecting parts. Because of this variety of configurations, joint models may range from a simple three-component model as in a welded beam-to-column connection, shown in Figure 2, to a complex extended end-plate multi bolt-row beam-to-column connection, illustrated in Figure 3.

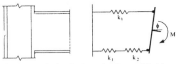

Figure 2. Mechanical model for welded beam-to-column steel connection.

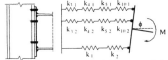

Figure 3. Mechanical model for extended end-plate beam-to-column steel connection.

Figure 4 shows the mechanical model for a beam-to-column composite joint where a single layer of reinforcing bars resists the tensile force in the joint.

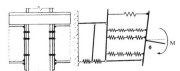

Figure 4. Mechanical model for a composite connection (COST 1996).

The evaluation of the maximum available rotation of a steel connection, essential to enable the safe utilisation of partial-resistant joints in steel construction, is currently not covered by the code specifications of EC3 (CEN 1997). There is a consensual opinion in the literature that any general approach to deal with this problem requires characterisation of the various connection components that extends well into the non-linear range.

Wide the component model to the evaluation of the full force-displacement response of each component is the object of a research project on the dynamic behaviour of composite structures. This project, comprises three major tasks, namely the characterisation of the dynamic behaviour at the structure level (Calado et al. 2000), the evaluation of the dynamic response of major and minor axis composite connections (Simões da Silva et al. 2000), and the investigation of the behaviour of various connection components. This last topic constitutes the main objective of this work.

3.2 Cyclic behaviour

Following the Northridge (1994) and Kobe (1995) earthquakes, the confidence of structural engineers in welded moment resisting connections was strongly compromised due to the extensive brittle damage detected in several frames. Starting form these observations, a great deal of theoretical and experimental research activity is presently being developed in USA, Japan and Europe on the cyclic behaviour of both welded and alternative configurations of beam-to-column connections. Since recently bolted connections, in particular top and seat with web angles connections have not been considered appropriate in seismic applications, due to the partial strength and semi-rigidity characteristics.

In this research framework, a wide experimental program on different types (welded and bolted) of beam-to-column steel and composite connections has been carried out at the Structures Test Laboratories of the Instituto Superior Técnico of Lisbon (Calado et al. 1999) and of the University of Coimbra, respectively (Simões et al. 1999). The experimental tests have been performed on specimens representative of frame structure beam-to-column joints close to the ones typical of European design practice.

In addition the possibility of extrapolating the theoretical prediction of the stiffness and the strength to the cyclic range was examined.

The experimental results obtained in this research allow todefine the collapse modes, the rotation capacity and the ultimate bending strength of bolted and welded beam-to-column connections.

4. T-STUB BEHAVIOUR

The T-stub is one of the most significant components to model adequately parts of a connection under tension, as shown in Figure 5.

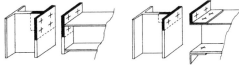

Figure 5. T-stub idealisation.

When adjacent bolt rows in the steelwork connection are subjected to tension forces, various yield mechanisms may form in the connected end plate and column flange: individual mechanisms which develop when the distance between the bolt rows are sufficiently large (Fig. 6a); group mechanisms including more than one adjacent bolt row (Fig. 6b). To these mechanisms are associated equivalent lengths of T-stubs and, through specific formulae, design resistances (CEN 1997).

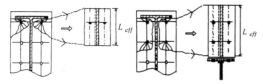

a) Individual mechanism.   b) Group mechanism.
Figure 6. Effective lengths.

According to the EC3, there are three different modes of failure. These modes can be defined as:

- Mode 1: Complete yielding of the flange (Fig. 7a);
- Mode 2: Bolt failure with yielding of the flange (Fig. 7b);
- Mode 3: Bolt failure (Fig. 7c).

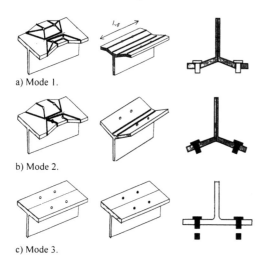

a) Mode 1.

b) Mode 2.

c) Mode 3.
Figure 7. Plate behaviour, equivalent T-stub and failure modes.

Figure 8. T-stub instrumentation (Gebbeken et al. 1998).

The knowledge and the requirements to perform reliable finite element analysis of a T-stub can be transferred to general bolt end plate connections.

At the Federal Armed Forces University of Munich some experiments on the T-stub behaviour under monotonic loading conditions were done (Gebbeken et al. 1998). The tests results have enough accuracy for a detailed and exact numerical

study. Figure 8 shows an example of a T-stub instrumentation.

Figure 9. Deformed configuration (Swanson et al. 1998).

Recently a series of 40-50 T-stubs were tested in the Georgia Institute of Technology by Swanson and Leon (1998), under cyclic tensile and compressive loads. Figure 9 illustrates one of the tested specimens.

5. FINITE ELEMENT ANALYSIS

The non-linear finite element model *DIANA* (1996) was used to simulate the non-linear behaviour of the T-stub tested by Bursi & Jaspart (1997). Three different models where considered: (a) "uni-dimensional" plane model, with beam elements, (b) "bi-dimensional" plane model, with plane stress elements (c) tri-dimensional model, with solid elements.

To reduce the complexity of the finite element model double symmetry was considered. In fact, only 1/4 of the T-stub was modelled in the plane models and 1/8 of the T-stub was modelled in the tri-dimensional model. The vertical displacements at the web plane of symmetry and the horizontal displacements at the bolt plane of symmetry are restricted.

The interaction between the two flanges was modelled with interface elements. These elements were considered by having an infinite stiffness under compression and a very small stiffness under tension. The border between the bolt and the profile has the same kind of interface elements. Both materials (for the bolt and for the profile) were modelled using a Von Mises plasticity model, considering the stress-strain values presented in Tables 1-2.

Table 1. Strain-stress values for the rolled profile steel.

$\varepsilon (\times 10^{-3})$	19	42	64	125	198	333
$\sigma (MPa)$	434.9	451.0	483.2	539.6	571.8	591.0

Table 2. Strain-stress values for bolt's steel.

$\varepsilon (\times 10^{-3})$	19	37	64	100
$\sigma (MPa)$	894.0	966.4	974.0	974.0

5.1 Uni-dimensional mesh

The mesh in Figure 10 includes 71 nodes and 51 beam elements, and as can be observed it is refined near the bolt and near the web and the flange intersection. The interaction was modelled with 4-nodes interface elements.

In order to make compatible the displacements of the different elements in contact (flange-flange and bolt-flange), the freedom degrees of the interface elements are related with the freedom degrees of the rolled section, of the bolt and of the supports. This dependence was imposed with convenient relationships between the displacements and the rotations at the axis level:

$$u_d = u_i + \theta_i \times d \qquad (1)$$

where u_d = displacement at the top or bottom levels; u_i = displacement at the axis level; θ_i = section rotation; and d = the section height.

Figure 10. Uni-dimensional mesh.

5.2 Bi-dimensional mesh

The FE type used in the analyses is the 8-nodes plane stress element. The mesh in Figure 11 includes 788 nodes and 262 elements.

Figure 11. Bi-dimensional mesh.

The connection between the two tees was simulated by overlapping two types of elements, one with the profile properties in the hole zone and the other with the bolt properties.

5.3 Tri-dimensional mesh

The FE type used in the analyses is the 20-nodes plane stress element. The mesh in Figure 12 includes 5316 nodes and 1120 elements. The interaction was modelled with 16-nodes interface elements.

Figure 12. Tri-dimensional mesh.

As could be expected this complex model was too time-consuming.

5.4 Result analysis

Figure 13 illustrates the response curves in terms of force-displacement. The comparison between the results obtained by the FEM code *DIANA* and the experimental results obtained by Bursi and Jaspart (1997) showed that the numerical models are accurate enough to analyse the behaviour of a T-stub.

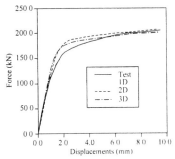

Figure 13. Force displacement curves.

The number of degrees of freedom (NDF), the precision and the CPU time of these models are compared in Table 3. The ratio between the integral of the experimental and numerical curves and the product of F_u and d_u was used to measure the precision of each model. F_u and d_u represents the force and the displacement of the last increment. This coefficient gives an adimensional measure of the error, which does not depend on the maximum displacement obtained in the analysis. For what concerns the CPU is convenient to refer that we used a personal computer having a PENTIUM II processor with 450 MHz and a RAM memory of 128 Mbytes.

Table 3. Comparison of precision and CPU times.

	NDF	CPU (min)	Error
1D Model	213	5	-0.2%
2D Model	1576	12	3.1%
3D Model	15948	960	1.1%

6. RELIABILITY ANALYSIS

6.1 Monotonic behaviour

6.1.1 Introduction

After some preliminary studies the relevant variables considered were the thickness of the flange (t_f), the flange width (b), the diameter of the bolt (d) and the size of the bolt head (e) (Fig. 14). It was considered that all these variables had a normal distribution and

159

the upper and lower limits of the parameters were 105% and 95% of the nominal value.

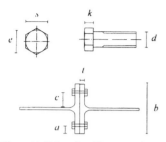

Figure 14. Bolt and profile geometries.

Afterwards the scatter of these variables was considered taking into account the dimensions and the tolerances given by the standards EN 10034 and ISO 4014. For the rolled section an upper and a lower value of the tolerances are given. It was considered that these limits were only exceeded in 0.1% of the times. The scatter of the variables was simulated by a normal distribution, $N(\mu, \sigma^2)$, where the mean, μ, and the standard deviation, σ, are given by:

$$\mu = \frac{U+L}{2} \quad ; \quad \sigma = \frac{U-L}{2 \times 3.09} \quad (2)$$

being U the upper value and L the lower value.

For some of the bolts dimensions only the nominal values and a minimal value were given in the standards. For these variables a normal distribution was also admitted, with upper tolerance equal to the lower tolerance. The other values were calculated as before.

Table 4. Dimensions of the T-Stub.

Dimension	Minimum	Nominal	Maximum
t_f (mm)	9.20	10.70	13.20
b (mm)	148.00	150.00	154.00
d (mm)	11.73	12.00	12.00
k (mm)	7.32	7.50	7.68
s (mm)	17.73	18.00	18.27

Firstly an IPE300 profile and two rows of M12 bolts were considered. The M12 bolts are grade A type. Table 4 summarises the standard dimensions adopted for bolts and rolled section. Taking into account the values defined at Table 4, the parameters which describes the distributions of the variables are evaluated according expression (2). The values obtained are indicated in Table 5.

Using the Latin hypercube method, the sample space was divided in sixty-two intervals and the representative values were considered at the centroid of each interval. Since the extreme intervals are unlimited the centroid value is too high or too low to be taken into consideration. To override this feature extreme intervals were not taken into account;

therefore, a truncated normal distribution was considered.

Table 5. Mean and standard deviation and relative dispersion of the geometric properties.

Dimension	μ	σ	σ/μ
t_f (mm)	11.20	0.32	2.89%
b (mm)	151.00	0.49	0.32%
d (mm)	11.87	0.02	0.18%
e (mm)	20.78	0.05	0.24%
k (mm)	7.50	0.03	0.39%
s (mm)	18.00	0.04	0.24%

The results stored from the sixty runs were the following: the initial stiffness, the ultimate force and the ultimate deformation. Failure occurs when the Von Mises deformation exceeds the value 0.01 in any Gauss point of the mesh.

The variability of the initial stiffness and the ultimate force was quite significant. The statistical analysis performed showed that these two variables depend mostly on the flange thickness. The failure occurs due to the yielding of the flange.

Taking into account these results, it would be interesting to analyse the same T-stub considering only changes in the bolt measures. Therefore, additional analyses were performed, considering the same properties for the profile and changing only the bolt characteristics (Table 6). The results obtained with these new values lead to other modes of failure.

Table 6. Description of the different T-stubs.

	d (mm)	e (mm)	k (mm)	A (mm)
T-stub 1	12.00	20.78	7.50	25.00
T-stub 2	8.00	14.70	5.30	15.00
T-stub 3	4.00	9.24	2.80	15.00

Figure 15 illustrates the strain patterns near failure for each model. In Fig 15a the formation of a plastic hinge in the flange without significant deformation of the bolt can be observed (T-stub 1). This phenomenon corresponds, according to EC3, to the first mode of failure. In Figure 15b the second mode can be observed, the formation of a plastic hinge with yielding of the bolt (T-stub 2). In Figure 15c the failure of the bolt occurs without significant deformation of the flange (T-stub 3).

6.1.2. Statistical analysis

6.1.2.1. Distributions with equal probability

The first results analysed correspond to the distributions with equal variability for all the parameters. The linear correlation coefficient between the parameters considered and the responses observed in finite element code were calculated. The values obtained are expressed in Table 7.

Table 7. Correlation between the results and some parameters

	Stiffness			Ultimate Load		
	t_f	b	d	t_f	b	d
T-stub 1	0.508	-0.941	0.210	0.672	-0.846	0.128
T-stub 2	0.650	-0.894	-0.167	0.612	-0.159	-0.104
T-stub 3	0.311	-0.900	0.231	0.563	-0.752	0.414

(a) – T-stub 1   (b) – T-stub 2   (c) – T-stub 3
Figure 15. Strain patterns near rupture.

After analysing these values, it was concluded that the most significant parameters were the width of the flange (b), the thickness of the flange (t_f) and the diameter of the bolt (d).

6.1.2.2. *Tolerances proposedby the ISO standards*

Performing the same type of analyses and considering the tolerances proposed by the standards mentioned before, the results obtained are quite different. The only parameter which has a significant correlation with the observed quantities is the flange thickness and, in a lower degree, the flange width. In fact, as can be observed in Table 8, there is not any significant correlation between the bolt diameter and the response variables.

Table 8. Correlation between the results and some parameters.

	Stiffness			Ultimate Load		
	t_f	b	d	t_f	b	d
T-stub 1	0.964	-0.269	0.065	0.988	-0.153	0.045
T-stub 2	0.971	-0.219	-0.091	0.988	-0.135	-0.064
T-stub 3	0.977	-0.217	0.057	0.982	-0.126	0.074

In table 9 and 10 the mean, standard deviation and relative dispersion of the results are presented. The main conclusion that can be taken from these results is that the variability of the dimensions is amplified. As a result both the ultimate force and the stiffness show a higher relative dispersion than any of the dimensions considered.

Table 9. Average, standard deviation and relative dispersion of the stiffness.

	σ	μ	σ/μ
T-stub 1	4.12	124.86	3.30%
T-stub 2	2.77	73.09	3.79%
T-stub 3	2.02	59.27	3.41%

Table 10. Average, standard deviation and relative dispersion of the ultimate force.

	σ	μ	σ/μ
T-stub 1	3.55	89.10	3.99%
T-stub 2	1.79	60.57	2.96%
T-stub 3	1.14	40.70	2.81%

6.1.2.3. *Conclusions*

Two conclusions can be taken from these results. Firstly, the variability of the response is highly conditioned by the scatter of the T-Stub dimensions. Secondly, the variability of the ultimate force increases with the diameter of the bolt. In fact, when the importance of t_f increases, the scatter of the results also increases.

6.1.3. *Proposed design rules*

As shown before the ultimate force depends on the failure mode. Expressions for the ultimate force for each of the three modes are herein presented. These expressions depend on the flange thickness, the flange width, the bolt diameter and the bolt position. The expressions obtained are quite close to the FEM results (square linear correlation factor above 98%). For the current three modes of failure, the ultimate force is given by the following equations:

- T-stub 1:

$$F_U = -40.2 \cdot b + (0.09 \cdot d)^2 + (0.32 \cdot t_f)^3 + 98 \cdot a + 78 \cdot c \quad (3a)$$

- T-stub 2:

$$F_U = 2.4 \cdot b - (0.35 \cdot d)^2 + (0.25 \cdot t_f)^3 - 5.5 \cdot a \quad (3b)$$

- T-stub 3:

$$F_U = 3.3 \cdot b + (0.81 \cdot d)^2 + (0.22 \cdot t_f)^3 - 4.8 \cdot a - 7.2 \cdot c \quad (3c)$$

Due to prying action the sum of the bolt forces, F_{Bolt}, is higher than the applied force, F_U. For the three models considered, the quotient between these two forces varies from 1.9 to 2.5. Expressions for this quotient, F_{Bolt}/F_U, are proposed.

- T-stub 1:

$$F_{BOLT}/F_U = -1.2 + \left(\frac{d}{12.8}\right)^2 - \left(\frac{t_f}{14}\right)^4 + \frac{28 \cdot c}{3 \cdot b} \quad (4a)$$

- T-stub 2:

$$F_{BOLT}/F_U = -2.1 + \left(\frac{d}{13.2}\right)^2 - \left(\frac{t_f}{13.4}\right)^4 + \frac{25 \cdot c}{2 \cdot b} \quad (4b)$$

- T-stub 3:

$$F_{BOLT}/F_U = -4.2 + \left(\frac{d}{6.3}\right)^2 - \left(\frac{t_f}{6.7}\right)^4 + \frac{16.7 \cdot c}{b} \quad (4c)$$

These expressions consider the influence of the bolt diameter, the flange thickness and the relation between the position of the bolt and the flange width (c/b). As stated before the results obtained through the expressions and FEM are quite close.

6.2 Cyclic behaviour

The T-stub 1 was modelled considering cyclic loading. After the results obtained for the monotonic model only the unidimensional mesh was considered. The properties of the T-Stub, including the dispersion of each variable, were the same used for the monotonic analysis. To model the cyclic behaviour of the T-stub a cyclic displacement was applied according to Figure 16.

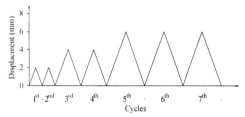

Figure 16. Loading sequence in cyclic modelling.

For each of the sixty models a force-displacement diagram was obtained. One of such diagrams is presented in Figure 17.

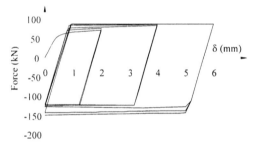

Figure 17. Example of force-displacement diagram

Two measures of the T-stub behaviour under cyclic load were considered: deformation energy and the residual force. In each cycle the dissipated energy is given by the area of the force-displacement diagram, the residual force is the force applied for at the end of a cycle (null displacement).

Table 11. Average, standard deviation and relative dispersion of the dissipated energy.

	μ (N.m)	σ (N.m)	σ/μ
1st cycle	294.4	6.3	2.12%
2nd cycle	239.7	3.0	1.26%
3rd cycle	674.7	21.0	3.11%
4th cycle	692.9	21.2	3.06%
5th cycle	1175.8	39.7	3.37%
6th cycle	1210.5	39.7	3.28%
7th cycle	1235.1	39.1	3.17%

From the analysis of all the runs made it was concluded that the energy presented a quite significant variability but the plastic force was almost constant for all the sixty models.

In table 11 the mean, standard deviation and relative dispersion of the dissipated energy for each cycle are presented.

Comparing the results presented in table 11 with table 5 it can be concluded that the deformation energy presents a higher relative dispersion and consequently a higher variability than any of the geometric parameters. Moreover, this variability increases with the amplitude of the cycle being reduced by the repetition of each cycle.

6.2.1. Proposed design rules

From the FEM results obtained is possible to produce an expression for the energy dependent of the flange thickness, the flange width and the bolt diameter. This expression is quite close to the numerical results obtained (square linear correlation factor above 99%).

$$E = 324.64 - \left(\frac{b}{39.9}\right)^3 - \frac{d_H}{3.44} + \left(\frac{t_f}{4.99}\right)^4 \tag{5}$$

7. CONCLUSIONS

A parametric study of the T-Stub component is presented. The importance of the geometrical T-Stub properties for the ductility, resistance and stiffness were verified.

It was concluded that the flange thickness and the bolt thread diameter are the most significant parameters to describe the T-Stub behaviour. It was also confirmed that the profile characteristics play an important role to the ultimate response, as their allowed tolerances are higher than the tolerances of the bolt dimensions.

As a result of the statistical analysis performed, some practical rules for design applications were proposed.

8. ACKNOWLEDGMENTS

Finantial support from "Ministério da Ciência e Tecnologia" - PRAXIS XXI research project PRAXIS/P/ECM/13153/1998 is acknowledged.

9. REFERENCES

Anderson D.: Steel-concrete composite connections, in Jan Breckelmans and Ton Toma (eds.) Proceedings of International Conference on Steel and Composite Structures, Delft, The Nederlands, pp 5.1-5.10, 1999.

Bursi O.S. & Jaspar J.P. 1997. Calibration of a finite element model for isolated bolted end-plate steel connections, *Journal of Constructional Steel Research*, 44, 3, 225-262.

Calado, L., Simões da Silva, L.A.P. & Cruz, P.J.S. 2000. Tests results on composite frame under dynamic loading, *Behaviour of Steel Structures in Seismic Areas, Proc. intern. symp., Montreal, Canadá, 21-24 August 2000.* Rotterdam: Balkema.

Calado, L. and Mele, E. (1999) Test results on welded and bolted beam-to-column connections, 2nd Pontuguese Conference on Steel and Composite Construction, 18-19 November, Coimbra , pp. 565-575.

EN 10034. 1998. Structural steel I and H sections – Tolerances and dimensions.

CEN. 1997. Eurocode 3, Part 1.1. Rev. Annex J, Joints in building frames. Ed. Approved Draft: Jan. 1997, *European Committee for Standardization*, Document CEN/TC250/SC3. Brussels.

CEN. 1996. Eurocode 4, Part 1.1. Annex J on composite joints in building frames, *European Committee for Standardization*, Document CEN/TC250/SC4. Brussels.

CEN. 1994. ENV-1993-1, Eurocode 3: Design of steel structures - Part 1.1:General rules and rules for buildings, *European Committee for Standardisation*, Brussels.

ISO 4014. 1998. Hexagon head bolts – Product grades A and B.

COST C1. 1996. Composite steel-concrete joints in braced frames for buildings, Brussels, ISSN 1018-5593.

Cruz, P.J.S. 1999. Economic studies of steel building frames, *EPMESC VII – Computacional Methods in Engineering and Science*, eds. João Bento, E. Arantes e Oliveira and Eduardo Pereira, Elsevier, 2-5 August, Macau, pp. 615-623.

Cruz, P.J.S., Simões da Silva, L.A.P. Rodrigues, D.S. & Simões, R.A.D. 1998. Database for the Semi-Rigid Behaviour of Beam-to-Column Connections in Seismic Regions, *Journal of Constructional Steel Research*, 46:1-3, Paper No. 120.

DIANA. 1996. *Diana User's Manual, Non-linear Analysis*, Rel. 6.1; TNO Bouw.

Florian A. 1992 An efficient sampling scheme: updated Latin hypercube sampling, Probabilistic Engineering Mechanics; 7, 123-130.

Gebbeken, N., Wanzek, T. & Petersen, C. 1997. Semi-rigid connections, T-stub modelle – Versuchsbericht-, Report on experimental Investigations, Berichte aus dem Konstruktiven Ingenieurbau der Universitat der Bundeswehr Munchen, Nr. 97/2, ISSN 1431-5122.

Henriques A.A.R., Calheiros F. & Figueiras J.A. 1994. A structural safety approach applied to non-linear methods of analysis, Computational Modelling of Concrete Structures, EURO-C International Conference, Innsbruck, Austria, Pineridge; 2, 975-986.

Maciej, Z. 1997. Numeircal analysis of bolted Tee-Stub connectios, *TNO-report*, Report 97-CON-R-1123, Delft.

Neves, L.A.C., Cruz, P.J.S. and Henriques, A.A.R. 1999, Reliability analysis of steel connections components based on FEM, *Engineering Failure Analysis*, Pergamon, (submited for publication).

Simões, R. A. D., Simões da Silva, L. A. P., and Cruz, P. J. S., (1999). Experimental models of end-plate beam-to-column composite connections, 2nd European Conference on Steel Structures, 26-29 May, Praha, Czech Republic, pp. 625-629.

Simões da Silva, L.A.P., Cruz, P.J.S. & Calado, L. 2000. Dynamic behaviour of composite structures with composite connections, *Behaviour of Steel Structures in Seismic Areas. Proc. intern. symp., Montreal, Canadá, 21-24 August 2000.* Rotterdam: Balkema.

Swanson, J.A. and Leon R.T. 1998. T-stub connection componnent tests, http://www.ce.gatech.edu/~sac/.

Behaviour of Steel Structures in Seismic Areas, Mazzolani & Tremblay (eds) © 2000 Balkema, Rotterdam, ISBN 90 5809 130 9

Cyclic behavior of semi-rigid angle connections: A comparative study of tests and modeling

G. De Matteis
Department of Structural Analysis and Design, University of Naples Federico II, Italy

R. Landolfo
D.S.S.A.R., University of Chieti G. D'Annunzio, Italy

L. Calado
Department of Civil Engineering, Instituto Superior Técnico, Lisbon, Portugal

ABSTRACT: Current paper focuses on structural response of typical semi-rigid steel beam-to-column joints. In particular, bolted cleat angle connections are investigated under both monotonic and cyclic loading conditions. The study is based upon available test results coming from two important experimental campaigns. Such results are duly compared, aiming at assessing the possibility to use analytical procedures matching the cyclic moment-rotation characteristics of angle connections. In particular, a purposely developed analytical model is applied to describe the whole response of the specimen under cyclic constant deformation conditions. The main energetic parameters affecting such a response are also investigated. Finally, test results have been re-analyzed in order to identify the fatigue life endurance of investigated connections. Therefore, fatigue-life relationships for the assessment S-N lines are applied and the structural performance of tested connection in terms of low-cycle fatigue is evaluated.

1 INTRODUCTION

Beam-to-column joints are one of the most important structural component affecting the seismic behavior of steel moment resisting frames. Usually, rigid full-strength connections are employed aiming at maximizing the dissipative capacity of the structure. On the other hand, terrible earthquake events occurred at the end of last century have clearly emphasized that welded joints may exhibit brittle failure modes, detrimentally penalizing the seismic response of the whole as respect to design assumptions (De Matteis et al. 2000). As an alternative, some Authors suggest bolted connections be also used in high-seismicity zones as moment connections. In fact, their performance could be particularly favorable in some range of beam sizes, provided that adequate resistance and/or ductility are guaranteed (Leon, 1997).

Since the connecting elements are subjected to considerable plastic deformations, in case of semi-rigid joints, the connection itself constitutes the main dissipative source of the system. As a consequence, the influence of joint behavior on the structural response of the frame is more and more remarkable. In addition, such connections introduce a new source of flexibility within the structure, providing a variation of soil-to-structure interaction as respect to the ideal case where connections are fully rigid.

All the above considerations require accurate and reliable models be developed in order to correctly assess and predict the behavior of bolted semi-rigid joints. A threefold approach needs to be followed. On one hand, reliable mechanical models should be set up in order to faithfully provide strength and stiffness of connection, allowing for sophisticated static design analyses of frame structures. Then, deterioration phenomena of mechanical features arising from plastic excursions should be considered for the correct prediction of the structural behavior under seismic actions. Such an aspect is very important for performing accurate time-history dynamic analyses and may be finalized by developing suitable analytical models describing the whole cyclic response of the joint up to failure. Finally, collapse of joints must be predicted by means of befitting models, which are able to appraise the fatigue life endurance of the structural component. This is essential when joints constitute the weakest element class of the structure and collapse is based upon local criteria.

A definitive assessment of the problem has not been reached yet. In fact, while several studies have been performed allowing the moment-rotation behavior under monotonic loading to be characterized (Kishi & Chen 1990, Faella et al. 1996, 1997, EC3-Annex J 1997), the prediction of connection response under cyclic loading appears to be still difficult, appropriate constitutive relationships being not available in technical literature. Similarly, fatigue damage assessment is almost troublesome, low-cycle fatigue capacity being highly sensitive to many variables, such as number of loading reversals, amplitude of inelastic excursion and connection typology and geometry.

Therefore, at the time being, in order to be reliable enough, interpretative models should be directly developed and verified on the basis of experimental evidence. In the framing of Copernicus European project RECOS 'Reliability of moment resistant connections of steel building frames in seismic areas' (Mazzolani 1999), a co-operation between University of Naples and Instituto Superior Técnico de Lisbon has been stated. An experimental and analytical study dealing with top and seat angle connections with double web angle has been carried out aiming at assessing the influence of the column size on the cyclic behavior of joint as well as to identify the main parameters influencing the low-cycle fatigue endurance of tested beam-to-column connections (Calado et al. 1999abc, 2000).

In the current paper available test results are re-analyzed in comparison with similar experimental results carried out at State University of New York at Buffalo where the low-cycle fatigue behavior of top-and-seat angle connection has been investigated under several deformation amplitudes. This allows the direct comparison between typical American and European semi-rigid connection typologies to be stated. Then, analytical models concerned with determination of cyclic behavior and fatigue life are applied *tout-court* at both European (EU) and American (USA) results. Therefore, the reliability of such models has been verified and the possibility to set some reference values of the main parameters controlling the behavior of such kind of connections assessed.

2 RELATED WORKS

In recent years, a number of research projects were carried out aiming at investigating the structural response of this and other semi-rigid connection typologies, including flush-end-plate and extended-end-plate connections.

Main experimental tests on top and seat angle connections, with or without web angles, were carried out by Marley & Gerstle (1982), Ballio et al. (1987), Azizinamini and Radziminiski (1989), Calado & Ferreira (1995), Mander et al. (1994), Bernuzzi et al. (1996, 1997), Kukreti and Abolmaali (1999), Calado et al. (1999b, 2000). Also, De Luca et. al (1995) reviewed available experimental data, analyzing monotonic test results from several data banks. The main findings of these studies are the following: (1) Connection behaves as non-linear starting from very small deformation amplitudes; (2) Connection strength is reasonable if compared with the one provided by connected member; (3) Connection stiffness is variable in a large range, even

due to difficulties and scatters arising from its evaluation; (4) Strength and stiffness of connection is strongly affected by the position of bolts, ruling the patterns of yielding lines in cleat legs; (5) Connection hysteretic behavior is characterized by remarkable pinching effects; (6) Cycles are rather stable, showing a not-negligible energy dissipation capacity; (7) Fatigue-based plastic rotation capacity is quite large, it being higher than the one normally expected in steel frames subjected to seismic actions; (8) Angle size and number of bolts may be properly selected in order to vary the moment capacity of the joint.

Mechanical models to determine initial stiffness and moment capacity of top and seat angle connections under statically increasing monotonic loads were developed, among others, by Kishi and Chen (1990) and Faella et al. (1996, 1997). A simplified procedure based on the component method approach for this type of connections has been also included in Eurocode 3 Annex J (1997). Such methods have been applied in several circumstances. Generally, they provide adequate results, but appear to be very sensitive to some parameters, whose relevant values are not easily computable. Also, Mander et al (1994) fitted their results with mechanical models, suggesting some alternative procedures as respect to the ones cited above.

Several curve-fitting equations were worldwide applied to interpreting the non-linear monotonic response of semi-rigid connections. Thus, bi-linea, piece-wise lineal, polynomial, exponential mathematical functions were proposed, as fairly reported by Kishi and Chen (1990) and De Stefano et al (1994).

Analytical approaches describing the moment-rotation hysteretic behavior of semi-rigid connections are rather complicated. Only few models have been therefore proposed. For the sake of example, mechanical model interpreting the cyclic behavior was proposed by De Stefano et al. (1994) for double-angle connections and by Bernuzzi (1998) for top –and-seat angle connections. Instead, Kukreti and Abolmaali (1999) applied four different types of mathematical models, going from the elastic-perfectly plastic one to the fully non-linear one, to top-and-seat angle connections. But all the above models do not account for the actual degradation of mechanical properties due to repeated cyclic actions through specific procedures. A hysteretic model adaptable to several connection typologies was also presented by De Martino et al. (1984). This is based on a numerical curve of Ramberg-Osgood type and damage is assumed to be dependent on dissipated energy through some empirical rules. The procedure was cumbersome and difficult to be implemented in general purpose computer programs. Instead, a more versatile analytical model interpreting the actual behavior of beam-to-column connections was presented by Della Corte et al (1999a). Such a model allows all the main experienced phenomena to be correctly described. In particular, non-linearity, strain hardening, isotropic hardening, damage of mechanical properties, pinching of hysteretic loops were considered. The model has been successfully applied to experimental tests carried out by the authors on top-and-seat angle connections with double web angles, showing an excellent agreement in terms of both moment capacity and energy dissipated per cycle (Calado et al. 1999a). Such a model has been also implemented in computer program for dynamic frame analysis, allowing the effect of joint hysteretic behavior on the global response of the structure to be investigated (Della Corte et al.1999b). The limit of the model is that some parameters have to be necessarily set up on the basis of experimental results. But such parameters could be fixed once at all for some connection typologies.

As far as the assessment of fatigue endurance of beam-to-column connections is concerned, several methods mainly based on the S-N line approach were proposed (Krawinkler & Zohrei 1983, Mander et al, 1995). Usually these methods refer to the connection rotation (either plastic or total) to develop damage accumulation functions, which rules the fatigue performance of the component. In the main, such methods assume linear damage accumulation hypothesis (Miner's rule) and provide special counting techniques for the number of equivalent excursions in case of variable cyclic loading. The method herein adopted is the one developed in Ballio & Castiglioni (1995), by assuming as failure criterion the one stated in Calado & Castiglioni (1995).

3 TEST RESULTS

3.1 General

In the current study reference is essentially made to full scale experimental tests on top-and-seat angle connections with double web angle recently carried out by the authors at Material and Structures Test Laboratory of the Instituto Superior Técnico in Lisboa and herein indicated as EU tests. Then, as useful comparison, tests carried out by Mander et al. (1995, 1995) at Laboratory of State University of New York at Buffalo, concerning top-and seat angle connections have been considered. The latter, herein referred to as USA tests, have been also used for a further calibration of the mathematical hysteretic model presented in Della Corte et al (1999b).

3.2 EU tests

Tests dealt with different types of semi-rigid, partial-strength connections, whose geometry and size is typical of European design practice (Calado et al. 1999b, 2000).

Tested specimens consisted of IPE300 beam section joined to three different column sizes: HEB160 (specimen type BCC9), HEB200 (specimen type BCC7) and HEB240 (specimen type BCC10). As connecting angles, L section 120x120x10 was adopted for all cases. Steel grade was S 235. Both top and bottom cleats were connected by means of 4 bolts (8.8-M 16), located on two rows, on both column and beam flanges. Similarly, 3 bolts (8.8-M16), located on one bolt row only, were used for double web cleat angles, both for beam web and column flange. Continuity plates (12 mm in thickness) were used in column panel zone. Bolts were tightened with a pre-loading equal to 88 kN. Connection set up used for tests and connection geometry are depicted in Figures 1 and 2, respectively.

Both monotonic and cyclic behavior were examined. In particular, in case of cyclic loading, both fatigue constant amplitude deformation tests and variable amplitude deformation tests were performed. Three different amplitude (v_b = 37.5, 50 and 75 mm) were considered, where v_b is the displacement at the load application point on the beam. In case of BCC7 series, these amplitudes corresponded to a drift Δ = v_b / L_b equal to 9.6, 6.4 and 4.8 %, respectively, where L_b is the distance of the load application point to the external face of the upper flange of the column (Fig. 1). Due to the variation of L_b, some little scatters respect to the above values of Δ should be considered for the other two series of tested specimens.

Obtained results shown the negligible effect of the column size on the behavior of the joint. In fact, the only component influencing the whole behavior was the tensioned cleat angle in bending. Failure was always due to cracks occurred on the leg of either top or seat angle connected to the beam flange, near the toe of the fillet. A hinge line also formed at the same location, but on the leg attached to the column. Finally, hinge line also developed at the leg attached to the column in correspondence of bolt location. The double web angle did not suffer any plastic deformation, whereas, due to bearing actions,

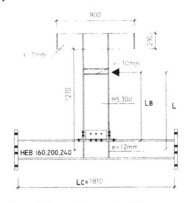

Figure 1. Connection setup for EU tests.

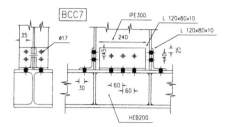

Figure 2. Geometry of EU connection.

web bolts produced the ovalization of beam web holes. Also, neither column web panel in shear, nor column web in compression, nor beam and web flange in compression provide any significant contribution to both deformability and resistance of the connection.

Hysteretic behavior was characterized by almost stable cycles, which provide a rather significant energy dissipation capability. Deterioration of stiffness and moment capacity due to repeated action was quite limited. On the other hand, connections exhibited a remarkable pinching effect, which limited the amount of energy absorbed per cycle. For the sake of example in Figure 3, the connection behavior of BCC7 specimen series for $v_b = 50$ mm is depicted. In such a diagram, according to the conventions adopted in (Calado et al. 1999b, 2000), moment M is evaluated starting from reaction force (F) with respect to center axis of the column (arm equal to L). Similarly, equivalent rotation (ϕ) is computed as ratio between applied displacement (v_b) and the arm L.

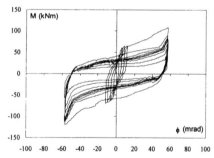

Figure 3. Typical response of EU connection (specimen BCC7-V_b=50).

1.3 USA tests

Tests performed at the University of Buffalo are concerned with top-and-seat angle connections. A series of 19 identical specimens, with the same nominal geometrical and mechanical characteristics were tested under several different cyclic loading histories. Column section was W8x31 (h_c = 203 mm, b_c = 203 mm); beam section was W8x21 (h_b = 210 mm, b_b = 134 mm); top and seat cleats were

L6x4x3/8'' (152x102x9.5 mm). All the sections were of ASTM A36 steel. Both top and bottom cleats were connected by means of 19 mm diameter A325-SC high strength bolts. Such bolts were located on two rows on the leg attached to the beam, while on one row only on the leg attached to the column. Proper flat hardened washers under both the head and the nut of the bolt were used, which, through a load indicator washers, allowed the required bolt tightening to be chosen. For this reason no slip was observed to occur in any of the tests. Local bending of the top flange of the column was prevented by means of a thick steel plate placed under the column flange at the connection. The connection test set up is shown in Figure 4, while the joint detail is depicted in Figure 5.

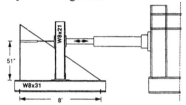

Figure 4. Connection setup for USA tests (Mander et al. 1994).

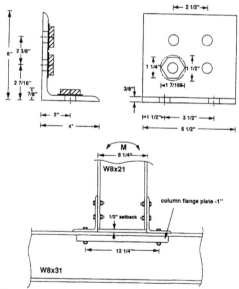

Figure 5. Geometry of USA connection (Mander et al. 1994).

Both monotonic and cyclic behavior were performed. In particular, in case of cyclic loading, both fatigue constant amplitude deformation tests and variable amplitude deformation tests were carried out. Several deformation levels (0.6%, 0.8%, 1%, 1.5%, 2%, 2.5%, 3%, 4%, 5%, 6%, 7%, 8%) were considered for constant amplitude tests, they being measured as drift $\Delta = v_b / L_b$. Also in this case, obtained results shown that the only component influencing the whole behavior was the tensioned cleat angle in bending. Failure was always due to cracks occurred on the leg of either top or seat angle connected to the beam flange, with a mechanism identical to the one described above.

Hysteretic behavior was characterized by fat cycles, which provide a more significant energy dissipation capability as respect to EU connections. In fact, pinching effect was limited, quite lacking. A somewhat moment degradation due to repeated action was noticed. For the sake of example in Figure 6, the connection behavior of R1-05 specimen, which is related to a drift equal to $v_b / L_b = 6\%$, is depicted. In such a diagram, following the conventions assumed in Mander et al. (1994), the moment M is evaluated starting from reaction force (F) with respect to the connection plane (arm equal to L_b), while the rotation (ϕ) is computed with reference to the only contribution of connection itself, i.e. without considering beam, column and web panel elastic deformation contributions.

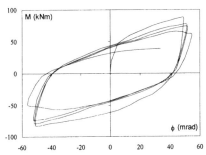

Figure 6. Typical response of USA connection (specimen R1-05-V_b/L_b=6%).

1.4 Comparison of results

The above test campaigns are related to different connection geometry, the first one being concerned with top-and-seat angles with double web angle, while the second one with top-and-seat angles only. Nonetheless, they behaved very similarly, due to the fact that in both of them either top or seat angle was definitely the weakest component. Obviously, web angles contributed to both stiffness and strength, but the hysteretic behavior of the joint was practically ruled by top-and-seat angles only. Another important distinction to be evidenced is the different bolt pretensioning: in the first case, it was determined empirically, following code provisions for bolt connections; in the second case, it was field-determined, resulting higher and more efficient. This difference was reflected on the cyclic response of the joint, the hysteretic behavior of USA connections being more highly dissipative than EU connections. Nevertheless, the comparison or results is very useful, because it allows the possibility to apply existing interpretative models to two widespread connection

typologies, exhibiting a very similar collapse mechanism, to be stated.

In Table 1, the main characteristics determined from monotonic tests are reported (Fig. 7). In such a table, results for EU connections are both refereed to each tested specimen and to the average of specimens with different column section. In order to be consistent to each other, results have been homogenized. Thus, in both cases the same conventions for both moment and rotation have been referred to. In particular, the ones chosen in Mander et al. (1994) have been taken into account, because these characterize better the sole specific behavior of the connection, avoiding to include the deformation contributions of other components.

Therefore, the displacement contribution due to connection, $v_{b,con}$ has been evaluated as:

$$v_{b,conn} = v_b - v_{b,bm} - v_{b,cl} \qquad (1)$$

where v_b is the total displacement at a given reaction force value F, $v_{b,bm}$ and $v_{b,cl}$ are the displacement contribution due to elastic deformation of beam and column, respectively. They have been determined according to the following formulas:

$$v_{b,bm} = \frac{F \cdot L_b^3}{3EI_{bm}}; \qquad v_{b,cl} = \frac{F \cdot L_c \cdot L^2}{16EI_{cl}} \qquad (2)$$

where, L_c is the total column length and EI_{bm} and EI_{cl} are the bending stiffness of the beam and column, respectively. The contribution of column web panel to the overall deformation of the joint has not been considered because panel zone was properly stiffened.

It is worthy to observe that the determination of

Table 1. Main results from monotonic tests.

TEST		M_y (kNm)	k_j (kNm/rad)		M_u (kNm)	ϕ_u (rad)
			$k_{j,sec}$	$k_{j,det}$		
EU	BCC9	44.5	39200	15170	122.5	0.115
	BCC7	45.2	44300	14710	127.5	0.115
	BCC10	42.3	30700	7150	127.6	0.110
	Average	*44.0*	*38100*	*12340*	*125.9*	*0.112*
USA		30.4	13750	8520	84.2	0.10

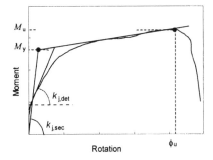

Figure 7. Conventions used for interpreting monotonic results.

stiffness from tests is a very difficult task. This is mainly due to the fact that the initial tangent is very sensitive to the irregularities of first data points, which generally provide very low deformations even for not-negligible applied loads. Therefore a lower and an upper bound may be estimated as respect to the adopted criterion. Herein two methods have been applied, they giving rise to very different results. As it is illustrated in Figure 7, the first method ($k_{j,sec}$) is related to the initial stiffness determined as secant of the first data points, as usually made in similar conditions (for instance in Mander et al. 1994). This criterion gives rise to the highest stiffness value. As an alternative, stiffness may be determined as tangent slope of the M-ϕ curve at the detachment from the moment axis, this giving rise to lower results ($k_{j,det}$) but sometimes more representative of the actual behavior of the connection, mainly when the global analysis of frame structure is aimed at. Therefore, in Table 1, the results due to both these two methods have been set out. It can be observed that they are rather different to each other, providing a wide range of possible values of connection stiffness.

The comparison between European and American connections emphasizes several analogies in global behavior, especially concerning the ultimate rotation (ϕ_u) of the connection. Stiffness and strength of EU connections are higher than the USA one. This is partially due to the contribution of web double angles, but mainly to the higher depth of beam for the EU connection ($h_b = 300$) as respect to USA connection ($h_b = 210$). In fact, beam depth is directly related to the moment arm of the resisting components of the connection. As a consequence, the ratio between beam depths of the two connections (equal to 1.43) is very similar to the ratio between corresponding elastic moment M_y (equal to 1.46) and ultimate moment M_u (equal to 1.49).

In order to compare the hysteretic behavior of the two connection typologies as well as their potential with respect to low-cycle fatigue behavior, in Figures 8 and 9, the performance obtained from cyclic constant amplitude tests is depicted. In particular, in Figure 8 the energy dissipated per cycle, normalized as respect to the total energy dissipated up to failure (W_f), is diagrammed versus the cycle number, normalized as respect to the number of cycles to failure (N_f). In case of EU connections, results are referred to specimen BCC7, i.e. the one with intermediate column section. Corresponding curves are traced for different values of deformation amplitude. Drift values are expressed as ratio between total displacement v_b and moment arm L_b.

Results are useful for inspecting the energy dissipating capability of the specimens as far as the residual fatigue life of connections is reducing. It is important to observe that the absorbed energy is rapidly decreasing after very few cycles. Then, it is almost constant up to incipient failure. This is a general trend, especially for low drift tests. Instead, for high drift tests, the dissipated energy seems to decrease linearly and rapidly up to failure. But this is due to the fact that in the latter circumstance, the cycles to failure (N_f) are few (two or three) and therefore the number of points constituting the curves is very low, i.e. the flat part of the curve is now missing.

Comparison between UE and USA connection typologies shows a very similar trend from the qualitatively point of view. This suggests the possibility to use for both cases the same model to predict the low cycle fatigue life of the connection. But, a deeper examination of relevant graphs evidences a strong similarity in quantitative terms as well. In fact, the percentage of energy dissipated for each cycle respect to the total energy W_f is similar for the two typologies. In Figure 8, results for EU connections appear lower, because it is higher the corresponding number of cycle to failure N_f for similar values of drift (Tab. 2).

Total amount of dissipated energy versus deformation amplitude is depicted in Figure 9. For both typologies, dissipated energy is higher for low drift values respect to higher drift values, i.e. in case of longer fatigue life. Furthermore, now it is apparent

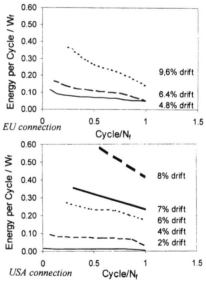

Figure 8. Energy vs. cycle diagrams.

Table 2. Number of cycles to failure (N_f)

EU connection		USA connection	
Drift (Δ %)	N_f	Drift (Δ %)	N_f
4.8	14	2	99
6.4	9	4	13
9.6	4	5	7.75
		6	4
		7	3
		8	1.75

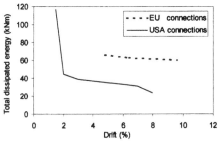

Figure 9. Dissipated energy (W_f) vs. deformation amplitude.

that the amount of energy dissipated by EU connection is higher than the one of USA connection. The ratio between the flat parts of the corresponding curves is about 1.41, which is practically correspondent to the ratio of beam depth of the two joint typologies. Anyway, also in this case, the analogy between the two connection typologies in the trend of dissipated energy versus drift amplitude is confirmed. Finally, Figure 10 shows the energy ratio α versus the number of cycle, the latter being normalized respect to N_f. α is defined as ratio between dissipated energy per cycle and the energy that might be absorbed in the same cycle if the connection had an elastic perfectly plastic behavior. Elastic displacement and plastic plateau have been obtained from corresponding cyclic tests, according to conventions indicated in Figure 7. The higher energy dissipation capability of USA connection as respect to EU connection is evident. This is due to the effect of pinching that is much more evident in the former connection typology.

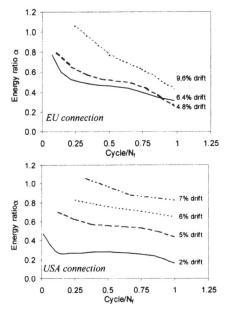

Figure 10. Normalized energy ratio (α) vs. cycle diagrams.

Such an effect is also reflected in the high slope of curves of EU connection in the range of first cycles, which means that after a first plastic excursion, subsequent loops are pinched and therefore the hysteretic energy is reduced. Eventually, it is also confirmed that in all the cases energy dissipative capability is decreasing as far as the number of cycle increases as well as that energy absorption potential is higher in case of small deformation amplitudes.

4 THE APPLICATION OF CYCLIC MODEL

A versatile analytical model interpreting the cyclic behavior of beam-to-column connections was developed by Della Corte et al. (1999). With reference to moment (M) and rotation (ϕ), the non-linear model is based on the analytical expression proposed by Richard and Abbott (1975). It is defined on the basis of four independent parameters characteristic of the monotonic curve: the initial stiffness k_0, the hardening of the system k_h, a reference force level M_0, which is assumed as the intersection of the hardening asymptotic line with the load axis, and a shape factor n, modeling the sharpness of the curve. It makes allowance for strength deterioration due to low-cycle fatigue, by means of a damage index, which is a function of the loading history in terms of number of reversals and amplitude of plastic excursions. The deterioration rate is ruled by the parameter β and the ultimate deformation ϕ_u, measured on the monotonic test. Besides, the effect of pinching is accounted, for by defining a lower bound curve, ruled by the parameters $k_{0,p}$, $M_{0,p}$, $k_{h,p}$ and n_p, and a transition curve, ruled by the parameters t_1, t_2 and λ.

The above model has been successfully applied to EU connections in Calado et al. (1999a). It appeared to be able to reproduce correctly the behavior of examined connections under different deformation history conditions. In the following, the same model is calibrated with reference to tests results of Mander et al. (1994). Qualitative comparison between experimental (dashed line) and numerical (full line) results are shown in Figure 11, for R1_03 and R1_05 tests, which are referred to a specimen drift equal to 5% and 6%, respectively. Adopted values for all model parameters are reported in indicated in Table 3. The model appears to be adequate to account for all the characteristic phenomenological aspects of examined connections. In fact, the virgin curve, the slight pinching behavior and the strength deterioration phenomenon are satisfactorily reproduced. In Figure 12, the comparison is shown in terms of both peak moment and energy dissipation per cycle. This allows us to emphasize the model capability to interpreting correctly the actual connection response in terms of the main mechanical features for each cycle and all over the deformation history. In fact, scatters are always in a reasonable limit (about 10%), indi-

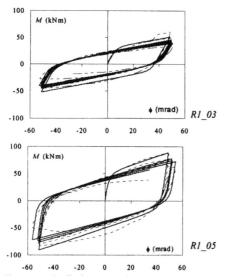

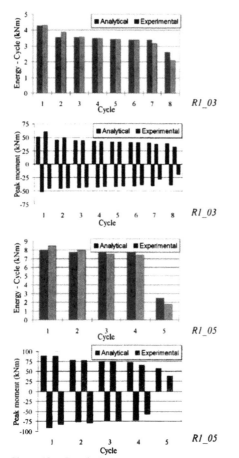

Figure 11. Qualitative comparison between experimental results (dashed line) and analytical model (full line).

Figure 12. Quantitative comparison between analytical and experimental results.

Table 3. Parameters used for the analytical model calibration

Test	M_0 (kNm)	k_0 (kNm/rad)	k_{inc} (kNm/rad)	$M_{0,p}$ (kNm)	$k_{0,p}$ (kNm/rad)	$k_{inc,,p}$ (kNm/rad)
R1_03	28.4	4260	450	24.7	4260	450
R1_05	50	16840	785	45.4	16840	675

Test	n	n_p	T_1	t_2	λ	β	$\phi_{u\,(rad)}$
R1_03	3.5	1	50	0.5	10	0.02	0.1
R1_05	1.9	1	50	0.5	10	0.03	0.1

cating the correct energy dissipation capability prediction of the model up to the collapse of the system. All the above conclusions were drawn out also for the simulation of EU connection behavior (Calado et al. 1999a) for different condition of pinching behavior. Eventually, the comparison of parameters used to calibrate the model for EU and USA connections is useful to set out some possible realistic values that could be used for the characterization of the cyclic behavior of this kind of connections for global dynamic frame analyses. In fact analyzed connection practically correspond to the two extreme conditions in terms of hysteretic behavior for top-and-seat angle typology, the EU connection being the one with utmost pinching effect and the USA connection the one with lowest amount of pinching. In particular, it can be observed that the strength deterioration rate parameter β is 0.02-0.03 for USA connection, at least for the examined tests, while it corresponds to 0.065-0.1 for EU connection, in the range of investigated deformation amplitude.

5 DETERMINATION OF FATIGUE LINES

As it is well known, one of the general purpose of the cyclic tests is to quantify the low cycle fatigue behavior of a structural element. Most common approaches to the fatigue behavior modeling can be classified into three categories, depending on the fatigue failure prediction function adopted: (1) the $S-N$ line approach, which assumes S to be the nominal stress range $\Delta\sigma_0$; (2) the local strain approach, which considers the local non-linear strain range $\Delta\varepsilon$; (3) the fracture mechanic approach, which adopts the stress intensity factor range. It is generally accepted that a modern methodology to assess the low-cycle fatigue endurance of civil engineering structures should adopt parameters related to the global structural behavior, such as displacements, rotations, bending moments, etc. Taking this into account, the $S-N$ curve approach may be adopted, by considering S parameter related to the global structural ductility (e.g., interstory drifts or joint rotations).

In order to establish the fatigue-life relationship, tests under study were re-elaborated following the Ballio & Castiglioni (1995) proposal (Eqn. 3a,b):

$$N(\Delta\sigma^*)^{m_{BC}} = K_{BC}; \quad \Delta\sigma^* = \frac{\Delta\delta}{\delta_y} f_y \qquad (3a,b)$$

where δ represents a generalized displacement component (e.g. a displacement v, or a rotation θ), Δ is the range of variation in a cycle, the subscript y identifies yielding of the material as well as conventional yielding with reference to the generalized displacement component (δ_y) and N represents the number of cycles to failure at constant stress (strain) range S. Parameter m identifies the slope $(-1/m)$ of the line interpreting, in a *Log-Log* scale, the relationship between number of cycles to failure (N) and the stress (strain) range (S). By using the Ballio & Castiglioni method it is suggested to adopt a value of $m=3$. In case of variable amplitude loading, it appears convenient to make reference to an equivalent value (S_{eq}) that can be derived from Miner's rule as the value of S at which the steel connection collapses for $N = n_{TOT}$ constant amplitude cycle at S_{eq}:

$$S_{eq} = \left(\sum_{i=1}^{L} n_i \cdot S_i^m \Big/ n_{TOT} \right)^{1/m} \tag{4}$$

The failure criterion adopted was the one proposed by Calado et al. (1998) and is based on parameters (i.e. stiffness, strength or dissipated energy) associated with the response of the component. This failure criterion has a general validity for structural steel components under both constant and variable amplitude loading histories and can be written in the following form:

$$\eta_f / \eta_0 \le \alpha \tag{5}$$

In this equation η_f represents the ratio between the absorbed energy of the considered component at the last cycle before collapse and the energy that might be absorbed in the same cycle if it had an elastic-perfectly plastic behavior, while η_0 represents the same ratio with reference to the first cycle in plastic range. The value of α, which depends on several factors, should be determined by fitting the experimental results. As it is particularly interesting to identify α *a priori*, in order to define a unified failure criterion, a value of $\alpha=0.5$ is recommended for a satisfactory and conservative appraisal of the fatigue life. In the case of variable amplitude loading histories the same criterion remains valid but should be applied on semi-cycles, which can be defined in the plastic range as the part of the hysteresis loop under positive or negative loads or as two subsequent load reversal points.

In Tables 4 and 5 the results of cyclic tests performed on the connections under study are reported in details. Best fitting straight line are depicted in Figure 13. They allow the values of slope (m) and the intercept $(\log K)$ to be properly defined for both typologies. Finally, in Figure 14 the design curves obtained according to the procedure proposed in the JWG XIII-XV - Fatigue Recommendations (1994) are shown. From these results it is evident that the low-cycle fatigue is mainly dependent of three variables: number of cycles, amplitude of deformation

and connection detail. However, as both types of connections can be considered as semi-rigid and plastic deformation is mainly concentrated in the top and seat angle connection, the fatigue strength lines S-N are similar to each other.

Table 4. Results for EU connections

EU Tests	Case	f_t (Mpa)	n_{tot}	v_y (mm)	Δv (mm)	Seq (Mpa)
	A	252.2	6	4.37	100	5771.2
BCC7	B	252.2	2	4.39	150	8617.3
	C	252.2	13	4.26	ECCS	4647.5
	D	252.2	6	4.21	75	4492.9
	A	252.2	5	4.73	100	5331.9
BCC9	B	252.2	3	4.67	150	8100.9
	C	252.2	6	4.88	75	3876.0
	D	252.2	12	4.90	ECCS	3832.6
	A	252.2	5	3.43	100	7352.8
BCC10	B	252.2	4	3.72	75	5084.7
	C	252.2	12	3.68	150	10280.0
	D	252.2	2	3.50	ECCS	9057.0

Table 5. Results for USA connections

USA tests	Case	f_t (Mpa)	n_{tot}	ϕ_y (m rad)	$\Delta\phi$ (m rad)	Seq (Mpa)
	R3	291.7	7	6.5	100	4487.7
	R4	291.7	25	2.7	500	5401.9
R	R5	291.7	4	4.0	104	7584.2
	R9	291.7	3	4.0	122	8896.9
	R10	291.7	6	3.75	65	5056.1

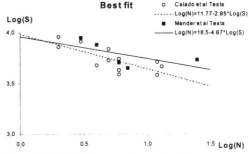

Figure 13. S-N lines for connections under study.

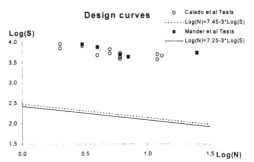

Figure 14. Design S-N lines for connections under study.

6 CONCLUSIONS

Current study allows several important conclusions to be drawn out. (1) Analyzed experimental tests have evidenced that the behavior of steel angle connections is mainly affected by bolt preloading, which determines the amount of pinching. (2) Web angles scarcely influence the energy adsorption capacity, because the major plastic deformation is concentrated into flange angles. (3) Proposed cyclic model appears to be adequate and pre-defined sets of parameters could be determined for assessing the hysteretic behavior of connections aiming at global dynamic analyses. (4) Fatigue strength lines are not influenced by the using of web angles, failure mode being similar for the two analyzed connection typologies. Therefore unique design S-N curves may be assumed.

ACKNOWLEDGMENTS

The current study has been developed with the support of EU-founded research. Besides, some activities have been carried out in the framework of the INCO-COPERNICUS Joint Research Project "*Reliability of Moment Resistant Connections of Steel Building Frames in Seismic Areas*" (*RECOS*).

REFERENCES

Azizinamini, A., & Radziminiski, J.B. 1989. Static and Cyclic Performance of Semirigid Steel Beam-to-Column Connections. *Journal of Structural Engineering*, ASCE, 115(12): 2979-2999.

Ballio G., & Castiglioni C.A. 1995. A Unified Approach for the Design of Steel Structures under Low and High Cycle Fatigue. *Journal of Constructional Steel Research*, 34: 75-101.

Ballio, G., Calado, L., De Martino, A., Faella, C. & Mazzolani, F.M. 1987. Cyclic Behavior of Steel Beam-to-Column Joints Experimental Research. *Costruzioni Metalliche*, 5: 69-88.

Bernuzzi C., Calado, L., & Castiglioni, C.A. 1997. Ductility and Load Carrying Capacity Prediction of Steel Beam-to-Column Connections under Cycling Reversal Loading. *Journal of Earthquake Engineering*, Imperial College Press, 1(2): 401-432.

Bernuzzi, C. 1998. Prediction of the behavior of top-and-seat cleated steel beam-to-column connections under cyclic reversal loading. *Journal of Earth. Eng.*, Imperial College Press, 2(1): 25-58.

Bernuzzi, C., Zandonini, R. & Zanon, P. 1996. Experimental Analysis and Modelling of Semi-Rigid Steel Joints under Cyclic Reversal Loading. *J. of Constructional. Steel Research*, 38(2): 95-123.

Calado L., & Castiglioni C.A. 1995. Low Cycle Fatigue Testing of Semi-Rigid Beam-to-Column Connections In *Connections in Steel Structures, Proc. inter. work.* Trento, 28-31 May: 371-380.

Calado, L. & Ferreira, J. 1995. Cyclic Behavior of Steel Beam-to-Column Connections – An Experimental Research, in F.M. Mazzolani and V. Gioncu (eds.) *Beh. of Steel Struc. in Seis. Areas*, STESSA'94, Timisoara, Romania, August 1994: 381-389.

Calado, L., Castiglioni, C. A. & Bernuzzi, C., 1997, Cyclic behavior of structural steel elements: method for re-elaboration of test data, *1st National Conference on Steel and Composite Construction*, ENCMM, Porto, pp 633 – 659.

Calado, C., Bernuzzi, C. e Castiglioni, C. A., 1998, Structural steel components under low-cycle fatigue: design assisted by testing", *Struct. Eng. World Congress*, SEWC,, San Francisco, CD-ROM.

Calado, L., De Matteis, G., Landolfo, R. & Mazzolani, F.M. 1999a. Cyclic Behavior of Steel Beam-to-Column Connections: Interpretation of Experimental Results. In D. Dubina and M. Ivanyi (eds.), *Stability and Ductility of Steel Structures*, SDSS'99, Timisoara, Romania, 9-11 September. Elsevier, 211-220.

Calado, L., De Matteis, G. and Landolfo, R., 1999b, "Experimental Analysis on Angle Beam-to-Column Joints under Reversal Cyclic Loading. In *Settimana della Costruzione in Acciaio*, XVII C.T.A., Vol. 1, Napoli, Italy, 3-5 October: 103-114.

Calado, L.,. Landolfo, R. and De Matteis, G. 1999c. Fracture resistance design of bolted joints. In *Construcao Metalica e Mista, Proc. nat. conf.*, Coimbra, Portugal, 18-19 November: 577-588.

Calado, L., De Matteis, G., & Landofo, R. 2000. Experimental Response of Top and Seat Angle Semi-Rigid Steel Frame Connections. Submitted for publication in *Materials and Structures*, RILEM.

Della Corte, G., De Matteis, G. & Landolfo, R. 1999a. A mathematical model interpreting the cyclic behavior of steel beam-to-column joints. In *Settimana della Costruzione in Acciaio*, XVII C.T.A., Vol. 1, Napoli, Italy, 3-5 October: 115-126.

Della Corte, G., De Matteis, G., & Landolfo, R. 1999b. Modellazione di nodi trave-colonna e risposta sismica di telai di acciaio. In *L'ingegneria Sismica in Italia*, 9° ANIDIS, Torino, 19-22 sept.

De Luca, A., De Martino, A., Pucinotti, R. & Puma, G. 1995. (Semirigid?) Top and Seat Angle Connections: Review of Experimental Data and Comparison with Eurocode 3. In *Settimana della Costruzione in Acciaio*, XV C.T.A., Riva del Garda, Italy, Oct. 1995: 315-336.

De Martino, A., Faella, C. & Mazzolani, F.M. 1988. Simulation of beam-to-column joint behavior under cyclic loads. *Costruzioni Metalliche, 6.*

De Matteis, G., Mazzolani, F.M. & Landofo, R. 2000. The behavior of connections in steel MR-frames under high intensity earthquake loading. In *Abnormal Loading on Structures*, London, 17-19 April, Rotterdam: Balkema.

De Stefano, M., De Luca, A., & Astaneh-Asl, A. 1994. Modeling of Cyclic Moment Rotation Response of Double-Angle Connections. *Journal of Structural Engineering*, ASCE, 120(1): 212-229.

Eurocode 3 - ENV 1993-1-1. 1997. *Design of Steel Structures. Annex J, Joint in Building Frames.* CEN/TC250/SC3-PT, Brussels.

Faella, C, Piluso, V & Rizzano, G. 1996 Prediction of the Flexural Resistance of Bolted Connections with Angles. In *Semi.Rigid Structural Connections*, IABSE Colloquium, *Proc. int. conf.*, Isatnbul, 309-318.

Faella, C, Piluso, V & Rizzano, G. 1997. Rotational Stiffness Prediction of Flange and Web Angle Connections with Preloaded Bolts. In *Settimana della Costruzione in Acciaio*, XVI C.T.A., Vol. 1, Ancona, Italy, 2-5 October: 316-327.

International Welding Institute, IIW, JWG XIII-XV. 1994. Fatigue Recommendations, Doc. XIII-1539-94/XV-845-94.

Kishi, N. & Chen, W.F., 1990. Moment-Rotation Relations of Semirigid Connections with Angles. *Jour. Struct. Eng.*, ASCE, 116(7): 1813-1834.

Krawinkler, H. & Zohrei, M., Cumulative Damage in Steel Structures subjected to Earthquake Ground Motions. *Journal of Computer and Structures*, Elsevier, 16 (1-4): 531-541.

Kukreti, A.R & Abolmaali, A.S. 1999. Moment-Rotation Hysteresis Behavior of Top and Seat Angle Steel Frame Connections. *Journal of Structural Engineering*, 125(8): 810-820.

Leon, R.T., 1997. Seismic Performance of Bolted and Riveted Connections. In *Program to Reduce the Earthquake Hazards of Steel Moment Frames Structures*, FEMA-288, Report No. SAC-95-09, Sacramento, California.

Mander, J.B, Chen, S.S & Pekcan, G. 1994. Low-Cycle Fatigue Behavior of Semi-Rigid Top-and-Seat Angle Connections. *Engineering Journal*, AISC, Third quarter: 111-122.

Mander, J.B, Pekcan, G & Chen, S.S. 1995. Low-Cycle Variable Amplitude Fatigue Modeling of Top-and-Seat Angle Connections. *Engineering Journal*, AISC, Second quarter: 54-62.

Marley, M.J. & Gerstle, K.H. 1982. Analysis of Tests of Flexibility-Connected Steel Frames. *AISC Project Report n. 199*, Dept. of Civil Engineering, University of Colorado, Boulder.

Mazzolani, F.M. 1999. Reliability of moment resistant connections of steel building frames in seismic areas. In *Seismic Engineering for Tomorrow, Proc. Intern. Semin.* in honour of Prof. Hiroshi Akiyama, Nov. 26, Tokyo, Japan.

Richard, R.M. & Abbott, B.J. 1975. Versatile elastic-plastic stress-strain formulas. *J. Eng. Mech. Div.*, ASCE, 101(4): 511-515.

Influence of loading asymmetry on the cyclic behaviour of beam-to-column joints

D. Dubina, A. L. Ciutina & A. Stratan
Faculty of Civil Engineering, 'Politehnica' University of Timisoara, Romania

ABSTRACT: The paper summarises the main results of the full scale tests on beam-to-column joints, carried out within the experimental programme of COPERNICUS "RECOS" European Research Project, in the Laboratory of Department of Steel Structures and Structural Mechanics of the "Politehnica" University of Timisoara. Two series of six double-sided joints with three different beam-to-column connection typologies have been tested under symmetrical and anti-symmetrical cyclic loading, respectively.

1. INTRODUCTION

The global performance of MR frames in seismic areas is strongly influenced by the beam-to-column joint properties, i.e. resistance, rigidity and plastic rotation capacity. For this reason, in university laboratories and research centres a large number of beam-to-column joints with different typology and constructional detailing, have been tested around the world (SAC 1995). But usually, these joints were either single sided or double-sided tested under symmetrical loading.

However, the anti-symmetrical loading, which is the natural action type in case of seismic motion, generates, compared with symmetrical one, important behavioural differences in case of double-sided beam-to-column joints.

The present paper summarises the results of the experimental programme carried out in the Laboratory of Steel Structures from the Department of Steel Structures and Structural Mechanics from the "Politehnica" University of Timisoara, during the year 1999, within the European Programme COPERNICUS: "Reliability of Moment Resistant Connections of Steel Building Frames in Seismic Areas" (Dubina & al., 2000). The tests were aimed to study the influence of loading asymmetry on double-sided beam-to-column joints with different connection typology.

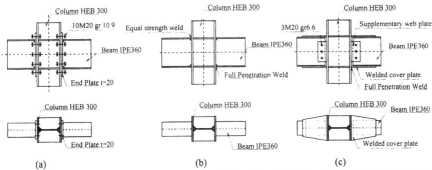

Figure 1. Connection configurations: (a) bolted with extended end-plate (EP), (b) welded (W) and (c) welded with cover and web plate (CWP).

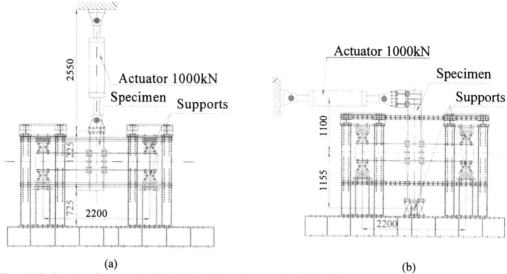

(a)   (b)

Figure 2. Testing set-up for symmetrical loading (a) and for anti-symmetrical loading (b).

2. SPECIMENS AND TESTING SET-UP DESCRIPTION

Two series of specimens, including three different connection typologies (see Figure 1) have been tested under symmetrical and anti-symmetrical loading, respectively. The design steel grade was S235.

Mechanical characteristics of joint components have been determined on samples extracted from: beam web and flange, column web and flange, end plate, web plate, cover plate and stiffeners. The tensile tests were conducted in accordance with SR EN 10002-1 (1990) code. The testing set-up for symmetrical joints (XS series) and anti-symmetrical ones (XU series) are shown in Figure 2.

The detailing for full penetration welds is shown in Figure 3. However, at a closer visual inspection on the welds of the joints, there was observed that not all the welds have been really full penetrated.

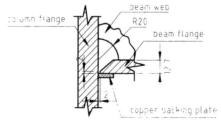

Figure 3. Detail of edge preparation.

In the case of bolted connection, the bolts have been prestressed to a moment of 620 Nm as according to Romanian Standard for prestressed bolts.

To see the influence of the welding procedure, for the XS-W1 specimen, that presented visible unconformities compared to design detailing, the sealing run of the full penetration weld between beam flanges and column was re-welded in the laboratory. The other specimen from the same series (XS-W2) remained as it was initially manufactured. Apparently, in case of XU-CWP1 specimen, the sealing run of the beam flange – to – column flange full penetration welds were rewelded by the manufacturer.

3. DATA PROCESSING

3.1 Symmetrical loading case

Since the plastic rotation was mainly expected to occur in the connections, the bending moment was computed at the column face. Starting from the actuator force P, the theoretic span between supports L and the depth of the column cross-section h_c, the bending moment can be found by the formula:

$$M=P(L-h_c)/4 \qquad (1)$$

To determine the global rotation of the joint, transducers nr. 1 and 2 are used -see Figure 4 (a), in

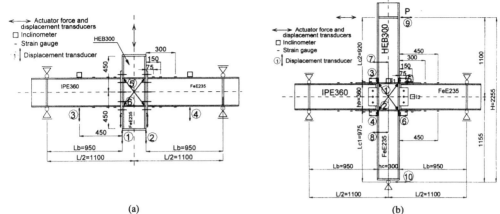

(a)   (b)

Figure 4. Instrumentation used in symmetrical (a) and anti-symmetrical (b) tests, respectively.

order to measure the displacements δ_1 and δ_2. The global rotation of the joint is given by (Grecea, 1999):

$$\phi_G = \frac{1}{L_b}\left(\frac{\delta_1 + \delta_2}{2} - \frac{PL_b{}^3}{6EI_b} - \frac{PL_b}{2Gh_b t_{wb}}\right) \qquad (2)$$

where:

L_b – the clear span of the beam, between the beam support and column flange.

E – modulus of elasticity

G – shear modulus

I_b – moment of inertia of the beam

P – actuator force

$h_b t_{wb}$ – the beam shear area

Practically, the term: $\dfrac{PL_b{}^3}{6EI_b} + \dfrac{PL_b}{2Gh_b t_{wb}}$ accounts for the elastic rotation of the beam on the length L_b, considered clumped at one end and free at the other.

3.2 Anti-symmetrical loading case

Geometry and basic instrumentation for anti-symmetrical loading is presented in Figure 4 (b). Rotations and moments are considered at the column face for all configurations. The bending moment is computed as follows:

$$M = \frac{H}{L} \cdot P \cdot L_b \qquad (3)$$

where: M – bending moment at the column face

H – column height

L – horizontal distance between beam supports

P – actuator force

L_b – clear length of the beam

Shear forces in the column web panel result from the combined action of forces in the tension and compression zones of the joint, and the shear forces resulting from the moment distribution on the adjacent sides of the column web panel. Due to these shear forces, additional deformations occur in the panel zone; even they do not represent real rotations, they lead to a change in angle between the axis of the column and the axis of the connected beam, as shown in Figure 5. Rotations that may be defined in the case of anti-symmetrically-loaded joints are:

(1) Panel zone rotation γ has two components, (γ_1 and γ_2). Due to the fact that it is difficult to estimate separately these components, they are considered equal for further calculations ($\gamma_1 = \gamma_2 = \gamma\,/2$). Overall panel zone rotation angle, γ is determined from displacement transducers 1 and 2:

$$\gamma = \frac{\sqrt{a^2 + b^2} \cdot (\delta_1 - \delta_2)}{2 \cdot a \cdot b} \qquad (4)$$

where: δ_1 and δ_2 are displacements measured by transducers 1 and 2 respectively

a and b are dimensions of the panel zone (see Figure 5a):

$$(a = h_c - t_{fc};\ b = h_b - t_{fb}) \qquad (5)$$

h_c – column depth

t_{fc} – thickness of the column flange

h_b – beam depth

t_{fb} – thickness of the beam flange

177

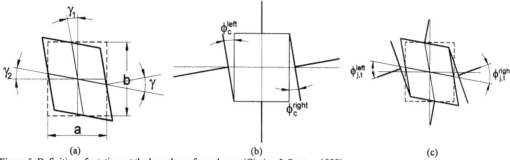

(a)   (b)   (c)

Figure 5. Definition of rotations at the boundary of panel zone (Ciutina & Stratan, 1999).

(2) Connection rotation, ϕ_c is determined from transducers 3 and 4 for the left side, and from transducers 5 and 6 for the right side (see Figure 5b):

$$\phi_c^{left} = \frac{\delta_3 - \delta_4}{b} \qquad \phi_c^{right} = \frac{\delta_6 - \delta_5}{b} \qquad (6)$$

The rotation determined in this way includes the contributions in the portion of the beam components adjacent to the column. The positive rotation is considered to be clockwise.

(3) Total joint rotation, $\phi_{j,t}$ (see Figure 5c) is the rotation between beam and column at the panel zone boundary, and is given by:

$$\phi_{j,t} = \gamma + \phi_c \qquad (7)$$

3.3 Testing Procedure

The testing procedure was made according to the Recommended Testing Procedure for Assessing the Behaviour of Structural Elements under Cycling Loading (ECCS 1996). Prior the plastic cycles, the simplified ECCS procedure was used in order to find the conventional yielding displacement e_y and the corresponding force F_y. The loading history for symmetrical and anti-symmetrical specimens is shown in Table 1.

The end of the experiment was considered when, the final force applied to the joint reached half of the maximum load. In some cases, due to premature failure of joints (XS-W1) or due to unexpected events during the test (one support felt during the XU-W2 testing), the experiment was stopped earlier.

4. TESTING RESULTS

4.1 Symmetrical loading case

XS-EP specimens

The moment vs. global rotation curve for specimen XS-EP2 is shown in Figure 6 (a), where the rotation represents the global rotation of the joint. Figure 7 (a) shows the envelope curve moment-rotation for the same specimen.

The XS-EP specimens are bolted connections with extended end-plate. Until collapse, the plastic energy was dissipated through the plastic cracking of the end-plate or beam flanges, plastic deformation of the end-plate (visible) or from the local buckling of the beam flanges, a fact that indicates that the connection components are very closely designed.

The specimen collapsed asymmetrically, due to different types of manufacturing imperfections.

Table 1. Loading history of symmetrical and anti-symmetrical specimens.

Plastic range	Number of plastic cycles (XS series)						Number of plastic cycles (XU series)					
	EP1	EP2	W1	W2	CWP1	CWP2	EP1	EP2	W1	W2	CWP1	CWP2
e_y-2e_y	4	4	5	3	3	**	7	-	4	-	4	1
±2e_y	3	3	3	3	3	**	3	3	3	3	3	3
±4e_y	4	3	3	4	3	**	3	3	3	3	3	3
±6e_y	-	6	3	-	3	**	3	3	3	3	3	3
±8e_y	-	-	2	-	9	**	18	28	17	14	3	6
±10e_y	-	-	-	-	-	**	-	-	-	-	34	15
TOTAL	11	16	16	10	21	**	34	37	30	23	50	31
Total Energy KNm rad	76.74	120.15	125.20	64.69	390.00	**	661.5	924.6	721.0	611.7	1666.8	1051.2

** accidental failure of the loading column end-plate

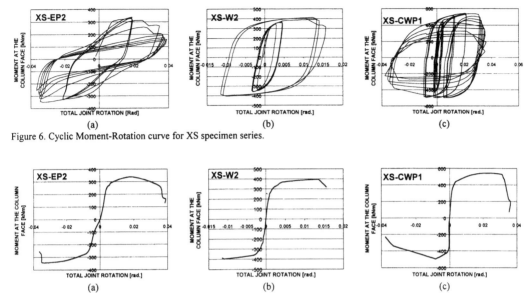

Figure 6. Cyclic Moment-Rotation curve for XS specimen series.

Figure 7. The envelope curve for XS specimen series.

The failure started by the cracks initiated from the weld of beam-to-end-plate or from the Heat-Affected Zone (HAZ), near the welds. Can also be observed a slippage due to the bolts, which permit a larger rotation of the joint. Generally, this kind of connections has good rotation capacity and energy dissipation.

XS-W specimens

The welded specimens resisted to a greater bending moment compared to the end-plate connection specimens (XS-EP), but the collapse was of brittle type. The cracks initiated in the HAZ of beam flange - to - column flange welding. It was then sudden developed on the beam depth. However, small plastic deformations have been observed during the first cycles. The behavioural curve moment-rotation of the joint is shown in Figure 6 (b), where the moment is computed at the column face, while the rotation represents the global rotation of the joint. The first cycles were asymmetrical, the actuator being unable to provide the necessary force in negative displacements (due to the limit capacity reached by the actuator). This effect was present during the entire testing of the specimen XS-W1 and can be observed in the first plastic cycles on the behavioural curve moment-rotation for specimen XS-W2.

Although the moment at the column face was significantly greater than the one obtained in the case of XS-EP specimens, the total dissipated energy remains only about half of the one in the case of end-plated joints. The envelope moment-rotation curve shows a very small softening branch, also due to premature and brittle failure - see Figure 7 (b). There have been important differences between the specimens XS-W1 and XS-W2 as follows:
- XS-W1 resisted to a larger numbers of cycles, supporting a greater maximum moment and showing a decreased stiffness
- the total energy dissipated by specimen W2 was considerable bigger (it resisted to a large number of cycles) than that of specimen W1 even if the energy dissipated per cycle was about the same.

These differences are due to the two different welding technologies used for the two specimens (see paragraph 2). The welding procedure seems to be the key-point for this type of joints, but not only.

XS-CWP specimens

These types of connections provide an increased stiffness and strength. The behavioural moment-rotation curve is shown in Figure 6 (c). At the first plastic cycles, there was also an asymmetry of loading, the actuator attaining its maximum capacity in compression.

The plastic hinge formed in the beam, at the end of the cover plate. It should be said that the connection plastic rotation is very small, the connection having behaved practically mainly within elastic range. The great values in the plastic rotation show the ductility of the beam element rather than the joint behaviour. The cyclic energy for these specimens are greater than the ones obtained for XS-EP and XS-W specimens, but this case presents the plastic energy dissipated in the beam.

For this specimen, the envelope curve - presented in Figure 7 (c) - shows a behaviour that is closed to an ideal moment-rotation curvature, with the mention that in the negative range the cut envelope curve obtained was affected by actuator limit to work in the negative range of $6e_y$ cycles

The beam flanges buckled each at the time they were in compression. It should be mentioned that the beam-to-column connections have no plastic deformations.

4.2 Anti-symmetrical Joints

Loading history for anti-symmetrical joints is presented in Table 1.

XU-EP specimens

The first signs of inelastic deformations were observed in the panel zone, where paint already started to blister at the $\pm e_y$ cycles. Plastic deformations in the panel zone increased progressively with the number of cycles. Deformations of end plate were observed starting with cycles of $\pm 2e_y$, a gap being formed between the end plate and column flange in the tension zone. First cracks in the welds between the beam bottom flange and end plate appeared at the $\pm 6e_y$, respectively $\pm 4e_y$ for the XU-EP1 and XU-EP2 specimens. Limited plastic buckling of the beam flanges was also observed. Deformation of the end plate was also given by the loosening of bolts, which decreased much the stiffness of the connection. Cracking of welds has shown at the top flange only at $\pm 8e_y$ displacement levels. After a number of plastic excursions at $\pm 8e_y$, the complete rupture of the extended part of the end plate occurred.

Panel zone showed stable hysteresis loops over the entire loading history, with an important strain hardening. It was the main source of deformation and resistance up to the point when important degradation of moment and stiffness occurred due to rupture of the end plate.

The dissipated energy per cycle began to be smaller after attainment of displacement levels of 8 e_y, and this mainly due to rupture of the extended end plate. The cyclic behaviour of the joint in terms of total joint rotation and its envelope are shown in Figure 8 (a).

XU-W specimens

Panel zone was again the weakest component. It showed important deformations (blistering of paint) at deformations levels exceeding $\pm e_y$. First cracks appeared at $\pm 6e_y$ in the sealing run of the welds between beam bottom flange and column flange. Top flange welds cracked only at the first $\pm 8e_y$ cycles. In the case of XU-W1 specimen, starting with the 9th cycle at $\pm 8e_y$ cracks in beam bottom flanges propagated progressively into the column flange, which was ruptured on the beam flange width. Column flanges were pulled out, together with the column web tearing. This phenomenon occurred first for the right beam, and continued in a smaller extent on the left side.

During successive cycles, top flanges cracked at near the weld access hole. At the end of the 16th cycle of $\pm 8e_y$ top beam flanges were completely ruptured, and bottom ones caused extensive pull out of the column flange and tearing of the web. Minor buckling of beam flanges occurred during the test.

The XU-W2 specimen showed a similar behaviour, but it was stopped in the 14th $\pm 8e_y$ cycle due to problems at one of the supports.

The shear of the panel zone brought the main contribution to the energy dissipation in the joint. Only minor plastic deformations occurred at the beam end, mainly during the last cycles, when effective rupture of the beam top flanges and column pullout at the bottom beam flanges occurred.

Cyclic behaviour of the XU-W1 specimen in terms of total joint rotation and its envelope are shown in Figure 8 (b) and Figure 9 (b) respectively.

XU-CWP specimens

Paint in the panel zone began to blister at deformation levels of $\pm e_y$. Through the loading history up to $\pm 10e_y$ panel zone continued to show increasing distortion, without any visible cracks. At the 14th cycle of $\pm 10e_y$ first cracks appeared in the welds at the lower part of bottom cover plates for the XU-CWP1 specimen. Similar cracks appeared at the XU-CWP2 specimen already at the 2nd cycle of $\pm 8e_y$.

The XU-CWP1 specimen showed a very stable energy dissipation capacity up to the 30th cycle of $\pm 10e_y$. For the other one, it begins to degrade already at the 4th cycle of $\pm 10e_y$, but the degradation is not so steep.

Cyclic behaviour of the XU-CWP1 specimen in terms of total joint rotation and its envelope are shown in Figure 8 (c) and Figure 9 (c) respectively.

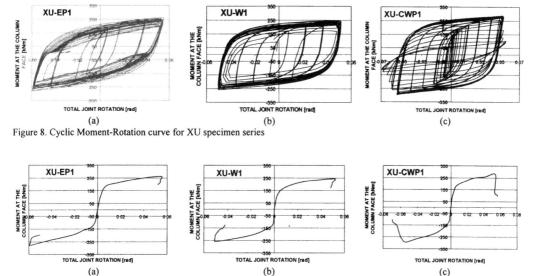

Figure 8. Cyclic Moment-Rotation curve for XU specimen series

Figure 9. The envelope curve for XU specimen series

It has to be underlined that for this type of joint and loading, plastic deformations concentrated in the panel zone. Plastic hinge did not form in the beam. Practically, the total joint rotation was the same with the rotation given by the panel zone.

5. COMPARISON BETWEEN THE EC3 AND EXPERIMENTAL RESULTS

Table 2 comprises the results of the experimental tests compared to that of EC 3 – Annex J, in terms of joint bending moments, rotational stiffness and ultimate rotation attained. It should be noted that for this comparison, the joint characteristics are computed with the measured strengths and dimensions of the joint components. The yielding bending moment $M_{Rd}^{(exp)}$ is computed according to the ECCS procedure.

Comparing the experimental and computed values of joint resistant moment, it can be observed that generally, close values are obtained for the XS series. An exception is the XS-CWP joint, which showed considerably lower experimental value. In the case of XU series, all experimental values are lower than the ones computed by Annex J of EC3. The difference between the computed and measured yielding bending moment could be explained by several causes:

- Annex J of EC3 does not consider cyclic loading neither strain hardening

- On the other hand, the procedure applied for determining the experimental yielding moments is a conventional one and is greatly influenced by the initial stiffness of the joint

In what concerns the initial stiffness, numerical and experimental results agree fairly well for the XU series, while significant differences are noticed for XS series. Anyway, stiffness is much lower for the anti-symmetrical joints both from experimental and computed stiffness values. This fact is again given by the deformability of the panel zone.

6. CONCLUDING REMARKS

As expected, the loading type (symmetrical or anti-symmetrical) significantly affects the response parameters of beam-to-column joints. The main component that introduces the difference is the panel zone in shear. The most important consequences on the cyclic behaviour of beam-column joints are the reduced moment capacity and, (in general) increased ductility with more stable hysteresis loops in the case of anti-symmetrical loading.

These tests were conducted under limiting cases of load asymmetry. The two loading types affect significantly joint properties in terms of initial stiffness, moment and rotation capacities. Therefore, when modelling the joint for structural analysis, different characteristics should be used for gravitational and lateral loading.

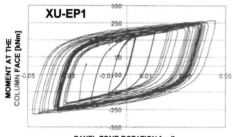

PANEL ZONE ROTATION [rad]

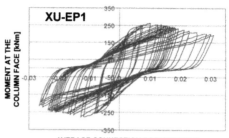

AVERAGE CONNECTION ROTATION [rad]

Figure 10. Moment-Rotation relationship for panel zone of specimen XU-EP1

Figure 11. Moment-Rotation relationships for connection of specimen XU-EP1

Table 2 Comparison between computed and experimental joint characteristics

SPECIMEN	$M_{j,Rd}^{(exp)}$	$M_{j,Rd}^{(th)}$	$S_{j,ini}^{(exp)}$	$S_{j,ini}^{(th)}$	$\phi_y^{(exp)}$	$\phi_u^{(exp)}$
	kNm	kNm	kNm/rad	kNm/rad	rad	rad
XS-EP 1	252.12	262.7	69978	142932.2	0.0036	0.031
XS-EP 2	277.05	261.3	63985	140886.8	0.0043	0.039
XS-W 1	343.14	319.4	220473	∞	0.0016	0.028
XS-W 2	281.58	320.2	291077	∞	0.001	0.017
XS-CWP 1	382.83	468.1	261697	∞	0.0014	0.036
XS-CWP 2	**	464.0	**	∞	**	**
XU-EP 1	144.4	169.2	44017	43727.2	0.0033	0.060
XU-EP 2	132.2	169.1	46713	43718.2	0.0028	0.062
XU-W 1	145.6	163.6	59429	68792.1	0.0026	0.052
XU-W 2	123.3	164.1	47794	69062.1	0.0026	0.052
XU-CWP 1	148.3	178.6	62939	75597.2	0.0024	0.064
XU-CWP 2	145.0	177.4	55411	74963.1	0.0026	0.060

Investigation of the different joint typologies revealed the importance of detailing of the connection and the welding procedure, as well as its quality. Defective welding was responsible for such phenomena as crack initiation and early cracks in welds or heat-affected zone.

Bolted end-plate joints showed an increased rotation capacity and more ductile behaviour with respect to welded joints. Extended end plate connections should be designed so as to prevent brittle failure by bolt rupture. Loosening of bolts during cycle reversals has lead to stiffness degradation. Another aspect characteristic to anti-symmetrical bolted joints is the distribution of ductility demands between the end-plate (connection) and the panel zone.

Generally, failure was brittle for welded joints and ductile for the other ones in the case of symmetrical loading. The ductile behaviour was due to connection (bolted joint) and due to shifting of the plastic hinge away from the column face in the case of CWP joint. Participation of panel zone to the plastic mechanism significantly increased the

ductility of the anti-symmetrically-loaded joints. Anyway, welded joints failed in a brittle manner in this case, too.

Generally, the joint with cover and web plates showed a good behaviour. Anyway, care should be taken when designing such joints due to potential problems caused by increased moment at the column face.

REFFERENCES

Ciutina A. & Stratan A. 1999. Cyclic behaviour of beam to column connections. *Buletinul Stiintific UPT, Tom. (44/58) /1999 (in print)*

Dubina D. & al. 1999. Influence of Connection Typology and loading asymmetry. *Final Report Copernicus "RECOS", Timisoara, Romania*

Grecea D. 1999. Caracterisasion du comportamnt sismique des ossatures metalliques – Utilisation d'assemblages a resistance partielle. *These de Doctorat, INSA Rennes, France*

SAC Joint Venture 1995. Connection test summaries. *Report No. SAC-96-02, Sacramento, California, USA*

Cyclic behaviour of bolted T-stubs: Experimental analysis and modelling

C. Faella, V. Piluso & G. Rizzano
Department of Civil Engineering, University of Salermo, Italy

ABSTRACT: The criteria and the formulations for predicting the rotational stiffness and the flexural resistance, under static loading conditions, of the most common connection typologies have been recently codified by Eurocode 3 in its Annex J (CEN, 1997) which is based on the so-called component approach. In order to extend the component approach to seismic applications, the modelling of the cyclic response of the joint components is necessary. Starting from the observation that the main sources of deformability and plastic deformation capacity of bolted connections can be modelled by means of an equivalent T-stub, an experimental program devoted to the cyclic response of the most important component of bolted connections has been planned aiming at the modelling of the cyclic force-displacement curve of bolted T-stubs. This paper presents a summary of the performed experimental tests and the description of the proposed models for predicting the cyclic behaviour of bolted T-stubs starting from the knowledge of their geometrical and mechanical properties.

1. INTRODUCTION

On the basis of the component approach, Eurocode 3 provides the rules for predicting the rotational stiffness and the flexural resistance of welded and bolted beam-to-column connections. These provisions allow the design of semirigid steel frames, under monotonic loading conditions, taking into account the actual rotational behaviour of beam-to-column joints. Conversely, regarding cyclic loading conditions, such as those occurring under seismic motion, the state of knowledge is still not sufficient to allow the prediction of the beam-to-column joint behaviour starting from their geometrical and mechanical properties.

Regarding the ability of a beam-to-column joint in providing sufficient plastic rotation supply and energy dissipation capacity for seismic design applications, the use of full-strength connections is commonly suggested. In such a case, the energy dissipation occurs at the beam ends by means of cyclic inelastic bending, so that dynamic inelastic analyses require the modelling of the cyclic response of the beams where plastic hinges develop.

However, an alternative design approach is based on dissipating the earthquake input energy through the cyclic response of the connecting elements of beam-to-column joints. In fact, an appropriate joint semirigid design can lead to a plastic rotation supply compatible with the plastic rotation demand under seismic motion. To this scope, the designer has to be aware that the overall joint ductility and energy dissipation capacity is provided by the contribution of all the components engaged in plastic range (Gobhorah et al., 1992).

It is clear that the knowledge of the joint cyclic behaviour and its modelling represents a fundamental point when the frame design is based on the dissipation of the seismic input energy in the connecting elements.

Many research programs on the cyclic behaviour of beam-to-column connections have been carried out worldwide (Bernuzzi et al., 1996; Calado, 1995; Pekcan et al., 1995; De Martino et al., 1981; Ballio et al., 1986) aiming at the identification of the behavioural parameters governing the cyclic response and at the modelling of the hysterethic behaviour (De Martino et al., 1984; De Stefano et al., 1994; Bernuzzi et al, 1996; Bernuzzi and Serafini, 1997). In addition, many efforts have been spent to analyse the low cycle fatigue (Krawinkler et al., 1971; Engelhardt and Husain, 1992, 1993; Mander et al., 1994; Bernuzzi et al., 1997; Calado et al., 1998; Bursi and Galvani, 1997; Faella et al., 1998).

Most of the mentioned works deals with the overall joint behaviour and its modelling. This approach does not allow the quantitative identification of the contribution of each component and, as a consequence, of the role played by the geometrical and mechanical parameters. A different approach can be based on the observation that the cyclic behaviour of beam-to-column joints can be predicted by properly combining the cyclic response of its basic components. This approach represents the extension to the cyclic behaviour of the component approach widely investigated in the case of monotonic loading conditions.

An accurate prediction of the joint rotational behaviour under cyclic loads, based on the component approach, requires the preliminary characterisation of the cyclic behaviour of the joint components. For this reason, an experimental program devoted to the analysis of the cyclic behaviour of the most important component of bolted connections, i.e. bolted T-stubs, has been planned. In this work, the results of this experimental program are summarised and models for predicting the cyclic response of such fundamental component are proposed.

2. EXPERIMENTAL RESULTS

The specimens of the experimental program have been designed to represent the basic components (column flange in bending and end plate in bending) of two beam-to-column joints, whose overall rotational behaviour has been tested at the University of Trento. In particular, the aim of the research program, which the University of Salerno is carring out, consists in the modelling of the cyclic behaviour of the most important joint components aiming at the prediction of the overall joint rotational behaviour starting from the knowledge of their geometrical and mechanical properties. The experimental program, herein presented, consists in the analysis of the cyclic behaviour of the components namely "column flange in bending", "end plate in bending" and "bolts in tension". The behaviour of these components has been examined by testing specimens constituted by the coupling of T elements connected through the flanges by means of two high strength bolts (class 8.8) as shown in Fig. 1. The bolt diameter is equal to 20 mm. The T elements can be divided into two groups. The first group is constituted by elements obtained from rolled profiles of steel grade Fe430 by cutting along the web plane, while the second group is based on T elements composed by welding.

The experimental program has required the testing of 28 specimens, 7 derived from an HEA180 profile (series HEA180), 7 from an HEB180 profile (series HEB180), 7 composed by welding with flange thickness equal to 12 mm (series W12) and 7 composed by welding with flange thickness equal to 18 mm (series W18). With reference to each series of specimens, 1 monotonic test, 5 constant ampli-

tude cyclic tests and 1 variable amplitude cyclic test have been carried out. All the tests have been performed under displacement control. With reference to the notation of Fig.1, the measured values of the geometrical properties of the specimens are given in Table 1.

The experimental tests have been carried out at the Material and Structure Laboratory of the Department of Civil Engineering of Salerno University. Under displacement control, all the specimens have been subjected to a tensile axial force which is applied to the webs tightened by the jaws of the testing machine, a Schenck Hydropuls S56 (maximum test load 630 kN, piston stroke ± 125 mm). In addition, coupon tensile tests have been performed to establish the mechanical properties of the material. The values of the main mechanical properties are given in Table 2.

The main purpose of the monotonic tests is the

Table 1. Gemetrical properties of specimens

Series	Test	B (mm)	b (mm)	t_f (mm)	t_w (mm)	r (a) (mm)	m (mm)	n (mm)
A: HeA 180	A1	181.25	158.75	9.71	6.78	15.0	37.39	37.85
	A2	181.75	158.25	9.68	6.83	15.0	37.23	38.24
	A3	181.25	158.25	9.79	7.05	15.0	37.15	37.95
	A4	181.50	158.25	9.85	6.70	15.0	37.58	37.82
	A5	181.00	158.50	9.80	6.75	15.0	37.13	38.00
	A6	182.00	158.25	9.74	6.63	15.0	37.74	37.94
	A7	181.50	158.75	9.79	6.63	15.0	37.51	37.93
B: HeB 180	B1	180.00	159.00	14.14	8.10	15.0	36.76	37.19
	B2	180.50	158.50	14.21	8.28	15.0	36.78	37.34
	B3	179.75	158.75	14.01	8.05	15.0	36.81	37.04
	B4	180.25	158.75	14.10	8.15	15.0	36.76	37.29
	B5	180.25	159.00	14.29	8.13	15.0	36.84	37.22
	B6	179.75	158.25	14.03	8.15	15.0	36.54	37.26
	B7	180.00	158.25	14.19	8.15	15.0	36.71	37.21
C: W12	C1	231.00	90.50	12.43	12.25	7.5	52.78	50.59
	C2	231.00	91.58	12.44	12.23	7.5	52.86	50.53
	C3	230.50	90.20	12.39	12.25	7.5	52.62	50.51
	C4	230.75	91.03	12.39	12.23	7.5	52.74	50.53
	C5	230.25	91.53	12.45	12.23	7.5	52.35	50.66
	C6	231.25	90.65	12.46	12.20	7.5	52.97	50.56
	C7	231.25	90.85	12.40	12.23	7.5	53.12	50.39
D: W18	D1	231.00	90.25	18.64	12.25	7.5	52.31	51.07
	D2	231.25	90.23	18.56	12.33	7.5	52.69	50.78
	D3	230.75	90.00	18.56	12.35	7.5	52.55	50.65
	D4	231.25	90.40	18.60	12.30	7.5	52.57	50.91
	D5	230.75	90.28	18.61	12.40	7.5	52.35	50.83
	D6	231.25	90.13	18.68	12.50	7.5	52.51	50.86
	D7	230.50	90.63	18.59	12.30	7.5	52.36	50.74

Table 2. Material mechanical properties

Series	A_o mm^2	ε_u %	f_y N/mm^2	f_u N/mm^2	C
A: HeA180	207.82	98.28	334.67	530.62	0.371
B: HeB180	106.28	109.92	280.10	464.56	0.457
C: W12	92.72	99.73	346.50	460.77	0.384
D: W18	373.13	98.32	307.34	464.94	0.364

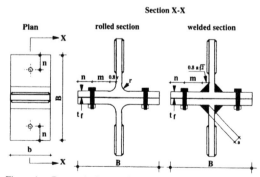

Figure 1. Geometrical properties of specimens

evaluation of the plastic deformation capacity of the specimens whose value has been used to establish the range of the amplitude values to be adopted in cyclic tests. The results of the monotonic tests of all the series of specimens are presented in Fig.2.

It is important to underline that specimens of series HEA180 and W12 have collapsed, under monotonic loading condition, according to a mechanism characterised by a significant yielding of the flanges at the flange-to-web connection zone and at the bolt axis. Despite of the collapse is due to the bolt failure which prevents the complete development of a second plastic hinge close to the bolts, it can be stated that such specimens exhibit a type-1 collapse mechanism, i.e. flange yielding, according to Eurocode 3. On the contrary, the other series of specimens, i.e. HEB180 series and W18 series, are characterised by minor yielding of the flanges and by a significant plastic engage of the bolts. In particular, with reference to the series W18, Fig. 2 shows a monotonic curve with a softening branch which corresponds to the necking of the bolts.

The application of the formulations suggested by Eurocode 3, for predicting the resistance and the collapse mechanism of bolted T-stubs, provides for such specimens a type-2 collapse mechanism, i.e. flange yielding with bolt fracture. However, it is important to underline that, according to experimental evidence, W18 specimen exhibits a type-3 collapse mechanism, i.e. bolt fracture only. This disagree is probably due to the fact that Eurocode 3 formulations do not account for the flexural engage of the bolts resulting from the deformation of the T-stub flanges which is followed by the bolts, due to compatibility.

Regarding the cyclic tests, the differences between the behaviour of the different groups of specimens are emphasized in the following. In fact, with reference to the specimens of HEA180 and W12 series, all specimens exhibited the same failure mode independently of the imposed displacement amplitude. Cracking of flanges initially developed in the central part of the flange at the flange-to-web connection zone. The number of the cycles corresponding to the development of the first cracking was dependent on the displacement amplitude of the

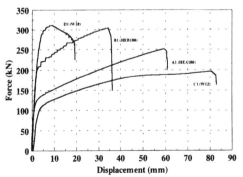

Figure 2. Monotonic Tests

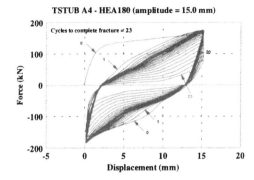

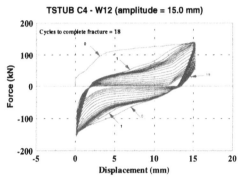

Figure 3. Examples concerning the behaviour of specimens belonging to series A and C

cyclic test, being as much greater as smaller is the displacement amplitude. By increasing the number of cycles these cracks progressively propagated towards the flange edges up to the complete fracture of one flange which produces the complete loss of the load carrying capacity. This behaviour gave rise to a progressive deterioration, up to failure, of axial strength, stiffness and energy dissipation capacity, as it is testified in Fig.3. It is important to underline that, for these series of specimens, the collapse mechanism under cyclic loading condition is different from that shown under monotonic tests, where the yielding of the flanges was accompanied by the bolt fracture. On the contrary, with reference to specimens of HEB180 and W18 series, due to relevant plastic deformations of the bolts, the cyclic behaviour is characterised by horizontal slips before reloading. During these slips the axial force is equal to zero up to the recovery of the bolt plastic deformation before reloading (Fig. 4). The failure mode of specimens HEB180-7 and W18-4 was characterised by the fracture of the bolts which, conversely, never occurred in HEA series and W12 series subjected to cyclic loading. W18-3 specimen prematurely failed due to the cracking of the web-to-flange connection which prevented the development of the typical collapse mechanism. For this reason, this specimen has not been included in the following analysis. Finally, all the other specimens failed as already described with reference to HEA and W12 series.

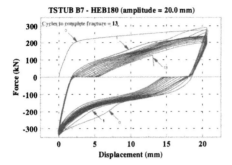

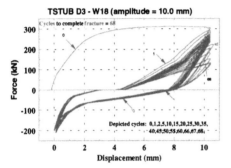

Figure 4. Examples concerning the behaviour of specimens belonging to series B and D

3. ENERGY DISSIPATION CAPACITY

With reference to the identification of the failure condition, there is not a criterion universally recognized. Generally reference is made to the deterioration of strength, stiffness or energy dissipation capacity. The influence of the failure criterion on the low cycle fatigue curve of bolted T-stubs has been analysed in a previous work by the same authors (Faella et al., 1998). Herein, a degradation of the energy dissipation capacity equal to 50% of the energy dissipated during the first cycle has been assumed as failure condition.

It can be observed that the prediction of the T-stub energy dissipation capacity under cyclic loads cannot be based on the energy dissipated in monotonic tests. This is due to the fact that the failure mode under monotonic loading conditions can be different from that occurring under cyclic loads. For this reason, the energy dissipation corresponding to the conventional failure condition has been related to the energy E_o dissipated in the monotonic test up to the achievement of a displacement amplitude corresponding to that of the cyclic test.

As the failure under cyclic loads is generally characterised by the complete fracture of one of the T-stub flanges, the corresponding monotonic failure mode is a collapse mechanism involving the T-stub flanges only. With reference to this mechanism, it has been recognized (Faella et al., 1997; 1999a) that

the theoretical value of the ultimate plastic displacement of a couple of bolted T-stubs is given by:

$$\delta_{p.th} = \frac{m^2}{t_f} \ C \tag{1}$$

where m is the distance between the bolt axis and the section corresponding to the flange-to-web connection (Fig.1), t_f is the flange thickness and C is a constant depending on the true stress-true strain curve of the material. Starting from the results of the coupon tensile tests, the values of C have been computed for each series of specimens and are given in Table 2.

The plastic part δ_p of the displacement amplitude δ of the cyclic tests can be properly expressed in nondimensional form by means of the parameter:

$$\overline{\delta} = \frac{\delta_p}{\delta_{p.th}} = \frac{\delta_p \ t_f}{2 \ C \ m^2} \tag{2}$$

Regarding the energy dissipation capacity E_{cc} exhibited under constant amplitude cyclic tests, it can be properly nondimensionalised considering the parameter:

$$\overline{E} = \frac{E_{cc}}{E_o} \tag{3}$$

On the base of these considerations and by means of a regression analysis of the experimental results regarding the specimens constituted by rolled profiles and considering also the tests on HEA160 and HEA220 performed by the same authors in a previous work (Faella et al., 1999b), the following relationship has been obtained (Fig.5):

$$\frac{E_{cc}}{E_o} = a_1 \left(\frac{t_f \ \delta_p}{2 \ C \ m^2} \right)^{-b_1} \tag{4}$$

where the coefficients a_1 and b_1 are equal to 2.081 and 1.212 with a correlation coefficient equal to 0.85.

A similar relationship could be derived for welded specimens, but these deserve further investigations due to the limited amount of experimental information.

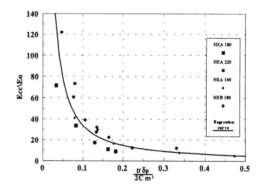

Figure 5. Nondimensional relationship between energy dissipation capacity and displacement amplitude

4. BEHAVIOURAL PARAMETERS AND MODEL

The aim of the model presented in this paper is the prediction of the cyclic behaviour of bolted T-stubs starting from the knowledge of their geometrical and mechanical properties. To this scope the monotonic force-displacement curve has to be preliminarily predicted. Such prediction can be developed by means of a theoretical approach presented by the same authors in a previous work (Faella et al., 1999a).

As soon as the monotonic behaviour has been predicted, the cyclic behaviour of bolted T-stubs can be modelled provided that the rules for strength and stiffness degradation and for the pinching of the hysteresis loops are derived. It can be observed from experimental results (Figs. 3-4) that the point corresponding to the load inversion remains practically unchanged during the loading process. These points (A and D in Fig. 6) can be identified starting from the maximum load achieved in the first cycle and by the initial stiffness. Therefore, the unloading branch is strictly identified for all the cycles, provided that the strength degradation law is known.

As already underlined in the previous sections, the cyclic behaviour of specimens collapsed, under monotonic loading, according to mechanism type-1 is different from that of specimens collapsed, under monotonic loading, according to mechanism type-2 or mechanism type-3. For this reason, the degradation laws of resistance and stiffness and the model have to be appropriately defined for each mechanism typology.

Mechanism type-1

On the basis of a regression analysis of the experimental results, for each cycle the load degradation has been related to the corresponding cumulated energy and displacement amplitude. The following relationship has been obtained:

$$\frac{F_i}{F_{max}} = 1 - a_1 \left(\frac{\delta_{max}}{2\,\delta_y}\right)^{a_2} \left(\frac{E_{ic}}{E_{cc}}\right)^{a_3} \quad (5)$$

where E_{ic} is the energy cumulated up to the i-th cycle and δ_{max} is the displacement amplitude (Fig. 6). The coefficients a_1, a_2 and a_3 are given in Tab.3.

The displacement δ_y corresponds to the limit of the elastic range and it is equal to the ratio F_y/K_o between the force corresponding to first yielding and the initial stiffness without bolt preloading which is exhibited in the experimental curve during unloading. Eq.(5) is characterised by a coefficient of correlation equal to 0.86 for rolled specimens and 0.96 for welded specimens.

It can be observed that the stiffness degradation and the pinching phenomenon are promoted by the detachment of the flanges at the bolt axis due to the plastic flexural and extensional deformation of the bolts. The reloading branch can be approximated by means of two straight lines with a different slope (Fig.6). The point C, corresponding to the intersection of the two straight lines, is approximately lined

Table 3. Coefficients of degradation laws for type-1 mechanism

	a_1	a_2	a_3	b_1	b_2	b_3
rolled sections	0.086	0.716	3.029	0.693	0.126	0.099
welded sections	0.345	0.158	3.595	0.849	0.053	0.137

up with the point A which corresponds to the inversion of the load sign and with the point B (Fig.6). Therefore, the slope of the straigth line connecting the above mentioned points (A, B and C) can be assumed equal to:

$$tg\,\alpha = \frac{F_{max}}{\delta_{max} - 2\,F_{max}/K_o} \quad (6)$$

In addition, in order to completely describe the pinching phenomenon, the knowledge of the stiffness K_i of the first part of the reloading branch is necessary (Fig. 6). On the basis of a regression analysis of the experimental results, the following relationship has been derived:

$$\frac{K_i}{K_o} = 1 - b_1 \left(\frac{\delta_{max}}{2\,\delta_y}\right)^{b_2} \left(\frac{E_{ic}}{E_{cc}}\right)^{b_3} \quad (7)$$

where the coefficients b_1, b_2 and b_3 are given in Table 3.

It is useful to observe that $2\,\delta_y$ is the threshold amplitude of δ beyond which degradation phenomena begin (Fig.6). Eq.(7) is characterised by a coefficient of correlation equal to 0.86 for rolled specimens and 0.81 for welded specimens.

As a conclusion, the modelling of the cyclic behaviour of bolted T-stubs requires the following steps:

- prediction of the monotonic force-displacement curve (Faella et al., 1999a);
- computation of the energy E_o dissipated under monotonic conditions up to the displacement δ_{max};
- estimation, through Eq. (4), of the energy dissipation capacity under cyclic action for the imposed displacement amplitude δ_{max};
- computation of the force F_{max} corresponding on the monotonic $F-\delta$ curve to the displacement amplitude δ_{max} of the imposed cyclic action;
- definition of the strength degradation rule by means of Eq. (5);

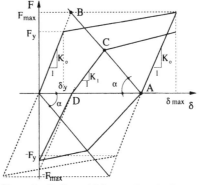

Figure 6. Cyclic model for type-1 mechanism

- definition of the stiffness degradation rule and of the pinching phenomenon by means of Eq.(7) and of the parameter α given by Eq.(6).

It can be recognized that the empirical part of the above modelling of the T-stub cyclic behaviour is constituted only by the strength and stiffness degradation rules given by Eqs. (5) and (7). The parameters K_o, F_{max} and α can be theoretically predicted starting from the theoretical prediction of the monotonic $F - \delta$ curve (Faella et al., 1997; 1999a).

In order to verify the reliability of the proposed model, the comparison with the experimental results has been performed. From the qualitative point of view, Fig.7 shows an example of the degree of accuracy of the model in predicting the cyclic behaviour. However, the reliability of the model can be better verified by means of a comparison in terms of energy dissipation. In Fig. 8, the results of this comparison are shown. It can be observed that, for all the experimental tests, the scatters between the experimental values of the energy dissipation and the ones predicted by means of the proposed model, are not particularly significant and always on the safe side.

With reference to cyclic tests with variable displacement amplitude, each test is characterised by several series of three cycles having constant amplitude. In the modelling, the passage from a series of three cycles to the following one, corresponding to a new value of the displacement amplitude, requires the evaluation of the maximum load by means of the predicted monotonic envelope and of the energy dissipation capacity by means of Eq.(4). Fig. 9 shows, from the qualitative point of view, the degree of accuracy of the model in predicting the cyclic behaviour while Fig.10 provides the comparison between the model and the experimental tests in terms of energy dissipation. Also in this case, it can be observed an acceptable degree of accuracy.

Mechanism type-2 or type-3

As already underlined, the specimens failing under monotonic loading conditions according to either mechanism type-2 or mechanism type-3 exhibit, under cyclic loading, a typical horizontal slip.

For this case, the cyclic model, depicted in Fig.11, is proposed. It is similar to that corresponding to type-1 mechanism, where the horizontal slips have been introduced and new stiffness and strength degradation laws are considered. To this scope, Eq.(4) for stiffness and Eq.(7) for strength degradation can still be adopted with the coefficients given in Table 4. The coefficients of correlation of strength degradation laws are equal to 0.76 and 0.93 for rolled and welded specimens, respectively. The corresponding coefficients of correlation of stiffness degradation laws are equal to 0.72 and 0.95, respectively. With reference to the horizontal slip, it can be observed that in the case of rolled specimens, exhibiting a type-2 mechanism, the horizontal slip increases as far as the number of cycles increases while, in the case of welded specimens, exhibiting a type-3 mechanism, the horizontal slip is practically constant. For this reason, with reference to rolled specimens, by means of a regression analysis, the following relationship has been obtained:

$$\delta_{slip} = a + b \, \ln\left(\frac{E_{ic}}{E_{cc}}\right) \qquad (8)$$

where, the coefficients a and b are given, for the unloading range, by:

$$a = 0.143 \, (\delta_{max})^{1.008}$$
$$b = 0.001 \, (\delta_{max})^{2.549} \qquad (9)$$

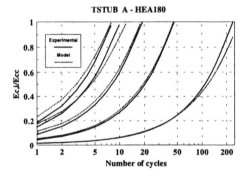

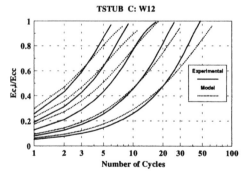

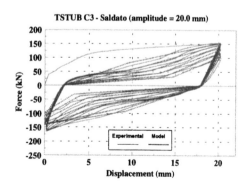

Figure 7. Modelling of the cyclic behaviour for type 1 mechanism: comparison with experimental tests

Figure 8. Accuracy of the proposed model in terms of energy dissipation capacity

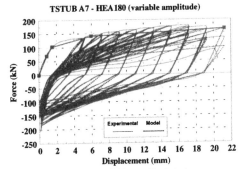

TSTUB A7 - HEA180 (variable amplitude)

Figure 9. Modelling of variable amplitude tests

while, for the reloading range, are given by:

$$a = 0.0075\ \delta_{max}^2 - 0.0306\ \delta_{max} - 0.0069 \quad (10)$$

$$b = 0.0079\ \delta_{max}^2 - 0.0989\ \delta_{max} + 0.3374$$

With reference to welded specimens, the value of the horizontal slip, independently of the energy dissipation, is given, for the unloading range, by:

$$\delta_{slip} = 0.3375\ \delta_{max} - 1.9373 \quad (11)$$

while, for the reloading range, are given by:

$$\delta_{slip} = 0.0072\ \delta_{max}^2 - 0.1373\ \delta_{max} + 0.6239 \quad (12)$$

From the qualitative point of view, the model appears sufficiently accurate (Fig.12). Moreover, the reliability of the model is also pointed out by the comparison in terms of energy dissipation (Fig.13). It can be observed that, for all the experimental tests, the scatters between the experimental values of the energy dissipation and the ones predicted by means of the proposed model are not significant.

5. CONCLUSIONS

In this work, the results of an experimental program devoted to the analysis and modelling of the cyclic behaviour of the most important components of bolted beam-to-column joints have been presented. In particular, on the basis of 28 experimental tests on

Table 4. Coefficients of degradation laws for type-2 and type-3 mechanisms

	a_1	a_2	a_3	b_1	b_2	b_3
rolled sections	0.380	0.025	1.867	0.872	0.008	0.037
welded sections	0.483	-0.168	1.089	0.805	0.098	0.020

T-stub assemblages (20 in cyclic loading conditions with constant amplitude, 4 in cyclic loading condition with variable amplitude and 4 in monotonic loading conditions), the stiffness and strength degradation laws have been derived both in the case of specimens failing, under monotonic loading conditions, according to type-1 mechanism and in the case of specimens failing, under monotonic loading conditions, according to either type-2 or type-3 mechanism. In addition, as the failure mode under monotonic loading conditions can be different from that occurring under cyclic loads, the correlation between the energy dissipation corresponding to the failure condition and the energy dissipated in monotonic conditions up to a displacement amplitude equal to that of the cyclic test has been provided. On the basis of the above analysis, semi-analitical models for predicting the cyclic behaviour of the T-stub assemblages starting from their geometrical and mechanical properties have been developed. Finally, the reliability of the proposed models has been testified by the good agreement with the expe-

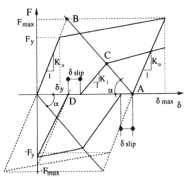

Figure 11. Cyclic model for type-2 and type-3 mechanisms

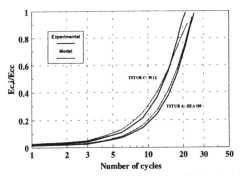

Figure 10. Accuracy of the proposed model in simulating variable amplitude tests

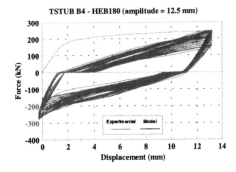

TSTUB B4 - HEB180 (amplitude = 12.5 mm)

Figure 12. Modelling of the cyclic behaviour for type-2 and type-3 mechanism: comparison with experimental tests

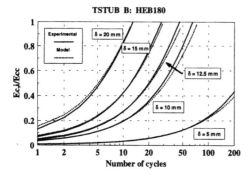

TSTUB B: HEB180

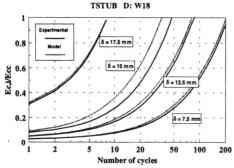

TSTUB D: W18

Figure 13. Accuracy of the proposed model in terms of energy dissipation capacity

rimental results in terms of energy dissipation capacity.

The above results are encouraging regarding the possibility of predicting the cyclic behaviour of bolted beam-to-column joints, starting from the knowledge of their geometrical and mechanical properties, by means of a component based approach.

AKNOWLEDGEMENTS

This work has been supported with Research Grants MURST 40% 1997 and MURST 60% 1999.

6. REFERENCES

Ballio, G., Calado, L., De Martino A., Faella, C. and Mazzolani, F.M. (1986):" Steel Beam-to-Column Joints under Cyclic Loads: Experimental and Numerical Approach", European Conference on Earthquake Engineering, Lisbona, 7-12, July.

Bernuzzi, C. and Serafini, N. (1997):" A Mechanical Model for the Prediction of the Cyclic Responce of Top-and Seat Cleated Steel Connections", CTA, Italian Conference on Steel Construction, Ancona, 2-5, October.

Bernuzzi, C., Zandonini, R. and Zanon, P. (1996):"Experimental Analysis and Modelling of Semi-Rigid Steel Joints under Cyclic Reversal Loading", Journal of Constructional Steel Research, Vol. 38, No. 2, pp. 95-123.

Bursi, O.S. and Galvani, M. (1997):"Low Cycle Behaviour of Isolated Bolted Tee-Stubs and Extended End Plate Connections", CTA, Italian Conference on Steel Construction, Ancona, 2-5, October.

Calado, L. (1995):"Experimental Research and Analytical Modelling of the Cyclic Behaviour of Bolted Semi-Rigid Connections", Steel Structures, Eurosteel '95, Kounadis ed., Balkema.

Calado, L., Castiglioni, C.A., Barbaglia, P. and Bernuzzi, C. (1998):"Seismic Design Criteria Based on Cumulative Damage Concept", 11th European Conference on Earthquake Engineering, Paris, 6-11 September.

CEN (1997): Eurocode 3, Part. 1.1: Joints in Building Frames (Annex J), Approved Draft, CEN/TC250/SC3-PT9, Comitè Européen de Normalisation.

De Martino, A., Faella, C. and Mazzolani, F.M. (1984): "Simulation of Beam-to-Column Joint Behaviour under Cyclic Loads", Costruzioni Metalliche, No. 6.

De Martino, G., Sanpaolesi, L., Biolzi, L., Caramelli, S., Tacchi, R. (1981):"Indagine Sperimentale sulla Resistenza e Duttilità di Collegamenti Strutturali", Ricerca Italsider - Comunità Europea, Monografia No. 5.

De Stefano, M., De Luca, A. and Astaneh, A. (1994): "Modelling of Cyclic Moment-Rotation Response of Double-Angle Connections", Journal of Structural Engineering, ASCE, Vol. 120, No. 1.

Engelhardt, M.D. and Husain, A.S. (1992):"Cyclic Tests on Large Scale Steel Moment Connections", Tenth World Conference on Earthquake Engineering, Madrid, Balkema Rotterdam, pp. 2885-2890.

Engelhardt, M.D. and Husain, A.S. (1993):"Cyclic-Loading Performance of Welded Flange-Bolted Web Connections", Journal of Structural Engineering, ASCE, Vol. 119, pp. 3537-3550, N.12, December.

Faella, C., Piluso, V. and Rizzano, G. (1997): "Plastic Deformation Capacity of Bolted T-Stubs", Second International Conference on Behaviour of Steel Structures in Seismic Areas, Kyoto, Japan, 4-7 August.

Faella, C., Piluso, V. and Rizzano, G. (1998): "Cyclic Behaviour of Bolted Joint Components", Journal of Constructional Steel Research, Vol.46, No.1-3, paper number 129.

Faella, C., Piluso, V. and Rizzano, G. (1999a): "Structural Steel Semirigid Connections", CRC Press, Boca Raton, Florida

Faella, C., Piluso, V. and Rizzano, G. (1999b): "Modelling of the Cyclic Behaviour of Bolted Tee-Stubs", Fourth International Conference on Steel and Aluminium Structures, ICSAS '99, Espoo, Finland, 20-23 June

Ghoborah, A., Korol, R.M. and Osman, A. (1992):"Cyclic Behaviour of Extended End Plate Joints", Journal of Structural Engineering, ASCE, Vol. 118, No. 5, pp. 1333-1353.

Krawinkler, H., Bertero, V.V. and Popov, E.P. (1971):" InelasticBehaviour of Steel Beam-to-Column Subassemblages", Report N. UBC/EERC-71/7, Earthquake Engineering Research Center, University of California, Berkeley.

Mander, J.B., Stuart, S.S. and Pekcan, G.:"Low-Cycle Fatigue Behaviour of Semi-Rigid Top-and-Seat Angle Connections", Engineering Journal, American Institute of Steel Construction, Third quarter, pp. 111-122.

Pekcan, G., Mander, J.B. and Chen, S.S. (1995):"Experimental and Analytical Study of Low-Cycle Fatigue Behaviour of Semi-Rigid Top-and-Seat Angle Connections", Technical report NCEER-95-0002, State University of New York at Buffalo, January.

Behaviour of Steel Structures in Seismic Areas, Mazzolani & Tremblay (eds) © 2000 Balkema, Rotterdam, ISBN 90 5809 130 9

Hysteretic behavior of steel moment resisting column bases

Mohamed Fahmy & Subhash C.Goel
Department of Civil and Environmental Engineering, G.G.Brown Building, University of Michigan, Ann Arbor, USA

Božidar Stojadinovic
Department of Civil and Environmental Engineering, University of California, Berkeley, USA

ABSTRACT: This paper reports on parts of the experimental investigation undertaken at the University of Michigan on the seismic behavior of typical exposed moment resisting steel column base connections designed following the US practice. The purpose of these tests are to examine the influence of detailing on potential brittle failure modes, and to verify that formation of a mechanism in the column and base plate can deliver the required capacity and ductility necessary to meet the imposed demands.

1 INTRODUCTION

Steel column bases designed following the US standards were damaged during the recent 1994 Northridge earthquake, as well as in the 1964 Anchorage earthquake. It is hard to evaluate the causes of this damage because the seismic response of US steel column bases is not well understood. Most of the available experimental data pertain to monotonic load tests or small-scale specimens. There are only a few analytical studies. More importantly, the design provisions in the US codes are limited to the axial load case allowing the designers to use any of several textbook procedures to design steel column bases for moment and for combined loads. A recent study at The University of Michigan was undertaken to investigate and quantify the seismic behavior of steel moment resisting column bases.

The main objective of this study was to investigate the seismic behavior of full-scale moment-resisting column base connections designed following the US practice. The research goal was to establish the conditions under which column base connections can reliably deliver the required strength, stiffness, ductility, and plastic rotations. The performance of the connections was examined through experimental and analytical parametric investigations. The experimental part of this investigation will be the main focus of this paper.

The objectives of the experimental investigation presented in this paper are to: 1) examine the influence of detailing on potential brittle failure modes such as: fracture of welds; lamellar tearing of the plate; fracture of rods; and grout crushing; 2) recommend specific weld detail and design considerations, which enhance the safety of the connection

against brittle fracture; and 3) verify that formation of a mechanism in the column and base plate can deliver the required capacity and ductility necessary to meet the imposed demand.

2 LITERATURE REVIEW

Design procedures for steel column bases used in the US design practice today are limited. Design for bending alone and for combined bending and axial loads is done following the AISC Steel Design Guide 1 (DeWolf, 90) and different procedures suggested in various textbooks. The design procedures suggested in the AISC Steel Design Guide 1 are based on the experimental results of small scale test from column base connection subjected to axial load only (Dewolf 78, Murray 83, Hawkins 68) and to combined monotonic bending and axial loads (Dewolf 80, Stephenson 81, Thambiratnam 91). Other monotonic lateral loading experiments on exposed column bases (Picard 85, Picard 87, Targowski 93, Carrato 91) provided more insight in the real connection behavior, but have not been used to support the AISC design provisions.

Experimental database on the seismic behavior of column bases constructed following the US practice is very limited (Burda 99). Experimental database on the behavior of column bases constructed following the Japanese practice is much more developed (Nakashima 92, Masuda 80, Ohtani 96). Unfortunately, column shapes, bolt configurations, connection details, and material properties used in these experiments are substantially different from those used in US practice. Therefore, it seems that the column base design produced according to the current US design practice have not been examined either experimentally or analytically in a strong seismic demand setting.

European efforts to predict the bolted beam column connection stiffness, strength and ductility have been focused on a component based approach. In this approach the connection is considered as an assembly of individual components. The mechanical properties of these components define the overall connection properties. By introducing normal force, the component-based concept used for bolted beam column connection was extended to include the column base connection (Wald, 95). The components considered in this method include: 1) concrete block in compression; 2) base plate in bending; 3) anchor bolt in shear and tension; and 4) column web and flange in compression. A series of monotonic load tests on individual components and complete column base connections have been conducted (Wald 96, Colson 87). These analytical and experimental results were used to develop the new Eurocode design provisions.

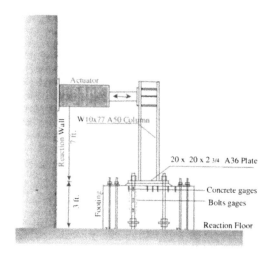

Figure 1. Test setup (side view).

3 TEST SPECIMENS

The experimental parametric study was performed by altering the weld detailing of two column base specimens (Sp.1 & Sp.2). These specimens had a W10X77 Gr50 column welded to a 20x20x2 ¾ in. A36 base plate. The base plate was anchored with four 2.0 in. A354BD rods (Fig. 1). The only difference between the two specimens was in the weld metal used to weld the column to the base plate. E70T-7 weld metal was used in the first specimen (Sp.1), while the E70TG-K2 weld metal was used in the second (Sp.2). Both specimens had identical weld details. The difference between these weld metals is that E70TG-K2 has a rated Charpy V-notch toughness of at least 20 lb-ft at -20 F while E70T-7 has no such rating. The web of the column was also fillet welded. The flanges of the column were beveled to leave a 1/8 in. wide gap for the partial penetration weld (Fig. 2). The 1/8 in. wide gap was the practical minimum weld gap that can be achieved in shop. This minimum weld gap was used to minimize the crack like effect inherent to partial joint penetration. By reducing the weld gap from 50% of the flange width to 13% of the flange width and by reinforcing the weld gap from the back by fillet weld, the weld detail represented an improvement over conventional partial joint penetration. This modification helped increase the fracture toughness of the weld. The use of complete joint penetration welds was considered to remedy this situation. However, the need for an access hole in the column web was thought to be more detrimental, with respect to the fracture toughness of the weld than the existence of the weld gap.

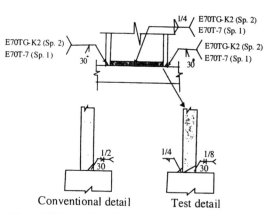

Figure 2. Weld detail for Specimens 1 and 2.

The specimens were designed to reach the full plastic moment capacity of the column. The connection design and details were reviewed by an advisory group for this study.

The dimensions of the footing (74x74x36 in.) were selected to accommodate attachments to the reaction floor. The mean compressive strength of the foundation concrete was 6000 psi. Concrete strain distribution was measured by twelve concrete strain gages. Non-shrink construction grout with 8000 psi strength was packed between the base plate and the concrete footing. The final grout thickness was about 2 in. Right angle strain rosettes together with linear strain gages were attached to the base plate and to the column at location where strains were expected to be critical. In addition to the strain gages, the column was equipped with LVDTs to measure the connection rotations.

The anchor rods in all tests were 48 in. long and fully anchored to the foundation so as to develop their tensile capacity. Each row of rods was anchored by means of an embedded steel bearing plate, as shown in Figure 1. The specified yield strength of the bearing plates was the same as that of base plates. The test configuration for all specimens is shown schematically in Figure 1. The specimen represented the column base and the column up to its first inflection points at mid-height.

Two angles mounted over a steel frame were used as a restraining system to prevent column twisting caused by local buckling of the column flanges during the large displacements. This bracing system is shown in Figure 3.

4 LOADING SEQUENCE

Earthquake load conditions for the column base models were simulated by applying a horizontal displacement at the top of the specimen column. Thus the column base was exposed to uniaxial bending only along the strong axis. Horizontal displacements were applied in a reversed cyclic quasi-static manner. The magnitude of the cycle amplitude was incremented in steps until failure. The number and the magnitude of cycles were determined according to the SAC Venture Protocol (SAC 99) (Fig. 4). No axial load was applied to the column.

5 TEST RESULTS FOR SPECIMEN 1

The normalized force-displacement plot for Specimen 1 obtained from the experiment is shown in Figure 5. The shear force V, was normalized with respect to the shear force V_p needed to form a plastic hinge in the column just above the base plate. The displacement was normalized with respect to the column height. The normalized displacement represents the drift of the test subassembly. Examination of the hysteretic loops shows that the column base behavior was brittle. The connection barely reached 90% of its ultimate capacity at 3% drift level. It had practically no energy dissipation. The hysteretic response of the specimen was stable, almost elastic up to 2% drift (Fig. 5). Upon reaching 2% drift level, the hysteretic loops began to widen, marking the beginning of the inelastic behavior of the connection. The inelastic behavior was initiated by the yielding of the flanges. With further loading and before finishing the first cycle of 3% drift, the weld fractured at 2.5 % drift and the connection strength dropped suddenly to almost zero. The weld fractured at the level of the weld gap of the partial joint penetration weld connecting the column flange to the base plate.

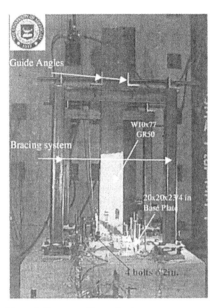

Figure 3. Test setup (front view).

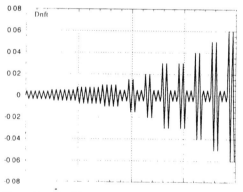

Figure 4 Loading history.

The fracture propagated from the edges through the whole flange width. The fractured specimen is shown in Figure 6.

5.1 *Failure Mode for Specimen 1*

Failure was caused by premature fracture of the weld. From the observation of the cracked surface, it appears that the weld cracked along one of the flange edges and then propagated toward the opposite edge quickly. This was concluded from 1) the non-broken crack line that extends from one edge to the other with a slight inclination (Fig. 7), and 2) the flaking of the white wash of this flange (Fig. 7). Apparently, the low notch toughness of the weld mate-

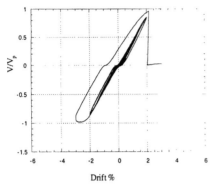

Figure 5. Normalized force-displacement hysteretic response of Specimen 1.

Figure 6 Failure mode for Specimen 1.

Figure 7 a) Cracks in the non-fractured flange b)Fracture surface in the weld.

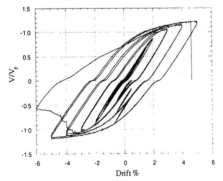

Figure 8. Normalized force-displacement hysteretic response of Specimen 2.

rial, the unintended increase in the root face length caused by the workmanship, the high residual stresses generated by the restrained joint configuration and the unusual increase of stresses at the edges of the flange contributed to this severe cracking. Figure 6 shows the fracture surface in the root pass of the weld together with details of the partial joint penetration weld. It shows that the weld detail was carried out by the fabricator as recommended except for the unwelded length. A thin crack was observed at the end of the non-fractured flange. The crack was originating from the far end of the flange weld (Fig. 7a). The several notches observed through the weld surface show a trend of dimple fracture (Fig. 7 b).

6 TEST RESULTS FOR SPECIMEN 2

The normalized force-displacement loops obtained from the experiment are shown in Figure 8. Examination of the hysteretic loops shows that the column base behavior was ductile. At drift level of 2% the whitewash started to flake near the bottom of the flanges, signifying that yielding and non-linearity in the force-displacement relations commenced, thus marking the end of the elastic response phase. As cyclic displacement progressed, whitewash flaking in the flange propagated vertically away from the base plate and horizontally toward the center of the flange.

The hysteretic loops were stable and exhibited good energy dissipation characteristics up to 5% drift level. The hysteretic loops overlapped each other and appeared to show no degradation even after several cycles. This shows the durability and stability of the yielding mechanisms. Before the end of the first cycle in the 6% drift level, fractures in the weld started to develop at the flange edges at 3.5 % drift and the load carrying capacity of the connection decreased. When the load was reversed, the crack closed and the other non-damaged flange fractured abruptly at the weld level. Subsequently the load dropped to zero and the test was stopped at 5.5 % drift limit. The fractured Specimen is shown in Figure 9. The drop of the connection strength was caused by flange fracture.

6.1 Failure Mode for Specimen 2

In this connection failure occurred after excessive yielding at the weld interface between the base plate and flanges. The column flanges apparently started to yield very close to the weld. With further loading, yielding spread through 1) edges of the column flange, 2) centers of the column flange, 3) web edges, and 4) parts of the plate located along the column flanges, respectively. During the first cycle of 6% drift, fractures in the weld started to develop

Figure 9 Origination of the cracks at the flange edges.

Figure 10 Flange fracture at the end of loading.

at the flange edges (Fig. 9). With further displacement, the fracture propagated towards the middle of the flange (Fig. 10). Subsequently, the weld in the web adjacent to the damaged flanges fractured and the column started to lose its load carrying capacity. When the load was reversed, the crack closed and connection carrying capacity picked up till the other flange fractured abruptly at the weld level. At that time, the load dropped to zero and the test was stopped.

7 COMPARISON OF TEST RESULTS

7.1 Strength

The measured yield and ultimate strengths of the two specimens are compared. The experimental values of yield and ultimate strengths are shown in Table 1. The strengths are expressed as normalized shear force applied at the tip of the specimen.

Both specimens shared the same yield value of $0.8 V_p$ at drift level of 2%. The inelastic response was initiated by yielding of the column flanges. The connection ultimate strength differed from one specimen to the other depending on the connection details. As shown in Table 1 the maximum load for the first Specimen was $0.8V_p$ and the corresponding drift was 2.5%. This load limit is lower than expected. This can not be considered satisfactory due to the generally expected drifts for seismic applications. The second specimen reached a load of 1.2 M_p at a drift level of 5.5%. The improved performance

Table 1 Ultimate and yield strength for tested connections.

	Yield		Ultimate		Yield		Ultimate	
	Disp.	Load	Disp.	Load	Drift	V/Vp	Drift	V/Vp
	in.	Kip	in	Kip	%		%	
Sp1	1.68	53.9	2.1 5	3.9	2	0.8	2.5	0.8
Sp2	1.68	53.9	4.2	67.4	2	0.8	5.0	1.0

Table 2 Ductility of the tested connections.

	Displacement Ductility	Energy Dissipated (E_d)	Plastic Rotation (θ_p)
Sp1	1.25	218 (K-in.)	0.5(radian)
Sp2	2.75	2670 (K-in.)	3.0(radian)

of this connection over the first one can be attributed to the use of notch tough weld metal.

7.2 Ductility

The estimated ductility measures for the two Specimens are compared using three ductility indices. These are deflection ductility, energy dissipation, and plastic rotation. For each connection, Table 2 lists the three ductility indices. The displacement deflection ductility factor μ_d is expressed as the ratio the tip deflection at the maximum value attained, δ_{max}, to the tip deflection at first yielding, δ_y. This parameter is useful for indicating the overall ductility that can be achieved during the most severe cycle of reversed deformation. Note that this concept of ductility ratio is based on the expected steel properties and does not account for residual stresses.

The energy dissipated by each specimen represents the accumulated strain energy dissipated at each load. It was computed by integrating the hysteresis loops step-by-step.

The plastic rotation capacity of the specimens is compared using maximum plastic rotation, θ_p. The quantity θ_p is equal to $|\delta_{max}-\delta_y|/L$ where δ_{max} is the maximum tip deflection in any loading direction (pull or push) and L is the column height. Physically, θ_p corresponds to the plastic rotation of the column end when the column is deformed in a double curvature deflection curve under lateral seismic loading.

The ductility factors and energy dissipation values for the first specimen indicate that the specimen did not have adequate energy dissipating mechanism due to the brittle failure of the connecting welds at the end of the elastic limit. The second specimen showed much higher ductility indices and energy dissipation capacities because of improvement in the connection details. The displacement ductility ratio for the first specimen shows that the connection exhibited some nonlinear behavior that might have been triggered by residual stresses.

The normalized energy dissipation values can be

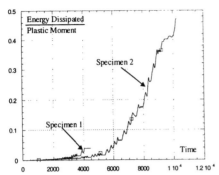

Figure 11 Normalized dissipated energy for the two specimens.

seen in Figure 11. The vertical axis in this figure represents the accumulated strain energy dissipated at each load step normalized with respect to the expected plastic moment of the connection, while the horizontal axis represents the accumulated time steps. The dissipated energy could be considered a better index than other ductility factors simply because its magnitude is affected by the history of force-deformation response before failure. The curve shows that the second specimen dissipated more energy than the first at the same displacement level. The total dissipated energy by Sp.2 exceeds 0.4 Mp. In these curves, some downward trends in the energy are noticed. These are caused by the recovery of elastic energy upon load reversal.

8 ACKNOWLEDGMENTS

The authors gratefully acknowledge the support of the American Institute for Steel Construction, specimens donation by the Great Lakes Fabricators and Erectors Association of Michigan and the National Science Foundation (Grant No. CMS 9810 / 595).

9 CONCLUSION

The partial joint penetration (PJP) weld in conjunction with notch tough weld metal detail used in these experiments proved to be resilient and a good substitute for a complete penetration weld detail with access holes and backing bar. This PJP weld detail differs from the conventional PJP weld detail in the length of the weld root toe. The root toe in this weld is about 13% of the flange thickness while in a conventional PJP weld it is about 50% of the flange thickness.

The use of E70T-7 weld metal, remains a potential hazardous issue since most of the pre-Northridge earthquake connections were designed and detailed similar to Specimen 1. Unfortunately, some weld fractures in column base connection at the weld level might be hard to detect, particularly after the closure of the crack under gravity loading and because of dust and paint covering the weld.

The test results of Specimen 2 verified that formation of a mechanism in the column instead of the anchor rods and/or the base plate can deliver the required capacity and ductility necessary to meet the imposed demand. The effect of the local behavior of the connection on the seismic resistance of buildings was studied in (Stojadinovic 98, Fahmy 98)

10 REFERENCES

AISC, "Manual of Steel Construction: Load and Resistance Factor Design.". AISC, 1994.

Astaneh A., Bergsma G. and Shen J.H., "Behavior and Design of Base Plates for Gravity, Wind and Seismic Loads." In Proceeding of the National Steel Construction Conference, AISC, Las Vegas, June 1992.

Burda J. and Itani A., "Studies of Seismic Behavior of Steel Base Plates." Report No CEER 99-7, Center for Civil Engineering Research, Department of Civil Engineering, University of Nevada, Reno, 1999.

Carrato, P., "Testing and Analysis of Base Plate Connections." Anchors in Concrete Design and Behavior, G. A. Senkiw and H. B. Lancelot III, editors, ACI Special Publication 130, ACI, 1991.

Colson A., "Three-dimensional physical and mathematical modeling of connections." In Connections in Steel Structures: Behavior, Strength and design, Elsevier Applied Science, 1987.

Dewolf J. and Sarisley E, " Column Base Plates with Axial Loads and Moments." Journal of the structural divisions, Proceedings of the American Society of Civil Engineers, Vol 106, No. ST11, Nov. 1980.

Dewolf J., "Axially loaded column base plates" ASCE Journal of the Structural Division, 104(ST5), May 1978.

Dewolf, J. and Ricker D., "Column Base Plates." Steel Design Guide 1, AISC, 1990.

Drake R. and Elkin S.," Beam-Column Base Plate Design-LRFD Method." AISC Engineering Journal, first quarter, 1999.

EUROCODE 3. ENV 1992-1-1, Part1.1: Design of Steel Structures, European Prenorm, Commission of the European Communities, Brussels, 1992.

Fahmy, M., Stojadinovic, B., S. C. Goel, "Analytical and Experimental Behavior of Steel column Bases." Proceedings of the 8th Canadian Conference on Earthquake Engineering, 1999.

Fahmy, M., Stojadinovic, B., S. C. Goel, and Sokol T., "Load Path and Deformation Mechanism of Moment-resisting Steel Column Bases." Proceedings of the Sixth U.S. National Conference on Earthquake Engineering, EERI, 1998.

Fahmy, M., Stojadinovic, B., S. C. Goel, and Sokol T., "Seismic Behavior of Moment-resisting Steel Column Bases." Proceedings of the eleventh European Conference on Earthquake Engineering, AFPS 1998.

Hawkins H. M., "The Bearing Strength of Concrete Loaded through Flexible Plates." Magazine of Concrete Research (June 1968): 30(62): 95-102.

Igarashi S, Kadoya H., Nakashima S., and Suzuki m., "Behavior of Exposed-type Fixed' column base connected to rise foundation." Earthquake Engineering, Tenth World Conference, Balkema, Rotterdam, 1992.

Masuda K. and Hirasaka T., "Connections in Steel Reinforcing

Concrete Structures-Column Bases, Part 2" Transactions of AIJ, 290, January 1980.

Murray, T., "Design of Lightly Loaded Column Base Plates" AISC Engineering Journal, Vol. 20, 4rth quarter, 1983.

Ohtani Y., Fukumoto Y., and Tsuji B. "Performance of mixed connections under Combined Loading." In connections in Steel Structures 3:Behavior, strength and Design, Pergamon, 1996.

Picard A., and Beaulieu D., "Behavior of a Simple Column Base connection." Canadian Journal of Civil Engineering, 12 (1), p 122-136, 1985.

Picard A.and Beaulieu D., "Rotational Restraint of a Simple Column Base Connection." Canadian Journal of Civil Engineering, 14: 49-57, 1987.

SAC Joint Venture, "Analytical and Field Investigations of Building Affected by the Northridge Earthquake of January 17,1994." Report SAC 95-04, part 2,1995

Stojadinovic, B., E. Spacone and S. C. Goel. 1998, "Influence of Semi-Rigid Column Base Models on the Response of Steel MRF Buildings." Proceedings of the Sixth U.S. National Conference on Earthquake Engineering, EERI, 1998

Thambiratnam D. and Parmasivan P., "Base Plates under Axial Load and Moments." ASCE Journal of structural engineering, 112(ST5):1166-1181, May 1986.

Thambiratnam D.P., "Strain distribution in Steel Column Base Plates Subjected to Eccentric Loads" Proceedings of International conference on Steel and Aluminum Structures, S.L. Lee and H.E. Shanmugam, editors, pages 601-609, Singapore, May 1991.

Wald F., "Column Bases." CeskÈ vysokÈ Uceni technike, Edicni stredisko CVUT, Zikova 4, 16635 Paha 6, The Check Republic, 1995.

Wald F., Sokol Z., and Steenhuis M., "Proposal of the stiffness design model for the column bases." In connections in Steel Structures 3: Behavior, Strength and design, p441-452. Pergamon, 1996.

Behaviour of Steel Structures in Seismic Areas, Mazzolani & Tremblay (eds) © 2000 Balkema, Rotterdam, ISBN 90 5809 130 9

Experiment and analysis of bolted semi-rigid beam-column connections Part I: Cyclic loading experiment

Kazuhiko Kasai & Yanghui Xu
Tokyo Institute of Technology, Yokohama, Japan

Arum Mayangarum
Thornton-Tomassetti Engineers, N.J., USA

ABSTRACT: Enormous failures of moment frame welded connections in the earthquakes of Northridge and Kobe gave impetus for renewed research into the use of bolted connections for high seismic zones. Test results of 24 full-size beam-column assemblies employing bolted semi-rigid connections are summarized in this paper. The beam is 604 mm deep, larger than typical sizes used in the prior studies on semi-rigid connections. For connecting the beam flange to column, either an angle or tee section was used, and they were designed to produce relatively high stiffness, strength, and ductility. For bolt-connecting the beam web to column, instead of conventionally used double angle, a shear tab with slotted holes was used. Analysis of the connection is explained in a companion paper (Part II, Xu et al. 2000).

1 INTRODUCTION

A moment frame consisting of relatively stiff semi-rigid connections can show similar lateral stiffness and vibration period, as compared with a frame of rigid connections. Due to yielding of the semi-rigid device, the frame dissipates seismic energy earlier, thereby being able to control the building drift as long as the device is ductile (Mayangarum & Kasai 1997, Maison et al. 2000).

It is important to look into feasibility of using the semi-rigid connection as an alternative scheme against a major seismic attack. The purpose of this study is to create the bolted semi-rigid connections having high stiffness, strength, and ductility. The device used is a portion of either angle or tee section which is designed to yield while beam remaining elastic.

2 SPECIMENS AND TEST SET-UP

2.1 *Material properties*

Totally 24 specimens were tested. Unlike the prior studies that had typically used A36 steel angle (L) or tee (WT) with nominal yield stress of 248 MPa, this project used higher strength steel A572 Grade 50, in order to obtain reasonably high moment capacity of the connection. According to the device sizes and types, the specimens are categorized into 4 groups as follows:

A1: L8 x 6 x 3/4 (8 specimens)
A2: L8 x 6 x 1/2 (6 specimens)
T1: WT8 x 33.5 (6 specimens)
T2: WT12 x 34 (4 specimens).

Specific details of group A1, A2, T1, and T2 are summarized in Table 1. Note that the width of angle or tee is measured in the direction of the beam or column flange width.

Table 2 shows the tensile coupon test results of yield stress σ_y, ultimate stress σ_u, and elongation. These values are average values of longitudinal and transverse

Table 1. Groups of specimens.

Specimen	Number	Designation	Size (mm)	Width (mm)
A1-1 to A1-8	8	L8 x 6 x 3/4	203 x 152 x 19	381
A2-1 to A2-6	6	L8 x 6 x 1/2	203 x 152 x 13	381
T1-1 to T1-6	6	WT8 x 33.5	207 mm deep	279
T2-1 to T2-4	4	WT12 x 34 *	207 mm deep	279

*Web trimmed from 301 to 207 mm.

Table 2. Tensile coupon test results.

Specimen group	σ_y (MPa)	σ_u (MPa)	Elongation (%)
A1	340.60	533.24	30.3
A2	400.17	533.30	40.6
T1	399.55	521.24	33.5
T2	440.43	531.03	38.1

Table 3. Charpy V-notch test results.

Specimen group	Energy Absorbed (J) *	
	Standard Location	Fillet Area
A1	95	50
A2	243	43
T1	138	57
T2	184	77

* At temperature = 23°C.

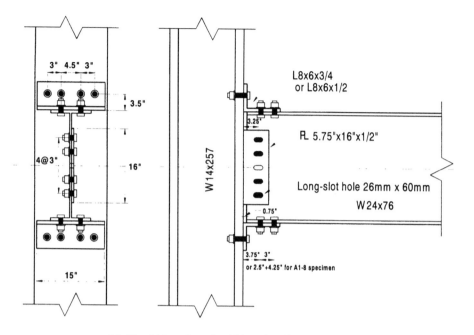

(a) A1 and A2 specimens (total 14 specimens).

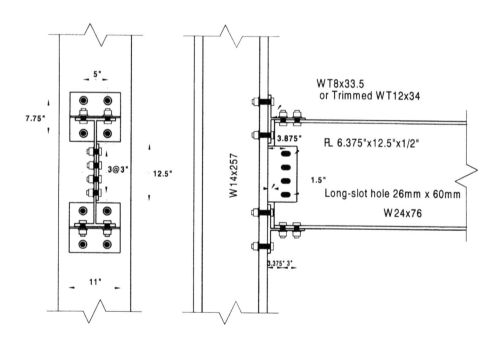

(b) T1 and T2 specimens (total 10 specimens).

Figure 1. Semi-rigid connection using either angle or tee section (unit: inch, 1 inch = 25.4 mm).

directions for each group. Charpy V-notch test results are listed in Table 3. Each value of the energy absorbed was calculated by averaging the test results of 3 specimens. Rockwell hardness tests showed that in case of angle, the average hardness was about 10 to 15% higher near the fillet area than those elsewhere. In case of tee, the higher value was recorded at the stem near the fillet.

2.2 Test set-up

Typical test connections using angle or tee sections are shown in Figures 1 (a) and (b), respectively. The

Table 4. Semi-rigid connection specimens and test results.

Specimen	Loading	Surface	θu (%rad.)	Nf	ΣE (kJ)	$\Sigma E/My$ (%)
(1)	(2)	(3)	(4)	(5)	(6)	(7)
A1-1	ATC	W	2.80	22	195	42
A1-2	L	W	3.00	6	180	38
A1-3	M	W	1.00	52	299	64
A1-4	S	W	0.39	350	N.A.	N.A.
A1-5	S	W	0.42	435	565	119
A1-6	ATC	B	3.00	22	245	52
A1-7	ATC	C	3.90	26	336	71
A1-8	ATC	W	2.80	22	160	34
A2-1	ATC	W	3.30	25	200	67
A2-2	L	W	3.40	5	106	35
A2-3	L	W	3.00	7	182	61
A2-4	M	W	2.20	52	322	107
A2-5	S	W	0.70	216	440	147
A2-6	ATC	W	3.30	25	190	63
T1-1	ATC	W	2.70	20	155	34
T1-2	L	W	3.10	3	114	25
T1-3	M	W	1.20	15	160	35
T1-4	S	W	0.55	158	195	43
T1-5	ATC	B	2.40	20	160	35
T1-6	ATC	C	2.25	18	174	38
T2-1	ATC	W	4.20	20	234	51
T2-2	L	W	3.30	11	214	47
T2-3	S	W	0.60	253	299	66
T2-4	EQS	C	3.50	158	460	101

Note:
ATC = ATC-24 loading;
L, M, S = constant displacement cyclic loading of
   large, medium, and small magnitude;
W = small fillet weld at the tip of the device;
B = blast-cleaned surface at the device-flange interface;
C = mill-scale surface at the device-flange interface;
θ_u = peak rotation at failure;
Nf = cycle number at failure;
Σ E = total energy dissipated up to failure;
My = experimental yielding moment of the connection, 471, 300,
   456, and 455 kN-m for group A1, A2, T1 and T2, respectively;
Σ E / My = equivalent cumulated rotation.

beam and column were W24×76 (609 mm deep) and W14×257 (355 mm deep) sections. Measured from the load point to the colum center line, the beam length is 3048 mm, and the column length is 1829 mm. The beam size was considerably larger (stronger) than those used for the prior experimental studies on semi-rigid connections. Moment capacities of the connections are provided by the flange connecting devices only, and shear resistance by the shear tab with long-slotted holes. Note that in the angle connection, four column bolts were used along the column width. Four beam bolts were used to attach either angle or tee to the beam flange. Four shear tab bolts are used for groups A1 and T1, while three bolts were used for groups A2 and T2. All bolts were Japanese F10T bolts of 15/16 in. (24 mm) diameter. The hole was 1/12 in. (2 mm) over the bolt diameter.

In some specimens of group A1 and A2, the beam bolts were located very close to the column face in order to examine their effects on the plastic mechanism of the angle. In many specimens, small fillet welds were provided to connect the tip of the angle or tee device to beam flange. The reason was to eliminate the slip at the interface for effectively measuring the deformation demand and corresponding low cycle fatigue strength of the device. For some specimens, the welds were not provided; the device and flange had either clean mill-scale surfaces or blast-cleaned surfaces. The above variations for the specimens are summarized in Table 4 (Column (3)), where W = small fillet weld; B and C = blast-cleaned surface and clean mill-scale surface.

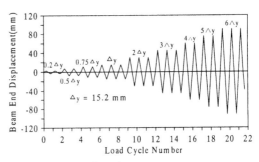

Figure 2. ATC-24 loading scheme.

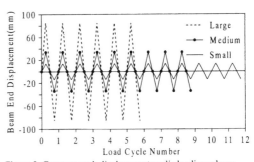

Figure 3. Constant peak displacement cyclic loading scheme.

201

2.3 Loading scheme

The specimens were tested by applying displacements at the cantilever end of the beam. There are three types of displacement histories. The first type is called ATC-24 protocol that has been the standard scheme in California since the 1994 Northridge earthquake. A typical ATC-24 loading is shown in Figure 2. Note that Δ_y is the displacement when the system lost stiffness considerably. The second type consisted of repeated cyclic displacements with a constant peak magnitude, which was selected to ensure about ±3% rad. (large), ±1% to ±2% rad. (medium), or ±0.4% to ±0.6% rad. (small) peak rotation of the connection (Fig. 3). The third type consisted of random earthquake simulated loading, which was applied to the specimen T2-4. Specific details of this loading will be explained later. The loading schemes applied to each specimen are summarized in Table 4 (Column (2)), where ATC = ATC-24 loading; L, M, S = constant peak displacement cycles of large, medium, and small amplitudes; EQS = earthquake simulated loading.

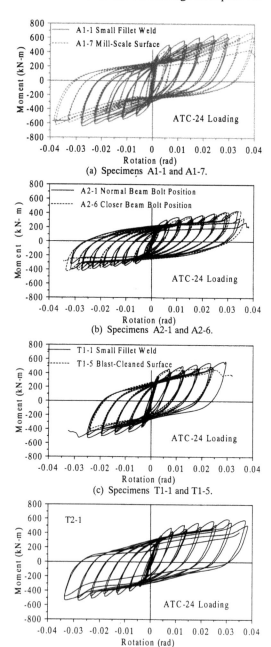

(a) Specimens A1-1 and A1-7.

(b) Specimens A2-1 and A2-6.

(c) Specimens T1-1 and T1-5.

(d) Specimen T2-1.

Figure 4. Moment vs. rotation for specimens under ATC-24 loading.

3 TEST RESULTS

3.1 ATC-24 loading

In Table 4, the peak rotation θ_u at failure (Column (4)) is given for each specimen. It can be seen that at failure under ATC-24 loading, most of them developed 2.5 to 4 % rad. rotation capacities. Figure 4 shows the moment vs. rotation $(M - \theta)$ curves under ATC-24 loading. Figure 4 (a) compares specimen A1-1 having small fillet welds between the tips of the angles and the beam, with specimen A1-7 which has no fillet welds. Figure 4 (b) compares specimen A2-1 with A2-6 for which the beam bolts was made closer to the vertical leg of the angles. Figure 4 (c) compares specimen T1-1 having fillet welds provided and T1-5 having blast-cleaned surface between the angles and the beam without weld. Figure 4 (d) gives the curve for specimen T2-1. The same type specimen T2-4 subjected to random loading will be described later.

The hysteresis was stable with large post-yielding stiffness. The specimens with the device tip fillet-welded to beam flange showed insignificant pinching of the hysteresis. This was also true for the case without welds due to the excellent slip resistance of the beam bolts. Absence of the weld leads to somewhat smaller moment capacity (see Fig. 4 (a)).

Absence of the weld also affected rotation capacity: For the connection using angle, it leads to larger rotation capacity (Fig. 4 (a)). This would be due to the slip betwen the angle and beam flange which caused less deformation demand to the angle. On the other hand, for the connection using tee, it leads to slightly smaller capacity (Fig. 4 (c)), the reason of which is currently investigated.

3.2 Constant peak displacement cyclic loading

$M - \theta$ curve for specimen A1-2 under large constant peak rotation (3% rad.) is shown in Figure 5. Regardless of peak rotation magnitude, typically the peak load stabilizes after 1 to 2 cycles. Figure 6 (a) shows $M - \theta$ curves when the peak load stabilized: Three identical

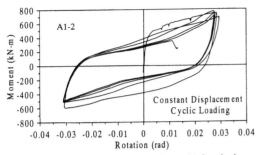

Figure 5. Moment vs. rotation for specimen A1-2 under large constant displacement cyclic loading.

specimens of A1 group are subjected to different peak rotations of 3%, 1%, and 0.4% rad. The number of cycles to failure Nf varies significantly over these specimens. Bernuzzi et al.'s criterion (1997) was used to determine the low-cycle fatigue failure and corresponding number (Nf) of the cycle shown in Table 4 (Column (5)).

Figure 6 (b) shows the similar plots for specimen group A2, having thinner angle than group A1. The thinner angle has lead to smaller stiffness and strength, but almost the same Nf as the thicker angle case. Figure 6 (c) shows the similar plots for T1 group which had smaller Nf among all 4 groups.

3.3 Earthquake simulated loading

An existing 5 bays by 5 bays 13-story steel frame building in Northridge (Maison & Kasai 1997, Mayangarum & Kasai 1997) was analyzed by using various strong earthquakes, and the critical connection rotation histories were obtained. The histories were

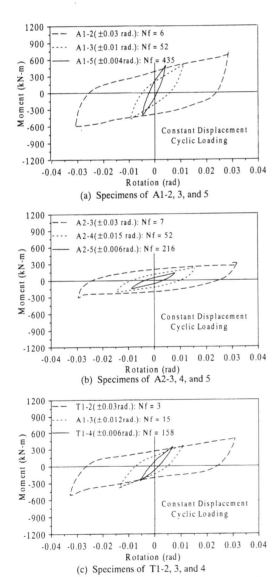

(a) Specimens of A1-2, 3, and 5

(b) Specimens of A2-3, 4, and 5

(c) Specimens of T1-2, 3, and 4

Figure 6. Comparison of moment vs. rotation under constant peak displacement cyclic loading with different magnitudes.

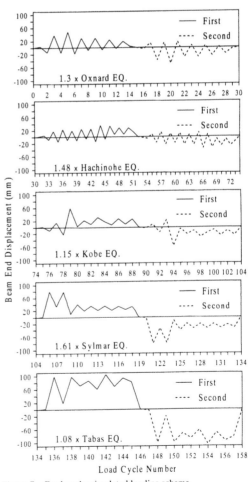

Figure 7. Earthquake simulated loading scheme.

203

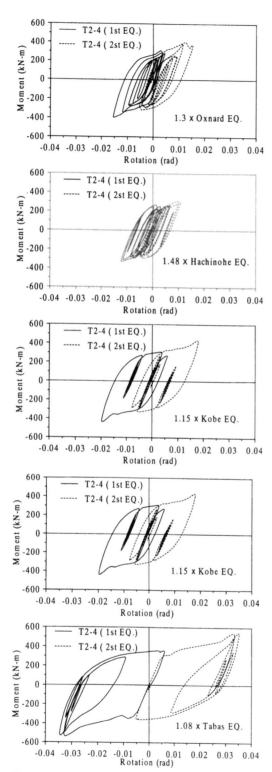

Figure 8. Moment vs. rotation curves under earthquake loading.

then applied to specimen T2-4 twice per earthquake. The earthquakes used were 1.3 times Oxnard record (1994 Northridge), 1.48 times Hachinohe record (1962 Tokachi-Oki), 1.15 times JMA Kobe record (1995 Kobe), 1.61 times Sylmar record (1994 Northridge), and 1.08 times Tabas record (1974 Iran), as shown in Figure 7.

The Oxnard record was obtained from the 13-story building's accelerometers at the basement. Also, the last three records have the spectrum comparable to the 1994 NEHRP design basis earthquake spectrum of probabilities of exceedence of 2 % in 50 years (Sommerville et al. 1997). Therefore, they are much stronger than the conventionally considered design earthquake (probabilities of 10 % in 50 years).

It should be noted that T2-4 specimen is identical with T2-1 specimen which survived up to 3% rad. under ATC-24 loading and could satisfy the post-Northridge requirement for connection ductility (Fig. 4 (d)). Figure 8 compares moment-rotation curves of specimen T2-4 under the earthquake simulated loading. It can be seen that the same type of specimen T2-4 sustained many major and catastrophic earthquakes. It sustained each of the first four earthquakes twice, and did not fail. Further, it could sustain 13 repeated histories from the catastrophic Tabas ground motion before the failure. It is important to understand that the bolted semi-rigid specimen that passed the 3% requirement above-mentioned could sustain more than 17 (= 2 + 2 + 13) earthquakes of the 2 % probabilities in 50 years. This may also mean the excessive conservatism of the US Federal Emergency Management Agency requirement, at least for the present specimen.

4 DISCUSSIONS

4.1 Stiffness and strength

The elastic rotational stiffness of all the tested semi-rigid connections were about 10 times or more than EI/L of the beam, where E = Young's modulus, I = 8.74×10^8 mm^4 of the W24×76 beam, and a typical building span length L = 9144 mm. This indicates that the connections are relatively stiff semi-rigid connections. Also, yield moments of specimen A1, T1, and T2 were about 0.5 to 0.6 times the plastic moment Mp of the W24×76 beam (A36 steel of nominal yield stress 36 ksi (248 MPa) assumed). Ultimate moments were about 0.7 to 0.8 times the Mp of the beam. These indicate relatively large strengths of the semi-rigid connections tested.

4.2 Ductility and energy dissipation

Energy dissipations by the specimens and equivalent cumulated plastic rotation are summarized in Table 4 (Column (6) and (7)). The equivalent cumulated plastic rotation (Σ E / My) is the total energy dissipated divided by the experimentally observed yield strength of the connection. The quantity indicates the cumulated plastic rotation of an equivalent elastic-perfectly plastic connection. Note that the cumulated rotations vary from 25% to 120% rad., depending on the loading history. In the case of T2 group specimens, the smallest value is 47% rad. when subjected to large constant peak displacement, and 101% rad. when subjected to the numerous simulated earthquakes. This suggests that even a specimen which shows the lowest rotation of 25% rad. could sustain a major earthquake without premature failure.

4.3 Responses of components.

In the tests, all of the specimens of group A1 and A2 fractured at the fillet area of the horizontal leg of the angle device. Figure 9 shows The connection moment vs. the deformation of the top angle for specimen A1-1. Figure 9 indicates the maximum ductility demand for the angle deformation is very high (tension direction only).

The specimens of group T1 and T2 fractured at the top portion of the flange of top tee, or bottom portion of the flange of bottom tee. The devices still had relatively large reserved strength after the fracture of one portion due to the remaining portion.

Figure 10 shows that the beam translated away from the column face as loading increased, and the maximum translation was 8.9 mm in case of specimen A1-1. The behavior is a key source of "softening" response of the connection during large displacement cycles. After the devices were stretched significantly, it requires large excursion of reversed displacement until it contacts and bears the column face. Before the contact of the compressed device, the connection is essentially soft, and this is why the stiffness and strength become progressively small as the number of cyclic excursions is increased.

The beam moment vs. the stain of the column bolt is shown for specimen A1-1 (Fig. 11). Up to the loading of $\pm 0.75\Delta y$, the bolt force remained essentially the same (Fig. 11 (a)). At 3 cycles of $\pm\Delta y$, the bolt force increased in proportion to the beam end load, indicating on-set of separation between the column face and the angle (Fig. 11 (b)). At these cycles, however, the bolt pretension force loss was not significant, as evidenced by the stable bolt force when the beam load was negative. The bolt pretension loss became significant stating from cycles of $\pm 2\Delta y$, suggesting that the pretension loss occurred when the devices yielded noticeably. This trend became more significant under the larger displacement (Fig. 11 (c)). In some case, the bolt strains were beyond yielding strain, as is believed to be caused by bending of the shank. All the bolts developed permanent bending deformations (except for specimen group A2), as observed after the tests.

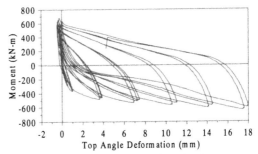

Figure 9. Top angle deformation (A1-1 specimen).

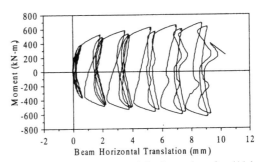

Figure 10. Beam horizontal translation from column face (A1-1 specimen).

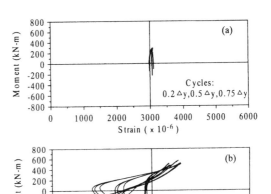

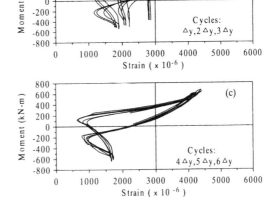

Figure 11. Moment vs. column bolt strain (A1-1 specimen).

205

5 CONCLUSIONS

The semi-rigid connections showed relatively high stiffness and strength as well as excellent ductility. The connection design which was experimentally proved to satisfy the 3 % rad. requirement under the ATC-24 loading demonstrated significantly ductile performance against the catastrophic earthquake simulated excitations. It sustained more than 17 earthquakes having the probability of 2 % in 50 years. This may indicate the need to re-examine the US post-Northridge requirement for the connection deformation capacity.

6 ACKNOWLEDGEMENT

The project explained in this paper were funded by Nippon Steel Corporation, and ATLSS Center of Lehigh University. The support is gratefully acknowledged. However, the opinions expressed in this paper are those of the authors, and do not necessarily reflect the views of the sponsors.

REFERENCES

Xu, Y., Kasai, K. & Mao, C., 2000. Experiment and analysis of bolted semi-rigid beam-column connections (Part II: 3-D finite element analysis of the connections using angles). STESSA.

Maison, B.F., Rex, C., Lindsey, S., & Kasai, K., 2000. Performance of PR moment frame buildings in UBC seismic zones 3 and 4. *Journal of Structural Engineering*, January.

Mayangarum, A., & Kasai, K., 1997. Design, analysis, and application of bolted semi-rigid connections for moment resisting frames, Report No. 97-15, Center of Advanced Technology for Large Structural Systems (ATLSS), Lehigh University.

Kasai, K., Mao, C., & Mayangarum, A., 1998. Feasibility of bolted rigid and semi-rigid connections for seismic regions. Summary report of third US-Japan workshop on steel fracture issues. Building Research Institute (BRI) of Japan, Kouzai-club, 469-481.

ATC-24, Guidelines for the cyclic testing of components of steel structures, No 24, *Applied Technology Council*.

Maison, B.F., & Kasai, K., 1997. Analysis of Northridge damaged thirteen story welded steel moment frame building, *Earthquake Spectra*, 13(3), 451-473.

Sommerville, P., Smith, N., Punyamurthula, S., & Sun, J., 1997. Development of ground motion time histories for phase 2 of the FEMA/SAC steel project, Report No. SAC/BD-97/04, SAC Joint Venture, Sacramento, CA, 1997.

AISC 1992. *Seismic provisions for structural steel buildings.* American Institute of Steel Construction, Chicago, IL.

Bernuzzi, C., Calado, L., Castiglioni, C.A., 1997. Ductility and load carrying capacity predictions of steel beam-to-column connections under cyclic reversal loading, *J. of Earthq. Eng.*, 1(2) 401-432.

Gross, J., Engelhardt, M.D., Uang, C.M., Kasai, K., & Iwankiw, N., 1998, Modification of exsisting welded steel moment connections for seismic resistance, *National Institute of Standards and Technology*, American Institute of Steel Construction, June.

Kasai, K., & Bleiman, D. 1996. Bolted brackets for repair of damaged steel moment frame connections, 7th US-Japan workshop on the improvement of structural design and construction practices: Lessons learned from Northridge and Kobe, Kobe, Japan, January 18-20.

Kasai, K., Hodgson, I., & Mao, C., 1997, Bolted repair methods for fractured moment connections, *Proc. of Internat. Cof. On Behavior of Steel Strs. in Seismic Areas* (STESSA), Kyoto 939-946.

Kasai, K., Hodgson, I., & Bleiman, D., 1997. Rigid-bolted repair method for damaged moment connections, *Engineering Structures*, 20(4-6), 521-532.

NEHRP, 1994. Recommended Provisions for seismic regulations of new buildings: Part 1, Provisions, Publication 222A, Federal Emergency Management Agency, Washington, D.C.

Behaviour of Steel Structures in Seismic Areas, Mazzolani & Tremblay (eds) © 2000 Balkema, Rotterdam, ISBN 90 5809 130 9

Experiment and analysis of bolted semi-rigid beam-column connections
Part II: 3-D finite element analysis of the connections using angles

Yanghui Xu & Kazuhiko Kasai
Tokyo Institute of Technology, Yokohama, Japan

Changshi Mao
Lehigh University, Bethlehem, USA

ABSTRACT: A sophisticated three-dimensional finite element model simulating the experimental specimens of bolted semi-rigid beam-column connection using angle sections (Part I, Kasai et al. 2000) is presented, and it is followed by 9 hypothetical models for parameter study. The connection stiffness, strength, and ductility as well as plastic mechanism of the top angle are discussed based on these models. Three yield zones are identified for the top angle, and the extent of yielding in each zone appears to be affected by the varied parameters. The parameters studied include: vertical leg length of angle, locations of column bolt and beam bolt, distance of beam to column, thickness and width of angle, strain hardening as well as column bolt pretension.

1 INTRODUCTION

Many experiments on bolted semi-rigid connections using angle sections have been performed to-date. However, local responses such as stress and strain distributions and interactions among the connection components are not easy to observe in the experiments. In order to closely examine the local responses and to investigate the effects of topological and material parameters of the connection, finite element analysis is employed in the present study.

A three-dimensional finite element model is formulated, simulating the experimental specimens explained in the companion paper (Part I, Kasai et al. 2000), and its accuracy demonstrated via correlative analysis. Then, a parameter study follows by hypothetically varying parameters of the model.

2 CORRELATIVE FINITE ELEMENT MODEL

The 3-D model (named as Correlative Model) simulates semi-rigidly connected full-scale beam-column subassembly experimented by Kasai et al. (1998, 2000). The subassembly is characterized in Table 1 (Beam length measured from load point to column center line is specified. Length up to the column face is 2815 mm).

In the test set-up, the beam was connected to the column by the top and bottom angles. The shear tab with slotted bolt holes was welded to the column flange and bolted to the beam web. Small fillet welds were provided at the tips of the horizontal legs of the angles.

Correlative Model is formulated by using ABAQUS. Only half of the subassembly is modeled with respect to the middle plane of the beam and column webs (Fig. 1). The beam, column, angles, shear tab, 4 beam bolts, and 4 column bolts as well as fillet welds are modeled by first-order 8-node brick elements. 4 shear tab bolts are simulated by truss elements. Portions of the beam and column near the angles are densely meshed, while other portions coarsely meshed, in order to save computational cost. A technique to treat mesh

Table 1. Components of subassembly.

Component	Designation	Size (mm)
Angles	L8 x 6 x 3/4	203 x 152 x 19, 381 wide
Beam	W24 x 76	609 deep, 3048 long (c-to-c)
Column	W14 x 257	355 deep, 1829 long (c-to-c)
Bolts	F10T	24 diameter
Shear tab		162 x 318 x 13

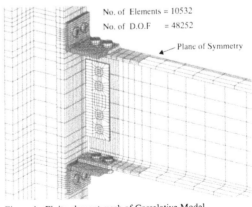

No. of Elements = 10532
No. of D.O.F = 48252
Plane of Symmetry

Figure 1. Finite element mesh of Correlative Model.

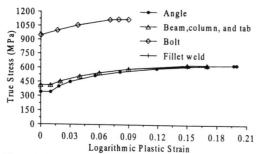

Figure 2. Input curves defining strain hardening behavior.

discontinuity will be explained later. Element numbering for each component is independent such that it can be freely varied.

Material nonlinearity is considered for all the components. Young's modulus is specified as 29500 ksi (203400 MPa), and Poisson's ratio 0.3. Strain hardening is defined by true stress vs. logarithm plastic strain, as shown in Figure 2. The curves are converted from either tensile coupon test or mill test results. The column and shear tab are assumed to have the same material as the beam. They remain essentially elastic in the test. Geometric nonlinearity is also considered during the analysis procedure.

Contacts and interactions among components are generally simulated by the option called "Surface" provided in ABAQUS. There are three types of Surfaces, namely Tied, Small-sliding, and Finite-sliding. Tied Surface ensures the compatibility of the neighboring elements who do not share nodes, thus, overlapping and separation of the elements are prevented. In contrast, either Small-sliding or Finite-sliding Surface prevents overlapping, but permits separation as well as sliding along the interface of the elements. The locations of Surfaces are depicted in Figure 3: Tied Surface is used for simulating weld between the shear tab and the column flange, and the fillet welds between the angles and beam (①, ②). It is also applied to the area of mesh discontinuity where the dense and coarse mesh meet (③, ④). Small-sliding Surface is used at the surface between either

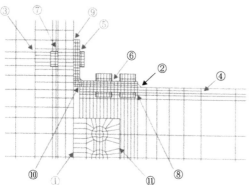

Figure 3. Locations of surface-based interaction simulations.

the bolt head or nut and the base metal (⑤ to ⑧). It is also used for the interface between the angles and either column or beam flange (⑨, ⑩), except for the portion of the beam flange beneath the top angle heel where Finite-sliding Surface is used due to potentially large slip between them. Finite-sliding Surface is also used for interactions between the shear tab and the beam web (⑪), and between the bolt shank and the base metal in order to prevent the bolt from penetrating into the metal and to permit application of the pretension force. The maximum slide distance is specified so as to reduce the size of the wavefront in finite-sliding simulations. The friction coefficient for all interactions is 0.33, except that 0.2 is specified at the interface between the shear tab and beam web (⑪). In addition, the nodes of the shear tab and the beam web at the shear tab bolt positions are constrained by equations to ensure they have the same vertical displacements.

Analysis is divided into two steps: First, all 12 bolts are pretensioned by specifying pretension forces at the pretension nodes. Second, monotonic downward load is applied at the cantilever beam end.

In the present analysis where surface-based interaction simulations as well as material and geometric nonlinearities are considered, the challenge is to obtain a convergent solution in the least possible computational time. For the interfaces between the bolt nuts or heads and the column and beam flanges(⑦, ⑧), the use of Tied Surface instead of Small-sliding is much less expensive, and virtually leads to the same result. Therefore, this technique is used in the subsequent analyses for the parameter study.

3   VALIDATION AND DISCUSSIONS OF CORRELATIVE MODEL

3.1 Validation

Figure 4 compares the connection moment vs. rotation curves (M-θ) of Correlative Model and the test. The moment is calculated at the column face. For the analysis result, M is twice the value obtained from actual analysis which used only half of the structure by considering symmetry (see Fig. 1). The rotation of the connected beam end relative to the column face is

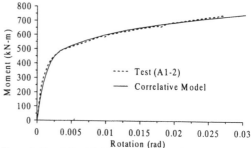

Figure 4. Correlative Model: moment vs. rotation compared with the test result.

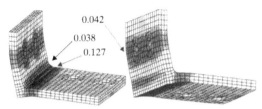

Figure 5. Correlative Model: equivalent plastic stain of the top angle (front and back view).

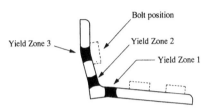

Figure 6. Plastic mechanism of the top angle.

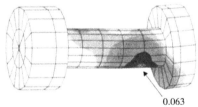

Figure 7. Correlative Model: maximum equivalent plastic strains for three yield zones.

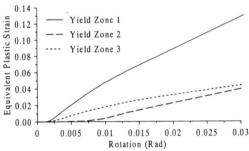

0.063

Figure 8. Correlative Model: equivalent plastic strain of the top column bolt.

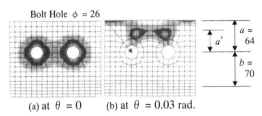

(a) at $\theta = 0$ (b) at $\theta = 0.03$ rad.

Figure 9. Correlative Model: contact pressure at the interface between the top angle and column (unit: mm).

considered. The clockwise moment and rotation are considered for the model. The two curves match well, indicating accuracy of the model.

3.2 Discussions

Like the test specimens, Correlative Model shows significant yielding of the top angle and moderate yielding of the column bolts under large rotation, while other components remain nearly elastic, as intended. Based on the contour of equivalent plastic strain (ε^{pl}) of the top angle (Fig. 5), three yield zones are identified as follows:

    Yield Zone 1: at the fillet area of the horizontal leg,

    Yield Zone 2: at the fillet area of the vertical leg,

    Yield Zone 3: through the column bolt holes,

which are depicted in Figure 6.

Figure 7 shows the maximum equivalent plastic strain (ε^{pl}) recorded for each yield zone. At $\theta = 0.03$ rad., the maximum ε^{pl} is about 13% at Zone 1, 4 % at Zone 2 and 3 (see also Fig. 5). This analytical result agrees with the test in which Zone 1 fractured due to large inelastic deformation.

Top column bolt also develops plastic bending deformation (Fig. 8), with the maximum ε^{pl} of 6.3% at the shank near its head. This also agrees with the permanent deformation of the bolt observed in the test.

The contact pressure between the top angle and the column face is shown in Figure 9. At $\theta = 0$, just after the column bolt pretension, the effective contact area due to the bolt pretension is circular and its diameter about 3 times the bolt diameter. As the moment and rotation develop, prying force center gradually shifts to the edge of the vertical leg. At $\theta = 0.03$ rad., its distance a' measured from the hole center is 40 mm, about 2/3 times the edge-distance.

4 PARAMETER STUDY

Parameter study is designed to examine the effects of topological and material parameters of the semi-rigid connection on its behavior, by hypothetically varying each parameter in Correlative Model.

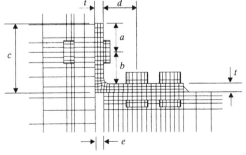

Figure 10. Topological parameters.

Table 2. Summary of 3-D finite element models.

Model No.	Identity	Description	ε^{pl} *	Yield Zone No.	M** (kN-m)
0	Correlative Model	Closely simulate the test specimens	0.127	1	747
1	Longer Vertical Leg Model (a = fixed)	Increase c from 152 to 203 mm b from 70 to 121 mm	0.075	1	516
2	Longer Vertical Leg Model (b = fixed)	Increase c from 152 to 203 mm a from 64 to 115 mm	0.128	1	748
3	Closer Beam Bolt Model	Reduce d from 95 to 64 mm	0.110	1	766
4	Closer Beam Model	Reduce e from 19 to 5 mm	0.111	1	778
5	Closer Beam and Beam Bolt Model	Reduce d from 95 mm to 64 mm e from 19 to 5 mm	0.118	2	803
6	Thinner Angle Model	Reduce t from 19 to 13 mm	0.095	1 and 3	461
7	Narrower Angle Model	Reduce w from 191 to 168 mm	0.131	1	707
8	No Hardening Model	Angles are elastic-perfectly plastic	0.269	1	607
9	No Pretension Model	Column bolts are not pretensioned	0.127	1	744

* Maximum equivalent plastic strain among three yield zones at θ = 0.03 rad.
**Connection moment at θ = 0.03 rad.

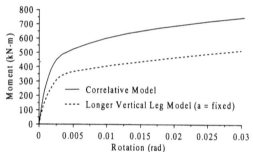

Figure 11. Longer Vertical Leg Model (a =fixed): moment vs. rotation.

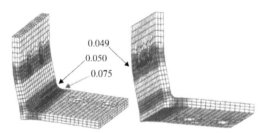

Figure 12. Longer Vertical Leg Model (a =fixed): equivalent plastic strain of the top angle.

The parameters studied include (see Fig. 10):
  a = edge-distance of column bolt;
  b = distance of column bolt to horizontal leg;
  c = length of vertical leg;
  d = distance of beam bolt to vertical leg;
  e = distance of beam to column;
  t = thickness of angle;
  w = width of the angle; and
  strain hardening of angle;
  pretension force of column bolt.
Note that the width of angle w is the dimension defined in the direction of the flange width of either column or beam. The parameter study models are summarized in Table 2.

4.1 Effect of column bolt position

In Model 1 named as Longer Vertical Leg Model (a = fixed), column bolts are shifted away from the beam, while maintaining its edge-distance a (Table 2). This causes weaker and more flexible performance of the connection, as shown by M- θ curve in Figure 11.

The distance b is 1.73 (= 121 mm / 70 mm) times that of Correlative Model. Accordingly, the moment at θ = 0.03 rad. is 0.69 times (516 kN-m / 747 kN-m) (Table 2). Ductility demand significantly decreases: maximum ε^{pl} of the top angle is 7.5%, occurring at Zone 1 (Fig. 12), 0.6 times that of Correlative Model. The column bolt inelastic deformation is also significantly less.

It is also noted in Figure 12 that Yield Zone 2 is more uniform along the angle width. Thus, with a longer distance b, the force from the bolts more uniformly spreads at Zone 2.

4.2 Effect of column bolt edge-distance

In Model 2, named as Longer Vertical Leg Model (b = fixed), the edge-distance a is increased 1.8 times (= 115 mm / 64 mm) that of Correlative Model. M- θ curve (Fig. 13) shows increasing the edge-distance a brings little change on the performance of the connection.

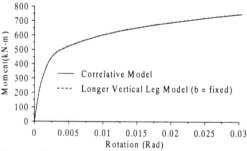

Figure 13. Longer Vertical Leg Model (b =fixed): moment vs. rotation.

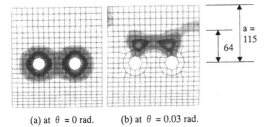

(a) at θ = 0 rad.   (b) at θ = 0.03 rad.

Figure 14. Longer Vertical Leg Model (b = fixed): contact pressure at interface between the top angle and column (unit: mm).

Figure 14 shows the contact pressure between the top angle and the column face for θ being 0 and 0.03 rad., respectively. The prying force appears to be almost the same as that of Correlative Model. In Figure 14, the edge-distance (= 115 mm) of Model 2 is compared with that (64 mm) of Correlative Model. In Model 2, there is only a slight pressure in the extended portion of the vertical leg, giving the reason why increasing the edge-distance has little effect on the connection performance.

4.3 Effect of beam bolt position

In Model 3 (Closer Beam Bolt Model), the beam bolt is moved closer to the column, with d reducing from 95 to 64 mm. M - θ curve (Fig.15) shows that making the beam bolt closer slightly increases the strength. Figure 16 shows von Mises stresses of the top angle and beam, as compared with Correlative Model. The

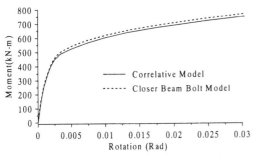

Figure 15.   Closer Beam Bolt Model: moment vs. rotation.

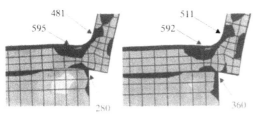

(a) Correlative Model   (b) Closer Beam Bolt Model

Figure 16. Von Mises stress of the top angle and beam. (viewed from the symmetric plane, unit: MPa)

beam end pushes up harder the heel of the angle, as evidenced by the greater stress in the beam end (i.e. 360 MPa), and smaller gap between them.

4.4 Effect of beam distance from column face

Model 4 (Closer Beam Model) is used for studying the effect of the distance e of the beam to the column face. e is reduced from 19 mm (thickness of the angle) to 5 mm. M - θ curve (Fig. 17) indicates that both stiffness and strength are increased by reducing e.

What is more important is that, the top angle develops less plastic deformation at Yield Zone 1, but more at Zone 2 (Fig. 18), compared with Correlative Model. In Closer Beam Model plastic deformation of the top angle is undertaken by both the vertical leg and horizontal leg rather than concentrating at the horizontal leg as in Correlation Model, suggesting more ductile performance.

Note, however, with the beam closer to the column a large contact force occurs between the beam end and the heel of angle (Fig. 19, at θ = 0.03 rad.). The compression force plays an important role in forcing the top angle to yield at Yield Zone 2, as well as increasing the stiffness, strength, and ductility. However, it may cause the beam end to yield: Figure 20 shows maximum ε^{pl} = 1.02% at the local area of the beam.

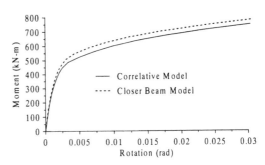

Figure 17. Closer Beam Model: moment vs. rotation.

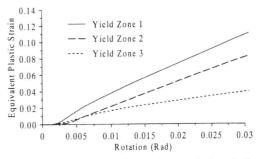

Figure 18. Closer Beam Model: maximum equivalent plastic strains for three yield zones.

211

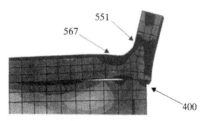

Figure 19. Closer Beam Model: von Mises stress of the top angle and beam (viewed from the symmetric plane, unit: MPa).

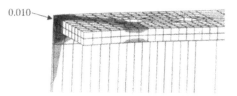

Figure 20. Closer Beam Model: equivalent plastic strain of the beam.

4.5 Effect of closer beam and beam bolt

The parameters examined from Model 3 and Model 4 are combined in Model 5 (Closer Beam and Beam Bolt Model), where *e* is reduced from 19 to 5 mm, and *d* from 95 to 64 mm. *M* - *θ* curve (Fig. 21) shows the moment at *θ* = 0.03 rad. is 1.07 times, and the elastic stiffness is about 1.08 times those of Correlation Model (see also Table 2).

The vertical leg develops more plastic deformation than horizontal leg (Fig. 22). This mechanism, in contrast to the others, indicates that yielding would be most significant at Yield Zone 2. Closer Beam and Beam Bolt Model shows higher stiffness and strength, and it could be more ductile than Correlative Model.

The beam yields at the beam end beneath the angle heel, as in Closer Beam Model (Figs.19 and 20). And due to the increased moment the column bolt exhibits 7.5% of plastic strain, as compared with 6.3% of Correlative Model.

4.6 Effect of angle thickness

In Model 6 (Thinner Angle Model) the angle thickness is reduced from 19 to 13 mm. Obviously, this will significantly weaken the connection. The reduction additionally leads to increase in *b* (distance of the column bolt to the horizontal leg) and *d* (distance of the beam bolt to the vertical leg). Further, the angle thickness (13 mm) becomes smaller than *e* (distance of the beam to the column). Based on discussions in Sections 4.1, 4.3, and 4.4, these further make Model 6 much weaker and less stiff than Correlative Model (see Fig. 23 and Table 2).

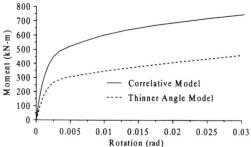

Figure 23. Thinner Angle Model: moment vs. rotation.

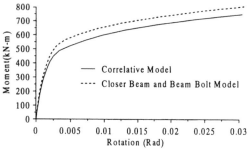

Figure 21. Closer Beam and Beam Bolt Model: moment vs. rotation.

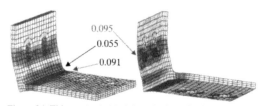

Figure 24. Thinner Angle Model: equivalent plastic strain of the top angle.

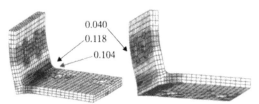

Figure 22. Closer Beam and Beam Bolt Model: equivalent plastic strain of the top angle.

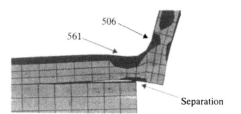

Figure 25. Thinner Angle Model: von Mises stress of the angle and the beam (viewed from the symmetric plane, unit: MPa).

Figure 24 depicts equivalent plastic strain of the top angle when θ = 0.03 rad. It shows large plastic deformation occuring at Yield Zone 1 and 3. The model uses the same column bolts as other models in spite of the relatively weak and flexible angle. Thus, the column bolt's restraining effect against the vertical leg of the angle increases. The strain gradient of the vertical leg (at Zone 3) becomes significant at the area immediately below the bolt head (Fig. 24).

Note also that the beam separates from the angle heel when θ = 0.03 rad.(Fig.25), which implies that the beam exerts mainly tension force on the angle.

4.7 Effect of angle width

In Model 7 (Narrower Angle Model), the width of the angle is reduced from 191 to 168 mm in the half subassembly model. Although the angle width is reduced to 0.88 times (= 168 / 191), the moment at θ = 0.03 rad. is reduced to 0.95 times, and the stiffness remains essentially unchanged (see Fig. 26 and Table 2).

Equivalent plastic strain of the top angle and beam at θ = 0.03 rad. (Fig. 27) shows the deformation of Model 7 is more uniformly distributed in the width direction, compared with Correlative Model. This stems from reduction of unsupported width of the angle. It implies that small unsupported width

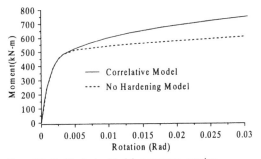

Figure 28. No Hardening Model: moment vs. rotation.

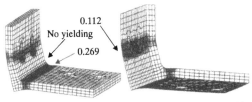

Figure 29. No Hardening Model: equivalent plastic strain of the top angle.

promotes 2-dimensional deformation: that is, the deformation pattern of the angle remains almost unchanged in the width direction. In this case, the moment becomes proportional to the angle width. The contribution of the unsupported part becomes less when the width is larger, since the angle exhibits 3-dimensional deformation pattern of having smaller bending deformation at unsupported part of the angle.

4.8 Effect of strain hardening of angle

In Model 8 (No Hardening Model) the angle is assumed to be elastic-perfectly plastic. M - θ curve in Figure 28 shows the stiffness remain unchanged compared with Correlative Model, but the moment at θ = 0.03 is reduced by the factor of 0.81. Figure 29 shows equivalent plastic strain of the top angle, where only two yield zones are found, and Yield Zone 2 remains elastic. Furthermore, spread of yielding in the two zones is modest compared with that of the other models. Thus, the strain demand is high and maximum ε^{pl} reaches 26.9% at Yield Zone 1 when θ = 0.03rad. This means the angle without strain hardening may be much less ductile.

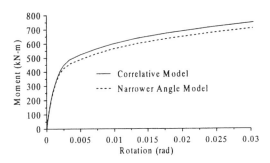

Figure 26. Narrower Angle Model: moment vs. rotation.

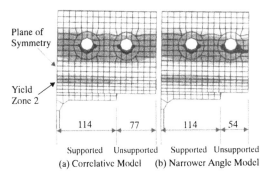

(a) Correlative Model   (b) Narrower Angle Model

Figure 27. Narrower Angle Model: equivalent plastic strain of the top angle and beam compared with Correlative Model.

4.9 Effect of column bolt pretension

In Model 9 (No Pretension Model) the column bolts are not pretensioned. M - θ curve in Figure 30 demonstrates reduction of the connection stiffness at small deformation, but at large deformation, the model response becomes almost the same as Correlative Model. The contact pressure between the top angle and the column flange at θ = 0.0001 and 0.03 rad. are shown in Figure 31, respectively, and the prying force

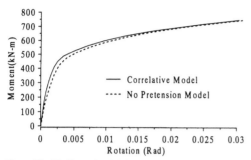

Figure 30. No Pretension Mode: moment vs. rotation.

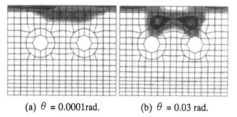

(a) $\theta = 0.0001$ rad.　　　(b) $\theta = 0.03$ rad.

Figure 31. No Pretension Model: contact pressure between the top angle and the column flange.

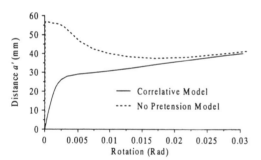

Figure 32. No Pretension Model: prying force center distance vs. rotation compared with Correlative Model.

center distance a' (defined in Fig. 9) against rotation θ is plotted in Figure 32. When the column bolt is not pretensioned, the prying force center is located at the edge of the angle when loading starts. Consequently, the angle starts separating from the column face from onset of loading. However, it is interesting to note that the prying force center shifts inward as loading continues, and its location becomes similar to that of Correlative Model.

5 CONCLUSIONS

Three-dimensional finite element models of bolted semi-rigid beam-column connections using angle sections have been presented in the paper. The connection stiffness, strength, and ductility as well as plastic mechanism of the top angle have been discussed

based on Correlative Model and 9 hypothetical models with varied parameters. The findings are summarized as follows:

1. Three yield zones are identified for the top angle when the connection is under negative moment. The maximum plastic strain typically occurs at the horizontal leg (e.g. Correlation Model), or at the vertical leg (Closer Beam and Beam Bolt Model). In some cases, only two yield zones take place (e.g. No Hardening Model).

2. Column bolt position (parameter b) and thickness of the angle significantly affect stiffness and strength of the connection. Width of the angle (measured in the direction of beam or column flange width) proportionally affects strength when its unsupported part is within a certain distance. Otherwise, the additional width has much less effect.

3. Distance of the beam to the column sensitively affects plastic mechanism of the angle, and hence affects the connection performance, especially ductility. Closer distance promotes similar magnitude of strains between Zones 1 and 2, and the connection would be more ductile. Beam bolt position (parameter d) also affects the plastic mechanism.

4. Location and magnitude of the prying force are considerably affected by column bolt pretension when loading is moderate. In contrast, they are hardly affected at ultimate loading. When bolt pretension is provided, the center gradually shifts towards the edge of the top angle. Otherwise, separation of the top angle from the column flange face will occur when loading starts, which causes less initial stiffness of the connection.

5. When edge-distance of column bolt is large, above-mentioned shifting of prying force center will terminate at a certain location defined by the parameter a'. Therefore, the edge-distance will bring little effect if it exceeds a certain limit.

The detailed analyses here are being furthered currently at Tokyo Institute of Technology. The extensive analysis results are also being utilized in order to propose the simplified analysis and design method of the semi-rigid connections.

REFERENCES

Kasai, K., Xu, Y., & Mayangarum, A., 2000. Experiment and analysis of bolted semi-rigid beam-column connections (Part I: cyclic loading experiment). STESSA.

Maison, B.F., Rex, J., Lindsey, S., & Kasai, K., 2000. Performance of PR moment frame buildings in UBC seismic zones 3 and 4. *Journal of Structural Engineering*, January.

Mayangarum, A., & Kasai, K., 1997. Design, analysis, and application of bolted semi-rigid connections for moment resisting frames. Report No. 97-15, ATLSS, Lehigh University.

Kasai, K., Mao, C., & Mayangarum, A., 1998. Feasibility of bolted rigid and semi-rigid connections for seismic regions. Summary report of third US-Japan workshop on steel fracture issues. Building Research Institute (BRI) of Japan, Kouzai-club.

Manual of ABAQUS. Hibbitt, Karlsson & Sorensen, Inc. 1997.

Behaviour of Steel Structures in Seismic Areas, Mazzolani & Tremblay (eds) © 2000 Balkema, Rotterdam, ISBN 90 5809 130 9

Fracture of beam-to-column connection simulated by means of the shaking table test using the inertial loading equipment

Y. Matsumoto
Yokohama National University, Yokohama, Kanagawa, Japan

S. Yamada
Tokyo Institute of Technology, Yokohama, Kanagawa, Japan

H. Akiyama
Nihon University, Tokyo, Japan

ABSTRACT: The structural performance of a steel member concerned with the fracture is influenced by the member's scale and loading speed . Thus, behaviors of full scale members under real seismic motion need to be clarified. The aim of this study is to simulate the fracture at the end of beam used in middle-rise moment resistant frames under severe earthquake. Specimens composed of H-shaped beams and box columns were subjected to seismic motions by means of shaking table tests, in which the inertial loading equipment was used. In this paper , two series of experiments are reported. The first aims at examining the effect of details of welded beam-to column connections on the deformation capacity. The second aims at examining the transition of fracture modes, from ductile ones to brittle ones, according to temperature.

1 INTRODUCTION

For ultimate design of moment resistant steel frames against severe earthquakes, it is important to estimate plastic deformation capacities of the members. However, the prediction method of the ultimate performance of fractured steel member is not established. In both the Northridge earthquake (1.17. 1994. U.S.A) and the Hyogoken-nanbu earthquake unexpected fractures were found at beam-to-column connections and it became necessary to clarify the mechanism of fracture.

The fracture modes are categorized into ductile and brittle ones and the latter can caused deterioration in ultimate performance of members. The geometrical notch effect is recognized as a factor of brittle fracture and the many improved details of beam-to-column connections have been suggested in order to reduce the stress concentration (e.g., Tabuchi, M. et.al. 1993). Equally, brittleness of material is another important factor. Therefore two series of experiments were performed in this research. The first aims at examining the effect of detail of welded connections on the deformation capacity. The second aims at examining the transition of fracture mode and deformation capacity according to the temperature.

In the case of fracture, ultimate performance of member is influenced by both the geometrical scale and loading speed. Thus, behaviors of full scale members under real seismic motions need to be examined. For this purpose, specimens were subjected to seismic motions by means of full scale shaking table tests, in which the inertial loading equipment was used (Yamada, S. et.al. 2000).

2 METHOD OF EXPERIMENT

2.1 *Specimens*

Each specimen is a partial frame taken out from middle-rise building structure. It was composed of a H-shaped beam and a column with rectangular hollow section (Fig. 1). The column and panel zone had enough strength and stiffness to eliminate the fracture other than fracture at the end of the beam.

2.2 *Parameters*

(1) Series 1, Details of Connections
Major parameters were concerned with details of the welded connections and the thickness of column's flanges, which governs the efficiency in transmitting the stress in the web of the beam. The list of specimens is shown in Tabel.1 and the details are shown in Fig. 2. The specimen No. 1, No. 2 and No. 6 had ordinary scallops. The two kinds of improved scallop (Tabuchi, M. et.al. 1993) and non-scallop (Yabe, Y. et.al. 1992) were employed in this research. The tests were performed at 13~18°C in temperature and the material was tough enough as mentioned below.

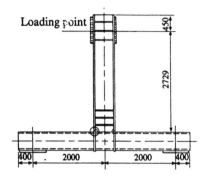

Fig.1 Shape of Specimen

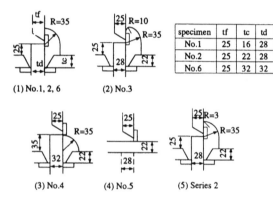

specimen	tf	tc	td
No.1	25	16	28
No.2	25	22	28
No.6	25	32	32

(1) No.1, 2, 6　　(2) No.3

(3) No.4　　(4) No.5　　(5) Series 2

Fig.2 Details of connections

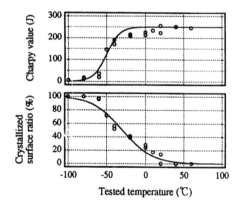

Fig.3 the result of Charpy impact test
for series 1

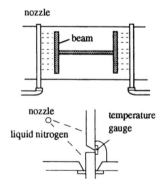

Fig.4 the result of Charpy impact test
for series 2

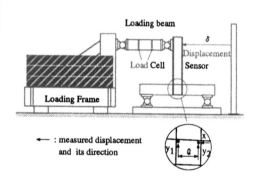

← : measured displacement
and its direction

Fig.5 Test set-up

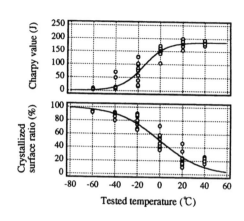

Fig.6 Refrigerating method

216

(2) Series 2 , Transition of fracture modes

The parameter was the test temperature which governs the material toughness. The range of temperature was set from -50℃ to +15℃, including -15℃, which was the transition temperature obtained by Charpy impact test. The list of specimens is shown in Table.2. All specimens had welded beam-to-column connections with ordinary scallops shown in Fig.2 (5).

2.3 Material properties

The mechanical properties of steel used in beam's flanges are shown in table 3. The results of Charpy impact tests are shown in Fig.3 and Fig.4.

2.4 Welding conditions

The CO_2 gas shield arc semi-automatic welding was employed for beam-to-column connections and the wire ranks as JIS Z3312 YGW11. The welding conditions are shown in Table 4.

2.5 Test set-up and loading processes

In this research the inertial loading equipment was used in order to apply seismic motions to the specimens. The test set-up is shown in Fig.5. The column of specimen was supported by a pin at each end, and the beam was connected to the loading beam by a pin. The specimen was subjected to inertia force which generated through the vibration of the shaking table.

Load cells were installed to the loading beam and displacement sensors were set as shown in Fig.5. These instruments recorded the shear force of the beam and displacements at the loading point and the end of beam. Using these records, the moment at the end of beam, M, and deflection of the beam, ϑ, can be obtained.

For the series 2, the specimens were covered by an insulation box and refrigerated by sprays of liquid nitrogen as shown in Fig.6. The temperature gauges were installed in the backing plates of welds and the temperature at the connections were measured.

The seismic wave used in the tests was the NS component of the 1995 Kobe(Japan) record and the energy input level was adjusted by multiplying amplification factor to the accelerogram. The three loading processes were applied to each spacemen as follows.

Pulse excitation: Pulse wave was inputted to the shaking table and subsequent free vibration was observed in order to examine the natural period and the damping factor of test set-up.

Elastic excitation: Seismic wave was inputted in order to examine the elastic properties of the specimen. The target peak velocity of shaking table was adjusted to be 5 kine or 10 kine.

Ultimate excitation: Seismic wave was inputted in order to realize fracture at the beam end. The target peak velocity of shaking table was 100 kine.

Table 1. The list of specimens (Series 1)

Specimen	Beam	Column	Scallop
No.1		Box-500×500×16	ordinary
No.2		Box-500×500×22	ordinary
No.3	H - 600×300	Box-500×500×22	improved
No.4	×12×25	Box-500×500×22	improved
No.5		Box-500×500×22	none
No.6		Box-500×500×32	ordinary

Table 2. The list of specimens (Series 2)

Specimen	Temperature	Beam	Column	Scallop
SC+15	15℃			
SC-10	-10℃	H-600×300	Box-500	
SC-20	-20℃	×16×25	×500×22	ordinary
SC-30	-30℃			
SC-50	-50℃			

Table 3. Mechanical properties of materials

Series	Specimen	σy (ton/cm²)	σu (ton/cm²)	Y.R.	εu (%)
1	No.1~6	3.30	5.15	0.64	20.4
2	SC+15	3.45	5.14	0.67	
	SC-10	3.60	5.31	0.68	
	SC-20	3.66	5.37	0.68	17.9
	SC-30	3.72	5.44	0.68	
	SC-50	3.86	5.58	0.69	

σ_y : yield point
σ_u : tensile strength
Y.R. : yield ratio
ε_u : uniform elongation capacity

Table 4. Welding conditions

Series	Pass	Current (A)	Voltage (V)	Heat Input (kJ/cm)	Interpass Temperature (℃)
1	11~15	300~380	32~37	11~27	50~180
2	15~22	300	35~37	5~24	60~140

217

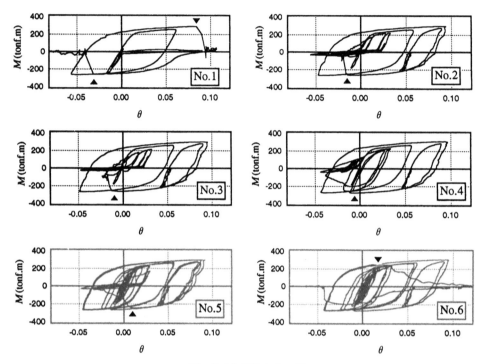

Fig. 7. *M*-θ Relationships

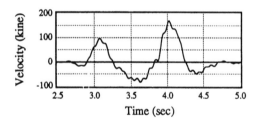

Fig. 8. Velocity at the loading point

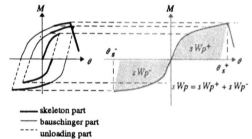

- ▬▬ skeleton part
- ─── bauschinger part
- - - - - unloading part

Fig.10 Hysterisis loop and Its Decomposition

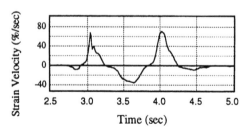

Fig. 9. Strain Velocity

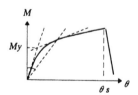

Fig.11 Definiton of experimental
full yield moment

3 RESULTS OF SERIES 1

The $M - \vartheta$ relationships for the ultimate excitations are shown in Fig. 7. The mark ▼ indicates the point where the fracture occurred. For all specimens, fractures occurred at the beam's end connected to the column. The fracture path propagated in the base metal of the flanges.

For the ultimate excitation of No. 2, the velocity at the loading point of beam is shown in Fig. 8. The peak velocity was 170 kine. For the same seismic excitation, the strain velocity at the fractured beam's flange is shown in Fig. 9. The peak velocity was 70%/sec. The similar response velocities occurred for the other specimens.

Hysteresis loop can be decomposed and skeleton curve can be extracted as shown in Fig. 10. (Kato, B. and Akiyama, H. 1969) The definition of experimental properties of the beam are shown in Fig. 10 and Fig. 11. The results are summarized in Table . 5.

For all specimens, crystallized surface ratio of torn flanges were below 55% and beam's plastic zone extended about 100 cm in length from the column surface. Thus, all fractures in this series can be

regarded ductile. Unfavorable effects of dynamic loading which causes brittle fractures were not observed.

The deformation capacity of the beam is evaluated in terms of the equivalent cumulative inelastic deformation ratio defined as follows.

$$_E\eta = W_p / (M_y \cdot \vartheta_y) \qquad (2)$$

$$_E\eta_S = {}_sW_p / (M_y \cdot \vartheta_y) \qquad (3)$$

$_E\eta$ and $_E\eta_S$ for all specimens are shown in Fig. 12. The specimens with ordinary scallops showed high deformation capacities, above 35 in $_E\eta_S$, comparable to those in improved connections. This result leads to the conclusion that high toughness of material reduces the sensitivity to geometrical notch effects and is effective to prevent brittle fracture. No. 5 and No. 6 showed higher $_E\eta$ than the others. Since No. 5 and No. 6 collapsed under the second application of ultimate excitation, these two specimens had a chance to develop a large energy absorption due to Baushinger effect. But, it is important to note that

Table . 5 Results of Series 1

Specimen	Loading Direction	Crystallized Surface ratio	T (sec)	My (tonf•m)	Mu (tonf•m)	$\eta\,s$	Wp (tonf•m)	sWp (tonf•m)
No.1	+	55%	0.66	194	290	15.1	71.3	50.8
	-	30%			265	11.5		
No.2	+	-	0.64	202	298	16.2	99.3	53.0
	-	12%			267	10.1		
No.3	+	-	0.64	206	301	16.3	100.2	53.5
	-	0%			268	10.2		
No.4	+	-	0.63	209	312	16.3	121.8	53.3
	-	0%			277	10.2		
No.5	+	-	0.63	193	299	15.4	204.7	54.2
	-	0%			271	12.5		
No.6	+	0%	0.61	206	307	16.3	176.2	55.8
	-	0%			275	11.9		

T : natural period of test set-up

My : full-yield moment obtained by means of the general -yield method(Fig.10)

Mu : maximum moment

η_s : cumulative inelastic deformation ratio for skeleton curve defined as follows

$$\eta_s = (\vartheta_s - \vartheta_y) / \vartheta_y \qquad (1)$$

where ϑ_s : maximum cumulative deflection for skeleton curve

ϑ_y : elastic deflection correspond to My

W_p : cumulative inelastic strain energy for hysteresis loop from the elastic excitation to the fracture point of the ultimate excitation

$_sW_p$: cumulative inelastic strain energy for skeleton curve

$_E\eta_S$-values are not so different from other specimens and $_E\eta_S$ can be regarded to be the most important parameter as an indicator of fracture.

4 RESULTS OF SERIES 2

The $M-\vartheta$ relationships for SC+15, SC-30 and SC-50 at the ultimate shakings are shown in Fig. 13. The mark ▼ indicates the point where fracture occurred. The $M-\vartheta$ relationships for SC-10 and SC-20 were similar to SC+15's. For theses three specimens, ductile fracture occurred at the end of the beam and fracture path propagated in the base metal of the flange (Photo. 1). For SC-30, brittle fracture was triggered by a crack initiated at the scallop of the beam and fracture path propagated in the base material of the flange (Photo. 2). For SC-50, brittle fracture was triggered from the weld toe near the end tab and fracture path propagated through the weld metal and the diaphragm (Photo. 3).

The response velocities were similar to series 1. The peak velocities at loading point were approximately 170 kine and the peak strain velocities were approximately 70%/sec.

The results are summarized in Table. 6. The column of 'temperature' indicates the temperature measured just before the ultimate excitation. However, the increases of temperature according to the inelastic strain energy absorption were observed for SC+15, SC-10 and SC-20. The temperature surge was 40℃ in the ultimate excitation of SC+15.

The transition of $_E\eta$ and $_E\eta_S$ according to the temperature is shown in Fig. 14. SC-50, in which the fracture took place through the welding metal and diaphragm, is not shown in this diagram. When the temperature was above -20℃, the specimens showed high deformation capacities which were above 20 in $_E\eta_S$ and above 25 in $_E\eta$. On the other hand, SC-30

fractured at an early stage of plastic deformation and showed the lower $_E\eta$ and $_E\eta_S$ than the others. It can be seen that the transition temperature of deformation capacity is around -20 ℃, which can be well correlated to Charpy impact test.

As mentioned above, the increases of temperature were observed for the specimens tested above -20℃ in temperature. It is conceivable that this temperature growth had an advantageous effect in reducing the transition temperature and exaggerated the transition of $_E\eta$ and $_E\eta_S$ from -30℃ to -20℃.

5 CONCLUSIONS

In order to simulate the fractures at the end of beam under severe earthquake, the full scale shaking table tests were performed. The specimens composed of H-shaped beams and box columns were subjected to seismic motions and the ultimate performances were investigated. The results were obtained as follows

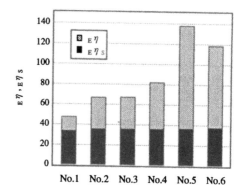

Fig.12.　$_E\eta$ and $_E\eta_S$ for series 1

Table 6. The results of series 2

Specimen	Temperature (℃)	Loading Direction	Crystallized Surface ratio	T (sec)	My (tonf•m)	Mu (tonf•m)	η_S	Wp (tonf•m)	sWp (tonf•m)
SC+15	15.1	+	0%	0.64	209	289	12.5	55.6	43.4
		-	-			286	9.4		
SC-10	-10.5	+	0%	0.65	230	311	12.6	52.4	42.0
		-	-			281	8.3		
SC-20	-20.0	+	5%	0.64	234	311	10.5	47.7	37.1
		-	95%			285	7.7		
SC-30	-29.0	+	100%	0.64	241	272	2.7	5.7	4.8
		-	100%			219	0.2		
SC-50	-44.9	+	100%	0.65	-	144	-	-	-
		-	100%			188	-		

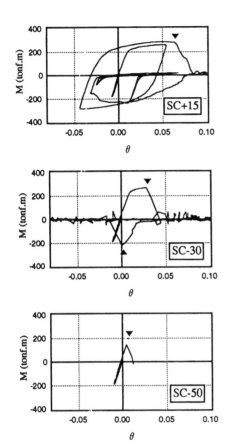

Photo.1. Fracture of SC+15

Fig.13. *M-θ* Relationships

Photo.2. Fracture of SC-30

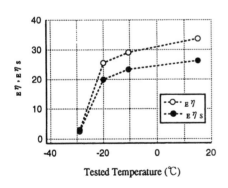

Fig.14. Transition of $_E\eta$ and $_E\eta$ s according to temperature

Photo.3. Fracture of SC-50

Series.1

In case of tough material, the specimens with ordinary scallops developed high deformation capacities, comparable to those in improved connections. The fracture mode were ductile ones and unfavorable effects of dynamic loading which causes brittle fracture were not observed. High toughness of material reduces the sensitivity to geometrical notch effects and is effective to prevent brittle fracture.

Series 2

When temperature was above -20℃, ductile fractures occurred and all specimens showed high deformation capacities. In this case, the increase of specimen's temperature according to the plastic deformation was observed. When temperature was below -30℃, brittle fracture occurred and the specimen showed the deterioration in deformation capacity according to the temperature.

REFERENCES

Kato,B. and Akiyama,H. (1969), Inelastic bar subjected to trust and cyclic bending, Journal of the structural division, Proc. of ASCE, Vol.95, No.ST1, pp.33-56

Tabuchi,M., Kanatani,H., Tanaka,T. and Sonoda,R. (1993), Improvement on scallop shape in RHS column to H beam connections by welding, Journal of constructional steel, Vol.1, pp.191-198, (in Japanese)

Yabe,Y., Sakamoto,S. and Nakagomi,T. (1992), Effect of scallops at beam end on mechanical behaviors of H-shaped steel beams connected to box column, Trans. of A.I.J., No.440, pp.125-132, (in Japanese)

Yamada,S., Matsumoto,Y. and Akiyama, H. (2000), Experimental Method of the Full Scale Shaking Table Test Using the Inertial Loading Equipment, Proc.of STESSA2000.

Behaviour of Steel Structures in Seismic Areas, Mazzolani & Tremblay (eds) © 2000 Balkema, Rotterdam, ISBN 90 5809 130 9

Quality assurance for welding of Japanese welded beam-to-column connections

M. Nakashima
Disaster Prevention Research Institute, Kyoto University, Japan

ABSTRACT: This paper presents an overview of post-Kobe research efforts on the ductility capacity of welded beam-to-column connections. Comparisons were made between the U.S. and Japan with respect to the damage patterns and post-earthquake actions. Quality assurance regarding the design and fabrication of post-Kobe welded beam-to-column connections is discussed in terms of the cost estimate based on the weight of steel, adoption of box columns and all rigid connections, employment of gas-shielded metal arc welding, and environment of fabrication industry.

1. INTRODUCTION

The Hyogoken-Nanbu (Kobe) earthquake shook Kobe and surrounding areas on January 17, 1995. In the long history of large Japanese earthquakes, the Kobe Earthquake was the first to cause widespread and serious damage to modern steel buildings (AIJ 1995a, b, AIJ 1997a). Many suggest that ground motions in Kobe were much larger than those experienced during previous earthquakes in Japan. Moreover, the Kobe area is one of the earliest urban developments in Japan and, consequently, contained a large inventory of older steel buildings designed and constructed with obsolete practices. Whatever reasons may be given, the fact remains that modern steel buildings experienced significant damage, refuting the popular myth in Japan that steel buildings are immune to strong earthquakes. Among various damage patterns, the damage to welded beam-to-column connections was considered most serious in that it occurred in steel frame buildings designed and constructed with the most recent practices. Since that time, extensive research efforts have been made to reevaluate the ductility capacity of the pre-earthquake connections, identify the causes of the damage, and improve the capacity by various modifications.

This paper presents an overview of post-Kobe steel research on welded beam-to-column connections and attempts to evaluate the ductility performance of post-Kobe welded connections, particularly form the perspectives of quality assurance. To expedite the discussion, comparison is made throughout the paper with the post-Northridge welded beam-to-column connections being developed in the U.S. after experiencing the serious damage to their welded beam-to-column connections in the 1994 Northridge Earthquake. This paper consists of the following four parts. In the first part, similarities and differences in the damage patterns between Japan and the U.S. are summarized. Then, significant differences in post-earthquake actions between the two countries are noted. These parts serve to highlight the unique features of Japanese post-Kobe connections. In the third part, a few design and fabrication practices characteristic of Japan are introduced, and their interaction with quality assurance is discussed. The issues discussed are: the design practice of all rigid connections in steel moment resisting frames, the design practice with box columns, the practice of the gas-shielded metal arc welding, and the environment of steel fabrication.

2. COMPARISON IN DAMAGE BETWEEN U.S. AND JAPAN

In both the Northridge and Kobe Earthquakes, steel buildings sustained significant damage, and many similarities in damage patterns were disclosed. Some notable similarities are summarized as follows.

2.1 *Damage Similarities*

(1) Steel buildings in Japan and the U.S. had not experienced much damage in previous earthquakes. These two earthquakes exposed for the first time in each country the significant

damage in welded steel moment resisting frame buildings.

(2) Many modern building structures designed and constructed with present practices were damaged. Thus, damage was not exclusively associated with old technologies and design practices.

(3) Much damage was found, but no building constructed using the most recent design and construction practices collapsed.

(4) Many welded beam-to-column connections failed by fracturing, indicating that welded connections were one of the weakest locations in steel moment frames.

2.2 Differences in Damage, Design, and Construction

Differences in damage patterns and sources were also observed. Notable differences are summarized as follows.

(1) Beam-to-column connections fractured, but in many instances fractures in Japanese structures were preceded by significant plastification and local buckling, meaning that the beams dissipated some energy before fracture. The vast majority of fractures in the U.S. involved virtually no plastification in either beams or columns. Thus, the degree of plastic rotation capacity of steel beams-to-column connections constructed using pre-Northridge and pre-Kobe practices may have been significantly different.

(2) Steel materials used may also be different. It appears that Japan has placed relatively more attention to the importance of material strength and strain hardening for securing beam plastic rotation capacity. For example, before the Kobe Earthquake, Japan developed a new type of steels (named SN steels) having a good margin between the yield and ultimate stresses, a smaller variation in these specified stresses, and larger fracture toughness. The use of so-called dual-certified steels in the U.S prior to the Northridge Earthquake suggests less concern in this regard.

(3) Welding processes and procedures are significantly different between the two countries. Japan almost exclusively employs gas-shielded metal arc welding with solid wires, whereas self-shielded flux-cored welding is commonly used in the U.S. Japanese welding is often conducted in the shop, whereas the critical welding of beams to columns in the U.S. is commonly done in the field.

(4) Connection details are also different. Japan construction typically uses square tube (box) columns, whereas wide-flange columns are usually employed in the U.S. This difference is accompanied by many differences in local connection details, such as through-diaphragm connections in Japan versus through-column connections in the U.S., and welded web to column joints in Japan versus bolted web to column in Northridge connections in the U.S (Figure 1).

(5) The redundancy of the moment frame system is not the same in the two countries. All beam-to-column connections are rigidly connected in Japan, whereas in the U.S. rigid connections are commonly assigned only to selected locations. In addition to the degree of redundancy, this difference affects member sizes, the importance of gravity loads relative to seismic loads, and the stress condition (bi- versus uni-directional bending) in columns.

More details on comparisons between the U.S. and Japan are presented elsewhere (Nakashima et al. 1998a, Nakashima 2000).

3. DIFFERENCES IN POST-EARTHQUAKE APPROACH

Damage to steel buildings in Northridge and Kobe was believed to have occurred because of a mixture of various sources related to design, materials, welding, connection details, and structural systems.

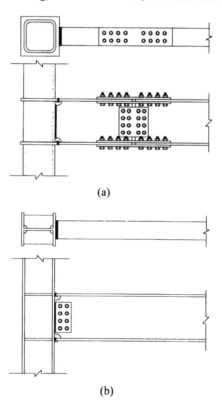

(a)

(b)

Figure 1. Typical welded beam-to-column connections: (a) through-diaphragm connection in Japan ; (b) web bolted connection in U.S.

This understanding is shared between the U.S. and Japan. However, solutions being provided after few years of post-earthquake efforts appear to be significantly different in many aspects between the two countries. Some examples of differences, particularly related to beam-to-column connections, are summarized below.

3.1 *Materials*

Use of materials with larger ductility can be a solution toward higher seismic performance of steel buildings. Japan developed SN steels before the Kobe Earthquake. Use of the new steels is still optional at the present time, but their use has been increasing significantly. The U.S. also introduced a new steel grade and revised specifications for testing of materials to be used in seismic details. It appears that the utilization of special high toughness steels is gaining acceptance faster in Japan than in the U.S.

3.2 *Welding*

Fractures at weld metals were very serious in the U.S., and use of different electrodes having a larger toughness and controlled deposition rate is now mandatory. In Japan, fractures at weld metals were also disclosed in many instances, and welding with stringer bead placement to avoid too large heat input is strongly recommended. Efforts to develop tougher electrodes are also underway in Japan. In general, however, the U.S. is more explicit as to the changes in welding and inspection practices.

3.3 *Connection Details*

Regarding connection details, the U.S. has pursued three courses: moving the plastic hinge away from the beam end, improving the local details and *in situ* material properties for conventional unreinforced connections, and substitution of welded connections by bolted connections. Many believe that moving the plastic hinge region is the most secure way to improve the ductility capacity of beam-to-column connections. Many new details have been developed along this line, such as strengthening of beam ends by cover plates, haunches, ribs, etc. or trimming beam flanges at a location away from the column face (named the Reduced Beam Section (RBS) connection). Such strengthening is considered as a possible solution also in Japan, but the general sentiment is that sufficient ductility capacity can be ensured by modifying connection details combined with good welding. Many efforts have been made to modify details by changing the size and shape of weld access holes, etc. Figure 2 shows some examples (AIJ 1996). After five years of extensive studies, it has been felt that connection details

without any weld access holes (shown in Figure 2c) can ensure the most ductile performance among the various post-Kobe connection details considered.

3.4 *Structural Systems*

As to the structural system employed, it is not likely that Japan will abandon square tube (box) columns and switch to wide flange columns at least for the foreseeable future. Similarly, the practice of using rigid connections for all beam-to-column connections will also likely remain unchanged. By the same token, the U.S. practices of using wide-flange shapes for columns and providing moment resisting connections in a small portion of the structure will likely remain unchanged as well.

More details on the differences in post-earthquake actions between the U.S. and Japan are found in Nakashima (2000) and Nakashima et al. (2000).

4. PRACTICIS CHARCTERISTIC OF JAPAN

4.1 *Market Share of Steel Buildings*

Steel is a very popular structural material in Japanese building construction. Figure 3a compares the total floor area of steel buildings constructed each year with that employed in construction using other structural materials. Wood has ranked first for years, but it is used almost exclusively for residential houses. Steel ranks second, followed by reinforced concrete (RC) and steel-encased reinforced concrete (designated SRC in Japan). Figure 3b shows the total floor area of steel buildings constructed each year with respect to the number of stories in each building. This figure suggests that the vast majority of steel buildings are shorter than five stories in height. In fact, most of steel buildings in Japan are

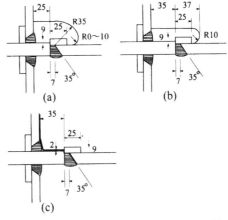

(a)

(b)

(c)

Figure 2. Weld access hole details employed in Japan: (a) Conventional detail ; (b) Modified detail with a smaller hole; (c) No hole detail.

low-rise, used for offices, shops and mixed occupancies, as well as for industrial and manufacturing structures.

4.2 Design Practice of All Rigid Connections

In Japan, the initial estimate of structural cost for steel frame buildings has been based on the weight of steel, and maximum efforts are made to reduce the total steel weight in structural design. When subjected to vertical loads, the beam connected rigidly to the columns sustains smaller bending moment than does the beam supported by shear connections, which makes it possible to reduce the required beam size and accordingly the cost (Figure 4a, b). This appears to be the strongest reason for adopting the design practice of all rigid connections. It is rather difficult to achieve rigid connections for the weak axis of wide-flange columns, which leads us to use box columns in that they can readily develop rigid connections in two horizontal directions. In moment resisting frames, the bending moment capacity of wide flange columns is smaller about the weak axis than about the strong axis. This is another motivation to use box columns.

When using closed sections like box columns, we need special considerations to join beams to the columns. The through-diaphragm connection, shown in Figure 1a, is a type of connections developed to this end. In this connection, the square-tube (often cold-formed in Japan) used for the column is cut longitudinally into three pieces: one used for the column of the lower story, one for the connection's panel zone (a short piece, often called a "dice" in Japan), and one for the column of the upper story. Two diaphragm plates are inserted between the three separated pieces and shop-welded all around. Then, short segments of beams are welded to the panel zone: the beam flanges are welded directly to the diaphragms and the web to the side of the dice. The entire piece (often called a Christmas tree) is transported to the construction site and connected to the mid-portion of the beam by high-tension bolts.

Comparison between the two types of connections shown in Figure 1 indicates that the through-diaphragm connection with a box column requires significantly more volume of weld in establishing one connection than the flange welded web bolted connection with a wide flange column. In the web bolted connection, welding is limited only to beam flanges. On the other hand, in the through diaphragm connection, the weld locations are: (1) the

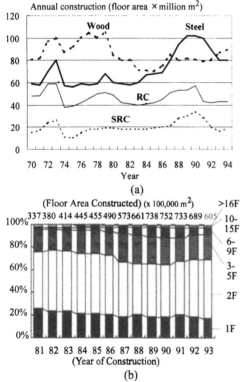

Figure 3. Adoption of steel in Japanese building construction: (a) Market share with respect to structural material; (b) Volume of construction with respect to number of stories.

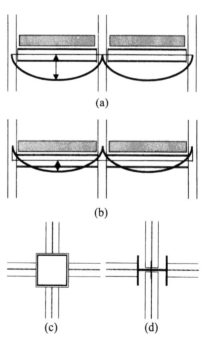

Figure 4. Design practice of all rigid connections:
(a) Beam moment distribution with rigid connection ;
(b) Beam moment distribution with shear connection;
(c) Two-way moment connection with box column;
(d) One-way moment connection with wide flange column.

column bottom with the top diaphragm, (2) the panel zone with the top diaphragm, (3) the panel zone with the bottom diaphragm, (4) the column top with the bottom diaphragm, (5) the beam top flange with the top diaphragm, (6) the beam bottom flange with the bottom diaphragm, and (7) the beam web to the column. This comparison suggests that Japanese practices are more weld-intensive. (It is notable that welding robots have been used increasingly lately, particularly for the welding of diaphragms to columns.)

4.3 *Practice of Gas Shielding Metal Arc Welding*

Welded flanges have been used since about 1970 in both the U.S. and Japan. U.S. practice started with shielded metal arc welding (SMAW) and rapidly progressed to self shielded flux core metal arc welding (FCAW). Japanese practice also started with SMAW, and gas shielded metal arc welding (GMAW) was introduced in the late 1960's. FCAW was also introduced in the late 1960's (Bada et al. 1969, Funabashi & Maruoka 1970) and used in the early 1970's primarily for field welding, with electrodes developed and marketed in Japan. FCAW lost favor (Morita et al. 1988) because of reatively low performance at that time, and GMAW became the process of choice.

Figure 5 shows deposition rates that can be achieved by GMAW with solid wires of 0.8, 1.2 and 1.6 mm in diameter (JWES 1994). With standard

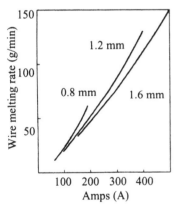

Figure 5. Deposition rate in GMAW process.

values of the voltage (30V) and amperage (300A), the deposition rate is about 50 kN per hour. Table 1 shows the deposition rates that can be achieved in typical FCAW process (Lincoln 1999). If compared for 1.2 mm solid wire in Figure 5 and E70TG-K2 in Table 1, those typically used for downhand welding, GMAW requires approximately two times to deposit the same amount of weld metal. Steel shop fabrication consists of multiple stages of preparation and work, and the actual arc time is a portion of the entire fabrication process. According to Yamamoto (1999), the time required for welding work is about 40% of the whole shop fabrication process. About half the welding time is exhausted by actual arcing, meaning that the actual arc time is about 20% of all. If the deposition rate were increased by two times, the fabricator would save about 10% of the total fabrication time.

Discussion on the design and fabrication practices of all rigid connections, box columns, and GMAW suggests that fabrication of Japanese connections requires a larger amount of weld and welding time than that of web bolted connection with FCAW, commonly used in the U.S.

4.4 *Environment of Fabrication*

Two fabricators' organizations are notable in Japan, commonly called with the acronyms of "Tekkenkyo" and "Zenkoren." As of September 1999, fifty-two and 3,933 fabricators join the "Tekkenkyo" and the "Zenkoren" respectively. In addition to the fabricators that join the organizations, there are many more unorganized, independent fabricators, whose exact number is not known. Market distribution among these three groups is approximately 100 ("Tekkenkyo"), 600 ("Zeonkoren"), and 300 (independent) in terms of the tonnage of annual production. Figure 6 shows the size of the fabricators in terms of the capital. The distribution is 5% for more than 100 million yen, 45% for 10-100 million yen, and 50% for smaller than 10 million yen (Tekkenkyo 1996). These statistical values indicate that Japanese fabrication industry is composed of many, small fabricators. Their income is not necessarily so appealing as compared to other industrial sectors, and furthermore, many of the fabricators now have difficulties to recruit young labor, because the image of the industry

Table 1. Deposition rate in typical FCAW processes.

Designation	Lincoln	Wire (inch)	Feed Speed (m/min)	Voltage (volts)	Current (amps)	Deposition Rate (kg/hr)
E70T-4	NS3M	0.12	355	38-40	600	18.0
E70TG-K2	NR311Ni	7/64	240	29-31	520	9.2
E70T-6	NR305	3/32	400	33-35	525	12.7
E71T-8	NR203MP	3/32	130	23-24	390	4.1
	NR232	5/64	180	22-23	365	3.9
E71T-8-Ni1	NR203Ni 1%	3/32	130	23-24	385	4.4

does not look fashionable (Zenkoren 1990, Zenkoren 1996, Nagano 1998).

Figure 7 shows the price of steel fabrication (in terms of yen per ton) with respect to the year (Steel 1999). Here, evidenced is the Japanese practice of estimating the cost based on the steel weight. In early 1990's when Japan enjoyed its bubble economy, the price was over 300,000 yen per ton, but under severe recession in recent years, it decreased to smaller than 200,000 yen per ton. This figure clearly shows large fluctuation of fabrication price and its strong dependency on economy. In its nature, the fabrication industry has to keep the fixed assets and expenses like machines, welders, engineers, etc., and therefore it is extremely difficult for them to adjust their production flexibly in accordance with economy. This often forces them to receive orders even when they cannot expect reasonable profit out of the orders.

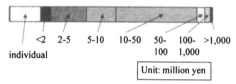

| | <2 | 2-5 | 5-10 | 10-50 | 50-100 | 100-1,000 | >1,000 |

individual

Unit: million yen

Figure 6. Distribution of steel fabricators in terms of capital.

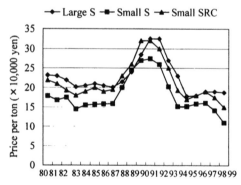

→ Large S → Small S → Small SRC

Figure7. Price of steel fabrication.

5. COMMENTS

Japanese welded beam-to-column connections generally require more volumes of weld than connections used in the U.S. This suggests that structural performance of Japanese connections is likely to be more dependent on the quality of welding. Then, how easy or difficult it is to ensure the quality of welding? Figure 8 shows the ductility capacities (expressed in terms of cumulative plastic ductility: η) of 86 full-scale beam-to-column connections, tested by the Steel Committee of the Kinki Branch of the Architectural Institute of Japan. The tests were carried out after the Kobe Earthquake, and ductility capacities of a few types of post-Kobe welded connections were examined. Details are provided in AIJ (1997b) and Nakashima et al. (1998b). In Figure 8, $\eta=40$ corresponds to about 0.03 radian of the maximum plastic rotation and about 0.3 radian of the cumulative plastic rotation. Those exhibiting $\eta>40$ are considered to possess sufficient plastic rotation capacity in light of the present Japanese seismic design code. According to Fig.8, 23 specimens (27% of the total) failed before reaching $\eta=40$. The vast majority of those exhibiting smaller rotation capacities were found to have defects associated with welding, such as large heat input by one-path per layer in weld bead placement and lack of fusion (disclosed in post-fracture observation). Structural steels with good quality were used in all specimens, and all welding had passed the UT inspection before the test. Nevertheless, 27% of the specimens showed the performance that was not necessarily encouraging. This figure clearly demonstrates the importance and at the same time difficulties of quality assurance of welding. The writer should note that the fabricators that fabricated the specimens were well qualified, sincere in their work, and did their best. The statement here is not against the fabricators by all means.

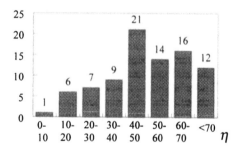

Figure 8. Distribution of plastic rotation capacity of welded beam-to-column connection.

Japanese welding is more time consuming (in terms of the disposition rate) than U.S. welding. The steel fabrication cost is estimated based on the weight of steel, which tends to blur the actual cost related to welding. The Japanese fabrication industry is fragmented, consisting of many, small fabricators. This makes it difficult for them to form a common voice on behalf of their profession. Their income fluctuates very significantly with respect to economy. Their labor condition is rather grim, with difficulties to recruit young labor. All of these observations suggest serious, potential difficulties for ensuring the quality of welding. It is very simple to state that we should ensure quality, but leaving the quality assurance only with the hands of fabricators is utterly unfair, unless the society respects the industry and offers reasonable rewards to them. The present situations indicate a strong doubt if the society is taking the right action.

In summary, the writer is not optimistic about the structural performance of Japanese post-Kobe welded beam-to-column connections, particularly from the viewpoint of quality assurance. Improvement of the environment of fabrication industry itself is no doubt an ultimate solution. It is a critical issue that the society has to face. It is very important but requires a long venue to resolve. From the perspective of structural engineering, the writer contends that the steel engineering community should place more efforts on developing steel connections that can be friendlier to fabrication.

6. CONCLUSIONS

This paper presented unique features of design and fabrication of Japanese welded beam-to-column connections. Steel is a very popular material for building structures. Japanese connections require more volumes of weld. The structural performance of welded beam-to-column connections is dependent heavily on the quality of welding, but it is not easy to achieve. Japanese welding is relatively slow. Cost of steel fabrication is estimated based on the weight of steel, which makes it difficult to reflect the actual welding cost. Japanese fabrication industry consists of many, small fabricators, and their work condition is not necessarily good. The fabrication cost fluctuates very significantly with economy, while the industry has difficulties in adjusting their orders according to the cost. Under these circumstances, it is unfair to leave the quality assurance of welding only to the fabrication industry. The writer wishes to emphasize that the society has to face the realism more positively and seek for solutions that do not torture the Japanese fabrication industry.

ACKNOWLEDGEMENT

The writer expresses his sincere appreciation to Prof. S. A. Mahin of the University of California at Berkeley for his precious input and comments regarding the U.S. post-Northridge actions about steel moment resisting frames.

DISCLAIMER

Any opinions expressed in this paper are exclusively those of the writer and do not necessarily reflect the views of Japan's steel engineering community.

REFERENCES

AIJ 1995a. *English edition of preliminary reconnaissance report of the 1995 Hyogoken-Nanbu Earthquake*, The Architectural Institute of Japan. Tokyo: AIJ.

AIJ 1995b. *Reconnaissance report on damage to steel building structures observed from the 1995 Hyogoken-Nanbu Earthquake*, The Kinki Branch of the Architectural Institute of Japan, Steel Committee (in Japanese with attached abridged English version). Osaka: AIJ.

AIJ 1996. JASS-6: *Technical recommendations for steel construction for buildings (Part 1: Guide to steel-rib fabrications)*, The Fourth Edition, The Architectural Institute of Japan (in Japanese). Tokyo: AIJ.

AIJ 1997a. *Report on the Hanshin-Awaji earthquake disaster: Structural damage to steel buildings*, Building Series Volume 3, The Architectural Institute of Japan (in Japanese). Tokyo: AIJ.

AIJ 1997b. *Full-scale test on plastic Rotation capacity of steel wide-flange beams connected with square tube steel columns*, The Steel Committee of the Kinki Branch of the Architectural Institute of Japan (in Japanese). Osaka: AIJ.

Baba, T., et al. 1969. Fuji nongas wire NA-50, *Fuji Technical Report*, 14(2):250-265 (in Japanese).

Funabashi, I. and Maruoka, Y. 1970. Non-gas semi-automatic welding method, *Report of Takenaka Technical Research Laboratory*, 97:1-17 (in Japanese).

JWES 1994. *Manual for semi-automatic CO2 shielded metal arc welding*, The Japan Welding Engineering Society, Tokyo: Sanpo Shuppan.

Lincoln Electric 1999. *Innershield Catalog*, Lincoln Electric, Cleveland.

Morita, K. et. al. 1988. Discussion on new technologies regarding field welding, *Journal of Steel Structural Technologies*, 20: 19-30, Tokyo: Kokozo Shuppan (in Japanese).

Nagano Prefecture Fabricators' Association 1998. *Action plan for steel fabrication* (in Japanese).

Nakashima, M., et al. 1998. Full-scale test on beam-column subasemblages having connection details of shop-welding type, *Proceedings of the Structural Engineers World Conference*, Paper#:T158-7, New York: Elsevier.

Nakashima, M., Inoue, K., and Tada, M. 1998. Classification of damage to steel buildings observed in the 1995 Hyogoken-Nanbu Earthquake", *Journal of Engineering Structures*, 20(4-6): 271-281, Essex: Elsevier.

Nakashima, M. 2000. Overview of damage to steel building structures observed in the 1995 Hyogoken-Nanbu (Kobe) earthquake, *The Final Report of SAC Joint Venture* (to appear).

Nakashima, M., Roeder, C. W., and Maruoka, Y. 2000. Steel moment frames for earthquakes in the United States and Japan, *Journal of Structural Engineering*, ASCE (tentatively accepted).

Steel Structure Journal 1999. July Issue, Tokyo: Kokozo Shuppan.

Tekkenkyo 1996. Toward establishment of building steel production system, *Tekkenkyo Report*, Tekkenkyo (in Japanese).

Yamamoto, Y. 1999. Personal communication with M. Nakashima.

Zenkoren 1996. Vision of building steel fabrications, creation of refined steel fabrication, *Zenkoren Report*, Zenkoren (in Japanese).

Zenkoren 1990. Report of committee on employment problems, *Zenkoren Report*, Zenkoren (in Japanese).

Behaviour of Steel Structures in Seismic Areas, Mazzolani & Tremblay (eds) © 2000 Balkema, Rotterdam, ISBN 90 5809 130 9

Experimental studies on post-tensioned seismic resistant connections for steel frames

J. M. Ricles, R. Sause, M. M. Garlock, S. W. Peng & L. W. Lu
Department of Civil and Environmental Engineering, Lehigh University, Bethlehem, Pa., USA

ABSTRACT: A series of experimental tests were conducted to investigate the behavior of an innovative post-tensioned (PT) top-and-seat-angle wide flange (WF) beam-to-column moment connection for steel moment resisting frames subjected to seismic loading conditions. Nine large-scale specimens were tested. Each specimen represented an interior connection and consisted of two WF beams attached to a column. The parameters investigated in the study include the angle size, angle gage length, beam flange reinforcing plates, connection shim plates, and post-tensioning force. The results of the test program demonstrate that post-tensioned connections possess exceptional cyclic strength and ductility. Energy dissipation occurs in the angles while other structural members remain elastic. The initial elastic stiffness is comparable to that of a welded connection, and following severe inelastic cycles of drift the connection has little permanent deformation. The angle parameters are shown to influence the connection moment capacity and energy dissipation. Therefore, experimental investigations were carried out on the angles alone to better evaluate the effects of angle size and angle gage length on the PT connection behavior.

1 INTRODUCTION

Structural steel has been widely used in moment resisting frame (MRF) systems for buildings. The connections in steel MRFs are either welded or bolted, with welding becoming common during recent decades. A typical welded moment connection detail consists of a bolted shear tab with full penetration beam flange welds. During the 1994 Northridge earthquake, over 130 steel-framed buildings suffered unexpected connection fractures (Youssef et al. 1995, FEMA 1995). The cyclic strength and ductility of these connections were diminished as a result of the fractures. Several alternative moment connection details have been proposed since the Northridge earthquake in an attempt to develop ductile response under earthquake loading. These details are intended to avoid weld failure and force inelastic deformation to develop in the beams away from the welds. Consequently, after the design-level earthquake the beams with these connections will have permanent damage caused by yielding and local buckling. This damage can result in a significant residual drift of a MRF.

As an alternative to welded construction the authors developed a post-tensioned (PT) moment connection for use in seismic resistant steel MRFs. There are several advantages of a PT connection. These advantages include: (1) field welding is not

required; (2) the connection is made with conventional materials and skills; (3) the connection has an initial stiffness similar to that of a typical welded connection; (4) the connection is self-centering without residual deformation, thus the MRF will not have residual drift after an earthquake if significant residual deformation does not occur at the base of the columns; (5) the beams and columns remain essentially elastic while inelastic deformation of the connection provides energy dissipation; and (6) the angles are easily replaced. The connection utilizes high strength steel strands that are post-tensioned after bolted top and bottom seat angles are installed (Figure 1). The post-tensioning strands run through the column, and are anchored outside the connection region (Figure 2). Although the top-and-seat angles are easily replaced, experimental studies conducted by the authors show that the angles have a sufficient low-cycle fatigue life to perform well over several earthquake loading events.

This paper presents experimental studies of PT connection subassemblies subjected to cyclic inelastic loading. Nine large-scale specimens were tested. Each specimen represented an interior connection and consisted of two WF beams attached to a column. The parameters investigated in the study include the angle size, angle gage length, beam flange reinforcing plates, connection shim plates, and post-tensioning force. The angle parameters were shown to influence the connection's behavior, therefore, experimental investigations were carried out on the angles alone to evaluate the effects of angle size and angle gage length on the PT connection.

The development of the PT steel connection utilizes research from prior studies of PT precast concrete construction and partially restrained steel connections. For more details on this prior research see Garlock et al. (2000) and Ricles et al. (1999). The paper by Garlock et al. includes analytical studies performed on PT MRFs and compares the behavior of these PT frames to the behavior of the same frame with conventional welded moment connections.

2 POST-TENSIONED CONNECTION

2.1 Connection Details

A PT steel MRF connection consists of bolted top and seat angles with post-tensioned high strength strands running parallel to the beam and anchored outside the connection (see Figure 1). The strands compress the beam flanges against the column flanges to resist moment, while the two angles and the friction at the beam and column interface resist shear. The proposed details are shown in Figure 2 for a connection to an exterior column. The angles' primary purpose is to dissipate energy. However, they also provide redundancy to the force transfer mechanisms for transverse beam shear and moment. The beam flanges are reinforced with reinforcing plates to control beam yielding. Also, shim plates are placed between the column flange and the beam flanges so that only the beam flanges and reinforcing plates are in contact with the column.

2.2 Flexural Behavior

The idealized moment-rotation (M-θ_r) behavior and the corresponding load-deflection (H-Δ) behavior of a PT steel connection are shown in Figure 3, where θ_r is the relative rotation between the beam and column. The M-θ_r behavior of a PT connection is characterized by gap opening and closing at the beam-column interface under cyclic loading (Figure 3 insert). The moment to initiate this separation is called the decompression moment. The connection initially behaves as a welded connection, but following decompression it behaves as a partially restrained connection. The initial stiffness of the connection is the same as that of a welded moment connection when θ_r is equal to zero until the gap opens at decompression (event "a" in Figure 3). The stiffness of the connection after decompression is associated with the stiffness of the angles and the elastic axial stiffness of the post-tensioned strands. With continued loading, the tension angle of the connection yields

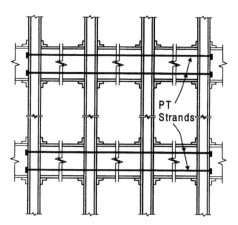

Figure 1. Elevation of a post-tensioned frame.

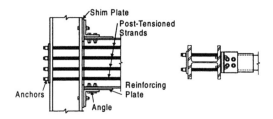

Figure 2. Post-tensioned connection detail at anchor point.

(event "b"), with full plastic yielding of the tension angles occurring at event "c". The compression angles yield at event "d". Until the load reverses at event event "e", the M-θ_r relationship has a nearly linear response where the connection stiffness is primarily due to the axial stiffness of the post-tensioned strands. Upon unloading, the angles will dissipate energy (between events "e" and "h") until the gap between the beam flange and the column face is closed at event "h" (i.e., when θ_r is equal to zero). A complete reversal in applied moment will result in a similar connection behavior occurring in the opposite direction of loading, as shown in Figure 3 where the M-θ_r and H-Δ relationships are symmetric.

As long as the strands remain elastic, and there is no significant beam yielding, the post-tensioning force is preserved and the connection will self-center upon unloading (i.e., θ_r returns to zero rotation upon removal of the connection moment and the structure returns to its pre-earthquake position). The level of the decompression moment, the flexural strength of the angles, and the elastic stiffness of the post-tensioning strands control the strength of the connection. The energy dissipation capacity of the connection is related to the flexural strength of the angles. To ensure that the strands remain elastic,

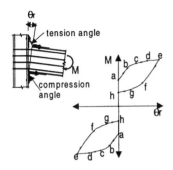
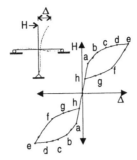

Figure 3. Moment-rotation and load-deflection behavior of a post-tensioned connection.

strands with lengths equal to one multiple of the beam clear span for each connection that the strands pass through are used. Thus, the deformation due to gap opening results in a small strain over a long strand length. The strands are post-tensioned to a stress level that is sufficiently below the yield stress.

3 EXPERIMENTAL STUDIES

3.1 *Subassembly Tests*

Nine large-scale cruciform-shaped subassemblies with PT connections were tested. The test matrix is shown in Table 1. The main parameters of this study were angle size, angle gage, the presence of reinforcing plates, shim plates, and post-tensioning force. All specimens except PC1 had shim plates and reinforcing plates. The angle gage length, g, is the distance along the column leg of the angle from the fillet to the near edge of the washer plate. The gage length was expressed as a non-dimensional gage length-to-thickness (g/t) ratio. All angles were 254 mm long. More detailed information on these experimental studies can be found in Peng et. al (2000).

Each cruciform-shaped subassembly with PT connections simulated an interior joint of a MRF. The beam size for all the test specimens was W24x62 with 248 MPa nominal yield strength (F_y). The column size was W14x311 with 350 MPa F_y for Specimens PC1 through PC5. A concrete filled tube (CFT) consisting of a 406 x 406 x 13 mm steel tube with 74 MPa compression strength concrete was used for Specimens PC6 through PC8. Figure 4 shows the test setup. Each PT connection had a total of 8 strands, with 4 to each side of the beam web. Commercially available strands were utilized in the PT connection. Each strand had an area of 140 mm^2 and an ultimate strength of 1860 MPa. The spacing between strands was 127 mm. The strands were ini-

Table 1. Specimen Test Matrix

Specimen	Angle Size	t (mm)	g/t
PC1	L152x152x7.9	7.9	9.0
PC2	L152x152x7.9	7.9	4.0
PC2-A	-	-	-
PC3	L203x203x15.9	15.9	7.2
PC4	L203x203x15.9	15.9	4.0
PC5	L203x203x25.4	25.4	4.0
PC6	L203x203x15.9	15.9	4.0
PC7	L203x203x15.9	15.9	4.0
PC8	L203x203x15.9	15.9	4.0

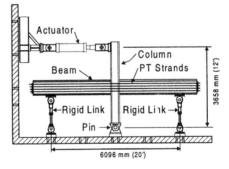

Figure 4. Subassembly test setup.

tially stressed to approximately 35% of their ultimate strength (about 93 kN per strand). Each strand was anchored at the end of the beams in the cruciform subassembly just beyond where the roller reactions (rigid links) were located.

Shim plates were placed between the end of the beam and column face for construction fit-up and to avoid contact of the beam web with the column. The shim plates were 275 x 254 x 9.5 mm (F_y=690 MPa) for Specimens PC2 through PC5, and 275 x 254 x 19 mm (F_y = 248 MPa) for Specimens PC6 through PC8. Reinforcing plates were welded to the beam

233

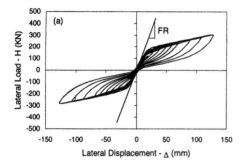

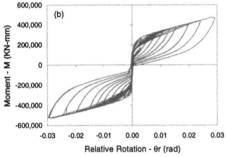

Figure 5. Specimen PC4 test results.

increasing amplitude. The initial cycles involved elastic response below the decompression moment of the connection, with the latter cycles corresponding to a maximum story drift of 3% to 3.5%.

3.2 Subassembly Test Results

Figure 5 shows the load deformation (H-Δ) and moment rotation (M-θ_r) plot of Specimen PC4. Notice that the initial stiffness of this connection is nearly that of a fully restrained (FR) welded connection. Also, at zero load, the relative rotation and displacement are zero. This demonstrates the self-centering capability of the connection. The decompression moment $M_{d,exp}$ was about 47% the plastic moment capacity of the beam, M_p. M_p was calculated as 576 kN-m using measured beam dimensions and tensile coupon values of the beam flange and beam web.

Figure 6 shows the effects of shim plates and reinforcing plates. Specimen PC1 did not have shim plates or reinforcing plates and therefore the region of the beam flanges in contact with the column yielded when the gap was open, but the beam web did not. When the gap closed, the flanges were no longer in good contact with the column. This caused

flanges. The reinforcing plates were 254 x 57 x 12.7 mm (F_y = 690 MPa) for Specimens PC2 through PC5, 610 x 203 x 12.7 mm (F_y = 690 MPa) for Specimens PC6 and PC8, and tapered 610 x (203 to 356) x 12.7 mm (F_y = 690 MPa) for Specimen PC7. Specimens PC2 through PC5 had two reinforcing plates on the inside of each flange, one on each side of the web. PC6 through PC8 had one reinforcing plate on the outside of each flange as shown in Figure 2. Specimen PC1 had neither shim plates nor reinforcing plates.

All connection bolts were 25 mm diameter. Two A325 bolts were used for column bolts (the bolts connecting the angle to the column) and beam bolts (the bolts connecting the angle to the beam) in the connection of Specimens PC1 and PC2. For Specimen PC5 two A490 column bolts and four A325 beam bolts were used. For the remaining specimens two A325 column bolts and four A325 beam bolts were used. The column bolts in each connection were snug tightened prior to applying the strand post-tensioning force. These bolts were then tightened to their required pretension following the application of the post-tensioning to the strands. This procedure avoided deforming the column leg of the top and seat angle. Strain gages were installed in the column bolts to verify the pretension as well as to monitor the bolt force during testing.

In each experiment the top of the column was subjected to a lateral displacement history consisting of a series of symmetric displacement cycles with

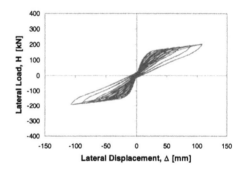

Figure 6. Specimen PC1 test results.

Table 2. Response quantities of subassembly test specimens.

Specimen	$\dfrac{M_{d,exp}}{M_p}$	$\dfrac{M_{max,exp}}{M_p}$	$\dfrac{T_{max,exp}}{T_u}$	$\theta_{r,\,max}$ (rad)
PC1	0.37	0.60	0.46	0.025
PC2	0.37	0.68	0.52	0.026
PC2-A	-	0.59	-**	0.030
PC3	0.40	0.72	0.53	0.021
PC4	0.47	0.89	0.51	0.025
PC5	0.47	0.59*	0.42*	0.014*
PC6	0.45	0.93	0.54	0.026
PC7	0.45	0.95	0.54	0.025
PC8	n.a.	0.29	n.a.	0.031

*Specimen PC5 was stopped at 1.75% maximum drift.
**Post-tensioning force in Specimen PC2-A not monitored during testing.

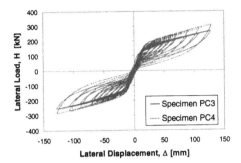

Figure 7(a). Effect of g/t ratio.

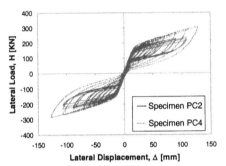

Figure 7(b). Effect of angle thickness.

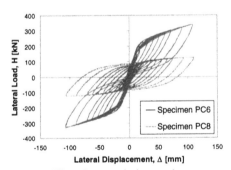

Figure 8. Effect of post-tensioning strands.

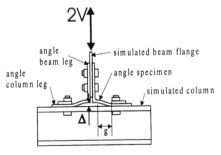

Figure 9. Isolated angle test setup.

a loss of stiffness and some self centering capability. The load displacement curves of all other specimens with shim and reinforcing plates look similar to the curve for PC4 shown in Figure 5.

Figure 7(a) and 7(b) show the effects of g/t ratio and angle size. Specimen PC4 had a smaller g/t ratio than PC3 and a larger thickness than PC2. The results show that connections with larger angle sizes and smaller g/t ratios dissipate more energy and achieve a greater moment for a given drift level.

By comparing the test results of Specimens PC6 and PC8, the beneficial effects of the post-tensioning strands are seen. Figure 8 shows that the post-tensioning strands (present in Specimen PC6 but not PC8) allow the connection to retain its self-centering capability. Also, PT strands increase the stiffness and moment capacity of the connection. Both specimens dissipate about an equal amount of energy.

A summary of the test results is shown in Table 2. Up to $0.95M_p$ was achieved in the connection (at the column face) at 3% drift ($M_{max,exp}$). The strand force at 3% drift, $T_{max, exp}$ never exceeded 54% of the ultimate strength of the strand, T_u, leaving a good margin of safety. A relative rotation, θ_r, of 0.021 to 0.030 radians at 3% drift indicates that the connections have excellent ductility since θ_r is analogous to the plastic rotation θ_p developed in a welded moment connection.

Test PC5 was terminated early because the bolts connecting the angles to the column developed large prying forces. Test PC2 was the only test in which the angles fractured. At the completion of this test, the angles were removed and the specimen was once again cyclically loaded as Specimen PC2-A. Notice that without angles, this connection develops $0.59M_p$. Since the angles provide most of the energy dissipation, Specimen PC2-A did not dissipate much energy.

3.3 Isolated Angle Tests

The angle parameters greatly influenced the connection behavior. Therefore, experimental studies were performed on isolated angles. For more details on these tests, see Garlock et. al (1999).

In the test setup, two angles were placed back-to-back as shown in Figure 9. The test matrix is shown in Table 3. All angles were 178 mm long. The L203x203 and L152x152 angles had a nominal F_y of 345 MPa and 248 MPa, respectively. Specimens with 203 mm angles had four 32 mm A325 beam bolts. Specimens with 152 mm angles had only two of these bolts. The column bolts were 25 mm A325 bolts for all specimens except L8-34-4 and L8-34-6 which had 25 mm A490 bolts. All specimens except L8-58-4-NW had 178 x 57 x 12 mm plates as washers instead of standard bolt washers. The purpose of these washer plates was to force a plastic hinge to occur in the angle leg at the washer plate edge and to reduce the bolt prying force.

The angles were cyclically loaded to specific levels of uplift, Δ, corresponding to SAC's testing protocol (SAC, 1997). The uplift values corresponded to estimated PT connection gap opening values at given story drifts for a connection with a W36 beam. For example, a 25 mm uplift, Δ, (i.e. gap opening) corresponds to approximately 4% story drift for a connection with a W36 beam.

Figure 10 shows a plot of V vs. Δ for Specimen L8-58-4. V and Δ are the shear force in the angle (assumed to be half the actuator force) and the angle uplift, respectively, as shown in Figure 9. Notice the similarity of the hysteresis loops of this angle to the hysteresis loops of PT connection PC4 shown in Figure 5. It was found that the angles can develop significant strength beyond the point at which a mechanism forms in the "column leg" (the leg of the angle that is connected to the simulated column). Figure 11 plots the points of peak V/V_p ratio value in each cycle against corresponding Δ for each specimen. V_p is the shear required to form a plastic mechanism, with the experimental value of V_p determined when a clear change in stiffness occurs in the V - Δ plot. A linear regression of V/V_p vs. Δ is also drawn. Figure 11 shows that the post-yield strength appears to be independent of gage length and angle size, and is a nearly linear function of Δ. The value of V/Vp is equal to β, which is an angle overstrength factor that accounts for geometric hardening (due to large deformations) and material strain hardening. The β factor is used to predict the moment capacity of a PT connection as shown in the next section.

The angle test results indicate that specimens with smaller g/t ratios dissipate greater energy, however, specimens with smaller g/t ratios have a shorter low cycle fatigue life. Therefore, one must find an optimum gage length that will produce adequate energy dissipation and sufficient fatigue life. The results also show that the washer plate had a negligible influence on the angle behavior.

4 PREDICTING PT CONNECTION MOMENT CAPACITY

The level of post-tensioning force in the strands controls the decompression moment. Decompression occurs when the contact force resultant in the beam tension flange is zero. The theoretical decompression moment is equal to:

$$M_{d,th} = d_c \frac{T_o}{2} \qquad (1)$$

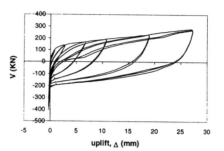

Figure 10. Isolated angle test results for Specimen L8-58-4.

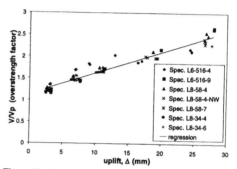

Figure 11. Angle overstrength factor.

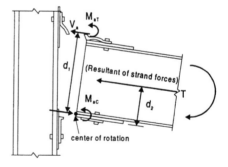

Figure 12. Free-body of PT connection.

T_0 and d_c are equal to the total initial post-tensioning force and the distance over the depth of the beam between the centroids of the contact areas between the beam flanges and column. This equation neglects the contribution of the bearing of the angle against the column face. To avoid excessive drift under gravity and wind load, $M_{d,th}$ should exceed the beam end moment due to gravity and wind load.

The moment capacity M_{max} of a PT connection is estimated considering the free body diagram shown in Figure 12. Assuming that the frame has undergone drift causing a relative rotation of θ_r between the end of the beam and column face, $M_{max,th}$ is determined as:

$$M_{max,th} = d_1 V_a + d_2 T + M_{aC} + M_{aT} \qquad (2)$$

where d_1, V_a, d_2, T, M_{aC}, M_{aT} are equal to the depth from the center of rotation to the fillet of the tension angle column leg, the tension angle force resultant, the height from center of rotation to the resultant force of the post-tensioning strands, the resultant force of the post-tensioning strands, the moment developed in the compression angle beam leg, and moment developed in tension angle column leg, as shown in Figure 12.

The tension angle force resultant is equal to:

$$V_a = \beta \frac{2M_g}{g} \qquad (3)$$

where β and M_g are the overstrength factor and angle moment developed at the end of the gage length g, respectively. The angle moment M_g is equal to the angle leg flexural capacity M_{pa}, whereby the angle shear V_p corresponds to $2M_{pa}/g$. β is based on test results of isolated angles [Garlock et al., 1999] described earlier. The PT connection specimens had a gap opening of about 17 mm at 3% story drift. Looking at Figure 11, this corresponds to a β value of about 2.

Considering the effect of strand elongation and beam shortening following decompression of an interior connection, the total force of the post-tensioning stands can be shown to be equal to:

$$T = T_0 + \frac{AE}{L} d_2 \theta_r - \Delta T \qquad (4)$$

where A, E, L, T_0 and ΔT are equal to cross-sectional area of the strands, modulus of elasticity of the strands, length of the strands, total initial post-tensioning force, and total decrease in post-tensioning force due to beam shortening after decompression of the PT connection occurs. The values of θ_r and ΔT vary with the story drift.

Figure 13(a) shows a plot of the average value of strand force versus the connection moment during the test of Specimen PC4. Notice that as the magnitude of the moment increases, the strand force increases. This is due to the elongation of the strands following decompression of the connection. Figure 13(b) compares the maximum value of the strand force in each cycle of the test of Specimen PC4 to the theoretical value based on Equation (4). This good correlation was also found for other specimens.

Table 4 shows that the correlation is good between $M_{max,th}$, $M_{d,th}$ and $T_{max,th}$, and the respective experimentally obtained values, where $T_{max,th}$ is based on Equation (4). Both $M_{max,th}$ and $T_{max,th}$ are determined at the maximum story drift. For Specimen PC1 the difference between the maximum theoretical strand force, $T_{max,th}$, and the maximum experimental strand force is larger than for the other

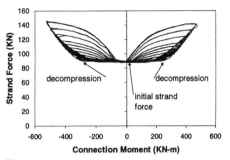

Figure 13(a). Average force per strand for Specimen PC4.

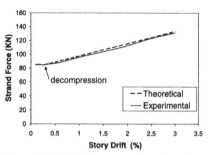

Figure 13 (b). Maximum experimental vs. theoretical strand force (Specimen PC4).

Table 3. Isolated angle test matrix.

Specimen	Angle Size (mm)	t (mm)	g / t
L6-516-4	L152x152x7.9	7.9	4.0
L6-516-9	L152x152x7.9	7.9	9.4
L8-58-4	L203x203x15.9	15.9	4.0
L8-58-4-NW	L203x203x15.9	15.9	4.0
L8-58-7	L203x203x15.9	15.9	7.2
L8-34-4	L203x203x19.0	19.0	4.0
L8-34-6	L203x203x19.0	19.0	5.8

*All specimens used washer plates except Specimen L8-58-4-NW which used a standard bolt washer.

Table 4. Theoretical response compared to experimental response.

Specimen	$\dfrac{M_{d,exp}}{M_{d,th}}$	$\dfrac{M_{max,exp}}{M_{max,th}}$	$\dfrac{T_{max,exp}}{T_{max,th}}$
PC1	0.97	0.90	0.84
PC2	1.00	0.90	0.93
PC2-A	-	0.94	-**
PC3	1.14	0.92	0.98
PC4	1.38	1.00	0.98
PC5	1.38	0.70*	0.96*
PC6	1.15	0.95	0.97
PC7	1.18	0.99	0.99
PC8	n.a.	0.89	n.a.

*Specimen PC5 was stopped at 1.75% maximum drift.
**Post-tensioning force in Specimen PC2-A not monitored during testing.

specimens, due to the beam yielding leading to a shortening of the beam.

5 SUMMARY AND CONCLUSIONS

The results of nine large-scale tests show that the PT steel connections can provide adequate strength and stiffness for a MRF system subjected to cyclic loading. Post-tensioning a top and seat angle connection results in a connection with an initial stiffness and strength similar to that of a fully restrained welded connection. The post-tensioning also provides a self-centering capacity, resulting in minimal story drift in a building after severe cyclic inelastic loading. Yielding occurred primarily in the top and seat angles. Since inelastic deformations are concentrated in the angles, it is easy to repair the building by replacing the angles.

It was found that connection shim plates and beam flange reinforcing plates are required in order to prevent the beam flanges from yielding under the large bearing forces developed at the zone of contact of the beam flanges and the column face. The size and gage length of the angles are shown to have an effect on the moment capacity and energy dissipation of the connection. Angles with either a larger thickness or smaller gage length produce a larger connection capacity and energy dissipation. The angle thickness and gage length, however, must be limited. An increase in angle thickness leads to larger bolt tension forces, and therefore requires more bolts. Shorter gage lengths result in larger accumulated plastic strain in the angles. This reduces the low cycle fatigue life of the angles. To maintain the self centering capability of the connection the post tensioning strands must be designed to remain elastic.

Expressions for predicting the connection moment capacity as a function of story drift were presented and found to agree well with the experimental results. These expressions are useful for design, where the required moment capacity must be provided while the self centering capability during an earthquake must be maintained.

6 ACKNOWLEDGEMENTS

The research reported herein was supported by the National Science Foundation (NSF) and a Lehigh University Fellowship. Dr. Ken Chong and Dr. Ashland Brown cognizant NSF program officials. The research was also supported by a grant from the Department of Community and Economic Development of the Commonwealth of Pennsylvania through the Pennsylvania Infrastructure Technology Alliance (PITA). PITA is co-directed by Dr. Pradeep Khosla and Dr. John Fisher. The opinions expressed in this paper are those of the authors and do not necessarily reflect the views of the sponsors.

7 REFERENCES

FEMA (1995), "Interim Guidelines: Evaluation, Repair, Modification and Design of Welded Steel Moment Frame Structures," Bulletin No. 267, FEMA, Washington, D.C.

Garlock, M, Ricles, J., Sause, R., Zhao, C., and Lu, L-W. (2000), "Seismic Behavior of Post-Tensioned Steel Frames," STESSA Proceedings, Montreal, Canada.

Garlock, M, Ricles, J., Sause, R., (1999), "Experimental Evaluation and Analytical Modeling of Bolted Angles Subject to Cyclic Loading", (in preparation for Journal of Structural Engineering).

Peng, S.W., Ricles, J., Sause, R., Chen, T.W., and Lu, L-W. (2000), "Experimental Evaluation of a Post-Tensioned Seismic Resistant Connection", (submitted to Journal of Structural Engineering).

Ricles, J., Sause, R., Zhao, C., Garlock, M., and Lu, L. (1999), "Post-Tensioned Seismic Resistant Connections for Steel Frames," (submitted to Journal of Structural Engineering).

SAC Joint Venture (1997) "Protocol For Fabrication, Inspection, Testing, and Documentation of Beam-Column Connection Tests and Other Experimental Specimens", Report No. SAC/BD-97/02, Version 1.1, October.

Youssef, N., Bonowitz, D., and Gross, J. (1995), "A Survey of Steel Moment Resisting Frame Buildings Affected by the 1994 Northridge Earthquake," NIST, *Report No. NISTIR 5625*, Gaithersburg, Md.

Behaviour of Steel Structures in Seismic Areas, Mazzolani & Tremblay (eds) © 2000 Balkema, Rotterdam, ISBN 90 5809 130 9

Inelastic seismic response of frame fasteners for steel roof decks

C.A. Rogers
Department of Civil Engineering and Applied Mechanics, McGill University, Montreal, Canada

R. Tremblay
Department of Civil, Geological and Mining Engineering, École Polytechnique, Montreal, Canada

ABSTRACT: Single-storey steel structures represent the vast majority of buildings that are constructed for light industrial, commercial and recreational uses in North America. It is often more cost effective to specify a non-ductile structural system, despite the fact that this type of structure and its occupants would be more vulnerable to seismic ground motions. The overall objective of this research is to investigate the possibility of allowing the metal roof deck diaphragm to absorb earthquake induced energy through plastic deformation. This paper provides preliminary information on the inelastic cyclic response, including load *vs.* displacement hysteresis and energy absorption capacity, of screwed, nailed (powder-actuated fasteners) and welded steel deck-to-structure connections. The results of monotonic, cyclic and quasi-static tests revealed that the type of connection influences the ultimate capacity and ability to dissipate energy.

1 INTRODUCTION

1.1 *General*

This paper highlights the results of a research project concerning the seismic performance of the steel deck-to-structure connections that are typically found in the roofs of single-storey steel buildings. Structures of this type are commonly used for light industrial, commercial and recreational buildings in North America. In Canada, a large proportion of these structures are located in the St-Lawrence and Ottawa River valleys, as well as along the Pacific coast, the most active seismic regions in the country. Seismic provisions have been recently included in Canadian design standards to ensure that an adequate level of seismic performance for steel structures exists. However, these required changes have increased construction costs, thus making it more attractive (comparing cost to safety increases) to use a non ductile structural system; despite the fact that this type of structure and its occupants would be more vulnerable to seismic ground motions.

The aim of this research project is to evaluate alternative design and construction methods that can be used to increase the cost efficiency of the bracing system while improving the seismic behaviour of the structure. One possible solution is to account for the inelastic response of the steel roof deck diaphragm in energy dissipation calculations.

In Canada, the seismic design of steel structures must conform to the National Building Code (NBCC) (1995), which refers to the CSA-S16.1 (1994) and CSA-S136 (1994) Standards for steel design related issues. In design, it is possible to use NBCC specified lateral seismic loads that are significantly lower than the maximum forces that would be expected under the design level earthquake, provided that the lateral load resisting system exhibits a stable and ductile cyclic inelastic response.

Under seismic ground motion, lateral inertia forces develop at the roof level due to the horizontal acceleration of the roof mass. To transfer and resist these lateral loads, the structure generally includes a metal roof deck diaphragm and vertical steel bracing. The roof diaphragm is made of steel deck units that are fastened to the supporting steel roof framing to form a deep horizontal girder capable of transferring lateral loads to the vertical bracing elements. The vertical bracing then transfers these loads from the roof level to the foundations. The lateral load resisting system of the building also includes all structural elements and connectors located along the lateral load path, *e.g.* the steel deck and its connections, collector beams at the roof level, bracing member connections, anchorage to the foundations, *etc.*

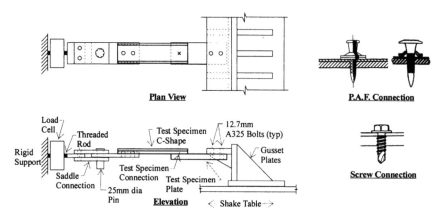

Figure 1. Deck-to-structure connection test set-up

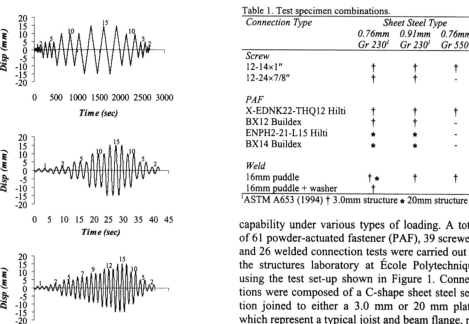

Figure 2. Quasi-static, 0.5 Hz and 3 Hz load protocol

Table 1. Test specimen combinations.

Connection Type	Sheet Steel Type		
	0.76mm Gr 230[1]	0.91mm Gr 230[1]	0.76mm Gr 550[1]
Screw			
12-14×1″	†	†	†
12-24×7/8″	†	†	-
PAF			
X-EDNK22-THQ12 Hilti	†	†	†
BX12 Buildex	†	†	-
ENPH2-21-L15 Hilti	*	*	-
BX14 Buildex	*	*	-
Weld			
16mm puddle	† *	†	†
16mm puddle + washer	†		

[1]ASTM A653 (1994) † 3.0mm structure * 20mm structure

2 CONNECTION TESTS

2.1 *Load protocol and test set-up*

The main objective of this phase of the investigation was to measure the performance of different deck-to-structure connections with regards to: stiffness, capacity, ductility and energy dissipation capability under various types of loading. A total of 61 powder-actuated fastener (PAF), 39 screwed, and 26 welded connection tests were carried out in the structures laboratory at École Polytechnique using the test set-up shown in Figure 1. Connections were composed of a C-shape sheet steel section joined to either a 3.0 mm or 20 mm plate, which represent a typical joist and beam flange, respectively, as listed in Table 1. Welds were formed using a 410-10 MPa welding rod for a duration of 3-4 sec at 250 V. Screws were installed using an adjustable torque drill with the appropriate driver, and powder-actuated fasteners were nailed using the corresponding Buildex Unidek, Hilti DX A41 or DX 750 tools.

The test frame was attached to an MTS shake table, which was used to displace the connections as required. Initial monotonic tests were completed, followed by quasi-static tests using the ATC 24 (1992) seismic testing guidelines, and finally cyclic tests at 0.5 and 3.0 Hz. The load protocol defined for the quasi-static and 0.5 Hz cyclic tests required displacements that ranged from ±1.0 to ±15 mm, with 5 increments of 3 cycles at the same amplitude and 3 decrements of 2 cycles, as illustrated in Figure 2. The number of incremental steps was increased to 7 for the 3.0 Hz cyclic tests to limit local buckling of the steel deck section when the test specimens were subjected to large displacements over a short period of time. Galvanised sheet steels with a specified base metal thickness of 0.76 and 0.91 mm, meeting ASTM A653 (1994) specifications, were obtained for fabrication of the specimen C-shapes.

2.2 Sheet steel material test properties

Material properties of the three different sheet steels were determined from triplicate tests of coupons fabricated according to ASTM A370 (1994) requirements (see Table 2). The 0.76 and 0.91 mm Grade 230 steels meet the requirements for ductility and ultimate strength to yield stress ratio contained in both the CSA S136 (1994) and American Iron and Steel Institute (AISI) (1997) design standards. The 0.76 mm Grade 550 material has a much higher ultimate strength and yield stress, although its ductility is significantly reduced compared with the other mild steel products.

Table 2. Sheet steel material properties.

Property	Steel Type		
	0.76 mm Gr 230	0.91 mm Gr 230	0.76 mm Gr 550
t base metal (mm)	0.74	0.84	0.69
f_y (MPa)	317	321	697[1]
f_u (MPa)	370	401	737
f_u / f_y	1.16	1.25	1.06
% elong (50 mm gauge).	42	25.6	2.5

[1] f_y calculated using 0.2% offset method.

2.3 Connection test results and observations

Information regarding the displacement versus load hysteresis for connections that were composed of a 0.76 mm sheet and a 3.0 mm plate, cyclically tested at 0.5 Hz, and fastened with either a Hilti EDNK22-THQ12 PAF, Buildex BX12 PAF, Buildex 12-14×1″ screw, Hilti 12-24×7/8″ screw, 16 mm puddle weld, or 16 mm puddle weld *with* washer (14.3 mm inner diameter, 26.5 mm outer diameter and 2.4 mm thickness), is provided in Figure 3, as well as Tables 3-6. Similar information for connections composed of a 0.91 mm sheet and a 20.0 mm plate connected with a Buildex BX 14 or Hilti EHPH2-21-L15 PAF tested at 3.0 Hz can

be found in Figure 4. Finally, the displacement versus load hysteresis for a welded specimen composed of a 0.76 mm sheet and a 20.0 mm plate, tested at 3.0 Hz, is shown in Figure 5. In all cases, the entire load *vs.* displacement hysteresis is provided, as well as graphs for the 1 and 5 mm incremental displacement cycles. The displacement at the connection was less than that imposed by the table due to elastic deformation outside of the fastener region. Typically, the PAF and screwed connections failed by either bearing distress of the sheet steel, tilting of the fastener with its eventual pullout, or a combination of the two. The welded connections failed through fracture of the heat affected zone at the edge of the weldment.

The graphs in Figure 3 illustrate that the ultimate capacity of the 0.76 mm sheet steel to 3 mm plate connections, tested at 0.5 Hz, depends on the type of fastener used, where the puddle welds *with* washers provided the highest shear resistance (13.3 kN). The puddle welds *without* washers and both the Hilti and Buildex powder-actuated fasteners were able to carry similar ultimate loads. However, the welds *without* washers were unable to resist shear loads past the first cycle of the 2 mm displacement stage. It was extremely difficult, if not impossible, to adequately fabricate welded connections of this type because the thinness of the sheet steel did not allow for consistent full perimeter welds (see Tables 5 and 6). The use of a washer provided the test specimen with the ductility needed to survive displacements of up to ±15 mm. The washer acted as a heat sink, which improved the weld quality and made fabrication of the connection much less operator dependent. The two screw types, with the same nominal diameter but different thread patterns (see Table 1), performed similarly with ultimate loads that ranged between 5 and 6 kN, less than those of the other fastener types.

The 3 mm plate, which was assumed to act as a flange section for a typical open web steel joist member, had an effect on the behaviour of the connections. In the case of the PAF and screwed test specimens, significant tilting of the fastener was observed for displacements greater than 5 mm because of the eccentric shearing force from the sheet and plate. A connection that is subjected to a monotonic load will typically have a higher capacity if the fastener remains upright and failure is due to bearing instead of tilting. In addition, when a structure is subjected to cyclic loads the tilting action of the fastener tends to cause the hole in the plate section to elongate, sometimes to an extent where the fastener either pops out or can be easily pulled out by hand. When the fastener is imbedded into a thicker plate structure the extent of tilting is reduced and the possibility for failure by loss of the fastener is decreased.

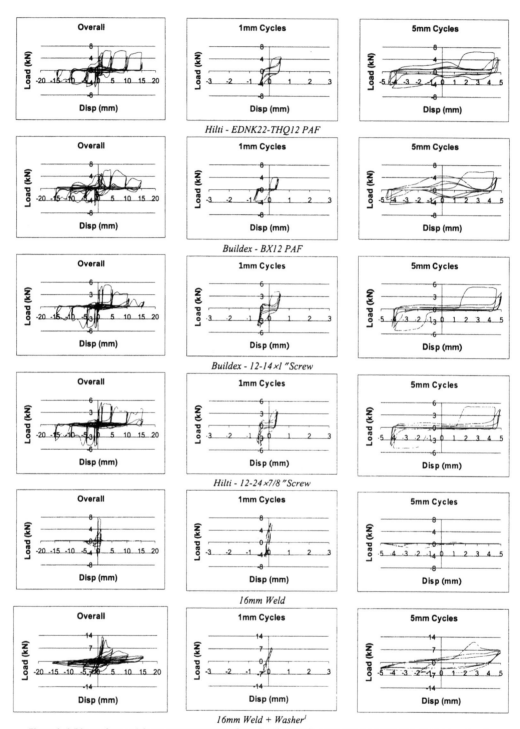

Hilti - EDNK22-THQ12 PAF

Buildex - BX12 PAF

Buildex - 12-14×1″Screw

Hilti - 12-24×7/8″Screw

16mm Weld

16mm Weld + Washer[1]

Figure 3. 0.76 mm sheet to 3.0 mm structure test specimen 0.5 Hz overall, 1 mm and 5 mm load vs. displacement cycles
[1] Displacement of table used for 16mm weld + washer figures in place of connection displacement

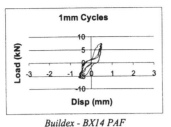
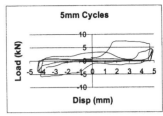
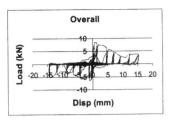

Buildex - BX14 PAF

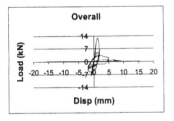
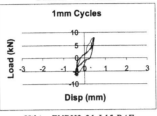
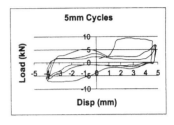

Hilti – ENPH2-21-L15 PAF

Figure 4. 0.91 mm sheet to 20.0 mm structure PAF test specimen 3.0 Hz overall, 1 mm and 5 mm load vs. displacement cycles

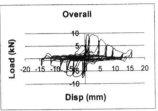
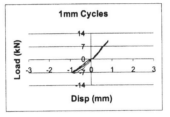
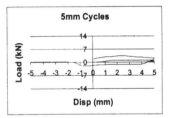

Figure 5. 0.76 mm sheet to 20.0 mm structure welded test specimen 3.0 Hz overall, 1 mm and 5 mm load vs. displacement cycles

The load *vs.* displacement results for the PAF and screwed connections exhibit a pinched curve, more evident in the 5 mm cycle graphs of Figure 3. This was caused by the permanent deformation of the sheet steel in the first cycle at each amplitude and was exacerbated by the tilting of the fastener. The second and third cycles of each incremental load step provide a load curve that is even more pinched because the sheet steel has been previously deformed.

The 0.91 mm sheet steel to 20 mm plate test specimens were connected with either a Hilti ENPH2-21-L15 or Buildex BX14 powder-actuated fastener. The load *vs.* displacement hysteresis graphs in Figure 4 show that their behaviour is similar, with increased loads compared to the 0.76 mm sheet steel to 3 mm test specimens because of the thicker sheet steel, the larger diameter fasteners, and the elimination of fastener tilting. The load *vs.* displacement curves are pinched due to the deformation of the sheet steel in the initial cycle of each incremental amplitude. The embedment length of the fasteners, and larger diameter washers used for the PAFs helped to eliminate the tilting action observed in the

test specimens attached to a 3 mm plate.

The welded connections *without* washers that were connected to a 20 mm plate (see Figure 5) showed no significant change in ductility behaviour, *i.e.* failure occurred in the first 2 mm cycle, however the ultimate load increased due mainly to the improvement in ability to fabricate the connection with the thicker plate acting as a more effective heat sink.

2.4 *Connection test energy dissipation*

Energy dissipation curves for representative test specimens are provided in Figures 6-10. These curves illustrate the plastic deformation characteristics for each type of connection with respect to time, as well as the cumulative displacement of the specimen while subjected to cyclic loading.

The results for the 0.76 mm sheet to 3 mm plate connections that were tested at 0.5 Hz indicate that the welds *with* washers had a higher capacity to absorb energy ($\approx$ 550 kN mm) in comparison with all other fasteners. In contrast, the welds *without* washers had the lowest energy dissipation level ($\leq$

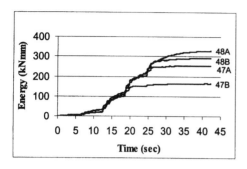
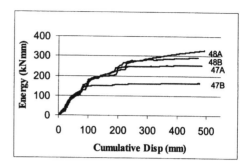

Figure 6. 0.76 mm sheet to 3.0 mm structure PAF test specimen 0.5 Hz energy dissipation history

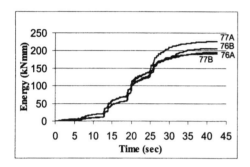
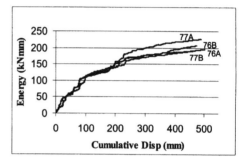

Figure 7. 0.76 mm sheet to 3.0 mm structure screw test specimen 0.5 Hz energy dissipation history

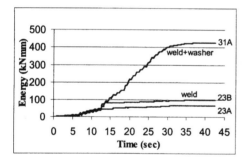
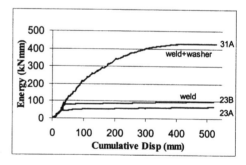

Figure 8. 0.76 mm sheet to 3.0 mm structure weld test specimen 0.5 Hz energy dissipation history

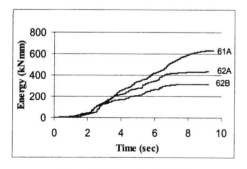
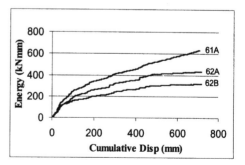

Figure 9. 0.91 mm sheet to 20.0 mm structure PAF test specimen 3 Hz energy dissipation history

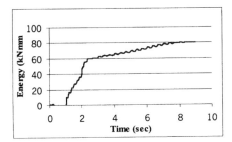 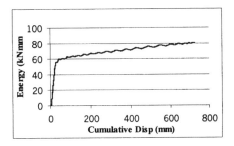

Figure 10. 0.76 mm sheet to 20.0 mm structure weld test specimen 3 Hz energy dissipation history.

Table 3. 0.76 mm PAF P_u and energy dissipation results.

Test Spec-imen	Connection Type	Load Type	P_u kN	$\Sigma Energy$ kN mm	$\Sigma Energy / P_u$ kN mm / kN
39a	EDNK22-THQ12	Mono	6.37	-	-
39b	EDNK22-THQ12	Mono	6.11	-	-
39c	EDNK22-THQ12	Mono	6.74	-	-
40a	BX12	Mono	6.59	-	-
40b	BX12	Mono	5.91	-	-
40c	BX12	Mono	6.07	-	-
43a	ENPH2-21-L15	Mono	7.42	-	-
43b	ENPH2-21-L15	Mono	6.57	-	-
44a	BX14	Mono	5.95	-	-
44b	BX14	Mono	6.66	-	-
47a	EDNK22-THQ12	0.5 Hz	6.56	253	38.6
47b	EDNK22-THQ12	0.5 Hz	6.04	165	27.3
48a	BX12	0.5 Hz	6.12	328	53.6
48b	BX12	0.5 Hz	6.86	291	42.4
		Avg.	*6.40*	*259*	*40.5*
51a	ENPH2-21-L15	0.5 Hz	7.32	405	55.4
51b	ENPH2-21-L15	0.5 Hz	7.33	445	60.8
52a	BX14	0.5 Hz	5.98	300	50.2
52b	BX14	0.5 Hz	6.30	163	25.8
		Avg.	*6.73*	*328*	*48.1*
55a	EDNK22-THQ12	3 Hz	6.55	182	27.8
55b	EDNK22-THQ12	3 Hz	6.98	512	73.3
56a	BX12	3 Hz	6.72	504	75.0
56b	BX12	3 Hz	6.32	361	57.1
		Avg.	*6.64*	*390*	*58.3*
59a	ENPH2-21-L15	3 Hz	7.41	512	69.1
59b	ENPH2-21-L15	3 Hz	7.56	425	56.2
60a	BX14	3 Hz	6.26	411	65.6
60b	BX14	3 Hz	6.91	376	54.5
		Avg.	*7.04*	*431*	*61.4*
63a	ENPH2-21-L15	Q-S	6.47	222	34.4
64a	ENPH2-21-L15	Q-S	6.72	260	38.6

Note: EDNK22-THQ12 and BX12 fasteners were used with 3 mm plate. ENPH2-21-L15 and BX14 fasteners were used with 20 mm plate

Table 4. 0.76 mm screw P_u and energy dissipation results.

Test Spec-imen	Connection Type	Load Type	P_u kN	$\Sigma Energy$ kN mm	$\Sigma Energy / P_u$ kN mm/kN
72a	12-14×1″	Mono	7.52	-	-
72b	12-14×1″	Mono	5.78	-	-
72c	12-14×1″	Mono	5.40	-	-
73a	12-24×7/8″	Mono	8.83	-	-
73b	12-24×7/8″	Mono	7.60	-	-
73c	12-24×7/8″	Mono	7.58	-	-
73d	12-24×7/8″	Mono	5.23	-	-
76a	12-14×1″	0.5 Hz	5.35	196	36.6
76b	12-14×1″	0.5 Hz	5.43	207	38.1
77a	12-24×7/8″	0.5 Hz	5.54	227	40.9
77b	12-24×7/8″	0.5 Hz	5.49	191	34.8
		Avg.	*5.45*	*205*	*37.6*
80a	12-14×1″	3 Hz	5.71	293	51.3
80b	12-14×1″	3 Hz	5.20	230	44.3
81a	12-24×7/8″	3 Hz	6.01	293	48.8
81b	12-24×7/8″	3 Hz	6.26	334	53.4
		Avg.	*5.79*	*288*	*49.5*
84a	12-14×1″	Q-S	4.91	141	28.8
85a	12-24×7/8″	Q-S	5.05	153	30.4

Note: 3 mm plate used for all screw specimens.

Table 5. 0.76 mm weld P_u and energy dissipation results.

Test Spec-imen	Connection Type	Load Type	P_u kN	$\Sigma Energy$ kN mm	$\Sigma Energy / P_u$ kN mm/kN
21a	½ 16.2mm[1]	Mono	7.43	-	-
21b	full 16.5mm	Mono	12.0	-	-
21c	full 16.2mm	Mono	13.6	-	-
23a	½ 16.5mm	0.5 Hz	7.61	67.3	8.8
23b	full 16.8mm	0.5 Hz	13.3	99.2	7.4
25a	¾ 15mm	3 Hz	8.52	48.8	5.7
25b	⅛ 15.5mm	3 Hz	2.02	82.4	40.7
27a	½ 14.5mm	Q-S	9.53	15.6	1.6
30a	⅝ + wash[2]	Mono	8.47	-	-
31a	full + wash	0.5 Hz	13.3	428	32.1
32a	full + wash	3 Hz	14.0	541	38.6
33a	full 16mm[3]	Mono	11.4	-	-
34a	full 16mm[3]	0.5 Hz	16.3	247	15.1
35a	full 15.4mm[3]	3 Hz	13.0	80.5	6.2

Note: 3mm plate used unless otherwise noted
[1]½ perimeter 16.2mm weld [2]Washer used [3]20mm plate used

100 kN mm), which can be attributed to the early failure of these specimens, typically in the 2 mm displacement range. The 0.76 mm to 20 mm welded welded connections that were tested at 3 Hz were also not able to dissipate as much energy (< 80 kN mm) (see Figure 10). The two types of

Table 6. 0.76 mm Gr 550 P_u and energy dissipation results.

Test	Connection	Load	P_u	$\Sigma Energy$	$\Sigma Energy$ $/P_u$
Spec- imen	Type	Type	kN	kN mm	kN mm / kN
36a[1]	¼ 17.4mm	Mono	3.48	-	-
37a[1]	½ 16.2mm	0.5 Hz	6.38	27.3	4.3
38a[1]	½ 14.5mm	3 Hz	1.42	54.7	38.4
69a[2]	EDNK22-THQ12	Mono	7.61	-	-
70a[2]	EDNK22-THQ12	0.5 Hz	6.97	151	21.6
71a[2]	EDNK22-THQ12	3 Hz	7.61	224	29.5
90a[3]	12-14×1″	Mono	7.34	-	-
91a[3]	12-14×1″	0.5 Hz	7.52	253	33.6
92a[3]	12-14×1″	3 Hz	7.62	381	50.0

Note: 3 mm plate used for Gr 550 tests. [1]Weld [2]PAF [3]Screw

screwed connections behaved consistently, which indicates that the different thread patterns (14 and 24 threads per inch) did not have a significant effect on performance (≈ 200 kN mm) and that these connections are the least dependent on installation technique. In most cases, both of the powder-actuated fasteners that were used with the 3 mm plate connections, Buildex BX12 and Hilti EDNK22-THQ12 (see Figure 6), were able to absorb more energy than the 12 gauge screw connections. The energy dissipation curves for the 0.91 mm sheet to 20 mm plate connections that were tested at 3 Hz reveal that the Hilti ENPH2-21-L15 and Buildex BX14 fastener powder-actuated fasteners were able to reach higher energy absorption levels (see Figure 9) in comparison with the other test specimens, again due to the thicker sheet steel, larger diameter fastener and the reduction of extensive tilting action.

Overall, the powder-actuated fasteners were typically able to absorb equal or more energy per kN of load than the screwed and welded connections, although the screwed connections provided the most consistent results. On the other hand, the welded connections *without* washers were the poorest performers, due to their inability to carry load when large displacements were induced.

The strain rate at which the specimens were tested had an effect on the ultimate load, P_u, the ability to dissipate energy E_H, and the E_H / P_u ratio. Specimens that were subjected to displacements at 3 Hz had higher values in all areas in comparison to the 0.5 Hz tests (see average results in Tables 3 and 4).

3 CONCLUSIONS

Screw fasteners provided the most consistent ultimate load and energy absorption results of all the deck-to-structure connections that were tested. The powder-actuated fasteners could carry higher loads and dissipate greater amounts of energy, but were not as consistent. The use of screws or PAFs with thin flange material in joists and beams can lead to premature failure of a connection when a structure is subjected to repetitive earthquake loading because the fasteners tend to pull out when tilted. The welded connections with washers can be easily and adequately fabricated and also are able to carry higher loads with significant ductility in comparison to welds without washers. If the plastic behaviour of deck-to-structure connections is to be accounted for in seismic design, then powder actuated and screwed connections imbedded in a material thick enough to limit fastener tilting would be recommended. If a deck is to be welded to the structure, then it is necessary for washers to be used at all times. Further full-scale seismic cantilever tests of steel roof deck assemblies would be beneficial to better understand the relative performance of these deck-to-structure connections when subjected to earthquake loading.

4 ACKNOWLEDGEMENTS

The authors would like to thank the Natural Sciences and Engineering Research Council of Canada, the Canadian Institute of Steel Construction, the Canadian Sheet Steel Building Institute, the Canam Manac Group, Hilti Limited, ITW Buildex and the Steel Deck Institute for their support. The authors would also like to acknowledge the assistance of the laboratory technicians at École Polytechnique, G. Degrange, P. Bélanger, and D. Fortier.

REFERENCES

American Iron and Steel Institute. (1997). "1996 Edition of the Specification for the Design of Cold-Formed Steel Structural Members", Washington, DC, USA.

American Society for Testing and Materials, A370. (1994). "Standard Test Methods and Definitions for Mechanical Testing of Steel Products", Philadelphia, PA, USA.

American Society for Testing and Materials, A653. (1994). "Standard Specification for Steel Sheet, Zinc-Coated (Galvanized) or Zinc-Iron Alloy-Coated (Galvannealed) by the Hot-Dip Process", Philadelphia, PA, USA.

Applied Technology Council. (1992). "ATC24 – Guidelines for Cyclic Seismic Testing of Components of Steel Structures", Redwood City, CA, USA.

Canadian Standards Association, S16.1. (1994). "Limit States Design of Steel Structures", Etobicoke, Ont., Canada.

Canadian Standards Association, S136. (1994). "Cold Formed Steel Structural Members", Etobicoke, Ont., Canada.

National Research Council of Canada. (1995). "National Building Code of Canada" 11th Edition, Ottawa, Ont., Canada.

Behaviour of Steel Structures in Seismic Areas, Mazzolani & Tremblay (eds) © 2000 Balkema, Rotterdam, ISBN 90 5809 130 9

Tests on the strain rate effects on beam-to-column steel connection

L. Sanchez & A. Plumier
Institute de Génie Civil, Université de Liège, Belgium

ABSTRACT: In this paper, test results on the strain rate effects on the cyclic behaviour of steel beam-to-column joints in moment-resisting frames are presented. Twelve full scale specimens have been tested under constant amplitude cyclic loading, for three connection typologies, for three different steel grades and for two different loading speeds. Qualitative conclusions are drawn in term of strength, energy dissipation and degradation. Test results are finally compared in term of low-cycle fatigue endurance.

1 INTRODUCTION

Post earthquake investigations, promoted by the unexpected structural damage to steel moment resisting frame connections, have identified the strain rate effect as one, between other, of the possible reasons to these damages.

After the Northridge (1994) and Kobe (1995) earthquakes, numerous tests have been made in the USA and in Japan on welded connections, since it is the typical detail considered in these countries.

These tests have been bearing on all the aspects of the problem: characteristics of base and weld material, design of connection, welding preparation and technology, strain rate effects. Many conclusions have been drawn and other research work under way will bring more, but it is already clear that the strain rate influences the connection behaviour:
1 It increases fy and fu, bringing higher demand on the connection.
2 It increases fy more than fu, which generates a localisation of the plastic strains subsequently reducing the apparent ductility.
3 It can propagate cracks in a brittle way.
However these results have been obtained for typical U.S. connection design and site welding, which does not cover European practice which favours shop welding and on site bolting.

The research work in the University of Liege bear on a not investigated aspect of the European connections, the influence of the strain rate in the capacity to cyclically dissipate energy by plastic mechanism for various typical beam-to-column European steel connection. The dimensions of the tested specimens correspond to typical size of real European steel structures used in common design of offices or apartment house buildings. The behaviour of some of these connection typology has been studied in a previous larger test programme under quasi-static loading condition, see reference [3].

As the aim is to study conditions similar to those produced in a real earthquakes, the problem is complex (stress gradient; material in the plastic range; local instabilities; brittle and ductile fracture; welded zones; low cycle fatigue; dependency of the material properties on the strain range but also on the strain rate time history). The approach to assess the influence of the strain rate in steel connections is thus the experimental testing of full-scale specimens representing the most common practice in Europe.

2 DESCRIPTION OF TESTED SPECIMENS AND TEST SET UP.

Three typologies of beam to column connection have been tested for a total of 12 specimens tested (see Table 1). These three typologies are the following:
☐ A1 is a typical European connection, involving welding at the shop and bolting on site (see Figure 1). It corresponds to a rigid full strength bolted connection. The end plate of the beam is welded with K preparation. A welding access hole is made in the beam web. Three different structural steel grades for beams have been used with this connection. Nine specimens with connection A1 have been tested.

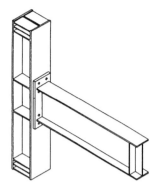

Figure 1. Connection A1.

□ C3 is the typical US frame connection with the beam flanges welded to the column, but the flat backing bar used in the weld is replaced by a new design of bar which should have created conditions for a good first full penetration weld (see Figure 2). In current practice, this bar would be an extruded steel wire with the designed cross section. In the present research program, the bar is machined out of a rectangular bar. Previous testing have shown that this backing bar give bad welds and a poor behaviour of the connection. In this research a dog-bone has been done in the beam in order to prevent brittle failure in the welded connection. Only one specimen with connection C3 has been tested.

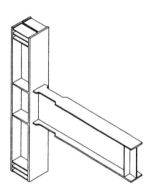

Figure 2. Connection C3.

□ B is a semi-rigid partial strength connection in which bolts are perpendicular to the bent beam (see Figure 3). It does not transfer the beam plastic moment to the column. It is intended to develop energy dissipation through the ovalization of the bolt-holes and through friction between the web of the UPN beam and the flange of the HEM column. Both mechanisms work in the following manner:

-friction gives a constant resistance throughout the displacement.

-bearing resistance provides the increases in resistance at large displacements.

The interest of connection B is an extremely low cost in the preparation of the connection and during the erection phase, as well as a good support surface to floor slabs or metal decking. Two specimens with connection B have been tested.

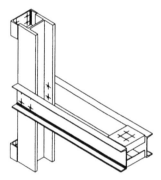

Figure 3. Connection B.

The imposed displacements are sinusoidal functions, with periods of 2.5 and 40 seconds respectively for the fast and slow loading speed, and with peak to peak amplitude of 80 mm., that correspond to two times the yield displacement vy. This corresponds approximately to strain rates equal to 0.1 and 0.001 1/s.

The amplitude and frequency of the imposed displacement have been chosen in order to compare the performance of the connections under quasi-static condition with those of real dynamic ones. The period of the dynamic test has been chosen similar to the period that could have a building with the beam-to-column connection tested.

Table 1. Brief description of the tested specimens.

Beam profile.	IPE 450	IPE 450	IPE 450	IPE 450 A	IPE 450 A	IPE 450 A	2 UPN 300
fy beam (N/mm2)	316	273	273	405	405	405	290
Connection type.	A1	A1	A1	A1	A1	C3	B
Slow loading speed.	LS1	LS2	LS3	HS1			US
Fast loading speed.	LF1	LF2	LF3	HF1	HF2	DOGF	UF

IPE 450 beams are likely to be used for spans about 8 m. Given the shape of the bending moment diagram under a combination of static and seismic loading the point of contra-flexure is about 2 m away from the connection. This defines the axis for the load application.

The test program intends to concentrate on beam plastic phenomenon. From this, it is derived that the column size and fixing in the test set up can be such that they minimise the column deformation and facilitate the testing work.

The measurements during the cyclic tests have been the following (Figure 4):
- □ The load P applied by the actuator is measured by means of a load cell.
- □ The load P is measured indirectly by the measurement of the strains in the two flanges of the steel profile at distance from the applied load point of the middle of the beam profile length.
- □ Displacement v under load is measured by a LVDT. It measures the relative displacement between the point of the beam where the load is applied and an external reference.
- □ The displacement used to pilot the test is the relative displacement of the actuator.
- □ Additional displacement transducers are installed in order to take into account settlements in supports.

→I DISPLACEMENT TRANSDUCTER
▮ STRAIN GAUGES
↔ ACTUATOR
∅ LOAD CELL

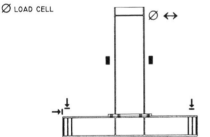

Figure 4. More relevant measured data.

For detailed description of specimen dimensions, see reference [4].

3 TESTS ON STEEL BASE MATERIAL.

The three different steel grades of the beams used in the tests on rigid connection A1 have been establish by standard tensile test. The results are shown in Table 2.

Table 2. Results tensile test.

Specimens	fy	fu	fu/fy	A%
LS1-LF1	316	429	1.36	35
LS2-LF2-LS3-LF3	273	406	1.49	38
HS1-HF1-HF2-DOGF	405	500	1.23	31

Three samples with V notch of the beams base material have been tested for each of the three following specimens (LS1, LS2, HS1), that represent the three different steel used in the IPE beams tested. The cuts have been made in the lamination direction, with a net area of 0.8 cm2 (see Table 3).

4 TEST RESULTS.

4.1 Results of Tests on Specimens with Low Strength Steel Grade (Group L).

The results of this group are summarised in Table 4. The failure mode for the six specimens of this group is the same, fracture in the flange close to the weld, with a stable propagation of the crack. Some local buckling of the beam flange can be observed. No differences can be observed between the failure mode for slow or fast strain rates. The number of cycles withstood in average up to collapse is 34 % lower when the higher load speed is applied and the total energy dissipated up to collapse is 37 % lower in average. Results for strength capacity against the number of cycles are presented in Figure 5 for the six specimens of the group and in average in Figure 6.

In Figure 7, the results are presented in term of dissipated energy against the number of cycles.

Table 3. Charpy V-notch test.

N° specimen	Temperature (°C)	Absorbed energy (Joules)			Resilience (J/cm2)			Notch toughness quality
LS1-1	+20	154	151	176	192.5	188.8	220	
LS1	0	25	95	35	31.25	118.8	43.75	C
LS1	-20	10	9	13	12.5	11.25	16.25	
LS2-2	+20	163	143	145	203.8	178.8	181.3	
LS2	0	155	125	140	193.8	156.3	175	DD
LS2	-20	154	100	140	192.5	125	175	
HS1-1	+20	160	169	155	200	211.1	193.8	
HS1	0	160	155	145	200	193.8	181.3	DD
HS1	-20	135	165	156	168.8	206.3	195	

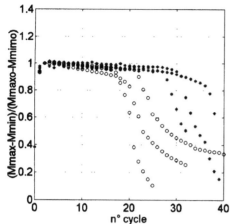

Figure 5. Six specimens group L. (*) Slow. (o) Fast.

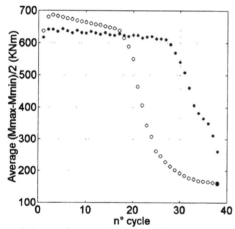

Figure 6. Average for group L. (*) Slow. (o) Fast.

pletely brittle; local buckling cannot be observed. HF1 and HF2 failure mode is the propagation of the flange cracks close to the weld toes, but the crack propagation has been more ductile than for the specimen HS1.

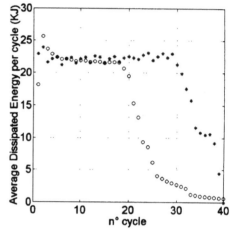

Figure 7. Average results for group L. (*) Slow. (o) Fast.

The average number of cycles withstood up to collapse in group H (6.6 cycles) is 76% lower than in the group L (27.5). This worse behaviour was in some manner expected, because of the lower fu/fy value of the group H base steel material (1.23 instead of 1.36 or 1.49) and because of the higher steel grade used for the base material while the weld material was the same.

Table 5. Results for the high steel grade. Group H.

Loading speed	Slow	Fast	
Specimen	HS1	HF1	HF2
N° cycles up to collapse	4	8	8
Mean	4	8	
Dissipated energy per cycle (KJ)	21	16.9	18.4
Mean (KJ)	20.8	17.6	
Total dissipated energy (KJ)	83	135.0	147.0
Mean (KJ)	83	141.0	

4.2 Results of tests on specimens made with high strength steel (H group).

The test results for the specimens HS1, HF1 and HF2 are shown in Table 5. In the HS1 specimen, the flange cracks propagation has been unstable, producing much noise and the failure has been com-

Table 4. Results for the low steel grade rigid connection A1. Group L.

Loading speed	Slow			Fast		
Specimen	LS1	LS2	LS3	LF1	LF2	LF3
Specimen	LS1	LS2	LS3	LF1	LF2	LF3
N° cycles up to collapse	33	38	29	24	21	20
Mean		33.3			21.7	
Dissipated energy per cycle (KJ)	18.8	25.5	22.1	18.8	21.4	25.0
Mean (KJ)		22.1			21.7	
Total dissipated energy (KJ)	620	970	640	450	450	500
Mean (KJ)		743.3			466.7	

The average number of cycles withstood up to collapse in H group (8 cycles) is a 100% higher when the higher load speed is applied than when the slow load speed is applied (4 cycles) and the total energy dissipated up to collapse is a 70 % higher. The number of specimens (one specimen tested at low speed and two tested at high speed) is too small to obtain definitive conclusions. It is necessary to perform more test for this steel grade, at least 3 specimens for each loading speed.

4.3 Dog-Bone connection.

The specimen has been able to withstand only 4 cycles up to the loss of the 50 % of its strength capacity and the failure took place in the weld, so that it can be concluded that the dog-bone did not protect the connection in this case. The upper flange that work in tension in the first cycle is detached, showing that this new welding detail design does not realise a full penetration weld. See Table 6.

Table 6. Results for specimen DOGF.

Loading speed	Fast
Specimen	DOGF
N° cycles up to collapse	4
Dissipated energy per cycle (KJ)	18.7
Total dissipated energy (KJ)	75

4.4 Partial strength connection.

Friction gives almost all the energy dissipation. For the imposed displacement, the increase in resistance due to plastic deformation of the bolt-holes is small and ovalization of the bolt-holes cannot be observed.

Degradation in maximum strength capacity with the number of cycles is faster for the specimen tested at high loading speed (the maximum moment has not an important influence on the energy dissipation, because of the relatively small imposed displacement). See Figure 8.

On the contrary, no important influence of the loading speed can be observed in the degradation of dissipated energy capacity per cycle, (see Figure 9).

For the imposed displacement, the energy dissipation is basically due to friction and loading speed does not influence the reduction of friction throughout the number of cycles. For the first 18 cycles, the reduction in Ec is a little higher for the specimen tested at low speed; it becomes lower for the following cycles.

The total energy dissipated is always greater for the specimen tested at low loading speed. It corresponds to the idea that an increment in velocity brings a reduction in friction (see Figure 10).

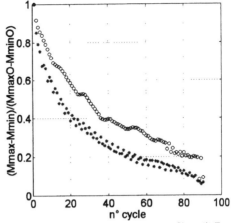

Figure 8. Strength degradation. for group H. (o) Slow. (*) Fast.

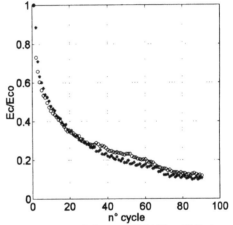

Figure 9. Energy per cycle for group H. (o) Slow. (*) Fast.

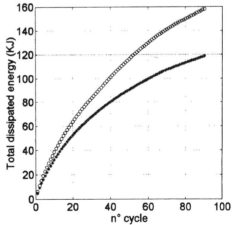

Figure 10. Total dissipated energy for group H. Slow (o) and fast (*)

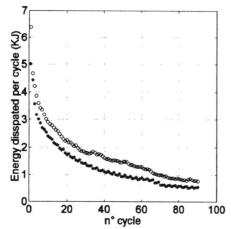

Figure 11. Slow (o) and fast (*). Group H.

Loading speed has a negative effect on the number of cycles sustained up to collapse. The reduction is 40%, when defining collapse as a 50% of strength capacity reduction (18% for a 75% strength reduction). See Table 7.

The dissipated energy average per cycle is lower for the high speed loading (2.8 kJ for each cycle); friction decreases 7 % in average during the loss of 50 % of strength capacity and 16 % in average during the loss of 75 % of strength capacity.

5 ANALYSIS OF RESULTS.

The strain rate has increased the yield moment, My, and the ultimate moment in all the specimens tested, thus the increase in loading speed has incremented strength demand in the welded zone for all specimens.

The results show that My value of the group H has been more sensitive to the strain rate than the group L. This result was not expected, because the higher steel grade, the less sensitive must be My to the strain rate. Further tests are necessary for group H.

The ratio of ultimate moment to yield moment must be higher for the specimens tested at high loading speed, because yield moment is more sensitive than ultimate moment to the strain rate. For all the specimens tested at high loading speed, this ratio is higher, see Table 8. It means that when loading speed is increased, the plastic hinge length and the beam ductility are reduced.

For group L, there is an average reduction in the number of cycles of 35 % when the load period is decreased from 40 s to 2.5 s. For group H, there is an increment of 100 % (it must be remembered that in this group the only specimen tested at low speed has a very brittle failure mode, and this result needs confirmation).

As it was expected, the steel with fu/fy=1.49 is more sensitive to loading speed than the steel with fu/fy=1.36. The steel with fu/fy=1.23 is the steel of group H.

All the former comments are summarized in Tables 9 and 10.

Table 7

Loading speed Specimen	slow US	fast UF
N° cycles up to 50% strenght reduction	25	15
N° cycles up to 75% strenght reduction	49	40
Average dissipated energy per cycle (50 % strenght reduction), (KJ).	3.0	2.8
Average dissipated energy per cycle (75 % strenght reduction), (KJ).	2.4	2.0
Total dissipated energy (50 % strenght reduction), (KJ).	74	42
Total dissipated energy (75 % strenght reduction), (KJ).	116	80

The total energy dissipated up to collapse is 43 % and 31 % lower when the loading speed is increased, respectively, when defining collapse as a 50% and 75% of strength capacity reduction.

Table 9. Results average ratio (fast/ slow loading speed for the three specimen groups)

Specimen Group	Group L	Group H	Group U
N° cycles failure	0.65	2.00	0.60
Total dissipated energy	0.63	1.70	0.57
Mu	1.04	1.11	
My	1.13	1.21	
Mu/My	0.925	0.919	

Table 10. Results average ratio (fast/ slow loading speed for the three steel grades)

fu/fy steel	1.23	1.36	1.49
N° cycles failure	2.00	0.73	0.61
Total energy dissipated	1.70	0.73	0.59

Table 8. Results for the low steel grade rigid connection A1. Group L.

Loading speed Specimen	Slow			Fast			Slow	Fast	
	LS1	LS2	LS3	LF1	LF2	LF3	HS1	HF1	HF2
Mu/My (*)	1.25	1.24	1.36	1.15	1.20	1.21	1.30	1.19	1.20

(*) Mu and My come not from a monotonic test but from the first cycle of the cyclic test.

The results of the tests have also been processed in term of S-N fatigue curves, see Figure 12. The EC3 curve, for high cycle fatigue, for detail category 112 that correspond to transverse butt welds, is far away from the test results. The results are relatively close to the detail category curve 50 for the low steel grade.

The test results have very good agreement with the S-N curves proposed in reference [2] for beam-to-column welded connections, with butt weld in K grooves connecting the flange of the beam to the column flange. This proposed curve depends on the welding details and imposed displacement (a threshold displacement is defined that governs the failure mode). Three types of failure mode are defined for welds with K preparation (B1= brittle, M = mixed and D = ductile).

B1 is a failure mode due to a crack formed in the centre of the weld between the beam flange and the end plate and propagated versus the edges, that correspond to the failure mode of all tested specimens of H group. Some specimens, where initiation of local bucking can be observed with a ductile crack propagarion (L group),could be in between failure mode B1 and M.

Using 0.9 for the parameter related to the weld quality, the threshold displacement limit between brittle and mixed failure is 112 mm for L group and 124 mm for H group. As the applied displacement is 80 mm, the S-N curve to be used is the one corresponding to B1 failure. If larger displacements were applied, the local buckling of the beam flange would protect the welding zone from higher strength demands, see reference [1].

From Figure 11, it can be seen that this approach is conservative for all specimens tested (included the one that collapsed in only four cycles with a very brittle and fast crack propagation). From the specimens of group L, it can be seen that the influence of loading speed in the connection behaviour in terms of low cycle fatigue is not negligible, with an influence on the joint behaviour as important as the failure mode.

6 CONCLUSIONS.

1 For the specimens tested there is no evidence that the loading speed does affect the failure mode of the steel beam-to-column connection. Nevertheless more tests are necessary for the rigid connection with high steel grade.

2 However, the number of cycles withstood up to collapse as well as the total absorbed energy decrease when the loading speed is increased. This conclusion contrasts with Japanese studies after Kobe earthquake on beam-to-column steel connections (for two connection typologies: flange welded to the column and bolted web and fully welded, see reference [5]), where it was concluded that the loading speed does not affect the number of cycles withstood up to collapse.

3 Loading speed decreases the ratio ultimate to

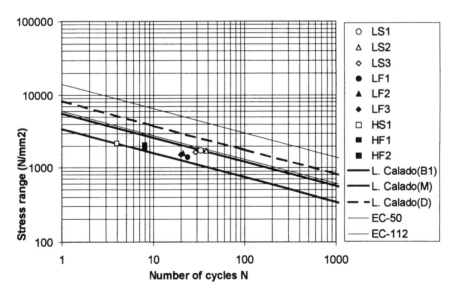

Figure 12. Fatigue strength curve.

yield moment, reducing apparent ductility, arising plastic localisation; and thus, for the same imposed displacement it increases the local strain demand and reduces the fatigue life and the absorbed energy.

4 For the semi-rigid connection,the loading speed also reduces the energy dissipated per cycle. The reduction in the number of cycles, although evident, could be related with the definition of collapse used, which is based on strength reduction (one half of the maximum strength capacity), but strength does not play an important role in the energy dissipation for the imposed displacement. Further tests are necessary for higher displacements.

7 ACKNOWLEDGEMENTS.

The current tests has been carried out in the framework of the INCO-COPERNICUS Joint Research Project "Reliability of Moment Resistant Connections of Steel Building Frames in Seismic Areas (RECOS) with the financial support of the DG XII of the European Commission.

REFERENCES.

[1] Bernuzzi, C., Castiglioni, C.A. & Vanjna de Pava, S., 1999, Behaviour of beam-to-column joints in Moment-Resisting Steel Frames, SDSS'99.

[2] Calado, L., C.A. Castiglioni, P. Barbaglia & C. Bernuzzi, 1998, Seismic design criteria based on cumulative damage concepts, Proc. of 11th European Conference on Earthquake Engineering, Paris, 6-11 September.

[3] Plumier, A., Agatino, M.R., Castellani, A., Castiglioni, C.A. & Chesi, C., 1998, Resistance of steel connections to low-cycle fatigue, Proc. of 11th European Conference on Earthquake Engineering, Paris, 6-11 September.

[4] "Moment Resisting Connections of Steel Buildings in Seismic Areas: Design and Reliability" E&FN SPON.

[5] Japanese Society of Steel Construction, July 1997, Kobe Eqrthquake Damage to Steel Moment Connections and Suggested Improvement, JSSC Technical Report N° 39.

Dynamic behavior of composite structures with composite connections

L. Simões da Silva
Department of Civil Engineering, University of Coimbra, Portugal

P.J.S. Cruz
Department of Civil Engineering, University of Minho, Guimarães, Portugal

L. Calado
Department of Civil Engineering, Instituto Superior Técnico, Lisboa, Portugal

ABSTRACT: Steel-concrete composite structures probably provide the best compromise for the building of the city of tomorrow, combining speed of construction, environment friendly solutions and optimising the advantages of steel and concrete in construction. Guidance for connection design of composite structures and its influence on the global behaviour of the structure under dynamic conditions is still scarce.

A research project between the University of Coimbra, University of Minho, Instituto Superior Técnico, University of Beira Interior and Martifer aiming at producing guidelines for design of steel and composite semi-rigid frames under static and dynamic loading, with major focus on the influence of the connections is presented. This project comprises three major tasks:

- The characterisation of the dynamic behavior at the structure level,
- The evaluation of the dynamic response of major and minor axis composite connections and
- The investigation of the behavior of various connection components.

Focussing on task 2, results from a series of tests on extended end-plate beam-to-column major and minor axis composite connections are presented and compared with analytical predictions.

1. INTRODUCTION

Steel-concrete composite structures combine the advantages of steel and reinforced concrete structures, making them a major building solution. In Portugal, composite structures have had so far limited application, mostly because of the conservative tradition of the construction industry and cheap labour. However, the increasing labour costs, the need for specialisation to achieve minimum quality standards and the required speed of construction will result in a widespread use of composite structures.

In spite of the intensive research done in this wide topic in the past years, the behaviour of composite and steelwork connections and its influence on the global behaviour of structure is still an open topic for investigation, particularly with regard to the seismic response and damage assessment. The southern countries of the EC have medium to high seismic risk. Portugal is one of such countries - needless to remind the destructive earthquake of 1755 in Lisbon and, more recently, the Azores earthquakes of 1980 and 1998 - and a steel-concrete composite code is required that adequately precludes adverse behaviour under seismic action. The so-called "capacity design" for structures in seismic zones requires the selection of the energy dissipation mechanisms, which can be characterised by the combination of stable hysteretic response with the possibility of controlling simply, but reliably, the key parameters of the relevant parts of the frame. With reference to steel frames, the energy dissipation capabilities of the whole system depend, basically, on the hysteretic behaviour of its individual components, such as members and joints. Therefore, in addition to the material ductility, overall and local buckling as well as low-cycle fatigue appear as very important phenomena able to affect significantly the seismic performance of the structural system. As a consequence, the design phase should be carried out by choosing an efficient structural configuration with respect to the inelastic response and selecting adequate details in which concentrate inelastic deformations. Moreover it appears necessary to ensure, through suitable overstrengths, that inelastic deformations do not occur at undesirable locations or by unexpected structural modes.

In the recent earthquake events of Northridge and Kobe, a significant number of steel structures suffered extensive damage. With reference to steel buildings that did not collapse, although exhibiting significant damage, local failures of steel elements and beam-column joints took place that did not result in severe overall deformations, remaining hidden behind undamaged architectural panels. Therefore, great efforts have been devoted to a complete understanding of the unexpected brittle behaviour of joints, which were traditionally considered able to satisfy ductility demands associated with capacity design (Zandonini et al. 1997).

Trying to provide additional insight in this area, a research project was established between the University of Coimbra, University of Minho, Instituto Superior Técnico, University of Beira Interior and Martifer aiming at producing guidelines for design of steel and composite semi-rigid frames under static and dynamic loading, with major focus on the influence of the connections. This project comprises three major tasks:

- the characterisation of the dynamic behavior at the structure level;
- the evaluation of the dynamic response of major and minor axis composite connections;
- the investigation of the behavior of various connection components.

Focussing on task 2, analytical models aimed at predicting the overall behavior of the connection are presented, together with results from a series of tests on extended end-plate beam-to-column major and minor axis composite connections. Tasks 1 and 3 are not referred here in detail, being described in two companion papers (Cruz et al. and Calado et al. 2000).

2. STATIC AND DYNAMIC BEHAVIOR OF STEEL AND COMPOSITE CONNECTIONS

Predicting the behavior of steel and composite connections under static and dynamic loading still remains a difficult task, despite the major advances in recent years (COST C1 1999, Jaspart 1998). Experimental and numerical calibration and verification is still required to allow the safe utilisation of analytical procedures in a practical range of connection configurations, sizes and loading types. A major advance in this area was achieved with the so-called "component method", that consists of idealizing a connection as a mechanical model composed of extensional springs and rigid links, whereby the springs (components) represent a specific part of a joint that, depending on the type of loading, make an identified contribution to one or more of its structural properties (Weynand et al. 1995).

This simpler approach provides good enough approximations to the strength and initial stiffness of steel and steel-concrete beam-to-column and beam-

to-beam connections under static loading, as specified in Annex J of Eurocodes 3 and 4 (1998 and 1996), while maintaining enough simplicity to be used by the practitioner. Also, the underlying philosophy of the method is capable of straightforward extension to new configurations by simply chosing an appropriate assembly of components or, whenever necessary, adding additional components.

However, the method is currently limited to predicting the strength and initial stiffness of the connection, basically because no information on post-limit response is available for the various basic components. While for some design conditions under static loading this limitation may not impose many restrictions, it prevents proper and safe application of the semi-rigid design concept, as well as further developments with respect to behavior under seismic loading.

3. ANALYTICAL MODELS

3.1 Introduction

As stated in the previous paragraph, the issue of predicting the monotonic moment-rotation response of steel and composite joints under static loading up to failure was first attempted. To achieve this objective, the proposed methodology (Silva et al 2000) is shown in Figure 1 for an extended end-plate beam-to-column steel connection, comprising all active components. Because of the complexity of the model, a simplifying transformation is introduced that replaces all associations of springs in series and paralel in equivalent springs at the tensile and compressive levels (Silva and Coelho 2000). Finally, the nonlinear response of the joint is obtained using the equivalent elastic model of fig. 1c), that yields identical results to the original elastic-plastic model of Figure 1b).

Here, an extension to composite joints is presented in detail, described next.

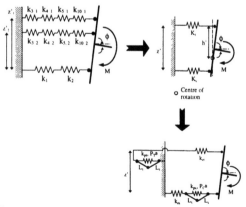

Figure 1. General substitute model for steel connections.

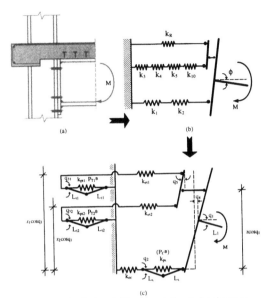

Figure 2. Mechanical model for extended end-plate beam-to-column composite connection.

3.2 General nonlinear model for composite joints

Composite joints require an additional row to adequately model the reinforcement. Therefore, the equivalent elastic model of Figure 2c) comprises the following six degrees-of-freedom: total rotation, q_1 ($=\phi$), rotation of the rigid links, q_2, q_{3_1} e q_{3_2}, axial displacement of the joint, q_4 and rotation of the tensile components (assembled in parallel as an eqivalent component), q_5, where indices t and c refer to the tensile and compressive zones.

(i) Elastic solution

$$
\begin{cases}
M = \theta \sin(2q_1) \\
\sin q_2 = 0 = \sin q_{3_1} = \sin q_{3_2} \\
q_4 = \dfrac{\xi \sin q_1 + \gamma_1 \sin q_5}{C} \\
\sin q_5 = -\dfrac{z k_{ec} \gamma_1}{A} \sin q_1
\end{cases}
$$

(ii) Non linear solution in q_2

$$
\begin{cases}
M = \theta \sin(2q_1) - \beta_c \cos q_1 (1 - \cos q_2) \\
1 - \cos q_2 = \dfrac{\beta_c \sin q_1 - P_C^B L_c}{2\alpha_c} \\
\sin q_{3_1} = 0 = \sin q_{3_2} \\
q_4 = \dfrac{\xi \sin q_1 + \gamma_1 \sin q_5 + 2 L_c k_{ec}(1 - \cos q_2)}{C} \\
\sin q_5 = \dfrac{k_{ec} \gamma_1}{A}\left[-z \sin q_1 + 2 L_c (1 - \cos q_2)\right]
\end{cases}
$$

Using an energy approach, the following eight analytical solutions (eqs. 1a) to 1h)) are obtained, P^B denoting twice the limit load of the equivalent component (Silva *et al.* 2000):

(iii) Non-linear solution in q_{3_1}

$$
\begin{cases}
M = \theta \sin(2q_1) - \beta_{t_1} \cos q_1 (1 - \cos q_{3_1}) \\
\sin q_2 = 0 = \sin q_{3_2} \\
1 - \cos q_{3_1} = \dfrac{\beta_{t_1} \sin q_1 - P_{T1}^B L_{t_1}}{2\alpha_{t_1}} \\
q_4 = \dfrac{\xi \sin q_1 + \gamma_1 \sin q_5 - 2 L_{t_1} k_{et_1}(1 - \cos q_{3_1})}{C} \\
\sin q_5 = \dfrac{-z k_{ec} \gamma_1 \sin q_1 + 2 L_{t_1} k_{et_1} \gamma_2 (1 - \cos q_{3_1})}{A}
\end{cases}
$$

(iv) Non-linear solution in q_{3_2}

$$
\begin{cases}
M = \theta \sin(2q_1) - \beta_{t_2} \cos q_1 (1 - \cos q_{3_2}) \\
\sin q_2 = 0 = \sin q_{3_1} \\
1 - \cos q_{3_2} = \dfrac{\beta_{t_2} \sin q_1 - P_{T2}^B L_{t_2}}{2\alpha_{t_2}} \\
q_4 = \dfrac{\xi \sin q_1 + \gamma_1 \sin q_5 - 2 L_{t_2} k_{et_2}(1 - \cos q_{3_2})}{C} \\
\sin q_5 = \dfrac{-z k_{ec} \gamma_1 \sin q_1 + 2 L_{t_2} k_{et_2} \gamma_3 (1 - \cos q_{3_2})}{A}
\end{cases}
$$

(v) Non-linear solution in q_2 and q_{3_1}

$$
\begin{cases}
M = \theta \sin(2q_1) - \beta_c \cos q_1 (1 - \cos q_2) - \beta_{t_1} \cos q_1 (1 - \cos q_{3_1}) \\
1 - \cos q_2 = \dfrac{2\alpha_{t_1}(\beta_c \sin q_1 - P_C^B L_c) - \chi_1(\beta_{t_1} \sin q_1 - P_{T1}^B L_{t_1})}{4\alpha_c \alpha_{t_1} - \chi_1^{2}} \\
1 - \cos q_{3_1} = \dfrac{\beta_{t_1} \sin q_1 - P_{T1}^B L_{t_1} - \chi_1(1 - \cos q_2)}{2\alpha_{t_1}} \\
\sin q_{3_2} = 0 \\
q_4 = \dfrac{\xi \sin q_1 + \gamma_1 \sin q_5 + 2 L_c k_{ec}(1 - \cos q_2) - 2 L_{t_1} k_{et_1}(1 - \cos q_{3_1})}{C} \\
\sin q_5 = \dfrac{-z k_{ec} \gamma_1 \sin q_1 + 2 L_c k_{ec} \gamma_1(1 - \cos q_2) + 2 L_{t_1} k_{et_1} \gamma_2(1 - \cos q_{3_1})}{A}
\end{cases}
$$

(vi) Non-linear solution in q_2 and q_{3_2}

$$
\begin{cases}
M = \theta \sin(2q_1) - \beta_c \cos q_1 (1 - \cos q_2) - \beta_{t_2} \cos q_1 (1 - \cos q_{3_2}) \\
1 - \cos q_2 = \dfrac{2\alpha_{t_2}(\beta_c \sin q_1 - P_C^B L_c) - \chi_2(\beta_{t_2} \sin q_1 - P_{T2}^B L_{t_2})}{4\alpha_c \alpha_{t_2} - \chi_2^{2}} \\
\sin q_{3_1} = 0 \\
1 - \cos q_{3_2} = \dfrac{\beta_{t_2} \sin q_1 - P_{T2}^B L_{t_2} - \chi_2(1 - \cos q_2)}{2\alpha_{t_2}} \\
q_4 = \dfrac{\xi \sin q_1 + \gamma_1 \sin q_5 + 2 L_c k_{ec}(1 - \cos q_2) - 2 L_{t_2} k_{et_2}(1 - \cos q_{3_2})}{C} \\
\sin q_5 = \dfrac{-z k_{ec} \gamma_1 \sin q_1 + 2 L_c k_{ec} \gamma_1(1 - \cos q_2) + 2 L_{t_2} k_{et_2} \gamma_3(1 - \cos q_{3_2})}{A}
\end{cases}
$$

(vii) Non-linear solution in q_{3_1} and q_{3_2}

$$
\begin{cases}
M = \theta \sin(2q_1) - \beta_{t_1} \cos q_1 (1 - \cos q_{3_1}) - \beta_{t_2} \cos q_1 (1 - \cos q_{3_2}) \\[4pt]
\sin q_2 = 0 \\[4pt]
1 - \cos q_{3_1} = \dfrac{2\alpha_{t_2}\left(\beta_{t_1} \sin q_1 - P_{T_1}^B L_{t_1}\right) + \chi_3\left(\beta_{t_2}\sin q_1 - P_{T_2}^B L_{t_2}\right)}{4\alpha_{t_1}\alpha_{t_2} - \chi_3^2} \\[8pt]
1 - \cos q_{3_2} = \dfrac{\beta_{t_2}\sin q_1 - P_{T_2}^B L_{t_2} + \chi_3(1 - \cos q_{3_1})}{2\alpha_{t_2}} \\[8pt]
q_4 = \dfrac{\xi \sin q_1 + \gamma_1 \sin q_5 - 2L_{t_1}k_{et_1}(1 - \cos q_{3_1}) - 2L_{t_2}k_{et_2}(1 - \cos q_{3_2})}{C} \\[8pt]
\sin q_5 = \dfrac{-zk_{ec}\gamma_1 \sin q_1 + 2L_{t_1}k_{et_1}\gamma_2(1 - \cos q_{3_1}) + 2L_{t_2}k_{et_2}\gamma_3(1 - \cos q_{3_2})}{A}
\end{cases}
$$

(viii) Non-linear solution in q_2, q_{3_1} and q_{3_2}

$$
\begin{cases}
M = \theta \sin(2q_1) - \beta_c \cos q_1(1 - \cos q_2) - \beta_{t_1}\cos q_1(1 - \cos q_{3_1}) - \beta_{t_2}\cos q_1(1 - \cos q_{3_2}) \\[4pt]
1 - \cos q_2 = \dfrac{\beta_c \sin q_1 - P_c^B L_c - \chi_1(1 - \cos q_{3_1}) - \chi_2(1 - \cos q_{3_2})}{2\alpha_c} \\[8pt]
1 - \cos q_{3_1} = \dfrac{2\alpha_c\left(\beta_{t_1}\sin q_1 - P_{T_1}^B L_{t_1}\right) - \chi_1\left(\beta_c \sin q_1 - P_c^B L_c\right)}{4\alpha_c\alpha_{t_1} - \chi_1^2} + \dfrac{\chi_1\chi_2 + 2\alpha_c\chi_3}{4\alpha_c\alpha_{t_1} - \chi_1^2}(1 - \cos q_{3_2}) \\[8pt]
1 - \cos q_{3_2} = \dfrac{\left(4\alpha_c\alpha_{t_1} - \chi_1^2\right)\left(4\alpha_c\alpha_{t_2} - \chi_2^2\right)}{\left(4\alpha_c\alpha_{t_1} - \chi_1^2\right)\left(4\alpha_c\alpha_{t_2} - \chi_2^2\right) - (\chi_1\chi_2 + 2\alpha_c\chi_3)^2} \times \\[8pt]
\qquad \times \left[\dfrac{2\alpha_c\left(\beta_{t_2}\sin q_1 - P_{T_2}^B L_{t_2}\right) - \chi_2\left(\beta_c \sin q_1 - P_c^B L_c\right)}{4\alpha_c\alpha_{t_2} - \chi_2^2} + \right. \\[8pt]
\qquad \left. + \dfrac{2\alpha_c\left(\beta_{t_1}\sin q_1 - P_{T_1}^B L_{t_1}\right) - \chi_1\left(\beta_c \sin q_1 - P_c^B L_c\right)}{4\alpha_c\alpha_{t_1} - \chi_1^2}\dfrac{\chi_1\chi_2 + 2\alpha_c\chi_3}{4\alpha_c\alpha_{t_2} - \chi_2^2}\right] \\[8pt]
q_4 = \dfrac{\xi \sin q_1 + \gamma_1 \sin q_5 + 2L_ck_{ec}(1 - \cos q_2) - 2L_{t_1}k_{et_1}(1 - \cos q_{3_1}) - 2L_{t_2}k_{et_2}(1 - \cos q_{3_2})}{C} \\[8pt]
\sin q_5 = \dfrac{-zk_{ec}\gamma_1 \sin q_1 + 2L_ck_{ec}\gamma_1(1 - \cos q_2) + 2L_{t_1}k_{et_1}\gamma_2(1 - \cos q_{3_1}) + 2L_{t_2}k_{et_2}\gamma_1(1 - \cos q_{3_2})}{A}
\end{cases}
$$

(1)

where the various parameters are defined as follows:

$$\rho_1 = (z_1 - z)^2 k_{ec}k_{et_1}$$

$$\rho_2 = (z_2 - z)^2 k_{ec}k_{et_2}$$

$$\rho_3 = (z_1 - z_2)^2 k_{et_1}k_{et2}$$

$$\gamma_1 = (z_1 - z)k_{et_1} + (z_2 - z)k_{et_2}$$

$$\gamma_2 = (z_1 - z)k_{ec} + (z_1 - z_2)k_{et_2}$$

$$\gamma_3 = (z_2 - z)k_{ec} + (z_2 - z_1)k_{et_1}$$

$$A = \rho_1 + \rho_2 + \rho_3$$

$$B = k_{ec}k_{et_1}k_{et2}$$

$$C = k_{ec} + k_{et_1} + k_{et2}$$

$$\chi_1 = 4L_cL_{t_1}\dfrac{(z_1 - z_2)(z - z_2)B}{A}$$

$$\chi_2 = 4L_cL_{t_2}\dfrac{(z_1 - z_2)(z_1 - z)B}{A}$$

$$\chi_3 = 4L_{t_1}L_{t_2}\dfrac{(z_1 - z)(z_2 - z)B}{A}$$

$$\alpha_c = 2L_c^2\left[\dfrac{(z_1 - z_2)^2 B}{A} + k_{pc}\right]$$

$$\alpha_{t_1} = 2L_{t_1}^2\left[\dfrac{(z_2 - z)^2 B}{A} + k_{pt_1}\right]$$

$$\alpha_{t_2} = 2L_{t_2}^2\left[\dfrac{(z_1 - z)^2 B}{A} + k_{pt2}\right]$$

$$\beta_c = \dfrac{2z(z_1 - z_2)^2 L_c B}{A}$$

$$\beta_{t_1} = \dfrac{2z(z_1 - z_2)(z - z_2)L_{t_1}B}{A}$$

$$\beta_{t_2} = \dfrac{2z(z_1 - z_2)(z_1 - z)L_{t_2}B}{A}$$

$$\theta = \dfrac{z^2}{2}\dfrac{(z_1 - z_2)^2 B}{A}$$

$$\xi = -\dfrac{z}{2}\left(k_{ec} - k_{et_1} - k_{et2}\right)$$

(2)

3.3 Ductility assessment

The evaluation of the ductility of steel connections in the context of the component method requires the characterisation of the ductility of each component. Here, assuming a bilinear idealisation of component behaviour and referring to Figure 3,

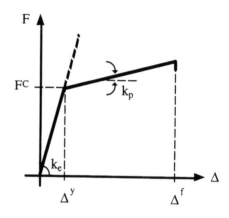

Figure 3. Component characterization.

where k^e, k^p, F^p, Δ^y and Δ^f denote, respectively, the initial elastic stiffness, the post-limit stiffness, the strength, the yield displacement and the collapse displacement of the component, a ductility index ϕ_i is proposed for each component i, defined as,

$$\phi_i = \dfrac{\Delta_i^f}{\Delta_i^y} \tag{3}$$

together with the adimensional post-limit stiffness,

$$\kappa_i^p = \dfrac{k_i^p}{E} \tag{4}$$

The component ductility index ϕ_i allows a direct classification of each component in terms of ductility, using, for example, the three ductility classes proposed by Kuhlmann et al. (1998):

Class 1 – Components with high ductility ($\phi_i \geq \alpha$)

Class 2 – Components with limited ductility ($\beta \leq \phi_i < \alpha$)

Class 3 – Components with brittle failure ($\phi_i < \beta$)

α and β representing ductility limits for the various component classes. In design terms, it seems reasonable to assume, for Class 1 components, a ductility index $\phi_i = \infty$. On the other end, for Class 3 components, a safe estimate can be obtained with a ductility index of $\phi_i = 1$. For Class 2 components, lower bounds for the ductility indexes must be established for each component type, as a result of experimental and analytical research to be carried out. As a crude indication, from the experimental results obtained by Kuhlmann (1999), a ductility index in the range of 4 to 5 seems reasonable for the component web in compression, if a negative plastic stiffness is used.

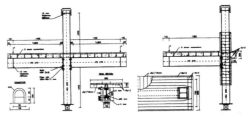

Figure 4. Experimental models for internal and external nodes

4. EXPERIMENTAL BEHAVIOUR

To provide insight into the behavior of composite joints under static and cyclic loading, a series of tests were undertaken on end-plate beam-to-column composite joints, described next.

4.1 Static tests

The test program performed at the Civil Engineering Department of the University of Coimbra included 8 prototypes, being 4 in internal nodes and 4 in external nodes. The description of each model includes the geometric definition, the material properties and the testing and instrumentation procedures. The prototypes, covering internal and external nodes, were defined such that they could reproduce the connections in a common framed structure, with spans of about 7m, 4m spacing between frames, live loads up to 4 kN/m^2 and a high energy dissipation capacity and a good fire resistance. According to the objectives of this study, the steel connection is the same in all prototypes, corresponding to a beam connected to the column by one end plate, welded to the beam and bolted to the column.

In all cases, the beams consist of an IPE 270, rigidly connected to a reinforced concrete slab (full interaction) by 8 shear connectors. The slab, 900 mm wide and 120 mm thick, is reinforced with 10ɸ12 longitudinal bars and 10ɸ8 transversal bars per meter, with 20 mm cover. The steel connection consists of a 12 mm thick end plate, welded to the beam and bolted to the column flange through 6 M20 bolts (class 8.8). The end-plate is flushed at the top and extended at the bottom, in order to achieve similar behaviour under positive and negative moments. The steel column is the same in all the tests (HEA 220), being envolved by concrete (300 × 300 mm) in tests E3 and E4, with longitudinal reinforcement of 4ɸ12, with one bar in each corner of the section and stirrups consisting of ɸ6 bars 0.08 m apart. The following materials were chosen: S235 in the steel components, steel class 8.8 in the bolts, steel A400 NR in the reinforcing bars.

4 tests were performed in internal nodes, tests E1 and E2 corresponding to the prototype arrangement between composite beams and a steel column shown in Figure 4a) and tests E7 and E8 between composite beams and a composite column. The

Figure 5. Photos of prototypes after testing

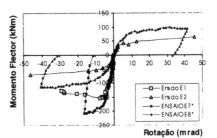

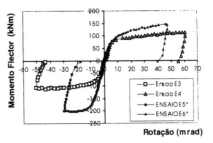

Figure 6. Moment-rotation response for tests on internal nodes.

loads were applied to the beams 1.40 m from the steel column face with two dynamic actuators with a capacity of 200 kN and 600 kN, and maximum displacement of 20 cm and 10 cm, respectively. In tests E1 and E7, loading is symmetric (both loads downward) while in tests E2 and E8 loading is asymmetric (one load upward and the other downward).

As for the internal nodes, four tests were performed in external nodes, tests E3 and E4 corresponding to a steel column, and tests E5 and E6 corresponding to a composite column and illustrated in Figure 4b). Loading is applied similarly to the internal node

Figure 7. Photos of external nodes after testing

tests. In tests E3 and E5 the load is applied downward, while in tests E4 and E6 it acts upward.

Comparative results for the tests on internal columns (E1, E2 and E7, E8) and on external columns (E3 to E6) are briefly presented. The resulting moment-rotation curves are illustrated in Figure 6. All tests were performed about 16 days after the slab was cast. Initially (elastic behaviour), loading was increased at 5 kN increments; close to the plastic resistance of the connection (plastic behaviour) displacement control was imposed (at the load application locations), with 3 mm increments. During the elastic response, the connection was unloaded to 5 kN, loading proceeding subsequently until collapse.

Failure ocurred by different reasons, varying from failure of the horizontal shear of the column web, yielding of the steel reinforcement closer to the column with concrete cracking mostly at the tension zone, instability of the column web to yielding of the lower part of the beam web and lower beam flange. At failure, for several of the tests, the dominant component of rotation was deformation induced by horizontal shear on the column web panel.

Comparison of experimental results with analytical predictions (for actual material and geometrical

data) based on EC4 and other authors, as described in detail in Simões (2000) is summarized in Tables 1 and 2.

4.2 Cyclic tests

The cyclic tests were performed under similar conditions as the static tests, according to the methodology proposed by the ECCS (1986). The evaluation of results was based on the moment rotation response of the connection, the various elastic parameters being summarized in Table 3

For the tests on internal nodes (E11 and E12), the joints presented high ductility (ductility index close to unity), with similar response for hogging and sagging moment. Because the maximum amplitude was not very high, the strength degradation was low.

Table 3. Elastic parameters for tests E9 to E12.

Elastic	E9		E10	
Parameters	Env. +	Env. -	Env. +	Env. -
K_y (kNm/mrad)	24.57	26.81	36.95	40.83
ϕ_y (mrad)	4.64	4.29	4.60	3.89
M_y (kNm)	114.0	114.9	169.8	158.9
Elastic	E11		E12	
Parameters	Env. +	Env. -	Env. +	Env. -
K_y (kNm/mrad)	16.50	18.83	34.44	36.22
ϕ_y (mrad)	5.05	4.28	3.39	3.25
M_y (kNm)	83.4	80.7	116.8	117.8

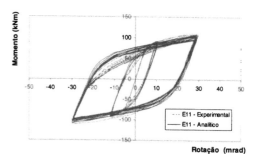

Figure 8. Moment-rotation response for test E11.

Table 1. Experimental and analytical moment resistance.

Ensaio	M_{exp} (kNm)		M_{anal} (kNm)		M_{anal}/M_{exp}	
E1		-153.6		-140.8		0.92
E2	62.8	-70.5	65.5	-73.0	1.04	1.03
E7		-219.8		-220.8		1.00
E8	126.1	-160.2	124.5	-168.0	0.99	1.05
E3	-111.3		-108.6		0.98	
E4	113.2		128.0		1.13	
E5	-217.6		-218.8		1.01	
E6	168.2		155.1		0.92	

Table 2. Experimental and analytical initial stiffness.

Ensaio	$S_{i,exp}$ (kNm/mrad)		$S_{i.anal}$ (kNm)		$S_{i,anal}/S_{i,exp}$	
	L	R	L	R	L	R
E1		62.5		57.8		0.92
E2	14.3	17.2	22.1	16.8	1.54	0.98
E7		81.3		77.6		0.95
E8	22.2	27.2	33.6	25.2	1.51	0.93
E3	27.0		25.3		0.94	
E4	28.7		37.0		1.29	
E5	43.1		36.1		0.84	
E6	44.4		50.5		1.14	

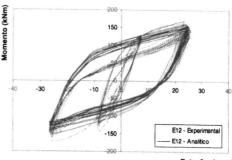

Figure 9. Moment-rotation response for test E12.

However, for the stiffness and energy degradation indices, the corresponding reductions were severe (Simões 2000). Figures 8 and 9 illustrate the experimental results for the internal node tests, shown superimposed with results from the Richard-Abbott analytical model, fitted with the parameter N taken as constant for all cycles from the envelope of the experimental hysteretic curve (Simões 2000). Parameter k_p is taken as 5 % of the initial k value. Parameters k and M_0 are evaluated for every cycle, M_0 being obtained from the straight line of slope k_p, going through the extreme point for each cycle, obtained experimentally.

For the tests on external nodes (E9 and E10), the ductility indices remained high (above 0.8), except for test E10 under negative hogging moment. The strength degradation was also severe, while the energy degradation indices have shown that the joints reached collapse. Figures 10 and 11 illustrate the moment-rotation response for these two tests, again superimposed with the Richard-Abbott analytical model.

5. COMPARISON WITH ANALYTICAL PREDICTIONS

5.1 *Monotonic results*

To assess the validity of the analytical model described in section 3, it was applied to test E7

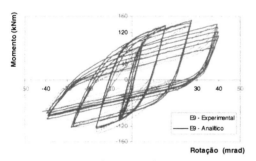

Figure 10. Moment-rotation response for test E9.

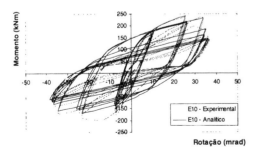

Figure 11. Moment-rotation response for test E10.

Table 4. Component characterization.

Component	F^C (kN)	k_e (kN/m)	k_p(kN/m)
Column web in compression	1383.85	3 712 800	10
Column web in tension	506.07	940 440	10
Column flange in bending	346.44	2 982 000	10 000
End-plate in bending	295.63	2 321 900	10 000
Beam flange and web in compression	582.21	∞	20 000
Beam web in tension	463.41	∞	∞
Bolts in tension	444.53	2 257 200	10 000
Reinforcement	511.86	1 025 000	1 200

(internal node, composite column). Table 4 reproduces the properties of the relevant components for this connection. Figure 12 shows the various moment-rotation curves, obtained experimentally and analytically (Annex J and eqs. (1)). It is noted that similar results are obtained in terms of resistance, while significant discrepancies are observed in terms of initial stiffness, as illustrated in Table 5. The proposed model identifies the sequence of components to fail: beam flange and web in compression, followed by yielding of the reinforcement. The test results (Simões *et al.*, 1999) confirm these predictions.

Careful examination of the experimental results of Figure 6 highlight a decrease of stiffness for relatively low values of bending moment. This

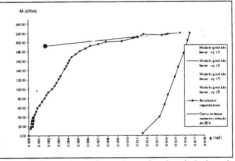

Figure 12. Comparison of experimental and analytical results for test E7.

Table 5. Strength and initial stiffness.

	Exp. results	Annex J of EC4	Analytical model
$M_{j.Rd}$ (kNm)	219.8	220.8	213.1
$S_{j.ini}$ (kNm/mrad)	81.3	77.6	110.7

tendency is even more marked in joints with composite columns (tests E5 and E7). According to the readings from extensometers applied to the rebars and the cracking history of the concrete slab, this reduction in stiffness is caused by cracking of the concrete in tension. The component 'slab reinforcement (R)', as characterized by Annex J of EC4, takes this effect into consideration in a semi-empirical way, assuming a constant initial stiffness for this component, deemed to remain valid up to yielding of the reinforcement. As a result of this crude assumption, the analytical results clearly deviate from the experimental evidence. To improve on this situation, it was decided to characterize the bilinear spring for the component R so that the cracking axial force of the concrete slab was identified, with due account of the post-cracking stiffness of the component, almost exclusively resulting from the rebars. The results of Figure 13 were thus obtained for test E7, clearly improving on the previous situation and rigourously reproducing the experimenatl results. It is noted, however, that the possibility to identify yielding of the reinforcement is lost, thus explaining the discrepancies for rotations larger than 12 mrad.

5.2 Cyclic results

The analytical predictions for the cyclic response of the composite joints presented in the previous section were no other than curve-fitting exercises that required the knowledge of the actual experimental results. Here a significant improvement is proposed (Simões 2000), also taking the Richard-Abbott model as the underlying mathematical description and exclusively deriving all results from the static monotonic response, as illustrated in Figure 14 for test E10. As before, the parameter N is kept constant for all cycles and obtained as described before. Parameter k_p is taken as 5 % of the intial elastic stiffness (K_y). A linear decaying law is assumed for parameters k e M_0, based on the values obtained for each cycle in the previous section; these parameters are

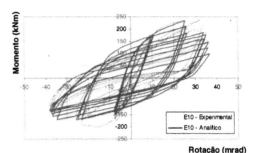

Rotação (mrad)

Figure 14. Comparison of experimental results and analytical prediction based on the Richard-Abbott model

adimensionalised with respect to the elastic parameters (M_y, K_y e ϕ_y), defined in the envelope of the hysteretic curves and related to the static monotonic response.

Given that the chosen composite joints are asymmetric with respect to the flexural axis, their behavior under negative and positive bending moments is distinct. Thus, the parameters for the Richard-Abbott formula are named k_a, k_{pa}, M_{0a} and N_a for the ascending branch and k_d, k_{pd}, M_{0d} and N_d for the descending branch.

As mentioned above, the parameters k_a, k_d, M_{0a} and M_{0d}, were obtained from adjusted curves to the values calculated for each cycle in the previous section. As the various cycles were performed for different amplitudes, it is important to define the variation of these parameters with the number of cycles and the amplitude. In general, both k_a and k_d, and M_{0a} and M_{0d} (initial values for a given amplitude) are kept constant for low amplitudes; as the amplitude of the cycles increases, these parameters reduce. With respect to the variation with the order of the cycles, the tendency is slightly different: k_a and k_d start to reduce right from the first cycle, even for low amplitudes, while M_{0a} and M_{0d} show a tendency to remain stable for cycles of low amplitude. These conclusions were obtained from the cyclic tests performed in this research program, considering 3 cycles for each amplitude only, except for the last.

6. CONCLUDING REMARKS

The developments presented in this paper were directed at the daunting task of predicting the dynamic behavior of steel and composite joints. Here, two directions were pursued: (i) full analytical evaluation of the moment-rotation response of composite joints under static monotonic loading leading to closed-form non-linear solutions, and (ii) analytical prediction of the cyclic moment-rotation response of end-plate beam-to-column composite joints from the knowledge of its static monotonic response.

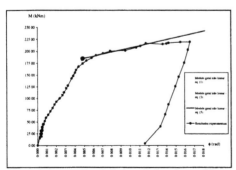

Figure 13. Improved analytical predictions for test E7.

An analytical model was presented that has shown good potential to fullfill the first objective. The assumption of bilinear characterization of the basic components was shown to be insufficient in some situations. Current work on the generalization of this methodology for arbitrary component (full non-linear) force-displacement response seems promising.

The curve-fitting prediction based on the Richard-Abbot model was mostly based on the maxima moments reached at the end of each half-cycle. Consequently, analytical and experimental results for this parameter were good, the maximum difference being obtained for test E12 with average ratios of $M_p^{anal} / M_p^{exp} = 1.10$ for positive bending and $M_a^{anal} / M_a^{exp} = 0.88$ for negative bending. Surprisingly, the analytical prediction based on the envelope of mnotonic response gave even better average results.

As for the energy dissipation, agreement was not as good, as expected, because slipage occurred in most tests and the Richard-Abbott model is not able to reproduce this phenomenon. The Mazzolani model (1988), with more parameters and able to simulate slipage, should provide better agreement and is currently the object of research.

7. ACKNOWLEDGMENTS

Finantial support from "Ministério da Ciência e Tecnologia" - PRAXIS XXI research project PRAXIS/P/ECM/13153/1998 is acknowledged.

8. REFERENCES

Calado, L., Simões da Silva, L.A.P. & Cruz, P.J.S. (2000). Tests results on composite frame under dynamic loading, *Behaviour of Steel Structures in Seismic Areas, Proc. intern. symp., Montreal, Canadá, 21-24 August 2000.* Rotterdam: Balkema.

COST C1. (1999). Recent advances in the field of structural steel joints and their representation in the building frame analysis and design process, Brussels, Luxembourg.

Cruz, P.J.S.; Calado, L.and Simões da Silva, L.A.P., (2000). Characterization of connection components in steel and composite connections under cyclic loading, *Behaviour of Steel Structures in Seismic Areas, Proc. intern. symp., Montreal, Canadá, 21-24 August 2000.* Rotterdam: Balkema.

ECCS (1986), "Recommended Testing Procedure for Assessing the Behaviour of Structural Steel Elements under Cyclic Loads" - N°45.

Eurocode 3, ENV - 1993-1-1:1992/A2 (1998), "Annex J", Design of Steel Structures – Joints in Building Frames, CEN, European Committee for Standardization, Ref. No. ENV 1993-1-1: 1992/A2:21998 E, Brussels.

Eurocode 4, ENV 1994-1-1 (1996), Proposed Annex J for EN 1994-1-1, Composite joints in building frames, CEN, European Committee for Standardization, Draft for meeting of CEN/TC 250/SC 4, Paper AN/57, Bruxelas.

Jaspart, J-P. (1997) Contributions to recent advances in the field of steel joints. Column bases and further configurations for beam-to-column joints and beam splices. Thèse d'agrégé de l'enseignement superieur, Université de Liège

Kuhlmann, U. (1999). Influence of axial forces on the component "web under compression", COST C1/WG2/99-01.

Kuhlmann, U., Davison, J. B. and Kattner, M. (1998). Structural systems and rotation capacity, *COST Conference*, Liège, Belgium, Sep. 18th-19th 1998.

Mazzolani, F.M. (1988). Mathematical Model for Semi-Rigid Joints Under Cyclic Loads. In Connections in Steel Structures: Behaviour, Strength and Design (ed. R. Bjorhovde et al.. Elsevier Applied Science Publishers, London, pp. 112-120.

Simões da Silva, L.A.P., Girão Coelho, A. & Neto, E.L. (2000). Equivalent post-buckling models for the analysis of steel connections, *Computers & Structures*, (in print).

Simões da Silva, L.A.P., Girão Coelho, A., (2000). A ductility model for steel connections, *Journal of Constructional Steel Research*, (submitted for publication).

Simões, R. A. D., Simões da Silva, L. A. P., and Cruz, P. J. S., (1999), Experimental models of end-plate beam-to-column composite connections, *2nd European Conference on Steel Structures, EUROSTEEL '99*, 26-29 May, Praha, Czech Republic, pp. 625-629.

Simões, R.A.D., (2000), Behaviour of beam-to-column composite joints under static and cyclic loading (in portuguese), PhD Thesis, Civil Engineering Department, Universidade de Coimbra, Coimbra, Portugal.

Weynand, K., Jaspart, J-P. and Steenhuis, M. (1995) "The stiffness model of revised Annex J of Eurocode 3", *Proceedings of the 3rd International Workshop on Connections*, Trento, Italy, May 8-31.

Zandonini, R., Bernuzzi, C. and Bursi, O., (1997). Steel and steel-concrete composite joints subjected to seismic actions - general report, in *Behaviour of Steel Structures in Seismic Areas, STESSA'97*, Kyoto, Japan, 1997.

Improvement of seismic behaviour of beam-to-column joints using tapered flanges

P. Sotirov, N. Rangelov & J. Milev
Faculty of Civil Engineering, UACEG, Sofia, Bulgaria

ABSTRACT: The good performance of steel moment-resisting frames under seismic motions depends strongly on the behaviour of their beam-to-column connections. The recent earthquakes generated an urgent need to increase the ductility of the connections normally used in practice. This paper presents a study of beam-to-column joints suitable for the case of column-tree configuration of moment-resisting frames, where tapered flanges are used to improve the seismic behaviour. Two types of joints are proposed and tested: type A, considered as an improved 'conventional' joint, and type B, that may be regarded as a 'new-generation' detail. Experimental investigation on six (3+3) full-scale specimens was carried out. The tests were performed in the Steel Structures Research Laboratory of the Faculty of Civil Engineering, University of Architecture, Civil Engineering and Geodesy (UACEG), Sofia, within the framework of the European RECOS project. Both specimen types exhibited stable hysteretic behaviour with adequate ductility and energy dissipation capacity. Though both variants were quantitatively comparable, the most significant advantage of type B was found in its predictable behaviour.

1 INTRODUCTION

Steel moment-resisting frames have been popular in many regions of high seismicity not only for their architectural versatility, but also for their good seismic performance as highly ductile systems. However, the seismic response of a ductile moment frame is strongly dependent on the available strength, stiffness and plastic deformation capacity of the connections between the framing members. The traditional strong-column – weak-beam design concept implies that the input energy is absorbed and dissipated primarily by plastic hinges formed near the beam-to-column connections. The Northridge earthquake definitely demonstrated a lack of deformation capacity of the site welded flange – bolted web connections, normally used in the American practice. The column-tree configuration, used in Japan and other countries, relies on the higher quality of the shop welded beam-to-column connections. However, the Kobe earthquake refuted to a great extent that belief. Thus an urgent need for improved detailing has been generated.

Two key strategies have been developed: (i) strengthening the connection, and (ii) weakening the beam that frame into the connection (Bruneau et al. 1998). Within the framework of (i), cover plates, upstanding ribs, side plates, and haunches are implemented. The weakening strategy (ii) is based on the idea of shaving beam flanges to intentionally weaken the beam at a predetermined location, originally proposed by Plumier (1990). Besides the classical "dog bone" profile and circular cuts, Chen et al. (1996) as well as Iwankiw & Carter (1996) proposed to taper flanges according to a linear profile so as to approximately follow the varying bending moment diagram. Both strategies aim at effectively moving the plastic hinge away from the column face, thus avoiding the problem of poor ductile behaviour and potential fragility of the welds. All the proposed and tested connections have some merits and disadvantages. It seems that no detail can be assumed to be perfect and new solutions are still worth exploring. For example, recently Popov et al. (1996) suggested a combination between the two strategies, namely a cover plates connection with circular "dog bone" cuts in the beam flanges.

In the study reported herein, an improvement of the hysteretic behaviour of the beam-to-column connections in moment-resisting frames is pursued by using tapered beam flanges in two variants. Both of them are more suitable for the case of built-up beams which have the advantage to be proportioned more flexibly to attain a better balance between internal forces and resistance, and the column-tree configuration with fully shop-welded rigid beam-to-column connections and field-welded or bolted erection splices in the beams.

2 PROPOSED BEAM-TO-COLUMN JOINT TYPOLOGIES

The two types of joints are shown in Figure 1. In type A the flanges are tapered to follow approximately the varying bending moment diagram. Thus, on one hand the solution is more economical, and, on the other hand, under an earthquake load, the yielded zones in flanges are expected to spread over the entire portion that follows the moment diagram. As a result, the energy dissipation capacity is believed to be higher than in the conventional non-tapered flanges due to the larger volume of plasticity. Nevertheless, the expected dissipative zone is mostly near the column face.

Detail B is of the 'new generation' joints. The shifting of the plastic hinge from the column flange is achieved by a specific flange profile, following the idea illustrated in the Figure. Thus both the above-mentioned strategies are incorporated in one detail. On one hand, the connection is strengthened without additional cover plates or ribs, keeping a constant beam depth, and, on the other hand, the beam is weakened at a predetermined location to provoke the development of a plastic hinge. This solution seems more economical than those mentioned in the literature and reserves the main advantage of possessing predictable behaviour.

3 SPECIMENS, TEST SETUP AND LOADING HISTORY

The test beam-to-column assemblies were full-scale, "extracted" from an especially designed two-bay four-storey regular moment-resisting frame. The specimens were fabricated by a Bulgarian industrial manufacturer. The column sections were produced by automatic welding under flux, whereas the beam sections, small in size for the welding equipment, were built-up by semiautomatic welding under CO_2 shield. The full penetration welds between beam and column flanges were also made under gas shield. The erection splices were hand-welded with covered electrodes in the laboratory, thus simulating the field conditions. Mild steel was used with an average yield stress measured $f_y = 252$ MPa. A total of 6 specimens were tested: 3 identical ones (labeled A-1, A-2 and A-3) of type A, and 3 identical specimens (B-1, B-2 and B-3) of type B. The layout and main dimensions are shown in Figure 2.

A total of 14 inductive displacement transducers were installed (Fig. 2) to measure the global behaviour of the specimens: beam end displacement, joint rotation, panel zone shear deformation, and possible movement of the supports. All the instruments were connected to a PC-based data acquisition system with software developed by the laboratory staff.

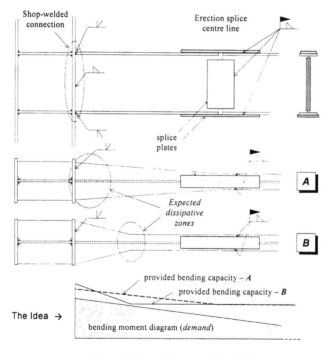

Figure 1. Proposed beam-to-column joints.

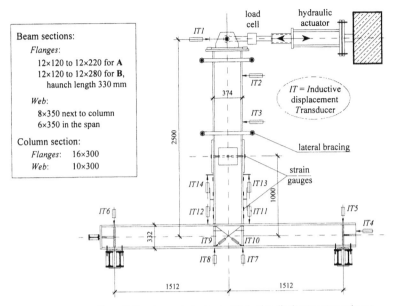

Figure 2. Specimens, test setup and inductive displacement transducers.

A *Keithly DAC-02* card was used to send the displacement command signal to the actuator servovalve. Strain gauges were used to investigate the local response of the beam flanges and splice plates.

The testing programme was based on the recommendations of ECCS (1986). The specimens were tested under displacement control, following a loading history consisting of stepwise increasing deformation cycles. Initially, four cycles in elastic range were applied with amplitudes of $\pm 0.25 v_y$, $\pm 0.50 v_y$, $\pm 0.75 v_y$, and $\pm v_y$, where v_y is the expected first-yield displacement of the beam end. Eventually, correction of v_y was made. Then the testing continued in the plastic range with three full cycles at each amplitude level $\pm 2 v_y$, $\pm 3 v_y$, and so on (instead of the recommended by ECCS (1986) $\pm (2+2n)v_y$; the latter was only applied to specimen A-3).

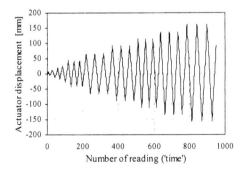

Figure 3. Typical loading history.

A typical loading history is shown in Figure 3. For specimens A-3 and B-3 an initial displacement corresponding to approximately 30% of the plastic bending capacity was applied to simulate the effect of the gravity loading.

4 EXPERIMENTAL RESULTS AND DISCUSSION

In type A specimens plastic deformations started to develop first in the panel zone and later in the flanges near the column. Satisfactory stable hysteretic behaviour was observed. The first tested specimen, A-1, failed by lateral-torsional buckling due to improper bracing. In specimen A-2 flange local buckling occurred, followed by web buckling. Shear deformations of the panel zone were apparent, demonstrating its very significant contribution. This was more pronounced in A-3, where no local buckling appeared. Both A-2 and A-3 failed due to fracture of the upper flange splice plate. The data from the strain gauges showed that considerable plastic deformations had developed there. Due to some lack of overstrength and different areas of the two splice plates, the top one became overloaded. This fact, together with the stress concentration, led to low-cycle fatigue. However, no other cracks were observed elsewhere. The good behaviour of the shop-welded full penetration welds was clearly proven.

All type B specimens exhibited a very stable cyclic hysteretic response, and behaved exactly as predicted. Plastic zones developed at the designed loca-

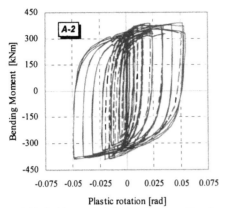
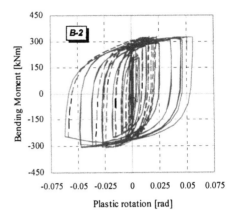

Figure 4. Typical hysteretic behaviour in terms of bending moment versus total (—) and beam (– –) plastic rotation at the 'reference' cross-section.

tion, accompanied by flange local buckling in the three specimens. Finally, the resulting secondary bending stresses at the crest of the buckles led to a low-cycle fatigue rupture of the buckled flange. No other cracks were detected either in the web-to-flange fillet welds in that region, or elsewhere. Both in A-3 and B-3, the effect of initial gravity loading was negligible and vanished well within the plastic range.

The cyclic response of the joints is illustrated in Figure 4 in terms of moment versus plastic rotation. Only the results for specimens A-2 and B-2 are given as typical. Both the bending moment and the rotation are determined at the cross-section where the dissipative zone is expected ('reference' cross-section), as shown in Figure 1. The plastic rotation is obtained by 'clearing' the measured values from their elastic components using the elastic stiffness determined from the experimental curve by a linear fit. The plastic rotation in the beam is separated to assess the relative contribution of the plastic zone in the beam and the panel zone. Generally, all joints

exhibited adequate ductility and in both types A and B, total plastic rotations of 0.05 rad were reached without a brittle fracture, which is well above the adopted demand of 0.03 rad (Bruneau et al., 1998). However, due to the different contribution of the panel zone, beam-only plastic rotations reached in type A were 0.02–0.025 rad, whereas in type B they were up to 0.03–0.04 rad. It means therefore that in case of type A joints, the ductility demand to the panel zone will be higher to provide the same rotational capacity.

Typical plastic hinges formed by local buckling in both specimen types. In type A specimens the notional plastic hinges developed near the column face and were not well pronounced. In contrast, in type B specimens classical plastic hinges took place exactly at the designed location, accompanied by very strong flange and web local buckling.

The idea of the two joint types is best illustrated by the strain distributions. Typical strain distributions in the flanges of both specimen types are shown in Figure 5 where only specimens A-2 and

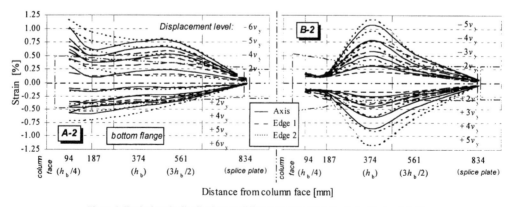

Figure 5. Typical strain distributions at different displacement levels for the two joint types.

268

B-2 are selected, and only the strains measured in bottom flanges are plotted. It is well seen that in case of type A, yielding has spread over the entire portion between the column and splice, thus illustrating the idea of more volume in plasticity (and therefore more dissipated energy). However, the plastic deformations apparently tend to concentrate near the column face. In contrast, the Figure definitely proves the concept of type B joints, clearly demonstrating a concentration of plasticity at the designed location.

This above was confirmed also by FEM simulations (Entchev 1999) as illustrated in Figure 6.

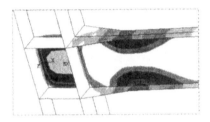

Figure 6. Typical strain picture obtained by FEM simulations of type B joint.

A very important parameter that characterizes the structural behaviour of a joint under seismic loading is its energy dissipation capacity. Herein, it is determined as the plastic work based on the hysteretic area of the experimental loops (those of Figure 4), and is plotted in Figure 7 versus the relevant cumulative beam end displacement. The contributions of the beam on one hand, and the column and panel zone on the other, are separated. Thus it is seen that the contribution of the plastic zones in the beams of type A specimens is only about 30%.

In contrast, in type B, the contribution of the beam plastic zones is approximately 50% and only for B-1 it is about 60%. Note the very close agreement of the curves for the total dissipated energy, demonstrating the similar behaviour of the three specimens.

To compare the dissipation capacity of the different specimens, the non-dimensional dissipated energy is defined as

$$E^* = \frac{E}{F_{pl}\, v_{tot}} \qquad (1)$$

where E is the dissipated energy (plastic hysteretic work), F_{pl} is the force at the beam end corresponding to the theoretical plastic resistance moment at the 'reference' cross-section, and v_{tot} is the total cumulative beam end displacement. Thus, referring the dissipated energy to that of a relevant perfectly plastic system, it is believed that not only the different bending resistance of the two types of specimens,

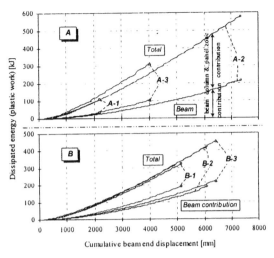

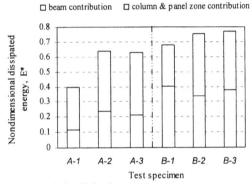

Figure 7. Dissipated energy.

Figure 8. Relative dissipation capacity.

but also some variations in the loading history may be accounted for.

The comparative results are shown in Figure 8. In this format, firstly, the different contributions of the two components – beams and panel zones – are well seen. Secondly, the relative dissipation capacity of type B specimens is on average some 20% higher. Moreover, the portion of the energy dissipated only by the beam plastic zones in specimens B is almost twice higher, which undoubtedly is an additional advantage.

5 CONCLUDING REMARKS

Not only the extensive research carried out, but also recent earthquakes in Japan and California demonstrated a lack of deformation capacity of the connections typically used in practice and thus generated an urgent need for improved detailing. Though

269

new strategies have been developed aiming at effectively moving the plastic hinge away from the column face, it still seems that no detail can be assumed perfect and new solutions are still worth exploring.

Haunching has been found to be an efficient tool for solving the problem of poor ductile behaviour and potential fragility of welded beam-to-column connections. In this paper, a type of haunching through variable-breadth flanges is used to develop two improved joint solutions. In joint type A, the flanges are tapered to follow approximately the varying bending moment diagram. The goal is not only to save material, but to spread yielding over the entire portion that follows the moment diagram and thus to dissipate more energy than in the case of conventional non-tapered flanges where the plastic zone is more localized. Detail B belongs to the 'new generation' joints, in which a shift of the plastic hinge from the column flange is achieved by a specific flange profile. Both variants are considerably suitable for the case of column-tree configuration.

An experimental study is reported on six specimens — 3 of type A and 3 of type B — following the ECCS (1986) Recommendations. Both specimen types exhibited stable hysteretic behaviour with adequate ductility. The plastic rotation capacity of type A was however largely due to the flexible panel zone. The energy dissipation capacity of both types was high, type B being some 20% better. Though the two variants were quantitatively comparable, the most significant advantage of type B lies in its predictable behaviour. The lack of a structural 'fuse' in type A leads to some variable behaviour and imposes higher demands to the panel zone.

It should be concluded that both joint typologies considered exhibited satisfactory cyclic behaviour and high dissipation capacity. However, type B, having behaved exactly as predicted and designed, possesses some definite advantages, therefore is suggested for the design practice.

ACKNOWLEDGEMENTS

The present study has been carried out in the framework of the INCO-COPERNICUS Joint Research Project "Reliability of Moment Resistant Connections of Steel Building Frames in Seismic Areas" (RECOS). Thanks are therefore due to all the partners for their comments, and especially to Professor F. M. Mazzolani, the coordinator of the Project. The financial support of the EU is also gratefully acknowledged.

REFERENCES

Bruneau, M., Uang, C.-M. & Whittaker, A. 1998. *Ductile Design of Steel Structures*, New York: McGraw-Hill.

Chen, S.-J., Yeh, C.H. & Chu, J.M. 1996. Ductile steel beam-to-column connections for seismic resistance. *Journal of Structural Engineering, Proc. of ASCE*, 122(11): 1292-1299.

ECCS 1986. *Recommended Testing Procedures for Assessing the Behaviour of Structural Elements under Cyclic Loads*. Brussels: European Convention for Constructional Steelwork (ECCS), Publication No. 45.

Entchev, A. 1999. *Analysis of ductile steel beam-to-column joints using FEM*, MSc Thesis, Faculty of Civil Engineering, UACEG, Sofia (in Bulgarian).

Iwankiw, R.N. & Carter, C.J. 1996. The dogbone: a new idea to chew on. *Modern Steel Construction* 36(4): 18-23.

Plumier, A. 1990. *New Idea for Safe Structures in Seismic Zones*. University of Liège, IABSE Symposium, Brussels.

Popov, E.P., Blondet, M. & Stepanov, L. 1996. *Application of "dog bones" for improvement of seismic behavior of steel connections*. Report No. UCB/EERC 96/05, University of California, Berkeley, USA.

Behaviour of Steel Structures in Seismic Areas, Mazzolani & Tremblay (eds) © 2000 Balkema, Rotterdam, ISBN 90 5809 130 9

Comparison of seismic capacity between post-Northridge and post-Kobe beam-to-column connections

Keiichiro Suita
Department of Architecture and Environmental Design, Kyoto University, Japan

Masayoshi Nakashima
Disaster Prevention Research Institute, Kyoto University, Japan

Michael D. Engelhardt
Department of Civil Engineering, The University of Texas at Austin, USA

ABSTRACT: This paper presents the results of full-scale tests of welded beam-to-column subassemblies having two types of improved connection details: i.e., the reduced beam section (RBS) detail adopted in the US and the no weld access hole detail adopted in Japan. Major findings obtained from the tests are summarized as follows. (1) Both types of improved connection details successfully prevented premature fracture of the welds and the surrounding base metal regions, indicating significantly larger plastic rotation capacity compared to the conventional details. (2) Both types of connection details exhibited nearly identical energy dissipation capacity, although the strength of the RBS detail is about 20 % smaller than the strength of the no weld access hole detail. This observation is derived from plastic strain distributions along the beam length as well as from contributions of the beam plastic rotation to the story drift.

1 INTRODUCTION

The 1994 US Northridge Earthquake and the 1995 Japan's Hyogoken-Nanbu (Kobe) Earthquake disclosed weaknesses of welded beam-to-column connections of steel moment frames. Reevaluation of the plastic rotation and energy dissipation capacity of the connections and development of more reliable connections are a key for providing better seismic performance of steel building structures in future earthquakes.

In light of the lessons learned from the Northridge Earthquake, various modifications were proposed in the US, including the use of higher toughness welding electrodes and other improved welding practices, and the reinforcement of the connection by cover plates, ribs, or haunches. Although these connection modifications showed good performance, new problems were introduced concerning fabrication, inspection, cost, and reliability. After extensive studies, the reduced beam section (RBS) design was accepted as one of the more reliable and least costly improvements (Engelhardt et al. 1996, Tremblay et al. 1997, Popov et al. 1998). This design intends to avoid the development of a plastic hinge in the beam adjacent to the face of the column, thereby reducing the likelihood of fracture occurring at the weld. On the other hand, in Japan, fractures of beam flanges typically initiated from the toe of a weld access hole, frequently accompanied by clear signs of local buckling and considerable plastification of the beam. Accordingly, modification of the size and shape of weld access holes, aimed at reducing the stress concentration at the toe, was discussed extensively in Japan. A variety of new shapes have been proposed and tested, with the no weld access hole (NWAH) detail as the ultimate modification (Tateyama & Inoe et al. 1988, Nakagomi et al. 1992, Nakashima & Suita et al. 1998).

The US and Japan have developed significantly different details for improving the seismic performance of steel moment connections. However, the performance of these details has not been compared to one another because of differences in US and Japanese practices with regard to materials, beam and column sections, welding practices, testing protocols, and other factors. Nonetheless, it is instructive to compare the seismic performance of these two connection improvements as well as to consider their relative merits and drawbacks. Such comparisons are intended to promote greater mutual understanding of steel moment connection performance.

For this purpose, cyclic loading tests were conducted using full-scale beam-column subassemblies having different types of improved connection details: i.e., the radius cut RBS detail adopted in the US and NWAH detail adopted in Japan. All specimens were fabricated with identical materials, sections, welding conditions, and workmanship except for the connection detail. The controlling failure modes and the overall seismic resistant performance of the specimens were compared to one another.

2 TEST SPECIMENS

2.1 Dimensions of test specimens

A total of 6 full-scale T-shaped beam-column subassemblies were prepared for testing as shown in Figure 1. A connection type referred to as the "through-diaphragm connection" was used for joining the column and beam. The major test variables were the methods used to improve the plastic deformation capacity of the connections, i.e. details of the weld access holes and the RBS.

• Column : Cold-formed square tube section of 350mm ×350mm× 12mm; material grade BCR295.
• Beam : Rolled wide-flange section of 500mm× 200mm× 10mm×16mm (d × b$_f$ × t$_w$ × t$_f$); material grade SN400B.
• Through diaphragm plate : 410mm×410mm×19mm; material grade SN490B.

Three types of connections were tested (Table 1). (1) Connections designated as Type S were constructed using the no weld access hole detail (NWAH) and did not include an RBS. (2) Connections designated as Type R were constructed with a radius cut RBS and the weld access hole recommended in AIJ JASS6 (1996). (3) A connection designated as Type SC was included as a baseline specimen. This connection included neither an RBS nor the NWAH detail. Connection Type SC most resembles typical pre-Kobe details used in Japan, except for one change described below.

2.2 Welding details

The weld access hole recommended in AIJ JASS6 (1996) was used for Type R and Type SC specimens. The basic shape of this weld access hole (Figure 2) is similar to a conventional access hole commonly used in Japan before the Kobe earthquake, except that the toe of the access hole is modified to have a 10 mm radius intended to reduce the stress concentration at the toe. The access holes were machine cut. The toe of the access hole was machined by a minor change of the cutter blade.

A welding detail having the no weld access hole (NWAH) was used for the type S specimens. During the process of cutting the weld bevel and the web cope, three blades are commonly used, i.e. (1) web cutting, (2) access hole cutting, and (3) bevel cutting. The second blade for cutting the web cope is eliminated from a machine in preparation for the NWAH details. In order to keep the root surface for the CJP weld flat along the flange width, the flange-web junction was not cut away by the machine. The thick lines in Figure 2 and Figure 3 indicate the profile of the beam end after the cut. For the NWAH design, the backing bar is split into two pieces, and is placed from both sides of the beam web. One end of the split backing bar is formed into a round shape to conform to the web-fillet curve of the wide flange shapes.

Table 1. Test variables and specimens.

Type	Weld Access Hole	RBS	Number of Tests	Specimen Designation
S	none	none	3	S1, S2, S3
R	JASS6 1996 recommended	radius cut	2	R1, R2
SC	JASS6 1996 recommended	none	1	SC1

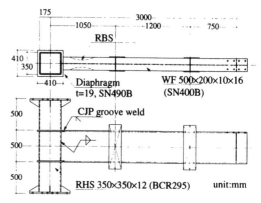

Figure 1. Dimensions of T-shaped test specimen.

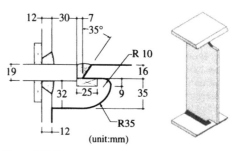

Figure 2. Weld access hole details recommended in AIJ JASS6.

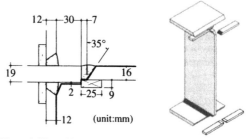

Figure 3. No weld access hole details.

Welding details for the CJP groove welds were as follows: gas metal arc welding (GMAW) with CO_2 shielding, YGW-11 (JIS) solid electrode with a diameter of 1.2 mm, stringer beads, 5 layers and 7 to 10 passes, welding current: 270-310A, arc voltage: 34-36V, travel speed: 235-667 mm/s, heat input: 9.5-26.9 kJ/mm, and interpass temperature: 135-355 °C. Tack welds for backing bars were avoided in the vicinity of the flange-web junction and beam flange sides, backing bars were left in place after welding, flux weld tabs were used for run-off tabs, and these tabs were removed after welding.

2.3 Material

Material properties of steels used for the test specimens obtained from coupon tests are summarized in Table 2. The transition temperature characteristics of both absorbed energy and crystallinity, obtained from CVN impact tests for the specimen beams and the deposited metals of CJP groove welds, are presented in Figure 4. These measured values were much larger than the minimum requirement stipulated in Japanese specifications for SN steel and YGW-11 electrode.

2.4 RBS design

The procedure for proportioning the radius cut RBS is based on the design recommendations suggested by M. D. Engelhardt 1999. According to these recommendations, the dimensions a, the distance from the CJP welds to the start of the RBS and b, the length of the RBS cut, were chosen as 100mm ($0.5b_f$: beam flange width) and 400mm ($0.8d$) respectively, where the length $a + b$ corresponds to the beam depth d (Figure 5).

The maximum moment expected at the center of the RBS was estimated as follows.

$$M_{RBS} = {}_wM_p + A_f(d - t_f)_f\sigma_u \qquad (1)$$

where, ${}_wM_p$ is the full-plastic moment of beam web, A_f is beam flange sectional area, t_f is beam flange thickness, and ${}_f\sigma_u$ is tensile strength of beam flange. As the depth of the RBS cut c is chosen as 45mm, M_{RBS} is 592 kN-m. When the moment at the center of the RBS reaches M_{RBS}, the moment at the CJP welds M_f is,

$$M_f = \frac{L}{L - (a + b/2)} M_{RBS} \qquad (2)$$

where L is beam length. Since $M_f/M_p = 0.97$, the moment at the CJP welds is limited to a value below the plastic moment of the beam (Figure 6).

Table 2. Material properties of steels used for test spesimens.

Element	Yield stress N / mm^2	Tensile strength N / mm^2	Y.R. * %	Elongation %
Beam Flange	305	462	66.0	30.8
Beam Flange-Web junction	294	460	64.0	30.0
Beam Web	362	490	73.8	28.8
Diaphragm	330	512	64.4	28.7

*YR : Ratio of yield to tensile strength

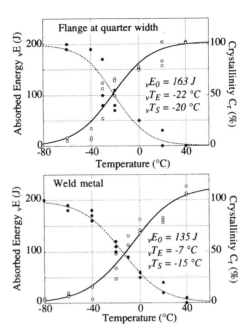

Figure 4. Results of CVN impact tests

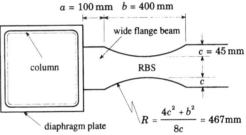

Figure 5. Radius cut RBS dimensions.

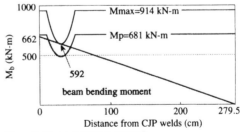

Figure 6. Moment diagram for RBS.

3 TEST PROGRAM

3.1 Loading Setup

The test specimen was placed in the loading setup as shown in Figure 7, oriented horizontally at a level of 1.0 m from the test bed. The column ends were con-

273

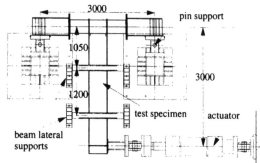

Figure 7. Load setup. (unit:mm)

nected to wide-flange column stubs, which were fastened to pin supports. The center-to-center length between the pins was 3.0 m. The beam end was clamped to an actuator at a distance of 3.0 m from the center line of the column. A swivel joint was attached at the head of the actuator to permit free motion of the beam end in the horizontal plane. Two sets of lateral supports were placed on the beam at points located 1.05 m and 2.25 m measured from the column center line to restrict out-of-plane deformation of the beam.

3.2 Loading Program

The specimens were subjected to displacement controlled cyclic loading with increasing amplitudes, in which $2\theta_y$, $4\theta_y$, $6\theta_y$ and $8\theta_y$ of the beam rotation amplitude were applied. For each amplitude, the specimen was loaded for two complete cycles (Figure 8). The specimens were considered to have failed when the beam flange fractured or the resistance at the peak rotation was reduced to less than 90% of the maximum resistance. If the specimen did not fracture after the completion of two cycles at the $8\theta_y$ amplitude, loading was continued to fracture with as many cycles as needed at the $8\theta_y$ amplitude. θ_y is the elastic rotation corresponding to the full-plastic moment of the beam at the face of the column for Type S (Figure 9). This value was used for all specimens (Types S, R and SC). The applied beam rotation (θ_m) was the net rotation of the beam only and was calculated using displacements measured at various locations of the test specimen. The load was applied in a quasi-dynamic

manner by a servo-controlled actuator at the beam tip. The beam rotation velocity for the $8\theta_y$ amplitude loadings was 0.00058 rad/s, corresponding actuator ram velocity was 10 mm/s.

4 TEST RESULTS

4.1 Global behavior

Moment versus rotation relationships obtained from the tests are presented in Figure 10. The abscissa is the net beam rotation (θ_m), and the ordinate is the bending moment (M_m) applied at the beam end. In this figure, M_m is normalized by the calculated full plastic moment M_p of full beam section. The general hysteretic behavior of the test specimens up to failure can be summarized as follows.

Type S (NWAH and non-RBS): During the cycles of $4\theta_y$ amplitude, slight local buckling of the beam was observed. During the cycles of $6\theta_y$ amplitude, a crack initiated from the toe of the CJP groove weld at the outer edge of the beam flange. Load reduced below 90% of the previous maximum load caused by local flange buckles at the first cycle of $8\theta_y$ amplitude, and the specimen achieved the ultimate state. After several cycles of $8\theta_y$ amplitude, local buckling of the beam flanges and web significantly progressed, and the crack penetrated through the thickness of the beam flange in the region just outside of the groove weld. The half width of beam flange fractured in a ductile manner, followed by complete brittle fracture. The crack extended into the beam web up to one-third of the beam depth.

Type R (RBS and JASS6 weld access hole): Initial local buckling of the beam web occurred at the midlength of theRBS during the cycle of $4\theta_y$ amplitude. Local buckling of the beam flanges occurred at the midlength of the RBS at the cycle of $6\theta_y$ amplitude. Load reduced below 90% of the previous maximum load caused by local flange buckles at the first cycle of $8\theta_y$ amplitude, and the specimen achieved the ultimate state. At the same time, a ductile crack initiated on the outer surface of the beam flanges at the location of a flange local buckle at midlength of the RBS. After several cycles of $8\theta_y$ amplitude, this crack caused the beam flange to fracture in a ductile

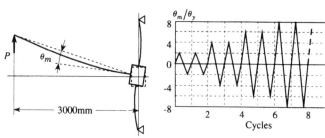

Figure 8. Loading program adopted in test.

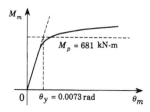

Figure 9. Definition of the beam rotation.

manner at midlength of the RBS. No cracks were detected in the vicinity of the CJP welds and the weld access holes.

Type SC (JASS6 weld access hole and non-RBS): During the cycle of $4\theta_y$ amplitude, a ductile crack initiated from the toe of the weld access hole and slight local buckling of beam flanges was observed. At the first cycle of $6\theta_y$ amplitude, this crack penetrated through the flange and propagated along the flange width, resulting in complete ductile fracture.

4.2 Beam deformation capacity

The normalized cumulative plastic rotation η was used to assess the deformation capacity of the test specimens. $\eta = \sum \theta_p / \theta_y$, where $\sum \theta_p$ is the cumulative plastic rotation (Figure 11). The value of η is useful to compare ductility among specimens having similar re-

storing force characteristics. In this project, however, it was necessary to evaluate the energy dissipation capacity for specimens having different maximum strength and hysteresis characteristics, i.e. the RBS design and the NWAH design. Since the value of η is concerned with only beam rotation, it is not, by itself, completely adequate to evaluate these tests. Another index of deformation capacity was used, namely, the normalized energy dissipation $\eta_e = E_p / E_y$, where $E_p = \sum \Delta E_p$ is the cumulative energy dissipation obtained from the loop areas of beam moment M_m vs. beam net rotation θ_m relations and E_y is the elastic energy absorption calculated by $M_p \theta_y / 2$ (Figure 11). The value of η_e is expected to better characterize real energy dissipation.

Table 3 summarizes the maximum moment and the deformation capacity indices η and η_e, obtained from the tests. The maximum moments of Type R (RBS)

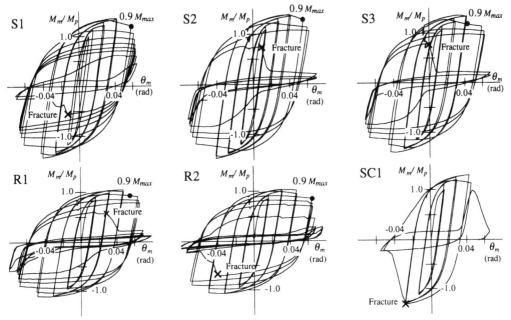

Figure 10. Moment versus beam rotation relationships obtained from tests.

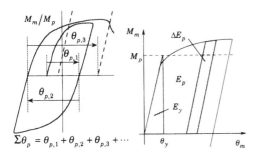

Figure 11. Definitions related to ductility indices

Table 3. Maximum moment and deformation capacity

Type	Moment Mmax/Mp	$\eta = \sum \theta_p / \theta_y$ Ulti.*	Fracture	$\eta_e = E_p / E_y$ Ulti.*	Fracture	End
S1	+1.36/-1.42	69	235	151	396	423
S2	+1.37/-1.45	69	153	152	308	344
S3	+1.35/-1.40	70	161	152	310	339
S ave.	+1.36/-1.43	69	183	152	338	367
R1	+1.05/-1.11	82	177	144	276	327
R2	+1.03/-1.07	73	165	127	255	302
R ave.	+1.04/-1.09	78	171	136	267	315
SC1	+1.28/-1.31	42	42	84	84	122

*ultimate state: 90%Mmax for Type S&R, fracture for Type SC

specimens were close to M_p (the full plastic moment for full beam cross-section). This result indicates that the maximum stress at the CJP groove welds was successfully reduced, and this is the main reason why cracks were not found in the vicinity of the CJP grooved welds for Type R specimens. Both Type S and Type R exhibited larger η and η_e compared with Type SC. At the ultimate state for each specimen, the value of η for Type S and R connections were 1.64 and 1.86 times larger than for the Type SC connections. The value of η_e for Type S and R connections were 1.81 and 1.62 times larger than for the Type SC connection. Except for Type SC, each test exhibited stable deformability after the ultimate state. Each test was continued up to fracture of the flange to evaluate the additional energy dissipation capacity after the ultimate limit state. The η and η_e at the instant of fracture for both Type S and Type R was about four times as large as for Type SC.

Figure 12 shows the history of energy dissipation E_p obtained in each loading cycle. In spite of lower flexural strength for Type R, both Type S and Type R show almost identical histories of energy dissipation E_p in the cycles of $2\theta_y$, $4\theta_y$ amplitude and the first cycle of $6\theta_y$ amplitude. As can be seen from the M_m/M_p vs. θ_m relations of Figure 10, the beam rotation amplitude of each cycle for Type R specimens is

lager than those for Type S specimens. Since E_p is calculated as the loop area obtained from M_m vs. θ_m relations, the larger values of θ_m for the Type R specimens will increase E_p. These results show that the energy dissipation capacity of Type R specimens is comparable to that of Type S specimens.

5 DISCUSSION

5.1 Distribution of strain

Figure 13 shows the distribution of cumulative plastic strain along the beam flange width at the beam end and at the end of the cycles of $2\theta_y$ and $4\theta_y$ amplitude. The abscissa is the distance from flange midwidth. The Type SC specimen experienced the largest cumulative plastic strains at the beam end because of the strain concentration at the toe of weld access hole. Type S specimens experienced smaller cumulative plastic strains at the beam ends, because they did not have weld access holes and strain concentrations in their vicinity. Though Type R specimens have weld access holes identical with those of the Type SC specimen, those cumulative plastic strains were not so large as those of Type SC, limited to the level similar to those of Type S specimens. This is because the maximum moment of Type R specimens at the beam end was controlled below the full plastic moment.

Figure 14 shows the distribution of cumulative plastic strain along the beam length at the ends of the cycles of $2\theta_y$ and $4\theta_y$ amplitude, and the end of the first cycle of $6\theta_y$ amplitude. For Type S and SC specimens, larger cumulative plastic strains were experienced at the beam end, and strains were reduced in proportion to the distance from the beam end. These diagrams reflect the

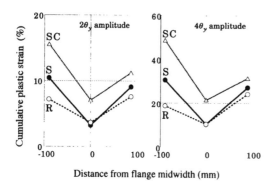

Figure 12. Energy disipation in each loading cycle.

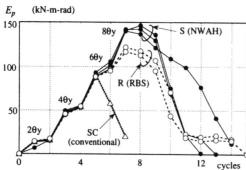

Figure 13. Distribution of cumulative plastic strain of the flange at beam end along flange width.

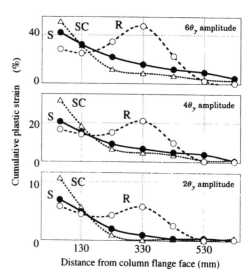

Figure 14. Distribution of cumulative plastic strain along beam length.

276

linear distribution of bending moment along the beam. On the other hand, for Type R specimens, cumulative plastic strains distributed somewhat more uniformly along the length of the RBS cut. The extent of plastification along the beam length was promoted by the RBS, with the consequence that larger beam rotations and corresponding amount of energy dissipation were obtained in the RBS.

5.2 Energy dissipation due to beam plastification

In order to discuss the energy dissipation capacity of the beam, the energy balance between the work done by the external applied force and the work absorbed as strain energy of beam plastification was evaluated.

The external work done by the applied force corresponds to the area of the loop of M_m vs. θ_m relationships E_p as shown in Figure 12. The shape of the loop obtained from the second cycle of $2\theta_y$ amplitude is illustrated in Figure 15 for each type of specimen. Both the maximum and minimum value of each loop was normalized by the absolute of each maximum value. The Type R specimens exhibited somewhat fuller loops than the Type S and SC specimens. This suggests that the Type R connection possesses hysteretic characteristics somewhat closer to an elastic-perfectly plastic relationship and a higher efficiency to dissipate energy.

The internal work obtained from cumulative plastic strain $\Sigma\varepsilon_p$ was calculated by integrating absorbed energy of each beam section along the beam length assuming that the distribution of $\Sigma\varepsilon_p$ along beam length is given by Figure 14 and the distribution of $\Sigma\varepsilon_p$ in a beam section is given as shown in Figure 16. In this calculation, the portion of the beam removed due to the RBS and the weld access hole were considered. Comparisons of the internal work and the external work E_p in each loading cycle obtained from tests are shown in Figure 17. This figure demonstrates that the external work and internal work were equivalent in each loading cycle, and that each type specimen exhibited almost identical energy dissipation capacity up to the $6\theta_y$ amplitude (0.044 rad) despite different extent of plastification and restoring force characteristics among specimen types.

5.3 Beam lateral instability

For Type R specimens, the radius-cut of the beam flange sections reduces flexural and torsional stiffnesses of the beam, and discussion continues whether or not additional lateral braces are needed for this type of specimen. In the present study, the beam was restrained against out-of-plane deformations by the two supports shown in Figure 7. To examine the forces resisted by the lateral supports, strains of their steel posts were measured from strain gages glued on the posts. The measurement was done for the lateral supports closer to the column, located 1.05 m from the column center line. Figure 18 shows examples of the forces estimated from the recorded strains during the cycles of $2\theta_y$ to $6\theta_y$ (Type R) or $8\theta_y$ (Type S).

Figure 18 shows that the forces exerted to the lateral supports are minimal up to the amplitude of $4\theta_y$ for both Type R and S specimens, which is understandable, because the beams remained nearly straight in

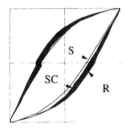

Figure 15. Shape of loops of M_m vs. θ_m relationship.

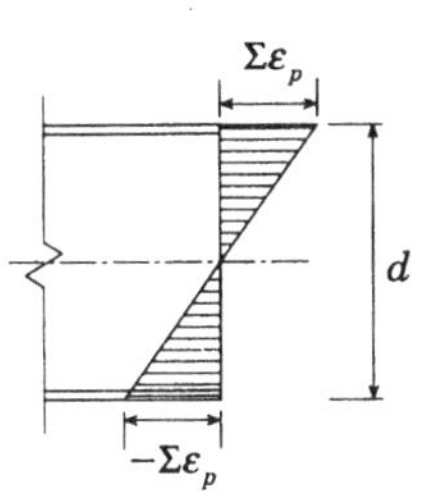

Figure 16. Cumulative plastic strain distribution.

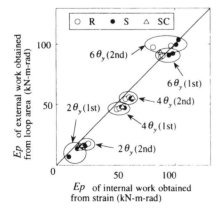

Figure 17 Comparison of E_p between external work and internal work.

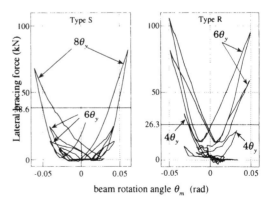

Figure 18 Lateral bracing force obtained from tests.

277

those cycles. For the amplitude of $6\theta_y$, there is a visible difference in the forces between the two specimens, showing larger forces for the Type R specimen. This is because the beam's unrestrained region between the column face and lateral support sustained more significant flange local buckling and out-of-plane deformations for the Type R specimen. The results indicate that the Type R specimen is more susceptible to out-of-plane deformations.

According to the AIJ Recommendation (1998), the required strength for bracing members to restrict notable out-of-plane deformations of flexural beams is 3% of the compressive force applied to the beam bottom flange when the beam carries the full-plastic moment. The required strength is estimated as 38.6 kN and 26.3 kN for Type S and R specimens, respectively (Figure 18).

For the Type S specimen, the force induced to the braces is smaller than 38.6 kN during the cycles of $6\theta_y$ amplitude (0.033 rad of plastic rotation), while it exceeded 38.6 kN during the cycles of $8\theta_y$ amplitude (0.054 rad of plastic rotation). This suggests that the AIJ requirement for bracing strength is sufficient for plastic rotations not greater than 0.033 rad. For the Type R specimen, the force induced to the braces is also smaller than the required strength of 26.3 kN up to the cycles of $4\theta_y$ amplitude (0.024 rad of plastic rotation) but exceeds the value for larger amplitudes. This observation suggests that the AIJ requirement is not sufficient to restrict out-of-plane deformations for beams with the RBS section adopted in this study.

The discussion thus far, however, is not sufficient to provide specific design requirements or procedures for the lateral bracing of beams with RBS sections. First, the tests presented in this study did not include floor slabs, which are known to retard the beam's out-of-plane deformations rather significantly. Second, the tests did not give information on how much the beam's in-plane resistance would have been reduced if lateral bracing had been absent. More study is needed to explore the effect of floor slabs as well as the possible reduction of in-plane resistance of unrestrained beams. However, the present study has provided information on the strength required for bracing members located near the plastic hinge region needed to restrict out-of-plane deformations of the beam.

Stability of beams with reduced beam sections has recently been evaluated by Uang and Fan 1999. This study concluded that the ductility of RBS specimens is also significantly affected by the beam's web slenderness. This study has also suggested web slenderness limits needed to develop large levels of ductility in beams with RBS connections.

6 CONCLUSION

Key observations from these tests were as follows:
1. Both types of improved connection details suc-

cessfully prevented premature fracture of welds and the surrounding base metal regions, indicating significantly larger plastic rotation capacity compared to the conventional details.
2. Both types of connection details exhibited nearly identical energy dissipation capacity, although the strength of the RBS detail is about 20 % smaller than the strength of the no weld access hole detail. This observation is reasoned from plastic strain distributions along the beam length as well as from contributions of the beam plastic rotation to the story drift.

ACKNOWLEDGMENT

This study was carried out as part of the US-Japan Cooperative Research Program on Urban Earthquake Disaster Mitigation, and financially supported by Grant-in-Aid for Scienctific Research of the Ministry of Education of Japan. T. Tamura and S. Morita, graduate students of Kyoto University, provided substantial help in carrying out the tests conducted in this study. These supports are gratefully acknowledged.

REFERENCES

AIJ (Architectural Institute of Japan), 1996. *Japanese Architectural Standard Specification JASS 6 Steel Work*, Tokyo (in Japanese).

AIJ, 1996. *Technical Recommendations for Steel Construction for Buildings*, Tokyo (in Japanese).

AIJ, 1998. *Recommendation for limit state design of steel structures*, Tokyo (in Japanese).

Engelhardt, M. D., 1999. Design of Reduced Beam Section Moment Connections, *Proceedings - 1999 North American Steel Construction Conference*, AISC.

Engelhardt, M.D., Winneberger, T., Zekany, A.J. and Potyraj, T.J., 1996. The Dogbone Connection, Part II, *Modern Steel Construction*, Vol.36, No.8, 46-55.

Nakagomi, T., Yabe, Y., Sakamoto, S., 1992. Effect of scallops at beam-end on mechanical behaviors of H-shaped steel beams connected to H-shaped column, *Journal of Structural and Construction Engineering, AIJ*, No.432, 51-59.

Nakashima, M., Suita, K., Morisako, K., and Maruoka, M., 1998. Tests of Welded Beam-Column Subassemblies. I: General Behavior, II: Detailed Behavior, *Journal of Structural Engineering, ASCE*, Vol.124, No.11, 1236-1244, 1245-1252.

Tateyama, E., Inoue, K., Sugimoto, S., Matsumura H., 1988. Study on ultimate bending strength and deformation capacity of H-shaped beam connected to RHS column with through diaphragm, *Journal of Structural and Construction Engineering, AIJ*, No.389, 109-121.

Popov, E.P., Yang, T.S. and Chang, S.P., 1998. Design of Steel MRF Connections Before and After 1994 Northridge Earthquake, *Engineering Structures*, 20(12), 1030-1038.

Tremblay, R., Tchebotarev, N. and Filiatrault, A.,1997. Seismic Performance of RBS Connections for Steel Moment Resisting Frames: Influence of Loading Rate and Floor Slab, *Proceedings, Stessa '97*.

Uang, C.M. and Fan, C.C., 1999. Cyclic Instability of Steel Moment Connections with Reduced Beam Section, Dept. of Structural Engineering, University of California at San Diego, Draft report prepared for the SAC Joint Venture.

Behaviour of Steel Structures in Seismic Areas, Mazzolani & Tremblay (eds) © 2000 Balkema, Rotterdam, ISBN 90 5809 130 9

Seismic responses of steel reduced beam section to weak panel zone moment joints

K.C.Tsai
Center for Earthquake Engineering Research, College of Engineering, National Taiwan University, Taipei, Taiwan

Wei-Zhi Chen
Kwan-Der Construction Company, Taipei, Taiwan

ABSTRACT: Five welded beam-to-column moment connections consist of reduced beam sections are cyclically tested. Key parameters include the amount of radius cut and the strength of the beam-column panel zone. Test results indicate that the average strain hardening ratio of the reduced beam section is 1.20. The total plastic rotation capacity of each specimen exceeds 0.04 radian. For strong panel zone joints, most of the plastic rotation concentrates in the reduced beam section. For weak panel zone joints, test results confirm that the amount of beam strength reductions can be reduced thereby delaying the initiation of the inelastic local buckling at the reduced beam section. Test results illustrate that properly proportioned and constructed moment connections employing the reduced beam section and the weak panel can possess excellent inelastic deformation capacity. The paper concludes with recommendations for the design of the reduced beam section and the weak panel in order to achieve well balanced inelastic deformations.

1 INTRODUCTION

The 1994 Northridge earthquake caused unprecedented damage to welded beam-to-column moment connections in more than 100 modern steel moment resisting frames (MRFs). Most common were non-ductile fractures near the weldments in the bolted web and welded flanges beam-to-column joint. Since then, extensive experimental investigations have been carried out in the U.S. in order to find possible causes of and solutions to these connection failures (SAC 1995). These efforts include the development and testing of various connection details. Effects of welding material, concrete floor slab and dynamic loading rates on the ductile performance of welded moment connections have also been investigated. These research and professional developments on many aspects of the problem have found that the connection employing the reduced beam section (RBS) details is a cost-effective choice for the construction of ductile moment connections (Tremblay et al 1997, Engelhardt et al 1998). However, the weak panel (PZ) designs are still possible following the model building code provisions even for RBS moment connections (UBC 1997). In addition, research results on the behavior of the RBS connections framed into a weak panel zone are rather limited. In order to obtain additional data, cyclic tests of five full-scale beam-column subassem-blies constructed with RBS-to-weak and strong panel zone joints were conducted at the National Taiwan University.

2 EXPERIMENTAL PROGRAM

2.1 Design of specimens

Each beam-column subassembly consists of one H600x200x11x17mm beam (SM400) framing into the flanges of one H300x300x16x25mm (A572, GR50) column (see Table 1). Detailed dimensions of the radius cut are given in Table 2 and Fig. 1. The design of the radius cut for the RBS connections is primarily based on the design recommendations published by Engelhardt et al (1998):

$$c \geq \frac{Z}{2t_f(d - t_f)}\left[1 - \frac{\beta_E(1 - a - 0.5b)}{\alpha L_b}\right] \quad (1)$$

where L_b is the distance between the applied load the column face, Z is beam plastic section modulus, t_f and d are beam depth and flange thickness, respectively. It is suggested that a strain hardening factor α of 1.15 be considered at the cut of the beam while the flexural demand imposed at the column face be smaller than $\beta_E M_P$ of the beam section. The recommended value for β_E should be limited between 0.85 and 1.0. The depth of the cut ranges from a 30% of the flange width for RB5, 37.5% for Specimens RB1 and RB3, and 50% for RB2 and RB4. The amount of the flange cut for RB5 is smaller than that suggested by Engelhardt. The design capacity of beam-to-column panel zone joint considers the ultimate strength of the panel zone including the contribution of the column flange at a panel zone deformation of $4\gamma_y$ (Krawinkler 1978, UBC 1997).

The design demands for the panel zone are 1.3MP for RB3 and RB4, but only 1.0MP for RB1, RB2 and RB5. The required doubler plate thickness is 6mm (A572 GR50) for the column in RB3 and RB4. Based on the actual material strength (Table 3), the design parameters for the panel zone of all specimens are given in Table 4. It can be found in the table that when the bending moment at the RBS reaches $1.15M_p^{RBS}$, the flexural demand imposed on the column face is $1.05\ M_p$ of the RB5 beam. The estimated maximum panel zone shear demand V_m given in Table 4 is computed from the estimated maximum beam moment $\beta_E M_p$. Values in column 5 of the table suggest that the panel zone in RB3 and RB4 should remain elastic when the maximum moment is developed in the RBS. Values in column imply that the panel zone deformations should exceed $4\gamma_y$ when the maximum moment is developed in RB1 and RB5.

2.2 Fabrication of specimens

The details of the panel zone follow the recommendations provided by SEAOC (1996). For Speicmens RB3 and RB4, the groove and fillet welds were employed to attach the doubler plate to the column flanges and the web, respectively. Additional two 25-mm plug welds were adopted between the column web and the doubler plate. Each doubler plate extended beyond the continuity plates by 100 mm. Each specimen consisted of one beam and one column. The beam web of all specimens was attached to the 15-mm thick shear tab using two 24-mm diameter A325X bolts. The connection between beam web and column was made by fully welding the beam web to the edge of the shear tab using 10-mm fillet welds as shown in Fig. 2. All beam flange welds were made at the Structural Laboratory at the National Taiwan University by using the shielded metal arc welding (SMAW) procedures.

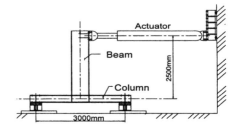

Figure 2. Schematic of test setup

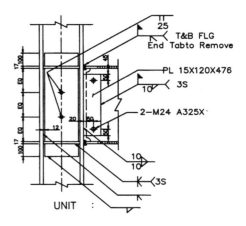

Figure 3. Details of beam-to-column

The welding rods are of E7016 grade with a minimum specified tensile strength of 483 Mpa and a specified CVN value greater than 20 ft-lbs at –20° F. The weld running tabs for the top and bottom flange welds were removed but both backing bars remained. No additional seal weld was made between the column flange and the backing bars. Beam flange welds for all specimens were made by the same welder and the welds were inspected by the same inspector.

3 EXPERIMENTAL RESULTS

3.1 Overview of test results

The test setup and the length of the column and the beam are shown in Fig. 3. Lateral supports were provided near the RBS and the point of applied loads as shown in Fig. 4. All specimens were loaded by imposing slowly increasing cyclic displacements at the tip of the cantilever beam according to the deformation history as shown in Fig. 5. This deformation history is identical to the standard cyclic loading protocol specified by SAC phase 2 tests (Clark

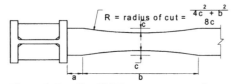

Figure 1. Details of RBS moment connection

Table 1 Schedule of specimens

Specimen	Beam Section (SM400) (mm)	Z_f/Z	Column Section (A572 Grade50) (mm)	Doubler Plate (A572 Grade50) THK (mm)
RB1	H600X200X11X17	0.69	H300X300X16X25	0
RB2	H600X200X11X17	0.69	H300X300X16X25	0
RB3	H600X200X11X17	0.69	H300X300X16X25	6
RB4	H600X200X11X17	0.69	H300X300X16X25	6
RB5	H600X200X11X17	0.69	H300X300X16X25	0

Table 2 Dimensions of radius cut

Specimen	a (mm)	b (mm)	c (mm)
RB1	100	400	37.5
RB2	100	400	50
RB3	100	400	37.5
RB4	100	400	50
RB5	100	400	30

Table 3 Material strength

Strength (MPa)	Beam		Column		Doubler Plate
	Flange	Web	Flange	Web	
F_y	280	308	389	370	419
F_u	412	425	281	503	521

Table 4 Expected maximum beam moment at column face and panel zone strength parameters

Specimen	M_p (kN-m) (1)	β_E (kN-m) (2)	$\beta_E M_p$ (kN-m) (3)	V_m (kN) (4)	$\dfrac{V_m}{V_y}$ (5)	$\dfrac{V_m}{V_{u,Code}}$ (6)
RB1	825	0.98	809	1132	1.22	1.05
RB2	825	0.87	718	1005	1.09	0.93
RB3	825	0.98	809	1132	0.84	0.78
RB4	825	0.87	718	1005	0.75	0.70
RB5	825	1.05	866	1212	1.31	1.13

$$V_m = \beta_E M_p \left(\frac{1}{0.95 d_b} - \frac{L}{L - d_c} \times \frac{1}{H} \right)$$

$$V_y = 0.55 F_y d_c t \text{ , } V_{u,Code} = 0.55 F_y d_c t \left(1 + \frac{3 b_c t_{cf}^2}{d_b d_c t} \right)$$

1997), except the number of the first few elastic cycles are reduced from 6 to 3. Important results, including the peak loads, the maximum plastic rotations attained, and the failure modes are summarized in Table 5. Except the bottom flange of Specimen RB2 fractured at the RBS after a total plastic rotation exceeded 0.041 radian, all other specimens sustained a total plastic rotation greater than 0.04 radian. All the specimens tested in this program performed quite well. All the beam plastic deformations concentrated in the RBS. Cyclic force versus total plastic deformation relationships are shown in Fig. 6. The primary failure mode is the local buckling of the beam web at the RBS followed by the flange buckling after a very large plastic rotation has developed.

Figure 4. Experimental setup

3.2 *Ultimate strength of specimens*

The maximum moments attained at the column face, M_{max}, and at the RBS, M^{RBS}_{max}, are given in the columns 4 and 5 of Table 6. In the table, the flexural strength considering the ultimate strength of the beam flange only, $Z_f F_u$, of all specimens is also given. Because of the sectional properties and actual material strength of the beam in this series of specimens, the ultimate flange flexural strength, $Z_f F_u$, is very close to the beam plastic moment capacity, M_p. Columns 6 and 7 of Table 6 indicate the ratio of the maximum moment developed at the colum face to the flexural strengths, $Z_f F_u$ and M_p, respectively. If the flexural demand imposed at the column face is to be kept below $\beta Z_f F_u$ ($\beta < 1.0$), then the minimum depth of the flange cut given in Fig. 1 should be:

Table 5 Failure modes, peak loads and plastic deformations

Specimen	P_u (kN) (1)		θ_p (%rad) (2)		Failure Mode (3)
	+	-	+	-	
RB1	332	379	5.23	5.09	buckled
RB2	310	327	4.10	4.12	buckled & BF cracked
RB3	331	367	4.25	4.23	buckled
RB4	309	309	4.47	4.03	buckled
RB5	379	394	4.05	4.98	buckled

Table 6 Experimental and theoretical maximum moment attained at the cut and column face

Spec.	$Z_f F_u$ (kN-m) (1)	M_p (kN-m) (2)	M^{RBS}_p (kN-m) (3)	M_{max} (kN-m) (4)	M^{RBS}_{max} (kN-m) (5)	$\beta = \dfrac{M_{max}}{Z_f F_u}$ (6)	$\beta_E = \dfrac{M_{max}}{M_p}$ (7)	$\alpha = \dfrac{M^{RBS}_{max}}{M^{RBS}_p}$ (8)
RB1	829	825	614	848	740	1.02(1.02)	1.02(0.98)	1.21
RB2	829	825	544	752	656	0.91(0.91)	0.91(0.87)	1.21
RB3	829	825	614	823	718	0.99(1.02)	1.00(0.98)	1.17
RB4	829	825	544	734	640	0.89(0.91)	0.89(0.87)	1.18
RB5	829	825	656	899	822	1.08(1.09)	1.09(1.05)	1.25
							average	1.20

Table 7 Maximum shear force and deformation attained in RB1, RB2 and RB5

Spec.	γ_{max} (%rad) (1)	γ_y (%rad) (2)	$\dfrac{\gamma_{max}}{\gamma_y}$ (3)	$V_{u,exp}$ (kN) (4)	$V_{max,exp}$ (kN) (5)	$V_{y,exp}$ (kN) (6)	$\dfrac{V_{u,exp}}{V_{y,exp}}$ (7)	$\dfrac{V_{max,exp}}{V_{y,exp}}$ (8)
RB1	1.71	0.28	6.1	1030	1082	812	1.27	1.33
RB2	1.12	0.28	4.0	996	1002	766	1.30	1.31
RB5	1.90	0.28	6.8	1073	1220	778	1.38	1.57

$$c \geq \frac{b_f}{2}\left[1 - \frac{\beta}{\alpha}\frac{F_u}{F_y}\frac{L_b - a - 0.5b}{L_b}\right] + \frac{t_w(d - 2t_f)^2}{8t_f(d - t_f)} \quad (2)$$

where t_w is the beam web thickness. In general, ultimate flange flexural strength $Z_f F_u$ is greater than the nominal beam plastic moment capacity M_p (Tsai et al. 1995, Tsai and Popov 1997). Test results shown in column 8 indicate that the average strain hardening factor of all specimens is 1.20, slightly larger than the value 1.15 estimated in the design of the specimens. Applying a β value between 0.9 and 1.0, the depth of the radius cut computed from Eq. 2 is generally smaller than that computed from Eq. 1 using an α value of 1.20, especially for A36 steel where F_u is substantially greater than F_y.

Using a strain hardening factor $\alpha = 1.20$ and the dimensions of the cut given in Table 2, the expected maximum flexural demand factor β values are computed from Eq. 2 and given in the parenthesis in column 6 of Table 6 for all specimens. Adopting a strain hardening factor of $\alpha = 1.15$ as described in the design of specimens, the values of the expected maximum moment factor, β_E, given in the column 2 of Table 4 are also shown in the parenthesis in column 7 of Table 6. Comparing the results shown in columns 6 and 7, it is evident that a strain hardening factor of $\alpha = 1.20$ gives better estimates of the maximum flexural demand. Bending moment factors given in column 6 for RB1 (1.02) and RB5 (1.08) also suggest that a slight exceeding of $Z_f F_u$ is tolerable when the beam web has been fully welded to the shear tab. Table 5 indicates that RB5 with a least amount of cut in all specimens sustained larger cyclic loads than other specimens did. A reduction in flange cut in RB5 also resulted in a larger restraining stiffness against the beam web buckling at the RBS as evidenced in Fig. 7, in which the plastic beam rotations developed before the onset of buckling are given for each specimen.

3.3 Panel zone responses

The maximum shear deformation and the maximum shear force imposed on the column panel zone joints are given in columns 1 and 5 of Table 7, respectively

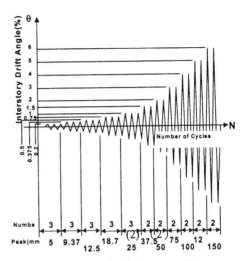

Figure 5. Cyclic displacement history

for Specimens RB1, RB2 and RB5. Test results confirmed that the panel zone possesses significant reserved strength (column 8) after the first yield. The maximum deformations achieved in RB1 and RB5 are greater than 4 times the yield deformation (column 3) as expected in the design of the specimens. The maximum shear force imposed on the panel zone in the last cycle (column 4) is smaller than the maximum shear (column 5) attained during the entire test. This is due to the local buckling of the beam web and flange at the RBS. The cyclic beam force versus panel zone ductility relationships are shown in Fig. 8 for Specimen RB 1 and RB5. The beam moments developed at the column face when the panel zone reached a rotation of $4\gamma_y$ are tabulated in column 1 of Table 8 for RB1, RB2 and RB5. The ratios of these moments to the expected maximum bending moment $\beta Z_f F_u$ are given in column 3. The ratios of the plastic deformation developed in the panel zone and the beam in each cycle are shown by dots in Fig. 9 for RB1 and RB5. In the same figure, the contributions of the panel zone and the beam in participating the plastic deformations in each cycle are also given. Column 4 of Table 8 gives the average ratios between the plastic rotations devel-

Table 8 Beam moment at specific panel zone deformation and ratio of panel zone and beam plastic deformations for Specimens RB1, RB2 and RB5

Specimen	$\Delta M_{4\gamma_y, exp}$ (kN) (1)	$\beta Z_f F_u$ (kN-m) (2)	$r = \dfrac{\Delta M_{4\gamma_y, exp}}{\beta Z_f F_u}$ (3)	$(\dfrac{\theta^p_{PZ}}{\theta^p_{Total}})_{avg}$ (4)
RB1	736	846	0.87	0.40
RB2	712	754	0.94	0.23
RB5	767	904	0.85	0.47

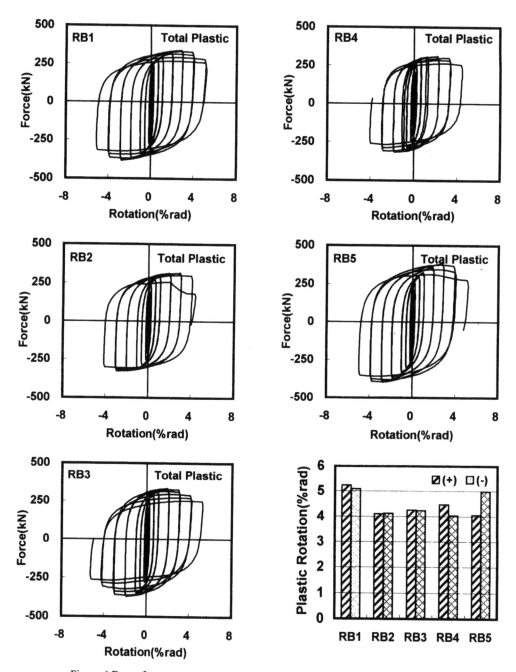

Figure 6 Beam force versus total plastic rotation relationships for all specimens

oped in the panel zone and the beam in all cycles for RB1, RB2 and RB5. Comparing the results shown in columns 3 and 4 in Table 8, it appears that the design moment for the panel zone can be taken about 85% (RB1 and RB5) of the expected maximum moment at the column face. In this manner, a somewhat balanced (0.40 and 0.47 in column 7 of Table 8) nonlinear deformation can be developed in both the panel zone and the beam.

4 CONCLUSIONS AND RECOMMENDATIONS

Based on this series of tests, conclusions and recommendations can be drawn as follows:

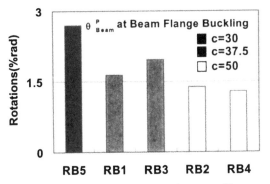

Figure 7. Beam plastic rotations at the onset of beam flange buckling

1. Welded moment connection employing RBS details suggested by Engelhardt performed very well. All specimens sustained a plastic rotational demand greater than 0.04 radian.

2. Well fabricated RBS moment connections can sustain large cyclic inelastic rotations without failure even when both backing bars are not removed or seal-welded to the column flange.

3. Test results indicate that the strain hardening factor developed in the RBS is about 1.20. The beam maximum flexural demand imposed at the column face should be kept below the ultimate beam flange flexural capacity. Fully welded web details should be considered as it can provide additional beam flexural capacity at the column face.

4. In order to achieve a somewhat balanced distribution of nonlinear deformations in both the panel zone and the RBS beam, the design moment for the panel zone can be taken as 85% of the expected maximum bending moment at the column face. In this manner, the panel zone will develop moderate inelastic deformations before RBS reach its peak

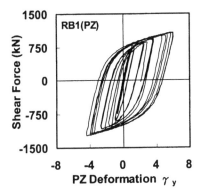

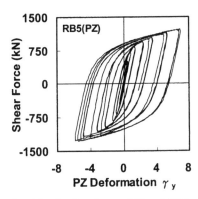

Figure 8. Panel zone shear versus deformation relationships for Specimens RB1 and RB5

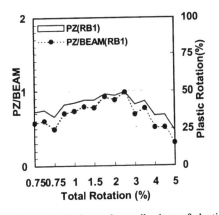

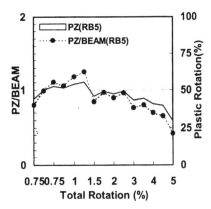

Figure 9. Ratios and contributions of plastic deformations between panel zone and beam

strength and plastic hinge rotations.

5. In most cases, applying Eq. 2 in the design of RBS can reduce the amount of flange cut thereby delaying the beam web local buckling during the development of the beam plastic hinge rotations. A reduction in flange cut also results in an increase on overall flexural strength as evidenced in Specimen RB5.

ACKNOWLEDGEMENTS

The authors gratefully acknowledge the financial support provided by the National Science Council and the Sinotech Foundation for Research and Development of Engineering and Technology.

REFERENCES

Clark, P. 1997. Protocol for fabrication, inspection, testing and documentation of beam column connection tests and other experimental specimens. *Report No. SAC/BD-97-02*. SAC Joint Venture.

Engelhardt, M.D., T. Winneberger, A.J. Zekany & T.J. Potyraj 1988. Experimental investigation of dogbone moment connections. *Engineering Journal*, Fourth Quarter.

Krawinkler, H. 1978. Shear in beam-column joints in seismic design of steel frames. *Engineering Journal*, American Institute of Steel Construction, 3rd quarter.

SAC 1995. *Interim Guidelines*-- Evaluation, repair , modification and design of welded steel moment frame structure, *Report No. SAC-95-02*, SAC Joint Venture.

SEAOC 1996. *Recommended Lateral Force Requirements and Commentary*, Sixth Edition, Structural Engineers Association of California.

Tremblay, R., N. Tchebotarev & A. Filiatrault 1997. Seismic performance of RBS connections for steel moment resisting frames: influence of loading rate and floor slab. *Proceedings: STESSA' 97*.

Tsai, K.C., S. Wu & E.P. Popov 1995. Experimental performance of seismic steel beam-column moment joints, *Journal of Structural Engineering*, ASCE, Vol.121, No.6, 1995.

Tsai, K.C. & E.P. Popov 1997. Seismic steel beam-column moment connections, Background Report, SAC Joint Venture, *Report No. SAC-95-09*, FEMA Publication No. 288.

Behaviour of Steel Structures in Seismic Areas, Mazzolani & Tremblay (eds) © 2000 Balkema, Rotterdam, ISBN 90 5809 130 9

Experimental method of the full scale shaking table test using the inertial loading equipment

S. Yamada
Tokyo Institute of Technology, Yokohama, Kanagawa, Japan

Y. Matsumoto
Yokohama National University, Kanagawa, Japan

H. Akiyama
Nihon University, Tokyo, Japan

ABSTRACT: Under the recent severe earthquakes, i.e. Northridge Earthquake (1.17.1994,U.S.A) and Hyogoken- Nanbu Earthquake (1.17.1995, JAPAN), many moment resistant steel frames showed typical mode of fracture at the beam-to-column connections. It is strongly requested to clarify the mechanism of fracture at beam-to-column connections. The most effective method to examine the process of fracture in the structural element of the building structures under severe earthquakes is the full scale shaking table test. But, it is not able to carry out the shaking table test on full-scale structures because of the capacity of the shaking table. When reduced-scale model is used as the specimen, the scale effect and the strain rate can't be reflected. Both of the scale effect and strain rate are the primary factors of fracture. In this paper, a new experimental method of the full scale shaking table test is proposed. The main feature of this experimental method is characterized by the use of the inertial force equipment. The inertial force equipment consists of a loading frame, 200 ton of counter weight and isolators. The loading frame was set on the shaking table supported by isolators. Specimens used in this experimental method were partial frames taken out from middle-rise building structures. Using partial frame as the specimen, full scale and real time dynamic loading test was realized. The test set-up was composed by a specimen, the inertial force equipment and loading beam which transmits the horizontal force to the specimen from the inertial force equipment. This test set-up, regarded as a single degree of freedom system, makes it easy to understand the dynamic behavior of the test set-up including a specimen. The shaking table used in this test is one of the largest shaking tables in the world, and it is can apply 100 kine velocity to 200 ton of weight. So, dynamic loading test of structural elements up to their collapse becomes possible.

1 INTRODUCTION

Under the recent severe earthquakes, i.e. Northridge Earthquake (1.17.1994,U.S.A) and Hyogoken- Nanbu Earthquake (1.17.1995, JAPAN), many moment resistant steel frames showed typical mode of fracture at the beam to column connections. It is strongly requested to clarify the mechanism of fracture at beam to column connections. The most effective method to examine the process of fracture in the structural element of the building structures under severe earthquakes is the full scale shaking table test. But, it is not able to carry out the shaking table test on full-scale structures because of the capacity of the shaking table. When reduced-scale model is used as the specimen, the scale effect and the strain rate can't be reflected. Both of the scale effect and strain rate are the primary factors of fracture. In this paper, a new experimental method of the full scale shaking table test is proposed.

2 CONCEPT OF THE EXPERIMENTAL METHOD

Carrying out a shaking table test on full scale building structures and reproducing dynamic behavior under severe earthquake is the most advisable way to investigate the fracture process of structural element and dynamic collapse process of building structures. However, the capacity of the existing largest shaking tables are limited ot be applied to the low-rise building structures with light weight. To test middle-rise building structures, a powerful test set-up including a large scale safety device against the collapse must be made. On the other hand, when a reduced-scale model is used as a specimen, the scale effect that must be an important control factor of fracture can't be reflected. Furthermore, strain rate can't be reproduced, because time scale must be reduced to fit it to the law of similarity.

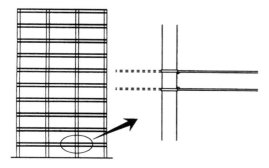

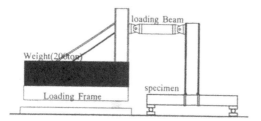

Fig.1 Partial frame

Fig.2 Outline of test set-up

In this experimental method, full-scale partial frames were taken out from middle-rise building structure as shown in Fig.1 and used as specimen in dynamic loading test. Then, dynamic fractures of structural elements and connection part of members were reproduced. Outline of the test set-up is shown in Fig.2 and concept of this experimental method is shown in Fig.3. The test set-up consisted of three parts: the inertial loading equipment, the loading beam and the specimen. The inertial loading equipment consisted of a counter weight, loading frame and isolators. The counter weight was mounted on the loading frame. The loading frame was set on the shaking table supported by isolators. The inertial loading equipment generates the inertial force according to the input from the shaking table.

The counter weight and loading frame can be regarded as a single mass. Isolators and the specimen can be regarded as a springs. So, the test set-up can be regard as a single degree of freedom system. Though the idea of the device of this type has already existed (Hagiwara, H., Akiyama, H., Kokubo, K. and Sawada, Y. 1991), its application to a huge model is a first experience.

Photo.1 Overview of the set-up

3 SHAKING TABLE

The shaking table, used in this shaking table test, is one of the largest shaking tables in the world. It is the equipment of the National Research Institute for Earth Science and Disaster Prevision (NIID) in Tukuba, JAPAN. The table has 15x14.5 meters. The maximum loading weight is 500 ton and the maximum displacement amplitude is ± 22 cm. Limit performance of the shaking table is shown in Fig.4.

4 EXPERIMENTAL EQUIPMENT

4.1 *Counter weight*

As for the weight of the counter weight, which generates the inertial force, the bigger one is desirable, because a big power is necessary to make strong members collapse. And, it is necessary to input the maximum acceleration of about 1G so as to reproduce the dynamic fracture of structures under the severe earthquake. Furthermore, the maximum velocity of the shaking table, thus the energy input, must be as big as possible to make a specimen to reach the ultimate state. Referring to the limit performance of the shaking table, shown in Fig.4, it was concluded that the most suitable weight and natural period are 200ton and 0.6 seconds respectively. So, the weight of the counter weight, which occupies the most part of the weight of the inertial loading equipment, was decided 200ton. Then, it was decided that 20 sheets of steel slab of about 10ton are used for the counter weight.

Input energy to the test set-up on the shaking table expressed in the equivalent velocity, V_E, was assumed to be two to three times as large as the maximum velocity of table motion. Equivalent velocity of energy is defined by Equation 1. (Akiyama, H. 1985)

$$V_E = \sqrt{2E/M} \qquad (1)$$

where V_E : Equivalent velocity of energy

E : Energy

M : Mass

An input energy spectrum to the test set-up with 200ton of the loading weight was supposed. Supposed input energy spectrum is shown in Fig.5. From Fig.5, it is predicted that the equivalent velocity of input energy to the test set-up becomes 300 (cm/s) from 200 (cm/s) in the range of the natural period of 0.6 second to 1.2 second.

4.2 *Loading frame*

Maximum shaking ability of the table is 1G in the acceleration. If the restoring force of isolators is ignored, the maximum inertial force of 200ton can be

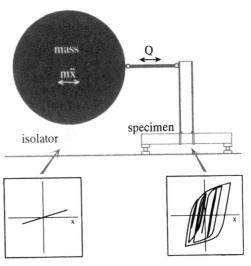

Fig.3 Concept of experimental method

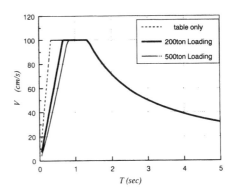

Maximum Velocity

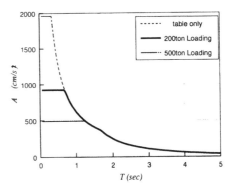

Maximum Acceleration

Fig.4 Limit performance of the shaking table

supplied to the specimen. The loading frame, which support the counter weight and transmits the inertial force to the loading beam, must be kept in elastic range when the inertial force of 200ton is supplied to the specimen. Furthermore, deformation of the loading frame, which causes the loss of the input energy, must be restrained. So, the loading frame was designed to keep the horizontal displacement at the loading position within 4mm against the maximum horizontal load of 200ton. As a result, the weight of the loading frame became 20ton and the weight of the counter weight and loading frame in total became about 220ton.

4.3 *Loading beam*
The loading beam, which transmits the inertia force from the inertial loading equipment to the specimen, had pins at both ends. Thus, the loading beam transmitted only axial force. Strain gauges were put on the loading beam, and it is used as a load cell measuring shear force acting on the top of the specimen.

4.4 *Lateral supports*
The shaking table used in this research is a shaking table of one dimension movement. Therefore, a set of lateral support was installed to restrict the out of plane deflection of the test specimens. The loading frame was supported at 4 point of both side by roller supports, as shown in the Fig.6, and it could move only in the same direction of the shaking table. And, the loading point of the specimen was prevented its out of plane behavior by the arm of the L shape projected from the loading frame as shown in Fig.7. Furthermore, the pins used in both end of the loading beam and support of the specimen were simple, and they allowed only rotational movement.

4.5 *Isolators*
Isolators must have enough deformation capacity to follow the ultimate deformation of the specimen at the same time must support the weight of the loading frame and counter weight. Isolators were arranged at the 4 corners of the loading frame. It was chosen that the natural period of the inertia loading equipment was 2 second. Elastic stiffness of the isolator was 0.50 (t/cm). And, allowable displacement of the isolator was 25cm.

4.6 *Stopper*
When a specimen is broken by fracture in the dynamic experiment, elastic energy of the specimen is released in a moment. If the released elastic energy of the specimen and the kinetic energy of the test set-up can't be absorbed by isolators within the allowable displacement, the loading frame might be broken. So, the stopper, which collides with the loading frame when the displacement of the loading

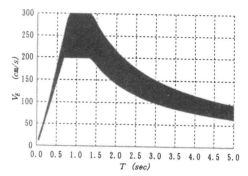

Fig.5 Supposed input energy spectrum

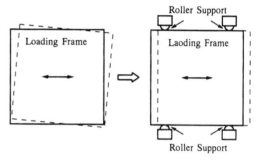

Fig.6 Lateral support of the loading frame

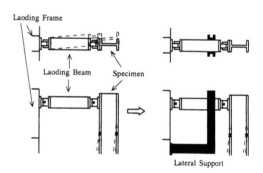

Fig.6 Lateral support of the loading beam

frame exceeds the allowable displacement of isolator, is set under the loading frame. Lead plates were installed on colliding surface of the stopper so as to absorb a shock at the instant of collision.

4.7 *Measuring device*
LVDT, acceleration meters, strain gauges and temperature gauges were installed on suitable points of the test set-up. The deformations of the specimen and the positions of each part of the test set-up measured by these sensors. Furthermore, the dis-

placement, the velocity and the acceleration of the shaking table in each direction were measured by the sensors installed in the shaking table. These data were measured by the dynamic strain meter with the sampling of 1000 Hz, and was recorded on the hard disk of the workstation.

5 TEST SET-UP AS A VIBRATION SYSTEM

The test set-up can be modeled as a single degree of freedom system. The equation of motion is given by Equation 2.

$$m\ddot{x} + (C_l + C_s)\dot{x} + (k_l + k_s)x = -m\ddot{z} \qquad (2)$$

where m : mass of the counter weight and the loading frame

k_l : stiffness of the isolators

k_s : stiffness of the specimen

C_l : damping of the loading equipment caused by isolators, loading frame, loading beam and connection.

C_s : damping of the specimen caused by the pin support

x : displacement of the mass relative to the shaking table

z : horizontal motion of the shaking table

Multiplied by $dx = \dot{x}dt$ on both side and integrated over the entire duration of motion t_0, Equation 2 is reduced to Equation 3.

$$\int_0^{t_0} \{m\ddot{x} + (C_l + C_s)\dot{x} + (k_l + k_s)x\}\dot{x}dt = -\int_0^{t_0} m\ddot{z}\dot{x}dt$$
$$= E \qquad (3)$$

E is the total input energy of the system. Equation 3 is reduced to Equation 4.

$$\int_0^{t_0} \{(C_l + C_s)\dot{x} + (k_l + k_s)x\}\dot{x}dt = -\int_0^{t_0} m(\ddot{z} + \ddot{x})\dot{x}dt$$
$$= E_D \qquad (4)$$

E_D is the consumed energy by the system.

6 PLAN OF THE EXPERIMENT

6.1 Test series
In this research, following test series were planned and carried out.

1) Beam to column connection series 1. Beams had wide flange section and columns had rectangular hollow section. Main parameters of this series were concerned with details of the connection. Beam of the representative specimen had a section of 600x300x12x25 (mm). (Mastumoto, Y. ,Yamada, S. and Akiyama, H. 2000)
2) Beam to column connection series 2. Beams had wide flange section and columns had rectangular hollow section. Main parameter of this series was the test temperature corresponding to the material property. Beam of the specimen had a section of 600x300x16x25 (mm). (Mastumoto, Y. ,Yamada, S. and Akiyama, H. 2000)
3) Beam to column connection series 3. Section of beams and columns were wide flange section. Main parameters of this series were concerned with details of the connection. Beam of the specimen had a section of 600x300x12x25 (mm).
4) Cold formed column series. Columns had rectangular hollow section and made by cold forming. Main parameter of this series was manufacturing process of the column. Column of the specimen had a section of 500x500x22x 22 (mm).
5) Column base series. Specimens were exposed type column base with different thickness of base plate.
6) Damper series. Partial frames with damper were tested in this series.
7) RC series. Some RC frames were tested and compared with steel frames.

6.2 Process of experiment
In all test series, experiment was carried out in the following process. Initially, pulse waves were applied in order to examine natural period and damping coefficient of the test set-up. Then, several seismic excitations of different peak velocity were applied to the test set up. Acceleration record used in this research is the NS component of the JMA Kobe record, which is recorded in the Hyogoken-Nanbu earthquake. And, the input level of the shaking table adjusted by multiplying a scaling factor to the applied acceleration record. In the preliminary excitation, the peak velocity was selected to be small enough to prevent plastic deformations of the specimens. The second excitation was scaled to make some plastic deformations on the specimens. The third excitation was intended to be large enough for the specimens to reach their collapse state.

7 CONCLUSION

Full-scale dynamic experimental method using inertial force equipment is proposed. The main feature of this experimental method is characterized by the use of the inertial force equipment. The inertial force equipment consisted of a loading frame, 200 ton of counter weight and isolators. The loading frame was set on the shaking table supported by isolators. Specimens used in this experimental method were partial frames taken out from middle-rise building structures. Using partial frames as specimens, full scale and real time dynamic loading tests were realized.

The test set-up was composed of a specimen, the inertia force equipment and the loading beam which transmits the horizontal force to the specimen from the inertial force equipment which generates the inertial force according to the input from the shaking table. This test set-up can be regard as a single degree of freedom system. So it is easy to understand the dynamic behavior of the test set-up including a specimen.

The shaking table, used in this test, is one of the largest shaking tables in the world, and it is possible to apply 100 kine velocity to 200 ton of weight. So, the dynamic loading test of structural elements up to their collapse is made possible to carry out.

REFERENCES

Akiyama, H. (1985), Earthquake resistant limit state design for buildings, University of Tokyo Press.

Hagiwara, H., Akiyama, H., Kokubo, K. and Sawada, Y. (1991), Post-buckling behavior during earthquakes and seismic margin of FBR main vessels, Int. J and Press & Piping 45, pp.s59-271.

Mastumoto, Y., Yamada, S. and Akiyama, H. (2000), Fracture of beam to column connection simulated by means of the shaking table test using the inertial loading equipment, Proc. of STESSA 2000.

Codification, design and application

Behaviour of Steel Structures in Seismic Areas, Mazzolani & Tremblay (eds) © 2000 Balkema, Rotterdam, ISBN 90 5809 130 9

Parametric analysis of the dynamic behaviour of ENEL towers in Naples by means of 'minimum' model

P. Belli & F. Guarracino
Dipartimento di Scienza delle Costruzioni, Università di Napoli 'Federico II', Italy

ABSTRACT: The dynamic behaviour of two towering buildings in Naples (ENEL headquarters) is analysed by means of a "minimum" model, which allows to obtain approximate and quick indications about the static and dynamic behaviour of the real structure. The equations are integrated analytically in absence of plastic deformations in the dissipating devices and by means of a standard Runge-Kutta procedure otherwise. Useful information about design parametrs can thus be readily obtained.

1 INTRODUCTION

In the following pages the authors wish to contribute some considerations about the design parameters of the ENEL towers in the Centro Direzionale of the city of Naples, which terminated in May 1995, by means of a simplified dynamic analysis.

The actual design of the buildings was developed by STIPE, an engineering firm based in Rome. The structural design revision and the supervision of the constructing process has been carried out by the first of the present authors.

Each building is composed of two towers realised in reinforced concrete (Figure 1). These towers are approximately 106 metres high and rest on a caisson foundation which, in turn, is supported by a group of very long piles (30 metres) embedded in the underground bank of rocks. The cross section of each tower is shaped like a trapezium and the major side measures about 18 metres. The distance between the towers is approximately 21 metres.

At their top the two towers support an horizontal steel frame by means of spherical hinges. Twenty-nine steel-concrete floors are suspended from the top frame by means of steel cables (Figure 2). The horizontal connection of each floor to the adjacent towers is obtained by means of several damping constraints AISI 304, as shown in Figure 3.

The mechanical behaviour of each connection is meant to be linearly elastic under the action of a shear

Figure 2 - Mounting of the suspended floors

Figure 1 - One of the ENEL buildings under construction

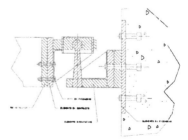

Figure 3 - Damping devices outline

force in the range of $0 \div 120$ KN for the group of floors 1-3 and 17-29 and $0 \div 180$ KN for the group of floors 4-16. For higher values of the forces the behaviour of the connections is meant to be perfectly plastic. These forces correspond to the actions which can be transmitted as a consequence of the wind and ordinary seismic shocks.

When earthquakes of exceptional magnitude take place the devices are driven in the perfectly plastic range so that the dynamic behaviour of the towers results to a certain extent disjointed from that of the suspended floors. In this manner a reduction of the eigenfrequencies of the whole building is attained (Mazzolani and Serino, 1997), together with some dissipation of the dynamic energy as a consequence of the hysteresis damping of the constraints in the plastic range (Belli and Guarracino, 1997). Therefore the value of the stresses in the towers are prevented to exceed a certain magnitude.

In the present work, in order to analyse the dynamic behaviour of the buildings and the influence of some very basic design parameters on their overall response, the authors make reference to a "minimum" model, which allows to obtain approximate and quick indications about the static and dynamic behaviour of the real structure.

2 STRUCTURAL DATA IDENTIFICATION

The first step towards the definition of a minimum model capable of giving reasonable information about the dynamic behaviour of the real structure consists in establishing the number of independent masses corresponding to the suspended floors. In the present case we assumed this number equal to two, so that at least the possibility that at a certain time some floors are moving in one direction and others in the opposite one is preserved.

Therefore, with reference to the scheme of Figure 4, we set the masses M_3 and M_4 to represent each half of the total mass of the twenty-nine suspended floors, i.e. 177500 kgm.

Conversely, the masses M_1 and M_2 correspond to the total mass of the concrete towers, i.e. 6.35×10^5 kgm, and are located at the respective centres of gravity.

Given that in the construction of the building two different types of horizontal constraints were employed, i.e.
• for the group of floors 1-3 and 17-29 a system characterised by a linearly elastic behaviour with stiffness 20 KN / mm in the displacement range ± 6 mm and by a perfectly plastic behaviour otherwise (the limit load being equal to 120 KN);
• for the group of floors 4-16 a system characterised by a linearly elastic behaviour with stiffness 30 KN / mm in the displacement range ± 6 mm and by a perfectly plastic behaviour otherwise (the limit load being equal to 180 KN);
we assumed a mean value of 180 KN /mm for the stiffness in the linearly elastic range of each of the springs K_1, K_2, K_3 and K_4. In case of axial displacements outside the range ± 6 mm the springs are thought to behave as perfectly plastic, with a limit load of 1080 KN.

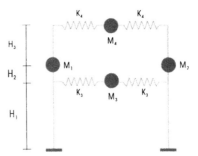

Figure 4 - The "minimum" model of the structure

Despite the fact that the Runge-Kutta procedure employed for integrating the differential equations would allow an immediate encorporation of such a term, but given that the examination of the effects of the AISI 304 hysteresis cycle were not whithin the scope of the present study, for the sake of simplicity no damping was taken into account in the present analysis.

The concrete modulus of the lateral towers is assumed to be 33.6 KN / mm², the second moment of area of the cross-sections is 290 m⁴.

Finally, with respect to the geometry of the structure we have: $H_1 = 42.8$ m , $H_2 = 18.7$ m and $H_3 = 23.3$ m.

At this point, it is goes without saying that the free elastic vibrations of the system in Figure 4 can be studied very easily with reference to an undamped mass spring system of equations of the type (Newmark and Rosenblueth, 1971)

$$[M]\{\ddot{x}(t)\}+[K]\{x(t)\}=0 \qquad (1)$$

where the degrees of freedom vector $\{x(t)\}$ encompasses the absolute horizontal displacements of the masses M_1, M_2, M_3 and M_4 from the initial position, i.e.

$$\{x(t)\}=\begin{Bmatrix} \Delta_1(t) \\ \Delta_2(t) \\ \Delta_3(t) \\ \Delta_4(t) \end{Bmatrix} \qquad (2)$$

However, it must be noticed that the system (1), when particularised with the insertion of the above introduced geometrical and mechanical data, leads to the following set of eigenfrequency values

$$\{\lambda\}=\begin{Bmatrix} 0.594 \\ 1.488 \\ 2.137 \\ 2.538 \end{Bmatrix} \quad (Hz) \qquad (3)$$

and, consequently, to the following set of periodic times

$$\{T\} = \begin{Bmatrix} 1.684 \\ 0.672 \\ 0.468 \\ 0.394 \end{Bmatrix} \text{ (sec)} \qquad (4)$$

These values turn out to be somewhat different from the output of the analysis performed by the designers of the buildings by means of the A.D.I.N.A. (Automatic Dynamic Incremental Non-linear Analysis) commercial code, which are (the sets are limited to the first four modes)

$$\{\lambda^*\} = \begin{Bmatrix} 0.327 \\ 0.347 \\ 0.802 \\ 1.482 \end{Bmatrix} \text{ (Hz)} \qquad (5)$$

$$\{T^*\} = \begin{Bmatrix} 3.058 \\ 2.882 \\ 1.247 \\ 0.675 \end{Bmatrix} \text{ (sec)} \qquad (6)$$

This is not surprising, given the extremely coarse discretization adopted for the minimum model under analysis. In fact, if we examine the behaviour of one of the concrete towers only, we have that the highest periodic time of the system considered as an elastic cantilever with evenly distributed mass μ results

$$T = 1.787 \, H^2 \sqrt{\frac{\mu}{EI}} = 1.632 \text{ sec} \qquad (7)$$

while, if the same tower is considered carrying the whole mass M_1 concentrated at the quote $H/2$, the highest periodic time results

$$T = 2\pi / \sqrt{3 \frac{EI}{M_1 (H/2)^3}} \qquad (8)$$

$$= 1.171 \text{ sec}$$

with a difference of about 40%.

On account of this fact, in order to make the highest value of the periodic time of the model close to that deriving from the A.D.I.N.A. analysis, we decided to assume a conventional value for H_1 equal to 67.8 m. This operation implicitly also accounts for the additional flexibility of the foundation.

Thus it results

$$\{\lambda\} = \begin{Bmatrix} 0.365 \\ 1.305 \\ 2.033 \\ 2.506 \end{Bmatrix} \text{ (Hz)} \qquad (9)$$

$$\{T\} = \begin{Bmatrix} 2.738 \\ 0.766 \\ 0.492 \\ 0.399 \end{Bmatrix} \text{ (sec)} \qquad (10)$$

3 INTEGRATION OF THE DIFFERENTIAL EQUATIONS

The vibrations of the minimum model introduced in the preceding section have been studied in presence of an external periodic force deriving from a sinusoidal ground motion.

The amplitude of the acceleration of the ground was set equal to 0.16 g and the periodic time of the ground motion was assumed equal to 0.3 sec. These values aim

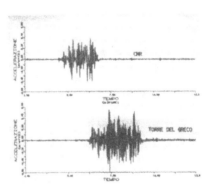

Figure 5 - Acceleration plots from CNR codes and Torre del Greco earthquake

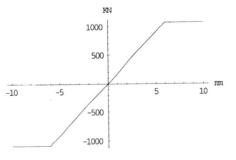

Figure 6 - Load-deformation law for the connecting springs K_1-K_4

to reproduce a situation not very different from the data gathered from the Torre del Greco Earthquake in 1980 and from the indications of the National Council of Research of Italy, shown in Figure 5.

Consequently, the system of differential equations under analysis became

$$[M]\{\ddot{x}(t)\} + [K]\{x(t)\} = a(t)\{I\} \qquad (11)$$

with

$$a(t) = A\sin(\omega\, t) \qquad (12)$$

Thanks to the extreme simplicity of the system (11), its integration has been performed analytically in the case of purely elastic vibrations (i.e. when the response of the connecting springs K_1-K_4 is considered linearly elastic for any value of the applied deformation) and by means of a straightforward Runge-Kutta numerical procedure (Press et al., 1992) when the load-displacement relationship of the springs K_1-K_4 is considered obeying to the elastic-perfectly plastic law plotted in Figure 6.

4 RESULTS FROM THE ANALYSES

The results from the performed analyses over a time interval of 20 sec are plotted in Figures 7-34. In order to investigate the effects of the stiffness and of the limit load of the connecting devices, the analyses were first performed for the effective number of AISI 304 constraints employed, and subsequently for half and double this number. This corresponds, of course, to halving and doubling, respectively, the value of the stiffness and of the limit load of the connecting devices. For the sake of the numerical stability of the procedure, a dummy hardening of .1 % was considered in the plastic range.

With reference to the plots 7-9 and 13-15, it is evident that the effect of considering the plastic deformation of the connecting devices results in an expected reduction of the frequency of the motion for all of the masses M_1, M_2, M_3 and M_4. Of course this is a beneficial effect, as it tends to increase the difference between the periodic time of the building motion and that of the ground motion.

Moreover, with reference to the plots 10-12 and 16-18, it can be noticed that the mean square acceleration of the masses M_3 and M_4 reduces in presence of plastic deformations of the connecting devices.

As said before, in order to investigate the effects of the stiffness and of the limit load of the connecting devices on the motion of the structure, the cases shown in the group of plots 19-26 (number of AISI 304 constraints halved) and 27-34 (number of AISI 304 constraints doubled) were analysed.

First it must be noticed that the eigenvalue analysis relative to the free elastic vibrations of the model

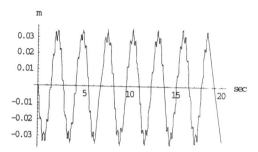

Figure 7 - Elastic analysis: absolute displacements relative to the masses M_1 and M_2

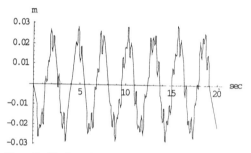

Figure 8 - Elastic analysis: absolute displacements relative to the mass M_3

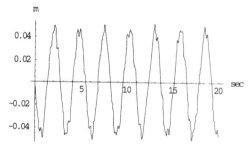

Figure 9 - Elastic analysis: absolute displacements relative to the mass M_4

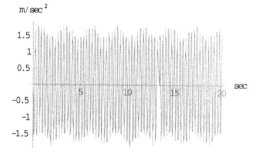

Figure 10 - Elastic analysis: acceleration relative to the masses M_1 and M_2

298

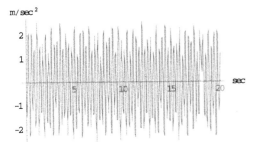

Figure 11 - Elastic analysis: acceleration relative to the mass M_3

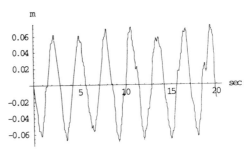

Figure 15 - Elastic-plastic analysis: absolute displacements relative to the mass M_4

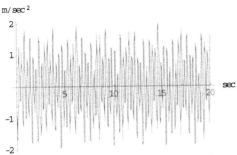

Figure 12 - Elastic analysis: acceleration relative to the mass M_4

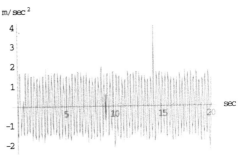

Figure 16 - Elastic-plastic analysis: acceleration relative to the masses M_1 and M_2

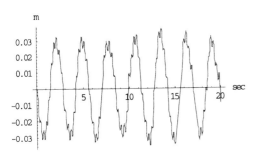

Figure 13 - Elastic-plastic analysis: absolute displacements relative to the masses M_1 and M_2

Figure 17 - Elastic-plastic analysis: acceleration relative to the mass M_3

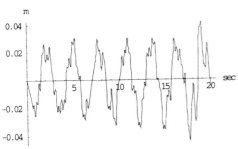

Figure 14 - Elastic-plastic analysis: absolute displacements relative to the mass M_3

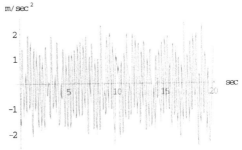

Figure 18 - Elastic-plastic analysis: acceleration relative to the mass M_4

299

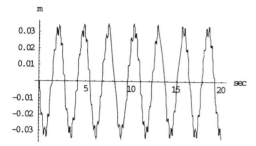

Figure 19 - Elastic analysis (number of AISI 304 constraints halved): absolute displacements relative to the masses M_1 and M_2

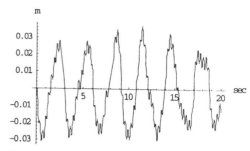

Figure 23 - Elastic-plastic analysis (number of AISI 304 constraints halved): absolute displacements relative to the masses M_1 and M_2

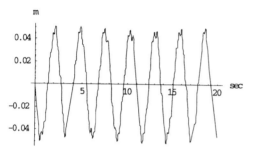

Figure 20 - Elastic analysis (number of AISI 304 constraints halved): absolute displacements relative to the mass M_4

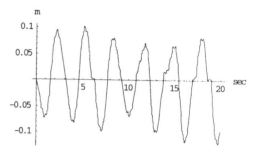

Figure 24 - Elastic-plastic analysis (number of AISI 304 constraints halved): absolute displacements relative to the mass M_4

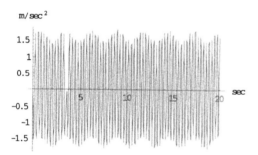

Figure 21 - Elastic analysis (number of AISI 304 constraints halved): acceleration relative to the masses M_1 and M_2

Figure 25 - Elastic-plastic analysis (number of AISI 304 constraints halved): acceleration relative to the masses M_1 and M_2

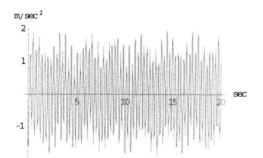

Figure 22 - Elastic analysis (number of AISI 304 constraints halved): acceleration relative to the mass M_4

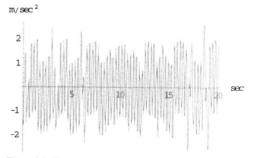

Figure 26 - Elastic-plastic analysis (number of AISI 304 constraints halved): acceleration relative to the mass M_4

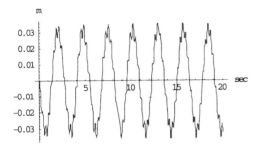

Figure 27 - Elastic analysis (number of AISI 304 constraints doubled): absolute displacements relative to the masses M_1 and M_2

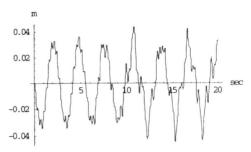

Figure 31 - Elastic-plastic analysis (number of AISI 304 constraints doubled): absolute displacements relative to the masses M_1 and M_2

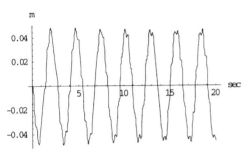

Figure 28 - Elastic analysis (number of AISI 304 constraints doubled): absolute displacements relative to the mass M_4

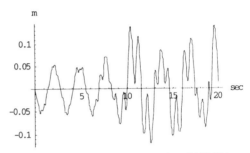

Figure 32 - Elastic-plastic analysis (number of AISI 304 constraints doubled): absolute displacements relative to the mass M_4

Figure 29 - Elastic analysis (number of AISI 304 constraints doubled): acceleration relative to the masses M_1 and M_2

Figure 33 - Elastic-plastic analysis (number of AISI 304 constraints doubled): acceleration relative to the masses M_1 and M_2

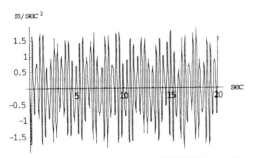

Figure 30 - Elastic analysis (number of AISI 304 constraints doubled): acceleration relative to the mass M_4

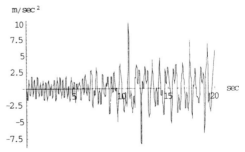

Figure 34 - Elastic-plastic analysis (number of AISI 304 constraints doubled): acceleration relative to the mass M_4

yields the following sets of periodic times for the modes of the first and the second of the above mentioned cases, respectively,

$$\{T^a\} = \begin{Bmatrix} 2.748 \\ 1.005 \\ 0.658 \\ 0.552 \end{Bmatrix} \text{ (sec)} \qquad (13)$$

$$\{T^b\} = \begin{Bmatrix} 2.734 \\ 0.584 \\ 0.385 \\ 0.290 \end{Bmatrix} \text{ (sec)} \qquad (14)$$

It follows that the periodic time corresponding to the first mode of the model is practically not affected by the stiffness of the springs K_1-K_4, while the other three obey to a substantially linear relationship.

Secondly, it can be noticed that in case of plastic deformations the reduction of the frequency of the motion is more marked when the number of AISI 304 constraints decreases.

Finally, it is important to notice that the value of the mean square acceleration relative to the masses M_1 and M_2 remains practically the same when the number of of AISI 304 constraints is halved, but increases when this number is doubled. On the contrary, the value of the mean square acceleration relative to the mass M_4 decreases when the number of of AISI 304 constraints is halved, but increases decidedly when this number is doubled. This behaviour is much more market for values of the time t>7-8 sec.

It is also important to notice that the absolute displacements of the suspended mass M_4 results sensibly increased in both of the last two cases under analysis, i.e. when the number of of AISI 304 constraints is halved or doubled with respect to the actual design of the building.

5 CONCLUSIONS

The adoption of a "minimum" model for the analysis of the dynamic behaviour of two towering buildings in Naples (ENEL headquarters) allowed to obtain approximate and quick indications about the dynamic behaviour of the real structure with the variation of some very basic design parameters, i.e. the stiffness and the limit load of the devices connecting horizontally the suspended floors to the adjacent concrete towers.

In the opinion of the writers, apart from constituting an effective and precious preliminary design tool, this approach allows a clear identification of the qualitative influence of some structural elements which may not result so immediate from other types of finite element analyses which are routinely performed by means of commercially available design codes.

ACKNOWLEDGEMENTS

This work was supported by a grant from the Italian Ministry of University and Scientific Research (M.U.R.S.T.), 1998.

REFERENCES

Belli, P. & F. Guarracino 1997. Experimental testing and damage evaluation for a typology of damping devices. *Proceedings STESSA '97, Kyoto, 3-8 August 1997.*

Mazzolani, F.M. & G. Serino 1997. Top isolation of suspended steel structures: modelling, analysis and applications. *Proceedings STESSA '97, Kyoto, 3-8 August 1997.*

Newmark, N. M. & E. Rosenblueth 1971. Fundamentals of earthquake engineering. *Prentice Hall.*

Press, W.H. et al. 1992. Numerical Recipes in Fortran 77, vol. I, 2nd ed. *Cambridge University Press.*

Behaviour of Steel Structures in Seismic Areas, Mazzolani & Tremblay (eds) © 2000 Balkema, Rotterdam, ISBN 90 5809 130 9

Serviceability and stress control in seismic design of steel structures

B. Calderoni
Department of Structural Analysis and Design, Faculty of Engineering, University of Naples 'Federico II', Italy

A. De Martino
Department of Construction and Mathematical Methods in Architecture, Faculty of Architecture, University of Naples 'Federico II', Italy

ABSTRACT: The majority of current structural codes for seismic design requires to calculate the structure for a strong earthquake under reduced horizontal loads, allowing to experience plastic deformations. The serviceability control on inter-story drift has to be performed too, but without any check on stress level. Then is not ensured that structural elements be not damaged. In the present paper, for steel structures, a proposal for modifying the design procedure is made, in order to obtain the structure remain in the elastic field, in case a moderate earthquake occurs.

1 INTRODUCTION

During the last earthquakes, significant structural damages have been found also in recent buildings, leading to a negative judgement of some design procedures, as stated by the codes. Actually seismic design is still oriented particularly to make the structure able to withstand a "strong" earthquake without collapse. Then the most important seismic codes do not show a great interest in controlling structural damages when a moderate earthquake occurs.

For these reasons a new approach is in development in the last few years. The so-called "performance-based design" (Bertero 1997) is going to become the conceptual base of the present seismic engineering: different levels of structural performance are associated with different intensities of ground motions. The objective of the design should be not only to ensure the no-collapse requirement for the most severe earthquake, but particularly to control the damage level (both in structural and non-structural elements) for earthquakes with higher probability of occurrence.

The current codes have not yet incorporated this new concept, even if, since some years, the researchers highlighted this need (Mazzolani et al. 1994). Nowadays only two verification levels are required (De Martino 1997): the first one refers to the severe earthquake the structure must survive, also suffering wide structural damages; the second one refers to a generic moderate event (often not well defined) which must be faced without significant damages to the whole building. Also the recent EC8 (Eurocode 8 1994) explicitly requires the ful-

fillment of two requirements: "no-collapse" and "damage limitation".

Unfortunately the second verification concerns only a displacement control, performed on inter-story drifts evaluated "a posteriori" from the ones obtained with reference to the "strong" earthquake action. No clear description of the moderate earthquake is given and no stress control in such condition is required.

On the contrary, the stress control is necessary for evaluating the damage level in the structural elements. Furthermore, the writers think that for the serviceability limit state (i.e. when the moderate ground motion is considered) the yield stress has not to be reached or (theoretical speaking) over-passed in a significant way, so to ensure the building to not experience spread structural damages.

In the present European code, similarly to many others (like the Italian one), even if two verification levels are considered, only the limitation of displacements is really concerned. In this paper, the criteria adopted by EC8 for the seismic flexibility check in steel structures are analyzed, highlighting its correlations with the tensional state in the serviceability conditions.

2 THE EC8 SEIMIC DESIGN PROCEDURE

Elastic analysis of the structure, loaded by conventional horizontal forces, is required by EC8 for performing the seismic design (force-based design) with reference to the "strong "earthquake (the one having a return period of 475 years). The safety verification requires the internal forces in beams and columns not exceed the ultimate resistance of the used sections. More in detail, the acting bending moment must be not greater than the full plastic moment, which results for the beams:

$$M_{pl,b} = \frac{f_y}{\gamma_m} \alpha W_{el} \qquad (1a)$$

and for the columns:

$$M_{pl,c} = 1.11 \left(1 - \frac{N_D}{N_{PL}}\right) \frac{f_y}{\gamma_m} \alpha W_{el} \qquad (1b)$$

where f_y = steel yield stress; γ_m = material safety factor; αW_{el} = section plastic resistance modulus, being W_{el} the elastic one; N_D = design axial force; and N_{PL} = axial full plastic resistance.

As well known, the horizontal forces, representative of the "strong" motion, are obtained starting from an elastic response spectrum reduced by a specific reduction factor or behavior factor (named q in EC8). This coefficient accounts for ductility and overstrength of the adopted structural scheme and ranges (according to EC8) from 5 to 6 for steel MRF, in line with other international codes.

To come to the point, the resistance verification (fulfillment of the "no-collapse" requirement) is made in the elastic field with reference to low values of design acceleration, trusting on the plastic capacity of the structural elements. And then, in order to ensure a satisfactory global plastic behavior, the adoption of a capacity-design criterion is required: the columns strength is increased with respect to the beams, so that plastic hinges occurs particularly in the beams, reducing the ductility demands in the columns. No check for displacements or for structural and non-structural damages is performed in this phase.

The serviceability control is required too. It consists in checking that inter-story drift ratio (δ_f/h) be less than 0.4% (for brittle non-structural elements) or 0.6% (for elements not fixed to the structure in a rigid way). The drifts are not specifically evaluated: they are the ones obtained from the resistance verification, increased by the adopted q-factor and reduced by another factor ($\nu = 2$), which accounts for the lower intensity of the earthquake to be considered for a serviceability limit state.

This kind of verification (fulfillment of "damage limitation" requirement) is practically related to the first one, even if it has no conceptual correlation with it. Anyway, the considered displacements correspond to an half ground motion level with respect to the "strong" earthquake, and they are independent of remaining the structure in the elastic field, due to the validity of the elastic-inelastic displacement equivalence (ductility theory).

3 REMARKS ON SERVICEABILITY LIMIT STATE

The displacements check concerns with the serviceability limit state. In the seismic case the actions are defined only in a probabilistic way, and then this limit state refers to an earthquake which have an high probability of occurrence (i.e. a low return period). When this event occurs, the whole building should preserve its integrity, i.e. to not suffer any significant damage to structural and non-structural elements. On the contrary, the limit values fixed by EC8 for the inter-story drift ratio are clearly defined and differentiated for saving from failure just the non-structural elements.

However, the EC8 limit values ($\delta_f/h = 0.4$-0.6%) correspond, in the SEAOC prescriptions (SEAOC 1995), to the performance level "operational" or "immediate occupancy", which is characterized by "light damage" for an "occasional" earthquake (having about a 50% probability of exceedence in 50 years). Instead, the performance level named "fully operational", which is characterized by "negligible damage" for a "frequent" earthquake (having a 100% probability of exceedence in 50 years), sets the inter-story drift ratio to 0.2%.

It seems to be confirmed that the EC8 serviceability control is oriented to limit the displacements rather than to limit the damages in the structure, in spite of that declared in the code.

As an example, in Figure 1 the capacity curve - normalized Base Shear (V_b/W, where W = weight of building) versus Roof Drift Angle (δ_r/H, where H = total height) - for a six floors and two bays MRF with rigid connections is drawn (Calderoni & Rinaldi 2000). The frame was designed according to EC8, for $q = 5$ (value lower than the maximum allowed), applying the capacity design criterion and giving to the sections no further design overstrength ($\Omega_d = 1$).

The sloping straight line represents the elastic behavior, while the three vertical dashed lines point out

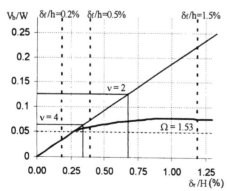

Figure 1. Capacity curve for a MRF designed for q=5.

the roof displacements related to the attainment of an inter-story drift ratio equal to respectively 0.2%, 0.5% and 1.5%. These are the limit values set by SEAOC for the following three performance levels: "fully operational", "operational" and "life-safe".

For $V_b/W = 0.05$, corresponding to the reduced design forces evaluated for the "strong" earthquake, the frame is correctly at its elastic limit. Furthermore it exhibits an adequate scheme overstrength ($\Omega \cong 1.5$) and a satisfactory plastic behavior, considered that it has experienced a significant roof displacement before to reach the 1.5% inter-story drift ratio. Finally, the structure is able to withstand the design earthquake (no-collapse requirement).

The force level corresponding to EC8 serviceability limit state is $V_b/H = 0.125$ (i.e. half of the elastic forces due to the "strong" earthquake – $v = 2$). At this level the structure does not fulfill the inter-story drift limit ($\delta_r/H > 0.5\%$) while is widely in the plastic field. Since no prescription is given about stress level, for coming back into the rules, it should be necessary only to stiffen the structure. Note that to stiffen does not necessarily mean to strengthen the frame, at least from a global point of view.

It is worthy to note that the structure is still beyond the elastic limit, even if the ground motion level is reduced by four ($v = 4$), representing, in the opinion of the authors, the "frequent" earthquake. Then, also in this case, the frame could suffer structural damages.

According to the above given remarks, the writers think that the code should have to require a damage control also for the structural elements, at least for the "frequent" earthquake ($v = 4$). The serviceability verification would consist in both displacement and stress control, so that the yield stress (or a lesser one) would be not over-passed (Uang & Bertero 1991) for a well defined moderate seismic event.

4 STRESS EVALUATION IN SERVICEABILITY LIMIT STATE

Suppose to design a steel frame, adopting a proper q-factor (e.g. $q = 6$). If no design overstrength is given to the sections ($\Omega_d = 1$), when acting the conventional horizontal loads (evaluated for the design "strong" earthquake) together with the design vertical loads (V = G + 0.3Q), the global bending moment (M_S) has to be equal to the ultimate bending strength (M_U). A fraction of this resistance faces the vertical loads (M_V), while the remaining part faces the horizontal seismic loads (M_E).

Setting $X_V = M_V/M_E$ we have:

$$M_S = M_E + M_V = M_E + X_V M_E = \\ = M_E (1 + X_V) = M_U \quad (2)$$

$$M_E = \frac{M_U}{1 + X_V} \quad (3)$$

With reference to the beams, see (1a),

$$M_U = M_{pl,b} = \frac{f_y}{\gamma_m} \alpha W_{el} \quad (4)$$

and then:

$$M_E = \frac{f_y}{\gamma_m} \alpha W_{el} \frac{1}{1 + X_V} \quad (5)$$

On the other hand the serviceability verification requires to evaluate the elastic displacements due to horizontal loads q/v times greater than the ones adopted for the ultimate limit state. The corresponding bending moment (M_{SER}) is:

$$M_{SER} = M_E(q/v) + M_V = M_E(q/v) + X_V M_E \quad (6)$$

and, by considering (5) and (6), it becomes:

$$M_{SER} = \left(\frac{q/v + X_V}{1 + X_V} \right) \frac{f_y}{\gamma_m} \alpha W_{el} \quad (7)$$

Then, the maximum stress (σ_{el}), due to M_{SER}, if evaluated in the elastic field, will result:

$$\sigma_{el} = \frac{M_{SER}}{W_{el}} = \left(\frac{q/v + X_V}{1 + X_V} \right) \frac{f_y}{\gamma_m} \alpha \quad (8)$$

This equation should represent the stress in steel at the serviceability limit state, in case the structure keeps elastic. Instead, as easily can be seen, the stress always exceeds the yield stress (f_y) and M_{SER} is greater than the full plastic moment of the section (M_U).

Setting $v = 2$ and $\gamma_m = 1.1$, as stated by EC8, and also adopting a low value for the shape factor α ($\alpha = 1.15$, suitable for common H sections), for $q = 6$ we have in fact:

$$\sigma_{el} = 1.045 \left(\frac{3 + X_V}{1 + X_V} \right) f_y > f_y \quad (9a)$$

$$M_{SER} = \left(\frac{3 + X_V}{1 + X_V} \right) \frac{f_y}{\gamma_m} \alpha W_{el} > M_U \quad (9b)$$

whatever be X_V.

Note that X_V indicates the amount of internal forces due to vertical loads (valid for the seismic load combination) with respect to the ones given by horizontal actions. It can range from about 0 (corresponding to negligible vertical load) to a maximum value (X_{Vmax}) dependent on the limit state verification performed for the non-seismic load condition.

As stated in EC3 (Eurocode 3 1992), when considering the factorized vertical loads (VL = 1.35G + 1.5Q), the corresponding design bending moment

(M_{VL}) has to be not greater than the section resistance $(M_{VL} \leq M_U)$.

Setting $\varepsilon = M_V/M_{VL}$ (always < 1), and remembering that $X_V = M_V/M_E$, we can write:

$$M_{V\,max} = X_{V\,max} M_E = \varepsilon M_{VL\,max} \qquad (10)$$

and, being

$$M_{VL\,max} = M_U = \frac{f_y}{\gamma_m} \alpha W_{el} \qquad (11)$$

the limit value X_{Vmax} is given by:

$$X_{V\,max} = \frac{\varepsilon}{1-\varepsilon} \qquad (12)$$

Consider that ε depends on the ratio (Q/G) of live load (Q) above dead load (G). In Table 1, for different Q/G ratios, the ε values and the related upper limits of X_V are given. Obviously X_{Vmax} tends to 0 as Q/G grows.

Table 1. Limit values of X_v for beams.

Q/G	0.00	0.25	0.50	1.00	1.50	2.00
ε	0.74	0.62	0.54	0.46	0.40	0.37
X_{Vmax}	2.86	1.65	1.16	0.84	0.68	0.58

As shown in equation (9a), the stress in material at the serviceability limit state exceeds (theoretically speaking) the yield stress, independently of fulfilling or not the inter-story drift limitations. Contemporary the acting bending moment results greater than the ultimate one.

This means that, even though the structure is acceptable as regards the resistance condition for the "strong" earthquake, the fulfillment of the EC8 displacement control does not assure the absence of plastic deformation for the "serviceability" earthquake, as Figure 1 showed too.

Therefore it seems to be reasonable to consider the strong ground motion reduced by two ($v = 2$, as stated by EC8) as the "occasional" earthquake rather than as the "frequent" one. The intensity of this low return period motion should be derived more properly by adopting a reduction by four ($v = 4$).

Equation (8) is drawn in Figure 2, normalized by f_y, for both "occasional" ($v = 2$) and "frequent" ($v = 4$) earthquakes. The ratio σ_{el}/f_y and the parameter X_V are reported respectively on vertical and horizontal axis, when $\alpha = 1.15$ and $\gamma_m = 1.1$. Note that σ_{el} tends to $\alpha f_y/\gamma_m$ (i.e. $1.045 f_y$ in this case) as X_v grows.

The upper decreasing curve corresponds to the serviceability control prescribed by EC8.

On the same diagram, the elastic stress due to the action of only vertical loads is also drawn (curve V). The horizontal solid line at 1.045 (named E+V), instead, corresponds to the seismic condition, since the

structure has not got any design overstrength and then the stress is always equal to the maximum allowed, whatever be the amount of gravity loads. The vertical dashed lines point out the limit of validity of the curves (i.e. X_{Vmax}) for different values of Q/G ratio.

The figure shows the frame exceeds the elastic limit ($\sigma_{el}/f_y = 1$) in a significant way also for the "frequent" earthquake ($v = 4$). In this case the stress should be from 25% to 50% greater than f_y, when X_V ranges within realistic values, i.e. for Q $\geq$ G/4.

However, both the decreasing curves lay above $\sigma_{el}/f_y = 1$ and then represent unacceptable conditions without a physical meaning, as they was obtained in the elastic field, which is whereas largely exceeded.

It is then clear that the serviceability control, as prescribed by EC8 (i.e. for the "occasional" earthquake), is more severe than the safety verification against the collapse, with reference to stress, whatever be the amount of vertical loads. The same conclusion is drawn also if we consider the "frequent" earthquake, as proposed by the writers.

Obviously the displacements has to be controlled too; but this is a problem of stiffness, which (at least in theory) is not directly correlated with the strength.

Finally the flexibility control, performed according to EC8, cannot be considered as a serviceability verification, since the elastic limit is always overpassed in a significant way.

Analogous results may be obtained if we refer to the columns. In fact, using equations (1b), (2) and (3), we have:

$$M_U = M_{pl,c} = 1.11\left(1 - \frac{N_{DS}}{N_{PL}}\right)\frac{f_y}{\gamma_m}\alpha W_{el} \qquad (13)$$

$$M_E = \frac{1.11\left(1 - \frac{N_{DS}}{N_{PL}}\right)\frac{f_y}{\gamma_m}\alpha W_{el}}{1 + X_V} \qquad (14)$$

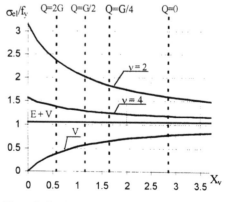

Figure 2. Serviceability stress (σ_{el}) in the beams versus $X_V = M_V/M_E$ for $q = 6$, $\alpha = 1.15$ and $\gamma_m = 1.1$.

where N_{DS} = axial force in seismic design situation.

Then, at the serviceability condition, also remembering the (6), it will result:

$$M_{SER} = \left(\frac{q/v + X_V}{1 + X_V}\right) \cdot \left(1 - \frac{N_{DS}}{N_{PL}}\right) 1.11 \frac{f_y}{\gamma_m} \alpha W_{el} \quad (15)$$

$$N_{SER} = N_E (q/v) + N_V \quad (16)$$

$$\sigma_{el} = \frac{N_{SER}}{A} + \frac{M_{SER}}{W_{el}} \quad (17)$$

where N_E = axial force due to horizontal seismic loads; N_V = axial force due to vertical loads ($V = G + 0.3Q$); N_{SER} = global axial force for the serviceability load combination; and A = cross-section area.

If we set:

$$\tilde{N}_{DS} = \frac{N_{DS}}{N_{PL}} = \frac{N_E + N_V}{N_{PL}} \quad \text{and} \quad \tilde{N}_V = \frac{N_V}{N_{PL}}$$

being :

$$N_{PL} = A \frac{f_y}{\gamma_m}$$

we can write:

$$N_E = \tilde{N}_{DS} N_{PL} - N_V \quad (18)$$

$$\frac{N_{SER}}{A} = \left[\tilde{N}_{DS}(q/v) - \tilde{N}_V(1 - q/v)\right] \frac{f_y}{\gamma_m} \quad (19)$$

By considering (17), (15) and (19), finally results:

$$\sigma_{el} = \frac{f_y}{\gamma_m}\left[\tilde{N}_{DS}(q/v) - \tilde{N}_V(1 - q/v)\right] +$$
$$+ 1.11\alpha \frac{f_y}{\gamma_m}\left[(1 - \tilde{N}_{DS}) \cdot \left(\frac{q/v + X_V}{1 + X_V}\right)\right] \quad (20)$$

which should represent the stress in the columns at the serviceability limit state, if the structure should keep elastic.

Equation (20) is drawn in Figure 3, when $q = 6$, $\alpha = 1.15$ and $\gamma_m = 1.1$, like in Figure 2, setting $\tilde{N}_V = 0.2$. For both $v = 2$ and $v = 4$ ("occasional" and "frequent" earthquake) the three decreasing curves are drawn for three different values of $\tilde{N}_{DS}$. The (E+V) and (V) curves have the already explained meaning.

The values X_{Vmax} are not reported in the picture, because of their change as $\tilde{N}_{DS}$ varies. Anyway, in Table 2, the upper limits for X_V, evaluated as for the beams, are listed for different Q/G ratios.

Figure 3 confirms that, also for columns, the yield limit is largely exceeded, even if the "frequent" earthquake is considered ($v = 4$).

Table 2. Limit values of X_v for columns ($\tilde{N}_V = 0.2$).

Q/G		0.00	0.25	0.50	1.00	1.50	2.00
	ε	0.74	0.62	0.54	0.46	0.40	0.37
X_{Vmax}	$\tilde{N}_{DS} = 0.1$	1.51	0.89	0.60	0.40	0.29	0.23
	$\tilde{N}_{DS} = 0.2$	2.09	1.12	0.72	0.47	0.33	0.27
	$\tilde{N}_{DS} = 0.3$	3.40	1.53	0.92	0.58	0.40	0.32

The variation of axial force due to seismic load ($\tilde{N}_{DS}$) does not significantly affect the stress values. Furthermore the trend of curves is very similar to that previously obtained for beams. So, the already made remarks are substantially effective also for the columns.

Nevertheless, the adoption of the capacity design criterion leads to a strengthening of the columns only, with an improvement of their stress condition at the serviceability state. Therefore the beams remain the structural elements more subjected to suffer structural damages with plastic engagement also during a moderate ground motion.

5 A NEW PROPOSAL

The fulfillment of the flexibility control, stated by EC8, does not ensure that all the structural elements are within the elastic field, if the allowed q-factors are adopted in designing the steel structure, also when considering the "frequent" earthquake. Then this check cannot be considered as an effective serviceability verification.

It was demonstrated – see eqn.(8) – that for a reduction factor still equal to 4, the structure (especially in the beams) exceeds the yield stress, when no design overstrength is given. Besides, for steel MRF, is really usual to adopt q-factors well higher than 4.

Then, just for MRF, the design procedure can be simplified, giving back the priority to the elastic analysis at the serviceability condition.

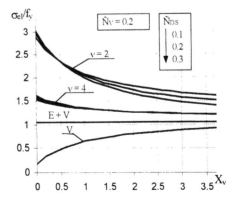

Figure 3. Serviceability stress (σ_{el}) in the columns versus $X_V = M_V/M_E$ for $q = 6$, $\alpha = 1.15$ and $\gamma_m = 1.1$.

Stated that the "frequent" earthquake is equal to one fourth (or rather, to one third, as proposed by others) of the severe one, and the corresponding performance level of the structure must be "fully operational" (i.e. exhibiting "negligible damage" in both structural and non structural elements), the authors propose the following design procedure, suitable for steel MRF (for which q-factor ≥ 4):

- elastic analysis of the frame loaded by vertical loads and horizontal seismic forces obtained from the elastic spectrum (not reduced by q) scaled according to the "frequent" earthquake;
- check on the above obtained elastic displacements and, if needed, stiffening of the structure;
- resistance verification of beams and columns in the elastic field, i.e. limiting the stress values to the yield (or a lower) one;
- adoption of a suitable capacity design criterion, in order to ensure a proper plastic behavior for the severe earthquake, so to avoid the collapse;
- possible structural performance evaluation (f.i. by means of a push-over analysis), in order to check, a-posteriori, whether the displacement, expected for that earthquake, can be experienced by the frame without suffering fatal ruptures, that is to not exceed fixed deformation limits, also with reference to different levels of ground motion.

This design procedure seems to be very clear and simple and it does not require to perform the serviceability control on the basis of results obtained for the ultimate limit state.

Moreover it stays within the performance-based design concepts: the project is made starting from the performance level required for the serviceability limit state, and, only after, the structure ability to ensure other performance levels for different ground motions is checked.

However, still according to the remarks already written, it is possible also to propose just a partial modification to the actual code.

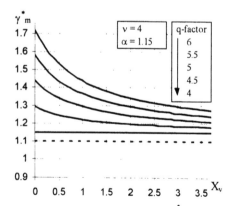

Figure 4. Increased material safety factor (γ^{*}_{m}) for beams.

In the practice, the fulfillment of inter-story drift limitation (which are quite severe) leads often to stiffen the structure and then to increase its strength; the use of commercial sections gives further additional resistance; the application of capacity design criteria is a strengthening for the columns. Then the structure exhibits quite always some design overstrength ($\Omega_d > 1$), even if its distribution among the elements is not easily predictable. Nevertheless, to give more resistance to the sections is like to design adopting an higher material safety factor.

Since our objective is to remain in the elastic field, when a moderate earthquake occurs, at the serviceability condition it shall be result $\sigma_{el} <= f_y$ whatever be the amount of vertical loads with respect to seismic actions (represented by X_V). If we refer to beams, this means:

$$M_{SER} = M_E(q/\nu) + M_V = \\ = M_E(q/\nu) + X_V M_E \leq f_y W_{el} \quad (21)$$

and, setting exactly f_y as limit value, the maximum allowed for M_E is given by:

$$M_E = \frac{f_y W_{el}}{q\cdot\nu + X_V} \quad (22)$$

At the same time, we have to fulfill the resistance condition for the severe earthquake, as expressed by equation (5):

$$M_E = \frac{f_y}{\gamma^*_m}\alpha W_{el}\frac{1}{1+X_V} \quad (23)$$

Comparing (22) with (23), we have:

$$\gamma^*_m = \alpha\frac{q\cdot\nu + X_V}{1+X_V} \quad (24)$$

which is the safety factor, which ensure the yield stress will be never exceeded at the serviceability limit state (i.e. for the moderate earthquake), if it is adopted in strength verification at the ultimate limit state (i.e. for the severe earthquake).

The here proposed code modification consists in adopting, just for the seismic load condition, an increased material safety factor (γ^*_m), variable with the ratio (X_V) of bending moment due to vertical loads above the one due to seismic actions.

This new factor, given in (24), should be anyway greater than γ_m (which is 1.1), as ensured being $\alpha \sim 1.1$ and $q\cdot\nu \ge 1$. It is easy to be evaluated, since X_V is always known in any structural elements.

In Figure 4, the relationship γ^*_m versus X_V is drawn, with reference to the "frequent" earthquake ($\nu = 4$), for different values of the reduction factor q.

Note that for q = 4, the safety factor is a constant (equal to α), being $q\cdot\nu = 1$, while, for usual values of q (5-6) and X_V(<1.5), it needs significant increasing.

The horizontal dashed line points out the safety factor fixed by European code ($\gamma_m = 1.1$), which always has to be increased.

6 CONCLUSION

The seismic serviceability control for steel MRF, as prescribed by European code, refers to a moderate earthquake, whose intensity is halved with respect to the severe one, and has to be performed only with reference to inter-story drift. Then is not ensured that, in such condition, steel be within the elastic field, so that structural elements can suffer widespread and significant damages also for the low return period ground motion.

Even if a further lower earthquake (like the "frequent" one) is considered, as the authors think right, the EC8 procedure leads still to unacceptable results, when the usual q-factors are adopted.

On the contrary, the design procedure proposed in this paper, in line with the performance based-design concepts, can ensure to preserve from structural damages for the moderate earthquake, as well as to withstand the severe one. The same objective can be reached without substantially changes in the present code, as also proposed, by using an increased material safety factor dependent on the amount of vertical loads (with respect to the horizontal seismic actions), when the no-collapse limit state verification is performed.

REFERENCES

Bertero, V. 1997. Codification, design and application - General Report. In F.M.Mazzolani and H.Akiyama (eds), *Proc. Behaviour of Steel Structures in Seismic Areas – STESSA '97, Kyoto-Japan, August 1997*: 189-205. Salerno (Italy): Edizioni 10/17.

Calderoni, B. & Rinaldi, Z. 2000. Inelastic dynamic and static analyses for steel MRF seismic design. In F.M.Mazzolani and R.Tremblay (eds), *Proc. Behaviour of Steel Structures in Seismic Areas – STESSA 2000, Montreal-Canada, August 2000*. Rotterdam: Balkema.

De Martino, A. 1997. State of the art report on basic problems of seismic behaviour of steel structures - General Report. In F.M.Mazzolani and H.Akiyama (eds), *Proc. Behaviour of Steel Structures in Seismic Areas – STESSA '97, Kyoto-Japan, August 1997*: 3-20. Salerno (Italy): Edizioni 10/17.

Eurocode 8 1994. *Design provisions for earthquake resistance of structures, ENV 1998*. CEN, European Committee of Standardization.

Eurocode 3 1992. *Design of steel structures, ENV 1993*. CEN, European Committee of Standardization.

Mazzolani, F.M., Georgescu, D. & Astaneh, A. 1994. Safety Levels in Seismic Design. In F.M.Mazzolani and V.Gioncu (eds), *Proc. Behaviour of Steel Structures in Seismic Areas – STESSA '94, Timisoara-Romania, June 1994*: 495-506. London: E & FN SPON.

SEAOC 1995. *Vision 2000, A framework for Performance Based Design*. Sacramento CA, USA: Structural Engineering Association of California.

Uang, C.M. and Bertero, V. 1991. UBC Seismic Serviceabilty Regulations - Critical Review. *Journal of Structural Engineering* 117/7: 2055-2068.

Behaviour of Steel Structures in Seismic Areas, Mazzolani & Tremblay (eds) © 2000 Balkema, Rotterdam, ISBN 90 5809 130 9

Romanian new Code for the design of steel structures subjected to seismic loads

C. Dalban, P. Ioan, S. Dima & St. Spanu
Technical University of Civil Engineering, Bucharest, Romania

ABSTRACT: The authors of this paper had the task to elaborate a new Code with the main purpose to cover the practical problems of the seismic structural design in agreement with the principal provisions of the general Seismic Code P100-92, existing in Romania. The new Code concerns mainly the multi-storey steel frames. New aspects taken into account are referring to the composite slab, connected to steel beams, and composite or steel shear walls acting as vertical bracing systems. For some basic provisions of existing Code P100–92 improvements are proposed as well.

1. INTRODUCTION

In Romania the last version of the Seismic Code P100–92 was published in 1992. The provisions of the Code concerning the general aspects of the Earthquake resistance rules applicable to buildings were in good agreement with those of similar codes from USA, Japan, Italy, France and from other seismic countries. The provisions of European Recommendations for Steel Structures in seismic zones (1988) and Eurocode 8 (1988) have been also taken into account.

In order to assess the proposals to improve the Code P100–92 and to elaborate the new Code for the design of steel structures various aspect have been taken into consideration:

- The national experience on the field of seismic design of steel structures.
- The recently published seismic codes and the theoretical and practical experience of the specialists of the seismic countries (USA, Japan, Italy, France and other), together with ECCS Manual – 93, EUROCODE 3-92, EUROCODE 4-92 and EUROCODE 8-94.

2. SEISMIC LIMIT STATES. EARTHQUAKE HORIZONTAL LOADS. LOAD COMBINATIONS LATERAL DEFLECTIONS, DRIFTS

2.1 Seismic limit states

According to the ECCS-proposals based on US tendencies, three limit states for buildings with steel structures were initially taken into consideration: a) ultimate limit states, b) damageability limit states, c) serviceability limit states. Our team agreed that the acceptance of a third (damageability) limit state seams to be more convenient for multi-storey structures. A proposal was presented to the board of authors of the general seismic Code P100-92, because it implies detailed knowledge on the behaviour of in-filled materials of lateral and partition walls. Till the agreement, the up to-date overall applied two limit states (ultimate respectively serviceability) remain available.

2.2 Earthquake horizontal loads. Load combinations

The horizontal seismic loads acting on a structure are determined using the relations:

$$S_r = c_r G \qquad c_r = \alpha \, k_s \, \beta_r \, \psi \, \varepsilon_r$$

where:
S_r - horizontal shear force due to the gravity loads of the whole building

G - resultant of gravity loads for the whole building

α - importance coefficient of the building (α=1.4; 1.2; 1.0; 0.8)

k_s - coefficient depending on the seismic zone (k_s=0.32 ; 0.25 ; 0.20 ; 0.16 ; 0.12)

β_r - dynamic amplification coefficient in the "r – th" vibration mode (max β_r=2.5)

ψ - coefficient of reduction seismic effects (equivalent to 1/q in EC 8)

ε_r - coefficient of equivalence between the actual system and a system with a single degree of freedom

The load combinations are given by the following general formulae:

$$\Sigma \quad P_i \quad + \quad \Sigma \quad C_i \quad + \quad \Sigma n_i^d V_i \quad + \quad E$$
(2.3)

$$\Sigma P_i + \Sigma C_i + \Sigma n_i^d V_i + \Omega_0 E \qquad (2.4)$$

where:

Pi - dead loads

Ci - quasipermanent live loads

Vi - live loads; n_i^d - represents the fractile of long duration ($n_i^d = 0.4$), wind / snow included

E - seismic load

Ω_0 - coefficient of amplification for special design combinations ($\Omega_0 = 2.0...3.0$)

Relation (2.3) is used for calculation of the dissipative zones and active links; relation (2.4) is used for calculation of the other members (columns, vertical bracings, beam segments outside the link) required to remain in elastic range during earthquake.

2.3 Combination of the components of the seismic action

The action effects due to the combination of the horizontal components of the seismic action may be computed using the following combinations:

(a) E_{Edx} "+" $0.30 E_{Edy}$ (2.5)

(b) $0.30 E_{Edx}$ "+" E_{Edy} (2.6)

where :

"+" implies "to be combined with"

E_{Edx} - action effects due to the application of the seismic action along the chosen horizontal axis "x" of the structure

E_{Edy} - action effects due to the application of the same seismic action along the chosen orthogonal horizontal axis "y" of the structure

The sign of each component in the above mentioned combinations should be taken as the most unfavourable ones for the effects under considerations. For the buildings, satisfying the regularity criteria in plan and in which the shear walls or vertical bracings are the only horizontal resisting components, the seismic action may be assumed to act separately along the main orthogonal horizontal axis of the structure.

2.4 Lateral deflections. Drifts

The following limitations for storey drift are provided in P 100-92: in general case 1/200 of the storey height; if non-structural members are not damaged by displacements, the drift condition is limited to 1/120 of the storey height.

With regard to the above mentioned drift limitations, some improvements have been provided in the new version of the Code, considering that the drift verification belongs to the serviceability limit state.

For buildings having non-structural elements of brittle materials attached to the structure, the drift condition is limited to:

$d_r/v = 1/200h$
(2.7)

For buildings having non-structural elements of brittle materials non affected by the structural deformation, the drift condition is limited to

$d_r/v = 1/120 h$
(2.8)

where:

d_r - elastic storey drift (calculated with $\psi = 1.0$)

h - storey height

v - reduction factor, taking into account the lower return period of the seismic event, associated with the serviceability limit state; the values of "v" depending on importance coefficient α are:

v=1.8 for α=1.2 - 1.4 and v=1.45 for α=0.8 - 1.0

3. DESIGN OF STRUCTURAL MEMBERS

3.1 Moment resisting frames (MRF)

MRF are widely used for low-and-medium-rise buildings. The dissipative zone is recommended to be located in the beams near the connections. The yielding in the panel zone and in the column are to be avoided. In order to fulfil these conditions, relationships similar to those of Chapter 9, 10 and 11 of AISC – 97 shall be satisfied.

The concept of strong-column weak-beam for the steel frame is applied.

These relationships may be used only in the case where the beam may be considered as "a bare steel member". If a concrete slab contributes to the resistance of beams, the effective strength and rigidity of the beams are higher and adequate modifications are to be operated in the formulae. Details of design, both elastic and plastic, have been provided in the Chapter 4 of this Code.

Taking into account the experience of seismic events in Japan (1995) and mainly in California (1994), in the Romanian Specification improved welded beam to column connections are proposed; there are the aim to increase the strength and to avoid untimely collapse of welds subjected to cyclic seismic forces. A ductile safe connection behaviour may be achieved by using electrodes capable to ensure tough weld metal, in conjunction with improved detailing. The trimming of beam flanges near the column is an effective way of enhancing ductility and reducing the defavourable effect of residual stresses in the flange welds.

Other type of connection consists either of designing beam-to-column connections with high strength bolted extended end-plate, or of providing columns with ramifications, all shop welded, with floor girders site connected (bolted or fillet welded) at a point away from face of column.

3.2 Concentrically braced frames (CBF)

The design provisions of CBF allows to apply the typologies with simple diagonal bracing and with "X" bracing which ensure the dissipation of the seismic input-energy by the plastic deformation of the tension diagonals. The design provisions are in good agreement with existing P100-92, and with EC8-94 as well. Some supplementary data belonging to AISC-97 (Chapter 14) and concerning the compressed diagonals were taken also into account.

The others typologies (V and inverted V) exhibiting a poor dissipative capacity are presented in the AISC formulation, different from the other. General behaviour factor q=4 (respectively $\psi=1/q=0.25$) is the same as for the previous mentioned bracing with tensioned diagonals, but considering an enhanced compression strength with

about 100%, provided for the bracing members in comparison with the required strength for normally load combination. (see also Table 1)

Such computation procedure for this type of seismic structure seems to be more convenient, especially in the design of dual systems. The difference between the initially computed design base shear, considered in the calculation of frames belonging to dual systems, both braced and unbraced is not significant. These frames shall differ only by lateral stiffness, which may be calculated by structural mechanics formulae.

A mention is to be done concerning the structures consisting of only CBF (V or inverted V) applied to steel towers. In this case, it may be more convenient to beware the existing value of ψ ($\psi = 1/q = 0.5...0.65$) in the design of the whole structure.

3.3 Eccentrically braced frames (EBF)

EBF may be considered a structural system possessing the stiffness of CBF with the ductility and energy dissipation capacity of MRF. EBF shall be designed so that under earthquake loads, yielding shall not occur in members like diagonal braces, columns, beam segments under the maximum forces that will be generated by the fully yielded and strain hardened links. Limitations on width-to-thickness ratios of the links correspond to class 1 in P100 and EC8-94 respectively "plastic" in AISC-97.

A classification of links integrated in beam of the frame corresponds to the classification assessed by ECCS Manual - 93.

$$\text{Short link} : e \leq 1.6 \frac{M_p}{V_p} \; ; \; \theta_{max} = \pm 0.08 \text{ rad} \quad (3.1)$$

$$\text{Long link} : 3 \frac{M_p}{V_p} < e \leq 5 \frac{M_p}{V_p} \; ;$$
$$\theta_{max} = \pm 0.02 \text{ rad} \quad (3.2)$$

$$\text{Intermediate link} : 1.6 \frac{M_p}{V_p} < e < 3 \frac{M_p}{V_p} \; ;$$
$$\theta_{max} = \text{by linear interpolation} \quad (3.3)$$

For the short link, the plastic rotation angle is dependent on the distance between the stiffeners. If the stiffeners are spaced at $30t_w - d/5$ a link rotation angle of ± 0.08 rad is ensured; if the distance is $52t_w - d/5$ the rotation angle is ± 0.02 rad. Linear interpolation shall be used for intermediate values.

The use of short, shear yielding link is to be preferred because this provides for the maximum strengths, both stiffness and energy dissipation capacity of the frame. Link length exceeding 1.6Mp/Vp should only be used when links are not attached to column. The use of link-to-column weak axis connections should be avoided.

Lateral support of link shall be provided at both top and bottom flanges of link at the ends of the link. End lateral support of link shall have a design strength of 6 percent of the link flange nominal strength ($F_y b_f t_f$).

Concerning the diagonal braces and beams outside the link, some requirements are to be fulfilled. The required combined axial force and moment strength of the diagonal brace shall be the axial forces and moments generated by 1.5 times the nominal shear strength of the link. The design strength of the diagonal brace shall be calculated in elastic range.

The required strength of the beam outside the link shall be calculated at the forces and moments generated by at least 1.5 times the nominal shear strength of the link; the beam segment outside the link shall be provided both top and bottom flanges with lateral supports possessing a strength of 0.02 $F_y b_f t_f$ (it shall be calculated in elastic range).

The coefficient 1.5 takes into account a higher steel strength than nominally specified and the extremely high degree of strain hardening in the shear web of the link.

Remark: In the case of concrete slab connected to the beams it is necessary that the link (especially the short one) to work inelastic, independently of slab, as a "bare steel member". An adequate proposed detail is presented in Figure 1, a gap being provided between the slab and the link.

4. MULTISTOREY STEEL FRAMES WITH REINFORCED CONCRETE SLABS. DESIGN OF COMPOSITE BEAMS AND BEAM-TO-COLUMN CONNECTIONS.

The present chapter deals only with multi-storey buildings where the slabs are performed by casting reinforced concrete on steel trapezoidal sheetings

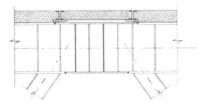

Figure 1: Improved detail for concrete slab supports in zone of short link

supported by double T steel beams. Shear studs ensure the connection between steel beams and concrete slab. In this cases, the slab ensure the horizontal diaphragm effect which allows to rigidly connect in plan the vertical bracings or structural walls ensuring the transfer of horizontal seismic forces.

Concerning the composite beams (steel beam connected to reinforced concrete slab) design provisions are contained by EUROCODE 4. These provisions concern only the statically acted composite beams simply supported and/or continuous supported on several columns.

EUROCODE 8 did not present any information concerning the cyclic seismic action, especially the composite behaviour near the beam to column joints, where cyclic plastic deformations develop in inelastic range of structural behaviour.

Concerning the damages observed after the Californian seismic events (January 1994), the pre-Northridge type of beam-to-column connections belonging to MR-Frames were submitted to an intensive programme of experimental investigations. The aim of the investigations was to explain the failures produced to the beam flanges of the composite beams and especially to the bottom flanges. Generally, the tested models of connections (full scale ones) consist of "bare steel" beam-to-column connections. Some of tests (Haijar and Leon) were performed on composite beams specimens consisting of double T rolled shapes connected to concrete slab (Fig.2); the column specimen was a "bare steel double T rolled profile". During the tests, it has been observed that the strains in the beam flange not connected to slab (the bottom flange) were significant higher than those in the top flange (connected to slab). That situation corresponds to the actual predominance of the bottom flange failure.

An improved solution in the case of rigid and full strength connections able to ensure an ultimate value of plastic rotation up to 2.5% is represented by an adequate unsymmetrical double T steel section obtained by supplementary welded plate to the bottom flange of the beam, in order to lower the neutral axis and the stresses in bottom flange. The general aim is to avoid the local and lateral buckling of the bottom flange (Bursi, 1997).

In order to ensure the formation of plastic hinge near the column, special longitudinal rebars (reinforcing bars) are provided in the slab. The rebars may carry tensile force after reaching of the tensile concrete strength, ensuring the rotational capacity and the stiffness of the joints. The shear studs ensure

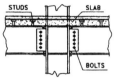

Figure 2: Typical North American rigid joint

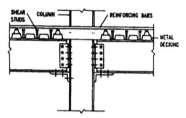

Figure 3: Partially restrained joint

the transfer of tensile forces from cracked concrete to the rebars.

Other theoretical / experimental studies were performed by different authors (Aribert 1997), (Leon 1997), (Nethercot 1997), (Plumier 1993).

Concerning the existing provisions in the field of composite joints some tendencies may be outlined.

1. In the case of nonsway composite frames (including also MR-Frames components of dual systems and coupled with braced frames or shear walls) the beam-to-column connections may be semirigid (Leon 1996), see also (Figure 3). The frames with semirigid connection may be designed applying a procedure based on the revised Annex J of EUROCODE 3 (Dubina 1998).

2. With regard to sway MRF the rigid Joint (Figure 2) may be applied. The bottom flange may be enlarged by an supplementary welded plate (cover plate), or by substitution of the bottom flange near the column by an enlarged plate butt welded to flange and fillet welded to web.

5. STRUCTURAL SHEAR WALLS

The following typologies of shear walls are proposed to be used for structural dual systems.
• Concrete shear walls.
• Steel-concrete composite shear walls.
• Steel panel shear walls with or without concrete covering.

The design of concrete shear walls is performed by applying the provisions of the Romanian special code. The connections between the adjacent walls

located in the same vertical plane may be performed by dissipative steel members (short links-semirigid connected) acting in shear.

Structural shear walls consisting of steel-concrete walls applied to medium and high rise buildings are mainly based on the steel structures (concentrically braced frames) erected before the other components of the dual structure and ensuring the strength for other steel members, profiled sheetings for casting the concrete slab and for concrete cast in structural walls.

Steel panel shear walls erected on the West Coast of USA (Xue 1994) and Japan (Sugii 1996) behaved well during the seismic events from 1994 and 1995. For example a tall building in Kobe (132m high, 32 stories) with central core steel panel shear walls shown very good resisting behaviour during the 1995 Hanshin Awaji - Earthquake (Yamada 1997).

Experimental and theoretical studies are in development on steel panel shear walls both with and without concrete covering (USA, Canada, Japan).

6. DUAL SYSTEMS. POSSIBILITY TO USE SEMIRIGID FRAMES IN DUAL SYSTEMS SUBJECTED TO SEISMIC LOADS.

The concepts presented in US codes are considered the most appropriate for the design of a complex dual system. This system consists of moment resisting frames and concentrically or eccentrically braced frames or shear walls (concrete, steel or composite).

All frames support the gravity loads. Resistance to lateral loads is provided both by MRF, which are capable of resisting at least 25 percent of the general base shear and braced frames or shear walls. The frames are interconnected by horizontal rigid concrete diaphragms (slabs) or steel braces, which allow that the total seismic shear force be transmitted to each frame in proportion to its lateral rigidity.

The necessity of dual systems is required to ensure the strength and lateral rigidity of all frames unbraced and braced frames (including shear walls) and to provide the imposed drift limitation to all component vertical structures.

The semirigid frames with both full-strength or partial-strength semirigid joints may be coupled with braced frames or shear walls in dual systems.

The braced frames, respectively shear walls, ensure the required lateral rigidity (drift) necessary to each frame of the dual system.

A remark is to be done for the frames with semirigid joints not sufficiently "safe" in cyclic loading; the existence of "old" braced frames represents a supplementary factor of safety.

If the dissipative capacity of the semirigid joint is not fully (experimentally) verified, the use of eccentrically braced frames is to be preferred in the dual system.

7. STRUCTURAL ANALYSIS OF SEISMICALLY LOADED MULTI-STOREY FRAMES

7.1 Horizontal Shear Force Method

This method is compulsory for all the structures presented in Romanian code.

This is a complete analysis, whiech considers in a simplified way the aspects of dynamic and post-elastic response of the structure. It implies the main operations according to point 2.2, 2.3. and 2.4.

In special cases, it may be suitable to use a linear dynamic analysis. The method is compulsory for all structures.

7.2 Refined Design Methods

The method refers to the pushover analysis (non-linear analysis) and to the dynamic non-linear analysis, using accelerograms of real earthquakes, in some cases accelerograms of simulated earthquakes may be used.

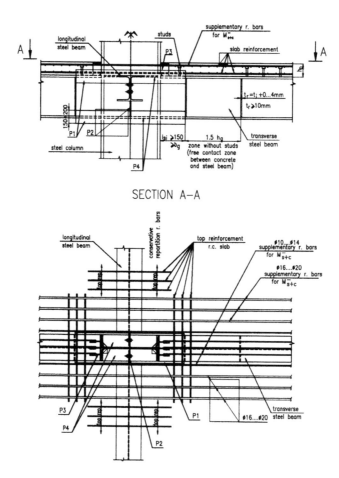

SECTION A–A

Fig.4 Connection of the Composite Beam to Steel Column of the ten- stories MR frames of fig.5

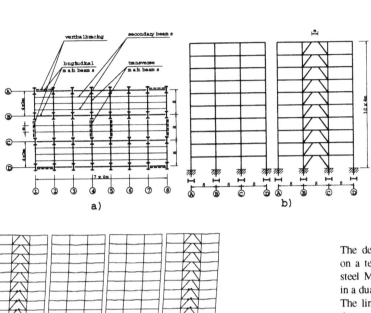

a)

b)

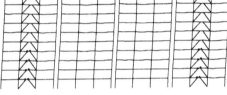

• $_0$=1.75

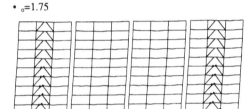

• $_0$=2.0

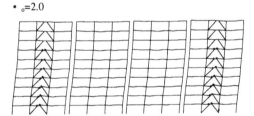

• $_0$=2.5

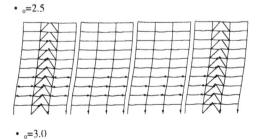

• $_0$=3.0

c)

Fig. 5. Ten-story building . a) layout; b) transverse frames
c) time history analysis.

The design example was performed on a ten-story building composed of steel MRF and EBF acting together in a dual system.

The links have been sized based on the maximum efforts given by the spatial linear analysis, according to the relationship 3.0 and to the provisions of item 3.3.

The columns, bracing and adjacent beam elements have been sized according to eq. 2.4, where • $_0$=2.0.

The beams have composite cross sections, with 15 cm thickness r.c. slab. The ultimate negative moment resistance (which is tensioning the r.c. slab) is greater or equal to the plastic moment of a homogenous cross section made by steel, which has sized the composite cross section.

In order to carry out the negative ultimate moment resistance, supplementary r-bars have been provided in the r.c. slab, along and parallel to the steel beam (see fig. 5).

As a result of the non-linear analysis, for • $_0$=2.5, the plastic hinges have occurred in links only and the maximum values of the displacements and rotations have not been exceeded. For • $_0$=3.0, the plastic hinges are occurred in the beams, located in the spans which are not braced and at the columns bases, as well. The rotations of the links have exceeded the maximum value of 0.08 rad.

317

REFERENCES

Astaneh ASL. A.(1995). Seismic Design of composite structures in the United States *Behaviour of steel Structures in Seismic Areas STESSA'94* E§FN.SPON p. 448

Aribert J.M. (1997). Modelisation par elements finis adoptée aux poutres et assemblages de bâtiments mixtes acier-beton. *Proc. of the 8th Intern. Conf. Steel Struct.* Timisoara (Romania)

AISC-97 (1997). *Seismic Provision for Structural Steel Buildings.*

ASCE.7-93 (1993). *Minimum design loads for Buildings and other Structures.*

Bernuzzi C., Chen S., Zandonini R., (1997). Modelling the Nodal Zone Behaviour in a Composite Frame. *IABSE , Intern Conf. on Composite Construction.* Innsbruck (Austria).

Bursi O.S., Gramola G., Zandonini R., (1997). Quasi-static cyclic and pseudo-dynamic composite substructures with softening behaviour in *SDSS'97* Nagoya (Japan).

Code for aseismic design of residential buildings, agrozootechnical and industrial structures (P100-1992).*Romanian ministry of Public Works and Teritory Planning.*Bucharest. (Romania).

Dalban C., Ioan P., Dima S., Betea St. & Spanu St. (1995). Proposals for Improving the Romanian Seismic Code. Provisions concerning multy-storey steel frames. *IABSE Int. Conf. of Steel Struct. Final Report.* Budapest.

Dalban C.,Ioan P., Dima S., Spanu St.,Betea St. & Tudorache Cr. (1997). Aspects concerning the behaviour and design of steel frames accordiong to various seismic Codes. *STESSA'97* Kyoto. (Japan).

Dubina D., Stratan A.& Dinu F. (1998) Suitability of semi–rigid joint steel building frames in seismic areas. *11th European Conference on Earthquake Engineering* Paris.

Eurocode 3 – ENV 1993 (1992). Design of steel structures. *Eur. Comitee for Standardisation.*

Eurocode 4 – ENV 1994 (1992). Design of composite Steel and Concrete Structures. Part 1.1 General rules for Buildings. *European Committee for Standardisation* (1992)

Eurocode 8 – ENV 1998 (1994). Design provisions for Earthquake resistance of structures. *European committee for Standardisation* (1994).

ECCS Manual of Design of Steel Structures located in Seismic Zones (1993). (Mazzolani F., Piluso V)., ECCS-TC 1.3 ; *Seismic Design.*

Haijar F.J., Leon R.T. (1996). Effect of floor slabs on the performance of SMR connections. *Proc. 11. World Conf. on Earth. Eng. Paper 656 Elsevier.*

JBLS (1981). Japan Earthquake resistance Regulation for Building Structures.

Japan Standard (1990) Standard for limit state of steel structures. Draft- *Architectural Institute of Japan (in english).*

Leon R.T. (1997). Seismic design for composite semi-continous frames. . *IABSE , Intern Conf. on Composite Construction.* Innsbruck (Austria).

Leon R.T. (1996). Design of partially restrained frames for seismic Loads. *Proc. 11. World Conf. on Earth. Eng. Paper 621 Elsevier.*

Nethercot D.A., (1997). Behaviour and design of composite connections. *IABSE , Intern Conf. on Composite Construction.* Innsbruck (Austria).

Norme francaise.(1995) Règles de construction parasismique appliquées aux bâtiments. Règles PS-92 *A.F.N.O.R.* (FRANCE).

Plumier A. § Schleich J.B. (1993). Seismic resistance of steel and composite frame structures. *J. Constr. Steel Research. 27.*

Plumier A.,Agatino M.R., Castellani A.,Castiglioni C.A., Chesi C., (1998) Resistance of steel connection to low – cycle fatigue. *11th European Conference on Earthquake Engineering* Paris.

Sugii K., Yamada M., (1996). Steel panel shear walls with and without concrete covering. *Proc. 11. World Conf. on Earth. Eng. Paper 436 Elsevier.*

Tirca L. & Gioncu V. (1999) Ductility demands for MRFs and LL-EBFs for different earthquake types. *Stability and ductility of steel structures. International Colloquium Timisoara. Proceedings.*

UBC-97 (1997) Uniform Building Code; *International Conference of building Officials, Withier* California.

Yamada M., Yamakaji T., (1997).Steel panel shear wall (as single and centre core type aseismic element) p. 477. *STESSA'97* Kyoto Japan.

Xue M., Lu Le-Wu (1994). Monotonic and cyclic behaviour of infilled steel shear panels. *Seventh Czech and Slovak Int. Conf. on Steel Structures.* Bratislava.

Behaviour of Steel Structures in Seismic Areas, Mazzolani & Tremblay (eds) © 2000 Balkema, Rotterdam, ISBN 90 5809 130 9

Lions' Gate Bridge North Approach – Seismic retrofit

David J. Dowdell & Bruce A. Hamersley
Klohn-Crippen Consultants Limited, Vancouver, B.C., Canada

ABSTRACT: The Lions Gate Bridge in Vancouver, British Columbia is one of the city's signature structures, bridging the entrance to the harbour. The bridge is currently undergoing a replacement in the main span deck and a seismic upgrade. Klohn-Crippen is providing consulting engineering services for the seismic retrofit of the North Approach Viaduct to the Contractor, American Bridge/Surespan Joint Venture. This is believed to be the first design-build seismic retrofit of a major structure. The seismic retrofit strategy is to permit the 24 steel bents to rock on their concrete pedestals. This rocking mechanism limits the seismic forces transmitted to the superstructure and allows a robust and cost effective retrofit. Non-linear time history analysis indicated member forces were larger than predicted by a static pushover analysis, primarily due to the horizontal/vertical coupling of the response mode at the time of impact. In this paper, key aspects of the retrofit design are discussed.

1 INTRODUCTION

The Lions Gate Bridge is one of Vancouver's signature structures bridging the first narrows at the entrance to the harbour. It provides a vital link to Vancouver's North Shore communities.

The bridge is composed of a 1550 ft (473m) cable suspended main span, two 614 ft (187m) suspended side spans, and a 2197 ft (670m) viaduct structure that approaches from the north. Banks (1942) detailed the initial design and construction of the bridge in a series of four papers.

The bridge, which was originally built to accommodate three narrow lanes of traffic, was opened in 1938 with just two lanes. Increased demands in the 1960s necessitated the expansion to three traffic lanes and the exclusion of trucks from the traffic stream. Currently the bridge carries 60,000 to 70,000 vehicles cross the bridge daily, far in excess of the originally conceived capacity of 45,000 vehicles per day.

The main suspended structure remains as originally designed, and has been in dire need of repair for at least 10 years. The North Approach Viaduct structure however underwent a complete deck replacement in 1975.

Now, after nearly 60 years of service, the main span structure is receiving a deck replacement and the North Approach Viaduct (NAV) is undergoing a seismic upgrade. The sidewalks on the NAV structure are also being widened to provide 2m of width

throughout the length of the bridge. The contract for the project has been awarded to American Bridge/Surespan Joint Venture. Klohn-Crippen Consultants Limited are provided consulting engineering services to the contractor for the seismic upgrade portion of the NAV. It is believed that this is the first major bridge structure to use the design-build contract format for a seismic retrofit.

2 VIADUCT STRUCTURE DESCRIPTION

The North Approach Viaduct, shown in Figure 1, is a steel structure composed of 25 composite plate girder spans. Span lengths vary from 81.5 ft (24.9m) at the north abutment end of the viaduct to 123 ft (37.5m) adjacent to the main span structure. The updated orthotropic plate girder deck is designed to act compositely with the original steel plate girders. Deck expansion joints are provided at each end of the viaduct and at four intermediate locations.

Longitudinal stability is derived by connecting adjacent bents at 4 locations with a longitudinal mid-height girder to form longitudinal portals. These 4-column bents are referred to as "H-Bents." Figure 2 shows a typical H-bent. At the north end of the viaduct, no H-Bent was provided. Rather, the shortest two bents in this segment (23 and 24) were designed to carry the longitudinal forces as fixed base cantilevers.

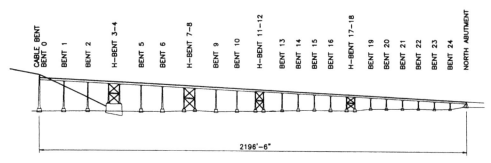

Figure 1. North approach viaduct.

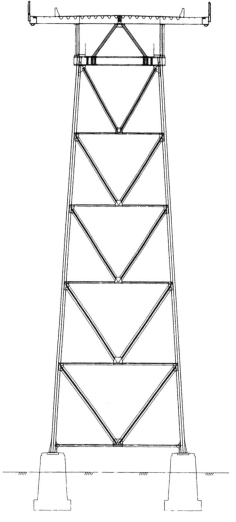

Figure 2. Typical bent.

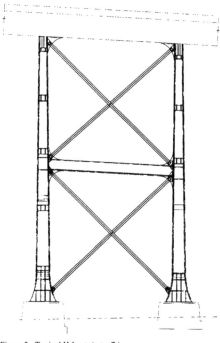

Figure 3. Typical H-bent. (retrofit)

Figure 3 shows a typical bent. Columns are built up of a 7/16-inch (11mm) stiffened web plate, with channel section flanges riveted to the web using pairs of angles as shown in Figure 4(a). The columns are braced by chevron braces built up of laced angles and struts. Struts are composed of two angles latticed in a single plane as shown in Figure 4(b). Diagonal braces are built up of 4 angles connected together by a single small angle bent in a 60-degree zigzag pattern. Figure 4(c) illustrates this detail. Welding was used for column and for plate girder web stiffeners and many connections particularly in the H-bent structures.

320

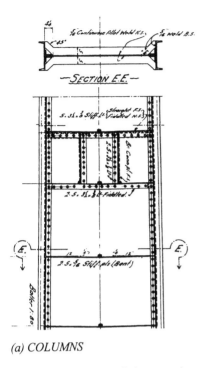

(a) COLUMNS

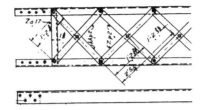

(b) STRUTS

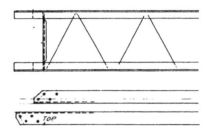

(c) DIAGONALS

Figure 4. Bent members typical cross-sections.

Throughout the viaduct the columns are rigidly connected at their bases. Individual spread footings support the loads of the superstructure.

3 FOUNDATIONS

The entire North Viaduct structure is founded on deep sediment deposits at the mouth of the Capilano River. The soils are characterised as coarse sands and gravels. Borehole data indicate the existence of pockets of sediments prone to liquefaction. A detailed liquefaction assessment and ground deformation analyses were carried out. Large settlements were predicted at seven bents. Smaller settlements were predicted at other bents. The locations where liquefaction can be expected include bents 8-10, 12-14 and 18. Limited horizontal ground movements were also predicted at these bents.

Bent columns are founded on individual spread footings. At piers where liquefaction is not predicted the footings are to be tied together with reinforced concrete tie beams to prevent differential rotations of footings.

Piles are to be added to the existing footings at the piers where large settlements were predicted. The piles are designed to be driven through the liquefiable layers and carry the full bent loads. Lique-faction will not be prevented but the settlements will be controlled. The effect of differential settlements and horizontal ground movements on the structure was evaluated by imposing settlements on the structural model to find the imposed loads on the steel structure.

4 SEISMICITY

The design earthquake for the bridge was specified as the 1/475-year event, having a 10% chance of ex-ceedence in 50 years. The response spectrum for the design earthquake is shown in Figure 5. Since steel structures typically exhibit low inherent damping, the 2% response spectrum was specified.

The Pseudo Displacement spectrum determined based on the given design spectrum exhibits the characteristic that the displacements increase and then level off following the "corner frequency". More commonly, the displacement spectrum increases indefinitely. The case offered with the Lions' Gate Bridge spectra is more realistic.

For more detailed time history analysis, 3 pairs of horizontal time histories were provided. Records were scaled to the target spectra and amplified through the soils to the surface. These histories were based on the following records described in Table 1.

Table 1. Earthquake record description.

Earthquake Record
1971 San Fernando Eq – CalTech record NS & EW
1989 Loma Prieta Eq – Capitola Fire Stn. NS & EW
1971 San Fernando Eq – Griffith Park NS & EW

Rather than utilise a horizontal record in the vertical direction, it was opted to use the associated vertical records uniformly scaled to the target peak ground acceleration. This was felt to be more realistic since vertical accelerograms tend to contain higher frequencies than do horizontal records.

Since the orientation of the faulting with respect to the bridge is unknown, orthogonal records were used in both longitudinal and transverse directions for a total of 6 time history analyses.

In Figure 5, time history acceleration and displacement spectra are superimposed over the design spectra and the pseudo displacement spectra. The acceleration spectra of the input records appear to be a good fit of the chosen design spectra. The displacement spectra appear to provide higher displacements than the specified design spectrum in the long period range. Since this inconsistency occurs at longer periods than those of interest in the NAV, it was not considered to be significant.

5 RETROFIT STRATEGY

The design life of the retrofit structure was prescribed to be 100 years. The retrofit was designed in

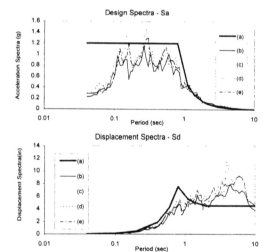

Figure 5. Design acceleration spectra and associated displacement spectra. *(a) Target Design Spectra; (b)&(c) San Fernando EQ – CalTech NS and EW respectively; (d)&(e) Loma Prieta EQ Capitola Fire Station NS and EW respectively*

accordance with the CAN/CSA S6-88 Design of Highway Bridges and the latest draft of the Canadian Highway Bridge Design Code (CHBDC) for the design earthquake. Following the design event limited service must be available immediately, and damage must be repairable. The bridge is required to re-open to traffic within a few days of the earthquake.

Early investigations with the 1/475-year earthquake response spectra revealed that the structure was prone to rocking. The spectral accelerations were found to be high enough such that the columns would pull the concrete foundations and a cone of soil surrounding them out of the ground. Such large forces, however, cannot be carried by unretrofit bents. Further, the magnitudes of such forces are uncertain, depending on the strength of the surrounding soils. Rather than rely on rocking at the base of the footing level, it was decided to allow uplift at the column base plates at the top of the footing as a means of limiting forces.

The aim of rocking in the design is to reduce the forces to a level where the bents can remain elastic. This is desirable since buckling of any of the bent members will lead to the formation of a mechanism and a subsequent collapse.

A simplistic explanation of rocking is that loads in the structure are limited to the forces required to uplift one column. Larger accelerations produce larger uplifts without an increase in the force. Seismic kinetic energy is transferred to potential energy through uplift of the centre of gravity of the structure

The advantages of using this strategy are:
- Forces are lower and more predictable as the bent is not required to carry the weight of the footing plus an unknown soil load.
- The area subjected to the uplift is easily inspectable and repairable in a post earthquake period.
- Work that is concentrated in the tower bases can be performed more economically than work in elevated portions of the structure.
- Rocking is a force limiting mechanism, therefore, force demands should not be substantially increased by larger seismic accelerations resulting from a larger magnitude earthquake.
- Foundation differential settlements can more easily be accommodated.
- Rocking at H-Bents 3 and 4, which are located on top of the massive north cable anchor block, can only occur at the base plate level.

Although rocking is an advantageous response to a seismic excitation, there were two areas of concern that needed to be addressed before the mechanism could be utilised, including:
- Dynamic impact of columns on the foundations
- Horizontal/Vertical interactions

ADINA

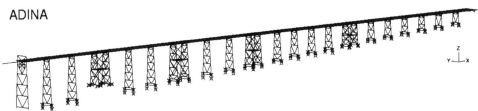

Figure 6. ADINA global model

Impact and modal coupling do increase loads, therefore member forces must be carefully evaluated using time history analysis.

6 ANALYSIS

6.1 Simplified Analysis

Static pushover analysis and linear response spectrum analysis were undertaken. These results were then compared with the results from an empirical energy balance and a substitute structure technique.

The energy balance technique according to Anderson (1993) is based on equating the maximum energy that the earthquake can transfer to the potential energy of the centre of gravity of the bent moving upwards as the bent rotates as a rigid body. The maximum kinetic energy is calculated using the maximum spectral velocity.

With the substitute structure technique, according to Priestly, Seible & Calvi (1996), the rocking deformation is approximated using an iterative technique. Rocking deflection is first estimated, then the compound rocking period and associated damping are derived using empirical formulae based on the assumption of completely inelastic collisions. Using the displacement response spectrum the deflection estimate is successively refined.

Static rocking loads and associated member forces were derived by calculating the overturning of the individual bents.

6.2 Non-linear Time History Analysis

The program ADINA 7.3.0 was used to perform the non-linear dynamic time-history analysis necessary to study the rocking behaviour of the structure. The global model used is illustrated in Figure 6. This model used beam-column elements to represent the bent members and the foundation elements.

The deck diaphragm was modelled using plate elements. The plate girders were modelled by beam elements and incorporated into the model deck level. The stiffness and material properties of both the girder elements and the deck plate were chosen to match the stiffness of the composite deck. The deck

spans were connected longitudinally at each bent using non-linear elastic truss elements having the force deformation curve of the proposed deck restrainers.

The interaction of the NAV with the main span suspended structure was taken into account by modelling a beam element with the same mass per unit length as the suspended side span with stiffness determined so as to match the frequencies of the fundamental side span vertical and horizontal modes.

The rocking behaviour was incorporated in the model by using contact surfaces at each of the column bases. The contact surface defines a constraint whereby the nodes of the contactor, in this case the base of the column cannot pass through the plane of the contact surface. The contact surface is connected to the foundation elements and is free to move as the foundation deflects. Although the columns were allowed to lift vertically, they were prevented from translating horizontally or rotating about a vertical axis.

The structural modifications necessary to ensure a controlled structural response were incorporated into the model. These included the H-bent cross bracing, girder end diaphragms and lead-core rubber bearings at the north abutment; lateral tie beams between the individual footing elements and longitudinal restrainers in the deck.

The lead-core bearings proposed for use at the north abutment were modelled using a rectangular beam element with a bilinear hysteretic material chosen to replicate the hysteresis of the lead core rubber bearing. ADINA enabled the incorporation of the non-linear material properties directly in the model.

Foundation elements with horizontal tie beams were included. Foundation springs and dampers were used to model the soil structure interaction. Longitudinal tie beams were not included in the model.

For each run the three orthogonal (two horizontal and one vertical) acceleration time histories were applied simultaneously. For convenience, relative displacement co-ordinates were chosen so that general mass proportional loading was applied rather than using displacement time histories at the more than one hundred foundation contact points.

6.3 Result correlation

The different analysis techniques used provided a reasonable correlation in describing the rocking behaviour. Maximum calculated baseplate uplifts at three locations are compared in Table 2. The uplifts compared favourably. Non-linear time history analysis at Bents 1 and 24 were found to be smaller than that indicated by the other two methods likely due to the lateral restraint of the Cable Bent and the North Abutment respectively.

The absolute maximum of each force in each member was extracted from the time history data. These forces were then used in the assessment of the existing and later the retrofit structural elements. The following, Table 3 compares for the ratio of the dynamic axial force to the static pushover force experienced by the members for the structure retrofit without member strengthening.

6.4 Impact Effects

The rocking mechanism causes the columns to uplift followed by an impact when the column drops back onto the footing. This results in two different effects that increase bent loads.

6.4.1 Pressure Wave
A pressure wave is propagated through the column when the base plate impacts the concrete footing. See Lindberg (1965) for a discussion of the theory.

Table 2. Maximum baseplate uplift (mm).

Bent Number	Non-Linear Time History Analysis	Energy Balance Calculations	Substitute Structure Methods
1	13*	31	48
5	39	25	44
24	24*	41	97

*Bents 1 and 24 are influenced by the cable bent and the abutment respectively

Table 3. Dynamic vs. static pushover force ratios.

Bents	Scale Factor	
	Axial Force	Strong Axis moment
1, 2	1.4	0.5
3, 4	1.5	0.6
5, 6	1.35	0.4
7, 8	1.4	0.4
9, 10	1.3	0.5
11, 12	1.4	0.5
13, 14, 15, 16	1.5	0.7
17, 18	1.4	0.75
19, 20, 21, 22	1.6	0.5
23, 24	1.2	0.5

Stresses are proportional to the column velocities at the time of impact. The maximum velocity observed in the model was approximately 0.5 m/s. Based on this observed velocity, the pressure wave impact related stresses are estimated as approximately 3.5Mpa.

6.4.2 Horizontal/Vertical Coupling
The impact of the column with the foundation provides a sudden vertical deceleration through the column and into the deck as the deck trajectory shifts from rotation about the rocking point to horizontal lateral shear deformation. The deceleration causes a rebound effect through the vertical vibration modes of the deck. The same is true in the opposite direction as the deck is forced to lift vertically at the point of rocking. The latter case (lift-off), although less obvious, appears to be more important as the vertical modes are excited at a time when many of the bracing members are experiencing their peak loads, whereas in the former case (impact) the loading in the members is dropping or reversing. In order to capture these effects the time step chosen for the non-linear time history analysis was small compared to the periods of vibration of the fundamental vertical deck modes.

6.5 Load Duration Effects

Axial forces can exceed the elastic buckling force providing that the time duration of application remains less than a fraction of the natural period of the member. The natural periods of the bent columns range from a approximately 1.3 to 3.3 Hz, and therefore the period below which significant axial force increases can be expected is .08 to .03 seconds. Since the time duration of the impact forces was observed to be in the order of 0.1 seconds, it was concluded that an increase in the load capacity due to short duration loads may not be reliable. Therefore increases in column capacity due to duration effects were not considered in this assessment.

6.6 H-Bents and Corner Rocking

The H-Bents with the addition of longitudinal X-bracing become very stiff space frames. When subjected to combined horizontal and vertical rocking, three legs uplift and the force that was carried in those members is then transferred to the remaining leg. Time history output clearly shows that under the given earthquakes corner rocking was induced in all four H-Bents.

During corner rocking, axial and bending forces induced in the columns are much higher than in the regular bents since a single leg must carry the dead weight plus dynamic load effects of four legs.

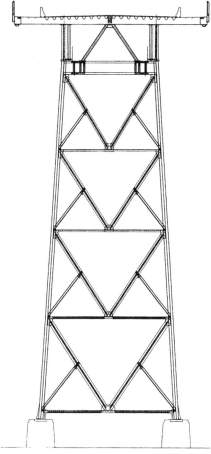

Figure 7. Diagonal brace retrofit.

In order to improve the response two options were investigated: crane buffers and yielding hysteretic elements. The aim of these devices is to provide some damping and improve the response characteristics so as to minimise the forces in the whole of the H-Bent structure.

A crane buffer is a hydraulic device that has the characteristic that it extends easily under its own internal spring but, on compression, the device reacts with a large damping force. The disadvantage of using a hydraulic device is that they are expensive and a monitoring program must be implemented to ensure that the devices are always operable.

A hysteretic element was developed utilising a triangular mild steel plate. Such an element provides coulomb damping in both directions of movement. Whereas a crane buffer does not apply downward forces on lifting legs, a hysteretic element will resist the leg that tries to uplift, potentially increasing the overturning moments and bent forces. A

steel hysteretic device has a definite cost advantage both in initial cost and in ongoing maintenance.

Both devices were investigated using a simplified model. The results indicated that relatively small damping forces were sufficient to improve the dynamic response characteristics of an H-Bent. The hysteretic dampers performance was equal to the more expensive crane buffers. It was concluded that a small 220kN per leg hysteretic damping force would be beneficial. This force was provided by a pair of triangular 40mm plates connected at the base of the H-Bent legs.

7 RETROFIT DESIGN

Several retrofit items were provided initially to effect the change in the boundary conditions and to modify the load path. These items are:

- Remove all nuts from the existing anchor bolts at the bases of the columns and provide lateral and longitudinal guides at the baseplate level to prevent translation or torsion in the bent legs.
- Add longitudinal bracing to the H-Bents to carry longitudinal loads.
- Tie the individual footing pedestals together to prevent differential movement that may interfere with the ability of the bent legs to lift freely
- Drive piles through the liquifiable soil layers and connect the pile cap to the existing footing to prevent large differential settlements
- Provide longitudinal and transverse restrainers at the deck level and new end diaphragms to circumvent the existing bearings
- Install new lead core rubber bearings at the north abutment to provide longitudinal restraint and damping to the spans from Bent 19 to the abutment.

Analysis of the structure was based on envelope forces extracted from the responses for the six time history combinations. Parameters that were not well understood such as the apparent foundation springs and the level of restraint in the existing deck were varied and the most critical combination was used in the assessment of the members.

7.1 Evaluation of the Existing Structure

Steel member capacities were determined based on the provisions in CAN/CSA S16.1.

Several areas were identified for which retrofit was necessary for structural integrity:

- H-Bent 3-4 struts and diagonals plus struts and diagonals in bents 19-24.
- Columns in all H-Bents plus ordinary Bents at expansion joints (4, 9, 13 and 19) plus Bents 14, 15, and Bents 20-24 (short bents).

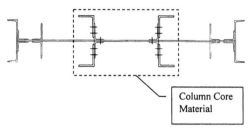

Column Core
Material

Figure 8. Column retrofit.

7.1.1 *Horizontal Struts*

Horizontal struts, where necessary were retrofit by bolting a small angle to the flange to prevent local buckling. The additional material was necessary to boost the elastic buckling load. Bolting was the preferred method of connection in order to avoid problems with the release of toxins when welding with lead paint coating the original members. Strut strengthening was required at H-Bent 3-4 and Bents 19-24.

7.1.2 *Diagonal Members*

Diagonal members generally were retrofit simply by providing support at the mid-point as shown in bold in Figure 7. Brace strengthening was required at H-Bent 3-4 and Bents 19-24.

7.1.3 *Columns*

The magnitude of the strong direction moments induced in the columns during longitudinal rocking created significant overloading of the columns. The moment demand, particularly in the short bents at times exceeded 65% of the bending strength.

The columns were strengthened by bolting angles to the web, essentially adding new internal flanges to the columns. It was found that material added in the centre region of the column, as shown in Figure 8, attracted substantially less bending moment.

In the most heavily overloaded columns the material added to the core of the section is proportioned such that the core is able to carry the entire load of the bent if the outer flanges buckle. If the bending capacity of the columns is exceeded during the earthquake, failure of the outer flange will result in a shedding of the load to the core material.

Adding angles to the web provides the additional benefit that the retrofit columns can be analysed as Class-2 sections, therefore the configuration makes better use of the existing material as well as providing additional strength. The existing lines of vertical stiffeners were utilised to facilitate the connection of the new material. Bolted connections were used to avoid problems associated with welding of members coated with lead paint.

8 SUMMARY AND CONCLUSIONS

Rocking was adopted as the retrofit strategy for the NAV to take advantage of the load limiting mechanism to provide a robust, cost-effective retrofit.

Non-linear analysis was undertaken to provide an understanding of the expected seismic response of the structure. The effect of impact and other structural interactions accounted for significant increases in static rocking loads.

H-bents, after retrofit with cross-bracing in the longitudinal direction were subjected to high loads during corner rocking. Response was improved, somewhat, by the addition of small hysteretic dampers at the bases of the H-Bents.

Retrofit of struts and diagonals was still necessary primarily at H-Bents 3-4 and in the shorter Bents 19-24.

Many of the columns were found to be deficient due to high moment demands induced by longitudinal rocking on the flared column bases. Column capacities were increased through the addition of interior flanges forming a strong core such that the outer flanges may fail, but the central core of column will still be able to carry the axial forces.

9 ACKNOWLEDGEMENT

The authors gratefully acknowledge the active participation of engineers Lev Bulkovshetyn, Yuming Ding, Dan Jennings, Leung Seto, Alex Sy and Li Yan, and drafters Ron Werner and Philomena Kwan without whose expertise the project could not be completed.

10 REFERENCES

Anderson, D.L. 1993. Buildings with Rocking Foundations, *Seismic Soil/Structure Interaction Seminar, Vancouver, B.C.*, May 29.

Banks, S.R. 1942. The Lions' Gate Bridge parts I, II, III and IV, *The Journal of the Engineering Institute of Canada*, April, May June and July.

Peter G Buckland 1981. The Lions' Gate Bridge – Renovation, *Can J Civ. Eng.* 8, 454-508

Lindberg, H.E., 1965. Impact Buckling of a Thin Bar, *Transactions of the ASME Journal of Applied Mechanics*, June.

Priestly, M.J.N., Seible F. & Calvi, G.M., 1996. *Seismic Design and Retrofit of Bridges*, John Wiley and Sons.

Behaviour of Steel Structures in Seismic Areas, Mazzolani & Tremblay (eds) © 2000 Balkema, Rotterdam, ISBN 90 5809 130 9

Seismic design and response of a 14-story concentrically braced steel building

L. Martinelli, F. Perotti & A. Bozzi
Dipartimento di Ingegneria Strutturale, Politecnico di Milano, Italy

In the paper some topics relevant to the design procedure and the seismic performance of steel concentrically braced frames (CBFs) are addressed. The design proposal of Eurocode 8 for CBFs is analysed in view of the application to fairly complex structural systems. Application of the "tension only" concept, to be used in strength verifications against ultimate-limit-state seismic effects, is applied employing a new strategy based upon the definition of a set of static equivalent seismic forces, computed from response spectrum analysis. Energy dissipation due to interaction between beams and diagonals in macro-bracing systems is also addressed.

The dynamic response of a complete 3-D structural system, encompassing several frames with different bracing solutions, is studied by means of a step-by-step numerical procedure implemented in an in house developed computer code. Non-linear response of bracing bars due to element buckling and inelastic material behaviour, as well as non-linear response of beams interacting with macro-bracing systems is considered.

Results, given either in terms of displacements and interstory drift envelopes or braces and beams hysteretic behavior, show higher modes and torsional effects importance. Effects of connection detailing, which may jeopardize bracing ductility is also considered through the introduction of a "death" option for bracing diagonals exceeding a predeterminate ductility in tension.

1 INTRODUCTION

The seismic behaviour and design of concentrically braced steel framed (CBF) structures has been the object of several research programs during the last decades; one of these started in the first 80's at DIS-Politecnico di Milano. At a first stage (Ballio and Perotti, 1987) the experimental behaviour of diagonal bracing elements was analysed and a numerical model of the brace was calibrated; subsequently, attention was devoted to the seismic behaviour and design (Perotti and Scarlassara, 1991) of one-story X-braced frames. The main object of this (numerical) study was the influence of the brace slenderness on the non-linear dynamic behaviour of the structure; a rule to account for this aspect in the design was proposed as an alternative to the Eurocode 8 provision, which states that the resistance of the compressed diagonal member must be neglected, regardless of its slenderness, in the ultimate limit states (ULS) checks.

In a later study (Perotti et al, 1994) the behaviour of different bracing solutions, for one-story systems as well, was considered; during this phase of the research the development of a new computer code

(NONDA) aimed at the simulation of the non-linear response of civil-engineering structural systems was started.

The availability of the NONDA code, encompassing at a subsequent stage non linear flexural elements as well, allowed for the study of the seismic behaviour of combined systems (Martinelli et al, 1996, 1998), trying to exploit the advantages offered by moment-resisting frames (MRFs) and CBFs; here too an alternative design strategy was proposed, within the EC8 context, with particular reference to the problem of serviceability limit states (SLS) for MRFs under moderate seismic actions.

More recently, attention has been focused on the problems related to the design and analysis of complete 3-D frames. In this light, the EC8 provisions for CBFs are analyzed, in the work here described, in view of their application to fairly complex structural systems; a test application is presented regarding a 14-story building structure encompassing several frames with different bracing solutions. The research addresses the following aspects.

– Assessment of the seismic behaviour of the original structure, which was designed against low seismic design forces.

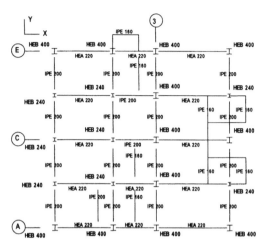

Figure 1. Location of bracing frames A, C, E and 3.

- Redesign of the structure according to the Euro-code 8 proposal and to higher seismic loads.
- Assessment of the performance of the new structure.

2 THE STRUCTURE AT STUDY

The analysed structure is a 14-story residential tower and can be deemed as typical for a high-rise building in the Italian practice. The building has a rectangular plan 12.68 x 13.02 m in X and Y direction respectively (see Figure 1). Interstory height is 3.2 m except for the basement and the first story which are 4.05 and 4.00 m respectively.

Floor construction consists of simply supported girders carrying a 13 cm concrete slab composite with profiled steel. The concrete slab is connected with floor girders using stud connectors.

Columns are obtained from hot-rolled profiles of compact section. Even though the vertical load carrying system can be regarded as a pinned frame, columns are continuous over more than one interstory; thus changes in column cross-section take place at 14.33 m, 23.93 m and 33.53 m from ground level (4.05 m above footing level).

Different bracing system types are used in the X and Y direction (see schemes at figures 4-6). In the X direction five frames act as vertical load carrying systems; three of them are used for horizontal loads as well (frames A, C and E). The latter are St. Andrew's cross CBFs, with diagonal braces spanning one interstory at basement and ground level and two interstories elsewhere. In the latter case (macro-bracing) the horizontal beams where diagonal bars intersect are flexurally loaded in the non-linear range; in fact, when buckling of compressed diagonals occurs, equilibrium of the midspan joint results in the transmission of a vertical force to the beam.

In the Y direction a single frame (#3) is designed to carry horizontal loads; in this frame usual X-bracing systems are present at outmost bays and are connected, at four different levels, by means of truss-type girders one interstory high ("belted-truss" system). Note that, due to non symmetric location of Frame #3, the structure plan possess a single symmetry axis.

The original structure was designed against a floor dead load of 4.5 kN/m^2 and a nominal live load of 2 kN/m^2. Base wind pressure q_{20}=0.8 kN/m^2 was considered in determining horizontal wind loading.

Allowable-stress verification of seismic resistance was performed against the following design spectrum representative of an area of intermediate seismic hazard (design PGA of about 0.25g):

$S(T) = 0.07$ g   for T<0.8 s
$S(T) = 0.0603/T^{2/3}$ g   for T>0.8 s

Reference codes were Italian regulations D.M. 14.02.1992, D.M. 12.02.1982, D.M. 24.01.96 and CNR-UNI 10011/88. As it can be easily verified, application of the above spectrum implicity results in the adoption of a behaviour factor of the order of six.

Column cross-section, from foundations to roof in the original structure, changes through HEB 400, HEB 300, HEA 280 and HEA 220 at the upper three stories. Floor girders are IPE 200 for Frame 3, acting in Y direction, and HEA 220 for the frames acting in X direction. Bracing diagonals are built-up with double-angle profiles. At the basement level section is 110 x 10 mm for all frames. At intestories two to six sections are 100 x 10 mm for frames A and E and 110 x 10 for frame C. Finally, for all frames section is 90 x 9 at intestories seven to ten and 80x8 above. In Frame #3 diagonals cross-section in outmost bays match those of Frame C.

3 DESIGN UPDATE ACCORDING TO EC8

The structure was subsequently redesigned in order to fulfil the Eurocode 8 provisions for a site of higher seismicity (design PGA = 0.35 g) and for a behaviour factor q=4. According to these provisions, seismic design of CBFs is mainly governed by some basic requirements, namely:

1 slenderness limits for bracing elements;
2 neglect of the compressed diagonal bars contribution in computing the design resistance;
3 application of capacity design rules for determining design actions for columns;
4 ductility and over-strength requirements for braces end-connections.

Here interest is mainly focused on structural analysis aspects of the design process: in this perspective, application of limits at point (1) does not pose any particular problem, while provisions at point (4) can be fulfilled in the detailing phase. Points (2) and (3), on the other hand, will be discussed in some detail.

3.1 Computation of design forces in brace elements

Based upon the results of the above quoted research activity, it can be said that requirement (2) can be objected when imposed to one-story CBFs; for multi-story systems, however, the conservatism implied by its application is, in the writers' opinion, adequate. Therefore, a consistent procedure for applying the requirement to the case of more complex structural schemes was devised; the procedure can be summarized as follows.

1 Performance of dynamic response spectrum analysis (DRSA) of the complete 3-D structural model encompassing all diagonal members; a reduced area of the braces is introduced in this phase, accounting for element buckling (see Martinelli et al, 1996).
2 Evaluation, for each CBF and for each significant normal mode, of a set of static equivalent forces; at each floor level the force is obtained as the difference between total shear forces in the upper and lower interstories.
3 Performance, for each CBF and mode, of static analysis under the equivalent forces computed in (2). At each interstory the compressed brace is removed, at this stage, and the axial force in the tensioned one is assumed as the modal contribution to the design seismic forces used in sizing each diagonal.
4 Combination of modal contributions via the SRSS rule.

It can be noted that, since compressed diagonals are removed from the static analysis, the procedure allows as well for a physically consistent estimation of seismic design forces in the girders interacting with the macro-bracing system.

3.2 Computation of columns design axial forces

To prevent columns failure, EC8 prescribes that the columns seismic axial force (point 3 above) should be multiplied by a capacity design coefficient (CDC) equal to the minimum value (among all braces pertaining to the same CBF of the column) of the ratio α of the design resistance to the design seismic force. The provision appears to be questionable, since yielding of a brace only limits the seismic forces at floor levels above the brace itself; in this light an "upward" interpretation of the rule was applied. To this end each CBF was loaded with a set of static forces having the following properties:
– vertical distribution as computed at section 3.1 for the lowest mode of interest for the frame;
– resultant equal to the total base shear as computed for the frame via the DRSA.

Under such force distribution the first brace, at level i, attaining its design resistance was detected; the α value computed, as prescribed by EC8, for this brace was taken as CDC for all columns above the i-th level. Subsequently, the static forces above this level were kept constant, while the ones below were increased until the design resistance in a new brace, at a lower level j, was reached. The α value computed for this new brace was assumed as CDC for all columns between the j-th and i-th level; the procedure went on until the brace resistance was reached at the first interstory. As a result of the redesign process the cross-sections of Figure 2 were obtained. Label HEM 400B denotes boxed cross-section built up from HEM 400 hot-rolled profiles and full penetration butt welded 10 or 15 mm plates.

4 SEISMIC BEHAVIOUR ASSESSMENT

4.1 Structural models and seismic excitation

The dynamic response of the complete 3-D structural systems, both the original and the re-designed one, were thoroughly studied by means of the NONDA numerical code, taking account of:
– non-linear response of bracing bars, beams and, when deemed necessary, columns;
– 2-D horizontal seismic excitation;
– torsional effects due to non-symmetric plan configuration of the building;

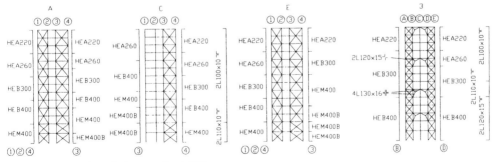

Figure 2. Structure after redesign according to EC8.

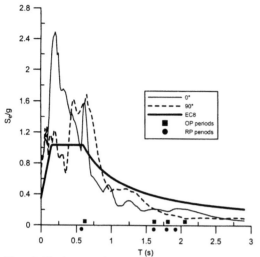

Figure 3. Elastic spectra for Corralitos 0 and 90 degree components. Periods of first to fourth normal modes of the original structure (OP) and redesigned structure (RP) are also shown.

– refined modelling of viscous damping forces, accounting for dissipation effects other than the hysteretic behaviour of non-linear elements.

Diagonal bars were modeled using the previously quoted non-linear brace element (Ballio Perotti, 1987) while floor girders and columns, when subjected to flexural damage due to seismic loads, were modeled by means of non linear one-component elements (Martinelli et al, 1996). The rigid-diaphgram hypothesis was adopted for floor systems.

A banded damping matrix corresponding to 3% of critical for significant elastic modes was introduced, according to the refined procedure described by Fogazzi & Perotti (2000).

Three structural cases were analyzed: the original structure was first considered as having truss-type columns (OP model). Columns structural continuity-from footing to roof was then considered (OC model); for the redesigned structure only pinned columns configuration was considered (RP model).

The Corralitos October 17, 1989 recordings of the Loma Prieta earthquake were chosen as seismic input for preliminary analysis. In Figure 3 the elastic pseudo-acceleration spectra for a damping factor of 3% of the East-West recording (0°) and North-South one (90°) are reported and compared against the EC8 elastic spectrum computed for the same damping factor, a type "B" soil and PGA = 0.35 g.

In the same figure natural periods of normal modes one to four are also shown for the original structure (OP) and the redesigned one (RP).

Corralitos earthquake horizontal component 0° and 90° were, respectively, first applied along X and Y direction; pertaining results will be referred to as 0-90 case. Directions of application of the two components were subsequently switched; these later results will be referred to as 90-0 case.

4.2 Results of the analyses

Figures 4 (OP model), and 5 (OC model) for the original structure and Figure 6 for the redesigned structure (RP model) give overall pictures of the damage patterns for the analysed systems. Frame drawings carry the indication of the collapsed braces and of the activation of flexural inelastic hinges. Graphs show extreme value of maximum (and minimum) frame lateral displacements at each story level and maximum absolute values of non dimensional interstory drifts.

The results of Figure 4 can be regarded as representative of the damage scenario for a structural system that was designed for seismic loads significantly lower than the ones introduced into the assessment analysis. If we look at the case of low-ductility braces, which is likely to occur for older designs, non-dimensional intestory drifts can be of the order of 10%, due to excessive damage in the diagonal braces. For high-ductility braces too, complete collapse of braces at one interstory can occur, with dramatic activation of soft-story behaviour. The most severe damage is found in the "belted" CBF at Frame #3, where high structural irregularity is present.

In Figure 5 the results obtained, for the original structure but taking account of columns flexural behaviour, apparently show an overall improvement of the system performance; intestory drift extremes, in fact are 20 to 50% lower, with one exception (excitation 0-90, μ=5) where concentration of inelastic damage at a single level led to significant increase in the intestory drift. On the other hand, examination of extremes of plastic-hinge (PH) rotations in the columns show unacceptable values, often exceeding 0.08 rad, clearly pointing out the insufficient design strength of the structural system against the upgraded seismic loading.

The redesigned structure (Figure 6) obviously shows a very different damage scenario, with large benefits especially gained from the bracing system at Frame #3, where deeper modifications were introduced.

When the high-ductility braces case is examined very limited and reasonably well-distributed interstory drifts are detected. For the low-ductility brace case, activation of very unfavourable torsional behaviour is detected in some cases due to diagonal failure. Even though this aspect merits further consideration, it can be observed that the ductility value μ=2 is clearly very pessimistic within the context of a new seismic design.

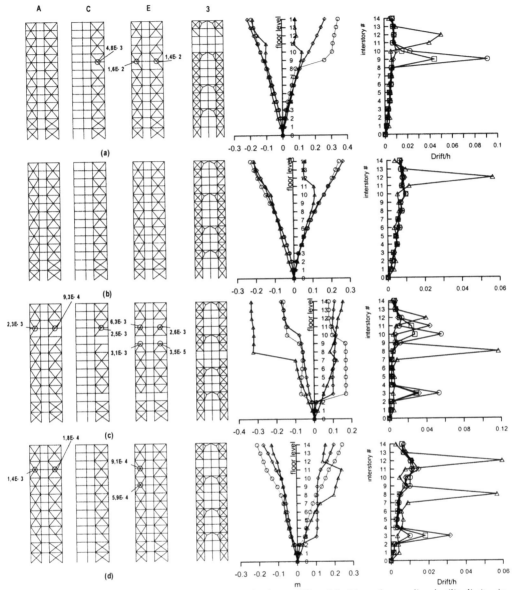

Figure 4. Damage patterns for original structure with pinned columns (OP model). Diagonals exceeding ductility limit μ in tension have been deleted and plastic rotation locations have been circled: a) 0-90 μ=2; b) 0-90 μ =5; c) 90-0 μ =2; d) 90-0 μ =5. Extreme values of lateral displacements and non dimensional interstory drifts are also shown for: $\Diamond$ Frame A, $\square$ Frame C, $\bigcirc$ Frame E, $\triangle$ Frame #3.

As an example of torsion effects, in Figure 7, Frame A and Frame E interstory drifts at the 5th interstory are reported for the RP model and μ=2. In fact, for this case most of the damage is located at this interstory. Since Frame A and E act parallel to each other, drift difference is a measure of torsional effects.

As it is possible to note, torsional effects start to become evident after time T=5 seconds, but it is only around T=8,30 seconds that this interstory braces reach their ductility limit. After braces failure, torsion remains evident showing notable effects even around T=19 seconds.

331

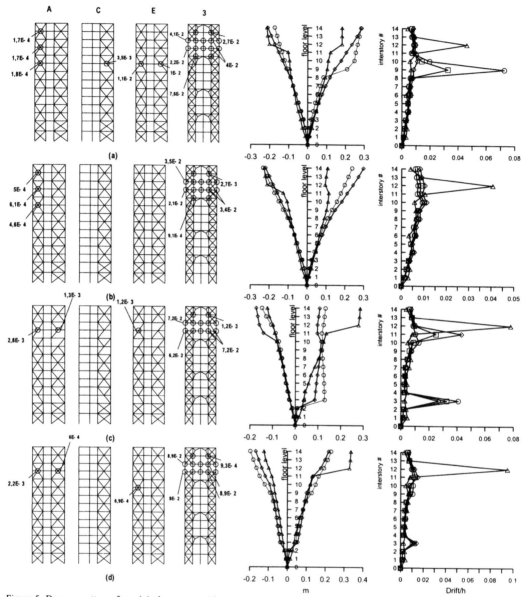

Figure 5. Damage patterns for original structure with continuous columns (OC model). Diagonals exceeding ductility limit μ in tension have been deleted and plastic rotation locations have been circled: a) 0-90 μ=2; b) 0-90 μ=5; c) 90-0 μ=2; d) 90-0 μ=5. Extreme values of lateral displacements and non dimensional interstory drifts are also shown for: $\Diamond$ Frame A, $\Box$ Frame C, $\bigcirc$ Frame E, $\triangle$ Frame #3.

Slenderness effect on diagonals response is shown in Figure 8. The graph on the left shows the response of one of the diagonals of the 9th interstory for Frame A and the OP model. On the right, the response, for the same structure and loading case, for the bracing at the 12th interstory in Frame #3 is given. Note that diagonal on the left has a slender-ness of 153.84 and the one on the right has slenderness of 74.084; this difference influences both the peak loading in compression and the cyclic inelastic behaviour. Both responses were obtained for the 0-90 loading case.

As an example of difference in inelastic flexural loading in beams and columns, response for an HEA 220 profile is reported in Figure 9.

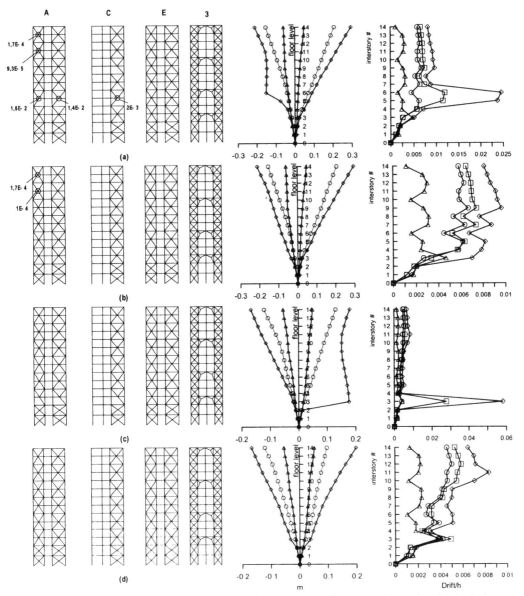

Figure 6. Damage patterns for redesigned structure with pinned columns (PC model). Diagonals exceeding ductility limit μ in tension have been deleted and plastic rotation locations have been circled: a) 0-90 μ=2; b) 0-90 μ 5; c) 90-0 μ=2; d) 90-0 μ =5. Extreme values of lateral displacements and non dimensional interstory drifts are also shown for: ◊ Frame A, □ Frame C, ○ Frame E, △ Frame #3.

On the left of Figure 9 the plastic cycles computed for the beam at the 11th floor of Frame A (lines 1-2) of the OC model are shown. On the right, the plastic rotation history at the bottom of the 11th interstory in column of line D of Frame #3 is shown for the same structural case. Both histories were computed for the 0-90 seismic input case. Differently than in columns, in beams plasticization occurs mostly for loading in one direction; plastic rotation extremes are much smaller and an important portion of the dissipated energy comes from low-amplitude inelastic cycles, which can take place after first yielding.

333

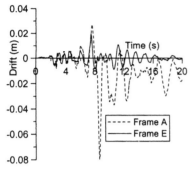

Figure 7. RP model, interstory drifts at fifth interstory for 0-90 loading case and μ=2.

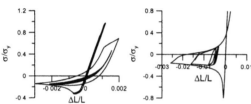

Figure 8. Left: OP model, 0-90 case, μ=2, slenderness=153.84. Right: slenderness=74.08.

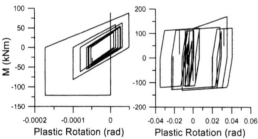

Figure 9. Left: plastic cycles in HEA 220 (beams). Right: plastic cycles in HEA 220 (columns).

configuration nor in terms of regularity in elevation. Subsequent work will address the behaviour under different accelerograms, both real and synthetic, the effect of other sources of eccentricity (uneven mass and/or resistance distributions) and the influence of second order effects.

REFERENCES

Ballio G., Perotti F. 1987. Cyclic behaviour of axially loaded members: numerical simulation and experimental verification. *J. of Construct. Steel Research*, 7: 3-41.

CEN, European Committee for Standardization, 1994. *Eurocode 3. Common unified rules for steel structures.*

CEN, Tech. Comm. 250/SC8, 1994-95. *Eurocode 8: Earthquake Resistant Design of Structures - Part 1: General Rules and Rules for Buildings*, (ENV 1998-1-1/2/3), Berlin: CEN

Fogazzi P., Perotti F. 2000 The dynamic response of seabed anchored floating tunnels under seismic excitation. Accepted for publication on *Earthquake Engineering and Structural Dynamics.*

Martinelli L., Mulas M.G., Perotti F. 1996. The seismic response of concentrically braced moment-resisting frames. *Earthquake Engineering and Structural Dynamics*, Vol. 25: 1275-1299.

Martinelli L., Mulas M.G., Perotti F. 1998. The seismic behaviour of steel moment-resisting frames with stiffening braces. *Engineering Structures*, Vol. 20, No. 12: 1045-1062.

Perotti F., Scarlassara G. 1991. Concentrically braced steel frames under seismic actions: non-linear behaviour and design coefficients. *Earthquake Engineering and Structural Dynamics*, Vol. 20: 409-427

Perotti F., De Amici A., Venturini P. 1994. Numerical analysis and design implications of the seismic behaviour of one-storey steel bracing systems. *Engineering Structures*, Vol. 18, No. 2: 162-178.

5 CONCLUSIONS

In the paper the design and seismic behaviour of fairly complex concentrically braced structural systems have been addressed.

The results given for a significant example, though preliminary and incomplete, seem to indicate that a satisfactory seismic behaviour can be achieved given that consistent and reasonably simple design rules are applied.

In this respect, a procedure for consistent applying the Eurocode 8 provisions is proposed.

It can be noted that, provided that braces possess adequate ductility in tension, the above considerations seem to hold true even for a system which does not meet strict regulations neither in terms of plan

Behaviour of Steel Structures in Seismic Areas, Mazzolani & Tremblay (eds) © 2000 Balkema, Rotterdam, ISBN 90 5809 130 9

Seismic behavior of steel lattice telecommunication towers

G. McClure & M. Lapointe
Department of Civil Engineering and Applied Mechanics, McGill University, Canada

M. A. Khedr
ROHN, Peoria, Ill., USA

ABSTRACT: Steel lattice self-supporting towers usually provide an economical alternative for telecommunication towers of heights up to 150 m located in areas where aesthetics are not a design constraint. Seismic analysis of such structures is not systematically done in practice, and satisfactory performance in past earthquakes has confirmed that it is perhaps not often necessary. In the context of the increasingly competitive market of tower design, simple formulas are proposed for the prediction of seismic response indicators such as maximum base shear, vertical dynamic reaction, and base overturning moment. These simplified predictors are applicable to towers of regular geometry and mass distribution, and for which serviceability limits can be exceeded during strong motion.

1 INTRODUCTION

It is generally recognized that in latticed telecommunication towers wind effects, and combinations of wind and ice effects, are more likely to govern the design than are earthquake effects. There have been very few documented reports of tower failures in the last 50 years, and none of which has been a direct threat to life safety. Not a single tower failure or serviceability defect has ever been reported in connection with earthquakes in Canada. However, tower damages observed after the Kobe earthquake of January 1995 (ASCE/TCLEE 1998) have indicated that broadcast towers and large building-supported microwave towers are the most vulnerable types. Microwave towers, in particular, must obey very stringent serviceability criteria, usually specified in terms of tilt and twist limits. Although the tower structure may appear sound after an earthquake, localized permanent deformations, especially in attachments of heavy antennas to mounts, may render it unserviceable. Apart from situations in which such damages are evident, serviceability-related defects largely go unreported by the expert reconnaissance teams that visit devastated areas after earthquakes.

There is an increased interest in North America in establishing earthquake-resistant design guidelines specifically for telecommunication towers. Seismic analysis of such structures is not systematically done in practice, and satisfactory performance in past earthquakes has demonstrated that it is not always necessary. In the context of the increasingly competitive market of tower design, simplified seismic response indicators are proposed for the prediction of maximum base shear, vertical dynamic reaction, and base overturning moment.

2 SEISMIC PERFORMANCE LEVELS

2.1 General

Earthquake-resistant design precautions vary depending on the seismicity of the tower location, including the potential amplification effects of special geotechnical conditions, and the tower performance level defined by the owner for the tower. It is assumed in the rest of the discussion that seismic hazard maps and corresponding earthquake data are available to tower designers, usually in a format similar to that used for building design, i.e. design spectra and/or horizontal peak ground velocities and accelerations. Also, in order to focus on the structural response itself, it is further assumed that the towers are on firm ground.

Although life safety is a first and foremost design concern, as in buildings, it is not the only performance objective appropriate to telecommunication towers. Depending on the economical value and the function of the structure, and the consequences of its failure or lack of serviceability, the owner should decide on the appropriate performance level among the following three: life safety, interrupted serviceability, and continuous serviceability.

2.2 Life safety

A telecommunication tower designed for life safety should not collapse in a failure mode that is a direct threat to life safety. This performance objective should apply without exception to all towers located in areas of human occupancy, with special attention paid to towers supported on building rooftops. A similar performance may be expected of towers whose failure would not pose a threat to life safety but would jeopardize the integrity of other nearby structures or equipment.

One could argue that linear elastic behavior is not required for life safety, as long as the failure modes are not dangerous: this is the rationale used for allowing inelastic response in earthquake-resistant design of buildings. However, probably because most telecommunication towers are governed by serviceability, current tower design codes are based on strickly elastic response. This conservative approach appears to have served the industry well since, as mentioned in the introduction, no loss of life has ever been directly attributed to a tower failure during an earthquake. This also means that there is some strength reserve in existing towers, which could be taken into consideration when evaluating their earthquake resistance.

2.3 Interrupted serviceability

A telecommunication tower designed for interrupted serviceability is not required to be fully serviceable during the strong motion, but it should not sustain any damage that would make it unserviceable immediately or shortly after the earthquake has occurred. The tower owner should establish the tolerable delay to resume full functionality, largely in view of economical considerations.

Most existing self-supporting steel towers are also likely to qualify for this performance objective, in view of the discussion in the previous section. As a matter of fact, deformed antenna mounts and adjacent members and connectors are easy to replace provided that some emergency stock is readily available.

2.4 Continuous serviceability

This performance level is the most stringent of all three and may be required only for exceptional towers. An example would be a strategic microwave tower used in the automatic control of an electric power grid. Full functionality must be ensured during the strong motion, which implies that serviceability criteria be met at all time.

3 PREDICTION OF SEISMIC RESPONSE

3.1 Methodology

The detailed methodology used in the derivation of the simplified indicators for base shear and vertical reaction described below, is presented in details in Khedr (1998) and summarized in Khedr & McClure (1999). Ten three-legged steel lattice towers (60° angle legs) with heights ranging from 30 m to 121 m were analyzed using modal superposition under the effect of 45 horizontal earthquake accelerograms. No further sophistication was required given the essentially linear elastic behavior of the towers. The number of modes considered varied with each tower so as to ensure that at least 90% of the total mass participates in the flexural modes. This requirement was somewhat relaxed in the study of vertical effects with a minimum participating mass of 85% of the tower mass. The viscous damping ratio was assumed to be 3% of the critical value in all modes, which is common practice for bolted steel lattice structures. Commercial software was used.

This study was later extended to include numerical simulations with the same data base of earthquakes on eight more towers with heights ranging from 30 m to 100 m, with solid round leg members. Results for the total set of 18 towers are used to derive the simplified equation for the base overturning moment presented in section 3.3.

3.2 Frequency coincidence

Basic structural dynamics principles (Chopra 1995) dictate that the seismic sensitivity of a telecommunication tower is influenced by the coincidence between its dominant natural frequencies and the frequency content of the strong motion. Past earthquake records have typical dominant frequencies in the range of 0.1 to 10 Hz, with a concentration in the 0.3 Hz to 3 Hz for horizontal motion, while vertical motion typically involves the higher frequency band.

The very first step in the assessment of the earthquake sensitivity of a tower is therefore the evaluation of its dominant natural frequencies. This can be done approximately using simplified or empirical expressions available in the literature, as summarized in the ASCE Guide (in preparation), or using formal eigenvalue analysis. The towers studied by Khedr (1998) had their fundamental flexural frequency between 0.8 and 4.4 Hz and their fundamental axial frequency between 8 and 30 Hz. The towers with round legs used in the extended study by the first two authors had their fundamental flexural frequency between 1.2 and 2.7 Hz and their fundamental axial frequency between 11 and 35 Hz.

It is clear from the above that all the towers studied would normally be subjected to dynamic amplification in the horizontal direction. On the other hand, only their fundamental axial mode is likely to be significantly excited by vertical ground motions.

3.3 Horizontal effects

The following simplified equation is suggested (Khedr & McClure 1999) as an upper bound for the maximum base shear, V_f:

$$V_f = MA_h(1.91 - 0.66T_f) \qquad (1)$$

where the base shear is expressed in N, M is the tower mass in kg, A_h is the horizontal peak ground acceleration in m/s^2, and T_f is the period of the fundamental sway mode of vibration of the tower in s.

Although base shear is the most common seismic response indicator used for buildings, this may not be the best one for self-supporting towers whose leg members develop axial forces to resist the overall overturning moment generated by the horizontal inertial forces. Using a similar methodology as for the derivation of the above formula, the following simplified equation is suggested for the maximum overturning moment, M_f:

$$M_f = MA_hL(1.16 - 0.93T_f) \qquad (2)$$

where the moment is expressed in N-m and L is the tower height in m.

It is suggested that the above two equations be used together as a rough check against the force response to other environmental design loads (wind, ice, and combinations). If seismic effects govern, most likely in active seismic zones, then a more detailed analysis is necessary to evaluate the individual member forces. Such an analysis may be static, using an equivalent horizontal force profile as suggested by Khedr (1998) and summarized in Khedr & McClure (2000), or dynamic. Whenever continuous serviceability is a requirement, however, dynamic analysis is necessary.

3.4 Vertical effects

Vertical effects have not been studied yet on the rounded leg towers, and the only simplified equation available to predict the maximum vertical dynamic reaction, P, at the tower base is from Khedr & McClure (1999):

$$P = MA_v(0.97 + 10.97T_a) \qquad (3)$$

where the axial reaction is in N, A_v is the peak vertical acceleration in m/s^2 specified for the site, and T_a

is the fundamental axial period of vibration of the tower in s. It is noted that self-weight must be combined with this vertical reaction.

3.5 Antennas and ancillary components

Antennas and tower ancillary components such as ladders, climbing platforms and transmission lines, must be properly attached to the primary structure. Their inertia effects can be taken into account in the total reacting mass of the structure, M, in the above equations. However, detailed simulations carried out by Khedr (1998) have shown that these effects are not significant unless the additional lumped masses are in the order of more than 5% of the total mass of the primary structure. Nonetheless, precautions are necessary for towers with heavy payloads and with performance objectives above the life safety level in active seismic zones.

4 CONCLUSIONS

This brief paper has presented three simple formulas to predict the overall seismic force response of self-supporting steel lattice towers, namely the tower base shear, overturning moment and vertical reaction. Such formulas are restricted to towers of relatively regular shapes and mass distributions, and should be used only as a rough design check against the maximum effects due to wind (and combined wind and ice if appropriate). More refined analysis is necessary whenever the results show that seismic effects may be more important than those due to the environmental design loads. Dynamic analysis is recommended for all towers that must remain fully functional during the design earthquake.

REFERENCES

ASCE Task committee on the dynamic response of lattice towers. Guide for the Dynamic Response of Lattice Towers. (in preparation.)

ASCE Technical council on lifeline earthquake engineering (TCLEE), 1998. Hyogoken-Nanbu (Kobe) Earthquake of January 17, 1995 Lifeline Performance. Ed. A.J. Schiff, ASCE/TCLEE Monograph No. 14, Reston, VA.

Chopra, A.K. 1995. Dynamics of structures Theory and Applications to Earthquake Engineering, Prentice Hall.

Khedr, M.A. 1998. Seismic analysis of lattice towers. Ph.D. thesis. Department of Civil Engineering and Applied Mechanics, McGill University, Canada.

Khedr, M.A. & G. McClure. 1999. Earthquake amplification factors for self-supporting telecommunication towers. Can. J. of Civ. Eng. 26(2):208-215.

Khedr, M.A. & G. McClure. 2000. A simplified method for seismic analysis of lattice telecommunication towers. Can. J. of Civ. Eng. In press.

Behaviour of Steel Structures in Seismic Areas, Mazzolani & Tremblay (eds) © 2000 Balkema, Rotterdam, ISBN 90 5809 130 9

Seismic upgrading of existing steel frames by a bracing system installed with fully mechanical interfaces

K.Ohi
Institute of Industrial Science, University of Tokyo, Japan

ABSTRACT: Effectiveness of steel bracing system installed with fully mechanical interfaces is experimentally demonstrated through a series of monotonic and cyclic loading tests. Also, sub-structuring pseudo-dynamic earthquake response tests are performed on a possible situation of a two-story braced frame upgraded by the proposed bracing system. An earthquake record and theoretical impulses are adopted as input excitations. The results show that an impulse excitation acts more stringently on the occurrence of brace breaking.

1 INTRODUCTION

In recent years after the Kobe earthquake, seismic diagnosis, retrofitting, and upgrading projects for existing buildings in public use have been widely carried out in Japan to prepare a destructive earthquake in the future. A steel gymnasium in school is one of such buildings and particularly important, because they are expected to serve as a refuge after a destructive disaster(Ohi, 1998). A main earthquake resisting element in the longitudinal direction of a gymnasium is a vertical bracing system, and its brace joints usually have too small strength to enable yielding in body portion of the bracing member. Then, a possible upgrading technique is to replace an old bracing member by a new one or to install an additional bracing system. This paper proposes an installation technique of such a bracing system with fully mechanical interfaces without use of site welding, and examines the performance of steel buildings upgraded by such a system experimentally through quasi-static and dynamic loading tests and sub-structuring hybrid simulations under seismic loading.

2 LOADING TESTS ON BRACING MEMBERS WITH A PROPOSED INTERFACE

A test setup commonly used in the loading tests here-in as well as pseudo-dynamic tests is shown in Figure 1. A tested frame is simply supported on a rigid base block. The detail of proposed interface is shown in Figure 2. The interface between a bracing member and an existing steel frame consists of a tee stub and a vertical stiffening plate, which are both directly bolted to the H-shaped column web. The tee stub is cut from a hot-rolled H-shaped cross-section, H-500 x 200 x 10 x 16 (depth, width, web thickness, and flange thickness) made of JIS SS400 grade steel. A brace member is a single angle cross-section, L-65 x 65 x 6(side lengths, thickness) made of JIS SS400 grade steel, and it is connected to the tee web through five high-strength bolts (JIS S10T, 20mm in diameter). This joint design just meets the condition specified in the AIJ recommendation (Architectural Institute of Japan, 1998). The mechanical properties derived from tension tests on coupons are summarized in Table 1.

Table 1 Mechanical properties of material used

Position	Yield stress (N/mm^2)	Tensile strength (N/mm^2)	Elongation (%)
Angle	323	464	24
Tee-web	322	460	27
Tee-flange	281	445	32
Column-web	319	451	28
Bolt	892	1060	14

3 QUASI-STATIC AND DYNAMIC LOADING TESTS

Figure 3 shows the results of loading tests on the braced frame specimen. Quasi-static and dynamic loading tests are performed under a monotonic

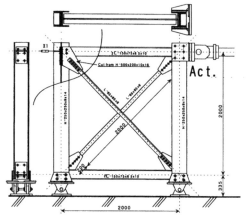

Act.

Figure 1 Test setup of loading tests

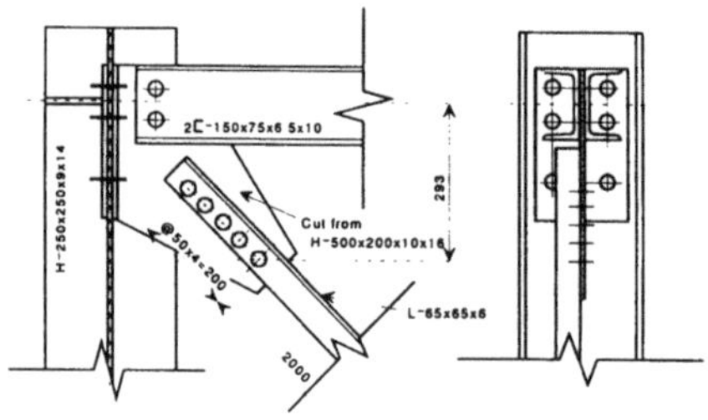

Figure 2 Details of interfaces proposed

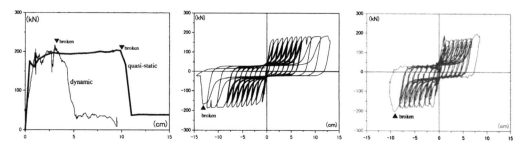

Figure 3 Loading test results (left: monotonic, middle: quasi-static and cyclic, right : dynamic and cyclic)

loading program as well as a cyclic loading program. All the braces were broken at the bolt-hole deducted portion of the angle member. Commonly in these loading tests, the column web plates are not damaged at all, and the proposed interface behave fairly well until the final failure.

In the dynamic tests, the velocity of loading apparatus is kept constant as fast as 50 cm/sec (0.25 radian/sec in the speed of story drift angle). The yield resistance of the bracing system under dynamic loading was found to be slightly greater than that observed under quasi-static loading, while the deformation capacity before breaking was considerably smaller. This is resulted from the increase of yield stress due to strain rate, which prevents yielding along the body-portion of angle member. Of course, the tensile strength at the bolt-hole deducted portion may increase due to strain rate, but it is not so great compared with the increase of the yield stress.

Under the cyclic reversals, the final deformation at breaking was slightly greater than the deformation under monotonic loading. Once a buckled angle member undergoes considerable flexural yielding in the middle portions, the plastic deformation does not recover to the straight line even if it is again pulled

back to the former unloaded point. This phenomenon generates a small delay of recovering to the plastic tensile resistance after it reaches to the former unloaded point. Summation of these small delays slightly postpones the final breaking at the bolt-hole deducted portion in total.

4 HYBRID RESPONSE SIMULATION COMBINED WITH LOADING TESTS

Hybrid responses of a 2-story fictitious frame were simulated by use of sub-structuring technique, where a computer-controlled loading test is performed only on an additional bracing portion at the second story, where a bracing system is installed by use of the interface proposed. The profiles of the remaining portions of the frame other than the specimen are assumed to stay in possible ranges that is experienced in ordinary school gymnasiums. The overall hybrid model is illustrated in Figure 4. This 2-story frame represents a structural system in the longitudinal direction of an ordinary school gymnasium in Japan. The first story is a reinforced concrete wall structure,

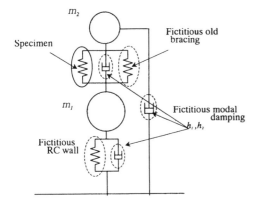

Figure 4 Hybrid model analyzed

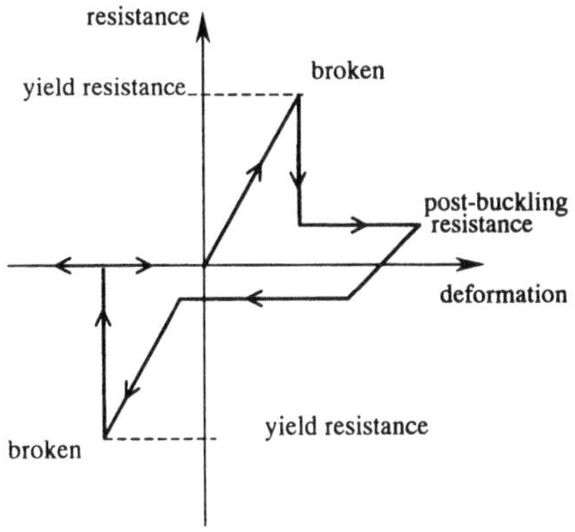

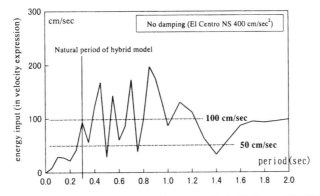

Figure 5 Hysteresis model of fictitious old bracing

Figure 6 Response spectrum of energy input (in velocity expression), El Centro NS 1940 400 cm/sec²

and the second story is steel bracing system. As for the first story, the amount of the reinforced concrete wall is enough and assumed to remain in elastic range. An existing bracing system at the second story has insufficient strength at its joint, and premature breaking will occur at the brace joint. To upgrade the seismic performance of the second story, an additional bracing system is installed with the interface proposed. Only a pair of additional new braces and its surrounding frame are loaded by computer-controlled actuator, and the restoring force measured from the test is reflected in the response analysis of the overall structural system after combined with the fictitious restoring forces of the remaining portions. The behavior of the fictitious old pair of braces is assumed to follow the hysteresis rule illustrated in Figure 5. At the moment that the resistance of the tension-side brace reaches to the yield strength, the brace is broken, and after braking the compression-side brace carries only small post-buckling resistance. The yield resistance and the elastic stiffness are arranged as the same values with those of speci-

men bracing system. The situation is that the existing story resistance and stiffness are doubled by adding a new ductile bracing system. The dynamic

Table 2 Dynamic properties of hybrid model

Mode	Natural period (sec)	Participation vector		Modal damping
1	0.3	(1.125	0.067)	0.02
2	0.1	(-0.125	0.933)	0.02

k_1/k_2	m_1/m_2	k_{ES} (kN/m)	Q_{YS} (kN)	Q_{Y2}/m_2G
15.65	2	19600	163	0.4

where k_1/k_2 :Story stiffness ratio, m_1/m_2 :Inertial mass ratio, k_{ES} :Specimen elastic stiffness, Q_{YS} :Specimen yield resistance, Q_{Y2}/m_2G : Yield shear coefficient of the second story

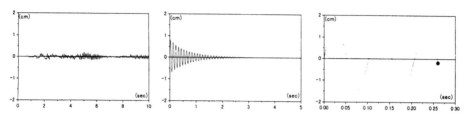

Figure 7 Time histories of story drift at the first story
(left: El Centro NS 1940 400 cm/sec², middle: impulse 50 cm/sec, right: impulse 100 cm/sec)

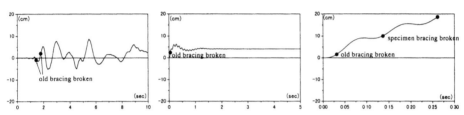

Figure 8 Time histories of story drift at the second story
(left: El Centro NS 1940 400 cm/sec², middle: impulse 50 cm/sec, right: impulse 100 cm/sec)

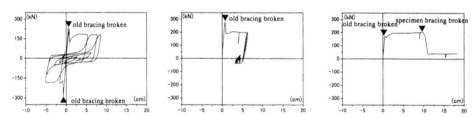

Figure 9 Hysteresis loops at the second story
(left: El Centro NS 1940 400 cm/sec², middle: impulse 50 cm/sec, right: impulse 100 cm/sec)

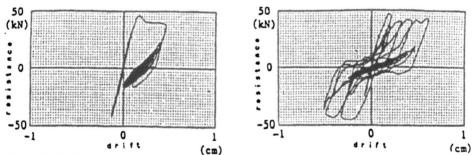

Figure 10 Real inelastic responses of a steel braced frame to two natural earthquakes
(left: near-field event, October1984, right: far-field event, June 1986)

342

Table 3 Input excitation used in the hybrid tests

Excitation	Magnitude
El Centro NS	400 cm/sec^2
Acceleration impulse I	50 cm/sec
Acceleration impulse II	100 cm/sec

properties of the overall system after upgrading is summarized in Table 2.

As shown in Table 3, two kinds of excitations are applied to the hybrid system: a theoretical impulse and an earthquake ground motion, which is the N-S component of El Centro 1940. The peak ground acceleration of El Centro 1940 is modified to 400 cm/sec^2 and the record of 10 sec in duration is used in the hybrid test. No scaling in time axis is considered. Figure 5 shows the response spectrum of energy input (in velocity expression) exerted by the modified earthquake record. Two theoretical acceleration impulses are also arranged to generate kinetic energy as much as the half of the average velocity level (50cm/sec) and almost the same average level (100 cm/sec) of the earthquake energy input. In the hybrid tests to theoretical impulses, free vibration is simulated from the condition that the initial velocity at each floor mass is set to the value, 50cm/sec or 100cm/sec.

Figures 7 and 8 show the time histories of the story drift at each story. Figure 9 shows the overall hysteresis loops at the second story. During these responses, the reinforced concrete wall at the first story remains elastic. Commonly in all the tests, fictitious old braces are first broken. In the response to El Centro 1940, both side of old braces are broken, while one of fictitious braces is broken to the impulse cases. Impulse responses are similar to monotonic loading process, but some higher (the second) mode vibration is mixed as observed in the impulse response to 50 cm/sec. Additional bracing system works well in the response to El Centro 1940, and the upgraded system did not collapse, while the system collapsed due to the breaking of the additional brace against the impulse input of 100 cm/sec. The reasons may be as follows: First, in the response to impulse, the energy is absorbed in a similar way to the monotonic loading, then the peak displacement becomes greater than the peak response to an earthquake when the amount of total energy to absorb is in the same level. Second, as experimentally shown in the previous section, apparent deformation capacity of bracing system under cyclic reversals is slightly greater than the deformation capacity under monotonic loading. Thus, impulse excitation acts more stringently on the occurrence of brace breaking.

It is difficult, however, to discuss how much impulsiveness shall be considered in the earthquake resistant design of steel frames. It depends on the characteristics of earthquake motions considered in the design. For instance, Ohi(1991) observed two moderate events that causes two different inelastic responses in a braced frame model for seismic monitoring, as shown in Figure 10. The peak ground acceleration is almost the same in the two events, but one epicenter is located near to the observation site, while the other is located offshore in the ocean and relatively far from the observation site. In the former event, the hysteresis behavior seems almost a monotonic loading test result. With the high possibility of such an earthquake as the former event, we could not rely on the advantage of cyclic reversals.

5 CONCLUDING REMARKS

(1) Fully mechanical interface for additional bracing system is possible and behaves well to seismic loading. Such a choice of details will avoid a premature failure in the vicinity of site welded portions.
(2) The loading test results and the hybrid simulation results show that the occurrence of the breaking at the bolt-hole deducted portion is affected by loading speed and by the impulsiveness of the excitations as well. Further studies are needed to clarify these effects.

REFERENCES

Architectural Institute of Japan: Recommendation of limit state design for steel structures, 1998

K.Ohi, K.Takanashi, Y.Homma: "Energy input rate spectra of earthquake motions," Journal of structural and constructional engineering, Transaction of AIJ, No.420, 1991.

K.Ohi and K.Takanashi: "Seismic diagnosis for rehabilitation and upgrading of steel gymnasiums," Engineering Structures, Vol.20, Nos.4-6, 1998.

Behaviour of Steel Structures in Seismic Areas, Mazzolani & Tremblay (eds) © 2000 Balkema, Rotterdam, ISBN 90 5809 130 9

Investigation and analysis of damage to expressway tollgate structures caused by the Hyogoken-Nanbu earthquake

Keiichiro Suita
Department of Architecture and Environmental Design, Kyoto University, Japan

Kozo Onikawa & Yukio Mitani
Hanshin Expressway Public Corporation, Japan

Shigeya Yamada & Masaki Arashiyama
Kume Sekkei, Japan

ABSTRACT:This paper presents the characteristics and the mechanism of the damage to the tollgate structures on the Hanshin Expressway caused by the Hyogoken-Nanbu Earthquake, through statistic and analytical study. The major findings obtained from the investigations and analysis are as follows, (1) the majority of damage to tollgate structure was observed in column footing system of base plate connection type, set up on the roadway bridges, especially at the anchor bolts, (2) the yielding and fracture of anchor bolts are caused by excessive tensile forces due to the rotation of column bases which is not considered in design practice, (3) and new design standards for horizontal seismic intensity and column base structures for tollgates on a roadway bridge were proposed.

1.INTRODUCTION

The Hanshin Expressway is an urban motorway that runs through the Osaka and Kobe areas. The 1995 Hyogoken-Nanbu Earthquake (Kobe Earthquake) caused major structural damage to the expressway (Watanabe et al. 1997). At the same time, heavy damage was sustained to tollgate structures set up on the roadway. A standard tollgate is a one-story steel-frame structure, with square tube columns and wide flange beams, of approximately 200 square meters in building area, and is located over the traffic lanes at the entrance to the expressway. It is important to ensure that damage to tollgates does not become an obstacle to the traffic on the expressway in case of a major earthquake because urban motorways carry out a significant role in the rescue and recovery operations after an earthquake. Because of the damage sustained to tollgates during the Kobe Earthquake, it is recognized that the design method should be revised.

2.STATISTICAL STUDY

2.1 Object of Investigation

The tollgate is composed of a booth for the billing personnel and a roof frame, which covers the booth. The typical figure of tollgate structure is shown in Figure 1.
There are a total of 122 tollgates on the Hanshin Expressway. Among these, detailed investigations have been performed at 112 locations.

2.2 Relation Between Damage to Tollgates and Peak Ground Acceleration

Three zones were selected where the peak ground acceleration was at approximately the same level, and the state of damage at each of these zones was investigated. The distribution of maximum acceleration is according to Obayashi Corp.,1995.
The zones were selected as follows.

- Zone I : over 400cm/s^2
- Zone II : 200 to 400cm/s^2
- Zone III : less than 200cm/s^2

The locations of these tollgates and the three zones mentioned above are shown in Figure 2.
The state of the damage to tollgates in each peak ground acceleration zone is shown in Figure 3.

The damage to tollgates is classified into "Minor" and "Severe" levels as shown below.

Minor: Destruction of column base mortar, damage to finishing material, residual story drift less than 1/100, etc.

Severe: Fracture or elongation of anchor bolt, welding crack in steel frame, residual story drift larger than 1/100, etc.

For zones I, II and III, the areas with larger peak ground acceleration sustained a higher incidence and rate of damage.

2.3 Relation between tollgate damage and the base structure

1)The effect of the base structure on the rate of damage to tollgates
The tollgates are divided into the following two types according to the base structure on which the tollgates are erected.

Figure 1. Typical tollgate structure.

• Tollgates on the ground
• Tollgates on bridges

Damages to tollgates on the ground are shown in Figure 4 and on bridges are shown in Figure 5, respectively.
The results show that the rate of damage to tollgates on bridges is higher than to tollgates on the ground.

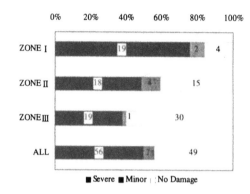

Figure 3. Zoning by peak ground acceleration and rate of damage.

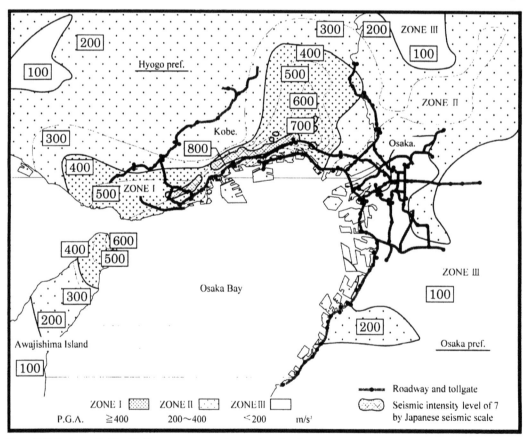

Figure 2. Tollgates on the Hanshin Expressway over peak ground acceleration contours.

2)The effect of pier structure
There are 4 types of pier structures that support roadway bridges, namely, reinforced concrete (RC) single-column pier, RC rigid frame pier, steel single-column pier, and steel rigid frame pier.

The relationship between pier structure of the bridges under the tollgates and the damage to tollgate was investigated, and no correlation was observed.

3)The effect of height of the bridge from the ground
The relationship between height of the bridge above ground and the damage to tollgates was investigated, and no correlation was observed.

2.4 Details of Damage to Tollgates

The characteristics of the damage sustained to tollgate frames by the earthquake can be shown as follows.
1)The damage locations
The damage locations of the tollgate frames can be roughly classified as follows.
• Column base
• Beam-to-column connection
• Other structures
• The finish, etc.

The ratios of the incidence of damage at these locations to the overall incidence of damage is investigated and the results are shown in Figure 6. When there are more than one damaged locations in a single frame, these are added up in the total figure, and therefore there will be overlaps in this figure. From the results, the column bases were the most outstandingly damaged location of all.

2)Type of column base
The type of column base connection is divided into the following two types.
• Standard base plate connection
• Concrete encased column base
(There is no embedded column base because the tollgate is placed directly on the bridge structure.)
From the results of investigation, all the damage occurred only at the standard base plate connection type, and there was no damage at all to concrete encased column bases.

3)Types of damage to standard base plate connection
The types of damage to standard column bases are as follows.
• Elongation of anchor bolt
• Fracture of anchor bolt
• Damage to mortar cover
(There were no significant cases of damage to base plates.)
The ratios of the incidence of each type of damage to the

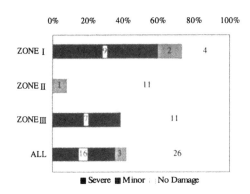

Figure 4. Damage ratio of tollgates on the ground.

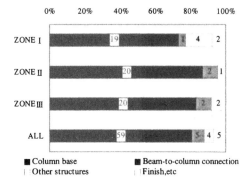

Figure 6. Damaged locations.

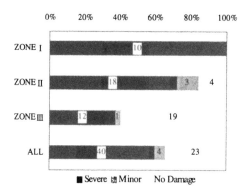

Figure 5. Damage ratio of tollgates on bridges.

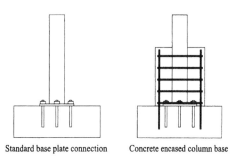

Figure 7. Type of column base connection.

347

total incidence of damage in standard base plate connection have been investigated as shown in Figure 8.

When there are more than one type of damage in a single column base these are added up in the total figure and therefore there will be overlaps in this figure. From these results it is apparent that elongation and fracture of anchor bolts were the most common types of damage.

3. ANALYTICAL STUDY

3.1 Subject

An analytical study was conducted using the non-linear earthquake response analysis method. A typical tollgate which sustained damage to the column base have been chosen as the subject of this study. It is Ikutagawa tollgate (Chuo Ward, Kobe), located in the area that registered a seismic intensity level of 7 by Japanese seismic scale.

The Ikutagawa tollgate is a single-story steel structure with a floor area of approximately 140 square meters, and a 1 span X 2 span frame layout, which has a size and a shape typically used on the Hanshin Expressway. The structure has square tube columns connected by wide flange beams, and all column bases are connected to the bridge using standard base plate connection. Among the 6 columns, 4 columns are fixed to the concrete deck on the bridge girder by anchor bolts. The remaining 2 columns are connected by high strength bolts to steel beams fixed to the bridge girder. The roadway bridge structure under the tollgate is a reinforced concrete deck on steel girders supported by independent columns made of reinforced concrete.

The following damage was sustained to the column bases: among the 12 anchor bolts, 5 bolts fractured at the screw, and the other 5 bolts were elongated by 3 - 14 mm. (The condition of the remaining two anchor bolts is not known as we were unable to examine them.) There was no damage to the column bases connected by high strength bolts. There was no deformation of the base plate. Further, the upper frame was also free from residual deformation. Though there was no damage to the columns or beams, some of the facility equipment on a roof were damaged.

3.2 Method

Since the incidence of damage observed at the column bases was higher, the study was performed according to the following three different analysis methods, focusing on the rotational rigidity of column base.

(1) Static analysis using a model in which the column base is pin supported and a design lateral seismic coefficient of 0.3 is used as in the original design

(2) Static analysis using a model in which the column base is partially fixed (spring supported) and the same lateral load as above is applied

(3) Dynamic analysis using a lumpmass type vibration model

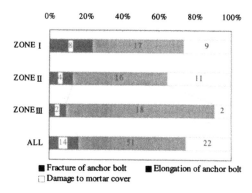

Figure 8. Types of damage to standard base plate connection.

Figure 9. Typical damage to base plate connection.

3.3 Results

1) Results from static analysis

The design seismic load in the horizontal direction was 45.9kN . The 'allowance' shown in the figure indicates the ratio (R/S) between the yield strength (R) of the members and actual stress (S).

The beams and columns with pin support at the column base, had 2 - 10 times more yield strength allowance against stress from the designed seismic load (this is because the members are designed according to the maximum story drift limitation). The allowance for the anchor bolt was more than 7 times as well. On the other hand, when the column base was regarded as being semi-rigid, the yield strength allowance was smaller 1.1 to 1.3, due to tensile forces acting on the anchor bolts. From this allowance, it is estimated that the anchor bolts reach the allowable yield strength at the shear coefficient of 0.3 to 0.4, and fracture at 0.7 to 0.9.

2) Results from dynamic analysis

The vibration model was a two-lump-mass model, where a tollgate and a bridge structure were modeled into a single mass respectively. The weight of the tollgate was 15.6 tons.

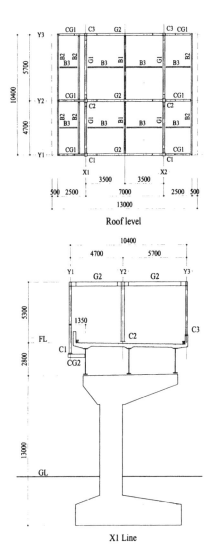

Figure 10. Ikutagawa tollgate.

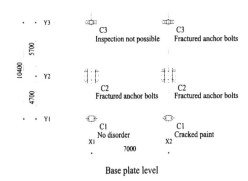

Figure 11. Damage of column base of Ikutagawa toll gate.

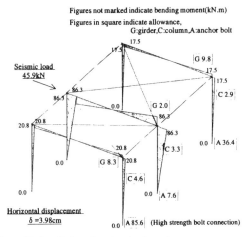

Figure 12. Results from the pinned column base.

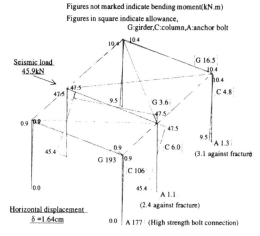

Figure 13. Results from the partially fixed column base.

The restoring force characteristic was regarded as a tri-linear model, determined by the yield strength of the anchor bolts and the ultimate strength of the steel frame. The damping factor was 2%.

The weight of the bridge structure was 573 tons. The restoring force characteristic was regarded as a tri-linear model, determined by the cracking strength and the ultimate strength of bridge pier. The damping factor was 5%. For the applied seismic record, a maximum acceleration of 400cm/s² was used as the standard value which was taken from the value observed in the vicinity.

From the result of dynamic analysis, the seismic load that acted on the tollgate in terms of shear coefficient, was 0.85 to 1.0, and compared to that of the static analysis, it was large enough to fracture the anchor bolts. This corresponds to the observation of the actual damage.

3.4 Damage Mechanism

The column bases of the tollgates have been designed as a pinned support, but if this is really the case, the anchor bolts should have a greater strength allowance than the beams and columns. However this does not correspond to the actual state of the damage. On the other hand, if the column base is regarded as being a semi-rigid support, bending moments will develop at the column base and excessive tensile forces will act on the anchor bolts. Consequently, the anchor bolts yield and fracture before the beams and columns yield. According to the dynamic analysis, it is estimated that the seismic load on the anchor bolts was sufficient to sustain damage, which roughly corresponds to the actual state of damage. From the above, it is believed that the reason for the more extensive damage to the column base (anchor bolts especially) can be referred to as the followings:

(1)   Additional tensile forces acting on the anchor bolts due to rotational rigidity of column base which were not anticipated in the design.

(2)   The tensile forces exceeded the strength of the anchor bolts.

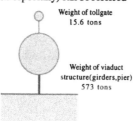

Weight of tollgate
15.6 tons

Weight of viaduct
structure(girders,pier)
573 tons

Figure 14. Vibration model.

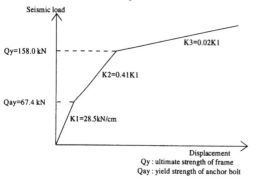

Figure 15. Restoring force characteristic of tollgate.

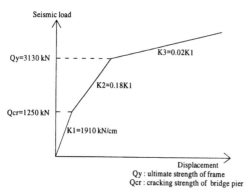

Figure 16. Restoring force characteristic of bridge pier.

4. SUMMARY OF DAMAGE TO COLUMN BASE

From the results of the statistical and analytical study, damage to column base can be summarized as follows.

(1)   Greater damage was sustained to the tollgates located in areas with greater peak ground acceleration. There is a clear correlation between the two.

(2)   The damage to tollgates was concentrated at the exposed column base and especially at the anchor bolts. This was a result of the difference in design assumptions and the reality, and greater forces acting on the anchor bolts. On the other hand, concrete encased column bases which had originally been designed to be fixed did not sustain any damage at all.

(3)   The rate of damage incidence was higher with tollgates on bridge than those on the ground. This is because the acceleration at bridge tollgates was amplified by the bridge structure.

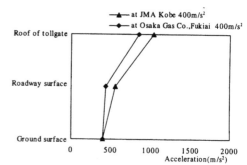

Figure 17. Maximum acceleration.

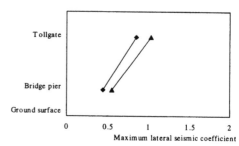

Figure 18. Maximum shear coefficient.

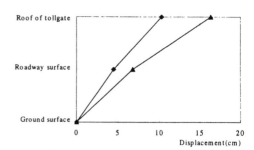

Figure 19. Maximum displacement.

(4)   The height, structural type and classification of the bridge structures were not found to have any correlation with the damage sustained to tollgates on bridges.

5.STORY SHEAR COEFFICIENT FOR DESIGN

The shear coefficients for seismic design was reviewed giving consideration to the fact that the incidence of damage to tollgates on bridges was higher than to those on the ground.

5.1 Method

Dynamic analysis was performed using a two-lump-mass shear type model, with the tollgate and the bridge structure each counted as a lump-mass point.

A standard type tollgate was chosen and then mass and stiffness were obtained from typical design. The bridge structure was modeled according to the structural type of the bridge pier and the natural period as the parameter. There are two structural types of bridge piers, reinforced concrete and steel structures, with five different natural periods, namely, 0.4, 0.6, 0.8, 1.0 and 1.2 seconds. The intensity of input acceleration were 100 cm/s^2 for medium earthquakes and 500 cm/s^2 for strong earthquakes.

5.2   Results

The results obtained from the analysis are shown in Figures 20 to 23. It can be observed that the longer the natural period of the pier the smaller the response of the tollgate. The reason for the large response shown in Figure 20, at 0.6 seconds, is due to the closeness of the natural period of the tollgate being 0.35 seconds, and the initial natural period of the bridge pier being 0.33 seconds. The maximum figures of shear coefficients were 1.72 for medium earthquakes and 2.5 for strong earthquakes. A residual deformation of several centimeters was produced by the strong earthquake.

5.3 The Proposed Shear Coefficient for Design

The seismic demand for a tollgate structure is to be able to use without any repairments after strong earthquakes. From the analysis, the structure will be roughly within the elastic region with a medium earthquake, if the design shear coefficient is 1.5. With a strong earthquake, the structure will not be in the ultimate state and even if it was, the deformation will not be residual, therefore satisfying the demand.

From the results of the analysis and by giving consideration to the architectural regulations and standards, the

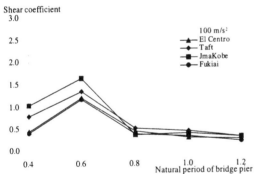

Figure 20. Analysis results of RC pier under medium earthquake conditions.

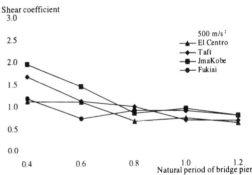

Figure 22.Analysis results of RC pier under strong earthquake conditions.

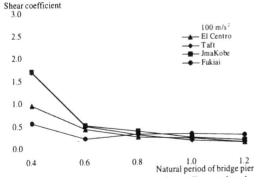

Figure 21.Analysis results of steel pier under medium earthquake conditions.

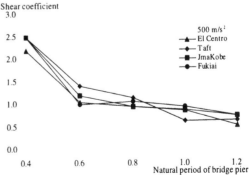

Figure 23. Analysis results of RC pier under strong earthquake conditions.

351

design shear coefficient of 1.5 shall be proposed for the allowable stress design of tollgates on a bridge. There was little damage to tollgates placed at ground level and therefore the design shear coefficient (for allowable stress design) in this case shall be 0.3 , which is as before.

6. PROPOSAL FOR COLUMN BASE DESIGN

The reason for the severe damage to the anchor bolts of standard column base connection comes from the fact that column bases have a certain level of rotational rigidity but the design have not taken into consideration. The following two improvement measures are important.
(1)   To have the column base perfectly fixed.
(2)   To be designed giving consideration to the rigidity together with the deformation capacity without fracturing under strong earthquakes.

Although bending moments have acted on a roadway bridge structure the rigid column base can be said to be the desirable method because of the clarity of stress transfer and the possibility of increasing rigidity without much change to the amount of steel presently used. The methods of making the column base rigid are as follows, either to encase the column base with reinforced concrete or to use prefabricated steel products. However, considering the possible deterioration of durability due to rain and exhaust gas, or nuts becoming loose because of heavy traffic, the concrete encased column base is desirable. For these reasons the concrete encased column base will be proposed as the type of column base to be used for tollgates on a bridge.

7. CONCLUSION

Through investigations of the damage to the tollgates on the Hanshin Expressway caused by the Kobe Earthquake, the characteristics as well as the mechanism of the damage were analyzed. From the results of the investigations and analysis, two major causes of the sustained damage have been identified. They are, (1) there was a gap between the design specifications and the actual requirements with respect to the estimation of rigidity for the column base connection and, (2) the actual seismic load applied to the tollgates by the earthquake was greater than the designed load level. The design base shear coefficient and the type of the column base of the tollgate have been reviewed for the measures to be taken in future.

ACKNOWLEDGEMENTS

The writers are grateful to Dr. Kiyoshi Kaneta, the chairman of the Committee on Seismic Design of Kobe Route 3 Tollgate Structure (Hanshin Expressway Public Corporation, and Hanshin Expressway Management Technology Center), the committee members, the secretary and all those involved for their valuable cooperation and guidance in writing this paper.

REFERENCES

AIJ(Architectural Institute of Japan),1990.*Recommendations for the design and fabrication of tubular structures in steel*, Tokyo(in Japanese)

Kaneta,K., et al.,1996. Kobe Route 3 Tollgate Structure Seismic Design Study Report, *Hanshin Expressway Public Corp., and Hanshin Expressway Management Technology Center* (in Japanese)

Obayashi Corp.,1995."Damage Investigation Report of the 1995 Hyogoken-Nanbu Earthquake" *Obayashi Corporation Technology Research Center* (in Japanese)

Watanabe,E.,Sugiura,K.,Nagata,K.and Kitazawa,M.,1997. Seismic Behavior of Highway Bridge Systems, *Behavior of Steel Structures in Seismic Areas (STESSA`97)*,885-896

Behaviour of Steel Structures in Seismic Areas, Mazzolani & Tremblay (eds) © 2000 Balkema, Rotterdam, ISBN 90 5809 130 9

Inelastic behaviour of a 52-storey steel frame building

Carlos E.Ventura & Yuming Ding
Department of Civil Engineering, University of British Columbia, Canada

ABSTRACT: This paper presents the results of a study on the seismic behaviour of a well-instrumented 52-storey steel frame building in downtown Los Angeles, California. This building has been subjected to ground motions from several earthquakes. The records obtained during the 1991 Sierra Madre earthquake and the 1994 Northridge earthquake were selected to calibrate a detailed 3-D dynamic computer model of the building. Non-linear dynamic computer analyses were then employed to investigate the possible inelastic responses of the structure during more severe ground shaking than that experienced by the building so far. A number of distinctive ground motions from recorded earthquakes were applied to the computer model and the responses were evaluated and compared. The results of the comparison between maximum inelastic responses in terms of story displacement, story drift ratio, story shear and overturning moments are presented here.

1 INTRODUCTION

Recent earthquakes like the 1994 California Northridge and the 1995 Kobe, Japan events have provided valuable lessons to the earthquake engineering profession, especially on the behaviour of steel structures. Many steel frame buildings with welded connections suffered brittle fractures during the Northridge earthquake, and several steel structures collapsed during the Kobe earthquake. Because of this, the evaluation of the inelastic behaviour of steel frame structures has become an important research topic in recent years.

The objective of the study leading to this paper was to use a recently developed nonlinear analytical model (Ding 1999), incorporated into a computer program, to investigate the possible 3-D non-linear response of a tall steel structure during severe earthquake shaking. The study was focussed on the non-linear global response and behaviour of the structure, rather than on the detailed performance of any of its elements. To this end, emphasis was placed on investigating the story displacement, story drift ratio, story shear and overturning moments of the building subjected to severe ground motions of different characteristics.

This paper presents a comparison of the results obtained from such analyses and helps elucidate the sensitivities on a nonlinear model to the characteristics of the input ground motions.

2 STUDY BUILDING

The study building, identified here as the FWT, is a 52-storey steel frame building located in downtown Los Angeles, California. This building was designed in 1988, constructed in 1988-90. The FWT consists of a 52-storey steel frame office tower and five levels of enlarged s underground parking. A picture of the building is shown in Fig. 1 and framing details are shown in Fig.2.

Due to architectural reasons, the floor plans of the tower are not perfectly square. On each floor, the tip of every corner is clipped and the middle third of each side is notched. Above the 36th story, in groups of about five stories, the corners of the floors are clipped further to provide a setback view to the exterior of the building.

The FWT was instrumented by the California Strong Motion Instrumentation Program (CSMIP) in 1990. Twenty accelerometers were installed on different levels of the building to measure its global translational, torsional, and vertical motions during significant shaking. To measure torsional motion of a floor, pairs of sensors were placed in the N-S

Figure 1 Overview of FWT building in Los Angeles

The interior core is concentrically braced. The sizes of mechanical ducts and door openings into the core dictated the configuration and sizes of braces (Fig. 2). The outrigger beams have to perform three functions (Banavalkar, 1991): 1) they have to support the design floor loads; 2) the outrigger beams along with core and perimeter columns have to act as a ductile moment resisting frame to carry a minimum of 25% design code level forces without the presence of interior core bracing; and 3) the stiffness of the beam should be such so as to create effective linkage between the interior core and the perimeter columns to provide effective overturning resistance to the seismic loads.

In high-rise buildings, it is always cost effective to minimize floor-to-floor height. The depth of outrigger beams is therefore dictated by the restriction on the ceiling cavity. To achieve these goals, the beams were offset into the floor and were notched at mid-span to allow for the passage of the mechanical ducts.

Typical floor plan dimensions are 47.45 m square. Typical story heights are 3.96 m. The type of building foundation is concrete spread footings (2.74 to 3.35 m thick) supporting the steel columns with 127 mm thick concrete slab on grade. The supporting soil is very stiff shale or sandstone and has an allowable bearing pressure of 718.2 kPa (15,000 lbs/sq ft). All structural steel framing including columns use ASTM A-572 (grade 50).

direction rather than the E-W direction because the floor eccentricity is larger in the E-W direction.

3 STRUCTURAL SYSTEM

The structural system of the FWT has three main components (see Fig. 2): a braced core, twelve columns (8 on the perimeter and 4 in the core), and eight 914mm (36 inch) deep outrigger beams at each floor connecting the inner and outer columns. The core, which is about 17.37m (57 feet) by 21.34 m (70 feet), is concentrically braced between the "A" level (the level just below the ground level) and the 50th story. Moment resisting connections are used at the intersection of beams and columns. The outrigger beams, about 12.19 m (40 feet) long, link the four core columns to the eight perimeter columns to form a ductile moment resisting frame. The outrigger beams are laterally braced to prevent lateral torsional buckling and are effectively connected to the floor diaphragm by shear studs to transmit the horizontal shear force to the frame.

4 CALIBRATION OF COMPUTER MODEL

In this study, computer program CANNY-E was used (Li 1996). CANNY-E is a general purpose computer program for 3-D non-linear static and dynamic analyses of reinforced concrete and steel building structures. The first step of the modelling process was to develop and calibrate a linear model of the building. Analytical studies were carried out to form a basis for the development of the non-linear model. The building was modelled as a combination of braced frames and moment frames consisting of 14 main column lines. The floor diaphragms were assumed to be rigid. Under this constraint, each diaphragm can be shown to consist of three in-plane degrees of freedom, one rotational and two translational. The building was assumed to be fixed at the ground level. All the beam and column centre lines were considered to coincide with column line co-ordinates.

The masses of the beam and column members associated with each floor were calculated from the

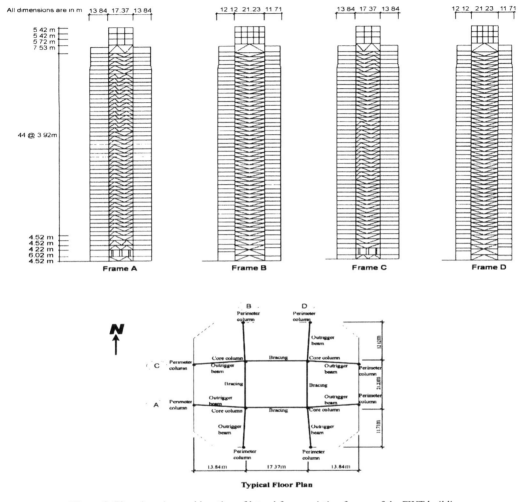

All dimensions are in m

13 84 17.37 13 84

5 42 m
5 42 m
5 72 m
7 53 m

44 @ 3 92m

4.52 m
4.52 m
4 22 m
6.02 m
4.52 m

Frame A

12 12 21.23 11 71

Frame B

13 84 17 37 13 84

Frame C

12 12 21 23 11 71

Frame D

N

B
Perimeter
column

D
Perimeter
column

Outrigger
beam

C Perimeter
column

Core column   Bracing   Core column

Outrigger
beam

Outrigger
beam

Perimeter
column

Bracing

Bracing

Perimeter
column

A Perimeter
column

Outrigger
beam

Outrigger
beam

Core column   Bracing   Core column

Outrigger
beam

Outrigger
beam

Perimeter
column

Perimeter
column

13.84 m   17 37 m   13.84 m

Typical Floor Plan

Figure 2 Elevation view and location of lateral force resisting frames of the FWT building

specified geometric dimensions of the elements and combined with the mass of the slab to determine the total story mass at corresponding floor. The estimated final mass and mass moment of inertia were lumped at the centre of mass at each floor.

A separate study (Ding 1999) showed that the computed responses fit well with the measured responses from two major earthquakes that the building experienced. The model was calibrated to optimize the match between computed and measured responses. The calibrated model was considered as a reliable representation of the structure system of the building during typical earthquake excitations. The second step in the modelling process was then to used the calibrated linear elastic model as the starting point for developing the non-linear model.

One of the most important tasks of non-linear analysis is the selection of the variables defining the inelastic properties of independent elements. In this study, all flexural members were modelled by a bilinear hysteresis model of the CANNY program. Yielding strengths were calculated based on the expected steel yielding strength, Fy = 380 Mpa, and the post-yielding stiffness factor was chosen as 2% (SAC 95-04). It was assumed that there would be 50% flexural strength remaining at a ductility level of 3. Shear stiffness of the beams and the axial stiffness of columns were assumed to be remain linear elastic.

355

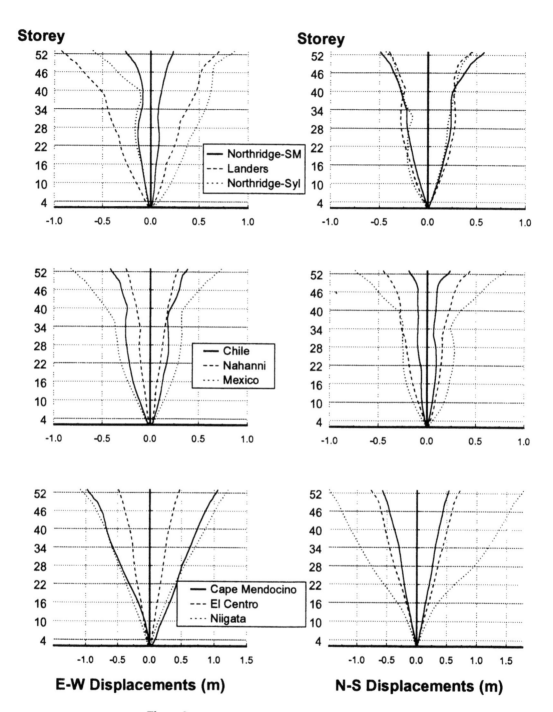

Figure 3 Comparison of lateral displacement envelopes

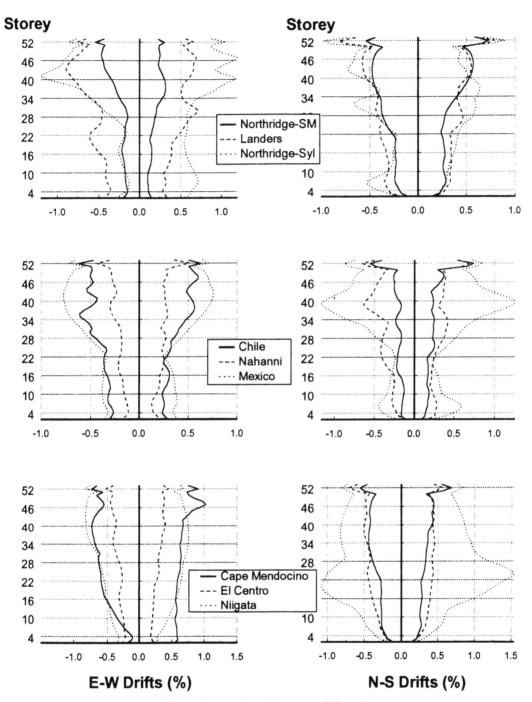

Figure 4 Comparison of interstorey drift envelopes

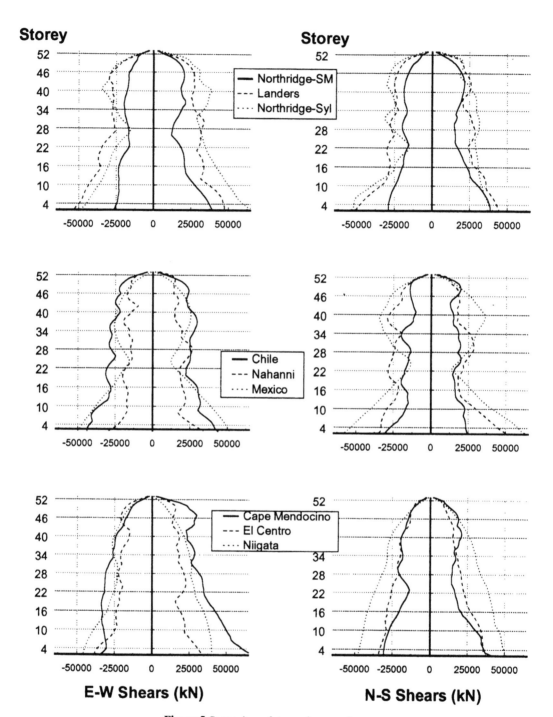

Figure 5 Comparison of storey shear envelopes

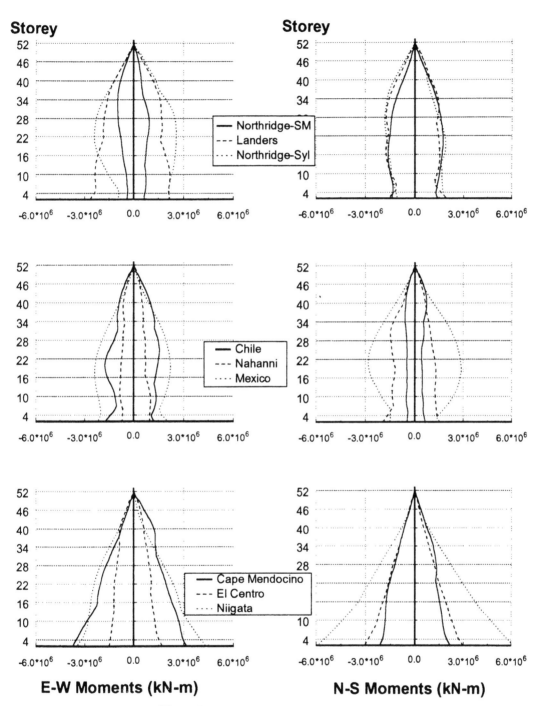

Figure 6 Comparison of overturning moment envelopes

5 TIME HISTORY ANALYSES

A set of 9 records from past earthquakes was used to conduct time history analyses. These records were divided into three groups, each with three records, based on their sites and earthquake characteristics. The first group includes the records from the 1992 California Landers earthquake measured at the Joshua Tree Fire station, the 1994 California Northridge earthquake records measured at the Santa Monica (SM) station and the Sylmar Hospital (Syl) free field station. The second group includes the 1985 Chile earthquake measured at the Llolleo station, the Canada Nahanni earthquake measured at Site 1, and the 1985 Mexico earthquake measured at the SCT station. The third group includes the 1992 California Cape Mendocino earthquake, the 1940 California El Centro/Imperial Valley earthquake, and the 1964 Japan Niigata earthquake measured at JMA station 1.

The results of the inelastic, time-history analyses are shown in Figs. 3, 4, 5 and 6 in terms of envelopes of lateral displacement, inter-story drift, story shear and overturning moment. The following observations are pertinent to the findings of this study.

Displacement envelopes of the third group of strong motion show the primary participation of the first modes in each principal of the building, while envelopes of the other two groups show the participation of higher modes. An evaluation of the corresponding response spectra of each of the ground motions of the third group also confirmed that great amount of energy is concentrated near the 6 seconds period, which is in the vicinity of the first modes of the structure in each principal direction for. The Niigata motion results in the highest demand of roof displacement at more than 1.5 metres.

The Mexico and Niigata ground motions result on the largest drift ratio demands to the building. However, the maximum drift ratios happen at different levels of the building. These large drift ratios indicate high ductility demands on structural members at these stories and the possible occurrence of structural damage.

In general, story shears are uniformly distributed between story 50 to story 20. This unique distribution of story shears is not predicted by building codes. The Cape Mendocino record results in modest roof displacements and drift ratios, but it provides the largest base shear demand. For this high base shear demand, it is worth to note the shear distribution at the upper stories (40 to 50) is of somewhat irregular shape.

The envelopes of the overturning moment show drastic differences for different ground motions. Group 3 responses are mainly controlled by the lowest modes of the structure. It is interesting to note the drum-shape distribution of moments for the Sylmar and Mexico records. The Sylmar record is a near-source, strong impulse ground motion, while Mexico record is for a station far away from the epicentre of the earthquake, but of very long duration and rich on long-period components.

6 CONCLUSIONS

A three-dimensional non-linear model of the FWT building was developed and analysed. Envelopes of building responses to various types of severe ground excitation were presented and discussed. Salient features of these responses were discussed. With the information generated from this type of study it is possible to gain a better understanding of the most probable sequence of damage to a high-rise building when subjected to severe ground shaking.

7 ACKNOWLEDGEMENTS

The financial support for this study was provided by a research grant from the Natural Sciences and Research Council of Canada awarded to the first author. The valuable information about the building provided by Dr. Banavalkar in 1993 is also acknowledged. CANNY-E program was kindly provided by its developer, Dr. Kang-Ning Li.

REFERENCES

Banavalkar, P.V. 1991. Spine structures provide stability in seismic areas. *Modern Steel Construction, January*, p.13.

Banavalkar, P.V. 1992. Personal communication with the first author. June.

Ding, Y. 1999. A study on the seismic behaviour of a 52-storey steel frame building. *M.A.Sc. thesis, Department of Civil Engineering, The University of British Columbia, Canada.* 126 pages.

Li, K.N. 1996. *CANNY-E users' manual.* Canny Consultants Pte Ltd. Singapore.

SAC 95-04 1995. *Analytical and field investigations of buildings affected by the Northridge earthquake of January 17, 1994.*

Base isolation and energy dissipation

Behaviour of Steel Structures in Seismic Areas, Mazzolani & Tremblay (eds) © 2000 Balkema, Rotterdam, ISBN 90 5809 130 9

A new ADAS system for seismic retrofitting of framed structures by means of the new hysteretical 'I' shaped device

Laura Anania, Antonia Badalà & Sebastiano Costa
Istituto di Scienza delle Costruzioni, University of Catania, Italy

ABSTRACT:

The seismic retrofitting of the structures can be carried out by strengthening some of the main structural elements so as to give a greater resistance to the earthquake thanks to the increasing of the structural stiffness as well as its energy dissipation capability. The ADAS systems are characterized by the employment of some non–structural elements with high dissipation capability in order to preserve the main structure. In the paper a hysteretical steel device, created by the same authors is used; it is capable of dissipating the energy transmitted by the seismic event, by means of repeated hysteresis cycles are used. This device was already the object of other publications but in this paper we study a different way of coupling between the frame and the proposed device. Namely, in the previous works the load was transferred from the frame to the device by means of tie rods acting only under traction. In this case the tie rods have been replaced by a rigid braces acting both in traction and compression. The system of load transferring was changed too, as well as the coupling frame-device which now is capable of deleting the vertical component of the load, transferred from the device to the mid-span of the frame, thus avoiding any damage along the beam frame.

1 INTRODUCTION

One of the faults of the most updated structural design, based on the ductility concept or, better, on the dissipation obtained by means of the plastic hinges development, consists on the damaging of both structural and non structural elements after a seismic event of high intensity. This damage derives from the great relative displacement, which occur during the event, between different floors which are necessary to activate the hinges formation in the plastic zones so very expensive retrofitting operations are required.

One of the most suitable solutions consists in the introduction in the braces system of some special devices whose efficiency of dissipating the energy transmitted by the earthquake, can limit the global response thus avoiding either the collapse or the mayor damaging of the main structure. That kind of operation uses ADAS (added damping and stiffness) elements, which have the advantage of introducing the mechanical device, accurately designed in order to dissipate the energy and capable of carrying great plastic deformations even after numerous disastrous earthquakes and, eventually, replaced in a very simple way. Commonly, an ADAS system can be described as a connection between two element which move in relative motion in respect to each other. The direct contact structural elements are very rigid and the device is inserted between them. In this way all the relative displacement is greatly limited and the improvement of

the characteristics of structural nodes ductility seems to be less important. In fact it is possible do work on the main structure in the elastic domain with a very low request of ductility at times. Besides the introduction of very rigid braces as well as the dissipating device give to the structure an increase of both the stiffness and the dissipation.

In order to have the efficiency of the ADAS system the device must start to work within the field of the elastic displacements of the frame structure. This condition especially in the case of the reinforced concrete structure gives some problem during the dimensioning phase of both the device and the system of load transferring from the main structure to the damper. The authors have realized a new hysteretical steel "I" shaped device, obtained from the cutting of a I steel series profile, and object of other works [Badala', Cuomo, Anania, Costa (1998); Anania (2000)]. The original device had been previously coupled with an r.c. frame by means of the "inverted V" braces. The braces represented the system used for transferring the load to the device [Anania, Badala' and Costa (1999)]. Although the system has worked well the application of the device by means of "inverted V braces" might produce some undesirable effects on the frame span because of the vertical component of the load transferred by the braces with subsequent increasing of the bending moment solicitation, on the beam of the frame, whose high values might cause the plastic hinge formation. Other problems are connected to the great geometrical deformation of the damper due to the variation of the angle of the load application. The new system concerns a new methodology of coupling the "I" hysteretical device to the frame by avoiding the employment of the braces.

2 THE NEW ADAS SYSTEM

As said before, in the previous system the connection between braces and device was performed by means of a hinge capable of transmitting both the horizontal an vertical component of the load from the main structure to the device. As shown in figure 1 this solution produces some undesirable effects in the mid span of the frame beam.

In order to avoid the transmission of the vertical component of the load to the beam, different kind of constrain between frame and braces-device must be used. This constrain must not transmit the shear force on the frame beam. That aim can be reached by means of either a horizontal pendulum or a roller bearing at horizontal axes Between the two solutions the most simple in terms of construction seems to be that of the roller bearing which also avoids of not changing the direction of the action swapped between frame beam and system brace-device even in the great displacements field. The latter advantage should disappear in the case of the use of a pendulum. From the experimental point of view, the roller bearing can be realized by an element capable of sliding along a guide. To this aim, the original prototype must be modified by assuming a different, although equivalent, statical scheme as reported in figure 2.

Furthermore, in order to limit the relative displacement between different storeys , a very rigid structure replaces the braces present in the previous analysis. In this way, the coupled system obtained can be represented as inn figure 3.

Finally, in the new coupled system the vertical restrain between frame span and device has been completely deleted and the braces, now, become very rigid.

3 DIMENSIONING OF THE HYSTERETICAL DEVICE

The design of the device is treated by referring to the elastic displacement of the tested r.c. frame equal to mm 1,3 for a corresponding load of 47 kN [Anania, Badala' and Costa (1999)]. In fact, the hysteretical device must start to work before the first plasticization hinge occur in any section of the frame so as to leave the frame always in elastic domain thus avoiding any damage.

During the dimensioning of the device, the frame displacement was assumed equal to mm 1, much lower than the true yielding one. To this aim a profile belonging to IEB 100 series was employed. A depth of mm100 was considered. The steel beam's stock is represented in figure 4.

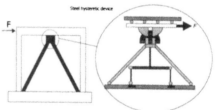

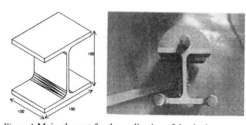

Figure 3. New system of coupling between frame and device

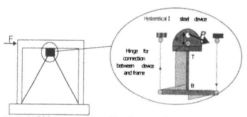

Figure 1. Old system of coupling between frame and device

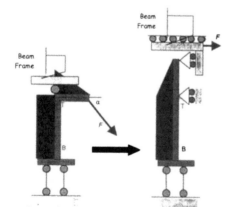

Figure 2. Comparison between old and new static scheme

Figure 4. Main element for the realization of the device

The characteristic parameters of the device are computed in the procedure described in Anania, Badala' and Costa (1999) . So, once given a strain ε_e equal to 2‰ the elastic rotation of the device ω_e, evaluated by neglecting the thickness of the profile's wings is equal to $\omega e = 0,0533$ [rad]. If the yielding device rotation should be reached at the elastic displacement of the frame of $\delta = 1$ mm, the lever to which the moment is connected, might be:

$$a \ sin(\omega_e) = \delta \qquad (1)$$

or more exactly:

$$a = \frac{\delta}{sin(\omega_e)} = \frac{1}{sin(0,0533)} = 18.77 \, mm \ (2)$$

In order to leave the frame in elastic domain, the real lever of the device was assumed equal to 12 mm.

In this way the strain level obtained by changing the imposed displacement is reported in table 1. It is possible to observe that the transition between the elastic and the plastic phase occurs in the gap defined by the δ frame displacements of 0,6 and 0,8 mm. Form this point the device start to dissipate the stored energy.

At a displacement equal to 8,49 mm the rotation in the device of $\pi/4$ is reached. This value represents the maximum limit of the corresponding deformation because we aim to investigate the efficiency of the device avoiding the damaging of the frame structure.

4 NUMERICAL INVESTIGATION OF THE COUPLED SYSTEM

The study of the coupled system is carried out by looking at two elastic- plastic springs set parallelly and connected rigidly at their extremities as shown in figure 5.

The system reported below, can be studied by increasing either the applied load or the displacement. That is preferable, because of its not linearity, to fix the displacement δ and evaluate the corresponding load capable of equilibrating the current configuration. Thus, in the analytical model, the congruence must be fixed among the displacements to find the forces acting on each spring and adding to each other. Namely, the force's component absorbed by the device is computed by the following procedure reported in [2] .

Table 1. Characteristic parameters of the device in function of the imposed displacement

δ [mm]	ω [rad]	χ [$\mu\mu^\wedge$-1]	ε [‰]	M [Nmm/10cm]	M/Mp
0,2	0,017	0,0002	0,63	77254	0,2083
0,4	0,033	0,0004	1,25	154529	0,4167
0,6	0,050	0,0006	1,88	231847	0,6253
0,8	0,067	0,0008	2,50	291813	0,7870
1,0	0,083	0,0010	3,13	320291	0,8638
2,0	0,167	0,0021	6,28	358261	0,9662
3,0	0,253	0,0032	9,48	365294	0,9851
4,0	0,339	0,0043	12,4	367756	0,9918
5,0	0,429	0,0053	16,1	368897	0,9949
6,0	0,524	0,0065	19,6	369518	0,9965
7,0	0,623	0,0078	23,3	369894	0,9976
8,0	0,663	0,0083	24,8	370000	0,9978

The force absorbed by the frame structure, can be, instead, computed in two different ways:

by using the data obtained from a experimental test in the case of absence of the device;

by carrying out a step by step analysis on the frame.

The following figure reports the curve Force- displacement obtained from the experimental investigation carried out on the 1:2 scaled frame model (studied in Anania, Badala' and Costa (1999)) and the curve force displacement obtained from the theoretical investigation carried out on the device. Finally the curve force displacement obtained by summing the previous contributes is also reported. It can be observed that the theoretical curve of the device presents a stiffness much greater than that of frame. So a great variation of the global stiffness of the system with respect to the original frame one is obtained. That is not a good result because, generally, is preferable, to keep the global stiffness in the ADAS systems, very close to the original structure, but, it must be underlined, how it is possible to obtain a great gain in terms of structural response. On the other hand we must note that the interview was carried out on a scaled model and that has produced a strong limitation in the design choices for the device.

5 THE EXPERIMENTAL PHASE

5.1 *The assembling procedure of the tested system*

The model frame studied is exactly the same object as the previous research reported in Anania, Badala' and Costa (1999) and represented in figure 7. It is a 1:2 size scaled r.c. frame where in this case a new type of braces was inserted. The junction points among the various elements such as: frame-braces and frame-device have been carried out by the employment of special chemical binding.

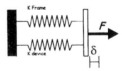

Figure 5. Schematization of the coupling of the system

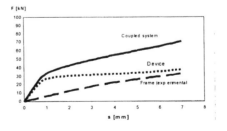

Figure 6. Load vs. displacement for the coupled system

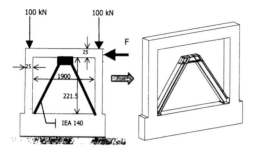

Figure 7. Frame model scaled 1:2 size

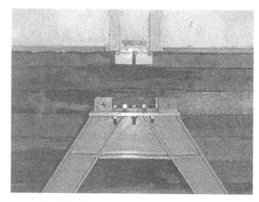

Figure 10. View of the fork for the connection of the device

Figure 8. Plate for the union of the steel brace

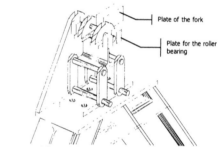

Figure 11. View of the connection braces-device

Figure 9. Global view of the frame with the braces

Figure 12. View of the device inserted in the frame mesh

The braces have been built in steel profile IEA 140 type located only after the binder plates had benn fixed (figures 8,9).

In order to realize the roller bearing a kind of "fork" was performed (figures 10,11). It was built by means of two plates placed along the shortest side and locked with a plate glued to the mid span of the frame. The transmission of the load occurs just in that element The distance between the plate of the fork is equal to the diameter of the hinge which represents the extremity of the lever of the force. The device was restrained to the upper extrem-

ity of the braces by means of a plate where the hinges of the double pendulum were located and to two lateral plates where the sliding guides for the roller bearing were placed. A global view is reported in figure 12.

5.2 Testing equipment for the measure of the parameters into play

The system was instrumented for the measuring of the displacements, deformations and strength acting on it. Namely, a couple of electronic potentiometers were placed for the displacements measure while the strain gauges were used to evaluate the strains into play. The strain gauges were located on each extremity of the frame member (columns and beam) apart from a couple of potentiometers located at the bottom of the column and capable of recording the deformations even after some fractures occur. Once the strains are known it is possible to determine the rotation and the axial deformation of the frame.

The vertical load was applied by following the same procedure adopted in Anania, Badala' and Costa (1999) and represented also in figure 11. Furthermore, a couple of potentiometers were placed along the diagonal of the frame in order to read both the horizontal displacement of the frame beam and the rigid rotations of the other frame elements directly connected to the elongation of these diagonals.

Finally the device was instrumented with strain gauges placed along the web profile and a potentiometer was put between the frame beam and the guide of the roller bearing so as to record the real relative movement.

The horizontal load was applied by means of a particular equipment, INSTRON, constituted by a digital controlled hydraulic actuator with ±250 kN capacity and a ±125 mm stroke.

Both the actuator and the loading jack have been erected on a very rigid testing frame (Figure 13).

The system has also been constrained out of plane by a double pendulum system located on the frame top .

All the instruments described above are connected to a dynamic data acquisition system.

The sampling frequency has been fixed equal to 140Hz while the excitation frequency of the test was equal to 0.2 Hz.

In addition to the electronic equipment, some mechanical displacement transducers were also located in order to better control some peculiar parameters during the test.

5.3 The test

A test on the coupled system was carried out by applying a series of sinusoidal loading cycles to the frame mesh at the frequency of 0.2 Hz and amplitude variable from to 0.2 mm up to 7 mm.

At this point, it is possible to observe that in the small amplitude cycles a good correspondence is recorded between the actuator displacements and the potentiometer placed for the reading of the relative displacement between frame beam and device while for the cycles of greater amplitude that correspondence disappears because of some inevitable problems in the connections. The applied load was increasing with a step of one tenth of millimeter up to a global imposed displacement of 3 mm; then, of two tenth of millimeter up to an imposed displacement of 7 mm. If we look at the diagrams for an imposed displacement equal to ±0.5 mm it can be seen that the loading and unloading curve coincide because of the re is not damaging.

The diagrams related to an imposed displacement of ± 2 mm show, instead, a little area under the curves which define a hysteretical cycle. The decreasing of the stiffness just near the origin is probably due to the faults in the

Figure 13.   Testing equipment of the frame

Figure 14.   Global view of the instrumented system

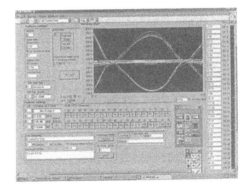

Figure 15.   Software for the data acquisition

screw jacks as well as in the hysteretical device. In the diagrams related to an imposed displacement of 4 and 6.5 mm the dissipation in much more evident and in addition, in the diagrams related to 6.5 mm, some very little damage appears on the frame, in fact a variation of the stiffness can be observed. Everywhere it is possible to note that the device dissipates a great amount of energy in plasticization domain, energy that otherwise should be absorbed by the frame with subsequent its damaging .

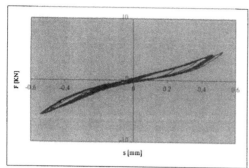

Figure 16. Load vs. displacement in the frame for a ± 0.5 mm of imposed displacement

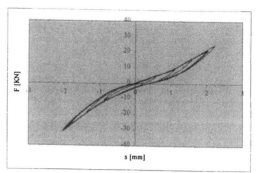

Figure 19. Load vs. displacement in the frame for a ± 2 mm of imposed displacement

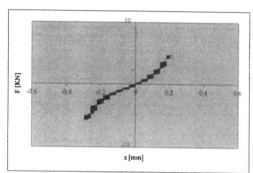

Figure 17. Relative displacement between frame – device

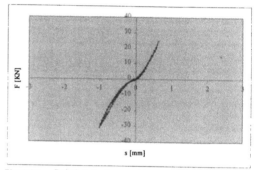

Figure 20. Relative displacement between frame - device

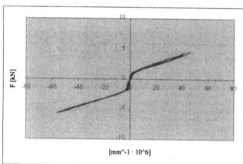

Figure 18. Load vs. average curvature in the device for a ± 0.5 mm imposed displacement on the frame

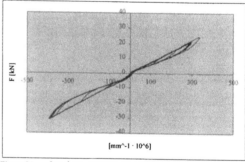

Figure 21. Load vs. average curvature in the device for a ± 2 mm imposed displacement on the frame

6 ANALYSIS OF THE RESULTS

The tests carried out on the A.D.A.S. system show both the values and the faults of the new system.

Compared to the common passive seismic protection systems the proposed system proves to be much simpler in terms of the realization of the hysteretical device.

The proposed system, does not require the use of tapered spindle profile for the fulfillment of the uniform plasticization condition, thanks to its bending moment whose constant distribution is guaranteed by the adopted statical scheme which is valid both in small and great displacement theory. It must be underlined that the realization of the hysteretical device requires the employment of skilled workers and equipment; otherwise we might get some undesirable plays among the assembling parts of the device that could lead to rigid displacements or rotations which sensibly decrease, the efficiency of the device. The deformation control of the hysteretical device by means of the right dimensioning of the lever permits us to couple very rigid systems such as reinforced concrete structures.

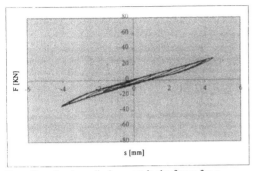

Figure 22. Load vs. displacement in the frame for a
± 4 mm of imposed displacement

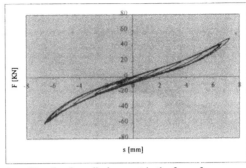

Figure 25. Load vs. displacement in the frame for
a ± 6.5 mm of imposed displacement

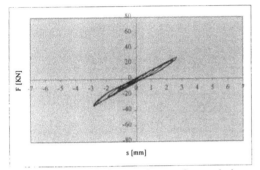

Figure 23. Relative displacement between frame – device
(± 4 mm)

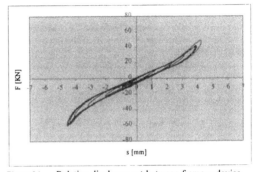

Figure 26. Relative displacement between frame – device
(± 6.5 mm)

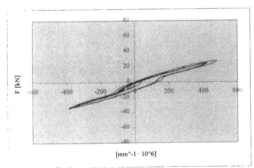

Figure 24. Load vs. average curvature in the device for a
± 4 mm imposed displacement on the frame

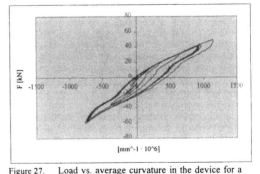

Figure 27. Load vs. average curvature in the device for a
± 6.5 mm imposed displacement on the frame

The assembling braces system as well as the position of the device give the possibility of employing the proposed device both in exiting and new buildings.

7 CONCLUSIONS

Compared to the previous tested system, many problems were overcome such as that of the damaging of the frame beam in the node to which the device was connected. In fact, the previous experience has shown, the possibility that a plasticization hinge occur on the mid span frame beam to which the device was connected. In previous case, in addition, the settling of the beam's center line due to the action of the tie rods leads to a loss of the efficiency of the hysteretical device in terms of its deformations.

The new system does not create any damage to the beam because there are no more extra movements . Later tests confirmed the validity of the new system because in

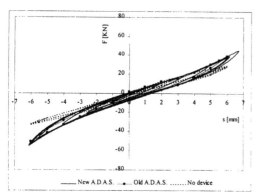

Figure 28.   Comparison between the old and the new ADAS system

effect, no damage was caused to the beam. The tests also showed other important aspects, necessary for the development of this system:

The notching made in the flanges to allow the rotation, produced a decrease of stiffness that created an unwanted deformation in the top plate. A greater attention in the notching plus the right connections will eliminate this problem

the settling due to the beam deformation limited the effects caused by the longitudinal bending moment of plasticization.

In figure 28 the new ADAS system is compared to the previous model; the difference between the two is hardly worth noting. This simply means that the imperfect functioning of the device had no effect on the final results, but the performance was not as good as expected, probably because of the problems mentioned before. The deformations and curvatures shown during the testing of the new system, can be compared to the previous model; besides, thanks to the protection system, no damage was caused to the frame beam by the new device. We must note that no damage occur to the device even after more than 500 loading cycles with 5.000 μm/m of strain. The testing stopped at this point because of the out- of- roundness (ovality) to the pendulum hinges.

As a conclusion after testing, we can see that it is important to improve the constraining system of the device, to avoid rigid rotation and displacements of the frame in respect to the dissipator. To obtain this result, we could consider taking some technical precautions such as the use of bronzing in steel elements in reciprocal rotation. This means the planning and the building of constraining groups that will be different form dissipating element that will be changed after serious damage to it caused.

Now, another interesting aspect to be studied is the possibility of the parallel coupling of various elements to obtain both a greater energy dissipation and a stiffness modulation of the passive group of protection depending on the number of parallel elements..

8 REFERENCES

Anania L, (2000)
   "Experimental analysis and analytical modelling of a innovative elastic-plastic dissipation device for seismic protection of the structures" Proceedings of 12WCEE - Auckland, New Zealand, 30 Jan.- 4 Feb. 2000

Anania L., Badala' A., Costa S.(1999)
   "Analytical and experimental analysis on a r.c. braced frame coupled to a new hysteretic dissipation device" Proceedings of ERES99 - Catania, 15-17 June 1999

Badala' A., Cuomo M., Anania L., Costa S., (1998)
   "A New elastic-palstic device for seismic protection of the structures: experimental analysis and analytical modelling of the hysteretic responce" Proceedings of Eleventh European Conference on Earthquake Engineering, Paris France, 6-11 Sett. 98

Behaviour of Steel Structures in Seismic Areas, Mazzolani & Tremblay (eds)© 2000 Balkema, Rotterdam, ISBN 90 5809 130 9

Active base isolation of m.d.o.f. structures by a time-convolution algorithm

A. Baratta & O. Corbi
Department of 'Scienza delle Costruzioni', Naples, Italy

ABSTRACT: In the paper an hybrid control system is designed; it consists of the coupling of a base isolation device and an active vibrator, able to bound the isolation absolute acceleration and, consequently, the inter-story drifts of the m.d.o.f. superstructure. The procedure implemented and numerically tested on a m.d.o.f. structural system has shown to be able to significantly mitigate the structural response with a minimum employ of energy.

1 INTRODUCTION

As well known the main purpose of adopting an isolation device for a structure built in a seismic area consists of increasing the structure main period and rendering the higher modes orthogonal to the ground motion. This ideal behaviour can be better pursued by combining the passive device with an active control system, able to further reduce the isolator absolute acceleration and, therefore, the super-structure drifts.

As a possible strategy one can adopt a procedure previously developed by the authors (A.Baratta & O.Corbi, 1999a,b,c) for the design of an active control algorithm acting in the frequency domain, and try to rearrange the whole for the particular case. The result is a pretty effective and low energy active base isolation device, working like a closed-loop active control system, based on the knowledge of the history of the isolator response variables up to the current instant and able to instantaneously counteract the action of the occurring seismic load, possibly elaborating the forcing spectrum on line.

2 THE M.D.O.F. BASE ISOLATED SYSTEM

2.1 The controlled equations of the motion for the m.d.o.f. base isolated system

Let's consider the dynamic equilibrium equations for a base isolated n-d.o.f. system, subject to an arbitrary forcing function; one can combine the passive isolation device with an active vibrator, contributing to the control action with a force $w_1(t)$, whose objective is to keep bounded the absolute acceleration of the isolator. The vectorial equation ruling the motion of the isolated frame is

$$\mathbf{M}\left[\ddot{\mathbf{u}}(t) + \ddot{\mathbf{u}}_g(t)\right] + \mathbf{C}\dot{\mathbf{u}}(t) + \mathbf{K}\mathbf{u}(t) + \mathbf{w}(t) = \mathbf{0} \qquad (1)$$

where $\mathbf{M}$ is the mass diagonal matrix, $\mathbf{C}$ and $\mathbf{K}$ the symmetric positive definite matrixes of damping and stiffness, $\mathbf{u}(t)$, with its time derivatives marked by the superimposed dots, respectively the vector of drift, velocity and acceleration, $\mathbf{w}(t)$ the control vector, whose only non zero element $w_1(t)$ concerns the (first) isolated floor. The vector of base acceleration, with the linked base velocity and displacement, is

$$\mathbf{u}_g(t) = u_g(t)\mathbf{1}; \quad \dot{\mathbf{u}}_g(t) = \dot{u}_g(t)\mathbf{1}; \quad \ddot{\mathbf{u}}_g(t) = \ddot{u}_g(t)\mathbf{1} \quad (2)$$

being $\mathbf{1}$ the unit $n+1$ vector. Equation (1) can be rewritten in terms of absolute displacements $\mathbf{y}(t) = \mathbf{u}(t) + \mathbf{u}_g(t)$

$$\mathbf{M}\left[\ddot{\mathbf{u}}(t) + \ddot{\mathbf{u}}_g(t)\right] + \mathbf{C}\left[\dot{\mathbf{u}}(t) + \dot{\mathbf{u}}_g(t)\right] + \mathbf{K}\left[\mathbf{u}(t) + \mathbf{u}_g(t)\right] + \\ + \mathbf{w}(t) = \mathbf{C}\dot{\mathbf{u}}_g(t) + \mathbf{K}\mathbf{u}_g(t) \qquad (3)$$

and therefore

$$\mathbf{M}\ddot{\mathbf{y}}(t) + \mathbf{C}\dot{\mathbf{y}}(t) + \mathbf{K}\mathbf{y}(t) + \mathbf{w}(t) = \mathbf{C}\dot{\mathbf{u}}_g(t) + \mathbf{K}\mathbf{u}_g(t) \quad (4)$$

2.2 The gain matrix of the m.d.o.f. system

Let's consider an harmonic forcing function $\mathbf{M}\ddot{u}_g(\omega,t) = \mathbf{M}\mathbf{1}\ddot{u}_{g_o}(\omega)e^{j\omega t}$ and a linear control law

$$w_1(\omega,t) = p(\omega)y_1(t) + q(\omega)\dot{y}_1(t) \qquad (5)$$

with $p(\omega)$ and $q(\omega)$ the control parameters; a particular integral $\mathbf{y}(\omega,t)$ of Equation (4), whatever the initial conditions are, should satisfy

$$\mathbf{M}\ddot{\mathbf{y}}(\omega,t) + \mathbf{C}\dot{\mathbf{y}}(\omega,t) + \mathbf{K}\mathbf{y}(\omega,t) + \mathbf{w}(\omega,t) = \mathbf{1}\ddot{u}_{g_o}(\omega)e^{j\omega t} \atop \text{with} \quad \mathbf{w}(\omega,t) = p(\omega)\mathbf{B}\mathbf{y}(\omega,t) + q(\omega)\mathbf{B}\dot{\mathbf{y}}(\omega,t) \qquad (6)$$

being $\mathbf{B}$ the $(n+1)$-square matrix, with $B_{11}=1$ the only non-zero element; as it can be expressed, with its time derivatives, in the form

$$\mathbf{y}(\omega,t) = \mathbf{y}_o(\omega|p,q)e^{j\omega t};$$
$$\dot{\mathbf{y}}(\omega,t) = j\omega\mathbf{y}_o(\omega|p,q)e^{j\omega t}; \ddot{\mathbf{y}}(\omega,t) = -\omega^2\mathbf{y}_o(\omega|p,q)e^{j\omega t} \qquad (7)$$

$\mathbf{y}_o(\omega|p,q)$ must satisfy

$$-\omega^2\mathbf{M}\mathbf{y}_o(\omega|p,q) + j\omega\mathbf{C}\mathbf{y}_o(\omega|p,q) + \mathbf{K}\mathbf{y}_o(\omega|p,q) + \\ + \mathbf{w}_o(\omega|p,q) = j\omega\mathbf{C}\mathbf{u}_{g_o}(\omega) + \mathbf{K}\mathbf{u}_{g_o}(\omega) \qquad (8)$$

with $\mathbf{w}_o(\omega|p,q) = p(\omega)\mathbf{B}\mathbf{u}_o(\omega|p,q) + j\omega q(\omega)\mathbf{B}\mathbf{u}_o(\omega|p,q)$,

that's to say

$$\left\{[\mathbf{K} + p(\omega)\mathbf{B} - \omega^2\mathbf{M}] + j\omega[\mathbf{C} + q(\omega)\mathbf{B}]\right\}\mathbf{y}_o(\omega|p,q) = -\left(\frac{\mathbf{K} + j\omega\mathbf{C}}{\omega^2}\right)\ddot{u}_{g_o}(\omega)(9)$$

whence

$$\mathbf{y}_o(\omega|p,q) = \bar{\mathbf{H}}_c(\omega|p,q)\mathbf{1}\ddot{u}_{g_o}(\omega);$$
$$\dot{\mathbf{y}}_o(\omega|p,q) = j\omega\mathbf{y}_o(\omega|p,q); \ddot{\mathbf{y}}_o(\omega|p,q) = -\omega^2\mathbf{y}_o(\omega|p,q) \qquad (10)$$

with
$$\bar{\mathbf{H}}_c(\omega|p,q) = -\frac{1}{\omega^2}\left\{[\mathbf{K} + p(\omega)\mathbf{B} - \omega^2\mathbf{M}] + \\ + j\omega[\mathbf{C} + q(\omega)\mathbf{B}]\right\}^{-1}(j\omega\mathbf{C} + \mathbf{K}) \qquad (11)$$

being $\bar{\mathbf{H}}_c(\omega|p,q)$ the gain matrix between the controlled structure absolute displacement vector and the base acceleration vector, and

$$\bar{y}_{o1}(\omega|p,q) = \bar{H}_{1c}(\omega|p,q)\ddot{u}_{g_o}(\omega);$$
$$\dot{y}_{o1}(\omega|p,q) = j\omega y_{o1}(\omega|p,q); \ddot{y}_{o1}(\omega|p,q) = -\omega^2 y_{o1}(\omega|p,q) \qquad (12)$$

with $\bar{H}_{1c}(\omega|p,q) = \sum_{j=1...5}\bar{H}_{1jc}(\omega|p,q) \qquad (13)$

the gain function between the isolator absolute displacement and the base acceleration.

It is clear that the structural absolute displacements/base displacement gain matrix $\mathbf{H}_c(\omega|p,q)$ and the structural absolute accelerations/base acceleration gain matrix coincide; actually

$$\mathbf{y}_o(\omega|p,q) = \bar{\mathbf{H}}_c(\omega|p,q)\mathbf{1}\ddot{u}_{g_o}(\omega) = -\omega^2\bar{\mathbf{H}}_c(\omega|p,q)\mathbf{1}u_{g_o}(\omega)$$
$$\rightarrow \mathbf{y}_o(\omega|p,q) = \mathbf{H}_c(\omega|p,q)\mathbf{1}u_{g_o}(\omega);$$
$$\ddot{\mathbf{y}}_o(\omega|p,q) = -\omega^2\mathbf{y}_o(\omega|p,q) = -\omega^2\bar{\mathbf{H}}_c(\omega|p,q)\mathbf{1}\ddot{u}_{g_o}(\omega) \qquad (14)$$
$$\rightarrow \ddot{\mathbf{y}}_o(\omega|p,q) = \mathbf{H}_c(\omega|p,q)\mathbf{1}\ddot{u}_{g_o}(\omega)$$

with
$$\mathbf{H}_c(\omega|p,q) = -\omega^2 \bar{\mathbf{H}}_c(\omega|p,q) = \\ = \left\{[\mathbf{K} + p(\omega)\mathbf{B} - \omega^2\mathbf{M}] + j\omega[\mathbf{C} + q(\omega)\mathbf{B}]\right\}^{-1}(j\omega\mathbf{C} + \mathbf{K}) \qquad (15)$$

and for the isolator

$$y_{o1}(\omega|p,q) = \bar{H}_{1c}(\omega|p,q)\ddot{u}_{g_o}(\omega) = -\omega^2\bar{H}_{1c}(\omega|p,q)u_{g_o}(\omega)$$
$$\rightarrow \frac{y_{o1}(\omega|p,q)}{u_{g_o}(\omega)} = H_{1c}(\omega|p,q);$$
$$\ddot{y}_{o1}(\omega|p,q) = -\omega^2 y_{o1}(\omega|p,q) = -\omega^2\bar{H}_{1c}(\omega|p,q)\ddot{u}_{g_o}(\omega) \qquad (16)$$
$$\rightarrow \frac{\ddot{y}_{o1}(\omega|p,q)}{\ddot{u}_{g_o}(\omega)} = H_{1c}(\omega|p,q);$$

with $H_{1c}(\omega|p,q) = -\omega^2\bar{H}_{1c}(\omega|p,q) = \sum_{j=1...5}\mathbf{H}_{1jc}(\omega|p,q) \quad (17)$

3 REDUCTION OF THE REAL STRUCTURE TO A SIMPLIFIED "DESIGN" MODEL

3.1 Introduction of the simplified model

In order to design the control algorithm, it could be convenient to reduce the original problem of analysis of the isolator at the base of n-d.o.f. structural system to the simplified one of analysis of the only s.d.o.f. isolation device; this purpose can be pursued by individuating the modified forcing function, that, acting on the isolator- simple oscillator (with the same mechanical characteristics of the original isolation device), induces the same response in the isolator alone as it was placed at the base of the real structural system. By using what exposed in the previous paragraph for the s.d.o.f. case, with $y_1(\omega,t) = y_{o1}(\omega|p,q)e^{j\omega t}$, and by referring to the harmonic modified forcing action $m_1\ddot{u}_g(\omega,t) = m_1\ddot{u}_{g_o}(\omega|p,q)e^{j\omega t}$, one gets

$$y_{o1}(\omega|p,q) = \bar{H}_c(\omega|p,q)\ddot{u}_{g_o}(\omega|p,q);$$
$$\dot{y}_{o1}(\omega|p,q) = j\omega y_{o1}(\omega|p,q); \ddot{y}_{o1}(\omega|p,q) = -\omega^2 y_{o1}(\omega|p,q) \quad (18)$$

with the isolator absolute displacement/base acceleration gain

$$\bar{H}_c(\omega|p,q) = -\frac{k_1 + j\omega c_1}{\omega^2\left\{\left[k_1 + p(\omega) - m_1\omega^2\right] + j\omega\left[c_1 + q(\omega)\right]\right\}} \quad (19)$$

Again the gain function between the isolator absolute displacement and base displacement, and between isolator absolute acceleration and base acceleration are given by

$$H_c(\omega|p,q) = -\omega^2\bar{H}_c(\omega|p,q) \quad (20)$$

3.2 Equivalence of the isolator response in the simplified and real model

In order to make the response $y_1(\omega,t)$ of the isolator inserted in the real structure [Eq.(12)] coinciding with the one of the isolator extracted from it [Eq.(18)], the following has to be verified

$$\left|\bar{H}_c(\omega|p,q)\right|\ddot{u}_{g_o}(\omega|p,q)e^{j\omega t} = \left|\bar{H}_{1c}(\omega|p,q)\right|\ddot{u}_{g_o}(\omega)e^{j\omega t} \quad (21)$$

and then

$$\ddot{u}_g(\omega,t) = \ddot{u}_{g_o}(\omega|p,q)e^{j\omega t} = \chi(\omega|p,q)\ddot{u}_{g_o}(\omega)e^{j\omega t}$$
$$\text{with} \quad \chi(\omega|p,q) = \frac{\left|\bar{H}_{1c}(\omega|p,q)\right|}{\left|\bar{H}_c(\omega|p,q)\right|} = \frac{\left|H_{1c}(\omega|p,q)\right|}{\left|H_c(\omega|p,q)\right|} \quad (22)$$

Therefore, the control parameters $p(\omega)$ and $q(\omega)$, chosen by the preferred optimality criterion, can be determined by referring just to the simple oscillator representing the real isolation device, as long as the forcing function, assumed to act on the s.d.o.f., is converted in the equivalent shaking $\ddot{u}_g(\omega,t)$ in accordance with equation (22).

In the following one will assume $p(\omega)=0$, as it is well known that the active control algorithm efficiency mainly relies on the dissipative term $q(\omega)\dot{y}_1(\omega,t)$.

4 ALGORITHM DESIGN

4.1 Search of the optimal control parameter

From the above, one deduces that one can refer to the simplified "equivalent" model for the design of the control algorithm, with no lack of accuracy.

A possible strategy for the individuation of the control parameters for the oscillator-isolator, can be obtained keeping in mind that, for $p(\omega) = 0$

$$w_{o1}(\omega|q) = q(\omega)\dot{y}_{o1}(\omega|q) = -j\frac{q(\omega)}{\omega}H_c(\omega|q)\ddot{u}_{g_o}(\omega|q)$$
$$\text{with} \quad H_c(\omega|q) = \frac{k_1 + j\omega c_1}{\left[k_1 - m_1\omega^2\right] + j\omega\left[c_1 + q(\omega)\right]} \quad (23)$$

The definition of $q(\omega)$ can be pursued by solving the optimum problem, where one minimizes the employed control force, while keeping the isolator absolute acceleration under a prefixed percentage (defined by means of the function $\alpha(\omega)\in[0,1]$) of the uncontrolled isolator acceleration $\ddot{y}_{o1}(\omega|0)$.

$$\begin{cases} w_{o1}(\omega|q) = min \\ sub \quad \ddot{y}_{o1}(\omega|q) \le \alpha(\omega)\ddot{y}_{o1}(\omega|0) \end{cases} \quad (24)$$

By squaring Equation (24) and writing a^2 for $|a|^2$, with $H_c(\omega|0)=H_o(\omega)$, one gets the final optimum problem

$$\begin{cases} F^2(\omega|q) = w_{o1}^2(\omega|q) = \frac{q^2(\omega)}{\omega^2}H_c^2(\omega|q)\ddot{u}_{g_o}^2(\omega|q) = min \\ sub \ G^2(\omega|q) = H_c^2(\omega|q)\ddot{u}_{g_o}^2(\omega|q) - \alpha^2(\omega)H_o^2(\omega)\ddot{u}_{g_o}^2(\omega) \le 0 \end{cases} \quad (25)$$

By the Kuhn-Tucker conditions, one gets

$$\begin{cases} \dfrac{\partial F^2(\omega|q)}{\partial q} = -\lambda\dfrac{\partial G^2(\omega|q)}{\partial q} \\ \lambda \ge 0 \\ \lambda G^2(\omega|q) = 0 \end{cases} \quad (26)$$

As the case $\lambda=0$ corresponds to a negative form of the control coefficient (that would result in a further worsening of the isolator response), the final solution to (24) is obtained for $\lambda\neq0$ and, in detail, it derives from the condition $G^2(\omega|q)=0$. As

$$H_c^2(\omega|q) = \frac{k_1^2 + \omega^2 c_1^2}{\left[k_1 - m_1\omega^2\right]^2 + \omega^2\left[c_1 + q(\omega)\right]^2} \quad (27)$$

the optimal form of the control parameter is then

$$q(\omega) = -c_1 + \frac{1}{\omega}\sqrt{\frac{k_1^2 + \omega^2 c_1^2}{\alpha^2(\omega)H_o^2(\omega)}\frac{\ddot{u}_{g_o}^2(\omega|q)}{\ddot{u}_{g_o}^2(\omega)} - \left(k_1 - m_1\omega^2\right)^2} \quad (28)$$

where $\alpha(\omega)$ is a function of reduction of the uncontrolled isolator absolute acceleration, that should be suitably chosen; actually Equation (28) represents the solution to the optimum problem (24), referred to the isolator modeled like a simple oscillator subject to the excitation $\ddot{u}_g(\omega,t)$, that induces the same response that the first (isolated) floor of the complete structure would have when the whole $n+1$ d.o.f. structural system is subject to the action of the real forcing function $\ddot{u}_g(\omega,t)$. Therefore $\alpha(\omega)$ has to be such that both the constraint in (24) and Equation (22) are satisfied. A possible criterion for the choice of the function $\alpha(\omega)$ is to keep constant the bound on the isolator absolute acceleration in the constraint condition

$$H_c^2(\omega|q)\ddot{u}_{g_o}^2(\omega|q) \le \alpha^2(\omega)H_o^2(\omega)\ddot{u}_{g_o}^2(\omega) = \gamma^2\left[H_o^2(\omega)\ddot{u}_{g_o}^2(\omega)\right]_{max}$$

$$\to \alpha^2(\omega) = min\left[1, \gamma^2\frac{\left[H_o^2(\omega)\ddot{u}_{g_o}^2(\omega)\right]_{max}}{H_o^2(\omega)\ddot{u}_{g_o}^2(\omega)}\right] \quad (29)$$

being γ a previously defined percentage. The analysis of the frequential composition of the excitation leads to the choice of a shape for $\alpha(\omega)$ able to mitigate the gain function $H_c(\omega|q)$ in the critical individuated frequency range. On the basis of the knowledge of the reduction function $\alpha(\omega)$, one can now numerically solve Equation (28), determining the form of $q(\omega)$, optimal with reference to the whole real system.

4.2 Control algorithm

Once the optimal control parameter for the real system has been determined, one can define the control action. For a generic bounded-support forcing function (like in case of seismic action) the control force can be expressed (A.Baratta & O.Corbi, 1999a,b,c) as

$$w(t) = \frac{1}{\sqrt{2\pi}}\int_{-\infty}^{+\infty}q(\omega)\dot{y}_1(\omega,t)d\omega = \frac{1}{\sqrt{2\pi}}\int_{-\infty}^{+\infty}j\omega q(\omega)y_{o1}(\omega|q)e^{j\omega t}d\omega \quad (30)$$

and, therefore, its instantaneous definition requires the instantaneous frequential decomposition of the structural response. As the only data available trough the current instant t are those monitored as far as that instant is reached, by considering the Fourier transforms of the whole time vector length and its restriction to the range $[0, t]$, one can obtain the final control force expression (A.Baratta & O.Corbi, 1999a,b,c)

$$w(t) = \frac{1}{\sqrt{2\pi}}\left[\int_0^t \dot{y}_1(\omega,\vartheta|t)Q(t-\vartheta)d\vartheta\right] \quad (31)$$

with $Q(x)$ the inverse Fourier transform of $q(\omega)$

$$Q(x) = \frac{1}{\sqrt{2\pi}}\int_{-\infty}^{+\infty}q(\omega)e^{j\omega x}d\omega. \quad (32)$$

5 NUMERICAL INVESTIGATION

One has considered a 4-d.o.f. frame with:

$m_2 = m_3 = m_4 = m_5 = m = 6\times10^4$ kg;
$k_2 = k_3 = k_4 = k_5 = k = 3\times10^8$ kg/m;
$c_2 = c_3 = c_4 = c_5 = c = 2\zeta\sqrt{\frac{k}{m}}$; $\zeta=2\%$;

and the two cases where the structure first floor corresponds to a:

■ **isolation device:**

$m_1 = m$; $k_1 = 0.05\,k$; $c_1 = 2\zeta_1\sqrt{\frac{k_1}{m_1}}$; $\zeta_1 = 1\%$.

■ **soft floor:**

$m_1 = m$; $k_1 = 0.4\,k$; $c_1 = 2\zeta_1\sqrt{\frac{k_1}{m_1}}$; $\zeta_1 = 5\%$.

For he isolated system the modal frequencies are $\omega = 6.8, 44.7, 83.5, 114.5, 134.4$, while for the soft floor case $\omega = 16.1, 51.3, 86.6, 115.75, 134.7$.

For both cases, the whole structural system is considered to be subject to a white noise base acceleration with zero mean and unitary variance, scaled in such a manner to have a peak acceleration of 0.4g (Fig.1, 2).

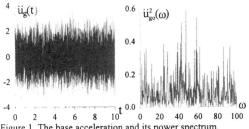

Figure 1. The base acceleration and its power spectrum.

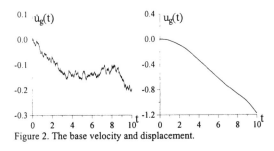

Figure 2. The base velocity and displacement.

In the first case, the presence of the rather flexible isolation device makes the super-structure mobile with respect to the ground, by deeply mitigating its sensitivity to the higher modes, as one can observe,

for example, from the gain function between the isolator floor absolute acceleration and the ground acceleration of Figure 3, that shows a strongly dominant peak. Through Equation (22) the gain function $H_{1c}(\omega|q)$ produces an "equivalent" excitation acting on the s.d.o.f. system able to take into account also the action transmitted to the isolator from the super-structure.

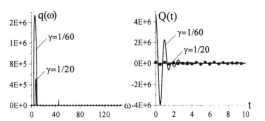

Figure 4. The optimal control parameter and its inverse FFT for the base isolated system.

From Figure 3 one can observe how the isolator gain function between the absolute acceleration and the base acceleration reduces when applying the control forces related to the optimal control coefficient of Figure 4. In detail two shapes of $q(\omega)$, derived respectively for $\gamma=1/20$ and for $\gamma=1/60$, have been considered: for $\gamma=1/20$, $q(\omega)$ has just one peak in correspondence of the main modal frequency of the isolated m.d.o.f. system (this is due to the constant power spectrum of the assumed forcing

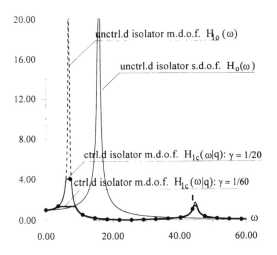

Figure 3. Controlled and uncontrolled gain functions between isolator absolute acceleration and base acceleration.

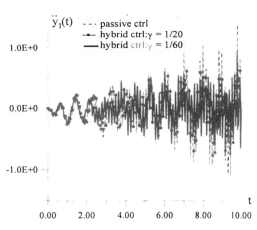

Figure 5. Controlled and uncontrolled isolator abs. acceleration.

action, that doesn't influence, with its own critical frequency range, the position of the peak of $q(\omega)$), while, for $\gamma=1/60$, $q(\omega)$ shows two peaks, respectively corresponding to the first and second structural system modal frequencies; as it is clear from the figure, the intensity of the second peak is, anyway, negligible with comparison to the first one.

It is important to emphasize that the choice of the shape of the reduction function $\alpha(\omega)$ is really significant; in this case, for example, it is such that the gain reduction is performed only at some particular frequencies. From Figure 5, 7 one can observe that the isolator absolute acceleration and drift is consistently reduced when adopting the control force corresponding to the case $\gamma=1/60$.

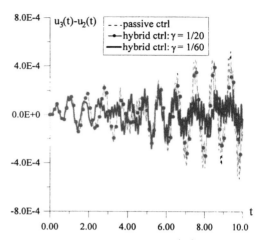

Figure 8. Controlled and uncontrolled 3rd-2rd floor inter-story displacement for the base isolated system.

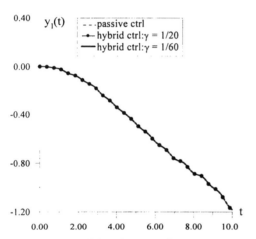

Figure 6. Controlled and uncontrolled isolator absolute displacement.

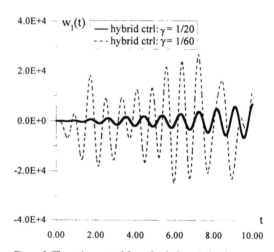

Figure 9. The active control force for the base isolated system.

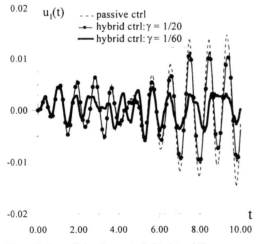

Figure 7. Controlled and uncontrolled isolator drift.

The inter-story structural displacements are deeply reduced as well, as one can notice from Figure 8 where, as an example, the relative displacement between third and second floor is reported.

The results seem to show a very good performance of the hybrid system with comparison to the simple passive isolation device, especially, because it employs very low energy (Fig. 9): actually for $\gamma=1/20$ the maximum value of the control action is equal approximately to 1/150 of the maximum value of the forcing function, while for $\gamma=1/60$ it is almost 1/30 of the maximum intensity reached by the occurring excitation.

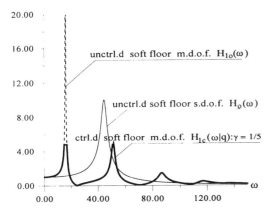

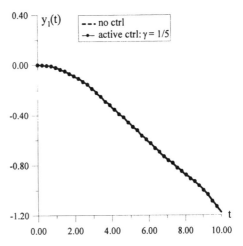

Figure 10. Controlled and uncontrolled gain functions between soft floor absolute acceleration and base acceleration.

Figure 13. Controlled and uncontrolled soft floor absolute displacement.

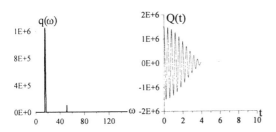

Figure 11. The optimal control parameter and its inverse FFT for the system with the soft floor.

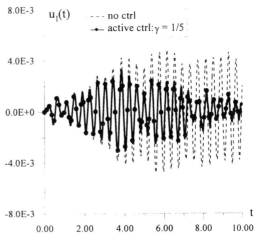

Figure 14. . Controlled and uncontrolled soft floor drift.

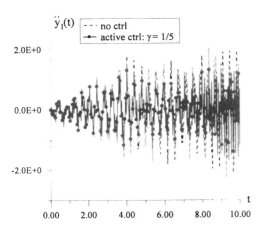

Figure 12. Controlled and uncontrolled soft floor absolute acceleration.

Also, in order to test the performance of the designed active control algorithm, on a m.d.o.f. structural system, a numerical investigation has been done on the same structure, in the hypothesis that the first floor consists, instead of the base isolation device, of a soft floor, with a certain flexibility and dissipative skill.

Even in this case, the flexibility of the introduced soft floor is such to render the first mode consistently dominant with respect to the others (Fig.10); however for $\gamma=1/5$ one gets a two-peaks shape of the control parameter $q(\omega)$ (Fig.11) and a maximum intensity of the control force (Fig.16) that is approximately 1/30 of the maximum value of the forcing function.

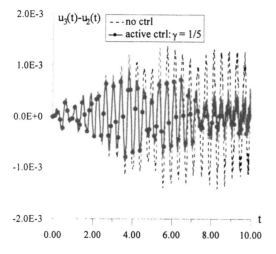

Figure 15. Controlled and uncontrolled 3rd-2rd floor inter-story displacement for the soft floor system.

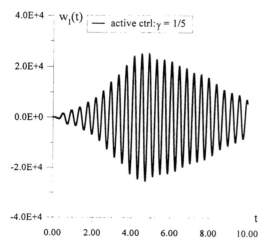

Figure 16. Active control force for the soft floor system.

Again the control action provides a good reduction of the soft floor absolute acceleration and drift (Fig. 12, 14) and, also, the inter-story drifts are consistently reduced (Fig. 15).

These results show that the control strategy results to be rather effective even when directly applied to the not base isolated m.d.o.f. structure.

6 CONCLUSIONS

In the paper a linear control algorithm, based on the harmonic decomposition of the forcing function has been applied for stress mitigation in strauctures endowed with a soft floor or even base-isolated.

A procedure to design the relevant control algorithm has been presented, and the result have been numerically tested, proving that significant improvement in the structure response can be achieved by a small control action.

The main advantage of the procedure is that the control force is engaged only at frequencies where the system exhibits a high dynamic magnification, thus allowing to save control energy where the passive device acts efficiently.

ACKNOWLEDGMENTS

Paper supported by grants of the Italian C.N.R. and co-financed by M.U.R.S.T.

REFERENCES

Baratta, A. & Corbi, O., 1999a. Improved Control of Structures by Time-Delayed Algorithms. *5th Intern. Conf. on the Application of Artificial Intelligence to Civil and Structural Engineering*, Oxford.

Baratta, A. & Corbi, O., 1999b. Algoritmo di Controllo per Forzanti Armoniche nel Dominio delle Frequenze. *12° Convegno Italiano AIMETA di Meccanica Computazionale, GIMC*, Napoli.

Baratta, A. & Corbi, O., 1999c. Controllo Attivo di Sistemi Strutturali Mediante Algoritmo Frequenziale Combinato. *14° Congresso Italiano AIMETA di Meccanica Teorica ed Applicata*, Como.

Buckle, I.G. & Mayes, R., 1990. Seismic Isolation: History, Application, and Performance- a World View. *Earthquake Spectra*, Vol.6.

Kelly, J.M., 1990. Base Isolation: Linear Theory and Design. *Earthquake Spectra*, Vol.6.

Behaviour of Steel Structures in Seismic Areas, Mazzolani & Tremblay (eds) © 2000 Balkema, Rotterdam, ISBN 90 5809 130 9

Effectiveness of viscous damped bracing systems for steel frames

A. Chiarugi & G. Terenzi
Department of Civil Engineering, Florence, Italy

S. Sorace
Department of Civil Engineering, Udine, Italy

ABSTRACT: A retrofitting trial design analysis of a thirteen-story steel moment frame building damaged by the 1994 Northridge earthquake is presented. The structural rehabilitation hypothesis consisted of introducing bracing members equipped with fluid viscous dampers in each one of the four perimeter frames constituting the lateral load resisting system of the building. A special energy-based design criterion developed within the context of a nonlinear dynamic procedure approach, previously proposed and utilized for a new construction project, was applied to this case study. The design analysis, complying with FEMA 273 - NEHRP Guidelines for the Seismic Rehabilitation of Buildings, was aimed at obtaining Immediate Occupancy Performance Level for the building. Feasible dimensions and costs came out for the dampers ensuring attainment of this notably demanding objective. The substantially upgraded seismic performance of the structural system was confirmed by the undamaged response derived from a more severe register of Northridge event than the one actually recorded at the basement of the building.

1 INTRODUCTION

The most recent international vision permeating earthquake engineering, essentially related to the concepts of performance-based and damage-controlled design, stresses the use of passive energy dissipation (PED) systems as a preferential strategy for new building structures as well as rehabilitation of existing ones. Among the various PED technologies conceived for both fields (Housner et al. 1997, Soong and Dargush 1997), growing attention is currently paid to the one concerning the insertion of fluid viscous (FV) dampers incorporated in steel bracing members within the load carrying skeleton of framed structures.

Experimental research (Pekcan et al. 1995, Reinhorn et al. 1995) on this innovative seismic protection strategy, and a first series of actual applications (Soong and Dargush 1997) have been noticed, essentially within the context of rehabilitation projects. Along this line, a trial design analysis aimed at assessing the effectiveness of a possible retrofitting intervention on a steel moment resisting frame building, damaged by the Northridge earthquake of 17 January 1994, is presented in this paper.

A nonlinear dynamic procedure, complying with FEMA 273 - NEHRP Guidelines for the Seismic Rehabilitation of Buildings (FEMA 1997), is followed by applying a special energy-based design criterion, initially proposed in an one-step version (Chiarugi et al. 1997), and then modified into an improved multi-step iterative one (Sorace & Terenzi 1999). By considering type and use of the building — the Blue Cross Headquarters facility located in the S. Fernando Valley — an Immediate Occupancy Performance Level was pursued. By meeting this objective level (named 1-B, FEMA 1997), determined by combining the Structural (S-1) and Nonstructural (N-B) Immediate Occupancy ones, the building is expected to sustain minimal or no structural damage, and only minor damage to the nonstructural components. Attaining this objective level in a high seismicity zone, such as the one in which the building is located, and for a structural system originally designed with the 1973 Uniform Building Code, can result in impracticable solutions if traditional rehabilitation strategies are followed. As it will be shown in the following, a feasible intervention from both a structural and an economic viewpoints can on the other hand be obtained, once repaired the damaged beam-column welded joints of the moment resisting frames, by applying the advanced retrofitting technique herein dealt with.

2 ADOPTED DESIGN CRITERION

The design procedure of fluid viscous dampers incorporated in steel braces consists of assigning

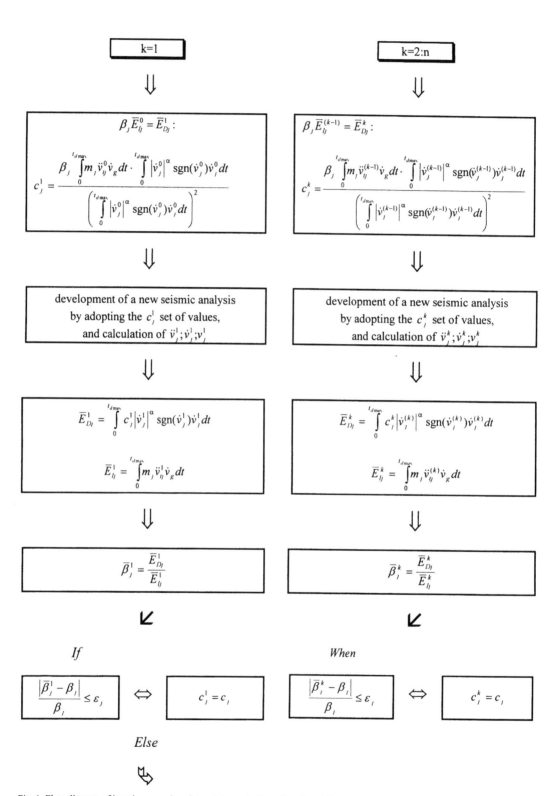

Fig. 1. Flow diagram of iterative procedure determining c_j design values from (4).

them the capability of dissipating a predefined share of the total input energy computed at each floor for the frame to be designed — as the result of a preliminary conventional seismic analysis — or to be retrofitted (Sorace & Terenzi 1999).

By recalling the expressions of the "absolute" input energy E_{Ij} (Uang & Bertero 1988) of the j-th floor and the energy E_{Dj} dissipated by the dampers placed at the same floor of a n-degree-of-freedom structural system:

$$E_{Ij} = \int_0^{t_f} m_j \ddot{v}_{tj} dv_g \qquad (1)$$

$$E_{Dj} = \int_0^{t_f} c_j |\dot{v}_j|^\alpha \operatorname{sgn}(\dot{v}_j)\dot{v}_j dt \qquad (2)$$

where t_f = time duration of input ground motion; m_j = mass associated to j-th floor; $\ddot{v}_{tj}$ = total j-th floor acceleration (here and in the following, one and two upper dots will denote first and second time derivatives); v_g = ground displacement; c_j = global damping coefficient characterizing j-th floor dampers; $\dot{v}_j$ = relative j-th floor velocity; $|\cdot|$ = absolute value; $\operatorname{sgn}(\cdot)$ = signum function; and α = fractional exponent defining the j-th floor damping reaction force F_{Dj} ($F_{Dj} = c \cdot \operatorname{sgn}(\dot{v}_j) \cdot |\dot{v}_j|^\alpha$), the adopted criterion imposes that

$$\beta_j E_{Ij} = E_{Dj} \qquad (3)$$

where β_j is the energy ratio prescribed for the j-th floor.

The c_j coefficients satisfying condition (3) are determined by means of the multi-step procedure summed up in Figure 1, which is started (k = 1) from the $\bar{E}_{Ij}^0$ values representing the work made by the inertia forces $m_j \ddot{v}_{tj}^0$ on the n floors of the frame — already including undamped braces — over the [0, t_{dmax}] time interval (where t_{dmax} = instant at which maximum j-th interstory drift is recorded).

As observed in Sorace & Terenzi (1999), the choice of the β_j ratios guiding c_j calculation should be based upon the mean elastic response of the frame structure without dampers, under the ensemble of input ground motions assumed for the dynamic design enquiry, suitably scaled to keep the response below the nominal yielding threshold. The relative distribution and the maximum values of the β_j ratios

can afterwards be adjusted by analyzing response obtained from first iteration, with the view of reaching nearly uniform member seismic demand along the height of the structure as well as overall optimization of the device sizes.

3 CASE STUDY BUILDING

The thirteen-story frame building selected for this design analysis, located approximately three miles from the epicenter of Northridge earthquake, was instrumented at the base, 6-th floor and roof levels. This circumstance made the building a largely investigated case study in the relevant post-seismic damage assessment phase (SAC 1995, Maison 1996, Uang et al. 1997, Gross 1997, Maison & Kasai 1997).

The lateral load resisting system of the structure, typical elevation and plan views of which are drawn in Figure 2, consists of four identical perimeter steel moment frames whereas internal column alignments and beams were designed to carry only gravity loads. Moment frame member sizes are recapitulated in Table 1. As shown in Figure 2, at the first floor above the ground the plan broadens to form a plaza; this architectural layout comprises basement perimeter reinforced concrete walls, which provide a high degree of lateral restraint at this level.

Table 1. Member sizes for perimeter frames.

Level	Beams	Column Rows 5-6-7-8 C-D-E-F	Column Rows 4-9 B-G (*)
Roof	W27x84	—	—
Fl. 12	W33x118	W14x176	371.5x25.4
Fl. 11	W33x118	W14x176	371.5x25.4
Fl. 10	W33x130	W14x255	384.2x31.8
Fl. 9	W33x130	W14x255	384.2x31.8
Fl. 8	W33x141	W14x283	390.5x34.9
Fl. 7	W33x141	W14x283	390.5x34.9
Fl. 6	W33x152	W14x311	403.2x41.3
Fl. 5	W33x152	W14x311	403.2x41.3
Fl. 4	W33x152	W14x398	409.6x44.5
Fl. 3	W33x152	W14x398	409.6x44.5
Fl. 2	W33x152	W14x426	422.3x50.8
Fl. 1	W36x230	W14x426	422.3x50.8
Plaza	W36x194	W14x500	447.7x63.5
Ground	—	W14x500	447.7x63.5

(*) Box sections (side x plate thickness - dimensions in millimeters

381

The welded beam-column connections belonging to the lower seven levels of the north-south oriented frames, which were subjected to the most severe local motion component (characterized by a peak ground acceleration of 0.411 g), suffered from light to considerable damage. More specifically, the observed damage configurations included fracture of the column or beam flanges, and fracture of the beam-to-column welds. The results of the careful checks carried out, similar to the ones of post-quake surveys developed in several other moment-resisting steel structures affected by the Northridge event, are sketched in the schematic of Figure 3, referred to most damaged line 4. In this figure circles represent inspected joints, whereas shaded quarters indicates the fractured sides of the connections.

The retrofitting hypothesis proposed in the following section would obviously include a preliminary repair intervention of all degraded joints, intended to restore their original performance; the relevant technical aspects will not be discussed within the present paper.

4 TRIAL DESIGN AND EVALUATION OF RETROFITTING INTERVENTION

The design analysis was developed by means of a series of eight artificial seismic accelerograms generated from a horizontal response spectrum constructed by complying with rules provided by FEMA 273. Spectral ordinates were scaled in accordance with the acceleration contour maps reproduced in the map package distributed with the same Guidelines, so as to obtain a ground shaking hazard corresponding to Base Safety Earthquake 2 (this last being characterized by a 2% probability of exceedance in 50 years). The resulting peak ground acceleration A_d was equal to 0.48 g. Spectral abscissas were calibrated on the fundamental vibration period of the frame structure including undamped braces, equal to 2.66 s. It should be noted that the first three vibration periods of the bare frame resulted to be equal to 3.03 s, 1.07 s and 0.65 s, that is, nearly coincident with the values calculated in Uang et al. (1997). This was the consequence of the assumed similar elastic model, which comprises rigid-end offsets, and rotational springs simulating panel zone flexibility — with elastic stiffness computed by the Krawinkler relation proposed for semi-rigid joints (Krawinkler 1978) — in correspondence with each beam-column connection.

The preliminary elastic enquiry led to determine the maximum interstory drift limit envelope shown in Figure 4, obtained from the median response to the ensemble of eight accelerograms. By following suggestions of other authors (Gross 1997, Maison & Kasai 1997), a reduced yielding strength compared

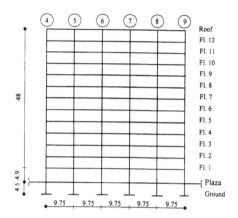

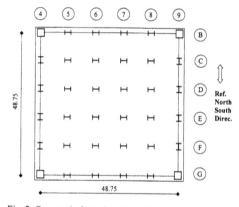

Fig. 2. Case study frame building (dimensions in meters).

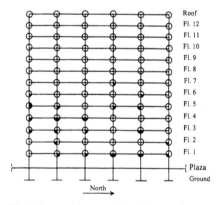

Fig. 3. Surveyed damage pattern in beam-column connections of line 4.

to the value provided by current standards was assumed (precisely equal to 276 Mpa), by considering the results of direct surveys conducted on similar structures built in the early Seventies.

382

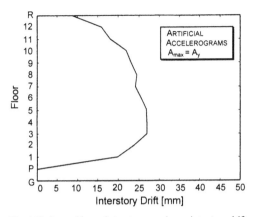

Fig. 4. Undamped braced structure: maximum interstory drift envelope at elastic limit state (median over eight design accelerograms).

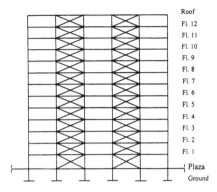

Fig. 5. Positioning of damped braces within perimeter frames.

A mean peak acceleration A_y of 0.173 g caused achievement of the limit elastic (yielding) threshold. By analyzing the graph in Figure 4, which points out that the maximum elastic demand is situated between the second and sixth floors (matching the actual post-seismic damage configurations), and is however quite comparable till to the eighth floor, the following β_j tentative distribution was adopted: 0.5 for levels from first to eighth; and 0.3 for remaining levels. The maximum 0.5 value emerged as a reasonable upper limit from an economic viewpoint; at the same time, further increased damping shares were not considered at this stage, with the aim of avoiding undesired contributions of superior vibration modes to global dynamic response.

The intervention hypothesis consisted of placing the brace elements within the two intermediate bays of perimeter frames (Figure 5). Jarret elastomeric devices — constituting the subject of a global research program begun by the authors several years ago — were adopted as fluid viscous dampers for this study. By applying the iterative procedure discussed in the second section, the aforementioned β_j values were attained after only two iterations. The corresponding mean envelopes of the input and damping energies calculated for any single floor at instant t_{dmax}, plotted in Figure 6, confirm the achievement of the target β_j values.

The energy time-histories relevant to one of the three most stressed floors (the fifth one) are reported in Figure 7, which highlights the effects of transition from the first to the second iteration on structural performance.

The c_j values derived from second iteration varied from 150 kN(s/mm)$^\alpha$ to 18 kN(s/mm)$^\alpha$, which correspond to medium through small size JE units in current production. In order to optimize the device dimensions, the anyhow quite similar c_j values

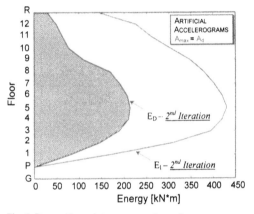

Fig. 6. Damped braced structure: maximum floor energy envelopes at t_{dmax} (median over eight design accelerograms).

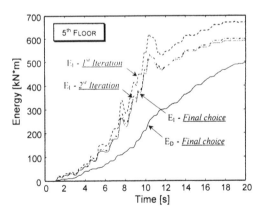

Fig. 7. Damped braced structure: 5-th floor energy time-histories (median over eight design accelerograms).

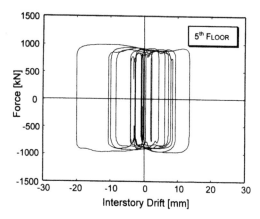

Fig. 8. Damped braced structure: 5-th floor damping force-interstory drift response loops corresponding to final c_j choice, obtained from one design accelerogram.

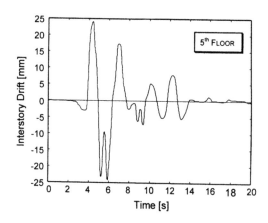

Fig. 10. Damped braced structure: 5-th floor interstory drift time-history corresponding to final c_j choice, obtained from Newhall record.

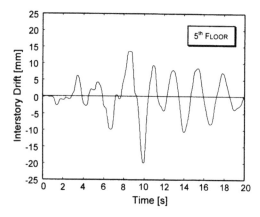

Fig. 9. Damped braced structure: 5-th floor interstory drift time-history corresponding to final c_j choice, obtained from same design accelerogram in Fig. 8.

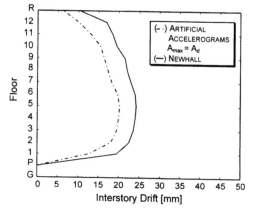

Fig.11. Damped braced structure: maximum interstory drift envelopes corresponding to final c_j choice (median over eight design accelerograms, and Newhall).

obtained for the lower nine floors were unified to 150 kN(s/mm)$^\alpha$. Damping coefficients equal to 70 kN(s/mm)$^\alpha$ for 9-th and 10-th floors; 55 kN(s/mm)$^\alpha$ for 11-th and 12-th floors; and 20 kN(s/mm)$^\alpha$ for under-roof level, were furthermore adopted. The results produced by the c_j design set correspond to the third E_I curve in Figure 7, where the related E_D curve is traced out as well; moreover, they are also demonstratively illustrated, for one selected motion and again the fifth floor, by the damping force-interstory drift response loops and interstory drift time-history drawn in Figures 8 and 9.

In particular, the derived 19.8 mm drift corresponds to a 0.049 interstory drift ratio that precisely conforms to the maximum 0.05 value of this parameter suggested by FEMA 273 for 1-B

Performance Level, as regards to braced frames. Although this indication is not intended as an explicit acceptance criterion for the rehabilitation design, but rather as a useful reference for a quantitative control about protection of nonstructural components, the outcome of the design process underlines a successful implementation of the considered technology within the frame building.

A numerical check about this aspect was carried out by subjecting the rehabilitated scheme to the N-S component of the Newhall record, characterized by a peak ground acceleration of 0.583 g. This represents one of the most severe registers of the Northridge earthquake, with amplitude equal to 1.21 times the peak design acceleration, and 1.42 times the peak acceleration of the ground motion actually recorded at the basement of the building.

The fifth floor interstory drift time-history, and the envelope of the same quantity over the entire frame, both corresponding to the final design c_j choice (Figures 10 and 11), confirm the effectiveness of adopted solution. In fact, the maximum drifts tolerably exceeds by no more than 15% the reference limit previously introduced, which ensures totally elastic structural response as well as a substantial preservation of nonstructural elements, in spite of the very significant hazard level expressed by the considered strong motion record.

5 CONCLUDING REMARKS

The trial design analysis carried out confirmed the effectiveness of the rehabilitation strategy based on insertion of fluid viscous dampers incorporated in bracing steel members within frame structures. This was highlighted by the substantial control of seismic response of the thirteen-story building assumed as the reference for this study, numerous moment-resisting connections of which were severely damaged by the Northridge earthquake.

An elastic response complying with FEMA-273 limit requirements for 1-B Immediate Occupancy Performance Level was in fact numerically surveyed under more severe input motions than the one actually recorded at the basement of this monitored building.

At the same time, the demanded damping capabilities could be satisfied by means of medium-size fluid viscous Jarret devices in current production, which underlines the economical as well as technical feasibility of the formulated intervention hypothesis.

REFERENCES

Chiarugi, A., S. Sorace & G. Terenzi 1997. Impiego di dispositivi siliconici in tecniche di controllo passivo generalizzato. *Proc. 8th Italian Conf. on Earthquake Engineering, Taormina, Italy, Sept. 1997*: 533-540.

FEMA 1997. *NEHRP Guidelines for the seismic rehabilitation of buildings*. Report No. 273, Federal Emergency Management Agency, Washington, D.C.

Gross, J. L. 1997. A connection model for the seismic analysis of welded steel moment frames. *Engineering Structures*, 20(4-6): 390-397.

Housner, G. W. et al. 1997. Structural control: past, present, and future. *Journal of Engineering Mechanics, ASCE*, Special Issue 123(9): 897-971.

Krawinkler, H. 1978. Shear in beam-column joints in seismic design of steel frames. *Engineering Journal, AISC*, 15(3): 82-91.

ICBO 1994. *Uniform Building Code*. International Conference of Building Officials, Whittier, CA.

Maison, B. F. 1996. VE dampers for upgrade of Northridge damaged steel building. *Proc. 11th World Conf. on Earthquake Engineering, Acapulco, Mexico, June 1996*: Paper No. 1437.

Maison, B. F. & K. Kasai 1997. Analysis of Northridge damaged thirteen-story WSMF building. *Earthquake Spectra* 13(3): 451-473.

Pekcan, G., J. B. Mander & S. S. Chen 1995. The seismic response of a 1:3 scale model RC structure with elastomeric spring dampers. *Earthquake Spectra* 11(2): 249-267.

Reinhorn, A. M., C. Li & M. C. Constantinou 1995. *Experimental and analytical investigation of seismic retrofit of structures with supplemental damping, Part 1: Fluid viscous damping devices*. Technical Report NCEER-95-0001, Univ. of New York at Buffalo, Buffalo, NY.

SAC 1995. *Analytical and field investigations of buildings affected by the Northridge earthquake of January 17, 1994*. Report No. SAC 95-04, SAC Joint Venture.

Soong, T. T & G. F. Dargush 1997. *Passive energy dissipation systems in structural engineering*. New York: John Wiley & Sons.

Sorace, S. & G. Terenzi 1999. Iterative design procedure of fluid viscous devices included in braced frames. *Proc. EURODYN '99 - 4th European Conf. on Structural Dynamics, Prague, Czech Republic, June 1999*: 169-174.

Uang C.-M. & V. V. Bertero 1988. Use of energy as a design criterion in earthquake-resistant design. Report No. UCB-EERC 88/18, Univ. of California at Berkeley, Berkeley, CA.

Uang, C.-M., Q.-S. Yu, A. Sadre, D. Bonowitz, N. Youssef & J. Vinkler 1997. Seismic response of an instrumented 13-story steel frame building damaged in the Northridge earthquake. *Earthquake Spectra* 13(1): 131-149.

Behaviour of Steel Structures in Seismic Areas, Mazzolani & Tremblay (eds)© 2000 Balkema, Rotterdam, ISBN 90 5809 130 9

Non-invasive passive energy dissipation systems for the seismic design and retrofit of steel structures

C.Christopoulos & A.Filiatrault
Department of Structural Engineering, University of California at San Diego, USA

ABSTRACT:

A passive axial elasto-plastic device is introduced locally near beam-to-column joints of steel moment resisting frames to enhance the energy dissipating capacity during earthquakes. The geometry, stiffness and slip or yield load are chosen to maximize the energy dissipated during a monotonic push. A typical 6-story moment resisting frame is first retrofitted with haunch type connections, with no slipping allowed. The building is then retrofitted with the same geometric configuration of the haunch but with an increased stiffness. Finally a third retrofit strategy consisting of replacing the haunch with energy dissipating devices is implemented. Non-linear time-history analyses where the fracture of welds is modeled with a strength degrading element are performed under different intensities of seismic ground motions. Results indicate that an increased haunch stiffness is more effective in protecting the weld fractures under large seismic loading. The energy dissipating haunch reduced significantly the response of the structure while still protecting the welded connections.

1 INTRODUCTION

Following the unexpected failure of several steel moment-resisting-frames (MRFs) during the 1994 Northridge, California earthquake the large SAC research program was launched in the U.S. to investigate the causes of these failures that typically occurred at beam-to-column weld connections. As findings quickly indicated that the major cause of these failures was a large rotational ductility demand at the level of the beam-to-column connections coupled with unexpectedly low capacity of the weld connections, focus was set on, first the retrofitting of existing MRFs that could potentially exhibit similar problems during a subsequent earthquake, and second on design provisions for new MRF structures. While the latter centered attention on a better understanding of the materials (weld and base steel) and a strict redefinition of welding practices, the former sparked a series of innovative concepts to improve the seismic resistance of existing MRFs. Three different possible connection retrofit schemes have been identified (Gross et al.1999), and design guidelines for each of these retrofit schemes have been elaborated. One of these retrofit schemes, the welded haunch connection, protects the welded section of MRFs by migrating the plastic hinge some distance away from the face of the column and by

redirecting the beam shear forces to the column through axial straining of the haunch. While focus was mainly placed on limiting the maximum stress developed at the weld location, the geometry, stiffness and strength of the haunch have been limited to practical values.

Passive energy dissipating devices, on the other hand, traditionally installed in buildings using braced configurations have proven to effectively raise the level of seismic safety. Nonetheless, the important disruption costs as well as loss of open space in commercial facilities have limited their actual implementations.

In this paper, a novel connection for MRFs, combining these two seismic retrofit strategies, is investigated numerically. This new connection is achieved by incorporating the haunch with an elasto-plastic energy-dissipating device.

2 OPTIMIZATION PROCESS

The energy dissipation capability of each connection configuration (geometry, stiffness of haunch, slip or yield load) is investigated first by computing the maximum energy dissipated when a vertical monotonic push is applied at mid span of the beam until either the maximum plastic rotation of the beam is reached, or yielding of the beam at the face of the column is instigated. Figure 1 illustrates a typical beam-to-column welded connection retrofitted with a haunch type device. The distance L' and the angle α characterize the geometry of the device, while the stiffness, K_{dev}, and the slip or yield load, F_s, define the properties of the device. If F_s is chosen to be very large, the connection is identical to the haunch connection.

To compare the effectiveness of each possible configuration and to determine the optimal values of geometry and desirable properties of the device, a computer program was developed to compute the amount of dissipated hysteretic energy for a mono-tonic push at midspan of the beam, i.e. the point of inflexion under lateral seismic loading of the MRF. Figure 2 illustrates the computation of the dissipated hysteretic energy. The force-deflection curve is inte-grated and the full elastic portion of the deflection is subtracted to compute the amount of dissipated hys-teretic energy. Two limit states were set to stop the monotonic push. The first limit state corresponds to the maximum moment that can be induced in the beam at the face of the column, while the second limit state corresponds to the maximum plastic rota-tion that can be achieved in any section of the beam. Although it is not clear which values are the most suitable, it was decided in this study to associate the first limit state to the yielding moment of the beam (M_y) and the second limit state to a plastic rotation of 0.03 rad. Both these values are consistent with current philosophy on welded MRF connections. Limiting the moment at the face of the column to M_y is believed to guarantee that the stress at the weld level will not exceed 80% of the maximum allow-able stress value, F_{EXX}. This limit is conservative on the amount of energy dissipated and favors, in the subsequent optimization schemes, solutions that achieve a large amount of energy dissipation without reaching yielding at the face of the column.

The beam is assumed to have a bilinear hysteretic behavior, with a post yielding stiffness equal to 2% of the initial stiffness. The yield moment of the

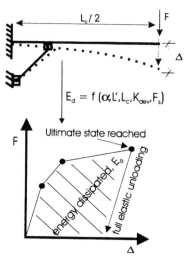

Figure 2: Computing energy dissipated in monotonic push

beam is set equal to the plastic moment (M_p). The length of the plastic hinge is set equal to 90% of the depth of the beam. When yielding or slipping occurs in the supplemental device, it is assumed that the post yielding stiffness is zero, i.e. that the device ex-hibits an elastic-perfectly-plastic hysteretic behavior. Elastic shear deformations are included in the analy-ses although no inelastic shear-flexure interaction is accounted for. The column is assumed to be fixed since according to capacity design principles it is expected to remain elastic, thus not contributing to the energy dissipation of the connection. Although numerous hysteretic behaviors can be achieved by varying the device parameters, the non-linearity is introduced only by, either the beam reaching the plastic moment at the desired plastic section or the haunch device reaching the predetermined axial force F_s.

3 OPTIMIZATION OF MOMENT-RESISTING CONNECTIONS

3.1 Definition of typical connection

To investigate the optimization of moment-resisting connections to increase the amount of energy dissi-pation that can occur before one of the limit states is reached, a typical connection is chosen. This con-nection is taken from the first story of the 6-story MRF that is defined later and used for non-linear time-history analyses. Since the column is consid-ered fixed, the beam section properties and clear length fully define the system that must be retrofit-ted. In this case, the beam is a W30x99 and the clear

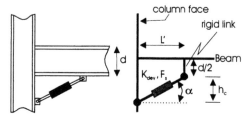

Figure 1: Configuration of retrofitted connection

Length, L_c, is 7000 mm. It is assumed that the steel is of A36 grade. According to the scheme presented in the previous paragraph, the bare connection exhibits zero energy dissipation since the yield moment will occur at the face of the column before any other section yields.

3.2 Proposed design of haunch connection

Following steps proposed by Gross et al. (1999) the chosen connection is retrofitted first according to the haunch scheme. The geometry of the haunch is determined by choosing L'= 0.55·d, where d is the depth of the beam, and the angle α is set equal to 30°. The flange area of the haunch is set equal to 4050 mm² to respect the strength and stiffness requirements. For the given geometry (L'= 410 mm and α = 30°) the stiffness of the haunch becomes K_{dev}=1711 kN/mm. Computing the force displacement hysteretic loop of this connection (Fig. 3) reveals that the critical limit state is the yielding of the beam at the face of the column. At this point, a small amount of energy is dissipated ($2.02 \cdot 10^4$ kN.mm) by plastic rotation at the section located a distance L' from the face of the column. When the stiffness of the haunch is increased to 2600 kN/mm, however, the critical limit state is defined by a plastic rotation of 0.03 rad at the section located at L' from the face of the column and the energy dissipated is increased to $4.94 \cdot 10^4$ kN.mm.

3.3 Optimization of haunch connection

Parametric analyses were carried out to determine the optimum configuration of the haunch to maximize the energy dissipation, E_d, as defined in the previous paragraph. For the case of the haunch ($F_s = \infty$), the maximum energy that can be dissipated by a given connection is determined by the energy dissipated by reaching a plastic rotation of 0.03 rad at the section of plastic deformation. Optimizing the haunch connection then becomes equivalent to providing sufficient stiffness (K_{min}) to the haunch such

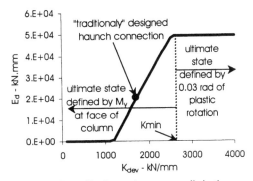

Figure 4: Defining K_{min} for maximum energy dissipation

that the maximum energy dissipation can be reached.

For the case of the W30x99 beam considered herein, keeping the geometry proposed by Gross et al. (1999) (L' = 410 mm and α = 30°) and plotting E_d as a function of K_{dev}, the minimum value, K_{min}, to reach the maximum value of E_d for this beam, is found to be K_{min} = 2600 kN/mm (Figure 4). The value of K_{min} was computed for different values of L' and α. Each point in Figure 5 represents the value of K_{min} as a function of L', for a value of α = 30°. The best-fit curve shows the relationship between these two parameters.

Using a multiple of the best fit curve to bound the data points, we get the following equation, for the stiffness of the haunch that maximizes the energy dissipation for a given value of L':

$$K_{dev} \geq \left(\frac{L'}{50000} \right)^{1.74}$$

Figure 6 shows that the angle α (K_{dev} = 1000 kN/mm) has little effect on the total amount of energy dissipated.

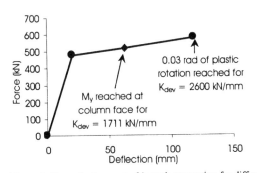

Figure 3: Force-displacement of haunch connection for different values of haunch stiffness

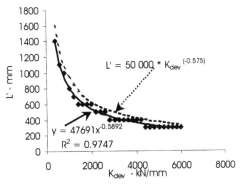

Figure 5: Values of K_{dev} for optimal energy dissipation

389

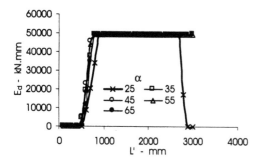

Figure 6: Effect of varying angle α on energy dissipation, E_d

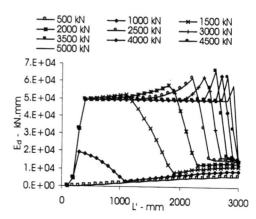

Figure 7: Effect of increasing slip load on energy dissipated, E_d

3.4 Optimization of haunch with elasto-plastic energy dissipating device

The elasto-plastic properties of the haunch further increase the energy dissipation capability of the connection. In this case, the force-displacement curve of the connection can assume linear, bi-linear or tri-linear characteristics. The limit states are kept similar to those previously defined, i.e. M_y at the face of the column or 0.03 rad of plastic rotation at the predetermined plastic section. When the slip load is too low, the beneficial effect of the haunch is lost, and the yielding moment is reached at the face of the column before large plastic deformations can occur in the plastic hinge. As the slip load is increased, the energy dissipated becomes a combination of hysteretic plastic hinging in the beam and axial elasto-plastic yielding or slipping of the haunch. As the slip load exceeds a certain value, the connection becomes similar to the haunch connection ($F_s = \infty$). Figure 7 shows the values of E_d as a function of L' for different values of the slip load F_s (500 kN to 5000 kN). The haunch stiffness is set equal to 3000 kN/mm and an angle of 30° is considered to remain consistent with the initial traditional haunch design. It can be seen that increasing the distance L' and adjusting the slip load F_s to different values increases significantly the amount of energy dissipated, E_d. The optimum haunch, computed in the previous paragraph resulted in a maximum energy dissipated of $4.94 \cdot 10^4$ kN.mm, whereas the optimum solution, for this case, results in a maximum value of E_d of $6.68 \cdot 10^4$ kN.mm, which represents an increase of 35.2 %. This occurs when the slip load is set to 3500 kN and for L'= 2600 mm.

In the previous paragraph, it was shown that for the haunch connection a minimum value of the haunch stiffness, K_{dev}, was required to assure that the maximum amount of energy dissipation could be reached. It is assumed that this value also applies for the slipping device.

In fact, the stiffness of the device is important before the device slips, and setting it equal to the

minimum required for the optimal haunch will assure the largest amount of energy dissipation by hysteretic hinging of the beam at the section of plastic deformation before the device slips. Table 1 shows the maximum value of E_d for different values of the angle α, as well as the slip load, F_s, and value of L' at which this value occurs. The stiffness of the device, K_{dev}, was set equal to 3000 kN/mm. The value of h_c presented in Table 1 shows how different choices of geometry for the device result in a various required heights on the column.

Table 1. Maximum values of E_d for different values of α

Angle α	Max. E_d	% incr in E_d	L'	h_c	F_s
Degrees	kN.mm		mm	mm	kN
15	$5.23 \cdot 10^4$	5.8	2100	562.7	3500
30	$6.68 \cdot 10^4$	35.2	2600	1501.1	3500
45	$8.31 \cdot 10^4$	68.2	2600	2600.0	4000
60	$10.7 \cdot 10^4$	116.6	2600	4503.3	4000

At this point, the designer is brought to weigh the advantages and disadvantages of each geometric configuration and chose one that is non-invasive (according to imposed restraints) while still providing a significant amount of supplemental energy dissipation. For example, if there is a strict restraint on h_c, a smaller angle α can be chosen and the stiffness K_{dev}, and slip load, F_s, of the device can be determined to assure a good amount of energy dissipation, while still protecting the welded section. More so, in cases where a week column problem is suspected as a result of the addition of a haunch type device, a larger angle α can be preferred, as it results in large axial forces rather than large shears and moments in the column where the device is attached. Although an increased amount of energy dissipation is not necessarily synonymous to a reduction of structural response (as the input energy is a function of the structural properties) it can be expected that a stable increase in the energy dissipating capability of each connection in the MRF will result in a reduction in the structural response. Furthermore, adding an energy-dissipating device allows for a tri-linear hysteretic behavior of the connections, thus increasing the number of natural periods that the structure can assume during important dynamic loading. This enables the structure to move away from resonant conditions that can occur during the earthquake.

4 CASE STUDY: 6-STORY MRF

The optimization techniques presented in the previous section are applied to retrofit all connections of a 6-story MRF.

4.1 Structure analyzed

The 6-story structure studied herein was first designed by Tsai and Popov (1988), and modified by Hall (1995). As shown in Figure 8, the structure is rectangular in shape, and spans 37 m by 22 m. Lateral loads in the North-South direction are resisted by two exterior moment-resisting frames. The structure is designed according to the 1994 edition of the

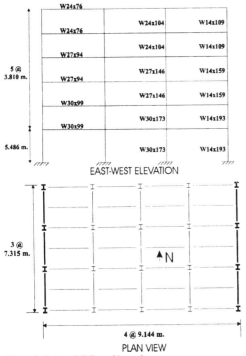

Figure 8: 6-story MRF used in analyses

Uniform Building Code (ICBO 1994) for a building in Zone 4 and on soil type S2.

The 2D analyses were performed using the Ruaumoko nonlinear dynamic analysis computer program (Carr 1996). Only one exterior frame was modeled since the interior frames were simply connected.

4.2 Non-linear modeling assumptions

A bilinear inelastic model was introduced to model the rotational hysteretic hinging at beam and column ends. Each member was assigned a plastic hinge length of 90% of its depth at each end, based on parametric analyses that showed that on average such a plastic hinge length yields reasonable values of the maximum computed moment at section ends. A strain hardening ratio of 2 % is used and, and a plastic rotation of 0.03 rad (after which local buckling occurs), is considered a failure criterion for the base metal of the steel sections. At failure, a value of M_{ult} = 1.2 M_p is considered reasonable, and is used here to obtain the strain hardening of 2 %. Furthermore, to account for the poor performance of pre-Northridge weld connections, a strength degradation model is introduced. This model reduces the strength of the welded section to 1% of its initial value after the value of M_{cr} = 1.05 M_y is reached two times during the time-history. The value of M_{cr} is difficult to define as it depends highly on the quality of the connection, nonetheless as suggested by Gross (1998) a value between M_y and M_p is reasonable. Shear deformations in panel zones were ignored, and only the response of the bare frame was included. The columns were assumed fixed at the ground level. A Rayleigh type damping of 2% critical was assigned based on the first and fifth elastic modes of vibration of the structure. To account for the combined effect of axial loads on the flexural strength, P-M interaction curves, such as defined by the LRFD (1993) code are introduced in the model.

4.3 Retrofit strategies

Each of the retrofit schemes is applied to all the connections in the building for comparison purposes although subsequent analyses showed that not all the connections reached yielding. First, the original MRF structure (OMRF), is retrofitted using the traditional haunch connection (HMRF). Then an optimized haunch (no slipping or yielding in the haunch) is designed by setting K_{dev} equal to K_{min} for each beam (OHMRF). Finally, an optimum energy-dissipating haunch, or optimum slip haunch (OSHMRF) is designed following the procedure presented in the previous paragraph. The optimum slip haunch is designed such that the value of h_c does not exceed 1500 mm. Table 2 summarizes for each of the connections the values of L', α, K_{dev} and F_s chosen for the different retrofit strategies.

Table 2. Definition of retrofitted structures

Model	Story	Beam	α	L'	h_c	F_s	K_{dev}
			Deg.	mm	mm	kN	kN/mm
HMRF	1	W30x99	30	410	237	∞	1711
	2	W30x99	30	410	237	∞	1711
	3-4	W27x94	30	370	214	∞	1451
	5-6	W24x76	30	330	191	∞	1233
OHMRF	1	W30x99	30	410	237	∞	2600
	2	W30x99	30	410	237	∞	2600
	3-4	W27x94	30	370	214	∞	2700
	5-6	W24x76	30	330	191	∞	400
OSHMRF	1	W30x99	30	2600	1500	3500	2700
	2	W30x99	30	2400	1386	2900	2700
	3-4	W27x94	30	2400	1386	2600	2700
	5-6	W24x76	30	2500	1443	2100	2700

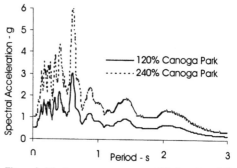

Figure 9: Elastic response spectra at 2% damping of Canoga Park records

5 NON-LINEAR TIME-HISTORY ANALYSES

5.1 Choice of earthquake records

The 1994 Northridge earthquake recording at Canoga Park is used for the time history analyses. The larger component of this recording amplified 1.2 times yields a 5% damping elastic response spectrum, which is reasonably compatible with the UBC 1997 design spectrum for a soil type D in a seismic zone 4.

The record presents peak ground accelerations of 0.5 g and a peak spectral acceleration of 3 g at a period of 0.6 s. The same record amplified by 2.4 times is also used for some analyses to further investigate differences between retrofit schemes as the demand on the system is increased to levels typical of larger seismic events. Figure 9 shows the elastic response spectra computed for 2 % of critical damping for both 1.2 and 2.4 times the Canoga Park record.

5.2 Results from analyses

5.2.1 Story deflections

Figures 10 and 11 show the maximum inter-story drifts predicted by the time-history analyses under the 120% and 240% Canoga park record, respectively. The original structure undergoes significant damage under both time-histories, especially under the 240% record where the model predicts collapse of the OMRF. The haunch connection is effective in reducing the inter-story drift except for the first story, where a seemingly soft story behavior is observed.

The OSHMRF is significantly more effective then the other retrofit schemes to reduce the inter-story drift. In fact this is due to both an increased amount of energy dissipation along with a considerable stiffening of the structure caused by an increase

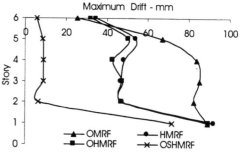

Figure 10: Maximum inter-story drifts under the 120% Canoga Park record

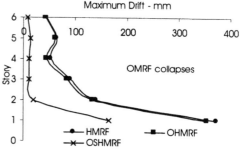

Figure 11: Maximum inter-story drifts under the 240% Canoga Park record

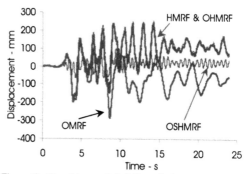

Figure 12 : Time-history of sixth floor displacements under the 120% Canoga Park record

392

in L' as well as K_{dev}. Figure 12 shows the time history of relative displacements for the sixth floor under the 120% Canoga Park record. Both the HMRF and OHMRF improve the response of the bare frame, while the OSHMRF is even more effective in reducing the displacements.

5.2.2 Moments at face of columns and maximum ductility demand in beams and columns

Since one of the major issues in retrofit strategies for pre-Northridge steel MRFs is the protection of the connections, each welded connection is examined to determine if fracture occurs during the earthquake loading. Furthermore, since the base metal can sustain plastic rotations of 0.03 rad for plastic sections before local buckling of the web, the maximum rotational ductility demand on each member is also computed.

The original structure (OMRF) sustained considerable damage under the 120% Canoga Park record, since all the welds in the first five floors fractured. The structure, however, did not collapse, as can be seen in figures 10 and 12.

Under the 120% Canoga Park record, all three of the retrofit strategies prevent any welds from fracturing by limiting the maximum moment at the face of the column below the value of M_{cr}.

Under the 240% Canoga Park record, however, the OMRF collapsed after almost all the welded connections fractured. Figure 13 shows the yielded sections, the maximum rotational ductility demand, the fractured welds and the activated energy dissipating devices under the 240% Canoga record for the HMRF, OHMRF and OSHMRF respectively. A rotational ductility demand of 11 corresponds to a plastic rotation of 0.03 rad for a 2% strain hardening ratio. All three structures predict serious problems at the base of the columns since the ductility levels exceed this value.

Five welds fractured for the HMRF while no welds fractured for the OHMRF and OSHMRF. This can be explained by the fact that the OHMRF has, for the same haunch geometry as the HMRF, an increased value of K_{dev}, computed using the approach described previously, which limits the maximum moment at the face of the column to M_y.

Figure 14, shows the time-history of the moment at one end of the first story central beam for both the HMRF and OHMRF. It can be seen that the reduction of moment caused by the increase of K_{dev} in the OHMRF protects the weld from fracturing.
Although some welds fracture for the HMRF, the global response of the structure is not greatly af-

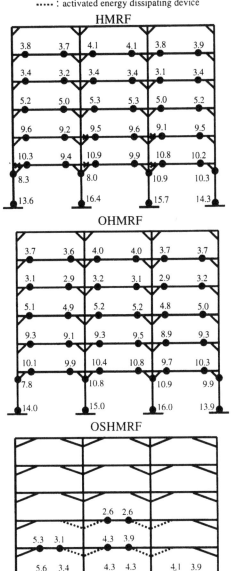

● : yielded section
✗ : fractured weld
····· : activated energy dissipating device

Figure 13: Rotational ductility demand, fractured weld sections and activated energy-dissipating devices under the 240% Canoga Park record

fected, since the presence of the haunch provides significant moment carrying capacity even after the weld fractures.

393

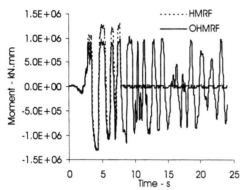

Figure 14: Time-history of moment at one end of the first story central beam under the 240% Canoga Park record.

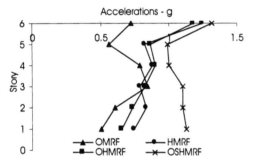

Figure 15: Maximum story absolute accelerations under the 120% Canoga Park record

5.2.3 *Maximum story accelerations*

Figure 15 shows the maximum story absolute accelerations under the 120% Canoga park record. The maximum accelerations are increased with all three retrofit strategies, especially with the OSHMRF. The significant increase of the structure's stiffness when the OSHMRF is used as well as the reduced damage and beam plastic deformation cause the increase in accelerations as the structure responds nearly elastically to the earthquake record. It may be desirable to adequately retrofit the non-structural elements of the structure if such a retrofit scheme would be used, to avoid any significant damage to equipment as well as injury to occupants.

6 CONCLUSIONS

A simple optimization procedure based on the amount of energy dissipated by a retrofitted beam-to-column connection under monotonic push is presented to compare different retrofit configurations for moment-resisting steel frames. The limits of the

pushover are set as the yielding moment at the face of the column where the welded connection is located, and a plastic rotation of 0.03 rad at any other plastic section. The haunch connection, an optimized haunch with increased stiffness, as well as an elasto-plastic passive energy-dissipating haunch were considered. Based on the time history analyses the following conclusions can be drawn:

The stiffness of the haunch connection should be increased to insure that when the plastic rotation reaches a value of 0.03 rad at the section of plastic rotation the moment is less then M_y at the level of the connection. Analyses showed that the angle of the haunch does not limit the total amount of energy dissipation that a connection can dissipate. Any angle, coupled with the corresponding haunch stiffness (K_{dev}) can achieve an optimum design. For cases where week column problems occur as a result of the added haunch, a large angle can be used to reduce the shear and moments induced in the column. The substitution of the haunch by an elasto-plastic energy-dissipating device is an efficient way to increase the seismic safety by increasing the initial stiffness of the structure, and by providing supplemental energy dissipation under seismic loading. The OSHMRF scheme proposed herein to meet non-invasive constraints reduced the structural response by more then 3 times. Further experimental studies are required to develop actual energy-dissipating devices with the required properties for haunch connections.

REFERENCES

Gross, J.L. 1998. A connection model for the seismic analysis of welded steel moment frames. Engineering Structures, Vol.20, Nos 4-6.

Gross, J.L, M.D. Engelhardt, C.-M. Uang, K. Kasai & N.R. Iwankiw 1999. Modification of Existing Welded Steel Moment Frame Connections for Seismic Resistance. American Institute of Steel Construction.

Carr, A.J. 1996. Ruaumoko - Inelastic Dynamic Analysis program. Department of Civil Engineering, University of Canterbury, New Zealand.

Hall, J. F. 1995. Parameter study of the response of moment-resisting steel frame buildings to near-source ground motions. Technical Report SAC95-05.

Tsai, K. C., & E.P. Popov 1988. Steel Beam-Column Joints in Seismic Moment Resisting Frames. Report No. UCB/EERC-88/19, Earthquake Engineering Research Center, University of California, Berkeley, Ca.

Behaviour of Steel Structures in Seismic Areas, Mazzolani & Tremblay (eds) © 2000 Balkema, Rotterdam, ISBN 90 5809 130 9

Earthquake protection of buildings and bridges with viscous energy dissipation devices

D. Di Marzo
Dipartimento di Analisi e Progettazione Strutturale, University of Naples 'Federico II', Italy

A. Mandara
Dipartimento di Ingegneria Civile, Second University of Naples, Italy

G. Serino
Dipartimento di Progettazione e Scienze dell'Architettura, Third University of Rome, Italy

ABSTRACT: The behaviour of a particular class of SDOF and MDOF systems provided with viscous damping devices is studied in detail. The examined models respectively represent isolated bridges and multi-storey buildings equipped with distributed floor isolation and energy dissipation devices. From the examination of the frequency response functions of the analysed systems, a new design procedure is developed which allows optimal selection of isolator stiffness and damping characteristics of viscous devices. The proposed procedure is finally applied and its efficacy is demonstrated through time history analyses to an isolated (Menshin) bridge and a multi-storey steel building.

1 INTRODUCTION

Many different types of viscous dampers have been proposed and applied for earthquake and wind protection of structures. These devices are usually based on the energy dissipation capabilities of silicon oils or bituminous fluids and are usually characterised by a damping force proportional to a non-unit power of the velocity. The typical location of such devices in earthquake protected structures is within energy dissipating braces in steel buildings or at the isolation interface in base isolated structures. Design procedures have been developed for both structural schemes and are commonly adopted (Serino 1994, Imbimbo & Serino 1994). Another structural configuration consists of building or bridge slabs supported through rubber pads on columns or piers with viscous dampers inserted between the slabs and the vertical supporting elements in order to limit the relative displacements and reduce the overall structural response. Such a scheme, largely adopted for seismic isolated bridges (Kawashima et al. 1991) and now starting to be utilised also in the case of highly earthquake protected buildings (Mazzolani & Mandara 1994, Mazzolani & Serino 1997, Mazzolani 1999), has the advantage of freely allowing the slow movements under thermal variations and quasi-static loading and at the same time can provide a significant energy dissipation capacity under a rapid dynamic excitation like that occurring during an earthquake. These features have proven to be rather effective for the minimisation of heat induced actions, e.g. the temperature changes on roof structures or long span bridge girders, providing at the same time a good performance also under seismic actions, due to the possibility to increase the effective degree of redundancy under dynamic loads.

A design approach adopted in the past for this class of structural systems has been to choose a column-to-deck interaction force threshold in order to limit the shear force in the supporting columns, adopting at the same time special viscous devices providing low energy dissipation below the thereshold and a high constant dissipative force – similar to that of an elastic perfectly plastic or Coulomb friction element – when the threshold level is attained (oleoplastic dampers). Nevertheless, such an approach does not allow to fully exploit the potentials of viscous dampers. Assuming a plastic threshold implies that almost the whole of the energy dissipation occurs only after that the plastic limit has been reached, but below that level the device must be very stiff in order to increase redundancy, and this corresponds to very high values of damping. Main scope of this paper is to demonstrate that a more effective optimisation in terms of forces transmitted to the supporting columns can be reached independently of any device plastic limit but only through an appropriate choice of device damping constant.

In what follows, a model with one dynamic degree of freedom and two kinematic parameters, representing a massive slab simply supported on top of vertical elements and connected in the horizontal direction through linear springs and dashpots denoting the bearing pads and the damping devices, respectively, is studied in detail. The investigation is also extended to the case of MDOF systems, representing the case of a steel multi-storey building in which the bearing pads and damping devices are located at each storey between the main girders and the columns. As a suitable design criterion for both kinds of structural system the minimisation of shear in the columns has been assumed.

2 ANALYSIS AND DESIGN OF SDOF SYSTEM

2.1 Structural model and equations of motion

The structural system is shown in Figure 1a. It consists of a portal frame in which the columns, having a lateral stiffness equal to k_c, are connected to the floor deck of total mass m by means of elastomeric bearings with lateral stiffness k_i and viscous dampers with damping constant c. This scheme can be used to represent a wide class of constructional types and has been adopted for isolated bridges (Kawashima et al. 1994) and in the rehabilitation of monumental buildings (Mazzolani & Mandara 1994, Mazzolani 1999). In the present study, the following assumptions are made:

1) infinite axial stiffness of columns and bearings;
2) infinite flexural stiffness of floor deck;
3) column mass negligible compared to deck mass;
4) linear viscous damping devices;
5) synchronous motion at column bases.

Due to anti-symmetry, the reduced system of Figure 1b has been analysed, whose corresponding mechanical model is represented in Figure 2, where $x_g(t)$ is the ground motion, $x_c(t)$ and $x(t)$ are the column top and the deck displacements, respectively, both relative to ground. For the system represented in Figure 2 the equations of motion can be written as:

$$\begin{cases} c(\dot{x}_c - \dot{x}) + (k_c + k_i)x_c - k_i x = 0 \\ \dfrac{m}{2}\ddot{x} + c(\dot{x} - \dot{x}_c) + k_i(x - x_c) = -\dfrac{m}{2}\ddot{x}_g. \end{cases} \quad (1)$$

For the sake of generality, the following nondimensional parameters are introduced:

$$\omega_i^2 = \frac{2k_i}{m}; \quad \kappa = \frac{k_c}{k_i}; \quad \upsilon = \frac{c}{m\omega_i} = \frac{c}{\sqrt{2k_i m}};$$

From the physical point of view ω_i and ν represent the system undamped frequency and damping ratio in the case of infinitely stiff columns,

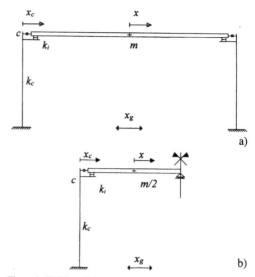

Figure 1. SDOF model: a) complete system; b) reduced system.

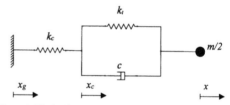

Figure 2. Mechanical model of the SDOF structure.

while κ is the column to isolator lateral stiffness ratio. With such assumptions, the equations of motion become:

$$\begin{cases} 2\upsilon\omega_i(\dot{x}_c - \dot{x}) + \omega_i^2(\kappa + 1)x_c - \omega_i^2 x = 0 \\ \ddot{x} + 2\upsilon\omega_i(\dot{x} - \dot{x}_c) + \omega_i^2(x - x_c) = -\ddot{x}_g. \end{cases} \quad (2)$$

2.2 Frequency response functions

Analysis of system response in the frequency domain allows to highlight the influence of the above defined non-dimensional parameters, independently of the specific dynamic input, and the choice of their optimal values to attain maximum response reduction. For a generic harmonic ground motion $x_g(t) = X_g(\overline{\omega})e^{i\overline{\omega}t}$ the steady-state solution can be written in the form:

$$x_c(t) = X_c(\overline{\omega})e^{i\overline{\omega}t}; \quad x(t) = X(\overline{\omega})e^{i\overline{\omega}t}. \quad (3)$$

and thus from Equations (2) one may obtain:

$$\begin{cases} AX_c - BX = 0 \\ -BX_c + CX = \beta^2 X_g \end{cases} \quad (4)$$

396

where $\beta = \overline{\omega}/\omega_i$ and:

$$A = (\kappa + 1) + i \cdot 2\upsilon\beta$$
$$B = 1 + i \cdot 2\upsilon\beta \qquad (5)$$
$$C = (1 - \beta^2) + i \cdot 2\upsilon\beta$$

It can be noted that system response depends on κ, υ and β only. Equations (4) represent a linear algebraic system with complex coefficients having $X_c(\overline{\omega})$ and $X(\overline{\omega})$ as unknowns. The solution is:

$$\begin{cases} X_c = \dfrac{B}{AC - B^2}\beta^2 X_g \\[4mm] X = \dfrac{A}{AC - B^2}\beta^2 X_g \end{cases} \qquad (6)$$

which, after algebraic manipulation on the complex terms, can be put in form:

$$\begin{cases} X_c(\overline{\omega}) = \left[\dfrac{\xi}{\Delta} + i\dfrac{\eta}{\Delta}\right]\beta^2 X_g(\overline{\omega}) \\[4mm] X(\overline{\omega}) = \left[\dfrac{\xi'}{\Delta} + i\dfrac{\eta'}{\Delta}\right]\beta^2 X_g(\overline{\omega}) \end{cases} \qquad (7)$$

where the following positions have been made:

$$\xi = \kappa - (\kappa + 1 - 4\upsilon^2\kappa)\beta^2 - 4\upsilon^2\beta^4$$
$$\eta = -2\upsilon\kappa\beta^3 \qquad (8)$$
$$\xi' = \kappa(\kappa + 1) - \left[(\kappa + 1)^2 - 4\upsilon^2\kappa\right]\beta^2 - 4\upsilon^2\beta^4$$
$$\eta' = -2\upsilon\kappa^2\beta$$
$$\Delta = \kappa^2 - \left[2\kappa(\kappa + 1) - 4\upsilon^2\kappa^2\right]\beta^2 + \left[(\kappa + 1)^2 - 8\upsilon^2\kappa\right]\beta^4 + 4\upsilon^2\beta^6 .$$

Equations (7) represent the complex frequency response functions of the system, which can be used for obtaining the expression of the transfer functions characterising the system steady-state response. These can be defined as follows:

$$\frac{X_c(\overline{\omega})}{X_g(\overline{\omega})} = W_1(\overline{\omega}) \qquad (9a)$$
representative of the shear in the column;

$$\frac{X(\overline{\omega}) - X_c(\overline{\omega})}{X_g(\overline{\omega})} = W_2(\overline{\omega}) \qquad (9b)$$
representative of the shear force in the isolator;

$$\beta\frac{X(\overline{\omega}) - X_c(\overline{\omega})}{X_c(\overline{\omega})} = W_3(\overline{\omega}) \qquad (9c)$$
representative of the force in the damper;

$$\beta^2\frac{X(\overline{\omega}) + X_g(\overline{\omega})}{X_g(\overline{\omega})} = W_4(\overline{\omega}) \qquad (9d)$$
representative of inertia forces on the mass;

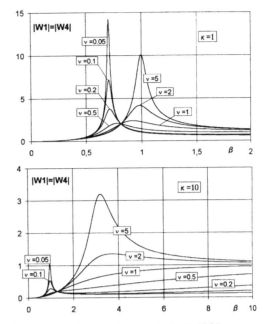

Figure 3. $|W_1| = |W_4|$ transfer functions for the SDOF system.

Note that $\beta W_2 = W_3$ and from Equations (1) it follows $W_2 + i \cdot 2\upsilon W_3 = \kappa W_1$ and $W_1 = W_4$. The modulus of W_1 as a function of β is diagrammed in Figure 3 for $\kappa = 1$, 10 and for several values of υ.

2.3 Discussion of results and design considerations

By analysing the transfer functions it is possible to observe the strong influence of the device damping factor on the system frequency response. Changing the viscous constant of dampers involves, in fact, a deep modification in the dynamic behaviour of the structure. To quantify this influence, the following limit cases can be considered:

- $c = 0$ no dampers;
- $c = \infty$ infinitely stiff dampers.

The corresponding mechanical schemes are shown in Figure 4. The free oscillation frequency for $c = 0$ is:

$$\omega_0^2 = \omega_i^2 \frac{1}{1 + 1/\kappa} \qquad (10)$$

and it can be easily demonstrated that for $\overline{\omega} = \omega_0$, i.e. for a β value equal to:

$$\beta_0 = \frac{\omega_0}{\omega_i} = \sqrt{\frac{\kappa}{1 + \kappa}} \qquad (11)$$

mathematical resonance occurs, in other words

$W_1(\omega_0) \to \infty$. It can be seen that when $\kappa \to 0$, $\beta_0 = 0$ but when $\kappa \to \infty$, $\beta_0 = 1$: therefore, in absence of any damping effect ($c = 0$), mathematical resonance always occurs in the frequency range $0 \le \beta_0 \le 1$.

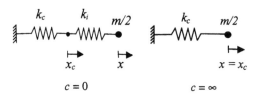

Figure 4. Mechanical schemes for limit cases ($c = 0$ and $c = \infty$).

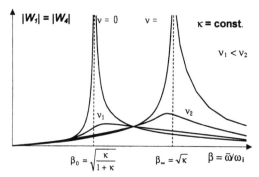

Figure 5. Influence of v on $|W_1| = |W_4|$ for SDOF system.

On the contrary, when $c = \infty$ the free oscillation frequency is:

$$\omega_0^2 = \frac{2k_c}{m} = \omega_\infty^2 \kappa \tag{12}$$

with mathematical resonance occurring when $\overline{\omega} = \omega_\infty$, i.e. a β value of:

$$\beta_\infty = \frac{\omega_\infty}{\omega_i} = \sqrt{\kappa} > \sqrt{\frac{\kappa}{1+\kappa}} = \beta_0 \tag{13}$$

It may be concluded that for very small values of the damping constant, the frequency response functions $W_1 = W_4$ have a peak for a frequency corresponding to the one of the undamped structure ($c = 0$), and its amplitude decreases as far as the damping coefficient increases, as usually occurs in a classical SDOF system. However, a further increase of the damping constant produces a shift of the resonance frequency toward the value corresponding to a rigid deck-to-column connection ($c = \infty$), increasing at the same time the peak amplitude (Figure 5). This particular behaviour indicates that it is possible to determine an optimal value of the damping coefficient, depending only on the non-dimensional parameter corresponding to the column-

pads lateral stiffness ratio κ, able to reduce the shear force in the columns to a minimum value.

Table 1. Values of v_{opt} as a function of κ.

κ	0.1	1	10	50
v_{opt}	2	0.5	0.2	0.1

The above results can be profitably used for setting up a design procedure for isolated bridges and buildings provided with viscous dampers and elastomeric isolators. In such a case, the acceleration and displacement spectra included in the relevant seismic codes may be adopted. The main design task is the determination of both the stiffness k_i of isolators and the damping factor c of viscous devices, compatibly with an assumed value of the relative deck-to-column displacement. In general the column stiffness k_c and the deck mass m are supposed to be known in advance. This is the case, for example, when the vertical elements have been previously designed according to the dead load, or simply they already exist as for example in the case of a building to be seismically upgraded. All what is needed is to determine the mechanical characteristics of the seismic protection devices.

Starting from a codified design spectrum, e.g. the elastic response spectrum of EC8, the value of k_i is normally evaluated in such a way that the actual resonance period T_{act} of the structure equipped with elastomeric bearing and viscous dampers, always between $T_\infty = 2\pi/\omega_\infty$ and $T_0 = 2\pi/\omega_0$ falls within the decreasing branch of the acceleration design spectrum, so as to take full advantage of the benefits deriving from seismic isolation. In other words, k_i is chosen in such a way that $T_{act} > T_B$, where T_B is the period value corresponding to the start of the decreasing branch of the spectrum. The optimal value of c can then be determined from the $|W_1| = |W_4|$ transfer function, so as to have the smoothest frequency response with a minimum value of the resonance peak. This because the displacement of column top is assumed as a leading parameter for the choice of the optimal damping value: minimising this displacement means, in fact, to reduce both bending moment and shear in the column at a minimum extent. The values to be assigned to the nondimensional damping constant v for obtaining the most damped response in terms of top-column displacement are shown in Table 1 for four different values of the nondimensional stiffness κ. Nevertheless, in some cases too large values of κ, that is an excessive reduction of the isolator stiffness k_i, would involve too large relative displacements in the isolators. This aspect has to be checked after the optimal value of c has been calculated.

2.4 Comparison with a worked example

The efficacy of the procedure outlined above is proven by a direct comparison with the analysis carried out by Kawashima (1994), referring to a girder bridge supported on vertical piers through rubber isolators. The methodology followed by Kawashima is based on time-history analyses adopting a wide frequency range simulated earthquake. The value of the optimal viscous constant is sought by assuming a given value of the isolator stiffness and finding through time history analyses the required minimising value of the damping ratio.

The assumed values of the relevant parameters are k_c = 17350 kN/m, k_i = 6276 kN/m, m = 241.4t. They correspond to κ = 2.76, T_∞ = 0.52s and T_0 = 1.02s. From Table 1, by suitable interpolation it is possible to estimate an optimal value for the damping ratio ν_{opt} = 0.45, able to minimise the shear force in the piers. This corresponds to a damping constant $c = \nu\sqrt{2k_i m}$ = 783 kN/(m/s).

Kawashima introduces the damping parameter:

$$h = \frac{c}{\sqrt{k_p m}}, \qquad (14)$$

$k_p = 2k_c k_i/(k_c + k_i)$ being the equivalent stiffness of columns and isolators arranged in series, and finds h = 0.5 as the damping value corresponding to the minimisation of column shear. Being the parameters ν and h linked by the relationship:

$$\upsilon = h\sqrt{\frac{k_c}{k_c + k_i}} \qquad (15)$$

h = 0.5 corresponds to ν = 0.43, very close to the ν_{opt} = 0.45 value obtained from the above proposed design procedure. Figure 6 shows the results in terms of maximum global shear force in the columns attained from time history analyses performed on the example bridge structure. The accelerogram recorded at Calitri during the earthquake which stroke Southern Italy in 1980, scaled to a PGA = 0.15g, has been considered has been considered as seismic input. This record has been chosen for its rich frequency content and wide flat amplification region of the acceleration response spectrum, so as to be a rather general seismic input for the evaluation of the optimal damping ratio. The time history response has been obtained using the SAP 2000 NL F.E.M. code (CSI 1999), which allows the use of dashpot elements for modelling the viscous dampers.

3 ANALYSIS AND DESIGN OF MDOF SYSTEM

3.1 Structural model and equations of motion

The above procedure can be formally extended to MDOF systems, by means of which the case of multi-storey building structures can be represented (Figure 7). First of all, it is worth pointing out that some preliminary assumptions are necessary if the same number of nondimensional parameters as in the SDOF system is wanted to be kept. To this purpose, the following positions are made: $m_1 = m_2 = ... = m_n = m$ and $k_{i,1} = k_{i,2} = ... = k_{i,n} = k_i$, $c_1 = c_2 = ... = c_n = c$, in which by m_i, $k_{i,i}$ and c_i have been denoted for the i-th storey the floor mass, bearing stiffness and damping constant, respectively. In addition, the assumption of constant cross section of the columns and constant inter-storey height is made. Based on these assumptions, it is possible to find a single value of the damping constant which minimises structural response. The analysis can be carried out in a way similar to the one followed for the SDOF system, by assuming the same basic assumptions and setting nondimensional parameters similar to those representing the SDOF structure. The vertical elements are represented by columns with flexural stiffness EI. The slab horizontal displacements x_1, x_2, ..., x_n and the top column horizontal displacements $x_{c,1}$, $x_{c,2}$, ..., $x_{c,n}$, all relative to the base, are assumed as kinematic unknowns of the problem.

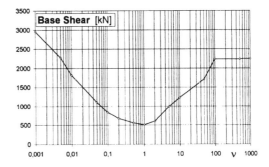

Figure 6. Base shear in the SDOF system as a function of ν.

Under the above assumptions the equations of motion can be expressed by a system of 2n 2nd order differential equations with constant coefficients. In matrix form:

$$[A]\{\ddot{q}\} + [B]\{\dot{q}\} + [C]\{q\} = -[A][I]\ddot{x}_g \qquad (16)$$

where:

$$\{q(t)\}^T = [x_{c,1}(t) \quad x_1(t) \quad ... \quad x_{c,n}(t) \quad x_n(t)] \qquad (17)$$

is the vector of the unknowns,

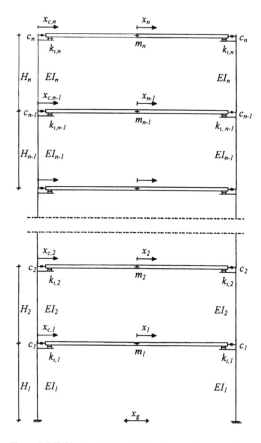

Figure 7. MDOF structural model (cantilever column case).

$$[A] = \begin{bmatrix} 0 & 0 & ... & 0 & 0 \\ 0 & m/2 & ... & 0 & 0 \\ ... & ... & ... & ... & ... \\ 0 & 0 & ... & 0 & 0 \\ 0 & 0 & ... & 0 & m/2 \end{bmatrix} \quad (18)$$

is the mass matrix,

$$[B] = \begin{bmatrix} c & -c & ... & 0 & 0 \\ -c & c & ... & 0 & 0 \\ ... & ... & ... & ... & ... \\ 0 & 0 & ... & c & -c \\ 0 & 0 & ... & -c & c \end{bmatrix} \quad (19)$$

is the viscous damping matrix and

$$[C] = \begin{bmatrix} k^{1,1} + k_i & -k_i & ... & k^{1,n} & 0 \\ -k_i & k_i & ... & 0 & 0 \\ ... & ... & ... & ... & ... \\ k^{n,1} & 0 & ... & k^{n,n} + k_i & -k_i \\ 0 & 0 & ... & -k_i & k_i \end{bmatrix} \quad (20)$$

the stiffness matrix, where the terms $k^{i,j}$ depend on the type of the supporting elements, which can be either cantilever columns or shear-type frames.

Under a generic harmonic input, the steady-state solution can be written as:

$$\{q(t)\} = \{Q(\overline{\omega})\}\, e^{i\overline{\omega}t} \quad (21)$$

with:

$$\{Q(\overline{\omega})\}^T = [X_{c,1}(\overline{\omega})\ X_1(\overline{\omega}) ... X_{c,n}(\overline{\omega})\ X_n(\overline{\omega})] \quad (22)$$

By substituting Equation (21) into (16), after simple algebraic manipulations one may obtain:

$$([C] + i\overline{\omega}[B] - \overline{\omega}^2[A])\{Q(\overline{\omega})\} = [A]\{I\}\overline{\omega}^2 X_g(\overline{\omega}) \quad (23)$$

where the matrix at the right hand of the equation represents the so called dynamic stiffness matrix. Equation (23) represents a system of complex linear equations in the unknown $\{Q(\overline{\omega})\}$, whose solution provides the structural response as a function of $\overline{\omega}$:

$$\{Q(\overline{\omega})\} = ([C] + i\overline{\omega}[B] - \overline{\omega}^2[A])^{-1}[A]\{I\}\overline{\omega}^2 X_g(\overline{\omega}) \quad (24)$$

On the basis of the assumptions made before the following nondimensional parameters can be set:

$\omega_i^2 = \dfrac{2k_i}{m}$ representative of the oscillation frequency in case of infinitely stiff columns;

$\nu = \dfrac{c}{m\omega_i} = \dfrac{c}{\sqrt{2k_i m}}$ representative of the system damping ratio in case of infinitely stiff columns;

$\tilde{\kappa} = (EI/H^3)/k_i$ representative of the column-to-isolator stiffness ratio;

The nondimensional expression of the response function is obtained by dividing equation (24) by $m/2$. This yields, after suitable manipulations:

$$\{Q(\overline{\omega})\} = ([\tilde{C}] + i \cdot 2\nu\beta[\tilde{B}] - \beta^2[\tilde{A}])^{-1}\{\tilde{I}\}\beta^2 X_g(\overline{\omega}) \quad (25)$$

where $\beta = \overline{\omega}/\omega_i$ and $[\tilde{A}]$, $[\tilde{B}]$ and $[\tilde{C}]$ are the nondimensional mass, stiffness and viscous damping matrix, respectively.

The solution of system (24) provides the transfer functions of the system in terms of top of columns displacements relative to base $X_{ci}(\overline{\omega})/X_g(\overline{\omega})$. From these, it is easy to evaluate the interstorey drifts and, hence, the shear force in the columns at each level. It has been shown by Di Marzo (1998) that, if the above assumption are fulfilled, it is possible to find a single value of the viscous damping ratio ν_{opt} which minimises the shear force at all levels. Based on the analysis carried out by Di

Marzo, the optimal values of ν have been evaluated for several number of storey n and for the two different types of supporting elements. The results are shown in Figure 8.

3.2 Analysis of an example case

The system depicted in Figure 9 is investigated, consisting of a five-storey scheme with vertical elements made of shear-type frames. The deck mass at each storey is 1032t, while the lateral isolator stiffness and the column EI flexural stiffness are 10.92kN/mm and 2.873kN·mm², respectively. These values correspond to a non dimensional stiffness ratio $\tilde{\kappa} = 4.80$. As respect to the case outlined in the previous chapter, the model under consideration is characterised by a greater height for the first storey (5.30m) compared to the upper storeys (3.80m). A similar scheme has been investigated by Mazzolani & Serino (1997), referring to the steel structures of the new Mobile Fire Brigade Building of Naples, in which viscous devices have been used to improve the seismic performance. Apart from the different inter-storey height, all the previous assumptions are fulfilled, as a result the nondimensional parameters defined above can be used. The results, summarised in Table 2, appear to be only slightly different from those referring to the general case: as expected, the most significant difference occurs for low values of $\tilde{\kappa}$, where the softer first storey of the analysed case involves higher values of the damping ratio.

Table 2. Values of ν_{opt} for the analysed scheme.

$\tilde{\kappa}$	0.1	1	10	50
ν_{opt}	1.5	0.5	0.35	0.30

From $\tilde{\kappa} = 4.80$ an optimal value of $\nu_{opt} = 0.44$ is obtained from Table 2 by linear interpolation. This is the value which produces the minimisation of shear in the columns.

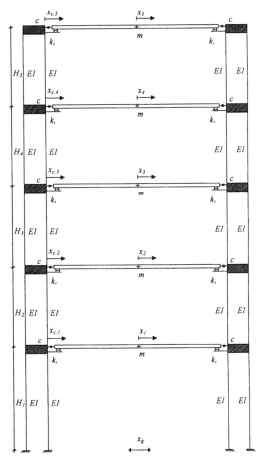

Figure 9. Model of investigated building (shear-type frame case).

3.3 Verification through time history analysis

A confirmation of the results obtained by means of the steady-state analysis in the frequency domain can be achieved through time history analyses carried out starting from a given earthquake. Again, the input accelerogram was the 1980 EW Calitri record scaled to 0.15g and the analyses have been performed using SAP 2000 NL (CSI 1999). The results are presented in Figure 10, where the plot of maximum global base shear versus damping ratio ν is shown. The optimal value $\nu_{opt} = 0.44$ is the same as the one evaluated through the steady-state

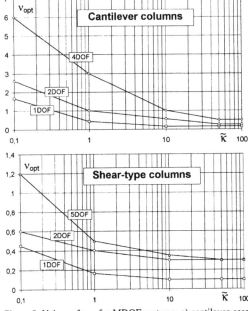

Figure 8. Values of ν_{opt} for MDOF systems: a) cantilever case; b) shear-frame case.

analysis. The beneficial effect of damping devices with c = c_{opt} compared to the case c = 0 (no dampers) and c = ∞ (rigid connection) is also shown in Figure 11, where the maximum deck and top column displacements relative to ground are plotted. It is worth to note the significant reduction of both displacements, which are attenuated by a factor equal to 3.8 as respect to c = 0 and to 6.2 as respect to c = ∞.

4 CONCLUSIVE REMARKS

The basic steps of a new design procedure able to determine the optimal values of damping and lateral stiffness characteristics of innovative earthquake protection devices in deck-isolated bridges and floor-isolated multi-storey buildings equipped with linear viscous dampers has been presented above. The procedure stems from the critical examination of the response functions of the analysed dynamic systems and has been verified through time history analyses carried out in two example cases. In this paper, only multi-storey buildings characterised by equal mass, isolator stiffness and damping constant at each floor, in addition to equal inter-storey stiffness and height, have been considered. Anyhow, the procedure could be easily generalised and applied also when these hypotheses are not valid, as well as to the case of non-linear viscous or fractional derivative energy dissipation devices.

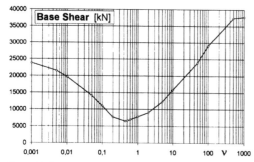

Figure 10. Base shear in the MDOF system as a function of v.

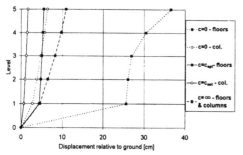

Figure 11. Effect of damping on displacements at each storey.

ACKNOWLEDGEMENTS

This work has been developed within the framework of "C.N.R. - Progetto Finalizzato Beni Culturali". Prof. F.M. Mazzolani, chief of the working group, suggested the authors to study the problem treated in this paper, as a preliminary step to the application of dissipative special devices to monumental buildings. His guidance, encouragement and precious advises are gratefully acknowledged.

REFERENCES

CSI 1999. SAP 2000 Integrated finite element analysis and design of structures, Version 7.1, Comp. and Structures Inc., Berkeley (California).

Di Marzo, D. 1998. Modelling and analysis of structural systems with viscous seismic protection devices (in Italian). Diploma thesis, University of Naples Federico II.

Imbimbo, M. & G. Serino 1994. Design of base isolation devices for steel structures. In F.M. Mazzolani & V. Gioncu (eds), *Behaviour of Steel Structures in Seismic Areas* STESSA '94: 669-680. London: E & FN SPON.

Kawashima, K., Hasegawa, K. & Nagashima, H. 1991. A perspective of Menshin design for highway bridges in Japan. *Proc. U.S. – Japan Workshop on Earthquake Protective Systems for Bridges: Tech. Rep. NCEER 92-0004*, Nat. Ctr. for Earthquake Engrg. Res., State Univ. of New York, Buffalo, N.Y.

Kawashima, K. 1994. Seismic response control of bridges by variable dampers. *ASCE, J.E.M.* vol. 129-9.

Mazzolani, F.M. 1999. Design and construction of steelworks in seismic area (in Italian). Keynote Lecture, Settimana della Costruz. in Acciaio, XVII Congresso C.T.A., Napoli.

Mazzolani, F.M. & A. Mandara 1994. Seismic upgrading of churches by means of dissipative devices. In F.M. Mazzolani & V. Gioncu (eds), *Behaviour of Steel Structures in Seismic Areas* STESSA '94: 747-758. London: E & FN SPON.

Mazzolani, F.M. & G. Serino 1997. Viscous energy dissipation devices for steel structures: modelling, analysis and application. In F.M. Mazzolani & H. Akiyama (eds), *Behaviour of Steel Structures in Seismic Areas* STESSA '97: 724-733. Salerno: Ed. 10/17.

Serino, G. 1994. Design methodologies for energy dissipation devices to improve seismic performance of steel buildings. In F.M. Mazzolani & V. Gioncu (eds), *Behaviour of Steel Structures in Seismic Areas* STESSA '94: 703-713. London: E & FN SPON.

Behaviour of Steel Structures in Seismic Areas, Mazzolani & Tremblay (eds) © 2000 Balkema, Rotterdam, ISBN 90 5809 130 9

Seismic performance of moment resistant steel frame with hysteretic damper

Y.H. Huang & A. Wada – *Structural Engineering Research Center, Tokyo Institute of Technology, Japan*

H. Sugihara & M. Narikawa – *Tokyo Electric Power Company, Japan*

T. Takeuchi – *Building Construction Division, Nippon Steel Corporation, Japan*

M. Iwata – *Department of Architecture and Building Engineering, Kanagawa University, Japan*

ABSTRACT: After the Hyogoken-Nanbu earthquake (1995) in Japan, unbonded brace as a kind of hysteretic damper has been widely used in earthquake-resistant building structures. The authors of this paper have carried out a series static and dynamic loading tests for the moment resistant steel frames with or without hysteretic damper during the past year. This paper is reporting the outlines of the experiments, test methods and results. For the purpose of comparison with the experiment results, dynamic response analysis for a 3 story steel frame structure has also been carried out. It has been confirmed that hysteretic damper can help the moment resistant steel frame to absorb a majority of input energy from earthquake so that the damage to the main steel frame can be greatly relieved. Due to the contribution of hysteretic dampers to the lateral stiffness of the structural system, the main steel frame can be manufactured relative slender and to achieve good economic performance.

1 INSTRUCTIONS

The conventional moment resistant steel frames are usually designed with weak beam and strong column. It means that the two end parts of weak beam are allowed to yield during an extreme earthquake disaster. However, it has learned from the Northridge Earthquake in the USA 1994 and the Hyogoken-Nanbu Earthquake in Japan 1995, that the extraordinary large plastic strain induced in the beam end parts is the key reason to result in the collapse of conventional moment resistant steel frames at the connection between beam and column.

Some improvement methods for the conventional moment resistant steel frames, such as, using semi-rigid connection joint at beam ends; shifting the yielding part from the welds at beam ends to the inside of the beam, have been proposed. Additionally, to attach supplementary energy dissipation devices to the steel frame is one of the most effective methods to reduce the damage of moment resistant steel frames. The supplementary energy dissipation devices help the steel frame to absorb/dissipate a great amount of earthquake input energy and protect the beam end parts from large plastic yielding or collapse.

This paper presents the seismic performance of such moment resistant steel frames with unbonded brace that acted as hysteretic damper (HD) through the static cyclic loading tests and dynamic

loading tests. Numerical response analysis for the test model has also been carried out for the purpose of comparison.

2 OUTLINES OF EXPERIMENTS

2.1 Motivation of the experiments

The bending moment distribution of a moment resistant frame is shown in figure 1 under the action of horizontal loads such as earthquake ground motion. The target of this experimental research is to see the ductility of beam end parts and the seismic performance of such frame.

Due to the limitation of the test facilities, only half span of the frame shown inside part of the circle of figure 1 is used for the test. For the convenience of loading method, the test specimen was rotated 90 degree in clockwise direction shown in right side of figure 1.

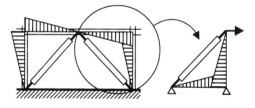

Figure 1 Part frame model

2.2 Installation of test specimens

Basically, two different types of specimens are used for this test series. MRF1 is a pure frame model without installation of hysteretic damper during the test shown in Photo 1. MRF2 is a frame model with hysteretic damper whose installation is shown in Photo 2.

In the photo pictures, the horizontal members are columns and the vertical members are beams. The column bottom (left side in the picture) is pin-supported on the reaction frame. The panel zone part between beam and column is supported by a pin-roller bearing so that allows horizontal displacement is allowed.

2.3 Test parameters

In order to compare the seismic performances of two different types of test specimens, the section size of MRF1 and MRF2+brace are adjusted so that they have the same strength approximately. The load dis-

placement relationships of these test models obtained from the static structural analysis are shown in Figure 2 and 3 respectively. Due to the contribution of the unbonded brace, the section of MRF2 frame is manufactured smaller than that of MRF1 frame. Therefore, the weight of MRF2 is only about 60% of that of MRF1. The stiffness of MRF2 frame is smaller than that of MRF1 so that MRF2 has larger elastic deformation capacity than that of MRF1 and MRF2 can keep elastic behavior even under large inter story deformation. The distribution ratio of the shear forces subjected by the hysteretic damper and the main frame is about 4:6 shown in Figure 3. Two different steels SM490 (mild steel) and HT590 (high strength steel are used for the test.

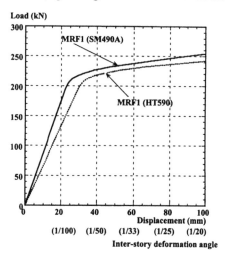

Figure 2 Load and deformation (MRF1)

Photo 1 Specimen of pure frame (MRF1)

Photo 2 Specimen of frame with hysteretic damper
(MRF2+brace)

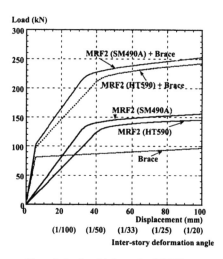

Figure 3 Load and deformation (MRF2)

Totally, 12 specimens shown in Table 1 are used for this test series. The meanings of the symbols used in Table 1 are: SM and HT stand for the steel type SM490A and HT590 respectively; w means with and w/o means without; S_cyclic means static cyclic loading and D_cyclic means dynamic loading test. HD here means hysteretic damper or unbonded brace. End taper means that the width of end flange of beam section extends gradually just like a taper. The end taper is expected to reduce the large plastic strain concentration at the welded part at beam ends.

Table 1. List of test specimens

Specimen	Steel	HD	End taper	Loading	Number
SM-A-Sta	SM490A	w	w/o	S_cyclic	1
SM-C-Sta	SM490A	w/o	w/o	S_cyclic	1
SM-A-Dyn1	SM490A	w	w/o	D_cyclic	1
SM-A-Dyn2	SM490A	w	w/o	D_cyclic	1
SM-A-Dyn3	SM490A	w	w/o	DC_cyclic	1
SM-B-Dyn	SM490A	w	w	D_cyclic	1
SM-C-Dyn	SM490A	w/o	w/o	D_cyclic	1
SM-D-Dyn	SM490A	w/o	w	D_cyclic	1
HT-A-Dyn	HT590	w/o	w/o	D_cyclic	1
HT-B-Dyn	HT590	w	w	D_cyclic	1
HT-C-Dyn	HT590	w	w/o	D_cyclic	1
HT-D-Dyn	HT590	w/o	w	D_cyclic	1

The sectional size of the specimens used for the experiments are shown in Table 2.

Table 2. Section shape and size of the specimens

	MRF 1		MRF 2	
Beam	200 / 380 / SM490A	160 / 380 / WT590	160 / 300 / SM490A	160 / 300 / WT590
Column	300 / 320 / SM490A WT590		160 / 260 / SM490A WT590	

2.4 Loading cycles

The horizontal load acted on the top of the test specimen is controlled by the inter story deformation which can be transferred into inter story deformation angle through dividing the story height. The inter story deformation can be calculated from the displacement measured from the loading point.

Loading cycles for static loading test is illustrated in Figure 4. After 2 cycle loading test at the level of 1/400 inter story deformation angle, 4 cycle loading test were carried out for each level 1/200, 1/100, and 1/67 of inter story deformation angle respectively. At last, about 100 cycle loading test were continuously carried out at the level of 1/50 inter story deformation angle until the specimen collapsed.

Loading cycles for dynamic loading test is illustrated in Figure 5 which is a kind of sinusoidal wave whose frequency is 0.65Hz. At the beginning, 2 cycles dynamic loading test were carried out at the level of 1/400 inter story deformation angle. After that, 5 cycles dynamic loading test were carried out for each deformation level shown in Figure 5. For the specimen MRF2+brace, the unbonded brace was first broken. The dynamic loading test was continued for the frame only until the frame collapsed.

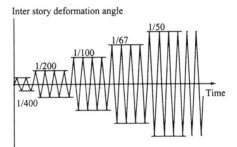

Figure 4 Static loading cycles

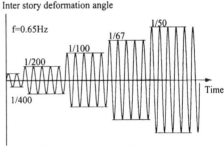

Figure 5 Dynamic loading cycles

3 ANALYSIS OF THE TEST RESULTS

3.1 Relationship of force and deformation

Assume the half span length of the beam to be L and the height of the column to be H as well as the horizontal displacement at top of the beam to be δ_L, then, the inter story deformation can be approximately obtained by $\delta_L * H/L$.

The relationships between the horizontal force and the inter story deformation obtained from the test results are illustrated from Figure 6 to 9 respectively for different specimens.

405

It is obviously understood that MRF2+brace specimen has much larger energy dissipation capacity compared with MRF1 specimen. Especially when the inter story deformation is very small like 12.5 mm (about 1/200 inter story deformation angle), MRF2+brace specimen has obvious energy dissipation capacity while the specimen of MRF1 has no any plastic loop.

When the specimen experiences large deformation like 50mm (about 1/50 inter story deformation angle), the frame MRF1 has large plastic behavior while the main frame MRF2 keeps in almost elastic region. It can be concluded that the hysteretic damper or unbonded brace helped the main frame of the specimen to reduce plastic deformation or damage while frame MRF1 that has no hysteretic damper subjected large damage.

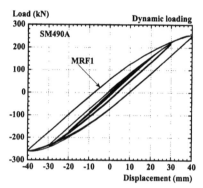

Figure 6 Load and displacement of MRF1 (SM490)

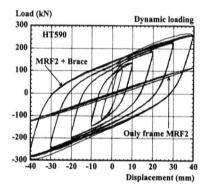

Figure 7 Load and displacement of MRF2+brace (SM490)

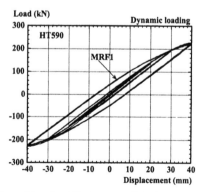

Figure 8 Load and displacement of MRF1 (HT590)

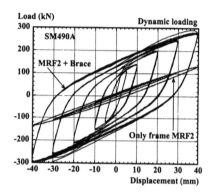

Figure 9 Load and displacement of MRF2+brace (HT590)

3.2 Energy dissipation capacity

Seen from the hysteresis loop shown in Figures 6 to 9, Specimen MRF2+brace model has much larger energy dissipation capacity. Integrating the load and displacement for each loop, the energy dissipated by each loop can be evaluated in quantity. Figures 10 and 11 show the energy dissipated at the second cycle for different types of specimen.

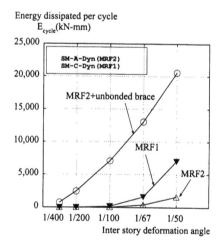

Figure 10 Energy dissipated per one cycle (SM490A)

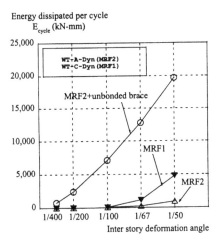

Figure 11 Energy dissipated per one cycle (HT590)

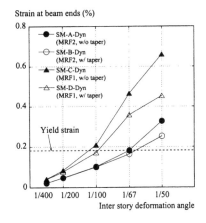

Figure 12 Local plastic strain at beam end part (SM490)

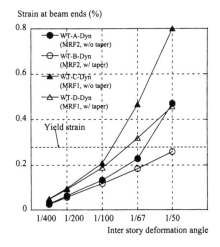

Figure 13 Local plastic strain at beam end part (HT590)

The curve denoted as MRF2 in the figures indicates the MRF2 frame only without hysteretic damper. Both the specimen manufactured of mild steel SM490A and the specimen manufactured of high strength steel have the same trend.

Under the little deformation whose inter story deformation angle is smaller than 1/100, both MRF1 frame and MRF2 frame has almost no energy dissipatd. However, under large deformation whose inter story deformation angle is about 1/50, MRF1 frame has much energy dissipated than MRF2. It means that MRF1 frame subjects much plastic deformation or damage than MRF2. MRF2 is much safer than MRF1. However, the specimen of MRF2+brace has much larger energy dissipation capacity than single MRF1 frame even under small deformation like 1/200. Due to the contribution of unbonded brace or hysteretic damer, MRF2 can keep in safe condition even under extreme large deformation.

3.3 Local plastic strain at beam end part

The trends of plastic strain at the center of beam end part increased with the increase of inter story deformation angle are shown in Figures 12 and 13 for the mild steel specimen and high strength steel specimen respectively. The plastic strain of MRF1 frame is about double larger than those of MRF2 frame. MRF1 frame begin to yield at the beam end part when the inter story deformation angle is 1/100 while MRF2 has not begun to yield even the inter story deformation angle is as large as 1/67. Taper shape flange at beam end can reduce the strain concentration at the end welded part. For the specimen MRF2+brace using high strength steel HT590 and with taper shaped flange at beam end, the largest strain at beam end is still kept in elastic even under large deformation with 1/50 inter story deformation angle.

4 DYNAMIC RESPONSE ANLAYSIS

4.1 Structural models for response analysis

In the static and dynamic cyclic loading test mentioned above, both the specimen MRF1 frame and MRF2+brace are subjected the same amplitude displacement. Since MRF2+brace model has much larger energy dissipation capacity, MRF2+brace model should has smaller displacement response than that of MRF1 frame. In order to confirm the test results, a few cases of dynamic response analysis have been carried out using 2 three story steel frames. One simulates MRF1 frame without any energy dissipation members. Another simulates MRF2+brace model with 2 unbonded braces in each story. The span length is 10 meter and story height is 6 meter. These 2 structural analysis models are shown in Figure 14. The member sizes are adjusted based on the results of static structural analysis under the action of A_i distributed external lateral force which are specified by the Japanese Building Construction Law. In the response analysis, 2% structure natural damping has been considered.

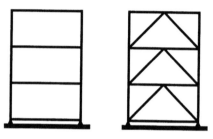

Figure 14 Structural model for dynamic response analysis

The relationship between the horizontal force and the inter story deformation at each story obtained from the static structural analysis are illustrated in Figure 15.

The solid curves in Figure 15 denote the results of MRF1 frame model while the dotted curves denote the results of MRF2+brace model. Due to the contribution of unbonded brace, the initial elastic lateral stiffness of MRF2+brace model is larger than

that of MRF1 frame. After the unbonded brace yields, the lateral stiffness of MRF2+brace become smaller than that of MRF1.

4.2 *Maximum response of inter story deformation*

The ground motion used for the time history response analysis is a kind of artificial ground motion that was generated to fit for a reasonable design spectrum using the phase filtered from the real earthquake ground motion EL Centro NS component. The strength of input ground motion was adjusted according to the level of induced maximum inter story deformation angle for MRF1 model. The distribution of the maximum inter story deformation of MRF1 and MRF2+brace models are illustrated in Figure 16 where the maximum inter story deformation angle is just equal to 1/100 for MRF1 frame. It is understood that the maximum inter story deformation angle induced in MRF2+brace model is only about 40% to 60% of that of MRF1 frame.

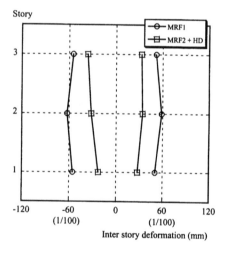

Figure 16. Maximum inter story deformation.

4.3 *Time history of energy dissipated*

Energy dissipated by the structural system was calculated in each time step. The time history of energies for MRF1 frame and MRF2+brace model are illustrated in Figure 17 and 18 respectively.

Seen from Figure 17, about 67% of total input energy has to be dissipated by the main frame MRF1. It means that MRF1 has to subject so much plastic deformation or damage. The rest 33% of total input energy was dissipated by the 2% natural structure damping.

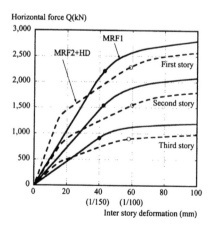

Figure 15 Relationship between force and inter story deformation for the analytical structural models

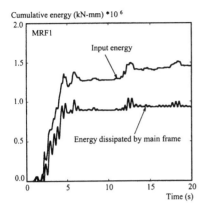

Figure 17 Time history of energy dissipated by MRF1 frame

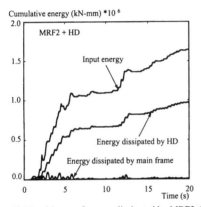

Figure 18 Time history of energy dissipated by MRF2+brace frame

While in Figure 18, about 70% total input energy was dissipated by the hysteretic damper or unbonded brace. The rest 30% of total input energy was dissipated by the 2% natural structure damping. Therefore, the main frame MRF2 is almost not necessary to dissipate any input energy. It means that the MRF2 frame can be protected from the major damage.

5 CONCLUSIONS

The following conclusions can be obtained from the experimental studies of this paper:

1) A pure steel frame without any energy dissipation members should absorb the majority of input energy and subject large damage itself. While the frame with supplemental hysteretic damper does not need to absorb the input energy by itself so that the main frame can be kept in elastic.

2) Due to the contribution of unbonded brace to the lateral stiffness of the whole structural system, the main frame can be manufactured relative slender with small section compared with that of regular pure steel frame. The main frame with supplemental hysteretic damper will keep in elastic behavior even it subject large inter story deformation angle. It can also reach good economic effect because it can be made lightly.

3) If the flange at beam end part is made in taper shape, it will relief the strain produced in the welds and reduced the damage to the weld parts during large earthquakes.

4) Although high strength steel has large elastic deformation capacity compared with mild strength steel, if the supplemental hysteretic damper is used in the steel frame, even mild steel can make the steel frame to have large elastic deformation capacity. Therefore, mild steel frame will have approximately the same seismic performance with that made of high strength steel.

6 REFERENCES

Akiyama H., Earthquake-Resistant Limit-State Design for Buildings, University of Tokyo Press, 1985.

Engelhardt M. D. and Sabol T. A. (1995), Lessons Learned from the Northridge Earthquake: Steel Moment Frame Performance, Proc. of Symposium on a New Direction in Seismic Design, Tokyo, pp.1-14.

Huang Y. H., Wada A. and Iwata M., Damage Tolerant Structures with Hysteretic Dampers, Journal of Structural Engineering, AIJ and JSCE, Vol.40B, pp.221-234, Mar. 1994.

Iwata M. (1995), Applications of Various Structural Steels to Seismic Design, Proceeding of Symposium on A New Direction in Seismic Design, Tokyo, pp. 171-194.

Iwata M., Huang Y. H., Kawai H. and Wada A., Study on The Damage Tolerant Structures, Journal of Technology and Design, Architectural Institute of Japan, No.1, pp.82-88, 1995.

Onishi, Y., Hayashi, K., Huang, Y. H., Iwata, M. and Wada, A., Cyclic Behaviors of Welded Beam-Column Connections Used in The Damage Tolerant Structures, Journal of Structure and Construction Engineering, AIJ, No. 501, 143-150, Nov. 1997.

Soong T.T. & Dargush G.F., Passive Energy Dissipation Systems in Structural Engineering, John Wiley & Sons, 1997.

Soong, T. T. and Constantinou, M. C., Passive and Active Structural Vibration Control in Civil Engineering, Springer-Verlag, New York, Inc., New York, N.Y., 1994.

Wada A., Connor J.J., Kawai H., Iwata M. & Watanabe A., Damage Tolerant Structures, Fifth US-Japan Workshop on the Improvement of Building Structural Design and Construction Practices, pp.1-12, 1992.

Behaviour of Steel Structures in Seismic Areas, Mazzolani & Tremblay (eds) © 2000 Balkema, Rotterdam, ISBN 90 5809 130 9

Seismic redesign of steel frames by local insertion of dissipating devices

M.G. Mulas & M. Arcelaschi
Dipartimento di Ingegneria Strutturale, Politecnico di Milano, Italy

J.E. Martinez-Rueda
Facultad de Ingenieria, Universidad Autònoma del Estado de México

ABSTRACT: The possibility of extending to steel moment resisting frames a retrofitting technique, previously developed for RC structures, is investigated in this work. This technique is based on the incorporation of energy dissipating devices around the regions where inelastic behaviour due to a strong earthquake is expected. This extension, however, is not straightforward, due to the deeply different nonlinear behaviour of steel and RC members under cyclic flexure. Therefore, in a first step, a numerically efficient analytical beam model has been developed, to represent the nonlinear behaviour of the devices, and has been implemented in the computer code STEFAN for the nonlinear analysis of steel plane frames, which adopts realistic models for both beam elements and joint panel zones. Secondly, making use of the code STEFAN, an 8-story, 5-bay frame, designed according to Eurocodes 3 and 8 prescriptions, has been analysed, both in its original and redesigned state. The results of the nonlinear analyses confirm the positive effects of dissipating devices for the case at study.

1 INTRODUCTION

Many old and some modern structures possess inadequate deformation capacity under earthquakes of moderate to high intensity and pose a severe risk to society. As a result of this, significant research effort has been devoted to the development of methods to enhance the seismic performance (*i.e.* seismic redesign methods) of vulnerable structures. Although conventional and innovative redesign techniques have proved effective in reducing the seismic vulnerability of existing structures, significant amount of construction work and severe disturbance to building and its occupants are the undesired side effects of these techniques. In many cases, expensive rehabilitation work results in larger seismic forces induced by the substantial increases of strength and stiffness and by the addition of considerable reactive mass. For this critical scenario, the redesign technique becomes structurally invasive with extreme cases requiring costly retrofitting of the foundation system.

To provide an alternative solution to the structural invasion and drastic increase of base shear of conventional and some innovative redesign techniques, a new redesign technique has been proposed and studied for RC structures by Martinez-Rueda (1992, 1997, 1998). As shown in Figure 1, the proposed technique incorporates hysteretic rotational devices around expected plastic regions.

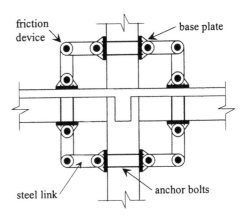

Figure 1. Proposed redesign technique applied to RC framed buildings using friction devices.

Under a major event, the main sources of energy dissipation are located in these regions; hence, the approach of the proposed technique is to deliberately activate complementary sources of energy dissipation around these inelastic regions. The main findings of the above studies can be summarised in the following terms:
- the installation of the devices results in significant reductions of seismic response in terms of inter-story displacements, residual displace-

ments and damage accumulation.
- the main parameter influencing the efficiency of the proposed redesign technique is the rotational strength of the devices. For a given original structure, the optimum device strength is independent of the seismic input and corresponds to the maximum strength preserving the location of the plastic hinges in the original structure.
- redesigned structures, with the devices tuned at optimum strength, experience mild increments of strength and stiffness, and the major source of seismic response improvement comes from the added hysteretic damping.

Based on the satisfactory results obtained for RC structures, in this research work the applicability of the proposed redesign technique is evaluated for the case of steel-framed buildings. For steel structures it is anticipated that the incorporation of the proposed device assemblies, using either bolted or welded connections, is virtually straightforward, when compared to the case of RC structures. A rigorous and efficient numerical evaluation of the performance of the devices, when installed in steel buildings, must be based on the adoption of realistic models for the behaviour of all the elements experiencing nonlinear behaviour. Therefore, a new beam model, describing the rigid perfectly plastic behaviour of the rotational devices, has been developed and implemented in the existing computer code JOINT (Mulas & Fogazzi, 1998) for the nonlinear analysis, both static and dynamic, of steel plane moment resisting frames (MRFs) having deformable joints panel zones. Beam elements are represented with the nonlinear lumped plasticity model developed by Martinelli, Mulas & Perotti (1996, 1998). A new code has been developed, named STEFAN (STEel Frames ANalysis); a series of nonlinear analyses, both static in displacement control and dynamic, have been performed on a 9-story, 5-bay MRF designed according to Eurocodes 3 and 8, both in its original and redesigned state. The results, even though limited to a single frame, show a substantial decrease of the structural response and indicate the satisfactory performance of the devices also for steel MRFs.

2 ANALYTICAL MODELLING OF ELEMENTS

2.1 Existing nonlinear elements

The analytical modelling of the MRF is based on an accurate, yet simple, representation of the elements undergoing nonlinear deformations during the earthquake. Due to capacity design, these are mainly concentrated in the beams regions around the beam-column joints and in the panel zone of the joint itself. In fact, this, even strengthened with

doubler plates, can evolve in the nonlinear range; moreover, some recent seismic codes allow yielding of the web panels prior to the fully developing of beams moment capacity. For the sake of completeness, the main features of the elements adopted in the present work will be presented in the following.

The beam element adopted in this work (Martinelli et al. 1996, 1998) is based on a lumped plasticity approach: in fact, the experimental evidence points out that, due to local buckling and low-cycle fatigue phenomena, the nonlinear effects due to bending do not spread along the beam. The beam model, represented in Figure 2, is composed of an elastic central beam, connected in series at the ends with two nonlinear springs, representing the plastic hinges. These follow separate and independent moment-rotation hysteretic relations, having a rigid-plastic primary curve, since in the linear range the beam flexibility is due entirely to the central beam.

The springs hysteresis law has been directly derived from the results of a series of experimental tests on cantilever beams. The primary curve, symmetric, is defined by the plastic moment M_y and the slope of the plastic branch. The generic hysteresis curve is trilinear, composed of three branches: a rigid unloading branch, a stiffness degraded reloading branch and a plastic branch. The damage accumulated in the beam, expressed as a function of the maximum semi-cycle amplitude previously experienced, causes a decrease in the stiffness of reloading and yielding branches and reduces the flexural strength. The latter effect is counteracted by isotropic strain hardening, which, modifying the moment level at which yielding onset occurs, leads to an expansion of the area subtended by the cycles.

A flexibility approach is adopted to derive the (2*2) stiffness matrix of the element; this is assembled in the structure stiffness matrix through a transformation matrix T, taking into account rigid body modes and the rigid offset zones resulting from the finite dimensions of the joints. When the latter are modelled as deformable elements, the matrix T is accordingly modified. Finally, an event-to-event strategy, applied at the element level, is adopted to solve the problem of state determination (determination of internal restoring forces) arising from the adoption of a flexibility based model in the

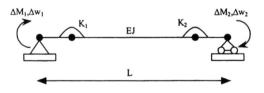

Figure 2. The nonlinear beam element.

framework of the direct stiffness method usually adopted by structural codes.

The joint model applies to a welded joint, with column flange stiffeners, where two beams and two columns frame, as indicated in Figure 3. It is supposed that the cross-sections of the bars do not change across the joint. The derivation of the model is based on two different assumptions, one concerning the static and the other the kinematics of the joint. The first hypothesis, usual in I-beam design, assumes that bending moment and axial force are transmitted mainly by the profiles flanges, and that shear is transmitted mainly by the panel zone, with a constant stress distribution. Therefore, two different resisting mechanisms can be identified for the joint, one for normal stresses and the other for shear stresses, independent in terms of both stiffness and strength. If the aspect ratio of the joint h/d is not too different from unity it can be furthermore assumed that the web panel edges deform maintaining rectilinear. Consistently with these two hypotheses, the plane truss depicted in Figure 3 represents the resisting mechanism for normal stresses, accounting for the stiffness of beam and column flanges; an ideal shear-resisting element is connected in parallel to the plane truss to represent the panel zone shear stiffness and to constrain a possible relative sliding among parallel bars. By further assuming that each bar of the truss is subjected to constant axial strain, the displacements u_i and v_i, depicted in the same figure, completely identify the deformed configuration of both mechanisms.

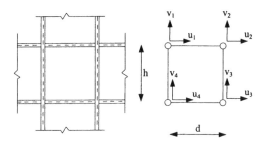

Figure 3. The joint at study and the adopted model.

Through a change in the coordinate system based on geometric considerations, as it is shown in Mulas (1996), these displacements can be resolved into two separate displacement systems, representing the rigid body modes of the joint and its deformation modes respectively. The former describes the nodal degrees of freedom (dofs) usually adopted in the analysis of plane frames (the in-plane translations and the rotation). The latter represents the joint deformations, namely the extension and the flexure in two perpendicular directions and the angular distortion due to shear, as depicted in Figure 4.

From the analytical relation governing the change in the coordinate system it is possible to derive both the stiffness matrix of the beam element referred to the new coordinate system and the generalised components of the beam internal forces deforming the panel. The joint stiffness for its deformation modes can be easily derived making use of the virtual work principle.

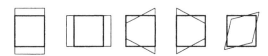

Figure 4. The joint deformation modes.

A constitutive relation for the panel zone is adopted, in terms of the average shear stress τ - shear strain γ, having a bilinear elastic-plastic primary curve and a trilinear hysteresis cycle (Mulas & Garavaglia, 1996). This, based on experimental data, is characterised by a degrading-evolutive behaviour in terms of stiffness for the reloading and plastic branch, and on a mixed kinematic-isotropic hardening. As for the beam element, the amount of stiffness degradation and isotropic hardening is a function of the maximum semi-cycle amplitude previously experienced.

2.2 The new element for the dissipating devices

The experimental moment-rotation relation of the friction devices is practically rigid-perfectly plastic (RPP): each dissipating device therefore can be thought as composed of two beam elements, one horizontal and one vertical, where an ideal plastic hinge can form, when the corresponding end friction device is in the plastic range. Due to the numerical problems arising when modelling branches having infinite or zero stiffness, an alternative approach, with respect to that adopted for the nonlinear beam element, has been followed, to avoid the explicit definition of the experimental hysteresis relation. The (2*2) element stiffness matrix (see Fig. 2 for reference) is assembled taking directly into account the possible end situations: no plastic hinges, one plastic hinge (at the right or left end) or two plastic hinges; this last situation leads to a null stiffness matrix. The finite dimensions of the friction device connection to the beam or the column is accounted for by a matrix T, as it was done for the beam element. The friction device rotation, that cannot be directly derived due to the null slope of the plastic branch, is computed from compatibility conditions. The transition from one stiffness matrix to another is not governed by a unique variable. In fact, the yielding onset happens when the moment exceeds, in absolute value, the yielding moment of the friction device; the rigid branch is entered when unloading is detected, due to a change in sign in the rotation increment. An event-to-event strategy applied at the element level is adopted also for this element, to exactly determine the restoring internal forces acting at the end of each load increment.

2.3 *The structural analysis code STEFAN*

The model proposed for the dissipating devices has been implemented in the computer code JOINT. The upgraded version of the code, named STEFAN, can perform nonlinear static and dynamic analysis of steel plane MRFs, both in the original and redesigned state, accounting for joint deformability. Nodes are defined at the intersections of horizontal and vertical bars: as it has been described, for each node up to 8 dofs can be separately and independently activated. A master-slave option is implemented to model in-plane rigid-floors; a geometric stiffness matrix, referred to the horizontal displacements of columns ends, takes into account $P\text{-}\Delta$ effects.

The nonlinear equations of motions are integrated by means of the Newmark's method; Newton-Raphson method, associated to a path-independent state determination, is adopted to eliminate unbalanced loads at the end of each time (or load) step. An energy criterion is adopted to check convergence.

To correctly model the effect of devices, which are supposed to be applied to the structure when gravity loads are already acting, the assembly of structure stiffness matrix is made twice: in a first step internal forces due to gravity loads are computed on the original structure. Then, the stiffness matrix is re-assembled accounting for the device presence, and the nonlinear incremental analysis, either static or dynamic, is performed.

A further problem is posed by the device modelling: a null term on the diagonal of the stiffness matrix can arise when the plastic hinge form on the intermediate node of the device. In this situation a zero rotational stiffness is found in both the elements (horizontal and vertical) facing into this node. The problem can be simply overcome in dynamics, by accounting for the rotational inertia of the device. In static, an infinite strength is forced by the code in one of the cross-sections: the exact value of the internal forces (corresponding to the null rotational stiffness) is found by the iterative Newton-Raphson process enforcing equilibrium.

Finally, it must be observed that, also due to an appropriate usage of the master-slave option, the *ad hoc* developed rigid-perfectly plastic flexural element allows for keeping to a minimum the total number of additional dofs needed by the modelling of the devices.

3 BEHAVIOUR OF A REDESIGNED MEMBER

As a first check on the device effectiveness the behaviour of an original (without devices) and redesigned (with devices) HE400B column, shown in Figure 5, has been numerically tested under cyclic displacements applied at the column top. The applied peak displacements were $1d_y$, $-2d_y$, $3d_y$, $-4d_y$, d_y being the yield displacement of the original column.

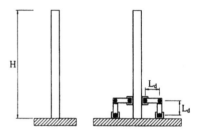

Figure 5. The analysed cantilever column.

Figure 6 compares the hysteretic response for the case of H = 4000 mm, L_d = 550 mm; the device strength M_d is equal to $0.02M_p$, M_p being the plastic moment of the column cross-section. For the selected device strength an efficient activation of the devices is found, since the location of the plastic hinge is the same, irrespective to the devices presence. The comparison of the hysteresis loops indicates an increase in the hysteretic damping of the redesigned column; the moment-rotation diagrams of the plastic hinge, not reported here, denote a decrease in ductility requirements.

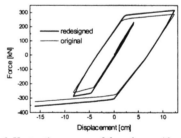

Figure 6. Hysteretic response of the column with and without devices.

4 THE FRAME AT STUDY

The 9-story 5-bay welded frame analysed in this work is the central frame, parallel to the long dimension, of a residential building structure,

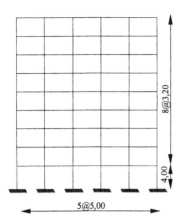

Figure 7. The frame at study.

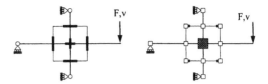

Figure 8. The redesigned story subassemblage.

The subassemblage extends up to the contra-flexure points due to horizontal loads; in Figure 8, left, the end constraints (rollers), the point of application of the load and the rigid ends are shown. Figure 8, right, shows the nodes and the elements adopted for the numerical analyses. The black squares denote the ends of the RPP elements where an infinite strength was forced; the central node is modelled with the joint element. The imposed history of reversed cyclic displacements includes the following cycles applied once: $\pm0.25d_y$, $\pm0.5d_y$, $\pm1d_y$, $\pm2d_y$, $\pm4d_y$, $\pm6d_y$, d_y being the beam vertical tip displacement producing yielding at the interface with the column in the original subassemblage. The dimension L_d of the device assembly (Fig. 5) was selected equal to the column depth and the device strength was 0.02 of M_p, the column plastic moment. Figures 9 and 10 compare the cyclic response and the stiffness degradation of the subassemblages with and without devices, respectively for different modelling and design assumptions for the joint panel.

composed of three parallel frames spaced 7.5m apart. The overall dimensions of the frame are depicted in Figure 7 and the adopted profiles are shown in Table 1.

The frame, built in Fe430 steel (f_y= 275MPa), has been designed for gravity loads and seismic actions as specified by the Eurocode 8 for a soil profile type B, damping ratio 0.03, behaviour factor q=6 and PGA equal to 0.40g. The design of the frame is governed by the requirements on interstorey drifts at serviceability limit state and by capacity design according to the strong column-weak girder design philosophy. The Eurocode 3-Annex J prescriptions have been followed in checking panel zones strength in shear. Plates doubling the panel zone thickness are required in all the interior joints. Further details on the frame design can be found in Mulas and Fogazzi (1998).

Table 1. European steel profiles adopted in the frame

Level	Beams	Interior columns	Exterior columns
1-2	IPE500	HEB500	HEB450
3-4	IPE450	HEB500	HEB450
5	IPE450	HEB450	HEB400
6-7	IPE400	HEB450	HEB400
8	IPE360	HEB400	HEB360
9	IPE360	HEB400	HEB360

5 BEHAVIOUR OF A REDESIGNED BEAM-COLUMN SUBASSEMBLAGE

To provide a first insight on the behaviour of the frame retrofitted with the friction devices, a beam-column subassemblage, representing an interior joint at the 2-*nd* floor of the frame at study, has been analysed, both in the original and redesigned form, under a cyclic load applied in displacement control.

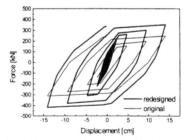

(a) Deformable joint panel without doubler plates.

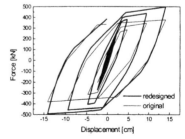

(b) Rigid joint panel.

Figure 9. Comparison of hysteretic response for the beam-column subassemblage.

As expected, the peak-to-peak secant stiffness of the hysteresis loops is higher in the model with a rigid joint panel. However, the stiffness degradation rate is virtually the same for both models. The initial stiffness of the original model with rigid joint panel is about 45% higher than that with the deformable panel, whereas for the redesigned model the difference in stiffness is about 10%.

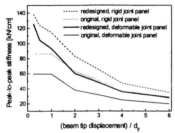

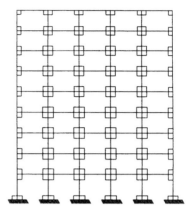

Figure 11. Redesigned frame model.

Figure 10. Evolution of stiffness degradation in the beam-column subassemblage.

It is also interesting to notice that the more realistic model, *i.e.* that modelling the inelastic response of the joint panel, predicts fatter hysteretic loops under short displacements. These findings pinpoint the need of modelling the joint panel deformations not only for the original model, but also for an adequate prediction of the device contribution to the hysteretic damping in the redesigned structure under short and large displacements. It must be finally observed that the devices presence does not modify the structural behaviour, showing a large damage in the beams and a limited amount of damage in the joint panel, if the latter is reinforced with doubler plates, and an opposite situation when the joint panel is not reinforced. The device presence causes, in both cases and for both the elements, a decrement in the excursion in the inelastic range; however, an undesired plastic hinge appears in the beams, immediately outside the device connection.

6 BEHAVIOUR OF A REDESIGNED FRAME

The frame presented in section 4 has been redesigned by adding the friction devices in all of the joints, as shown in Figure 11. The devices strength has been calibrated to 2% of the plastic strength under pure bending of the largest column connected to the node. Dynamic analyses of the original and redesigned frame have been performed with the following criteria. The frames under study were modelled as presented in the previous section, considering different modelling and design assumptions for the joint panels; only the joint shear defor-

mation has been accounted for. In-plane rigid floors have been considered, and inertia forces are associated with the dof describing the horizontal translation of each floor. Rayleigh type viscous damping have been considered, assuming a modal damping ratio equal to 0.03 for the first two modes of the original elastic structure. The model for the original frame was made of 99 inelastic beam elements, 54 inelastic joint panel elements and 60 nodes which, accounting for boundary conditions, correspond to 171 dofs. The redesigned frame was modelled with 364 RPP elements, 297 inelastic beam elements, 54 inelastic joint panel elements and 452 nodes, corresponding to 1221 dofs. Two artificial accelerograms (named *Eurb1* and *Eurb11*), compatible with the curve B of the elastic spectrum of Eurocode 8, PGA=0.40g, have been adopted to perform dynamic analyses; these were identified, in a wider set, as those imposing the most severe ductility demands for the original frame.

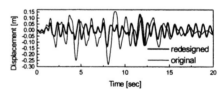

(a) Top story displacement.

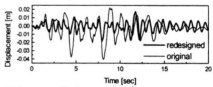

(b) Interstory displacement of 5th story.

Figure 12. Response history of displacements of frame model with deformable joints.

The redesigned model exhibited significant displacement reductions when compared to its original counterpart. An example of this reduction is shown in Figure 12, illustrating the time history, due to *Eurb11*, of the top story displacement and of the interstory displacement at the *5th* level, where the peak value was found for the original structure. Figures 13 and 14 illustrate the envelopes, due to *Eurb11*, of story and interstory displacements of the frames studied, respectively. Similar results, not reported here, are found for *Eurb1*, differing only in the fact that the scatter due to the joint modelling practically disappears, as it happens when the average of the envelopes, over several accelerograms, is performed. From these figures the dramatic reduction, at global and local level, in the frame response is apparent, with a maximum difference, due to the incorporation of the devices, of the order of 50%.

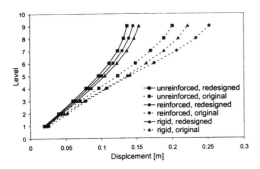

Figure 13. Maximum story displacements.

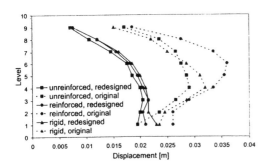

Figure 14. Maximum interstory displacements.

The above results can be better understood by analysing the distribution of the plastic deformations inside the frame; this is illustrated in Figures 15 and 16, for the cases of the frame with and without doubler plates respectively. In these figures circles denote yielding in the panel zones and squares in the beam elements, the largest amount of plastic deformations corresponding to the black symbols and the

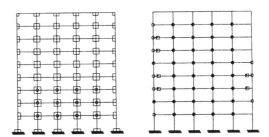

Figure 15. Distribution of plastic deformations, frame without doubler plates (*Eurb11*).

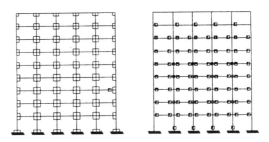

Figure 16. Distribution of plastic deformations, frame with doubler plates (*Eurb11*).

smallest amount corresponding to the white symbols. In the original frame without doubler plates inelastic deformations are mainly found in panel zones yielding in shear, while in the original frame with doubler plates are mainly due to beams yielding in flexure. The redesigned frames, irrespective to the presence of doubler plates, are practically in the linear range, showing only a very limited amount of inelastic deformations in structural elements. Differently from the results of the static analyses over the story subassemblage, only one (Fig. 16) plastic hinge is found in an undesired location.

The frame behaviour is further clarified by the analysis of Figures 17 and 18, showing the amount of energy dissipated by the hysteresis cycles of ine-

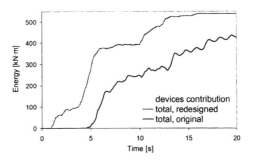

Figure 17. Dissipated energy, frame without doubler plates (*Eurb11*).

417

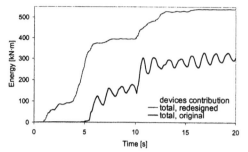

Figure 18. Dissipated energy, frame with doubler plates (*Eurb11*).

lastic elements for the frames at study. The dissipated energy in the redesigned frames, practically entirely due to the dissipating devices, shows a significant increase, consistent with the observed substantial reduction in displacements, when compared to the original frame.

7 CONCLUSIONS

In this paper the possibility of extending to steel moment resisting frames a retrofitting technique previously developed for RC frames has been analysed. To this purpose a numerically efficient analytical beam element, modelling the response of the rotational friction devices, has been developed and implemented in an existing computer code.

A series of nonlinear analyses, both static in displacement control and dynamic, of a prototype steel frame have been performed, with the twofold purpose of testing the proposed element and verify the devices effectiveness. The results confirm the adequacy of the element; it must be noted, however, that the adoption, in the same structural model, of nonlinear flexural elements having strength which differs by two orders of magnitude, requires a careful choice of the time (or load) step.

From the structural point of view, it can be surely concluded that the proposed technique has a highly positive effect on the seismic behaviour of the frame. It is interesting to notice that the static analysis in displacement control of the subassemblage predicts the formation of plastic hinges in beams in undesired locations, which has not been found with the dynamic analyses. This fact points out that, while monotonic static analyses are necessary to exactly calibrate the devices strength, a dynamic analysis is necessary to exactly evaluate the devices effect on the frame seismic behaviour.

The work presented here can be viewed as a necessary first step toward the application of the proposed redesign technique to steel frames. The results can be further validated by the analysis of a large number of frames, for several different accelerograms. The development of a design procedure to determine the optimal number, strength and location of the devices to be used appears to be very important. In fact, it is not necessary to maintain the frame in the linear range, and the application of so many devices can be costly and unpractical.

ACKNOWLEDGEMENTS

The work herein presented has been performed as a part of a thesis (presented at Politecnico di Milano in partial fulfilment for the bachelor's degree in Civil Engineering) developed by the second author under the guidance of the first author. The contribution of C. Graneroli, co-author of the thesis, is gratefully acknowledged.

REFERENCES

CEN, European Committee for Standardization, 1994. *Eurocode 3. Common unified rules for steel structures.*

CEN, Techn. Comm. 250/SC8, 1994-95. *Eurocode 8: Earthquake Resistant Design of Structures - Part 1: General Rules and Rules for Buildings*, (ENV 1998-1-1/2/3), CEN, Berlin.

Martinelli L., Mulas M.G., Perotti F. 1996. The seismic response of concentrically braced moment-resisting frames. *Earthquake Engineering and Structural Dynamics*, Vol. 25, 1275-1299.

Martinelli L., Mulas M.G., Perotti F. 1998. The seismic behaviour of steel moment-resisting frames with stiffening braces. *Engineering Structures*, Vol. 20, No. 12, 1045-1062.

Martinez-Rueda, J.E. 1992. A novel technique to upgrade existing RC structures with soft first storey. *MSc Dissertation.* Imperial College University of London.

Martinez-Rueda, J.E. 1997. Energy dissipation devices for seismic upgrading of RC structures. *PhD Thesis.* Imperial College University of London.

Martinez-Rueda, J.E. 1998. Seismic redesign of RC frames by local incorporation of energy dissipation devices. *Proceedings of 6thUSNCEE.* Seattle.

Mulas M.G. 1996. An analytical model for beam-to column joints behaviour. Costruzioni Metalliche, No. 2, 15-31.

Mulas M.G. & Fogazzi, P. 1998. Seismic behaviour of steel MRF with inelastic joint panels. *Proceedings of 11thECEE*, Paris. Balkema.

Mulas M.G. & Garavaglia E. 1996. The effect of joint deformations on steel plane frames behaviour. *Int. Conf. On Advances on Steel Structures*, Hong-Kong, Vol. I, 189-194.

Behaviour of Steel Structures in Seismic Areas, Mazzolani & Tremblay (eds) © 2000 Balkema, Rotterdam, ISBN 90 5809 130 9

Seismic dampers utilization and design for steel structures

Wenshen Pong
School of Engineering, San Francisco State University, Calif., USA

ABSTRACT: A special steel moment frame building with seismic dampers was designed to meet a enhanced structural performance level in the Bay area. In order to protect the valuable computer data center, passive control with steel moment frame system were selected as a most cost-effective solution to control the lateral acceleration and displacement for earthquakes. The story drift ratio was limited to 1% of the story height during the Design Basis Earthquake (DBE), which has a mean return period of 475 years. Seismic dampers were mounted on Chevron braces in selected locations each floor. Approximately 20% of critical damping were provided. The primary seismic resistant frames, without dampers, were designed to meet the Uniform Building Code of 1997. In addition, the seismic resistant frames, with dampers, were designed to remain elastic under a Design Basis Earthquake. 3 site-specific DBE time-history analyses were performed and the maximum values were selected for design. The goal is to limit the lateral drift to 1% of the story height and, the Demand-to-Capacity ratio (DCR) of the earthquake resistant members within 1 under the DBE event. As a result, collector forces were found to be much higher than originally expected. More piles with larger lengths and bigger pile caps were designed to meet the DBE demands.

1 INTRODUCTION

1.1 Energy Dissipation System

The addition of Energy Dissipation System ("EDS") in structures provides a new and innovative option to improve structural performance. This is an emerging technology that enhances structural performance by implementing supplemental damping to the building. With such an implementation, the structural displacement can be significantly reduced. Force demands would also be reduced provided the structure is responding elastically. The effectiveness of seismic dampers is reduced when the structure experiences inelastic deformations. Therefore, force might be either increased or reduced if the structure is responding beyond elastically.

Currently, the Uniform Building Code ("UBC") has not recognized the use of passive energy dissipation systems while it does address the design procedures for base isolation. Due to the complexity of structural seismic response and the uncertainty of ground motion, the effects of inelastic behavior of frames with a EDS are important. In addition, the supplemental damping becomes insignificant in comparison to that generated by the structural inelastic deformations.

This study investigates the behavior of steel moment frame structures, with and without seismic dampers, as they are subjected to various site-specific earthquake records. The EDS chosen for this study was a Fluid Viscous Damper because the damper forces are out of phase with the axial loading of the columns, and the structural period does not alter due to the addition of fluid dampers.

1.2 Structural Performance Level

Methods and design criteria to achieve different levels and ranges of seismic performance are well defined in FEMA-273 (FEMA-273, 1997) document. The four building performance levels are collapse prevention, life safety, immediate occupancy and operational. Probabilistic hazard levels used in this design and their corresponding mean return periods (the average number of years between events of similar severity) are shown in Table 1. The Basic Safety Objective (BSO) for the building design is to meet the life safety building performance level of 10%/50-year earthquakes and to meet the collapse prevention performance level of 2%/50-year earthquakes. Any objective intended to provide performance superior to that of BSO is termed an Enhanced Objective (FEMA-273, 1997).

In order to meet immediate occupancy performance level, structural damage shall be limited to the following situations which include; (1) minor local yielding at a few places; (2) no fractures; and (3) minor buckling or observable permanent distortion of members. The use of energy dissipating devices in the structures, or the isolation of a structure from earthquake are commonly used to achieve that performance level because application of such seismic protective systems can effectively reduced the demand on structures.

Table 1. Earthquake Definition

Definition	Earthquake Having Probability Exceedance	Mean Return Period (years)
Lower Level Earthquake (LLP)	50%/50 year	72
Design Basis Earthquake (DBE)	10%/50 year	474
Maximum Credible Earthquake (MCE)	10%/100 year	950

2 BUILDING DESCRIPTION

2.1 Steel Moment Frames

A 4-story 11,150 m^2 office building with supplemental damping devices was designed for better seismic performance. Typical floor height is 4.3 m and floor mass is 16,400 KN. The beam members are typical W24 and Columns are W27.

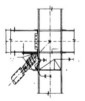

Figure 1. Brace Connection

Approximately 20% of critical damping were provided using viscous dampers that are mounted on diagonal braces in selected locations on the each floor. The brace with a damper and the brace connection are shown in figure 1. Due to the change from a moment frame to a braced configuration, load paths were altered to produce substantial axial loads in the columns. As a result, the number of piles and the size of pile cap were significantly increased. Foundation cost is one of major issues for structures with seismic dampers.

2.2 Site Specific Ground Motions

Site specific ground motion records are required for time-history dynamic analyses. Development of site-specific response spectra shall be based on the geologic, tectonic, seismologic, and soil characteristics associated with a specific site (UBC 1997). Response spectra should be developed for an equivalent viscous damping ratio of 5%. Additional spectra are also developed for other damping ratios. Ground motion time histories developed for the specific site shall be representative of actual earthquake motions. One pair of LLE and three pairs of DBE and MCE were prepared by geo-technical engineers (Treadwell & Rollo 1997). The time histories used in spectral matching are shown in Table 2, and the peak acceleration for three level earthquakes is shown below in figure 2. Figure 3 shows the time histories used in spectral accelerations at various damping ratios.

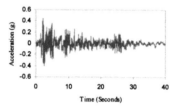

Figure 2. Ground Motion (DBE)

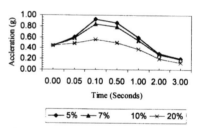

Figure 3. Spectral Accelerations

Table 2. Ground Motion Used in Spectral Matching

Design Earthquake	Earthquake	Time History	Dist. (km)
LLE	Loma Prieta	Santa Teresa	18
DBE	Imperial Valley	El Centro	12
	Landers	Joshua Tree	15
	Loma Prieta	Saratoga	3
MCE	Imperial Valley	El Centro	12
	Landers	Joshua Tree	15
	Kern County	Taft	56

3 ENERGY DISSIPATION SYSTEM

3.1 Fluid Viscous Dampers

The simplest model to simulate the mechanical behavior of fluid viscous damper is the Maxwell model (Bird, Armstrong & Hassager 1987) given by

$$P + \lambda \dot{P} = C_0 \dot{U} \qquad (1)$$

Where λ is the relaxation time, C_0 is the damping constant at zero frequency, P is the damping force, and U is the damper position velocity. Constantionu & Symans 1992 and Pong, Tsai & Lee 1994 also investigated the mechanical behavior of fluid viscous dampers.

Since fluid viscous damper force is a function of the damping constant and velocity, the damper force can be defined as $F = CV^\alpha$ where C is damping constant, V is velocity, and α is velocity exponents. The cyclic response of the damper is dependent on the velocity of motion. The mechanical behavior of the damper is dependent on the frequency and amplitude of the motion (Contantionu & Symans 1992). The addition of fluid viscous dampers should be detailed so that the period of damped structure does not change in comparison to that of bare frames without dampers.

3.2 Design Criteria

Special steel moment frames without dampers were designed to conform to the UBC of 1997 requirements for strength and drift. The seismic design parameters are as follows:

Important Factor, I	1.0
Site Profile Type	S_D
Ductility Factor, R	8.5
Seismic Force Amplification Factor, Ω_0	2.8

The maximum inelastic response displacement under the UBC earthquake design force was limited to 0.02 times the story height. The structural period was found to be about one second. As a result, the static base shear force equals 7.2% of total story weight. The stress of members was relatively small using the Load and Resistance Factor Design method because structural design for moment frames is typically controlled by story drift. However, the final design of moment frame members was governed by the time-history analyses for the damped structure. Steel moment frames with dampers were designed to remain elastic under a DBE. The goal was to limit the lateral drift to 1% of the story height and the Demand-to-Capacity ratio ("DCR") of the moment frame members and connections to within 1 under the DBE events. In case of a larger earthquake such as a MCE, plastic hinge formations and some structural damage would be expected. Therefore, the post-Northridge moment frame connections were designed to provide the ductility to minimize the structural and nonstructural damage in the event of a MCE. Reduced beam section connections (FEMA 167A 1997) were selected to produce an intended plastic rotation hinge zone.

Steel moment frames with supplemental damping were designed to have a DCR of less than 1 under site-specific time-history dynamic analyses. Three DBE time-history analyses were performed and the maximum response of the parameter of interest was used for the final design. Each pair of site-specific time histories was applied simultaneously to the computer model, considering the most disadvantageous location of mass eccentricity (UBC 1997).

Diagonal braces were designed to have a DCR remain within 1/2 so that the braces would have higher safety factors. As a result, 10-inch extra strong pipes were selected to meet the design and construction requirements. This will ensure that the braces function properly under higher earthquake demands. Although, the seismic dampers can reduce the story drift and, thus reduce the column bending moment, the load paths are also changed. This alteration of load paths leads to substantial axial loads in the columns. As a result, demands on the foundation were significantly increased with the addition of dampers. More piles and larger pile caps were designed to keep the DCR to be less than 1 at strength level.

3.3 Structural Seismic Response

The DCR for beams and columns at LLE, DBE and MCE are tabulated in the table 3 and 4. Table 5 to 10 show that both the story drift and story lateral acceleration are reduced significantly with the use of dampers during earthquakes.

Table 3. Demand/Capacity Ratios on a Beam Member

	LLE	DBE	MCE
Roof	0.58	0.78	0.93
4th Floor	0.66	0.92	1.00
3rd Floor	0.74	0.94	1.05
2nd Floor	0.75	0.99	1.10

Table 4. Demand/Capacity Ratios on a Column Member

	LLE	DBE	MCE
Roof	0.44	0.65	0.75
4th Floor	0.72	0.92	1.10
3rd Floor	0.52	0.67	0.78
2nd Floor	0.82	0.98	1.15

Table 5. Story Displacement for LLE (cm)

	Without Dampers	With Dampers
Roof	14.6	7.8
4th Floor	11.2	6.3
3rd Floor	7.3	4.3
2nd Floor	3.4	2.0

Table 6. Story Displacement for DBE (cm)

	Without Dampers	With Dampers
Roof	23.4	13.8
4th Floor	17.8	11.1
3rd Floor	11.6	7.2
2nd Floor	5.2	3.4

Table 7. Story Displacement for MCE (cm)

	Without Dampers	With Dampers
Roof	27.4	14.4
4th Floor	20.8	11.5
3rd Floor	13.6	7.5
2nd Floor	6.23	3.6

Table 8. Story Lateral Inertial Acceleration under LLE (g)

	Without Dampers	With Dampers
Roof	0.533	0.225
4th Floor	0.371	0.237
3rd Floor	0.405	0.223
2nd Floor	0.375	0.186

Table 9. Story Lateral Inertial Acceleration under DBE (g)

	Without Dampers	With Dampers
Roof	0.833	0.420
4th Floor	0.648	0.322
3rd Floor	0.563	0.296
2nd Floor	0.463	0.338

Table 10. Story Lateral Inertial Acceleration under MCE (g)

	Without Dampers	With Dampers
Roof	0.999	0.480
4th Floor	0.750	0.396
3rd Floor	0.631	0.350
2nd Floor	0.566	0.390

3. 4 Nonlinear Structural Analysis

A two-surface plasticity model (Tesng & Lee, 1983) was adopted to model structural nonlinear behavior. The nonlinear time-history analysis was made and the results are shown in figures 4 and Table 11 below.

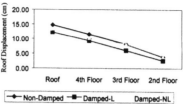

Figure 4. Floor Displacement at MPE

Figure 4 shows that the roof displacement is reduced about 15% during MPE. This result indicates that the structural members are able to resist lower level earthquakes efficiently with or without supplemental damping. The structure was designed to have enough stiffness and ductility for lower level earthquakes since they were designed per the UBC codes. In addition, the structural members were revised to meet the strength requirements for DBE. This figure also shows the response between linear and nonlinear model is insignificant. This indicates that structural members remain elastic for MPE.

Table 11 shows that the roof displacement is reduced about 55% for structural linear model and 45% for structural nonlinear model during DBE. This result indicates that the damping devices play a major role in reducing story drift significantly during the 475-year-return-period earthquake. The story drift ratio is limited to 1 during DBE. The immediate occupancy performance level is therefore achieved. The structural responses both floor displacement and story acceleration are greatly reduced because the added damping reduces the demands on the structure effectively. On the other hand, the non-damped structure was only designed for Basic Safety Objectives. During stronger earthquakes, such as a DBE, the non-damped structure will be forced to dissipate energy through structural inelastic yielding behavior and joints rotations. These demands are great on the building and will eventually cause the primary earthquake resistant members fractured. The results verify the advantages of using seismic dampers to prevent the primary structural members fractured by reducing the demands on the building.

The result also shows that the structural linear model tends to give a liberal result by ignoring the inelastic behavior of structural member. This model does not take the change in structural stiffness into account so that the displacement is smaller compared to that of the structural nonlinear model. When taking the structural nonlinear behavior into account, the structural stiffness will change. Depending on the frequency contents of the input ground motion,

Table 11. Floor Displacement (cm) at DBE (2 Surface Model)

	Without Damper	W/Damper Linear Analysis	W/Damper Nonlinear Analysis
Roof	38.2	17.1	21.0
4th Floor	30.5	13.4	18.8
3rd Floor	20.7	8.9	15.1
2nd Floor	9.6	4.2	6.6

the induced base shear may either increase or decrease. Therefore, the result based on a structural, linear and nonlinear, could be significantly different. That is the reason why it is recommended to run time-history analyses by using at least three pairs of site-specific ground motion for design.

4 CONCLUSIONS

Since the story drift ratio is limited to less than 1% during a DBE, the structural and nonstructural damage will be minimal. In the event of a MCE, the post-Northridge moment connections will yield to provide additional structural damping to reduce the demand on primary structural frames. More piles with increasing length and larger pile caps were designed to meet higher performance demands. In addition, collector forces were found to be much higher than originally expected. Therefore, larger connections and beam collector members were redesigned to meet this requirement.

Structural behavior is not much different during lower demand earthquakes such as a LLE or DBE, since the structure remains elastic under those events. However, structural behavior is quite different under a MCE because demands on the joints increase significantly. Table 3 shows that the DCR is larger than 1 under a MCE. Under MCE events, the primary structural members contribute larger energy dissipation, which reduces the role of the dampers as a major mechanism for energy dissipation. Due to the complexity of structural seismic response and the uncertainty of ground motion, the effects of inelastic behavior of frames with a EDS become important. The effectiveness of seismic dampers is reduced when the structure experiences inelastic deformations. In such cases, force might be either increased or reduced when the structure is responding beyond elastically. The complexity of structural behavior during earthquakes makes it difficult to present a clear design method for engineering professionals. Therefore, extensive verification of structural dynamic behavior is required.

In this study, the results show that the intensity of an earthquake is the primary factor effecting structural performance. For lower level earthquakes, the structure has enough stiffness and strength to resist earthquake demands. Therefore, the addition of supplemental damping is relatively insignificant. At the DBE level, the added damping becomes a very important means of reducing structural response assuming the structure remains elastic. At the most severe level earthquakes, such as a MCE, the added damping may become insignificant if the structure undergoes inelastic action. Because the characteristics of ground motion, such as its amplitude and frequency content, which will effect the amount of energy imparted to a structure, the time history analyses should be done with many different ground motions. As a result, design parameters should be based on the various characteristics of ground motions. It is also recommended the structural stiffness and strength are proportionally distributed to ensure the structural regularity. The dampers should be arranged so that structural integrity and strength remain uninterrupted. Structural connections should be detailed properly so that they provide significant ductility in case of inelastic deformation during severe seismic excitations.

Design criteria should be written in a much stricter form so that larger safety factors will ensure the success of a project. In addition, both the structural linear and nonlinear behavior should be investigated for stronger level earthquakes. In this study, the results show that the intensity of earthquake is the primary factor effecting structural performance. For the lower level earthquakes, the structure has enough stiffness and strength to resist earthquake demands. Therefore, the addition of supplemental damping is relatively insignificant. At the stronger level earthquakes, the added damping becomes a very important means of reducing structural response. Because the characteristics of ground motion, such as its amplitude and frequency contents, which will effect the amount of energy imparted to a structure, the time history analyses should be done with many different ground motions. As a result, design parameters should be based on the various characteristics of ground motions. Such approach is essential.

Although it is not required to design a structure remain elastic during stronger earthquakes, it is recommended that the structural behavior be checked for a MCE. Under this earthquake level, the structural connections will form plastic hinges to release energy and furthermore complicate the structural analysis. The importance of structural nonlinear model is recognized under these circumstances.

Energy dissipation devices should be designed considering environmental conditions such as wind, fatigue, ambient temperature, operating temperature, and other damaging substances. A good quality control program should be implemented to ensure that the dampers will consistently function properly after installation. A well-established maintenance document is a useful tool to ensure better long-run service. Supplemental damping devices are still new to many professionals. The seismic behavior of structures with dampers requires more research and testing to ensure reliability. Therefore, research efforts and professional education and training are urgently needed in this field.

5 REFERENCES

Bird, R.B., Armstrong, R.C, & Hassager, O. 1987. *Dynamics of Polymeric* Liquids, Wiley and Sons, New York.

Constantionu, M.C. and Symans, M.D. 1992 Experimental and Analytical Investigation of Seismic of Structures with Supplemental Fluid Viscous Dampers. Report No. NCEER-92-0032, National Center for Earthquake Engineering Research, SUNY at Buffalo, New York.

FEMA-267A, 1997, Interim Guidelines Advisory No. 1. Federal Emergency Management Agency, Report No. 267-A, 1997.

FEMA-273, 1997, NEHRP Guidelines for the Seismic Rehabilitation of Buildings. Federal Emergency Management Agency, Report No. 273.

Pong, W.S., Tsai, C.S. and Lee, G.C. 1994. Seismic Study of Building Frames with Added Energy-Absorbing Devices. Report No. NCEER-94-0016, National Center for Earthquake Engineering Research, SUNY at Buffalo, New York.

Tseng, N.T. and Lee, G.C. 1983. *Simple Plasticity Model of Two-surface J. Engrg. Mech,*, ASCE, Vol. 109, No.3, 795-810.

Treadwell & Rollo 1997. Geo-technical Investigation 3COM Phase IV, Santa Clara, California, Report No. 1854.04, 1997.

Behaviour of Steel Structures in Seismic Areas, Mazzolani & Tremblay (eds) © 2000 Balkema, Rotterdam, ISBN 90 5809 130 9

Earthquake resistant performance of moment resistant steel frames with damper

M. Yamaguchi, S. Yamada & A. Wada
Tokyo Institute of Technology, Japan

M. Ogihara & M. Narikawa
Tokyo Electric Power Company, Japan

T. Takeuchi & Y. Maeda
Nippon Steel Corporation, Japan

ABSTRACT: In this paper, an experimental method and results of a shaking table test on the steel frames designed by the method of "Performance-Based Design" are discussed. This method has a system of proper estimation for the response of partial frame against Real Time Speed Earthquake, also specimens are scale down models. This system consists of a Weight, a Spring, a Loading Beam, a Specimen and a shaking table. Specimens are half size scale model of medium rise steel buildings, and the types of specimens are Moment Resistant Frame and Moment Resistant Frame with Hysteretic Damper. The natural period of this system is set at about 0.7-0.8 second that is nearly the same as middle rise steel buildings. A visualized performance makes easy to understand the effect of Moment Resistant Frame with Hysteretic Damper compared with Moment Resistant Frame, because of real time speed. The system was subjected to excitation with real time speed, the reaction was observed and the data was obtained. This data proved the superiority of "Performance-Based Design". From a series of shaking table tests, the following results were obtained. A Moment Resistant Frame with Hysteretic Damper demonstrated sufficient seismic performance for the demand.

1. INTRODUCTION

After two big earthquakes, Northridge and Hyogoken-Nanbu occurred in 1994 and 1995, it became popular to require much seismic performance for structures.
Moreover, in recent years steel materials for structures varied form low yield strength steel to high yield strength steel and these materials have been adapted to strengthen structures as well as viscosity and viscoelastic materials.

With this background, A. Wada is suggested the concept of "Damage tolerant structures"(A. Wada 1991). This structure consists of main frames and seismic members. Main frame only support the vertical load and remains elastic during earthquakes. While input earthquake energy is concentrated and absorbed by seismic members. This kind of structure has a lot of advantages such as: "better seismic performance", "more economic", "large life span", environmentally friendly and so on.

Recently, many studies for various damping devices have been taken(such as Y.MAEDA 1998 and E.SAEKI 1996 A.WADA 1997). However, member tests by means of fixing within the frame is not so many (E.SAEKI 1996 and K.KASAI 1997). A behavior of an actual building during earthquake that is most important to grasp is normally done by means of analytic tools. Though real behavior of frames and damp-

ing devices with elastoplastic ranges are very complicated. Therefore, it needs to understand the behavior of frames under shaking table test to simulate real earthquake.

2. METHOD OF EXPERIMENT

In common cases, shaking table tests are carried with specimens of partial frame model, input wave is modified a lot on time history domain. It is compressed on time history in order to fit in frequency spectrum for that natural period shortened. In this case, with such method, strain rate is different from that of real behavior in structures. Especially in the case with damping device that depends on velocity, the response behavior is a completely different matter.

In order to solve these problems, The experiment system using shaking table tests is preposed in this paper. The concept of this system is show in Fig 2-1. and outline of this system is shown in Fig.2-2 and Photo.2-3. This experiment system correspond to real buildings. This system consists of the weight, a spring, a loading beam, a specimen and a shaking table. the weight simulates the weight of the upper part of the building. A spring connecting the weight and the loading beam by series methods modeles the stiffness of upper part of the building. The whole system including a specimen is compared to a real building. And the natural period of this system is about 0.7-0.8 seconds, which is nearly the same as the natural period of a medium rise steel building. Such a system decreases a scale down effect of a partial frame specimen by setting a natural period. Visualized behavior makes easy to understand the superiority of damage tolerant structures because of real time speed.

The major part of weight is a row slab whose weight is about 10 tonf. In addition, balance weights and safety devices are included. Total weight of the system is about 16.0 tonf. The weight is hanging like a pendulum with wire ropes. This hanging technique makes shaking table release from a restriction under a gravity load, especially for small and medium size shaking tables. The weight produces inertia force while the shaking table moves the specimens.

As for the elastic spring, this is modeled as the stiffness of upper part of buildings. This spring consists of two isolators that arranged in parallel. Shear deformations are indicated in Fig.2-4. The stiffness of elastic spring is 10.0kN/mm. Deformation limit is 25.0cm.

The specimens is a 1/2 span and half size scale partial frame model, which is picked up from a building. The beam (literal member) end is supported on a pin condition and near beam-to-column connection is supported on a pin-roller condition. A connection between a specimen and loading beam which is wide flange beam is used split-T that connected with bolts.(Fig. 2-5) The split-T is connected to web member of loading beam. The rotation stiffness is much smaller than that of other part of experiment system. Therefore, this connection is dealt with as a pin support. Another end of the loading beam is in connected too.

In this test, shaking table is used as single degree of freedom system. There are some devices to prevent out-plate displacements in the experiment system. These are shown in follows.

1. Two pins are arranged in parallel at the end of a specimen. This technique is a role to enhance stiffness against another degree of freedom.

2. Middle part of the column, steel plate with a teflon sheet is welded. If it occurs out-plate displacement , it plays a role to restrict extra deformations.

3. The loading beam is restricted by an equal technique .

4. The wight is supported with four rollers, which are very tough and restricts extra deformations.

Attention to a safety of this shaking table tests, it is constructed a safety devices. Its outline is indicated in Fig.2-6. The mechanism is that O-section bars absorb energy of the system to be deformed by a wide flange beam connected to be the weight, if it occurred extra deformations.

3.SPECIMENS (shapes and performance)

Tests were taken with two types of specimens, Moment Resistant Frame 1 (M.R.F.1),show in Fig.3-1, using ordinary so far and Moment Resistant Frame 2 with Hysteretic Damper (M.R.F.1 with damper),show in Fig.3-2.

Specimens are partial frame models that are divided from a verticl steel building. Considering symmetry of buildings, 1/2 span half-size scale model is used. The length of beam member (literal member, Lb) is 2000mm; the length of column (vertical member, Lc)

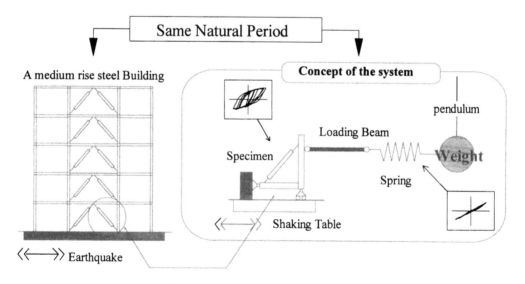

Fig.2-1 Concept of the experiment system

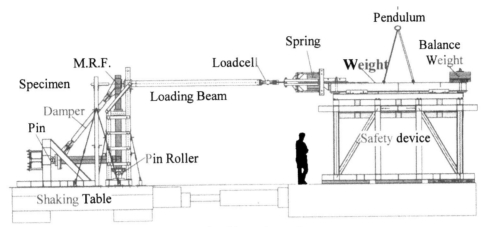

Fig.2-2 Outline of the experiment system

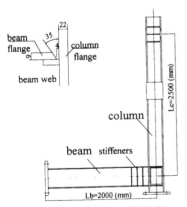

Photo.2-3 Experiment system

Fig.2-4 Spring　　　　Fig.2-5 Pin joint

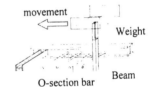

O-section bar
Fig.2-6 Outline of a safety device

Fig.3-1 Specimen of M.R.F.1

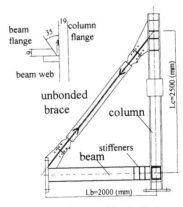

Fig.3-2 Specimen of M.R.F.2 with damper

is 2500mm. Both members have H-section (shown in Table.3-3)

The connection panel is reinforced with double plates. In order to prevent local buckling, three pieces of stiffeners are arranged with equal distance in beam-to-column connection.

Table.3-3 sections of the elements

	M.R.F.1	M.R.F.2(with damper)
Beam	BH-340x200x 6x 9	BH-240x170x 6x 9
Column	BH-280x280x12x22	BH-220x220x12x19
Yielding Steel Core	-	25x16

Non-scallap detail is applied for beam-to-column connections. Because it is paid no attentiom to great earthquakes that it occurs fractures at beam-to-column connections. A damper is a Hysteretic and brace type damper. It is called "unbonded brace". The mechanism of unbonded brace is shown in Fig.3-3. The unbonded brace used in this experiment is shown in Fig.3-4. This is to prevent Euler buckling of central steel core by encasing it over its length in a steel filled concrete or mortar. Therefore unbonded brace provides a stable behavior in both sides of tension and compression.

Table.3-5 Shows the mechanical properties of material of specimens obtained from tension tests.

In this test, upper limit of resisting force is as 15 ton for performance of shaking table.For a yielding steel core of the unbonded brace, it is selected low yield steel (BT-LYP100) which is able to expect an energy absorption in ranges of small amplitudes, so its yield strain is very small.

Fig.3-6 shows resisting force-displacement relation of both specimens. The curves are obtained from analysis.

Design points of specimens to which we pay attention in the case designed cross section are shown as follows.

1.Shear forces at yield point of both specimens are designed equally (about 90 kN).

2.Shear ratio with horizontal resisting force between M.R.F.2 and damper is about 1:1.

3.M.R.F.2 is designed to have large elastic ranges, compared with M.R.F.1, therefore its section depth beam and column are slender.

4-1. SHAKING TABLE

The Shaking table, used in this test, is medium size and belongs to the Disaster Prevention Research Institute, Kyoto University. This table has six degree of freedoms. In this test, however, it is used as single degree of freedom system by means of fixing other 5 degrees of freedom. Fig.4-1 and Fig.4-2. show the performance curve of this shaking table concerned with Acceleration- Period relation and with Velocity-Period relation.

At first, we assume the weight of a Mass to be15 tonf. In attention to ranges 0.7-0.8sec (in Fig.4-1. gray zone) which is natural period of this system considering an expansion of period by plasticity. We recognize that this shaking table has sufficient capacity to shake table so far as 1.0 G. In order to inflict much damage to specimens, it is required to have capacity to shake with sufficient velocity. Fig.4-2. shows that the shaking table has enough capacity to shake table as fast as 100 kine.

4-2.TYPES OF INPUT WAVES

Considering the variety of real ground motions, we choose three earthquake records shown in Table 4-3.

Table4-3.input waves

Year	Earthquake	Record	component	description
1940	Imperial Valley	El_Centro	NS	El_Centro
1968	Tokachi	Hachinohe	EW	Hachinohe
1995	Hyogoken-nanbu	Kobe JMA	NS	Kobe

Shaking time is determined at 30.0 sec, since amount of Energy input with elastoplastic analysis during first 15.0 sec is more than 90 percents of total input energy, and very little over 30.0 sec with these input waves. It is an acceleration record to control shaking table. However, those input waves are little reversed. Because integration of original acceleration records of earthquake gives phenomena such as remaining velocity of shaking table or divergence of displacement. Compensation approachs of input waves are shown in the follows.

1. Removal of long period components more than 10.0 sec.

2. Base line compensation.

Through these compensations, numerical integration was taken again. Input waves were generalized with Maximum-velocity of ground motion (shaking table), after being confirmed maximum displacement.

4-3.INPUT LEVELS

Generally, in the case of designs of real buildings,the input level was established at first as a design criteria. On this experiment, the process order is reversed, decided with considering relative evaluation between performance of specimens, experiment system and intensity of input wave. Design criteria and maximum-velocity of ground motion are shown in table 4-4.

Table.4-4 Input Level

Input Level	Design Criteria	Maximum-velocity
Level_1	Protection of Faculty	25.0kine
Level_2	Protection of Property	50.0kine

In ordinary input level is depend on the structural designer's requirment. According to "the theory of Damage Tolerant Structure", the purpose of structure demanded, if it occur equivalent Level 1 earthquakes, is "Protection of Faculty". And in Level 2 is "Protection of Property". These are an outline of required seismic performance by the structures.

Deciding an approach of concrete input level is as follows: It was determined by means of elastoplastic analysis of SDOF model that is this experiment sytem. Fig.4-5 shows correspondence relation between input level and force-displacement relation which is derivered from pre-analysis of specimen. Input levels are shown with maximum response displacement which is the results of elastoplastic analysis using El Centro earthquake.

428

Assumptions of behavior in both input levels are as follows: On Level 1 shaking, frames of both specimens are remained within elastic ranges, although the steel core of the damper is already yields and absorbs energy.On Level 2 shaking ,M.R.F.1 is yielding a lot, although the frame of M.R.F.2 with damper remained within elastic ranges.

4-4.TEST PROCEDURES

The procedure is planed and shown as follows: concerned with section 4-4-1~4-4-3, these are defined as tests to compare the seismic performance of both specimens, M.R.F.1 and M.R.F.2 with damper.

4-4-1.Pulse-wave test

Shaking table test using pulse wave with small amplitude is taken in order measure the damping ratio and natural period of this system.

4-4-2. Level 1 tests

Maximum-velocity of input wave is 25.0 kine. Using this Input level, seismic performance of M.R.F.2 with damper is compared with M.R.F.1. Behaviors of both specimens are maintained within elastic ranges. Procedure of input waves are El Centro, Hachinohe and Kobe.

4-4-3. Level 2 tests

Maximum-velocity of Input wave is 50.0 kine. Using this Input level, seismic performance of M.R.F.2 with damper is compared with M.R.F.1. on elastoplastic ranges. Only M.R.F.1 received a lot of damage.

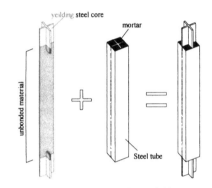

Fig.3-3. mechanism of unbonded braces

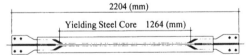

Fig.3-4. unbonded braces

Table3-5. Mechanical properties

Steel	Yield stress Mpa	Tensile strength Mpa	Yield strain %	Elongation %	Yield ratio
SN400A(web)	311	435	0.148	31.7	0.71
SN400A(flange)	292	451	0.142	30.8	0.65
LYP100(damper)	96	258	0.240	59.6	0.37

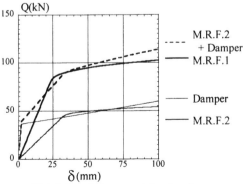

Fig.3-6. force-displacement relation of specimens

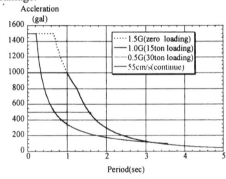

Fig.4-1. Performance curve of the shakig table (acceleration-period ralation)

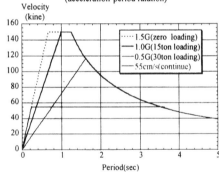

Fig.4-2. Performance curve of the shakig table (volocity-period ralation)

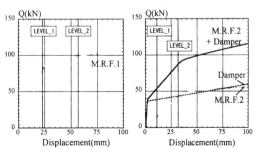

Fig.4-5. relationship between specimens and input levels

5. MEASURED DATA

Equipments for measuring data is used some types of instruments such as laser type displacement gauges, a spring type displacement gauges, a wire type displacement gauges, and accelerografs, strain gauges, and loadcell with 500kN, these are total 41 measuring points.

Strain gauges are put on an edge of a beam at beam-to-column connection and flanges in two section of a column, which is estimate a share force on the M.R.F.2.

Data sampling frequency is 200 Hz and the time measuring is 81.96 seconds. The data is initialized by means of each initial data of each test, and modified by ten-point moving average.

When it occurs range over in displacement gauges, acceleration records by accelerograph on the same place at the displacement gauge is integrated and the result of integration is used for compensation.

A value of relative story displacement(δ) is defined as equation 5-1, and measuring location concerned with relative story displacement(δ) are shown in Fig 5-2. An explanation about shear force(Q) used is shown in follows too.

$$\delta = \delta_{top} + \frac{1}{2}(\delta_{panel1} + \delta_{panel2})\qquad(5\text{-}1)$$

δ: relative story displacement
Q: shear force (value of the loadcell)

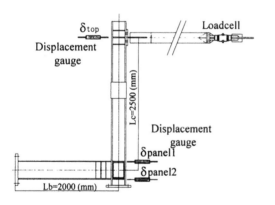

Fig.5-2. locaton of displacement gauges

6. RESULTS AND CONCLUSIONS.

Natural period of this test system with both specimens and damping ratio are shown in table6-1, and shown in Eqs.6-1 and 6-2

Table.6-1 result of pulse wave test

	Period	damping ratio
	(sec)	(%)
M.R.F.1	0.82	1.70
M.R.F.2 with damper	0.70	2.30

$$T = 2\pi\sqrt{\frac{M_{mass}}{K_{system}}}\quad(6\text{-}1)\qquad K_{system} = \frac{K_s \cdot K_{spe}}{K_s + K_{spe}}\quad(6\text{-}2)$$

T: natural period of the system, M_{mass}: mass of the weight
K_{system}: stiffness of the system, K_s: stiffness of the spring
K_{spe}: stiffness of the specimen

It is recognized that natural period of this system depends on Ks (stiffness of spring). A difference in natural period is about 10%. Therefore, it is anticipated that amount of input energy to the system is equal order for both specimens. Damping ratios in both tests were 1.7%(M.R.F1) and 2.3% respectively. We judged its values for proper. Tests for confirming seismic performances was continuously taken.

All experimental results in force-displacement relations are shown in Table.6-3 and Fig.6-4

In the beginning, with level 1 tests, it is recognized that M.R.F.1 performed almost elastically from a macroscopic view When it was taken with three kinds of input waves. Maximum absolute Values of shear force and displacement are recorded in a test with Kobe. On the other hand, the steel core of M.R.F.2 with damper yields already and the histeresis loop is described, and it absorbed the energy.

In the results in Level 2 tests, M.R.F.1 largely yields a lot and large hysteresis loops are described. This is because the plasticity occurs on the beam member close to beam-to-column connection. The maximum value of force and displacement are not so large in El_Centro in both positive and negative sides, although M.R.F.1 was experienced some medium amplitudes

Tble 6-3 result of tests

Test Results		M.R.F.1			M.R.F.2 with damper			reduction ratio		
Input wave	Input level	|Q|max(kN)	|δ|max(mm)	strain(%)	|Q|max(kN)	|δ|max(mm)	strain(%)	|Q|max	|δ|max	strain
El_Centro	LEVEL_1	93.0	36.0	0.245	54.6	7.7	0.025	0.59	0.21	0.10
Hachonohe	LEVEL_1	84.3	32.0	0.153	56.6	7.3	0.023	0.67	0.23	0.15
Kobe	LEVEL_1	93.5	36.3	0.192	69.8	13.0	0.045	0.75	0.36	0.23
El_Centro	LEVEL_2	112.1	52.5	1.024	91.6	23.2	0.085	0.82	0.44	0.08
Hachonohe	LEVEL_2	126.4	93.9	1.499	112.2	31.1	0.164	0.89	0.33	0.11
Kobe	LEVEL_2	128.9	*93.8	1.380	121.6	44.4	0.236	0.94	0.47	0.17

reduction ratio = Maximum Value of M.R.F.1 / Maximum value of M.R.F.2 with Damper

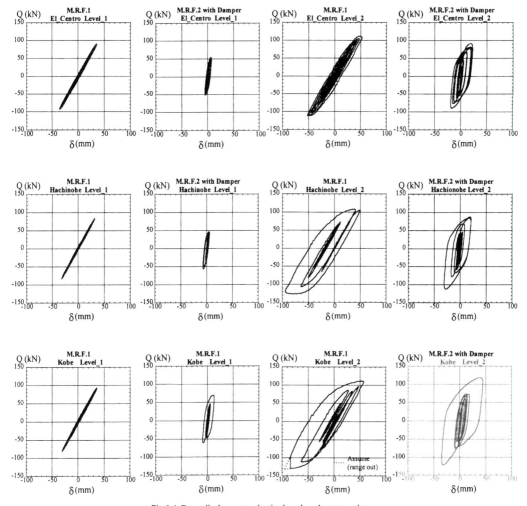

Fig.6-4 Force-displacement relatoins based on the test results.

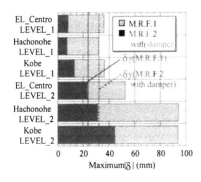

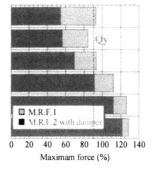

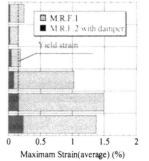

Fig.6-5 comparison with Maximum relative displacement

Fig.6-6 comparison with Maximum force

Fig.6-7comparison with Maximum strain at beam edge

again and again. The damage is accumulated at beam-to-column connection. In the case using Hachinohe ground motion, the specimen experienced a large deformation in the negative side and described a large hysteresis loop and left the residual deformation (about 15mm) after the test. In the case using Kobe ground motion, since large displacement occurred in the negative side, a displacement gauge produced the range over, while residual deformation was left. In this part, it was used for the interpolation by means of integrating acceleration record of same point with the displacement gauge. Maximum force and deformation are calculated based on the measured results.

Concerned about M.R.F.2 with damper, in the case using El Centro ground motion, the specimen deformed similarly in positive and negative sides, however, the displacement is small compared with another tests. In the case using Hachinohe ground motion, comparatively with large amplitude is occurred, and the yielding steel core greatly deformed plastically. However, M.F.F2 has still large secondary stiffness as whole frame itself, since M.R.F.2 does in elastic range. Therefore, it is thought that M.R.F.2 with damper has much seismic performance than M.R.F.1. In the case using Kobe ground motion, the reduction of relative story displacement decreases less than other input tests, since that input is rapidly type.

From the viewpoint of relation maximum displacement and each yield deformation value of M.R.Fs, it is compared in Fig.6-5. Though M.R.F.1 has greatly been yielded, M.R.F.2 remained within elastic range. The value of M.R.F.2 are under 1/2 of the yield displacement of the specimen. The reduction ratio of maximum displacement, M.R.F2 / M.R.F1, is greatly reduced with 0.21 (El_Centro), 0.23 (Hichinohe), 0.36 (Kobe).
In level 2 tests, only in Kobe, M.R.F.2 deformed over its yield displacement, however it does not so have large damage. The reduction ratio are 0.44, 0.33, 0.47. Therefore it was also effectively able to reduce the response displacement in level 2.

The maximum shear force in each test is shown in Fig.6-6. Qy is a value of shear force at yield point of frames. The reduction ratio of maximum force is 0.67 in three averages in level 1 input, and 0.88 in level 2 input. It was able to reduce the force as well in both input levels as well.

The following conclusion seem from Fig.6-7: Largest average strain of the beam edge in each test and yield strain of the material testing. It is shown the maximum absolute value, on three averages, of strain gauges attached in the beam edge. It is proven that frame of M.R.F.2 with damper has positioned in the elastic range except for Kobe ground motion. The value of M.R.F.2 with damper is 20% or less than that of M.R.F1. However, the beam edge of M.R.F1 is greatly plasticized in Level 2 tests. The value of about 10 times of the yield strain has been experienced. It

was able to confirm that strain of beam edge can be greatly reduced at both input levels.

REFERENCE

Akira WADA: Damage Tolerrant Structure, Proseeding of Fifth U.S.Japan Workshop on The Inprovement of Structural Design and Construction Practices, Applied Technology Council, 27-39, 1991

Yasushi MAEDA, Yasuhiro TANAKA, Mamoru IWATA and Akira WADA: Fatigue properties of axial-yield type hysteresis dampers, J. Struct. Constr. Eng., AIJ, No.503,109-,Jan., 1998

Eiichiro SAEKI, Yasushi MAEDA, Kouichi IWAMATSU and Akira WADA:Analytical study on the unbonded brace fixed in a frame, J. Struct. Constr. Eng., AIJ, No.489,95-,Nov., 1996

Eiichiro SAEKI, Mitsuru SUGISAWA, Tanemi YAMAGUCHI, Haruo MOCHIZUKI and Akira WADA:A study on htsteresis and hysteresis energy characteristics of low yield strength steel, J. Struct. Constr. Eng., AIJ, No.473,159-,Jul., 1995

Eiichiro SAEKI, Yasushi MAEDA, Hideji NAKAMURA, Mitsumasa MIDORIKAWA and Akira WADA: Experimental study od practical-scale unbonded braces, J. Struct. Constr. Eng., AIJ, No.476,149-,Oct., 1995

Hiroshi AKIYAMA, Satoshi YAMADA, Chikahiro MINOWA, Takayuki TERAMOTO, Fumio OTAKE and Yoshitaka YABE: Experimental method of the full scale shaking table using inertial loading equipment, J. Struct. Constr. Eng., AIJ, No.505,139-,Mar., 1998

Global behaviour

Behaviour of Steel Structures in Seismic Areas, Mazzolani & Tremblay (eds) © 2000 Balkema, Rotterdam, ISBN 90 5809 130 9

Effects of repeated seismic events on structures

C. Amadio, M. Fragiacomo, S. Rajgelj & F. Scarabelli
Civil Engineering Department, University of Trieste, Italy

ABSTRACT: The repetition of strong seismic actions during a seismic sequence has been observed in many parts of the world and requires new assessments of the effects on structures. In this paper, the responses of five types of single degrees of freedom structural systems and a 3-storey, 2-bay moment resistant steel frame have been analysed when subjected to repeated seismic events of medium-strong intensity. Obtained results show that for SDOF and steel frame seismic responses there is a significant increase in the required ductility or strength, particularly for structures with a period less than about 1.5s with an elastic-perfectly plastic behaviour.

1 INTRODUCTION

The repetition of medium-strong earthquakes during a seismic sequence is very frequent and characteristic of many areas (Elnashai et el. 1998, Murià & Jaramillo 1998). This type of seismic sequence has been observed in Romania, Italy, Mexico, Japan, Turkey, California, Taiwan, etc. The evaluation of cumulative damage caused by repeated medium-strong seismic actions is very important because experience points out that structures with an adequate behaviour and a tolerable level of damage under the first earthquake, generally become irreparably damaged under the second or third seismic event. It is then evident the importance to explore in a parametric manner the features of structural response under repeated seismic actions and provide some preliminary information about the vulnerability of structures subjected to this type of event.

This necessity is evident also if we consider design rules proposed by actual codes that in general require satisfying with two fundamental criteria:

• Under non-destructive earthquakes no structural damage should occur and only limited non-structural damage is acceptable (serviceability limit state). To this aim the structure should remain elastic with limited interstory drift under a moderate earthquake (period of return about 35 to 50 years)

• For a severe earthquake (period of return 475 years), significant damage in both structural and non-structural elements is acceptable but any collapse with subsequent loss of life must be avoided (ultimate limit state). To this aim the structure requires a significant energy dissipative capacity under a destructive earthquake to avoid collapse without excessive costs. High q factors are, in general, adopted in the design to achieve this.

The effects of repeated seismic events of medium-strong intensity, characteristic of some countries, are therefore not adequately evaluated by these criteria since the accumulation of damage, certainly present if the structure is designed adopting a high ductility or q factor, is not considered. It is therefore evident that a balance between strength, stiffness and energy dissipation represents the goal of a correct seismic design for repeated events to avoid damage increases that are not economic for structural repair.

In order to reach these results and to understand the structural behaviour under this type of loading, a series of single degree of freedom structural systems (SDOF) with different hysteretic laws is analysed for recorded or generated ground motions repeated one, two or three times. In particular, to simulate the behaviour of typical multi-storey frames under seismic actions, various behavioural models are

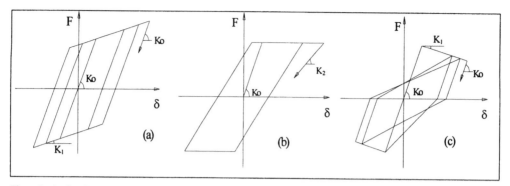

Figure 1. Analysed SDOF systems

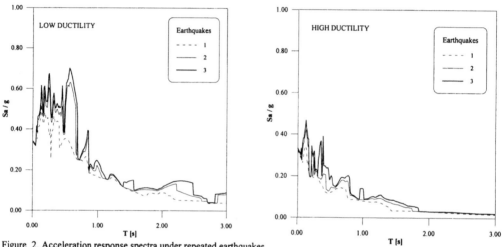

Figure 2. Acceleration response spectra under repeated earthquakes

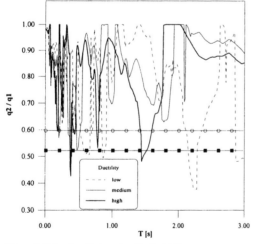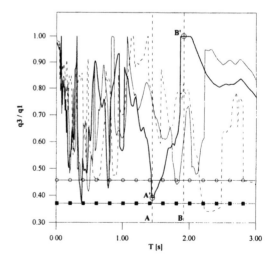

Figure 3. Behaviour factor ratios for the EPP system

436

assumed, including: elastic-perfectly plastic model (EPP), elastic-plastic with hardening model, EPP degrading model, Clough's model and a degrading with softening model. The analyses were performed with a vibration period from 0.05s to 3s, a damping ratio of 5% and a displacement ductility equal to 2, 4 and 6, representative of structures with low, medium and high ductility respectively.

The most important parameters analysed are been: displacement, velocity, strength (elastic and inelastic spectra), behaviour factor q, energy spectra and ratios of elastic and inelastic spectral values.

To evaluate the cumulative damage history under repeated seismic actions, time dependent charts of the damage index D_μ (ratio between plastic displacement level reached and maximum plastic displacement available) and Park & Ang index $D_{P\&A}$ (Park & Ang 1985) are investigated.

At last, to validate the results obtained by the SDOF, a series of analyses of 3-storey, 2-bay moment resisting steel frames were also performed.

Obtained results show that analyses performed with SDOF under repeated seismic events require a significant increase in the ductility or strength, in particular for structures with a period less than about 1.5s. Also moment resisting steel frames, designed according to EC8, are sensitive to these events and in general highlight an accumulation of damage greater than the equivalent SDOF systems.

2 NUMERICAL ANALYSES

2.1 SDOF systems

It is known how SDOF can, in general, be considered representative of the seismic behaviour of a structure when this vibrates according to the first mode. Generally, more commonly used models are:

• the elastic-perfectly plastic (EPP) model (fig.1a where $K_1=0$), representative for example of a moment resisting frame where all plastic hinges are simultaneously formed under seismic forces and cyclic behaviour is stable (similar to steel frames with rigid joints);
• the elastic plastic with hardening (EPH) model (fig.1a), characteristic of frames where plastic hinges are not formed simultaneously;
• the elastic plastic degrading (EPD) model (fig.1b), characteristic for example of composite frames with semi-rigid joints;

• the Clough model (fig. 1c where $K_1=0$), representative in general of concrete frames or masonry structures;
• the degrading with softening (DS) model (fig. 1c), representative for example of steel frames where second order effects are important.

For this preliminary analysis, earthquakes adopted are the SOOE El Centro 40 component and two generated earthquakes named G1 and G2, compatible with Eurocode 8 spectra for stiff and soft soil. Any earthquake, normalised to a maximum ground acceleration of 0.35g, is applied one, two or three times (a sequence of different earthquakes give similar results), considering a time gap of about 40 seconds between two consecutive events for the structure to cease moving. These inputs have been selected because the first is very common and typically adopted in the seismic analyses to characterise the response, whereas the others are representative in average of limit seismic inputs adopted by design codes.

For the EPP model and EC40 earthquake, a comparison between acceleration response spectra obtained by one, two or three events is shown in figure 2 varying the ductility level. Analysing these responses, it is evident that repeated events require a strength increase of the system with respect to a single event, mainly for period T=0.1 to 1.5 seconds. For longer periods, T>2 s and particularly for high ductility systems, the response under repeated earthquakes is very similar to that of only one earthquake.

In order to best appreciate the strength capacities under more earthquakes in series with respect to only one event, using the same symbols, ratios q_2/q_1 and q_3/q_1 between the q factor evaluated for 2 or 3 events (denoted as q_2 and q_3) and the one evaluated for only one event (q_1), varying the natural vibration period T, are plotted in figure 3 for a fixed value of ductility. These q ratios can be seen also as the peak ground acceleration ratios at the collapse $a^c_n/a^c_1 = q_n/q_1$ or the ratios between shear base design forces $Q_1/Q_n = q_n/q_1$. This figure shows that multiple events can require a very high change of the q factor and its variability with the period is elevated.

To best understand this response, two systems with different periods, denoted in figure 3 as A and B, that for high ductility show high decay (A' point) or no decay (B' point), are analysed in detail. In the figures 4, varying the ductility μ, the plot of q_1, q_3 are reported. In figures 5, 6, the plastic dissipated energy at the collapse is reported for a single earthquake and

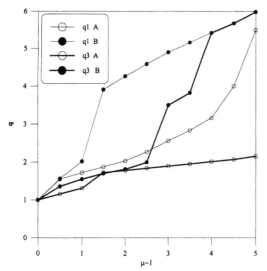

Figure 4. Behaviour factor for A and B structures

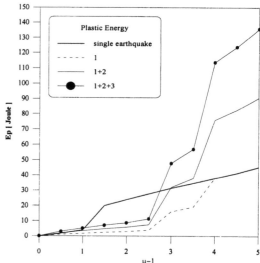

Figure 6. Plastic energy for B structures

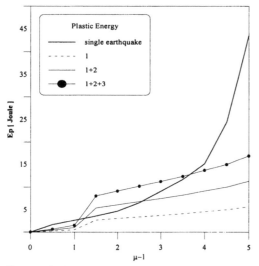

Figure 5. Plastic energy for A structures

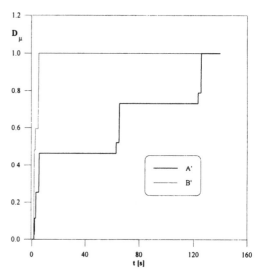

Figure7. Ductility index in time, for structures A' and B'

for three events in series (denoted as 1+2+3). For this last case also the plastic energy under first event (denoted as 1) or first and second event (denoted as 1+2) are reported.

It can be observed a behavioural analogy between q factor and plastic dissipated energy, justified by fact that q is a measure of strength reserve in plastic field. For structures A the q loss increases with μ beginning from $\mu-1=2$ whereas for

structures B the q loss is limited in the range $\mu-1=1$ to 3.

For three or n events structures A dissipate under each event the same energy and this is practically that of a structure with a ductility equal to $\mu/2$ for one event. Energetically a structure type A with $\mu=6$, presents for example under three earthquakes about the same behaviour of a structure with $\mu=3$ under one earthquake. Structures B present

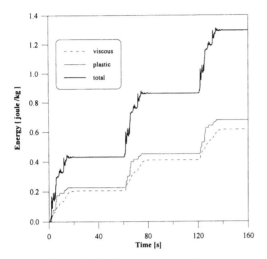

Figure 8. Energies in time for A' structure at collapse

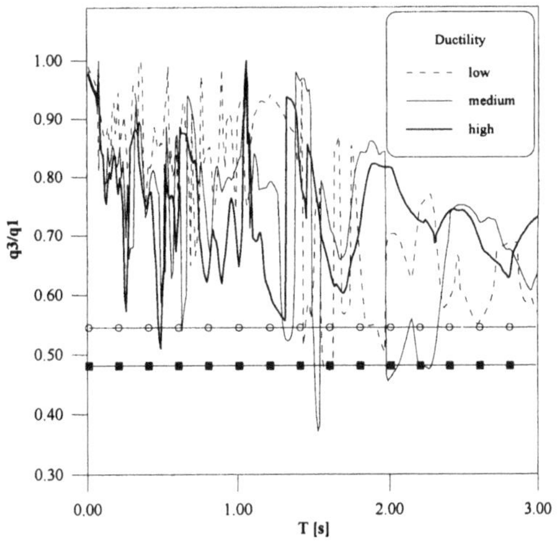

Figure 9. Behaviour factor under G2 repeat earthquakes

instead behaviour opposite of the structures A, ductile structures are not penalised under repeated events whereas structures with low and medium ductility are penalised. For $\mu=2$ to 4, it can be seen that plastic energy dissipated under more earthquakes is less than that dissipated with a single earthquake and in this zone we have a high decrement of q factor.

Diagrams in time of damage index D_μ, relates to the ductility μ by the equation:

$$D_\mu = \frac{\mu - 1}{\mu_0 - 1} \qquad (1)$$

where μ_0 is the available ductility and $0 \leq D_\mu \leq 1$, are shown in figure 7 under three events type EC40.

It can be observed as for the structure A', the damage is on average distributed in time whereas structure B' accumulates the damage essentially during the first event. Park & Ang damage index follows the same behaviour of D_μ. Absorbed energy in time, shown in figure 8, highlights how structure A' has in terms of energy, practically the same trend in time at any event (the total energy behaviour is similar also for structure B').

By remarks above seen, it is evident that the q factor is tied to the ductility μ in a very complex way; this link depends in fact on the structure period and frequency content of the earthquake. For the G2 earthquake, for example, the ratio q_3/q_1 is shown in figure 9: its comparison with that of figure 3, points out the variability of this parameter with the earthquake.

For safety it is then opportune to adopt a reduction of the q factor independently of the period T of the structure. The average value minus 1.5 or 2 times the standard deviation can then be adopted, obtaining limit curve a or b of figure 3 and 9, denoted with hollow circles and solid squares lines respectively. For the EPP system the average value of limit curves a and b, obtained considering all examined earthquakes and a prefixed ductility are reported in table A. This shows that the variability with the ductility is very limited.

The same comparison carried out for the EPP model in terms of q factor is reported in figure 10 for the EPH model, considering a hardening branch with a stiffness K_1 equal to 0.1 times the initial stiffness

Table A. Limit curves for EPP System

		Ductility=2	Ductility=4	Ductility=6	Average
q2/q1	Curve a	0.610	0.661	0.646	0.639
	Curve b	0.540	0.604	0.592	0.579
q3/q1	Curve a	0.494	0.475	0.511	0.493
	Curve b	0.418	0.396	0.443	0.419

of the system K_0 and EC40 earthquake. This shows that the presence of the overstrength reduces the decrement of the q factor under repeated events, especially for high ductility, significantly. Parameters of average limit curve a, b for this model are reported in table B.

Also the presence of stiffness degrading, typical of EPD and Clough systems, produces a favourable effect on the q factor. Indeed, analysing figures 11 and 12, the ratio q_3/q_1 is on average greater than that of EPP system. For the case when T>0.8 to 1 s, this is due to the increment of the period in plastic phase

and to the consequent decrement of required strength (the acceleration spectra decreases with T in this range). For the EPD and Clough model, parameters of limit curves a and b are calculated considering all the examined earthquakes, and are reported in table B. For the EPD system, the stiffness degrading was considered linear. The limit value k_2 of the stiffness when $\mu=\mu_0$ is assumed equal to (Sulaimani &. Roessett 1985):

$$k_2 = \left(\frac{1}{\mu_0}\right)^e k_0 \qquad (2)$$

where k_0 is the initial stiffness of the system. In table B the values e=0.35, e=0.70 and e=1 are considered.

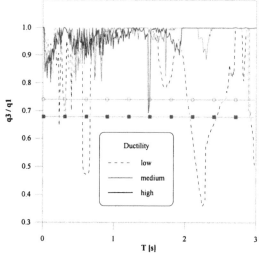

Figure 10. q3/q1 response for EPH system

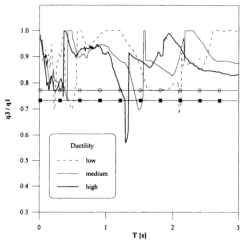

Figure 12. q3/q1 response for Clough Model

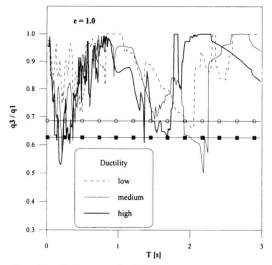

Figure 11. q3/q1 response for EPD system

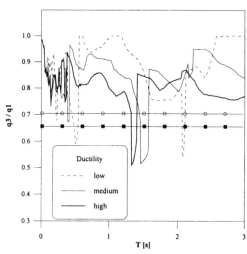

Figure 13. q3/q1 response for DS system

Table B. Limit curves

		q3/q1
EPH	Curve a	0.799
	Curve b	0.756
EPD e=0.35	Curve a	0.607
	Curve b	0.533
EPD e=0.70	Curve a	0.712
	Curve b	0.603
EPD e=1.00	Curve a	0.738
	Curve b	0.650
Clough	Curve a	0.812
	Curve b	0.746
DS	Curve a	0.719
	Curve b	0.678

Using these values, it is evident that when the degrading of stiffness is high (e=1), the limit curves of q_3/q_1 ratio are elevated with respect to the EPP model and similar to EPH model. At last, considering as DS a Clough model with softening (a stiffness of softening branch K_1 equal to 10% of the initial stiffness k_o has been considered), obviously an increment of the strength demand (fig. 13) under repeated earthquakes with respect to the classical Clough model (fig. 12) can be observed. In table B, average values of a and b curves for all analysed earthquakes are also reported.

Comparing these results with those of the EPP model, it is possible to observe how the EPP model is characterised by maximum decay of the q factor. Working in advantage of safety, this last can then be considered as the reference model for the problem of repeated events and the limit curves a or b can define therefore in a significant way these effects. They should be determined as a function of the site using a large number of possible events. First results obtained by these analyses show however how the repetition of more events can produce a meaningful reduction of the q factor to utilise in the design. It is evident that the use of a high q factor (5 or 6 for a steel frame for example) cannot be acceptable, since under a limited number of medium-high events (3 to 5), limiting conditions for some structural elements are certainly reached.

2.2 Moment resisting frames

In order to verify the corrispondence between SDOF and frame response, some results obtained from analysis of moment resisting steel frames are here reported. In particular, the response of the three storey, two bay rigid steel frame of figure 14, characterised by Italian compact profiles of steel Fe360, is shown. The frame was designed according to Eurocode 3 and 8, considering the same as an unbraced transversal frame of a parallel frames system posed at a distance of 6 m. Periods are: T_1=0.934 s, T_2=0.305 s and T_3=0.167 s. Distributed loads in the beams and second order effects have been considered in the analyses. Static and dynamic

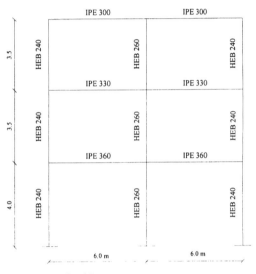

Figure 14. Analysed Frame

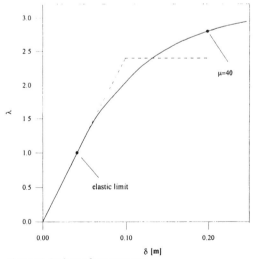

Figure 15. Pushover frame response

analyses were performed by Abaqus code using beam elements type B33, characterised by three Gauss points along the beam and thirteen points in the section (Hibbit et al. 1994). In this way, the interaction between the normal force and bending moment is automatic, with control achieved directly on the stress-strain material law. For limit strain the value $\varepsilon_{lim} = 40\varepsilon_y$ was assumed, with ε_y equal to the yielding steel strain, which corresponds to a ductility $\mu_e \cong 16$ for the beams. The frame pushover response in terms of load multiplier λ versus the top displacement δ, under equivalent seismic forces, is shown in figure 15 (continuous line). It is characterised by a strong over-strength because plastic hinges are not formed simultaneously. Behaviour factors q, considered as ratio between the ultimate and the yielding maximum ground acceleration, are reported in table C for the El Centro earthquake applied one, two or three times. Also in this case, multiple events involve a strong reduction of the q factor (analogous results were obtained for G1 and G2 earthquakes). Comparing these values with those of equivalent SDOF systems, obtained by static response, has been observed that in general they are not able to simulate the global frame response if they are considered of type EPH. Only the EPP model shown in figure 15 (dashed line), characterised by the same total energy of the real frame, has provided sufficiently adequate behaviour factors and ratios q_2/q_1 and q_3/q_1 (see table C).

In general, we have observed that SDOF are not able to correctly predict the actual response of the frame. A SDOF system in fact does not include the complex interaction between plastic hinges during an earthquake. In particular, the interaction with high modes and the effect of normal force in the external columns cannot be considered by a SDOF system. It is then evident that a large parametric analysis should be developed for a complete comprehension of the problem.

3 CONCLUSIONS

In this work some preliminary results regarding the behaviour of SDOF systems and real moment resisting steel frames under repeated earthquakes are shown. The analysis cannot be considered exhaustive but first results indicate that multiple events of medium-high intensity can imply a considerable accumulation of damage and a consequent reduction of the behaviour factor. Many parameters influence the response, the principal are the type of earthquake, the behaviour of the structure and the level of available ductility.

The EPP model is the most vulnerable and significant under this type of event. Indications reported here show how the q factor adopted in the design for a single event cannot be acceptable.

For the design of structures in areas historically characterised by multiple seismic events, a decrement of the q factor is in fact necessary to guarantee a seismic behaviour with limited damage and a consequent reduced vulnerability for the structure.

REFERENCES

G.J. Al Sulaimani & J.M. Roessett 1985. Design spectra for degrading systems. *Journal of Structural Engineering ASCE Vol. 111, December.*

Commission of the European Communities. *Eurocode n° 3* 1994. Design of steel structures.

Commission of the European Communities. *Eurocode n° 8* 1994. Design provisions for earthquake resistance of structures.

A.S. Elnashai, J.J. Bommer & A. Martinez-Pereira 1998. Engineering implications of strong-motion records from recent eartquakes". *11th European Conference on Earthquake Engineering, Paris.*

Hibbit, Karlsson and Sorensen. 1994. Abaqus version 4.9. User Manual.

D. Murià Vila & A.M. Toro Jaramillo 1998. Effects of several events recorded at a building founded on soft soil. *11th European Conference on Earthquake Engineering, Paris.*

Y.J. Park & AH-S. Ang 1985. Mechanistic seismic damage model for reinforced concrete. *Journal of the Structural Engineering.*

Table C. Behaviour factor for the analysed frame

	Abaqus	EPP
q1	4.291	3.496
q2	3.031	2.270
q3	2.610	2.136
q2/q1	0.706	0.649
q3/q1	0.608	0.611

Behaviour of Steel Structures in Seismic Areas, Mazzolani & Tremblay (eds) © 2000 Balkema, Rotterdam, ISBN 90 5809 130 9

Toward a consistent methodology for ductility checking

A. Anastasiadis & V. Gioncu
Department of Architecture, University 'Politehnica', Timisoara, Romania

F. M. Mazzolani
Department of Structural Analysis and Design, University of Naples, Italy

ABSTRACT: An accurate design of structure subjected to seismic loads has to consider the verification of the stiffness, resistance and ductility triad. Unfortunately, in the present codes only the direct checking for stiffness and resistance is required, the ductility demands being ensured just by detailing rules. During the last great earthquakes this provision has been proved to be inadequate, the damage of steel structures being very important. Thus, a consistent methodology for direct ductility checking is required by design practice. This paper presents a proposal for such methodology, which considers the interaction between local and global ductility.

1 INTRODUCTION

For an efficient seismic design is necessary to use plastic analysis in which ductility plays an important role. The behaviour of a structure depends on ductility requirements, comprising both the earthquake characteristics and the available ductility of the individual members, which is limited by buckling of compression plates or fracture of tension parts. Therefore, for a proper design of steel structures subjected to seismic loads, the ductility checking should be quantified at the same level as for stiffness and strength. Unfortunately, in the present codes there are only vague provisions "...when plastic global analysis is used, the members shall be capable of forming plastic hinges with sufficient rotation capacity to enable the required redistribution of bending moment to develop..." (EUROCODE 3, 5.3.1), "...sufficient local ductility of members or parts of members in compression shall be assured..." (EUROCODE 8, 3.5.3.1). These two examples show the very rough definitions given by codes. For the structural designer is essential to have a clear definition of what " sufficient rotation capacity" or " sufficient local ductility" means and how these terms can be quantified.

The EC 8 considers that sufficient ductility for members shall be assured by limiting the width-to-thickness ratio of compression parts, according to the cross-sectional classes specified in EC 3. For plastic global analysis, EC 3 requires that all members developing plastic hinges shall have class 1 cross-sections, and under special conditions also

class 2. EC 8 gives limitations for the q-factor value in relation with three behavioural classes, being the use of class 4 sections not allowed in dissipative zones. For joints the provisions given in Annex of EC3 considers only some constructional details, without any explicit ductility determination. This methodology to assure a sufficient ductility by means of constructional rules only, contains many shortcomings and in some cases is proved to be not effective because:

(i) The local ductility of members depends not only on width-to-thickness ratios, but also on the flange and web interaction, member length, moment gradient, level of axial forces, etc. As a consequence of such additional factors, the concept of cross – section behavioural classes should be substituted by the concept of member behavioural classes (Gioncu & Mazzolani, 1994, Gioncu and Petcu, 1997, Mazzolani & Piluso, 1993, Anastasiadis, 1999).

(ii) The provisions of EC 3 concerning the ductility of members and joints refer to the static loads. In case of seismic actions, the local behaviour of cross-sections is considerably different due to cyclic and high velocity characteristics of loads (Gioncu, 2000, Gioncu et al, 2000a, Anastasiadis et al, 2000). These new factors reduce the local ductility and, in some cases, can transform the plastic deformations in a brittle fracture (for instance, the connection failure during Northridge and Kobe earthquakes). In addition, the ductility demand is strongly influenced by the earthquake type (near or far-source) (Gioncu et al, 2000b).

(iii) The code imposes that plastic deformations occur only at the beam ends and at the column bases, but without considering the joints, which under some conditions can show a stable behaviour. But in reality, the required overstrength of connections (the joint capacity must be 20% stronger than the adjacent member) does not assure the elastic behaviour of joints. As a consequence, the joint could be the weakest component of the node and its ductility cannot be ignored (Gioncu, 1999a, Gioncu et al, 2000a).

For these reasons, it is strongly required by design practice to have a comprehensive methodology for ductility checking. The present paper presents such a method in which all the above mentioned factors are considered.

2 DUCTILITY CHECKING IN SEISMIC DESIGN

Building in seismic areas requires the development of a particular design philosophy. The basic principle of this philosophy consists in considering that it is not economically justified that, in a seismic active area, all structures should be designed to survive the strongest possible ground motion without any damage. In the rare event of very strong ground motion, damage would be tolerated as long as the structure collapse is prevented. The main goal of seismic design and requirement is to protect life and structure collapse. However, the last earthquakes have been characterized by element collapses, interruption of functionality for many buildings, evacuation of people, losses in work places for varying periods, monetary losses and, therefore, they have shown that the above mentioned goal is not sufficient for a proper design methodology. So, in the last time the concept of multi-level design approach is proposed as a basic design philosophy. In the Vision 2000 Committee of SEAOC (Bertero, 1996) four levels of structural performance are proposed: *fully operational, operational, life safety* and *near collapse* for frequent, occasional rare and very rare earthquakes. Mazzolani and Piluso (1996) propose three levels: *serviceability, damageability* and *survivability* limit states. Contrary EC 8 proposes two levels verification: *serviceability,* and *ultimate* limit state. Among these proposals, the verification for three levels seems to be more reasonable for design practice. To be effective for design, these performance levels must be translated into seismic action values in term of design accelerations. In this context, it is necessary to decide the return period for each level. For three performance levels it is admitted that 10, 50 and 450 years correspond for the above mentioned limit states, respectively. The adequate accelerations result from recurrence relations established for each seismic area (Figure 1a).

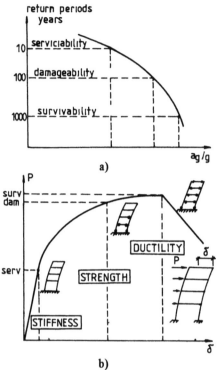

a)

b)

Figure 1 Stiffness, strength and ductility triad.

In the capacity design method a proper seismic design must consider the verification of structure stiffness, strength and ductility (Bachmann et all, 1995). Because the verification of this triad for each above limit states is too cumbersome, it seems that it is more rational to perform the stiffness, strength and ductility checks at different limit states: stiffness for serviceability, in case of frequent and weak earthquakes, strength for damageability, for rare and moderate earthquakes, and ductility for survivability, in case of very rare and strong earthquakes (Figure 1b).

The designer must verify the stiffness in elastic range (linear analysis), the strength by elasto-plastic analysis, using one of the well known methods (equivalent static analysis, push-over analysis, time-history analysis) and the ductility with the collapse kinematic mechanisms of structure (local and global mechanisms).

3 REQUIRED AND AVAILABLE DUCTILITIES

Ductility assessment of a structure is provided by satisfying the limit state criterion:

$$\gamma_r D_{req} \leq \frac{D_{av}}{\gamma_a} \qquad (1)$$

where D_{req} is the required ductility, obtained from the global plastic behaviour of structure, and D_{av} is the available ductility determined from the local plastic deformation, while γ_r, γ_a are the partial safety factors for required ductility and available ductility, respectively. These two safety factors must be determined considering the scatter of data with a mean plus one standard variation. Values $\gamma_a = 1.3$ and $\gamma_r = 1.2$ are proposed for this verification, if the available ductility is determined by plastic deformation. If the available ductility results from local fracture, a greater value of γ_a must be used ($\gamma_a = 1.5$). The relationship (1) is presented in Figure 2a. In the range where this relation is not satisfied, the inelastic force redistribution is not assured and the structure may collapse. Another indicator of structure behaviour is the ductility index (Figure 2b):

$$I_D = \gamma_r \gamma_a \frac{D_{req}}{D_{av}} \qquad (2)$$

The elastic limit with minor damage corresponds to the ductility index of 0.1, superficial and repairable damage to the value of 0.4 and collapse limit to the

damage index of 1.0. Values of the ductility index greater than 0.6 show unrepairable damage and values over 1.0 correspond to the extensive damage and progressive collapse state. This ductility index can be used as indicator of survivability limit state.

4 GLOBAL DUCTILITY AS REQUIRED DUCTILITY

The global ductility is directly related to the earthquake characteristics. In the last time a great amount of information concerning the feature of earthquakes is collected and important databases are operative. Important activity in macro and microzonation has been carried out all over the World to identify and characterize all the potential sources of ground motions. For the structural engineers the interest of these results is focused in the source characteristics with direct influence on seismic action. Source depth has a considerable influence on the earthquake behaviour and may be classified as (Figure 3):
-*surface sources*;
-*deep sources*.

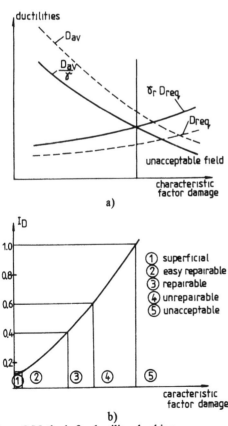

a)

b)

Figure 2 Methods for ductility checking.

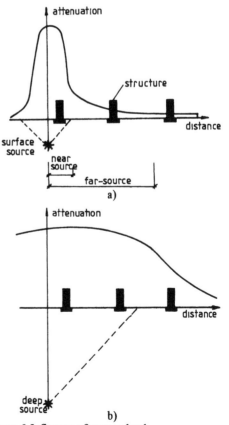

a)

b)

Figure 3 Influence of source depth.

Generally the surface sources are more frequent, over 85 percent of recorded earthquakes being ranged within 15Km. The importance of source depth is underlined by the attenuation low, which is very important for surface earthquakes (Figure 3a). So, the surface earthquakes have a great influence on the reduced area around the epicenter. In the last time, two main regions with different ground motion characteristics are considered (Figure 3a) (Iwan, 1996):

-*near–source region*, which can be defined as the region within few kilometres from either the surface rupture or the projection on the ground surface of the fault rupture zone. This region is also referred as near field region;

-*far-source region*, situated at some hundred kilometres far from the source.

For deep sources, the attenuation is reduced and the affected areas are very large (Figure 3b).

Unfortunately, the ground motions and the design methods adopted in the majority of codes are mainly based on records obtained from intermediate or far-source fields, being unable to describe in a proper manner the earthquake action in near-source field. Only the last UBC 97 has introduced some supplementary provisions concerning the near-source earthquakes, considering the lessons learned from the last dramatic events (Northridge, Kobe).

Another very important factor influencing the ground motions is the source mechanism, which may be:

-*interplate mechanisms* (Figure 4a) produced by sudden relative movement of two adjacent tectonic plates of their boundaries. Very large magnitude and large natural periods and duration characterize such earthquake events. The amplification of ground motions is strongly influenced by the nature of the soil under the site, and the corner periods are very large.

-*intraplate mechanisms* (Figure 4b) associated with relative slip across geological faults, within a tectonic plate. Such earthquake types generally gives smaller values of magnitude, natural periods, duration and corner periods. An important amplification occurs for rigid structures with small natural periods.

By coupling of these two aspects, source depth and mechanism, some differences in earthquake characteristics can be observed, which must be considered in design (Gioncu, 1999):

-*directionality* of wave propagation, very important in the case of near-source earthquakes;

-*soil influence*, as a result of travelling path and local site stratification;

-*velocity pulse*, which is one of the main characteristics of near-source earthquakes, where ground motions have distinct low-frequency pulses in accelerations and coherent pulse in velocity and displacement;

-*cyclic movements*, characteristic for far-source earthquakes, where the number of high value cycles is essential for the determination of ductility demands;

-*vertical components*, very high in the near-source region, being in many cases greater than the horizontal components;

-*velocity of ground motions*, with very important values in near-source regions, giving rise to very high strain-rates and impending the formation of plastic hinges in the structure members.

Without considering all these aspects in the evaluation of the required ductility, every design methodology should be incomplete. But this is a very difficult task, which oversteps the possibilities of structural engineers. The co-operation with the seismologists, geologists and geotechnical engineers is necessary. The interaction between ground motion types and ductility demand requires to pay attention to some important aspects concerning the interaction between local-source conditions and structures (Table1):

-*Seismic macrozonation*, which is an official zoning map, at the level of a Country, based on a hazard analysis elaborated by seismologists and geologists. This map divides the national territory in different categories and provides for each area the minimum values of earthquake intensity. At the same time, this macrozonation must characterize the possible ground motion type, as a surface or deep source, interplate or intraplate fault, etc.

-*Seismic microzonation*, which considers the possible earthquake sources at the level of region or town, on the basis of common local investigation of geologists and seismologists. The result of this study is a local map, which indicates the positions and the characteristics of sources, together with general information about the soil conditions.

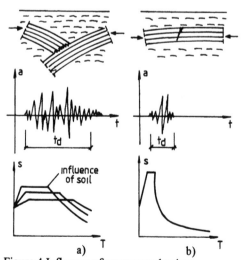

Figure 4 Influence of source mechanisms.

Table 1. Steps in control of required ductility.

Activity	Scheme	Specialits	Informations
Macrozonation		• seismologist • geologist	• earthquake types • intensities
Microzonation		• seismologist • geologist	• source position • intensities • attenuation • duration
Site conditions		•· geologist • geotechnical eng.	• soil stratification • soil type • amplification • duration • time-history records • spectrum
Structure characteristics		• geotechnical eng. • structural eng. • architect • builder • owner	• level of protection • general configuration • materials • foundation type • structural system

-*Site conditions*, established by geologists and geotehnical engineers, from the examination of the stratification under the proposed structure site. The changing in ground motions (amplification of accelerations, modification of natural vibration periods, increasing of duration, etc) due to soil conditions must be specified as a result of site examination.

-*Structure characteristics*, which result from the collaboration among geotehnical engineers and structural engineers, architects, builders and owners. At this step the level of seismic protection is established and the ductility demand is fixed as a function of this level. General configuration, structural materials, foundation and elevation types, technology of erection, etc. are the results of this activity.

The definition of required ductility inevitably calls for a series of engineering judgements of seismology, safety policy as well as structural matters. For this reason, the required ductility should be established in close collaboration between seismologists and structural engineers. Unfortunately, there are some difficulties in communication between these professionals. Seismologists break their research works at the level of spectra without being interested in structure behaviour. In contrast, structural engineers have no sufficient knowledge in the seismological problems. So, an important gap exists between the view points

of these two specialists category, impeding a reliable definition of seismic actions.

In order to establish the required ductility, the available methods for the designer are: monotonic static linear analysis (equivalent static analysis), monotonic static nonlinear analysis (push-over analysis) and dynamic nonlinear analysis (time-history analysis), presented in Table 2 with all the determinant factors:

-*Equivalent static analysis*, based on the assumption that the structural behaviour is governed by the first vibration mode. The characteristics of ground motions are described by means of linear elastic spectrum. For the inelastic deformations the design spectra are obtained by means of a reduction factor, namely q-factor. In this method the required ductility, D_{req}, is directly related to q-factor. These values are given by (Figure 5a):

$$D = \frac{\Delta_{pu}}{\Delta_e} \quad ; \quad q = \frac{Q_e}{Q_p} \qquad (2a,b)$$

In the literature there are some proposals for the relationship between D and q:

• Veletsos and Newmark (1960) for SDOF systems:

$$q = \sqrt{2(D+1)-1} \qquad (3a)$$

resulting

$$D_{req} = \frac{q^2 - 1}{2} \qquad (3b)$$

447

Table 2. Available methods for determining the required ductility

Method	Structure actions	Loading type	Structural response	Required ductility
Equivalent static analysis				
Push-over analysis				
Time-history analysis				

- Shinozouka and Moriyama (1989) for MDOF systems:

$$q = \varepsilon\sqrt{2(D+1)-1} \qquad (4a)$$

resulting:

$$D_{req} = \frac{\left(\dfrac{q}{\varepsilon}\right)^2 - 1}{2} \qquad (4b)$$

where ε is determined taking into account the scattering of numerical tests, using the average ± one standard deviation. For buildings with 3, 5 and 10 levels, after the examination of 711 cases, results $\varepsilon \approx 0.85$.

- Mazzolani and Piluso (1993) for MDOF:

$$q = \frac{2}{3}D + 1 \qquad (5a)$$

resulting:

$$D_{req} = \frac{3}{2}(q-1) \qquad (5b)$$

The proposals coming from Veletsos & Newmark and Mazzolani &Piluso correspond very closely with the medium values for required ductility. The Shinozouka & Moriyama relation gives the maximum values of ductility demands (Figure 5b).

-*Push-over analysis.* The structure is subjected to incremental lateral loads, using one or more predetermined local patterns of horizontal forces. These load patterns are supposed to describe the lateral load distributions which occur when the structure is subjected to earthquakes (Mazzolani & Piluso, 1996). The determination of these patterns is a very difficult task, because it depends on the influence of superior vibration modes and the progressive plastic hinge formation. Mazzolani and Piluso (1997) develop a simplified methodology based on the rigid plastic collapse mechanism by substituting the actual curve with a tri-linear one (Figure 6). The first part corresponds to a linear behaviour, while the equilibrium curve of collapse is determined by second-order rigid-plastic analysis and can be described by the following relationship:

$$\alpha = \alpha_0 - \gamma_s\delta \qquad (6)$$

where is α_0 is the collapse multiplier of the horizontal forces, obtained by rigid-plastic analysis

448

and γ_s is the slope of the linearized mechanism curve, determined in function of mechanism type. The cusp produced by the intersection of elastic curve and mechanism equilibrium curve is cutted by a horizontal straight line, corresponding to a point of mechanism equilibrium curve with a sway displacement equal to 2.5 times the elastic displacement.

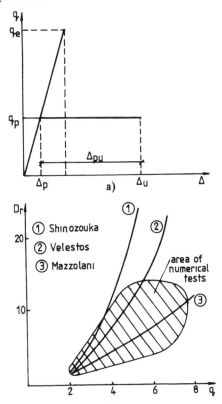

Figure 5 Equivalent static analysis.

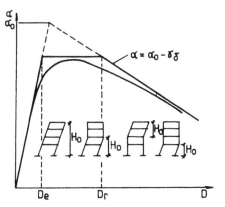

Figure 6 Push-over analysis.

The required rotation of plastic hinges can be determined by the relationship:

$$\theta_{req} = \frac{1}{H_0}(\delta_u - \delta_y) \qquad (7)$$

where H_0 is the sum of the intrestorey heights of the storeys involved in the collapse mechanism. The ultimate displacement value can be determined corresponding to near collapse criteria (Gioncu 1999b). Using this methodology, the required ductility for each storey must be determined.

The push-over analysis is relatively simple to be implemented, but contains a great number of assumptions and approximations that may be reasonable in some cases and unreasonable in other ones. Especially, when the superior vibration modes have important effects, the obtained results can be very far from the actual behaviour of structure.

-*Time history analysis.* The structure is subjected to an artificial or recorded accelerogram and the structure response is determined by considering the nonlinear elasto-plastic deformation of structure. The result of this analysis is an envelop of required ductility for each structure levels. The maximum demands may occurs at different levels along the structure height, in function of the earthquake natural period. In the case of short periods, the structure top is more affected, while for long periods the maximum required ductilities occurs at the first levels (Figure 7) (Gioncu et al., 2000b). Due to the development of computer science, today is not a problem to perform such a complex analysis. But the real problem of this method is the option for an accelerogram, which adequately represents the earthquake at the structure site.

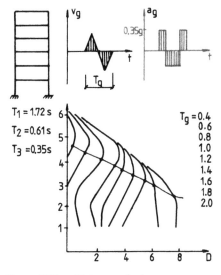

Figure 7 Time history analysis.

449

Table 3 Collapse modes of members

Member type	In-plane buckling	Out-of plane buckling	Flange induced buckling	Flange fracture

Table 4 Collapse modes of joints

Joint type	Collapse types		
Welded joint	Panel buckling	Panel crushing	Weld fracture
Bolted joint	Bolt fracture	End plate fracture	Weld fracture

Table 5 Collapse types under seismic loads

Earthquake type	Local behaviour
Pulse loads	M, θ, fracture
Cyclic loads	M, θ, fracture

Table 6 Classification criteria of joints

Properties	Behaviour	Joint type
Rigidity	M, θ	• rigid joint • semi-rigid joint
Strength	M, θ	• full strength joint • partial strength joint
Ductility	M, θ	• ductile joint • semi-ductile joint • brittle joint

450

The choice of an accelerogram is a very complex task due to the fact that at the same site, as a result of the same source, the ground motions may be very different in characteristics for different events. Therefore, the method of amplification of the peak ground acceleration without changing other characteristics (periods, duration, velocities, etc), what in generally done according to this method, is very disputable.

All the above mentioned methods contain many assumptions which can introduce some errors in the evaluation of required ductility. Thus, the interpretation of results must be done within the context of the used assumptions. The determination of a realistic ductility demands is one of the most complex problem because contains many uncertainties and discussions beyond the current knowledge of a structural engineer. This may be an explanation why today the verification of structure ductility is more an exception than a rule. But these problems do not differ very much from the ones concerning the strength and rigidity verifications. However, in order to minimize the assumed risk in the prediction of the ductility requirements, it is necessary to estimate the seismic activity, to evaluate the local soil conditions as well as to assess the structural behaviour under the estimated and predicted conditions.

5 LOCAL DUCTILITY AS AVAILABLE DUCTILITY

The determination of local ductility is more related to the structural engineer judgements than to the required ductility and contains less uncertainties (Gioncu, 1997). Beams, columns and joints compose a framed structure. In seismic design some critical sections are chosen to form a suitable plastic mechanism able to dissipate an important amount of the input energy. Generally, it is considered that these sections are located at the beam ends, where plastic hinges occur during a strong earthquake. But the beam is joined to a node, which connects also the column. Furthermore, the local plastic mechanism in the structure can be located not only at the beam or column ends, but also at joints, or at both member ends and joints. Consequently, the local ductility has to be defined at the level of node, composed by panel zone (column web), connection elements (bolts or welds, plates, angles, etc) and member ends (Figure 8) (Gioncu, 1999, Gioncu et al, 2000a).

The collapse modes of the members are presented in Table 3: in-plane, out-of-plane, flange induced buckling types and flange fracture. The collapse modes for welded or bolted joints are shown in Table 4. It is interesting to notice that in case of welded joints the collapse mode is governed by the panel collapse, while for bolted joints, by connection elements. The main aspects of these collapse modes are presented in Gioncu and Petcu (1997), Gioncu et al (2000a), as well as, in the companion papers Anastasiadis et al (2000) and Gioncu (2000).

The types of local failure must be considered for available ductility under seismic loads (Table 5). In the case of pulse loads, characteristic for near-source earthquakes, the great velocity induces very high strain-rate and fracture of members or joints occurs at the first or second cycle. Contrary, if the action is characterized by cyclic loads, especially for far-source earthquakes and soft soils, an accumulation of plastic deformation occurs, producing a degradation in behaviour and the fracture takes place after a high number of cycles.

In order to establish the weakest component of a node, the joint properties must be compared with the properties of connected members in terms of rigidity, strength and ductility (Table 6). So, the joints may be classified according to their capacity to restore the properties of beams and columns. Based on the method of components, the overall behaviour of the node is dictated by the behaviour of the weakest component (Tschemmernegg, 1998), which is determined by the comparison of the two plastic moments. The node ductility is given by the component with the smallest value.

6 DUCTILITY CHECKING

The capacity design method is based on the concept that the available ductility, determined from local ductility, is greater than the required ductility, obtained from global ductility. A chart for determining the global and local ductilities, as well as, the conceptual ductility checking is illustrated in Figure 9.

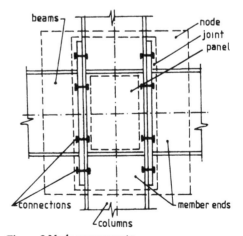

Figure 8 Node components

For global ductility the hierarchy is at the level of source, epicentral distance, site and structure, while for local ductility material, cross-section, member and connections are the main factors. The comparison between required and available ductilities, D_{req}, D_{av}, can be performed in two ways as follows:
-*direct verification* using the equation (1);
-*calculation of ductility index* given by the equation (2).
The use of direct verification has the purpose to assure that the redistribution of forces after the formation of plastic hinges, in some predetermined sections, is going to be under stable conditions in order to prevent the structure collapse. Contrary, the use of ductility index has the advantage to limit the member and joint damage below an acceptable level, in order to allow for an easy repairing. The reserve

of ductility in frames subjected to some repeated earthquakes may be determined by using the latter approach.

7 CONCLUSIONS

In seismic analysis the most difficult problem is to predict in a proper manner the earthquake type and the seismic actions, because a great variability of these characteristics exists. The code provisions are normaly very poor, being based only on a reduced number of design parameters, which cannot cover the possible seismic actions. Due to this code lack, in many cases the structure behaviour is studied by structural engineers starting from a wrong distribution of lateral forces. Consequently the obtained results are far from the reality. Only the introduction in code provisions of more reliable methods to establish the seismic actions can solve this situation. Therefore, the co-operation with the seismologists must be enlarged. As usually the seismologists have a limited knowledge on the structure behaviour, it is the duty of structural engineers to fill the existing gap.

Recent developments of advanced design concepts, as the ones introduced in the capacity design method, are based on the scope to provide the structure with sufficient ductility, in the same way as for strength and rigidity, in order to minimize the aforementioned problems. For these reasons, a consistent, comprehensive and transparent methodology is developed here which considers the required and available ductilities determined at the levels of the overall structure as well as at the local levels of the structural components. The main factors influencing these ductilities are presented. One can consider that today the accumulated knowledge allows to elaborate a sufficiently simple, but consistent methodology, which can be implemented in the modern codes.

For instance in EC8, instead to refer to the provisions of EC 3 concerning the ductility of cross-section under statical conditions, it should be more useful to elaborate an Annex, in which the bases of ductility checking in seismic conditions are presented. At the same time, some constructional details, very important to assure an adequate seismic behaviour, preventing the local damages, are required to be introduced in this Annex.

8 REFERENCES

Anastasiadis, A. 1999. Ductility problems of steel moment resisting frames. Ph.D Thesis. University Politehnica Timisoara.
Anastasiadis, A., V. Gioncu & F.M. Mazzolani

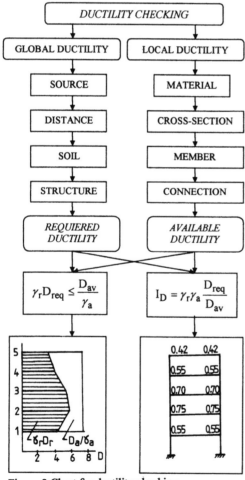

Figure 9 Chart for ductility checking

2000. New trends in the evaluation of available ductility of steel members. In *Behaviour of Steel Structures in Seismic Areas. STESSA '2000.* Eds. F.M. Mazzolani &R. Tremblay, *Montreal, 21-24 August 2000.* Rotterdam: Balkema.

Bachmann, H., P. Linde & Th. Werik 1995. Capacity design and nonlinear dynamic analysis of earthquake-resistant structures. In *10th European Conference on Earthquake Engineering.* Vienna, 28 August- 2 September 1994: 11-20. Rotterdam :Balkema.

Bertero, V.V. 1996. State-of the-art report on design criteria. In *11th World Conference on Earthquake Engineering, Acapulco, 23-28 June 1996.* CD-ROM No. paper 2005.

Gioncu, V. & F.M. Mazzolani 1994. Alternative methods for assessing local ductility. In *Behaviour of Steel Structures in Seismic Areas. STESSA '94.* Eds. F.M. Mazzolani & V. Gioncu,*Timisoara, 26 June-1July 1994:182-190* London: E&FN Spon.

Gioncu, V. 1997. Ductility demands. General report. *Behaviour of Steel Structures in Seismic Areas. STESSA '97.* Eds. F.M. Mazzolani & H. Akiyama, *Kyoto, 3-8 August 1997*: 279-302. 10/17 Salerno.

Gioncu, V. & D. Petcu 1997. Available rotation capacity of wide-flange beams and beam-columns. Part 1: Theoretical approaches. Part 2: Experimental and numerical tests. *J. of Constr. Steel Research,* Vol. 43, 1-3, 161-217, 219-244.

Gioncu, V. 1999a. Design criteria for seismic resistant steel structures. CISM course on *"Seismic Resistant Steel Structures: Progress and Challenge": Udine, 18-22 October, 1999.*

Gioncu, V. 1999b. Framed Structures. Ductility and seismic response. General report. In *Stability and Ductility of Steel Structures, SDDSS'99,* Eds. D. Dubina & M. Ivanyi. *Timisoara, 9-11 September.*

Gioncu, V. 2000. Influence of strain-rate. In F.M. Mazzolani & R.Tremblay, In *Behaviour of Steel Structures in Seismic Areas, STESSA 2000, Montreal.* Rotterdam: Balkema.

Gioncu, V., G. Mateescu, D. Petcu, A. Anastasiadis. 2000b. Prediction of available ductility by means of local plastic mechanism method: DuctRot computer program. In "Moment Resistant Connections of Steel Building Frames in Seismic Areas". Ed. F.M. Mazzolani. London: E &FN Spon.

Gioncu, V., G. Mateescu, L. Tirca, A. Anastasiadis. 2000a. Influence of the type of seismic ground motion. In "Moment Resistant Connections of Steel Building Frames in Seismic Areas". Ed. F.M. Mazzolani. London: E &FN Spon.

Iwan, W.D. 1995. Drift demand spectra for selected Northridge sites. Technical Report SAC 95-05, 2.1-2.40.

Mazzolani, F.M. & V. Piluso 1993. Member behavioural classes of steel beams and beam columns. *XIV Congresso CTA,* Viareggio, Italy, Ricerca Teoretica e sperimentale, 405-416.

Mazzolani, F.M. &V. Piluso 1996. *Theory and Design of Seismic Resistant Steel Frames.* London: Champan & Hall.

Mazzolani, F.M. &V. Piluso 1997. A simplified approch for evaluating performance levels of moment-resisting steel frames. In *Seismic Design Methodologies for the Next Generation of Codes* .Eds. P. Fajfar & H.Krawinkler. *Bled, 24-27 June, 1997.* Rotterdam: Balkema.

Shinozouka, M. & Moriyama K. 1989. An assessment of uncertainties in seismic response. In *Earthquake Hazards and the Design of Constructional Facilities in the Eastern United States.* Eds. K.H. Jacob & C.J. Torkstra. Annuals of the New York Academy of Sciences, Vol 558, 234-250.

Tschemmernegg. F., D. Rubin & A. Pavlov 1998. Application of the component method of composite joints. In *Control of Semi-Rigid Behaviour of Civil Engineering Structural Connections.* COST C1 Conference, *Liege, 17-18 September 1998,* 145-154.

Veletsos, A. & Newmark N.M. 1960. Effect of inelastic behaviour on response of simple system to earthquake motions. Proc. of *the 2nd World Conference on Earthquake Engineering, Tokyo,* 895-912.

Behaviour of Steel Structures in Seismic Areas, Mazzolani & Tremblay (eds) © 2000 Balkema, Rotterdam, ISBN 90 5809 130 9

Numerical investigation of the q-factor for steel frames with semi-rigid and partial-strength joints

J. M. Aribert
Laboratory of Structural Mechanics, INSA of Rennes, France

D. Grecea
Department of Steel Structures and Structural Mechanics, Politehnica University of Timisoara, Romania

ABSTRACT: A new definition of the q-factor is presented briefly relating this factor to the maximum inelastic base shear force of the structure during its time-history. Such a definition is well applicable to steel frames with semi-rigid and partial-strength joints. In this case, assuming a design elastic response spectrum to characterize the seismic action, an elastic-plastic static global analysis should be used to check the structural resistance of frames which generally are partially dissipative. Then, adopting an elastic-perfectly plastic cyclic behaviour of the beam-to-column joints and considering different limited rotation capacities of these joints, a parametric investigation of the q-factor by means of non-linear dynamic analysis is performed for different types of multi-bays multi-storeys frames subject to severe ground accelerograms. In addition to the instructive conclusions deduced from this investigation, conservative values of the q-factor are proposed in low seismicity zones as a function of both the joint rotation capacity and the nominal ground acceleration. Finally, a design example is given to illustrate the new concepts and results of this paper.

1 INTRODUCTION

Generally European codes are based on elastic static global analysis when designing a steel structure subject to seismic actions. Nevertheless, advantage of the very significant dissipative phenomena in the structure is taken by means of the q-factor reducing the seismic forces which would be obtained assuming a perfectly elastic behaviour of structural steel. As reminder, a steel frame with elements (beams and columns) of high rotation capacity (for example when the cross-sections are Class 1) can reveal values of q-factor greater than 6 (CEN 1995) provided that their joints in dissipative zones are rigid and show sufficient overstrength. So, Clause 3.5.3.2(1) in Eurocode 8-1-3 requires clearly that: «Connections in dissipative zones should have sufficient overstrength to allow for yielding of the connected parts...». However, Clause 3.5.3.2(4) accepts a possible alternative, namely: «The overstrength condition for connections needs not apply if the connections are designed in a manner enabling them to contribute significantly to the energy dissipation capability inherent to the chosen q-factor». In fact, Eurocode 8 does not provide any application rule for the latter Clause so that the procedure to evaluate the effect of the behaviour of partial-strength joints on the q-factor remains to be established totally.

An attempt of procedure is proposed in this paper adopting a new definition of q-factor related essentially to the maximum inelastic shear force at the base of the structure during its time-history response. Several advantages may be assigned to this definition; in return, an elastic-plastic static global analysis (including 2nd order geometrical effects if necessary) should be used to check the frame resistance/or stability. A design example is treated at the paper end to demonstrate the efficiency of the so-proposed procedure.

2 PROCEDURE TO CHARACTERIZE AND TO DESIGN FRAMES WITH PARTIAL-STRENGTH JOINTS

2.1 Determination of the q-factor

As clearly illustrated in a chapter of Mazzolani and Piluso's book (1996), comparisons of different methods existing in the scientific literature to evaluate the q-factor show a large scattering of results. That may be explained largely by the conventional nature of the adopted definitions, which generally are not consistent with the «directions for use» of the q-factor.

Though somewhat laborious, the approach adopted here consists in performing a series of non-linear dynamic analyses, which are able to take

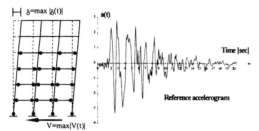

Figure 1. Dynamic response of a structure subject to ground acceleration λa(t) (horizontal component)

account of elastic-plastic behaviour of the structural steel and possibly 2nd order geometrical effects in a MDOF structure. For a given ground accelerogram (natural or artificial), a(t), multiplied by some positive factor λ (Figure 1), the maximum response of the structure during its time-history should be determined.

Contrary to the choice of the literature (Ballio & Setti 1985, Sedlacek & Kuck 1993, Mazzolani & Piluso 1996), the present authors (Aribert & Grecea 1997, 1998, 1999) decided to characterize the maximum response of the structure, not by a displacement (generally the upper horizontal deflection δ is considered significant), but by the horizontal base shear force V of the structure. Increasing λ step-by-step, such a determination should be repeated systematically up to a certain ultimate value $λ_u$ which corresponds to the attainment of the rotation capacity of a particular joint (it is worth noting that other criteria as rotation capacity and stability of beam and column elements should be also considered at ultimate limit state if more unfavourable, specially when their cross-sections are Class 2 or 3). Curve OEU in Figure 2 represents the variation of V versus the multiplier λ where the shear-force values $V^{(e)}$ and $V^{(inel)}$ correspond to the first yielding state and the ultimate limit state, respectively. In the same figure, straight line OEU* would be the curve (V, λ) assuming an ideal elastic behaviour of the structure up to $λ_u$.

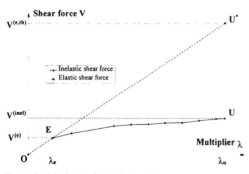

Figure 2. Maximum base shear force versus accelerogram multiplier

Whereas the q-factor in the literature is generally defined by the ratio $λ_u/λ_e$, the new definition of the present authors is given by the ratio between the elastic theoretical base shear force, $V^{(e,th)}$, and the real inelastic base shear force, $V^{(inel)}$:

$$q = V^{(e,th)}/V^{(inel)} = \left(V^{(e)}/λ_e\right)/\left(V^{(inel)}/λ_u\right) \qquad (1)$$

The main advantage of this definition is the correct calculation of the reduced horizontal force applied to the foundation, which allows to expect a suitable evaluation of the internal forces and moments in the structure provided that an appropriate global analysis is used taking account of the dissipative phenomena.

Another advantage of definition (1) is the fact that there is no need to know the value $λ_e$ corresponding to the first yielding stage; it is only necessary to know the slope $V^{(e)}/λ_e$. Besides, determination of $λ_e$ is sometimes difficult (due to a not always clear bifurcation of curve segment EU), and even significance of $λ_e$ in dynamic behaviour remains questionable.

It is important to mention a complementary aspect of definition (1) which has been neglected likely in the literature; as an explicit result of the determination procedure itself, the q-factor should be associated with a precise value of the maximum ground acceleration, namely:

$$a_N^{(u)} = λ_u \max|a(t)| \qquad (2)$$

Practically, $a_N^{(u)}$ may be considered as a nominal acceleration.

2.2 Evaluation of equivalent static forces

Considering only the case of a regular structure in elevation according to Clause 2.2.3 of Eurocode 8-1-2 (CEN 1994) and assuming the relevant ground accelerogram to be well representative of a given normalized elastic design response spectrum in pseudo-acceleration, $R_e(T)$, where T is the vibration period, the reduced inelastic base shear force corresponding to a certain doublet $(a_N^{(u)}, q)$ can be evaluated by:

$$V^{(inel)} = V^{(e,th)}/q = MR_e(T_1)a_N^{(u)}/q \qquad (3)$$

where T_1 is the fundamental period and M the total mass of the structure. It is clear that relationship (3) may be approximative because the inelastic response spectrum is never deduced perfectly by anamorphosis from the elastic one, specially when the latter is very irregular. Then, the distribution of the static horizontal seismic forces is given by relationship (4) where x_{jl} represents the displacement of storey j in the fundamental mode of vibration:

$$F_j^{(inel)} = \frac{m_j x_{jl}}{\sum_j m_j x_{jl}} V^{(inel)} \qquad (4)$$

In fact, the above distribution is adjusted a little so that the participant mass M_1 of the fundamental mode is equal to the total mass M of the structure. If the structures were not regular, it should be taken account of several vibration modes x_{ji} so that the sum of the participant modal masses exceeds 90% of the total mass M; but such a generalization will be developed no more in the present paper.

2.3 *Checking of the structural resistance and stability*

As often as not, partial strength joints have got a limited rotation capacity (in comparison with Class 1 elements) so that the frame cannot be fully dissipative under seismic actions. That means that under the distribution (4) of forces the frame does not collapse by a global mechanism consisting of plastic hinges developed in the joints at all the beam ends and in the columns at the base of the first storey. Consequently the most appropriate global analysis to check the structural resistance appears an elastic-plastic static one which has to be performed step-by-step up to reach the horizontal forces $F_j^{(inel)}$. Before the final step, neither a global mechanism nor a storey local mechanism (meaning a plastic hinge mechanism concentrated on one storey or a few ones including plastic hinges at column ends) should occur. In addition, the required rotation of each joint should be less than its available rotation capacity. Also, if plastic hinges occur at the ends of some columns, specially towards the structure base, any column local mechanism should be avoided due to the occurrence of a third plastic hinge between the column ends; for that, checking may consist in limiting the axial force and/or the slenderness of the concerned columns, for example in accordance with empirical relationships given in Clause 13.7.1.1 of French code PS 92 (AFNOR 1995). Finally, it is worth underlining the undeniable advantage of the step-by-step elastic-plastic static analysis providing directly a suitable evaluation of the inelastic displacement, $\delta_j^{(inel)}$, of any storey j (possibly with 2nd order geometrical effects), as confirmed later on by numerical examples where the static evaluation will be compared to the non-linear dynamic solution.

Once $\delta_j^{(inel)}$ is known, the level of the 2nd order geometrical effects may be evaluated by means of the interstorey drift sensitivity coefficient θ_j defined as follows:

$$\theta_j = \frac{\Delta\delta_j}{h_j} \frac{W_j}{H_j} \qquad (5)$$

where $\Delta\delta_j$ is the horizontal displacement at the top of the storey j relative to the bottom of the storey, h_j is the storey height, H_j is the total horizontal shear force at the bottom of the storey and W_j is the total vertical axial force acting at the bottom of the storey. When $\theta_j \leq 0.1$ for all the storeys, 2nd order effects may be neglected.

3 PARAMETRIC INVESTIGATION

As already mentioned, the structure should be considered partially dissipative when the rotation capacity of dissipative members is limited (due to cross-sections in Class 2 or 3) or partial strength joints are used in dissipative zones.

In this case, the maximum values (q, $a_N^{(u)}$) should be determined as a function of the allowable rotation capacity directly from dynamic analyses.

3.1 *Data*

A parametric investigation (Grecea 1999) was carried out on the structures presented in Figure 3 (structures A, B, C with characteristics of Table 1), using partial strength joints with different properties and considering the two accelerograms of Figure 4 in order to evaluate their influence on the q-factor. The moment resistance $M_{j,R}$ and initial rotational stiffness $S_{j,ini}$ of the studied joints are given in Table 2 where $M_{b,pl,R}$ is the plastic resistance moment of the connected beam to the joint and K_{sup} is the limit rotational stiffness corresponding to the distinction between rigid joint behaviour and semi-rigid one according to Annex J of Eurocode 3. Figure 5 shows the skeleton moment-rotation curves of the joints (for a quarter space) assuming here perfect parallelograms without stiffness degradation when repeated cyclic bending moments are applied. This type of joint behaviour was introduced in the inelastic dynamic analyses performed using DRAIN-2DX computer code (Prakash 1993). As often as not, partial strength joints have got a limited rotation capacity ϕ_u. Here three values of rotation capacity were selected, namely 0.015 ; 0.030 and 0.045 radians which seem realistic to cover most of applications.

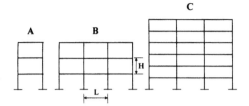

Figure 3. Investigated frames

Table 1. Characteristics of analysed frames

Frame	L[m]	H[m]	w[kN/m]	m_j[kg]	Beams	Columns
A	5.0	4.0	22	11000	IPE300	HEB240
B	4.0	4.0	32	38400	IPE330	HEB240
C	4.5	3.0	35	47250	IPE330	HEB360 (1,2)
						HEB300 (3-5)
						HEB260 (6)

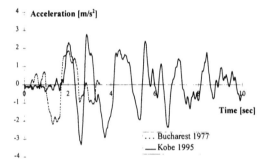

Figure 4. Accelerograms of Bucharest (1977) and Kobe (1995)

Table 2. Characteristics of analysed frames

Moment resistance $M_{j,R}$	Rotational stiffness $S_{j,ini}$
$1.0 M_{b,pl,R}$	K_{sup}
	$0.8 K_{sup}$
	$0.6 K_{sup}$
$0.8 M_{b,pl,R}$	K_{sup}
	$0.8 K_{sup}$
	$0.6 K_{sup}$
$0.6 M_{b,pl,R}$	K_{sup}
	$0.8 K_{sup}$
	$0.6 K_{sup}$

This type of joint moment-rotation cyclic curve based on an elastic-perfectly plastic model, but with limited values of the rotation capacity ϕ_u can be considered significant as proved by a research program developed at INSA Rennes (Aribert & Grecea 1998).

This research program dealt with 8 beam-to-column welded joints of different sizes under monotonic and repeated cyclic loading (Table 3).

The specimens were major axis joints with a symmetrical cruciform arrangement comprising a H or I column connected to two cantilever beams by full penetration butt welds with double bevel in the beam flanges. No transverse stiffener was welded in the compression zone of the column web, so that the static partial resistance of the joints was governed by local buckling of the column web.

The tests were performed according to the Recommended Testing Procedure of ECCS (1985). Each type of joint was subject first of all to a monotonic loading (CPP11, CPP13, CPP15 and CPP17), which allowed to determine the main static characteristics as the moment resistance, the initial rotational stiffness, the maximum elastic rotation and the rotation capacity.

After the determination of these characteristics, the elements were subject to cyclic reversal loading (CPP12, CPP14, CPP16 and CPP18). The cyclic moment-rotation curves for the four tested joints are presented in Figure 6.

The cyclic behaviour of this type of joint may be characterized by a good regularity in the loops shape with more or less deterioration between loops. This regularity could be explained by the sufficient continuity of internal forces and the reasonable value of the local buckling slenderness of the column web subject to transverse compression (non-dimensional slenderness $\bar{\lambda} \le 0.8$). Generally the failure mode occurred by local buckling of the column web. Nevertheless, due to plastic deformations developed in the column web under alternate compression and tension, a crack in the web panel near the flange was observed in a few cases. Simultaneously, in the column flange subject to alternate bending, another crack started close to the weld connecting the column flange with the beam flange. Experimental values of ultimate resistance moment and rotation capacity obtained in both monotonic and cyclic reversal loadings can be compared in Table 4.

>From the moment-rotation curves, it is observed that the ultimate moment and the initial stiffness of

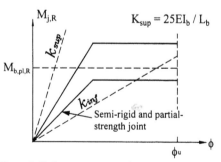

Figure 5. Skeleton moment-rotation curves

the joints are not strongly influenced by the repeated cyclic loading, so that in seismic design the corresponding formulae given in Eurocode 3 for the case of static loading can be used, as reasonable approximations.

On the other hand it appears clearly that the rotation capacity of the joints is systematically reduced by a factor about 2. As in the literature there are no formulae to evaluate the rotation capacity of the relevant joints, the present authors have proposed the following one for monotonic loading :

$$\Phi_u = 0.030 h_b / h_c \qquad (6)$$

and the following one for cyclic loading :

$$\Phi_u = 0.015 h_b / h_c \qquad (7)$$

where h_b is the beam depth and h_c the column depth (it should be noted that ϕ_u would be proportional to

ratio h_b/h_c, and not to ratio h_c/h_b as mentioned in Clause J.5(5) of Annex J of EC3).

In addition the failure mode may be different under repeated cyclic loading in comparison with the static one ; for example fracture of the column flange and web may occur instead of local buckling of the column web, which can be controlled only by design methods including low-cycle fatigue phenomena and damage models.

3.2 Numerical results

Tables 5 and 6 collect, for example the results of the q-factor for structure C subject to the accelerogram of Vrancea (1977) and Kobe (1995), respectively.

3.3 Interpretation of the parametric investigation

From examination of Tables 5 and 6 it is clear that the initial rotational stiffness $S_{j,ini}$, even for the lowest value $0.6K_{sup}$, has no influence practically on

Table 3. Experimental elements

Test specimens	Column	Beam	Loading type	Failure mode
CPP 11	HEB 200	IPE 360	Monotonic	Buckling of column web
CPP 12	HEB 200	IPE 360	Cyclic Reversal	Buckling of column web with fracture of column flange and web
CPP 13	IPE 360	IPE 360	Monotonic	Buckling of column web
CPP 14	IPE 360	IPE 360	Cyclic Reversal	Buckling of column web
CPP 15	HEB 300	IPE 360	Monotonic	Buckling of column web
CPP 16	HEB 300	IPE 360	Cyclic Reversal	Buckling of column web
CPP 17	HEB 300	IPE 450	Monotonic	Buckling of column web
CPP 18	HEB 300	IPE 450	Cyclic Reversal	Buckling of column web with fracture of column flange and web

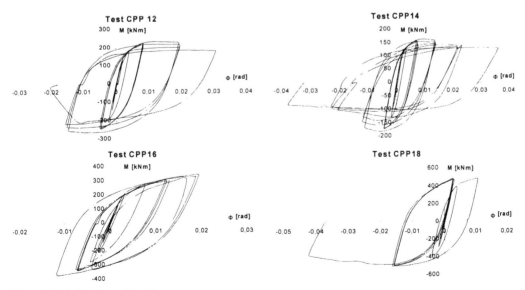

Figure 6. Cyclic behaviour of the joints

Table 4. Comparison between joint characteristics under cyclic and monotonic loadings

	CPP11	CPP12	CPP13	CPP14	CPP15	CPP16	CPP17	CPP18
M_u [kNm]	230.0	253.0	166.9	180.3	349.2	368.5	467.5	486.2
Φ_u [rad]	0.064	0.031	0.045	0.023	0.045	0.020	0.052	0.030

Table 5. Values of q-factor for accelerogram of Bucharest (1977)

$\dfrac{M_{j,R}}{M_{b,pl,R}}$	Criterion	K_{sup}			$0.8K_{sup}$			$0.6K_{sup}$		
		q	a_N	θ_i	q	a_N	θ_i	q	a_N	θ_i
1.0	$\phi_{0.015}$	1.5	1.30	0.042	1.6	1.30	0.039	1.6	1.30	0.043
	$\phi_{0.030}$	2.0	2.20	0.061	2.0	2.20	0.061	2.1	2.20	0.065
	$\phi_{0.045}$	2.8	3.00	0.090	2.9	3.00	0.094	3.0	3.00	0.096
0.8	$\phi_{0.015}$	1.8	1.20	0.044	1.8	1.20	0.049	1.8	1.20	0.049
	$\phi_{0.030}$	2.3	2.10	0.074	2.3	2.00	0.074	2.4	2.00	0.075
	$\phi_{0.045}$	2.9	2.90	0.097	2.9	2.80	0.101	3.0	2.80	0.102
0.6	$\phi_{0.015}$	2.0	1.10	0.060	2.0	1.10	0.060	2.0	1.10	0.058
	$\phi_{0.030}$	2.4	2.00	0.086	2.4	2.00	0.087	2.6	2.00	0.090
	$\phi_{0.045}$	2.9	2.60	0.111	3.0	2.60	0.112	3.0	2.60	0.115

Table 6. Values of q-factor for accelerogram of Kobe (1995)

$\dfrac{M_{j,R}}{M_{b,pl,R}}$	Criterion	K_{sup}			$0.8K_{sup}$			$0.6K_{sup}$		
		q	a_N	θ_i	q	a_N	θ_i	q	a_N	θ_i
1.0	$\phi_{0.015}$	1.5	0.50	0.044	1.6	0.60	0.046	1.8	0.80	0.054
	$\phi_{0.030}$	2.3	1.00	0.054	2.8	1.20	0.066	2.7	1.20	0.071
	$\phi_{0.045}$	3.4	1.60	0.069	3.5	1.60	0.073	3.2	1.60	0.079
0.8	$\phi_{0.015}$	1.5	0.50	0.043	1.7	0.50	0.047	1.9	0.70	0.052
	$\phi_{0.030}$	2.8	1.00	0.068	2.9	1.00	0.070	2.8	1.10	0.074
	$\phi_{0.045}$	3.5	1.40	0.069	3.8	1.50	0.072	3.5	1.60	0.078
0.6	$\phi_{0.015}$	1.9	0.50	0.042	2.3	0.50	0.049	2.3	0.60	0.056
	$\phi_{0.030}$	3.2	1.00	0.065	3.5	1.00	0.070	3.2	1.10	0.075
	$\phi_{0.045}$	4.2	1.40	0.075	4.6	1.50	0.079	4.0	1.70	0.082

Table 7. Average values of q-factor for accelerogram of Bucharest (1977)

$M_{j,R}$	$1.0M_{b,pl,R}$			$0.8M_{b,pl,R}$			$0.6M_{b,pl,R}$		
ϕ	$\phi_{0.015}$	$\phi_{0.030}$	$\phi_{0.045}$	$\phi_{0.015}$	$\phi_{0.030}$	$\phi_{0.045}$	$\phi_{0.015}$	$\phi_{0.030}$	$\phi_{0.045}$
q	1.6	2.0	2.9	1.8	2.3	2.9	2.0	2.4	3.0
a_N	1.30	2.20	3.00	1.20	2.00	2.80	1.10	2.00	2.60
θ_i	0.04	0.06	0.09	0.05	0.07	0.10	0.06	0.09	0.11

Table 8. Average values of q-factor for accelerogram of Kobe (1995)

$M_{j,R}$	$1.0M_{b,pl,R}$			$0.8M_{b,pl,R}$			$0.6M_{b,pl,R}$		
ϕ	$\phi_{0.015}$	$\phi_{0.030}$	$\phi_{0.045}$	$\phi_{0.015}$	$\phi_{0.030}$	$\phi_{0.045}$	$\phi_{0.015}$	$\phi_{0.030}$	$\phi_{0.045}$
q	1.6	2.6	3.4	1.4	2.8	3.6	2.2	3.3	4.3
a_N	0.60	1.20	1.60	0.60	1.00	1.50	0.50	1.00	1.50
θ_i	0.05	0.06	0.07	0.05	0.07	0.07	0.05	0.07	0.08

the q-factor. Moreover, decrease in the joint moment resistance tends to increase slightly the q-factor, maybe favouring the occurrence of global dissipative mechanism. So, these results incite to present the average values of the q-factor for the two accelerograms, as a function only of the joint resistance moment and of the rotation capacity. These average results are collected in Tables 7 and 8 and represented in Figure 7.

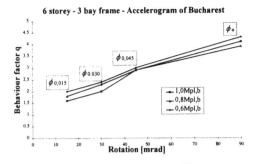

6 storey - 3 bay frame - Accelerogram of Bucharest

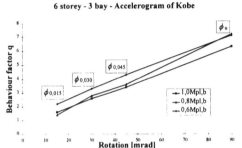

6 storey - 3 bay - Accelerogram of Kobe

Figure 7. Average values of q-factor as a function of the joint rotation capacity for accelerograms of Bucharest (1977) and Kobe (1995).

Average values of q for $\phi_{0.015}$ can be deduced from Tables 7 and 8, applicable to the three values of resistance moment of the joints $1.0M_{b,pl,R}$, $0.8M_{b,pl,R}$ and $0.6\ M_{b,pl,R}$. For Bucharest accelerogram, q can be taken equal to 1.8 provided that a_N and θ_j do not exceed 1.2m/s^2 and 0.050, respectively. Similarly for Kobe accelerogram, the values become q=1.7, with $a_N=0.6\text{m/s}^2$ and $\theta_j=0.050$.

For the criterion $\phi_{0.030}$ and Bucharest accelerogram, it is obtained q=2.2 with $a_N=2.1\text{m/s}^2$ and $\theta_j=0.075$, while for Kobe accelerogram, q=2.7, with $a_N=1.0\text{m/s}^2$ and $\theta_j=0.070$.

Lastly for the criterion $\phi_{0.045}$ and Bucharest accelerogram, it is obtained q=2.9 with $a_N=2.8\text{m/s}^2$ and $\theta_j=0.100$, while for Kobe accelerogram, q=3.8, with $a_N=1.5\text{m/s}^2$ and $\theta_j=0.075$.

Following to the previous comments, some more general conclusions can be expressed.

For q-values higher than q=3.0 it appears that the geometrical second order effects (or P-Δ effects) begin to influence the structural behaviour. So, for q-values higher than this value P-Δ effects cannot be neglected.

In accordance with all performed analyses, even for low q-values, most of the plastic hinges are developed in beams or joints but also someones in columns. Consequently, care must be taken to control the rotation capacity at the column ends with respect to the risk of local mechanism (due to

Table 9. Conservative values of q-factor for steel frames with partial-strength joints

	Accelerogram duration			
	≤ 4 sec.		≤ 20 sec.	
	a_N [m/s²]	q	a_N [m/s²]	q
$\phi_{0.015}$	1.20 - 2.20	1.3 - 1.8	0.60 - 1.70	1.7 - 1.8
$\phi_{0.030}$	2.10 - 2.90	1.7 - 2.2	1.10 - 2.20	2.0 - 2.9
$\phi_{0.045}$	2.80 - 3.70	2.0 - 2.9	1.50 - 2.70	2.3 - 3.8

occurrence of an intermediate plastic hinge).

For the chosen three rotation capacities, the joint moment resistance seems also not to influence the q-factor, noting that with the reduction of the moment resistance, the q-factor has a little tendency to grow up a little bit. In practice, it should be preferable to keep constant the value of q-factor till the value $0.6M_{b,pl,R}$ (Figure 7).

3.4 Proposal of conservative q-factors

According to the numerical study, an indicative proposal of q-factors could be promote for steel structures with semi-rigid and partial-resistant joints, for different categories of joint rotation capacity (ϕ_u), acceleration (a_N) and type of accelerogram, as in Table 9. Evidently, these values are imbued with the results deduced from the accelerograms of Bucharest and Kobe, which are quite severe in comparison with those given by most of the seismic codes, specially in Europe. Proposed values of Table 9 could be generalized and improved by further researches ; they are already giving a first answer to the design problem of steel frames with partial-strength dissipative joints.

4 EXAMPLE

An application of the design procedure explained in 2 is given for the 6 storey - 3 bay frame of Figure 3 including partial strength joints with resistance moment $M_{j,R} = 0.8M_{b,pl,R}$ and with rotation capacity $\phi_u = 0.030$ radians is presented bellow.

In this example, the frame is subject to the accelerogram of Bucharest-1977 (Figure 3). The elastic response spectrum associated to the accelerogram is presented in Figure 8.

The q-factor deduced from non-linear dynamic analyses is q = 2.1 with the associated acceleration $a_N = 2.1\ \text{m/s}^2$. An equivalent static second order elastic-plastic analysis is performed using PEP-micro computer code developed in CTICM by Galea & Bureau (1995) up to reach step-by-step the base shear force equal to $V^{(inel)} = 759$ kN. It leads to 6th storey displacement equal to $\delta_6^{(inel)} = 0.029$m and a maximum required rotation 0.029 radians.

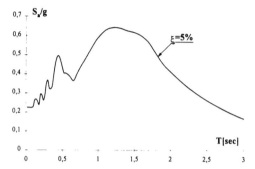

Figure 8. Elastic response spectrum associated to the accelerogram of Bucharest - 1977.

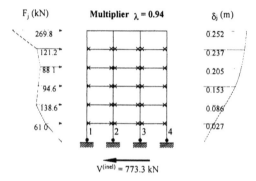

$V^{(inel)} = 773.3$ kN

Figure 9. Distribution of seismic forces and storey displacements from inelastic dynamic analysis for required rotation 0.030 radians

Table 10. Internal forces at the base columns

Base floor column	Static elastic-plastic analysis (PEP-micro)			Inelastic dynamic analysis (DRAIN-2DX)		
	M[kNm]	V[kN]	N[kN]	M[kNm]	V[kN]	N[kN]
1	737.0	171.2	68.5	737.0	177.0	72.0
2	711.9	207.5	931.6	712.2	211.0	927.2
3	701.4	208.8	947.9	700.2	210.8	960.6
4	714.2	171.0	887.0	713.8	174.5	895.2

From the inelastic dynamic analysis, the real solution is in fact: $V^{(inel)} = 773.3$ kN, with a 6th storey displacement equal to $\delta_6^{(inel)} = 0.025$m and a maximum required rotation 0.030 radians. These results are illustrated in Figure 9 for the accelerogram multiplier $\lambda = 0.94$ which corresponds to the required rotation 0.030 radians.

The comparison between the internal forces evaluated by the static elastic-plastic analysis and those calculated by the inelastic dynamic analysis, into the base columns is presented in Table 10 (bending moment M, horizontal shear V and axial force N).

5 CONCLUSION

Using the new definition of the behaviour q-factor, some important conclusions have to be underlined.

The joint rotation capacity ϕ_u has a very significant effect on the values of the q-factor.

The appropriate equivalent static elastic-perfectly plastic approach is very simple to apply even for steel frames with partial strength joints.

This method, together with proposed values of the q-factor still has a clear research character, which needs to be developed in the future to be applicable in design practice.

REFERENCES

Aribert, J.M. & Grecea, D. 1997. A new method to evaluate the q-factor from elastic-plastic dynamic analysis and its application to steel frames. Proceedings of STESSA'97, Kyoto, 3-8 August 1997.

Aribert, J.M. & Grecea, D. 1998. Experimental behaviour of partial-resistant beam-to-column joints and their influence on the q-factor of steel frames. The 11th European Conference of Earthquake Engineering, Paris, 6-11 September 1998.

Aribert, J.M. & Grecea, D. 1999. Dynamic behaviour control of steel frames in seismic areas by equivalent static approaches. Proceedings of SDSS'99, Timisoara, 9-11 September 1999.

AFNOR-Règles PS 92 appliquables aux bâtiments, NFP 06.013, Décembre 1995.

Ballio, G. 1985. ECCS approach for the design of steel structures against earthquakes. Symposium on Steel in Buildings, IABSE-AIPC-IVBH Report, Vol.48, pp. 373-380, Luxembourg, September 1985.

CEN 1994. Eurocode 8 (ENV): Design provisions for earthquake resistance of structures. Part 1-2: General rules for buildings.

CEN 1995. Eurocode 8 (ENV): Design provisions for earthquake resistance of structures. Part 1-3, Section 3: Specific rules for steel buildings.

ECCS TWG1.3. 1985. Recomanded testing procedure for assesing the behaviour of structural steel elements under cyclic loads.

Eurocode 3. Part 1.1. Revised annex J : Joints in building frames. ECCS Committee TC 10-Structural Connections. April 1996.

Galea Y. & Bureau A. 1995. PEP-micro. Analyse plastique au second ordre de structures planes à barres. Manuel d'utilisation, Version 2b, CTICM, France.

Grecea D. 1999. Caractérisation du comportement sismique des ossatures métalliques - Utilisation d'assemblages à résistance partielle. Thèse de Doctorat. INSA de Rennes, France.

Mazzolani, F.M. & Piluso, V. 1996. Theory and design of seismic resistant steel frames. E & FN Spon, London

Prakash, V. Powell, G.H. & Campbell, S. 1993. DRAIN-2DX, base program description and user guide. Version 1.10.

Sedlacek, G & Kuck, J. 1993. Determination of q-factors for Eurocode 8, Aachen, 31 August 1993.

Setti, P. 1985. Un metodo per la determinazione del coefficiente di strutura per la construzioni metalliche in zona sismica. Construzioni Metaliche n°3.

Behaviour of Steel Structures in Seismic Areas, Mazzolani & Tremblay (eds) © 2000 Balkema, Rotterdam, ISBN 90 5809 130 9

Composite frames with under dynamic loadings: Numerical and experimental analysis

L. Calado & J. M. Proença
Instituto Superior Técnico, Department of Civil Engineering, Portugal

L. Simões da Silva
Universidade de Coimbra, Department of Civil Engineering, Portugal

Paulo J. S. Cruz
Universidade do Minho, Department of Civil Engineering, Portugal

ABSTRACT: The pseudodynamic testing method is considered, with respect to the quasi-static system identification and the shaking table tests, to represent an optimum compromise between budget limitations and accuracy in reproducing the strongly non-linear earthquake structural behaviour. Nevertheless, some problems may arise that deserve a particular study. This paper reports some of the results already achieved in an ongoing research project devoted to the study of the dynamic behaviour of subassemblages and frames. A comparative study was initially performed to assess the accuracy and stability of some numerical integration schemes. The most revealing findings resulted of the subsequent development of a simulation algorithm, which served to study the efficiency of several numerical integration schemes, as well as to assess the effects of different error-compensation procedures. These conclusions led to the development of a pseudodynamic testing system, which accuracy will be compared with results obtained with shaking table.

1 INTRODUCTION

The structural dynamic experimental methods can, coarsely, be subdivided into the following categories:
- System identification (system-id) methods;
- Quasi-static tests;
- Shaking-table tests;
- PSD testing method.

The system-id methods aim the assessment of some of the most relevant modal parameters (modal frequencies, mode shapes and modal damping ratios) for a given structural model that, in numerous cases, is the structure itself. This model is subjected to a forced excitation (controlled or uncontrolled) and the correspondent response is monitored at specified points, allowing for the estimation of the transfer functions which, in turn, allow for the identification of some of the more relevant modal parameters. The fact that the excitation levels are generally low justifies one of its major advantages, since it is basically a non-destructive testing technique. On the other hand, this latter fact also limits the extension of the results so obtained to situations, like those found when structures are subjected to severe ground motions, where structural response is dominated by the non-linear, post-elastic, behaviour.

In the quasi-static tests, the structural model, generally a component or a sub-assemblage, is subjected to a predetermined displacement history. This loading history, although arbitrary, aims at the activation of the same non-linear effects that would otherwise determine the earthquake behaviour. The major disadvantage of this testing technique lies precisely in the arbitrary nature of the loading, which can hardly reproduce the damage accumulation processes customary to strong motion structural response. The most striking advantages result of the simplified testing apparatus which, in turn, allows for the derivation of extensive experimental data within more common budget restraints.

The shaking table tests, while being the most realistic testing procedures, present, also, some limitations as a consequence of the possible lack of similitude with respect to real, full-scale, situations. This testing technique is also, by far, the most expensive, which explains the relative lack of experimental results (materialised, for example, in design code prescriptions) achieved so far.

The PSD testing method, firstly presented by Takanashi (Takanashi et al. 1977), represents a compromise between the quasi-static tests, since it requires much simpler testing facilities than shaking-table tests, and these latter tests, since it aims at reproducing the behaviour of the specimen (generally a structural component or sub-assemblage) to a given (natural or artificial) earthquake record. It is, in essence, a hybrid method since the inertial, damping (and excitation) forces are modelled analytically, whereas the restoring forces are directly measured from the specimen. In contrast with the shaking-table tests, it allows for a close observation of the non-linear effects, like buckling and crack propagation, which determine the damage accumulation processes. Figure 1 depicts the overall configuration and flowchart for a 2DOF, Newmark-β explicit numerical integration scheme ($\beta=0$, $\gamma=0.5$), where variables q, F, Q, a and Δ respectively stand for displacements, external applied forces, restoring forces, ground acceleration and integration time step. Matrices M and C are, respectively, the mass and the viscous damping matrices assumed for the numerical model.

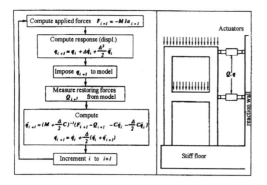

Figure 1. Flowchart and scheme for a PSD test. Adapted from Mahin & Shing (1985).

The major limitation of this testing procedure lies in the dual nature of the errors, both experimental, in terms of imposed displacements and measured restoring forces, and numerical, as a consequence of the adopted numerical integration method. The sources of errors can be subdivided into the following classes:

- Structure idealisation errors. The numerical idealisation of the test structure is, inevitably, a discrete-parameter system, such that the mass is considered to be concentrated at predetermined locations. The numerical model so developed may not reflect the real dynamic behaviour of the otherwise continuous test specimen (thus neglecting the higher mode effects and distorting the frequencies of lower modes). The structure idealisation also implies the arbitrary assumption of the damping and stiffness characteristics of the test specimen.
- Numerical integration errors. The numerical integration schemes yield only approximate solutions, as they are derived from arbitrary assumptions relative to the response variation within the time step. In general, these errors decrease with the time step. These errors are particular important since they are potentially cumulative.
- Experimental errors. The experimental errors are introduced into the numerical computation procedures during the feedback and the displacement imposition processes. In fact, as a conse-

quence of a variety of reasons (transducer miscalibration, digitising errors, etc.), not only the actually imposed displacement differs from the computed value, but also the measured restoring force is subjected to measurement errors. The strain-rate effects can also lead to errors since the test is conducted quasi-statically whereas the strong-motion response is essentially dynamic.

Some studies trying to assess the extent of the aforementioned errors have already been conducted. As to the structure idealisation errors, Mahin & Shing (1984, 1985), concluded that if more than 80% of the mass is effectively concentrated, the lumped-mass model can reliably be used. Moreover, the viscous damping errors have practically no effect whenever most of the energy dissipation is due to the strongly post-elastic behaviour. As to the numerical integration errors, some studies have already been conducted, namely Bathe & Wilson (1976), to establish the stability and accuracy limits for some of the most common numerical integration schemes. These studies show a tendency for the decline of the numerical errors below acceptable levels whenever the time step Δ to natural period ratio T is kept below 1/40 to 1/20. The experimental errors can be dealt with by using high-sensitivity transducers and, as to the strain-rate effects, these show little influence when testing steel specimens (where the strain-rate effects are negligible in the 0-10 Hz range, which dominates the strong-motion structural response). Anyway, the spectral response and the amount of plastic deformation values, evaluated considering the static structural behaviour, are generally higher than those based on the real dynamic behaviour, thus leading to conservative results. The use of the computed, in place of the measured, displacement, avoids the experimental error propagation along the test.

As a conclusion, while the majority of the errors can be dealt with by means of an adequate selection of the test specimen and of the testing apparatus, the effect of the prescribed time step deserves a particular study. This is particularly so, since the experimental errors show an opposite sensitivity when compared to the numerical errors. In fact, while the former errors decrease with an increasing time step (the number of hold periods decreases), the latter errors show a tendency to increase, independently of the adopted numerical integration scheme.

2 PSEUDODYNAMIC ALGORITHM

As stated previously, the pseudodynamic test results are partially affected by errors, both from experimental and numerical nature, whose extent should be investigated and contained. A simulation model (materialised in the SIMULATION program) was developed so as to compare its results with those, assumed correct, conveyed by the RESPONSE program. This simulation algorithm served also to study the effects of different error-compensation procedures to be later employed in the pseudodynamic tests to be performed.

The SIMULATION program was designed to closely reproduce the characteristics of the testing apparatus under development. An error-sensitivity analysis was furthermore conducted, considering the following control variables:

- Numerical integration time step Δ.
- Numerical integration scheme. Two distinct Newmark-β methods were considered, the constant average acceleration method ($\gamma=0.5$, $\beta=1/4$) and the linear acceleration method ($\gamma=0.5$, $\beta=1/6$).
- Damage potential of the accelerogram. The effect of the relative intensity of the accelerogram upon the performance of the different numerical integration schemes was investigated considering an auxiliary variable, the accelerogram magnifying factor γ_F, which, in turn, is directly correlated with the displacement ductility factor μ_q.

The simulation algorithm considered a state determination routine correspondent to a (SDOF) bilinear (elastoplastic with strain hardening) model.

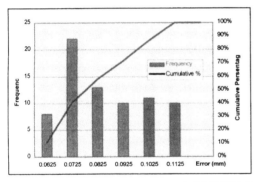

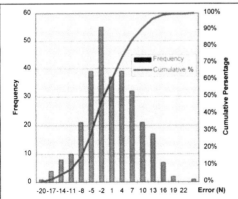

Figure 2. Experimental errors histograms.

One of the distinctive features of the SIMULATION program was the artificial generation, based upon sequences of pseudo-random numbers, of the experimental errors. This procedure was calibrated considering samples of experimentally measured errors, both in terms of the imposed displacement increment and of the measured restoring force. Figure 2 depicts the histograms obtained for the particular testing devices (actuator controller system, displacement and force transducers) to be used.

As seen above, the displacement increment error fits into an uniform distribution whereas the restoring force errors seem to follow a normal Gaussian distribution. The time lag intrinsic to the actuator controller system justifies the non-zero mean value for the displacement increment errors. This conclusion led to the implementation of an error-compensation procedure that consists in subtracting the mean value to the actually imposed displacement increment (obviously, this procedure only takes place when the displacement increment is large enough).

The RESPONSE program was developed in order to obtain potential exact solutions for each of the situations considered in the SIMULATION program. The comparison between equivalent situations, allowed for the assessment of the different error indexes, from which the error sensitivity to each of the considered variables, could be estimated. The main differences from the SIMULATION program are twofold: the error-generation procedures are deactivated; and the yielding point is determined exactly (when, for a given time step, the specimen suffers yielding, this time step is subdivided into two different sub-steps, correspondent to the elastic and to the strain-hardening segments). This latter difference, altogether with the exclusive consideration of very small time steps ($\Delta T = 1/40$), supports the assumption that the RESPONSE program results are unaffected by numerical errors (in accordance with the non-existence of experimental errors).

The results conveyed by these two programs were compared for each of the eighteen different situations described in Table 1.

Table 1. Control variables correspondent to the sensitivity analysis conducted.

γ_F	μ_q	ΔT			Integration Method	
0.35	<1.0	1/20	1/10	1/5	$\gamma=1/2$, $\beta=1/4$	$\gamma=1/2$, $\beta=1/6$
1.25	~2.5	1/20	1/10	1/5	$\gamma=1/2$, $\beta=1/4$	$\gamma=1/2$, $\beta=1/6$
2.00	~5.5	1/20	1/10	1/5	$\gamma=1/2$, $\beta=1/4$	$\gamma=1/2$, $\beta=1/6$

The first group of situations ($\gamma_F=0.35$) was designed so that the system remained in the elastic range, whereas in the remaining groups ($\gamma_F=1.25$ and 2.00) the system exhibited a response progressively in the plastic range. The ΔT ratios considered situations were intended to be representative of real-test situations where the optimum compromise between numerical and experimental errors would be found. Two distinct, implicit-type, numerical integration schemes were considered: the constant-average acceleration method (β=1/4); and the linear acceleration method (β=1/6).

All situations analysed with the SIMULATION (as well as the RESPONSE) program corresponded to a SDOF system with a natural (elastic) period of T=0.2 s. The N-S component of the 1940 El-Centro earthquake record was selected, being suitably scaled by means of the magnifying factor γF, so as to have different damage potentials.

Figure 3 matches the results, expressed in terms of the displacement, conveyed by the SIMULATION and the RESPONSE programs for the first 15 s, of one of the considered situations.

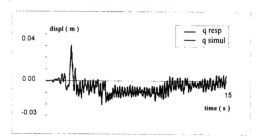

Figure 3. Match between SIMULATION and RESPONSE programs (γF=2.0; Δ/T=1/10; β=1/4)

The results conveyed by SIMULATION program agree favourably with the reference results, despite the fact that this latter results were obtained de-activating all kinds of experimental errors and considering a Δ/T ratio significantly smaller. The agreement shown above was, however, only obtained after the implementation of the error-compensation procedure already referred to (the results obtained before this error-compensation procedure were developed in a much less favourable manner).

Apart from the visual observation of the response *versus* time charts, the following error indexes were considered:

- Displacement ductility factor μq;
- Work performed by the restoring force W_Q.

Table 2 depicts the results obtained in this sensitivity analysis for one of the considered numerical integration schemes.

Table 2. Error indexes for the constant average acceleration method (β=1/4).

Δ/T	Displacement ductility μq			Work (kN.m) WQ		
	γF=0.35	γF=1.25	γF=2.00	γF=0.35	γF=1.25	γF=2.00
1/5	0.6356	1.7181	2.5493	-0.075	1.0974	5.1186
1/10	0.8578	2.2877	4.6318	-0.116	2.1040	9.5206
1/20	0.8537	2.3958	5.4669	-0.206	2.4635	11.172
*1/40**	*0.8525*	*2.4738*	*5.6080*	*0.0007*	*2.9885*	*12.200*

* *italic* values correspond to the reference (RESPONSE) results.

One of the most striking conclusions is that the SIMULATION program generally underestimates both damage indexes. This can be explained by the fact that, as a consequence of digitising both the action and the response, some of the local *maxima* and *minima* are lost (this loss is obviously more pronounced for high Δ/T ratios, which explains the attenuation of the errors for smaller integration time steps). In all situations, the error decreases with a decreasing time step, which indicates that the optimum compromise would, eventually, be achieved for Δ/T values lower than 1/20.

The above mentioned conclusions also hold true for the other numerical integration scheme, as shown in Table 3.

Table 3. Error indexes for the linear acceleration method (β=1/6).

Δ/T	Displacement ductility μq			Work (KN.m) W_Q		
	γF=0.35	γF=1.25	γF=2.00	γF=0.35	γF=1.25	γF=2.00
1/5	0.7653	2.2329	3.5308	-0.065	1.6279	6.6818
1/10	1.0816	2.2732	4.8247	-0.106	2.1979	9.6169
1/20	0.8695	2.3503	5.4643	-0.227	2.6041	11.632
*1/40**	*0.8534*	*2.4775*	*5.6130*	*0.0009*	*3.0078*	*12.233*

* *italic* values correspond to the reference (RESPONSE) results.

When comparing both numerical integration schemes, one could state that, with some exceptions, the linear acceleration method generally conveys better results (particularly for high Δ/T values).

3 EXPERIMENTAL RESULTS

A SDOF pseudodynamic testing system was assembled in the Instituto Superior Técnico with the following two main objectives:

- Start the implementation of a more complete pseudodynamic testing apparatus (two additional hydraulic actuators are expected to be mounted in the near future, allowing for up to 3DOF pseudodynamic tests);
- Check some of the conclusions attained by the sensitivity study already presented.

The data acquisition and control software was written in an object-oriented programming language and the results were stored in spreadsheet format (that, thanks to multiprocessing possibilities, allowed for real-time display of the more significant results in the spreadsheet environment).

A group of four HE180B cantilever steel specimens were tested with a common configuration as depicted in Figure 4.

The differences between the four pseudodynamic tests (PSD1 to PSD4) lay in the adopted time step vs. natural period ratio (Δ/T) and, also, in the adopted earthquake record magnifying factor (γF), as shown in Table 4. All tests were performed consid-

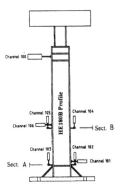

Figure 4. Specimen configuration and location of the displace-. ment transducers.

Table 4. Pseudodynamic tests distinctive features.

γ_F	$\Delta/T = 1/20$	$\Delta/T = 1/10$
2 ($\mu_q \sim 5.5$)	PSD1	PSD3
1.25 ($\mu_q \sim 2.5$)	PSD2	PSD4

ering the same earthquake record that had been previously adopted in the simulation algorithm (appropriately scaled by γ_F).

Under these circumstances, tests PSD1 and PSD3 (as well as PSD2 and PSD4) could be directly related, and the comparative analysis of PSD1 and PSD2 (the same for PSD3 and PSD4) would reflect the influence of the damage potential of the accelerogram alone.

The processing and post-processing stages of the test results were performed discriminating the so-called direct results, those directly obtained from the monitoring devices (i.e. tip force vs. tip displacement), from the indirect results, such as those relating indirect variables, like the ones computed by means of algebraic expressions based in direct variables (i.e. bending moment vs. relative rotation). Figures 5 and 6 depict some of the direct results.

The specimen clearly entered in the plastic range, although in a much less pronounced manner than what had been predicted by the simulation algo-

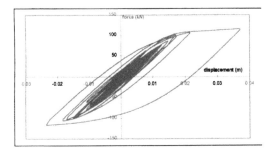

Figure 5. Restoring force vs. displacement chart for specimen PSD1 ($\Delta/T=1/20$; $\gamma_F=2$).

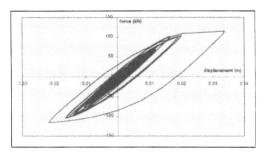

Figure 6. Restoring force vs. displacement chart for specimen PSD3 ($\Delta/T=1/20$; $\gamma_F=2$).

rithm. A close look at the test data showed that there were significant discrete rotations in what was otherwise considered the built-in section of the cantilever (this resulted of the lack of fit between the base plate and the supporting rigid beam, aggravated by the non-existence of pre-stressing of the connecting bolts). As a consequence, this specimen (and all others, as this defect was only identified in the end of the test campaign) presented elastic stiffness values much lower than what had been anticipated in the simulation algorithm, discrediting any further comparative study between the predicted and the measured results.

Despite the eventual lack of similitude in the base connection between models PSD1 and PSD3, the response (presented for the first 9 s, within which the maximum displacement was attained) agrees quite favourably. The only exceptions seem to be circumscribed to some of low amplitude cycles for which the lack of error compensation (the displacement increments were excessively small) notably increases the experimental errors.

The results presented by the other two tests (PSD2 and PSD4), both corresponding to a magnifying factor of $\gamma_F=1.25$, showed a less pronounced agreement. The reasons for this discrepant behaviour can be found in the more frequent occurrence of small amplitude, non-compensated, cycles, as well as in the stronger relative influence of the nearly rigid-body base rotations. This result confirms the assumption that the stronger the accelerogram damage potential, the smaller the numerical errors are bound to be. The decrease of the average stiffness leads to a decrease of the integration time step to average period ratio, increasing the accuracy of the particular numerical integration scheme chosen.

The energy that enters in the system, the input energy W_I, corresponds to the work of the force dues to the ground acceleration (ma_g) in the top displacement δ:

The total dissipated energy W_T is the sum of the following energies:

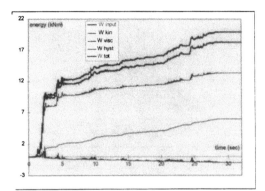

Figure 7. Energies diagrams of the specimen PSD1.

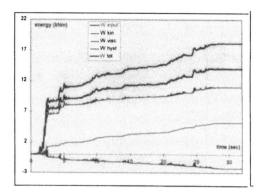

Figure 8. Energies diagrams of the specimen PSD3.

$$W_I = \int_{\delta_,}^{\delta_,} (m \cdot a_g) \cdot d\delta = \int_{\delta_,}^{\delta_,} (m \cdot a_g) \cdot d\delta \cdot \frac{dt}{dt}$$

$$W_I = \int_{t_,}^{t_,} (m \cdot a_g) \cdot \dot{\delta} \cdot dt$$

kinetic energy:

$$W_K = \int_{t_,}^{t_,} (m \cdot \ddot{\delta}) \cdot \dot{\delta} \cdot dt$$

viscous energy:

$$W_V = \int_{t_,}^{t_,} c \cdot \dot{\delta}^2 \cdot dt$$

energy dissipated by hysteresis:

$$W_H = \int_{\delta_,}^{\delta_f} F(\delta) d\delta = \int_{t_,}^{t_f} F \cdot \dot{\delta} \cdot dt$$

where m and c are the mass and the coefficient of damping of the specimen, while δ, $\dot{\delta}$ and $\ddot{\delta}$ are the displacement, the velocity and the acceleration of the top.

Therefore, the total energy in output of the model is:

$$W_T = W_K + W_V + W_H$$

From Figures 7 and 8 it is shown that the total energy that exists has the same behaviour of the one that enters, and the small difference between them can be explained as the energy dissipated in the connection.

The results obtained show that the methodology developed is suitable for pseudodynamic tests of other types of specimens namely frames.

This research project is hence aimed to the construction and testing sub-assemblages, representative of steel framed structures with concrete slab on steel decking, designed in accordance with damage control and failure limit states.

Reference is made to three different typologies of frames: moment resisting (i.e., frames with rigid joints) and semi-continuous (i.e., frames with semi-rigid joints) steel frames and braced frames. It is proposed to consider one of the following typologies of beam-to-column connections: bolted extended end plates, welded connections with weld access holes and backing bars, as well as full penetration butt welds for moment resisting frames, while for semi-continuous frames, top-and-seat and web cleated steel connections (Figure 9).

The presence of the concrete slab, remarkably modifies the strain distribution within the cross section (hence the seismic response) of the members. In the last two decades, however, a number of experimental research works were carried out all over the world, on the cyclic behaviour of steel connections, testing joint specimens without the concrete slab. These results presently form a large data base, that might be very useful to designers and Codes Drafting Panels. However, at present, one of the main issues being discussed in the research community is how these data can be relied on, when designing a connection in a composite concrete-steel structure. In order to try to give a first answer to this question, and to investigate the actual difference in behaviour caused in a steel joint by the presence of the concrete slab, in the case of fully welded connections, two specimens will be tested, the first one with the full concrete slab the second one with the concrete slab removed near the beam-to-column connections, but having a larger thickness, hence allowing the same overall mass.

Furthermore, the frame specimens will be designed to have an eccentric mass distribution, in one plane, so that they can be tested either under in-plane conditions or under strong torsional effects, depending on the direction of the input seismic ex-

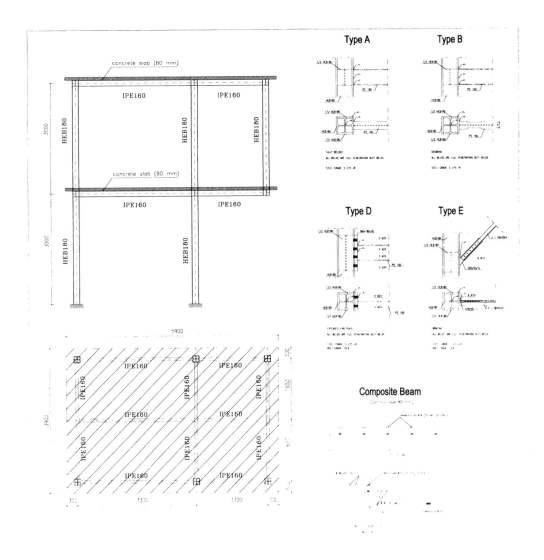

Figure 9. Composite frame configuration.

citation. A vertical excitation will also be considered, simultaneously with the horizontal one, in order to enhance the effect of the vertical loads.

It should be noted that this experimental programme will be carried out on frames composed by member types that have already been tested on isolated specimens under low-cycle fatigue. This allows to profit and to make reference to the available data-base related to the fatigue strength and endurance of both members and connections. The available data related to the low-cycle response of the key components of the specimen, tested under cyclic quasi static reversal loading, and the new data related to the behaviour of the whole structures as well as its significant components under seismic loading

will allow for a fully validation phase for the design method. Moreover, it will be possible to appraise both the influence of the concrete slab on the seismic performance of the subassemblages as well as the strain rate effect.

The attached drawings show the geometry of the proposed specimens as well as the details of the connections. In particular, it should be noticed that: a) a large number of connection typologies will be tested; b) very few data are available in the literature about full scale frame structures subjected to strong torsional effects, hence the results from the research project will be innovative; c) the proposed research will allow a comparison among the response of different types of framed structures (moment resisting,

Figure 10. Composite frame at NTUA.

braced and semi-continuous frames); d) in the case of fully welded connections (MR frames), the proposed research will allow an assessment of the influence of the concrete slab on the behaviour of the connections. Some of the proposed typologies of beam-to-column connections allow for reparability and/or replacement after being damaged during a seismic event. Hence, a further aim of the research program is to analyse possible repair and/or replacement procedures, for structures damaged during an earthquake.

However due to technical problems in the laboratory of Instituto Superior Técnico, the first composite frame was not yet tested. Meanwhile, a shaking table test on a similar frame was performed at the Laboratory of National technical University of Athens (NTUA) in order to assess the accuracy of the developed methodology. The shaking table of NTUA has six degrees of freedom control, offering control of the three orthogonal translational degrees of freedom and the associated rotational degrees of freedom. The steel platform measures 4 x 4 m^2 and weighs 10 tonnes. It can carry a maximum payload of 10 tonnes. Figure 10 shows the composite frame at the laboratory of NTUA.

The first test using the shaking table was performed last December and for that reason the experimental data are not yet re-elaborated. However latter one, as mentioned before it will allow to check the accuracy of the pseudodynamic methodology developed at IST.

4 PRELIMINAR CONCLUSIONS

The pseudodynamic simulation algorithm proved to form an effective development platform from which conclusions could be drawn relative to the accuracy of a particular numerical integration scheme, to its dependency upon the integration time step and, also, to the effect of different accelerogram damage potentials. This platform was also successfully used to try out different error-mitigation strategies from which an efficient, yet simple to implement, error-compensation procedure was derived. The results so obtained need, however, a more refined sensitivity analysis to determine, for each of the considered numerical integration schemes (and others, such as the piecewise linear and the Runge-Kutta family of integration methods) the optimum compromise between numerical and experimental errors.

The developed SDOF pseudodynamic testing system confirmed one of its most striking advantages: the simplicity (actually the testing apparatus requirements are similar to those corresponding to the more common quasi-static tests). The results obtained in the limited experimental campaign confirmed, partially, the conclusions that had been previously attained by means of the simulation algorithm. These results indicate that, for the considered class of numerical integration schemes (as well as for the actuator controller and force transducer specifications under use), the optimum compromise between the different sources of errors should correspond to a Δ/T ratio below 1/20. Moreover, it became evident that the developed error-compensation procedure, although effective for large displacement increments (such as those corresponding to higher damage potentials), leaves small displacement increments practically unaffected.

ACKNOWLEDGEMENTS

This work was funded by the JNICT. Alberto Ferrari developed his graduation thesis in close cooperation with the IST under the supervision of Prof. Carlo Castiglioni, of the Politecnico di Milano.

REFERENCES

Bathe, K. & Wilson, E. 1976. Numerical Methods in Finite Element Analysis. Englewood Cliffs, NJ: Prentice-Hall.
Calado, L. 1994. Amplificador linear de deflectómetros e de transdutores de deslocamentos (in Portuguese). INPI - Instituto Nacional da Propriedade Industrial. Lisbon.
Ferrari, A.. 1997. Pseudodynamic testing method for a single degree of freedom system. Tese di Laurea. Politecnico di Milano. Milan.
Mahin, S. A. & Shing, P. B. 1984. Pseudodynamic test method for seismic performance evaluation: theory and implementation. Report N° UCB/EERC-84-01. University of California, Berkeley.
Mahin, S. A. & Shing, P. B. 1985. Pseudodynamic method for seismic testing. ASCE, Journal of the Structural Engineering Vol. 11(7): 1482-1503.
Takanashi, K. & Udagawa, K. & Tanaka, H. 1977. A simulation of earthquake response of steel buildings. 6th. World Conference on Earthquake Engineering Vol. 11: 165-169.

Behaviour of Steel Structures in Seismic Areas, Mazzolani & Tremblay (eds) © 2000 Balkema, Rotterdam, ISBN 90 5809 130 9

Analytical study of buildings constructed with riveted semi-rigid connections

K.C.Chessman
Steel Structures Education Foundation, Toronto, Canada

M.Bruneau
Multidisciplinary Center for Earthquake Engineering Research, State University New York at Buffalo, USA

ABSTRACT: A summary of an analytical study of the seismic response of buildings incorporating riveted connections is presented. The study focuses on the moment resistance of two types of connections: stiffened seat angles and stiffened tee stubs. Four frames based on an existing 18-story building have been analyzed using a non-linear inelastic dynamic analysis program. The frames were subjected to earthquakes typical of Eastern and Western Canada. The four frames have structural periods much longer than similar modern buildings. These longer periods are caused by heavy building materials and flexible connections. The inter-story drifts for the frames located in Quebec City fall within the bounds set forth by the National Building Code. However, of the frames located in Victoria only the 18-story structure's drifts are less than the allowable values.

1 INTRODUCTION

The use of riveted semi-rigid connections was wide spread in building construction in the first half of the 20th century at a time when earthquake design was not required. When assessing the seismic resistance of these old buildings, engineers typically consider semi-rigid connections as having no moment resistance. As a result of this assumption, these historic buildings may need to undergo costly retrofits or simply be demolished.

Recent experimental work attests that these connections are able to develop stable hysteretic moment rotation behavior, but must undergo large rotations in order to reach a moderate moment capacity. It is therefore legitimate to question the seismic response of these buildings.

In that perspective, this study looks at the hysteretic behavior of semi-rigid connections as demonstrated by past experimental studies and at the seismic behavior of four frames, based on an existing building located in Eastern Canada. These frames incorporate either semi-rigid stiffened seat angle connections or more rigid stiffened tee-stub connections.

2 RIVETED CONNECTIONS

In recent years the connections that have been studied were riveted stiffened seat angles. The typical details of this type of connections are shown in Figure 1. These connections were widely used as rigid joints in old steel frames.

Although the building used as the basis for this study was built in the same era. The connections used in its steel frame are quite stiffer. Built at the end of the 1920's the designers used stiffened tee-stub connections. The details of this second connection are shown in Figure 2.

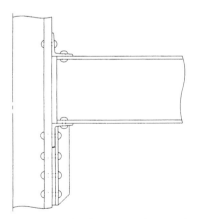

Figure 1. Riveted stiffened seat angle connection

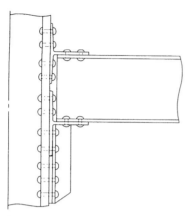

Figure 2. Riveted stiffened tee-stub connection

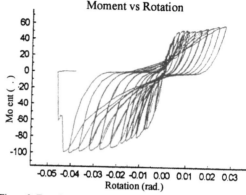

Figure 3. Experimental hysteresis curve for stiffened seat angle connection (Bruneau & Bisson 2000)

3 PAST EXPERIMENTAL STUDIES

A limited number of specimens where obtained from the Daly Building in downtown Ottawa when this building was demolished in 1992 after 83 years of service. These specimens of riveted stiffened seat angle connections where used in a series of experimental studies to determine their hysteretic behavior and possible retrofit strategies. (Sarraf & Bruneau 1996 and Bruneau & Bisson 2000). These studies demonstrated that although severely pinched the moment-rotation (M-θ) curves offer non-negligible seismic resistance. The hysteretic behavior of this type of connection is shown in Figure 3.

This curve gives a good descriptive and quantitative expression of the resistance of this type of connection. The pinching in the hysteretic loops are caused mainly by:
- Slippage at rivet holes due to lack of fit (diametric shrinkage after cooling) and low clamping force of the rivets
- Rocking of the vertical leg of the top angle (the angles deforms into a convex shape and rivets elongate)
- Lack of integrity between parts of stiffened seat, resulting in separation of the seat angle and the stiffener angles under reverse loading

4 ANALYTICAL APPROACH

In the present study both the aforementioned connection types have been used in determining the seismic behavior of steel frames.

In order to effectively determine the frames' behaviors, non-linear inelastic dynamic analysis was conducted using the Mehran Keshavarzian Degrading and Pinching Hysteresis model available in

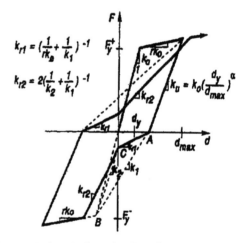

Figure 4. Mehran Keshavarzian Degrading and Pinching Hysteresis rule (Carr 1998)

Ruaumoko (Carr 1998) and shown in Figure 4. This model provides in a general sense the stiffness degradation and pinching that occur in stiffened seat angle connections.

The frames chosen for the analysis are all based on an existing building located in Eastern Canada. This 18-story building was built in 1929. The bay dimensions of the frames are the same as in the existing building. All column and beam sizes were also taken from the existing structure. Not all the structural drawings were available for this study. The missing drawings were of the connection details. It was thus necessary to re-design the connections using old literature.

Although the connections in the existing building were of a different type their hysteretic behavior is mostly the same as that described earlier. The pinching described above will also take place for much the same reasons (rivet shrinkage, lack of integrity between bottom tee and stiffener angles). It is important to note that the rigidity of the stiffened tee-stub will be an order of magnitude higher than that of the stiffened seat angle. This factor was used to approximate the flexural stiffness of the spring elements

Table 1. Earthquake records selected

Earthquake	Year	PGA (g)	Scaling factor
Western Earthquakes			
Kobe	1995	0.84	0.4
New Hall	1994	0.59	0.6
Olympia	1949	0.28	1.2
Pacoima Dam	1994	1.28	0.3
Parkfield	1966	0.49	0.7
San Fernando	1971	1.17	0.3
Eastern Earthquakes			
Atkinson long period *		0.09	2.0
Atkinson short period *		0.15	1.2
Nahanni**	1985		
Saguenay	1988	0.17	1.1

* Synthetic time-history (Atkinson 1996)

**Although occurred in Western Canada, time history similar to Eastern quakes, at time of writing analyses not completed

for the seat angle connections. Springs were used as the beam to column connecting elements. For this study the beams and columns have been selected as elastic elements.

Four frames have been analyzed representing an 18, 7, 4 and 2-story building. Figure 5 shows how which parts of the 18-story frame has been used to create the lower frames. The lower frames have been based on the upper stories of the existing building to simulate smaller buildings (smaller col

umns, more realistic wind forces).

In this study, the building was considered as if located in Quebec City or Victoria. These locations have been chosen to represent the seismic activity of both the Eastern and Western regions of North America. To further generalize the results, 10 earthquakes have been used in the analyses. Table 1 lists the earthquakes used, their original peak horizontal ground acceleration (PGA) and the scaling factors used to linearly scale the earthquakes to the PGA given by the National Building Code of Canada 1995 (Quebec City 0.19g and Victoria 0.34g). These maximum values of PGA represent the expected values for PGA having an exceedance probability of 10% in 50 years for both cities. Table 1 also gives the scaling factors used for each earthquake.

A total of seven series of analyses have been performed, three for each type of connection and one for non-scaled earthquakes, for each of the four frames considered. Analyses are referred to in the following as:

- As is: no modification to the existing configuration found in the building (heavy floors and stone cladding)
- P-Δ: same structure masses as above but including P-Δ effects.
- Modern mass: modified structure masses (thinner concrete floor slabs and glazed cladding as found in modern office buildings)

These three configurations have been applied to frames incorporating both the stiffened seat angle and stiffened tee-stub connections. The seventh set of analyses was applied to the frames corresponding to the building in its existing condition (tee-stub connections, large masses) without scaling the earthquakes.

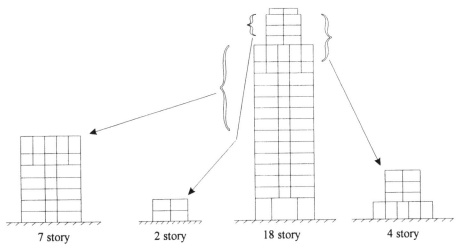

7 story   2 story   18 story   4 story

Figure 5. Analyzed frames

473

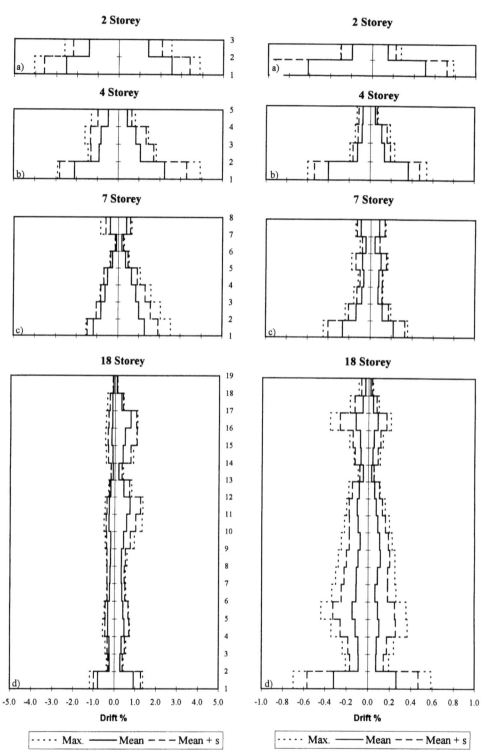

Figure 6. Inter-story drifts for tee-stub connections, Victoria: a) as is b) as is c) modern mass d) p-delta

Figure 7. Inter-story drifts for tee-stub connections, Quebec City: a) as is b) as is c) modern mass d) p-delta

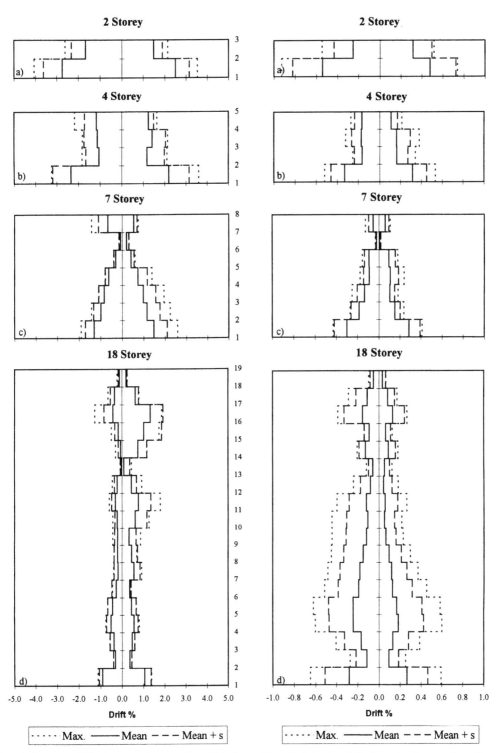

Figure 8. Inter-story drifts for seat angle connections, Victoria: a) as is b) as is c) modern mass d) p-delta

Figure 9. Inter-story drifts for seat angle connections, Quebec City a) as is b) as is c) modern mass d) p-delta

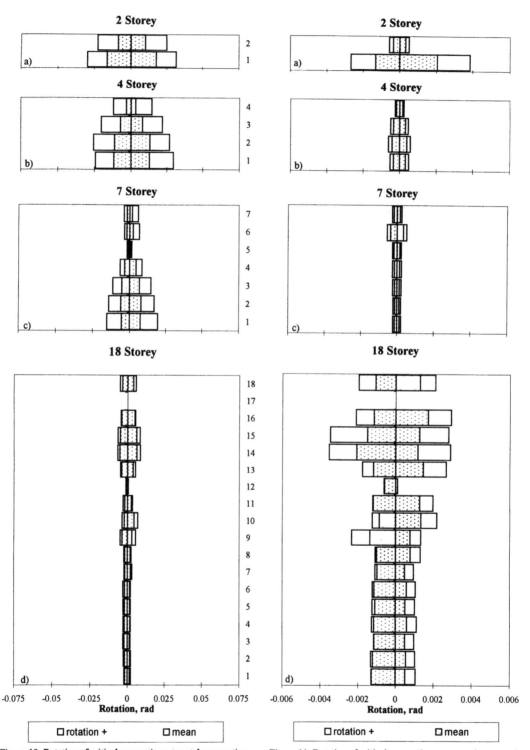

Figure 10. Rotation of critical connections, tee-stub connection
Victoria: a) as is b) as is c) modern mass d) p-delta

Figure 11. Rotation of critical connections, seat angle connection
Quebec City: a) as is b) as is c) modern mass d) p-delta

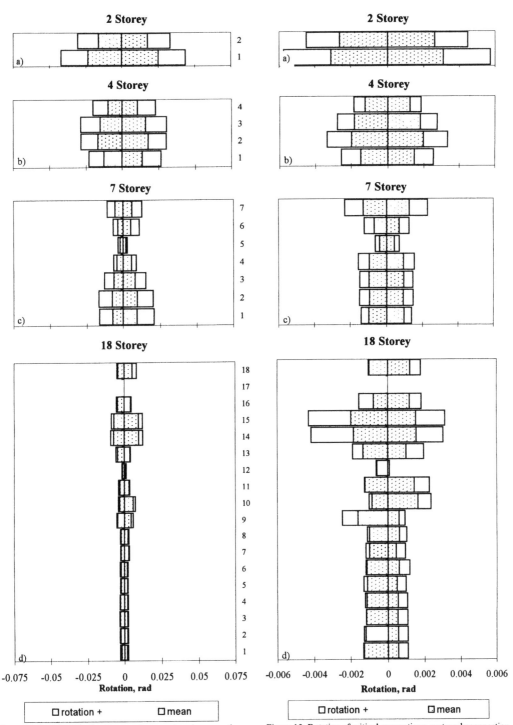

Figure 12. Rotation of critical connections, seat angle connection, Victoria: a) as is b) as is c) modern mass d) p-delta

Figure 13. Rotation of critical connections, seat angle connection, Quebec City: a) as is b) as is c) modern mass d) p-delta

5 ANALYTICAL RESULTS

Each analysis configuration described above gave a different period for each of the four frames. Table 2 shows the periods for each frame. All the periods given are longer than that for similar modern buildings. This characteristic is mainly caused by:

- Large mass of the existing building (the cladding is stone blocks, the steel elements are fireproofed with terracotta bricks and the concrete floor slabs are very thick).
- Low rigidity of the connections, and
- Particular flexibility of the columns and beams.

Table 2. Frame periods for each analysis configuration, in seconds.

Analysis Configuration	Frame			
	2 story	4 story	7 story	18 story
Stiffened tee-stub connections				
As is	1.23	2.30	2.33	5.04
P-delta	1.30	2.56	2.47	5.74
Modern mass	0.90	1.54	1.74	3.39
Stiffened seat angle connections				
As is	1.81	2.83	2.95	5.97
P-delta	2.00	3.23	3.22	6.65
Modern mass	1.38	1.93	2.20	4.01

The analyses have generated large amounts of data and in order to present this data in an acceptable format the maximum, mean and in some cases mean plus one standard deviation for each analysis set is presented. Figures 6 to 9 show inter-story drifts for both connection types and Figures 10 to 13 show rotations of the critical connection for each floor. Since all the output could not be included in this paper and to simplify the comparison of the connection types each figure repeats the same order of analysis configurations. Figure 6, 8, 10 and 12 a) are all of the 2-story frame located in Victoria in the as is configuration. When comparing the figures for the buildings located in Victoria and Quebec City it is important to note that the scales are not the same.

It is observed that Eastern earthquakes generate larger drifts in the lower levels while Western earthquakes seam to create larger deflections towards the top of the buildings.

The National Building Code of Canada 1995 limits the inter-story deflection of non-post-disaster buildings to 2% of the story height. As shown in the in the analysis results, all the frames located in Quebec City met this limit, but the shorter structures in Victoria did not. For the 18-story frame in Victoria, the connection type plays an important role. The stiffer connection allows the building to remain within the allowable limit while the seat angle permits more lateral movement. However the values obtained for the seat angle are still borderline ac-

ceptable. Note that exceeding the arbitrarily set code limit does not imply collapse.

As it would be expected, the rotation values for the seat angle connections are larger than for the tee-stub connections. In this study only the rigidity of the connections have been modified, the maximum rotation allowable for both types were the same.

6 CONCLUDING REMARKS

The behavior exhibited by the frames, caused by the longer period may be the saving grace of the larger frames. The longer period has pushed the frames further down the spectral acceleration curve thus lowering or increasing (this depends on the earthquake) the effects of seismic excitation. However this beneficial property may also be a detrimental for the shorter frames.

The mass of the structure obviously plays an important role in the overall performance of the buildings. A difficulty arises from the fact that the structure's initial use may have changed many times since its construction and with these renovations, masses may have changed significantly.

ACKNOWLEDGEMENTS

Funding from the Natural Sciences and Engineering Research Council is appreciated. The writers are grateful to Luc Jolicoeur and the Engineering Department of the City of Quebec for providing useful information on the type of buildings considered in this study. However, the opinions expressed in this paper are those of the writers and do not reflect the views of the aforementioned individual or sponsors.

REFERENCES

Bruneau, M. & M. Bisson 2000. *Cyclic Testing and Retrofitted Concrete-encased RSSA Connections*, Engineering Journal (in press)

Carr, A.J. 1998, *Ruaumoko*, University of Canterbury, New Zealand

Ketchum, M.S. 1924, *Structural Engineer's Handbook* McGraw-Hill Book company

Sarraf, M. & M. Bruneau 1996. *Cyclic Testing of existing and retrofitted stiffened seat angle connections*, ASCE Structural Journal, Vol. 122 No. 7, pp. 762-775

National Building Code of Canada 1995, Canadian Commission on Building and Fire Codes, National Research Council of Canada

Behaviour of Steel Structures in Seismic Areas, Mazzolani & Tremblay (eds) © 2000 Balkema, Rotterdam, ISBN 90 5809 130 9

Seismic response of steel columns under multi component seismic motion

M. Como & R. Ramasco
Department of Structural Analysis and Design, University of Naples 'Federico II', Italy

M. De Stefano
Department of Constructions, University of Firenze, Italy

ABSTRACT: This paper aims at examining interaction phenomena between axial and lateral forces arising in steel columns in the plastic range of behaviour under multi-directional earthquake motion. Analyses have been carried out by means of two different models: a fiber one and a mathematical one. It is believed that the fiber model is more useful for a detailed response description while the mathematical model is more useful to draw more general conclusions, being its response dependent on a few number of parameters. Both models have clearly shown how axial and horizontal force interact in the plastic field, with a consequent development of important vertical displacements, even when axial force is due to gravity loads only. When considering the vertical earthquake component in addition to the horizontal ones, system response does not vary significantly in the inelastic range of behaviour.

1 INTRODUCTION

Seismic behaviour of steel columns in framed structures is significantly affected by variations in axial force and bending moments which interact and modify yielding conditions; this interaction could lead to a reduction in bending capacity with a consequent increasing in ductility demands.

In order to investigate the effects due to interaction phenomena, a proper definition of the input ground motion is necessary, particularly with reference to the vertical earthquake component.

The topic of effects of vertical earthquake component was substantially dismissed at the beginning of 70's when Newmark and others (1973) concluded that peak ground vertical acceleration could be considered as a fraction of the horizontal one. Still today vertical seismic acceleration is usually not considered in seismic design except in special cases.

However, the analysis of records from recent major earthquakes (Bozorgnia et al. 1995) has shown that the peak vertical ground acceleration can be even higher than the horizontal ones at sites near to the epicentre (near field earthquakes). Previous studies (Elnashai 1997, Elnashai & Papazoglou 1996) have indeed shown that, under earthquakes with a strong vertical component, significant variations in axial forces in columns arise, which interact with bending components. It means that yielding conditions, and consequently the damage suffered by the system, could be affected to a large degree by the action of the vertical earthquake component, in addition to the horizontal ones.

The behaviour of a single steel column subjected to multi-directional earthquake excitation and to gravity loads is analyzed in this paper by means of a refined fiber model in order to investigate: i) effects due to interaction between axial force and bending moment in the presence of horizontal input ground motion only; ii) effects of vertical seismic component.

Further analyses have been conducted by means of a simplified three-degree-of-freedom mathematical model, in order to extend the above investigation considering a wide parametric range. Namely, short-to-long period systems designed to undergo different levels of inelasticity have been studied with special emphasis on damage evaluation.

It has been found that, when considering horizontal input ground motion and gravity loads only, significant interaction phenomena arise, leading to development of important plastic excursions in the vertical direction. In particular, such vertical plastic displacements and the corresponding energy dissipation cannot be neglected with respect to those obtained for the horizontal directions. Analyses including vertical earthquake component in the input ground motion show that, even if axial forces are significantly increased, horizontal and vertical displacement time-histories are not significantly modified by the presence of the vertical component. Consequently, damage suffered by the system is practically very similar, including or not the vertical seismic component in the loading condition.

2 SEISMIC INPUT

In this study records from "near field" earthquakes have been used as input ground motion, which are characterised by large vertical peak ground accelerations. In this way, the investigation covers system response under seismic excitations characterised by significant vertical accelerations. In the following, for the sake of brevity, results obtained for the three component Newhall record from the Northridge earthquake (17-1-1994) are shown. For such record, the ratio of peak ground vertical acceleration (A_V) to peak ground horizontal acceleration (A_H) is equal to 0.93.

3 NUMERICAL ANALYSIS: FIBER MODEL

Effects of multi-directional earthquake excitation have been investigated with reference to a single steel column with a mass at the tip.

The column is 3 m high with an *IPE270* cross section, and presents a mass $M = 30$ $KNsec^2/m$ at the tip. As a consequence, its fundamental periods along x - and z - directions are $T_x= 0.91$ sec and $T_z=0.062$ sec. The column cross section has been designed according to the following requirements: horizontal strength in the x direction matching that corresponding to the constant ductility spectrum S_μ for $\mu=4$, and a safety coefficient of 2 against vertical loads. The column is idealised as a fiber element (Prakash et al. 1993) allowing to monitor spread of plasticity throughout the cross section and the whole member volume (Fig. 1). Furthermore, refined material cyclic constitutive relationships, which in turn may affect interaction phenomena, can be implemented (Filippou et al. 1991). However, for the sake of simplicity, only bilinear elastic-perfectly plastic (E.P.P.) and bilinear strain hardening (Hard.) material constitutive relation have been considered to model steel cyclic behaviour.

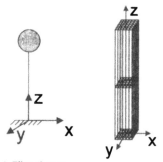

Figure 1. Fiber element

The column is divided into 7 elements in the vertical direction; while the cross section is discretized into 104 fibers. The element at the base of the column has the same length as the largest cross section dimension (i.e. depth), which can be taken as plastic hinge length (Mazzolani and Piluso 1996). For the isostatic nature of the system, there is no way of spreading plasticity outside this element.

In order to investigate effects of interaction between axial and horizontal forces, system response has been evaluated for two loading conditions: horizontal seismic component only (1H) and horizontal seismic component with gravity loads (1H+Gr. loads). Figure 2 shows time-histories of the acting horizontal force for the two loading conditions: it can be seen that maximum values for (1H +Gr. loads) are substantially lower than those obtained for the (1H) loading condition. Such reduction in horizontal strength is definitely due to interaction between axial and horizontal forces in the inelastic range of behaviour.

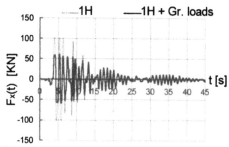

Figure 2. Horizontal force time-history.

When looking at horizontal top displacement time-histories (Fig. 3), it emerges that the presence of gravity loads does not lead to significant variations in the whole trend; however, peak displacements and, particularly, plastic residuals at the end of excitation increase due to reduction in horizontal capacity.

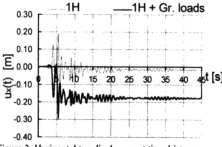

Figure 3. Horizontal top displacement time-history.

The most important finding is related to the displacement response in the vertical direction, as represented by the time-history of u_z (Fig. 4). Under

(1H+Gr. loads) important vertical top displacements (u_z) develop due to interaction phenomenon between the horizontal seismic force and the axial force due to gravity loads only; such displacements u_z are monotonically increasing with time even if no vertical seismic action is applied.

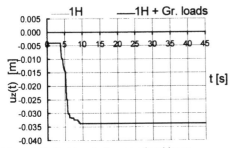

Figure 4. Vertical top displacement time-history.

Such behaviour has been already found by Popov & Bertero (1975) from cyclic tests on steel columns and by Lopez & Chopra (1978) from the study of a simplified mathematical model. Subsequently, the same column model has been studied including vertical earthquake component in the input ground motion (1H+V+ Gr. loads). Results have been compared to those obtained for (1H+ Gr. loads). As expected, the vertical seismic component does not substantially modify horizontal top displacements u_x with respect to the case of application of gravity loads only along the vertical direction (Fig. 5).

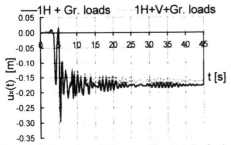

Figure 5. Horizontal top displacement time histories for the two loading conditions (1H+V+Gr. Loads) and (1H+Gr. Loads).

It is interesting to compare vertical top displacements obtained with and without the earthquake vertical component (Fig. 6).

Even if axial force F_z varies dramatically due to the high peak vertical ground acceleration (Elnashai & Papazoglou 1996, Elnashai 1997), the u_z time-histories are practically coincident as the vertical input ground motion induces a slight oscillation in u_z values about those computed for (1H+Gr. loads).

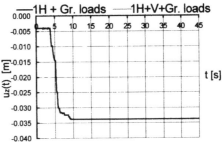

Figure 6. Vertical top displacement time histories for the two loading conditions (1H+V+Gr. Loads) and (1H+Gr. Loads).

Therefore, it can be concluded that effects of interaction due to gravity loads and horizontal seismic forces are very substantial as they lead to the development of large inelastic displacements in the vertical direction; furthermore, the inclusion of the vertical earthquake component does not lead to any significant change in the above described trends.

3.1 Damage evaluation

Characterization of plastic demands and of damage can be carried out as a function of maximum fiber plastic strains ε_{max}.

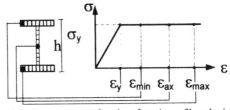

Figure 7. Damage as a function of maximum fiber plastic strain

Figure 8 shows how different loading conditions affect maximum plastic strain ε_{max}.

If only horizontal earthquake excitation is applied (1H), value of maximum strain is modest. When only gravity loads (Gr. loads) are applied, their effect is negligible, leading to very small axial deformations. However, when both horizontal earthquake component and gravity loads are applied, maximum strains become significant, thus implying that interaction phenomena between axial force due to gravity loads and horizontal seismic forces result in important amplifications in plastic excursions, as represented by ε_{max}. Finally, if earthquake vertical component is included in the input ground motion, ε_{max} is very similar to that obtained without the vertical ground motion, thus confirming results already shown with reference to vertical displacements time-histories.

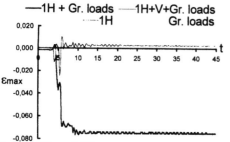

Figure 8. Effect of different loading conditions on the fiber maximum strain.

Further insight in the system response can be gained from analysis of cross section strain distribution, as shown in Figure 9 for the loading condition (1H+Gr. loads). Values of ε_{max} and ε_{min} have been derived as follows:

$$\varepsilon_{max} = \varepsilon_{ax} + |\chi|\frac{h}{2} \qquad \varepsilon_{min} = \varepsilon_{ax} - |\chi|\frac{h}{2} \qquad (1)$$

where ε_{ax} is the strain at the cross section centre, χ is the section curvature and h is the section depth (Fig.7).

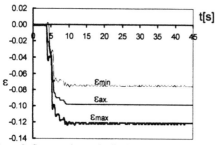

Figure 9. Cross section strain distribution.

Namely, maximum strain ε_{max} is primarily due to the axial deformation ε_{ax} arising from the interaction phenomena. Therefore, it appears that main cause of damage is development of vertical plastic displacements induced by interaction phenomena, while horizontal plastic displacements influence to a lesser degree fiber plastic deformation.

3.2 Hardening effects

Fiber model allows to implement refined material cyclic constitutive relationships, which in turn may affect interaction phenomena. A first analysis has been conducted to evaluate response when a simple elastic-strain hardening constitutive relationship is considered; hardening ratio has been selected a value

of 3%, which is considered to be realistic for steel structures. It can be seen that behaviour does not vary substantially, being displacements of the hardening model slightly lower than those computed for the elastic - perfect plastic one (Fig. 10-11).

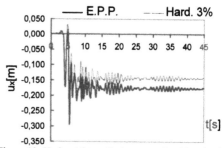

Figure 10. Horizontal top displacement time-histories for different material constitutive relationships.

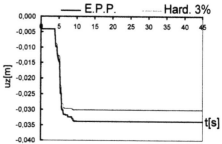

Figure 11. Vertical top displacement time-histories for different material constitutive relationships.

Namely, development of large inelastic vertical displacements is found again and very similar top displacement time-histories, both in the horizontal and in the vertical directions, are obtained.

4 NUMERICAL ANALYSIS: MATHEMATICAL MODEL

Herein, a mathematical model representing a single mass restrained by elastic-perfectly plastic springs is analysed (Fig. 12).

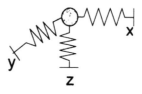

Figure 12. The mathematical model.

In the following, the model is presented as a system with three degrees of freedom, i.e. x, y (horizontal) and *z* (vertical) displacements; however in the initial analyses its behaviour has been investigated in the vertical *x-z* plane.

For such a model it has been assumed that the axial force-bending moment interaction in the plastic field is described by an ellipsoidal yield domain (Fig. 13) and the plastic behaviour follows the well-known associated flow rule.

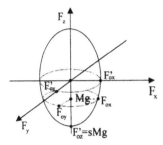

Figure 13. System plastic interaction domain.

This plastic condition is expressed by:

$$\left(\frac{F_x}{F'_{ox}}\right)^2 + \left(\frac{F_y}{F'_{oy}}\right)^2 + \left(\frac{F_z}{F'_{oz}}\right)^2 = 1 \qquad (2)$$

It is believed that mathematical model is very useful to draw more general conclusions since its response is governed by a few number of parameters such as fundamental periods and strengths along the three *x*, *y* and *z* directions, while it is worth using the fiber model to look for a detailed description of the system response and for numerical simulation of experimental tests.

This mathematical model has been adopted to investigate seismic response of the same single steel column previously examined by means of the fiber model. It should be noticed that the mathematical model is not able to capture: i) actual shape of yield domain, which is affected by the section shape and ii) cyclic behaviour in the inelastic range, which is represented in a very simplified manner by the assumption of elastic- perfectly plastic material, thus not accounting for hardening effects and deterioration in stiffness and strength arising during cyclic excitation.

In order to simulate response of the steel column, the mathematical model parameters (i.e. fundamental periods T_x, and T_z and strengths F'_{ox}, and F'_{oz} in the two principal directions *x* and *z*) have been set at the values characterising the column itself.

5 MODEL COMPARISON

First analyses have been aimed at comparing response trends obtained by using the two different models. If a horizontal earthquake component only is applied, the two models lead to practically coincident response predictions both in terms of forces and in terms of displacements (Fig.14-15).

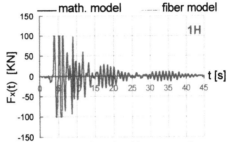

Figure 14. Horizontal force time-history comparison

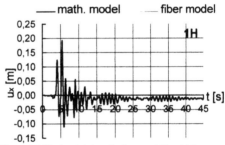

Figure 15. Horizontal top displacement time-history comparison.

Such behaviour results from the closeness of the force- displacement relationships obtained by using the two models, as shown by Figure 16.

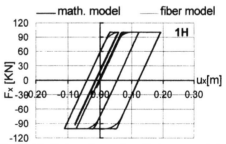

Figure 16. Horizontal top force-displacement relationship comparison.

When results obtained for the loading condition (1H+Gr. loads) are compared, displacements evaluated by means of the fiber model are greater than those obtained with the mathematical one (Fig. 17) while horizontal strengths are reduced (Fig. 18). This difference is due, as already said, to the difference between the ellipsoidal yielding domain adopted for the mathematical model and the actual one, which is affected by the section shape and is well represented by the fiber model.

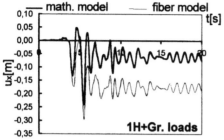

Figure 17. Horizontal top displacement time-histories comparison.

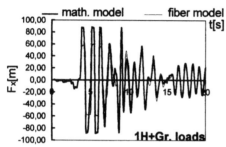

Figure 18. Horizontal force time-histories comparison

6 EFFECTS OF EARTHQUAKE VERTICAL COMPONENT

In the following, results obtained including vertical earthquake component in the input ground motion are presented and compared to those obtained without the vertical component itself.

As expected, the presence of the vertical seismic component (1H+V+Gr. loads), independently of the adopted model, does not substantially modify horizontal displacements with respect to the case of application of gravity loads only (1H+Gr.loads) along the vertical direction (Fig. 19).

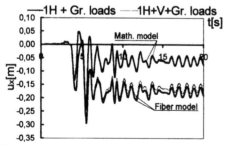

Figure 19. Horizontal top displacement time-history fort the two models in the two loading conditions.

But the most important findings come from the comparison between vertical top displacements obtained with and without the earthquake vertical component (Fig. 20).

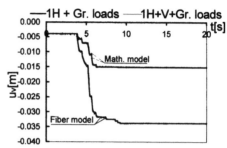

Figure 20. Vertical top displacement time-history fort the two models in the two loading conditions.

Both models, with a different degree of approximation because of their different way to capture interaction phenomena, show that in the inelastic range of behaviour there is no sensitivity of the system to the vertical component action, as it has already been noticed by means of the fiber model (Sect. 3). Vertical seismic component just induce oscillation of the vertical displacement time-history around values induced by gravity loads and horizontal seismic forces.

7 PARAMETRIC ANALYSES BY MEANS OF THE MATHEMATICAL MODEL

As previously pointed out, it is believed that mathematical model is suitable to investigate in a simplified manner seismic response of framed structures under multi-directional excitations. As an example, consider a one-storey steel framed building: if the building is symmetric, a storey interaction domain can be defined as a function of shear strengths along the two horizontal directions and of vertical strength. In such case, if steel sections are distributed along both x and y directions, the resulting yield domain approaches the ellipsoidal shape more closely than in the case of a single column.

In the following several parametric analyses have been carried out for two different loading conditions: gravity loads and bi-directional seismic action, gravity loads and all three seismic components.

7.1 Parametric field

Analysis have been conducted by varying parameters characterising elastic and inelastic behaviour of the system within ranges that are believed to cover most actual situations. Elastic periods in the two horizontal principal directions - T_x and T_y - have been assumed to be equal, in order not to introduce effects arising from different flexibility of the system in the horizontal directions, and to vary from 0.1 sec to 1.5 sec, to represent structures with medium and medium-low lateral stiffness. The vertical period T_z has been obtained with a simple formulation which relates it to the horizontal one; this relationship takes into account the limitation imposed by Eurocode 8 to the inter-storey drift Ψ, as shown elsewhere (Como et al. 1999). For steel structures, the proposed formulation satisfactorily correlates values reported by Elnashai and Papazoglou (1996) for real structures.

Inelastic system parameters define the interaction domain. Horizontal system strengths are dependent on the design ductility μ used to evaluate the inelastic spectra S_μ of the horizontal earthquake components; the vertical system strength is defined by the safety coefficient s against axial forces. For the safety coefficient s, it has been adopted $s=2$; this value is thought to be relevant to not very slender steel structures as the ones having lateral periods included between 0.1 sec and 1.5 sec.

7.2 Damage indices

For what concerns the assessment of the plastic demand experienced by the system, available studies are far from exhausting the topic of plastic demand evaluation under multi-directional excitations (De Stefano & Faella 1996).

It is easy to define plastic demands along each single direction, while combining plastic demands along the three x, y and z direction, when a multi-directional action is applied, is questionable as ductility demands along horizontal directions correspond to development of dissipative mechanisms very different from the ones related to ductility demands along vertical direction. Such inelastic response parameters have been defined extending the well-known concepts of hysteretic and kinematic ductility under uni-directional action. Particularly, the kinematic ductility demands μ_x, μ_y and μ_z along the three principal directions have been evaluated as:

$$\mu_x = \frac{u_{x,max}}{u_{xo}} \quad \mu_y = \frac{u_{y,max}}{u_{yo}} \quad \mu_z = \frac{u_{z,max}}{u_{zo}} \quad (3)$$

where $u_{x,max}$, $u_{y,max}$ and $u_{z,max}$ are the maximum displacements during excitation and u_{xo}, u_{yo} and u_{zo} are the yielding displacements along the three directions. In a previous study (Como et al. 1999), in order to represent plastic demand caused by multi-directional excitations, it has been considered the radial ductility μ_{rad} that is a direct three-dimensional extension of the well-known concept of kinematic ductility for uni-directional action. Namely, μ_{rad} is the ratio of the maximum displacement modulus to

the yielding displacement modulus in the same direction:

$$\mu_{rad} = \max \left| \frac{u_x(t)^2}{u_{xo}^2} + \frac{u_y(t)^2}{u_{yo}^2} + \frac{u_z(t)^2}{u_{zo}^2} \right|^{\frac{1}{2}} \quad (4)$$

It can be also defined a parameter of horizontal kinematic ductility μ_h obtained as the ratio of the maximum displacement horizontal projection modulus to the corresponding horizontal yielding displacement in the same direction:

$$\mu_h = \max \left| \frac{u_x(t)^2}{u_{xo}^2} + \frac{u_y(t)^2}{u_{yo}^2} \right|^{\frac{1}{2}} \quad (5)$$

The parameter μ_h is directly comparable to the widely used kinematic ductility to evaluate plastic demand under uni-directional horizontal action.

With an hysteretic approach to the definition of collapse, it has been introduced elsewhere (Como et al. 1999) the hysteretic ductility μ_{hyst} under multi-directional action:

$$\mu_{hyst.} = 1 + \frac{E_{diss}}{\max_{j=x,y,z}(F_{oj}u_{oj})} \quad (6)$$

This last parameter seems to be more adequate to describe the system response of the system under multi-directional action: hysteretic ductility, being an expression of dissipated energy, does not present, as the radial ductility, a problem of combining plastic demands in the three directions.

7.3 Effects of design ductility

By using the mathematical model, a parametric study has been carried out in order to evaluate how design ductility influences system inelastic response.

Namely system strength along vertical direction is proportional to gravity loads by means of a safety coefficient of 2, while strengths along horizontal directions have been calibrated by using the constant ductility spectra of horizontal ground motion components for μ ranging between 1 (elastic design) and 8. Kinematic and hysteretic parameters have been used to compare response under both horizontal components and gravity loads (2H+Gr. loads) and under all three seismic components and gravity loads (2H+V+Gr. loads). Figures 21-22 report increase in radial and in horizontal ductility demands, while Figure 23 reports increase in hysteretic demands.

It emerges that increments in radial ductility due to vertical earthquake component are significant only for systems designed to resist horizontal seismic action in the elastic range of behaviour ($\mu=1$). Otherwise, for larger values of design ductility, variation in the radial ductility is negligible.

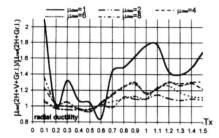

Figure 21. Increase in radial ductility demands due to the vertical earthquake component.

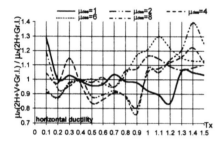

Figure 22. Increase in horizontal ductility demands due to the vertical earthquake component.

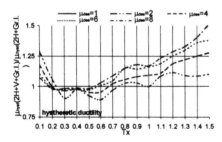

Figure 23. Increase in hysteretic ductility demands due to the vertical earthquake component.

Regarding horizontal ductility (Fig. 22), it is found that the seismic vertical component causes small variations which do not exceed about 20%, over the whole range of design ductilities.

In the same way, it can be seen that hysteretic ductility is only slightly modified when the vertical seismic component is included in the input ground motion, as the computed increment of $\mu_{hyst.}$ due to its action does not exceed about 25%.

In conclusion, both with a kinematic and with an energetic definition of the system collapse, vertical earthquake component does not affect in a significant manner damage parameters.

8 CONCLUSIONS

Interaction phenomena between axial and lateral forces developing in the plastic range of behaviour in a steel column have been investigated by means of two different models. Both of them confirm experimental results showing important plastic displacements in the vertical direction, also when only gravity loads and horizontal seismic forces are acting. The presence of the vertical earthquake component does not significantly modify such a behaviour.

The underlined phenomenon is very interesting and probably it should be taken into account in structural design.

REFERENCES

Elnashai, A.S. 1997. Seismic design with vertical ground motion. In Fajfar & Krawwinkler (eds), *Seismic design Methodologies for the next generation of codes*, 91-100. Rotterdam: Balkema.

Elnashai, A.S. & Papazoglou, A.J. 1996. Analytical and field evidence of the damaging effect of vertical earthquake ground motion. *Earthquake Engineering and Structural Dynamics.* Vol 25: 1109-1137.

Newmark, N.M., Blume, J.A. and Kapur, K.K. 1973. Seismic design spectra for nuclear power plants. *J. Power Div.* 99: 287-303.

Bozorgnia, Y., Niazi, M. and Campbell, K.W. 1995. Characteristics of free-field vertical ground motion during the Northridge earthquake. *Earthquake Spectra.* 11: 515-525.

Como, M., De Stefano, M. and Ramasco, R. 2000. Seismic response of yielding systems under three component earthquakes. *Proceedings of 12th World Conference of Earthquake Engineering*, Auckland, New Zealand.

De Stefano, M. & Faella, G. 1996. An evaluation of the inelastic response of systems under biaxial seismic excitations. *Engineering Structures*, vol. 18, n°9: 724-731.

Filippou, F., Spacone, E. and Taucer, F. 1991. A fiber beam-column element for seismic response analysis of reinforced concrete structures. *Report n° 91/17 EERC*, University of California at Berkeley.

Lopez, O. & Chopra, A. 1978. Studies of structural response to earthquake ground motion. *Report n° EERC 78/07*, University of California at Berkeley.

Mazzolani, F.M. & Piluso, V. 1996. *Theory and Design of Seismic Resistant Steel Frames.* London: E & FN Spon, an Imprint of Chapman & Hall.

Prakash, V., Powell, G.H. and Campbell, S. 1993. Drain-3dx, Base Program user guide.

Popov, E.P., Bertero, V.V. and Chandramouli, S. 1975. Hysteretic Behaviour of Steel Columns. *Report n° EERC 75/11*, University of California at Berkeley.

Behaviour of Steel Structures in Seismic Areas, Mazzolani & Tremblay (eds) © 2000 Balkema, Rotterdam, ISBN 90 5809 130 9

Establishing seismic force reduction factors for steel structures

R.G.Driver
Lafayette College, Easton, Pa., USA

D.J.L.Kennedy & G.L.Kulak
University of Alberta, Edmonton, Canada

ABSTRACT: Modern seismic design codes recognize that inelastic behaviour can reduce the earthquake forces generated in structures significantly and that this reduction is accompanied by an appreciable ductility demand. Currently, no generally accepted means exists of establishing the degree of force reduction suitable for use in design codes. A standardized means of establishing force reduction factors is therefore required. This paper describes a method for establishing force reduction factors for lateral load resisting systems based on large-scale physical tests. Lower and upper bounds to the force reduction factor are found from the hysteretic behaviour obtained through a cyclic test of the system to failure. Within these bounds, an appropriate value for code use is then selected based on properties that influence the behaviour of the structure in an actual earthquake, such as ductility, hysteretic energy dissipation, resistance to degradation, inherent redundancy, number of cycles resisted, and the failure mode.

1 INTRODUCTION

Seismic design codes have long recognized the ability of some structures to undergo significant inelastic deformation without reaching the point of collapse. This behaviour reduces the forces generated under seismic action to values significantly less than those consistent with elastic behaviour. However, the lateral deflection of a structure designed to behave inelastically can be several times that based on an elastic analysis under the reduced loads. Therefore, in order to survive a design earthquake, such a structure must possess good ductility and hysteretic energy dissipation capacity and be resistant to degradation under cyclic loads. Consequently, when designing structures using equivalent static forces, the level of force reduction, the level of available overstrength, and the anticipated deflection amplification due to inelastic action must all be assessed. Determination of the level of force reduction is the focus of this paper.

Most codes provide force reduction factors for common lateral load resisting systems that permit the calculation of the reduced seismic force. These values have been based historically on the collective judgement of code writers and the limited qualitative information available from previous earthquakes. In many cases there is little or no experimental evidence to substantiate the values used. An accepted standardized means of establishing force reduction factors experimentally is required; this will lead to more rational values for design, more confidence in the values used, and will also focus future research efforts. A standardized approach will also facilitate systematic comparisons of different systems.

The general principles of force reduction, overstrength and displacement amplification have been present in seismic codes for many years. Uang (1991) discusses explicitly their interrelationship for code development purposes. These concepts, based on a bilinear idealization of the behaviour, are depicted in Figure 1, where R = force reduction factor; Ω = overstrength; V = design shear; V_m = shear capacity; V_e = elastic shear; the elastic deflections that correspond to the base shears V, V_m, and V_e are Δ_V, Δ_y, and Δ_e, respectively; and Δ_m = the maximum deflection at which a shear of V_m can be sustained.

In defining R, it must be recognized that a structural system designed for the reduced forces must possess the required associated ductility and robustness in order to survive the design earthquake. Therefore, the force reduction factor is implicitly related to behavioural properties such as the ability to remain stable under inelastic lateral cyclic deflection, energy dissipation characteristics, resistance to cyclic degradation, inherent redundancy, and the failure mode, which can only be assessed reliably through large-scale physical tests.

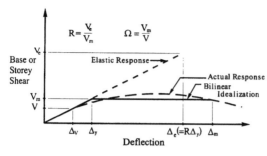

Figure 1. Elastic and inelastic response of a structural system.

The structural overstrength, Ω, is the ratio of the actual lateral resistance of the structure to that determined by conventional analysis. It can arise from several sources, some of which may be difficult to quantify such as the contribution to lateral resistance of non-structural elements. Others, such as the over-sizing of members to meet drift limitations or other load combinations and the ratio of the mean to nominal resistance of the members, can be quantified reasonably well either analytically or statistically. Uang (1994) notes that overstrength reduces the ductility demand.

2 FORCE REDUCTION FACTOR

As shown in Figure 1, the deflection Δ_m occurs at the intersection of the actual and idealized bilinear response curves. Defining the displacement ductility ratio, μ, as:

$$\mu = \frac{\Delta_m}{\Delta_y} \qquad (1)$$

the relationship between R and μ can be determined. Newmark and Hall (1982) demonstrate that for single degree of freedom systems with periods greater than about 0.5 seconds, the elastic deflection, Δ_e, for the elastic force, V_e, is approximately equal to the inelastic deflection, Δ_m, associated with the reduced force, V_m. Therefore:

$$R = \frac{V_e}{V_m} = \frac{\Delta_e}{\Delta_y} = \frac{\Delta_m}{\Delta_y} = \mu \qquad (2)$$

They also showed that for systems with periods from about 0.125 to 0.5 seconds, the deformation energies for the elastic and the elasto-plastic cases are approximately the same. Equating the areas under the two curves leads to:

$$R = \sqrt{2\mu - 1} \qquad (3)$$

Because $\mu \geq 1$, this implies that $R \leq \mu$ in this range of periods. Using a value of R equal to μ for

these low period structures is considered by many to be unconservative. Uang and Maarouf (1994) claim that the ratio of the building period to the predominant period of the ground motion is the important parameter. Tso and Naumoski (1991) propose a linear reduction from $R = \mu$ to $R = 1$ as the period reduces from 0.5 s to zero to account for the lower level of force reduction for short period structures. Nevertheless, a method of determining a reliable value of R for periods greater than 0.5 s is still required and can be achieved using the method described herein.

Uang and Maarouf (1994) demonstrate, using seismic analyses of four existing structures, that the value of the force reduction factor, R, for multi-storey buildings can in some individual storeys be less than μ (particularly for larger values of R), even for buildings with periods significantly longer than 0.5 s. However, the values of R that they presented correspond to the peak ductility demand that occurs in only one—or perhaps a few—cycles and therefore do not represent a sustained cyclic demand during the analysis. In contrast, the cyclic tests described subsequently provide many cycles at or near the maximum ductility demand and result in relatively severe cumulative damage. Therefore, in spite of the values of R presented by Uang and Maarouf corresponding to the peak ductility demand, based chiefly on Newmark and Hall (1982) and the large number of cycles at or near the maximum ductility demand in the proposed method for determining R experimentally, it is suggested that a value of $R = \mu$ be used for periods greater than 0.5 s. Thus, R is as given in equation (2).

The value of R is therefore determined from the lateral deflections Δ_y and Δ_m determined in tests, taking into consideration the observed behaviour. With multiple tests, the force reduction factor could be selected statistically. The method presented herein is based on an elasto-plastic idealization of the behaviour, to which the resulting value of R is sensitive. Thus, the method of establishing this bilinear curve is of paramount importance.

3 TEST STANDARDIZATION

A logical and consistent standardized approach is required to determine force reduction factors experimentally so that the relative performance of different structural systems can be assessed. It is proposed that the general procedures outlined in the *Guidelines for cyclic seismic testing of components of steel structures*, ATC–24 (ATC 1992), be followed. The method is particular to "slow cyclic" loading in which the test specimen is loaded in a controlled and predetermined manner. This method reveals many of the characteristics of a lateral load resisting system that are important for predicting seismic performance. Because dynamic tests are often unfeasible, except perhaps at a small scale, slow cyclic loading has been the mainstay of experimental seismic investigations and is likely to remain so. Krawinkler (1988) presents evidence that, as compared with dynamic testing, slow cyclic tests tend to lead to a lower measured strength and an increase in the rate of deterioration. They can therefore be considered to lead to conservative conclusions about the cyclic performance of structures. Some of the fundamental criteria for test design and execution outlined in ATC-24, and considered by the authors to be the most important for the purpose of determining the force reduction factor, are:

- use of a large-scale test specimen;
- fabrication to simulate field conditions;
- inclusion of gravity loads;
- provision of appropriate out-of-plane bracing;
- determination of material behaviour prior to testing;
- prediction of the specimen behaviour to develop a suitable loading protocol;
- selection of a deformation parameter such as the interstorey drift to control the test; and
- increase of cyclic displacements until severe strength deterioration is evident.

The first two items help to ensure that the specimen is representative of the actual structure. A scale on the order of 50% or more of the prototype structure is likely to be satisfactory. In some cases, smaller models may be adequate, but increased scrutiny of the detailing is required. Krawinkler (1988) discusses several details that at significantly reduced scale may produce misleading results. Testing multiple storeys is desirable because of the reduced dependence on the modelling of the boundary conditions adjacent to the storey of interest. Fabrication that simulates field conditions is essential to achieving reliable results and thus typical commercial fabrication is preferable to fabrication under "laboratory conditions."

The inclusion of gravity loads and appropriate out-of-plane bracing is necessary to ensure that the specimen is representative of the actual structure. Gravity loads, and the associated second-order effects, may have an effect on the hysteretic performance (e.g., the rate of deterioration or failure mode) and appropriate bracing ensures that the out-of-plane behaviour of the members is realistic.

The behaviour of the specimen must be predicted so that the point of significant yielding can be estimated before testing. This point is utilized in developing the entire loading regime and therefore must be based on measured material properties.

The deformation control parameter selected should be one that is easily measured and monitored and is representative of the overall behaviour. It should reveal the effect of damage in the structure from cyclic loading and the amount of energy dissipation.

The test is continued until the strength of the system is exhausted so that the full range of behaviour can be assessed. Stopping the test early may not lead to conservative conclusions, even though a reduced deformation capacity may be assumed, because the failure mode cannot be assessed.

ATC-24 describes a loading protocol that cycles the structure with gradually increasing peak deflections in each direction, primarily based on multiples of the deflection, Δ_y, the point of significant yielding. The specimen is subjected to multiple elastic cycles prior to reaching the yield deflection and then two or three cycles at each multiple of Δ_y thereafter until significant strength degradation is evident.

The SAC Joint Venture protocol (SAC 1997) supersedes ATC-24 for projects within Phase 2 of the SAC Steel Project. However, because much of the SAC report applies to moment-resisting frames in particular, ATC-24 is considered more suitable for general use.

It may not be possible to follow a comprehensive prescription of procedures such as those contained in ATC-24 in every detail. Hence, in the evolution of the standardization process for determining the force reduction factor experimentally, it is desirable to outline the few fundamental mandatory requirements, as well as those that are simply desirable. The onus is then on the researcher to report the particular methods used so that the results and conclusions can be assessed critically by others.

4 SHEAR VS. DEFLECTION CURVE

4.1 Fundamental Concepts

To determine a force reduction factor, an idealized bilinear elasto-plastic shear vs. deflection relationship is required. The sensitivity of R to the curve selected dictates that the various properties that influence seismic performance be carefully evaluated. These properties include ductility, hysteretic energy dissipation, resistance to degradation, inherent redundancy, number of cycles resisted, and failure mode. Each one has an effect on the process of se-

489

lecting a suitable force reduction factor, and is discussed in section 4.2.

A cyclic test in a symmetrical set-up with excursions of equal magnitude in the two loading directions provides two potential idealized load vs. deflection curves. Should the two directions not exhibit similar behaviour, the direction that results in the lower value of R should generally be utilized.

The unidirectional "actual response" depicted in Figure 1, is taken as the envelope of cyclic curves from a plot of storey shear vs. storey deflection for the most severely loaded storey. The envelope is established by connecting the data points at the extreme deflection of the excursions that give the highest storey shear for each deflection level. (Any deterioration in strength in subsequent excursions to the same deflection level is accounted for by assessing the resistance to degradation, as described below.) The envelope contains most of the relevant behavioural information, such as ductility and rate of strength degradation with increasing deflection levels. The other properties are assessed from the cyclic data and test observations and used with judgement in developing the particular bilinear idealization used to calculate R.

Identification of the lower and upper bounds of R demarcates the range of possible values, as shown in Figure 2. The lower and upper bounds occur when the plastic curve intersects the maximum load on the envelope and the failure load, respectively. The bilinear idealization is truncated at the point at which it crosses the descending envelope curve. At deformations greater than this, the structure can no longer resist the assumed inelastic force. This defines the maximum deflection, Δ_m. Selection of the slope of the elastic segment to approximate the linear (or near-linear) initial portion of the envelope curve establishes the value of Δ_y at the intersection of the two segments. The value of R is then be calculated from these two quantities using equation 5.

4.2 Properties of influence

4.2.1 Ductility

Although ductility is perhaps the most important property for resisting seismic loading, it is not a quantity that is easily assessed strictly through analytical means. Experimentation permits a quantitative assessment of ductility which is represented explicitly in the envelope curve of the storey shear vs. deflection behaviour. Of the many definitions of ductility that have been proposed, the measure that appears most suited to this application is the ratio of Δ_m to Δ_y based on interstorey drift. As this ratio is

the definition of R that has been proposed, the ductility of the system is directly taken into account.

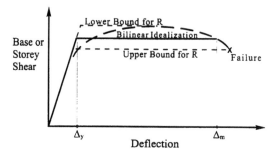

Figure 2. Lower and upper bounds for R.

4.2.2 Hysteretic energy dissipation

Hysteretic energy that is dissipated during cyclic loading contributes to the ability of a structure to survive an earthquake. The energy dissipated in one complete cycle is measured as the area enclosed by the storey shear vs. deflection curve during that cycle. Two characteristics are particularly desirable. The amount of energy dissipated should be as great as possible and it should not decline significantly— or preferably should increase—in each subsequent cycle of increasing deflection. Favourable energy dissipation characteristics support a more liberal value of R.

4.2.3 Resistance to degradation

Although a structure may be able to undergo many cycles of loading and withstand large deformations without collapse, its ability to resist an earthquake can be severely impaired if the capacity of the structure is not maintained. As can be seen in Figure 2, any increase in the value of R beyond the lower bound presumes that the concomitant degradation in strength can be tolerated. This interpretation should only be used if the degradation is gradual. Rapidly degrading behaviour should be seen as similar to collapse for the purpose of determining R.

Any decrease in capacity observed in second and subsequent excursions to a particular deflection level should also be assessed. Even if the envelope of cyclic curves shows gradual strength degradation with increasing deflection levels, significant decreases in strength from one excursion to the next with the same peak deflection warrants conservatism in the selection of R.

4.2.4 Inherent redundancy

Many lateral load resisting systems possess some structural redundancy, but large-scale experimental

490

work can reveal the degree to which the redundancy can be utilized. For example, if tests show that all but one of the multiple load paths degrade rapidly through damage to the structure, the favourable qualities of redundancy are largely lost. Conversely, if the multiple load paths all participate and maintain their integrity, a more liberal value of R could be considered.

4.2.5 *Number of cycles resisted*

Structures exhibiting similar performance based on the envelope of cyclic curves may have been subjected to significantly different numbers of cycles of loading. Because it is often the repeating aspect of the loading that leads to the degradation of performance, the number of load cycles to which the structure has been subjected at the time of failure should be considered.

Derecho et al. (1980) used 170 time history analyses of reinforced concrete structures for earthquakes of 20 second duration to determine that the number of fully reversed large amplitude cycles was fewer than four in 95% of cases, with an extreme value of six. Although this should not be seen as a guideline for establishing a threshold that represents adequate performance, it gives a means of comparing the number of test cycles to what might be expected during an earthquake. As the database of large-scale slow cyclic test results grows, relative comparisons may become more appropriate.

4.2.6 *Failure mode*

Although failure can occur at very large deflections and after a large number of cycles of load, the failure mode is also of importance in assessing the suitability of a structural system for seismic applications. Sudden modes of failure that result in rapid or complete loss of load carrying capacity are clearly undesirable and should indicate that a conservative value of R is appropriate.

However, because the test is to be continued until severe strength deterioration of the specimen is evident, a brittle failure mode does not indicate unsuitability of the system itself. Rather, as long as gradual degradation has taken place prior to the ultimate brittle failure, performance should be considered adequate, with the value of R selected accordingly. Observation of the failure mode can also give valuable insight into how the design could be modified to improve the performance.

5 EXAMPLE APPLICATIONS

5.1 *Moment-resisting frame*

Because of the qualitative nature of some of the parameters of influence and the judgement required in their interpretation, an illustrative example is presented to demonstrate the intent of the methodology. Due to the premature fractures of connections of moment-resisting frames in the Northridge earthquake of 1994, a frame configuration developed to circumvent this problem is selected: the reduced beam section (RBS), also known as the "dogbone." Since no large-scale complete moment frames with these improved details have been tested, the results of a large-scale connection test have been adapted to apply to a complete frame by making a few simple assumptions, as described subsequently.

The specimen selected is specimen DB4 tested by Engelhardt et al. (1998). It consists of a W920x289 (W36x194) beam connected using an all-welded moment connection to a 3.45 m long section of a W360x634 (W14x426) column. The beam was tested as a cantilever, 3.40 m long from the face of the column, bent about its strong axis by a concentrated force at the tip. The beam flanges were reduced near the columns by means of circular cutouts, which reduced the beam flange width by approximately 40% at the minimum section. Details of the test specimen and set-up can be found in Engelhardt et al. (1998).

The tests were conducted using the ATC-24 loading protocol. Displacement control was used, with three load cycles to each of ± 13 mm, ± 19 mm, ± 25 mm, ± 51 mm, and ± 76 mm measured at the load point at the cantilever tip. Thereafter, displacements were increased by 25 mm with two cycles conducted at each displacement level. For specimen DB4, a total of 21 complete cycles of load were applied. The tests were then terminated because of limitations of the set-up. The connections, therefore, did not experience total failure. In the final cycle, the total plastic rotation (measured to the column face) was 0.037 radians.

In order to estimate the force reduction factor, values for storey shear and storey deflection for a frame are required. To illustrate the proposed method, the connection test results are transformed to frame shear and deflection data by assuming that the storey height is 3.66 m and that the bay width is 7.26 m (centreline dimensions) and that the column bases are pinned. The bay width selected is twice the cantilever distance to the point of load application in the test (assumed to be the point of inflection in the frame). The plastic rotation observed in the test is assumed to occur in a discrete plastic hinge at the centre of the reduced beam section. (Engelhardt et al. noted that deformations within the column panel zone were negligible in the test.) The plastic portion of the implied storey deflection, Δ_{ps}, is determined from the plastic hinge rotation, θ_p, solely from geometry, as shown in Figure 3. (The plastic hinge rotation is defined differently from the total plastic rotation presented in Engelhardt et al.) For any particular plastic storey deflection, the storey shear is then taken as the horizontal force that would cause the associated moment as measured at the face of the column during the test. The elastic portion of the storey deflection is determined from an elastic analysis of the frame using this horizontal storey force, neglecting the reduced beam section. (Engelhardt et al. indicate that an investigation of frames with a localized 40% reduction in flange area revealed a reduction in elastic frame stiffness of only 4 to 5 percent.) The total deflection is the sum of the elastic and plastic contributions.

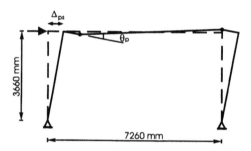

Figure 3. Assumed frame configuration.

A storey shear vs. storey deflection curve is established, as shown in Figure 4. The frame behaves in a virtually linear manner up to the end of cycle 9 at a peak storey deflection of about 31 mm. Thereafter, a gradual softening occurs until the maximum storey shear of 2230 kN is reached in cycle 15. The peak storey shear remained virtually the same through cycle 17 and then declined gradually. In the final cycle (the second to a deflection of 160 mm), the peak storey shear had declined to about 73% of the maximum achieved, but failure of the specimen did not occur, as described previously.

Figure 5 shows the envelope of cyclic curves presented in Figure 4, plotted in terms of the displacement ductility ratio based on the yield deflection for the particular elasto-plastic idealized curve shown (solid line). For this curve, the force reduction factor, R, is taken as the displacement ductility ratio where the bilinear curve intercepts the envelope. For the lower and upper bound curves the corresponding R values, based on their respective Δ_y values, are 2.4 and 4.9, respectively. Furthermore, recalling that the specimen did not fail, the true upper bound exceeds 4.9 by an indeterminate amount. The value of R selected must fall between the lower and upper bounds and is dependent upon the properties of influence, as discussed in the following.

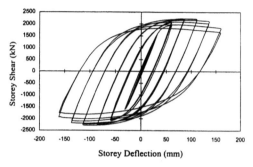

Figure 4. Experimental hysteresis curves.

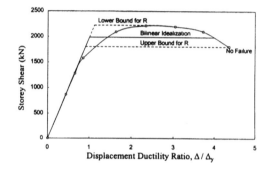

Figure 5. Envelope of cyclic curves and bilinear idealization.

Figure 4 shows that the moment-resisting frame is very ductile; large deformations were achieved. Even at the final cycle, the ductility had not been entirely exhausted.

The area enclosed by the hysteresis curves is a measure of the energy dissipated. As can be seen in Figure 4, the loops demonstrate favourable energy dissipation characteristics. Figure 6 shows the amount of energy dissipated during a complete cycle (one excursion in each direction) as a function of the displacement ductility ratio. The displacement ductility ratios shown, based on the yield deflection, Δ_y, as determined from the particular bilinear idealization shown in Figure 5, are not necessarily integer values. Figure 6 demonstrates that the amount of energy dissipated increases with increasing deflection levels. This increase continues until the final level, even though a slight decrease in the rate of increase can be observed at this point. This indicates that although collapse did not occur, the performance in the final cycles had begun to decline and therefore selecting a force reduction factor near the upper bound level is not recommended.

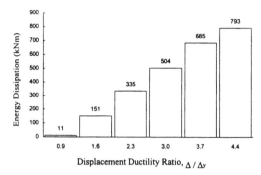

Figure 6. Cyclic energy dissipation.

The hysteresis loops in Figure 4 are not only wide, but are also uniform and stable. This indicates that there will be no sudden losses in stiffness under severe cyclic loading. Significant degradation begins only after a displacement ductility ratio of about 3.7. The test beam capacity declined gradually as a result of local and lateral buckling after a total plastic rotation (measured to the column face) of about 0.020 radians, but Engelhardt et al. (1998) point out that this buckling behavior was less severe than that observed in similar tests of other types of moment frame connection configurations.

In total, the moment frame was subjected to 21 cycles of loading. Twelve of these cycles carried the structure into the range of significant inelastic deformations. The failure mode was not observed in this test, indicating a need for conservatism in the selection of the force reduction factor. However, the degradation was gradual, with the hysteresis curves remaining wide and stable.

Although the upper bound value is somewhat greater than 4.9, because the test was not taken to failure the value of 4.9 should be used as the upper bound since the behaviour of the connection was not explored at greater deformations. Furthermore, lack of knowledge of the failure mode suggests that the value of R should not be selected near the upper bound. Selecting the value of R to be near the lower bound is also considered inappropriate because of the extremely favourable cyclic characteristics observed. Taking into account the excellent hysteretic behaviour, the increasing energy dissipation capacity and stable behaviour, and the long, nearly flat plateau of the envelope curve of Figure 5 that extends to a ductility ratio of about 3.5, a reasonable force reduction factor would fall in the range of about 3.5 to 4.5. It is suggested that a value of approximately 4 would be appropriate in this case. Selection of this value implies a decline in shear capacity to about 90% of the value at the lower bound for R.

5.2 Steel plate shear wall

The authors have recently reported the results of a cyclic test of a four storey steel plate shear wall (Driver et al. 1998). These results are used to provide a supplementary example of the application of the methodology described here. In this case, a full frame was tested and gravity loads were applied to the columns. The top three storeys were 1.83 m high and the first storey was 1.93 m high. The columns were 3.05 m on centre. Full moment connections, without special details to improve panel zone ductility, were used at all beam-to-column joints. The specimen was constructed entirely in a commercial fabrication shop using standard methods. A more complete description of the test specimen and loading regime is provided in Driver et al. (1998).

Figure 7 shows the envelope of cyclic response curves for the most severely loaded storey. Recalling that the value of the yield deflection, Δ_y, depends on the particular bilinear curve considered, it can easily be shown that the lower and upper bounds for R are 4.8 and 10.8, respectively. The actual value selected follows from an evaluation of the various properties of influence.

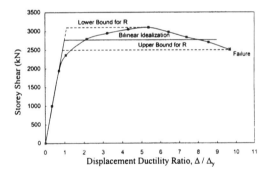

Figure 7. Envelope of cyclic curves and bilinear idealization.

Figure 7 demonstrates the exceptional ductility exhibited by the steel shear wall. The maximum interstorey deflection of 76.5 mm achieved prior to failure is nine times the value of Δ_y of the idealized bilinear curve shown in the figure (solid line).

As described in Driver et al. (1998), the hysteresis loops generated are relatively wide, indicative of significant energy dissipation during each cycle. The uniformity of the hysteresis curves indicates that the behaviour was also extremely stable and resistant to degradation. There was no sudden loss of stiffness and, even after the peak load had been reached, deterioration was slow and controlled. This resistance to degradation can be attributed partially to the inherent redundancy of the system. Observed tearing of the infill plates, which is a mechanism for dissipating energy, occurred in a gradual manner. The tearing does not result in a sudden decrease in stiffness because the continuous infill plate redistributes loads to areas unaffected by the tearing. Further redundancy exists because the moment-resisting frame also contributes to the lateral resistance.

Thirty cycles of loading were applied to the test specimen prior to failure, of which 20 were well beyond the point of significant yielding. The test specimen finally failed by fracture at the base of one column.

The envelope of the cyclic curves for the steel plate shear wall and for the moment frame described previously are quite similar. However, at large deformations the rate of deterioration for the former is considerably less than that of the latter. Moreover, the steel plate shear wall is a dual system in which the infill plates limit the deformations, resulting in reduced plastic rotational demand at the beam-to-column connections. Conservatively placing the bilinear idealization curve at a shear of about 90% of the peak, as was done for the moment-resisting frame, results in an R value of approximately 8.

6 SUMMARY AND CONCLUSIONS

For structures with periods greater than 0.5 s, it is proposed that the force reduction factor, R, be taken as equivalent to the displacement ductility ratio, μ. Based on this relationship, appropriate values of R for use in the design of lateral load-resisting systems can be determined from large-scale tests that use quasi-static cyclic loading. In order to attain consistency for different systems and for different materials, standardization of the test design protocol is essential. Lower and upper bound values can be established from the storey shear vs. storey deflection envelope curves. By examining the properties of influence critically, a specific value can be selected within these bounds. The properties considered in this evaluation include ductility, hysteretic energy dissipation, resistance to degradation, inherent redundancy, number of cycles resisted, and the failure mode. By using this method, values of the force reduction factor can be established that lead to a better representation of the effects of inelastic behaviour during an earthquake and that permit direct comparison among different systems.

Currently, there are few test results suitable for establishing R values for lateral load-resisting systems. In general, the values in use have been selected with little or no experimental substantiation. Therefore, more large-scale tests are needed that follow a generally accepted test design and loading protocol. Guidelines for such a protocol are presented here.

ACKNOWLEDGEMENTS

The authors thank M.D. Engelhardt of the University of Texas at Austin for generously providing the data from the reduced beam section test and C.-M. Uang of the University of California San Diego for providing insightful comments.

REFERENCES

ATC. 1992. *Guidelines for cyclic seismic testing of components of steel structures.* Report No. 24, Applied Technology Council, Redwood City, CA.

Derecho, A.T., M. Iqbal, M. Fintel & W.G. Corley 1980. Loading history for use in quasi-static simulated earthquake loading tests. ACI Publication SP63, *Reinforced Concrete Structures Subjected to Wind and Earthquake Forces.* American Concrete Institute: 329-356.

Driver, R.G., G.L. Kulak, D.J.L Kennedy & A.E. Elwi 1998. Cyclic test of four-story steel plate shear wall. *ASCE Journal of Structural Engineering.* 124(2):112-120.

Engelhardt, M.D., T. Winneberger, A.J. Zekany & T.J. Potyraj 1998. Experimental investigation of dogbone moment connections. *AISC Engineering Journal.* 35(4):128-139.

Krawinkler, H. 1988. Scale effects in static and dynamic model testing of structures. *Proceedings of the 9th World Conference on Earthquake Engineering* (Vol. VIII), Tokyo-Kyoto, Japan: 865-876.

Newmark, N.M. & W.J. Hall 1982. *Earthquake spectra and design,* Earthquake Engineering Research Institute, Berkeley, CA.

SAC. 1997. *Protocol for fabrication, inspection, and documentation of beam-column connection tests and other experimental specimens.* Report No. SAC/BD-97/02, SAC Joint Venture, Sacramento, CA.

Tso, W.K. & N. Naumoski 1991. Period-dependent seismic force reduction factors for short-period structures. *Canadian Journal of Civil Engineering.* 18(4):568-574.

Uang, C.-M. 1991. Establishing R (or R_w) and C_d factors for building seismic provisions. *ASCE Journal of Structural Engineering.* 117(1):19-28.

Uang, C.-M. 1994. Balancing structural strength and ductility requirements for seismic steel design. *Proc., AISC National Steel Construction Conference,* Pittsburgh, PA: 24-1 – 24-14.

Uang, C.-M. & A. Maarouf 1994. Deflection amplification factor for seismic design provisions. *ASCE Journal of Structural Engineering.* 120(8):2423-2436.

Behaviour of Steel Structures in Seismic Areas, Mazzolani & Tremblay (eds) © 2000 Balkema, Rotterdam, ISBN 90 5809 130 9

Seismic response of tied and trussed eccentrically braced frames

A. Ghersi, F. Neri & P. P. Rossi
Dipartimento di Igegneria Civile e Ambientale, Faculty of Engineering, Catania, Italy

A. Perretti
Dipartimento di Analisi e Progettazione Strutturale, Faculty of Engineering, Napoli, Italy

ABSTRACT: This paper analyses the inelastic behaviour of eccentrically braced frames by means of pushover and dynamic response analyses. The structural typologies considered include standard eccentrically braced frames and schemes provided with vertical or diagonal elements, which connect the ends of each link to the upper storey (tied or trussed braced frames). The length of the links has been varied so as to examine both short links (which yield for shear) and long links (which yield for bending moment). The number of storeys has been varied from 4 to 12, so as to point out its influence on the behaviour of the scheme. Pushover analyses have shown some differences in the behaviour of schemes with short, intermediate or long links, together with an increase of ultimate displacements when ties or trusses are added, independently of the number of storeys. On the contrary, dynamic analyses have pointed out the strong influence of the number of storeys in the case of standard eccentrically braced frames, which are dramatically penalised by the arising of a softstorey behaviour. The adoption of ties or trusses is really effective in preventing such behaviour and in granting the eccentrically braced frames a highly dissipative response also in the case of tall buildings.

1 INTRODUCTION

As it is well known, buildings located in seismic areas have to be designed in such a way to fulfil specific requirements. Although nowadays more detailed sets of structural performances are under discussion, the basic behavioural aspects to be analysed are those related to the occurrence of low or moderate seismic events and of very strong earthquakes. In the first case all structural elements should remain in the elastic range, while non-structural elements should be only slightly damaged; this aim is in practical applications achieved by limiting the storey drift, i.e. checking the stiffness. In the second case the structure may undergo large inelastic deformations; it is therefore fundamental in such conditions to assure to the structure the capacity to dissipate large amounts of energy by means of a stable hysteretic behaviour (i.e. granting proper values of local and global ductility). A Moment Resistant Frame (MRF) shows a really good dissipative capacity, thanks to the large number of plastic hinges developed when the structure fails in a global mechanism, but it is at the same time very flexible. For this reason its design is often governed by the limits imposed to the maximum displacements for low seismic events and its cross-sections are usually oversized respect to those strictly necessary in presence of the ultimate limit state loadings. On the contrary, a Concentrically Braced Frame (CBF) is very stiff but it is characterised by a quite poor inelastic behaviour: in occurrence of strong earthquakes this imposes the use of larger design forces in order to counterbalance the low levels of available ductility. Aiming at obtaining at the same time stiffness and ductility, a new typology has been proposed about 25 years ago: the Eccentrically Braced Frame (EBF). The presence of bracings, although not converging in the same point, provides the scheme enough stiffness (e.g. see Hjelmstad and Popov 1984), while the beam segment between the braces, named *link*, is able to undergo large plastic deformations and to dissipate a conspicuous amount of energy. The link is subjected to constant shear forces and linearly varying bending moments. When it is short, i.e. its length $e < 1.6\, M_p / V_p$, being M_p and V_p the limit values of bending moment and shear respectively, the shear yielding of the whole link dominates the inelastic response. When it is long, i.e. $e > 2.6\, M_p / V_p$, flexural yielding arises at its ends; in intermediate cases, the inelastic response is governed by a combination of shear and flexural yielding (see Kasai & Popov 1986a, AISC 1997).

The cyclic behaviour of short links is really stable, provided that the web buckling is prevented by means of proper web stiffeners (e.g. see Hjelmstad and Popov 1983). A very large monotonic plastic deformation (even 0.2 radiant) can be resisted without

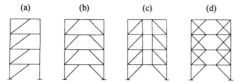

(a)   (b)   (c)   (d)

Figure 1. D-braced frames (a); split K-braced frames (b); tied braced frames (c); trussed braced frames (d)

a significant loss of capacity (Malley and Popov 1984), although a lower value (0.09 radiant) is suggested in order to limit strength deterioration in cyclic deformation (NEHRP 1994). Many tests have furthermore showed that, mainly because of steel hardening, the maximum shear force can be 1.5 times the plastic capacity V_p (Kasai & Popov 1986b, Ricles & Popov 1989).

Some technological aspects have to be considered in designing eccentrically braced frames. First of all, link-to-column connections in a D-braced scheme (Fig. 1a) may be subjected to deformation demands even larger than those of beam-to-column connections in a MRF, because the deformation is confined in a shorter portion of the beam. This requires more complex and costly connections, together with the necessity to carry out cyclic tests to confirm the inelastic behaviour. Furthermore the lack of symmetry of D-braced schemes may give rise to strongly different bending moments at the ends of the links, with the possibility of an early flexural yielding of the most stressed end. All these problems are bypassed when Split-K-braced schemes are used (Fig. 1b), which grant symmetry and require very simple beam-to-column pinned connections.

A second aspect to be considered is the influence of the connection of links to the floor slab. As a matter of fact, the vertical loads acting on the link modify both the value and distribution of the internal actions; this may give rise to an unexpected inelastic behaviour (flexural instead of shear yielding; partial shear yielding of the link). Furthermore, the large inelastic displacements of the link induce considerable deformation and damage in the floor slab. A suggested way to avoid this is to disconnect the floor slab from the lateral load resisting system, introducing, when necessary, an additional beam parallel to the one of the link (Perretti 1999).

The dissipative capacity of EBFs has been highlighted in many studies, mainly on the basis of pushover analyses but in some cases also by means of inelastic response analyses. Nevertheless it has been also shown that links may undergo very large deformation at a single floor, often the first one (Foutch 1989, Popov et al. 1989) or an intermediate or upper storey (Lu et al. 1997). AISC (1997) remarks that "in extreme cases this may result in a tendency to develop a soft storey" but in spite of this it gives no particular relevance to the negative consequences of such a possibility. A reliable method

for obtaining uniform link deformation at all storeys has been devised by Ricles & Bolin (1991), who suggest to introduce vertical elements (ties) to connect the corresponding ends of the links of contiguous floors. This scheme (Fig. 1c) is here referred as Tied eccentrically Braced Frame (TBF). For the same purpose, an alternative (Fig. 1d) has been recently proposed by Perretti (1999), who suggests to connect the ends of each link to the beam-to-column node of the upper storey. It is thus obtained a scheme that in some way recalls a truss and which is named for this reason TRussed eccentrically Braced Frame (TRBF). Such a solution allows furthermore, respect to the previous ones, larger architectonical flexibility in placing windows within the frame.

This study analyses EBF, TBF and TRBF schemes by means of both pushover and inelastic dynamic response analyses, in order to compare their dissipative capacity and to find out how much it may be penalised by large inelastic deformations at single storeys and by soft storey mechanisms.

2 STRUCTURAL SYSTEMS AND DESIGN CRITERIA

This paper analyses the seismic behaviour of three steel buildings, having a square plan (24×24 m²) and 4, 8 and 12 storeys. Their structure is constituted by pinned frames arranged along an orthogonal grid, with span length L equal to 8 m. The horizontal actions are withstood by eccentrically braced frames located along the perimeter of the system (Fig. 2).

Decks are made by sheetings and light concrete so as to limit the mass of the structure. For the same reason high quality and light non-structural elements have been used. Dead and live loads of the deck are therefore supposed to be in total 5 kN/m².

The structure firstly examined encloses standard split K-braced frames with a link length $e=0.1\,L$; all the columns are pinned at the base. The seismic design actions have been evaluated according to Eurocode 8, using static analysis with the design response spectrum proposed for subsoil class C with a peak ground acceleration $a_g=0.35$ g, a behaviour factor $q=5$ and a damping factor of 0.05. At this stage the

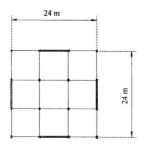

Figure 2. Plan of the structural schemes

Table 1. Parameters of design

n. storeys	Total seismic weight (kN)	Design period (s)	Design acceleration	Base shear for each EBF (kN)	Actual period (s)
4	11520	0.5	0.158 g	907.2	0.656
8	23040	1.0	0.136 g	1563.6	1.108
12	34560	1.5	0.104 g	1789.9	1.779

fundamental periods of vibration of the structures under analysis have been estimated in 0.5, 1.0 and 1.5 s for the systems having 4, 8 and 12 storeys respectively. Their actual values, calculated at the end of the phase of design, do not show great differences (Table 1).

The first step of the design consists in defining the cross-section of the links at the generic k^{th} storey. In presence of horizontal actions only, the vertical translation equilibrium of half of the frame from the top to the kth floor gives the vertical action N_k transmitted to the floor below

$$N_k = \sum_{j=k}^{n_s} V_j \tag{1}$$

being n_s the number of storeys and V_j the shear in the link at the jth storey. The rotational equilibrium of the portion of the frame above the level k gives the sum of the shear of the links from the top to the kth floor

$$\sum_{j=k}^{n_s} V_j = \frac{\sum_{j=k}^{n_s} F_j \, (h_j - h_{k-1})}{L} \tag{2}$$

where F_j is the horizontal action and h_j is the height of the floor j respect to the base. This formulation may be used to design their shear strength, proceeding from the top to the base of the scheme. If the strictly necessary strength is given to each link, the value of the design shear at each storey is obtained by the formula

$$V_k = \frac{\left(\sum_{j=k}^{n_s} F_j \right) h_k^s}{L} \tag{3}$$

being h_k^s the storey height at the storey under consideration.

The necessity of using commercial cross-sections implies anyway some overstrength for the links. Economic reasons, which suggest adopting the same cross-section at many storeys, lead to further increases in the overstrength. In the examined cases, aiming at limiting at the same time the overstrength and the number of different cross-sections of the links, the same cross-section has been adopted every two floors.

All the other elements (columns, bracings and ties) shall be defined according to the capacity de-

sign criterion, i.e. basing on the internal actions corresponding to the maximum capacity of the link. As previously mentioned a realistic relationship between the ultimate shear V_u and the yielding value V_y may be given by the following expression

$$V_u = 1.5 V_y \tag{4}$$

In order to take into account all other uncertainties in geometrical and mechanical characteristics, this value is furthermore increased by means of a factor 1.2 (i.e. multiplying V_y by 1.8) so as to effectively grant that yielding or buckling will never occur in bracings and ties before links have reached their ultimate shear.

The design value of the axial force in columns is sum of the share due to the vertical loads and of that due to the presence of the plastic shear in the links. The low probability that in tall buildings the links contemporarily reach the ultimate shear value at all the storeys allows the designer to decrease the amplification factor previously described, used for bracings and ties. In the analysed schemes, the overall amplification factor of the axial force deriving from the yielding shears has been assumed equal to 1.8 for the upper four storeys; it has been differently fixed equal to 1.5 and 1.2 for the stories starting from the top from the fifth to the eighth and from the ninth to the twelfth respectively. The cross-section chosen according to this procedure are shown in Table 2.

In order to analyse the influence of the link length on the seismic behaviour of the system, eccentrically braced frames with different values of the link length (0.1 L, 0.15 L, 0.2 L, 0.3 L and 0.4 L) have been considered. The same cross-sections previously selected have been used in all these structures. Furthermore, analogous schemes with the addition of tie or truss elements have been studied. Thus, a total of 3×5×3 schemes have been considered.

Table 2. Cross-sections

storey	link section	V_y (kN)	M_y (kNm)	bracing section	column section
		4 storey scheme			
3-4	HEA240	221.3	158.9	HEA160	HEA160
1-2	HEA320	357.9	347.8	HEA180	HEA240
		8 storey scheme			
7-8	HEA240	221.3	158.9	HEA160	HEA160
5-6	HEA360	449.0	446.1	HEA220	HEA240
3-4	HEA400	550.3	547.3	HEA240	HEA340*
1-2	HEA450	649.1	687.1	HEA260	HEA400**
		12 storey scheme			
11-12	HEA220	188.6	121.3	HEA160	HEA160
9-10	HEA320	357.9	347.8	HEA180	HEA240
7-8	HEA400	550.3	547.3	HEA240	HEA300*
5-6	HEA400	550.3	547.3	HEA240	HEA500*
3-4	HEA500	754.3	843.4	HEA280	HEA500**
1-2	HEA500	754.3	843.4	HEA280	HEA500**

* steel grade Fe430   all other sections:
** steel grade Fe510   steel grade Fe360

Table 3. First period of vibration of the analysed schemes (s)

Scheme; n. storeys		e/L 0.10	0.15	0.20	0.30	0.40
EBF	4	0.656	0.768	0.899	1.196	1.518
	8	1.108	1.206	1.329	1.630	1.980
	12	1.779	1.883	2.018	2.361	2.775
TBF	4	0.649	0.754	0.878	1.158	1.465
	8	1.103	1.196	1.312	1.590	1.928
	12	1.775	1.873	2.000	2.322	2.716
TRBF	4	0.616	0.726	0.854	1.138	1.447
	8	1.073	1.167	1.284	1.569	1.900
	12	1.739	1.839	1.967	2.291	2.683

In order to synthesise their dynamic characteristics, the values of their first period of vibration are reported in table 3.

3 PUSHOVER ANALYSES

The inelastic behaviour of all the above-mentioned structures has been firstly tested by means of pushover analyses. Vertical loads are directly applied to the columns, assuming that a proper framing of the deck or an additional beam are used to avoid to apply vertical loads to the link. Horizontal forces are linearly variable along the height, i.e. proportional to the design actions.

Both shear and flexural yielding have been checked at the ending cross-sections of the links. According to the experimental results previously referred to, the shear deformation of the link is described by a bilinear elastic-plastic model. The relationship between V_u and V_y is given by Equation (4) and the ultimate rotation is equal to 0.09 radiant; the plastic modulus of elasticity is supposed to be approximately 1/150 of the elastic one. The flexural yielding is schematised by means of plastic hinges; their ultimate flexural rotation is assumed equal to 0.06 radiant because of the presence of stiffeners, which reduce the effects of local buckling.

The possibility of flexural yielding of the columns has not been expressly taken into account in the analysis. Anyway a final check has confirmed that the columns would quite always remain in the elastic range up to the collapse of the structure, consistently with the capacity design criterion used in design.

The intensity of the lateral forces has been increased up to the failure of the frame, conventionally defined as the achievement of the above listed values of the ultimate shear or flexural rotation. The second order effects of the vertical loads have been not taken into account at this stage of the research.

Figures 3, 4 and 5 show the base shear versus top displacement relationship for EBF, TBF and TRBF respectively. In each figure the curves corresponding to different values of e/L are clearly distinct, because the strength and the stiffness of the schemes decrease as the length of the link increases.

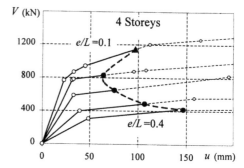

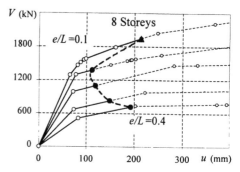

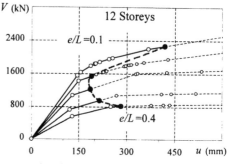

▲ Shear failure

● Flexural failure

○ Shear or flexural yielding

Figure 3. Pushover analyses of eccentrically braced frames

The presence of ties or trusses gives a moderate increase to the stiffness and in some cases also to the strength of scheme.

In all the EBFs (Fig. 3) the links do not yield in correspondence of a unique value of the multiplier of the lateral forces, because of the overstrength of some links respect to the design actions.

In the frames with long links (e/L=0.30-0.40), which experience flexural yielding, the increase of flexibility at first plastifications is so large as to bring the structure to collapse with only one or two links in the plastic range.

498

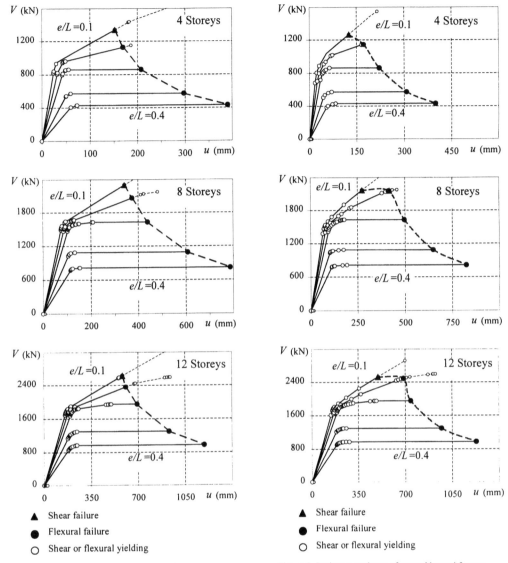

Figure 4. Pushover analyses of tied braced frames

Figure 5. Pushover analyses of trussed braced frames

This may be partially owed to the fact that flexural hardening is neglected in the model, because the stiffness given by it might in some cases modify the described behaviour. On the contrary, in the frames with short links (e/L=0.10) nearly all links yield for shear before failure. The ultimate horizontal displacement is influenced by the length of the links, being larger in systems with very short or long links. The number of storeys does not seem to substantially modify the above-mentioned observations.

Differently from standard eccentrically braced frames, the presence of ties (Fig. 4) or truss elements (Fig. 5) force links to present quite equal deforma-

tions at each level of the lateral load. Because of that in tied and trussed braced frames the first plastic hinge occurs at values of the lateral forces slightly higher than those of standard eccentrically braced frames. For the same reason, most plastifications of the links occur in correspondence of a narrow range of values of the horizontal forces. Also the collapse load is slightly higher than that of EBFs and it is achieved with a larger number of yielded sections.

Anyway, the most important differences may be noted in the values of horizontal displacements at collapse. In the case of schemes with short links (e/L=0.10) they are about 50% larger than the corresponding values of EBFs. The differences are even more relevant in the case of systems with longer links and seem to be emphasised in the buildings with a larger number of storeys. The P-Δ effect, neglected in the analyses, could slightly reduce the maximum displacements, but it would not change the overall behaviour of the schemes.

4 DYNAMIC ANALYSES

Each one of the systems described in section 2 has been subjected to a set of ten accelerograms, artificially generated by means of the procedure proposed by Falsone & Neri (1999). The mean value of their spectral pseudo-accelerations matches the response spectrum proposed by EC8 for soft soil (class C) and for a damping ratio equal to 5%. The accelerograms are enveloped by a trapezoidal intensity function and are characterized by a total duration of 35 s and by a stationary part of 22.5 s.

The shear and flexural yielding of the links has been modelled as described in the previous section; no strength or stiffness deterioration has been considered in the cyclic behaviour of the elements. The non-linear behaviour of the ending cross sections of the columns has been schematised by means of an elastic-perfectly plastic moment-curvature relationship, taking into account the influence of the axial force on the yielding value of the bending moment.

The inelastic response analyses, performed by means of the well-known DRAIN-2D code, have been repeated scaling each accelerogram to different values of the peak ground acceleration, so as to detect the intensity $a_{g,y}$ which causes within the structure the first plastic hinge and that one $a_{g,u}$ which provokes the structural failure. For each accelerogram, has thus been possible to evaluate the ratio

$$q = \frac{a_{g,u}}{a_{g,y}} \qquad (5)$$

The mean of the values computed for the ten accelerograms has been assumed as representative of the actual behaviour factor q of the scheme.

As already noticed for the pushover analyses, the structural failure generally occurs for achievement of the shear or flexural deformation capacity in systems with short and long links respectively.

The mean value of the behaviour factor in eccentrically braced systems is always lower than the values noticed in tied and trussed braced frames, both in presence of short and long links (Fig. 6). Furthermore, it is strongly conditioned by the number of storeys. In particular, it is quite comparable to the values suggested by Eurocode 8 in the case of low buildings (4 storeys), ranging from 4 to 5.5, but it rapidly decreases in the case of taller buildings, ranging from 2.4 to 3.6.

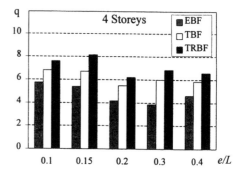

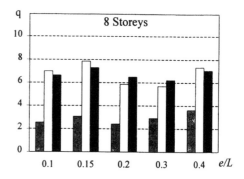

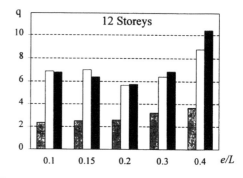

Figure 6. Behaviour factors of standard eccentrically braced frames, tied and trussed braced frames

On the contrary the values in tied and trussed frames are quite stable on varying the number of storeys, ranging from 5.5 to 8 in frames with four storeys and from 6 to 10 in systems with 12 storeys.

The difference between standard EBFs and tied or trussed braced frames and the influence of the number of storeys on it is confirmed by the amount of energy dissipated by the different schemes in the collapse configuration (Fig. 7). As expected, the dissipated energy is mostly owed to shear yielding in the case of short links and to flexural yielding in the case of long links; the amount of damping energy is always negligible. In the case of low buildings (4 storeys), the hysteretic energy dissipated by standard

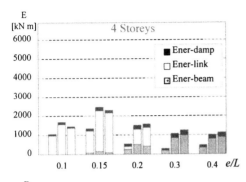

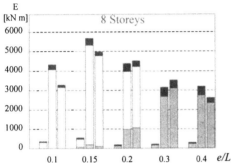

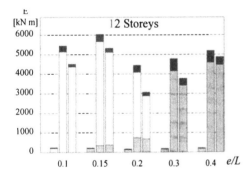

Figure 7. Dissipated energy of standard eccentrically braced frames, tied and trussed braced frames

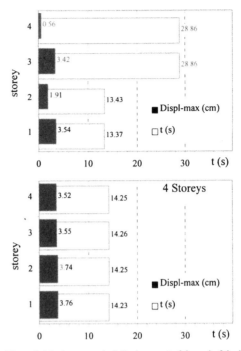

Figure 8. Maximum vertical displacement of the end of the link and time at which it is reached, for a standard EBF (a) and a tied braced frame (b) having short links and 4 storeys.

flexural plastic deformation) at the same time and show quite the same value at every storey (Fig. 8 b, 9 b). On the contrary, in EBFs the links present different values of the maximum plastic deformation, reached at different times. Anyway, in frames having a low number of storeys most links can reach a large plastic deformation (Fig. 8 a), thus providing a good dissipation, while in higher frames very few links are significantly yielded (Fig. 9 a) and the inelastic behaviour is very poor.

Such a behaviour, which is not dependent on the length of the link, may be explained by the fact that when a link is yielded the flexibility at that storey increases dramatically. This has minor importance in pushover analyses, in which horizontal forces increase proportionally independently of the plastification of any cross-section, while it has enormous influence on the dynamic behaviour of the scheme (soft storey) where the inertial forces change at different times according to the stiffness of the structure.

5 CONCLUSIONS

The performed analyses have first of all pointed out that the dynamic response of standard eccentrically braced frames may be really worst than what

EBFs is always smaller than that of TBFs or TRBFs, but the difference is not extremely large. As the number of storeys increases, the first one decreases while the other increase; the difference thus becomes really conspicuous.

Such different behaviour of standard eccentrically braced frames and tied or trussed braced frames may be understood by analysing in detail the step-by-step response of the schemes. As already noticed in pushover analyses, the deformation of the links is quite equal at all the storeys in tied and trussed frames while it is appreciably different in standard EBFs. Consequently, in tied and trussed schemes the links reach the maximum shear deformation (or

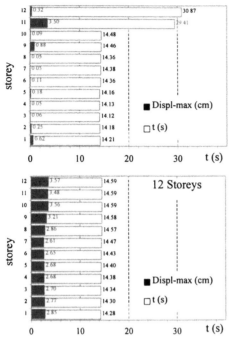

Figure 9. Maximum vertical displacement of the end of the link and time at which it is reached, for a standard EBF (a) and a tied braced frame (b) having short links and 12 storeys.

considered by many researchers and assumed by seismic codes. The large increase of flexibility at a storey when a link yields gives rise to a "soft-storey" behaviour, which modifies the dynamic response and does not allow the other links to develop their dissipative capacity. The suggestion of providing some overstrength to the links of lower storeys, given by AISC (1997), appears to be ineffective because in most cases this problem arises at intermediate or upper storeys. The location of possible soft storeys cannot be easily foreseen, because it depends on the distribution of strengths and masses and on the random characteristics of the seismic input.

On the contrary, the analyses have confirmed the good inelastic behaviour of eccentrically braced frames with vertical connections between links (tied braced frames). Independently of the number of storeys, the presence of ties forces all the links to the same deformation, sweeping away the risk of soft storey behaviour.

Finally, the analyses have shown that eccentrically braced frames with diagonal connections between links and nodes (trussed braced frames) have a dynamic inelastic behaviour quite similar to that of tied braced frames. This new typology thus appears to be a possible alternative for designing structures able to sustain strong seismic events.

REFERENCES

American Institute of Steel Construction – AISC. 1997. *Seismic Provisions for Structural Steel Buildings*. Chicago, IL.

Falsone, G. & Neri, F. 1999. Stochastic modelling of earthquake excitation following the EC8: power spectrum and filtering equations, *European Earthquake Engineering*, 3

Foutch, D.A. 1989. Seismic behavior of eccentrically braced steel building. *Journal of Structural Engineering*, ASCE, vol. 115, no. 8: pp. 1857-1876.

Hjelmstad, K.D. & Popov, E.P. 1983.Cyclic behavior and design of link beams. *Journal of Structural Engineering*, ASCE, vol. 109, no. 10: pp. 2387-2403.

Hjelmstad, K.D. & Popov, E.P. 1984. Characteristics of eccentrically braced frames. *Journal of Structural Engineering*, ASCE, vol. 110, no. 2: pp. 340-353.

Kasai, K. & Popov, E.P. 1986a. General behavior of WF steel shear link beams. *Journal of Structural Engineering*, ASCE, vol. 112, no. 2, pp. 362-382.

Kasai, K. & Popov, E.P. 1986b. Cyclic web buckling control of shear link beams. *Journal of Structural Engineering*, ASCE, vol. 112, no. 3.

Lu, L.W., Ricles, J.M. & Kasai, K. 1997. Global performance: general report. *Behaviour of Steel Structures in Seismic Areas*: pp. 361-381.

Malley, J.O. & Popov, E.P. 1984. Shear links in eccentrically braced frames. *Journal of Structural Engineering*, ASCE, vol. 110, no. 9: pp. 2275-2295.

National Earthquake Hazard Reduction Program – NEHRP. 1984. Recommended provisions for the development of seismic regulations for new buildings. *Bldg. Seismic Safety Council*, Wash., D.C.

Perretti, A. 1999. Comportamento sismico di telai in acciaio con controventi eccentrici (in Italian), Doctorate thesis, University of Naples.

Popov, E.P., Engelhardt, M.D. & Ricles, J.M. 1989. Eccentrically brace frames: U.S. practice. *Engineering Journal*, AISC, vol. 26, no. 2, pp. 66-80.

Ricles, J.M. & Bolin, S.M. 1991. Seismic performance of eccentricity braced frames. *Str. Sys. Research Project Rpt. 91-09*. Univ. of California, San Diego, CA.

Ricles, J.M. & Popov, E.P. 1989. Composite action in eccentrically braced steel frames. *Journal of Structural Engineering*, ASCE, vol. 115, no. 8.

Roeder, C.V. & Popov, E.P. 1978. Eccentrically braced frames for earthquakes. *Journal of Structural Engineering*, ASCE, vol. 104, no. 3.

Behaviour of Steel Structures in Seismic Areas, Mazzolani & Tremblay (eds) © 2000 Balkema, Rotterdam, ISBN 90 5809 130 9

Large scale tests of steel frames with semi-rigid connections under quasi-static cyclic loading

M. Iványi & G. Varga
Department of Steel Structures, Budapest University of Technology and Economics, Hungary

ABSTRACT: The paper presents a test series consisting of six large scale frames under quasi-static cyclic loading. The purpose of the tests is to examine the frame behaviour under such loads and to look at the behaviour of the connections as part of a frame. The frames are two-storey one-bay structures built up from welded I sections, which are connected by flush end plate type joints. The detailing of the joints is such that it ensures sufficient rotation capacity for a plastic hinge to occur within the joint; and the applied lateral restraint system prevents the premature occurrence of global buckling phenomena. Loading consists of vertical loads applied in a proportionally increasing manner, and a set of horizontal loads applied in load cycles at levels of vertical loads. The paper provides a detailed description of the testing methods and an overview of the test results. The presented results include behaviour curves representative to the behaviour of the frame as a whole. The presentation of the results is supplemented with specific observations related to the behaviour of connections.

1 INTRODUCTION

Semi-continuous framing, i.e. framing in which partial strength and semi-rigid joints according to the Eurocode 3 definitions (CEN 1992) are applied, is gaining importance in building structures due to the economy offered and the possibilities for modelling and calculation provided by a new set of design codes (Davies 1996). The connection models presented in these design standards and in the technical literature in general are based on the enormous number of tests, experimental and numerical, which were performed in the last decade or two (see e.g. Mazzolani 1992), most of them examining joints isolated from the whole structure in which they are to be incorporated. However, it is not evident how these joints behave when they become part of a real frame; it is recognised that this problem needs experimental investigations performed on full scale frames and realistic solutions for connection behaviour.

Experiments on full-scale or nearly full-scale specimens applying semi-rigid connections are rather rare, possibly because of the instrumentation demand of these tests. However, the results of some efforts are already available in the literature (El-nashai et al. 1998), although with little attention to the behaviour of the connections themselves.

The first author directed a series of tests in 1993/94 consisting of four, two-storey and two-bay frames (Iványi 2000), in which partial strength and semi-rigid joints were applied for column bases as well as beam-to-column connections. However, these joints were not designed to accommodate plastic hinges, therefore the failure of the frames in all instances was due either to loss of member stability or to connection failure in a brittle manner.

The new test series presented below consists of frames prepared with semi-rigid beam-to-column joints where special attention was devoted to the detailing of these joints in order to ensure a rotation capacity sufficient for a plastic mechanism to occur within the frame. Furthermore, the tests concentrated mostly on the behaviour of these frames under quasi-static cyclic loading with different types of load histories. This way two objectives were set at the start of the tests: (i) to examine the behaviour of the frame itself under the circumstances covered, and (ii) to look at the behaviour of the partial strength and semi-rigid connections within these frames.

The purpose of this paper is to present the testing methods and the testing apparatus including the specimen itself and the measuring systems, and to present some results of generic type, focusing on the behaviour of the frame as a whole and on the measured performance of joints. Specific observations related to the behaviour of joints are also presented.

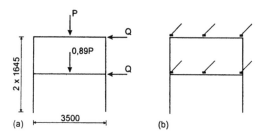

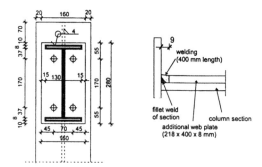

Figure 1. General layout and lateral restraints of the structure

Figure 2. The layout of the connections

Table 1. Beam and column sections applied in the tests

Frame	Beam flange/web	Column flange/web
1,2	130-8/260-8	200-12/260-8
3,4	160-10/260-8	200-12/260-8
5,6	130-8/260-8	160-10/260-8

2 TESTING METHODS

2.1 *Structure*

An overall view of the testing arrangement is shown in Figure 1. The frames examined are two-storey single-bay ones. Both the columns and the beams are welded I sections, see Table 1. Pairs of test frames are identical. Columns are connected to a rigid steel base element by two bolts through an end plate (layout generally regarded as pinned joint in the practice).

Beams and columns are connected with flush end plate joints (Fig. 2). In frames 1 to 5, the connections are strengthened with single-sided additional web plates. These were found to be necessary on the basis of an analysis of joint behaviour according to Revised Annex J of Eurocode 3 (CEN, 1992). According to the predictions, the joints were to fail in

their tension zone, by Mode 1 bending of the end plate, a failure mode generally recognised as one ensuring sufficient rotation capacity to accommodate a plastic hinge within the connection. These predictions also showed that omitting the additional web plates would have caused column web buckling to become the governing failure mode, which is regarded to be ineffective in terms of plastic hinge behaviour. Despite these predictions, joints in frame 6, in which these plates were indeed omitted, showed similarly ductile behaviour as those in frames 1 to 5.

In addition, these joints showed a type of asymmetric behaviour which, according to the best knowledge of the authors, is not treated in the technical literature, and which is not covered by design code regulations. This behaviour, characterised by excessive yielding of the column flange at one side and similarly excessive yielding of the end plate at the other, is due to the asymmetry of the joint because of the single-sided additional web plate, and

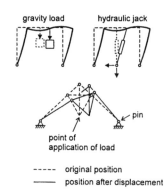

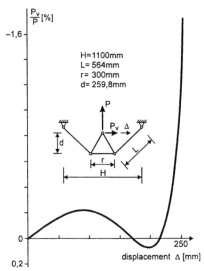

Figure 3. The gravity load simulator

504

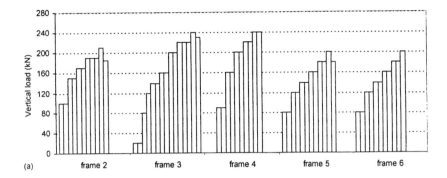

Figure 4. Loading programmes for frames 2 to 6. One column in the diagram represents one full cycle of horizontal loads

suggests that these web plates play a more sophisticated role in the joint behaviour than as implied in current European regulations. This issue is discussed in more detail in section 3.2 below.

In order to avoid lateral-torsional buckling, lateral restraints are applied to the frame at the beam-to-column joint locations and at the mid-spans of the beams, see Figure 1.

2.2 Loading

The frames are loaded by two vertical concentrated loads at the mid-spans of the beams, and two horizontal loads applied at one side of the frames in the levels of the beams (Fig. 1).

The two vertical loads are increased and decreased proportionally using three hydraulic jacks (one larger to the lower beam and two smaller and identical to the upper) connected into one oil circuit. Because of the slight difference between the pressure surfaces of the larger jack on one hand and the two smaller jacks on the other, the lower beam was loaded by a concentrated load 89% in magnitude of the load on the upper beam. The vertical loads are applied through so-called gravity load simulators (Halász & Iványi 1979), devices which ensure the verticality of the loads within certain limits of lateral displacements of the points of application of the loads (Fig. 3).

The horizontal loads are applied using one hydraulic jack through a simply supported vertical beam, which ensures the applied load to be equally distributed between the two beam levels. The direction of these horizontal loads is reversible.

The loading histories under which test frames 1 to 6 were examined are shown in Table 2 and Figure 4. Frame 1 was loaded by proportionally increasing loads; frames 2 to 6 were loaded by increasing vertical loading and cyclic horizontal loading at the levels of the vertical loads. In the case of frames 2, 3, 5 and 6, the amplitude of the horizontal loads was set as a ratio of the vertical loads, while in the case of frame 4, this amplitude was set to a constant value. In the case of all frames except frame 4, the ratio of the vertical loads to horizontal loads was set to 1:6 (value of the total horizontal load as compared to the vertical load acting on the upper beam); for frame 4, the amplitude of the horizontal load cycles was set to a constant value of 36 kN (value of the total horizontal load).

2.3 Measurements

During the tests beam mid-span deflections and lateral storey displacements were measured by using inductive transducers; forces, by pressure transduc-

Table 2. Loading programmes

Test frame	Application of horizontal load*	Ultimate vertical load kN
1	monotonic	193,4
2	cyclic / fixed ratio	208
3	cyclic / fixed ratio	240
4	cyclic / fixed value	240
5	cyclic / fixed ratio	196
6	cyclic / fixed ratio	200

*Vertical loads are always monotonic

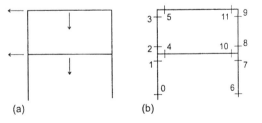

Figure 5. The locations of measurements: (a) displacements; (b) rotations and strains

505

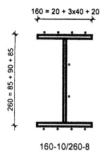

160 = 20 + 3x40 + 20

260 = 85 + 90 + 85

160-10/260-8

Figure 6. The measurement of strains in cross-sections

ers built into the oil circuit of the hydraulic system; strains, by strain gauges; and cross-section rotations, by purpose-designed devices (Fig. 5).

Strains were measured at the end cross-sections of beams and columns, by using ten gauges for each cross-section. The arrangement applied has been proposed by Gibbons et al. (1991), and makes possible to derive the four cross-sectional internal forces (the axial force, the two bending moments and the bimoment) even in the case of plasticity occurring within a part of the cross-section (Fig. 6). In order to properly interpret the measurements of these strain gauges, knowledge about the residual stresses present in the cross-section is needed. This issue is not discussed in more detail in this paper. One has also to note that we did not expect – and, effectively, did not measure – excessive plasticity in these cross-sections because of the fact that the beams and columns were connected by partial strength connections.

We applied a special purpose-designed device referred to as rotation indicator for the measurement of the absolute rotations of cross-sections (Fig. 7). This

device consists of a spring steel connected rigidly on its top to the cross-section being examined. While the cross-section and the top of the spring element

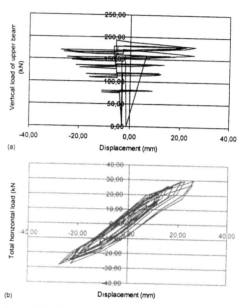

Figure 8. Load vs. top lateral displacement curves for frame 5 in terms of vertical and horizontal loads

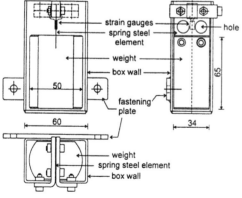

Figure 7. The rotation indicator

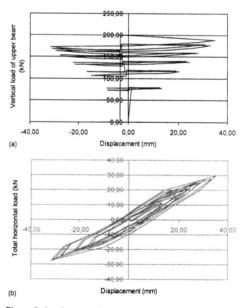

Figure 9. Load vs. top lateral displacement curves for frame 6 in terms of vertical and horizontal loads

506

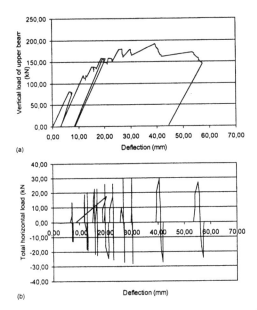

(a)

(b)

Figure 10. Load vs. top beam deflection curves for frame 5 in terms of vertical and horizontal loads

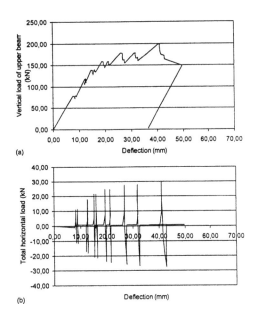

(a)

(b)

Figure 11. Load vs. top beam deflection curves for frame 6 in terms of vertical and horizontal loads

rotates, a weight connected to the bottom of this spring element tries to remain in place, which causes bending within the spring steel element. The deformations depend on the rotation to be measured, and are assessed by two strain gauges on the two surfaces of the spring steel element. The device is suitable for the measurement of rotations within the range of ±10 degrees. The device was calibrated for this range and was found to behave linearly; its coefficient, i.e. the ratio between the rotation and the difference of strains as measured by the two strain gauges, was in the range of 7.8 to 9.0×10^{-3} degrees per microstrains (the results come from the calibration of the fifteen rotation indicators constructed).

Material properties of the applied steel were determined on the basis of tension tests.

3 RESULTS AND DISCUSSION

3.1 Frame behaviour

Figures 8 to 15 present representative behaviour curves for frames 5 and 6. As described in section 2 above, these two frames are essentially identical and they are loaded by the same loading programme; the only difference is that the column webs in frame 5, unlike those in frame 6, are strengthened with single-sided additional web plates at the joint regions. The behaviour curves clearly show the resulting dif-

ference in overall frame stiffness, and they also show that the strength of the frames is not affected significantly. It is to be noted that the calculated joint initial stiffness according to Eurocode 3 Revised Annex J (CEN 1992) is 8087 kNm in the case of the beam-to-column joints of frame 5, and 5826 kNm in the case of joints of frame 6.

Joint strengthening was found necessary on the basis of an analysis of joint behaviour according to Eurocode 3 Revised Annex J (CEN 1992) in order to shift the failing zone from the compressed column web elsewhere within the joint where there is adequate ductility available, which in turn is necessary from the point of view of the rotation capacity of the joint. This analysis showed that these additional plates are needed for each of the frames 1 to 6. However, these plates were omitted in frame 6 to see what happens, and indeed what happened was that nothing happened, at least nothing that indicated that joint rotation capacity did drop. In fact, joints plastified in the same fashion as in frame 5, and the change in the overall behaviour curves is solely due to the decrease of the joint stiffness (and the fact that in one case ultimate failure was reached while increasing the horizontal load, and in the other, while increasing the vertical load).

3.2 Joint behaviour

Although the underlying phenomena of the observed behaviour of joints is not yet fully understood, one

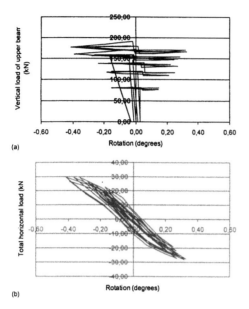

(a)

(b)

Figure 12. Load vs. absolute rotation curves in cross-section "0" (Fig. 5) for frame 5 in terms of vertical and horizontal loads

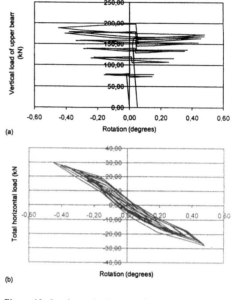

(a)

(b)

Figure 13. Load vs. absolute rotation curves in cross-section "0" (Fig. 5) for frame 6 in terms of vertical and horizontal loads

specific observation concerning the behaviour of end plate connections with one-sided supplementary web plates needs attention. The observation is illustrated in Figure 16, where the same joint is shown from its two sides, and it can be seen that at the loading stages near the ultimate failure of the joint, the two sides appear to behave fairly differently: in one (the strengthened) side it is the end plate, in the other (the non-strengthened), it is the column flange that ultimately fails. This leads to the conclusion that the approach presented in Eurocode 3 (CEN 1992) may not be capable to describe the behaviour of such joints, since that approach takes into account these supplementary web plates exclusively at the determination of the resistance of the column web.

It needs further study to decide whether this effect is worth taking into account, i.e. whether it significantly affects the overall joint response (including available rotation capacity). However, as it has been demonstrated elsewhere (Iványi & Varga 1999), it appears possible that staying within the logic of the component model as presented in Eurocode 3 one cannot fully explain the phenomenon.

4 CONCLUSIONS

This paper presents the testing and measurement methods and apparatus applied to testing of six frames of similar dimensions under alternating

quasi-static horizontal loading and monotonic vertical loading, and discusses some of the measurements and observations. In terms of results, the paper concentrates on two of the specimens loaded with the same loading programme, with the only difference being between them is that the beam-to-column joints are different, specifically, in one case, supplementary web plates are used, and in the other, they are omitted.

The presented results are preliminary and allow only rough conclusions to draw. The presented results clearly show, however, the effect of the presence of these supplementary web plates on the overall frame behaviour; while affecting significantly the overall stiffness through stiffening the beam-to-column joints themselves, they have little effect on the strength of the frames. In terms of joint behaviour and local effects within the joints, it is observed that a phenomenon, leading to the asymmetric behaviour of the joint, needs further study so that it could be described more appropriately than as offered by the model presently available in Eurocode 3.

Further work will concentrate on both examining more thoroughly some of the phenomena not discussed in this paper, and on analyzing the behaviour of the joints under the cyclic loading as provided by the experiments. Specifically, analytical work will be carried out to look at the behaviour of the joints under non-regular cyclic moment loading.

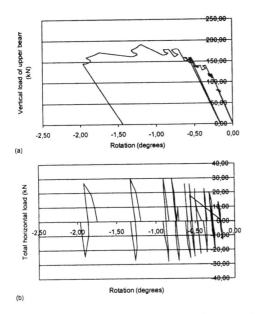

(a)

(b)

Figure 14. Load vs. absolute rotation curves in cross-section "11" (Fig. 5) for frame 5 in terms of vertical and horizontal loads

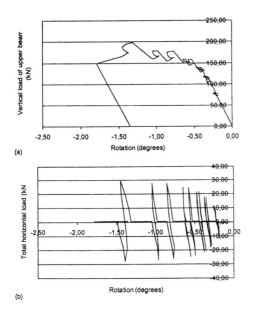

(a)

(b)

Figure 15. Load vs. absolute rotation curves in cross-section "11" (Fig. 5) for frame 6 in terms of vertical and horizontal loads

5 ACKNOWLEDGEMENT

This work is supported by the National Scientific Research Fund of the Republic of Hungary (OTKA)

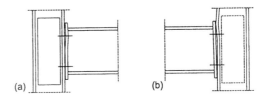

Figure 16: Illustration of the effect of the presence of the supplementary web plate on the joint behaviour near failure load

under grant No. T020358. The test arrangement and the device for the measurement of rotations was designed by Mr. László Kaltenbach, chief engineer at the Structural Testing Laboratory of the Budapest University of Technology and Economics. The measurement system was installed and operated by Dr. Miklós Kálló. The valuable contribution of both these persons, as well as other staff members of the Structural Testing Laboratory and the Department of Steel Structures, is gratefully acknowledged.

REFERENCES

CEN 1992. *ENV 1993-1-1:1992 Eurocode 3: Design of steel structures. General rules. General rules and rules for buildings.* As amended by *ENV 1993-1-1:1992/A2:1998 Amendment A2.* European Prestandard. Brussels: European Committee for Standardisation

Davies, J.M. 1996. The stability of steel frames: General report. *Stability of steel structures, 1995, Budapest*, ed. M. Iványi: Vol. 1, 377-392, Budapest: Akadémiai Kiadó

Elnashai A.S., A.Y. Elghazouli & F.A. Demsh-Ashtiani 1998. Response of semirigid steel frames to cyclic and earthquake loads. *J. Struct. Eng.* 124(8): 857-867. American Soc. of Civ. Eng.

Gibbons, C., P.A. Kirby & D.A. Nethercot 1991. The experimental assessment of the force components within structural steel 'H' sections. *Journal of Strain Analysis* 26(1): 31-37

Halász, O. & M. Iványi 1979. Tests with simple elastic-plastic frames. *Periodica Polytechnica Civil Engineering* 23(3-4): 151-182. Technical University of Budapest

Iványi, M. 2000. Full-scale tests of steel frames with semi-rigid connections. *Engineering Structures* 22: 168-179. Elsevier

Iványi, M. & G. Varga 1999. Full scale tests of steel frames under quasi-static cyclic loading. *Proc. Eurosteel '99*, Vol. 1:251-254 and CD-Rom supplement. Prague: Czech Technical University

Mazzolani, F.M. 1992. Stability of steel frames with semi-rigid joints. *Stability problems of steel structures: CISM courses and lectures No. 323*, ed. M. Iványi and M. Skaloud: 357-415, New York: Springer-Verlag

Behaviour of Steel Structures in Seismic Areas, Mazzolani & Tremblay (eds) © 2000 Balkema, Rotterdam, ISBN 90 5809 130 9

Influence of P-Δ effect on a proposed procedure for seismic design of steel frames

F. Neri
Dipartimento di Ingegneria Civil e Ambientale, Università di Catania, Italy

ABSTRACT: This paper presents a method to design moment resisting steel frames, so as to fail in global mode, which takes into account also geometrical second order effect. After having defined the beam cross-sections, dimensioned to resist only to vertical loads or also to some horizontal actions, the cross-sections of columns are the unknowns of the problem. The procedure proposed in this paper is an improvement of that described in previous papers, because it allows to define the strength of the columns by means of a direct approach also when the geometrical second order influence is considered, avoiding the necessity of an iterative procedure. A worked example is finally presented in order to show the practical application of the proposed method.

1 INTRODUCTION

The structural design classical approach in seismic zone should be based on the control of the triad: strength, stiffness and ductility. Nowadays the seismic codes impose the check of the strength and the stiffness, while, in force of the practical operative difficulty, they give only few indication about the ductility. Being the structures in seismic zone designed for internal action quite smaller than those the structure could experience during strong earthquake, the ductility at local and global level is a fundamental problem. At local level the elements are dimensioned in way that they can undergo large deformation in plastic range without substantial reduction of strength, so as to be able to dissipate high quantity of energy. At global level the structure should be designed in way to develop plastic hinges in a large number of sections before a mechanism is formed or the rotation capacity in any section is reached.

For the moment resisting frame (MRF) the last goal is reached when the collapse mechanism present plastic hinges at all the beams end and at the base of the first order columns so as to reach a mechanism called global.

The most modern seismic codes propose some criterions with the aim of avoiding partial collapse mechanisms; other methods for the same purpose are given by researchers.

The procedure proposed by the modern codes (Eccs 1988, Eurocode 8 1996, Gndt 1984) is based on the "capacity design criterion". According to this

method the designer should first define the suitable mechanism that the structure must reach and then design each cross-section which has to remain elastic using design actions greater than the resistant capacity of the plastic sections to which it is connected.

This criterion, based on the node equilibrium, could be effective if the joint connects only two elements or if the plastic hinges would develop all together. This last condition is never fulfilled in real structures, mainly for two reasons: the first is that the internal actions differ from the values used in design, also if the modal analysis is performed, and vary during the seismic event in a non proportional way; the second is that also if the actions increase proportionally a simultaneous plastification is not possible in consequence of the unavoidable overstrength of the sections. For the aforementioned reasons the code prescription quite never grants the global collapse mechanism, so as is shown in several works (Landolfo et al. 1988, Lee 1996).

Starting from these considerations and basing on the observation of the increase of bending moment in a frame subjected to fixed vertical load and increasing seismic action, Lee proposed a modification of the capacity design (Lee 1996). His method is based on the fact that when the collapse is incipient the ratios of the moment of the lower column to that of the upper column is about three.

Mazzolani and Piluso (Mazzolani & Piluso 1997) have proposed a different approach based on the kinematic theorem of plastic collapse. To grant that the frame collapses according to a global mechanism the authors impose, by means of an iterative procedure, that the collapse multiplier associated to this mechanism be the smallest one among the kinematic multipliers of all possible mechanism.

In section 2 a design procedure able to grant a global collapse mechanism, presented in previous paper, is shortly recalled; an improvement of the method, which allows to take into account P-Δ effects in a direct way, is described in detail in section 3; an example of its application is finally reported in section 4.

2 BASIC CONCEPTS

The method proposed in other works (Neri 1999), in which the author impose that the collapse mechanism is of global type, is based on the application of the limit analysis.

In a MRF subjected to a seismic horizontal forces the collapse mechanisms can be considered to belong to three main typologies (Fig. 1), and the global one is a particular case of these (Mazzolani & Piluso 1997).

For each collapse mechanism the collapse multiplier α_c can be evaluated equating the internal work owed to the plastic hinges to the external work produced by horizontal forces.

$$\mathbf{M_p}^T dr = \alpha_c \mathbf{F}^T ds + \Delta \mathbf{F}^T ds \qquad (1)$$

where
$\mathbf{M_p}$ is the vector of the plastic moments of the cross-sections where hinges shall arise according to the assigned mechanism
dr is the vector of the virtual rotations of the plastic hinges for the assigned mechanism
$\mathbf{F}$ is the vector of the horizontal seismic actions
ds is the vector of the horizontal virtual displacements for the considered mechanism
$\Delta \mathbf{F}$ is the vector of the deviation forces produced by the horizontal displacement of the application point of the vertical load. The k-th component relative to the k-th storey can be obtained as follow (Fig. 2):

$$\Delta F_k = \Delta V_{k+1} - \Delta V_k \qquad (2)$$

where ΔV_k is the shear at the k-th interstorey, produced by the second order effects:

$$\Delta V_k = \sum_{i=1}^{n_c} P_{ik} \frac{(s_k - s_{k-1})}{h_k} \qquad (3)$$

in which
P_{ik} is the axial force in the i-th column of the k-th storey
h_k is the k-th interstorey height

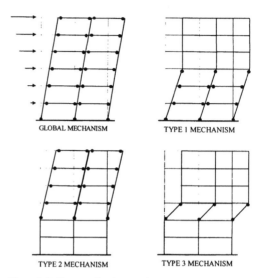

GLOBAL MECHANISM   TYPE I MECHANISM

TYPE 2 MECHANISM   TYPE 3 MECHANISM

Figure 1. Collapse mechanism typologies of MR frames under seismic horizontal forces

s_k is the horizontal virtual displacement at the k-th storey

2.1 The global mechanism

For a global type mechanism the aforementioned terms can be explicated, utilising the following notation.

n_c number of columns
n_b number of beams
n_s number of storeys
i column index
j bay index
k storey index
$M_{c,ik}$ plastic moment, reduced for the interaction with the axial force, in the i-th column at the k-th storey
$M_{b,jk}$ plastic moment in the j-th beam at the k-th storey

The plastic moment vector, with $[n_c + 2 n_s (n_c-1)] \times 1$ dimension, is composed by the reduced plastic moment of the bottom section of first order columns and the plastic moment of all the beams:

$$\mathbf{M_p}^T = \left\{ M_{c,11}^B, ..., M_{c,n_c 1}^B, M_{b,11}^L, ..., M_{b,jk}^L, M_{b,jk}^R, ..., M_{b,n_b n_s}^R \right\} \qquad (4)$$

where the superscript B, L and R refer to the bottom section of the columns and to the left and right ends of the beams.

The vector of virtual rotation, having the same dimension of $\mathbf{M_p}$, has all terms equal to the rotation at the base of the frame dr:

$$dr^T = \{1,1, ... 1,1\} dr = \mathbf{I} \, dr \qquad (5)$$

in which $\mathbf{I}$ is the unit vector.

512

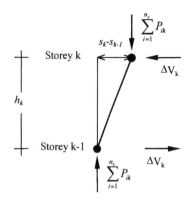

Figure 2. Definition of deviation forces

The vector of the virtual horizontal storey displacements, of $n_s \times 1$ dimensions, can be too related to the rotation at the base of the frame dr:

$$d\mathbf{s}^T = \left\{h_1, h_2, ... h_k ...., h_{n_s}\right\} dr = \mathbf{h}^T dr \qquad (6)$$

where $\mathbf{h}$ is the storey height vector.

The vector of horizontal seismic forces, with $n_s \times 1$ dimension, is composed by the horizontal forces F_k applied to the different storeys:

$$\mathbf{F}^T = \left\{F_1, F_2, ..., F_k, ..., F_{n_s}\right\} \qquad (7)$$

Finally, the k-th component of the vector of deviation force, with $n_s \times 1$ dimension, defined in the Equations (2-3), for the global mechanism assumes the following form:

$$\Delta F_k = N_k \, \vartheta \qquad (8)$$

where N_k is the total vertical load at the k-th storey; then the aforementioned vector becomes:

$$\Delta \mathbf{F}^T = \left\{N_1, N_2, ..., N_k, ..., N_{n_s}\right\} \vartheta = \mathbf{N}^T \vartheta \qquad (9)$$

The collapse multiplier can be obtained, in vectorial form, by the Equation 1 utilising the (2)-(9):

$$\alpha = \frac{\mathbf{M_p}^T \mathbf{I}}{\mathbf{F}^T \mathbf{h}} + \frac{\mathbf{N}^T \mathbf{h}}{\mathbf{F}^T \mathbf{h}} \vartheta \qquad (10)$$

Therefore the collapse multiplier can be defined as follows:

$$\alpha_c = \frac{\sum_{i=1}^{n_c} M_{c,i1} + \sum_{k=1}^{n_s}\left(\sum_{i=1}^{n_b} 2 M_{b,ik}\right)}{\sum_{k=1}^{n_s} F_k \, h_k} - \frac{\sum_{k=1}^{n_s} \Delta F_k \, h_k}{\sum_{k=1}^{n_s} F_k \, h_k} \qquad (11)$$

For an assigned structure the knowledge of the collapse multiplier permits to evaluate the sum of bending moments $M_{c,ik}$ in any horizontal section by

imposing rotation equilibrium of the upper part of the structure respect the considered section. In particular for the r-th storey

$$\sum_{i=1}^{n_c} M_{c,ir} = \alpha_c \sum_{k=r}^{n_s} F_k \, (h_k - h) - \sum_{k=r}^{n_s}\left(\sum_{j=1}^{n_b} 2 M_{b,jk}\right) + \\ + \sum_{k=r}^{n_s} \Delta F_k \, (h_k - h) \qquad (12)$$

where $h=h_r$ for the top section while $h=h_{r-1}$ for the bottom section.

For an assigned structure and for a prefixed value ϑ of the plastic rotation the Equations 11 and 12 give the collapse multiplier and the sum of bending moments acting in all horizontal sections of the frame.

Equilibrium condition does not allow to evaluate the bending moment at each column (Marino et al. 1999). Push-over analysis shows that when all columns at a storey have the same cross-section the bending moments in the proximity of the collapse are quite similar for the different columns; larger values are obtained in the internal columns but the differences are smaller then 20%. In order to design the columns it seems acceptable to assume for each column the average of the results of the limit analysis:

$$M_{sd} = \frac{1}{n} \sum_{i=1}^{n_c} M_{c,ir} \qquad (13)$$

The axial force in each column is equal to the sum of the shears transmitted by the beams framing into the column, which are easily evaluated because in the global collapse mechanism mode beams are subjected only to the vertical loads and the plastic bending moment at the ends. Then the shear transmitted by the beams in the collapse configuration is:

$$V = \frac{q \, l_j}{2} + \frac{2 \, M_{b,j}}{l_j} \qquad (14)$$

where q is the vertical load in seismic condition and l_j is the span of the j-th bay

The knowledge of the axial forces and the bending moments acting in the top and bottom section of the columns at different levels permits to define the cross-section of these elements.

In the next section is shown how the previous equations can be utilised for defining a design procedure for MR steel frame falling in global mode.

2.2 Design of structures with the proposed method

The application of the aforementioned equations to define the flexural moment at each cross-section of the columns require the knowledge of all the terms present in Equation 11. The only terms known in the

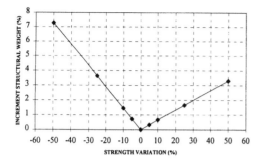

Figure 3. Columns weight varying the bending moment at bottom of first order columns respect the optimum value

design phase are the storeys height h_k and the total vertical load at the k-th storey; if the allowable plastic rotation ϑ is defined, also the deviation forces ΔF are known. Nevertheless the distribution of horizontal seismic forces can be defined under some hypotheses, while the plastic moments of the beams and the reduced plastic moments of the first order columns can be estimated by means of economic and technological considerations.

If the structure is regular, the distribution of the horizontal seismic actions can be assumed proportional to that of the fundamental mode, or more simply it can be considered triangular.

The cost of the structure depends on the beam and column cross-sections. In order to obtain a global mechanism, the design moment of the columns is function of the resisting moment of the beams. For minimising the cost of the structure the beams must be dimensioned with the strength strictly required by vertical loads in non-seismic condition. Lower level of strength would be not acceptable because inelastic deformations would arise just in occurrence of vertical loads, while higher level of strength in the beams would require greater strength in the columns and major cost of the structure.

After having defined the horizontal forces and the resisting moments of the beams, the last term of the Equation (11) to be defined is the strength at the base of the first order columns. This strength plays an important role in the design of steel frames which would comply with structural, economic and technologic requirements. The structural and technologic requirements impose that the columns decrease along the height of the structure, while the economic one imposes that the columns weight and then its cost is minimised.

For an assigned vertical load on the beams, which implyes an axial load in the columns, the weight of the columns is related to the plastic modulus and then to the bending moment acting on them. The value of this parameter is minimum when the absolute value of the global resisting bending moment at the base of the frame equals the maximum absolute value at the upper storeys. The structural weight in-

creases almost linearly when the strength differs from this optimum value, mainly when it decreases (Fig. 3).

The bending moment at the base of the frame which equals the maximum absolute value at the upper storeys can be evaluated by means of an iterative approach. Starting with a tentative value of the resisting reduced bending moment at the base of the columns, e.g. the sum of resisting bending moment at all ending cross-sections of the beams at the first storey, the bending moment acting at the ending cross sections of the columns can be evaluated by the Equations 11 and 12. As long as this condition is not verified, the absolute value of the global resisting bending moment at the base of the frame should be replaced with the greatest of the global resisting bending moments at the upper storeys. The previous approach can be applied with or without considering the P-Δ effects (Ghersi et al. 1999b).

If the geometrical second order effects are neglected the iterative procedure can be avoided, because the optimum value of the global strength of the base section is related to the total strength of first order beams and to the number of storeys (Neri 1999, Ghersi et al. 1999a).

In the following section is proposed an approach that allows to avoid the iterative procedure also when the geometrical second order effects are considered in the design of the steel frame.

3 IMPROVEMENT OF THE METHOD

In an frame collapsing according to a mechanism of global type the equilibrium rotation respect to the base can be written:

$$\sum_{i=1}^{n_c} M_{c,i1} + \sum_{k=1}^{n_s} \left(\sum_{j=1}^{n_h} 2 M_{h,jk} \right) - \alpha_c \sum_{k=1}^{n_s} F_k \cdot h_k +$$
$$- \sum_{k=1}^{n_s} \Delta F_k \cdot h_k = 0 \tag{15}$$

In an analogous way the equilibrium rotation respect to the top cross-section of the columns at the r-th storey is:

$$\sum_{i=1}^{n_c} M_{c,ir}^T + \sum_{k=r}^{n_s} \left(\sum_{j=1}^{n_h} 2 M_{h,jk} \right) +$$
$$- \alpha_c \sum_{k=r+1}^{n_s} F_k \cdot (h_k - h_r) - \sum_{k=r+1}^{n_s} \Delta F_k \cdot (h_k - h_r) = 0 \tag{16}$$

For the frame can be written a set of these equations equal to the number of storey.

The optimisation of structural weight imposes that the absolute value of the global resisting reduced bending moment at the base of the frame equals the maximum absolute value at the upper storey.

Being the sign of bending moment at the base opposite to that of the moment at the top of the other storeys the previous condition can be written for each storey as follows:

$$\sum_{i=1}^{n_c} M_{c,i1}^{B} = -\sum_{i=1}^{n_c} M_{c,ir}^{T} \qquad r = 1.....n_s \qquad (17)$$

By each one of these equations a collapse multiplier can be evaluated, obtaining a set of n_s different values:

$$\alpha_r = \frac{\sum_{k=1}^{n_s}\left(\sum_{j=1}^{n_h} 2 M_{h,jk}\right) + \sum_{k=r}^{n_s}\left(\sum_{j=1}^{n_h} 2 M_{h,jk}\right)}{\sum_{k=1}^{n_s} F_k \cdot h_k + \sum_{k=r+1}^{n_s} F_k \cdot (h_k - h_r)} +$$
$$- \frac{\sum_{k=r}^{n_s} \Delta F_k \cdot h_k + \sum_{k=r+1}^{n_s} \Delta F_k \cdot (h_k - h_r)}{\sum_{k=1}^{n_s} F_k \cdot h_k + \sum_{k=r+1}^{n_s} F_k \cdot (h_k - h_r)} \qquad (18)$$

It can be demonstrated that among the n_s multipliers the one that grants the optimisation of structural weight is that which assume the maximum value:

$$\alpha_c = \max\left\{\alpha_1...., \alpha_r,....\alpha_{n_s}\right\} \qquad (19)$$

The previous method enables to remove the iterative approach also when the second order effect is considered in the design.

With further conditions (interstorey height constant for the different storeys, i.e. $h_k - h_{k-1} = h$; beams with the same cross-sections, i.e. $M_{b,jk}$=const; total vertical load N_k constant at different storey) the Equation 15 can be rewritten as follows:

$$\sum_{i=1}^{n_c} M_{c,i1}^{B} = -n_s \sum_{j=1}^{n_h} 2 M_{h,j} + \alpha_c h \sum_{k=1}^{n_s} F_k \cdot k +$$
$$+ \Delta F h \sum_{k=1}^{n_s} k \qquad (20)$$

while Equation 16 becomes:

$$\sum_{i=1}^{n_c} M_{c,ir}^{T} = -(n_s - r + 1)\sum_{j=1}^{n_h} 2 M_{h,j} +$$
$$+ \alpha_c h \sum_{k=r+1}^{n_s} F_k \cdot (k - r) + \Delta F h \sum_{k=r+1}^{n_s} (k - r) \qquad (21)$$

Under these hypotheses Equation 18 can be rewritten as follows:

$$\alpha_r = \frac{(2 n_s - r + 1)\sum_{j=1}^{n_h} 2 M_{h,j} - \left[\sum_{k=1}^{n_s} k + \sum_{k=r+1}^{n_s}(k - r)\right]\Delta F h}{\left[\sum_{k=1}^{n_s} F_k\, k + \sum_{k=r+1}^{n_s} F_k\,(k - r)\right] h} \qquad (22)$$

The previous hypotheses imply that the horizontal seismic force at the k-th storey can be expressed in the following form:

$$F_k = \frac{k}{\sum_{k=1}^{n_s} k} \qquad (23)$$

thus obtaining

$$\alpha_r = \frac{(2 n_s - r + 1)\sum_{k=1}^{n_s} k}{\sum_{k=1}^{n_s} k^2 + \sum_{k=r}^{n_s} k^2 - r\sum_{k=r}^{n_s} k} \cdot \frac{\sum_{j=1}^{n_h} 2 M_{h,j}}{h} +$$
$$- \frac{\left[\sum_{k=1}^{n_s} k + \sum_{k=r}^{n_s}(k - r)\right]\sum_{k=1}^{n_s} k}{\sum_{k=1}^{n_s} k^2 + \sum_{k=r}^{n_s} k^2 - r\sum_{k=r}^{n_s} k}\Delta F \qquad (24)$$

This equation can be written:

$$\alpha_r = W \frac{\sum_{j=1}^{n_h} 2 M_{h,j}}{h} - R\,\Delta F \qquad (25)$$

where:

$$W = \frac{(2 n_s - r + 1)\sum_{k=1}^{n_s} k}{\sum_{k=1}^{n_s} k^2 + \sum_{k=r}^{n_s} k^2 - r\sum_{k=r}^{n_s} k} \qquad (26)$$

and:

$$R = \frac{\left[\sum_{k=1}^{n_s} k + \sum_{k=r}^{n_s}(k - r)\right]\sum_{k=1}^{n_s} k}{\sum_{k=1}^{n_s} k^2 + \sum_{k=r}^{n_s} k^2 - r\sum_{k=r}^{n_s} k} \qquad (27)$$

A large number of numerical tests has shown that for a frame with an assigned number of storey the value of r, that grants that the value of the bending moment at the top of columns of the r-th storey is maximum, is correlated to the number of storey n_s. In Table 1 are reported the value of r which satisfies the economic condition and the corresponding value of W and R, for an number of storey up to ten.

515

Table 1. Values of collapse multiplier for different number of storey

n_s	r	α_c	
		W	R
10	7	1.7460	7.6077
9	6	1.7462	6.8507
8	6	1.7444	6.1850
7	5	1.7500	5.4250
6	4	1.7500	4.6666
5	4	1.7500	4.0000
4	3	1.7647	3.2352
3	2	1.7647	2.4705
2	2	1.8000	1.8000

$\dfrac{\sum_{j=1}^{n_b} 2\,M_{bj}}{h}\ \Delta F$

Table 2. Values of the sum of reduced bending moment of the columns at the base of the frame for different number of storey

n_s	r	$\sum_{i=1}^{n_s} M_{c,i1}^{B}$	
		$\overline{W}$	$\overline{R}$
10	7	2.2222	1.7460
9	6	2.0597	1.6119
8	6	1.8855	0.9515
7	5	1.7500	0.8750
6	4	1.5833	0.7778
5	4	1.4167	0.3333
4	3	1.2941	0.2941
3	2	1.1177	0.2353
2	2	1.0000	0.0000

$\sum_{j=1}^{n_b} 2\,M_{bj}\ \ \Delta F\,h$

By equating the 20 and the 25 the maximum bending moment to be considered for the design of the columns for the first r storeys can be evaluated:

$$\sum_{i=1}^{n_c} M_{c,i1}^{B} = \left(W \sum_{k=1}^{n_s} F_k\,k - n_s \right) \sum_{j=1}^{n_b} 2\,M_{b,j} +$$

$$+ \left(-R \sum_{k=1}^{n_s} F_k\,k + \sum_{k=1}^{n_s} k \right) \Delta F\,h \tag{28}$$

Using Equation 23 it can be rewritten:

$$\sum_{i=1}^{n_c} M_{c,i1}^{B} = \left\{ \frac{\left[\sum_{k=1}^{n_s} k + \sum_{k=r}^{n_s}(k-r) \right] \sum_{k=1}^{n_s} k^2}{\sum_{k=1}^{n_s} k^2 + \sum_{k=r+1}^{n_s} k^2 - r \sum_{k=r+1}^{n_s} k} + \sum_{k=1}^{n_s} k \right\} \Delta F\,h$$

$$+ \left[-n_s + \frac{(2n_s - r + 1)\sum_{k=1}^{n_s} k^2}{\sum_{k=1}^{n_s} k^2 + \sum_{k=r+1}^{n_s} k^2 - r \sum_{k=r+1}^{n_s} k} \right] \sum_{j=1}^{n_b} 2\,M_{b,j} \tag{29}$$

or in the following form

$$\sum_{i=1}^{n_c} M_{c,i1}^{B} = \overline{W} \sum_{j=1}^{n_b} 2\,M_{b,j} + \overline{R}\,\Delta F\,h \tag{30}$$

where

$$\overline{W} = -n_s + \frac{(2n_s - r + 1)\sum_{k=1}^{n_s} k^2}{\sum_{k=1}^{n_s} k^2 + \sum_{k=r+1}^{n_s} k^2 - r \sum_{k=r+1}^{n_s} k} \tag{31}$$

and

$$\overline{R} = \frac{\left[\sum_{k=1}^{n_s} k + \sum_{k=r}^{n_s}(k-r) \right] \sum_{k=1}^{n_s} k^2}{\sum_{k=1}^{n_s} k^2 + \sum_{k=r+1}^{n_s} k^2 - r \sum_{k=r+1}^{n_s} k} + \sum_{k=1}^{n_s} k \tag{32}$$

In Table 2 are reported the value of r which satisfy the economic condition and the corresponding values of $\overline{W}$ and $\overline{R}$, for a number of storey up to ten. The proposed design procedure may be resumed in the following six steps.

1. definition of the beams strength, as the value necessary to resist to vertical loads or according to other prescriptions

2. evaluation of the sum of plastic bending moments at the end of all the beams of a storey and of the product of deviation forces for the interstorey height $\Delta F\,h$

3. evaluation of the collapse multiplier α_c by means of Equation 23 with r value reported in Table 1

4. evaluation of the design reduced plastic moments at each storey by means of Equations 12 and 13

5. evaluation, by means of Equation 14, of the axial forces in the columns considering the contribution of all the beams framing in the columns in the upper storeys

6. definition of the cross-section of the columns according to the reduced bending moments and the axial forces and taking into account the technological considerations which impose that the column sections can only decrease along the height.

4 EXAMPLE

The previously proposed method is here applied for the design of a set of steel frames. The results are compared to those reported in other papers (Ghersi et al. 1999b).

The first frame to be designed is shown in Figure 4. The bay span is equal to 6.50 m; the interstorey height is equal to 3.20 m. The characteristic values of distributed vertical loads acting on the beams are equal to 15 and 10 kN/m for permanent (G_k) and live (Q_k) actions respectively.

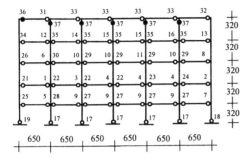

Figure 4. Geometry of analysed frame, yielding order and collapse mechanism

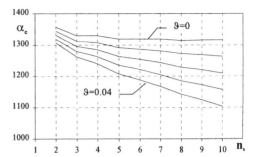

Figure 5. Collapse multiplier, varying the number of storeys and the plastic capacity rotation ϑ

The beams are designed with the maximum bending moments given by two limit conditions: resistance for vertical loads only, evaluated as $(\gamma_g\, g_k + \gamma_q\, q_k)\, l^2/10$ (M_{sd}=148.9 kNm) and value able to grant in seismic condition the formation of the plastic hinges at the end of the beams; this last condition corresponds to assure that the maximum positive moment is achieved at the end of the beam: $(g_k + \psi\, q_k)\, l^2/4$ (M_{sd}=190.1 kNm). The selected beam is a IPE 330 in Fe430 steel grade with a flexural strength $M_{b,rd}$=201 kN m. In this example a plastic rotation in the collapse configuration equal to 0.04 rad. at the base of the frame has been considered.

By the Table 1 we can see that for a five storey frame the storey r were the maximum bending moment is reached is the fourth.

With the Equation 25 or the Table 1, for a sum of bending moments at k-th storey equal to 2412 kN m and a deviation force equal to 28.8 kN, the collapse multiplier can be easily obtained (α_c=1206.74). Analogously the sum of bending moments, reduced for the interaction with the axial force, to be used for the design of the first r storeys columns can be evaluated by means of the Equation 30 or the Table 2 (3446.9 kN m).

In Table 3 the seismic horizontal forces in the collapse configuration are reported. In the same table are shown the bending moment and the axial force in the top and bottom sections of internal and external columns. The values utilised in the column design are report in bold. The table finally reports the cross-sections chosen for the columns.

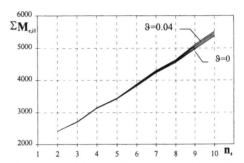

Figure 6. Sum of plastic bending moments at the base of the first order columns, varying the number of storeys and the plastic capacity rotation ϑ

The comparison of these values with those obtained for the same frame by means of the iterative approach, reported in a previous work (Ghersi et al. 1999b), shows that the same results have been obtained. In fact both the collapse multiplier and the reduced bending moment of the columns (3446.9/7= 492.4 kN m) are the same.

In Figure 4 are reported the plastic hinge sequence and the collapse mechanism that shows how the global one is full reached.

In the same paper (Ghersi et al. 1999b) a comparative analysis with the same frame designed according with other method (EC8, Lee, Mazzolani-Piluso) has been carried out. This comparison have shown that the weight of the structure given by the proposed procedure is quite larger than that obtained following the EC8 provisions, but the increase is slightly smaller than that suggested by other researchers (Mazzolani-Piluso). The inelastic behaviour of frames designed according to the proposed procedure is significantly better than that shown by frames designed following EC8 provisions and comparable to that of frames obtained by the more complex and costly procedure suggested by Mazzolani and Piluso.

For schemes having the same total vertical load, number of columns, beams section and interstorey

Table 3. Collapse horizontal force, bending and axial force in the columns and designed sections

Storey	F_{crit} kN	M_t kN m	M_p kN m	N_{ext} kN	N_{int} kN	Sections
5	401.9	**344.5**	147.8	**120.3**	117.0	HEB 280
4	321.7	**492.4**	135.7	**240.7**	234.0	
3	241.3	**480.3**	0.4	**361.0**	351.0	HEB 320
2	160.8	**345.0**	-221.1	**481.4**	468.0	
1	80.42	123.4	**-492.4**	**601.7**	585.0	

height than the one previously analysed, the collapse multiplier and the sum of design bending moments at the base of the first order columns are reported, (Figg. 5, 6), for different number of storeys and plastic rotation in the collapse configuration.

The Figure 5 shows that the collapse multiplier of the horizontal seismic force decreases with the increase of the storeys number. The slope of reduction of this parameter depends considerably on the plastic rotation value. The sum of reduced bending moments increases quite linearly with the number of storeys of the frame and it is quite independent on the available deformation in the collapse configuration (Fig. 6).

5 CONCLUSIONS

In the present paper a method for designing steel frames such in way to collapse in global mode has been presented. The method here proposed takes into account the geometrical second order effects that not always can be neglected in the design of moment resisting steel frame. The original proposal of this paper is that the design can be performed with a direct approach, avoiding any iteration.

The method applied to a five storeys-seven bays frame has shown the fulfilment of the design aim that consists in achieving the collapse according to a global mechanism.

The author thanks Marco Pistollato who co-operated to the method development within the preparation of his thesis

REFERENCES

ECCS 1988. European Recommendations for Steel Structures in Seismic Zones. European Convention for Constructional Steelwork, Brussels.

EUROCODE 8 1996. Design Provisions for Earthquake Resistance of Structures. CEN, European Committee of standardisation, Brussels.

GNDT 1984. Norme tecniche per le costruzioni in zone sismiche. Gruppo Nazionale per la Difesa dai Terremoti, (Italy).

Ghersi A., Neri F. & Rossi P.P. 1999a. A global approach to the design of steel structures, *Proceedings of the 6th International Colloquium on Stability & Ductility of Steel Structures, Timisoara (Romania)*.

Ghersi A., Neri F. & Rossi P.P. 1999b. A seismic design method for high ductility steel frames taking into account P-Δ effect, *Proceeding of XVII C.T.A. conference, 3-5 October, Napoli*.

Ghersi A., Marino E. & Neri F. 1999. A simple procedure to design steel frames to fail in global mode, *Proceedings of the 6th International Colloquium on Stability & Ductility of Steel Structures, Timisoara (Romania)*.

Landolfo R., Mazzolani F.M. & Pernetti M. 1988. L'influenza del criterio di dimensionamento sul comportamento sismico dei telai in acciaio, *Atti del IV Convegno Nazionale L'ingegneria sismica in Italia, Milano (Italia)*.

Lee H.S. 1996. Revised Rule for Concept of Strong-Column Weak-Girder Design. *Journal of Structural Engineering 122, 359-364*.

Marino E., Neri F. & Rossi P.P. 1999. A design procedure for steel frames with rigid connections, *Proceedings of the Conference Eurosteel '99, Maggio, Praga* .

Mazzolani F.M. & Piluso V. 1997. Plastic Design of Seismic Resistant Steel Frames. *Earthquake Engineering and Structural Dynamics 26, 167-191*.

Neri F. 1999. Comportamento sismico di telai in acciaio a nodi rigidi. Doctorate Thesis, Università di Catania.

Behaviour of Steel Structures in Seismic Areas, Mazzolani & Tremblay (eds) © 2000 Balkema, Rotterdam, ISBN 90 5809 130 9

Steel building response under biaxial seismic excitations

R. Ramasco & G. Magliulo
University of Naples 'Federico II', Italy

G. Faella
University of Perugia, Italy

ABSTRACT: In the paper the effects of the bi-directional input ground motion on a six story steel moment resisting frame building, designed according to EC3 and EC8 provisions, are evaluated. Some response parameters, obtained considering only the "primary" horizontal component of a real earthquake acting on a principal direction of the building, are compared to the ones obtained when the "secondary" horizontal component, acting on the orthogonal direction, is added. Nonlinear analyses are performed using five real earthquakes. The action of the secondary component does not increase the maximum top floor displacement and the interstory drifts in the direction of the primary component, while a low increment is obtained considering their vectorial values. On the contrary the column cross-section damage is much increased.

1 INTRODUCTION

In the last years the interest in the seismic response of the buildings subjected to bi-directional input ground motion is increased. It is proved by the high number of research studies concerning elastic and inelastic, static and dynamic bi-directional analyses and experimental tests: this is the subject of many papers presented at the recent 12th World Conference on Earthquake Engineering. The common aim of all these studies is to understand if uni-directional analyses and design methods can be used to simplify a reality which is three-dimensional both in terms of input ground motion and in terms of dynamic behavior and damage of the structures.

The most research studies on this subject are parametric studies on very simple structures. On the contrary, the authors of the paper believe that to gain reliable conclusions precisely modeled three-dimensional structures must be analyzed. The main reasons of this opinion are three: 1) recent studies (Faella et al. 1999, Faella et al. 2000) on reinforced concrete frame buildings show that local damage analyses are necessary for determining bi-directional input ground motion effects and these analyses are only possible if the 3D structural behavior is well reproduced; 2) nowadays experimental results to calibrate numerical models for reproducing the inelastic dynamic response of three-dimensional steel and reinforced concrete buildings are available (Taucer et al. 1998, Molina et al. 1999); 3) computer programs for nonlinear dynamic analysis of three-dimensional buildings, e.g. (Li 1996a, b) and (Pow-

ell & Campbell 1994, Prakash et al. 1993), are available too.

In the paper the seismic response of a steel building subjected to both the horizontal components of the earthquake is analyzed. The vertical component of the earthquake is not taken into account, both because it can be often considered negligible and to isolate the effects of the two horizontal components of the earthquake.

As regards the methodology, nonlinear dynamic analyses are performed, instead of the elastic dynamic ones carried out by other authors (Fernandez-Davila et al. 2000, Cruz & Cominetti 2000, López et al. 2000, Menun & Der Kiureghian 1998), because the aim of the paper, as already mentioned, is to analyse the bi-directional input ground motion effects in terms of global and local damage of the structure.

2 GEOMETRY AND DESIGN OF THE BUILDING

A six story steel frame building is analysed in the paper. It is doubly symmetric in plan (Fig. 1); four moment resisting frames are parallel to the XZ plane of the global reference system (Fig. 2), while six moment resisting frames are parallel to the YZ plane (Fig. 3). Secondary beams are used to decrease the span between the Y direction beams (Fig. 1).

The building is designed (Urbano et al. 1997) according to Eurocode 3 (CEN 1992) and Eurocode 8 (CEN 1994a, b, c) provisions. The design was carried out assuming PGA equal to 0.35g, soil type B, behavior factor equal to 6, φ coefficient equal to 0.8 for intermediate stories and Ψ_{2i} coefficient equal to 0.3. A steel grade Fe 360 was adopted.

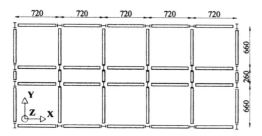

Figure 1. Plan of the analyzed building.

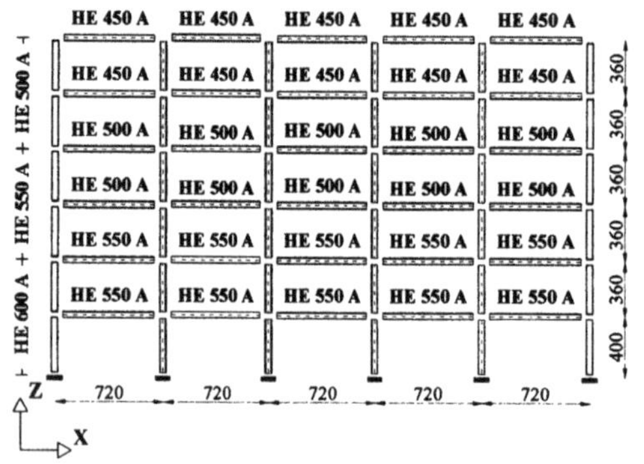

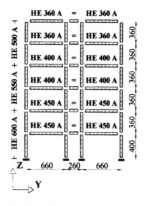

Figure 2. XZ plane oriented moment resisting frames.

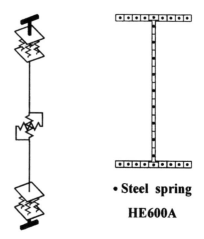

Figure 3. YZ plane oriented moment resisting frames.

3 NUMERICAL MODELING

The building response is analyzed by means of the computer program CANNY-E.

The beams are idealized as linear plane elements - with an elastic flexural and shear behavior - which have two nonlinear bending springs at the ends. The moment rotation relationship assigned to these springs is characterized by a bilinear skeleton curve, elastic until the section plastic moment and with an hardening equal to 3%. The stiffness of the unloading curve is assumed equal to the initial stiffness.

The columns are idealized by a linear elastic element with flexural and shear elastic behavior in both the cross-section principal directions and elastic axial deformations. At the ends of each column two multi-spring elements are placed, which are able to model the flexural and axial inelastic behavior (Fig. 4) of the end cross-sections taking into account biaxial bending-axial force interaction.

Each flange of all the column end cross-sections is divided into 9 springs, the HE550A and HE600A webs into 9 springs (Fig. 4), the HE500A web into 7 springs. A force-displacement relationship is assigned to each spring, assuming a plastic hinge length equal to the greatest of the two column section dimensions (Mazzolani & Piluso 1996) and multiplying the steel stress by the area of each spring and its strain by the plastic hinge length. A bilinear stress-strain relationship is assumed; it is elastic until the yielding stress and is characterized by a 3% hardening. An unloading stiffness equal to the initial one is assigned.

A 5% damping ratio and a damping matrix proportional to the instantaneous stiffness matrix are considered. Rigid floor slabs are assumed.

Figure 4. Column modeling.

Table 1. Earthquake records used as input ground motion.

Earthquake	Date	Station	Duration [sec]	Primary component PGA [g]	Secondary component PGA [g]
Imperial Valley	18.05.40	El Centro	53.40	0.348	0.214
Kern County	21.07.52	Taft	54.40	0.179	0.156
Montenegro	15.04.79	Petrovac	19.60	0.438	0.305
Valparaiso	03.03.85	El Almendral	72.02	0.284	0.159
Northridge	17.01.94	Newhall	59.98	0.590	0.583

4 PERFORMED ANALYSES

The nonlinear dynamic analyses are performed using five real earthquakes (Tab. 1); for each of them, as "primary" is called the component which have the largest peak acceleration value. These earthquakes are chosen because the Eurocode 8 elastic spectrum (PGA equal to 0.4g and soil type B) is well fitted by the average of the elastic spectra of their primary components (Fig. 5) and because they represent earthquakes having different intensity and different primary/secondary component intensity ratio.

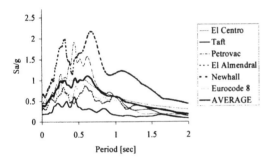

Figure 5. Primary components and Eurocode 8 elastic spectra.

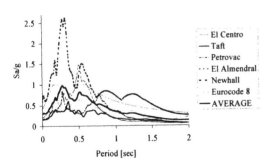

Figure 6. Secondary components and Eurocode 8 elastic spectra.

The effects of the bi-directional input ground motion are evaluated by a comparison with the results obtained under uni-directional excitation. The primary component of each earthquake is always assumed acting in the Y direction of the global reference system (Fig. 1); for the bi-directional analyses, the secondary component acting in the orthogonal direction is added. The comparison is carried out both in terms of building global seismic response and at the columns cross-section level. All the "directional" results, i.e. base shear, top story and frame displacements and interstory drifts, are evaluated in the direction of the primary component and calculating their maximum vectorial values.

5 RESPONSE COMPARISON UNDER UNI- AND BI-DIRECTIONAL EXCITATION

5.1 Comparison in terms of building global seismic response

As dynamic response global parameters, the base shear, the displacements of the top floor center of mass, the interstory drifts, the top floor rotations and the top displacements of the lateral Y-direction frames are evaluated. In the following figures "Uni-dir" means that the only primary component of each earthquake is applied in the global reference Y direction and, obviously, Y direction results are shown. "Bi-dir Y" means that Y direction results are shown when also the secondary component of each earthquake acts in the X direction. The latter analyses give the "Bi-dir vec" results too, which are computed by:

$$\max \sqrt{\left(E_x(t_i)\right)^2 + \left(E_y(t_i)\right)^2} \qquad (1)$$

where $E_x(t_i)$ and $E_y(t_i)$ are the X and Y direction considered effects at the i-th instant of the time history.

Under bi-directional excitation the Y direction maximum base shear (Fig. 7) is generally lower then

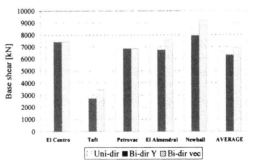

Figure 7. Maximum base shear.

521

the one obtained under uni-directional excitation, even though the decrement never exceeds 10%: for the five considered records, the average value of the reduction is equal to 4%. On the contrary, the "Bi-dir vec" base shears are generally larger than the "Uni-dir" ones, even though the increment, except under Taft record (27%), is lower than 10%: on average it is equal to 5%. With reference to this, it is to be noted that the building under Taft record remains in the elastic range of behavior, as it is also confirmed by the Y direction results, which are always equal under uni- and bi-directional excitation.

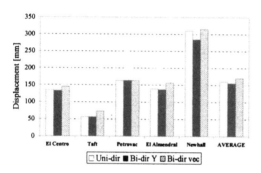

Figure 8. Maximum top center of mass displacement.

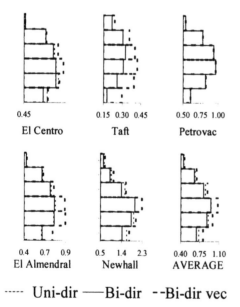

Figure 9. Maximum interstory drifts as interstory height percentage.

Figure 8 confirms that under bi-directional excitation also the Y direction maximum displacement of the top floor center of mass is generally lower than the one obtained under uni-directional excitation. The decrement never exceeds 10% and on average, for the five earthquakes, it is equal to 4%. Conversely, the "Bi-dir vec" values are larger than the "Uni-dir" ones, even though the difference, except in the case of Taft record (31%), does not exceed 13% and on average is equal to 6%.

Under both the components of El Centro, Petrovac, El Almendral and Newhall earthquakes, the maximum decrement in Y direction interstory drifts, plotted in Figure 9 as interstory height percentage, never exceeds 12% and, considering the average of all the earthquakes at each floor, does not exceed 6%. On the contrary, the maximum vectorial interstory drifts "Bi-dir vec" are generally larger than the "Uni-dir" ones, even though the increment is lower than 22% and, considering the average of the five earthquakes at each floor, does not exceed 10%.

Interesting remarks come out from Figures 7, 8 and 9: when the building behavior is not elastic, the presence of the secondary component of the earthquake generally decreases the maximum base shear, the maximum top floor center of mass displacement and the maximum interstory drifts in Y direction. The maximum vectorial values computed by formula (1) under bi-directional excitation are generally larger than the ones computed under uni-directional excitation, even though, when the building behavior is inelastic, the increment is low: it does not exceed 22%.

This is due to the accumulation - when only the primary component of the earthquake acts - of the damage and, above all, of the plastic displacement in Y direction. On the contrary, under bi-directional excitation, the direction of the input ground motion vectorial component (sum of the Y and X direction components) and of the building oscillations varies at every instant; consequently, the plastic displacement has a variable direction and it does not accumulate in Y one. Furthermore, in the latter case, the hysteretic damping forces, which decrease the displacement, are at each instant larger than in the former case. This effect seems to be greater than the opposite effect of the reduction of the flexural strength due to the interaction between the YZ and the XZ plane bending moments and the axial force, which in a doubly symmetric building, as the analyzed one, arises when the secondary component of the earthquake is applied.

The maximum XY plane rotations of the top floor of the building, due to the asymmetry generated in the inelastic range of behavior by the action of the secondary component of the earthquake are negligible. This is evident considering the maximum top floor rotation obtained under the action of both the components of Newhall record, the more violent of the five considered earthquakes. Its contribution to the top displacement of the Y direction external frames is about 3 mm, that is equal to 1% of the displacement of the top floor center of mass (Fig. 8).

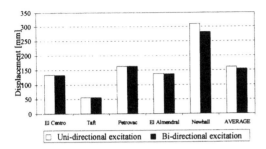

Figure 10. Y direction maximum top displacement of the left side frame.

This is confirmed in Figure 10 where the maximum top displacements of the Y direction left side frame are shown: the values obtained under bidirectional ground motion are not larger than the ones obtained under uni-directional ground motion. Consequently, it can be stated that the asymmetry in the Y direction generated, in the nonlinear range of behavior, by the variation of axial force due to the secondary component of the earthquake acting in X direction does not significantly affect the building torsional response. This conclusion was also drawn analyzing a symmetric reinforced concrete frame building (Faella et al. 2000).

In conclusion, by the analysis of the building global damage, it seems that the combination rules prescribed by Eurocode 8 between the effects due to the application of the seismic action along two horizontal and orthogonal directions, are conservative.

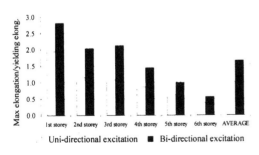

Figure 11. Top and bottom column cross-sections plastic demand.

5.2 *Comparison in terms of column damage*

As regards the damage analysis at the column cross-section level, four parameters are analyzed.

The first one is the ratio between the maximum and the yielding elongation of the springs belonging to the multi-spring elements, which is representative of the plastic demand. In Figure 11 this parameter is shown considering all the five earthquakes and all the column bottom and top cross-sections of each story; the last two histograms represent the average values considering all the stories. It is evident that under bi-directional excitation the plastic demand of the analyzed cross-sections is much larger than under uni-directional excitation; the increment is equal to 96% at the third floor and it is 64% considering the average of all the stories.

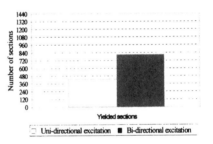

Figure 12. Number of yielded cross-sections.

In Figure 12 the number of yielded cross-sections, considering all the columns of the building and all the five earthquakes, are shown; the increment due to the action of the secondary component is equal to 88%.

These results clearly evidence that the secondary component of the earthquake much increases the damage at the column cross-section level. It is due to 1) the larger energy which has to be absorbed by the structure when two components instead of one strike it; 2) the increment of the range of variation of the axial force in the columns due to the secondary component; 3) the biaxial bending-axial force interaction at the column cross-section level.

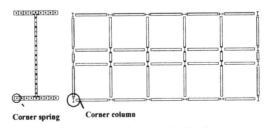

Figure 13. Corner column and corner spring location.

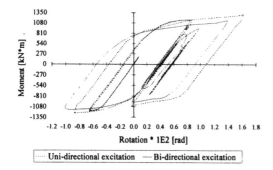

Figure 14. Corner column bottom cross-section YZ plane bending moment-rotation time history.

To analyze more in detail the local damage of the columns, one of them is chosen, i.e. a corner one of the first floor (Fig. 13) and the moment-rotation time history of its bottom cross-section is shown (Fig. 14); it is a HE600A section. The input ground motion is the Newhall one.

Table 2. Corner column maximum and minimum axial forces, bending moment and rotation maximum absolute values.

	Nmax [kN]	Nmin [kN]	Myzmax [kN*m]	Φyzmax [rad]
Uni-directional excitation	1423*	-552	1303	0.0162
Bi-directional excitation	1917*	-1000	1178	0.0139

* The positive values mean compression.

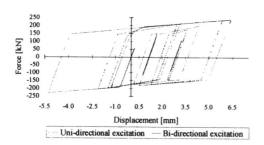

Figure 15. Corner spring force-displacement time history.

Figure 14 and Table 2 show that the maximum absolute values of the bending moment and rotation in YZ plane of the corner column bottom cross-section under bi-directional excitation are lower than those obtained under uni-directional excitation: the decrements are about equal to 10% and 15% respectively; on the contrary, the increment of the range of variation of the axial force due to the secondary component is about equal to 50%. It means that even

though the flexural strength of the section decreases because of the triaxial interaction, the maximum rotation in YZ plane is not increased. This fully confirms what already noted in terms of global parameters of the response of the building. This result is conditioned by the shape of the section, i.e. HE, which is not much influenced by the XZ bending moment, because its stiffness in XZ plane is lower; it also depends on the increment of the strength domain of the section when the section yields, because of the assigned hardening.

Table 3. Corner spring maximum (in compression) and minimum (in tension) forces and ductility demand.

	Fmax [kN]	Fmin [kN]	μmax	μmin
Uni-directional excitation	235	-203	8.46	-3.04
Bi-directional excitation	239	-230	9.21	-7.62

Figure 15 and Table 3 show the time history of the force-displacement relationship and the extreme values of the forces and ductility demand (maximum/yielding elongation ratio) of the corner spring (Fig. 13). Under bi-directional excitation the increment of the force and ductility demand in compression (positive values) is equal to about 2% and 9% respectively, the increment in tension is equal to 13% and 150% respectively. It is evident that the secondary component of the earthquake affects more the values in tension than those in compression, which are much conditioned by the gravity loads.

In conclusion, the column cross-section plastic demand and the axial force range of variation given in Table 2 show that the combination rules prescribed by Eurocode 8, for taking into account the contemporaneous action of both the horizontal component of the earthquake, could be not conservative; this is true also considering that the local damage, e.g. the joint local instability, is an important problem for the steel buildings.

6 CONCLUSIONS

Interesting results are obtained comparing the response of a steel building subjected to the primary component of real earthquakes to the response obtained when the secondary horizontal component of the same earthquakes is added.

The maximum base shear, the top displacement and the interstory drifts computed in the direction of the primary component are not increased by the contemporaneous action of the secondary component. Furthermore, the increment of their maximum vectorial values (square root of the Y and X direction square values sum) due to the action of the secondary horizontal component with respect to those obtained when the only primary component acts is low. This is due to the building swinging, under bi-directional excitation, in many directions; consequently, the damage and, in particular, the plastic displacement do not accumulate in one direction, as in the case of uni-directional excitation. This effect also covers the possible strength decrement due to the triaxial (bending moments and axial force) interaction.

On the contrary, the damage at the column end cross-sections, analyzed in terms of plastic demand, is much increased by the secondary horizontal component of the earthquake: in the examined case the increment reaches 96%. This is due to the increment of the axial force variation range, to the larger energy of the input ground motion and to the triaxial interation.

Considering the global response "directional" parameters of the examined case, it seems that the combination rules prescribed by Eurocode 8 for taking into account the bi-directional input ground motion effects are conservative. On the contrary, when the column cross-section damage is analyzed, it seems that they could be not.

REFERENCES

CEN 1992. Eurocode 3: Design of steel structures. Part 1.1 General rules and rules for buildings.
CEN 1994a. Eurocode 8. Design provisions for earthquake resistance of structures. Seismic Actions and General Requirements of Structures.
CEN 1994b. Eurocode 8. Design provisions for earthquake resistance of structures. General rules for buildings.
CEN 1994c. Eurocode 8. Design provisions for earthquake resistance of structures. Specific Rules for Various Materials and Elements.
Cruz, E.F. & Cominetti, S. 2000. Three-dimensional buildings subjected to bi-directional earthquakes. Validity of analysis considering uni-directional earthquakes. *Proc. 12th World Conf. on Earthquake Engineering, Auckland, 31 January – 4 February 2000.*
Faella, G., Magliulo, G., Ramasco, R. 1999. Seismic response of asymmetric r/c buildings under bi-directional input ground motion. In Karadogan & Rutemberg (eds), *Irregular structures* (vol. 2) – *Proc. of the 2nd European Workshop on the Seismic behavior of asymmetric and irregular structures, Istanbul 8-9 October 1999.*
Faella, G., Kilar, V. & Magliulo, G. 2000. Symmetric 3D r/c buildings subjected to bi-directional ground motion. *Proc. 12th World Conf. on Earthquake Engineering, Auckland, 31 January – 4 February 2000.*
Fernandez-Davila, I., Cominetti, S. & Cruz E.F. 2000. Considering the bi-directional effects and the seismic angle variations in building design. *Proc. 12th World Conf. on Earthquake Engineering, Auckland, 31 January – 4 February 2000.*
Li, K.N. 1996a. CANNY-E. Three-dimensional nonlinear dynamic structural analysis computer program package. Canny Consultants Pte Ltd. Technical manual.
Li, K.N. 1996b. CANNY-E. Three-dimensional nonlinear dynamic structural analysis computer program package. Canny Consultants Pte Ltd. Users' manual.
López, O.A., Chopra, A.K. & Hernández, J.J. 2000. The significance of the direction of ground motion on the structural response. *Proc. 12th World Conf. on Earthquake Engineering, Auckland, 31 January – 4 February 2000.*
Mazzolani, F.M. & Piluso, V. 1996. *Theory and design of seismic resistant steel frames.* London: E & FN Spon, an imprint of Chapman & Hall.
Menun, C.H. & Der Kiureghian, A.D. 1998. A replacement for the 30%, 40%, and SRSS rules for multicomponent seismic analysis. *Earthquake Spectra* 14:153-163.
Molina, F.J., Verzeletti, G., Magonette, G., Buchet, P. & Geradin, M. 1999. Bi-directional pseudodynamic test of a full-size three-storey building. *Earthq. Engg. Struct. Dynam.* 28:1541-1566.
Powell, G.H. & Campbell, S. 1994. DRAIN-3DX element description and user guide for element type 01, type 04, type 05, type 08, type 09, type 15, and type 17. Version 1.10. *Department of Civil Engineering, University of California, Berkeley, California.*
Prakash, V., Powell, G.H., Campbell, S. 1993. DRAIN-3DX base program user guide. Version 1.10.
Taucer, F., Negro, P. & Colombo, A. 1998. Cyclic testing of the steel frame. Final report.
Urbano, C., Mazzolani, F.M., De Luca, A., D'Amore, E. 1997. I nuovi DD.MM. '96 e gli Eurocodici. Strutture in acciaio. *Corso di aggiornamento in ingegneria sismica. Messina January-February 1997* (in Italian).

Behaviour of Steel Structures in Seismic Areas, Mazzolani & Tremblay (eds) © 2000 Balkema, Rotterdam, ISBN 90 5809 130 9

Influence of brace slenderness on the seismic response of concentrically braced steel frames

R.Tremblay
Epicenter Research Group, Department of Civil, Geological and Mining Engineering, Ecole Polytechnique, Montreal, Canada

ABSTRACT: This paper describes an analytical study performed to evaluate the influence of the brace slenderness ratio on the seismic performance of concentrically braced steel frames. Nonlinear time history dynamic analyses have been performed on typical building structures. The brace slenderness, λ, for the braces was varied from 0.35 to 2.65 for tension-compression brace design and from 0.85 to infinite for tension-only brace design. The building height was varied from 1 to 8 storeys. The results of the analyses indicate that the inelastic demand in the tension-compression bracing systems decreases as the brace slenderness is increased. For tension-only bracing, the inelastic demand generally increases when increasing the brace slenderness. For both bracing systems, it was also found that higher inelastic demand was generally associated with a concentration of inelastic damage over the height of the structures. The results suggest that a brace slenderness limit of λ = 2.65 is appropriate for tension-compression bracing systems. Tension-only design should be allowed, provided that the brace slenderness is kept below the same limit. For both systems, building height limitations are proposed to prevent the formation of a soft-storey mechanism.

1 INTRODUCTION

Seismic input energy in concentrically braced steel frames is essentially dissipated through inelastic deformation in the bracing members, i.e., stretching of the braces in tension and inelastic bending upon brace buckling and in subsequent brace straightening. Numerous experimental studies (e.g. Black et al. 1980, Jain & Goel 1978, Khan & Hanson 1976, Prathuangsit et al. 1978, Wakabayashi et al. 1977) have shown that the energy dissipation capacity of bracing members generally increases when reducing the brace slenderness. Ghanaat (1980) concluded from shake table tests on braced frame models with various brace slenderness ratios that braces with a slenderness ratio, KL/r, less than 100 could be very efficient in moderate-to-strong ground motions. Limited numerical analyses by Remennikov & Walpole (1997) indicated that reducing the brace slenderness ratio from 90 to 40 in chevron and X-braced frames resulted in smaller storey drifts.

Limits on the brace slenderness, which approximately correspond to the value proposed by Ghanaat, have then been introduced in several codes and guidelines to ensure minimum energy dissipation characteristics. For instance, in SEAOC (1990) and CSA (1989), the brace slenderness parameter, λ, as defined in Equation 1, must not exceed 1.35. In ECCS (1991), this limit is λ = 1.5.

$$\lambda = \frac{KL}{r} \sqrt{\frac{F_y}{\pi^2 E}} \tag{1}$$

In Japan (Tremblay et al. 1996), λ must be less than 1.41 but the higher energy dissipation provided by stockier braces is also recognized by allowing reduced design seismic loads when lower brace slenderness is maintained in the structure. For instance, the lateral loads specified for ductile moment resisting frames can be used when λ is less than 0.35. If λ exceeds 0.63, the loads are 1.6 times greater. The limit λ = 1.35 also applies in the New Zealand steel design code (SNZ 1997). In addition, the design seismic loads increase with both the building height and the brace slenderness. Moreover, building height limitations are imposed for braced frames and these limits are more stringent when slender braces are used. For example, for the

Category 1 (ductile) braced frames, the seismic loads are reduced by 1.6 and the maximum building height is raised from 2 to 8 storeys if λ of the braces is less than 0.35 instead of being greater that 0.90.

Experimental studies (e.g. Tang & Goel 1987) also revealed, however, that the low-cycle fatigue life of bracing members generally diminishes when brace slenderness is decreased. The limit on brace slenderness has been raised recently from 1.35 to 1.87 in the USA (AISC 1997) as an attempt to allow the use of braces with improved fracture life.

This paper presents the results of a parametric study performed to examine the influence of brace slenderness on the overall seismic response of concentrically braced steel frames. The objective was to determine brace slenderness limits as well as building height limitations for braced frames of the Ductile Braced Frame category described in the Canadian Standards. Nonlinear dynamic analyses of a typical building have been performed under seismic ground motions. The building height and the brace slenderness were varied. In addition, two brace design approaches were considered: tension-compression and tension-only.

2 BUILDING STUDIED

2.1 Description

The building studied by Robert & Tremblay (2000) was modified for this study. The structure is braced by two identical X-bracing bents in each direction (Fig. 1a) and the response in the N-S direction is examined herein. Five different building heights were considered: 1, 2, 4, 6, and 8 storeys. The storey height at every floor was 3.8 m.

The building was assumed to be in Vancouver, B.C., Canada, which is located in a moderate seismic zone. The design was made according to the NBCC 1995 (NRCC 1995). The following gravity loads were considered in the calculations: a roof dead load of 1.2 kPa for buildings up to 4 storeys and 3.4 kPa for the 6- and 8-storey buildings, a floor dead load of 4.5 kPa, a roof snow load of 1.48 kPa, and an occupancy floor live load of 2.4 kPa. The weight of the exterior cladding was taken equal to 1.2 kPa. A load factor of 0.5 was applied to the live loads when combined with seismic loads. The applicable NBCC reduction factor for occupancy floor live loads was considered in the calculations.

2.2 Seismic loads

The braced frames were designed for a seismic horizontal load $V_f = vSFIW(U/R)$, where v is the velocity ratio for the site, S is the seismic response factor, F is the foundation factor, I is the importance factor, W is the seismic weight of the structure, U is a calibration factor (U = 0.6), and R is the force modification factor.

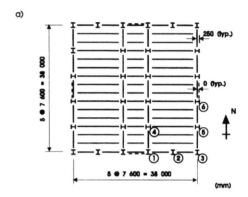

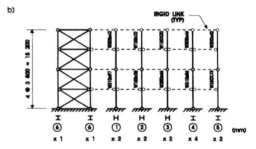

Figure 1. Building studied: a) Plan view; b) Analytical model.

The value of S varies with the fundamental period of the structure, T, and the seismic zone at the site. For Vancouver, v = 0.21 and S was taken as 3.0 for the 1- and 2-storey buildings and 2.11, 1.74, and 1.51 for the 4-, 6-, and 8-storey frames. The importance factor and the foundation factor were taken equal to 1.0 (structure of normal importance on stiff soil). The seismic weight, W, included the floor and roof dead loads, the weight of the exterior walls, and 25% of the roof snow load. W was then equal to 2680 kN, 10,050 kN, 24,790 kN, 42,800 kN and 57,540 kN for the 1-, 2-, 4-, 6-, and 8-storey buildings, respectively. The R factor for the Ductile Braced Frame category is equal to 3.0. For all structures, the fundamental period computed from free vibration analysis exceeded the NBCC value and the S factor could then be reduced to 80% of its initial value, as permitted in NBCC.

The base shear force was distributed over the height of the structures according to the NBCC static procedure. A concentrated lateral force equal to

5.2% and 6.9% of the base shear had to be applied at the top of the 6- and 8-storey buildings, respectively. In-plane torsion was neglected and each N-S frame was designed for half the total applied lateral loads.

2.3 Tension-Compression (T/C) Brace Design

In this design, the storey shear was assumed to be equally resisted by the tension- and compression-acting braces. Six different values of brace slenderness were considered: $\lambda = 0.35$, 0.85, 1.35, 1.60, 1.85, and 2.65. The first value corresponds to very stocky braces which have an hysteretic response approaching that of the bare steel material. The 1.35 and 1.85 values correspond to the current CSA limit and the newly introduced AISC limit, respectively. The $\lambda = 2.65$ slenderness corresponds approximately to the maximum slenderness ratio allowed for a compression member, KL/r = 200, for a steel with $F_y = 350$ MPa. The other values of λ were considered for completeness.

In design, all the braces in the building were assigned the brace slenderness under consideration, assuming that there would exist structural shapes having both the required axial resistance and brace slenderness. For a given slenderness, the brace cross-section at every floor, A, was then determined to develop a factored compressive resistance, C_r, as specified in CSA (1994) and given by Equation 2, equal to the seismic induced brace compression force acting at that level. Brace loads due to gravity loads were neglected in the calculations.

$$C_r = \phi C_u = \phi A F_y \left(1 + \lambda^{2n}\right)^{(-1/n)} \qquad (2)$$

where ϕ is the resistance factor ($\phi = 0.9$) and n = 1.34. The braces were assumed to be made of G40.21-350W steel with $F_y = 350$ MPa.

2.4 Tension-Only (T/O) Brace Design

In this design, the storey shear at any level is assumed to be entirely resisted by the tension-acting brace only. The contribution of the compression-acting braces is then ignored in design and, therefore, the cross-section area of the braces was determined to develop a factored tensile resistance, $T_r = \phi A F_y$ with $\phi = 0.9$, equal to the seismic induced brace tension loads. By comparing the equations for C_r and T_r, it can be shown that a tension-only design requires less steel material for λ greater than 1.17 in a symmetrical braced frame.

For this design, 7 values of λ were considered: 0.85, 1.35, 1.85, 2.65, 3.35, 4.00, and ∞. The second

last value corresponds to the maximum slenderness ratio generally permitted for tension members, KL/r = 300, for $F_y = 350$ MPa. The "infinite" brace slenderness was considered to examine the performance of structures braced with very slender elements such as rods, flat bars, etc. Note that contrary to tension-compression design, brace slenderness in tension-only design does not influence the size of the braces. The compression properties of the braces were included, however, in the numerical model, as described next.

3 NUMERICAL ANALYSES

3.1 Building Model

The nonlinear time history dynamic analyses were performed using the Drain-2D computer program (Powell & Kanaan 1973). The numerical model is shown in Figure 1b for the 4-storey building. It includes the bracing bent studied as well as all the gravity columns that are laterally stabilised by the bracing bent. The bracing members were modelled using the inelastic brace buckling element with pinned ends (element no. 9) developed by Jain & Goel (1978). The initial buckling strength of the braces was set equal to C_u, as given by Equation 2. The post-buckling resistance of a brace, C'_u, has a considerable influence on its energy dissipation capacity and C'_u was determined from an extensive survey of inelastic cyclic tests by Archambault et al. (1995), Black et al. (1980), Gugerli (1982), Lee & Goel (1987), Leowardi & Walpole (1996), Liu (1987), Perotti & Scarlassara (1991), Wakabayashi et al. (1977) and Walpole (1996). For this study, C'_u was taken as the compressive strength exhibited by the bracing members at a ductility of 3.0

Figure 2 shows the test results as well as the best fit curve obtained by regression analysis and given by Equation 3. This equation is similar to those

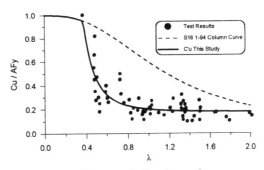

Figure 2. Post-buckling strength of bracing members.

proposed by others (Nakashima & Wakabayashi 1992, Remennikov & Walpole 1998).

$$C'_u = AF_y \left(0.176 + 0.024\lambda^{-3.51}\right) \le C_u \qquad (3)$$

All columns were made continuous over two storeys with zero-moment connection splices. They were modelled using beam-column elements with plastic hinges forming at their ends. A bi-linear hysteretic response with 2% strain hardening was considered for these elements. Beams were pin-connected to the columns and could then be ignored in the model. Rigid diaphragm behaviour was however assumed at every level. A Newmark constant acceleration integration scheme with a time step of 0.0005 s was used in the study. P-Δ effects were considered in the calculations with 100% of the dead load and 50% of the live load applied to the structure. Rayleigh damping equal to 5% of critical damping in the first two modes was adopted.

3.2 Ground Motions

A group of five historical ground motion records and one artificially generated time history was retained for this study. The historical records were: 1940 Imperial Valley, Ca (El Centro, S00E), 1971 San Fernando, Ca (Hollywood Storage, L.A., N90E & S00W), 1949 Western Washington, Wa (Olympia Highway Test Lab, N04W), and 1983 Coalinga aftershock, Ca (Oil Fields Fire Station, N270). The generated motion corresponded to a M_w 7.2 at 70 km scenario (Atkinson & Beresnev 1998). The ground motions were scaled to match, in average, the NBCC design spectrum for Vancouver (Fig. 3). The scaling factors for the records were respectively: 0.8, 1.3, 1.5, 1.3, 1.3, and 1.6.

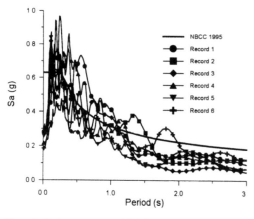

Figure 3. Design spectrum and 5% damped response spectra of the scaled ground motion records.

4 RESULTS

4.1 Results of the analyses

Figure 4 shows the storey drift ductility-base shear relationship computed for the single-storey building under record #2 when considering various brace slenderness ratios. The storey drift ductility corresponds to the peak storey drift divided by the storey drift at yield, the latter being the storey shear deformation which induces tension yielding in the braces. For the frame studied, it is equal to 16.63 mm, i.e., 0.44% of the storey height, h_s. Hence, a ductility of 4.57 was required to reach the maximum drift limit of 0.02 h_s specified in NBCC.

As illustrated, the storey drift in tension-compression (T/C) systems (Fig. 4a) decreases significantly when increasing brace slenderness. This behaviour is attributed to the larger storey shear resistance exhibited in the post-buckling range by the systems with slender braces. Such braces have much higher capacity in tension than in compression and, because design is governed by compression, a significant reserve lateral capacity is provided by the tension-acting braces which compensates for the strength degradation of the compression braces. In addition, slender braces have a larger cross-section area and are therefore stiffer.

For $\lambda = 0.35$, C'_u is the same as C_u which, in turn, is nearly equal to the tensile yield resistance of the braces (AF_y). Hence, for very stocky braces, yielding in the tension brace develops shortly after buckling of the compression brace and the system exhibits a nearly elastic-plastic response with overall yielding at $V/V_f \cong 1/\phi \cong 1.1$. When much slender braces are used, considerable lateral resistance must be mobilised before yielding is reached in the tension brace. Hence, the tension-acting brace in each direction remains essentially elastic and acts as a back-up stiffness preventing large storey drifts to develop. This is shown in Figure 4a for $\lambda = 1.85$ and 2.65 in which cases the peak ductility reaches approximately 1.0, as opposed to 4.0 for $\lambda = 0.35$. Such a better response is achieved, however, at the expense of much higher brace loads that must be accounted for in capacity design of the surrounding structural elements.

For intermediate brace slenderness ($\lambda = 0.85$ and 1.35), the degradation in compressive strength is relatively more important and the difference between the tensile resistance and C_u is smaller (Fig. 2). Yielding then developed in the tension braces followed by some pinching in the hysteresis due to the loss in compression capacity and the slackness experienced by the braces. Nevertheless, the ductility demand remained within acceptable limits and the tension-acting brace in each direction was still effective in limiting the storey drift.

Figure 4b shows that increasing the brace slenderness ratio in tension-only (T/O) design results in more severely pinched hysteresis and higher deformation and ductility demand. In this design, the cross section area remains unchanged when slenderness is varied. Hence, the better performance of the less slender braces is mainly attributed to their better load carrying capacity in compression, which provides additional lateral strength and stiffness to the system and, thereby, higher energy dissipation characteristics.

Figure 5 shows the mean plus one standard deviation over the earthquake record ensemble of the peak brace ductility computed at every floor of the multi-storey buildings. Brace ductility is obtained by dividing the computed brace elongation (or contraction) by the yield deformation of the braces (14.9 mm). In these multi-storey structures, a brace ductility of 1.0 translates into a storey shear deformation of 0.44% of h_s.

For the tension-compression systems, brace ductility generally decreases when increasing the brace slenderness, as observed in the single-storey frames. For the 6-storey structures, the use of braces with $\lambda = 0.35$ led to excessive inelastic demand in the top floors. Dynamic instability occurred under one ground motion record with the $\lambda = 0.35$ and 0.85 braces in the 8-storey building. Note that no results are displayed in the figures for the brace slenderness-building height combinations for which dynamic instability was observed.

Multi-storey tension-only bracing systems exhibited a behaviour similar to that of the single-storey buildings, with larger deformations being associated with higher brace slenderness. Storey drifts well in excess of the NBCC limit were computed in the uppermost floors of the 2-, 4- and 6-storey frames when using braces with KL/r greater than 300. For the 8-storey building, dynamic instability occurred under two ground motion records with the "infinite" brace slenderness.

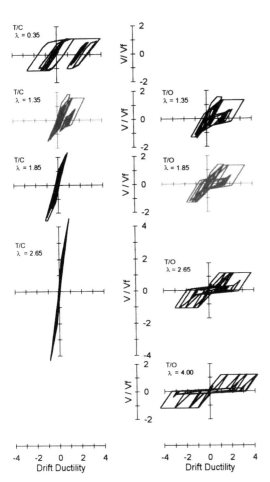

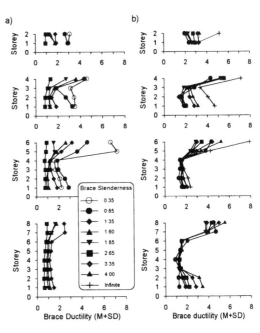

Figure 4. Influence of brace slenderness on the response of the single-storey structure under record #2: a) tension-compression design; b) tension-only design.

Figure 5. Mean plus one sigma of the peak brace ductility in multi-storey structures with 2-storey column tiers: a) tension-compression design; b) tension-only design.

The maximum values over the building height of the peak brace ductility are presented in Figure 6 for all structures. For the tension-compression systems, these curves clearly indicate that structures with brace slenderness λ greater that 1.35 generally exhibit a more stable response, with lower inelastic demand and storey drifts, than the frames built with stockier braces. If tension-only design is used, the inelastic demand gradually increases with the brace slenderness. Except for the 4-storey frames, the computed storey drift remained below the NBCC limit for a slenderness λ equal to or less than 2.65.

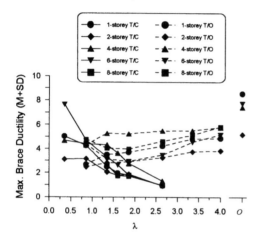

Figure 6. Influence of brace slenderness on the maximum peak brace ductility (2-storey column tiers).

Figure 5 shows that large inelastic deformations tend to concentrate in a few stories over the building height, which is attributed to the limited capacity of concentrically braced frames to redistribute vertically the inelastic demand (Jain et al. 1993, Perotti & Scarlassa 1991, Shibata 1988). This phenomenon was more pronounced at the top floor of the 4-storey structure and at the upper two levels of the 6- and 8-storey buildings when tension-only design was used. Similar results have been obtained by Tremblay et al. 1997.

As shown in Figure 7, continuity of the columns in the framework, even if limited to two consecutive stories, helps in mitigating this soft-storey response (Tremblay and Stiemer 1994). Bending moments that develop in the columns must be accounted for in design, however, to maintain the integrity of the gravity load carrying system. The flexural demand on the columns then becomes another parameter which could be examined to assess the performance of braces frames with various brace slenderness ratios.

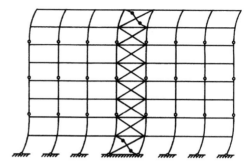

Figure 7. Bending of continuous columns due to concentration of inelastic demand in multi-storey braced frames.

Figure 8 gives the mean plus one standard deviation over the earthquake ensemble of the peak bending moments computed in the columns of the bracing bent. As expected, larger bending moments generally correspond to the regions of concentration of inelastic demand in Figure 5.

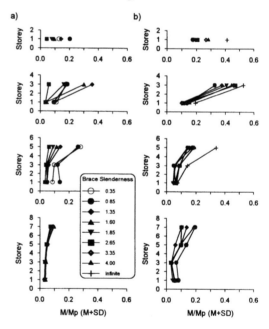

Figure 8. Mean plus one sigma of the peak bending moments in the bracing bent columns (2-storey column tiers): a) tension-compression design; b) tension-only design.

Relatively higher flexural demand also developed in the four-storey tension-only braced frame, even when a brace slenderness of 2.65 or less was used. For that structure, using four-storey column tiers can still represent an economical solution and could help in achieving a more uniform inelastic demand over the building height. Additional analyses have then been performed on this building, both for the T/C and T/O systems, assuming that the column members were continuous and of same cross-section over the full height of the structure. The results are given in Figures 9 and 10. As shown, this modification resulted in a less severe concentration of inelastic deformation at the top floor and a reduction in the bending moments acting in the columns.

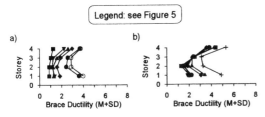

Figure 9. Mean plus one sigma of the peak brace ductility in the 4-storey structures with 4-storey column tiers: a) tension-compression design; b) tension-only design.

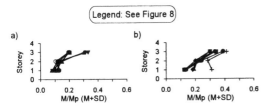

Figure 10. Mean plus one sigma of the peak bending moments in the bracing bent columns (4-storey column tiers): a) tension-compression design; b) tension-only design.

4.2 Discussion

The results indicate that using braces with λ greater than 1.35 in tension-compression bracing systems does not lead to inadequate seismic performance. On the contrary, larger inelastic demand was observed when very stocky braces were used. This behaviour is mainly attributed to the large lateral overstrength exhibited by systems with slender braces. Such overstrength appears to offset the poorer energy dissipation characteristics of these braces.

Excessive demand and occurrences of dynamic instability in the upper part of the taller tension-compression systems with stocky braces is explained by the nearly zero storey shear stiffness exhibited by these frames upon yielding of the braces and by the amplified storey shear demand due to higher modes. In actual braced frames, this phenomenon is expected to be less pronounced as braces in the upper floors would typically be more slender.

Tension-only braced frames were found to perform better when stockier braces were used, most likely due to the relatively higher contribution of the compression braces in resisting lateral loads (which was ignored in design) and dissipating energy. Four-storey and taller structures however exhibited significant concentration of inelastic demand at their top levels, even when made with less slender braces. For the 4-storey building, this undesirable behaviour can be mitigated if the columns are made continuous over the building height.

5 CONCLUSIONS

Nonlinear dynamic analyses have been performed on concentrically braced steel frames up to 8 storeys in height to evaluate the influence of brace slenderness on their seismic performance. The structures were assumed to be located in Vancouver, Canada, and were designed according to NBCC requirements. Tension-compression and tension-only design approaches were considered in the study and every column in the structures was assumed to be continuous over two storeys. The objective of this project was to determine maximum brace slenderness limits as well as building height limitations for the Ductile Braced Frame category specified in Canadian seismic design provisions. The results suggest that:

• Brace slenderness λ up to 2.65 should be permitted in tension-compression bracing systems.

• Tension-compression bracing systems should not be used in non-moment resisting structures beyond 8-storey in height, unless it can be shown that these structures exhibit a stable inelastic response.

• Tension-only design should be permitted in single- and two-storey buildings, provided that the brace slenderness λ is less than 2.65. Structures up to four storeys in height could also be allowed, provided that all columns in the frame are continuous over the full building height.

The study also revealed that bending moments develop in the columns which must be accounted for in design. These bending moments typically reach between 20% and 30% of the column plastic moment capacity.

In this study, the brace slenderness was assumed to be constant over the height of the structures. Further analyses should be performed to investigate the effects of having a brace slenderness increasing from the bottom to the top of the building, as this is typically the case in actual structures.

6 ACKNOWLEDGEMENTS

This research was supported by the Natural Sciences and Engineering Research Council of Canada.

7 REFERENCES

AISC. 1997. *Seismic Provisions for Structural Steel Buildings.* American Institute of Steel Construction Inc. (AISC), Chicago, Il.

Archambault, M.-H., Tremblay, R. and Filiatrault, A. 1995. *Étude du comportement séismique des contreventements ductiles en X avec profilés tubulaires en acier.* Rapport no. EPM/GCS-1995-09, Département de génie civil, École Polytechnique, Montréal, Canada.

Atkinson, G.M. and Beresnev, I.A. 1998. Compatible ground-motion time histories for new national seismic hazard maps. *Can. J. of Civ. Eng.*, 25, 305-318.

Black, R.G. Wenger, W.A.B. and Popov, E.P. 1980. *Inelastic Buckling of Steel Struts Under Cyclic Load Reversals.* Report no. UCB/EERC-80/40, Earthquake Engineering Research Center, Univ. of California, Berkeley, Ca.

CSA. 1989. *CAN/CSA-S16.1-M89, Limit States Design of Steel Structures.* Canadian Standard Association, Rexdale, Ont.

CSA. 1994. *CAN/CSA-S16.1-94, Limit States Design of Steel Structures.* Canadian Standard Association, Rexdale, Ont.

ECCS. 1991. *ECCS-TWG 1.3 "Seismic Design" European Recommendations for Steel Structures in Seismic Zones.* European Convention for Constructional Steelwork. Brussels, Belgium.

Ghanaat, Y. 1980. *Study of X-Braced Frame Structures under Earthquake Simulation.* Report no. UCB/EERC-80/08, Earthquake Engineering Research Center, Univ. of California, Berkeley, Ca.

Gugerli, H. 1982. Inelastic Cyclic Behavior of Steel Members. Ph.D. Thesis, Dept. of Civ. Engrg., Univ. of Michigan, Ann Arbor, Michigan.

Jain, A.K., Redwood, R.G., and Lu, F. 1993. Seismic Response of Dual Concentrically Braced Dual Steel Frames. *Can. J. of Civ. Eng.*, 20, 672-687.

Jain, A.K. and Goel, S.C. 1978. *Hysteresis Models for Steel Members Subjected to Cyclic Buckling or Cyclic End Moments and Buckling (User's Guide for Drain-2D: EL9 and EL10)*, Report UMEE 78R6, Dept. of Civ. Eng., University of Michigan, Ann Arbor, Mi.

Jain, A.K., Goel, S.C. and Hanson, R.D. 1978. Inelastic Response of Restrained Steel Tubes. *J. of the Struct. Div.*, ASCE, 104, 897-910.

Kahn, L.F. and Hanson, R.D. 1976. Inelastic Cycles of Axially Loaded Steel Members. *J. of the Struc. Div.*, ASCE, 102, 947-959.

Kanaan, A.E. and Powell, G.H. 1973. *DRAIN-2D, Reports no. EERC 73-6 and 73-22 (revised in 1975)*, Earthquake Eng. Research Center, University of California, Berkeley, Ca.

Lee, S. and Goel, S.C. 1987. *Seismic Behavior of Hollow and Concrete-Filled Square Tubular Bracing Members.* Research Report UMCE 87-11, Dept. of Civ. Engrg., Univ. of Michigan, Ann Arbor, Mich.

Leowardi, L.S.and Walpole, W.R. 1996. *Performance of steel brace members.* Research Report 96-3, Dept. of Civil Engrg., University of Canterbury, Christchurch., New Zealand.

Liu, Z., 1987. *Investigation of Concrete-Filled Steel Tubes Under Cyclic Bending and Buckling.* Ph.D. Thesis, Dept. of Civ. Engrg., Univ. of Michigan, Ann Arbor, Michigan.

Nakashima, M. and Wakabayashi, M. 1992. Analysis and Design of Steel Braces and Braced Frames in Building Structures. In Y. Fukumoto and G.C. Lee (ed.), *Stability and Ductility of Steel Structures under Cyclic Loading*, CRC Press, Boca Raton, USA, 309-321.

NRCC. 1995. *National Building Code of Canada (NBCC)*, 11th ed., National Research Council of Canada (NRCC), Ottawa, Ont.

Perotti, F. and Scarlassara, P. 1991. Concentrically Braced Steel Frames under Seismic Actions: Non-Linear Behaviour and Design Coefficients. *Earthquake Eng. and Struct. Dynamics*, 20, 409-427.

Prathuangsit, D., Goel, S.C. and Hanson, R.D. 1978. Axial Hysteresis Behavior with End Restraints. *J. of the Struc. Div.*, ASCE, 104, 883-895.

Remennikov, A.M. and Walpole, W.R.1997. Analytical prediction of Seismic Behaviour for Concentrically-Braced Steel Systems. *Earthquake Eng. and Struct.. Dyn.*, 26, 859-874.

Remennikov, A.M., and Walpole, W.R. 1998. A note on compression strength reduction factor for a buckled strut in seismic-resisting braced system. *Eng. Struct.*, 20 (8), 779-782.

Robert, N. and Tremblay, R. 2000. Seismic Design and Behaviour of Chevron Steel Braced Frames. *Proc. 12th World Conf. on Earthquake Eng.*, Auckland, NZ.

SEAOC. 1990. *Recommended Lateral Force Requirements and Commentary.* Structural Engineers Association of California, Seismology Committee, Sacramento. Ca.

Shibata, M. 1988. Hysteretic Behavior of Multi-Story Braced Frames. *Proc. of the 9th World Conf. on Earthquake Eng.*, Tokyo, Japan, IV, 249-254.

SNZ. 1997. NZS3404:Part1:1997 – Steel Structures Standard. Standards New Zealand, Wellington, New Zealand.

Tang, X. and Goel, S.C. 1987. *Seismic Analysis and Design Considerations of Concentrically Braced Steel Structures.* Report UMCE 87-4, Dept. of Civil Eng., Univ. of Michigan, Ann Arbor, Mi.

Tremblay, R. and Stiemer, S.F. 1994. Back-up Stiffness for Improving the Stability of Multi-storey Braced Frames under Seismic Loading. *Proc. 1994 SSRC Annual Techn. Session*, Bethlehem, PA, 311-325.

Tremblay, R., Bruneau, M., Nakashima, M., Prion, H.G.L., Filiatrault, A., and DeVall, R. 1996. Seismic Design of Steel Buildings: Lessons From the 1995 Hyogoken-Nanbu Earthquake. *Can. J. of Civ. Eng.*, 23, 727-756.

Tremblay, R., Robert, N. and Filiatrault, A. 1997. Tension-Only Bracing: a Viable Earthquake-Resistant System for Low-rise Steel Buildings? *Proc. SDSS '97 5th Intern. Colloquium on Stability and Ductility of Steel Structures*, Nagoya, Japan, 2, 1163-1170.

Wakabayashi, M., Nakamura, T., and Yoshida, N. 1977. Experimental Studies on the Elastic-Plastic Behavior of Braced Frames under Repeated Horizontal Loading. *Bull. Disaster Prevention Research Institute*, Kyoto University, 27 (251), 121-154.

Walpole, W.R. 1996. Behaviour of Cold-Formed Steel RHS Members under Cyclic Loading. Research Report 96-4, Dept. of Civil Engrg., University of Canterbury, Christchurch, New Zealand.

Behaviour of Steel Structures in Seismic Areas, Mazzolani & Tremblay (eds) © 2000 Balkema, Rotterdam, ISBN 90 5809 130 9

Inelastic seismic analysis of eccentrically loaded steel bridge piers

T. Usami & Q. Y. Liu
Department of Civil Engineering, Nagoya University, Japan

ABSTRACT: This paper presents a simple SDOF formulation for inelastic seismic response analysis of eccentrically loaded steel bridge piers, considering the vertical inertia associated with horizontal ground motion. Modeling of the unsymmetrical hysteretic behavior is verified by pseudodynamic test results. It is shown that influence of vertical inertia force can not be neglected when e/h (e = eccentric distance, h = pier height) is relatively large. A parametric study shows that under design accelerograms, steel bridge piers with intermediate eccentricity (e/h around $0.1 \sim 0.2$) are susceptible to highest maximum displacement; And eccentricity definitely leads to higher residual displacement level than centrally loaded steel bridge piers.

1 INTRODUCTION

Thin-walled steel bridge piers are widely used to support highway bridges in urban areas of Japan. According to different ground area conditions, steel bridge piers are built in different forms: load from the upper structure acts centrally on a T-shaped pier while it acts eccentrically on an inverted L-shaped pier. Neglecting the effects of vertical ground motions, a bridge pier is seen as under the combined action of a constant vertical force and cyclic horizontal loading during an earthquake. And the constitutional characteristic of steel bridge piers makes them susceptible to damage in the form of coupled local and overall buckling. According to the current *Design Specifications of Highway Bridges* (Japan Road Association 1996), state-of-practice in design of such steel bridge piers is carrying out time-history analysis with prescribed earthquake accelerograms to ensure seismic safety. With reliable modeling of the hysteretic behavior, a lumped-mass SDOF system proves to be adequate for predicting seismic response of centrally loaded steel bridge piers. Substantial research has been done on the horizontal resisting mechanism of centrally loaded steel bridge piers, and quite a few hysteretic models are now available for predicting their seismic response. But there seems to be little corresponding research on eccentrically loaded type so far.

This paper sets out to solve the problem of seismic response analysis of eccentrically loaded steel bridge piers. With a view to practical application, the problem is also handled with a simple lumped-mass SDOF model just like in the case of centrally loaded type. Because the system mass has an offset from axis of the pier, horizontal ground motion alone will bring about considerable vertical vibration, and the vertical inertia force is taken into account when formulating the equation of motion. Modeling of hysteretic behavior takes advantage of an approximate correlation of hysteretic behavior between the centrally loaded steel bridge piers and eccentrically loaded ones. And this approach to analytical modeling of hysteretic behavior is verified with pseudodynamic test results. To settle the dispute over whether it is necessary to consider vertical inertia, a numerical experiment is carried out covering the possible range of e/h (e = eccentricity, h = pier height). To answer concerns from practical design, a parametric study is carried out to investigate the effect of different e/h values on computed seismic response.

2 SDOF MODEL OF ECCENTRICALLY LOADED STEEL BRIDGE PIERS

2.1 *Formulation of equation of motion*

In practice, centrally loaded steel bridge piers are usually modeled as SDOF system with the mass concentrated at the top. Under a ground shake $\ddot{x}_g$, the standard equation of motion for centrally loaded steel bridge piers is:

$$M\ddot{u} + C\dot{u} + H(u) = -M\ddot{x}_g \qquad (1)$$

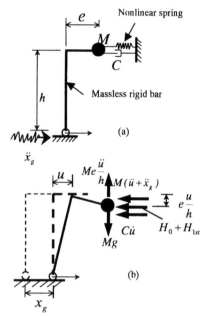

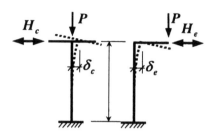

Figure 2. Centrally loaded and eccentrically loaded steel bridge piers

Figure 1. SDOF modeling of eccentrically loaded steel bridge piers

wherein $H(u)$ stands for the horizontal restoring force of the structure.

A similar lumped mass model is adopted in this study for seismic response analysis of eccentrically loaded steel bridge piers. In Figure.1a, the structure is equivalently modeled as a massless L-shaped rigid bar attached to ground by a hinge, and constraint to the rotation of the rigid bar is provided by the inelastic spring in the horizontal direction.

From static equilibrium, the eccentric vertical load Mg causes an initial reaction H_0 in the nonlinear spring:

$$H_0 = Mg \cdot e / h \qquad (2)$$

Due to eccentricity of the mass, horizontal ground motion during an earthquake will also cause vertical vibration of the mass in an eccentrically loaded pier, which is different from the case of centrally loaded piers. And considering the resulted vertical inertia is a point in formulating the equation of motion. Figure.1b illustrates the dynamic equilibrium under a ground shake of $\ddot{x}_g$. The vertical inertia force due to rotation of the rigid bar is $Me\ddot{u} / h$. Equilibrium of moment at the hinge yields:

$$\qquad (3)$$

Linearize Eq.(3) by omitting higher order terms and consider Eq.(2), one arrives at:

$$M(1 + \frac{e^2}{h^2})\ddot{u} + C\dot{u} + (H_{1st} - \frac{Mg}{h} \cdot u) - M\frac{e}{h}\frac{\ddot{x}_g}{h}u = -M\ddot{x}_g$$
$$\qquad (4)$$

in which $H_{1st} - Mgu / h$ represents the restoring force in second order theory. Since $P - \Delta$ effect is to be accounted for in modeling the horizontal restoring force, denoting the restoring force in an eccentrically loaded steel bridge pier as $H_e(u)$, it is obvious that

$$H_e(u) = H_{1st} - Mgu / h \qquad (5)$$

Introducing Eq.(5) into Eq.(4), one arrives at:

$$M\left(1 + \frac{e^2}{h^2}\right)\ddot{u} + C\dot{u} + H_e(u) - M\frac{e}{h}\frac{\ddot{x}_g}{h}u = -M\ddot{x}_g \qquad (6)$$

Compared with Eq.(1), Eq.(6) has on its left-hand side two extra terms: one is $Me^2 / h^2\ddot{u}$ (equivalent vertical inertia), the other is $-Me / h \cdot \ddot{x}_g / h \cdot u$. Under normal conditions, u / h is rather small, and the latter turns out to have little effect on the overall seismic response. Thus Eq.(6) can be simplified to:

$$M(1 + \frac{e^2}{h^2})\ddot{u} + C\dot{u} + H_e(u) = -M\ddot{x}_g \qquad (7)$$

From Eq.(7), it is clear that considering vertical inertia in the analysis has the effects of enlarging overall inertia force as well as elongating the natural period of the system; And the influence of vertical inertia is related to e / h.

2.2 Modeling of hysteretic behavior

The eccentric load from upper structure causes an extra moment M_0 and an initial displacement δ_0 in the eccentrically loaded pier, and its hysteretic behavior is different from that of an otherwise identical centrally loaded pier (Fig.2). But there exists an approximate correlation of hysteretic behavior between the two, which can be expressed by (Gao et al. 1998):

$$H_e(\delta_e) = H_c(\delta_c) - M_0 / h \qquad (8)$$

$$\delta_c = \delta_e + 2/3\delta_0 \qquad (9)$$

Through Eq.(8) (heretofore referred to as the translation equation), the restoring force of an eccentrically loaded steel bridge pier H_e can be derived from H_c — the restoring force of an otherwise identical centrally loaded pier. Thus any appropriate hysteretic model developed so far for centrally loaded steel bridge piers can also be used for predicting the hysteretic behavior of eccentrically loaded steel bridge piers; In this study, the damage-based hysteretic model (Kumar & Usami 1996) for centrally loaded steel bridge piers is adopted for all the seismic response analysis.

3 SEISMIC RESPONSE ANALYSIS

3.1 Discussion on the influence of inertia

A numerical example is used here to illustrate the influence of vertical inertia on the calculated seismic response under different e/h values. Input ground motion in this example is JR-Takatori record. The analysis is separately carried out based on Eq.(7) (considering vertical inertia) and Eq.(1) (neglecting vertical inertia). The results are compared in Figure.3 (with e/h =0.1. 0.2 and 0.4). It can be seen that when e/h is as small as 0.1, it is perfectly acceptable to leave out vertical inertia. However, when e/h is relatively large, it is necessary to include vertical inertia in seismic response analysis. It can also be seen from this numerical example that although vertical inertia will add to the overall inertia force in the equation of motion, including vertical inertia does not necessarily result in larger response: In Figure.3, the predicted response based on Eq.(7) is larger than that based on Eq.(1) in the case of e/h =0.2, while this comparison is reversed in the case of e/h =0.4 . This phenomenon indicates that variation in natural period resulted from the vertical inertia also has a considerable weight in determining the computed response.

3.2 Simulation of pseudodynamic tests

Simulation of two pseudodynamic tests of eccentrically loaded steel bridge piers (Usami et al. 1999) is carried out based on the SDOF model. Input ground motion in both tests is JR-Takatori accelerogram (N-S component). Test specimens are two stiffened box-section steel bridge piers, both have the e/h of 0.0726. With such a small e/h, vertical inertia is expected to have little influence on response.

Analysis results are compared with test results in Figure.4 and Figure.5. Good agreement between analysis results and test results indicates that the hysteretic model and the translation equation are valid and adequate for the analysis.

3.3 Parametric study

Based on Eq.(7), a parametric study is carried out on the effects of eccentricity on computed seismic response under design conditions. The analysis is carried out on six series of specimens with different e/h values (ranging from $0 \sim 0.5$) but identical stiffened box section (width-thickness ratio parameter of the flange plate R_f =0.35; stiffness ratio of the longitudinal stiffeners $\gamma / \gamma^* = 3$; and the aspect ratio of flange plate between diaphragms =0.5). These specimens are designed using seismic coefficient method specified in the current *Design Specifications of Highway Bridges* (Japan Road Association 1996). Damping ratio is assumed as 0.05, and the natural period is calculated by:

$$T = 2\pi\sqrt{M / K_e} \qquad (10)$$

Figure.6 shows the calculated responses (average responses to the three accelerograms contained in Ground Type II group) under Level 2 · Type II design accelerograms. It can be seen that the mildly eccentrically loaded series with e/h around $0.1 \sim 0.2$ have the largest maximum displacements; Among all eccentrically loaded specimens, the response tends to go lower with higher e/h; And the response of the centrally loaded series (e/h =0) comes between the highest level and the lowest level of all eccentrically loaded specimens. When it comes to residual displacement, it can be seen that among all eccentrically loaded groups, the response still goes lower with higher e/h, but the centrally loaded specimens tend to have the lower level of residual displacement than all eccentrically loaded specimens irrespective of e/h values. It is worthwhile to point out that the above trends in both maximum displacement and residual displacement are common under all the Level 2 accelerograms and for all the ground types[5].

4 CONCLUSIONS

In previous sections, seismic response analysis of eccentrically loaded steel bridge piers is tackled by a simple SDOF dynamic formulation which takes into account the vertical inertia associated with horizontal ground motion. Effect of eccentricity on the predicted responses under Level 2 design ac-

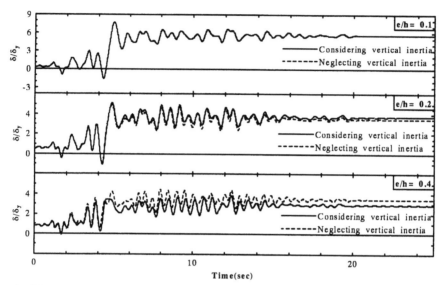

Figure 3.   Effect of vertical inertia on calculated responses

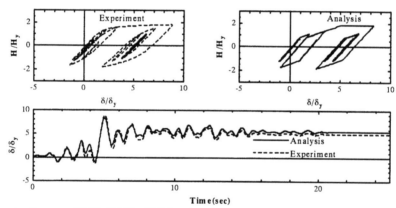

Figure 4.   Response of SE35-35H[A] to JR-Takatori accelerogram

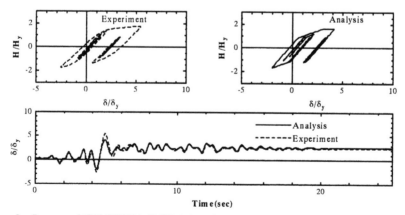

Figure 5.   Response of SE35-35H[B] to JR-Takatori accelerogram

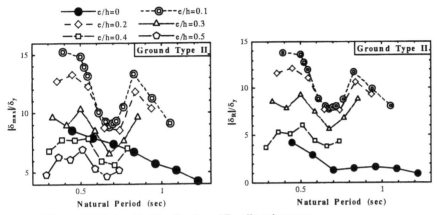

Figure 6. Responses under Level 2 · Type II · Ground Type II accelerograms

celerograms is investigated based on the proposed dynamic analysis approach. Major findings and conclusions are summarized as follows:

1. Through the translation equation, modeling of hysteretic behavior for eccentrically loaded steel bridge piers becomes equivalent to modeling for the centrally loaded steel bridge piers. And any appropriate SDOF hysteretic model for centrally loaded steel bridge piers can also be made use of in seismic response analysis of eccentrically loaded steel bridge piers.

2. When e/h is relatively small, including vertical inertia in the analysis almost makes no difference on the predicted response. However, in case of relatively large e/h, it is necessary to consider vertical inertia and use Eq.(7) for seismic response analysis of eccentrically loaded steel bridge piers.

3. Including vertical inertia will amplify the overall inertia force and elongate the natural period of the system. Effect of vertical inertia on the amplitude of predicted response varies from case to case; It does not necessarily lead to larger computed response.

4. Eccentrically loaded steel bridge piers with e/h around $0.1\sim0.2$ tend to have the highest level of maximum displacement response among all eccentrically loaded piers designed under the same conditions based on the seismic coefficient method. However, with higher value of e/h, the maximum displacement response tends to become lower due to the fact that load from the upper structure goes lower in the elastic seismic design by the seismic coefficient method. The maximum displacement response of centrally loaded steel bridge piers usually comes between the highest level corresponding to those with e/h around $0.1\sim0.2$ and the lowest level corresponding to those with e/h of 0.4 or 0.5 .

5. Designed under the same conditions, eccentrically loaded steel bridge piers will generally have higher level of residual displacement than centrally loaded piers, which may be a matter of concern in practical design.

REFERENCES

Gao, S., Usami, T. and Ge, H. (1998), Numerical study on seismic performance evaluation of steel structures, NUCE Research Report No. 9801, Dept. of Civil Engineering, Nagoya University.

Japan Road Association (1996), Design Specifications of Highway Bridges: Part V. Seismic Design (In Japanese).

Kumar, S. and Usami, T. (1996), An evolutionary-degrading hysteretic model for thin-walled steel structures, Engineering Structures, Vol.18, No.7, pp.504-514.

Liu, Q.Y, Usami, T., and Kasai, A.(1999), Inelastic Seismic Design Verification of Steel Bridge Piers, NUCE Research Report, Dept. of Civil Engineering, Nagoya University.

Usami, T., Honma, D., and Yoshizaki, K. (1999), Pseudodynamic tests of eccentrically loaded steel bridge piers, to appear in J. Strut. Mech. and Earth. Engrg, JSCE (In Japanese).

Behaviour of Steel Structures in Seismic Areas, Mazzolani & Tremblay (eds) © 2000 Balkema, Rotterdam, ISBN 90 5809 130 9

Steel panel shear wall – Analysis on the center core steel panel shear wall system

M. Yamada
Kansai University, Osaka, Japan

T. Yamakaji
Newjec Incorporated, Osaka, Japan

ABSTRACT: An analytical research on elasto-plastic sway process of steel rigid frame with center core steel panel shear wall are carried out and compared with test results. For Type B of steel rigid frame with 9 story-3 span, the relationship between the vectors of incremental sway angle and shear force at each story is derived from equilibrium and compatibility equations at each story and joint, respectively. For Type A, which is Type B added center core steel panel shear wall, the wall is idealized into shear springs differentiated by each story. Then restoring forces in opposite direction with shear forces are added at each story, so that fundamental matrix equation can be derived as well as Type B. As a result, the correspondences between tested and calculated values in both cases are fairly good qualitatively as well as quantitatively.

1 INTRODUCTION

1.1 Background

Recently it has been applied to design to control earthquake excitation by partially installed steel panel in buildings and resistance by tall steel panel throughout multistory as main anti-seismic element against overturning moment. Especially, on the second main anti-seismic element, it has already been reported that center core steel panel shear wall set in the New Kobe City Hall of 32 stories and 132 m high showed good resisting behaviour subjected to severe ground motion at the last Kobe earthquake on 17. Jan. 1995 in Kobe Japan (Yamada 1996). But there have been few basic researches on mechanical property of steel panel for building yet (Mimura & Akiyama 1977, Caccese & Elgaaly et al. 1990, 1993a,b, Yamada 1992, 1996).

The fact that ultimate strength of steel panel with frame subjected to shear forces is not determined by shear buckling of steel panel is well-known. It is originally in civil engineering, shipbuilding and aircraft practice field where steel panel had been used as main structural element. Before long steel panel that was used in aircraft became much thinner and at last in 1928 Wagner reported mechanical property of a thin shear web plate (Kuhn 1956). Wagner theoretically treated pure diagonal tension field and pure shear field in the limiting cases of the extremely thin and thick web, respectively, and incomplete diagonal tension field between the two limiting cases. However, the actual behaviour is so complex that the

formulation has largely and conservatively depended on experimental results and empirical judgements.

On the other hand, at the last STESSA'97, Kyoto, Japan, the authors had already presented their experimental researches on the fundamental mechanism of steel panel as both single and multi-story anti-seismic elements (Yamada & Yamakaji 1997). And they had proposed steel panel shear wall as one of the best anti-seismic elements because steel panel shear wall had been found to be beneficial on higher not only resistance but also initial stiffness despite lower ductility in combining steel panel shear wall with various kind of rigid frame. Furthermore steel panel is more available on construction than steel brace to have been used. And at the coming WCEE 2000, Auckland, New Zealand, it will be indirectly explained the behaviour of steel panel itself by changing various kind of parameters as the surrounding boundary conditions of steel panel and then directly comparing the whole behaviour with each other (Yamakaji & Yamada 2000). As a result, it had been found that the behaviour of steel panel depends on the surrounding conditions

1.2 Objectives and scopes

Under the above mentioned situation, also in the field of building engineering, to propose the formulation of the behaviour of the extremely thin steel panel for building more rationally and to give how to analyze the behaviour of structure with steel panel are objectives and scopes in this research.

2 EXPERIMENT

2.1 Test specimens and testing method

Tests were carried out on two specimens under the same loading condition. Two tested specimens are shown in Figure 1. In structural form, Type A and B are rigid frames with and without center core shear wall, respectively, and are with the same scale of about 1/30 of 9 story-3 span i.e. story height of 120 mm and colum span of 240 mm. Both Type A and B are pure steel structures which are build up of beams and columns made of combined 2-channels $19 \times 10 \times 10 \times 1.2$ mm and, furthermore, Type A combines center core steel panel of a thickness of 0.4 mm which is spot-welded to beams and columns. In the material property of members, steel profile and steel panel of Type A have yield strength of 336 MPa and 238 MPa, respectively, and steel profile of type B has yield strength of 328 MPa.

As shown in Figure 2, horizontal force P was loaded at 6th story i.e. 2/3 point from the base of overall height of structures. This loading level is considered as near the gravity of inverse triangular distribution of horizontal forces excited at each floor level and near the inflection point of center core frames under ground motions. The horizontal force P were loaded by measuring and controlling cyclic repeated incremental sway angle amplitude R of 0.001, 0.002, 0.003, 0.005, 0.007, 0.010, 0.015, 0.020, 0.030, 0.050 and 0.090.

Both Type A and B are plane structures, so because they are predicted to be laterally buckled out of the planes by a certain constant horizontal force, buckling protection facility were set with a system separated from the structures.

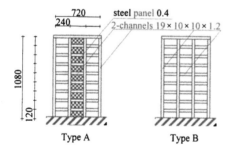

Figure 1. Specimens (unit: mm)

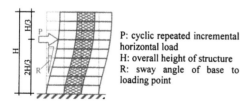

P: cyclic repeated incremental horizontal load
H: overall height of structure
R: sway angle of base to loading point

Figure 2. Loading method

2.2 Elasto-plastic sway characteristics

P-R relationship of Type A is described as follows. At the sway angle of +5/1000 and +15/1000 on the way to +20/1000, it starts to yield and it shows the ultimate strength with yield ratio of about 55 % and ductility factor of about 3, respectively. On steel panel, it tends to sound due to buckling itself from mainly 3rd and 4th story within the sway angle of about 3/1000 in each cycle. This can be analytically regarded as the formation of diagonal tension field in one direction and as vanishment of the diagonal compression field in the conjugate direction at the same time since the behaviour of unit steel panel shear wall has been already found (Yamada 1999). It was found as well as the unit wall that reversed loading make the buckling waves turned over, so that the structure shows actual behaviour unstable slightly and negligibly but wholly stable hysteresis loop with slipping characteristics of the P-R relationship. And formed state of buckled waves of steel panel is shown as Type A in Figure 3. This represents that the center core steel panel forms the buckled waves, which are with an inclination of 45 degrees, a pitch of one story height and anchors for running lineaments near beam-column panel zones and penetrate into neighbouring stories. Such buckled waves can be guessed to mainly result from shear deformation. Conversely, steel panel mainly resists shear forces.

On the other hand, Type B represents P-R relationship as follows. The sway angle reaching at +7/1000 after nearly elastic behaviour in the early loading, it starts to yield and after that it continues to show the remarkably stable behaviour. At the sway angle of +90/1000, it shows the ultimate strength with yield ratio of about 35 % and ductility factor of about 13. Whole hysteresis loop shows so-called spindle-shaped and *Baushinger's effects*. And as Type B in Figure 3 showing, it can be assumed that Type B had undergone shear deformation predominantly.

Namely steel panel and steel rigid frame are shear element and system, respectively, so that Type A that the element and system are compatible parallel with each other may be assumed to be shear system.

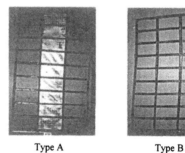

Type A   Type B

Figure 3. Final states of tested specimens

3 ANALYSIS

3.1 Modeling for structures

The discussion on the experimental results makes both Type A and B simplified into the same shear system. It is called Frame system model shown in Figure 4 and composed of rigid members on beam and column, elasto-plastic hinges on both ends of each member (Yamada and Kawamura et al. 1984) and shear springs on steel panel shear wall differentiated into each story (Yamakaji and Yamada 2000). For Type A, the stiffness of shear spring at each story can be put at a certain constant numerical value greater than zero as initial condition. While, for Type B, the stiffness of shear spring at each story must be put at zero as initial condition. Namely, the difference between two models applied to the abstract model is distinctly that shear springs are set or not at all story as initial condition.

3.2 Fundamental equation

Frame system model simplified diagrammatically as general story subjected to sway deformation shown in Figure 4 can be formulated and then the fundamental equation can be derived as follows. Each variable is expressed as the increment because of carrying out the incremental numerical analysis.

The equation expressed as 3-terms on story sway angle among the general story, its upper and lower story is to be derived. From equilibrium condition on moment at joint, increment of external joint rotation angle is given as follows :

$$d_e r_i = \frac{\left({}_{eo}K_i \cdot dR_{i+1} + {}_{eu}K_i \cdot dR_i \right)}{\Sigma_e K_i} \tag{1}.$$

From the same condition as Eq. (1), increment of internal joint rotation given as follows :

$$d_i r_i = \frac{\left({}_{io}K_i \cdot dR_{i+1} + {}_{iu}K_i \cdot dR_i \right)}{\Sigma_i K_i} \tag{2}.$$

From the relationship between story shear force at each story and external force, shearing equation is given as follows :

$$\frac{2\left({}_{eu}K_i + {}_{eo}K_{i-1} + {}_{iu}K_i + {}_{io}K_{i-1} \right) + h \cdot G_i}{h} \cdot dR_i -$$

$$\frac{2\left({}_{eu}K_i \cdot d_e r_i + {}_{eo}K_{i-1} \cdot d_e r_{i-1} + {}_{iu}K_i \cdot d_i r_i + {}_{io}K_{i-1} \cdot d_i r_{i-1} \right)}{h}$$

$$= \begin{array}{l} dP_i \ (i=2 \ to \ 6) \\ 0 \ (i=7,8) \end{array} \tag{3},$$

where $h = H - B - 2\lambda_c$, λ_c : hinging zone of column.

And Eqs. (1) and (2) are substituted into Eq. (3) and then the 3-term difference equation is derived as follows :

$$\left\langle \frac{{}_{eu}K_i \cdot {}_{eo}K_i}{\Sigma_e K_i} + \frac{{}_{iu}K_i \cdot {}_{io}K_i}{\Sigma_i K_i} \right\rangle \cdot dR_{i+1} +$$

$$\left\langle \frac{{}_{eu}K_i^2}{\Sigma_e K_i} + \frac{{}_{eo}K_{i-1}^2}{\Sigma_e K_{i-1}} + \frac{{}_{iu}K_i^2}{\Sigma_i K_i} + \frac{{}_{io}K_i^2}{\Sigma_i K_{i-1}} - \Sigma_c K_i - \frac{h}{2} \cdot G_i \right\rangle \cdot dR_i$$

$$+ \left\langle \frac{{}_{eu}K_{i-1} \cdot {}_{eo}K_{i-1}}{\Sigma_e K_{i-1}} + \frac{{}_{iu}K_{i-1} \cdot {}_{io}K_{i-1}}{\Sigma_i K_{i-1}} \right\rangle \cdot dR_{i-1}$$

$$= \begin{array}{l} -\frac{h}{2} dP_i \ (i=2 \ to \ 6) \\ 0 \ (i=7,8) \end{array} \tag{4},$$

where $\Sigma_c K_i = {}_{eu}K_i + {}_{eo}K_{i-1} + {}_{iu}K_i + {}_{io}K_{i-1}$.

And when i =1 and 9, put $d_e r_0$ and $d_i r_0$, ${}_{eo}K_9$, ${}_{io}K_9$ and dP_9 at zero, respectively, and by the same procedure as giving Eqs. (1) to (3) and deriving the Eq. (4), the 2-term difference equation of Eqs. (5) and (6), respectively, are derived as follows :

$$\left\langle \frac{{}_{eu}K_1 \cdot {}_{eo}K_1}{\Sigma_e K_1} + \frac{{}_{iu}K_1 \cdot {}_{io}K_1}{\Sigma_i K_1} \right\rangle \cdot dR_2 +$$

$$\left\langle \frac{{}_{eu}K_1^2}{\Sigma_e K_1} + \frac{{}_{iu}K_1^2}{\Sigma_i K_1} - \Sigma_c K_1 - \frac{h}{2} \cdot G_1 \right\rangle \cdot dR_1 = -\frac{h}{2} dP_1 \tag{5},$$

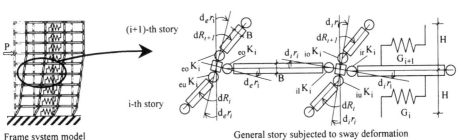

Frame system model   General story subjected to sway deformation

Prefix abbreviation of
e and i : exterior and interior, respectively
o, u, r and l : on, under, right and left, respectively

Suffix abbreviation of
(i+1), i and (i-1) : (i+1), i and (i-1)-th story, respectively

Variable increment of
dr and dR : joint rotation and translation angle, respectively

Stiffness of
K and G : hinge and shear spring, respectively

Figure 4. Frame system model

$$\left\langle \frac{_{eu}K_9{}^2}{\Sigma_e K_9} + \frac{_{eo}K_8{}^2}{\Sigma_e K_8} + \frac{_{iu}K_9{}^2}{\Sigma_i K_9} + \frac{_{io}K_8{}^2}{\Sigma_i K_8} - \Sigma_c K_9 - \frac{h}{2} \cdot G_9 \right\rangle \cdot dR_9$$

$$+ \left\langle \frac{_{eu}K_8 \cdot _{eo}K_8}{\Sigma_e K_8} + \frac{_{iu}K_8 \cdot _{io}K_8}{\Sigma_i K_8} \right\rangle \cdot dR_8 = 0 \qquad (6).$$

Then the coefficient of each term of Eqs. (4) to (6) is expressed by using simple sign as horizontal stiffness at each story, so that the equations are set as follows :

$$\left.\begin{array}{l} B_9 \cdot dR_9 + C_8 \cdot dR_8 = 0 \\ \quad \vdots \\ A_{i+1} \cdot dR_{i+1} + B_i \cdot dR_i + C_{i-1} \cdot dR_{i-1} = \begin{array}{l} 0 \;\; (i=7, 8) \\ dP_i \,(i=2 \text{ to } 6) \end{array} \\ \quad \vdots \\ A_2 \cdot dR_2 + B_1 \cdot dR_1 = dP_1 \end{array}\right\} \qquad (7),$$

and Eq. (7) can be expressed as matrix equation composed of band stiffness matrix of $[\mathbf{K}]$, stress and strain vector of $\{\mathbf{dP}\}$ and $\{\mathbf{dR}\}$, respectively. And then this matrix equation is solved on $\{\mathbf{dR}\}$, so that fundamental equation is obtained as follows :

$$\{\mathbf{dR}\} = [\mathbf{K}]^{-1} \{\mathbf{dP}\} \qquad (8).$$

On Eq. (8), when an arbitrary $\{\mathbf{dP}\}$ composed of numerical elements as small as possible is given, $\{\mathbf{dR}\}$ can be computed, so that $d_s r_i$ and $d_r r_i$ can be computed from Eqs. (1) and (2). After that, decisions on $\{\mathbf{dR}\}$, $d_s r_i$ and $d_r r_i$ give rotation angle increment of each hinge, sway angle increment of shear spring at each story and sway angle increment of base to load point. While each variable value can be obtained by adding each variable increment to itself continuously. When each variable value reaches at boundary points where hysteresis loops of structural elements are assumed as subsequent paragraph, the stiffnesses of the elements should be changed and Eq. (8) should be solved repeatedly the same procedure as above. This incremental numerical analysis was carried out with controlling the sway angle R of base to load point as experiments.

3.3 Hysteresis loops of structural elements

3.3.1 Hinges
The relationship between stress σ and strain ε of the material in steel profile is assumed as bilinear.

The cross section of steel profile is simplified into five lumped masses model, which is equivalent to the original cross section on sectional area, moment of inertia and plastic modulus of cross section. When hinging zone of columns and beams is assumed to be 1/6 of their length and pure bending moment acts on the model, the relationship between bending moment M and rotation angle θ of column and beam is assumed as trilinear hysteresis loop. The procedure as above is shown in Figure 5.

3.3.2 Shear springs
It has been already found that, in general, ultimate strength of extremely thin steel panel subjected to shear force is not determined by shear buckling of the steel panel but after that rises by secondary resisting mechanism. This fact has been explained that the panel resists by pure shear mechanism and, after that, buckles in the early deformation, so that resists by tension about in diagonal direction. This has been so-called *Theory of pure diagonal tension*. Taking account of more two hypotheses for cyclic hysteresis on the basis of this theory, the hysteresis loop of shear spring is assumed as follows.

At first, as shown in Figure 6, steel panel of 240 mm by 120 mm square in which the material has perfectly elasto-plastic relationship between stress σ and strain ε is divided into band braces A and B in the conjugate diagonal directions. The width of these band braces assumed as computed for domain closed with 1/2 of span and height. The ratio of 1/2 can be experimentally determined by thickness, shear span ratio and elastic modulus of unit steel panel shear wall. And consider the unit moment free frame with the band braces hinged in the ends. When the frame with the band braces is subjected to shear force, the band braces A and B describe axially perfectly elasto-plastic σ-ε relationships without compressive resistance due to their buckling out of plane. Therefore, the Q-R relationships on the band braces A and B are added wholly with each other, so that the slipping hysteresis loop of shear spring can be simply obtained. This indicates that the band braces A and B do not show conjugate resistance at the same time but the brace A and B resist one after the other. Namely, the idea of tension field can be simply and rationally explained as if only tensile stress essentially occurred in steel panel.

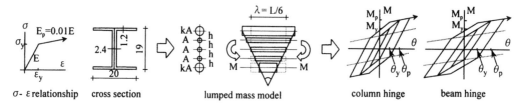

σ- ε relationship   cross section   lumped mass model   column hinge   beam hinge

Figure 5. Assumed hysteresis loop of hinge

Secondly, in order to consider practical panel as one body, hysteresis loop assumed as above should be modified. Moment of inertia on band braces as diagonal members having undergone tension should be consider increasing by buckled waves formed in rectangular direction to diagonal members. This is like anisotropy. Therefore, after diagonal members having undergone tension is unloaded, they continues to formed compression field till a certain constant stress, which is assumed as 1/6 of yielding stress in this case. And one more consideration is that the tensile residual strain at the crossing domain in one diagonal member hastens formation of tension field in the other diagonal member. Taking account of these two factors, modified hysteresis loop is shown in Figure 7. This diagram also shows the deformation process of simplified steel panel as follows. At the elastic stage, steel panel shear wall the forms the diagonal tension field in one direction after the shear buckling in early deformation and comes up to the yielding point of steel. Next, the previous tension stress is unloaded, and at last, this tension field is turned out to be compression field and comes up to the yielding point on the side of compression. From this point, steel panel shear wall starts to forms the diagonal tension field in the other conjugate direction and to vanish the compression field at the same time. After that, as the first elastic stage, the new diagonal tension field is formed. And the deformation processes of the band braces are symmetrical on loading direction. As above, when the slipping hysteresis loop of shear spring in Figure 6 is modified, more actual behabiour of steel panel shear wall can be explained.

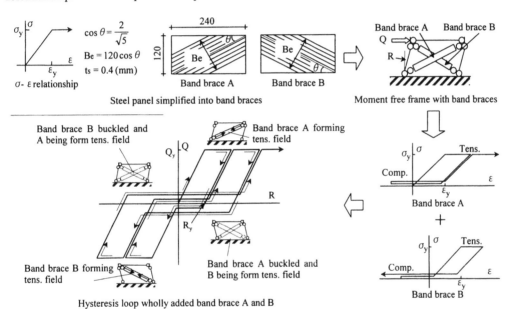

Figure 6. Hysteresis loop of shear spring and formation process of steel panel shear wall

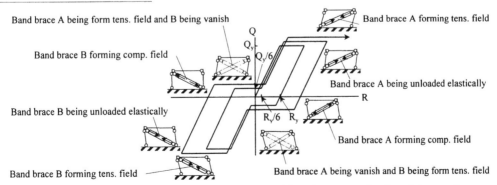

Figure 7. Modified hysteresis loop of shear spring and formation process of steel panel shear panel

3.4 *Analytical results*

3.4.1 *Hysteresis loops*

Dotted hysteresis loops for Type A and B are obtained by incremental numerical analysis. Hysteresis loops by solid lines indicates experimental results for comparison, such as shown in Figure 8.

On Type A, the hysteresis loop represents slipping characteristics which is due to assumed hysteresis loop of shear spring for steel panel shear wall. It is found that the analysis can follow the experimental result fairly well qualitatively as well as quantitatively. On Type B, the hysteresis loop represents spindle-shaped and *Baushinger's effects*. This loop is derived from the integration on all of hinges with trilinear relationship between bending moment and rotation angle. It is also found that the analysis can follow experimental result essentially. When this hinge is determined, the point where the bending yield on the original member starts is not considered. Therefore, yielding behaviour of whole hysteresis results in unnaturalness. This can be improved by increase of lumped mass in number. The incremental number must be equal to the incremental number of sectional condition i.e. yielding point, sectional area on web and flange in detail.

3.4.2 *Deflection curves*

Shear deflection curve can be obtained from horizontal displacements at the level of the center of gravity in each beam, such as shown in Figure 9. Both Type A and B similarly represent S-shaped curves nearly symmetrical between plus and minus on horizontal displacement, with inflection point at 3rd or 4th level.

3.4.3 *Distributions of story sway angles*

Distributions of story sway angles can be obtained from difference between horizontal displacement of neighbouring stories, such as shown in Figure 10. On Type A, it is well found that the larger the shear deformation becomes, the larger story sway angle in the middle of stories becomes than story sway angle at 1st and 6th stories. But on Type B, when the shear deformation becomes larger, the concentration of the shear deformation in the middle of stories is not as large as that on Type A. This indicates that shear spring do not resists uniformly at each story.

3.4.4 *Yield patterns of structural elements*

Yield pattern of structural elements can be obtained from assumed hysteresis loops which express mechanical states, such as shown in Figure 11. On

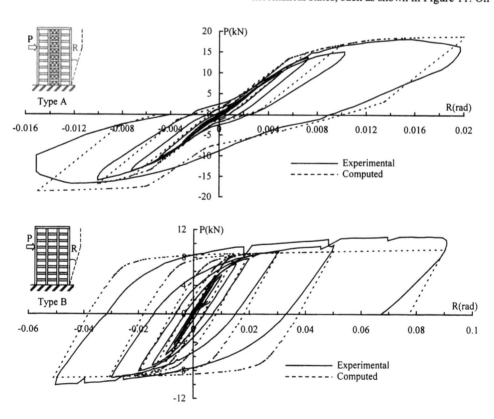

Figure 8. Comparison of P-R relationship between tested and calculated values

546

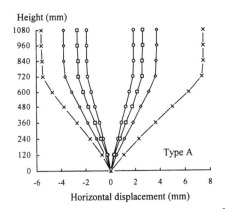
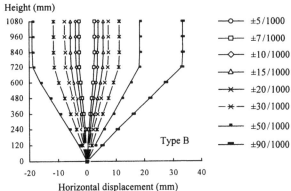

Figure 9. Deflection curve

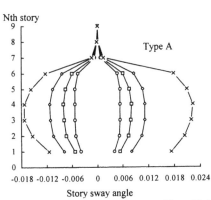
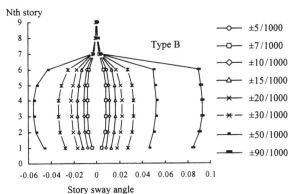

Figure 10. Distribution of sway angle at each story

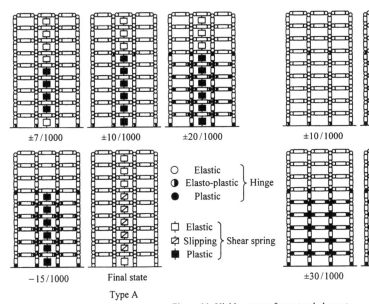

Figure 11. Yield pattern of structural elements

547

Type A, it is well found that shear springs for steel panel shear wall start to yield from the middle of stories and yield of shear spring go up to 6th story and down to 1st story. On the other hand, hinges start to yield from column hinge at the bottom and top within 1st to 6th story and, the next, the most outside beam hinges start to yield. After that, the yield of column and beam hinges go to the middle of story. This indicate that yield pattern of shear springs are a contrast to hinges in expanding yields of structural elements. On Type B, yield pattern of hinges are the same as Type A. Namely influence of setting shear spring.

4. CONCLUSION

In this paper, an analytical approach on the elasto-plastic cyclic sway behaviours of multistory-multispan steel rigid frame with center core steel panel shear wall system are developed and compared with test results. The coincidence between calculated results and tested results on the 9 story-3span steel rigid frame models with and without center core shear panel shear wall are very well. From this research the effect of center core steel panel shear wall on the cyclic sway behaviours are clarified.

5. REFERENCES

Caccese, V., Elgaaly, M., & Chen, R. 1993b. Experimental Study of Thin Steel Plate Shear Walls under Cyclic Loads. *J. Str. Engrg. ASCE.* 119 : 573-587.

Elgaaly, M., Caccese, V., & Du, C. 1990. Steel Plate Shear Walls Post-Buckling Behaviour under Cyclic Loads. *Proc. 4 U.S. National Conf. Earthquake Engrg.* Vol.2 :895-904.

Elgaaly, M., Caccese, V., & Du, C. 1993a. Postbuckling Behaviour of Steel Plate-Shear Walls under Cyclic Loads. *J. Str. Engrg. ASCE.* 119 : 588-605.

Mimura, H. & Akiyama, H. 1977. Load-Deflection Relationship on Earthquake Resistant Steel Shear Walls developed Diagonal Tension Field. *Trans. Architectural Institute of Japan.* 260 :109-114. (in Japanese)

Kuhn, P. 1956. *Stress in Aircraft and Shell structures.* New Yark, Toronto & London : McGRALL-HILL.

Yamada, M. 1996. Das Hanshin-Awaji-Earthquake. Japan. 1995-Schaden an Hochbauten (The Hanshin-Awaji-Earthquake. Japan. 1995-Collapses and damages od buildings). *Bauingenieur 71.* : 73-80. (in German)

Yamada, M. 1999. Steel Shear Panels for anti-seismic Elements. *Light-Weight Steel and Aluminum Structures (ed. Mäkelinen-Hassinen.).* Helsinki. : Elsevier Science. : 853-860.

Yamada, M. & Kawamura, H. et al. 1984. Elasto-Plastic Deformation and Fracture Behaviour Multi-Story. Multi-Span Steel Frame under Constant Axial Load. *Annual Meeting of AIJ. No. 24.* :453-456.

Yamada, M. & Yamakaji, T. 1997. Steel Panel Shear Wall-As single and Center Core Type Aseismic Element-. *STESSA '97. Kyoto.* : 477-484.

Yamakaji, T. & Yamada, M. 2000. Resisting Characteristics of Hybrid Center Core Shear Wall Systems. *Proc. 12. World Conf. Earthquake Engineering.* : Paper No.1479. (in printing)

Performance based design and moment resisting frames

Behaviour of Steel Structures in Seismic Areas, Mazzolani & Tremblay (eds) © 2000 Balkema, Rotterdam, ISBN 90 5809 130 9

Earthquake resistance of frames with unsymmetric bolted connections

D. Beg & P. Skuber
University of Ljubljana, Slovenia

C. Remec
Steel Construction Institute, Ljubljana, Slovenia

ABSTRACT: The paper deals with unsymmetric endplate bolted connection which is suitable to resist gravity loading, but does not provide adequate resistance and ductility in earthquake conditions, when fully ductile design is required. Two full-scale dynamic tests were conducted to evaluate connection behaviour. It was then modelled numerically as a nonlinear spring, taking into account unsymmetric behaviour, pinching effect and lower ductility for the tension at the weaker side. This connection model was used in nonlinear dynamic analysis of single storey frames and portal frames subjected to earthquake excitation. The obtained results show that the behaviour of unsymmetric connection was better than one would expect. Even after bolt rupture at the weaker side of the connection the carrying capacity of frames was not exhausted. This could be good news for some existing low rise steel buildings in Slovenia, where unsymmetric connections were used instead of symmetric ones, reminding on the fact that Slovenia is a seismic area.

1 INTRODUCTION

Our interest in unsymmetric bolted endplate connections with two rows of bolts at one flange and with only one row of bolts at the other flange (see Fig. 1) arises from the past misuse of this type of connection in Slovenian steel construction practice. Unsymmetric connections are suitable to resist gravity loading in non-sway frames and can be used also in sway frames when horizontal loading is not very important as in the case of moderate wind loading or in· non-seismic regions. They have been successfully used mainly in portal frame design and also in some multi-storey frames in countries with low seismicity such as United Kingdom and Germany.

Under the German influence this type of connections was introduced in Slovenia and sometimes used without sufficient precaution disregarding the fact that Slovenia is unlike Germany a seismic region. The real problem concerned more loose interpretation of the old Slovenian seismic design code than unsymmetric connections themselves. In the code the behaviour factor of 5 or even more was implicitly incorporated, but there were no detailed provisions for adequate ductile design of steel structures.

There was only a statement saying that ductile behaviour must be assumed. As before the Kobe and Northridge earthquakes steel structures were recog-

nised as ductile more or less automatically, also unsymmetric connections were regarded as satisfying as far as they were strong enough to carry the internal forces arising from seismic load combination. As a result a certain number of sway frames was designed and erected in Slovenia where unsymmetric end plate bolted beam-to-column connections were used with questionable resistance and ductility under the influence of bending moments causing tension on the weaker side of the connection.

To get an insight into the behaviour of unsymmetric endplate connections we performed experimental and numerical investigations. Two full-scale dynamic tests on connections were carried out, accompanied by non-linear dynamic analysis of simple

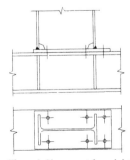

Figure 1. Unsymmetric endplate bolted connection

frames containing unsymmetric connections. The main purpose of this work was to assess the sensitivity to weaker side collapse of the connections in earthquake conditions and further behaviour of partly damaged frames.

2 EXPERIMENTAL ANALYSIS

Tee shaped beam-column assembly was chosen to represent a part of a real frame around the connection. The beam part of the assembly was made of IBE 300 hot rolled profile and the column part of HEB 200 hot rolled profile. The specimens were produced by a steel fabricator implementing standard workmanship quality. Both test specimens containing unsymmetric endplate bolted connection were nominally equal and made of steel with yield strength 315.5 MPa and ultimate tensile strength 460.1 MPa. The connections were made with M20 bolts of grade 10.9 preloaded to half of full preloading. The test set-up is described in Beg et al. (1999).

The tests were run under displacement control following the sinusoidal pattern with constant amplitude of approximately two times the yield displacement. The constant frequency of 0.5 Hz was applied. In each test the yield displacement was established in the first static part of the test according to the ECCS testing procedure.

To simulate the effect of the gravity loading the static displacement causing the bending moment equal to 30% of the beam plastic moment was introduced prior to dynamic loading. The preloading certainly caused tension at the weaker part of the connection, as in real frame situation.

The displacement v_y was estimated to 8-10 mm. The first specimen was subjected to the displacement amplitude of ±18 mm and the second one to ±20 mm. Unfortunately this increase caused tensile rupture of the two bolts at the weaker side of the connection at the beginning of the second test. New bolts were installed and the second test was run at the displacement amplitude of ±16 mm.

The test results for the first test are shown in Figures 2 and 3. In Figure 2 the relation moment - total relative rotation for the connection is plotted. Strong pinching effect mainly due to plastic extension of bolts producing larger deformability and decrease of absorbed hysteretic energy is evident. Relative moment amplitude in connection in relation to the number of cycles is shown in Figure 3.

In both tests the rupture by stable crack propagation occurred in the beam flange at the stronger part of the connection at the edge of the weld (HAZ). Due to preloading in those flanges plastic deformations were larger than at the weaker side of the connection.

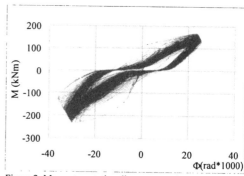

Figure 2. Moment – rotation diagram : test 1

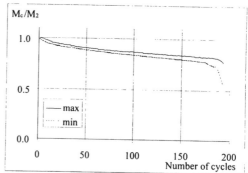

Figure 3. Moment amplitude throughout test1

The first specimen was able to withstand 195 cycles up to the loss of one half of its strength capacity and 93 cycles up to the loss of one half of energy dissipated in each cycle (in relation to the value in the second cycle).

The second specimen survived 118 cycles up to the loss of one half of its strength capacity.

Relatively large number of cycles was obtained because the imposed displacement amplitude of two times the yield displacement is relatively small and can be expected in moderate earthquakes.

3 NUMERICAL MODELLING OF THE CONNECTION

On the basis of the test results the connection behaviour was modelled numerically. A rotational spring element which was originally developed by Žarnić & Gostič (1997) for the analysis of masonry infilled RC frames and implemented into the computer program for nonlinear dynamic analysis DRAIN 2DX was adapted for our purpose.

The numerical model of the connection is in a way simple (bilinear behaviour in loading stage), but on the other hand represents all the important features of the behaviour of unsymmetric endplate connections. It includes unsymmetric hysteretic behaviour, pinching effect, low ductility for tension on the weaker side of the connection due to the possibility of rupture of bolts. It can also represent the connection behaviour after the possible rupture of the weaker side bolts. At bending moments causing tension on the stronger side the connection behaves like normal moment connection and at the reverse moments it acts as pinned connection opening a gap at the position of failed bolts.

The connection behaviour is described by the hysteretic envelope and the pinching effect and is shown in Figure 4.

4 NONLINEAR DYNAMIC ANALYSIS OF SIMPLE FRAMES

Numerical analysis of simple frames containing unsymmetric bolted connections was performed using computer program DRAIN D2X because there was a suitable spring element for connection modelling available. One bay single storey frame with horizontal beam and one bay portal frame were used in the analysis. In both cases pinned and rigid column bases were considered. The unsymmetric endplate bolted connections were placed at both beam-to-column connections and were represented by nonlinear springs. For beams and columns the bilinear elastoplastic moment-curvature relationship was assumed according to the plastic resistance of the corresponding cross-sections.

The basic behaviour of connection springs was taken from the tests (the shape of hysteretic envelope and individual hysteretic loops influenced by pinching effect). The actual strength was determined from the design requirements in each case and for the initial stiffness three characteristic values representing rigid, slightly semi-rigid and semi-rigid behaviour were chosen (S_{ini} = 50 K, 16.67 K, 5.55 K, K = EJ_b/L_b).

Rotational capacity for the rotations causing tension on the upper stronger side of the connection was chosen as 0.05, 0.055 and 0.06, respectively for the three initial stiffnesses. Rotational capacity for the opposite rotations causing tension on the weaker side of the connection were smaller (0.01, 0.015, 0.02), taking into account the possibility of premature bolt rupture. After the rotational capacity is exceeded, the corresponding strength in the numerical model quickly decreased and finally vanished completely.

Prior to the numerical analysis all frames were designed according to EC 3 (CEN 1992) and EC 8 (CEN 1994) design rules, taking into account the design ground acceleration of 0.3 g, which corresponds to the highest value adopted in Slovenia. For the

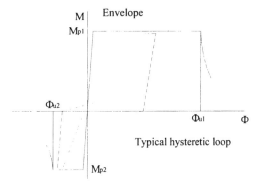

Figure 4. Modelling of the connection spring

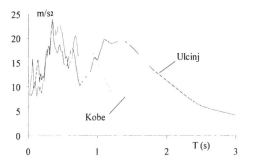

Figure 5. Acceleration spectra (Ulcinj adapted)

numerical analysis two actual accelerograms, Kobe (1995) and Ulcinj (1979) were used. The Ulcinj accelerogram was adapted to have the same peak ground acceleration as the Kobe accelerogram (0.836g). Then suitable scaling factors ranging from 0.5 to 2.0 were used in the analysis for both accelerograms. Acceleration spectra are shown in Figure 5.

To resume, all frames were designed according to EC 8 to resist strong earthquake. Only the beam to column connections were deliberately chosen to be unsymmetric and with satisfactory behaviour only for rotations producing tension on the stronger upper side of connections. The frames were then analysed numerically using accelerograms with fairly strong peak ground acceleration. The results of this analysis follow.

4.1 Single storey frame

The geometry and characteristic values of loadings are shown in Figure 6. Snow load was not considered in seismic load combination. In the numerical analysis the following parameters were varied:
- pinned and rigid column bases
- two accelerograms: Kobe, Ulcinj (K, U)
- scaling factors for accelerogram: 0.5, 1.0, 1.5, 2.0

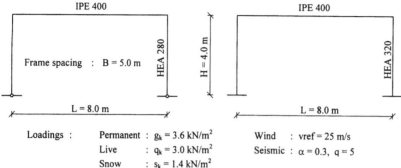

Figure 6. Geometry and loadings for single storey frame

- three different initial stiffnesses for connection spring (S1 - rigid, S2 – semi-rigid (stiffer), S3 – semi-rigid)
- influence of connection preloading (0%, 50%, 100%)

Gravity loading distributed along the beam produces favourable preloading of unsymmetric connections by initial compression at the weaker lower side. 100% of preloading means that all the gravity loading is distributed along the beam and no preloading (0%) means that all the gravity loading is concentrated as vertical forces on the top of columns, serving only as a mass in a dynamic analysis and producing no direct bending moments in beams. 50% of preloading means that the loading is equally shared between two previous cases.

Altogether 44 cases were analysed and the summary of results is shown in Tables 1 and 2 where

maximal horizontal displacement relative to storey height and observed damage is reported. *Elastic* means that the structure remains completely in elastic range, *Small* means that small plastification occurred with relative connection rotations below 0.01. *Large* damage indicates rotations over 0.01 and *Bolt Rupture* (B.R.) indicates that bolts at the weaker side of the connection failed in tension (L for left connection, R for right connection, B for both connections). *Collapse* indicates that the structure became unstable.

From the results it is clear that with increasing scaling factor the damage increases and that the difference between the results for the two different accelerograms are not very pronounced, despite the differences in their spectra. The initial stiffness influences the maximum displacements obtained, but has little influence on the damage developed in connections.

Table 1. Single storey frame results

Pinned column bases

Parameters	u_{max}/h	Damage
K-0.5, S1, 0%	0.023	Small
K-0.5, S1, 50%	0.023	Small
K-0.5, S1, 100%	0.023	Small
K-1.0, S1, 100%	0.030	B.R.-B
K-1.5, S1, 100%		Collapse
K-0.5, S2, 100%	0.025	Small
K-1.0, S2, 100%	0.040	B.R.-L
K-1.5, S2, 100%		Collapse
K-0.5, S3, 100%	0.026	Elastic
K-1.0, S3, 100%	0.058	B.R.-L
K-1.5, S3, 100%		Collapse
U-0.5, S1, 0%	0.025	B.R.-L
U-0.5, S1, 50%	0.029	B.R.-L
U-0.5, S1, 100%	0.020	Small
U-1.0, S1, 100%	0.041	B.R.-R
U-1.5, S1, 100%		Collapse
U-0.5, S2, 100%	0.053	B.R.-B
U-1.0, S2, 100%	0.040	B.R.-L
U-1.5, S2, 100%		Collapse
U-0.5, S3, 100%	0.018	Elastic
U-1.0, S3, 100%	0.067	B.R.-B
U-1.5, S3, 100%		Collapse

Table 2. Single storey frame results

Rigid column bases

Parameters	u_{max}/h	Damage
K-2.0, S1, 0%	0.021	B.R.-L
K-2.0, S1, 50%	0.016	Small
K-2.0, S1, 100%	0.015	Small
K-1.0, S1, 100%	0.010	Elastic
K-1.5, S1, 100%	0.013	Small
K-1.0, S2, 100%	0.010	Elastic
K-1.5, S2, 100%	0.013	Elastic
K-2.0, S2, 100%	0.017	Small
K-1.0, S3, 100%	0.009	Elastic
K-1.5, S3, 100%	0.013	Elastic
K-2.0, S3, 100%	0.020	Large
U-2.0, S1, 0%	0.025	B.R.-R
U-2.0, S1, 50%	0.017	Small
U-2.0, S1, 100%	0.017	Small
U-1.0, S1, 100%	0.011	Elastic
U-1.5, S1, 100%	0.014	Small
U-1.0, S2, 100%	0.010	Elastic
U-1.5, S2, 100%	0.014	Small
U-2.0, S2, 100%	0.017	Small
U-1.0, S3, 100%	0.008	Elastic
U-1.5, S3, 100%	0.013	Elastic
U-2.0, S3, 100%	0.017	Small

Preloading of connections due to gravity loading acts favourably. Decreasing of preloading causes bolt rupture, while at full preloading the damage is small. This can be seen from Figure 7, where maximum relative displacement and damage are shown for rigid column base, the largest initial connection stiffness S1 and adapted Ulcinj accelerogram with scaling factor 2. The displacement and damage are increasing with the decreasing preloading. Fully preloaded connection underwent small damage, while in the case of non-preloaded connections one of them failed by bolt rupture.

In the case of rigid supports there was no bolt rupture except in the case when no preloading was applied in the connections. The damage induced in these frames is shared between fixed column bases and beam-to-column connections. Pinned frames are more subjected to larger damages in the connections and bolt rupture on the weaker side of connections occurred regularly at scaling factor 1.0 and complete collapse of frames at scaling factor 1.5. Nevertheless these frames survived in the numerical simulation the Kobe earthquake with peak ground acceleration of 0.836 g.

Typical frame response is shown in Figures 8-11, where the results for the frame with pinned column bases, 100% preloading, intermediate initial stiffness S2 and the Kobe accelerogram (scaling factor 1.0) are presented. Moment-rotation diagrams for the left and right connection are plotted in Figures 8 and 9. Partial failure due to bolt rupture on the weaker side of the left connection can be noticed. The behaviour of both connections before and after bolt rupture is evident also from moment-time diagram in Figure 10. After bolt failure there are no positive moments in the left connection. The connection behaves as pinned for positive rotations. In Figure 11 the top displacement as a function of time is plotted and a residual displacement that opens the gap at the position of failed bolts can be observed.

4.2 Portal frames

In portal frame like structures the share of steelwork is the highest. Very often the only variable loading contributing to masses for seismic design is snow loading. In our analysis portal frames with span L=20 m and L=30 m were chosen and the column height was varied from 6 m to 10 m for fixed supports and from 4 m to 8 m for pinned supports. Three different initial stiffnesses of connections were taken into account. In the design the combination factor Ψ_2 for snow was taken as 0.3. In EC 1 Ψ_2 is basically set to zero, but non-zero values can be used in areas where snow loading is important. In dynamic analysis snow loading was varied (and consequently the masses) by applying three values of Ψ_2: Ψ_2= 0 (no snow), 0.5, 1.0 (full snow load). Ge-

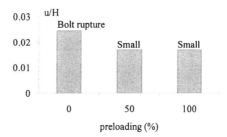

Figure 7. Influence of preloading on behaviour of connections

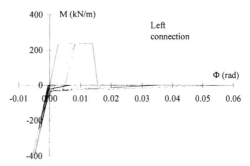

Figure 8. Moment - rotation diagram : left connection

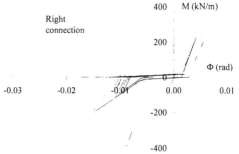

Figure 9. Moment – rotation diagram : right connection

ometry and characteristic loadings are shown in Figure 12. Deliberately large snow loading was chosen to intensify the influence of acting masses. Only the Kobe accelerogram was used with scaling factor 1.0 for pinned supports and 1.5 for fixed supports.

The results are summarised in Tables 3 and 4. Larger damages are observed at pinned frames for the same reason as at single storey frames.

The increase of snow load (masses) increases the damages, which is the case particularly at frames with pinned supports. When no snow was taken into account, all connections remained elastic and with larger snow loads the displacement and damages increased.

The increase of span and column height on frames behaviour is not very pronounced, because it

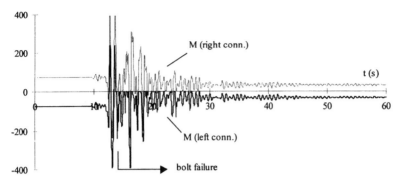

Figure 10. Moment – time diagram

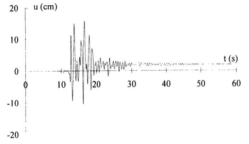

Figure 11. Top displacement – time diagram

is compensated by stronger cross-sections of frame members and stronger connections due to design requirements. From the results it is evident that portal frames that we analysed survived the strong Kobe earthquake with no damage in connections, when snow load was excluded. At rather large snow loads included, larger damages were produced, but only in two cases bolts on the weaker side of the connections failed and the collapse was observed only in the case of 10 m high columns and very large snow load.

The main reasons for fairly good behaviour of usymmetric connections under consideration besides low masses and corresponding inertia forces are:

– after the bolt rupture occurs at the weaker side of one or even both connections, the frame become more flexible and the fundamental period increases considerably, which usually means decrease of seismic loadings

– the change of the fundamental period is more pronounced in one storey frames than in more complex structures which are more redundant

– after the bolt rupture at the weaker side the connection can undergo further rotations, as it behaves as pinned for rotations producing tension at the position of failed bolts.

Table 3. Portal frame results

Pinned column bases		
Parameters	u_{max}/h	Damage
H = 4, L = 20, ψ_2 = 0, S1	0.021	Elastic
H = 4, L = 20, ψ_2 = 0.5, S1	0.029	Small
H = 4, L = 20, ψ_2 = 1.0, S1	0.027	Large
H = 4, L = 20, ψ_2 = 0.5, S2	0.028	Small
H = 4, L = 20, ψ_2 = 0.5, S3	0.024	Elastic
H = 6, L = 20, ψ_2 = 0, S1	0.020	Elastic
H = 6, L = 20, ψ_2 = 0.5, S1	0.028	B.R.-R
H = 6, L = 20, ψ_2 = 1.0, S1	0.036	Large
H = 6, L = 20, ψ_2 = 0.5, S2	0.035	Large
H = 6, L = 20, ψ_2 = 0.5, S3	0.051	B.R.-L
H = 8, L = 20, ψ_2 = 0.5, S1	0.038	Large
H = 8, L = 20, ψ_2 = 0.5, S2	0.028	B.R.-L
H = 8, L = 20, ψ_2 = 0.5, S3	0.041	Large
H = 6, L = 30, ψ_2 = 0.5, S1	0.036	Large
H = 6, L = 30, ψ_2 = 0.5, S2	0.039	Large
H = 6, L = 30, ψ_2 = 0.5, S3	0.044	Elastic

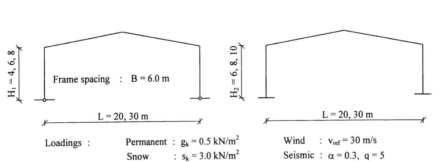

Frame spacing : B = 6.0 m

H_1 = 4, 6, 8

H_2 = 6, 8, 10

L = 20, 30 m

L = 20, 30 m

Loadings : Permanent : g_k = 0.5 kN/m²
Snow : s_k = 3.0 kN/m²

Wind : v_{ref} = 30 m/s
Seismic : α = 0.3, q = 5

Figure 12. Geometry and loading of portal frames

Table 4. Portal frame results

Rigid column bases

Parameters	u_{max}/h	Damage
H = 6, L = 20, ψ_2 = 0, S1	0.021	Elastic
H = 6, L = 20, ψ_2 = 0.5, S1	0.013	Elastic
H = 6, L = 20, ψ_2 = 1.0, S1	0.021	Small
H = 6, L = 20, ψ_2 = 0.5, S2	0.012	Elastic
H = 6, L = 20, ψ_2 = 0.5, S3	0.013	Elastic
H = 8, L = 20, ψ_2 = 0.5, S1	0.019	Elastic
H = 8, L = 20, ψ_2 = 0.5, S2	0.020	Elastic
H = 8, L = 20, ψ_2 = 0.5, S3	0.019	Elastic
H = 10, L = 20, ψ_2 = 0, S1	0.012	Elastic
H = 10, L = 20, ψ_2 = 0.5, S1	0.017	Elastic
H = 10, L = 20, ψ_2 = 1.0, S1		Collapse
H = 10, L = 20, ψ_2 = 0.5, S2	0.017	Elastic
H = 10, L = 20, ψ_2 = 0.5, S3	0.017	Elastic
H = 6, L = 30, ψ_2 = 0.5, S1	0.003	Elastic
H = 6, L = 30, ψ_2 = 0.5, S2	0.003	Elastic
H = 6, L = 30, ψ_2 = 0.5, S3	0.003	Elastic

5 CONCLUSIONS

On the basis of test results on connections and numerical simulations on frames it is possible to conclude that unsymmetric endplate bolted connections behave in earthquake conditions better than it would be expected considering low strength and ductility for the moments causing tension at the weaker side of the connections. It is certainly not our aim to encourage the use of such connections in earthquake resistant steel frames, but it is important to recognise that at least for the existing single storey frames and portal frames with unsymmetric connections there is a potential resistance to seismic actions. Even after rupture of bolts on the weaker side, connections (and frames) can sustain further earthquake shocks acting as rigid in one and as pinned in the other direction. This is particularly true for portal frames where small masses are involved and the structural response even at very strong earthquake is not far from elastic behaviour.

It is important to note that there are some additional aspects that require respect, such as the influence of low cycle fatigue (in our tests no important differences in comparison to normal symmetric connections in this respect were observed) and the influence of tensile and shear forces on the behaviour of partially ruptured connections.

ACKNOWLEDGEMENTS

The research presented in the paper was supported by grants from INCO-COPERNICUS Joint Research Project "RECOS" and from Slovenian Ministry for Science and Technology.

REFERENCES

Beg, D. et al. 1999. Behaviour of Unsymmetric Bolted Connections Subjected to Dynamic Loading. In D. Dubina & M. Ivanyi (eds), Stability and Ductility of Steel Structures; Proc. 6th inern. colloq., Timisoara, 9-11 September 1999. Amsterdam: Elsevier.

Bernuzzi, C. et al. 1997. Steel Beam-to-Column Joints: Failure Criteria and Accumulative Damage Models. In F.M. Mazzolani & H. Akiyama (eds), Behaviour of Steel Structures in Seismic Areas; Proc. 2nd intern. conf., Kyoto, 3-8 August 1997. Salerno: Edizioni 10/17.

Calado, L. et al. 1999. Cyclic Behaviour of Steel Beam-to-Column Connections: Interpretation of Experimental Results. In D. Dubina & M. Ivanyi (eds), Stability and Ductility of Steel Structures; Proc. 6th inern. colloq., Timisoara, 9-11 September 1999. Amsterdam: Elsevier.

CEN 1992. ENV 1993: Design of Steel Structures (EUROCODE 3).

CEN 1994. ENV 1998: Design Provision for Earthquake Resistance of Structures (EUROCODE 8).

Žarnić, R. & Gostič, S. 1997. Masonry Infilled Frames as an Effective Structural Sub-Assemblages. In P. Fajfar & H. Krawinkler (eds), Seismic Design Methodologies for the Next Generation of Codes; Proc. intern. workshop, Bled, 24-27 June 1997. Rotterdam: Balkema.

Behaviour of Steel Structures in Seismic Areas, Mazzolani & Tremblay (eds) © 2000 Balkema, Rotterdam, ISBN 90 5809 130 9

Inelastic dynamic and static analyses for steel MRF seismic design

B. Calderoni
Department of Analysis and Structural Design, University of Naples 'Federico II', Italy

Z. Rinaldi
Department of Civil Engineering, University of Rome 'Tor Vergata', Italy

ABSTRACT: The performance of steel MRF with rigid connections, proportioned by adopting different capacity design criteria, is evaluated in order to highlight the effectiveness of static non-linear procedure in predicting the structural seismic behavior. In the framework of the performance-based design, some considerations are made on the base of the results obtained by both dynamic time histories and push-over analyses, particularly with reference to the damage levels and the structure ability to withstand a strong earthquake.

1 INTRODUCTION

The unexpected high damages experienced during recent earthquakes by many buildings, even if designed according to the seismic codes, have driven the researchers to a conceptual revision of the methodologies currently adopted in the practice. The criteria until now used in seismic design of structures, particularly devoted to ensure the no-collapse requirement, are not considered any longer fully satisfactory, due to their inadequacy in predicting and limiting the damage levels. Since 1995 the so-called performance based design (SEAOC 1995) is considered the most appropriate way to face the problem of controlling the structure performance even with reference to different levels of seismic hazard (Bertero 1997, Hamburger 1997).

While this new approach is very attractive from a conceptual point of view, its practical application is not yet well defined and many researchers are now working for developing simple and reliable methodologies to be incorporated in the new codes (Gobarah et al. 1997). The prediction of the structure performance related to a specified level of ground motion, in fact, requires to evaluate displacements and forces distribution also beyond the yield limit, which cannot be derived from simplified elastic calculations (Priestley 2000).

The inelastic dynamic time history seems to be the most effective tool for evaluating the structural behavior, but it is too cumbersome and time consuming. Furthermore this methodology is strictly connected to the earthquake characteristics, and the use of a large ensemble of accelerograms is necessary for obtaining meaningful results. The static push-over analysis, on the contrary, is considered a suitable procedure for predicting the non linear behavior of the structure in a more accessible way (Tso & Moghadam 1998).

In this paper, the performance of moment resisting steel frames with rigid connections, designed according to different criteria, has been evaluated in order to judge: the effectiveness of the non-linear static analysis if compared to the dynamic time history; the influence of the overstrength given by different capacity criteria and the effectiveness of a displacement oriented approach.

2 MRF PERFORMANCE EVALUATION

Force-based design is nowadays the methodology used in proportioning seismic structures and required by present codes. Conventional horizontal forces are obtained by reducing the elastic ones by means of a factor (R_w, q-factor and so on), accounting for the plastic deformation experienced during the ground motion. The resistance of the system for this load is then verified.

Obviously the necessary ductility has to be assured and the application of capacity design criteria is required in order to improve the ductile behavior. Anyway, whatever be the adopted design criterion, it is effective only if the structure is really able to withstand severe earthquakes.

At this aim, the dynamic behavior of MRF, designed according to various criteria, has been yet analyzed by the authors (Calderoni et al. 1996, 1997b). The obtained results have pointed out that a judgement on the bearing capacity of structures, designed according to European seismic code - EC8 (Eurocode 8 1994), cannot be independent of the way the design has been really developed, with particular reference to the amount and distribution along the height of the design overstrength. It was highlighted that the capacity criterion, as proposed by EC8, sometimes can give also unsatisfactory results, and to provide the column with a given overstrength, "correctly" distributed, can be more effective and simple in many cases.

Frames designed in order to exhibit an ultimate global mechanism, were also analyzed (Calderoni et al. 1995): improvement of the structural inelastic performance has been shown, facing a significant overstrength given to the columns and the corresponding increase of structural weight.

Finally, the force-based design gives satisfactory results with reference to the collapse behavior, when an adequate capacity criterion is applied, even if some uncertainties are still related to the force-reduction factor. Anyway the present criticism to this procedure is the inadequacy in predicting and limiting the damage levels, particularly for the action of earthquakes having a larger probability of occurrence.

The displacement-based design, on the contrary, is not oriented to ensure the structures an assigned strength, but to control the damage level related to displacements expected to occur for different ground motion intensities. Practically the structural performance is evaluated in terms of deformation rather than of resistance.

The application of the "displacement-based design", however, requires the use of inelastic static analyses. Therefore the behavior of some MRFs has been evaluated by means of both static push-over and dynamic procedures, and the corresponding results are compared and discussed also with reference to the adopted design criteria.

3 ANALYSED FRAMES AND DESIGN CRITERIA

Four different steel MRFs (named 13-14-17-18) has been analyzed. Their geometrical and load schemes are reported in fig.1.

It must be noted that the masses have been assigned in such a way that all the frames exhibit the same first period, equal to 1.5 s, despite the differences in the element cross-sections. Furthermore two levels of vertical loads have been considered for the beams: a lower one in the frames 13 and 17 and a higher one in the other two (14 and 18).

All the frames have been designed according to EC8 requirements, for a seismic zone characterized by PGA=0.35 g and soil type A. Multi-modal response spectrum analysis has been performed, by adopting a reduction factor $q = 5$. P-Δ effects have been also accounted for in the approximate way allowed by the code.

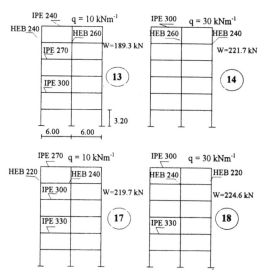

Figure 1. Analyzed frames: geometrical and load schemes.

Furthermore three different design criteria (named B, C and Z) (Calderoni et al. 1997a) have been considered in proportioning the frame elements, so that twelve different structures have been analyzed.

It must be pointed out that the necessary resistance has been given to the frame conventionally assigning to each section a proper (and different) value of the yielding stress, without any change of the prefixed cross-sections. In this way the structure is provided with the exact overstrength required by the used design criterion, and contemporary no variation affects the fundamental period. Furthermore, the inter-story drift limitation, stated by the code, has been deliberately not accounted for.

The first adopted design criterion (B) consists in not applying any capacity criterion, i.e. in giving the frame no design overstrength. The second one (C) respects the EC8 requirements, increasing the columns ultimate strength respect to the beams one.

The last design criterion (Z) gives the columns a prefixed overstrength, which presents a linear pattern along the height of the frame, and then is independent of the beams strength. More in detail, the columns strength increase is equal to 25% at the first floor, with a mean value of 50% for the whole frame. Note that this overstrength distribution is consistent with the column failure pattern, usually resulting from dynamic analyses of frames designed without any overstrength (Calderoni et al. 1995).

4 DYNAMIC ANALYSES

The evaluation of the seismic performance of the above described twelve frames has been performed

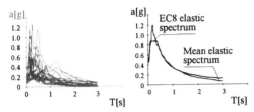

Figure 2. Elastic spectra of the thirty selected earthquakes.

Figure 3. Acceleration mean elastic spectrum.

by means of a non linear dynamic time history analysis. As the dynamic behavior is strongly affected by the ground motion, an ensemble of thirty real records, selected among a great number of Italian earthquakes (Fig.2), has been adopted as seismic input. The seismic records have been suitably scaled in order to improve their homogeneity and to obtain an average spectrum similar to the EC8 elastic one, used in the design phase (Fig.3), with a PGA=0.35 g.

The dynamic response of the frames subjected to the whole set of accelerograms has been evaluated by means of the Drain2DX code. The obtained dynamic quantities (displacement, rotation etc), when statistically interpreted, refer to their mean values, as the design spectrum is analogous to the mean spectrum of the selected records.

The performance of the frames has been then judged in terms of both collapse and damage, as described below.

Furthermore incremental analyses have been performed, for PGA values ranging from 0.1g to 0.8 g.

4.1 Collapse analysis

The dynamic behavior of the frames has been firstly evaluated in terms of ultimate resistance.

The percentage of earthquakes, under which the structure reaches the limit fixed by a proper collapse criterion, is assumed as failure index.

Since the dynamic frame performance is defined by different response parameters, the choice of the one related to the collapse is a first problem. Once chosen the parameter, the definition of its limit value is a problem too. A number of proposals on the matter can be found in literature. In this paper three different collapse criteria, frequently used, have been considered, which refer to the attainment, respectively, of:
- the maximum plastic rotation;
- the maximum accumulated plastic rotation;
- the maximum inter-story drift ratio.

According to many research results, the failure is supposed to occur when the following values of the selected parameters are reached:

- 4% for beams and 2.5% for columns as regards the plastic rotation;
- 10% for beams and 5% for columns as regards the accumulated plastic rotation (Akbas & Shen 1998);
- 1.5% for the inter-story drift ratio (life-safe limit in SEAOC 1995).

In Figures 4-5, for all the analyzed frames, the failure index is plotted with reference to the three adopted criteria.

When maximum plastic rotation is used as collapse parameter (Fig. 4), all the frames seem to exhibit a good behavior, as the percentage of failures is always very low, quite independently of the adopted design criteria.

On the contrary, if we refer to the maximum accumulated plastic rotation (Fig. 5) significant differences appear: in this case the adoption of a capacity design criterion (C and Z) proves to be effective in reducing the failure probability, up to 50%. In fact the accumulated plastic rotation is related also to the cyclic behavior and the energy absorption, which are necessarily influenced by the adopted design criteria. It is evident that in the C and Z cases the columns are less engaged in cumulative plastic deformation, due to the overstrength given to them with respect to the beams. This aspect is particularly highlighted by the damage analysis, reported in the following.

It is noteworthy that, as already pointed out, the Z criterion, very simple to be applied, is equivalent or even safer than the capacity criterion prescribed by EC8.

As far as the maximum inter-story drift is concerned, the corresponding failure index has proved to be higher if compared with the first collapse criterion, but with a similar pattern.

4.2 Damage analysis

The damage level of the frames has been evaluated by means of a damage index (D_I), defined as the ratio of the maximum accumulated plastic rotation attained during the earthquake above the fixed limit value of it.

In Figure 6, for all the examined cases, the mean value, obtained from the thirty earthquakes, of D_I, separately for beams, columns and the whole frames,

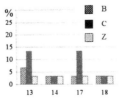

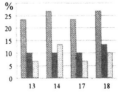

Figure 4. Failure index based on maximum plastic rotation.

Figure 5. Failure index based on maximum accumulated plastic rotation.

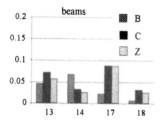 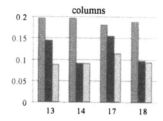 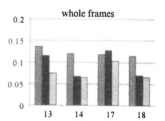

Figure 6. Damage index (mean value) for beams, columns and whole frames.

is plotted. It can be noted that the damage levels in C and Z schemes are higher for beams and lower for columns, if compared to the B schemes. Moreover, the D_I for the whole frame is quite always lower in cases C and Z.

Similar results (not reported) have been obtained with reference to the maximum plastic rotation.

The adoption of whatever capacity design criterion confirms to be quite effective in reducing distribute damages in the structure, shifting the plastic engagement from columns to beams.

It can be noticed that a low level of vertical loads on the beams (frame 13 and 17) reduces the effectiveness of the EC8 design criteria (C), both for failure and for damage, provided that the column strength is dependent on the beams one.

5 STATIC NON LINEAR ANALYSES

The behavior of the above considered frames has been studied also by means of static inelastic procedures. Push-over analysis has been performed on the structures by increasing a fixed pattern of lateral forces. Since different criteria have been proposed in order to define the proper load pattern for performing a significant push-over analysis, in this paper two different lateral load distributions along the frame height have been considered. The first one is the same of the first vibration mode shape, while the second one is similar to the modal forces envelope. Even if these two patterns are quite different, the obtained results are not significantly affected by them and then the reported results will be related only to the second adopted distribution.

The capacity curves (normalized base shear V_b/W versus roof drift angle δ_{top}/H) for all the frames and the considered design criteria are given in Figure 7.

Three horizontal lines are plotted in each diagram, which are related to significant strength levels of the frames. The lower one points out the design base shear (V_{DB}), obtained by means of the performed modal analysis. Since no overstrength was given to the beams, the first plastic hinge, for all the frames, should be developed at the attainment of that shear base value, as exhibited by frames 13 and 17,

independently of the adopted design criterion. Small shiftings are possible, due to the approximate way the P-Δ effects have been evaluated.

Instead, frames 14 and 18, which bear more considerable vertical loads, experience the first yielding for a higher base shear, when criteria C and Z are applied. For this frames the beams design is governed by the factorized vertical load condition rather then by the seismic one, so that the beams exhibit some overstrength respect to the seismic strength demand. Then, if no overstrength is given to the columns (case B), the first yielding at V_{DB} occurs just in these elements, while the elastic limit is obviously shifted up when a capacity design criterion is considered.

The higher horizontal line plotted in the diagrams points out the frame ultimate shear resistance (V_U), defined as:

$$V_U = \sum_{c=1}^{n} \left| \frac{Mu_b + Mu_t}{h} \right|$$

where Mu_b and Mu_t are the ultimate bending moments at the end sections of the column, h is the inter-story height and n is the number of the columns.

Practically V_U is the ultimate shear strength provided by the first floor at the mechanism development and it should be coincident with the design base shear V_{DB} when no overstrength is given (case B). Instead an additional strength respect to the design global base shear is due to gravity load presence, and it cannot be removed. In fact, while the bending moments for horizontal forces change in sign when inverting the action, the ones for vertical loads are fixed, so forcing to oversize the columns, when having, as usual, symmetrical cross-sections. This means that a frame exhibits an additional amount of horizontal shear strength just because it bears gravity loads, even if the global base shear due to vertical action is null.

The push-over curves show that, in all the cases, the frames exhibit a maximum strength practically equal to the theoretical one (V_U).

The intermediate horizontal line indicates the increase in shear strength with respect to V_{DB} correlated to the applied capacity design criterion (C, Z).

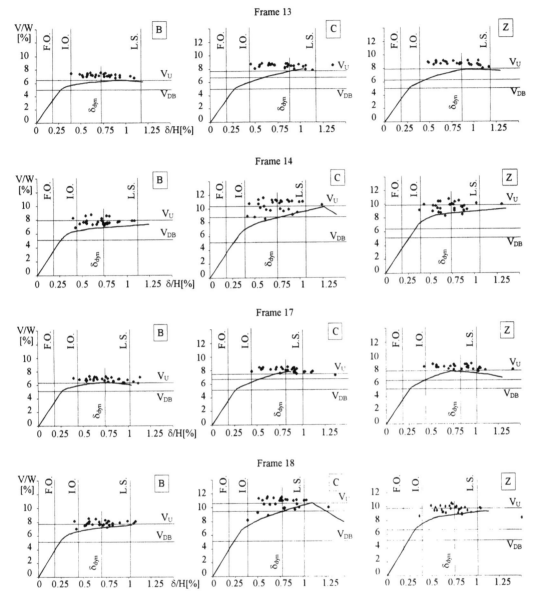

Figure 7. Capacity curves of the frames for the adopted design criteria.

Obviously, when no overstrength is given (B), this line is coincident with the lower one (V_{DB}).

Note that, while the increase of resistance in the Z case is quite constant whatever be the frame, being fixed a priori, in the C case, this increase is influenced by the beams strength, i.e. by the vertical loads level.

On each curve three vertical lines indicate the roof displacements related to the attainment of the maximum inter-story drifts ratio (δ/h), which define the performance levels stated by SEAOC, as follows:

- Fully Operational (F.O.) - $\delta/h = 0.2\%$
- Immediate Occupancy (I.O.) - $\delta/h = 0.5\%$
- Life Safe (L.S.) - $\delta/h = 1.5\%$

With reference to this last limit state, in Figure 8, only for two frames, but for all the adopted criteria, the plastic hinge patterns are depicted.

It can be noted that no mechanism has been already reached, even if the first floor failure is close to be attained. Anyway yielding is widely spread both in columns and beams, but not according to a global mechanism.

On the base of the push-over results, it can be observed that, contrary to the expectation, no significant differences among the behavior of frames emerged. In fact, despite the various adopted design criteria, which provide the structures with different resistance levels, the limit inter-story drifts, related to the above said performance levels, are reached in all the cases quite for the same roof displacements. Furthermore, also the maximum values of plastic rotation in the elements are quite similar and only slight differences have been shown in plastic engagement distribution.

Anyway it must be reminded that the analyzed frames have been designed without any additional design resistance, contrary to the most of cases reported in literature, which refers to real frames. The unavoidable overstrength of these last ones can, in fact, influence in a random way the structural behavior and can create some misleading in the interpretation of the results.

6 COMPARISON BETWEEN STATIC AND DYNAMIC BEHAVIOUR

On the push-over diagrams the results of the dynamic non linear analyses are reported too. The solid square dots represent the couples V_{bmax}-δ_{topmax} obtained from the time histories performed for each of the thirty used ground motions. The few cases, in which dynamic instability has occurred, have been not considered. A vertical solid line points out the mean dynamic roof displacement (δ_{dyn}).

A good agreement emerges between the dynamic

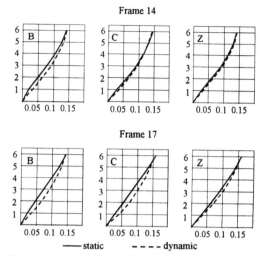

Figure 9. Static and dynamic floor displacements [m].

results and the static curves. As predictable, the dots are grouped close to the mechanism line (V_U), proving the development of several plastic hinges in the columns during the quakes.

A little higher scatter in the dynamic results is shown, when vertical loads are considerable and govern the beam design (frame 14 and 18). In these cases a number of dots can be found also well below the mechanism line, even if the push-over curves are always conservative.

In Figure 9 the dashed line outlines the dynamic deformed shape of the frames 14 and 17. For each floor the mean values of the maximum floor displacements obtained from the dynamic analyses are reported. The solid line, instead, represents the distribution of floor displacements from push-over static analyses at the roof deflection corresponding to the above defined dynamic one.

A satisfactorily agreement between static and dynamic displacement shape can be noted. The slight scatters are due particularly to the different behavior at the first floor. Anyway the figure confirms that the push-over method can provide lateral deflections similar to the ones obtained from inelastic dynamic analyses, and that the horizontal forces pattern used in performing the static analyses is quite suitable, at least for frames of such height.

Similar consideration can be made also with reference to the inter-story drift ratios. In Figure 10 the solid line refers to the values obtained from the push-over analyses, at the top displacement already reported in Figure 9, while the dashed line refers to the mean values of the maximum dynamic drifts.

Again, significant scatters are shown particularly at the first floor, where the dynamic results are quite always higher, highlighting a larger plastic engagement of columns.

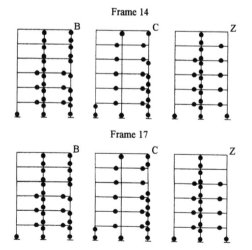

Figure 8. Plastic hinge patterns at Life-Safe limit.

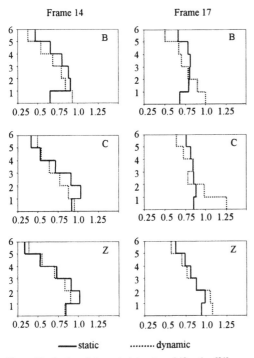

Figure 10. Static and dynamic inter-story drift ratios [%].

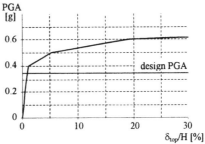

Figure 11. Frame 13 C: dynamic push-over curve.

Anyway, for both floor displacements and inter-story drifts, no meaningful differences of behavior have been pointed out among the frames, whatever be the design criterion adopted (B, C, Z).

It is also of interest to discuss about results obtained from the dynamic analyses performed by increasing the PGA from 0.1g to 0.8 g. In Figure 11 the relationship between the mean value of the roof drift angle and the increasing PGA is plotted with reference to frame 13 C. This kind of diagram can be considered as a sort of dynamic push-over.

In Figure 12 the zone closer to the origin is enlarged. The three vertical solid lines drawn in this figure indicate the roof displacements (from the static push-over analysis) related to the attainment of the maximum inter-story drift ratios (δ/h), which define the performance levels stated by SEAOC. The horizontal solid lines, instead, point out the PGA values corresponding to different levels of ground motion.

The higher line (PGA = 0.35g) represents the design strong earthquake, for which no failure has to occur in the structure (no-collapse requirement in EC8 or "life-safe" in SEAOC).

The intermediate line (PGA = 0.175g) represents the serviceability earthquake (the design one reduced by two), for which the maximum inter-story drift should be not greater than 0.4 ÷ 0.6%, as stated in EC8. This no-damage requirement is equivalent to the "operational" SEAOC performance level related to an "occasional" seismic event.

The lower line (PGA = 0.0875g) represents, in authors opinion, the "frequent" earthquake, for which, according to SEAOC, the structure must exhibit, at the most, negligible damages ("fully operational") with maximum inter-story drift ratio not greater then 0.2%.

The figures show that the limit value of inter-story drift ratio (1.5%), corresponding to the "life-safe" performance level, is not exceeded for the design PGA (0.35g) while the maximum ground acceleration bearable by the frame is about 0.60g, higher than the design value. The no-damage requirement is well fulfilled too, while the "fully operational" performance level is not respected, but just for a little.

Finally the global behavior of the frame can be considered quite satisfactory, if judged by using the dynamic results, both for serviceability and ultimate limit states.

If we refer to the static analysis, the values of normalized base shear (V_b/W), corresponding to the "occasional" and "frequent" earthquakes, are respectively 12% and 6.3%. From the push-over curve (frame 13 C in Fig. 7) it can be easily found that the above said serviceability requirements are, instead, not respected.

As regards the frame performance under the strong earthquake (no-collapse requirement), when adopting a push over analysis it is necessary to refer to a

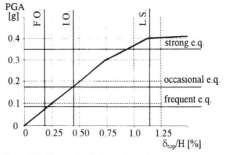

Figure 12. Frame 13 C: dynamic push-over curve (detail).

target displacement, in the framework of the displacement-based design.

The best way to evaluate this displacement should be to perform a number of time history dynamic analyses, as made in this work; but, in this case, the static push-over should be useless. Actually some simplified criteria, suggested in order to make feasible in the practice this kind of design procedure (Faijfar & Gaspersic 1996, Chopra & Goel 1999), can be used.

If we refer in particular to the SDOF equivalence approach, the target displacement is obtained from the elastic displacement spectrum, considering the first period of the frame increased in order to account for the plastic engagement of the structural elements. According to Mendis & Chandler (1998) and setting the ductility index equal to the adopted q-factor, the shifted period of all examined frames becomes 3 s and the corresponding target displacement is 15 cm (Fig.13).

Note that this value is quite the same of the mean displacements obtained from the dynamic analysis and already displayed, so highlighting a sufficient reliability of the adopted method, at least for frames of the analyzed typology.

The so evaluated target displacement is always lower than the one corresponding to 1.5% inter-story drift ratio, as can be seen in the push-over curves (Fig. 7). All the analyzed frames, whatever be the adopted design criteria, appears to be able to withstand the design strong earthquake without exceeding the SEAOC life-safe limit, as already found from the dynamic results.

On the base of these remarks, static and dynamic procedures might be considered as perfectly equivalent in giving information about the behavior of the frames at the ultimate seismic limit state. Nevertheless, some uncertainties can arise if we consider that no significant differences have been shown by the push-over analyses developed for the three adopted design criteria. In particular this result would lead to conclude that giving overstrength to the columns should be not necessary for improving the seismic performances. But, obviously, this cannot be true!

In fact, substantial differences among the design criteria are found when the damage index related to the accumulated plastic rotation is considered (see par.4). A better behavior for the frame provided with some overstrength in the columns (C and Z criteria) is clearly pointed out, when the dynamic cyclic behavior is accounted for. Note that this kind of information can be never given by the static push-over procedure.

With reference to the serviceability limit state, the static analyses have proved to be more severe than the dynamic ones. More in detail, the displacements related to moderate earthquakes, when obtained from the push-over curve, do not fulfill at all the co-

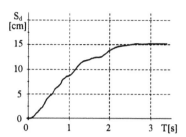

Figure 13. Mean displacement spectrum of the thirty real earthquakes.

de requirements, contrary to what highlighted by the dynamic analysis, at least for the reported case.

7 CONCLUSIONS

The new performance-based design concepts appear to be useful and interesting in evaluating the behavior of steel MRFs for different levels of ground motion.

The comparison of the static non linear approach, which is the most suitable tool for applying the new criteria, with the dynamic time history methodologies has shown a quite good agreement between the two procedures.

Anyway, the developed analyses have pointed out that the inter-story drift ratio is not sufficient to completely define the performance level of the structure, as instead stated in recent codes. This parameter, as well as the others related to a push-over curve, cannot account for the cyclic behavior and the accumulation of plastic deformations.

In fact, the static non linear analyses have been not able to highlight significant differences in the structural behavior of frames having different amount and distribution of overstrength in the columns. On the contrary, the cumbersome dynamic analyses have shown the effectiveness of adopting capacity design criteria: lower damage level has been obtained, as expected, for the frames provided with a more correct overstrength distribution.

In order to make the displacement-based design more reliable and effective for practical application, it is necessary to improve the push-over analysis with an additional parameter or procedure related to the cyclic behavior.

As regards the serviceability limit state, since, at this stage, the global behavior of the frame is not affected in a significant way by cyclic plastic deformations, the static non linear analysis seems to be suitable to control the damage level when a moderate earthquake occurs. In this case, moreover, the push-over results appear more conservative than the dynamic ones.

REFERENCES

Akbas, B. & Shen, J. 1998. Energy-based earthquake resistant design methodology for steel moment resisting frames. In *Proc. 11th European Conference on Earthquake Engineering, Paris, September 1998*. Rotterdam: Balkema.

Bertero, V. 1997. Codification, design and application - General Report. In F.M.Mazzolani and H.Akiyama (eds), *Proc. Behaviour of Steel Structures in Seismic Areas – STESSA '97, Kyoto-Japan, August 1997*: 189-205. Salerno (Italy): Edizioni 10/17.

Biddah, A. & Heidebrecht, A. C. 1998. Seismic performance of moment-resisting steel frames structures designed for different levels of seismic hazard. *Earthquake Spectra*. Vol.14 n.4: 597-625.

Calderoni, B., Ghersi, A. & Rinaldi, Z. 1995. Influenza dei criteri di progetto sul comportamento dinamico dei telai in acciaio (in italian). *Proceedings of 7th National Conference L'ingegneria sismica in Italia, Siena, Italy, 25-28 September 1995*:925-934.

Calderoni, B., Ghersi, A. & Rinaldi, Z. 1996. Statistical analysis of seismic behavior of steel frames: influence of overstrength. *Journal of Constructional Steel Research*. Vol.39 n.2: 137-161. Elsivier Applied Science.

Calderoni, B., Ghersi, A. & Rinaldi, Z. 1997a. Column ovestrength distribution as a parameter for improving the seismic behavior of moment resisting steel frames. In F.M. Mazzolani & H. Akiyama (eds.), *Proc. 2nd Int. Conf. Behaviour of Steel Structures in Seismic Areas, Kyoto, Japan, 3-8 August 1997*:402-409. Salerno (Italy):Edizioni 10/17.

Calderoni, B., Ghersi, A. & Rinaldi, Z. 1997b. Effective behaviour factor for moment resisting steel frames. In F.M. Mazzolani & H. Akiyama (eds.), *Proc. 2nd Int. Conf. Behaviour of Steel Structures in Seismic Areas, Kyoto, Japan, 3-8 August 1997*:410-417. Salerno (Italy):Edizioni 10/17.

Chopra, A. K. & Goel, R.K. 1999. Capacity-Demand-Diagram methods for estimating seismic deformation of inelastic structures: SDF systems. *Report no. PEER 1999/02, April*. Pacific Earthquake Engineering Research Center, University of California, Berkeley.

Eurocode 8 1994. *Design provisions for earthquake resistance of structures, ENV 1998*. CEN, European Committee of Standardization.

Fajfar, P. & Gaspersic, P. 1996. The N2 method for the seismic damage analysis of RC buildings. *Earthquake Engineering and Structural Dynamics*. Vol. 25: 31-46.

Gilmore, A. T. 1998. A parametric approach to performance-based numerical seismic design. *Earthquake Spectra*. Vol.14 n.3: 501-520.

Ghobarah, A., Aly, N.M. & El-Attar, M. 1997. Performance level criteria and evaluation. In Faifar & Krawinkler (eds), *Seismic Design Methodologies for the Next Generation of Codes*: 33-42. Rotterdam: Balkema.

Gupta, A. & Krawinkler, H. 1998. Quantitative performance assessment for steel moment frame structures under seismic loading. In *Proc. 11th European Conference on Earthquake Engineering, Paris, September 1998*. Rotterdam: Balkema.

Hamburger, R.O. 1997. Definig performance objectives. In Faifar & Krawinkler (eds), *Seismic Design Methodologies for the Next Generation of Codes*: 33-42. Rotterdam: Balkema.

Priestley, M.J.N. 2000. Performance-based seismic design. In *Proc. 12th World Conference on Earthquake Engineering, Auckland, New Zealand, 30 January-4 February 2000*.

Mandis, P.A. & Chandler, A. 1998. Comparison of force and displacement based seismic assessments. In *Proc. 11th European Conference on Earthquake Engineering, Paris, September 1998*. Rotterdam: Balkema.

Prakash, V., Powell, G.H. & Campbell, S. 1993. DRAIN-2DX: base program description and user guide, Version 1.10. *Technical Report UCB/SEMM-93/17, November*. Department of Civil Engineering, University of California, Berkeley .

SEAOC 1995. *Vision 2000, A framework for Performance Based Design*. Sacramento CA. USA: Structural Engineering Association of California.

Tso W.K. & Moghadam A.S. 1998. Pushover procedure for seismic analysis of buildings. *Construction Research Communication Limited*. ISSN 1365-0556.

Behaviour of Steel Structures in Seismic Areas, Mazzolani & Tremblay (eds) © 2000 Balkema, Rotterdam, ISBN 90 5809 130 9

Seismic performance of dual steel moment-resisting frames

D. Dubina & A. Stratan
'Politehnica' University, Timisoara, Romania

G. De Matteis
University of Naples 'Federico II', Italy

R. Landolfo
University of Chieti, G. D'Annunzio, Italy

ABSTRACT: The paper analyses the performance of dual configuration for steel Moment Resisting Frames, obtained by mixed use of rigid/ full strength and semirigid/ partial strength joints. Cases in which such configurations may be conveniently adopted for design are also highlighted. Seismic performance of dual MRF configurations is compared to the fully rigid one by means of dynamic inelastic analysis. Consequences of the design methodology on the structural response are also investigated. The ability of pushover analysis to reliably represent the structural response is finally analysed, in comparison with time history analysis.

1 INTRODUCTION

Steel moment resisting frames (MRFs) are widely used as structural systems for low to middle rise buildings. They provide flexibility in the use of interior spaces, as well as clear facades, which is an advantage from the architectural point of view. Also, steel MRFs are considered particularly suitable for seismic applications due to their inherent ductility and energy dissipation capacity.

This structural typology relies on flexural behaviour of beams and columns, and on rigid beam-to-column connections in order to resist horizontal seismic forces. Rigid beam-to-column connections are usually realised by welding. They are required to posses an overstrength with respect to connected members, aiming at promoting plastic hinges in beams rather than in the connection itself. In order to obtain really rigid and full-strength joints, significant amount of stiffeners (including panel zone) and workmanship may be necessary. This generally leads to complex and therefore expensive beam-to-column connections. Rigid and full-strength joints are necessary if minimum weight solutions are sought. Anyway, it has been shown that minimum weight solutions may be as much as 20% more expensive than solutions where also fabrication costs are considered for design optimisation (Steenhuis et al. 1998).

Nonetheless, the use of semirigid joints in seismic applications is prone to certain problems:
- There are relatively few analytical models that can be used to determine properties of beam-to-column joints.
- Reduced rigidity of semirigid joints will increase the flexibility of MRFs under seismic forces.

Accurate models for semirigid joints are available, but most of them have been developed for particular joint typologies, and the design procedure is quite cumbersome. A practical design tool has become available by the implementation of the so-called "component" method in the Annex J of Eurocode 3 (1997). Anyway, one of its main drawbacks is the limitation to static loading, model for cyclic behaviour of joints being not available.

Increased flexibility of semirigid frames is partially counterbalanced by lower seismic forces, but this effect is usually small. An increase of structural members is usually necessary in order to verify the inter-storey drift limitations of seismic design codes.

Traditional design adopts only the two limiting cases of rigid and pinned joints. In reality, most of practical beam-to-column joints will exhibit some sort of semirigid behaviour. This may happen due to disregard of some of the components (such as the panel zone), and will cause the assumed rigid joint to be classified as semirigid/partially resistant. On the other hand, some bolted joints that would be considered traditionally as pinned, can be classified as semirigid, especially when the effect of floor slab is considered.

The unexpected poor behaviour of welded connections during the earthquakes of Northridge (1994) and Hyogoken-nanbu (1995) has led to an increased interest in bolted connections, which are usually semirigid and partial resistant. Bolted connections have provided very good performance historically, and their distributed characteristics lead to tough and redundant structural systems, the latter two being a key to good seismic performance. It has to be considered that in most pre-Northridge steel MRFs erected in US, the number of rigid connections was minimised, this approach providing the most economical solution. In many common structural layouts, the ratio of pinned to rigid connections could be as high as six or even greater (Leon 1999).

In order to optimise lateral stiffness and system energy dissipation capability, traditional space MRFs present all beam-to-column connections of rigid and full-strength type. This solution should provide the greatest number of dissipative zones which, according to the strong column - weak beam (SCWB) design philosophy, have to be located at the beam ends (Eurocode 8 1994). On the other side, it is common in nowadays practice to design steel MRFs as perimeter frames and/or dual rigid-pinned frames. Aiming at optimising the structural configuration, building layouts where some frames (interior and/or perimeter) have a reduced number of bays with moment resisting connections are adopted. Main reason for using these alternative solutions is to reduce the number of three- and four-way rigid connections in space MRFs, which are relatively expensive and present a questionable mechanical performance, unless particular precautions in details are taken (Krawinkler 1995). Therefore, it is preferred to eliminate weak-axis moment connections, adopting pinned ones instead. Consequently, the number of moment-resisting connections is reduced as much as possible, resulting in structures with low redundancy. Reduced number of moment connections in perimeter MRFs, due to reduced redundancy, provides higher ductility demands to rigid beam-to-column connections, resulting in their premature failure.

2 WHY DUAL FRAMES?

It is recognised that semirigid joints can provide significant savings over fully rigid ones, due to simplification of the joint detailing and the fabrication process (Weynand et al. 1998). Sometimes dual frame configurations (with rigid and semirigid joints) could be considered as an alternative to homogeneous configurations (all joints semirigid). In order to satisfy the interstorey drift limits under seismic forces, homogeneous configurations with semirigid joints will lead to increased size of structural members in respect with frames with rigid joints. At the same time, certain types of dual frame configurations with mixed use of rigid and semirigid joints, will accommodate drift limits without any increase of structural weight (De Matteis et al. 1999). It means a reduction in the number of rigid joints in the structure without any change of structural ele-

ments, which is again economy.

For different reasons, such as design optimisation and elimination of troublesome three- and four-way beam-to-column connections, frames with only few bays with rigid connections have been used extensively in US prior to Northridge earthquake. It has been shown that the poor behaviour of this structural system was, among others, due to low redundancy and increased ductility demands of fewer rigid joints. A possible way of "upgrade" of these frames is to use semirigid joints instead of pinned ones. The semirigid joints will provide a backup system to the rigid and full strength ones, increasing at the same time the redundancy of the structure.

Most of semirigid/partial resistant joints are of bolted type. Some of these, such as joints with angle cleats or tee stubs, can be easily repaired after they are damaged during a seismic excitation. This is certainly an advantage from the point of view of seismic retrofit of the damaged structure, as it could be done at a minimum cost with no difficult interventions, and without changing the original joint properties. Semirigid joints are usually partially resistant as well. If both rigid and semirigid joints are used in dual structural configurations, the latter ones are expected to yield earlier under seismic forces, due to lower moment capacity. It means the possibility to control of damage distribution throughout the structure. However, taking into consideration the reduced stiffness of semirigid connections, their earlier yield is no more guaranteed. Therefore, care should be taken when selecting semirigid joints for damage control.

Akiyama (1999) demonstrated that a combined use of rigid and semirigid connections could yield a very preferable structural behaviour under earthquakes. The semirigid connections remain almost elastic, preventing excessive development of drifts, while the rigid connections absorb the seismic input energy by developing plastic deformations. Such a type of structure is called "flexible-stiff mixed structure".

It has been shown that the SCWB concept is very favourable for a good seismic performance of moment resisting structural systems. The goal of a design according to this concept is to prevent hinging in columns, except at the base of the frame and at the top floor of multi-storey buildings (Eurocode 8 1994). Plastic hinges in beams only are deemed to promote a plastic mechanism of global type. The global collapse mechanism is the most efficient for dissipation of seismic energy and minimises the structure P-Δ effects (Mazzolani et al. 1996). Provisions of modern seismic codes, through stating the SCWB requirement for steel MRFs, do not assure a collapse mechanism of global type, nor do they prevent column hinging.

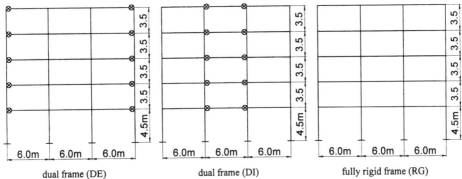

Figure 1. Considered structural schemes.

Nevertheless, it is believed that yielding of the beams will predominate and frames designed according to simplified code provisions will have adequate seismic performance. However, there are certain situations that could worsen the beam to column moment ratio. Some of this (shifting of beam plastic hinge away from the column face, yield limit of the beam different from the one assumed in design) are accounted by some codes (UBC 1997). In case of composite structures, the interaction between the steel beam and the concrete slab changes the beam-column moment ratio. If rigid joints are used in such cases, increased demands will be exerted on columns. Use of semirigid/ partial resistant joints in some bays will reduce the seismic demand in adjoining columns, reducing the risk of partial collapse mechanisms.

3 DESIGN OF FRAMES

Some of the reasons to use dual rigid/semirigid MRFs have been stated above. In order to investigate the influence of dual joint configuration on the seismic response of a steel frame, two different dual configurations have been considered, together with a reference fully rigid configuration (Figure 1). The structure is a 3-bay × 3-bay 5-storey steel MRF. It is supposed to be located in a high seismicity zone, with stiff subsoil conditions. Different column and beam sections are considered for levels 1-3 and 4-5, respectively. According to Eurocode 8, the design conditions are:
- dead floor load: $G_{FLOOR}=4.75$ kN/m^2
- dead load for exterior surface: $G_{CL}=1.70$ kN/m^2
- live load: $Q=3.0$ kN/m^2
- PGA=0.35g
- Subsoil class - A
- Behaviour factor $q_d=6$
- Interstorey drift limit $d_{lim}=0.006$ h

The actual behaviour of the semirigid/partial resistant joints shall be accounted for when designing

the structure. Therefore, the design procedure has been followed for each structural scheme considered, according to the actual joint characteristics. As it was expected, the design has been ruled by serviceability conditions (interstorey drift limitation) rather than by member strength requirements.

Bolted angle cleat connections have been chosen for semirigid joints. They are rather weak and sometimes are considered as nominally pinned. However, the connecting elements (cleats and bolts) of these connections present the advantage of being easily replaceable in case they are damaged under earthquake forces. Also, significant transfer of moment can be accomplished when detailed correspondingly. Connections with relatively low moment capacity have been chosen intentionally, in order to verify the possibility to control the damage distribution throughout the structure. Rigid connections have not been designed explicitly. A possible solution could be a welded joint with cover plates, which will move the plastic hinge away from the column face. It has to be stressed out that the column web panel will need to be stiffened by web plates or diagonal stiffeners, if the joint shall be considered rigid and full strength. The two possible configurations of joints for the DI and DE frames are shown in Figure 2a and Figure 2b, respectively. It can be observed that exterior joints are much simpler than the dual interior ones, because the panel zone does not need to be stiffened. Therefore, possible economy in the case of DI frames will be minimal.

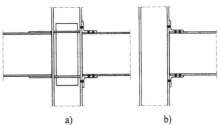

Figure 2. Possible joint configurations for DI and DE frames.

Mechanical properties of joints have been computed according to revised Annex J of Eurocode 3 (1997), even if it gives guidance for static loading only. Thick angle cleats have been chosen to assure a greater moment capacity, anyway, these have been the weakest component in all the cases. Fully pretensioned high-strength bolts have been considered, in order to eliminate some sources of deformation within the joint.

Table 1. Fundamental period of vibration and cross-sectional dimensions of the frames.

| Frame | T_1, s | Storeys 1-3 | | Storey 4-5 | |
		Beams	Columns	Beams	Columns
WB-DE	0.94	IPE450	HEB600	IPE450	HEB400
WB-DI	0.95	IPE450	HEB550	IPE400	HEB400
WB-RG	0.90	IPE450	HEB550	IPE400	HEB400
SB-DE	0.94	IPE550	HEB450	IPE450	HEB320
SB-DI	0.94	IPE550	HEB400	IPE450	HEB300
SB-RG	0.90	IPE550	HEB400	IPE450	HEB300

When designing the frames, two different strategies have been followed. The first one is to keep beam dimensions to the minimum required by the vertical loads and to chose appropriate column sections, in order to verify the serviceability condition under seismic load combination (interstorey drift limits). This design strategy will promote a collapse plastic mechanism of the global type. Frames designed according to this strategy are denoted by WB (weak beam). The second design strategy is to increase the beam sections, while complying with the hierarchy criterion (SCWB concept). This is more effective than increasing column sections, but may increase the risk of soft stories. Frames designed according to the second strategy have been denoted by SB (strong beam).

The fundamental period of vibration and the dimensions of the structural members can be followed in Table 1. Structural weight of the designed frames is shown in Figure 3. First, the rigid frame was designed, which served as the starting point for the dual ones. For both series of frames, no change of structural members was needed for the DI frames with respect to the RG ones. Instead, the increase of structural weight for the DE frame was about 5% for both series. Increase of beam dimensions was more effective than that of columns. Weight of WB series of frames is about 6% greater than that of SB series.

Table 2. Properties of the semirigid joints.

| Frame | Storeys 1-3 | | Storey 4-5 | |
	$S_{j,ini}/S_{beam}$	$M_j/M_{pl,beam}$	$S_{j,ini}/S_{beam}$	$M_j/M_{pl,beam}$
WB-DE	7.43	0.52	7.54	0.52
WB-DI	12.36	0.52	13.44	0.43
SB-DE	5.12	0.38	6.20	0.36
SB-DI	9.72	0.38	11.42	0.36

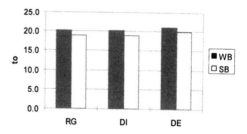

Figure 3. Structural weight for designed frames

Properties of the semirigid joints are given in Table 2. $S_{j,ini}$ stands for joint initial stiffness, S_{beam} is the flexural stiffness of the beam connected to the joint, M_j and $M_{pl,beam}$ are plastic moment capacities of the joint and beam, respectively. Angle cleats are 200×200×24 with M24 gr10.9 and 200×200×20 with M20 gr10.9 for the storey 1-3 and 4-5, respectively. Semirigid joints of the DE series are less stiff than for DI frames, due to the contribution of the panel zone, which was intentionally unstiffened, to minimise as much as possible fabrication costs for this joint typology. The reduced values of the joint moment capacity did not influence the design, since it was governed by the serviceability conditions under lateral loads.

4 DYNAMIC ANALYSIS

Inelastic time-history analysis has been carried out using the DRAIN2DX computer program. Lumped plasticity model has been adopted for structural members with elastic - perfectly plastic behaviour for plastic hinges and connections. A 2% viscous damping has been assumed in the analysis.

4.1 Ground motion records

Acceleration time histories have been selected so as to represent different types of ground motions. The following records have been used:
- El Centro: Imperial Valley earthquake, May 18, 1940, SE component, El Centro record.
- Kobe: Hyogoken-nanbu earthquake, January 17, 1995, NS component, Kobe JMA record.
- Northridge: Northridge earthquake, January 17, 1994, 360deg component, Sylmar Converter Station record.

The problem of ground motion scaling for dynamic analysis is subjected to severe criticism, as this will affect the characteristics of the seismic record. For a realistic representation of a ground motion at a specific site, this should include the effects of seismic source and travel path, as well as local site conditions. Anyway, taking into account the variability of seismic motion and that the goal of this study is primarily concerned with the response of different structural typologies under general seismic conditions, the exact modelling of seismic motion is not determinant. At the same time, since recorded earthquake spectra are quite different from the one used in design, considered ground motions have been scaled so that the mean spectral acceleration in the range from 0.8 to 1.0 seconds will match the corresponding design mean spectral acceleration (Eurocode 8 spectrum). The period range is close to that of the designed frames (0.90 – 0.95 seconds). This will assure initial forces approximately equal to the design ones, and roughly the same input of seismic energy into the structures. Acceleration response spectra of the scaled records, together with the design one, are shown in Figure 4.

With ground motion records scaled in this way, an acceleration multiplier $\lambda=1$ will correspond to the design earthquake, while $\lambda=0.5$ will correspond to the earthquake for the serviceability limit state ($v=2$).

ACCELERATION RESPONSE SPECTRA SCALED RECORDS

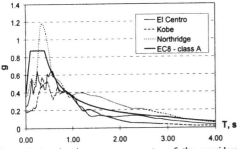

Figure 4. Acceleration response spectra of the considered ground motions.

4.2 *Collapse criteria*

Different collapse criteria could be used for the evaluation of structural performance under seismic actions. One frequently used criterion is the attainment of maximum rotation in structural members or connections. AISC 1997 specifies 0.03 radians plastic rotation capacity for connections in Special MRFs. It will roughly correspond to 0.03 interstorey drift normalised to the storey height. Both these criteria have been widely investigated and they have shown similar results. In the current study, 3% interstorey drift only is considered, being a more global and general parameter with respect to the other one.

A phenomenon corresponding to the effective structural collapse under seismic excitation is the

dynamic instability. Its essence is the change of response from vibration to drift in a single direction (Bernal 1997). The task of obtaining analytical predictions of dynamic instability for realistic buildings is an extremely complex one. It strongly depends on the modelling assumptions and available analysis techniques. In this study, the following procedure is adopted in order to determine dynamic instability: the structure is subjected to the increasing acceleration record and afterwards it is let to freely vibrate for a time increment under no external excitation. If the structure is stable, it will damp down, otherwise, displacements will increase indefinitely under no external loading. This latter situation is used to characterise the state of dynamic instability.

4.3 *Results*

The two series of frames have been subjected to acceleration records of increasing magnitude, in order to determine the different structural states. First yielding including beam-to-column connections and first yielding in structural members only are shown in Figure 5 and Figure 6, respectively. Eurocode 8 (1994) specifies only limitation of interstorey drifts for the serviceability limit state. However, this limitation prevents structural yield under serviceability earthquake, but only for rigid frames. Earlier yield of semirigid joints in the case of DI frames and SB-DE frame did not postpone yield in structural members (by dissipation of some of the seismic input energy); on the contrary, in some cases, it was accelerated (see Figure 6). The dual frame configuration with internal semirigid joints (DI) is characterised by earlier yielding of the semirigid joints compared to the DE frames. In the case of WB-DE frame, first yielding occurred in the structural members and not in weaker semirigid joints. It can be noted that structural members of WB series yielded at smaller acceleration levels than for the SB series of frames, even if their stiffness (fundamental period of vibration) is quite the same. It could be concluded that for low values of acceleration level, response of dual frames in terms of damage to structural members is very close to the one of rigid frame. Semirigid connections will undergo some plastic deformations and therefore it has to be verified that the damage to semirigid connections be in acceptable limit.

Level of seismic input for the attainment of 3% interstorey drift and that characterising dynamic instability are shown in Figures 7 and 8, respectively. It has to be observed that P-Δ effects for small displacements only have been considered in the analysis. Therefore, the very high values of acceleration multiplier (λ), in the order of 15 or greater, do not represent reliable results. However, they indicate that there is no risk of dynamic instability at levels of seismic input characterising 3% drift limitation.

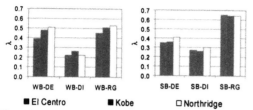

Figure 5. Acceleration multiplier at first yielding.

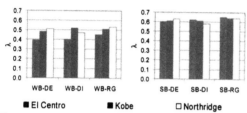

Figure 6. Acceleration multiplier at first yielding (structural members only).

First of all, it should be noted the higher performance of WB series of frames, with respect to the SB series, in terms of both 3% drift level and dynamic instability. This is definitely due to collapse mechanism, WB series being characterised by more favourable ones. The collapse mechanism was determined by investigating the deformed shape of the frame at the attainment of dynamic instability. The form of the plastic mechanism is given in Table 3, where Mij represents a tilted portion from level i to level j.

Table 3. Number of stories involved in the collapse mechanism

Frame	El Centro	Kobe	Northridge	Pushover
WB-DE	M04	M04	M04	M05
WB-DI	M05	M03	M05	M05
WB-RG	M05	M24	M04	M05
SB-DE	M03	M02	M02	M04
SB-DI	M01	M24	M01	M02
SB-RG	M01	M01	M01	M01

Response of dual frames from the WB series is very close to that of the rigid frame, with the exception of the results of Kobe record. Very strong columns of the WB series frames promoted a global type mechanism, the influence of semirigid joints being not important and the global response of the frames being practically the same. On the other hand, dual frames of the SB series present better behaviour than the rigid one. In particular, DE frame is the best of the SB series, but it should be recalled

that this case is characterised by the strongest columns (see Table 1). Anyway, analysis performed on a frame with dual joint configuration of DE type, but with member dimensions of the rigid frame showed results very close to the SB-DI frame, in terms of acceleration multiplier for both 3% drift limitation and dynamic instability. In the whole, the above results emphasise the importance of the collapse mechanism. For SB series, the governing pattern of the collapse mechanism for the rigid frame is a first soft storey. On the contrary, semirigid joints in dual frames imposed less moment demand on the columns enhancing the structural response, by promoting a higher level collapse mechanism.

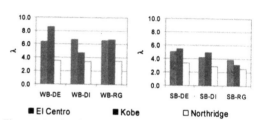

Figure 7. Acceleration multiplier at 3% interstorey drift.

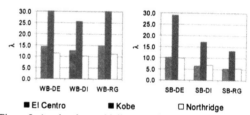

Figure 8. Acceleration multiplier at attainment of dynamic instability.

The beneficial influence of semirigid joints in the dual frame configurations of the SB series could be inspected by monitoring other parameters, such as maximum and residual interstorey drifts. Residual interstorey drift is measured after the structure damps down, and could be used to characterise structural sensibility to second-order effects and dynamic instability. In WB series, at a given level of seismic input, maximum interstorey drifts are similar for all the frames. In SB series, the interstorey drifts are close from one frame to another, but only for low values of seismic input ($\lambda \leq 1.5$). For higher values, interstorey drift for rigid frame is much larger than for the dual frames. The same picture could be also drawn in terms of residual interstorey drift.

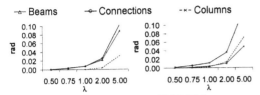

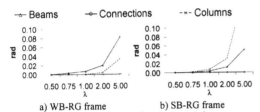

a) WB-DE frame   b) SB-DE frame

Figure 9. Cumulated plastic rotations in the DE frames, mean of dynamic analyses.

a) WB-RG frame   b) SB-RG frame

Figure 10. Cumulated plastic rotations in the RG frames, mean of dynamic analyses.

The maximum and cumulated plastic rotations in structural members and connections present substantially different patterns for the two frame series (see Figures 9 and 10). WB series is characterised by low values of plastic rotations in columns, as compared to those in beams and connections. Plastic rotations in beams are very close to those in connections for the dual frames. The opposite is valid for the SB frames. Plastic rotation demands in columns are greater than in beams for SB-RG frame. Dual joint configuration shifts in a certain extent the plastic rotation demand from columns to connections and beams. Ductility demand in connections is greater than that of beams only in the case of SB-DE frame, and these are quite similar for the SB-DI frame. The same situation is valid for cumulated plastic rotation demand, but the difference is more evident due to its cumulative character.

5 COMPARISON OF DYNAMIC AND PUSHOVER ANALYSIS

The inelastic time history analysis is the most accurate and reliable method for the prediction of seismic response of structures. It needs an appropriate structural model, realistic modelling of cyclic behaviour of elements and a representative set of ground motions. Time history analysis is very cumbersome and time-consuming and it is not feasible yet for practical design. Elastic static analysis is the common currently employed design tool, it being supported by seismic codes. Anyway, the elastic static analysis fails to characterise reliably the seismic response in some cases. This is especially true

for dual frame configurations, with mixed use of rigid and semirigid joints. Earlier yield of the weaker semirigid joints in dual frames, but ultimate performance equal to or even better than that of corresponding fully rigid frames will result in high and unrealistic values of seismic force reduction factor. This mainly happens because Eurocode 8 (1994) uses the first yield in a structural element instead of the effective global yield of the structure.

Static pushover analysis, if appropriately used, may overcome these problems. It is also much simpler to be performed than inelastic time history analysis, but may provide reasonable estimates for response of structures characterised by the first mode of vibration. Account for higher modes could be considered by adopting several constant load patterns (triangular, uniform, SRSS) or adaptive patterns (Krawinkler 1995). In the current study, the triangular distribution of lateral forces has been considered only.

In order to perform a pushover analysis, the structure has to be pushed to a target displacement, usually represented by the top displacement. Here the top displacement corresponding to the attainment of the 3% interstorey drift ratio has been used. It readily provides an evaluation of the ultimate limit state characterised by the drift limitation.

Table 4. Prediction of maximum plastic rotation in beams by means of dynamic and pushover analysis (rad).

Frame	El Centro	Kobe	Northridge	Pushover
WB-DE	0.028	0.028	0.028	0.029
WB-DI	0.029	0.029	0.029	0.029
WB-RG	0.028	0.030	0.028	0.028
SB-DE	0.022	0.026	0.024	0.025
SB-DI	0.022	0.022	0.023	0.024
SB-RG	0.015	0.012	0.012	0.010

Estimation of the maximum plastic rotation attained in beams by means of dynamic and pushover analyses is summarised in Table 4. It shows a very good correspondence. This is also true for maximum rotations attained in columns and connections.

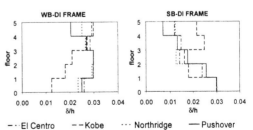

Figure 11. Comparison of storey ductility demands prediction according to dynamic and pushover analyses for the DI frames (δ – storey drift, h – storey height).

Storey ductility demand for some analysed frames are shown in Figure 11. First of all, the important differences in storey drift distribution along the height of the building for the three earthquakes have to be noticed. El Centro record is characterised by a uniform distribution of storey ductility demand along the height, Kobe record amplifies drifts in the two upper stories, while Northridge record in the middle stories. Higher modes amplification effects could be identified for the last two ground motions. However, this pattern is not so evident for the SB frames, especially for the SB-RG one, due to the soft storey mechanism forming at the first level. Pushover analysis predicts correctly the storey ductility demands for the El Centro record, which is characterised by a response governed by the fundamental mode of vibration. In any case, ductility demands in the upper stories are underestimated by the inelastic static analysis. Prediction of collapse mechanism could be examined in Table 3. The results of the dynamic and pushover analysis do not match exactly, but the tendency of the SB frames to form a partial collapse mechanism, and the beneficial effects of dual joint configuration for this series may be detected from inelastic static analysis as well.

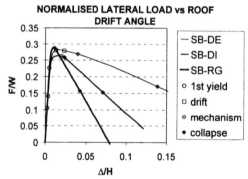

NORMALISED LATERAL LOAD vs ROOF DRIFT ANGLE

Figure 12. Capacity curves for the SB series of frames (F – lateral load, W – weight, Δ – roof displacement, H – roof height).

The capacity curves of the SB series frames, and the states corresponding to drift limit, plastic mechanism and collapse are given in Figure 12. The collapse is defined by the intersection of the capacity curve and the P-Δ curve (Bertero et al. 1999). Slope of the softening branch of the pushover curve used to characterise the sensivity to P-Δ effects clearly indicates the beneficial effect of semirigid joints in the dual joint configuration in case of frames of SB series. On the contrary, for the WB series, the slope of the softening branch of the pushover curve is the same for all three frames and the capacity of the DI frame only is lower that that of RG and DE frames. Therefore, results from pushover analysis confirm the main outcomes that could be identified by dynamic analysis.

6 CONCLUSIONS

Seismic performance of steel MRFs is greatly influenced by the design methodology. The global collapse mechanism is favoured by the SCWB design approach. However, there are some practical cases where this could not be feasible. Dual joint configuration of MRFs does not affect substantially the seismic behaviour with respect to the homogeneous fully rigid configuration, if the global mechanism is assured. On the contrary, dual frames reduce the risk of soft storey and may enhance the structural response in terms of dynamic P-Δ effects if the structure is sensible to partial collapse mechanisms.

Due to different moment capacities and stiffness of rigid and semirigid connections, elastic static analysis could be misleading. On the contrary, pushover analysis seems to be a useful and relatively simple tool for evaluation of seismic performance of dual frame configurations, to be used as an alternative to dynamic time history analysis. However, care should be taken when the interaction between the structure and the earthquake characteristics amplify higher modes of vibration.

Finally, it is worthy to remark that results and conclusions of this study have to be limited to the analysed cases only. They could be considered as a first attempt to generalise the behaviour of dual frame structures, that sustains and motivates the authors to continue their studies.

REFERENCES

Akiyama, H. 1999. Behaviour of connections under seismic loads. *Control of semi-rigid behaviour of civil engineering structural connections.* COST C1. Proc. inter. conf., Liege, 17-19 September 1998.

Bernal, D. 1997. Instability of buildings during seismic response. *Engineering structures,* 20 (4-6): 496-502.

Bertero, R.D. & Bertero, V.V. 1999. Redundancy in Earthquake-Resistant Design. *Journ. of Struct. Eng.,* 125 (1), ASCE.

De Matteis, G., Landolfo, R. Dubina, D. & Stratan, A. 1999. *Influence of the Structural Typology on the Seismic Performance of Steel Framed Buildings.* Chapter 7.3 of final report of UE COPERNICUS "RECOS" project.

Krawinkler, H. 1995. *Systems Behaviour of Structural Steel Frames Subjected to Earthquake Ground Motion,* Report No. SAC-95-09. SAC Joint Venture, California, USA.

Leon, R.T. 1999. Developments in the use of partial restraint frames in the United States. *Control of semi-rigid behaviour of civil engineering structural connections.* COST C1. Proc. intern. conf., Liege, 17-19 September 1998.

Mazzolani, F.M. & Piluso V. 1996. *Theory and design of seismic resistant steel frames,* Chapman & Hall, London, UK.

Steenhuis, M., Weynand, K. & Gresnigt, A.M. 1998. Strategies for Economic Design of Unbraced Steel Frames. *Journal of Constructional Steel Research,* 46:1-3, Paper No. 069.

Weynand, K., Jaspart, J.-P., Steenhuis M. 1998. Economy Studies of Steel Building Frames with Semi-Rigid Joints. *Journal of Constructional Steel Research,* 46:1-3, Paper No. 063

Behaviour of Steel Structures in Seismic Areas, Mazzolani & Tremblay (eds) © 2000 Balkema, Rotterdam, ISBN 90 5809 130 9

Comparative study on seismic design procedures for steel MR frames according to the force-based approach

B. Faggiano & G. De Matteis
University of Naples 'Federico II', Italy

R. Landolfo
University of Chieti G. D'Annunzio, Italy

ABSTRACT: Paper aims at examining the seismic performance of steel moment resisting frame structures with rigid connections, facing together the serviceability and damage control issues. Firstly, with reference to the force-based approach, a review of seismic design procedures provided by the most up-to-dated international codes is reported and some improvements are proposed. Then, a wide numerical investigation is carried out by means of pushover analysis, which has been chosen as a tool for evaluating the seismic performance of the examined frame typologies. The outcomes allow the reliability of design procedures in terms of strength, stiffness as well as ductility requirements to be assessed.

1 INTRODUCTION

The deformation capacity of dissipative elements of moment resisting (MR) frame structures is usually considered as large enough to tolerate inelastic seismic demands according to design assumptions. But, during the recent American and Japanese earthquakes, steel framed structures experienced severe damages, showing limited available ductility and energy dissipation capacity (SAC 96-03 1997, Bruneau et al. 1998).

Thus, it has to be underlined the urgent need of code provision revaluation, based on a more realistic assessment of assumed demands and available capacities, which firstly requires a clear understanding of the system behavior. The main objective is to define simplified design procedures, which are able to guaranty that the structure fulfils the hyphotesized design performance, with regard to displacement limitations, strength and energy dissipation capacity.

In this context, the paper moves within the force-based methodology in order to accomplish a critical survey of current seismic design procedures, paying attention to their efficacy in the fulfillment of serviceability and ultimate limit states requirements. Special regard is devoted to analyse the specific rules for ductile design of structures, such as the hierarchy resistance criterion, whose deficiencies and advantages are pointed out.

The consequences of different design methodologies are then investigated by comparing the seismic performance of corresponding structures. In this respect, inelastic pushover analyses, which can be considered as a possible valid alternative to more complicated and cumbersome dynamic time history analyses, are performed.

2 BASIC PRINCIPLES FOR DUCTILE DESIGN

2.1 General

Within recent codes, the seismic design of steel frames basically consists on the identification of two performance levels (EC8 1996): (1) limited damage, for frequent moderate seismic events; (2) no collapse, in occasion of rare major earthquakes. The attainment of the above performance levels requires a check against the serviceability and the ultimate limit states, respectively.

Since during a strong earthquake the elastic limit of the structure may be exceeded, in order to take into account for the energy dissipation due to cyclic plastic deformations, current seismic design methodologies prescribe a reduction of the design forces by means of a coefficient (called q-factor in Europe). This allows for a simplification of the structural analysis for design purpose without loosing the economical benefit due to a proper combination of strength and energy dissipation capacity.

In case of steel moment resisting frames (SMRFs), at local level, adequate ductility and dissipation capacity requirements can be conferred by an appropriate conception of constructional details, in order to avoid the occurrence of local failure. At this scope seismic provisions introduce a cross-section classification imposing the use of ductile sections in case of seismic design (EC8 1996), although a member classification is considered more correct and exhaustive (Gioncu & Mazzolani 1995, Mazzolani & Piluso 1996). Furthermore, connections have to be full-strength, so that plastic hinge is going to be formed in the adjacent member, preferably in the beam, instead of in the connection itself.

At global level, the ductile behavior should be

guaranteed by a design procedure aimed at controlling the collapse mechanism, which might be of global type in order to optimize the exploitation of the plastic resources of the structural scheme. Nowadays, for obtaining this result, international seismic codes recommend the capacity design methodology, represented by the hierarchy resistance criterion, which is based on the strong column-weak beam concept. Generally, due to hierarchy criterion, even if plastic hinging in columns are not completely avoided, undesirable story mechanisms are more rare and involve a major number of dissipative zones. This assumption has been stated by means of several inelastic numerical analyses of many typologies of SMRF structures (Roeder 1989, Landolfo & Mazzolani 1990, Krawinkler 1995, Faggiano & Mazzolani 1999)

Moreover, a minimum level of lateral stiffness is required for building structures, because the construction functionality must be guaranteed during a moderate earthquake. In this condition, seismic design codes prescribe that structures lateral displacements evaluated by means of an elastic analysis have to be contained within some deformation limits.

2.2 A revised amplification factor method for the hierarchy resistance criterion application

The methods existing in literature for applying the hierarchy resistance criterion can be subdivided in three groups (Faggiano & Mazzolani 1999):
1 the hierarchy condition can be achieved designing the columns to resist a bending moment obtained by an elastic analysis amplified by a proper α coefficient (CNR-GNDT 1984; ECCS 1988);
2 the hierarchy condition is checked a posteriori (EC8 1996; UBC and AISC 1997; Lee 1996);
3 the hierarchy condition is checked within more refined approaches based on the application of the limit design theory (Mazzolani-Piluso method 1997; Ghersi-Neri method 1999).

Within the first group the α coefficient is usually determined according to the column bending moments. For example in *ECCS* recommendations (1988) it is obtained by the following equation:

$$\sum M_{S,c,i,v} + \alpha \cdot \sum M_{S,c,i,s} = \sum M_{pl,b,i} \quad (1)$$

where $M_{S,c,i,v}$, $M_{S,c,i,s}$ and $M_{pl,b,i}$ are respectively: the bending moments acting in columns converging in the ith node due to vertical loads only (v); the ones due to seismic loads only (s); the plastic moments of the beams converging in the ith node. The coefficient α is interpreted as the amplification factor of the design horizontal loads leading the structure to the beam collapse, while vertical loads remain constant.

It can be evaluated by the following relationship:

$$\alpha \cdot = \frac{\sum M_{pl,b,i} - \sum M_{S,c,i,v}}{\sum M_{S,c,i,s}} \quad (2)$$

As an alternative it can be defined with reference to the beam, as the amplification factor of the bending moment due to design seismic loads, leading the most stressed beam section to the inelastic strain hardening deformation corresponding to the local instability. By considering a given column between node (i) and node (j) (Fig. 1), the following condition may be written at each beam end:

$$M_{S,b,v,i,i} + \alpha_{i,i} \cdot M_{S,b,s,i,i} = s_i \cdot M_{pl,b,i,i} \quad (3)$$

where $M_{S,b,v,i,i}$, $M_{S,b,s,i,i}$ and $M_{pl,b,i,i}$ are respectively: the bending moment acting in the ith beam converging in the ith node due to vertical loads only (v); the one due to seismic loads only (s); the plastic moment of the ith beam converging in the ith node. It gives:

$$\alpha_{i,i} = \frac{s_i \cdot M_{pl,b,i,i} - M_{S,b,v,i,i}}{M_{S,b,s,i,i}} \quad (4)$$

Then for both column ends (i) and (j)

$$\alpha_i = \max(\alpha_{i,1}, \alpha_{i,2}); \quad \alpha_j = \max(\alpha_{j,1}, \alpha_{j,2}) \quad (5)$$

Finally, for the column (i,j)

$$\alpha = \max(\alpha_i, \alpha_j) \quad (6)$$

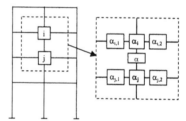

Figure 1. Characterisation of the amplification coefficient α.

The parameter s identifies the member ductility category as a function of: (1) the slenderness of cross-section flanges and web (λ_f and λ_w), (2) the material properties, (3) the bending moment distribution along the axis (Mazzolani & Piluso 1996). In particular, it is defined as the ratio between the stress leading to the local instability (f_u) and the yielding stress of the steel (f_y). Clearly, the more is $f_u > f_y$, the more is the available local ductility, which means that the available local ductility is greater as far as the local instability phenomenon is delayed in the plastic range. The parameter s is expressed by the following equation obtained from empirical results:

$$s = \frac{1}{0.695 + 1.632 \cdot \lambda_f^2 + 0.062 \cdot \lambda_w^2 - 0.602 \frac{b_f}{L^*}} \quad (7)$$

where b_f is the flange width and L^* is the distance between the section where bending moment is equal to zero and the one where the plastic hinge forms.

Depending on the value assumed by the parameter s, three member ductility categories can be defined (Tab.1). According to such classification, the member is ductile when the local instability appears in the plastic range after large inelastic deformation. It is plastic when the local instability appears quite soon in the plastic range, reducing the available rotational capacity. The member is slender when the local instability anticipates the yielding.

Table 1. Member classification

s >1.2	ductile
1< s <1.2	plastic
s <1	slender

An extensive investigation on the Italian standard profiles IPE, HE and welded sections has shown that all examined shapes belong to the first two classes and the majority to the first one.

The design internal force and moment for columns are then evaluated as follows:

$$M_{c,d,i} \geq \chi \cdot \left(M_{S,c,v,i} + \alpha \cdot M_{S,c,s,i} \right) \qquad (8)$$

$$N_{c,d,i} = \left(N_{S,c,v,i} + \alpha \cdot N_{S,c,s,i} \right) \qquad (9)$$

where the factor $\chi = 1/(1-COV) \cong 1.2$ accounts for the overstrength due to the random material variability, being COV the coefficient of variation of the beam yield strength (Mazzolani et al. 1998). It has to be underlined that, contrary to other methods, the design axial forces in columns are defined too, by using the α coefficient.

3 RESPONSE EVALUATION OF SMRF

3.1 General

As it is well known, the better understanding of structure behavior is paid by the complexity of sophisticated numerical analysis, which cannot be compulsory requested in common design practice. As a consequence, simplified but suitable procedures have to be developed.

In order to supply to designers the essential characteristics of the seismic response of buildings by means of a simple procedure, the static inelastic pushover analysis is used as numerical tool for the evaluation of the SMRF capability to sustain seismic action within the performance design requirements.

The basic assumptions and main approximations of such analysis are: (1) the dynamic response of a MDOF system is dominated by the first mode of vibration, neglecting higher modes effects, (2) this mode remains constant during the deformation pattern. Therefore, an invariant load pattern is usually used corresponding to the distribution of inertia forces along the height of the structure associated to the first mode of vibration. Parametrical studies (Krawinkler & Seneviratna 1998) have shown that for low-rise building (up to 5 floors) the above assumptions give quite accurate results.

The static inelastic pushover analysis provides the so-called "behavioral curve", which represents the lateral load multiplier α, or the base shear V as well, versus top displacement δ relationship. It has to be noted that the structural response is more properly characterized by means of the V-δ curve, because the base shear is actually an effect of external actions into the structure, whereas α represents only the external forces which structure can resist according to the deformation pattern. In fact the α-δ and V-δ curves are different, because due to the second order effects the structure can resist a maximum value of lateral forces (α_{max}), while the base shear force (V), determined as the sum of shear forces in the first story columns, is always increasing. However, as long as the structure remains in linear range base shear and global horizontal forces are coincident, they being different only in case of large deformations. But, usually, reference is made to the α-δ curve, because it is directly obtained from the pushover analysis (Fig.2). Such a curve represents the overall capacity of the structure, which is characterised by some behavioral parameters.

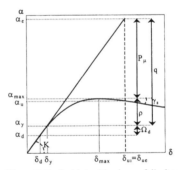

Figure 2. Typical lateral load – roof displacement relationship

3.2 Performance characterisation

A seismic performance evaluation aims at: verifying that structures are able to respect the design assumptions; identifying the structure capability, in terms of strength, stiffness and ductility; assessing the effectiveness of different design criteria.

According to force-based approach, the check of design requirements at the ultimate limit state mainly consists on verifying two conditions:

1 Structure should sustain without collapse an earthquake of a certain damage potential, which in code provisions is characterized by means of the elastic response spectrum amplified by the peak ground acceleration value $(PGA)_d$.

2 Structure should undergo inelastic deformations within ductility and energy dissipation capacities, which are quantified in the design phase by means of the q-factor (q_d).

Therefore, it has to be controlled that computed PGA

and q-factor values, deduced by the pushover analysis, are higher than the design ones:

$$\frac{PGA}{(PGA)_d} \geq 1 ; \qquad \frac{q}{q_d} \geq 1 \qquad (10a,b)$$

Condition (10a) can be assessed through the ratio α_e/α_{ed} between the maximum force multiplier that the structure, considered as indefinitely elastic, can resist (Fig. 2) and the elastic design one corresponding to PGA_d $(\alpha_{ed}=\alpha_d q_d)$. In fact, PGA is assumed as proportional to the horizontal seismic actions, as follows:

$$\alpha = M \cdot R(T) \cdot PGA \qquad (11)$$

α_e is determined according to the ductility factor theory, assuming that the top sway displacements at collapse of the inelastic and the corresponding elastic systems are equal $(\delta_{ui} = \delta_{ue})$ (Fig. 2).

In order to check the condition (10b), q-factor can be evaluated, as the ratio α_e/α_y, where α_y is the static seismic force corresponding to the first yielding (first plastic hinge in the structure). Moreover, special concern has been directed to the assumption of design q-factor values. In fact ECCS and EC8 European code provisions suggest quantifying q_d for SMRFs as 5 α_{max}/α_y, making reference to the response evaluated by means of the static pushover analysis. The ratio $\rho = \alpha_{max}/\alpha_y$ (Fig. 2) represents the inelastic redistribution coefficient. It is a function of the structural redundancy and synthesizes the structure capability to sustain increasing lateral load respect to the first yielding one. Without a specific analysis, a value in the range of 1.2-1.6 is suggested, so to have q_d ranging from 6 to 8. On the other hand, the factor 5 (P_μ in Figure 2) represents the ratio between the indefinitely elastic and the inelastic maximum overall bearing capacity of the structure. It depends upon the effective structural ductility. Both these parameters has been determined and checked by the behavioral curve.

The negative slope (γ_s, in Figure 2) of the lateral load-displacement relationship at large displacements serves to define the stability coefficient $\gamma=\gamma_s \cdot \delta_r$, where δ_r represents the roof displacement under the unit lateral force obtained with a linear elastic analysis (Mazzolani & Piluso 1996). It is a measure of the P-Δ effects.

Furthermore, the first yield point (δ_y, α_y), which permits to know how much the elastic range is extended, and the point corresponding to the design static seismic force α_d can be also identified. They allow for the estimation of the overstrength share due to drift limitations and to larger than necessary member sizes because of the availability of standard shapes and detailing requirements.

The check of design requirements at the serviceability limit state mainly consists on controlling the lateral displacements of the structure. In particular, EC8 recommends the following verification:

$$d_r/\nu \leq d_{lim} \qquad (12)$$

where $d_r = q \, d_e \, \eta$, is the interstory drift corresponding to the elastic response spectrum; q is the reduction factor; d_e is the interstory drift corresponding to the inelastic response spectrum; ν is a reduction coefficient which considers a lower return period for moderate earthquakes compared to severe earthquakes, also depending on the importance of the building (it is equal to 2 for residential buildings); d_{lim} is the assumed interstory drift limit value.

The performance evaluation is completed by observing the damage pattern, in order to detect undesirable characteristics in the building such as soft story conditions, strength and stiffness discontinuities. To this aim, interstory drift ratios, plastic hinge locations and plastic rotations can be examined.

4 NUMERICAL APPLICATIONS

4.1 Study cases and design criteria

The examined structural schemes are low-rise SMRFs with 2 and 4 bays, 2 and 5 floors (2B2F, 2B5F, 4B2F, 4B5F in Figure 3). The geometrical characteristics are given in Table 2. In all cases the frame spacing is equal to 6.00m.

Table 2. Scheme geometrical characteristics

Bays		Floors		
n°	L (m)	n°	h (m)	h_1 (m)
2	6	2	3.5	4.5
4		5		

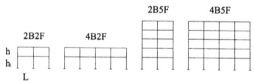

Figure 3. Examined structural typologies.

The structural members are designed according to EC3 and EC8, with reference to ultimate and serviceability limit states (ULS and SLS). Different design conditions have been assumed: q_d=4, 6, ν=2, 3, d_{lim}=0.004h, 0.006h and 0.008h. Steel grade is Fe 430. Dead and live loads are equal to 4.75 and 2.00 kN/m², respectively. The seismic action is defined assuming a soil profile A, a viscous damping factor equal to 5%, PGA=0.35g.

Structures are designed by means of a first order analysis, considering the second order effects in an approximate manner as indicated in EC8, but limiting structural member sizes to the condition $\theta \leq 0.3$, θ being the interstory drift sensitivity coefficient.

In order to assess the impact of ULS and SLS checks on the frame design, these criteria are considered separately. So, frames designed according to ULS have not to fulfil necessarily serviceability conditions too; frames designed according to SLS are checked also in respect to resistance, according to the serviceability equivalent static forces. Moreover, a reference case designed according to EC8, assuming q_d=6, v=2 and d_{lim}=0.006h is considered.

For structures designed according to the SLS, the reduction coefficient v is assumed equal to 2, as recommended by EC8 for residential buildings, and equal to 3, on the basis of a proper seismological study. d_{lim} is assumed equal to 0.004h and 0.006h (EC8 1996) for non-structural elements having more or less interference with the structure, and equal to 0.008h in case of flexible facades elastically connected to the structure. For structures designed according to the ULS, q_d is assumed equal to 4 and 6, which seems to be more suitable values, representing two structural ductility classes (low and high, respectively). It has to be noted that, in the range of periods of the study cases, v equal to 3 and 2 corresponds to q_d values equal to 5 and 3 for 5F structures, to 4 and 3 for 2F structures. In fact the design response spectrum shape at ULS is different from the elastic one, while at SLS the elastic response spectrum is simply reduced by v. The hierarchy resistance criterion has been also considered for structures dimensioned at SLS, so that the improvement of available ductility could be emphasised.

The capacity design is applied according to EC8 and the new proposal (NP). The cases in which the hierarchy resistance criterion are not used are labeled with N.

Summary of design data is presented in Table 3.

Table 3. Design data

Ultimate limit state (ULS)		
q-factor (q_d)	Hierarchy Resistance Criterion	
4		
6	N* EC8 NP**	
Serviceability limit state (SLS)		
v	d_{lim}	Hierarchy Resistance Criterion
	0.004h	
3	0.006h	
	0.008h	N EC8 NP
2	0.006h	

* N=None;   **NP=New Proposal

In total 72 different structures have been designed.

Static pushover inelastic analysis is performed by means of the computer program Drain-2DX (Prakash et al. 1993). P-Δ effects are taken into account and the inelastic behavior of dissipative zones is defined by means of concentrated elastic-perfectly plastic hinges.

As collapse criterion, the attainment of the maximum available plastic rotation θ_{pl} in the most engaged section is assumed. The rotational capacity, $R= \theta_{pl}/\theta_y$, is determined by Mazzolani-Piluso method (Mazzolani & Piluso 1996).

In the next, for each structural typology, the re-

sults of structure dimensioning and the corresponding performance parameters, obtained by means of the pushover analysis, are plotted. The influence of the adopted design criteria, in terms of limit state and hierarchy resistance criterion, are evidenced and some trends are highlighted, due to the large number of performed analyses.

4.2 The analysis of results

4.2.1 Structural weight
Figure 4 shows the influence of the design assumptions and the hierarchy resistance criterion on the structural weight (W) of the analyzed frames.

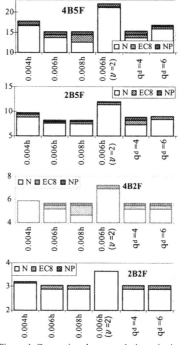

Figure 4. Comparison between design criteria in terms of W (t).

It can be observed that the influence of hierarchy criterion is quite important only in case of frames designed according to SLS with a drift limit d_{lim}=0.008h ($W_{NP}/W_N \cong 1.20$, $W_{EC8}/W_N \cong 1.15$ as average for all structures), which corresponds to the lightest solution. For heavier structures it does not provide significant effects ($W_{NP}/W_N \cong 1.03$, $W_{EC8}/W_N \cong 1.01$). In almost all the cases the NP hierarchy criterion gives rise to a little heavier structures as respect to EC8. This appears especially in case of 4B frames. As an average for all the structures, EC8 and NP cases provide a weigth increment as respect to the N-cases of about 5% and 10%, respectively.

SLS strongly conditions member sizes. In particular v factor has a remarkable influence, a considerable weight increment being between the 0.006h-

v=3 and −v=2 cases (for 5F and 2F N-structures it is about 50% and 30%, respectively, as average for all the structures). For v=3 cases the differences are attenuated when a hierarchy criterion is adopted, most of all for the 0.006h and 0.008h 5F structures. Moreover, frames designed according to d_{lim}=0.004h are heavier than the ones designed at ULS. For frames designed at the ULS, design q-factor has a somewhat influence only in case of 5F frames. In particular their weigth for q_d=6 is higher than for q_d=4, due to the influence of the II order effects, prevailing in the first case.

Moreover an evident weight increment, between q_d=6 and 0.006h-v=2 cases, has to be noticed for both 5F and 2F structures, it being of about 30% and 25% respectively. This confirms that SLS check is dominant in member sizing compared to SLU check.

For 2B2F typology the design criteria 0.006h, 0.008h, q_d=4 and q_d=6 give always the same member dimensions and for design criteria 0.004h and 0.006h-v=2 the hierarchy criterion is uninfluencing.

4.2.2 PGA

In Figure 5, the ultimate PGA is reported as normalized value respect to design value $(PGA)_d$.

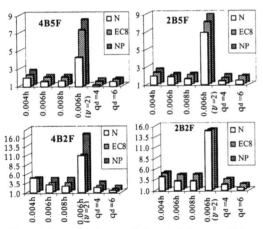

Figure 5. Comparison between design criteria in terms of PGA.

First it can be noted that whatever is the adopted design criteria it is always $PGA>PGA_d$. The application of the hierarchy resistance criterion is favorable, being the NP and EC8 PGA increments as respect to N-case of about 30% and 18%, respectively, as an average for all the structures. It is evident that NP gives better results compared to EC8.

For 5F frames the structures dimensioned with 0.006h and 0.008h have a better performance compared to the one designed with q_d=4 and q_d=6.

Generally, by comparing Figures 4 and 5, it seems that trend of increasing performance in terms of PGA is the same of increasing structural weight.

4.2.3 q-factor

In Figure 6 behavior factor q evaluated by pushover analyses are reported for all examined structures.

It can be noted that whatever is the adopted design criteria, q-factor values are always higher than 4, except for case 4B5F-0.006h-v=2. The influence of the hierarchy resistance criterion on q-factor value is very large, being the increments due to NP and EC8 as respect to the N-case of 50% and 30%, respectively, as an average for all the structures. In particular, in case of q_d=6, evaluated q-factors for structures designed applying the hierarchy resistance criterion are always greater than 6. Moreover, for 5F frames, the structures dimensioned with 0.006h and 0.008h have a better performance compared to the ones designed with q_d=4 and q_d=6, specially for NP frames. For 2F frames q-factor values are very large, the most favorable cases being the ones corresponding to the heavier structures (0.004h and 0.006h-v=2), with q-factor values higher than 9.

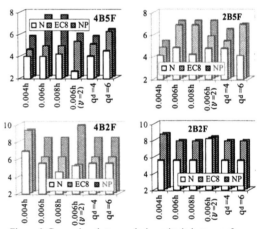

Figure 6. Comparison between design criteria in terms of q.

By considering that q-factor can be evaluated as $q=\rho \cdot P_{\mu}$, it has to be emphasized that the redundancy factor ρ always increases if NP and EC8 hierarchy conditions are used, being the increment almost of 25% and 18%, respectively, as an average for all the structures. It assumes always values higher than 1.2, and reaches values higher than 1.6 in the 88% and 63% of the study cases if NP and EC8 hierarchy resistance criteria respectively are applied. The P_{μ} factor also increases in case the NP and EC8 hierarchy conditions are used, being the increment of almost 22% and 12%, respectively, as an average for all the structures. Anywise, it never reaches the value P_{μ}=5 indicated by EC8.

4.2.4 Damage pattern

The benefit of the hierarchy resistance criterion on the ductility of the structure is confirmed by the evolution of the damage pattern. In fact it appears evident that, when EC8 and NP are applied, the plastic hinges are mostly distributed in the beams, involving all floors. On the contrary, if the hierarchy criterion is not applied, hinges are concentrated in some floors only, involving several columns. In case of NP and EC8 criteria the decrement of column plastic hinge number is about 50% and 23%, respectively, as an average for all the structures. Besides, as respect to the N case, the increment of beam plastic hinge number is equal to 85% and 50%, respectively. The distribution of the interstory drift along the height validates this trend.

4.2.5 *Stability coefficient*
As far as the II order effects are concerned, the trend of structural weight is followed inversely. In particular, the benefit due to the hierarchy resistance criterion is sensible, the factor γ being reduced of 44% and 24%, respectively, as an average for all the structures, if NP and EC8 are applied (Fig. 7).

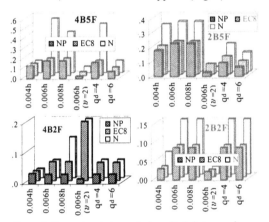

Figure 7. Comparison between design criteria in terms of γ.

4.2.6 *Overstrenght*
The overstrength factor trend among structures designed with different criteria is the same as for structural weight. The hierarchy criterion is favourable, because the overstrength reduces, mainly for 5F structures. It means that the structural member sizes better approach the strictly necessary ones.

4.2.7 *The performance coefficient*
In order to compare in a simple and synthetic way the behavior of designed structures, a performance index *PI* is defined in equation (10). It takes into account the most important performance parameters, namely the structural weight W, the q-factor, the stability coefficient γ, by assuming for each one a weigthing factor.

$$PI = 1 + 2\frac{q - q_0}{q_0} - 1.5\frac{W - W_0}{W_0} - 0.5\frac{\gamma - \gamma_0}{\gamma_0} \qquad (10)$$

The higher is *PI* the better the structure behaves, being the effect of q-factor prevalent on the negative influence of *W* and γ. The reference values (q_0, W_0, γ_0) are taken as the ones corresponding to the lighter study frame for each structural typology. In Figure 8 the *PI* values for all study cases are plotted. The reference case is the $0.008h$ structure for all the examined structural typologies.

It can be noticed that most of the designed structures present *PI* values higher than 1, except some cases, mainly belonging to the 5F-N-cases. The trends highlighted in the previous sections are generally confirmed. Therefore, such a coefficient can be considered as a useful aid for evaluating structural performance of frame structures.

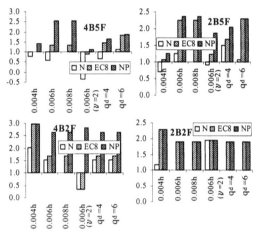

Figure 8. The performance coefficients.

5 CRITICAL ANALYSIS OF RESULTS

5.1 *Main outcomes for structures designed at ULS*

From the performed analyses, it is recognized that in all study cases designed with q_d=4, the structural response in terms of oserved q-factor is good enough (q = 4 has been always reached), even in N-case, and even if most of the structural typologies present a soft story collapse mechanism. If the NP and EC8 hierarchy conditions are applied, the q-factor increases of about 40% and 20%, respectively, as an average for all structures and the soft story is rather always avoided. It is interesting to observe that the weight increment as respect to the N-case is only of 12% and 7%, respectively, in average. On the contrary, structures designed with q_d=6 behave accordingly only in case hierarchy resistance criterion is applied.

In case the SLS check is performed, all structural typologies designed with q_d = 4 and 6 are able to provide interstory drift within the limits $0.006h$ and $0.008h$, with ν=3, but not in case of $0.004h$ and absolutely not for ν=2. It appears evident if q_d=6 and

0.006h-v=2 structures are compared. It means that in the latter cases SLS check is more restrictive than ULS one. In this circumstance the advantage of the inelastic design, coming from the use of reduced seismic forces, is partially lost. Besides, the application of capacity design rules, aiming at a ductile structure, is not fully exploited.

5.2 *Main outcomes for structures designed at SLS*

The careful examination of frames designed according to SLS allows for assessing the possibility to impose less restrictive ductility requirements when SLS conditions structural design as respect to ULS.

A first comment can be addressed to the effects of the assumed different values of v. Making reference to the N-case, because as a principle no ductility or energy dissipation capacity control is required by the SLS check, the 0.006h-v=3 case presents a considerable weight increment as respect to v=2 case (Section 4.2.2). The 2B2F-0.006h-v=2 case is better performing (P =1.94). In fact, N-structure is coincident with EC8 and NP structures, so its ductility is large enough, while for 0.006h-v=3 case it is equal to 1. For 4B2F, 2B5F and 4B5F the 0.006h-v=3 case is better performing as respect to corresponding 0.006h-v=2 structures (see factor *PI* in Figure 8).

In general, it can be stated that for strong interstory drift limitation, namely 0.006h-v=2 and 0.004h-v=3, the corresponding overstrength (intended as the ratio between α_y of structures designed according to SLS and ULS, respectively) assures a saving of the ductility requirements to structures for ULS check. In other words, the overstrength itself represents the ductility discount that can be exploited to the structure. For example, overstrength values of N-case structures designed according to SLS as respect to $N_{qd=4}$, $NP_{qd=4}$ and $NP_{qd=6}$ structures are always greater than 1.2.

6 CONCLUSIVE REMARKS AND FURTHER DEVELOPMENTS

The current study allows for several interesting conclusions to be drawn: (1) the value q_d=4 can be used for structures with low ductility; (2) in order to fulfill high ductility requirements, for structures designed according to ULS with $q_d \geq 4$ the application of the hierarchy resistance criterion appears to be convenient; (3) the proposed method for hierarchy resistance criterion is generally advantageous in terms of structural performance; (4) structures designed according to ULS satisfy the deformability check only in case of less restrictive limitation, namely v=3 and d_{lim}=0.006h-0.008h; (5) the assumption of v=3 for defining the serviceability forces distribution leads to a better performing structures as respect to v=2; (6) for strong d_{lim} (0.004h), it is convenient to design frames at the

SLS, imposing less restrictive conditions for ductility requirements at the ULS, due to the overstrength.

Anyway, further developments appear to be necessary. In fact, the possibility of using pushover analysis as a tool for investigating the seismic performance of a structure has not been definitively assessed yet. This is mainly due to the difficulty to correctly schematize the ground motion, which cannot be represented solely by the *PGA*, but other important parameters should be considered, like the duration, the energy and frequency contents, the prevalence of impulsive or pure cyclic characteristics. But all analyzed aspects constitute valid information for helping the understanding of the structure behavior. On the other side, dynamic time history analyses are needed in order to confirm the above outcomes as well as to assess the range of limitations for the applicability of static pushover analyses.

REFERENCES

Bruneau, M., Uang, C.M. and Whittaker, A., 1998. *Ductile Design of Steel Structures*. McGraw-Hill.

Gioncu, V. & Mazzolani, F. M., 1995. Alternative methods for assessing local ductility. In *Proc. Intern. Workshop, STESSA '94*, Timisoara, Romania. F.M. Mazzolani, and V. Gioncu edrs. London: E & FN Spon.

Faggiano, B. & Mazzolani, F.M., 1999. Interpretazione del criterio di gerarchia nel dimensionamento dei telai in acciaio. In *L'Ingegneria sismica in Italia, Proc of the IX National Conference,* Torino, Italy, 20-23 September 1999.

Ghersi, A., Marino, E. and Neri, F., 1999. A simple procedure to design steel frames to fail in global mode. In *Proc. of the SDSS Intern. Conf. 1999,* Timisoara, Romania, September.

Krawinkler, H., 1995. Systems behavior of structural steel frames subjected to earthquake ground motion. In *Background Reports SAC 95-09.*

Krawinkler, H. & Seneviratna, G.D.P.K., 1998. Pros and cons of a pushover analysis of seismic performance evaluation. *Engineering Structures,* Vol.20, Nos 4-6, pp.452-464.

Landolfo, R. & Mazzolani, F.M., 1990. The consequence of design criteria on the seismic behavior of steel frames. In *Proc. IX Intern. ECEE,* Moscow.

Lee, H-S., 1996. Revised rules for concept of Strong-Column Weak-Girder design. *Journal of Structural Engineering.* Vol.122, No 4, April, 359-364.

Mazzolani, F.M. & Piluso, V., 1996. *Theory and Design of Seismic Resistant Steel Frames.* London: E & FN Spon, an Imprint of Chapman & Hall.

Mazzolani, F. M. & Piluso, V., 1997. Plastic Design of Seismic Resistant Steel Frames. *Earthquake Engineering and Structural Dynamics,* Vol. 26, 167-191.

Mazzolani, F. M., Piluso, V. and Rizzano, G., 1998. Design of full-strength extended end-plate joints for random material variability. In *Proc. of COST C1 Congress,* Liège.

Prakash, V., Powell, G.H. and Campbell, S., 1993. DRAIN - 2DX Base Program Description and User Guide. Version 1.10. EERC, University of California, Berkeley, November.

Roeder, C. W., EERI M., Carpenter, J.E. and Taniguchi H. 1989. Predicted ductility demands for steel moment resisting frames. *Earthquake Spectra,* Vol. 5, No. 2.

SAC 96-03, 1997. Interim guidelines. FEMA 267/A, SAC Join Venture, California, USA.

Behaviour of Steel Structures in Seismic Areas, Mazzolani & Tremblay (eds) © 2000 Balkema, Rotterdam, ISBN 90 5809 130 9

Optimum bending and shear stiffness distribution for performance based design of rigid and braced multi-story steel frames

C.J.Gantes, I.Vayas & A.Spiliopoulos
Civil Engineering Department, National Technical University of Athens, Greece

C.C.Pouangare
Limassol, Cyprus

ABSTRACT: The preliminary design of multi-story building structures to resist lateral loads due to earthquake and wind is investigated. It is attempted to gain a qualitative understanding of the influence of bending and shear stiffness distribution on the response of such structures. It is observed that the conventional stiffness distribution, dictated by strength constraints, is not the best to satisfy deflection criteria in the case of flexible braced frames. This suggests that a new approach to the design of such frames may be appropriate when serviceability governs.

1 INTRODUCTION

1.1 *Design requirements and strategies*

Strength based, conventional design of multi-story buildings dictates higher bending and shear stiffness at the base of the building, which gradually decrease with height, following the variation of bending moments and shear forces, respectively. This approach, however, can be questionable for the design of flexible structures, which is very often governed by deflection criteria.

In such cases it may be appropriate to reverse the common approach of designing for strength and checking for stiffness. Instead, the required stiffness to satisfy deformation criteria is derived first, followed by checking for strength.

In the present work it is attempted to gain a qualitative understanding of the influence of bending and shear stiffness distribution on the lateral deflection response of such structures. This is viewed as a first step towards developing a preliminary design strategy for structures governed by deflection requirements.

1.2 *Stiffness requirements*

Frames subjected to seismic loading are designed for stiffness in order to limit their lateral deformations. This limitation refers to both the serviceability and the ultimate limit states, it applies, therefore, for wind as well as moderate and strong earthquakes. In the serviceability limit state lateral deformations have to be controlled primarily in order to limit the damage of non-structural elements, and also to prevent feelings of uneasiness among the occupants. In

the ultimate limit state deformations must be limited in order to avoid instabilities due to second order effects and to be consistent with the common, first order analysis methods that assume small displacements.

The structural response to lateral loads may be described by the base shear – top displacement curve as shown in Figure 1. The initial slope of the curve expresses the elastic stiffness of the structure. For the serviceability limit state, the lateral deformations calculated on the basis of the elastic stiffness correspond to the actual deformations of the structure, as the structural response is expected to be nearly elastic in the event of moderate earthquakes.

However, the actual deformations at the ultimate limit state are larger than those calculated on the assumption of an elastic structural response. This is due to the nonlinear behavior of the structure, parts of which yield during strong earthquakes. Nevertheless, in order to avoid nonlinear analysis, most seismic design codes, such as Eurocode 8 (1994), propose a simple relationship between the nonlinear elastic-plastic and the elastic deformations. The two types of deformations are related through the value of the behavior factor q, assumed in the analysis as follows:

$$\delta = q \cdot \delta_e \qquad (1)$$

where δ=lateral deformations for nonlinear response, and δ_e=corresponding deformations for linear response.

Although the above methodology constitutes an approximation of the real behavior in case of strong earthquakes, it is evident that the knowledge of the elastic lateral stiffness of a frame is important in structural design practice. Therefore, the analysis in the present paper will be based on the assumption of elastic response.

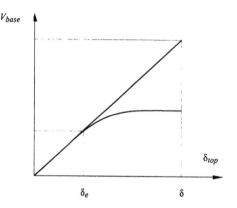

Figure 1. Base shear – top displacement curve of a frame.

Lateral deflection constraints for multi-story buildings consist of two requirements: (a) that the maximum inter-story drift is within a predefined percentage of the story height, and (b) that the total maximum drift does not exceed another predefined percentage of the total building height. It is reasonable to combine the two criteria into one, requesting a near-linear deflection line that satisfies the most stringent among the two constraints.

Consequently, the objective of this paper is to study the influence of bending and shear stiffness distribution on the deflection response of multi-story buildings subjected to lateral loading. Moreover, suitable stiffness combinations will be sought, which result in a linear or near-linear deflected shape.

For preliminary design purposes, the stiffness may be determined by substitution of the frame by an equivalent vertical cantilever beam having the same deformation properties, as shown in Figure 2.

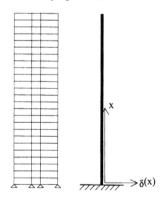

Figure 2. Frame and equivalent vertical cantilever beam.

This beam is subjected to deformations due to bending moments and shear forces. This is illustrated in Figure 3, showing that the total rotation at any cross-section is the sum of a flexural rotation ψ and a shear rotation γ. The beam has correspondingly a bending stiffness, EI, and a shear stiffness, S_v. In other words, it is not a Bernoulli beam but a Timoshenko beam, and its deflected shape must be determined using Timoshenko beam theory.

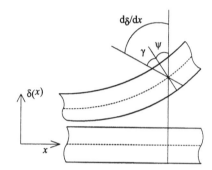

Figure 3. Bending and shear rotation of a Timoshenko beam.

This process consists of two steps: (i) substitution of the frame by the equivalent beam, and (ii) calculation of the deflections of this beam. These steps will be presented in sections 2 and 3 of the paper, respectively. In section 4 the results of this analysis are used to carry out a parametric study and draw conclusions on appropriate stiffness distributions for different frame structural systems.

2 STIFFNESS OF EQUIVALENT BEAM

2.1 Diagonally and X- braced frames

In this section, the expressions of bending and shear stiffness of the equivalent beam representing the simplest case of a diagonally or X-braced, single-bay frame will be derived. In the next section the corresponding expressions for other frame configurations will be given.

A typical floor of a diagonally braced multi-story frame is shown in Figure 4. When subjected to lateral loads, this structure behaves basically as a vertical truss. Bending moments are resisted by axial deformations of the columns acting as flanges of this truss. The bending stiffness of the frame may be accordingly determined from:

$$EI = EA_c h_0^2 / 2 \qquad (2)$$

where E is the Young's modulus of the material, A_c is the area of the cross-section of one column, and h_o is the width of the frame, equal to the distance between the centerlines of the columns.

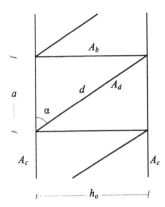

Figure 4. Typical floor of single-bay diagonally braced frame.

For the evaluation of the shear stiffness, a unit transverse force $V=1$ is applied and the axial forces in the members are determined, as shown in Figure 5. The resulting lateral deformation is determined by means of:

$$\delta_s = \Sigma \int \frac{N^2}{EA} dx \qquad (3)$$

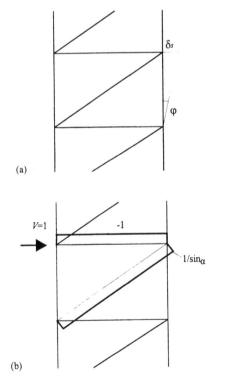

Figure 5. Deformation (a) and axial forces (b) of typical floor of single-bay diagonally braced frame subjected to unit shear.

where N are the axial forces in the truss members due to the unit shear load, and A are the areas of the corresponding cross-sections, namely A_d is the cross-sectional area of the diagonal and A_b of the beam. The application of equation (3) to the specific frame of Figure 4 gives:

$$\delta_s = \frac{d}{\sin^2 \alpha\, EA_d} + \frac{h_o}{EA_b} \qquad (4)$$

The shear deformation of the truss, defined by the angle of sway, is given by:

$$\varphi = \frac{\delta_s}{a} \qquad (5)$$

The shear stiffness is determined from:

$$S_v = V / \varphi = 1 / \varphi \qquad (6)$$

Substituting (4) and (5) into (6) and eliminating the lengths of the members leads to the following expression for the shear stiffness:

$$S_v = \frac{1}{\dfrac{1}{EA_d \sin^2 \alpha \cos \alpha} + \dfrac{1}{EA_b \cot \alpha}} \qquad (7)$$

For the X-braced frame of Figure 6, where two diagonals participate in the shear force transfer, the relevant expression may be written as:

$$S_v = \frac{1}{\dfrac{1}{2EA_d \sin^2 \alpha \cos \alpha} + \dfrac{1}{EA_b \cot \alpha}} \qquad (8)$$

However, if in X-braced frames the contribution of the compression diagonal is neglected due to buckling, equation (7) instead of (8) shall be used.

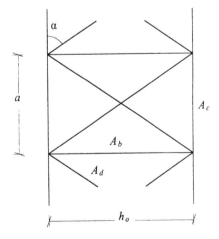

Figure 6. Typical floor of single-bay X-braced frame.

2.2 Other types of frames

The bending stiffness of other single-bay frames is determined like for diagonally or X-braced frames from equation (2). The shear stiffness is determined by applying similar procedures to the one adopted for X-braced frames. The results for several typical structural systems of braced frames and a single-bay rigid frame are summarized in Table 1. In the following relations, G is the shear modulus of the material, A_{bv} the shear area of the beam cross-section, I_b and I_d the moments of inertia of the cross-sections of beam and diagonal, respectively, and h_b is the height of the cross-section of the beam.

where

$$S_1 = EA_d \sin^2 \alpha \cos \alpha \qquad (9)$$

$$S_2 = EA_b \cot \alpha \qquad (10)$$

$$S_3 = GA_{bv} \tan \alpha \qquad (11)$$

$$S_4 = \frac{3EI_b \tan \alpha}{e^2} \qquad (12)$$

$$\varepsilon = \frac{e}{h_o} \qquad (13)$$

$$\phi = r(1 - \varepsilon) + \varepsilon \qquad (14)$$

$$r = \frac{1}{1 + \mu \dfrac{I_d}{I_b} \sin \alpha} \qquad (15)$$

$$\psi = 1 - \varepsilon \qquad (16)$$

where $\mu = 0$ if the braces are pin-connected at their ends, $\mu = 1$ if the braces are rigidly connected to the beam and pin-connected at the far end, and $\mu = 4/3$ if the braces are rigidly connected at both ends.

Table 1. Shear stiffness of typical single-bay frames.

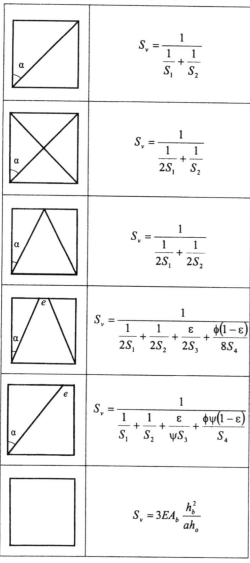

3 DEFLECTIONS OF EQUIVALENT BEAM

3.1 Governing equations

As outlined before, a multi-story frame can be substituted by a simple, cantilever, Timoshenko beam that is equivalent with respect to stiffness. The equivalent properties of the beam are its bending stiffness, EI, and its shear stiffness, S_v. The beam is subjected to a distributed lateral load, simulating equivalent static wind or earthquake excitation. For the sake of simplicity, the stiffness properties and the lateral loads are considered in this analysis to vary linearly along the height of the beam as shown in Figure 7.

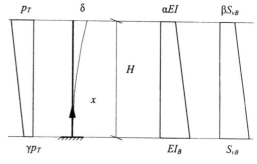

Figure 7. Loading, deflected shape, and stiffness distribution of equivalent cantilever beam.

The relevant expressions for the stiffness properties and the lateral loading may then be written as:

$$p(\xi) = p_T(\gamma + \xi - \gamma\xi) \tag{17}$$

$$EI(\xi) = EI_B(1 - \xi + \alpha\xi) \tag{18}$$

$$S_v(\xi) = S_{vB}(1 - \xi + \beta\xi) \tag{19}$$

where

$$\xi = \frac{x}{H} \tag{20}$$

The system of differential equations of equilibrium for this beam is given by:

$$-(EI\psi')'' = -p \tag{21}$$

$$S_v(\delta' - \psi) = -(EI\psi')' \tag{22}$$

where ' denotes differentiation with respect to x. For given stiffness distributions, these equations can be solved analytically for the deflections δ and the flexural rotations ψ, taking into account the appropriate boundary conditions:

• zero deflection at the base:

$$\delta(0) = 0 \tag{23}$$

• zero bending rotation at the base:

$$\psi(0) = 0 \Rightarrow \tag{24}$$

• zero bending moment at the top:

$$M(H) = 0 \Rightarrow \psi'(H) = 0 \tag{25}$$

• zero shear force at the top:

$$V(H) = 0 \Rightarrow \gamma(H) = 0 \Rightarrow \delta'(H) - \psi(H) = 0 \tag{26}$$

3.2 Calculation of deflections with the force method

Alternatively, the lateral deformations may be determined by application of the force method. The shear forces and bending moments due to the external loads and a unit horizontal force at the position x_o where the displacement is to be determined are, respectively, equal to:

• Due to external loads:

Shear force:

$$V(\xi) = \frac{1}{2}p_T(1 + \gamma + \xi - \gamma\xi)(1 - \xi)H \tag{27}$$

Bending moment:

$$M(\xi) = \frac{1}{6}p_T(2 + \gamma + \xi - \gamma\xi)(1 - \xi)^2 H^2 \tag{28}$$

• Due to a unit force at the position $\xi_0 = x_0/H$:

Shear force:

$$\overline{V}(\xi) = \begin{cases} 1, & \xi \leq \xi_0 \\ 0, & \xi > \xi_0 \end{cases} \tag{29}$$

Bending moment:

$$\overline{M}(\xi) = \begin{cases} \xi_0 - \xi, & \xi \leq \xi_0 \\ 0, & \xi > \xi_0 \end{cases} \tag{30}$$

The lateral deformations at position ξ_0 may be determined by summing up the bending and shear deformations, δ_b and δ_s, respectively:

$$\delta = \delta_b + \delta_s \tag{31}$$

Using the force method, δ_b and δ_s are obtained by appropriate integration:

$$\delta_b = \int_0^{x_0} \frac{M\overline{M}}{EI} dx \tag{32}$$

$$\delta_s = \int_0^{x_0} \frac{V\overline{V}}{S_v} dx \tag{33}$$

By executing the integration the following expressions are found for the deformation at any position ξ:

$$\delta_b = \frac{H^4 p_T}{6EI_B} D_B \tag{34}$$

$$\delta_s = \frac{H^2 p_T}{2S_{vB}} D_S \tag{35}$$

where

$$D_B = \frac{1}{12(\alpha - 1)^5}\{(1 - a)\xi\{\alpha^2[-36 - 18(3 + 2\gamma)\xi + (2 + 16\gamma)\xi^2 + 3(1 - \gamma)\xi^3] + \xi[-12 + 2\xi + \xi^2 - \gamma(6 - 4\xi + \xi^2)] + \alpha\xi[48 - 4\xi - 3\xi^2 + \gamma(24 - 14\xi + 3\xi^2)] + \alpha^3[24 + 18\xi - \xi^3 + \gamma(12 + 18\xi - 6\xi^2 + \xi^3)]\} + 12\alpha^2[-3 + \alpha(2 + \gamma)] \cdot [1 + (\alpha - 1)\xi]\log[1 + (\alpha - 1)\xi]\} \tag{36}$$

$$D_S = \frac{1}{2(\beta - 1)^3}\{(\beta - 1)\xi[2 + \xi - \beta\xi + \gamma(2 - 4\beta - \xi + \beta\xi)] + 2\beta(-2 + \beta + \beta\gamma) \cdot \log[1 + (\beta - 1)\xi]\} \tag{37}$$

The total deformation may be written as:

589

$$\delta = \frac{p_T H^2}{2 S_{vB}} \left(\rho_M \frac{D_B}{3} + D_S \right) \qquad (38)$$

where:

$$\rho_M = \frac{S_{vB} H^2}{EI_B} \qquad (39)$$

The base shear is equal to:

$$V_B = \frac{1}{2} p_T (1 + \gamma) H \qquad (40)$$

Inserting equation (40) into (38) the total lateral deformation is obtained in dimensionless form as:

$$\frac{\delta}{H} \cdot \frac{S_{vB}}{V_B} = \frac{1}{1+\gamma} \cdot \left(\frac{1}{3} \rho_M D_B + D_S \right) \qquad (41)$$

4 ANALYSIS RESULTS

4.1 *Performance criteria*

All seismic codes provide limitations for inter-story drifts in order to limit the damage in non-structural elements, as well as second order effects. In addition, limits are imposed on the maximum total drift on top of the building. The two criteria can be combined, so that the design target is a uniform, as far as possible, drift distribution over the height of the building, suggesting a linear variation of lateral deflections.

A fundamental assumption of the proposed approach must also be reminded here, namely that elastic deformations, as obtained by the above analysis, are considered to adequately represent the deformations of the real, elasto-plastic system. This assumption is in accordance to most codes, which suggest to obtain elasto-plastic displacements by multiplying elastic ones with an appropriate behavior factor.

4.2 *Numerical results*

Equations (41) and (39) indicate that for low values of ρ_M the shear deformations are prevailing, while for high values of ρ_M the bending deformations are more important.

It is reminded that values of the parameters α and β smaller than 1 correspond by definition to a reduction in stiffness, and therefore, normally, in strength too, along the height of the building.

The variation of the deformation along the height of the building for the case of triangular loading ($\gamma=0$) is presented in Figures 8, 9 and 10 for various values of the parameters ρ_M, α and β.

Figure 8 shows the influence of parameter ρ_M for a case of decreasing stiffness along the height of the building. It is observed that for low values of ρ_M,

where shear deformations govern the response, the design target of linear variation of the lateral deformations is best achieved by a reduction in both bending and shear stiffness over the height of the structure, defined here by values of the parameters α and β equal to 0.1. However, for increasing values of ρ_M the deformation line becomes increasingly non-linear if this stiffness distribution is maintained.

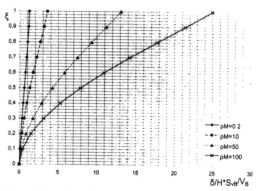

Figure 8. Variation of drift along the height of the building for $\alpha=\beta=0.1$ and values of ρ_M between 0.2 and 100.

This is also verified by Figure 9, where ρ_M is kept constant, equal to 0.2, so that shear deformations govern, while α is varied between 0.1 and 1, and β between 0.5 and 2. The influence of the bending stiffness distribution parameter α is insignificant in this case of prevailing shear response. Again, the design target is best served by small values of β, which suggests decreasing shear stiffness with height.

In the case of mostly flexural buildings, characterized by increasing ρ_M, better results are achieved for higher α values, as illustrated by Figure 10. As expected, the lateral deformations are mostly

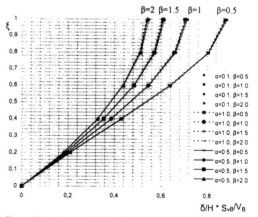

Figure 9. Variation of drift along the height of the building for ρ_M =0.2, α between 0.1 and 1, and β between 0.5 and 2.

expected, the lateral deformations are mostly influenced by the distribution of the bending stiffness (parameter α) for such buildings with prevailing bending deformations. On the contrary, the influence of shear stiffness distribution (parameter β) is practically insignificant.

tion of low α and β-values, while the deformation requirements, as discussed before, are served better by high α and β-values. This constitutes a disadvantage for slender bracing systems with large height-to-width ratios, in which the bending deformations prevail.

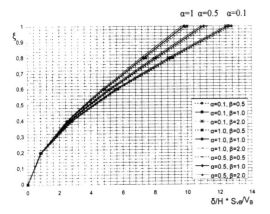

Figure 10. Variation of drift along the height of the building for ρ_M =50, α between 0.1 and 1, and β between 0.5 and 2.

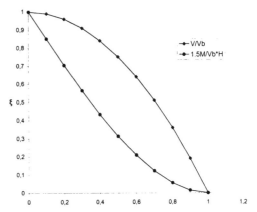

Figure 11. Variation of demand in shear and bending strength over the height of the building for triangular lateral loading.

4.3 Discussion of results

As already mentioned, strength based, conventional wisdom dictates higher bending and shear stiffness at the base of the building, which gradually decreases with height, following the variation of bending moment and shear force, respectively. Using equations (27) and (40) the applied shear force, related to the base shear, is given by:

$$\frac{V}{V_B} = \frac{(1 + \gamma + \xi - \gamma\xi)(1 - \xi)}{1 + \gamma} \qquad (42)$$

Similarly, the applied overturning moment, related to the overturning moment at the base of the structure is given by:

$$\frac{3M}{2V_B H} = \frac{(2 + \gamma + \xi - \gamma\xi)(1 - \xi)^2}{2(1 + \gamma)} \qquad (43)$$

Equations (42) and (43) are presented graphically for the case of triangular loading (γ=0) in Figure 11. The curves express the requirements of the building in shear and bending strength. For braced frames these requirements correspond to requirements for the dimensions of braces and columns, respectively. As expected, the requirements decrease from the top to the base of the structure.

Assuming that the variation of strength is not very much deviating from the variation in stiffness, a discrepancy is found for structures with high value of ρ_M, and accordingly prevailing bending deformations The strength requirements suggest the adop-

Figures 12 to 14 illustrate types of buildings for which the above discussion is relevant. The rigid frames of Figure 12 deform primarily as shear cantilevers, when subjected to lateral loads. Therefore, there is no discrepancy between the stiffness distributions resulting from strength and deflection requirements.

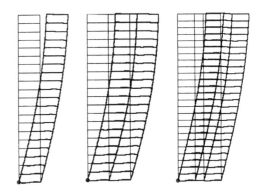

Figure 12. Rigid frames with prevailing shear mode of lateral deformation.

Slender braced frames with high aspect (height to width) ratio, such as the first example of Figure 13, deform primarily as flexural cantilevers. For such structures a design governed by deflection requirements will result in different stiffness distributions than one for which strength criteria prevail.

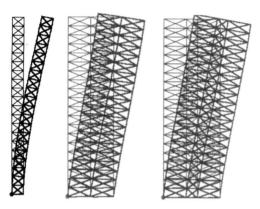

Figure 13. Braced frames with prevailing flexural or combined shear-flexural mode of lateral deformation.

The other two braced frames of Figure 13 with higher aspect ratios, as well as the combined rigid and braced frames like the one shown in Figure 14 constitute intermediate cases, for which further analysis is needed.

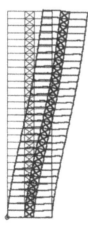

Figure 14. Mixed rigid-braced frames with combined shear-flexural mode of lateral deformation.

5 SUMMARY AND CONCLUSIONS

The preliminary design of multi-story building structures subjected to lateral loads has been investigated. First, a simplified analysis method has been suggested, based on substitution of the building by a stiffness-equivalent Timoshenko beam. Then, the results of this analysis have been used to gain qualitative understanding of the influence of bending and shear stiffness distribution on the response of such structures. A design criterion has been established that requires linear or near-linear deflection along the height. It has been observed that the conventional stiffness distribution, dictated by strength constraints, is not the best to satisfy this deflection criterion in the case of flexible braced frames. This suggests that a new approach to the design of such frames may be appropriate when deflection criteria govern.

REFERENCES

American Institute of Steel Construction 1995. *Load and resistance factor design (LRFD)*.

American Institute of Steel Construction 1997. *Seismic provisions of structural steel buildings*.

European Committee for Standardization (CEN) 1992. *Eurocode 3: Design of steel structures, Part 1.1: General rules and rules for buildings*, ENV 1993-1.1.

European Committee for Standardization (CEN) 1994. *Eurocode 8: Design provisions for earthquake resistance of structures, Part 1.3: General rules – Specific rules for various materials and elements*, ENV 1998-1.3.

Gionis, G.A. 1999. *A new method for the preliminary design of high-rise buildings*. Diploma thesis, Civil Engineering Department, National Technical University of Athens, Greece (in Greek).

Haris, A.A.K. 1977. Approximate stiffness analysis of highrise buildings, *ASCE, Journal of the Structural Division*, Vol. 104, No. ST4, August.

Heidebrecht, A.C., Stafford Smith, B. 1973. Approximate analysis of tall wall-frame structures, *ASCE, Journal of the Structural Division*, Vol. 99, No. ST2, February, 199-221.

Lu, L.-W., Ricles, J.M. and Kasai, K. 1997. Global performance, General report. In Mazzolani, F., and Akiyama, H., eds., *Behaviur of steel structures in seismic areas, Stessa '97*, Edizioni 10/17: 361-381.

Pouangare, C.C. 1990. *New design concepts for tall buildings*, Ph.D. thesis, Department of Civil Engineering, Massachusetts Instityte of Technology, Cambridge, MA.

Stafford Smith, B., Kuster, M. and Hoenderkamp, J.C.D. 1981. A generalized approach to the deflection analysis of braced frame, rigid frame and coupled wall structures, *Canadian Journal of Civil Engineering*. Vol. 8, January, 230-239.

Stafford Smith, B. and Coull, A. 1991. *Tall building structures: Analysis and design"*, John Wiley & Sons, Inc., New York.

Takewaki, I. 1997. Efficient semi-analytical generator of initial stiffness designs for steel frames under seismic loading – Part 1: Fundamental frame. *The structural design of tall buildings*. Vol. 6: 151-162.

Takewaki, I. 1997. Efficient semi-analytical generator of initial stiffness designs for steel frames under seismic loading – Part 2: Slender frame. *The structural design of tall buildings*. Vol. 6: 163-170.

Vayas, I. 1997. Stability and ductility of steel elements, *Journal of Constructional Steel Research*. 44(1-2): 23-50.

Vayas, I. 1999. Design of braced frames, *Seismic resistant steel structures: Progress and challenge*, Advanced professional training course, Coordinators: F.M. Mazzolani, V. Gioncu, International Center for Mechanical Sciences (CISM), Udine, Italy, October 18 – 22, Springer Verlag (to appear).

Behaviour of Steel Structures in Seismic Areas, Mazzolani & Tremblay (eds) © 2000 Balkema, Rotterdam, ISBN 90 5809 130 9

Seismic behavior of post-tensioned steel frames

M. M. Garlock, J. M. Ricles, R. Sause, C. Zhao & L. W. Lu
Department of Civil and Environmental Engineering, Lehigh University, Bethlehem, Pa., USA

ABSTRACT: A post-tensioned connection for earthquake resistant steel moment resisting frames (MRFs) is introduced. The connection has excellent ductility, limits inelastic deformations to the easily replaceable components of the connection, requires no field welding, and returns the structure to its pre-earthquake position. The connection includes bolted top and seat angles with post-tensioned high strength strands running parallel to the beam. Analyses were performed on a 6-story, 6-bay post-tensioned steel MRF to study its response to strong ground motions. Results show good energy dissipation, strength, and ductility in the post-tensioned system. The analyses indicate that the seismic performance of a post-tensioned steel frame can exceed that of a frame with conventional moment resisting connections.

1 INTRODUCTION

The inadequate performance of steel moment resisting frames (MRFs) in recent earthquakes has caused much concern. During the 1994 Northridge Earthquake, many steel-framed buildings suffered unexpected premature fractures in their welded beam-to-column connections (Youssef et al. 1995, FEMA 1995). The conventional earthquake resistant steel moment connection is composed of full penetration welds between the beam flanges and the column face, with the beam web bolted to a shear tab extending from the column. Standard design practice assumes that these connections are capable of withstanding repeated inelastic deformation cycles. However, the occurrence of premature fractures in the beam flange welds between the beams and columns indicates an immediate need to revise current design and construction standards. Recent research has provided new details for MRF construction (FEMA 1995, FEMA 1997) but these new details are more expensive to fabricate. Moreover, use of the standard MRF connection often results in damage and permanent drift in the building following a major earthquake.

To overcome the difficulties of the standard MRF connection, a post-tensioned (PT) connection for earthquake resistant steel MRF structures is introduced. This connection has excellent ductility, limits inelastic response to the replaceable components of the connection, requires no field welding, and returns the structure to its pre-earthquake position (i.e., exhibits a self-centering capability). A PT MRF connection uses high strength steel strands that are post-tensioned after the bolted top and seat angles are installed (Figure 1). Details of this connection are described below. After a major earthquake, repairs would be limited to the replaceable connection components, which are the top and seat angles. Although these angles are easily replaced, experimental studies conducted by the authors show that the angles have a sufficient low-cycle fatigue life to perform well over several earthquake loading events.

This paper describes the details and behavior of a PT steel MRF connection, and analytical studies of a MRF with PT connections. Experimental test results were used to calibrate a fiber element model that was developed to analyze a 6-story, 6-bay PT steel MRF. Non-linear static push-over and seismic time-history analysis were performed on the PT MRF model and compared to the response of the same MRF with standard fully restrained welded connections.

A companion paper (Ricles et al., 2000) discusses experimental studies of large-scale interior connection subassemblies of a PT MRF subjected to cyclic inelastic loading.

2 PRIOR RESEARCH ON PT AND SEMI-RIGID CONNECTIONS

The development of a PT connection utilizes knowledge acquired from studies of two areas of related research; PT precast concrete construction and partially restrained steel connections.

Choek and Lew (1993) conducted a series of cyclic loading tests of one-third scale post-tensioned concrete connections. An analytical study of the seismic behavior of unbonded post-tensioned precast concrete frames was conducted by El-Sheikh et al. (1997). The results from these and other studies of PT connections for precast concrete structural systems indicate that these connections are well suited for seismic resistant structural systems. Their adaptation to steel frames is therefore appealing.

Partially restrained (PR) steel connection details with top and seat angles have been studied extensively. Recent research has demonstrated the energy-absorption ability of PR connections, (Nader and Astaneh 1992, Leon and Shin 1994, Mander et al. 1994, Mayangarum and Kasai 1997).

More details on prior research related to precast construction with PT connections, and partially restrained steel connections can be found in Ricles et al. (1999).

3 POST-TENSIONED CONNECTION

3.1 *Connection Details*

The PT steel MRF connection consists of bolted top and seat angles with post-tensioned high strength strands running parallel to the beam and anchored outside the connection (Figure 1). The strands compress the beam flanges against the column flanges to resist moment, while the two angles and the friction at the beam and column interface resist shear. The proposed details are shown in Figure 2 for a connection to an exterior column. The angles are intended to dissipate energy. However, they also provide redundancy to the force transfer mechanisms for transverse beam shear and moment. The beam flanges are reinforced with reinforcing plates to control beam yielding. Also, shim plates are placed between the column flange and the beam flanges so that only the beam flanges and reinforcing plates are in contact with the column.

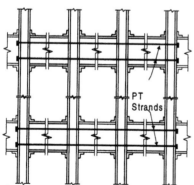

Figure 1. Elevation of a post-tensioned frame.

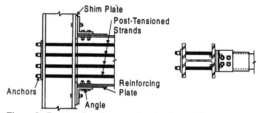

Figure 2. Post-tensioned connection detail at anchor point.

One erection sequence of the PT connection assembly could be as follows: (1) shop bolt the seat angle to the column, (2) shop bolt the top angle to the beam, snug tight, (3) erect the column and the beam, field bolting the remainder of the beam-to-column connection, snug tight, (4) plumb and level the building, (5) insert the PT strands into the beams of the post-tensioned frame, (6) tension the strands starting from the bottom and working to the top, (7) check and refine building for plumb and level; and (8) fully pretension the bolts of the beam-to-column connections. For a high rise building, this sequence of construction could be done in sets of floors.

3.2 *Flexural Behavior*

The idealized moment-rotation (M-θ_r) behavior and the corresponding load-deflection (H-Δ) behavior of a PT steel connection are shown in Figure 3, where θ_r is the relative rotation between the beam and column. The M-θ_r behavior of a PT connection is characterized by a gap opening and closing at the beam-column interface under cyclic loading (see Figure 3 insert). The moment to initiate this separation is called the decompression moment. The connection initially behaves as a fully restrained connection, but following decompression it behaves as a partially restrained connection. The initial stiffness of the connection is the same as that of a fully restrained welded moment connection when θ_r is equal to zero until the gap opens at decompression (event "a" in Figure 3). The stiffness of the connection after decompression is associated with the stiffness of the angles and the elastic axial stiffness of the post-tensioned strands. With continued loading, the tension angles will yield (event "b"), with eventual full plastic yielding of the tension angles occurring at event "c". The compression angles yield at event "d". Until the load reverses at event event "e", the M-θ_r relationship has a linear response where the connection stiffness is primarily due to the axial stiffness of the post-tensioned strands. Upon unloading, the angles will dissipate energy (between events "e" and "h") until the gap between the beam flange and the column face is closed at event "h" (i.e., when θ_r is equal to zero). A complete reversal in applied moment will result in similar behavior occurring in the opposite direction of loading, as shown in Figure 3 where the M-θ_r and H-Δ relationships are symmetric.

As long as the strands remain elastic, and there is no significant beam yielding, the post-tensioning force is preserved and the connection will self-center upon unloading (i.e., θ_r returns to zero rotation upon removal of the connection moment and the structure returns to its pre-earthquake position). The level of the decompression moment, the flexural strength of the angles, and the elastic stiffness of the post-tensioned strands control the strength of the connection. The amount of energy dissipation of the connection is related to the flexural strength of the angles. To ensure that the strands remain elastic, long strands (with lengths close to the beam span length)

are used so the deformation due to gap opening results in a small strain over a long strand length. The strands are post-tensioned to a stress level that is sufficiently below the yield stress.

4 ANALYTICAL MODELING OF A PT CONNECTION

4.1 Description of Model

The computer program DRAIN-2DX (Prakesh et al., 1993) was used to develop a model of the PT steel MRF connection and the associated beams and columns. This computer program is well suited for this

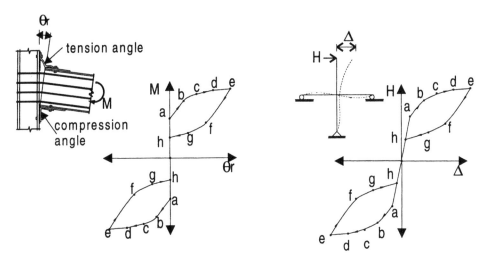

Figure 3. Moment-rotation and load-deflection behavior of a post-tensioned connection.

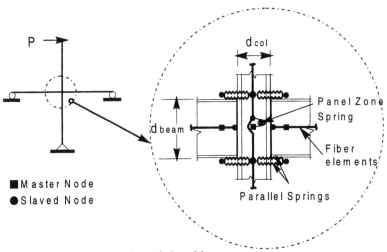

Figure 4. Post-tensioned connection analysis model.

purpose since it accounts for geometric and material non-linearities in static and dynamic two-dimensional structural analysis. A brief description of the model is presented. More detailed information on the modeling of PT connections is given in Ricles et al. (1999).

The beams and columns are modeled using fiber elements. Each fiber element is divided into a number of segments along the element length. The cross-section of each segment is comprised of several fibers. A material stress-strain relationship, a cross sectional area, and a distance from the longitudinal reference axis of the member characterize each fiber. The fibers in the segment of the beam adjacent to the column are used to model gap opening. The fibers of the beam cross section initially in contact with the shim plates are assigned a stress-strain relationship that has stiffness in compression, but none in tension. Fibers are omitted from the beam cross section not in contact with the shim plates.

To properly account for the depth of the beam and the size of the panel zone in the connection region, it was necessary to utilize sets of master-slave nodes as shown in Figure 4. The flexibility of the panel zone was modeled by placing a rotational spring with moment-rotational characteristics determined from the shear stiffness and strength of the panel zone. Each strand in the PT connection was modeled using a truss element. Each angle was modeled by two parallel spring elements resulting in a tri-linear force deformation relationship. The spring elements modeled the flexural behavior of the vertical leg of the angle.

4.2 Model Verification

Figure 5 compares the model's response with the experimental measured response of a PT connection (Specimen PC3) tested by Ricles et al. (2000). The predicted response is seen to correlate well with the experimental response. The predicted initial stiffness, however, is 10 to 15% higher than that of the test specimen. Analyses of the other test specimens had a similar accuracy to experimental results. These other analyses are not discussed herein.

5 SEISMIC BEHAVIOR OF A PT STEEL MRF

5.1 Description of Frame

To investigate the seismic behavior of steel frames with PT connections, a 6-story perimeter MRF with 6 bays was designed and then analyzed using the PT connection model described previously. The structure was designed as a special moment resisting frame in accordance with the 1994 NEHRP provisions [BSSC 1994]. The connections of this MRF

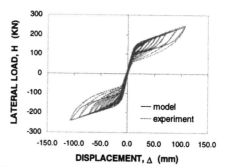

Figure 5. Comparison of analytical model with Specimen PC3 experimental results by Ricles et al. (2000).

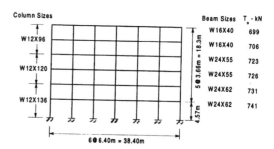

Figure 6. Frame used in analytical study.

were assumed to be rigid (i.e., a FR MRF). The design is consistent with the design of a typical office building in San Francisco or Los Angeles, where the structure is designed for a site with stiff soil conditions. The frame is shown in Figure 6 where T_o is the total initial force in the strands.

The rigid connections in the FR MRF were replaced with PT connections to create a second frame, thereby enabling a comparison to be made between the seismic behavior of a typical FR MRF and a PT MRF. The connection consisted of L203x203x15.9 A36 (i.e. 248 MPa yield strength) steel angles. The bolts connecting the angles to the column were located 63.5 mm from the fillet. This connection was consistent with Specimen PC4 described in Ricles et al. (2000). The post tensioned strands ran continuously through the 6 bays of the frame, and were anchored to the outside faces of the exterior columns. The resulting connection moment capacities, as determined from the dynamic time history analyses discussed below, are approximately $0.9M_p$, $1.0M_p$, and $1.3M_p$ at the first and second, third and fourth, and fifth and sixth floors, respectively, where M_p is the beam flexural strength.

5.2 Static Pushover Analysis

A nonlinear static pushover analysis of the moment resisting frames was conducted. The lateral loads

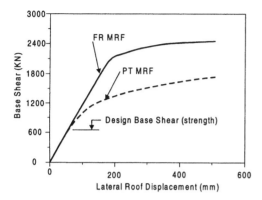

Figure 7. Static pushover results.

were distributed over the height of the structure in accordance with the 1994 NEHRP provisions. The relationship between the base shear and the roof lateral displacement for both frames is shown in Figure 7. Both frames have the same initial response before

some of the PT connections reach the decompression moment in the lower floors at a base shear of 578 kN. This magnitude of base shear is close to the design base shear, which is 650 kN. After decompression, the lateral stiffness of the PT MRF began to decrease relative to that of the FR MRF. At a roof displacement of about 165 mm, corresponding to 0.7% of the height of the building, the top and seat angles yield in the lower floors. Additional angles in the upper floors yield with further imposed lateral displacement. In the FR MRF, yielding occurs at the column base and in the beams.

5.3 Time History Analysis

To investigate the seismic behavior of the PT MRF and FR MRF, nonlinear dynamic time history analyses were conducted [Ricles et al, 1999], using an ensemble of four accelerograms. These accelerograms include: (1) the north-south component of the 1940 El Centro Earthquake; (2) the north-south component of the 1994 Northridge Earthquake recorded at Newhall; (3) the north-south El Centro Differential

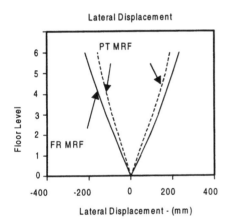

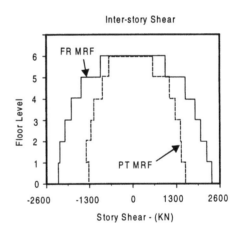

Figure 8. Envelop results of 1.5 x El Centro Earthquake.

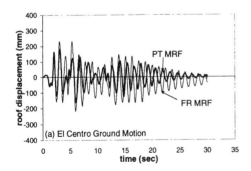

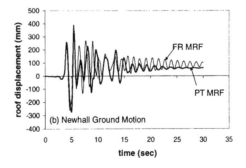

Figure 9. Time history results of 1.5 x El Centro Earthquake.

Table 1. Response of MRF with PT connections and MRF with welded connections.

Ground Motion	Max. Roof Displ. (mm)		Max. Story Drift (%)		Max. Base Shear (kN)	
	PT	Welded	PT	Welded	PT	Welded
El Centro	191	234	1.0	1.2	1579	2353
Newhall	379	391	2.2	2.4	2491	2914
Imperial Valley - El Centro	163	193	0.9	1.1	2456	2794
Artificial	226	282	1.4	1.9	2144	2714

Array recording of the 1979 Imperial Valley Earthquake; (4) and an artificially generated earthquake that was compatible with the NEHRP response spectrum for the seismicity conditions noted previously. The El Centro Earthquake was scaled to a peak ground acceleration of 0.5g in order to make it commensurate with the other records in the ensemble. The peak ground accelerations for the Northridge and Imperial Valley Earthquake records were 0.59g and 0.49g, respectively. The artificial earthquake had a peak ground acceleration of 0.5g.

The response envelop of lateral displacement and story shear for the PT MRFs and the FR MRFs to the scaled El Centro earthquake is shown in Figure 8. It can be seen that the PT MRF drifts less than the FR MRF. Also, the maximum story shears are shown to be less for the frame with PT connections.

A time history plot of the roof displacement response to the scaled El Centro earthquake is shown in Figure 9(a). The response of the PT MRF to the scaled El Centro earthquake consists of inelastic deformations that are confined to the top and seat angles of the first floor PT connections. The beams and columns of the frame remain elastic. Therefore the frame maintains its self centering capability, returning to its original undeformed position following the earthquake. The slight elongation of the period of the PT MRF occurs due to the softening of the structure when decompression and yielding occurs in the PT connections. For the FR MRF the first floor beams yielded at the face of the columns, with the other beams and all columns remaining elastic throughout the structure. Approximately 25 mm of residual lateral displacement existed at the end of the earthquake in the FR MRF.

Figure 9(b) shows the roof displacement time history response of both frames to the Newhall ground motion. The maximum roof displacement of the frame with PT connections is 379 mm, which is less than the maximum roof displacement of 391 mm for the MRF with welded connections. In the MRF with PT connections, the connections in the first through fourth floors yield. In addition, plastic hinges formed at the base of the first story columns, which results in a residual roof displacement of 61 mm. For the MRF with welded connections, plastic hinges formed in the first and second floor beams as well as at the base of the first floor columns. The

residual roof displacement of the MRF with welded connections is 87 mm.

In all cases studied the PT MRF developed a smaller magnitude of response (roof displacement, base shear, story drift) to the selected earthquake records. This is shown in Table 1. The Newhall ground motion produced the largest response for both frames compared to the other earthquakes.

6 SUMMARY AND CONCLUSIONS

An innovative connection for seismic resistant design of steel structures has been presented. Combining top and bottom seat angles with high strength post-tensioned steel strands results in a connection with initial rigidity that is similar to fully welded moment resisting connections. Full scale tests show that this new connection has the benefit of concentrating the inelastic deformations in the replaceable top and bottom seat angles, making it more easily repaired than a welded moment connection. In addition, the connection has a self-centering capability, resulting in minimal permanent story drift in a building following a severe earthquake. However, the frame must be properly designed to avoid permanent rotational deformation at the base of the columns. An analytical model was presented that was based on fiber elements. Analytical studies indicate that moment resisting frames with post-tensioned connections may perform better than moment resisting frames with conventional welded moment connections.

7 ACKNOWLEDGEMENTS

The research reported herein was supported by the National Science Foundation (NSF) and American Institute of Steel Construction (AISC). Dr. Ken Chong and Dr. Ashland Brown cognizant NSF program officials. Nestor Iwankiw cognizant AISC program official. The opinions expressed in this paper are those of the authors and do not necessarily reflect the views of the sponsors.

8 REFERENCES

Building Seismic Safety Council, (1994) "NEHRP Recommended Provisions for Seismic Regulation for New Building," Washington D.C.

Cheok, G., and Lew, H. (1993), "Model Precast Concrete Beam-to-Column Connections Subject to Cyclic Loading," *PCI Journal*, PCI, Vol. 38, No. 4, July-August, pp. 80-92.

El-Sheikh, M., Sause, R., Pessiki, S., Lu, L.W., and Kurama, Y. (1997), "Seismic Analysis, Behavior, and Design of Unbonded Post-Tensioned Precast Concrete Frames," *Report No. EQ-97-02*, Dept of Civil and Environmental Eng., Lehigh University, Bethlehem, PA

FEMA (1995), "Interim Guidelines: Evaluation, Repair, Modification and Design of Welded Steel Moment Frame Structures," Bulletin No. 267, FEMA, Washington, D.C.

FEMA (1997), "Interim Guidelines Advisory No. 1: Supplement to FEMA 267," Bulletin No. 267A, FEMA, Washington, D.C.

Leon, R.T. and Shin, K.J. (1994), "Seismic Performance of Semi-Rigid Composite Frames," Proceedings of the 5th U.S. National Conference on Earthquake Engineering, Chicago, Illinois.

Mander, J.B., Chen, S.S., and Pekcan, G. (1994), "Low-Cycle Fatigue Behavior of Semi-Rigid Top and Seat Angle Connections," Engineering Journal, AISC, Third Quarter.

Mayangarum, A. and Kasai, K. (1997), "Design, Analysis, and Application of Bolted Semi-Rigid Connections for Moment Resisting Frames," ATLSS Report 97-15, ATLSS Engineering Research Center, Lehigh University.

Nader, N.M., and Astaneh-Asl, A. (1992), "Seismic Behavior and Design of Semi-Rigid Steel Frames," Report No. UCB/EERC-92/06, Earthquake Engineering Research Center, Univ. of Calif., Berkeley.

Prakesh, V., Powell. G., and Campbell, S. (1993), "DRAIN-2DX Base Program Description and User Guide; Version 1.10," *Report No. UCB/SEMM-93/17 & 18*, Structural Eng Mechanics and Materials, Dept of Civil Eng, Univ. of Calif., Berkeley, CA, December.

Ricles, J., Sause, R., Zhao, C., Garlock, M., and Lu, L. (1999), "Post-Tensioned Seismic Resistant Connections for Steel Frames," (submitted to Journal of Structural Engineering).

Ricles, J., Sause, R., Garlock, M., Peng, S.W., and Lu, L. (2000), "Experimental Studies on Post-Tensioned Seismic-Resistant Connections for Steel Frames," STESSA Proceedings, Montreal, Canada.

Youssef, N., Bonowitz, D., and Gross, J. (1995), "A Survey of Steel Moment Resisting Frame Buildings Affected by the 1994 Northridge Earthquake," NIST, *Report No. NISTIR 5625*, Gaithersburg, Md.

Behaviour of Steel Structures in Seismic Areas, Mazzolani & Tremblay (eds) © 2000 Balkema, Rotterdam, ISBN 90 5809 130 9

The effect of connection hysteretic behaviour on seismic damage to moment resisting steel frames

R. Landolfo
University of Chieti Gabriele D'Annunzio, Italy

G. Della Corte & G. De Matteis
University of Naples 'Federico II', Italy

ABSTRACT: Current paper focuses on beam-to-column connections, which play a very important role in determining the seismic performance of moment resisting steel frames. In particular, the mechanical modeling of these structural components is investigated, aiming at correct dynamic analyses of this structural typology. On the basis of a previously developed mathematical model, which simulates the static cyclic inelastic behaviour of steel beam-to-column joints, a parametric study is herein presented, relating the effect of strength cyclic deterioration and pinching of hysteresis loops on the predicted values of plastic deformation and energy dissipation demand.

1 INTRODUCTION

A reliable prediction of seismic performance of structures is basically related to the accuracy of mechanical behaviour modeling and to the appropriate definition of earthquake characteristics.

Ordinary structures can survive a strong earthquake just relying on their own energy dissipation capacity, which is in turn related to the local mechanical behaviour of their components.

In moment resisting steel frames, beam-to-column joints are key components for the global performance of the structure, what is also testified by recent strong earthquakes occurred in different World Regions. In particular, in case of Northridge (1994) and Kobe (1995) earthquakes the inspection of steel-framed buildings evidenced many unexpected connection failures. The unforeseen bad performance of connections have stimulated the development of many research activities worldwide, aiming at improving the knowledge of connection behaviour in steel buildings, when large plastic deformation demand occurs. Among these activities, the European Research Project entitled "Reliability of Moment Resistant Connections of Steel Building Frames in Seismic Areas" has been developed in Europe through the co-operation of Universities and Institutions belonging to eight different countries, under the co-ordination of University of Naples (Mazzolani 2000). Both from the already available experimental test results and from the ones obtained within this project, it is apparent a complex cyclic inelastic behaviour of beam-to-column joints. Gradual plastification, kinematic hardening, cyclic hard-

ening, strength degradation due to both plastic fatigue and local buckling, pinching of hysteresis loops are the main phenomenological aspects which determine this complexity. On the contrary, hysteresis models usually adopted for describing joint behaviour are very simple, neglecting some or all of the above mentioned phenomenological aspects. In the framework of the above research project, a mathematical model interpreting the static cyclic inelastic behaviour of the most usual connection typologies has been developed. Also, some preliminary numerical investigations were performed, concerning the effect of the type of hysteresis model on the response of both single degree of freedom systems and multi-degree of freedom structures (Della Corte et al. 2000).

Current paper provides numerical results from a new and more comprehensive parametric analysis, aiming at verifying whether the simplification of connection mechanical behaviour is allowed or a more accurate hysteresis model is needed for a reliable prediction of deformation and energy dissipation demand to the frame structural components.

2 THE HYSTERESIS MODEL

The hysteresis model adopted in the study is semi-empirical and therefore its parameters are to be calibrated on the basis of experimental test results. It correlates the bending moment (M) transmitted by the beam to the column and the dual kinematic parameter, i.e. the relative rotation (ϕ) between the beam and the column. All the important features exhibited by usual connection typologies in static cyclic inelastic laboratory tests are taken into account.

The fundamental concept, upon which the model is based, is that the basic unit for the description of a generic loading history is the deformation excursion, which is constituted by a loading branch and by the subsequent unloading branch of the load-deformation path (ATC 1992) (see Figure 1). Once defined the 'perfect dissipative system' - the one whose mechanical properties are stable, i.e. independent from the deformation history - then the description of cyclic hardening and cyclic damaging phenomena is obtained by specifying the variation law of excursion's parameters, as a function of the dissipated hysteretic energy.

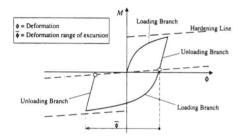

Figure 1. Basic concepts for the hysteresis model proposed.

In particular, strength degradation has been modeled by means of the following relationship:

$$M_{0,dam} = M_0(1 - \beta) = M_0(1 - E_h/M_0\phi_{ult})$$ (1)

where M_0 is the initial strength, $M_{0,dam}$ is its reduced (damaged) value, E_h is the hysteretic dissipated energy and, finally, ϕ_{ult} is the conventional deformation capacity under monotonic loading.

Figure 2 shows the methodology adopted for simulating pinching. A lower bound and an upper bound curve are introduced. The actual load-deformation path is then described by defining the transition law between these two limit curves.

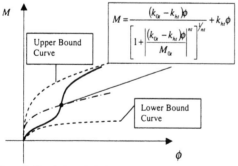

Figure 2. Simulation of pinching.

The formula reported in Figure 2 is due to Richard and Abbott (1975). It is based upon the definition of

four parameters, namely initial stiffness (k_0), reference moment level (M_0), stiffness of the hardening branch (k_h) and shape factor (n). The generic parameter defining the transition path (for example, M_{0t}) has been defined by means of the following linear convex combination:

$$p_t = p_l + t (p_u - p_l)$$ (2)

where p_l and p_u are the relevant parameters for the lower bound and the upper bound curve, respectively, and t is a parameter ranging in the interval [0,1]. In particular, the following expression has been assumed for the transition parameter t

$$t = \left[\frac{(\phi/\phi_{lim})^{t_1}}{(\phi/\phi_{lim})^{t_1} + 1} \right]^{t_2}$$ (3)

where t_1, t_2 and ϕ_{lim} are three empirical coefficients. An appropriate choice of them allows every type of behaviour, with or without pinching, to be described.

More detailed information about model assumptions (with reference also to other aspects such as the simulation of unstable branches in the load-deformation relationship) are given in Della Corte et al. (2000).

Figures 3 and 4 show the application of the model for simulating the experimental response of a beam section in bending (Fig. 3) and of a typical top and seat angle connection (Fig. 4).

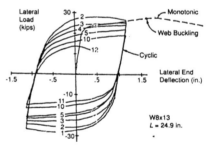

Figure 3a. Experimental behaviour of a beam (I section) in bending (Vann et al. 1973).

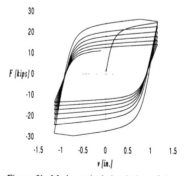

Figure 3b. Mathematical simulation of the experimental results related in Figure 3a.

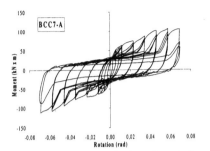

Figure 4a. Experimental behaviour of a typical top and seat angle connection (Calado et al. 1999).

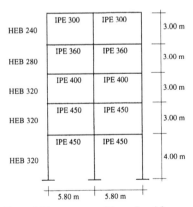

Figure 4b. Mathematical simulation of the experimental results related in Figure 4a.

3 NUMERICAL ANALYSES PLAN

3.1 Frame under study

Numerical analyses have been performed with reference to a regular frame scheme (Fig. 5).

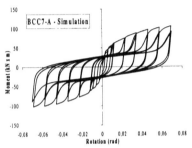

Figure 5. The geometry of the analyzed frame.

Beam-to-column joints were assumed to be rigid and full-strength for design purposes. Then, beams and columns were designed according to Eurocode 3 (1992) and Eurocode 8 (1994). Sub-soil class *B* (medium density sand), peak ground acceleration equal to 0.20*g* and design *q*-factor equal to 6 were

assumed in the design procedure (De Matteis et al. 1998). Vertical loads were computed by considering the frame extracted by a regular plan civil building.

It is useful to remark that beam and column cross sections were chosen taking in mind the serviceability limit state check imposed by Eurocode 8. In particular the maximum inter-story drift was limited to 0.006*h*, *h* being the inter-story height. This limit is very severe; consequently structural members much stronger than the ones strictly necessary to comply with ultimate limit state design requirements were obtained. However, it is important to observe that the following numerical analyses have been performed with reference to very strong earthquakes, whose magnitude is larger than the one concerning the earthquake scenarios upon which Eurocode 8 is based.

3.2 Joint modeling

In the current study column web panels have been assumed to be always rigid and resistant enough to remain elastic. Therefore, the only sources of joint deformation (elastic and/or plastic) were found into beam-to-column connections, which have been modeled by means of lumped elasto-plastic springs. Two types of connections have been considered:

1 Rigid full-strength connections equipped with elasto-plastic bi-linear strength degrading restoring force characteristics.

2 Semi-rigid partial-strength connections equipped with elasto-plastic fully non-linear restoring force characteristics with pinching.

Figure 6 synthesizes the assumed joint's mechanical model by means of qualitative drawings.

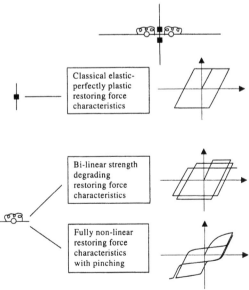

Figure 6. Mechanical modeling hypotheses.

Connection stiffness have been evaluated according to Annex J of Eurocode 3 (1994). Here, the ratio between connection stiffness and beam flexural stiffness is used as control parameter. The rigid connection has been characterized by a value of this ratio equal to twice the limit value that signs the boundary between the rigid and the semi-rigid joints. The semi-rigid partial-strength connection has been obtained by assuming the above ratio equal to the half of the limit value, while the moment capacity of the connection has been hypothesized equal to the half of the beam strength.

Strength deterioration rate and pinching amount have been controlled by means of appropriate model parameters, whose values have been selected according to the results obtained for the calibration of the model against a number of connection hysteretic experimental responses (Della Corte et al. 2000).

3.3 Seismic input

Dynamic analyses have been carried out with reference to three natural accelerograms. Two of them have been selected as near-fault recordings; the last one has been chosen as representative of long-duration type earthquakes. This criterion of subdivision derives from the need of evaluating the effect of strength deterioration on damage prediction to the structure. In fact, it might be foreseen a quite small sensitivity of experienced damage to strength deterioration in case of near-fault earthquakes, characterized by few strong pulses. On the contrary, strength deterioration could be important in case of long-duration earthquakes. Unfortunately, the number of recordings related to strong (large magnitude) long-duration earthquakes is very small. Just one accelerogram has been found as useful for the analyses being (Elgamal et al. 1998). Table 1 gives a summary of the fundamental characteristics of the selected earthquakes and relevant accelerograms. Predominant period was taken as the period corresponding to maximum value of the relative input energy.

Table 1. Fundamental characteristics of the chosen earthquakes.

Earthquake	Moment magnitude	Record	a_{max} (g)	Arias Intensity (cm/s)	Predominant Period (s)	Trifunac duration (s)
Northridge (17 January 1994)	6.7	Rinaldi 228	0.84	14.86	1.36	7.04
Kobe (17 January 1995)	6.9	JMA NS	0.83	71.48	0.69	8.36
Chile (3 March 1985)	8.0	Llolleo 10	0.71	46.16	0.53	35.91

Figures 7 and 8 give the pseudo-acceleration and equivalent velocity ($\sqrt{2E_i/m}$, E_i and m being the relative input energy and the system mass, respectively) spectra for the chosen acceleration records.

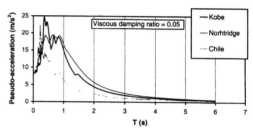

Figure 7. Pseudo-acceleration spectra for the acceleration time histories under study.

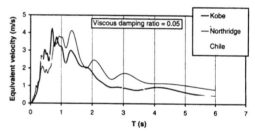

Figure 8. Pseudo-velocity spectra for the acceleration time histories under study.

4 OUTCOMES FROM THE ANALYSES

4.1 General

Earthquake induced frame damage is herein described by means of local and global parameters. For local damage evaluation different criteria have been chosen as a function of the type of hysteretic behaviour. In case of absence of pinching effects, local damage has been described by means of the maximum required plastic excursion. On the contrary, in case of presence of pinching effects, the maximum apparent deformation (i.e. deformation measured from the initial undeformed configuration) has been assumed as damage indicator. Moreover, local hysteretic energy dissipation demand has been monitored.

As for global damage evaluation, the definition of a 'story' by the union of the beams that constitute it and the below columns has been introduced. Then, the distribution of the story hysteretic energy (the sum of the hysteretic energy dissipation for all the elements of that story) along the frame height has been evaluated. Eventually, residual frame story displacements (i.e. displacements at the end of frame motion) have been computed.

Finally, in order to emphasize the modification of the local response when different hysteretic behaviours are assumed, load-deformation histories for the connection between the first column and the first beam on the left side of the first floor ('reference connection') have been computed.

4.2 The effect of strength degradation

The effect of strength degradation rate (β) has been studied with reference to rigid and full-strength connections. It is worthy to notice that the range of β values corresponding to usual strength degradation rate associated to plastic fatigue for steel members and connections is [0, 0.25], with a large predominance of small values (ranging from 0.025 to 0.1) (Della Corte et al. 2000). On the other hand, highest values of β could be considered equivalent to other sources of strength degradation, namely local buckling phenomena. In fact, laboratory tests usually evidence the existence in this case of unstable

branches in the load-deformation relationship, producing higher damage rate. Even if the mathematical model developed is able to explicitly consider unstable branches in the load-deformation relationship, the latter are not directly introduced in the present parametric analysis in order to limit the number of factors to be investigated. Therefore, in current analysis the range of β values investigated has been extended to [0.025, 0.4].

Figures 9a, 9b and 9c show the load-deformation histories computed for the reference connection, in case of Northridge, Kobe and Chile earthquakes, respectively. They have been obtained by assuming a very small value of the damage rate control parameter ($\beta = 0.025$), which practically corresponds to absence of strength deterioration. In figures 10a, 10b and 10c the analogous histories are depicted, but with reference to a greater value of the damage rate control parameter ($\beta = 0.40$).

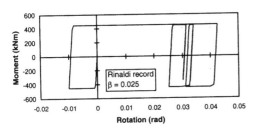

Figure 9a. Moment-Rotation history for the reference connection: Northridge earthquake – No strength degradation.

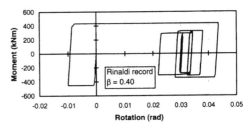

Figure 10a. Moment-Rotation history for the reference connection: Northridge earthquake – High strength degradation.

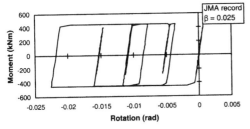

Figure 9b. Moment-Rotation history for the reference connection: Kobe Earthquake – No strength degradation.

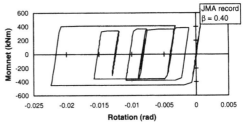

Figure 10b. Moment –Rotation history for the reference connection: Kobe earthquake – High strength degradation.

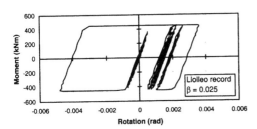

Figure 9c. Moment-Rotation history for the reference connection: Llolleo Earthquake – No strength degradation.

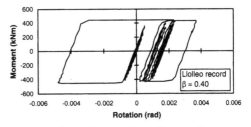

Figure 10c. Moment-Rotation history for the reference connection: Llolleo Earthquake – High strength degradation.

From figures 9 and 10 it can be observed that local behaviour is not very sensitive to strength degradation, at least in the range of values investigated for β and for the chosen ground acceleration records. This conclusion is confirmed by both graphs relating maximum plastic excursions (Fig. 11a, b) and hysteretic energy dissipation (Fig. 12a, b) to the damage rate control parameter. A small increase of maximum plastic deformation and hysteretic energy dissipation demand can be observed in the degrading components (connections), with a correspondent decrease of demands in columns. Moreover, it can be observed that the maximum values of required deformations are quite constant, while energy dissipation demand is increasing with β, meaning an increase of the number of plastic excursions with an unchanged value of the maximum deformation.

Figures 13a, 13b and 13c synthesize the story displacements resulting at the end of the earthquakes. Kobe earthquake only gives rise to appreciable influence of strength degradation.

Finally, Figure 14 illustrate the distribution of hysteretic energy among the five stories of the examined frame. Once again the effect of strength degradation seems to be not very important.

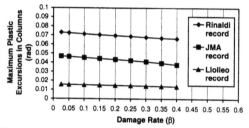

Figure 11a. Maximum required plastic excursion in columns.

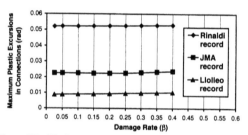

Figure 11b. Maximum required plastic excursions in connections.

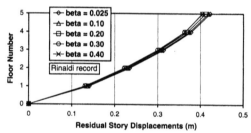

Figure 13a. Residual story displacements for Northridge earthquake.

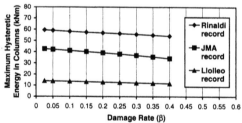

Figure 12a. Maximum required hysteretic energy in columns.

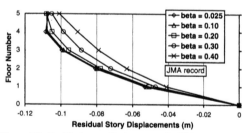

Figure 13b. Residual story displacements for Kobe earthquake.

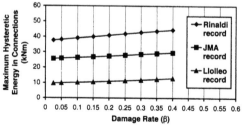

Figure 12b. Maximum required hysteretic energy in connections.

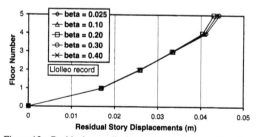

Figure 13c. Residual story displacements for Chile earthquake.

606

4.3 The effect of pinching

Pinching of hysteretic cycles can be more or less strong, depending essentially on the connection typology. For example, no pinching effects are present in the case of fully welded connections, while top and seat angle connections are generally characterized by strong pinching. On the other hand, extended and/or flush end-plate connections are characterized by moderate pinching. Moreover, pinching is usually characteristic of semi-rigid and partial-strength connections. Therefore, in order to investigate the sensitivity of damage prediction to the amount of reduction of energy dissipation capacity due to pinching of hysteresis loops, a parametric analysis has been performed with reference to semi-rigid and partial-strength connections. As basic parameter the ratio of t_1 and t_2, whose meanings have been briefly

described in Section 2, has been chosen. This ratio regulates the sharpness of transition from the lower bound to the upper bound curve and, therefore, the amount of pinching.

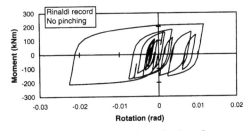

Figure 15a. Moment-Rotation history for the reference connection: Northridge earthquake – No pinching.

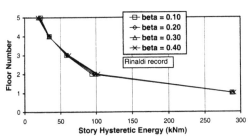

Figure 14a. Story hysteretic energy demand distribution along the frame height: Northridge earthquake.

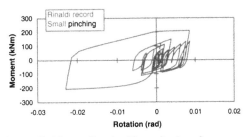

Figure 15b. Moment-Rotation history for the reference connection: Northridge earthquake – Very moderate Pinching.

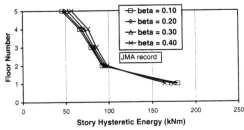

Figure 14b. Story hysteretic energy demand distribution along the frame height: Kobe earthquake.

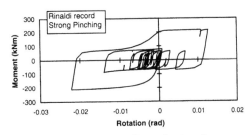

Figure 15c. Moment-Rotation history for the reference connection: Northridge earthquake – Strong Pinching.

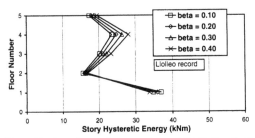

Figure 14c. Story hysteretic energy demand distribution along the frame height: Kobe earthquake.

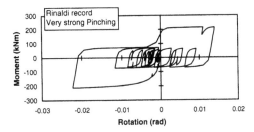

Figure 15d. Moment-Rotation history for the reference connection: Northridge earthquake – Very strong Pinching.

Figures 15 and 16 clarify, with reference to Northridge and Kobe earthquakes, respectively, the type of hysteretic behaviours considered in the study. In particular Figures 15a, b, c and d (the same is true for figure 16) are associated to decreasing values of the ratio t_1/t_2 (10000, 2500, 500 and 100), which correspond to an increase of pinching amount. It is interesting to observe that the first deformation

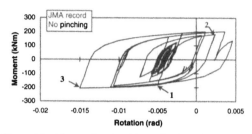

Figure 16a. Moment-Rotation history for the reference connection: Kobe earthquake – No pinching.

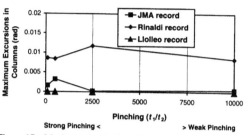

Figure 17a. Maximum required rotation excursions in columns.

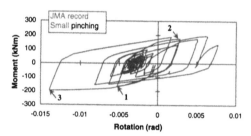

Figure 16b. Moment-Rotation history for the reference connection: Kobe earthquake – Very moderate Pinching.

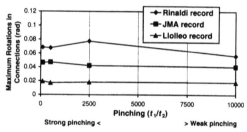

Figure 17b. Maximum required apparent rotations in connections.

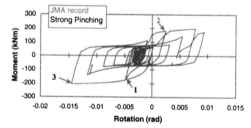

Figure 16c. Moment-Rotation history for the reference connection: Kobe earthquake – Strong Pinching.

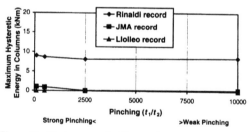

Figure 18a. Maximum required hysteretic energy in columns.

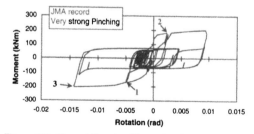

Figure 16d. Moment-Rotation history for the reference connection: Kobe earthquake – Very strong Pinching.

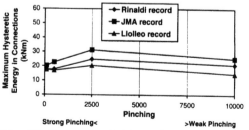

Figure 18b. Maximum required hysteretic energy in connections.

excursion is identical for all the examined systems, because they have been characterized by the same skeleton curve. This is visible in a very clear manner from Figure 15, due to the type of acceleration history characterized by few strong pulses. This aspect is also highlighted in Figure 16 by means of the indication of the first three excursions.

Figures 17a and 17b illustrate the sensitivity of the predicted values of maximum local deformations to pinching amount. Unfortunately, no clear trend can be identified. The same observation holds for the maximum local hysteretic energy dissipation demand, which is shown in Figures 18a and 18b for columns and connections, respectively. In any case, the variation of local response parameters is small.

Figure 19 shows the residual frame story displacements after the earthquakes occurred. In case of Northridge earthquake (Fig. 19a) there is no signifi-

cant difference between the system without pinching and that with strong or very strong pinching. However, for small pinching there is a singular behaviour with a significant increase of lateral residual displacements. In the case of Kobe earthquake (Fig. 19b), pinching had a beneficial effect on frame final configuration, leading to smaller residual lateral displacement. Finally, in case of Chile earthquake (Fig.19c), a singular behaviour of the system with small pinching can again be observed, with a significant increase of residual displacements. In any case, the effect of pinching on the frame final configuration seems to be not negligible.

In Figures 20a, b and c the distribution among the five frame stories of hysteretic energy dissipation demand is shown. It can be observed a quite uniform distribution of energy among the five stories, for all the three ground acceleration histories examined.

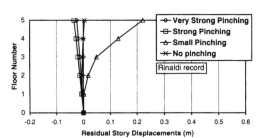

Figure 19a. Residual story displacements for Northridge earthquake.

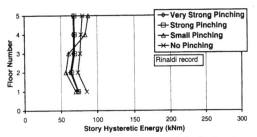

Figure 20a. Story hysteretic energy demand distribution along the frame height: Northridge earthquake.

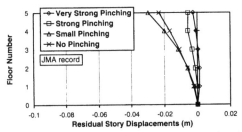

Figure 19b. Residual story displacements for Kobe earthquake.

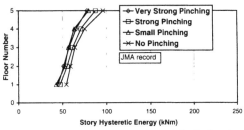

Figure 20b. Story hysteretic energy demand distribution along the frame height: Kobe earthquake.

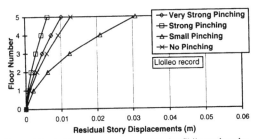

Figure 19c. Residual story displacements for Chile earthquake.

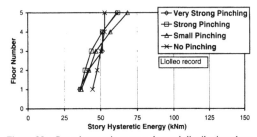

Figure 20c. Story hysteretic energy demand distribution along the frame height: Chile earthquake.

This uniform distribution of hysteretic energy among the frame stories is due to the scarce plastic engagement in columns. The shape of the residual frame configuration, which is like the deformed shape of a cantilever (Fig. 18), confirms this. It is interesting to compare this shape with that obtained in the case of rigid and full-strength connections, where the contribution of columns in dissipating energy (mainly of the column bases of the first floor) determined a different residual configuration (Fig.13) and energy dissipation path (Fig.14).

5 CONCLUSIONS

The effects of strength deterioration rate and pinching amount on damage prediction have been mainly investigated in the current paper. The following conclusions may be drawn:

1 Both local and global damage parameters are slightly influenced by connection strength deterioration, at least in the examined damage rate range and for the three ground acceleration histories considered. Further investigation is needed, looking at the possibility of using generated accelerograms for simulating long duration and large magnitude earthquakes. Moreover, the effect of strength degradation due to local buckling phenomena, which is often associated to unstable branches in the load-deformation relationship of structural components, should be investigated.

2 Pinching of hysteresis loops had a small but non-negligible influence on maximum values of local damage parameters, such as maximum plastic deformation and maximum hysteretic energy demand. The same conclusion may be drawn with reference to story hysteretic energy demand. Moreover, the effect of pinching on the residual lateral frame displacements seems to be important, leading to substantial differences when changing the type of model.

Comparing results obtained for rigid and full-strength connections with those related to semi-rigid and partial-strength connections can draw an additional interesting observation. The semi-continuous frame behaved quite better than the continuous one, with smaller residual lateral displacements and smaller hysteretic energy dissipation demand.

6 ACKNOWLEDGEMENTS

The results related in the present work were carried out in the framework of the Italian research project (COFIN '99) titled "Damage of joints in metal and composite construction". The financial support of MURST is gratefully acknowledged.

7 REFERENCES

Akiyama H. 1985. Earthquake-resistant limit state design for buildings. Tokyo: University of Tokyo Press.
ATC 1992. Guidelines for cyclic seismic testing of components of steel structures. Applied Technology Council, Publication N. 24.
Calado, L. De Matteis G., Landolfo R. 1999. Experimental analysis on angle beam-to-column joints under reversal cyclic loading. In Proc. of XVII Congresso CTA, Napoli 3-5 October: 103-114.
Della Corte G., De Matteis G., Landolfo R. 2000. Influence of connection modeling on seismic response of moment resisting steel frames. In F.M. Mazzolani (ed.), Moment resistant connections of steel building frames in seismic areas. London: E & FN SPON, in press.
De Matteis, G., Landolfo, R., Mazzolani, F.M. 1998. Dynamic response of infilled multistorey steel frames. In Proc. of XI European Conference on Earthquake Engineering, Paris la Defense 6-11 September CD-ROM.
Eurocode 3 1992. Design of Steel Structures. Commission of the European Communities.
Eurocode 8 1994. Structures in Seismic Regions. Commission of the European Communities.
Mazzolani F.M. (ed.) 2000. Moment resistant connections of steel building frames in seismic areas. London: E & FN SPON, in press.
Elgamal A., Ashford S., Kramer S. (eds.) 1998. Workshop conclusions and recommendations. In Proc. of the 1st PEER Workshop on Characterization of special source effects, University of California, San Diego 20-21 July 27-28.
Richard, R.M. & Abbott, B.J. 1975. Versatile Elastic-Plastic Stress-Strain Formula. J. Engrg. Mech. Div., ASCE, 101(4): 511-515.
Vann, W.P, Thompson, L.E., Whalley, L.E., Ozier, L.D. 1973. Cyclic behaviour of rolled steel members. In Proc. of the 5th World Conference on Earthquake Engineering, Rome.

Behaviour of Steel Structures in Seismic Areas, Mazzolani & Tremblay (eds) © 2000 Balkema, Rotterdam, ISBN 90 5809 130 9

Performance based design of seismic-resistant MR-frames

Federico M. Mazzolani
Department of Structural Analysis and Design, University of Naples, Italy

Rosario Montuori & Vincenzo Piluso
Department of Civil Engineering, University of Salermo, Italy

ABSTRACT: Within the framework of performance based earthquake-resistant design of buildings, the authors have recently proposed a simple approach for evaluating the performance levels of seismic-resistant steel frames. With reference to global moment-resisting steel frames (GMRFs), i.e. frames designed to assure a collapse mechanism of global type, this paper presents a refinement of the proposed approach.
The frame structural response is modelled by means of an equivalent SDOF system exhibiting a trilinear force-displacement behaviour. The force-displacement relationship is derived by the coupling of a simple frame elastic analysis with a second order rigid-plastic analysis. Therefore, the main feature of the approach is its ability in providing the designer with a quick evaluation of the performance objectives of moment resisting steel frames. In fact, neither push over inelastic analyses nor dynamic inelastic analyses are required.
The refinement presented in this paper regards the characterization of the state of damage, for global moment-resisting frames, corresponding to different performance levels which have to be guaranteed as far as the return period of the expected ground motion increases.

1 INTRODUCTION

The traditional design philosophy of seismic-resistant civil engineering constructions basically identifies three performance levels which should be achieved as the return period of the expected ground motion increases (Mazzolani et al., 1994). In fact, in case of frequent minor earthquakes, it is required that structures remain in elastic range and interstorey drifts are limited enough giving rise to minimum discomfort to the activities developed in the structures. In addition, the minimization of both structural and non-structural damage during occasional moderate seismic events is expected. Finally, serious damage in rare major earthquakes is accepted provided that collapse is prevented in order to assure the safeguard of the human lives.

The fulfilment of the above design objectives needs the use of multi-level design criteria, so that it can be stated that the traditional philosophy of earthquake-resistant design qualitatively agrees with the *performance based earthquake design* (Bertero, 1996).

It is the author opinion that the main difficulties in setting up performance based earthquake design is the quantitative definition of the degree of damage corresponding to each performance level and the prediction of the earthquake magnitudes leading to the attainment, for a given structure, of the pre-defined damage levels, by means of simple methods to be applied in everyday design practice.

In fact, a comprehensive but simple design procedure should correlate the resistance of a structure at various limit states to the probability that the earthquake action can reach the intensity required to achieve such limit states.

A detailed definition of performance based earthquake-resistant design has been provided within the activities of SEAOC Vision 2000 Committee (SEAOC, 1995). The aim of the performance based earthquake-resistant design is to provide the designers with the criteria for selecting the appropriate layout of the structural system and its layout and for proportioning and detailing both structural and non-structural components, so that at specified levels of earthquake intensity the structural and non-structural damage will be constrained within given limits. In fact, the state of damage has to be quantitatively defined for both structural and non-structural elements, so that performance levels can be divided in structural performance levels and non-structural performance levels. A building performance level is obtained by combining a structural performance level with a non-structural performance level (Hamburger, 1997).

The coupling of a performace level with a specific level of ground motion provides a performance design objective (SEAOC, 1995; Bertero et al., 1996).

In this perspective, the SEAOC Vision 2000 Committee proposes four performance levels: Fully Operational, Operational, Life Safe and Near Collapse. Fully Operational is a state in which the facility continues in operation with negligible damage to non-structural elements only. In the Operational state the facility continues in operation with minor damage to both structural and non-structural elements and minor disruption in non-essential services. Structures in the Life Safety condition are significantly damaged, but are expected to be repairable, although perhaps not economically. Structures in the Near Collapse condition still guarantee the safeguard of the human lives, but corresponding to potential complete economic losses.

The requirements related to the above performance levels for different civil engineering facilities and constructional materials are also suggested (SEAOC, 1995).

With reference to ductile moment-resisting reinforced concrete frames, it is useful to note that it has been proposed (Ghobarah et al., 1997) to characterize, from the quantitative point of view, the above performance levels by means of the maximum interstorey drift. In particular, values equal to 0.7%, 2%, 4.5% and 5.6% have been proposed for the Fully Operational, Operational, Life Safe and Near Collapse limite states, respectively. More stringent limitations are suggested in the SEAOC Vision 2000 Document (0.2%, 0.5%, 1.5% and 2.5% for the Fully Operational, Operational, Life Safe and Near Collapse limite states, respectively), but it is also recognized that *permissible drift levels are functions of the structural as well as nonstructural systems* so that the suggested limits should be applied with caution and judgement.

In addition, four earthquake design levels are specified: Frequent, Occasional, Rare and Very Rare (the corresponding return periods are equal to 43, 72, 475 and 970 years) (SEAOC, 1995; Bertero et al., 1996).

According to the framework, briefly outlined above, it can be recognized that the complete knowledge of the seismic performances of a structure requires sophisticated numerical procedures, because the quantitative evaluation of the structural damage for different earthquake design levels would require nonlinear dynamic analyses. In addition, in this context, also the random nature of loads and resistances deserves to be mentioned. As such analyses cannot be applied in common design practice, the setting up of simplified design procedures is a pressing need.

2 DRAWBACKS OF TRADITIONAL DESIGN

As already underlined, the traditional design philosophy qualitatively agrees with performance based earthquake-resistant design, but falls behind in providing the designers with the awareness of the actual performances which the structures are able to develop.

In fact, there is not a comprehensive investigation about the fulfilment of the seismic performance levels obtained for different magnitudes of the ground motion. The structural analysis is limited to elastic analyses under seismic horizontal forces which are defined by reducing the base shear, required to the structure in order to remain in elastic

range during the most severe design earthquake, by means of a coefficient, namely q-factor, which takes into account the structural ductility and the energy dissipation capacity. Under such horizontal forces, the structure has to possess sufficient strength and stiffness in order to guarantee the fulfilment of the requirements associated with the serviceability limit state, which basically corresponds to the Fully Operational performance level. The safety against the ultimate limit state, which corresponds to the Life Safe performance level, is considered automatically verified, provided that the detailing rules and design procedures, suggested by seismic codes in order to control the failure mode and, therefore, the energy dissipation capacity, are satisfied. In other words, it is practically assumed that the use of the suggested design procedures and detailing rules with the adoption of the design value of the q-factor given by the code leads to safe results. It means that seismic codes assume that the respect of the suggested design and detailing rules automatically leads to the fulfilment of the seismic design requirements.

This procedure leads to a great simplification because it allows to perform in a single shot both the check against the serviceability limit state and the one against the collapse limit state. At the same time, the limits of such procedure are evident; in fact not always the suggested design procedures lead to the foreseen failure mode and the expected ductility, so that the energy dissipation capacity of the structure can result less than the one required in order to prevent collapse under the most severe design earthquake. In addition, the designer is not aware about the collapse mechanism of the structure, the local and global ductility demands and the actual energy dissipation capacity, because his analyses are limited to the elastic range. As a consequence, the designer has not an exhaustive knowledge of the inelastic response of the structure and of the design measures to be adopted for its improvement.

The designers must be aware of the need to check the actual seismic inelastic behaviour of the structure. The main gap of modern seismic codes is that there are no rules to solve this problem. As a consequence, it seems to be necessary to assess general provisions which are always on the safe side. It means that the current design procedure should be integrated by simple methods for checking the fulfilment of the design performance objectives.

In this paper, a simplified approach for evaluating the performances of moment-resisting steel frames for different earthquake intensity levels (Mazzolani and Piluso, 1997a) is improved with reference to global moment resisting frames. The proposed approach is the generalization, within the framework of performance based earthquake-resistant design, of an approach already proposed for evaluating the q-factor (Cosenza et al., 1988). From the operative point of view, the procedure does not require any sophisticated analysis, because the inelastic behaviour is predicted by means of simple second order rigid-plastic analyses.

3 PERFORMANCE LEVELS OF MR-FRAMES

One of the difficulties in the development of performance based earthquake resistant design is the quantitative definition of performance levels and of the corresponding degree of damage. A quantitative definition is fundamental. In fact, the most delicate task the seismic engineer has to accomplish is the evaluation of the seismic performances corresponding to increasing levels of the earthquake intensity, so that the peak ground acceleration leading to the achievement of a given limit state can be computed.

A simple but rational proposal concerning the quantitative definition of performance levels has been recently developed (Mazzolani and Piluso, 1997a) for moment-resisting steel frames. The proposal is based on a trilinear approximation of the frame behavioural curve relating the multiplier of seismic horizontal forces (or the base shear) to the top sway displacement, which is the result of a static pushover analysis (Fig. 1).

The first branch of the trilinear approximation has a slope according to the linear elastic analysis. The second branch is an horizontal plateau corresponding to the maximum load carrying capacity. Finally, the softening branch corresponds to the mechanism equilibrium curve. As a result of the above trilinear modelling of the actual frame behaviour, four characteristic points can be recognized. Within the framework of performance based earthquake-resistance design, these points can be also used to identify the performance levels.

The first point, A, corresponds to the minimum value α_A between the first yielding multiplier of horizontal forces α_y and the multiplier corresponding to the attainment of the maximum interstorey drift compatible with serviceability requirements. Therefore, the *Fully Operational* performance level can be identified with the attainment of the corresponding top sway displacement δ_A.

The *Operational* performance level is associated with the point B of the trilinear modelling which corresponds to the first significant departure of the structure from linearity to the plastic range. Overstrength due to plastic redistribution capacity is exploited with limited plastic rotation demands θ_{pB}, as

those corresponding to the attainment of the top sway displacement δ_B.

Under rare earthquakes the frame can be engaged in plastic range with significant ductility demands up to the attainment of the point C which, in the trilinear modelling, corresponds to the complete development of a kinematic mechanism. Therefore, the attainment of the displacement δ_C can be assumed to be representative of the *Life Safe* performance level. The corresponding plastic rotation demands θ_{pC} are significant.

Finally, the mechanism equilibrium curve can be exploited in the case of very rare earthquakes. In fact, the dynamic equilibrium is still possible, thanks to the inertia forces, but significant local ductility demands are expected. Therefore, a new limit point, D, is identified on the basis of the available plastic rotation supply θ_{pu}. The corresponding top sway displacement δ_u can be computed as:

$$\delta_u = \delta_C + (\theta_{pu} - \theta_{pC})\, H_o \qquad (1)$$

where H_o is the sum of the interstorey heights of the storey involved in the collapse mechanism.

It is evident that no structural damage occurs when the point A of the frame behavioural curve is attained. On the contrary, the attainment of the point D corresponds to a plastic rotation demand defined *a priori* by considering the structural detail of beam-to-column joints, governing the plastic rotation supply in the case of partial-strength connections, and the width-to-thickness ratios of the plate elements constituting the member sections, governing the plastic rotation supply in the case of full-strength connections.

Regarding the plastic rotation demand corresponding to the attainment of the characteristic points B and C of the frame behavioural curve, it mainly depends on the plastic redistribution capacity, the frame sensitivity to second order effects and the collapse mechanism typology. Obviously, these plastic rotation demands could be computed by means of static pushover analyses, but these analyses are not usual in common design practice.

Within the framework of the approach proposed by the authors for evaluating performance levels of moment-resisting steel frames (Mazzolani and Piluso, 1997a), it is important to predict the above plastic rotation demands directly from the knowledge of the parameters characterizing the frame behavioural curve $\alpha - \delta$. In fact, the main feature of the proposed approach is that it does not require any nonlinear analysis, neither dynamic nor static, but only the use common analyses such as the elastic structural analysis and the classical plastic analysis, i.e. rigid-plastic analysis.

4 RIGID-PLASTIC ANALYSIS IN SEISMIC DESIGN

The design of moment-resisting steel frames can be commonly developed according to different criteria

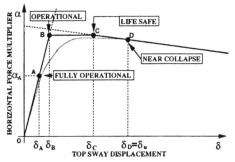

Figure 1. Trilinear modelling of the load-displacement curve

leading to two main typologies, namely ordinary moment-resisting frames (OMRF frames) and special moment resisting frames (SMRF frames).

As soon as the preliminary design has been carried out, elastic analyses can be performed by means of usual procedures (static or dynamic). The aim of these analyses is the check of the members (stability and resistance) and of the lateral displacements under the load combinations provided by the codes. If the above checks are satisfied, the preliminary design can be accepted. In fact, it is assumed that the safety against the ultimate limit state is automatically verified, provided that the detailing rules and design procedures suggested by seismic codes are satisfied.

As already stated, this procedure leads to a great simplification because it allows to perform in a single shot both the check against the serviceability limit state and the one against the collapse limit state, but evident limitations occur, because the actual inelastic behaviour of the structure is not investigated.

A simple approach to overcome the main drawback of the common design practice is the combination of elastic structural analysis with classical rigid-plastic analysis. In fact, by means of an appropriate use of such analyses, with a minimum increase of the computational effort the designer can be aware about the overstrength due to plastic redistribution capacity, the global ductility supply and the collapse mechanism typology.

The collapse mechanism can be obtained by evaluating the kinematically admissible multiplier of horizontal forces corresponding to the main collapse mechanism typologies of moment-resisting frames subjected to horizontal forces (Mazzolani and Piluso, 1996). According to the kinematic theorem of plastic collapse, the true collapse mechanism is that corresponding to the minimum kinematically admissible multiplier. In addition, the linearized mechanism equilibrium curve can be obtained according to second order rigid-plastic analysis:

$$\alpha = \alpha_o - \gamma_s \, \delta \qquad (2)$$

where α_o is the collapse multiplier of the seismic horizontal forces (first order rigid-plastic analysis) and γ_s is the slope of the linearized mechanism equilibrium curve.

The following notation is adopted:
- n_s is the number of storeys;
- n_c is the number of columns;
- n_b is the number of bays;
- k is the storey index;
- i is the column index;
- j is the bay index;
- F is the vector of the design horizontal forces;
- s is the shape vector of the storey horizontal virtual displacements ($du = s \, d\theta$, where $d\theta$ is the virtual rotation of the plastic hinges of the columns involved in the mechanism);
- V is the vector of the storey vertical loads whose element V_k is the total vertical load acting at the kth storey;

- B is a matrix of order $n_b \times n_s$ whose elements B_{jk} are equal to the plastic moments of beams (i.e. $B_{jk} = M_{b.jk}$);
- C is the matrix of order $n_c \times n_s$ whose elements C_{ik} are equal to the column plastic moments (i.e. $C_{ik} = M_{c.ik}$);
- R_b is the matrix (order $n_b \times n_s$) of the beam plastic rotations;
- R_c is the matrix (order $n_c \times n_s$) of the coefficients $R_{c.ik}$:
  $R_{c.ik} = 2$ when the column is yielded at both ends
  $R_{c.ik} = 1$ when only one column end is yielded
  $R_{c.ik} = 0$ when the column does not participate to the collapse mechanism;
- q is the matrix (order $n_b \times n_s$) of the vertical loads acting on the beams.
- D_v is a matrix (order $n_b \times n_s$) whose elements $D_{v.jk}$ represent the vertical displacements of the jth beam of the kth storey;

The collapse multiplier α_o, according to rigid-plastic analysis, is given by (Mazzolani and Piluso, 1997b):

$$\alpha_o = \frac{[tr(C^T R_c) + 2tr(B^T R_b) - tr(q^T D_v)]}{F^T s} \qquad (3)$$

where tr denotes the trace of the matrix.

The slope of the mechanism equilibrium curve is given by (Mazzolani and Piluso, 1997b):

$$\gamma_s = \frac{V^T s \, \dfrac{1}{H_o}}{F^T s} \qquad (4)$$

where H_o is the sum of the interstorey heights of the storeys involved in the collapse mechanism.

Due to second order effects, the ultimate multiplier (the true collapse multiplier) of seismic horizontal forces is less than that evaluated by means of rigid-plastic analysis. An accurate estimate of this multiplier can be obtained by means of the Horne's method (Fig. 2). The linear elastic analysis is represented by the straigth line whose equation is given by:

$$\alpha = \frac{1}{\delta_1} \delta \qquad (5)$$

where δ_1 is the top sway displacement corresponding to the design value of the seismic horizontal

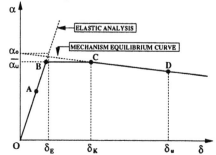

Figure 2. Method for deriving the trilinear model

forces (i.e. scaled down by means of the design value of the q-factor).

According to the Horne method, an accurate estimate $\bar{\alpha}_u$ of the true collapse multiplier α_u can be obtained through the point of the mechanism equilibrium curve corresponding to a top sway displacement δ_K equal to 2.5 times the elastic displacement δ_E (Horne and Morris, 1973) corresponding to the collapse value of the horizontal forces derived from rigid-plastic analysis:

$$\delta_K = 2.5\, \alpha_o\, \delta_1 \qquad (6)$$

The approximated value of the ultimate multiplier is given by the following relationship:

$$\bar{\alpha}_u = \alpha_o - \gamma_s\, \delta_K \qquad (7)$$

As a result of the Horne's method also the trilinear approximation of the curve relating the multiplier of the horizontal forces to the top sway displacement is obtained.

5 PLASTIC ROTATION DEMANDS

As already stated, within the framework of an approach for evaluating performance levels, which does not require any nonlinear analysis, it is important to predict the plastic rotation demands corresponding to the *Operational* and *Life Safe* limit states (points B and C of Fig. 1) directly from the knowledge of the parameters characterizing the frame behavioural curve $\alpha - \delta$. To this scope, a wide parametric analysis has been carried out with reference to global moment-resisting frames.

Global moment resisting frames (GMRFs) are designed by means of procedures able to assure a collapse mechanism of global type. This objective can be obtained by means of an iterative design procedure based on repeated push over static inelastic analyses which are used to determine the column sections required to control the failure mode. A more effective procedure based on second order rigid-plastic analysis and on the application of the kinematic theorem of plastic collapse has been recently proposed (Mazzolani and Piluso, 1997b) and has been applied within the present work.

The parametric analysis has regarded 168 low rise frames. The number of bays varies between 1 and 6; the number of storeys varies between 2 and 8. Four values of the bay span have been considered: 4.50 m, 6.00 m, 7.50 m and 9.00 m. The characteristic values of dead and live loads are equal to $3.0 \times L$ kN/m and $2.0 \times L$ kN/m, respectively (where L is the bay span in meters). Each frame has been designed according to the method of Mazzolani and Piluso (1997b). The subsequent static pushover analyses have confirmed in all cases the development of a collapse mechanism of global type and have allowed the computation of the plastic rotation demand θ_{pC} corresponding to the top sway displacement δ_C given by the intersection between the linearized mechanism equilibrium curve and the horizontal line ($\alpha = \alpha_u$) corresponding to the ultimate multiplier of horizontal forces. In addition,

Table 1. Relationship cor computing the ψ factor

COEFFICIENTS FOR COMPUTING PLASTIC ROTATION DEMANDS
$\psi_1 = 653.994 - 960.398\,\zeta$
$\psi_2 = 664.163 + 43.507\,\zeta$
$\psi_3 = -1.427 + 0.840\,\zeta$
$\psi_4 = -395.099 - 106.511\,\zeta$
$\psi_5 = 182.204 - 525.565\,\zeta$
$\psi_6 = 3.707 - 0.015\,n_s$
$\psi_7 = 1.284 + 0.064\,n_b$
$\psi_8 = 2.539 - 0.286\,n_s$
$\psi_9 = 3.014 - 0.252\,n_b$
$\psi_{10} = 1.504 \times 10^{-5}\,\zeta^{-1.533}$
$\psi_{11} = 160.544 - 152.888\,\zeta$
$\psi_{12} = 2.060 - 0.940\,\zeta$
$\psi_{13} = -3.359 + 20.700\,\zeta$
$\psi_{14} = -858.394 + 33.888\,\zeta$
$\psi_{15} = 235.530 - 1080.019\,\zeta$

the plastic rotation demand $\theta_{p.\alpha_u}$ corresponding to the attainment of the ultimate multiplier of the horizontal forces has been computed.

Successively, by means of a regression analysis, the above plastic rotation demands have been related to the main parameters α_y, α_u and γ_s characterizing the frame behavioural curve (α_y is the multiplier of horizontal forces corresponding to the formation of the first plastic hinge). The following relationships have been obtained:

$$\frac{\theta_{pC}\,H}{n_s\,\delta_y} = \psi_1 \left(\psi_2 \frac{\alpha_u}{\alpha_y} - 1 \right)^{\psi_3} \frac{1 - \psi_4\,\gamma_s}{1 - \psi_5\,\gamma_s}\,\psi_6\,\psi_7 \qquad (8)$$

$$\frac{\theta_{p.\alpha_u}\,H}{n_s\,\delta_y} = \psi_8\,\psi_9\,\psi_{10}\,\psi_{11} \left(\psi_{12} \frac{\alpha_u}{\alpha_y} - 1 \right)^{\psi_{13}} \frac{1 - \psi_{14}\,\gamma_s}{1 - \psi_{15}\,\gamma_s} \qquad (9)$$

where H is the total height of the frame ($H_o = H$ for frames failing in global mode) and δ_y is the top sway displacement corresponding to the formation of the first plastic hinge.

The coefficients from ψ_1 to ψ_{15} are given in Table 1 as a function of the number of storeys n_s, the number of bays n_b and the parameter ζ, referred to the first storey, given by:

$$\zeta = \frac{\sum E\,I_b/L}{\sum E\,I_c/h} \qquad (10)$$

where I_b and L are the beam inertia moment and the bay span, respectively; I_c and h are the column inertia moment and the column height, respectively; E is the modulus of elasticity.

Regarding the accuracy of the proposed relationships, it is useful to note that Eq. (8) provides an average error equal to -3.26% with a standard devia-

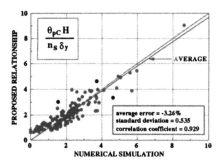

Figure 3. Accuracy of the formulation for computing the plastic rotation θ_{pC}

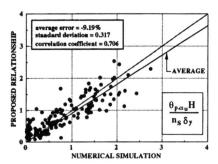

Figure 4. Accuracy of the formulation for computing the plastic rotation $\theta_{p.\alpha_u}$

tion equal to 0.535 and a correlation coefficient equal to 0.929 (Fig. 3); Eq. (9) leads to an average error equal to -9.19% with a standard deviation equal to 0.317 and a correlation coefficient equal to 0.706 (Fig. 4).

As described in the previous Section, in the original approach it was suggested an estimate $\overline{\alpha}_u$ of the true ultimate multiplier α_u by means of the Horne method (Horne, 1959). However, a more accurate prediction can be here obtained for global moment-resisting frames by properly exploiting the results of the numerical simulations. In fact, starting from the Merchant-Rankine formula relating the ultimate multiplier α_u to the collapse multiplier α_o obtained by means of rigid-plastic analysis and to the critical elastic multiplier α_{cr} of vertical loads:

$$\frac{1}{\alpha_u} = \frac{1}{\alpha_o} + \frac{1}{\alpha_{cr}} \tag{11}$$

and considering that for global moment-resisting frames (Mazzolani and Piluso, 1996):

$$\frac{1}{\alpha_{cr}} \approx \gamma = \gamma_s \, \delta_1 \tag{12}$$

the following relationship can be suggested:

$$\alpha_u = \frac{\alpha_o}{1 + (\alpha_o \, \gamma_s \, \delta_1)^\psi} \tag{13}$$

where δ_1 is the top sway displacement under the design horizontal forces ($\alpha = 1$) and ψ is a correcting factor ($\psi = 1$ corresponds to the Merchant-Rankine formula) given by:

$$\psi = 0.787 + 0.569 \, \zeta \tag{14}$$

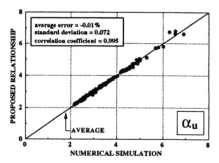

Figure 5. Accuracy of the formulation for computing the ultimate multiplier of seismic forces

Eqs. (13) and (14) allow the prediction of the true ultimate multiplier with an average error equal to -0.01% with a standard deviation equal to 0.072 and a correlation coefficient equal to 0.995 (Fig. 5).

6 PERFORMANCE OBJECTIVE EVALUATION

As soon as information about the intensity of the seismic actions with the corresponding occurrence probability is available, the designer has to check the fulfilment of the performance objectives. Therefore, the designer should be able to estimate the value of the peak ground acceleration corresponding to the attainment of any given limit state. Then, the recurrence period can be evaluated from seismic hazard studies.

The check of the design performance objectives requires the approximated evaluation of the values of the peak ground acceleration corresponding to the attainment of the characteristic points of the $\alpha - \delta$ curve.

The method proposed by the authors (Mazzolani and Piluso, 1997a) for checking the fulfilment of the design performance objectives is the generalization of a rational approach for estimating the q-factor (Cosenza et al., 1988). With respect to the original proposal, a refinement is herein presented. According to this new model, the peak ground acceleration corresponding to the *Fully Operational* limit state is predicted on the basis of the elastic frame overall response, while the ground motion intensities corresponding to the attainment of limit states involving the inelastic response of the structure are predicted by substituting the trilinear approximation of the $\alpha - \delta$ curve by means of a bilinear degrading model leading to the same energy absorption capacity. Therefore, the inelastic response of the structure is predicted by means of an equivalent bilinear degrading SDOF system (Fig. 6).

The design value of the seismic horizontal forces corresponds to a base shear given by:

$$V_d = \frac{M \, A \, R(T)}{q_d} \tag{15}$$

where A is the peak ground acceleration specified by the seismic code for the check against the ultimate limit state (which corresponds to the *Life Safe* limit state), M is the total mass of the building,

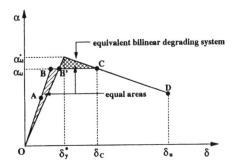

Figure 6. Equivalent bilinear degrading system

$R(T)$ is the elastic response spectrum normalized with respect to the peak ground acceleration and q_d is the design value of the q-factor.

It is useful to note that the design value of the q-factor is given by the ratio between the elastic design response spectrum $S_{ae}(T)$ and the inelastic design response spectrum $S_{ad}(T)$:

$$q_d(T) = \frac{S_{ae}(T)}{S_{ad}(T)} \qquad (16)$$

Therefore, the design value of the q-factor is coincident with the behaviour factor provided by seismic codes only in the constant acceleration range of the design spectra (Mazzolani and Piluso, 1996).

By denoting with α_{FO} the horizontal force multiplier corresponding to the attainment of the point A of the trilinear approximation of the $\alpha - \delta$ curve, the peak ground acceleration corresponding to the *Fully Operational* performance level is given by:

$$A_{FO} = \alpha_{FO} \frac{A}{q_d} \qquad (17)$$

As already stated, the values of the peak ground acceleration corresponding to the attainment of the limit states involving the inelastic behaviour of the structure are predicted by means of an equivalent SDOF system with a bilinear degrading response whose energy absorption capacity is equal to that of the trilinear representation of the $\alpha - \delta$ curve.

The mass of the equivalent SDOF system is given by:

$$M_{eq} = \sum_{i=1}^{n_s} m_i \, \Phi_i \qquad (18)$$

where m_i is the mass of the i-th storey and Φ_i is the i-th component of the fundamental mode of vibration.

The effective period of vibration of the equivalent SDOF system is given by:

$$T = 2\pi \left(\frac{M_{eq} \delta_y^*}{\alpha_u^* V_d} \right)^{1/2} \qquad (19)$$

where V_d is the design base shear and α_u^*, δ_y^* define the yielding point of the bilinear degrading model.

The point B' corresponding to intersection between the elastic branch of the bilinear degrading

model and the plateau of the trilinear representation of the $\alpha - \delta$ curve is assumed as representative of the *Operational* performance level. The peak ground acceleration corresponding to the attainment of the point B' can be estimated as:

$$A_O \approx \alpha_u \frac{A}{q_d} \qquad (20)$$

Regarding the plastic rotation demand corresponding to the attainment of the *Operational* performance level, with reference to global moment-resisting frames, its value can be estimated by means of Eq. (9).

The attainment of the point C requires the exploitation of the global ductility supply of the structure within the plateau of the $\alpha - \delta$ curve (trilinear representation). With reference to the equivalent bilinear degrading SDOF system, the point C corresponds to a global ductility demand $\mu_C = \delta_C / \delta_y$. The corresponding q-factor (reduction factor) can be expressed as (Cosenza et al., 1989; Mazzolani and Piluso, 1993):

$$q_C = \frac{q_o(\mu_C, T)}{\varphi(\mu_C, \gamma)} \qquad (21)$$

where $q_o(\mu_C, T)$ is the reduction factor for the bilinear elastic-perfectly plastic system (i.e. for $\gamma = 0$ and $\varphi(\mu_C, \gamma)$ is a factor accounting for the influence of second order effects.

The reduction factor $q_o(\mu, T)$ can be predicted by means of the formulation due to Nassar and Krawinkler (1992):

$$q_o = [c(\mu - 1) + 1]^{1/c} \qquad (22)$$

where:

$$c = \frac{T}{1 + T} + \frac{0.42}{T} \qquad (23)$$

Obviously, in the examined case, Eq. (22) has to be applied with $\mu = \mu_C$.

Regarding the effects of geometrical non-linearity, they can be taken into account by means of the following average reduction factor (Cosenza et al., 1989; Mazzolani and Piluso, 1996):

$$\varphi = \frac{1 + 0.62(\mu - 1)^{1.45} \gamma}{1 - \gamma} \qquad (24)$$

where γ is the stability coefficient ($\gamma = \gamma_s \, \delta_1$). Obviously, in the examined case, also Eq. (24) has to be applied for $\mu = \mu_C$.

Therefore, the peak ground acceleration corresponding to the "Life Safe" performance level can be computed as:

$$A_{LS} = q_C \alpha_u \frac{A}{q_d} \qquad (25)$$

It is now important to underline that, with reference to global moment-resisting frames, the plastic rotation demand corresponding to the attainment of the *Life Safe* performance level can be computed by means of Eq. (8).

In the case of earthquakes whose peak ground acceleration exceeds A_{LS}, additional plastic deformation demand occurs, so that the softening branch of the $\alpha - \delta$ curve is furtherly exploited. The kinematic mechanism can be considered completely developed, but dynamic equilibrium is still possible thanks to inertia forces. Failure can be prevented provided that plastic rotation demands are compatible with the local ductility supply.

The ultimate displacement can be computed by means of Eq. (1) where, with reference to global moment-resisting frames, $H_o = H$ and θ_{pC} can be computed by means of Eq. (8). The available global ductility is given by $\mu_u = \delta_u/\delta_y^*$. Therefore, the reduction factor q_D corresponding to a global ductility μ_u can be computed as:

$$q_D = \frac{q_o(\mu_u, T)}{\varphi(\mu_u, \gamma)} \qquad (26)$$

where $q_o(\mu_u, T)$ can be still computed by means of Eq. (22) with $\mu = \mu_u$, and also $\varphi(\mu_u, \gamma)$ can be still computed by Eq. (24) with $\mu = \mu_u$.

Therefore, the value of the peak ground acceleration A_{NC} corresponding to the attainment of the point D, i.e. at the *Near Collapse* performance level, can be estimated as:

$$A_{NC} = q_D \alpha_u \frac{A}{q_d} \qquad (27)$$

The above described procedure allows the prediction of the values of the peak ground acceleration corresponding to the occurrence of any given limit state. Therefore, by means of seismic hazard analysis, it is also possible to approximately predict the recurrence period of any limit state. It means that the check of the fulfilment of the design objectives can be carried out. If one or more design objectives are not satisfied the preliminary sizing of the structure can be modified aiming to the increase of its lateral strength and stiffnes and/or its global ductility and energy dissipation capacity. It has to be underlined that some iteration is necessary when the check of the obtained seismic performances shows that the seismic design goal has not been achieved.

7 CONCLUSIONS

A procedure, based on elastic and rigid-plastic analysis, for predicting the values of the peak ground acceleration corresponding to the attainment of different performance levels has been proposed. The main feature of the proposed approach is that it can be applied within the framework of performance based earthquake resistant design without requiring cumbersome inelastic analyses, neither dynamic nor static.

Within the proposed procedure, with reference to global moment-resisting frames, relationships for predicting the plastic rotation demands corresponding to the *Operational* and to the *Life Safe* performance level have been also suggested on the basis of a wide parametric analysis.

The proposed procedure represents just an example of how to address the problem and to codify its practical aspects. Of course, being the simplified procedure based on many approximations, several criticisms can be raised. Notwithstanding, the proposed approach represents a powerful design tool when compared with common design practice in which the designer is not aware of the seismic inelastic performances of the designed structure.

8 REFERENCES

Bertero R.D., Bertero V.V., A. Teran-Gilmore (1996): «Performance-Based Earthquake-Resistant Design Based on Comprehensive Design Philosophy and Energy Concepts», 11th World Conference on Earthquake Engineering, Acapulco.

Bertero V.V. (1996): «The State-of-the Art Report on Design Criteria», 11th World Conference on Earthquake Engineering, Acapulco.

Bertero V.V. (1996): «The Need for Multi-Level Seismic Seismic Design Criteria», 11th World Conference on Earthquake Engineering, Acapulco.

Cosenza E., De Luca A., Faella C., Piluso V. (1988): «A Rational Formulation for the q-Factor in Steel Structures», 9th World Conference on Earthquake Engineering, Tokyo, August.

Cosenza E., Faella C., Piluso V. (1989): «Effetto del Degrado Geometrico sul Coefficiente di Struttura», IV Convegno Nazionale, L'Ingegneria Sismica in Italia, Milano, 5-7 Ottobre.

Ghobarah A., Aly N.M., El-Attar M. (1997): «Performance Level Criteria and Evaluation», International Workshop on Seismic Design Methodologies for the Next Generation of Codes, Bled, Slovenia, June, Balkema, Rotterdam.

Hamburger R.O. (1997): «Defining Performance Objectives», International Workshop on Seismic Design Methodologies for the Next Generation of Codes, Bled, Slovenia, June, Balkema, Rotterdam.

Horne M.R., Morris L.J. (1973): «Optimum Design of Multi-Storey Rigid Frames», Chapter 14 of *Optimum Structural Design: Theory and Applications*, edited by R.H. Gallagher and O.C. Zienkiewicz, Wiley.

Krawinkler H., Nassar, A.A. (1992): «Seismic Design Based on Ductility and Cumulative Damage Demands and Capacities» in *Nonlinear Analysis and Design of Reinforced Concrete Buildings*, eds. P. Fajfar and H. Krawinkler, Elsevier, London.

Mazzolani F.M., Georgescu D., Astaneh-Asl A. (1994): «Safety Levels in Seismic Design», Proceedings of the 1st International Workshop on "Behaviour of Steel Structures in Seismic Areas", Timisora, Romania, June, published by E & FN Spon an Imprint of Chapman & Hall, London.

Mazzolani F.M., Piluso V. (1993): «P-Δ Effects in Seismic Resistant Steel Structures», SSRC Annual Technical Session and Meeting, Milwaukee, February, April.

Mazzolani F.M., Piluso V. (1995): «Seismic Design Criteria for Moment Resisting Steel Frames», *Steel Structures*, Proceedings of the 1st European Conference on Steel Structures, Athens, 18-20 May, published in "Steel Structures", Balkema, Rotterdam.

Mazzolani F.M., Piluso V. (1996): «Theory and Design of Seismic Resistant Steel Frames», FN & Spon, an Imprint of Chapman & Hall, London.

Mazzolani F.M., Piluso V. (1997): «Plastic Design of Seismic Resistant Steel Frames», Earthquake Engineering and Structural Dynamics, Vol.26, 167-191.

Mazzolani F.M., Piluso V. (1997) «A Simple Approach for Evaluating Performance Levels of Moment-Reisting Steel Frames», International Workshop on Seismic Design Methodologies for the Next Generation of Codes, Bled, Slovenia, June, Balkema, Rotterdam.

SEAOC, Structural Engineers Association of California (1995): «VISION 2000: Performance Based Seismic Engineering of Buildings»

Behaviour of Steel Structures in Seismic Areas, Mazzolani & Tremblay (eds) © 2000 Balkema, Rotterdam, ISBN 90 5809 130 9

Energy based methods for evaluation of behaviour factor for moment resisting frames

J. Milev, P. Sotirov & N. Rangelov
Faculty of Civil Engineering, UACEG, Sofia, Bulgaria

ABSTRACT: This paper presents the results of a series of parametric studies performed to evaluate the behavior factors of some typical for the Bulgarian construction practice steel moment resisting frames. Several frames are designed according to current Bulgarian codes and the capacity design procedure of EC8. Their behaviour factors are evaluated by the modified version of the Akiyama-Kato energy based method. Static pushover analysis with assumed load distribution is made for obtaining of story stiffness and yield shear forces. Time history response analysis for several acceleration records is performed aiming to obtain the critical value of cumulated story ductility ratio. Some comparison with experimental results of six full-scale specimens tested in the Steel Structures Research Laboratory of the Faculty of Civil Engineering, University of Architecture, Civil Engineering and Geodesy (UACEG), Sofia, are carried out. Parameters that influence q-factors are discussed and some conclusions are drawn.

1 INTRODUCTION

It is not efficient to design buildings that will not be damaged during any earthquake in the areas with high degree of seismic risk. The design philosophy of the current earthquake codes including Eurocode 8 is as follows:

1. Buildings should be prevented from collapsing during the most severe earthquakes. However some non-destructive damages may be permitted and some of the energy input during the severe earthquake is dissipated through inelastic deformations.

2. Building should be designed to remain almost within the elastic range for the earthquakes, which are expected to occur several times during the lifetime of the building.

Akiyama (1985) defined in that the "collapse" occurred when one story of a multi story frame loses its restoring force and collapses. And the term "almost elastic range" refers to a case when the displacements of all stories do not exceed the yield story displacement.

2 MODIFICATION OF AKIYAMA-KATO ENERGY METHOD

The evaluation of the total energy input from an earthquake contributing to the damage is made by Housner (1959) assuming that energy input attributable to the damage of an elastic-plastic system is the same as that producing damage in the relevant elastic system. Applying this idea Akiyama (1985) proposed the method for behaviour factor evaluation. The following assumptions are made on the basis of large amount of experimental data and parametric studies:

- The storey restoring forces of a multi-storey structure are assumed with the elastic-perfectly plastic characteristics (see Fig. 1^a);
- The coefficient of optimum distribution of story yield shear force, $\alpha_{opt,i}$, is established;
- The fundamental damage distribution law is introduced;
- The energy spectrum is derived.

The original Akiyama-Kato method does not require any non-linear analysis. However the evaluation of each story capacity needs several assumption concerning the simplified expression of the inelastic strain energy absorbed by the entire structure and the optimum distribution of this work between different storeys. A modification of the original Akiyama-Kato method has been already proposed by Milev (1998). The most important revisions of the original Akiyama-Kato method being as follows:

1. The i-th story stiffness K_i and yield shear forces Q_i are calculated by performing the non-linear pushover analysis on the considered structure. The distribution of horizontal forces along the height of the structure is considered on the basis of story shears obtained by elastic multi-modal analysis of the

structure. Then the story yield shear force coefficients α_i and the parameters c_i are calculated following the original Akiyama-Kato procedure as follows:

$$\alpha_i = \frac{Q_{yi}}{\sum\limits_{j=i}^{N} m_j g} \quad ; \quad c_i = \frac{4\pi^2}{K_i T_1^2} \frac{\left(\sum\limits_{j=i}^{N} m_j\right)^2}{\sum\limits_{j=1}^{N} m_j} \quad (1)$$

where: m_j = mass of the j-th story, T_1 = period of the fundamental mode, N = number of stories, g = gravity acceleration.

2. The cumulative inelastic strain energy W_p for the entire structure and the i-th story inelastic strain energy W_{pi} are calculated by performing the non-linear time history analysis with the scaled recorded or artificial accelerograms. Then the damage distribution coefficient γ_i is calculated as follows:

$$\gamma_i = \frac{W_p}{W_{pi}} \quad (2)$$

Then the original Akiyama-Kato procedure is followed:

3. The cumulated story ductility ratios η_i (see Fig. 2) are calculated as follows:

$$\eta_i = \frac{W_{pi} K_i}{Q_{yi}} \quad (3)$$

4. If the maximum η_i is equal (with some tolerance) to the assumed critical value of the cumulated story ductility ratio η_{lim} then the lower bound of the q-factor, derived by the assumption that the collapse in the i-th story of the building is prevented, is calculated by applying the following equations:

$$q_i = \sqrt{1 + 2c_i \gamma_i \eta_i \left(\frac{\alpha_i}{\alpha_1}\right)^2} \quad (4)$$

Therefore the value of the q-factor, which assures that all stories, will be prevented from the collapse is

$$q = \min(q_1, q_2, ..., q_N) \quad (5)$$

The modification of Akiyama-Kato method does not require the simplified expression of the inelastic strain energy absorbed by the entire structure and the optimum distribution of this work between different storeys. However it needs both non-linear static pushover and dynamic time history analysis.

Both the original and modified Akiyama-Kato methods require to judge the limit value of the story cumulated ductility ratio η_{lim}. The assumption made by Akiyama (1988) is that the collapse occurs when the energy dissipated at the observed story during the entire time history in each direction (positive or negative) is equal to the critical energy dissipated under monotonic loading. Under this hypothesis

$$\eta_{lim} = 2\mu_{lim} \quad (6)$$

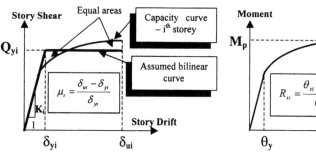

a) storey monotonic behaviour

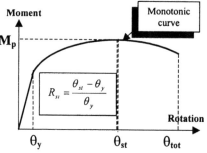

b)Plastic rotation capacity of beam-columns

Figure1: Monotonic behaviour

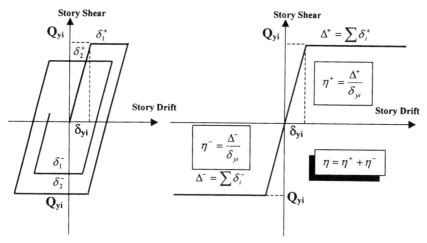

Figure 2. Cumulated ductility ratio definition

where μ_{lim} = limit value of storey monotonic ductility in one direction.

For the "weak beams-strong column" frames only the beams are deformed and according to Akiyama & Yamada (1997) the story ductility is related to the beam rotation capacity ($R_{st}=R_b$). In the case of shear type frame, the columns only are deformed and therefore the storey monotonic ductility μ_{lim} is related to the rotation capacity of the columns, ($R_{st}=R_{col}$). The following values for η_{lim} are assumed in this study (see Mazzolani & Piluso 1996 for more details):

$$\eta_{lim} = 2R_{st} \text{ for "weak beams-strong columns" frames}$$
$$\eta_{lim} = \frac{4}{3}R_{st} + 4 \text{ for shear type frames} \tag{7}$$

3 ANALYSED FRAMES

Some parametric studies are performed for evaluation of the behaviour factors of some typical for the Bulgarian construction practice steel-moment resisting frames with two types of haunching. The types of haunching are denoted as follows:
- "A" is linear haunching;
- "B" is haunching aimed at provoking the plastic hinge location at a specific point (see Sotirov et al. 2000 for more details).

In fact, the "B-type" haunching is an equivalent of the "dog bone" idea.

A total number of 12 studied frames has been carefully designed, having 1 to 3 bays and 2 to 6 stories.

Results for the frames of type B only are presented herein. The frames are denoted as follows:
- one span two stories frame – RF21B;
- two spans three stories frame – RF32B;
- two spans four stories frame – RF42B;
- three spans four stories frame – RF43B;
- three spans five stories frame – RF53B;
- three span six stories frame – RF63B.

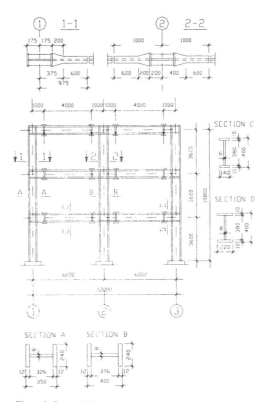

Figure 3. Frame RF23B layout

One of the frames (RF32B) is presented in Figure 3. The dimensions adopted for all frames are as follows: all bays = 6.0 m, all story heights = 3.6 m, all erection splices located at 1.0 m from column centre line, frames @ 6.0 m longitudinally.

Both columns and beams have been designed with welded built-up I-sections. The column cross-sections are assumed constant over the entire height, different for internal and for external columns. The beam-to-column connections are fully shop-welded with continuity plates (stiffeners) in the columns, and therefore should be classified as rigid full-strength connections. The column bases are considered as fixed. The erection splices are made with splice plates (single in the flanges and double in the web), site-welded by fillet welds (see Fig. 3). Such a detail may also be regarded as rigid connection, therefore no semi-rigidity is actually introduced.

The design has been carried out according to the current BG specifications: BG Steel design code (1987), BG Code for actions on structures (1991) and BG Seismic code (1987). Additionally, the provisions of the capacity design methodology have been taken into account. Thus all the frames designed fail into global plastic mechanism. Moreover all are applicable for real construction needs. Generally, the strength and stability criteria according to BG steel design code are similar to those of EC3 for Class 3 cross-sections. Additionally, the capacity design rules have also been taken into consideration.

No irregularities have been considered and all the frames are symmetrical. The most important feature is the column–tree configuration (erection splices located within the beam spans), as well as the horizontal haunching of beams (normally, there is no space for the more efficient vertical haunching at the joints).

Mild steel is considered of grade equivalent to grade S235, having $f_y = 235$ N/mm^2. The material safety factor adopted is $\gamma_m = 1.1$.

4 MODELLING OF THE FRAMES

Both static inelastic pushover analysis and dynamic inelastic time history analysis were performed with the frames described above. The general-purpose computer program DRAIN-2DX (see Parkash et al. 1994) is used.

Beams and columns are modeled with the plastic hinge beam-column element 02 (see Powel & Campbell 1994). The in-plane rigid floor is modeled as the cross sectional area of the beam is set to infinity. The damping in the structure is considered as 5% of the critical with damping matrix, which is linear combination of stiffness and mass matrix.

5 EARTHQUAKE RECORDS

Four earthquake records were selected in this study. They were normalised to the gravity acceleration (g) in order to eliminate the influence of the peak ground acceleration (PGA). The earthquakes are refereed as El Centro, Kobe JMA, Tolmezo and Vrancea77. Their elastic pseudo acceleration response spectra are presented in Figure 4.

6 ANALYSIS PROCEDURE

The applied procedure is as follows:
1. The SRSS of story shear are determined based on the modal analysis of the frames.

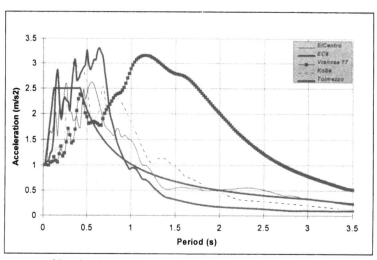

Figure 4. Response spectra of the selected earthquakes

Table 1. Story stiffness, story yield forces and horizontal force distribution (HFD) of the analyzed frames

Frame	Story	HFD	Story stiffness, K_i [kN/cm]	Story yield force, Q_{yi} [kN]
RF21B	1	1.00	83.64	215.77
	2	0.44	46.77	150.00
RF32B	1	1.00	137.30	324.74
	2	0.50	80.25	266.25
	3	0.33	73.16	177.50
RF42B	1	1.00	161.44	357.86
	2	0.44	85.99	309.72
	3	0.34	77.00	250.56
	4	0.28	74.21	174.00
RF43B	1	1.00	235.61	500.14
	2	0.45	129.18	434.97
	3	0.34	117.25	352.30
	4	0.27	112.36	243.00
RF53B	1	1.00	326.1	584.60
	2	0.41	159.6	522.00
	3	0.28	132.9	441.12
	4	0.31	128.2	358.13
	5	0.24	105.7	261.00
RF63B	1	1.00	377.98	685.35
	2	0.49	195.42	629.10
	3	0.29	168.27	550.80
	4	0.26	163.55	480.60
	5	0.26	162.29	402.30
	6	0.21	154.47	270.00

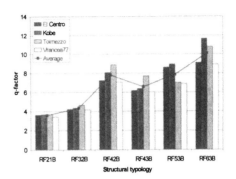

Figure 5. Calculated behaviour factor values

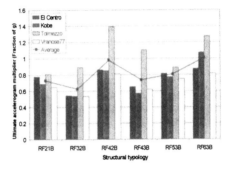

Figure 6. Calculated ultimate accelerogram multipliers

Table 2. Beam and columns plastic rotation capacity

Frame		R_{st}			θ_{st} [rad]		
		Mazzolani-Piluso	Mitani-Makino	Average	Mazzolani-Piluso	Mitani-Makino	Average
RF21B	Beam	7.59	5.18	6.40	0.0530	0.0360	0.0445
	Column	3.66	3.84	3.75	0.0176	0.0184	0.0180
RF32B	Beam	7.59	5.18	6.40	0.0530	0.0360	0.0445
	Int. col.	3.43	3.84	3.63	0.0147	0.0161	0.01556
	Ext. col.	3.02	3.03	3.03	0.007	0.007	0.007
RF42B	Beam	12.48	7.37	9.90	0.093	0.055	0.074
	Int. col.	6.73	5.81	6.3	0.0269	0.0311	0.029
	Ext. col.	8.57	7.13	7.85	0.0247	0.0206	0.0226
RF43B	Beam	7.59	5.18	6.40	0.0530	0.0360	0.0445
	Int. col.	5.84	6.27	6.05	0.0246	0.0264	0.0255
	Ext. col.	4.43	5.72	5.10	0.0108	0.0139	0.0124
RF53B	Beam	7.59	5.18	6.40	0.0530	0.0360	0.0445
	Int. col.	7.57	8.10	6.40	0.0263	0.0282	0.0273
	Ext. col.	11.35	8.49	9.92	0.0255	0.0191	0.0223
RF63B	Beam	12.00	8.41	10.20	0.0823	0.0577	0.0700
	Int. col.	13.42	8.50	10.96	0.0251	0.0159	0.0205
	Ext. col.	7.90	8.10	8.00	0.0257	0.0263	0.0260

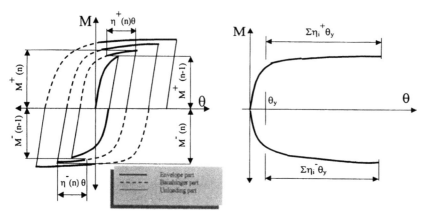

Figure 7. Decomposition of moment-rotation diagram

2. The distributions of fictional story forces (DFSF), which correspond to the obtained story shears, are calculated. Their normalized to unity distributions are listed in Table 1.

3. The procedure of modified Kato-Akiyama method, described above, is applied. Story stiffness and story yield forces, obtained by nonlinear static pushover analysis with assumed story forces distribution, are listed in Table 1.

The ultimate limit value of the story cumulated ductility ratio η_{lim} is calculated for the case of "weak beams-strong columns" frames.

The monotonic rotation capacity of beams and columns Rst and corresponding plastic rotation capacity θst are obtained by both the theoretical method of Mazzolani & Piluzo (1996) and the empirical method of Mitani-Makino (see Mazzolani & Piluzo 1996 for details). The stable part of the moment-rotation relationship (see Fig. 1^b) was taken as average of the values calculated by both methods. The results are listed in Table 2.

7 RESULTS OF ANALYSIS

The summary of the analysis results is presented in Figures 5-6. The values of the q-factor obtained by the energy method for different earthquakes show very small variation. The difference is in the reasonable limits. The behavior factors increase with increase of story number. That is reasonable because with the increase of story number the energy dissipation capacity of the frame increases.

8 COMPARISON WITH EXPERIMENTAL DATA

The hysteretic moment-rotation curves of columns, beams and panel zones can be decomposed

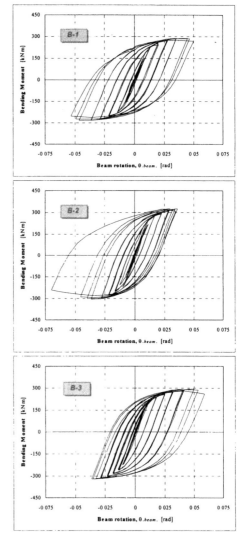

Figure 8. Moment–rotation relationship for specimens

into envelope parts, Baushinger parts and unloading parts (see Fig. 7). The bold solid lines in the Figure 7 represent the inelastic part in each loading cycle under assumption that the load at the beginning of unloading of the foregoing loading cycle corresponds to the load at the onset of plastification after one cycle has elapsed. Those inelastic parts in each cycle are connected in each direction of loading (positive and negative).

According to Akiyama (1985), the two curves obtained, which correspond to the positive and negative directions, coincide with the inelastic part of the moment-rotation curve under monotonic loading. The slope of the unloading parts agrees with the initial elastic slope. After developing of inelastic deformation in one direction, the initial path in the subsequent reverse loading is not linear. It takes a non-linear curve with considerable deterioration in stiffness (see the dashed parts in Fig. 7). This phenomenon is termed as Baushinger effect.

Six full-scale beam-column specimens were tested in the test laboratory of UACEG - Sofia. Those specimens were extracted from frames RF42B and RF42A (see Sotirov et al. 2000 for details). Only the specimens with B-type haunching are discussed herein. Experimental beam moment-rotation relationships of specimens B-1, B-2 and B-3 are presented in Figure 8. Moment-rotation diagram decomposition for all specimens in both direction of loading (positive and negative) are presented in Figure 9. The monotonic rotation capacity of tested beams Rst and corresponding plastic rotation capacity θst obtained with the theoretical method of Mazzolani & Piluzo (1996) and the empirical method of Mitani-Makino (see Mazzolani & Piluzo 1996 for details) as well as average experimental values of Rst and θst are listed in Table 3.

The analysis procedure, which is described above, is applied again with the frame RF42B employing all four earthquake records (El Centro, Kobe JMA, Tolmezo and Vrancea77). The ultimate limit value of the story cumulated ductility ratio η_{lim} is calculated for the case of "weak beams-strong columns" frames by using Equation 7. However the average experimental value for Rst=7.24 is assumed. Comparison, between behavior factor values obtained by assuming average experimental value for Rst=7.24 and previously considered value Rst=9.90 as average of the values calculated by both theoretical methods (Mazzolani-Piluso and Mitani-Makino), is presented in Table 4.

9 CONCLUSIONS

This study is focused on the energy approach for behaviour factor evaluation of steel moment resisting frames. A modification of Akiyama-Kato method is

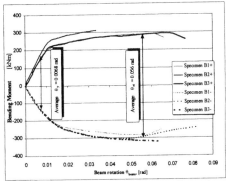

Figure 9. Decomposition of M-θ diagrams for specimens B-1, B-2 and B-3

Table 3. Theoretical and experimental plastic rotation capacity of the specimens

Frame RF24B	Mazzolani-Piluso	Mitani-Makino	Experimental (average)
R_{st}	12.48	7.37	7.24
θ_{st} [rad]	0.093	0.055	0.056

Table 4. Behaviour factors of frame RF42B based on analytical prediction and experimental results for Rst

Earthquake	Theoretical		Experimental	
	Rst	q-factor	Rst	q-factor
El Centro	9.90	7.25	7.24	6.85
Kobe JMA	9.90	8.08	7.24	7.02
Tolmezo	9.90	8.59	7.24	7.24
Vrancea 77	9.90	7.12	7.24	6.71
Average	-	7.76	-	6.96

employed. Some parametric study, with frames designed according to current Bulgarian codes and considering the capacity design approach, is performed. Besides, the theoretically calculated and experimentally obtained beam plastic rotation capacity is set to one of analysed parameters. The following conclusions could be drawn on the bases of the current study:

- The scatter of behaviour factor values, which are calculated by applying the energy approach, is in reasonable limit (within 15%) for different earthquake records;
- The behaviour factor values increases with increasing of story number of the frames. In the case of properly designed "strong columns-weak beams" frames, this trend could be explained by

increasing of frame energy dissipation capacity with increasing of story number;

- One of the most important assumptions in applying of energy approach is the limit value of member rotation capacity. The comparison between theoretical recommendation and experimental data shows that the member rotation capacity could be theoretically predicted with reasonable accuracy. The difference between analytical and experimental values of member rotation capacity affects the q-factor values in the limits of 10-15%. This scatter is in the same range as one influenced by the different earthquake records.

However it is necessary to underline that more experimental and theoretical studies are necessary for collapse state definition based on energy approach.

REFERENCES

Akiyama, H. 1985. *Earthquake Resistant Limit-State Design for Buildings*, University of Tokyo Press.

Akiyama, H. 1988. Earthquake Resistant Design Based on the Energy Concept, *Proceedings of the 9th World Conference on Earthquake Engineering*, Tokyo, Kyoto.

Akiyama, H. & Yamada S. 1997. Seismic Input and Damage of Steel Moment Frames, General Report of STESSA'97, *Behaviour of Steel Structures in Seismic Areas*, Kyoto, Japan.

Housner, G. 1959. Behaviour of Structures during the Earthquakes, *Journal of ASCE*, EM4.

Mazzolani, F., & Piluso, V. 1996 *Theory and Design of Seismic Resistant Steel Frames*, E&FN SPON, 1996

Milev, J. 1998. Modelling of Shear Walls for Seismic Analysis of Frame-Wall Structures, *PhD Thesis*, (in Bulgarian).

Parkash, V., Powell, G. H., & Campbell, S. 1994. DRAIN-2DX. Base Program Description and User Guide, *Report No UCB/SEMM/94/07*.

Powell, G. H., & Campbell, S. 1994. DRAIN-2DX. Element Description and User Guide for Element Type01, Type02, Type04, Type06 and Type15, *Report No UCB/SEMM/94/08*.

Sotirov, P., Rangelov, N. & Milev, J. 2000, Improvement of seismic behaviour of beam-to-column joints using tapered flanges, *Proceedings of STESSA 2000*, Montreal, Canada.

Behaviour of Steel Structures in Seismic Areas, Mazzolani & Tremblay (eds) © 2000 Balkema, Rotterdam, ISBN 90 5809 130 9

Plastic design of steel frames with dog-bone beam-to-column joints

Rosario Montuori & Vincenzo Piluso
Department of Civil Engineering, University of Salermo, Italy

ABSTRACT: The use of "dog-bones", i.e. the weakening of the beam section by properly trimming its flanges around the connection, has been proposed aiming at both the protection of the beam-to-column joints and the promotion of the plastic hinge formation in the beams rather than in the columns. In this paper the design issues regarding the use of "dog-bones" within a design procedure for failure mode control are dealt with. The design procedure, already proposed for rigid and semirigid frames, assures both the development of a collapse mechanism of global type and the fulfilment of the serviceability requirements. In order to stress the influence of the amount of reduction of beam flexural resistance obtained by means of "dog-bones", and the influence of their location, the application of the design procedure is presented with reference to a significant structural situation.

1 INTRODUCTION

According to the traditional design philosophy of seismic resistant structures, structures have to remain in elastic range during frequent seismic events having a return period comparable with the service life of the structure. Conversely, in the case of destructive earthquakes having low probability of occurrence (usually a 475 years return period is considered), it is accepted the damage of both structural and nonstructural elements which derives from the development of dissipative mechanisms. Therefore, the plastic reserves of the structure have to be exploited, only in the case of rare major earthquakes, to dissipate the earthquake input energy in some zones of the structure, namely dissipative zones, which have to be properly selected.

The column hinging has to be absolutely avoided, because, due to the action of axial forces and the premature occurrence of local buckling, they exhibit poor ductility. Moreover, the failure modes which can result from column hinging could involve a limited number of dissipative zones. For these reasons, aiming at the complete development of the plastic reserves of the structure, modern seismic codes provide simple design criteria whose goal is the prevention of local failure modes and, instead, promotion of a global mechanism, i.e. a collapse mechanism characterised by the hinging of all the beam ends and the hinging of the base sections of the first storey columns.

In the case of moment resisting frames, the design criterion suggested by seismic codes is the adoption of columns having a flexural resistance greater than that of the connected beams. However, the fulfilment of this design criterion, namely member hierarchy criterion, is only able to prevent the development of storey mechanisms, but is not sufficient to guarantee the formation of a collapse mechanism of global type.

Within this design framework, aiming at the safeguard of brittle elements and to the maximisation of the ductile elements engaged in dissipating the earthquake input energy, the idea of realising the so-called "dog-bones", i.e. the weakening of the beam at its ends by reducing the flange width, is born. This structural detail is aimed at the promotion of the beam end hinging to prevent yielding of columns and/or of connections. In addition, by means of this structural detail, it is possible to promote a collapse mechanism of global type.

The goal of the work herein presented is the setting up of design rules regarding the magnitude of the weakening to be realised and the location of the weakened beam sections. In particular, the location of the weakened section has to be selected in order to assure the development of the plastic hinges in "dog-bones" and/or in intermediate beam sections, while the yielding of the beam-to-column connections has to be prevented. Furthermore, by means of a design methodology already successfully adopted with reference to rigid (Mazzolani and Piluso, 1996; 1997) and semirigid (Montuori, 1997; Faella et al., 1998; 1999) frames, it is possible to design frames failing in global mode.

2 DOG-BONE LOCATION

The first problem to be faced within rigid-plastic analysis of seismic-resistant structures is the location of plastic hinges in the beams. Regarding this issue, considering that seismic action can be represented by means of an appropriate distribution of horizontal forces increasing with a common multiplier α, it is preliminarily necessary to observe that the bending moment diagram of a generic beam is due to the superposition of that due to vertical loads and of the one, increasing with α, due to the seismic forces (Fig. 1). Therefore, the bending moment diagram looks like that depicted in Fig. 2, where the sections corresponding to the beam ends (sections 1 and 5), those corresponding to the "dog-bone" locations (section 2 and 4) and that corresponding to the maximum sagging moment (section 3) have been pointed out.

It is evident that the design parameters are the location of the "dog-bones" (which is denoted with the distance a in Fig. 2) and the magnitude of the weakening characterising the "dog-bones". This second parameter can be expressed in non-dimensional form as:

$$m_{db} = \frac{M_{p.db}}{M_p} \qquad (1)$$

where $M_{p.db}$ is the plastic moment of the weakened beam section and M_p is the plastic moment of the complete beam section.

In this phase of the design procedure, similarly to the case of semirigid frames (Montuori, 1997; Faella et al., 1998; 1999), the m_{db} value is assumed to be fixed, while the location a of the "dog-bones" has to be properly selected.

Increasing the seismic horizontal forces, i.e. increasing the multiplier α, the conditions assuring that sections 1, 2, 3 and 5 remain in elastic range, while section 4 yields, have to be determined. These conditions correspond to the fulfilment of all the following relationships:

$$M_A < M_p \qquad (2)$$

$$M(a) < m_{db} M_p \qquad (3)$$

$$M(x') < M_p \qquad (4)$$

$$M(L - a) = -m_{db} M_p \qquad (5)$$

$$M_B < M_p \qquad (6)$$

The bending moment at the generic section x is given by:

$$M(x) = M_A + T_A x - q \frac{x^2}{2} =$$
$$= M_A + \left(q \frac{L}{2} - \frac{M_A + M_B}{L} \right) x - q \frac{x^2}{2} \qquad (7)$$

By imposing the fulfilment of relationship (5) regarding the yielding of the "dog-bone" in section 4, i.e. by combining Eq. (7) with $x = L - a$ and Eq. (5), and by solving with respect to M_B, the following relationship is obtained:

$$M_B = M_A \frac{a}{L-a} + q \frac{La}{2} + \frac{L}{L-a} m_{db} M_P \qquad (8)$$

which represents the relation occurring between the end moments when the first plastic hinge develops at section 4 corresponding to the right "dog-bone".

By means of Eqs. (8) and (7), it is possible to express the design requirements (2), (3), (4), and (6) as follows:

$$M_A < M_{A1} \text{ with } M_{A1} = M_P \qquad (9)$$

$$M_A < M_{A2} = \frac{m_{db} L}{L - 2a} M_P - q \frac{a(L - a)}{2} \qquad (10)$$

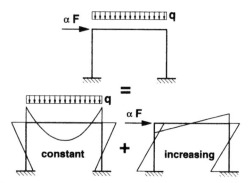

Figure 1. Superposition of bending moments due to vertical loads and seismic forces.

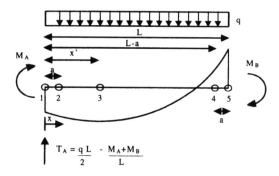

Figure 2. Typical beam bending moment diagram

$M_A < M_{A3}$ with

$$M_{A3} = \sqrt{2q(L-a)^2(1+m_{db})M_P}$$
$$-\left(\frac{q(L-a)^2}{2} + m_{db}M_P\right)$$

(11)

$M_A < M_{A5}$ with

$$M_{A5} = \frac{(1-m_{db})L-a}{a}M_P - q\frac{L(L-a)}{2}$$

(12)

Obviously, the first plastic hinge develops in the right "dog-bone" provided that Eq. (6) is satisfied. Under this condition, it is required that the second plastic hinge develops either in the left "dog-bone" or an intermediate beam section. On the contrary, the yielding of the beam ends close to the beam-to-column connections has to be prevented, because, as already stated, the use of "dog-bones" is also aimed at the protection of beam-to-column connections. It is easy to recognise that increasing the seismic horizontal forces, i.e. increasing M_A, relationships (9), (10), (11) and (12) allow to identify the section where the second plastic hinge develops. To this scope, it is sufficient to control what is the minimum limit value among M_{A1}, M_{A2}, M_{A3}, M_{A5}. In other words, it is sufficient to identify the first relationship to be unsatisfied as far as M_A increases.

Therefore, all the yielding conditions can be expressed by means of the limit values M_{Ai} (with i=1,2,3,5) of the bending moment M_A occurring at the first beam end. In particular, the condition:

$$M_{A3} < M_{A2} \quad \textbf{condition A} \quad (13)$$

identifies the a values assuring that the yielding of the beam in the section where the maximum sagging moment occurs (section 3) precedes the yielding of the left "dog-bone" (section 2); the condition:

$$M_{A3} < M_{A5} \quad \textbf{condition B} \quad (14)$$

identifies the a values assuring that the beam yielding (section 3) precedes the yielding of the connection B (section 5); the condition:

$$M_{A2} < M_{A5} \quad \textbf{condition C} \quad (15)$$

identifies the a values assuring that the left "dog-bone" yielding (section 2) precedes the yielding of the connection B (section 5); the condition:

$$M_{A3} < M_{A5} \quad \textbf{condition D} \quad (16)$$

identifies the a values assuring that the beam yielding (section 3) precedes the yielding of the left connection A (section 1); finally, the condition:

$$M_{A2} < M_{A1} \quad \textbf{condition E} \quad (17)$$

identifies the a values assuring that the yielding of the left "dog-bone" (section 2) precedes the yielding of the left connection A (section 1).

It is evident that conditions (14), (15), (16), (17) have to be absolutely satisfied, because they assure the development of the second plastic hinge either in the left "dog-bone" or in the intermediate beam section where the maximum sagging moment occurs, while the yielding of the connections at the beam ends is prevented. In other words, relationships (14), (15), (16) and (17) are the design requirements.

Conversely, condition (13), depending on its fulfilment or not, can be used to discern if the second plastic hinge develops in the left "dog-bone" or in intermediate beam section.

However, the five conditions are not all significant. In fact, condition D can be written as follows (by expressing relationship (16) by means of the M_{A1} and M_{A3} values given by Eqs. (9) and (11), respectively):

$$-\left(\frac{a}{L}-1+\sqrt{2(1+m_{db})\frac{M_P}{qL^2}}\right)^2 < 0$$

(18)

therefore, condition D is always satisfied.

Moreover, it is easy to check that, for $0<a/L<1/2$, condition E is automatically fulfilled when condition 3 is satisfied. Therefore, in the range $0<a/L<1/2$, which is the significant one from the design point of view, only the three conditions A, B and C remain to be analysed. Such conditions give rise to the following non-dimensional relationships:

- **condition A**:

$$4\left(\frac{a}{L}\right)^3 + \left(4\sqrt{2(1+m_{db})\frac{M_P}{qL^2}} - 8\right)\left(\frac{a}{L}\right)^2$$
$$+ \left(5 + 4m_{db}\frac{M_P}{qL^2} - 6\sqrt{2(1+m_{db})\frac{M_P}{qL^2}}\right)\left(\frac{a}{L}\right)$$
$$- 4m_{db}\frac{M_P}{qL^2} - 1 + 2\sqrt{2(1+m_{db})\frac{M_P}{qL^2}} < 0$$

(19)

whose solutions are:

$$\frac{a}{L} < \frac{a_3}{L} \quad \text{and} \quad \frac{a_2}{L} < \frac{a}{L} < \frac{a_1}{L}$$

(20)

where:

$$\frac{a_1}{L} = 1 \; ; \frac{a_2}{L} = \frac{1}{2} - \sqrt{\frac{(1+m_{db})}{2}\frac{M_P}{qL^2}} + \sqrt{\frac{(1-m_{db})}{2}\frac{M_P}{qL^2}}$$
$$\frac{a_3}{L} = \frac{1}{2} - \sqrt{\frac{(1+m_{db})}{2}\frac{M_P}{qL^2}} - \sqrt{\frac{(1-m_{db})}{2}\frac{M_P}{qL^2}}$$

(21)

- **condition B:**

$$-\left(\frac{a}{L}\right)^3 + \left(1 - 2\sqrt{2(1 + m_{db})\frac{M_P}{qL^2}}\right)\left(\frac{a}{L}\right)^2$$

$$+ \left(2(1 - m_{db})\frac{M_P}{qL^2} + 2\sqrt{2(1 + m_{db})\frac{M_P}{qL^2}}\right)\left(\frac{a}{L}\right) \quad (22)$$

$$- 2(1 - m_{db})\frac{M_P}{qL^2} < 0$$

whose solution are:

$$\frac{a_6}{L} < \frac{a}{L} < \frac{a_5}{L} \quad \text{and} \quad \frac{a}{L} > \frac{a_4}{L} \quad (23)$$

where:

$$\frac{a_4}{L} = 1 \quad ; \quad \frac{a_5}{L} = -\sqrt{2(1 + m_{db})\frac{M_P}{qL^2}} + 2\sqrt{\frac{M_P}{qL^2}} \, ;$$

$$\frac{a_6}{L} = -\sqrt{2(1 + m_{db})\frac{M_P}{qL^2}} - 2\sqrt{\frac{M_P}{qL^2}} \quad (24)$$

- **condition C:**

$$-2\left(\frac{a}{L}\right)^4 + 5\left(\frac{a}{L}\right)^3 - 4\left(1 + \frac{M_P}{qL^2}\right)\left(\frac{a}{L}\right)^2 \quad (25)$$

$$+ \left(1 + 2(3 - m_{db})\frac{M_P}{qL^2}\right)\frac{a}{l} - 2(1 - m_{db})\frac{M_P}{qL^2} < 0$$

whose solutions are:

$$\frac{a}{L} < \frac{a_8}{L} \quad \text{and} \quad \frac{a}{L} > \frac{a_7}{L} \quad (26)$$

where:

$$\frac{a_7}{l} = 1 \quad ; \quad \frac{a_8}{l} = \frac{1}{2} + \frac{1}{6}\sqrt[3]{T} - 6\frac{\frac{2}{3}\frac{M_P}{ql^2} - \frac{1}{12}}{\sqrt[3]{T}} \quad (27)$$

with:

$$T = -108 \, m_{db} \frac{M_P}{qL^2} +$$

$$3\left[-3 + 72\frac{M_P}{qL^2} - 576\left(\frac{M_P}{qL^2}\right)^2 \right.$$

$$\left. + 1296 \, m_{db}{}^2\left(\frac{M_P}{qL^2}\right)^2 + 1536\left(\frac{M_P}{qL^2}\right)^3\right]^{1/2} \quad (28)$$

Now it can be observed that, being $a/L < 1/2$ (which means that a "dog-bone" cannot be located beyond the midspan), the three examined conditions provide the following significant solutions:

condition A $\quad \frac{a}{L} < \frac{a_3}{L} \quad$ and $\quad \frac{a}{L} > \frac{a_2}{L} \quad (29)$

which is obtained from Eqs. (20) and (21);

condition B $\quad \frac{a}{L} < \frac{a_5}{L} \quad (30)$

which is obtained from Eqs. (23) and (24) and by observing that a_6 provides negative values which are not significant;

condition C $\quad \frac{a}{L} < \frac{a_8}{L} \quad (31)$

which is obtained from Eqs. (24) and (27). Therefore, taking into account that condition A has to be used only to recognise the location of the second plastic hinge which can develop either at the left "dog-bone" or at an intermediate beam section, it means that conditions B and C show the existence of an upper bound concerning the parameter a expressing the "dog-bone" location (this upper bound value is given by the minimum value between a_5 and a_8).

Conversely, there is not any lower bound value other than the technological one.

As the expression for computing a_8/L is particularly complex, in order to identify the governing limit value of a/L, a numerical analysis has been carried out. For any given value of m_{db}, by varying the non-dimensional parameter M_p/qL^2 in the range between 1/16 and 1/3, which covers all the possible design situations, the values of a_2, a_3, a_5 and a_8 have been computed.

As an example, in Table 1 the values corresponding to $m_{db}=0.7$ are given; however, the following consideration is valid independently of m_{db}.

Table 1. Values of a_2, a_3, a_5 e a_8 for $m_{db} = 0.7$

qL^2/M_p	a_2	a_3	a_5	a_8
16	0.3663	0.1727	0.0390	0.0325
12.12	0.3464	0.1239	0.0448	0.0409
9.75	0.3288	0.0808	0.0500	0.0485
8.16	0.3129	0.0418	0.0546	0.0552
7.01	0.2982	0.0058	0.0589	0.0614
6.15	0.2845	-0.0278	0.0629	0.0669
5.47	0.2716	-0.0593	0.0667	0.0718
4.93	0.2594	-0.0892	0.0702	0.0763
4.49	0.2478	-0.1176	0.0736	0.0803
4.12	0.2367	-0.1447	0.0769	0.0840
3.80	0.2261	-0.1708	0.0800	0.0873
3.53	0.2158	-0.1959	0.0830	0.0904
3.30	0.2059	-0.2201	0.0859	0.0932
3.10	0.1964	-0.2435	0.0886	0.0957
2.91	0.1871	-0.2662	0.0913	0.0981

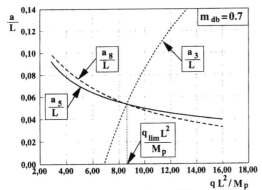

Figure 3. limit values a_3, a_5 and a_8 for $m_{db}=0.7$

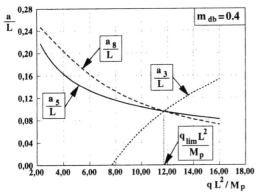

Figure 4. limit values a_3, a_5 and a_8 for $m_{db}=0.4$

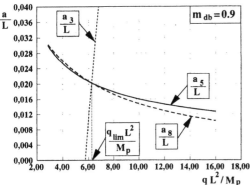

Figure 5. limit values a_3, a_5 and a_8 for $m_{db}=0.9$

From the analysis of the table, it is evident that a_2 is always the greatest value, so that the design solution concerning the "dog-bone" location can be expressed as follows: the smallest value between a_5 and a_8 is the upper bound of a, while the location of the second plastic hinge depends on the a_3 value; in

particular, according to Eq. (29), if $a<a_3$ the second plastic hinge occurs at the left "dog-bone", otherwise the second plastic hinge develops in the intermediate beam section, where the maximum sagging moment occurs.

From the analysis of Table 1 and Fig. 3, it is evident the existence of a limit value of qL^2/M_p for which a_3, a_5 and a_8 are coincident. Such limit value can be easily determined by equating the relationships (21) and (24) providing a_3 and a_5. The following relationship is thus obtained:

$$q_{lim} = \frac{4M_p}{L^2}\left(5 + \sqrt{8(1 - m_{db})} - 2\sqrt{2(1 + m_{db})} - \sqrt{1 - m_{db}^2}\right) \quad (32)$$

As a conclusion, the design solution concerning the "dog-bone" location and its influence on the location of the second plastic hinge can be expressed as follows:

- **case $q<q_{lim}$:** if $a<a_3$ the yielding of the beam occurs; otherwise ($a_3<a<a_5$) the yielding of the "dog-bone" develops;

- **case $q>q_{lim}$:** the design requirement is $a<a_8$; the second plastic hinge always develops at the intermediate beam section where the maximum sagging moment occurs.

It is interesting to note that the q_{lim} value already obtained by Mazzolani and Piluso (1996; 1997) for full-strength beam-to-column joints is obtained, as a particular case, from Eq. (32) for $m_{db}=1$, i.e. in absence of "dog-bones":

$$q_{lim} = \frac{4M_P}{L^2} \quad (33)$$

The results of the analyses concerning the appropriate location of "dog bones" are given in Figs. 4 and 5 also for $m_{db}=0.9$ and $m_{db}=0.4$, respectively.

From these figures it is evident that the range where "dog bones" can be located, assuring the protection of the beam-to-column connections, decreases as m_{db} increases. In particular, for high values of m_{db}, the width of such range is so limited that it could be incompatible with the length of the weakened beam zone required for an appropriate spreading of plasticity.

3 THE USE OF "DOG-BONES" FOR FAILURE MODE CONTROL

A the theoretical procedure for designing frames failing in global mode has been proposed with reference to full-strength rigid joints (Mazzolani and Piluso, 1996; 1997) and, successively, extended to the case of semirigid frames with partial-strength joints (Montuori, 1997; Faella et al., 1998; 1999). The theoretical bases are herein briefly recalled as the same design procedure is applied in this work with reference to moment-resisting frames with "dog-bone" connections.

The design procedure is based on the kinematic theorem of plastic collapse and on the concept of mechanism equilibrium curve. In particular, it is observed that the collapse mechanism of a frame subjected to horizontal forces can, basically, belong to three collapse mechanism typologies, so that the failure mode control can be obtained by analysing $3n_s$ collapse mechanisms, being n_s the number of storeys (Fig. 6).

It is assumed that the beam sections are known as they are designed to resist vertical loads and, in the case herein under examination, it is also assumed that the "dog-bone" locations and the corresponding magnitude of the beam weakening are known. Therefore, the unknowns of the design problem are the plastic moments of the column sections.

Moreover, the design procedure accounts for the influence of second order effects by extending the kinematic theorem of plastic collapse to the concept of mechanism equilibrium curve.

In fact, the plastic moments of the columns are derived by imposing that, within a given displacement range depending on the plastic rotation supply of members and connections, the mechanism equilibrium curve corresponding to the global mechanism has to lie below the mechanism equilibrium curves corresponding to all the remaining 3ns-1 kinematically admissible mechanisms.

In order to understand the results of the analyses herein presented, it is useful to remember that the main result of the design procedure for failure mode control is the sum, at each storey, of the plastic moments of the columns required to assure a collapse mechanism of global type. The plastic moments are reduced due to the influence of axial forces. The knowledge of both the axial forces at collapse and the sum of the reduced plastic moments, required at each storey, allows the dimensioning of the columns. The aim of the following example is to point out the influence of the parameter a, i.e. the "dog-bone" location (as it has been previously shown, the parameter a can vary within a range depending on m_{db}).

With reference to the case of a 6 storey-3 bay frame with a bay span equal to 10.5 m and interstorey height equal to 3.50 m for the first storey and 3.2 m for the upper storeys, dead load G equal to 21 kN/m, live load Q equal to 10.5 kN/m, IPE500 beams and "dog-bones" having a non-dimensional flexural

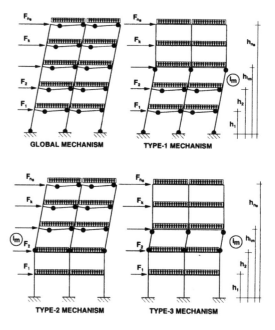

Figure 6. analysed collapse mechanism typologies

resistance $m_{db}=0.6$, Fig. 7 shows the influence of the "dog-bone" location a on the sum M_k of the plastic moments of the kth storey, required to assure a collapse mechanism of global type:

$$M_k = \sum_{i=1}^{n_c} M_{i,k} \qquad (34)$$

where $M_{i,k}$ is the plastic moment (reduced due to the influence of the axial force) of the ith column of the kth storey and n_c is the number of columns.

Figure 7 has been obtained by varying the "dog-bone" location a within the range determined by means of the procedure described in Section 2, considering the limit value of the uniform vertical load given by Eq. (32).

It can be observed that M_k is increasing with a. It means that the structural weight required to assure a collapse mechanism of global type increases as far as the distance of the weakened beam section from the beam ends increases. The values given in Fig. 7 can be computed by varying m_{db}, as an example Fig. 8 refers to the case $m_{db}=0.4$.

As soon as the column sections have been designed to assure a global failure mode, the control of the serviceability requirements is necessary. This requires an elastic analysis to perform the resistance and stability checks of members, connections and "dog-bones" under the design load combinations prescribed by the codes.

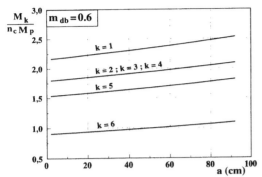

Figure 7. Influence of the "dog-bone" location a on the sum M_k of the plastic moments of the kth storey, required to assure a collapse mechanism of global type for m_{db}=0.6

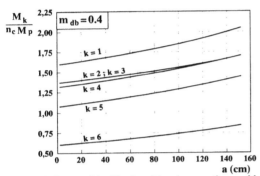

Figure 8. Influence of the "dog-bone" location a on the sum M_k of the plastic moments of the kth storey, required to assure a collapse mechanism of global type for m_{db}=0.4

As an example, with reference to Eurocode 3 and Eurocode 8, it is necessary to verify that the above checks are satisfied under the non-seismic load combination (1.35G + 1.5Q) and the seismic load combination (G + 0.3 Q + E, being E the design earthquake action). In addition the control of the interstorey drift has to be performed. In particular, it is important to stress that the minimum value of mdb which can be practically used is strongly affected by the need of fulfilling all the design requirements under the load combinations prescribed by the codes.

4 PARAMETRIC ANALYSIS

In order to investigate the structural configurations where the use of "dog-bones" can lead not only to the protection of the beam-to-column connections, but also to a significant saving in structural weight of frames failing in global mode, a wide parametric analysis has been carried out (Galdi, 1999). In particular, 27 frame configurations have been analysed by varying the number of storeys n_s (n_s = 2 ; 4 ; 6), the number of bay n_b (n_b = 2 ; 3 ; 4) and the bay span L (L = 9.00 m ; 10.50 m ; 12.00 m).

The interstorey height of the first storey is equal to 3.50 m, those of the upper storeys are equal to 3.20 m. Regarding the vertical loads, the characteristic values of the dead and live load are equal to 4.0 x L/2 kN/m and 2.0 x L/2 kN/m, respectively (being L the bay span). For each structural scheme the design solution leading to the minimum weight with the restraint of a collapse mechanism of global type has been searched by varying the "dog-bone" non-dimensional flexural resistance m_{db} and the corresponding location.

Therefore, the solutions leading to the minimum weight requires several analyses for each frame configuration. The structural solution leading to the minimum weight have been compared with the corresponding structural solutions assuring a collapse mechanism of global type with traditional connections, where the beam flanges are not weakened, i.e. without "dog-bones". As an example, with reference to a 3 bay-4 storey frame with a bay span equal to 12.00 m, Fig.9 shows the structural solution leading to a collapse mechanism of global type for the reference case of full-strength rigid joints. The corresponding structural weight is equal to 29.6 tons. Conversely, Fig. 10 shows for the same structural scheme the optimal solution adopting "dog-bone" connections which corresponds to m_{db} = 0.4 and to "dog bones" located at a distance from the column axis equal to 67.5 cm. The structural weight of the frame with "dog bone" connections is equal to 26.2 tons. The analysis of the "dog bone" location according to Section 2 points out that in order to protect the beam-to-column connections, preventing their yielding, the distance between the face of the column flange and the "dog bone" axis has to be less than 177 cm.

According to Carter and Iwankiw (1998) the "dog bone" axis has to be located at a distance from the face of the column flange greater than $(5/8)d_b$ (being d_b the beam depth) and the "dog bone" length l has to satisfy the limitation $3d_b/4 < l < d_b$. Therefore, taking into account that the column sections vary from an HEB500 at the first storey up to an HEB320 at the top storey, it is evident that the "dog bone" location (67.5 cm from the column axis) satisfies also these design requirements concerning the structural detail of the "dog bone".

Regarding the structural weight it is important to underline that both structural solutions are characterised by a total beam weight equal to 17.6 tons. Therefore, the total column weight is equal to 29.6 - 17.6 = 12 tons for the traditional frame.

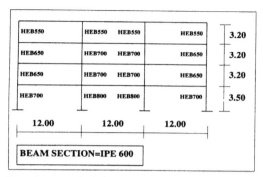

Figure 9. structural solution adopting traditional connections.

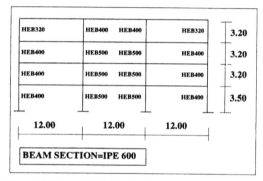

Figure 10. Structural solution adopting "dog bone" connections.

Conversely, the total column weight is equal to 26.2-17.6 = 8.6 tons in the case of the structural solution adopting "dog bone" connections. As a result the use of "dog bone" leads to a 28.3% saving concerning the column weight which corresponds to 11.5% saving in total structural weight.

With reference to the performed parametric analysis a maximum saving in total structural weight equal to 17% has been achieved.

5 CONCLUSIONS

The criteria for designing "dog-bone" connections have been analysed. In particular, depending on the magnitude of the weakening of the beam section, the "dog-bone" location has to be properly selected in order to protect the beam-to-column connections which is one of the goals to be achieved in designing frames adopting such a structural detail.

In addition, "dog-bone" connections can be properly used aiming to the design of frames failing in global mode. Under this point of view, it is recognised that the smallest m_{db} value provides the widest range where it is possible to locate the "dog-bones"; in fact, increasing m_{db} the range identified by a reduces up to the loss of meaning for per $m_{db}=1$.

Moreover, increasing m_{db} and a the column plastic moments required to assure a global failure mode increase; as a consequence the structural weight increases. By repeating the design procedure for different values of the "dog-bone" non-dimensional resistance m_{db} and for different "dog-bone" locations a, it is possible to identify the values of the design parameters m_{db} and a which lead to the minimisation of the structural weight within frames assuring both a collapse mechanism of global type and the fulfilment of the serviceability requirements.

In particular, the use of "dog bone" connections can lead to a saving concerning the total weight of columns up to 30% about which can correspond to a saving in terms of total structural weight up to 20% about.

ACKNOWLEDGEMENTS

This work has been supported with Research Grants MURST 40% 1997 and MURST 60% 1999.

REFERENCES

Ballio, G., Plumier, A. & Thunus, B (1990). The Influence of Concrete on the Cyclic Behaviour of Composite Joints, IABSE Symposium, Brussels.

Chen, S.-J., Chu, J.M., & Chou, Z.L. (1997). Dynamic Behaviour of Steel Frames with Beam Flanges Shaved Around Connection. *Journal of Constructional Steel Research*, Vol 42, No.1, pp.49-70.

Faella, C., Montuori, R., Piluso, V., Rizzano, G. (1998): Failure Mode Control: Economy of Semi-Rigid Frames, XI European Conference on Earthquake Engineering, Paris 6-13 September,1998.

Faella, C., Piluso, V., Rizzano, G. (1999). Structural Steel Semirigid Connections. CRC Press, Boca Raton, FL.

Ivankiw, N.R., Carter, C.J. (1998): Improved Ductility in Seismic Steel Moment Frames with Dogbone Connections, Journal of Constructional Steel Research, Vol 46:1-3, Paper No 253.

Mazzolani, F.M. & Piluso, V. (1996). *Theory and Design of Seismic Resistant Steel Frames*. London, New York, E & FN Spon, an Imprint of Chapman & Hall.

Mazzolani, F.M. & Piluso, V. (1997). Plastic Design of Seismic Resistant Steel Frames. *Earthquake Engineering and Structural Dynamics*, Vol. 26, pp.167-191.

Montuori, R (1997). Seismic-Resistant Semirigid Steel Frames: Design for Failure Mode Control (in italian). Thesis presented for obtaining the Civil Engineer Degree. Tutors: Faella, C., Piluso, V. & Rizzano, G. University of Salerno, Italy.

Galdi, M. (1999). Failure Mode Control of Seismic-Resistant Steel Frames with "Dog-bone" Connections (in italian). Thesis presented for obtaining the Civil Engineer Degree. Tutors: Piluso, V., Faella, C. & Montuori, R., University of Salerno, Italy.

Behaviour of Steel Structures in Seismic Areas, Mazzolani & Tremblay (eds) © 2000 Balkema, Rotterdam, ISBN 90 5809 130 9

Behaviour of MRFs subjected to near-field earthquakes

L. D. Tirca

Politehnica University, Civil Engineering and Architecture Faculty, Timisoara, Romania

ABSTRACT: The paper studies the behaviour of two MRFs subjected to velocity pulse ground motions. The main parameter examined in the numerical tests is the pulse period. Because the second and the third vibration mode strongly interacts with the first one, same irregularities in comparison with the code provisions occur in the bending moment and shear force diagram, displacement shape and distribution loads.

1. INTRODUCTION

Each strong earthquake is basically unique, offering new surprises in the vulnerability of the affected buildings and showing the great complexity of the phenomena. In the recent earthquakes, such Northridge earthquake in 1994 and Kobe earthquake in 1995 a lot of different failure modes which not to be predicted from the past experience, were observed. So, the code design philosophy fall behind these inadequate performances of modern buildings, because the structure damage arose also when both design and detailing have been performed in perfect accordance with the code provisions. Between these collapse modes, the damage of intermediate levels of moment-resisting frames (MRFs) were particularly a rule than an exception. Usually the failure of a MRF, is considered occur concentrated at the first levels due to the resonance of first structure vibration mode with the ground motion vibrations. The occurrence of failure at superior levels suggests that the resonance is produced due to influence of the superior structure vibration modes.

The both Northridge and Kobe earthquakes were recorded in near-source areas where many unusual ground motion characteristics are noted (Gioncu and all, 2000): (i) velocity pulse loading; (ii) important vertical components; (iii) high velocities for both horizontal and vertical components. Between these aspects, the velocity pulse is one of the main characteristics of near - source earthquakes. The ground motions has a distinct low-frequency pulse in acceleration and a pronounced coherent pulse in velocity and displacement (Iwan, 1995). So, it is very clearly that for very short period (< 0.5sec) which correspond with the second vibration mode of steel MRFs, the influence of superior vibration modes may be more important than the first vibration mode. Therefore, the structure failure types shift from the base levels to intermediate or top levels.

The paper deals with the study of MRFs subjected to velocity pulse ground motions. The main parameter of this study is the pulse period of the near-field earthquakes.

2. ANALYSED MRFs

Two MRFs (MRF1 and MRF2) with six stories and one bay were selected for analysis. For beams, IPE profiles are used, while for columns, HEB profiles are choosing. The column sections vary in two steps on the frame high. The MRF1 is an ordinary moment resisting frame, sized without any control of the plastic mechanism, so that story mechanism occurs. The MRF2 are special moment resisting frames at which global mechanism must be the collapse type. The characteristics of analysis frames and their first three fundamental periods are present in Table 1.

Because the Northdridge and Kobe records reveal the presence of relative large low-frequency pulses in the acceleration time-histories and these pulses are evident for the velocity and displacement (Figure1), one artificial pulse velocity seismogram was created. For the frame analysis it was considered one single

Table 1 Characteristics of analysed frames for pulse action

Frame type	Name	Columns		Beams		Periods [sec]			Collapse mechanism
		T1	T2	T1	T2	T_{s1}	T_{s2}	T_{s3}	
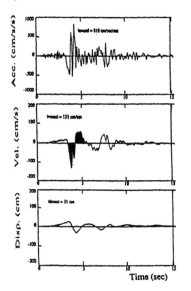	MRF1	HE240B	HE200B	IPE360	IPE300	1.81	0.65	0.40	Local
	MRF2	HE260B	HE220B	IPE360	IPE300	1.72	0.61	0.35	Global

pulse acceleration (Gioncu and al, 2000) with pulse period T_g=0.4 sec, which coincides with the period of the third vibration mode of the analysed frames (Figure 2).

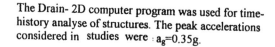

Figure 1 Velocity pulse characteristics of Sylmar record (Northridge)

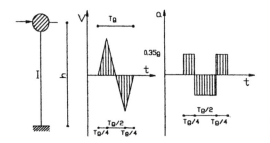

Figure 2 Artificial pulse loading

The Drain- 2D computer program was used for time-history analyse of structures. The peak accelerations considered in studies were : a_g=0.35g.

2.1 Displacements

Due to the short period of the artificial motion and to the pulse characteristics of this loads, the effects of higher vibration modes increases. For structures subjected to pulse actions, the impact propagates through the structure as a wave, causing large localized deformations (Iwan, 1995). Analyzing the time-history displacements, it can observe that till t=0.4sec, the frame movement is forced and after this time, the frame vibration is free. During the forced movement the shape of displacement for the both frames is the same (Figure 3). For the both frames, after the time t=0.4sec. when free vibration begin, the influence of higher vibration modes increases and the concentration of maximum deformations rises up to the frames upper levels.

So, at the six level a large displacement occurs and the first plastic hinge produces at the time t=0.6sec for MRF1 and t=0.66sec. for MRF2.

Between this time and the followings till t=2.0sec, the frame oscillates influenced by the second vibration mode and the maximum values of displacements on the frames level decrease in time due to damper effects.

2.2 Basic shear force

The time-history of shear forces variation for the base and top levels are shown in Figure 4. At the beginning the structures develops a maximum value of the shear forces at the first level, where the deformation shape is like in the first vibration mode. During the forced movement (till t=0.4sec)

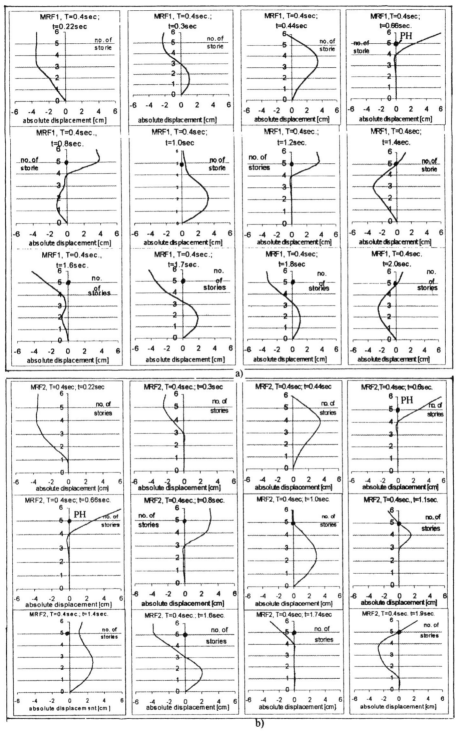

Figure 3 Displacements shapes a) MRF1, b) MRF2

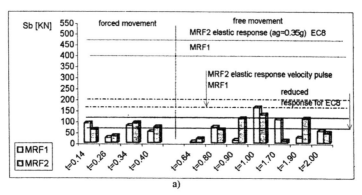

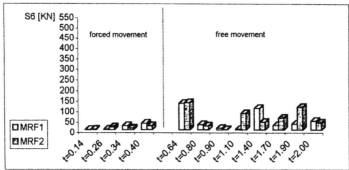

b)

Figure 4 Variation of shear forces; a) base shear forces; b) top shear forces

the value of the shear forces are approximately constant. After this step, the movement is free and the value of the shear forces increase at the top level, where the displacement reaches the maximum value at the time t=0.66sec for MRF1 and t=0.64sec for MRF2. The top shear forces for MRF1 is S_6=128KN and for MRF2 is S_6= 132KN. Till the time t=2.0sec, the maximum value of the shear forces oscillates between the top and the base.

When the base shear forces reach minimum values, the structure oscillates influenced by second or the third vibration mode.

So, for MRF1 at the time t=1.0sec, base shear force is S_1=160KN and for MRF2, S_1=128KN.

The maximum values of shear forces occur during the free movements.

The base shear force for a pulse load is considerable reduced than the elastic value resulted from EC8 prevision (considering the acceleration a_g =0.35g and Tc=0.7sec).

In exchange, one can see that the code behaviour factor q is too large. So, for pulse loading, special studies must be performed for determining proper values for q factors.

2.3 Distribution of loading

The simplified code analysis and the most performed method as the push over analysis consider that the structure is subjected to triangular pattern of horizontal forces. After a time history analysis under a pulse load, the diagram of distribution loading is very different and that load shape is shown in Figure 5. Because the pulse period Tg=0.4sec corresponds with the period of the third structure vibration mode, the distribution of the loading system is influenced by the shape of this mode. The maximum loading, which is in concordance with the maximum displacement, is localized at the top level for the step t=0.6sec, when the first plastic hinge appears. Beside this, the maximum value of distribution loading is correlated with the bending moment diagram (Figure 6). When the bending moment diagram for the columns has not inflection point, the load reaches maximum value at that level.

So, the loading distribution on the high of the structure must be choose in concordance with the effects of the higher vibration modes.

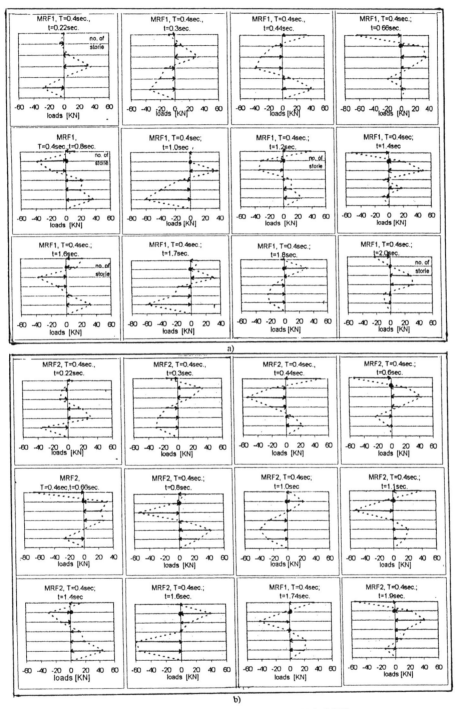

Figure 5 Seismic load distributions a) MRF1, b) MRF2

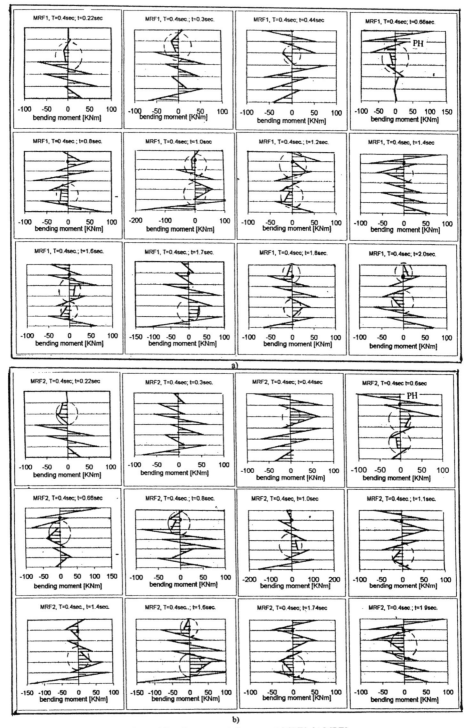

Figure 6 Bending moment diagrams a) MRF1, b) MRF2

2.4 Variation of moments

The bending moment diagrams resulted from a triangular load distribution, presents one inflection point at each level. In this case, the envelope of loading distribution is triangular and is in concordance with the first vibration mode.

Because the second and the third vibration modes interact with the first one, some irregularities occur in this diagram. Usually these perturbations appear at the middle or the top part of the frame height (Figure 6). These levels of the frame can be potential collapsing zones of the structure. The maximum value of the bending moment for the MRFs studied was reached when the value of the base shear forces is maximum. This is happened at the top level for the time t=0.6sec and at the base level for the time 1.6-1.7sec.

2.5 Ductility demands

Due to the pulse characteristics of the actions, developed with great velocity, the ductility demands could be high. Owing to the higher vibration mode, which caused irregularities in the bending moment diagrams, a dramatic reduction of the available ductility is recorded on the middle and the top part of the structure (Tirca and Gioncu, 1999). It is known (Gioncu and Petcu, 1997) that the minimum available ductility reaches, when the moment diagram is one curvature type. So in these critical zones, the value of ductility requirement can exceed the available ductility, and the collapse of the building may occurs at the mid part of structure (see Kobe structure damage).

For MRF1, the first plastic hinge was appeared in the beam of the 5th level, at the time t=0.66sec.

For MRF2, the first plastic hinge appear in the same beam at the time t=0.6sec The rotation of hinge increases till t=0.64sec and reaches the value $\theta= 1.6x10^{-2}$. After this time, till t=2.00sec the value of rotation hinge remain the same.

Because the require ductility increase at the top of structure, a solution can be the constant cross-section of the columns at the height frame.

2.6 Interstory drift

A plastic hinge appears in that level where the value of interstory drift exceed the available value.

Figure 7a shows for MRF1 a comparison between the interstory drift values at the time t=0.4sec (when forced movement stop) and at the time t=0.66sec

when all the studied parameter reach the largest values.

At this time, for a little increasing of action, a joint mechanism will be form. Because the return forces of pulse action are missing, the formed mechanism is in concordance with the structure failure. In this place value of require ductility exceeds the available ductility. One can see that, due to influence of superior vibration modes, the maximum interstorey drift are concentrated at the frame top.

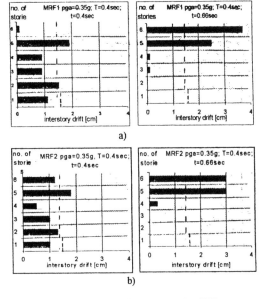

Figure 7 Interstory drift a) MRF1, b) MRF2

3. INFLUENCE OF PULSE LOAD PERIODS

Studying the behaviour of this two frames subjected at seismic actions, with pulse periods varying from 0.3sec to 2.0sec some important remarks can be specified.

For the short pulse period, which are near to second vibration mode, the maximum required ductility is concentrated at the top of structure, while for the pulse periods near to first vibration mode, the maximum ductility demands occurs at the first level. The required ductilities for beam and columns, resulting from the second vibration mode, are more reduced than the ones obtained from the first vibration mode (Figure 8). For MRF2, which was designed to formed plastic mechanism, the structure behaves elastically for pulse periods smaller than 0.3sec. For 0.4sec, the first plastic hinge occur at the frame top. With the increasing of the impulse period, the formation of plastic hinge moves dawn, till Tg=0.7sec. when two plastic mechanisms appear due to the second vibration mode. With the increasing of pulse period the collapse mode is changing.

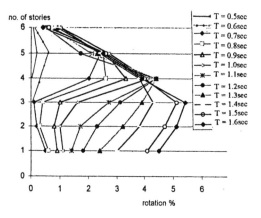

Figure 8 The require ductility for different pulse period

The variation of the base shear forces is present in Figure 9. The influence of vibration modes is evidently for MRF1 and for pulse period Tg=0.5÷0.8sec. In this case due to the fact that the structure period for second vibration mode corresponds to the pulse period, the frame collapses due to the resonance effect.

MRF1c is a frame with the same characteristics that MRF1, but has the cross-section of columns constant. In this case the collapse of structure is avoid.

For buildings, EC3 and EC8 recommend that the horizontal deflections as a structure as a whole does not exceed 0.2% H (H being the height of the structure). Considering this recommendation as a limit for operational level of seismic loads, the live safe level is given by the 0.4%H. Excepting the frame MRF1, the value of damage index defined as the ratio De/H% increases with the pulse period value (Figure 10).

4. CONCLUSIONS

The actual code provisions for regularly buildings are based mainly on the effect of first vibration mode. But for near-field earthquakes, characterized by velocity and displacements pulses, the influence of superior vibration modes is essentially, the displacements, load distributions, moment diagrams, ductility demands, interstory drifts being very different from the results obtained using code provisions. So, it is very important that these new aspects related to the effects of near-field earthquakes to be introduced in the last actions of code revisions.

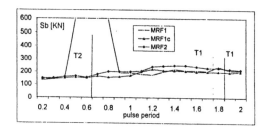

Figure 9 The variation of base shear forces for different periods of pulse loads

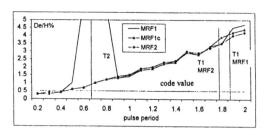

Figure 10 The variation of damage index for different periods of pulse loads

REFERENCES

Gioncu and Petcu, 1997. Available rotation capacity of wide-flange beams and beam-columns. *J. Construct. Steel Research*, Vol. 43, pp.161-217

Gioncu V., G. Mateescu, L. Tirca & A. Anastasiadis 2000. Influence of the type of seismic ground motions. *Moment resistant connections of steel building frames in seismic areas*. Ed. F.M. Mazzolani, E &FN Spon, London.

Iwan W.D. 1995. Drift demands spectra for selected Northdrige sites. *Technical report SAC 95-05*, 2.1-2.40

Tirca L.,V. Gioncu 1999. Ductility demands for MRFs and LL-EBFs for different earthquake types. *Stability and Ductility of Steel Structures*. Ed. D. Dubina & M. Ivany, Elsevier, 429-438

Behaviour of Steel Structures in Seismic Areas, Mazzolani & Tremblay (eds) © 2000 Balkema, Rotterdam, ISBN 90 5809 130 9

Evaluation of the response of moment frames in respect to various performance criteria

A. I. Vayas
Department of Civil Engineering, National Technical University of Athens, Greece

B. F. Dinu
Romanian Academy, Timisoara Branch, Romania

ABSTRACT: Criteria for the evaluation of the response of moment resisting steel frames subjected to earthquakes are proposed. The criteria are used for the definition of the ultimate state of the structure. They relate to serviceability and ductility requirements. A parametric study on the response of several frames is performed. The parameters under consideration are the frame geometry, the level of vertical loading and the type of seismic record. It may be seen that the current serviceability criteria set up in the seismic Codes are governing the design, as long as minimum ductility requirements for the members and the connections are met.

1 INTRODUCTION

Building frames are designed for stiffness and strength to comply with the criteria set up in the serviceability and ultimate limit states. For structures subjected to earthquakes, the design criteria have to be extended due to the fact that during strong earthquakes yielding is allowed to occur so that part of the energy input may be dissipated through inelastic deformations. The extension refers to the requirement for provision of ductility, which correspond to the capacity of certain parts of the structure to undergo under cyclic loading conditions large inelastic deformations without considerable reduction in stiffness and strength. However, the occurrence of large inelastic deformations is associated to a certain damage of the relevant structural elements, so that low-cycle fatigue may become critical for design. The criteria in respect to stiffness, strength, ductility and low-cycle fatigue resistance may be complementary in design. It is therefore to be examined if one of those is prevailing according to the rules of existing seismic Codes, under what conditions and to what extend.

In the present study, four limit state criteria for moment resisting steel frames are introduced. They relate to allowable drift limitations, allowable residual drift limitations, maximum rotation capacities of members or connections and low-cycle fatigue resistance of structural parts. After an introduction of the selected performance criteria, a parametric study on the non-linear response of several moment resisting steel frames under various types of seismic records is performed. The maximal accelerations un-

der which each one of those criteria is satisfied at the limit are determined. The comparison between the limit accelerations indicates which criterion is most unfavourable for each specific case of frame geometry and seismic record under consideration. This allows for conclusions in respect to the structural response and the Code provisions governing design.

2 PERFORMANCE CRITERIA

As outlined before, structures have to comply with several requirements and associated design criteria. Frames subjected to strong earthquakes are considered to behave satisfactory if during the seismic event the criteria are met. Alternatively, it is possible to evaluate the maximal acceleration of a specific seismic record, for which a specific frame under certain loading conditions meets at the limit the relevant performance criteria. Such criteria are either introduced in seismic design Codes or used during building inspections after an earthquake event. In the following, four performance criteria associated with relevant limit states are proposed for use for moment resisting steel frames.

2.1 Stiffness criterion (A)

The limit state is defined as the situation where the interstorey drift exceeds 2% of the relevant storey height.

Compliance with the above criterion roughly assures that the serviceability criteria proposed in seismic design Codes are met. For example, Euro-code 8 requires a drift limitation under serviceability conditions (moderate earthquakes) between 0.4% and 0.6% in order to limit damage in nonstructural elements. The lower value is valid for brittle non-structural elements, the higher value for flexible elements. This Code does not explicitly include characteristics for a moderate earthquake but relies instead on processing of the results of the analysis for the strong design earthquake. Accordingly, the lateral deformations for the moderate earthquake are determined by division of the corresponding defor-mations calculated for the strong record with a cer-tain factor that expresses the ratio of the peak ground accelerations between the two records. The value of this factor may be taken equal to 2.5 as proposed in the Code. The application of this factor on the above referred drift limits leads to drift limitations under a strong earthquake between 1% (=0.4% x 2.5) and 1.5%. The adopted limit value of 2% proposed be-fore covers accordingly with some safety the Code requirements.

2.2 Strength criterion (B)

The limit state is defined as the situation where the residual non-recoverable part of the interstorey drift exceeds 1% of the relevant storey height.

The residual permanent drift is generally accepted as a criterion for the assessment of a building's con-dition as it provides an indication of the structural damage during inspections after an earthquake. Based on the residual drift, the structural damage may be estimated as low, moderate or high. Al-though it is difficult to prescribe precise limit values, some figures are proposed and applied in the prac-tice (Ohi, Takanashi 1998). For a residual drift ex-ceeding roughly 3%, it is supposed that the structural damage is so heavy that the building must be de-molished. For values of the residual drift over 1%, the damage is considered to be moderate to heavy so that it is possible for the building to be repaired. This criterion is characterized as strength criterion due to the fact that if the structural strength is not sufficient, large inelastic deformations will occur that will po-tentially lead to unacceptably high residual drifts.

2.3 Ductility criterion (C)

The limit state is defined as the situation where the rotation at any developing plastic hinge becomes larger than 0.03 rad.

The above limit value of rotation is supposed to correspond to an exhaustion of the rotation capacity at beam-to-column joints and connections. In that sense this criterion expresses the limits imposed by local ductility. The proposed value of 0.03 radians

corresponds to the relevant requirements of the re-cent AISC-Code (1997) for special moment frames.

2.4 Low-cycle fatigue criterion (D)

The limit state is defined as the situation in which the low-cycle fatigue resistance at a structural ele-ment has been exhausted.

The low-cycle fatigue resistance curve is ex-pressed in terms of the plastic rotation according to:

$$\log N = \log a - m \log \Delta\varphi \qquad (2.1)$$

where $\Delta\varphi$ = fatigue deformability (plastic rotation); N = number of plastic rotation range cycles; m = slope of the fatigue plastic rotation range curves; and $\log a$ = a constant.

Reference values for the damage evaluation may be provided from the results of monotonic tests. Monotonic loading corresponds to ½ of a cycle of a specimen deformed to up to $\Delta\varphi = \varphi_{mon}$ and unloaded to zero plastic deformation. That means that mono-tonic loading corresponds to a pair $N_{mon}=1/2$ and $\Delta\varphi = \varphi_{mon}$. Accordingly, if the rotation capacity un-der monotonic loading φ_{mon} is known, the number of cycles for a certain range of plastic rotation may be determined from:

$$N = \frac{1}{2}(\frac{\varphi_{mon}}{\Delta\varphi_p})^m \qquad (2.2)$$

Studies from experimental investigations reveal that the value of the slope m is approximately equal to 2. Accordingly a value m = 2 was adopted for the fatigue curve. For the rotation capacity under monotonic loading φ_{mon}, a value equal to 0.05 radi-ans, higher as the corresponding value for cyclic loading, was adopted for the fatigue curve.

For variable ranges of plastic rotation, the damage assessment is performed in accordance to the linear Palmgren-Miner cumulative law in accordance to:

$$N = \frac{1}{2}(\frac{\varphi_{mon}}{\Delta\varphi_p})^m \qquad (2.2)$$

Studies from experimental investigations reveal that the value of the slope m is approximately equal to 2. Accordingly a value m = 2 was adopted for the fa-tigue curve. For the rotation capacity under mono-tonic loading φ_{mon}, a value equal to 0.05 radians, higher as the corresponding value for cyclic loading, was adopted for the fatigue curve.

For variable ranges of plastic rotation, the damage assessment is performed in accordance to the linear Palmgren-Miner cumulative law in accordance to:

$$D = \Sigma \frac{n_i}{N_i} \qquad (2.3)$$

where n_i = number of cycles of deformation range $\Delta\varphi_i$; and N_i = number of cycles of the same deformation range that cause failure.

For the determination of the design spectrum in the fatigue assessment, the *rainflow* or *reservoir* method for counting the cycles for a certain deformation history has been employed.

3 NONLINEAR ANALYSIS

In order to study moment resisting frames by means of nonlinear dynamic analysis, the general purpose DRAIN-2DX software package was used. In respect to earthquakes, two records from Aigion (1985) and Thessaloniki (1978), Greece and one from Kobe (1995), Japan has been considered. Their acceleration spectra scaled to the peak ground acceleration are shown in Fig. 1. The Figure shows also the design spectrum of the Greek seismic Code for moment resisting steel frames, adopting a value of the behaviour factor q = 4.

The geometric and other properties of the frames under investigation are presented in Table 1. They range from two- to eight-storey, one to three bays regular frames. The beam-to-column joints are considered as rigid, the level of vertical loading is varied between 40% and 60%.

This level represents the portion of the moment capacity of the beams that is consumed in resisting vertical loads, the rest remaining to resist seismic actions.

The frames cover a wide range of frequencies, the relevant fundamental periods range from 0.6 to 1.9 sec as shown in Table 1.

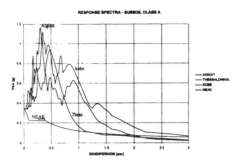

Figure 1: Acceleration response spectra of the records under consideration and design Code spectrum

For all frames, the yield and the ultimate acceleration α_y, respectively α_u, were determined by appropriate scaling of the relevant records. The yield acceleration corresponds to the appearance of the first plastic hinge, the ultimate acceleration to the approach of the limit state in accordance to one of the definitions described above.

4 RESULTS

The yield acceleration α_y for two levels of vertical loading is presented in Figure 2. It may be seen that both the type of record and the frame characteristics influence this acceleration. The Kobe record with the highest response values over a wide range of periods provides generally the lowest values of accelerations. On the contrary, the Aigion record whose peak

Table1. Data of investigated frames

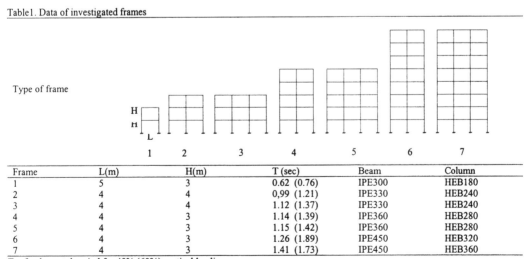

Frame	L(m)	H(m)	T (sec)	Beam	Column
1	5	3	0.62 (0.76)	IPE300	HEB180
2	4	4	0,99 (1.21)	IPE330	HEB240
3	4	4	1.12 (1.37)	IPE330	HEB240
4	4	3	1.14 (1.39)	IPE360	HEB280
5	4	3	1.15 (1.42)	IPE360	HEB280
6	4	3	1.26 (1.89)	IPE450	HEB320
7	4	3	1.41 (1.73)	IPE450	HEB360

T = fundamental period for 40% (60%) vertical loading
Yield stress for beams and columns f_y = 235 Mpa

response values concentrate over a narrow frequency range provides the highest yield accelerations. It is interesting to observe the low value of α_y for the frame 1 whose fundamental period is near the peak spectral acceleration value for this record. The yield acceleration is also dependent on the level of vertical loading, higher levels of vertical loading producing as expected lower acceleration values, as the structural reserves for resisting seismic loading are lower in this case.

The values of the ultimate accelerations in accordance to the previously defined limit states criteria are illustrated in Figures 3 to 5.

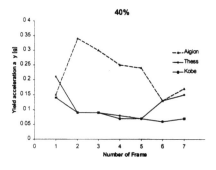

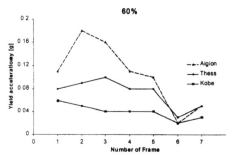

Figure 2: Yield acceleration a) for 40% and b) for 60% of vertical loading

By examining the results from Figures 3 to 5, following observations may be made:

a) Like the yield accelerations, the ultimate accelerations are strongly dependent on the characteristics of the structure and the type of record. For the frames investigated here, the Aigion record provides generally the highest acceleration values, almost twice as large as the Kobe record.

b) The serviceability criterion A provides for the Aigion record considerably lower values of α_u, compared to the other three criteria (Figure 3a). This is due to the fact that Aigion constitutes a shock-type earthquake, where the energy input occurs during a short time. The resulting deformations in this time may be high, but the structure is afterwards more or less oscillating freely as the ground accelerations are becoming small.

This results in low requirements for strength, ductility and low-cycle fatigue resistance. The other three criteria provide similar values of the limit acceleration.

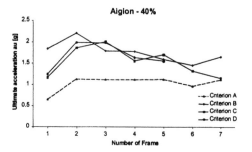

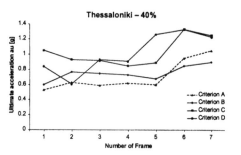

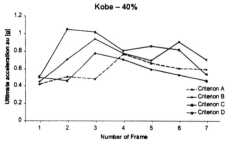

Figure 3: Ultimate accelerations for 40% level of vertical loading for a) Aigion, b) Thessaloniki and c) Kobe records

c) For the Thessaloniki record, the serviceability criterion A provides also generally lower limit acceleration values (Figure 3b). However, the distance to the limit values of the other criteria is not so high. This is due to the fact that the duration of this record is higher, imposing more cycles to the structure. Comparing the limit acceleration for the ductility and low-cycle fatigue criteria C and respectively D, it is interesting to observe that frames 1 and 2 are critical in low-cycle fatigue, frames 4 and 5 in ductility and frames 3, 6 and 7 in both. The strength criterion is critical for the highest period frames 6 and 7.

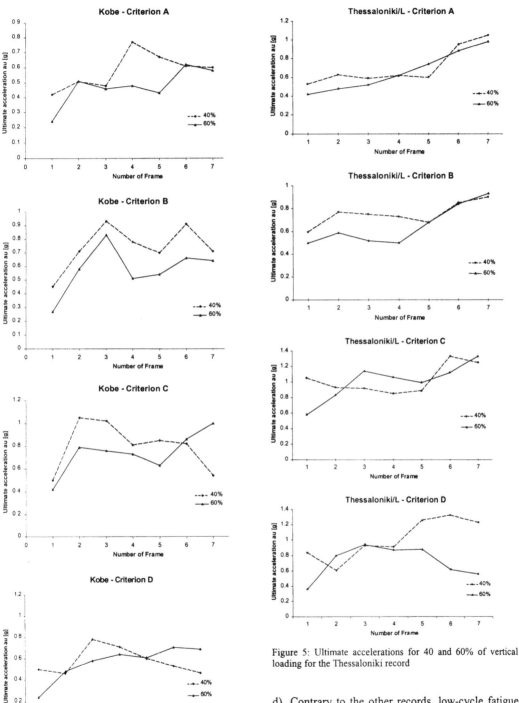

Figure 4: Ultimate accelerations for 40 and 60% of vertical loading for the Kobe record

Figure 5: Ultimate accelerations for 40 and 60% of vertical loading for the Thessaloniki record

d) Contrary to the other records, low-cycle fatigue appears to be the most critical, with one exception, for the Kobe earthquake (Figure 3c). This is confirmed by the actual behaviour of steel frames during the Kobe earthquake, where fatigue cracks were observed in beam-to-column joints. The fact that ductility provides larger critical values shows that many large magnitude deformation cycles have occurred during the Kobe earthquake, each contributing considerably to the structural damage. Stiffness and strength requirements were less severe for this record.

e) Figure 4 shows that for the Kobe record the limit accelerations are generally lower for higher levels of vertical loading. However, it is interesting to observe that the 8-storey frames 6 and 7 are more critical for lower levels of vertical loading, especially in respect to their low-cycle fatigue resistance.

f) Figure 5 shows that for the Thessaloniki record, the level of vertical loading appears not to have a considerable influence on the limit accelerations. The limit acceleration curves are generally very similar for all frames and performance criteria. An exception constitutes the low-cycle fatigue resistance, criterion D, where the higher frames are much more susceptible for higher vertical loading.

g) The above suggest, that if the normal ductility requirements are met, the decisive design criterion for moment resisting frames will be normally the drift limitation. This may be confirmed by practical design applications applying the current Eurocode 8 provisions. Low-cycle fatigue may become critical for certain types of earthquakes, like Kobe, that impose many loading cycles to the structure.

The ratio between the ultimate and the yield acceleration defines, according to Eurocode 8, the value of the behaviour factor. The mean value of this ratio for the three records ranges between 4.5 and 23 as shown in Table 3. It may be seen that the ratio increases with increasing height of the building and that criterion A provides the lowest values. However, as this factor is a ratio and not an absolute value, it does not necessarily express the capacity of the structure to resist higher seismic actions. This capacity depends also on the value of the yield acceleration that is highly influenced, as we have seen, by the type of record and structure under consideration.

Table 3. Mean value of the ratio between ultimate and yield acceleration (q-factor) for three records and two levels of vertical loading

Frame	1	2	3	4	5	6	7
A	4.50	6.40	6.30	8.60	8.10	16.80	13.00
B	7.20	8.50	9.70	9.50	9.40	18.2	14.10
C	6.90	10.50	10.20	11.80	11.20	22.70	18.00
D	5.55	8.91	8.90	11.01	9.16	17.90	9.64

5 CONCLUSIONS

Based on stiffness, strength, ductility and low-cycle fatigue resistance criteria, critical limit accelerations of several moment resisting frames subjected to various seismic records were determined. The results are strongly influenced by the type of frame and the type of record under consideration. However, it may be seen that stiffness criteria, expressed by limit values of interstorey drifts as proposed by seismic Codes are usually governing the design. Low-cycle fatigue criteria may be decisive for records of long duration with considerable energy input over a large range of structural frequencies.

REFERENCES

Eurocode 8 Design provisions for earthquake resistance of structures. CEN, European Committee for Standardization, ENV 1998-1-1, 1994

Kannan, A., Powel, G.: DRAIN-2D. A general purpose computer program for dynamic analysis of inelastic plane structures, EERC 73-6 and EERC 73-22 reports, Berkeley, USA, 1975

Ohi, K., Takanashi, K.: Seismic diagnosis for rehabilitation and upgrading of steel gymnasiums. *Engineering structures, 20,* No 4-6, 533-539, 1998

Seismic Provisions for Structural Steel Buildings, American Institute of Steel Construction, Chicago, Illinois, 1997

Vayas I. Interaction between local and global ductility. in F. Mazzolani (ed.), Moment resisting connections of steel building frames in seismic areas, E & FN SPON, 2000

Behaviour of Steel Structures in Seismic Areas, Mazzolani & Tremblay (eds) © 2000 Balkema, Rotterdam, ISBN 90 5809 130 9

Low-cycle fatigue behaviour of moment resisting frames

I. Vayas & A. Spiliopoulos
Department of Civil Engineering, National Technical University of Athens, Greece

ABSTRACT: The low-cycle fatigue behaviour of moment resisting steel frames is investigated. Damage is attributed to plastic rotations at members or joints. Fatigue curves are expressed by the relevant rotation capacity under monotonic loading and the value of the slope. A linear damage accumulation low and the rainflow method for counting of cycles are applied. Parametric studies for several frames subjected to various earthquake records are performed. The influence of a series of factors like the type of frame, the type of record, semi-rigidity of joints, level of vertical loading, slope of the fatigue curves all of which significantly affect the fatigue behaviour was investigated.

1 INTRODUCTION

It is generally accepted that structures subjected to strong earthquakes are allowed to yield in order to dissipate part of the energy input through inelastic deformations. For moment resisting frames, yielding is usually associated with the formation of plastic hinges developing at various positions of the structure. Depending on the type and the duration of the strong motion as well as on the characteristics of the structure, plastic hinges of variable rotation amplitudes may form one or several times at certain sections. The appearance of inelastic deformations results in structural damage that, as the number of cycles is low but the deformation amplitudes are high, may be evaluated by means of low-cycle fatigue procedures. Damages reported during the recent earthquakes in Northridge and Kobe where cracks in the regions of beam-to-column joints were observed indicate that fatigue resistance may indeed be a significant design criterion for steel building frames.

This paper presents a low-cycle fatigue procedure for the evaluation of the structural damage of steel moment frames. The fatigue curves are expressed in terms of the developing inelastic rotations. The response of several moment resisting frames by means of inelastic dynamic analysis is studied. The main parameters under consideration are the type of frame, the type of strong motion, the level of vertical loading, the flexibility of joints and the low-cycle fatigue resistance at members and joints.

2. FORMULATION OF LOW-CYCLE FATIGUE RULES

Steel structures subjected to dynamic loading are susceptible to fatigue. The evaluation in respect to fatigue is expressed by means of fatigue curves that provide a generalized "strength" as a function of the applied number of cycles. The selection of the appropriate type of fatigue curve depends on the characteristics of both the dynamic loading and the structural response.

When the structural response is elastic, the fatigue curves are expressed in terms of stresses. The fatigue strength represents the stress range $\Delta\sigma$ that may be resisted under a certain number of applied cycles. The number of cycles to failure is in such cases high. Structures prone to (high-cycle) fatigue are those frequently subjected to dynamic loading such as bridges, crane girders etc. loaded by heavy traffic or slender structures like chimneys, masts etc. excited by wind.

When the structural response is inelastic, stresses are substituted by deformations. The fatigue curves provide then the fatigue deformability, i.e. the range of deformation Δv, as a function of the applied cycles. The fatigue curves may include either total or only inelastic deformations, depending on the extent of the latter.

Structures in seismic regions are designed such that they exhibit large inelastic deformations during strong earthquakes. In such an event elastic deformations are generally small compared to the total

Table 1. Formulation of fatigue rules

Structural response	Elastic	Inelastic	
No of cycles to failure	$\sim 10^4 - 10^8$	$\sim 10^2 - 10^4$	$\sim 10^0 - 10^2$
Fatigue curves for ranges of	Stresses	Total deformations	Inelastic deformations
Usual fields of application	Bridges, crane girders, chimneys, masts etc.	Slender plate girders, tanks etc.	Seismic loading

deformations, so that only the inelastic part may be accounted for in the fatigue curves. Damage is therefore associated only with inelastic deformations, as predicted for example by the Japanese design practice. The above formulation constitutes a low-cycle fatigue problem, as the number of cycles to failure is small.

However when elastic and inelastic deformations are of similar order of magnitude, the fatigue evaluation should be based on total deformations. Such formulation may be applied e.g. in cases of post-critically designed slender structures, where under service loads the response may be inelastic.

The number of cycles to failure is in such cases somehow intermediate between high- and low-cycle fatigue. Example of such a response constitutes the problem of web breathing of slender plated girders, where secondary local bending due to the initiation of a tension field formation may lead to inelastic local stress and strain concentrations.

Table 1 summarizes the various possible formulations of fatigue rules as outlined above. Alternative formulations for low-cycle fatigue problems have also been proposed, e.g. by introduction of equivalent stresses that correspond to the total deformations if the response would be supposed to be elastic (Bernuzzi et. al. 1997).

3. LOW-CYCLE FATIGUE RULE FOR MOMENT FRAMES

In moment resisting frames, plastic hinges develop in structural members during strong earthquakes. The resulting inelastic rotations are accommodated by the ductility of the corresponding members. Ductility is understood in this case as the capacity of the member to develop plastic rotations without considerable reduction in stiffness and strength. The low-cycle fatigue curve expresses the amount of inelastic rotation that may be sustained under a certain number of applied cycles and may be written as:

$$\log N = \log a - m \log \Delta \varphi \qquad (3.1)$$

where $\Delta \varphi$ = fatigue deformability (plastic rotation); N = number of plastic rotation range cycles; m = slope of the fatigue plastic rotation range curves; and $\log a$ = a constant.

The values of the above referred parameters m and a may be determined experimentally by constant amplitude fatigue tests. Approximate fatigue curves may be derived once the rotation capacity under monotonic loading is known and the failure modes from monotonic and cyclic loading are the same (Vayas et al 1999). Monotonic loading corresponds to application of ½-cycle and a rotation amplitude equal to the rotation capacity φ_{mon}. Accordingly, the number of cycles to failure N for an applied range of rotation $\Delta \varphi_p$ may be determined from:

$$N = \frac{1}{2} (\frac{\varphi_{mon}}{\Delta \varphi_p})^m \qquad (3.2)$$

Experimental investigations indicate that if only inelastic deformations are accounted for, the value of the slope m is not differing very much from 2. A simple algebraic summation of the inelastic deformations, as foreseen by the Japanese practice (Akiyama 1985), corresponds to $m = 1$.

The damage assessment for variable ranges of plastic rotation is performed as for high-cycle fatigue in accordance to the linear Palmgren-Miner cumulative law:

$$D = \Sigma \frac{n_i}{N_i} \qquad (3.3)$$

where n_i = number of cycles for an applied range of plastic rotation $\Delta \varphi_i$; and N_i = number of cycles to failure for the same range of rotations.

For the determination of the design spectrum in the fatigue assessment, the *rainflow* or *reservoir* method for counting the cycles for a certain deformation history may be employed.

4. PARAMETRIC STUDIES

The fatigue behaviour of various moment resisting frames shown in Table 2 was studied. They range from three to eight storey, two-bay regular frames. Both rigid and semi-rigid beam-to-column joints are considered. Semi-rigidity in joints is expressed by means of the parameter $K=25$ EI_b/L_b as proposed by Eurocode 3 and shown in Table 2. Two levels of vertical loading, 40% and 60%, expressing the exploitation of the beam bending resistance due to vertical loading, are used. The rotation capacity under monotonic loading φ_{mon}, employed in the fatigue curves, is supposed to be 0.05 radians. The slope of the fatigue curve m is considered as a pa-

Table 2. Data of investigated frames

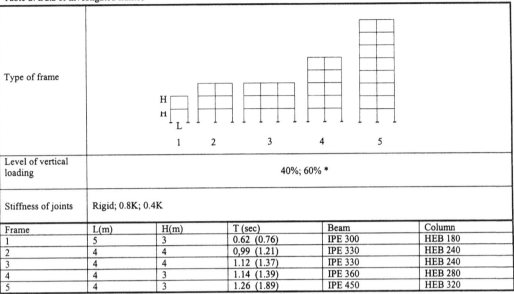

Type of frame					
Level of vertical loading	40%; 60% *				
Stiffness of joints	Rigid; 0.8K; 0.4K				

Frame	L(m)	H(m)	T (sec)	Beam	Column
1	5	3	0.62 (0.76)	IPE 300	HEB 180
2	4	4	0,99 (1.21)	IPE 330	HEB 240
3	4	4	1.12 (1.37)	IPE 330	HEB 240
4	4	3	1.14 (1.39)	IPE 360	HEB 280
5	4	3	1.26 (1.89)	IPE 450	HEB 320

* Percentage of the beam moment resistance exploited for vertical loading
EI_b = stiffness of the connected beam; L_b = length of the connected beam ; $K=25\ EI_b/L_b$
T = fundamental period for 40% (60%) vertical loading and rigid joints
Yield stress of members f_y = 235 Mpa

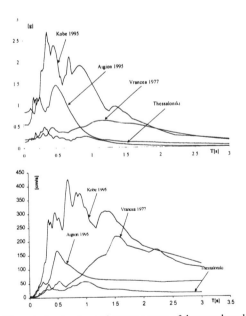

Figure 1: Acceleration and energy spectra of the records under consideration

rameter, taking values 1, 2 and 3. Nonlinear dynamic analysis was employed, by application of the general purpose software program DRAIN-2DX.

Records from earthquakes in Greece (Aigion 1985, Thessaloniki 1978), Japan (Kobe 1995) and Romania (Vrancea 1977) were considered. Their acceleration and input energy spectra are shown in Figure 1.

5. RESULTS

The frames were subjected to non-scaled records from earthquakes in Aigion, Kobe and Vrancea. For each frame, the damage index at the positions where plastic hinges appear was evaluated. Figure 2 illustrates results for two frames. It may be seen that plastic hinges appear not only directly at the nodes, but also at any position along the members.

The highest values of the damage index in the beams and the columns of the frames are summarized in Table 3. Values larger than 1 correspond to failure at the relevant section. Some frame collapsed during the seismic motion as no numerical convergence was achieved. For those frames no damage values could be determined. What the columns is concerned, the largest damage was always produced at their bases. Table 3 indicates that the Kobe record produces the highest damage.

The damage caused by the Vrancea record is larger than the damage of the Aigion record, although its PGA is smaller. This is due to the fact that the acceleration and energy spectral values of Vrancea are higher than the corresponding ones of Aigion.

Table 3. Data of investigated frames

F r a m e	vert loa din g	join t stif fne	Aigion (GR) PGA=0.54g						Kobe (JN) PGA=0.85g						Vrancea (RO) PGA=0.21g					
			beams			columns			beams			columns			beams			columns		
			slope m			slope m			slope m			slope m			slope m			slope m		
			1	2	3	1	2	3	1	2	3	1	2	3	1	2	3	1	2	3
2	40 %	0.4	.09	.01	0	.14	.01	0	3.4	8.6	25.	3.5	2.3	2.2	2.3	1.8	2.4	2.9	2.7	3.0
		0.8	.12	.01	0	.08	0	0	3.1	9.6	30.	3.6	2.4	2.5	2.4	1.5	1.4	2.4	2.3	2.6
		rig	.18	.03	.01	.05	0	0	3.4	11.	39.	3.2	2.6	2.6	2.4	1.6	1.4	2.2	2.1	2.1
	60 %	0.4	.31	.09	.03	.73	.30	.12	2.9	7.5	21.	3.2	3.2	4.1	1.7	2.8	4.8	5.6	4.3	4.3
		0.8	.34	.07	.04	.96	.26	.08	3.1	7.8	22.	3.3	3.7	5.2	2.0	3.8	7.4	5.2	4.0	4.0
		rig	.37	.09	.05	.44	.10	.02	5.7	11.	22.	3.6	4.2	6.3	2.2	4.8	10.	5.0	3.8	3.6
4	40 %	0.4	.29	.07	.02	1.0	.31	.10	3.5	2.5	3.5	3.3	2.6	2.6	2.2	1.6	1.4	5.4	4.4	4.5
		0.8	.31	.07	.02	.86	.21	.05	2.3	2.2	2.2	3.5	3.4	4.2	2.3	1.7	1.5	5.0	4.1	4.1
		rig	.38	.07	.02	.66	.12	.02	2.7	2.9	3.5	3.8	4.2	5.9	4.6	3.5	3.1	4.4	3.4	3.2
	60 %	0.4	.28	.08	.01	1.3	.63	.34	3.3	3.3	6.5	6.0	4.7	5.4	-	-	-	-	-	-
		0.8	.29	.09	.02	1.4	.64	.35	3.4	4.2	7.8	7.9	6.8	7.5	-	-	-	-	-	-
		rig	.30	.09	.02	1.3	.61	.32	3.5	5.3	12.	5.6	5.7	9.7	-	-	-	-	-	-
6	40 %	0.4	.15	.02	0	1.2	.53	.28	-	-	-	-	-	-	1.8	1.0	1.0	3.9	3.1	3.3
		0.8	.23	.03	0	1.1	.51	.27	-	-	-	-	-	-	1.8	2.4	3.8	4.4	5.7	8.7
		rig	.28	.04	.01	.98	.49	.26	-	-	-	-	-	-	2.2	4.8	10.	7.1	10.	20
	60 %	0.4	.18	.03	.01	3.6	1.1	.60	-	-	-	-	-	-	-	-	-	-	-	-
		0.8	.23	.05	.01	3.0	.79	.41	-	-	-	-	-	-	-	-	-	-	-	-
		rig	.26	.07	.02	3.0	.82	.47	-	-	-	-	-	-	-	-	-	-	-	-

- = structural collapse

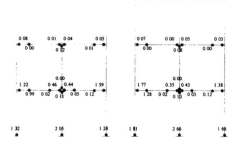

Figure 2: Damage index at the members of a 3-storey frame

The influence of semi-rigid joints is illustrated in Figure 3. Figure 3a presents the mean values of the damage index in beams and columns for frames with semi-rigid joints related to the corresponding values for frames with rigid joints. The joint rigidity is expressed by the parameter αK, as indicated in Table 2. For rigid joints the parameter α was set equal to 1.2, as proposed in Eurocode 3. It may be seen that semi-rigidity leads generally to less damage in the beams and higher damage in the columns. Flexible connections may affect positively the structural behaviour as they prevent high moment concentrations around the beam-to-column joints. However, with increasing flexibility the frame action gradually reduces stressing more the columns and especially their bases.

Figures 3b and 3c present the same quantities as before, differentiated in respect to the records and correspondingly to the types of frames under conside-ration.

The influence of the joint flexibility is low for the Kobe record and high for the Aigion record. In the latter case, more flexible joints result in an increase in column damage and a decrease in beam damage compared to rigid joints. However, considering that the beam damage for the Aigion record is lower than the column damage in case of rigid joints (Figure 3d), it may be concluded that semi-rigidity is beneficial in this case as the damage is more evenly distributed among the members. The influence of joint flexibility is also dependent on the type of frame as indicated in Figure 3c.

The influence of the level of vertical loading is illustrated in Figure 4. Figure 4a presents the mean values of the damage index in beams and columns related to the values for 60% vertical loading. It may be seen that higher vertical loading results in an increase in damage. However there exists a differentiation in respect to the records as shown in Figure 4b. Here again the lowest influence is for the Kobe record and the highest for the Aigion record. This may be explained by the fact that the damage due to the Kobe record is so high that it is insensitive to small variations of the frame properties. The frames should be therefore thoroughly redesigned to withstand the Kobe earthquake.

The influence of the slope of the fatigue curve is illustrated in Figure 5. Only results for the Aigion record are presented due to the fact that the other records produce higher damage index than 1, at which the relevant sections are supposed to fail. The curves present values for rigid and semi-rigid joints with stiffness 0.8K. It is reminded that m=1 corresponds to a simple addition of the positive and negative plastic rotations. The Figure shows that the damage

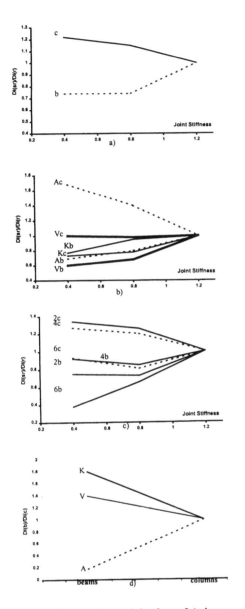

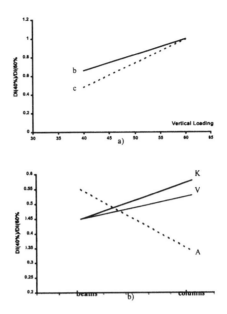

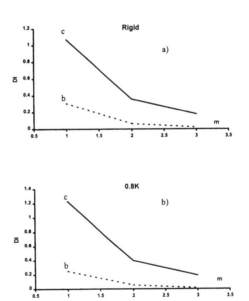

Figure 4: Relative damage index for m=2 in beams and columns vs. level of vertical loading a) mean value, b) dependence on the record

Figure 3: Relative damage index for m=2 in beams and columns vs. joint rigidity a) mean value, dependence b) on the record, c) on the frame d) on the member type for rigid joints

Figure 5: Damage index for slopes m = 1, 2 and 3 in beams and columns a) rigid, b) semi-rigid joints

for higher slopes of the fatigue curves is significantly lower. The reduction in the damage index depends on the number and the value of the occurring plastic rotations. A large numbers of low plastic rotation result in large reduction, while a small number of large rotations a low reduction. Figure 5 indicates that the first happens in the columns and the second in the beams. The difference between rigid and semi-rigid joints is not significant. The Figure indicates also that analyses in which the plastic rotations are simply added are conservative.

653

Table 4. PGA for no damage and complete damage and damage for non-scaled records

Frame	Vertical loading	Aigion (GR)			Kobe (JN)		
		D = 0 A	D = 1 A	A = 0.54g D	D = 0 A	D = 1 A	A = 0.85g D
2	40%	0.33g	1.99g	0.03g	0.09g	0.46g	11.6g
	60%	0.18g	1.69g	0.10g	0.05g	0.48g	10.9g
4	40%	0.25g	1.64g	0.07g	0.07g	0.71g	4.2g
	60%	0.11g	-	0.61g	0.04g	0.64g	5.7g
6	40%	0.12g	-	0.49g	0.07g	0.53g	-
	60%	0.02g	-	0.82g	0.02g	-	-

slope of fatigue curve m = 2   rigid joints   - = structural collapse

Another type of analysis with scaled records has been subsequently performed. For the Aigion and the Kobe records, the peak ground accelerations were scaled so that the maximal damage indexes in the structure are equal to 0 (no damage) and 1 (complete damage) for m = 2. The results, together with the damage index under the non-scaled records are convergence indicating structural collapse was observed prior to the attainment of complete damage. It may seen that in order to reach complete damage, the Aigion record has to be scaled in respect to PGA 3 to 4 times up and the Kobe record 1.5 to 2 times down.

The PGA of the Kobe record leading to damage is around 3 to 4 times larger than that of the Aigion record. A similar relationship holds for the PGA where damage starts. These observations confirm that PGA alone does not provide sufficient information for the damage potential of an earthquake. Figure 6 illustrates the increase in damage with increasing PGA, as mean values of the frames and records under consideration. An over-proportional increase in damage with increasing acceleration may be observed. This indicates that it is not possible to make a linear extrapolation of the results of a damage evaluation under a specific record for higher accelerations.

Another type of analysis was performed, in which the records of Aigion, Thessaloniki and Kobe have been scaled in respect to acceleration to result in a maximum damage index in the structure equal to 1. Frames 1 to 4 with semi-rigid and rigid joints were investigated, the results are summarized in Table 5. In this analysis, the fatigue curves are derived for a rotation capacity under monotonic loading equal to 0.05 radians and a slope m=2. As expected, the type

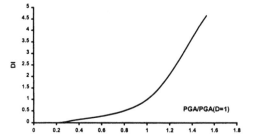

Figure 6: Damage index vs. peak ground acceleration

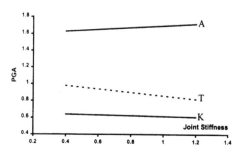

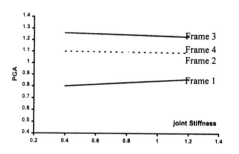

Figure 7: Limit PGA for frames with rigid and semi-rigid joints

Table 5. Limit PGA for complete damage (D = 1) at one section

Frame	Joint stiffness	Aigion	Thessaloniki	Kobe
1	0.4 K	1.28g	0.65g	0.46g
	rigid	1.25g	0.84g	0.50g
2	0.4 K	1.98g	0.93g	0.60g
	rigid	1.99g	0.61g	0.46g
3	0.4 K	1.87g	1.21g	0.70g
	rigid	1.98g	0.93g	0.78g
4	0.4 K	1.41g	1.08g	0.80g
	rigid	1.64g	0.91g	0.71g

of record largely influences the limit accelerations. These accelerations are for the Aigion record in the mean around twice as large as for the Thessaloniki record and three times as large as for the Kobe record. It may be observed that the limit accelerations for the Kobe record are for all the frames lower that the actual PGA of this earthquake (=0.85g), indicating the high damage potential of this earthquake.

The influence of semi-rigid joints is illustrated in Figure 7 that presents the mean values for all frames of the limit accelerations for rigid and semi-rigid joints. It may be seen that semi-rigidity does not significantly affect the structural response in respect to damage. However, Figure 7a shows that there exist differences for the three records under consideration. For the Aigion record the influence of semi-rigid joints seems as a mean to be negative, for the Thessaloniki record the influence is positive, while for the Kobe record almost no differentiation may be observed. In respect to the different frames, Figure 7b shows that frame 1 is negatively influenced by more flexible joints, while the other frames are more or less positively influenced.

6. CONCLUSIONS

This work presents a parametric study of the low-cycle fatigue behaviour of moment resisting steel frames. The results of this study indicate that:
a) The fatigue damage is strongly dependent on the type of seismic record expressed by its acceleration and energy spectra.
b) The fatigue damage grows over-proportional with increasing peak ground acceleration of a record.
c) A better fatigue behaviour, expressed by a larger slope of the fatigue curve may reduce significantly, especially for the columns, the damage.
d) Semi-rigidity in beam-to-column connections leads generally to less damage in the beams and higher damage in the columns. It may therefore be beneficial in cases where for rigid connections beams are more critical than columns.
e) Higher levels of vertical loading lead generally to more damage.

REFERENCES

Akiyama, H.: Earthquake-Resistant Limit-State Design for Buildings, University of Tokyo Press, 1985
Bernuzzi, C., Calado, L., Castiglioni, C.: Ductility and load carrying capacity predictions of steel beam-to-column connections under cyclic reversal loading. J. of Earthquake Engineering, 401-432, 1997
Kannan, A., Powel, G.: DRAIN-2D. A general purpose computer program for dynamic analysis of inelastic plane structures, EERC 73-6 and EERC 73-22 reports, Berkeley, USA, 1975
Vayas, I. Ciutina, A., Spiliopoulos. A.: Low-cycle fatigue gestützter Erdbebennachweis von Rahmen aus Stahl. Bauingeniuer 74, 448-457, 1999

SAC steel project

Behaviour of Steel Structures in Seismic Areas, Mazzolani & Tremblay (eds) © 2000 Balkema, Rotterdam, ISBN 90 5809 130 9

SAC program to assure ductile connection performance

C.W. Roeder
University of Washington, Seattle, USA

ABSTRACT: Steel moment frame buildings were damaged during the January 17, 1994, Northridge Earthquake. The SAC Steel Project was funded by the US Federal Emergency Management Agency to determine the causes of the earthquake damage and to provide solutions to the problems associated with this damage for new and existing buildings. This study included extensive analysis and testing of connections, which may be suitable for seismic design, and a summary of this connection performance research is provided in this paper. An overview of the research will be provided, and the strategies for assuring good connection performance will be discussed.

1 INTRODUCTION

During the January 17, 1994, Northridge Earthquake, a number of steel frame buildings sustained cracking and other damage. The damage occurred to the welded flange bolted web connection illustrated in Figure 1, and most damage consisted of cracking in the beams, columns and welds of these connections. The SAC Steel Project was funded by the Federal Emergency Management Agency (FEMA) to address this problem. Phase 2 of the SAC Steel Project included extensive analysis and testing of connections.

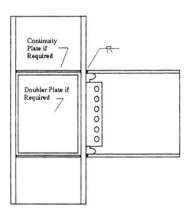

Figure 1. Pre-Northridge Connection

The ultimate goals of this work were the development of practical, simple models for predicting the resistance, stiffness, and ductility of steel moment frame connections. The methods used for developing these models emphasize the separation of ductile yield mechanisms from failure modes that limit the strength and ductility. The failure modes and yield mechanisms for different connections were reviewed and evaluated based upon analysis, experiments performed as part of the SAC research program, and earlier experimental investigations. The relative ductility of different yield mechanisms were compared for a number of different connection alternatives. The comparison led to recommendations for a range of connection types.

2 OVERVIEW OF ANALYTICAL RESEARCH

Eleven experimental tasks and four analytical tasks with funding of approximately 2.3 million dollars were included in the Connection Performance portion of the SAC Phase 2 research program. These studies include both bolted and welded steel moment frame connections. Some connections were direct extensions of Pre-Northridge connection illustrated in Figure 1, while others represent significant departures from past seismic design practice. A brief summary of this analytical and experimental research is appropriate..

Analytical studies (Chi et.al. 1997, Deierlein et al. 1999) examined crack growth based on elastic and inelastic crack propagation theory. Initial work focused on pre-Northridge connections and attempted to match the different failure modes and crack directions noted after the Earthquake and during early post-Northridge testing. The analyses showed good correlation between the measured and calculated force deflection behavior, and showed that the weld was the critical region in pre-Northridge connections because of the flaws in this region and the low weld toughness provided by the E70T-4 self shielded flux core arc welding electrodes. The work showed that early yielding of beam steel shields the critical regions of weld until redistribution of stress and strain hardening occur. This suggests that overmatched weld metal or reduced yield stress in the beam steel may reduce the potential for cracking and increase the plastic rotation capacity of the connection. Large inelastic shear strains in the panel zone cause large local deformations, and the analyses show that excessive panel zone yielding can reduce the ductility of some connections. The work showed that removal of backing bars was important for connection ductility, but analysis also indicated that continuous supplemental fillet weld on the underside of the backing bar benefits Pre-Northridge connections, since it translates an external initial crack into a less critical internal flaw. The analysis shows that if fracture of the welds is prevented through elimination of flaws in the welds and the use of tougher electrodes, other locations such as the weld cope region are susceptible to cracking at only slightly larger plastic rotations.

Another analytical study (El-Tawil et al. 1998) used the ABAQUS Computer Program to predict inelastic deformation of the connection and to examine the local stresses and deformations for cracking potential. The study examined the parameters, which affect yielding and crack potential. These parameters include shear yielding of the panel zone, beam and column flange thickness, beam depth, web and continuity plate thickness, size and geometry of weld cope hole, relative yield and ultimate tensile stress, shear tab thickness, and span to depth ratio. The results also suggested that panel zone yielding increases the potential for cracking. The results show that ductility of the connection is insensitive to yield/tensile ratios less than 0.8, but ductility was reduced by larger ratios.

Another analytical study (Coons 1999, Hoit 1997) developed an extensive database including hundreds of past connection test results for different connection types. This past data was combined with the SAC experimental results to develop simple analytical models for all connections used in seismic design. More than a thousand past connection tests were evaluated. The work established relatively simple but accurate models for establishing the strength and stiffness associated with each individual yield mechanism and failure mode for each moment frame connection. These models provide the

basis for determining the yield mechanism and failure modes for each connection.

Additional analyses were performed by each of the experimental research teams, but the details of these analyses are too lengthy to summarize here.

3 OVERVIEW OF EXPERIMENTAL RESEARCH

Most of the connection tests used a cyclic deformation pattern similar to that defined in ATC-24 (ATC-24 1992) for the displacement controlled experiments. However, the test program was modified to more closely simulate the inelastic deformation demands expected during real earthquakes.

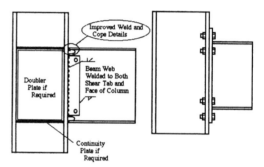

a) Welded Flange - Welded Web   b) Extended End Plate

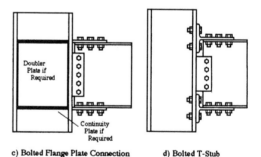

c) Bolted Flange Plate Connection   d) Bolted T-Stub

Figure 2. Typical Connections Examined in Experimental Research

Bolted web and welded flange connections such as the Pre-Northridge connection depicted in Figure 1 and similar connections with notch tough Post-Northridge weld details were tested at the University of Michigan. The tests were completed to fully understand the behavior of Pre-Northridge connections, to establish the extent that improved welding will improve connection performance, and to examine other effects. This work clearly showed that the welded flange bolted web connection is unable to achieve the required seismic ductility with the member sizes commonly used in today's steel moment frames even with notch tough electrodes and with backing bars and runoff tabs removed. A modified free flange connection proposal, which shows considerable promise for seismic design, was examined as part of this study, but is not illustrated in this paper.

Welded flange-welded web connections as illustrated in Figure 2a were tested at Lehigh University in response to the behavior noted in the Michigan tests. This later study examined the effects of improved cope details, continuity plates, and web welds on connection performance. Nonlinear computer analyses were performed to examine the effect of the cope transition angles and surface finish on connection performance. This work showed that connections with thoroughly welded webs provide increased ductility and plastic rotation over their bolted web alternatives. It should be noted that the beam web is welded both to the shear tab with fillet welds and to the column web with full penetration welds. Stiff, strong web connections combined with good cope finish and good cope geometry resulted in good inelastic rotational capacity even for connections with very large members.

The extended end plate connection as illustrated in Figure 2b was studied at Virginia Tech University. Extended End Plates are one of the more promising bolted connection options for seismic design, but failure modes are complex and methods for predicting the behavior required improvement. A large body of past test data on this connection was available, and this test program filled gaps in this data and further developed design models for the connection. This connection appears to provide significant resistance and rotational capacity, but its applicability for seismic design is limited to moderate sized members. Bolted flange plate connections as illustrated in Figure 2c were tested (Schneider et al. 1999) at the University of Illinois. This study examined failure modes of bolted flange connections, and worked toward improved design models for prediction of connection behavior. This connection provided among the largest plastic rotational capacity of all Post-Northridge connections in the study. It develops the full plastic capacity of the beam and can be used for relatively large members. However, the bolted flange plate connection requires balancing of three different yield mechanisms, and the resulting design equations are relatively complex. This study also examined the possibility of supplementing bolted connections with friction dampers, but the size of the connections required for friction damping appeared to render friction dampers impractical for modern steel moment frame construction.

Bolted T-Stub connections (Leon et al. 1999) as illustrated in Figure 2d were examined at Georgia Institute of Technology. The work addressed the complex behavior of these connections including prying forces, connection stiffness, and the large

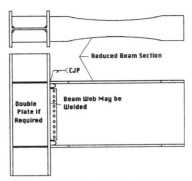

Figure 3. Sketch of Radius Cut RBS Connection

number of failure modes and variations in bolted connections. Many of these tests are relatively simple push-pull tests, since these simple tests were economical and helped to understand the complexities of bolted connection behavior.

The reduced beam section (RBS) connection with the radiused reduced section depicted in Figure 3 provides among the best inelastic seismic performance seen in recent tests. Many past tests were completed, and so only focused testing on the RBS system was completed at the U. of Texas, U. of Texas A&M, and the U. of California at San Diego. These tests focus on the effect of the composite slab, the effect of lateral-torsional buckling and unsupported length, the effect of beam depth and column orientation and the effect of panel zone yielding on the RBS system behavior. The research showed that the composite slab changed the strain distribution in the beam and increased the lateral stability of the specimen, but the slab did not adversely affect the rotational capacity of the connection.

Connection repair and retrofit methods were also examined. Coverplated connections and haunch connections are depicted in Figure 4a and 4b, and these are suitable for strengthening connections to develop the full capacity of the connection for very large connections. No haunch connections were tested during the SAC Phase 2 research, but a number of coverplated connections were tested at University of California, Earthquake Engineering Research Center in Richmond, CA. This research was intended to address several failure modes which have recently come to light with the coverplated connection. In general, the haunch and coverplated connections are relatively expensive connection modifications, but they can achieve large plastic rotations by moving the plastic hinge to the end of the haunch or coverplate. Several changes in the proposed design procedure were made for these connections to reduce the potential for premature connection fracture. Weld overlays are a very economical method for repair or retrofit as depicted in Figure 4c. These connections were studied at the University of Southern California. Tests showed that this method provided reasonable ductility for modest sized connections. The plastic rotational capacity of these modified connections was clearly less than that achievable with many other Post-Northridge connections, but nevertheless, it was a significant improvement over Pre-Northridge behavior. It is unclear that this modification will provide significant improvement to the performance of very large connections, but it appears promising for future use in modest size specimens. Unfortunately, the research did not adequately evaluate design procedures to permit direct use of the overlay method, and further research is needed to establish that design procedure.

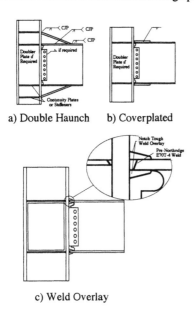

a) Double Haunch   b) Coverplated

c) Weld Overlay

Figure 4. Typical Repair or Retrofit Methods

Steel moment frame buildings constructed in the US commonly use simple bolted web shear tab connections at most gridline intersections. These connections are assumed to have no moment resistance, and they are not considered in the seismic design. The beams are designed for gravity loads, and the beams and the connections commonly have composite floor slabs. Despite normal design assumptions, the resulting shear tab connection can develop composite connection behavior with significant moment resistance as depicted in Figure 5. Research was completed on these connections to determine if this resistance and stiffness was adequate to reduce the levels of repair or retrofit required for damaged or undamaged steel frame buildings with Pre-Northridge connections. The research was completed at the University of California at Berkeley. The tests showed that these connections have significant rotational capacity, but the capacity is a function of the depth of the connection. Bending

moments develop at these connections, and the bare steel moments can be retained through very large rotations. Composite action significantly increases the rotational resistance, but this additional resistance is lost at rotations much smaller than those that can be developed by the bare steel connection. The stiffness and resistance of these connections benefit buildings with Pre-Northridge connections, and help these buildings remain stable after connection fracture occurs. However, the stiffness and resistance is small compared to that noted for frames with the more rigid connections commonly used in seismic design. Despite the large number of these connections, it is unclear that they provide significant benefit to undamaged buildings.

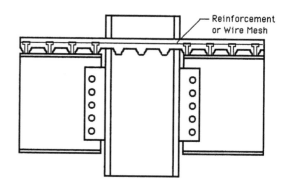

Figure 5. Composite Shear Tab Connection

4 DEVELOPMENT OF RECOMMENDATIONS

The goal of the SAC Connection Performance research effort was development of simple models, which can be used by structural engineers to accurately predict connection performance. This prediction must provide the connection strength (or resistance), stiffness and ductility. Connection stiffness is important, because it is needed to establish the predicted performance and seismic demands for the building. Some connections are sufficiently stiff that they may be treated as rigid, but others are not. The connection resistance is important, because it is used to establish the global structural resistance, and global structural resistance combines with the structural analysis to establish inelastic ductility demands. Inelastic demands must be compared to the connection ductility (or rotational capacity) to determine whether the connection is adequate for the given seismic loading and performance requirements. Therefore, ductility (or plastic rotational capacity) becomes the last and probably most critical element of the connection performance.

This discussion shows considerable interaction between the connection resistance, stiffness and ductility, the seismic demand and the performance evaluations of the system. Unfortunately, there is not a direct closed form solution for the strength, stiffness and ductility for any given connection type. Stiffness, resistance and ductility vary widely depending upon the connection type and the yield mechanisms and failure modes which control the connection behavior. Many yield mechanisms and failure modes occur for each connection type. The welded flange-welded web connection illustrated in Figure 2a is used as an illustration of the importance of yield mechanisms and failure modes for a given connection. Figure 6a shows the yield mechanisms, which can occur with this connection, and Figure 6b illustrates failure modes for this and comparable connection types.

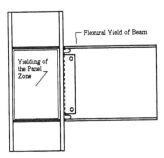

a) Yield Mechanisms

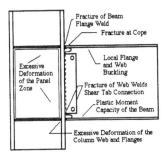

b) Failure Modes

Figure 6. Yield Mechanisms and Failure Modes for Welded Flange-Welded Web Connections

The yield mechanisms are primarily flexural yielding of the beam and shear yielding of the panel zone for the welded flange-welded web connection. Some plastic rotation may be achieved by local deformation of the columns, but the plastic rotations that are achieved by this local deformation are relatively small. In addition, these local deformations are a concern for local failure of the connection and are minimized to avoid potential problems of this type. Flexural deformation of the column may also contribute significant plastic rotation, but plastic ro-

tations in the columns are discouraged, because research has shown that weak column strong beam behavior concentrates inelastic deformation (Schneider et al. 1993) into individual stories or parts of the structure. It is difficult to achieve the large rotational capacities required by some weak column-strong beam structural systems, and so weak column behavior is restricted by seismic design provisions. As a consequence, this rotational capacity associated with flexural yield of the columns and local deformation of column web and flange can not be achieved in practice and are not included in the figure.

Fracture initiating from the weld and the heat affected zone around the weld were very common in Pre-Northridge connections and are also a potential mode of failure for welded flange-welded web connections. Fracture of the beam flange in the region of the cope has also been noted in a number of connection tests. Local web and flange buckling and lateral torsional buckling of the beam are potential failure modes for these connections. These buckling phenomenon have been noted in many tests, and they usually result in rapid deterioration of resistance and often lead to ultimate tearing and fracture of the metal in the deformed region. Excessive shear deformation of the panel zone has seldom resulted in a direct fracture of the column panel zone, but these large deformations frequently place large demands on the welds and contribute to local fractures in the connection. Further, large panel zone yield deformations may result in large inelastic shear buckling which leads to significant loss of resistance and stiffness. Plastic bending of the beams has been a fairly common failure mode for ductile connections and it is commonly associated the local buckling noted above. Finally, fracture of the web connection is an important concern with the welded web connection, because the stiff, strong web connection places greater demands upon this attachment and the welds accomplishing it.

The failure mode strongly influences the ductility of the connection regardless of the connection type. It is important that structural engineers have reliable but simple design models to predict the full range of connection behavior for all connection types. These design models must provide -

- accurate prediction of the yield mechanism and the mode of failure for each connection type,
- reliable estimation of the resistance associated with each mechanism and mode of failure, and
- and the ductility and deformational capacity expected for the connection.

These diverse goals were achieved (Roeder 2000) by first predicting the resistance associated with each yield mechanism and mode of failure for a given connection. The failure mode with the lowest resistance is expected to be the critical failure mode, and the resistance associated with that critical failure mode is taken as the best estimate of the maximum resistance of the connection. Yield mechanisms, which are capable of sustaining significant inelastic deformation, provide a ductile mechanism to the connection if the resistance associated with that given yield mechanism is significantly smaller than the resistance associated with the critical failure mode. The yield mechanism with the lowest resistance is the critical yield mechanism, although connections with closely space resistances for different yield mechanisms and different failure modes are likely to have coupled contributions from these different modes and mechanisms. The ductility and deformational capacity of the connection was then determined a function of the critical yield mechanism and critical failure mode for the connection, as well as the proximity of the resistances for these two conditions. In comparing failure modes and yield mechanisms, it is important to recognize that yielding or failure may occur at different locations, and thus the comparisons must be normalized to a standard location on the connection such as the face of the column. Many different yield mechanisms and failure modes are possible for each connection type, and so a number of simple yet thorough equations are needed for each connection. These equations were obtained by analysis of the connection and by comparison of predicted behavior with experimental results. Principles of elementary mechanics were used to develop individual equations, but the test of the accuracy of the resulting equations (Hoit 1997, Coons 1999) depended upon how well the simplified equations predicted the experimental results.

Other factors in addition to the yield mechanisms and failure modes affect connection performance. For example, the plastic rotation that can be achieved with the connection depends upon the column spacing and the beam depth, since these factors affect the strain distribution required to achieve a given connection rotation. These added factors were also considered in the evaluation, but their contribution to the ductility secondary to that of the yield mechanisms and failure modes and are not emphasized here.

5 EQUATIONS AND RECOMMENDATIONS

To illustrate the application of these methods, the equations and recommendations of the welded flange-welded web connection will be discussed.

Shear yielding of the panel zone and flexural yielding of the beam are the primary yield mechanisms for the welded flange-welded web connections and equations for these yield mechanisms are quite easily developed. Panel zone yielding occurs when the panel zone shear force, V_{pz}, equals the yield shear force, V_y, where

$$V_y = 0.55 \, F_y \, d_c \, t_{wc}. \tag{1}$$

In this equation, F_y is the tensile yield stress of the column, d_c is the depth of the column and t_{wc} is the panel zone thickness. This yield shear force has been noted for a large number of past experiments, and is an accurate indication of the shear force associated with the initiation of significant panel zone yielding. Initial yielding of the beam occurs when the bending moment at the face of the column equals the yield moment, M_y, where

$$M_y = S \, F_y \tag{2}$$

The elastic section modulus, S, is used because the yield mechanism is the initial point of significant yielding. It should be noted, that US building codes have a long history of estimating the shear yield stress as $0.6 F_y$ rather than the $0.55 F_y$ noted in past experiments and included in Equation 1, but this estimate is achieved only after substantial yielding and strain hardening have occurred.

The experiments showed that connections with either flexural yielding or panel zone yielding were capable of developing significant plastic rotations. However, very large strain hardening was noted in connections with good ductility, and even connections with strong panel zones developed large plastic deformations in the panel zone. As the panel zone connections became extremely large, the research (Roeder 2000, Deierlein et al 1999) showed an increased potential for early connection fracture. In general, panel zones do not fail, but the large deformation of the panel zone may cause failure in other locations of the connection. As a result, a significant change was proposed in the way panel zone yielding is addressed. This change attempts to balance panel zone yielding with flexural yielding. Panel zone yielding is encouraged so that the ductility provided by this important yield mechanism can readily be used in practice, but at the same time, the balance condition prevents overly weak panel zones, which result in overly large panel zone deformation and increased potential for connection fracture. This balance condition is

$$V_{pz} \leq V_y \tag{3a}$$

Where V_{pz} is the panel zone shear force computed by equilibrium with the yield moment as illustrated in Figure 7. Therefore, the equation may be restated as

$$\frac{\Sigma \, S \, F_{ybm}}{d_b} \left(\frac{L}{L - d_c}\right)\left(\frac{h - d_b}{h}\right) \leq 0.55 \, F_{yc} \, d_c \, t_{w-c} d_b \tag{3b}$$

Failure modes were dealt with in similar ways. The research showed that fracture of the weld is prevented if notch tough electrodes combined with removal of the bottom flange backing bar and the runoff tabs are employed. In addition, the underside of the top backing bar is reinforced with a notch tough fillet weld as depicted in Figure 8. The toe of the cope is a second location of potential fracture during large inelastic deformations, and fractures at this location were commonly noted with welded flange connections, which used tough electrodes and welding detail that inhibited weld fracture. Experiments showed that fracture at this location could be controlled if the cope was prepared to a high quality surface finish with appropriate transition angles of the cope. The improved finish and geometry were employed on the welded flange-welded web connections to eliminate the concern. The reader is referred to other publications (Roeder 2000, Ricles 2000) for details on this cope geometry.

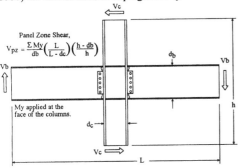

Figure 7. Computation of V_{pz}

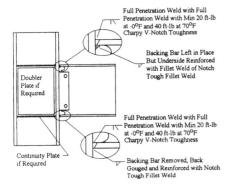

Figure 8. Flange Weld and Backing Bar Requirements

The SAC Phase 2 research clearly showed the benefits of having stiff, strong web attachments, since these reduced the demands on the weld and cope regions and provided significant increases in ductility of the connection. Cracking and fracture of the web attachment was the third location of concern in welded web connections. In general, cracking at this location appeared to occur at slightly larger rotations than initiation of cracking in the cope region if the cope did not have the optimized geometry and finish noted earlier. However, web connections that were welded as described earlier with these optimized cope details delayed this mode of failure until large plastic rotations were achieved for even large sized specimens. Nevertheless, cracking and frac-

ture in and near the web attachment remained the first occurrence of failure in some of these ductile connections.

The web and flange details noted earlier, resulted in quite large plastic rotations and ductility for welded flange-welded web connections. As a result, the failure mode that controls the behavior of these ductile connections is the plastic bending of the beams. These ductile specimens developed large strain hardening moments, since bending moments at the face of the column were typically 40% to 50% larger than the plastic moment capacity with measured yield stress of the steel. As a result, the equation

$$M_{pfailure} = Z \frac{F_y + F_t}{2} \qquad (4)$$

represents a conservative failure moment capacity for these ductile connections where F_t is the tensile strength of the steel. The experiments commonly achieved maximum bending moments at the face of the column 10% to 15% larger than Equation 4 for these ductile connection details.

The consequences of this large strain hardening are important to other issues of seismic design. In particular, strong column-weak beam connections are required because excessive plastic deformation in the column concentrates the inelastic deformation into a single story of the structure. This results in excessive local damage and increased likelihood (Schneider et al. 1993) of collapse during severe earthquakes. Strong column behavior is assured by balancing the summation of the plastic capacity of the beams to the summation of the plastic capacity of the columns, but the large strain hardening of the beams means that much larger beam moments develop in these ductile connections. As a result, a more appropriate balance connection for these ductile connections is

$$1.1 < \frac{\sum Z_c \left(F_{yc} - \frac{P_{uc}}{A_g} \right)}{\sum Z_b \frac{F_{yb} + F_{tb}}{2}} \cdot \qquad (5)$$

Connection ductility also requires control of the slenderness of the beam flange and web. The present limits are based upon intuitive evaluation of very old test results. The research (Roeder 2000) shows that control of these slenderness limits is important, but that the present limits are clearly imprecise. Lateral support is also required to assure that lateral torsional buckling does not result in early deterioration of the seismic performance of the connection. The research suggests the present support limits are overly conservative and significant improvements in these provisions are possible. There is considerable room for improvement in these equations. The reader is referred elsewhere (Roeder 2000) for further discussion on these issues.

Given the above behavior, the plastic rotational capacity, θ_p, and the statistical variation in this rotational capacity was estimated for this connection from the test results. For welded flange-welded web connections which satisfy the previously described requirements,

$$\theta_{pmean} = 0.041 \qquad (6a)$$

and the standard deviation expected for this plastic rotation is

$$\sigma_p = 0.003. \qquad (6b)$$

This limit can conservatively be applied for all connections with beams less than W36 in depth.

This brief discussion has attempted to illustrate the consideration of yield mechanisms and failure modes for a given connection. These considerations were applied (Roeder 2000) to all connection types. Much greater complexity was encountered for many of the other connections, but this brief description provides an overview of the issues addressed.

6 SUMMARY AND CONCLUSIONS

This paper has attempted to illustrate the concerns, focus and scope of connection evaluation work in the SAC Steel Project. This paper cannot present a comprehensive presentation of the research results, and so the reader is referred to other documents (Roeder 2000) for more detailed information on specific issues. In addition, individual project reports are referenced in the Connection Performance State of the Art Report (Roeder 2000), and these are another source of more detailed information on individual research programs.

7 ACKNOWLEDGEMENTS

The work forming the basis for this publication was conducted pursuant to a contract with the Federal Emergency Management Agency. The substance of such work is dedicated to the public. The funding is provided through the SAC Joint Venture and California Universities for Research in Earthquake Engineering (CUREe) Contract No. 7. The author is solely responsible for the accuracy of statements or interpretations contained in this publication. No warranty is offered with regard to the results, findings and recommendations contained herein, either by FEMA, the SAC Joint Venture, the individual joint partners, their directors, members or employees. These organizations and individuals do not assume any legal liability or responsibility for the accuracy, completeness or usefulness of any of the information, product or processes included in this publication.

REFERENCES

ATC-24 (1992), *Guidelines for Seismic Testing of Components of Steel Structures*, Applied Technology Council, Redwood City, CA.

Coons, R.G., (1999),"Seismic Design and Database of End-Plate and T-Stub Connections," A thesis submitted in partial fulfillment of the MSCE Degree, University of Washington, Seattle, WA.

Chi, W.M, Deierlein, G.G., and Ingraffea, A.R., (1997),"Finite Element Fracture Mechanics Investigation of Welded Beam-Column Connections," *Report No. SAC/BD-97/05*, SAC Joint Venture, Sacramento, CA.

Deierlein, G.G., and Chi, W.M., (1999) "SAC TASK 5.3.3 - Integrative Analytical Investigations on the Fracture Behavior of Wellded Moment Resisting Connections," Final Report, John A. Blume Earthquake Engineering Center, Stanford University, Palo Alto, CA.

El-Tawil, S., Mikesell, T., Vidarsson, El, and Kunnath, S.K., (1998),"Strength and Ductility of FR Welded-Bolted Connections," *Report No. SAC/BD-98/XX*, SAC Joint Venture, Sacramento, CA.

ICBO (1991), *Uniform Building Code*, International Conference of Building Officials, Whittier, CA.

Hoit, M. (1997) "An Investigation into the Seismic Design of Flange Plated Moment Resistant Connections," A thesis submitted in partial fulfillment of the MSCE Degree, University of Washington, Seattle, WA.

Leon, R.T., Swanson, J.A., and Smallidge, J.M., (1999), "SAC Steel Project - Subtask 7.03: Results and Data CD," Georgia Institute of Technology, Atlanta, GA, 1999.

Ricles, J., (2000), Report to SAC Steel Project, Final Report in preparation from Lehigh University, Bethlehem, PA.

Roeder, C.W., (2000), "State of the Art Report Connection Performance," FEMA report 355D, Federal Emergency Management Agency, Washington, D.C.

Schneider, S.P., Roeder, C.W., and Carpenter, J.E., (1993)," Seismic Behavior of Moment-Resisting Steel Frames", "Analytical Study" and "Experimental Investigation", ASCE, <u>Journal of Structural Engineering</u>, Vol. 119, No. 6, June.

Schneider, S. P., and Teeraparbwong, I., (1999),"SAC Task 7.09: Bolted Flange Plate Connections," Dept. of Civil Engineering, U. of Illinois, Urbana, IL, October.

Behaviour of Steel Structures in Seismic Areas, Mazzolani & Tremblay (eds) © 2000 Balkema, Rotterdam, ISBN 90 5809 130 9

Development of the free flange steel moment connection

Jaehyung Choi, Božidar Stojadinović & Subhash C.Goel
Department of Civil and Environmental Engineering, University of Michigan, Ann Arbor, USA

ABSTRACT: Premature failures of steel beam-to-column moment connections, consisting of welded flange-bolted web detail, shows that conventional beam theory may not be applicable in the vicinity of the connection. In addition, commonly used low notch tough weld metal and poor welding practice are another key reason of brittle fracture of these connections. Through extensive analytical work conducted at The University of Michigan, a new beam-to-column steel moment connection called "free flange connection" was developed based on understanding of the true stress distributions and force flows in the connection region as well as adopting notch tough weld material and improved welding details. Results of five free flange connection specimen tests presented in this paper show excellent ductile behavior with adequate plastic deformation without premature fracture in the connection region.

1 INTRODUCTION

It has been almost five years since the 1994 Northridge earthquake that we learned that traditional US-designed fully restrained steel beam-to-column connections fracture before they yield. This finding shook designers' confidence in steel moment-resisting frames. Much work has been done since then to understand the behavior of steel moment connections and to develop new earthquake resistant designs (Goel, 1996; Engelhardt, 1996; Yu, 1997; Allen, 1998).

This paper presents a new fully restrained steel moment connection, called the free flange connection. Development of this connection is the culmination of five-year long research work conducted at The University of Michigan.

The free flange design is based on a fundamental understanding of the flow of forces in fully restrained steel moment connections. Finite element analyses have shown that the principal stresses in the connection tend to flow toward the beam flanges. Such stress redistribution in the vicinity of the connection causes over-stressing of the flanges. Thus, flanges are required to carry the axial stresses produced by the beam moment as well as approximately half of the beam shear. Recently completed research at The University of Michigan shows that Saint Venant's Principle can be used to explain this redistribution of stresses (Lee, 1999).

The details of the free flange design are also based on a comprehensive analytical and experimental study of unreinforced moment connections conducted at The University of Michigan within the SAC steel program (Stojadinović, 2000). This study produced two fundamental findings. First, the use of notch-tough weld metal and the improvements in welding details are essential for increasing the fracture resistance of the beam flange welds. Second, beam flanges undergo severe local bending in the region between the weld and the end of the beam access hole. Strain concentrations produced by such local bending result in large-amplitude stress-strain reversals under reversed cyclic loading. Low-cycle fatigue caused by such strain reversals may explain fractures of the beam flanges observed in virtually all of the specimens tested in the SAC study.

The seismic behavior of free flange connection tests was investigated in a comprehensive parametric study. The results of five tests and some important factors for the free flange connection designs are summarized in this paper. They show that the free flange connection is resilient and ductile. It satisfies the new drift based rotational capacity requirement for moment connections in special moment resisting frames.

2 CONFIGURATION OF FREE FLANGE CONNECTION

The free flange connection has two unique characteristics, shown in Fig. 1: a certain length of the beam flange that is not connected with the beam web, and tall trapezoid shaped shear tab welded to the beam web.

First, the disconnected region of the flange, called the free flange, is intended to reduce the beam flange local bending in the vicinity of the connection. Extensive research on beam-to-column connection boundary region showed that the stress distribution and force flow do not follow the classical beam theory in the vicinity of the connection (Lee, 1997). Instead, the force flow in the connection region is directed towards the beam flanges, like diagonal truss action (Goel, 1997). According to that study, relatively strong column flange confines Poisson's effect and the shear deformation at the column face. To achieve a self-equilibrium in this region, the beam flange connection, which was designed to resist only normal force due to moment, is overloaded with additional shear. By leaving enough free flange length between the column face and the beam web end, local deformation of the beam flange is drastically reduced. In addition, the force flow is re-directed back into the web.

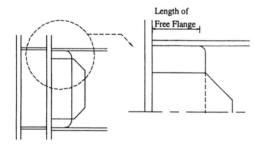

Fig. 1 Configuration of the free flange connection.

Clearly, using longer free flange length should improve the ductility of the free flange connection. The longer the free flange length, the lesser the local bending and the shear force in the beam flange. However, using too long a free flange length may also cause stability problem. The flange may buckle before yielding spreads into the beam web and a complete plastic hinge forms in the beam. Thus, an optimum free flange length needs to be determined by balancing the reduction of local flange bending with flange buckling strength. Aspect ratio of the free flange length in terms of flange thickness is used to define the length of free flange.

Second, using a strong shear tab is another element in reducing the stress concentration and local bending in the beam flange. It can be explained by considering a parallel shear spring model. The shear force acting on the connection is distributed between the beam web and the flanges according to their shear stiffness as shown in Fig. 2. The stiffness of tall shear tab is dominant in this system. Thus, it can attract more force into the beam web so as to relieve beam flange overloading. In addition, stiff shear tab also resists more normal force due to moment.

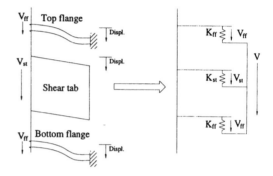

Fig. 2 Parallel spring model.

The shape and size of the shear tab is determined based on two main considerations. The shear tab is extended enough to protect the vulnerable juncture of the web and the flange. The trapezoidal shape helps the formation of the typical hourglass patterned yielding of beam plastic hinge. Another factor in determining the shear tab size is the weld length between the shear tab and the beam web designed to resist the required normal and shear forces. Complete design procedure for the free flange connection and more details of the study are presented in the SAC report Subtask 7.02a (Choi, 2000).

3 TEST SPECIMENS

Five free flange connection specimens were tested under SAC steel program Subtask 7.02a. All specimens were fabricated with A572 grade 50 steel. For Specimen 8.2, W24×68 beam and W14×120 column sections were used. For Specimens 9.1 and 9.2, W30×99 beam and W14×176 column sections were used. For Specimens 10.1 and 10.2, W30×124 beam and W14×257 column sections were used. Three parameters were used to study the behavior of the free flange connection.

Column panel zone strength effect: the strength of panel zone can be defined by using the shear strength formula specified in AISC seismic provisions for structural steel buildings (Eq. 9-1, AISC, 1997). The design shear strength satisfying this formula with resistance factor $\varphi_v = 0.75$ is defined as strong panel zone; Specimens 8.2, 9.2 and 10.2 are included in this category. Those with $\varphi_v = 1.0$ are considered as medium strength panel zone; Specimen 10.1 is included in this category. The specimens with $\varphi_v > 1.0$ are defined as weak panel zone; Specimen 9.1 is included in this category. For strong panel zones, 5/8", 3/4" and 1/2" doubler plates were added in the column panel zones of Specimens 8.2, 9.2 and 10.2, respectively.

Effect of beam flange thickness and size: Three different pairs of beam and column sections were used. Identical aspect ratio of free flange length in terms of flange thickness was used in all specimens. Based on the results of finite element analyses, free flange aspect ratio of 5.22 was selected. Specimen 8.2 is considered as small size beam and column section with 3.0" free flange length. Specimens 9.1 and 9.2 are considered as medium size beam and column section with 3.5" free flange length. Specimens 10.1 and 10.2 are considered as large size beam and column section with 4.5" free flange length.

Shape of web end cut: Two different beam web end cut out shapes were used to study its effect and to determine optimal free flange connection configuration as shown in Fig. 3. Specimens 8.2, 10.1 and 10.2 employed the 90-degree straight beam web cut, which is intended to eliminate redundant web portion. Specimens 9.1 and 9.2 employed the extended web end shape. The beam web end was extended to approximately 3 times the shear tab thickness from the column face. This shape was intended to add some stiffness in the shear tab area as it restrains beam web out-of-plane deformation during severe inelastic stage.

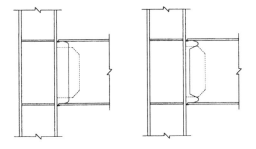

Fig. 3 Beam web end shape (left: Straight cut, right: Extended web end shape).

Fig. 4 shows general view of the test set up. The specimens were rotated 90 degrees from the actual position of the exterior beam and column in the building. Lateral bracing was provided in order to prevent out-of-plane movement of the specimens. The shear tab and doubler plate were placed on the same side. White wash was applied on the beam and column surface around the connection region in order to investigate yielding location and yielding pattern.

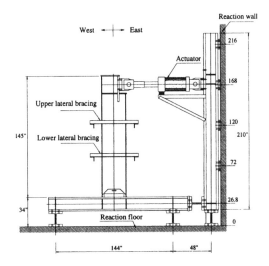

Fig. 4 General test specimen set up.

A total of 12 potentiometers were installed on each specimen to measure rotational components of deformation. The potentiometer readings were used to obtain information such as beam plastic rotation contribution, column plastic rotation contribution and panel zone plastic rotation contribution, as suggested in SAC Interim Guide lines (FEMA/SAC, 1995; FEMA/SAC 1996).

The applied load patterns follow the SAC Phase 2 Testing Protocol, which prescribes a displacement controlled quasi-static load pattern (SAC, 1997). Testing was continued until either reaching the maximum actuator loading capacity or occurrence of severe lateral-torsional instability problem due to excessive beam web and flange yielding.

For all specimens, E70TG-K2 notch tough welding wire was used between beam flange and column and E70T-7 welding wire was used for shear tab, doubler plate and continuity plate groove welding. E71T-8 wire was used for shear tab and beam web fillet weld and beam flange reinforcing fillet weld. To minimize the notch effect, back up bar at the beam bottom flange was removed, back gauged and reinforced with fillet weld.

Except for the beam flange weld to the column and shear tab weld to the beam web, other welds were done in the shop. The field welding was performed with beam and column in a position so as to simulate the field condition. All welding was done by a qualified welder and then certified by a qualified inspector through visual inspection and ultrasonic tests.

4 TEST RESULTS

Weak panel zone specimen (9.1) showed good behavior in terms of total plastic rotation. This specimen survived 6 cycles at 5% drift without any fracture. Panel zone yielding was significant. Small yielding occurred in the panel zone during the 1st cycle at the 0.5% drift level. Beam flange yielding started at the same time as panel zone yielding and then spread out towards the beam web during 2% drift cycle. Small local buckling was observed in the free flange portion but the strength degradation of the specimen was not observed. The test was stopped after extensive yielding in the beam flanges and the panel zone with ductile fracture near the access hole in the bottom flange. Panel zone yielding dominated the overall behavior. It provided 60% of total plastic rotation capacity of the specimen. Although this weak panel zone specimen showed excellent behavior, excessive panel zone yielding caused local column flange kink that may be a factor in beam flange fracture. In addition, excessive yielding in the panel zone prevented formation of "complete" plastic hinge in the beam. The yielding pattern of the panel zone and column flange kink due to excessive shear deformation in the panel zone are shown in Fig. 5.

Strong panel zone specimens (8.2, 9.2, and 10.2) did not fracture during the tests. Specimens 8.2 and 9.2 showed similar behavior. First, beam top and bottom flange yielding occurred at 0.75% drift. After sufficient yielding in beam flange, it spread out into the beam web during the 1.5% drift cycle, forming "hourglass" patterned plastic hinge located at approximately half the beam depth away from the column face. Panel zone yielding was not observed during the tests for both specimens. Typical mode of failure was lateral-torsional buckling of the beam. First, the beam flange yielded significantly and then free flange buckled slightly. Next, the beam flange local buckling and beam web buckling occurred at the location of plastic hinge followed by severe out-of-plane deformation of that region. Since panel zone did not yield during the test, all plastic rotation was provided only by the beam, which may have caused severe lateral-torsional buckling of the strong panel zone specimens. Unlike other strong panel zone specimens, behavior of Specimens 10.2 was somewhat different. It allowed some yielding in the panel zone. Also, more yielding spread into the beam web.

Fig. 5 Panel zone yielding and column kink in weak panel zone specimen (5% drift).

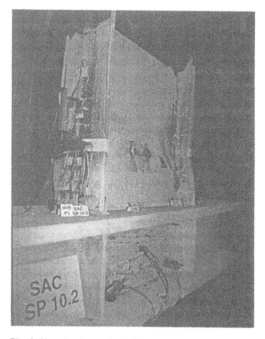

Fig. 6 "Hourglass" type plastic hinge for strong panel zone (4% drift).

The test was stopped at 4% drift level without any fracture in the specimen because the applied load reached the actuator capacity.

During the last cycle of the test, a small crack was observed in the beam top flange near the access hole, at the junction of the beam flange and beam web.

Potentiometer data analysis showed the panel zone strength with 1/2" doubler plate of Specimen 10.2 was not enough to totally prevent the shear yielding. The strain exceeded three times of the elastic yield strain. Consequently, Specimen 10.2 showed excellent behavior in terms of not only the total drift rotation but also some shared yielding of the beam and the panel zone. This balanced behavior is considered good. The formation of plastic hinge and slight panel zone yielding are shown in Fig. 6.

Medium panel zone specimen (10.1) also showed good behavior. It survived 4% drift without any fracture until applied load reached the maximum actuator capacity. Panel zone yielding started during the 1.0% drift cycle and beam flange yielding spread out to the corner of the beam web near the access hole during the 3% drift cycle. Small crack occurred at each corner of the shear tab connected to the column face during the 1st cycle at 4% drift. Those cracks propagated by 3" and penetrated slightly into the groove weld between the shear tab and the column flange. Despite cracking in the shear tab weld, beam flange fracture did not occur nor any significant degradation of the global stiffness was observed during the test. The yielding pattern of the medium panel zone was similar to the weak panel zone case rather than to the strong panel zone case. Significant panel zone yielding was observed and severe column flange deformation and local kink was observed in the connection region. This excessive column deformation may have caused the crack in the shear tab. Although beam web yielding occurred, it was not as much as in the strong panel zone case.

The size effect of beam depth or column was not obvious. Three different beam and column sections were used. Although some differences from the test results in the maximum rotational capacity or yielding pattern were observed, all free flange connection specimens obtained sufficient amount of inelastic deformation during the tests.

The beam web end cut effect can be found from the comparison of Specimens 9.2 and 8.2. Both specimens were designed as strong panel zone, but different shapes of the web end cut were used. For the 90 degree straight end cut shape, some yielding at the middle of the shear tab due to out-of-plane bending was observed. For the extended web end shape, yielding due to shear tab bending did not occur. Thus, the redundant portion of the beam web may add some stiffness in the out-of-plane direction. But, the results of both specimen tests showed that the trend of the lateral-torsional buckling was independent of this local reinforcing effect.

5 ROTATION CAPACITY OF SPECIMENS

Fig. 7 shows the maximum drift angle and plastic rotation of the free flange connection specimens. As can be seen in this figure, all free flange specimens exceeded 4% drift based rotation capacity that is specified in the new seismic provisions. Fig. 8 shows the comparison of strong and weak panel zone in terms of hysteretic loops.

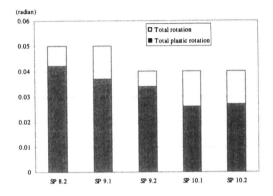

Fig. 7 Maximum drift angle and plastic rotation of the free flange connection specimens.

Specimen 9.1 showed the best result in terms of total rotational capacity that reached up to 5% drift angle without any fracture in the connection. Total rotation of Specimen 8.2 also reached 5% maximum drift. However, as can be seen in Fig. 8, this specimen suffered significant strength degradation after lateral-torsional buckling occurred. Fig. 9 shows the percentage of the beam and panel zone plastic rotation to the total plastic rotation as a function of the panel zone strength. For the weak panel zone case, contribution of the panel zone due to shear deformation is dominant. This result implies that the beam plastic hinge is located very near the connected region where the possibility of connection failure is high due to boundary effects. For the strong panel zone case, the beam contributed to all the inelastic behavior. This result indicates that beam can develop sufficient web yielding before reaching the maximum rotational capacity of the connection. Strong panel zone design of the beam-to-column connection is preferable and safe if sufficient lateral supports are provided. Generally, concrete slab provides enough lateral support to the moment connection in the building. Column panel zone is an excellent earthquake energy-dissipating device. Thus, some column inelastic deformation can be considered acceptable in the beam-to-column connection design.

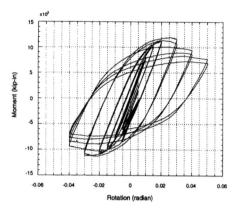

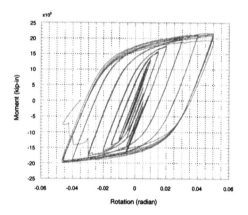

Fig. 8 Moment-rotation responses for strong and weak panel zone specimens (upper-strong: lower-weak panel zone).

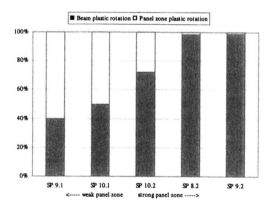

Fig. 9 Percentage of the beam and panel zone plastic rotation.

6 CONCLUSIONS

Important findings from this study and key recommended design parameters for the free flange connection are summarized as follows:

1. All free flange connection specimens satisfied the minimum required drift based rotational capacity of 4% for special steel moment resisting frames specified by the new seismic provisions and exhibited excellent ductile behavior without fracture.
2. The range of free flange aspect ratio between 5.0 and 6.0 is proved good for free flange length. It was shown from the test results that small free flange local buckling did not cause any stability problem nor did it decrease the stiffness of the connection.
3. The column panel zone strength was a significant factor for the free flange connection behavior. Using adequate shear resistance factor in the design equation can control panel zone yielding.
4. Although panel zone yielding is a good energy dissipation mechanism, excessive column deformation can cause some detrimental effects, such as column flange kink or permanent distortion of the column, and also lead to beam flange tearing, shear tab cracking and excessive repair costs.
5. The size effect (using different beam and column sections sizes) was not obvious in these tests.

REFERENCES

ABAQUS, User's Manual-Version 5.7 (1997), Hibbit, Karlsson, and Sorenson, Inc, 1080 Main Street, Pawtucket, RI 02860.

Allen, J., Richard, R. M., and Partridge, J. (1998), *"Seismic Connection Designs for New and Existing Steel Moment Frames Structures"* Proc., 2nd World Conf. On Steel in Construction, Elserier Science Ltd., Oxford, England.

Choi, J. H., Goel, S. C., and Stojadinovic, B. (2000), *"Parametric Tests on the Free Flange Connections"* SAC Joint Venture, Subtask 7.02, Final Report.

Engelhardt. M. D., Winneberger, T., Andrew J. Z. and Timothy J. P. (1996), *"The Dog-bone Connection: Part 2"*, Modern Steel Construction Aug.

Federal Emergency Management Agency. Interim Guidelines: *Evaluation, Repair, Modification and Design of Welded Steel Moment Frame Structures*, 1995, Report FEMA-267.

Federal Emergency Management Agency. Interim Guidelines advisory, FEMA/SAC Joint Venture,1996, No. 1 SAC-96-03

Goel, S. C., and Stojadinovic, B., and Lee, K. H. (1996), *"A New Look at Steel Moment Connections"*, Technical Report MUCEE 96-19 Dept. of Civil and Envir. Engrg. University of Michigan, Ann Arbor, MI.

Goel, S. C., and Stojadinovic, B., and Lee, K. H. (1997), *"Truss Analogy for Steel Moment Connections"*, AISC Engineering Journal, 2nd Qtr., pp 43 – 53.

Lee, K. H., Goel, S. C., and Stojadinovic, B. (1997), *"Boundary effects in Welded Steel Moment Connections"*, Technical Report UMCEE 97-20, Dep. of Civil and Environmental Eng., The University of Michigan, Ann Arbor, MI.

Lee, K. H., Goel, S. C., and Stojadinovic, B. *(1999). "Saint Venant's Principal in steel moment connection."* J. Struct. Engrg., ASCE, submitted for publication.

SAC Joint Venture, *Protocol for Fabrication, Inspection, Testing and documentation of Beam-Column Connection Tests and Other Experimental Specimens,* Report No. SAC/BD-97/02, 1997.

Seismic Provisions for Structural Steel and Building, 1st edition, AISC, Chicago, 1997.

Stojadinović, B., Goel, S. C., and. Lee, K. H. (2000), *"Parametric Tests on Unreinforced Steel Moment Connections",* J. of Struct. Engin. ASCE, Vol. 126, No. 1, 40-49, Jan.

Yu, Q. S., Noel, S., and Uang, C. M. (1997), *"Experimental and Analytical Studies on Seismic Rehabilitation of Pre-Northridge Steel Moment Frame Connection: RBS and Haunch Approaches",* Report No. SSRP –97/08, University of California, San Diego, La Jolla, CA.

Behaviour of Steel Structures in Seismic Areas, Mazzolani & Tremblay (eds)© 2000 Balkema, Rotterdam, ISBN 90 5809 130 9

Fracture mechanics analysis of moment frame joint incorporating welding effects

P. Dong, J. Zhang & F. W. Brust
Center for Welded Structures Research, Battelle, Columbus, Ohio, USA

ABSTRACT: Both the actual connection configurations of the moment resistant frames and local weld details are considered by means of advanced residual stress analysis techniques. Then, the effects of triaxial residual stress state on plastic deformation capacity are discussed based on finite element results on a wide-plate specimen loaded in tension. Finally, a series of fracture mechanics analyses were carried out to assess the effects of residual stresses and strength-mismatch on the fracture behavior. In addition, the current study also investigated several other factors that are commonly believed to affect the weld joint behavior, such as beam strength, weld discontinuity size, and backing plate. As a result, a great deal of insight has been obtained on the fracture behavior of the moment frame welds. Among other things, the results indicate that welding-induced residual stresses significantly increase fracture-driving force due to the presence of high tensile residual stresses. In the meantime, the high triaxial residual stress states identified in these joints can greatly reduce the plastic deformation capability, consequently promoting brittle fracture under dynamic loading conditions.

1 INTRODUCTION

Brittle weld fracture encountered in 1994 Northridge Earthquake damaged steel frames has been a major subject of investigation over the recent years. Various potential factors that could have contributed to the brittle fracture phenomenon are currently under investigation. These include low temperature, low weld metal toughness, weld defects, etc. However, high stress triaxiality due to severe structural restraint conditions, residual stresses, weld strength mismatch effects have not been adequately examined. For instance, there exists ample evidence that residual stresses can play dominant role in the fracture process of highly-restrained welded joints as discussed in Dong and Zhang (1999), Dong and Kilinski, et al (1999), Zhang and Dong (2000), and Brust, et al (1997a-c). The design of welded moment resistant frame connections presents perhaps the most severe mechanical restraint conditions both during welding and in service. Consequently, high weld residual stresses are present. In addition, the triaxiality of the residual stress state in these joints can be significant such that the anticipated plastic deformation may not develop before the fracture driving force reaches its critical value, resulting in brittle fracture.

In this paper, some of the important residual stress characteristics relevant to moment frame joints are first summarized. Then, fracture mechanics analysis results are presented by considering both welding-induced residual stresses and external loading conditions. In addition, weld strength mismatch and backing-bar effects are also investigated. Experimental testing results of laboratory specimens simulating some of the joint features are also discussed, based on a series of comprehensive studies under the auspices of SAC Steel Project (Dong, Kilinski, et al, 1999). Finally, the major findings are summarized in light of both finite element analysis and experimental results.

2 APPROACH

To facilitate the planning and interpretation of the mechanical and fracture mechanics testing, advanced finite element techniques for welded structures were used to assess some of the welding and specimen design effects. As discussed in (Zhang and Dong, 2000), a minimum specimen width requirement of $w/t = 6$ was established. where w is specimen width and t specimen thickness.

Under such conditions, both weld residual stress and structural constraint conditions on plastic deformation capacity can be fully contained in the specimens, as shown in Figure 1, in addition to standard small coupon testing, as discussed in Zhang and Dong (2000). The final designs of two types of test specimens are shown in Figures 2 and 3. The SWPS specimens were intended to capture the structural restraint effects in the width direction, while T-Stub specimens to capture additional load transfer effects. By the means of advanced weld residual stress modeling procedures (see Brust, et al, 1997d, and Dong, et al, 1998), the residual stress distributions in the SWPS specimen is shown in Figure 4. It can be seen that high tensile residual stresses in both the longitudinal and transverse directions of the weld at the mid-section of the specimen. As for the T-Stub specimens, due to the presence of the servere restraint conditions near the column web, significant tri-axial residual stresses are present, as shown in Figure 5.

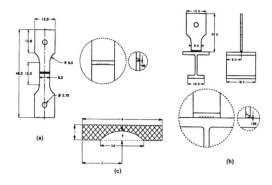

Figure 2. Test specimen geometry: (a) simple welded plate specimen (swps (b) t-stub specimen (c) crack geometry

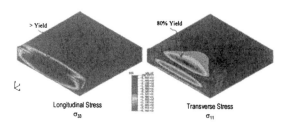

Figure 3. Welding-Induced Residual Stresses in SWPS

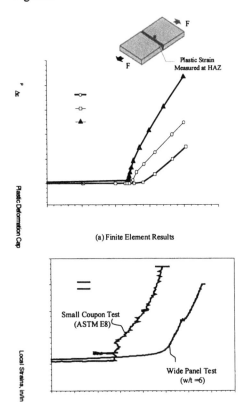

(a) Finite Element Results

(b) Tensile Test Results

Figure 1. Specimen size effects on structural behavior

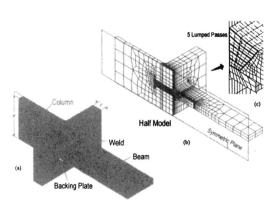

Figure 4. Weld Residual Stress Model for T-Stub Specimens: (a) Specimen geometry, (b) Finite element model, (c) Weld pass details

676

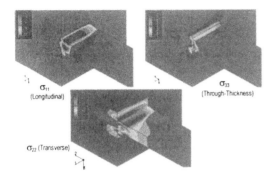

Figure 5. Predicted residual stress distributions in T-stub specimen

2.1 Experimental Procedures

To investigate the effects of weld strength and notch toughness on the fracture behavior of SWPS and T-Stub specimens, an artificially machined crack (circular-arc shaped) is introduced in both SWPS and T-Stub specimens, as shown in Figure 3. Various strength mismatch conditions between base material (BM) and weld metal (WM) were considered in the experimental study, as summarized in Table 1. The crack geometry and position are shown in Fig. 3 for both SWPS and T-Stub specimens, respectively. The crack mouth opening displacement (CMOD) was used to provide a *rough* estimation of crack initiation point from load-CMOD curves measured for calculating stress intensity factors.

Table 1. Base metal and weld metal strength and matching ratio for test specimens.

BM (Yield/Ult[*])	A572 Grade 50 (54/81)			
WM (Yield/Ult.)	NR203 NiC (65/79)	NR311 Ni (81/93)	NR203+ (78/68)	NR305 (88/75)
Marching Ratio: WM/BM	20%/-2%	50%/15%	25%/-4%	39%/9%

* Unit:ksi

2.2 Fracture Mechanics Analysis

A detailed experimental study was first conducted for a SWPS specimen with an artificial surface crack. Fracture toughness was inferred using the experimental data. Then, by considering a typical cross-section of the T-Stub specimen configuration, both strength-match conditions and weld residual stress effects were considered using finite element model. The residual stress effects on fracture driving force in term of both energy release rate and stress intensity factors are examined.

2.3 Test Results

A detailed example is summarized in Figures 6 and 7. The detailed discussions on all the rest results are given in (Dong, et al., 1999). The load-CMOD curve in Figure 7 was directly taken from the measured load-CMOD curve for A572 Gr. 50/NR203 NiC SWPS specimen. A rapid change ("pop-in") of the load-CMOD behavior can be observed, as indicated in Fig. 23b. The stress intensity factor calculations based on the pop-in load are shown in Figure 8. The K_{IC} under plane-strain conditions was inferred as a rough estimated value (Dong, et al., 1999) as a lower bound. The K_{IC} based on J_{IC} (taking into account of nonlinear effects) was inferred from an existing WM database covering a wide range of weld metal toughness for the nuclear industry as an upper bound. It can be seen that the current stress intensity factor solution (indicated by diamond symbols) based on the estimated crack initiation load was well bounded by both the lower bound K_{IC} under plane

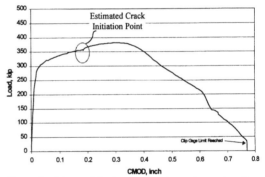

Figure 6. Measured load versus crack-mouth opening displacement (CMOD) – SEPS (A572 Gr. 50/NR203 NiC and fast loading.

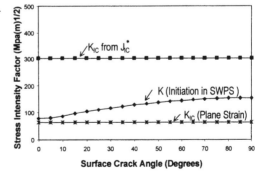

Figure 7. Stress intensity factor solutions along crack front-SWPS (A572 Gr. 50/NR203 NiC and fast loading).

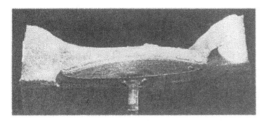

beam material. The external load (σ_p) was applied up to a level slightly beyond the initial yield stress of the beam material. The applied load plotted in the figure was normalized by the yield stress of the weld material (σ_{yw}). It is clear that the energy release rates are much greater for the case with residual stress effects than those without residual stresses. The presence of residual stresses leads to a non-zero initial value of the energy release rate and its more rapid increase as the load increases. However, the energy

strain conditions and the upper bound K_{IC} based on J_{IC}. The maximum stress intensity factor at the deepest position (surface crack angle = 90 degree) was estimated at about 160 MPa (m)1/2 at estimated "pop-in" load of 358 kip. The final fracture surface is shown in Fig. 8.

2.4 Fracture Mechanics Analysis Results

With the finite element weld model as shown in Figure 9, the transverse residual stress (perpendicular to the weld direction) distributions are summarized in Figure 10. The transverse residual stresses are highly tensile at the weld root and rapidly changed to compression within a short distance along line A-A, reversing to tension again near the top weld surface. Such a residual stress distribution is typical of multi-pass welds (e.g., Dong and Zhang, 1999) with a high gradient along the interface between due to the presence of the backing bar.

In the model, the weld flaw was represented by defining duplicated nodes along the crack plane assumed along line A-A. These duplicated nodes were tied (bonded) together in the weld simulation and debonded when the crack was introduced. As a result, the as-welded residual stresses are redistributed to achieve the stress-free condition at the crack surface. Without losing generality, a crack size of 2.5mm was assumed in this study. External loads were then applied at the end of the beam flange by using a displacement control approach. The fracture driving force was determined by the energy release rate at the crack tip. In calculating the method (Zhang and Dong, 2000) was used by releasing the pair of bonded nodes at the crack tip (virtual crack extension) and calculating the work required to close them back. The refined mesh in the weld area was used to ensure an adequate resolution.

The solutions of the energy release rate (G_{tip}) are shown in Figure 11a for the cases with and without residual stress effects. These solutions were obtained from a weld defect of 2.5mm and with A36-Low

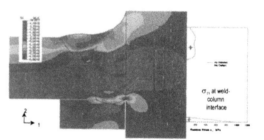

Figure 9. Detailed 2D cross-section model for T-stub specimen.

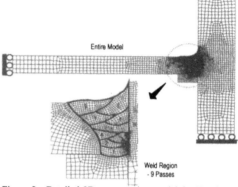

Figure 10. Predicted transverse weld residual stress (σ_{11}) – matched BM/WM: (a) contour plot; (b) distribution along Line A-A.

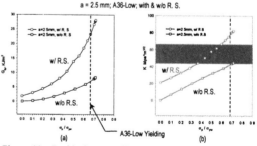

Figure 11. Residual stress effects on fracture driving force: (a) energy release rate; (b) stress intensity factor.

release rate ceases increasing (or increasing rather slowly) after the load reaches the initial yield stress (indicated by the dashed line) of the beam material. This is different from the phenomenon commonly observed in the elasto-plastic fracture behavior for homogeneous materials in which the crack driving force (e.g., J-integral) exhibits a sudden increase as the material is fully yielded. This can be explained by considering the yield stress mismatch effects. The weld joint analyzed represents an over-matched weld metal strength (60% higher than that of the beam material). The yield strength of the column material is relatively close to the weld strength. As the beam becomes fully yielded under the external tension, the further change in the stress state at the crack tip becomes negligible.

The corresponding stress intensity factor solutions are plotted in Figure 11b. Again, the effects of weld residual stresses are clearly seen. The fracture toughness (K_{IC}) of the weld material typically ranges from 44 MPa*m$^{1/2}$ to 66 MPa*m$^{1/2}$ from the testing results (see Figure 7), as shown in a shaded region. Thus, the present results considering the residual stress effects predict that brittle fracture can occur before the external load reaches the yielding stress of the beam material, while the results without considering residual stresses indicate that the crack remains stable even after the beam is fully yielded.

2.5 Beam Strength Effects

The effects of beam yield strength on the energy release rate are shown in Figure 12a. The corresponding stress intensity factor solutions are shown in Figure 12b. These results were obtained from a crack size of 2.5mm and with A36-Low and A36-High beam materials respectively. Weld residual stresses were considered in both cases. For A36-Low, the results are the same as those shown in Figure 11. These results indicate that the lower-slightly higher crack driving force before the beam

is yielded. The initial crack driving force is also higher for A36-Low due to higher residual stresses at crack tip. In the present case, over-matching conditions between beam and weld metal yield strengths exist. The increase in fracture driving force in terms of either energy release rate or stress intensity factor is consistent with an early systematic study (Dong and Zhang, 1999).

3 CONCLUDING REMARKS

Based on the present study, the following observations can be made:

a) Weld residual stresses can have significant effects on fracture behavior of moment frame connections due to the presence of severe restraint conditions both during welding and in service

b) The effects of weld residual stresses on fracture behavior of moment frame joints are primarily in the form of stress tri-axiality which can significantly reduce the plastic deformation capacity and elevate fracture driving force.

c) In developing the property requirements for both base material and weld metals, specimen size and joint restraint effects must be considered.

4 REFERENCES

Brust, F. W., Zhang, J., and Dong, P., 1997a. "Pipe and Pressure Vessel Cracking: The Role of Weld Induced Residual Stresses and Creep Damage during Repair", *Transactions of the 14th International Conference on Structural Mechanics in Reactor Technology (SMiRT 14)*, Lyon, France, Vol. 1, pp. 297-306.

Brust, F. W., Dong, P., and Zhang, J., 1997c. "Influence of Residual Stresses and Weld Repairs on Pipe Fracture", Approximate Methods in the Design and Analysis of Pressure Vessels and Piping Components, W. J. Bees, Ed., PVP-Vol. 347, pp. 173-191.

Brust, F. W., Dong, P., and Zhang, J., 1997d. "A Constitutive Model for Welding Process Simulation using Finite Element Methods", Advances in Computational Engineering Science, S. N. Atluri and G. Yagawa, Eds., pp. 51-56.

Dong, P., Zhang, J., and Li, M.V., 1998. "Computational Modeling of Weld Residual Stresses and Distortions – An Integrated Framework and Industrial Applications, ASME Pressure Vessel

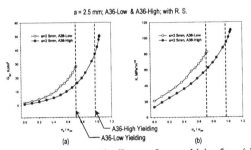

Figure 12. Beam strength effects on fracture driving force (a) energy release rate; (b) stress intensity factor.

and Piping Conference Proceedings, PVP-Vol. 373, *Fatigue, Fracture, and Residual Stresses,* pp. 311-335.

Dong, P., Kilinski, T., Zhang, J., and Brust, F.W., 1999. "Effects of Strength/Toughness Mismatch on Structural and Fracture Behaviors in Weldments (Sub-Tasks 5.2.1 and 5.2.2), *Battelle Final Report* to SAC Steel Project.

Zhang, J., Dong, P., and Brust, F.W., 1998. "Residual Stress Analysis and Fracture Assessment of Welded Joints in Moment Resistance Frames," Proceedings of International Conference on Advances in Computational Engineering Science, Vol. II: *Modeling and Simulation Based Engineering,* Atluri, S.N., and O'Donoghue, P.E., eds., pp 1894-1902.

Zhang, J., Dong, P., Brust, F. W., William J. Shack, Michael E. Mayfield, Michael McNeil, 2000. "Modeling Weld Residual Stresses in Core Shroud Structures", *Nuclear Engineering and Design,* 195, pp. 171-187.

Zhang, J. and Dong, 2000. "Residual Stresses in Welded Moment Frames and Effects on Fracture," *ASCE Journal of Structural Engineering,* March, No. 3

Behaviour of Steel Structures in Seismic Areas, Mazzolani & Tremblay (eds) © 2000 Balkema, Rotterdam, ISBN 90 5809 130 9

Reduced beam section welded steel moment frame connections

G.T. Fry & S.L. Jones
Texas A&M University, College Station, USA

M.D. Engelhardt
University of Texas, Austin, USA

ABSTRACT: Improved welding procedures used in combination with the reduced beam section (RBS) connection detail appear to perform much better in steel moment frames than details used prior to the 1994 Northridge, CA earthquake. As a result, the likely failure mode for RBS moment frames is beam instability during very large (> 3%) inelastic drift excursions rather than weld fracture at relatively small drifts (< 1.5%). The question now becomes what is an acceptable drift limit for design-level earthquakes? To address this question, one must consider the consequences of beam instability. This paper presents results from two tests of full-scale interior moment frame subassemblies with composite slabs. Measurements and observations from large post-buckling drift excursions are discussed.

1 REDUCED BEAM SECTION CONNECTIONS

Since the 1994 Northridge, CA earthquake, researchers have performed extensive testing on welded steel special moment resisting frames (SMRF). A subset of the testing involves experiments on SMRFs with reduced beam sections (RBS). Chen et al (1996), Iwankiw and Carter (1996), and Zekioglu et al. (1997a, 1997b) investigated the performance of tapered cut RBS details in single-sided beam-column subassemblages. Engelhardt et al (1996, 1997), Popov et al (1998), Tremblay et al (1997), Plumier (1990, 1997), and Uang (1998, 1999) performed tests on radius cut RBS details in single-sided beam-column subassemblage tests. Engelhardt and Venti (1999) and Fry et al (1999) performed tests on radius cut RBS details in double-sided beam-column subassemblages (Figure 1) to account for higher demands on the panel zone in interior frame joints. The tests on RBS connection details indicate that the RBS, when used in conjunction with improved post-Northridge welding specifications, helps reduce both the potential for brittle fracture of the welded joint and the plastic strain demand on the beam flange-column flange weldment.

The likely failure mode for RBS moment frames, which incorporate improved welding details, is beam instability during very large (> 3%) inelastic drift excursions rather than weld fracture at relatively small drifts (< 1.5%). The question now becomes what is an acceptable drift limit for design-level earthquakes? To address this question, one must consider the consequences of beam instability. Experiments indicate that beam local buckling precedes more serious global instabilities. Among the contributors to buckling initiation and global instability are beam torsion and weak axis bending. Fry et al. (1999) developed a method for measuring the beam-end torsion in a double-sided beam-column subassemblage experiment. This study presents the methodology for experimentally determining the beam-end torsion, in conjunction with beam-end vertical load, and discusses the meaning of measurement results from two tests.

2 EXPERIMENTAL PROGRAM

2.1 *Beam-end boundary conditions*

The beam ends in tests of double-sided beam-column subassemblage performed at Texas A&M University were restrained from moving vertically, laterally, and from rotating out of plane. The restraint was provided by roller bearings attached at the neutral axis location of the beams as shown in Figure 2. The roller bearings were restrained vertically by guides attached to vertical reaction frames (Figure 3). Each of the four roller bearings was instrumented with strain gauges (Figure 2) and subsequently calibrated against load in an MTS 500-kip testing machine.

The beam-end load determined from the calibrated roller bearings was verified during testing using a redundant measurement of column tip load from the

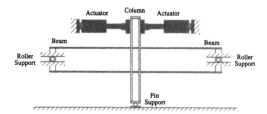

Figure 1. Double-sided beam-column subassemblage experiment set-up.

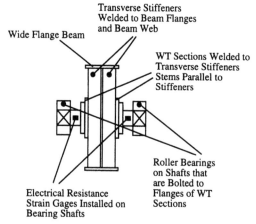

Figure 2. Schematic drawing of roller support detail at beam ends. The roller bearings are restrained from moving vertically. Guides are also provided to prevent the beam ends from moving laterally.

at the beam end when looking toward the end of the beam while facing the column. The equation for torque will depend on the beam end and on the direction of the applied column tip load. For purposes of the present discussion,

Figure 3. Photo of beam-end roller bearing, designed to resist vertical load and torsion, and its guides.

actuator load cells. First, beam-end load is calculated using the roller bearing readings.

Summing moments about the pin at the base of the specimen, column tip horizontal load is related to the beam-end vertical loads by the following equation:

$$P = \frac{(R2 - R1)L}{2h} \qquad (1)$$

where P is column tip load (positive to the right), R1 is the left-hand beam reaction (positive upward), R2 is the right-hand beam reaction (positive upward), L is the span between the ends of the beams, and h is the column height. Column tip load calculated using equation 1 may be compared to measured column tip load to verify the load cell accuracy of the roller bearings.

The beam-end torsion can be determined conveniently from the load cell readings using equilibrium of moments about the shear center at each beam end. For the purposes of this study, positive torsion will be defined as a clockwise-applied torque

the specimen will be loaded in the east-west direction and the bearings will be on the north and south sides of the beam. Figure 4 illustrates equilibrium of the west beam end with a westward column tip load. Pushing the column tip to the east results in downward loads acting on the roller bearings (Figure 5). The equations for beam-end torque resulting from moment equilibrium of the west beam end follow:

$$T = (|R_N| - |R_S|) * \frac{s}{2} \qquad P > 0 \qquad (2a)$$

$$T = (|R_S| - |R_N|) * \frac{s}{2} \qquad P < 0 \qquad (2b)$$

where R_N is the north bearing load cell reading, R_S is the south bearing load cell reading, and s is the center to center spacing of the roller bearings at each end of the beam. Column tip load, P, is defined as positive when directed toward the west. A similar analysis of the east beam also yields equations 2a and 2b.

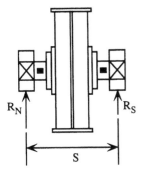

Figure 4. Positive torque viewing end of West beam when column tip load is pushing the specimen to the West ($R_N > R_S$).

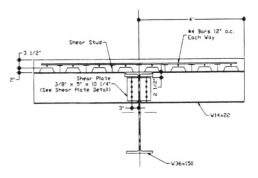

Figure 6. Composite slab details.

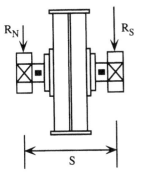

Figure 5. Positive torque viewing end of West beam when column tip load is pushing the specimen to the East($R_N < R_S$).

Figure 7. Photo of experimental set-up using double-sided moment frame subassemblage.

2.2 Test specimens

Two interior subassemblage test specimens, one without panel zone doubler plates (DBWWC) and one with panel zone doubler plates (DBBWSPZC), were constructed and tested at Texas A&M University. Both specimens comprised a wide-flange column with wide-flange beams attached to both sides of the column (Figure 1). The column was a W14x398 made with ASTM A572 Grade 50 material. Specimen DBBWSPZC had 19 mm (0.75 in.)-thick stiffener plates attached to both sides of the column web in the panel zone. Both beams were W36x150 made with ASTM A572 Grade 50 material. The beam webs were bolted to shear tabs that were shop welded to the column flanges in order to facilitate welding the beam flanges to the column flange. The beam flanges were field welded to the column flange with complete joint penetration (CJP) single bevel groove welds made with the self-shielded flux cored

arc-welding process using an E70T-6 electrode. The bottom flange backing bars were removed and a small reinforcing fillet (E71T-8) was placed at the root of the groove weld. The top flange backing bars were left in place and sealed along their bottom edges with fillet welds (E71T-8). In specimen DBWWC, the beam webs were field welded to the column flange with CJP single bevel groove welds made with the self-shielded flux cored arc-welding process using an E71T-8 electrode. In specimen DBBWSPZC The beam webs were bolted to the shear tab using ten 1" A490 bolts tightened using the turn-of-the-nut method. In order to square the beams with the columns, shim plates were placed between the beam web and the shear tab.

The 3-1/2-inch thick composite slab on both specimens was supported by 6 W14x22 beams attached to the center of the column web and to the beam 8 feet from the center of the column web (Figure 6). The slab spanned 8 feet perpendicular to the w36x150 beams and 20 feet parallel to the beams.

Figure 8. Verification of beam load cell readings for specimen DBWWC.

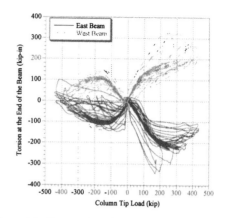

Figure 10. Beam-end torsion versus column tip load for specimen DBWWC.

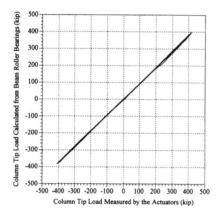

Figure 9. Verification of beam load cell readings for specimen DBBWSPZC.

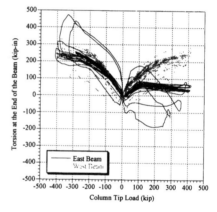

Figure 11. Beam-end torsion versus column tip load for specimen DBWWSPZC.

2-inch metal decking was attached to the beams using 5-inch tall shear studs spaced at 1 foot. No shear studs were attached between the column face and the RBS nor within the length of the RBS to avoid stress concentrations and embrittled heat affected zone material where extreme levels of inelastic strain were expected. The slab was reinforced in both directions using Grade 60 #4 bars spaced at one-foot on center. The compressive strength of the concrete on the day testing was 6800 psi for specimen DBWWC and 6100 psi for specimen DBBWSPZC.

The experiments were designed to emulate the deformation of a special moment resisting frame subjected to lateral loading. To this end, the column base was pin-connected to the lab floor and the beam ends were restrained from moving vertically by bearing guides supported by vertical reaction frames (see Figure 1). Two 220-kip actuators supported by horizontal reaction frames applied the lateral column tip load. Figure 7 is a photograph of the experimental set-up with a specimen in place. Both specimens were instrumented to measure column tip load, column tip displacement, and beam-end vertical load. Data readings were taken once every two to three seconds during the experiments. Torsion at each beam –end was calculated using equations 2.

Each specimen was loaded cyclically, with the lateral displacement amplitude increasing incrementally at pre-determined cycles, until either failure of one of the components occurred or limits of the testing equipment were reached. The tests were controlled by giving the actuators a deflection rate of one inch in 90 seconds and reversing the loads at prescribed story drifts.

3 EXPERIMENTAL RESULTS

The beam-end load cells gave reasonable readings during tests on both specimens DBWWC and DBBWSPZC. The verification of these numbers by comparison of column tip load calculated using equation 1 with measured column tip load is shown in figures 8 and 9. In both tests, specimen stiffness and strength degraded at the onset of local buckling of the beam webs and flanges in the RBS region. Peak load, or peak resistance threshold (PRT), occurred at a total (elastic plus plastic) story drift ratio of approximately 4%. Beam-end torsion was calculated for each test and plotted versus measured column tip load. Results from these measurements and calculations are presented as Figures 10 and 11.

Specimen DBWWC performed well under cyclic lateral column loading, remaining globally stable beyond local buckling of the beam in the RBS. Local buckling of the RBS initiated in the beam web near the flanges at a column tip load of 440 kips. As the column tip displacement was further increased, the maximum load for each cycle slowly decreased due to further web and flange local buckling and twisting of the shape. When the story drift reached approximately 6%, substantial local flange buckling and strength deterioration had occurred. Shear studs connecting the slab to the beams began to fracture at the 4% drift cycles. Nearly all of the studs connecting the slab to the main girder had fractured before the 7% drift cycles. At the first half-cycle to 7%, ductile tearing of the upper flanges in the RBS initiated. Out-of-plane movement of the column was negligible throughout the test, even when the specimen was pushed to 11%. By the 11% drift level however, the top beam flanges were completely fractured throught trhe middle of the RBS and the web was beginning to tear. The resistance of the specimen at 11% drift was approximately 20% of the resistance at the PRT.

Specimen DBBWSPZ also exhibited stable behavior in the early cycles, but testing was ultimately stopped due to global instability. Local buckling of the RBS initiated in the beam web near the flanges at a column tip load of 420 kips. Strength degradation, local buckling patterns, and shear stud fractures were all similar to the behavior described for specimen DBWWC. As the column was pushed east to a story drift of approximately 5%, out-of plane movement of the column became apparent. The test was terminated after the first half-cycle to 7% drift when it became apparent that the out-of-plane instability was becoming more severe. The beam flanges underwent substantial local buckling, but no ductile tearing had occurred by the end of the test.

4 CONCLUDING REMARKS

1) In the present tests, the peak resistance to column tip load occurred at approximately 3.5% to 4% total (elastic plus plastic) drift. This level of drift has been termed herein the Peak Resistance Threshold (PRT).

2) Improved connection detailing, combined with improved welding procedures allow a steel moment frame to be exercised well beyond the PRT – without weld fracture.

3) During testing through post-PRT drift levels, beam webs and flanges experience substantial local buckling deformations which in turn cause tearing in beam flanges and large twisting moments in the beams.

4) In the present testing, shear studs connecting the slab to the main girder began to fracture at PRT drift levels.

5) Combined with progressive loss of connectivity between girder and slab, the progressive increase in girder twisting moment could represent a significant destabilizing influence on the frame as a whole. Thus, it would appear that a rational design limit for drift, under the effect of design-level earthquakes, is a number smaller than the expected value of frame PRT.

5 ACKNOWLEDGEMENTS

This work was coordinated under the auspices of the SAC Joint Venture and supported financially by the Federal Emergency Management Agency and units within the Texas A&M University System (the Texas Engineering Experiment Station, the Department of Civil Engineering, and the Center for Building Design and Construction). The authors wish to thank the following individuals for invaluable assistance: James Malley, Stephen Mahin, Charles Roeder, Robert Dodds, Stanley Rolfe, John Barsom, Subhash Goel, Bozidar Stojadinovic, Loren Lutes, and Walter P. Moore, Jr.

6 REFERENCES

Engelhardt, M. D., T. Winneberger, A. J. Zekany & T. J. Potyraj 1996. The Dogbone Connection, Part II. *Modern Steel Construction*. American Institute of Steel Construction, Inc.

Engelhardt, M. D., T. Winneberger, A. J. Zekany & T. J. Potyraj 1997. Experimental Investigation of Dogbone Moment Connections. *Proceedings; 1997 National Steel Construction Conference*. American Institute of Steel Construction, Chicago.

M. D. Engelhardt & M. Venti 1999. Unpublished preliminary test reports for SAC Phase 2 tests. University of Texas at Austin.

Fry, G., S. L Jones & S. D. Holliday 1999. Unpublished preliminary test reports for SAC Phase 2 tests. Texas A & M University.

Plumier, A. 1990. New Idea for Safe Structures in Seismic Zones. *IABSE Symposium - Mixed Structures Including New Materials*. Brussels.

Plumier, A. 1997. The Dogbone: Back to the Future. *Engrg. J.*, American Institute of Steel Construction, Inc. 2nd Quarter.

Popov, E. P., T. S. Yang & S. P. Chang 1998. Design of Steel MRF Connections Before and After 1994 Northridge Earthquake. *Engrg. Struct.*, 20(12), 1030-1038, 1998.

Tremblay, R., N. Tchebotarev & A. Filiatrault 1997. Seismic Performance of RBS Connections for Steel Moment Resisting Frames: Influence of Loading Rate and Floor Slab. *Proceedings, Stessa '97*, Kyoto, Japan.

Uang, C. M. 1998. Unpublished preliminary test reports for SAC Phase 2 RBS tests. University of California at San Diego.

Uang, C. M. 1999. Unpublished preliminary test reports for SAC Phase 2 RBS tests. University of California at San Diego.

Behaviour of Steel Structures in Seismic Areas, Mazzolani & Tremblay (eds) © 2000 Balkema, Rotterdam, ISBN 90 5809 130 9

Relating ground motion spectral information to structural deformation demands

A. Gupta
Exponent Failure Analysis Associates, Menlo Park, Calif., USA

H. Krawinkler
Department of Civil and Environmental Engineering, Stanford University, Calif., USA

ABSTRACT: A procedure is presented for estimating the global and local seismic deformation demands for steel frame structures. In this procedure the ground motion spectral displacement at the first mode period of the structure is related to the roof drift demand for the multi-degree-of-freedom (MDOF) structure, which in turn is related to the story drift demands. The relationships are based on statistical values obtained from time history analyses of nine steel moment frame structures subjected to sets of ground motions representative of different hazard levels. The story drift demand is related to the beam and panel zone plastic deformation demands using strength and stiffness properties of the components. The procedure should prove useful for estimating strength and stiffness requirements in conceptual design and estimating deformation demands for performance assessment.

1 INTRODUCTION

A simplified procedure that provides quick and reasonable estimates of the structural deformation demands is attractive from many standpoints; 1) it permits estimation of parameters intrinsic to a deformation based seismic evaluation of the structure, including global (e.g., roof), intermediate (e.g., story), and local (component) deformation demands, 2) it provides estimates of seismic demands to serve as design targets in the conceptual design phase, and 3) it provides such estimates without requiring the effort associated with parametric nonlinear time history analyses and complex analytical modeling of structures. The demands obtained from the procedure can also serve as an indicator of the need for more detailed analysis in cases where the demand estimates exceed the boundaries of the simplified process.

Figure 1 illustrates a loop that relates the spectral displacement demand at the structure's first mode period to local element plastic deformation demands. Quantitative information for this loop usually is obtained from rigorous structural analysis. The objective of this paper is to explore simple relationships that can be established between individual parts of the loop for essentially regular steel moment resisting frame (SMRF) structures. The detailed background for arriving at these relationships is not discussed in this paper; only representative results are presented. For details the reader is referred to Gupta & Krawinkler (1999).

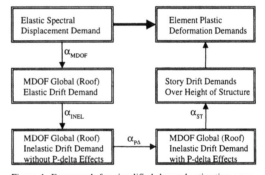

Figure 1. Framework for simplified demand estimation procedure

The loop illustrated in Figure 1 assumes that stable relationships can be found between the spectral displacement demand at the first mode period of the structure and the system and element deformation demands, using the following definitions:

1 *MDOF modification factor, α_{MDOF}:* a factor that relates the elastic spectral displacement demand at the first mode period of the structure to the elastic roof drift demand of the MDOF structure, neglecting P-delta effects.

2 *Inelasticity modification factor, α_{INEL}:* a factor that relates the elastic roof drift demand to the inelastic roof drift demand, neglecting P-delta effects.

3 *P-delta modification factor, α_{PA}:* a factor that accounts for the effect of P-delta on the inelastic roof drift demand.

4 *Story drift modification factor, α_{ST}:* a factor (or function) that relates the story drift demands to the roof drift demand.

5 *Component deformation modification functions:* a set of relationships that relate the story drift demand to the component plastic deformation demands.

Portions of an extensive study on the seismic performance of steel frame structures, which was carried out as part of the SAC steel program, were used to evaluate the potential of estimating global and local demands for essentially regular steel frame structures by means of the suggested modification factors and functions. Extensive linear and nonlinear dynamic analyses of nine SMRF structures form the basis for the quantitative information presented here. These structures are of various heights (3, 9, and 20 stories) and are located in three regions of different seismicity (Los Angeles, Seattle, and Boston). The first mode properties of the structures are presented in Table 1.

Table 1. First mode properties of model buildings.

		Period (seconds)	Participation Factor
LA	3-story	1.01	1.30
	9-story	2.24	1.38
	20-story	3.74	1.36
Seattle	3-story	1.36	1.29
	9-story	3.06	1.37
	20-story	3.46	1.42
Boston	3-story	1.97	1.28
	9-story	3.30	1.37
	20-story	3.15	1.45

The response of the structures is evaluated for sets of ground motions (each set consisting of 20 records) representative of three hazard levels (Somerville 1997). The different hazard levels have a probability of exceedance of 2% in 50 years (denoted as 2/50), 10% in 50 years (10/50), and 50% in 50 years (50/50, for Los Angeles only). The analytical models were 2-D representations of the perimeter frames in which inelastic behavior was modeled by means of point plastic hinges with 3% strain hardening. Details of the structures, analytical modeling, and the ground motions used are given in Gupta & Krawinkler (1999).

2 ESTIMATION OF ROOF DRIFT DEMAND

The MDOF systems inelastic roof drift inclusive of structure P-delta effect is estimated in three steps. First, the spectral displacement is related to the roof drift for an elastic MDOF system excluding P-delta

effects. The elastic roof drift is then modified to represent inelastic behavior of the structure. Finally, the inelastic drift demand is modified to include the P-delta effect.

Tables 2 and 3 summarize some of the time history analyses results that form the basis for the MDOF and inelasticity modifications factors discussed here. The statistical values are presented for the 10/50 set, the 2/50 set, and all 40 records combined (i.e., all records in the 10/50 and 2/50 sets). It should be noted that most of the LA 2/50 records are near-fault records (albeit rotated by 45 degrees with respect to the fault-normal direction) with pulse-type characteristics.

Table 2. Statistical values of MDOF modification factor.

			Ground Motion Sets		
			10/50	2/50	Combined
LA	3-story	Median	1.27	1.27	1.27
		Dispersion	0.05	0.05	·0.05
	9-story	Median	1.46	1.50	1.48
		Dispersion	0.10	0.13	0.12
	20-story	Median	1.58	1.51	1.54
		Dispersion	0.14	0.11	0.13
Boston	9-story	Median	1.80	1.69	1.74
		Dispersion	0.22	0.15	0.19
	20-story	Median	2.23	2.01	2.12
		Dispersion	0.25	0.17	0.22

Table 3. Statistical values of inelasticity modification factor.

			Ground Motion Sets		
			10/50	2/50	Combined
LA	3-story	Median	0.80	0.95	0.87
		Dispersion	0.21	0.56	0.43
	9-story	Median	0.78	0.69	0.73
		Dispersion	0.22	0.29	0.26
	20-story	Median	0.73	0.70	0.72
		Dispersion	0.21	0.18	0.19

The median values are calculated as the exponent of the mean of the natural log values of the data set. The associated dispersion values are computed as the standard deviation of the natural log values of the data set. All statistical values are obtained by first computing the modification factor for each earthquake record, and subsequently computing the median and dispersion for the set of modification factor values.

2.1 *Estimation of MDOF Elastic Roof Drift Demand*

The roof displacement demand for an elastic MDOF system can be computed using an appropriate combination of modal displacements. For systems whose roof response is not greatly affected by higher mode effects (i.e., systems subjected to ground motions that do not have a relatively small spectral displacement at the first mode period), the first mode participation factor, is expected to provide a good estimate for the relationship between the spectral displacement at the first mode period and the roof displacement demand.

Table 2 shows that for the 3-story structure in LA the displacement response is controlled by first mode vibrations, as the modification factor is very close to the first mode participation factor PF_1 given in Table 1. Furthermore, the associated dispersion in the data set is very small. These results are independent of the intensity of the ground motions.

The dispersion associated with the MDOF modification factor values increases with an increase in the number of stories to 9 and 20 for the LA structures. The median values also exceed the first mode participation factor, indicating increased higher mode participation. The influence of higher modes, over all the ground motions, is reflected in the median modification factor being about 10% larger than the first mode participation factor. The statistical values are rather insensitive to the severity of the ground motions, which is somewhat surprising since the LA 2/50 set of ground motions contains mostly near-fault records. Data points for the MDOF modification factor, plotted against the spectral displacement at the first mode period (normalized by story height) are shown in Figure 2 for the LA 20-story structure.

defined as the MDOF modification factor α_{MDOF}, is best evaluated from an appropriate modal analysis. A good estimate is the first mode participation factor PF_1 – unless higher mode spectral displacements are relatively very large. For structures with a period exceeding about 2 seconds it is advisable to use $1.1 \times PF_1$ for the MDOF modification factor. For ground motions with large higher mode spectral displacements, α_{MDOF} should be obtained from a modal analysis.

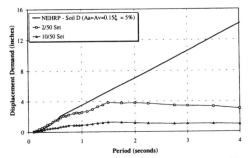

Figure 3. Median displacement spectra for Boston 10/50 and 2/50 ground motion sets.

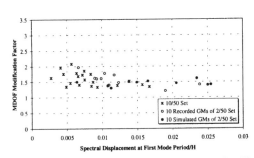

Figure 2. Data points for MDOF modification factor; LA 20-story structure.

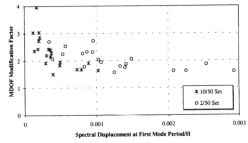

Figure 4. Data points for MDOF modification factor; Boston 20-story structure.

The influence of higher mode effects on elastic roof displacements becomes large in the response of the 9- and 20-story Boston structures. This is reflected in the statistical values shown for the Boston structures in Table 2. The median modification factors are much larger than the first mode participation factors and are associated with higher dispersion values. The reason is the relatively small spectral displacements of the Boston ground motions at the first mode period of the structures, see Figure 3. Figure 4 shows the data for the MDOF modification factor for the Boston 20-story structure. As expected, the MDOF modification factor is particularly large for records with a small first mode spectral displacement.

In summary, the relationship between the spectral displacement and roof elastic displacement demand,

2.2 Modification of MDOF Elastic Roof Drift Demand for Inelasticity

The effect of inelasticity on the roof drift demand is expected to depend on the ground motion and structural characteristics. Past work on this subject (Seneviratna & Krawinkler 1997) has revealed that (1) inelasticity significantly increases the roof drift for short period structures (first mode period is shorter than the transition period between the constant acceleration and constant velocity range of the spectrum), and in this range of periods the increase depends strongly on the period and the extent of inelasticity (global ductility demand); (2) inelasticity somewhat decreases the roof drift for structures whose first mode period is longer than the transition period, and in this range of periods the decrease is only mildly dependent on the global ductility demand; (3) the amount of decrease in not very sensitive to the period provided that it is clearly longer than the transition period; and (4) the amount of decrease can be evaluated from studies with inelastic SDOF systems.

The global effect of inelasticity is described here by the factor α_{INEL}, which relates the roof displacement of the inelastic structure to that of the elastic structure. The statistical values for the inelasticity modification factor given in Table 3 are consistent with and support the observations made in the previous paragraph. For the steel structures studied the first mode period is clearly longer than the transition period of the median 10/50 and 2/50 spectra, and therefore the inelasticity factor is consistently smaller than 0.8 and has an acceptable dispersion. The exception is the response of the LA 3-story structure ($T_1 = 1.0$ sec.) under the 2/50 set of ground motions. This anomaly arises due to use of 10 simulated ground motions that have peculiar characteristics at about 1.0 seconds due to the process used to generate these records. Removing the 10 simulated records from the 2/50 set reduces the median to 0.79 and the dispersion to 0.39. The response of the 9- and 20-story structures is not affected much by removal of these 10 records. With this modification, the medians of all sets are between 0.80 and 0.69; a rather narrow range.

An example of data points for the inelasticity modification factor, plotted against the elastic roof drift angle, is presented in Figure 5 for the LA 9-story structure.

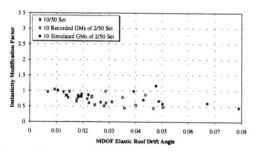

Figure 5. Data points for inelasticity modification factor; LA 9-story structure.

In summary, the median value of the inelasticity factor for regular structures with a first mode period greater than about 1.0 second is between 0.7 and 0.8, if significant inelastic behavior is expected (say, ductility greater than 2). An average value of 0.75 appears to be a good estimate for all long period structures and the sets of ground motions used in this study. The measure of dispersion is of the order of 0.3. Median MDOF values are somewhat smaller than median SDOF values, and the latter can serve as conservative estimates.

The MDOF and inelasticity modification factors can be combined to provide an estimate of the spectral displacement and the MDOF inelastic roof displacement. For long period structures the combined factor is estimated as $1.1PF_1 \times 0.75 = 0.825PF_1$. Using the LA 20-story structure as an example, the combined factor is $0.825 \times 1.36 = 1.12$. Using individual data points for the ratio of MDOF roof displacement to first mode spectral displacement, the results shown in Figure 6 are obtained. The concentration of higher than median values observed at low spectral displacement demands (around $S_d/H = 0.005$) in Figure 6 are attributed to two reasons; 1) the MDOF modification factor is more significantly affected by higher modes in this range, thus, $1.1PF_1$ underestimates the elastic roof drift demand, and 2) since the demands are low, the inelasticity factor value of 0.75 is non-conservative in this region, where the value is closer to 1.0.

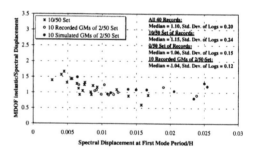

Figure 6. Combined MDOF and inelasticity factor; LA 20-story structure.

The estimates summarized here are based on case studies with SMRF structures and are supported by other studies on generic structures (Seneviratna & Krawinkler 1997). The results are for standard ground motions and should not be generalized to soft-soil or near-fault type ground motions. A study by Krawinkler & Alavi (1999) suggests that the inelasticity modification factor may be sensitive to the pulse characteristics of fault-normal ground motions.

2.3 Modification of Inelastic Roof Drift Demand to Include Structure P-Delta Effect

P-delta effect may result in a negligible influence on the inelastic response of the structure or may bring the structure to a state of dynamic instability. The latter may occur when the ground motion is strong enough to drive the structure into the range of negative post-yield stiffness, as seen from a global load-deflection pushover curve. For such cases the response becomes very sensitive to ground motion characteristics, and the potential for dynamic instability is not insignificant. Furthermore, only a detailed non-linear dynamic analysis would indicate the full impact of the P-delta effect on the response of the structure. The P-delta effect is discussed in detail in Gupta & Krawinkler (2000). An important design issue is to provide the structure with sufficient post-yield reserve strength so that the negative stiffness range is not entered under the ground motions the structure is expected to experience. The quoted reference provides recommendations how this can be accomplished.

In the simplified demand estimation process, the effect of structure P-delta on the inelastic roof drift demand is accounted for using the P-delta modification factor $\alpha_{P\Delta}$. Figure 7 presents data points for this factor, defined as the ratio of roof displacements with and without P-delta effects, for the LA 20-story structure. The LA 20-story structure attains a negative post-yield stiffness at about 0.02 roof drift angle. For ground motions that drive the structure to roof drift demands less than 0.02, the P-delta effect changes the roof drift demand by about −15% to +30%. The median increase in drift demand is only 2%. P-delta reduces the effective stiffness of the structure thereby increasing the fundamental period and thus, changing the interaction between the ground motion and the structure. The response of the structure is, however, greatly affected when the ground motion drives the structure past the 0.02 drift demand level. For example, as shown in Figure 7, at a roof drift demand without P-delta of about 0.022 the LA 20-story structure exhibits P-delta modification factors of 0.61, 0.96, and 1.95 for three different records.

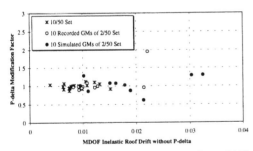

Figure 7. Data points for P-delta modification factor; LA20-story structure.

Thus, as long as the negative post-yield stiffness region is not entered (a very desirable design objective) the P-delta modification factor $\alpha_{P\Delta}$ is expected to be relatively small. A reasonable estimate may be $\alpha_{P\Delta} = 1/(1-\theta)$, where θ is the maximum elastic story stability coefficient, defined as $P_i\Delta_i/V_ih_i$. P_i is the total vertical load acting above story i, Δ_i is the elastic story drift associated with the story shear V_i, and h_i is the height of story i.

In summary, the various modification factors that relate the spectral displacement demand to the inelastic roof drift demand inclusive of P-delta effect have been quantified. These factors are applicable for regular SMRF structures whose roof displacement is not greatly affected by higher mode effects and for which the P-delta effect is sufficiently contained so that the structure is not driven into the post-yield negative stiffness range. The factors are developed for ground motions that do not have severe soft-soil or near-fault pulse type characteristics.

3 RELATIONSHIP BETWEEN ROOF DRIFT AND STORY DRIFT: α_{ST}

Many attempts are reported in the literature to determine the distribution of story drift demands over the height of the structure as a function of the roof drift. However, this relationship is strongly dependant on the ground motion and structure characteristics. Figure 8 presents, as an example, medians of the ratio of story drift demands to roof drift demand over the height of the 9-story structures for the 10/50 and 2/50 sets of ground motions. The results presented in this figure coupled with similar results obtained for other structures lead to the following conclusions: (1) the distribution of story drift demands over height is not strongly affected by the intensity of ground motions [except if P-delta effects become prominent or the ground motions have pulse type characteristics], (2) there is no evident pattern based on height, number of stories, or geographic location, that can be generalized, (3) structures with significant higher mode effects (Boston structures) exhibit very large ratios for the upper stories, and (4) the median of the maximum story drift in every story usually is larger than the median of the maximum roof drift.

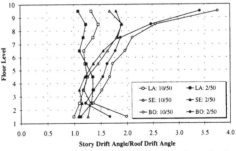

Figure 8. Variation over height of median values of ratio of maximum story drift to maximum roof drift.

An estimation of story drift demands over the height of the structure as a function of the roof drift is difficult to accomplish on account of the dependence of the relationship on a multitude of factors, including, among others, relative strength and stiffness of the stories, higher mode and P-delta effects, and characteristics of the ground motions. The pushover analysis procedure is useful in estimating these story drift demands for many cases. It will provide good predictions for low-rise structures, however, it is not a good means for predicting drift demand distributions in structures subjected to ground motions that cause significant higher mode effects.

The only ratio that can be generalized from the analyses is the ratio of maximum story drift over the height of the structure to the maximum roof drift. The median values for the different structures and ground motions are presented in Figure 9. The following conclusions can be drawn from this figure: (1) the median values increase with an increase in number of stories, (2) median values are not very sensitive to the ground motion intensity, and (3) structures whose response is controlled by higher mode effects (Boston 9- and 20-story structures) have very large median values (the same applies to the dispersion values not shown here).

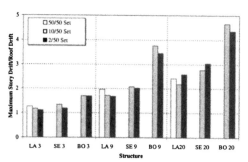

Figure 9. Median values of ratio of maximum story drift to roof drift.

The patterns that can possibly be generalized for other regular SMRF structures are that the maximum story drift to maximum roof drift ratio is small (about 1.2) for low-rise structures, about 2.0 for mid-rise structures, and about 2.5-3.0 for tall structures. These factors are valid for structures whose response is not controlled by higher mode effects, in which case the factor may become much larger.

In summary, the maximum story drift demand for an inelastic MDOF structure in which the P-delta sensitive range is avoided can be estimated as:

$$\theta_{i,ST,max} = \alpha_{MDOF} \times \alpha_{INEL} \times \alpha_{P\Delta} \times \alpha_{ST} \left(\frac{S_d}{H} \right) \quad (1)$$

where S_d, is the spectral displacement demand at the first mode period of the structure and H is the height of the structure.

4 RELATIONSHIP BETWEEN STORY DRIFT AND COMPONENT PLASTIC DEFORMATION DEMANDS

The final step of the process outlined in Figure 1 is to relate the story drift demand to the component plastic deformation demands. This step is carried out in two parts. First the story plastic drift demand is estimated by subtracting the elastic drift component from the total story drift demand, and then the story plastic drift demand is related to the beam and panel zone plastic drift demands, assuming that columns remain elastic. The relationships presented are for "standard" fully-restrained connections, but can easily be modified for different connection types, e.g., connections using the reduced beam section concept, cover plates, etc.

4.1 Estimation of Story Plastic Drift Demand

The following assumptions are made on the elastic behavior of the structure in order to facilitate estimation of the story plastic drift demand: 1) inflection points are at mid-height of columns and mid-span of beams, 2) gravity load moments are small compared to seismic moments, 3) a story has regular geometry and uniform section properties, and 4) second order effects and lateral deflection due to column axial deformations can be neglected. Detailed derivations and background for the process are given in Gupta & Krawinkler (1999).

The column shear force associated with first yielding (beam or panel zone) at the connection is given as the smaller of:

$$V_{col} = \frac{\Delta M}{h \left[1 - \dfrac{d_c}{l} \right]} = \frac{2 \times M_{pb}}{h \left[1 - \dfrac{d_c}{l} \right]}$$

and $\qquad (2)$

$$V_{col} = \frac{V_{y,pz}}{\dfrac{1}{d_b} \left[h \left(1 - \dfrac{d_c}{l} \right) \right] - 1}$$

where M_{pb} = beam strength, h = height between the two inflection points in the column, d_c = depth of column, and l = distance between the two inflection points in the beams (see Figure 10). The yield strength of the panel zone in shear, $V_{y,pz}$ is given as:

$$V_{y,pz} \approx 0.55 F_y d_c t_p \quad (3)$$

where F_y = yield strength of the column web material and t_p = thickness of the panel zone.

Using the computed value of the shear force in the column associated with first yielding at the connection, the various components of the elastic story drift can be computed as follows (see Figure 10):

$$\delta_b = \frac{1}{2} \times \frac{V_{col} \times h^2 \left(1 - \frac{d_c}{l}\right)}{\left(\frac{6EI_b}{l - d_c}\right)}$$

$$\delta_c = V_{col} \times \frac{(h - d_b)^3}{12EI_c} \qquad (4)$$

$$\delta_{pz} = V_{col} \times \frac{(h - d_b) \times \left(\frac{h}{d_b} - 1\right)}{Gt_p d_c}$$

where δ_b, δ_c, δ_{pz} = lateral drift from beams, columns, and panel zone, respectively, d_b = beam depth, E = elastic modulus of steel, and I_c and I_b = moment of inertia of the column and beam, respectively.

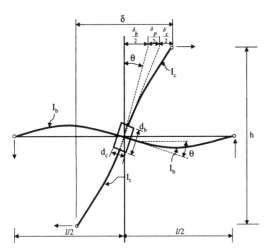

Figure 10. Components of lateral deflection in beam-column assembly.

The story plastic drift angle is then estimated as follows:

$$\theta_{p,st} = \theta_{total,st} - \frac{(\delta_b + \delta_c + \delta_{pz})}{h} \qquad (5)$$

where $\theta_{p,st}$ = story plastic drift demand and $\theta_{total,st}$ = estimated total story drift demand. The second term on the right-hand side is the elastic story drift, i.e., the story drift at which the first component at the connection reaches its yield strength.

The relationship between the story plastic drift demand and the component plastic deformation demands is based on the following premise:

$$h \times \theta_{p,st} = h \times \theta_{p,b} + (h - d_b) \times \theta_{p,pz} \qquad (6)$$

where $\theta_{p,b}$ and $\theta_{p,pz}$ = plastic deformation demand of the beam and panel zone, respectively.

The initial assumption is made that plastic deformation exists in both the beam and the panel zone. The moment in the beam, M_b, is written as:

$$M_b = M_{pb} + \alpha \left(\frac{6EI_b}{l}\right) \times \theta_{p,b} \qquad (7)$$

where α = strain-hardening in the moment-rotation relationship. Based on this moment in the beam, the shear force in the panel zone is calculated as:

$$V_{pz} = 2 \times M_b \left[\frac{1}{d_b} - \frac{1}{h\left(1 - \frac{d_c}{l}\right)}\right] \qquad (8)$$

This shear force in the panel zone can then be related to the panel zone plastic deformation demand as follows:

$$V_{pz} = V_{y,pz} + \beta K_{pz} \theta_{p,pz} \qquad (9)$$

where K_{pz} = panel zone elastic shear stiffness, given as:

$$K_{pz} = 0.95 d_c t_p G \qquad (10)$$

and β is the slope of the second segment of a tri-linear shear force-shear distortion relationship (Krawinkler 1978).

Equating Equations 8 and 9 and solving for the panel zone plastic deformation results in the following expression for the plastic panel zone deformation:

$$\theta_{p,pz} = \frac{2 \times M_b \left[\frac{1}{d_b} - \frac{1}{h\left(1 - \frac{d_c}{l}\right)}\right] - V_{y,pz}}{\beta K_{pz}} \qquad (11)$$

Substituting Equation 11 into Equation 6 results in the following expression for the beam plastic deformation demand:

$$\theta_{p,b} = \frac{\theta_{p,st} \times \beta K_{pz} \left(\frac{h}{h - d_b}\right) - 2 \times M_{pb} \left[\frac{1}{d_b} - \frac{1}{h\left(1 - \frac{d_c}{l}\right)}\right] + V_{y,pz}}{2 \times \alpha \left(\frac{6EI_b}{l}\right) \times \left[\frac{1}{d_b} - \frac{1}{h\left(1 - \frac{d_c}{l}\right)}\right] + \beta K_{pz} \left(\frac{h}{h - d_b}\right)} \qquad (12)$$

Once Equation 12 has been evaluated, Equation 6 or Equation 11 provides the plastic deformation demand for the panel zone. If only one component at the connection deforms plastically, then the equations will return a negative value for the other component. For such cases, Equation 6 will provide a complete solution once the non-yielding element term is set to zero.

If the plastic deformation demand in the panel zone reaches the third slope of the element force-deformation relationship, then Equation 9 needs to be modified as follows:

$$V_{pz} = V'_{y,pz} + \beta_1 K_{pz} \theta_{p,pz}$$

where (13)

$$V'_{y,pz} = V_{y,pz} + (\beta - \beta_1) \times K_{pz} \times 3\theta_{y,pz}$$

where β_1 = slope of third segment of the panel zone force-deformation relationship and $\theta_{y,pz} = V_{y,pz}/K_{pz}$.

This derivation is based on a trilinear shear force-shear distortion relationship for the panel zone, a bilinear relationship for the beams, and standard beam-to-column welded connections. The equations can be modified for special circumstances, as needed.

5 CONCLUSIONS

The seismic deformation demands of SMRF structures can be estimated using discernible patterns that permit the development of relationships between the spectral displacement demand at the first mode period of the structure and the global (roof), intermediate (story) and element level deformation demands. A simplified procedure is presented to relate the spectral deformation demand to the system and component deformation demands using a series of modification factors and functions.

The various modification factors are quantified, and the functions relating the story drift to element deformation demands are presented. The factors, which are derived from elastic and inelastic time history results of 3-, 9-, and 20-story structures subjected to various sets of ground motions, are in good agreement with similar factors given in the literature for simplified cases/scenarios.

For the structures and ground motions used in this study, it is found that the roof displacement demand of an inelastic structure can be estimated with good confidence, provided that the P-delta effect is sufficiently contained to avoid drifts associated with negative post-yield story stiffness.

The relationship between the roof and maximum story drift demands depends on the height of the structure. Relationships with reasonable dispersion were found, except for structures dominated by higher mode effects. The distribution of story drifts over height depends strongly on ground motion and structure characteristics.

The functions relating story drift demand to element plastic deformation demands are developed using basic geometric and member section properties.

The story yield drift is estimated based on the weakest element at the connection and then the plastic deformation demands are distributed amongst the beam(s) and panel zone in proportion to their strength and interaction based on the geometric properties of the sub-structure (story).

The simplified demand estimation process proposed in this paper should prove useful in the conceptual design phase, in estimating deformation demands for performance assessment, and in improving basic understanding of seismic behavior.

6 ACKNOWLEDGEMENTS

The work forming the basis for this publication was conducted pursuant to a contract with the Federal Emergency Management Agency (FEMA) through the SAC Joint Venture. No warranty is offered with regard to the results, findings and recommendations contained herein, either by FEMA, the SAC Joint Venture, the individual joint venture partners, their directors, members or employees, or the authors of this publication. These organizations and individuals do not assume any legal liability or responsibility for the accuracy, completeness, or usefulness of any of the information included in this publication.

REFERENCES

Fajfar, P. & Krawinkler, H. (eds), 1997. *Seismic Design Methodologies for the Next Generation of Codes*. Rotterdam: Balkema.

FEMA 273. 1997. *NEHRP guidelines for the seismic rehabilitation of buildings*. Federal Emergency Management Agency.

Gupta, A. & Krawinkler, H. 1999. Seismic demands for performance evaluation of steel moment resisting frame structures. *John A. Blume Earthquake Eng. Center Report No. 132*, Dept. of Civil Eng., Stanford University.

Gupta, A. & Krawinkler, H. 2000. Dynamic p-delta effects for flexible inelastic structures. *Jo. of Str. Eng., ASCE*, Vol. 126, No. 1.

Krawinkler, H. 1978. Shear design of steel frame joints. *Engineering Journal, AISC*, Vol. 15, No. 3.

Krawinkler, H. & Alavi, B. 1998. Development of improved design procedures for near fault ground motions. *SMIP98 Seminar on Utilization of Strong-Motion Data*, Oakland.

Seneviratna, G.D.P.K. & Krawinkler, H. 1997. Evaluation of inelastic MDOF effects for seismic design. *John A. Blume Earthquake Eng. Center Report No. 120*, Dept. of Civil Eng., Stanford University.

Somerville, P. et al. 1997. Development of ground motion time histories for phase 2 of the FEMA/SAC steel project. *Report No. SAC/BD-97/04*, SAC Joint Venture, 555 University Ave., Sacramento, CA 95825.

Behaviour of Steel Structures in Seismic Areas, Mazzolani & Tremblay (eds) © 2000 Balkema, Rotterdam, ISBN 90 5809 130 9

Reinforced steel moment-resisting joints

Taejin Kim, Andrew S. Whittaker & Vitelmo V. Bertero
Earthquake Engineering Research Center, University of California, Berkeley, USA

ABSTRACT: In an attempt to resolve outstanding questions related to the seismic design and detailing of steel moment-resisting connections reinforced with cover-or flange-plates, ten full-scale single sided joints were tested to failure to investigate the effects of key design parameters on their seismic response. One beam size (W30x99) and one column size (W14x176), both of grade 50 steel, were used for all ten tests. The results obtained show that the target value for the plastic rotation capacity of 0.03 radians without fracture was exceeded in all the cases studied. Although local buckling of the beam flanges started relatively early (at plastic beam rotation, θ_{pb}, of about 0.01 radian) no significant degradation of strength was noted until local buckling of the web was detected at a $\theta_{pb} < 0.03$ radians. The best performance was obtained from test on the following specimens; the one with partial restraint against web buckling; and the specimen with balanced yielding mechanisms at the panel zone and beam.

1 INTRODUCTION

1.1 Introductory remarks

The 1994 Northridge earthquake exposed the vulnerability of steel moment-resisting space frames to critical damage in moderate-to-severe *Earthquake Ground Motions (EQGMs)* (Bertero at all 1995). In the week and months following the earthquake, forensic engineering identified failed welded beam-column connections in more than 200 buildings. The damage to steel moments resisting frames raised many questions for the *Earthquake Engineering Community (EQEC)* (Whittaker at al 1998). Answering these questions involved consideration of many complex technical professional and economic issues including metallurgy, welding fracture mechanism, connection behavior, system performance and practice related design, fabrication, erection, and inspection. Thus to attempt to answer the questions a joint venture of design professionals and researchers was formed. Known by the acronym *SAC*, the joint venture partners are the Structural Engineers' Association of California *(SEAOC)*, the Applied Technology Council *(ATC)*, and the California Universities for Research in Earthquake Engineering *(CUREe)*. SAC formulate a comprehensive, coordinated problems focused programs of investigation, guidelines developing and professional training. A 5-year program sponsored by the U.S. *Federal Emergency Management Agency (FEMA)* was initiated in 1995. The program was conducted with the active involvement of engineers, researches, construction experts, and others through the U.S. The program was conducted in two phases. The second phase program, which is near completion, is divided into eleven major tasks spanning over 4 years (Mahin et al 2000). Among the technical studies undertaken during phase 2 were a series of studies devoted to examine the connection performance through coordinate analytical and experimental research program. The research reported herein is part of this program and had the following objectives and scope.

1.2 Objectives

The main objective of the research reported herein was to conduct detailed finite element analysis and tests of welded reinforced connections having (a) cover plated connections, and (b) flange plated connections, in order to investigate if it is possible to predict their behavior when they are subjected to seismic actions and to properly design them considering the following key parameters or variables: cover- and flange-plate geometry, weld geometry, web connection details (welded and/or bolted), panel zone strength (strong, balanced, or weak), loading or deformation history (cyclic versus monotonic), partial restraint against buckling (local and torsional).

1.3 Scope

This paper summarizes only the experimental research program that was conducted at the *Earthquake Engineering Research Center (EERC)* laboratory. Ten full-scale single side beam-column joint assemblies were tested to failure to investigate the effects on their seismic response (behavior or performance) of the above mentioned key variables. To simplify the interpretation of the data one beam size (W30x99) and one column size (W14x176), both of grade 50 steel were used for all ten tests.

Mechanical characteristics of the grade 50 steel were determined and are given, the applied displacement history is briefly discussed, and the experimental set-up is discussed and illustrated.

Detailed test results are given for four specimens (two cover plate, and two flange plated connection assemblies). These results are presented as; Actuator Force – Tip Displacement Relation; Total Moment – Beam Plastic Rotation Relation; and Panel Zone Plastic versus Moment at Column Face. The cumulative energy dissipated for each of these specimens is also computed and a discussion regarding the observed maximum displacement ductility is also presented.

An evaluation of the results obtained and main conclusions as well as recommendations for future research are presented.

For the purpose of this paper, a *beam-column assembly* refers to the entire test specimen, a *beam-column joint* is defined as the entire assemblage at the intersection of the beam and column, and a *beam-column connection* is the group of components (elements) that connect the beam to the column. As such the beam-column connection is part of the beam-column joint.

2 TEST SPECIMENS

Tests were performed on single-sided beam-column assemblies as shown in Figure 1. Such tests are representative of exterior beam-column connections. A total of ten specimens were designed, fabricated and tested. Five of these specimens were designed with cover plated connections and the other five were designed with flange plated connections.

2.1 Cover plated connection specimens

Detail of this type of connection is depicted in Figure 2. The purpose of cover plate is to reinforce (strengthen) the welded connection so that the yielding of the beam occurs at the end of the cover plate and not at the face of the column where the beam is welded to column. The five full scale specimens were welded in the shop. For the fabrication of this type of connections in the field it is sug-

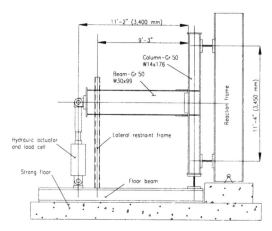

Figure 1. Test specimen and test set-up

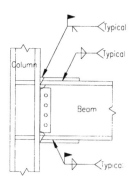

Figure 2. Cover plated connection

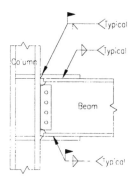

Figure 3. Flange plated connection

gested to weld in the shop the bottom cover plate and the web plate to the column and the top cover plate to the top flange at the beam. All the other needed welds should be done in the field. The main geometry characteristics of the cover plates used in the five specimens as well as the type of welds used and type of load applied are given in Table 1. The length of the reinforcing plate was 15 in. (380 mm).

696

Table 1. Characteristics of the test specimens[1]

Specimen No	RC01	RC02	RC03	RC04	RC05	RC06	RC07	RC08	RC09	RC10
Reinforcing plate[2]	CP	CP	CP	FP	CP	FP	FP	FP	FP	CP
Plate geometry[3]	Re	Re	Re	Sw	Tr	Re	Re	Re	Re	Re
Plate fillet weld (# sides)	2	2	3	3	3	3	3	3	3	3
Plate width (in.)	12.0	12.0	12.0	13.0	10.5	13.2	13.2	13.2	13.2	12
Plate thickness (in.)	0.62	0.62	0.62	1.12	0.69	1.00	1.00	0.78	0.78	0.62
Doubler plate (in.)	0.37	0.37	0.37	0.37	0.37	0.37	0.37	0.37	0.00	0.37
Continuity plates (in.)	1.12	1.12	1.12	1.12	1.12	1.00	1.00	0.87	0.62	1.12
Loading history	Cyclic	Cyclic	Cyclic	Cyclic	Cyclic	Cyclic	Mono	Cyclic	Cyclic	Cyclic
LTB restraint(# places)[4]	1	2	1	1	1	1	1	1	1	1

1. All specimens have welded beam web to column flange connections.
2. CP designates cover plate, FP denotes flange plate.
3. Re denotes Rectangular plate, Sw designates Swallow shape plate, Tr is trapezoid plate.
4. All specimens have lateral torsional buckling restraint near the actuator; specimen RC02 was also restrained against LTB at the theoretical plastic hinge.

2.2 Flange plated connections specimens

The details of the welded flange plated connections are illustrated in Figure 3. The five flange plated connections were completely welded in the shop. For the fabrication and erection of this type of connection in the field the web plate is welded to the column in the shop as it is also the back bar for the bottom flange plate. The flange plate (bottom and top) are welded to the flanges of the beam in the shop. In the field the top and bottom flange plates are welded to the face of the column and the web plate is bolted and in some cases also welded to the web of the beam in the field.

The flange plates are designed to develop the full plastic capacity of the beam at the ends of the flange plate. The desired yield mechanisms for this type of connections are primarily shear yielding of the panel zone and flexural yielding of the beam. The main characteristics of the plate (geometry, width, length, and thickness), fillet weld of the plate (number of sides), doubler plate thickness at the panel zone, continuity plate thickness and deformation (loading) history are given in Table 1.

3 FABRICATION OF TEST SPECIMENS AND EXPERIMENTAL SET-UP

3.1 Fabrication and material mechanical charateristics

The ten specimens were fabricated by the Gayle manufacturing Company of Woodland, California specifically for testing at the EERC from the steel grade 50 provided by the Nucor-Yamato Steel Company.

Coupon testing of the beams and columns were undertaken to determine the mechanical characteristics of the test specimens. Table 2 present the data obtained from the coupon tests. Also listed in Table 3 are the mill certificate data for the steel.

All local fabrication work was performed in accordance with SAC approved AWS procedures and monitored by inspectors from the Signet testing Labs. Ultrasonic testing (UT) was employed to detect weld imperfections for all specimens. All the specimens passed the UT inspection prior to shipment for testing.

3.2 Test setup

The test setup for the specimens is shown in Figure 1. The specimens were tested in the up-right position, axial loads were not applied to the column and a composite slab was not included in the test setup. The column were attached to the strong floor and to the strong reaction frame using short pieces of steel W section which reacting in the weak direction simulate simple support (pin) boundary condition. For all specimens, lateral restrain to the beam flange was provided adjacent to the tip of the cantilever beam; for specimen RC02, lateral restrain against lateral torsional buckling was also provided at the theoretical mid point of the plastic hinge.

Cyclic displacements were imposed on the beam at a distance of 134 in. (3.4 m) from the column face. The displacement-control test algorithm was used for the testing. For all specimens except specimen RC07, a cyclic loading history designated the standard ATC (SAC) cyclic loading was utilized. For specimen RC07, a type of monotonic loading history was used to simulate the effects of near-field EQGMs.

4 CYCLIC RESPONSE OF SPECIMENS

4.1 General

The protocol used to test the UCB specimens is similar to that set fourth in the ATC-24 report (ATC, 1992). The displacement history consisted of stepwise increasing displacement cycles. The displacement loading history was continued until either the

Table 2. Coupon test data

Member	Size	Section	Yield Stress (ksi)	Ultimate Stress (ksi)
Beam	W30x99	Flange	52.9	72.0
		Web	63.1	73.8

Table 3. Mill certificate data

Member	Size	Section	Yield Stress (ksi)	Ultimate Stress (ksi)
Beam	W30x99	Flange	52.5	68.5
		Web	52.5	68.5
Column	W14x176	Flange	53.5	71.5
		Web	53.5	71.5

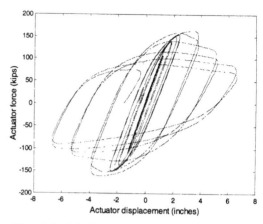

Figure 4. Load-displacement relation for RC01

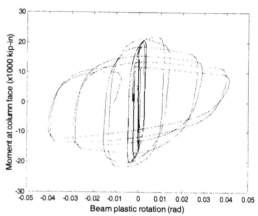

Figure 5. Beam plastic rotation - moment at column face relation for RC01

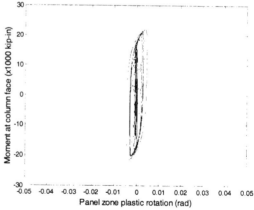

Figure 6. Panel zone plastic rotation – moment at column face relation for RC01

specimen reached its deformation capacity or the displacement limit of the test fixture (=±8 inches) was reached. Table 4 summarizes the results obtained from the tests of the 10 specimens. However this paper present only a detailed description of the response of the following four specimens, RC01, RC03, RC06, and RC09. The nominal yield displacement (δ_y) for all the specimens was assumed as 1.42 in. (36 mm), or equal to 1-percent inter-story drift ratio.

4.2 Response of specimen RC01 (Figures 4-6)

As illustrated in Figure 4 the specimen UCB-RC01 was tested to a tip displacement of 6.7 inches (4.8% inter-story drift). There was no brittle fracture either at the beam-column connection or at the cover plate-beam flange connections. The specimen showed signs of yielding in the panel zone at displacement cycles to $0.75\delta_y$. During displacement cycles to δ_y, beam flanges yielded outside the cover plate region. During displacement cycles to $1.5\delta_y$ and $2\delta_y$, the yielding spread along the beam flanges and the flanges started to buckle. During the first displacement cycle to $3\delta_y$, there was local buckling in the beam web and small tears in the fillet weld joining the top cover plate to the beam flange caused by high local strains due local buckling. The crack grew inward the beam flange during the second cycle and there was also evidence of beam lateral torsional buckling. During displacement cycles to $4\delta_y$ and $5\delta_y$, the amplitude of local buckling in the beam flanges and beam web increased and the tear propagated towards the beam web. The maximum joint plastic rotation was 4.15% radians, comprised almost exclusively from plastic deformation in the beam (Figures 5 and 6). These large plastic rotations occurred at a value of actuator force of approximately 52 kips which is 32% of the maximum observed value of actuator force (164kips) during the test. The cumulative dissipated energy was 10,400 kips-in.

Table 4. Summary data for ten specimens

Specimen No	RC01	RC02	RC03	RC04	RC05	RC06	RC07	RC08	RC09	RC10
Maximum moment at column face (k-in.)	22,000	22,500	22,500	21,200	21,800	22,200	21,300	22,500	21,600	25,800
Maximum beam plastic rotation (radian)	0.0415	0.0430	0.0430	0.0490	0.0522	0.0420	0.0590	0.0500	0.0500	0.0450
Recorded peak beam tip displacement (in.)	6.7	6.7	7.1	7.5	7.7	7.7	7.6	7.8	7.4	7.3
Specimen initial stiffness (k/in.)	89	87	90	88	84	84	89	91	83	93
Cumulative dissipated energy (k-in.)	10,400	8,700	10,200	10,900	16,100	11,700	8,800	11,800	16,500	12,400

4.3 Response of specimen RC03 (Figures 7-9)

As shown in Figure 7 the specimen UCB-RC03 was tested to a tip displacement of 7.1 in. (5% inter-story drift). There was no brittle fracture either at the beam-column connection or at the cover plate-beam flange connections. The specimen showed signs of yielding in the panel zone during displacement cycles to δ_y. During displacement cycles to $1.5\delta_y$, beam flanges outside the cover plate region yielded and during displacement cycles to $2\delta_y$, the beam flange outside the cover plate buckled and the length of the yielded zone in the flanges increased. During displacement cycles to $3\delta_y$, the beam web buckled and there was a small tear in the K-line (between the beam top flange and beam web) caused by high local strains from local buckling. During displacement cycles to $4\delta_y$ a similar crack formed in the K-line in the bottom flange. During displacement cycles to $5\delta_y$, these two cracks slowly propagated outward (away from beam web) and the specimen also experienced large lateral torsional buckling. The maximum joint plastic rotation was 4.3% radians, comprised almost exclusively from plastic deformation in the beam (Figures 8 and 9). These large plastic rotations occurred at a value of actuator force of approximately

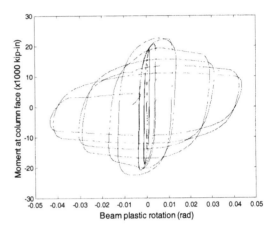

Figure 8. Beam plastic rotation - moment at column face relation for RC03

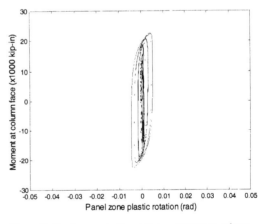

Figure 9. Panel zone plastic rotation – moment at column face relation for RC03

52 kips which is 31% of the maximum observed value of actuator force (168 kips) during the test. The cumulative dissipated energy was 10,200 kips-in.

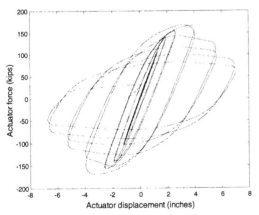

Figure 7. Load-displacement relation for RC03

4.4 Response of specimen RC06 (Figures 10-12)

As illustrated in Figure 10 the specimen UCB-RC06 was tested to a tip displacement of 7.7 in. (5.4% inter-story drift). There was no brittle fracture either at the flange plate-column connection or at the flange plate-beam flange connections. The specimen yielded in the panel zone during displacement cycles to $0.75\delta_y$. During displacement cycles to δ_y, the specimen yielded in the beam flanges both outside flange plate region and close to column flange and in the continuity plates. During displacement cycles to $2\delta_y$ beam flanges buckled and the yielding zone in the beam flanges outside the flange plate extended. During displacement cycles to $3\delta_y$ there was beam web buckling and lateral torsional buckling. During displacement cycles to $4\delta_y$, the amplitude of buckling was increased, In addition, ductile cracks appeared in both beam flanges due to large strains (caused by the buckling). These cracks grew during displacement cycles to $5\delta_y$ and during displacement cycles to $5.5\delta_y$ there was a crack at the beam top flange K-line in addition to large ductile tearing of beam flanges. The maximum joint plastic rotation was 4.2% radians, comprised almost exclusively from plastic deformation in the beam (Figures 11 and 12). These large plastic rotations occurred at a value of actuator force of approximately 52 kips which is 31% of the maximum observed value of actuator force (165 kips) during the test. The cumulative dissipated energy was 11,700 kips-in.

4.5 Response of specimen RC09 (Figures 13-15)

As shown in Figure 13 the specimen UCB-RC09 was tested to a tip displacement of 7.4 in. (5.3 % inter-story drift). There was no brittle fracture either at the flange plate-column connection or at the flange plate-beam flange connections. The specimen yielded in the panel zone during displacement cycles of $0.75\delta_y$. During displacement cycles to δ_y, the specimen yielded in the beam flanges. During displacement cycles to $1.5\delta_y$, flange plates and column flange opposite the continuity plates yielded. During displacement cycles of up to $3\delta_y$, there was minor beam flange buckling, yielding of beam web, large plastic deformation in the panel zone, and small yielding in the continuity plates. During displacement cycles to $4\delta_y$, there was large buckling of beam flanges and beam web, and small ductile cracks formed in the beam flanges. During displacement cycles to $5\delta_y$ the amplitude of beam local buckling increased, a crack was formed in the transverse weld between top flange plate and beam top flange. During the cycles to $5.5\delta_y$ a crack initiated at the beam bottom flange K-line and the fracture propagated through the width of the bottom flange. During the last set of cycles, a maximum joint plastic rotation of

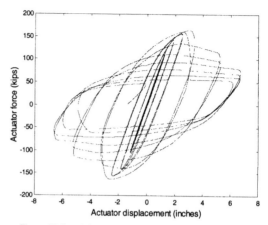

Figure 10. Load-displacement relation for RC06

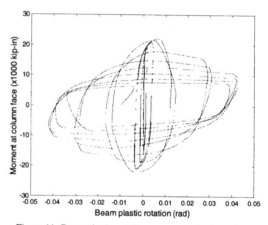

Figure 11. Beam plastic rotation - moment at column face relation for RC06

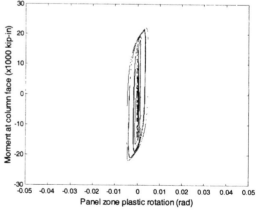

Figure 12. Panel zone plastic rotation – moment at column face relation for RC06

5.3% radians (comprised of 0.3% radian in the panel zone and 5.0% radian in the beam) was reached; earlier in the tests, a maximum panel zone rotation of 2.5% radian was obtained (Figure 15). The 5.0% radian large plastic rotations occurred at a value of actuator force of approximately 38 kips which is 24% of the maximum observed value of actuator force (161 kips) during the test. The cumulative dissipated energy was 16,500 kips-in.

5 SUMMARY, CONCLUSIONS AND RECOMMENDATIONS

5.1 Summary

A total of 10 full scale beam-column assemblies with reinforced welded connections, 5 with cover plates and 5 with flange plates, were tested at the EERC facilities, as a part of the SAC program phase 2. The ten test specimens were designed to develop the full plastic capacity of the beam and to have an improved seismic performance by strengthening the connections at the column face so that they avoid fracture at this face. This was achieved by moving the flexural plastic deformations from this face to the end of the strengthened segment (end of cover plate or end of flange plate). With the exception of specimen RC07 all the other specimens were tested to failure using a cyclic loading history consisting of stepwise increasing displacement cycles. Specimen RC07 was subjected initially to one cycle of deformation to a tip displacement of 2.8 in. and then the deformation was reversed and increased monotonically up to a tip displacement 8.0 in. This specimen was capable of resisting 2 full cycles of reversals of tip displacement at 8.0 in. No premature brittle fracture was observed in any of the specimen tested, and the following observations and conclusions can be drawn from the data obtained in this experimental study.

5.2 Observations and conclusions

1. All the tested specimens reached their maximum resistance (which exceeded the nominal full plastic flexural resistance of the beam) at a beam plastic rotation, θ_{pb}, of about 1% radian.

2. Beam flange local buckling was observed at a θ_{pb} of about 1% radian.

3. The lateral resistance (strength) drops to about 80% of the nominal full plastic flexural capacity of the beam at about a $\theta_{pb} \cong 3\%$ radian.

4. Significant drop in strength was observed in all specimens when local buckling of the web was detected, which occurred at a $\theta_{pb} < 3\%$ radian for all the specimens with the exception of RC10 which was provided with longitudinal stiffeners which

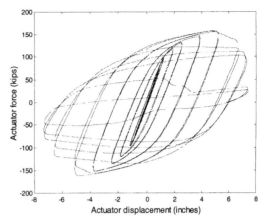

Figure 13. Load-displacement relation for RC09

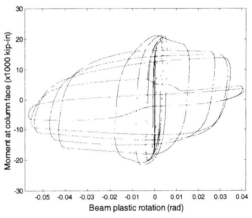

Figure 14. Beam plastic rotation - moment at column face relation for RC09

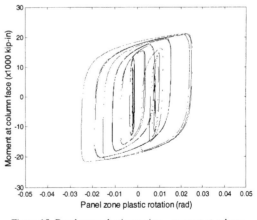

Figure 15. Panel zone plastic rotation – moment at column face relation for RC09

were welded to the beam web to delay its local buckling.

5. Fracture of the specimens occurred at a $\theta_{pb} \geq$ 4.15% radians. In the specimen RC07 which was subjected to a monotonically increasing deformation up to $\theta_{pb} = 5.9\%$ radians, the fracture occurred after 2 cycles with full reversals at $\theta_{pb} = 5.9\%$ radians being its resistance 26% of the maximum measured strength. This is considered to be usually larger than the resistance that can be demanded by the gravity loads that can be acting on the beam. Thus this $\theta_{pb} = 5.9\%$ radians can be considered as a safe rotation against failure to resist gravity loads.

6. Regarding the values of the displacement ductility, μ_δ, that has been attained they vary considerable depending on what is considered as the yielding displacement, δ_y. If δ_y is considered at the first observed yielding then the μ_δ at maximum resistance is about 3.3 and the μ_δ at the maximum measured displacement is larger than 5.6 and for specimen RC08 reached a value near 6.5. However if the first yielding is considered as the idealized elastic displacement at a load level equal to the maximum resistance, the above values are reached at the following values: 2.1, 3.6 and 4.1 respectively. In view of this variability of the value of the ductility is very difficult to estimate the accumulative ductility due to cyclic response. Thus it is better to estimate the dissipated energy due to plastic deformations.

7. The accumulative dissipated energy due to cyclic response vary from 8700 k-in. for specimen RC02 to 16,500 k-in for specimen RC09, which was designed and constructed without doubler plate at the panel zone.

8. The results obtained show the advantage of trying: (a) to achieve a balanced condition of yielding due to shear at the panel zone and of flexural yielding of the beam at the end of the strengthened segment of the beam; and (b) to delay as much as possible the local buckling of the web.

5.3 Recommendations

The results of the tests conducted using different histories of the deformations applied to the specimens clearly reveal that they affect the amount of energy dissipated due plastic deformation, thus a comprehensive experimental research program should be conducted using different deformation histories that vary from a pure monotonically increasing deformations up to collapse to the one that consist in applications of full cycles of reversal inelastic deformations having different intensities.

There is also a need to conduct integrated analytical and experimental studies on the relationship among local buckling of the flanges, local buckling of the web and lateral torsional buckling of the critical regions of the beam, attempting to find out how it is possible to detail these regions to delay as much as possible the local buckling of the flanges and of the web and so to delay lateral torsional buckling.

At least two bay frames should be study experimentally loading the beam with the expected gravity loads and the column with the expected axial forces due to gravity loads and then to subject them to increasing lateral displacement that simulate the effects of the horizontal component of EQGMs up to obtain the collapse of the frames (fall down of the beams).

6 ACKNOWLEDGEMENTS

Funding for the studies reported herein was provided by FEMA through the *SAC Joint Venture*. The test specimens were fabricated by Gayle Company of Woodland, CA. using the steel provided by Nucor-Yamato Steel Company and the inspection of the welds and the tests of the materials were conducted by Signet Testing Laboratory, Inc. The FEMA funding and the pro-bono support of Gayle, Nucor-Yamato and Signet are acknowledged and appreciate by the authors.

The help of Dr. A. Gilani and S. Takhirov during the instrumentation of the specimens and their testing are also acknowledged and appreciated.

7 REFERENCES

ATC 1992. Guidelines for cyclic seismic testing of components of steel structure, *Report No. ATC-24*, Applied Technology Council, Redwood City, CA

Bertero, V.V., J.C.Anderson & H. Krawinkler 1994. Performance of steel building structures during the Northridge earthquake, *Report No. UCB/EERC-94/09* College of Engineering, University of California at Berkeley

Mahin, S., J. Malley, R.Hamburger & M.Mahoney 2000. Overview of a program for reduction of earthquake hazards in steel frame structures. *Proceedings of the 12th world conference on Earthquake Engineering* Aukland, NZ.

Whittaker, A.S., A. Gilani & V.V. Bertero 1998. Evaluation of pre-Northridge steel moment-resisting frame joints. *Structural design of tall buildings* 7: 263-283

Behaviour of Steel Structures in Seismic Areas, Mazzolani & Tremblay (eds) © 2000 Balkema, Rotterdam, ISBN 90 5809 130 9

Cyclic modeling of T-stub bolted connections

R.T. Leon
Georgia Institute of Technology, Atlanta, USA

J.A. Swanson
University of Cincinnati, Ohio, USA

ABSTRACT: This paper presents the development of a model for cyclic behavior of T-stub connections. Beginning with experimental data generated from 48 T-stub component tests, a comprehensive monotonic model was developed. In this paper, the monotonic model is briefly described and its extension to the cyclic case is discussed. Comparisons with test data indicate that the model captures well most of the cyclic behavior features, with the exception of the varying location of the contact area between the T-stub flange and the column flange.

1 INTRODUCTION

In the aftermath of two severe earthquakes (1994 Northridge and 1995 Kobe) in which steel connections suffered unexpected failures, an improved understanding of cyclic connection behavior is clearly needed. Models possessing both sufficient accuracy and ease of use must be developed to allow designers to conduct evaluations of the effects of connection energy dissipating characteristics, strength deterioration, and loss of stiffness on overall frame behavior. Such studies require that the designer know the cyclic moment-rotation characteristics of the connections.

Two approaches are commonly used to generate cyclic moment rotation curves. The first is a strict mechanistic model of a connection and the second is a modification of existing monotonic curves. In this paper, modifications to the monotonic stiffness model discussed in Swanson and Leon (2001) will be presented. These modifications will lead to a model that accurately and efficiently predicts the cyclic response of T-stub connections (Fig. 1).

2 BACKGROUND

In response to the steel connection failures observed during the Northridge earthquake, the Federal Emergency Management Agency (FEMA) began a comprehensive research project (SAC) aimed both at clarifying the reasons for the failures and at developing new design guidelines and alternatives. Bolted connections are one of the alternatives investigated, and the authors conducted a large-scale experimental effort to assess the use of T-stub and clip angle connections in areas of high seismicity. A series of 48 T-stubs and 10 clip angles were tested individually under cyclic tensile and compressive loads to help develop expressions for quantifying the components' strength, stiffnesses, and deformation capacity. Additionally, 8-12 full scale beam-column tests were used to calibrate the overall connection model. Descriptions, drawings, plots, photos, and all test data for this work are available at http://www.ce.gatech.edu/~sac/.

Because the development of the models required careful measurement of the components of deformation, careful instrumentation of the specimens was required. Figure 2 shows the instrumentation employed in the component tests. Linear Variable Displacement Transformers (LVDTs) A monitored the connection slip, B measured the uplift of the T-stub flange from the face of the column, C measured the

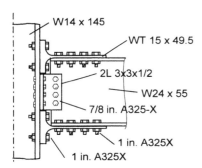

Figure 1 – Typical T-stub connection

uplift of the T-stub flange at the bolt line, D measured the elongation of the T-stem, and E measured the overall T-stub deformation. G indicates the use of bolts instrumented with strain gages to obtain the bolt forces, including the effects of prying. Strain gages were also used on the T-stub.

3 TYPICAL EXPERIMENTAL RESULTS

The typical specimen behavior will be illustrated with data from Test TA-25. The details of the specimen are shown in Fig. 3. Material tests indicated a yield stress of 51.1 ksi, an ultimate strength of 68.0 ksi, and a strain at fracture of 38.0%. Figures 4 shows the load vs. total deformation (LVDT E). It shows that the total deformation of a T-stub in compression is smaller than it is in tension. This is because in compression the T-stub is being pushed into the column flange forcing any deformation to occur in the stem of the T-stub. The T-stub exhibits a high initial stiffness followed by a region of low stiffness. The change in stiffness represents the transition from elastic behavior to plastic.

Figure 5 shows the load vs. slip (LVDT A). The slip load of the T-stub decreases throughout the test as the pretension in the shear bolts is lost. Figure 6 shows the load vs. flange uplift (LVDT B). For this specimen, the uplift is negligible. Figure 7 shows the load vs. stem deformation behavior. In the case of TA-25 the latter was a major component of deformation because the shear bolt spacing was only 2.67 times the bolt diameter. Thus bearing deformation and yielding of the stem dominated over the flange mechanism and uplift.

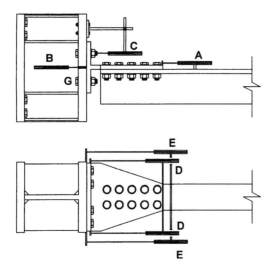

Figure 2 - Instrumentation for component testing.

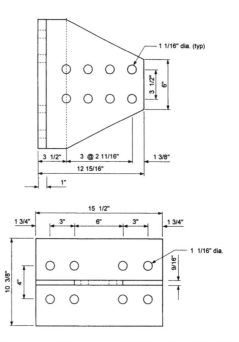

Figure 3 – Details of specimen TA-25 (cut from W16x100).

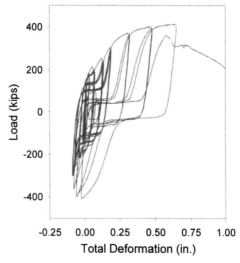

Figure 4 – Total force vs. displacement (TA-25)

4 MONOTONIC MODEL

The overall connection stiffness typically consists of the T-stub stiffness, web angle or web plate stiffness, panel zone stiffness, and the connection region of the beam. The model of the individual T-stub, which is the main issue addressed here, typically consists of contributions from 1) the tension bolts, 2) T-stub flange, 3) T-stub stem, 4) shear bolts, 5) bearing deformations and 6) connection slip. Each of these deformation modes is non-linear, and often two or more deformation modes interact with one another. The overall load-deformation curve is complex, and careful modeling and calibration are required to insure the model's robustness.

The first step in the model development was to transform the data from the cyclic component tests, which comprised a large variation in geometric parameters, into a series of monotonic, backbone curves. The data in this form were used to develop non-linear monotonic spring relationships for each of the deformation modes. The development of this model is described in Swanson and Leon (2001), and only the basics will be discussed next.

The monotonic model is comprised of three basic springs. The first considers the deformations of both the bolts and the flange. The model itself is based on the simple mechanics of the model proposed by Kulak which is used in current American practice (Kulak et al. 1987). A significant modification in this model, however, was the introduction of a semi-empirical method to allow for the gradual formation of the plastic hinges. There are three stiffnesses for each critical region (elastic, transition from elastic to plastic, and plastic). Since the cross section at the K-zone is different from that at the bolt lines, there are five separate combinations of elastic, transition and plastic conditions that are possible when modeling the steel flange.

Another innovative aspect of this model is the careful modeling of the bolt strength and stiffness. This feature is missing from many similar models, but is essential in determining the ultimate deformation capacity of the connection as many failures are governed by the tension (prying) capacity of the bolts. The bolt stiffness model is made up of four linear segments. The first segment models the bolt before its pretension is overcome, the second seg-

ment models the bolt during the linear-elastic portion of its response, the third segment models the bolt after initial yielding has started and the fourth segment represents the bolt after it has reached a plastic state.

The resulting spring model, which combines the steel flange and bolt submodels, can have as many as 9 different linear segments, and is iterative, as the interaction between the deformation of the bolt and flange do not permit a closed-form solution. Cali-

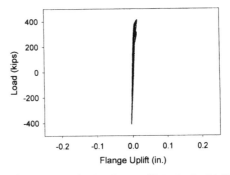

Figure 6 – Total load vs. flange uplift (LVDT B - TA-25)

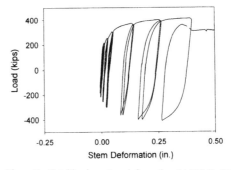

Figure 7 – Total load vs. stem deformation (LVDT D - TA-25)

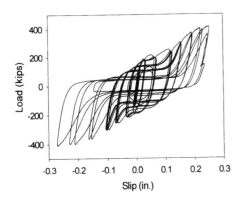

Figure 5 – Total deformation vs. slip (LVDT A - TA-25).

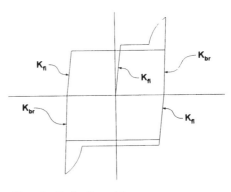

Figure 8 – Cyclic slip model.

705

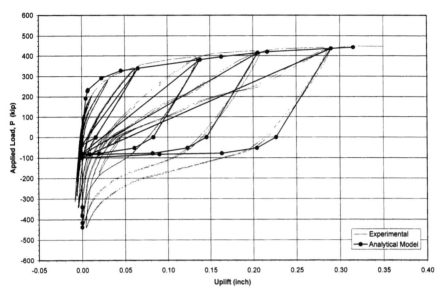

Figure 9 – Comparison of model and experimental cyclic flange response of T-stub TA-03.

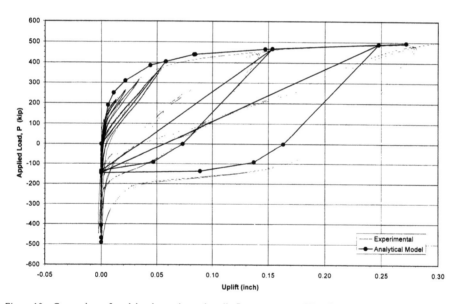

Figure 10 – Comparison of model and experimental cyclic flange response of T-stub TC-12.

bration with the test data indicated very good agreement with the mechanistic model. The experimental data were used primarily to calibrate the transition stiffness of the plastic hinges.

The second spring is used to model the stem deformations. These deformations include the shear deformation of the bolts and the yielding of the stem in tension. It was found that a rational bi-linear model satisfactorily predicts the initial stiffness, yield load, plastic or secondary stiffness, and ultimate deformation of the stem. The initial stiffness is predicted by considering a uniformly distributed axial load over a tapered member. The secondary stiffness was based on the assumption that the majority of the yielding in the stem takes place in the region directly between the last two shear bolts and

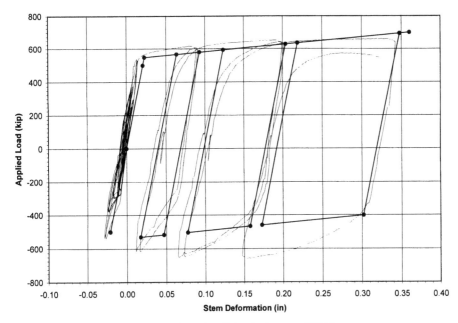

Figure 11 – Comparison of model and experimental prediction for the stem of TC-09.

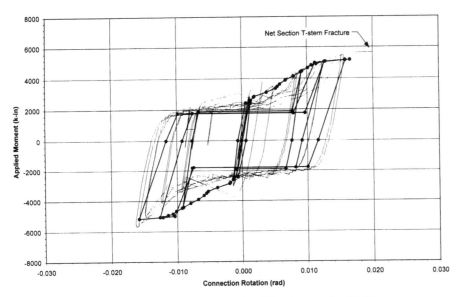

Figure 12 – Model and experimental moment-rotation curve for a full-scale connection (FS-03).

roughly accounts for the strain hardening characteristics of the T-stem.

The third spring is used to model the slip and bearing deformations around the shear bolts. The basic characteristics of this spring were based on work by Rex and Easterling (1996). Their model was further calibrated with the experimental data obtained from the component tests. The details of the development of this spring, which are too lengthy to be included in this paper, are given in Swanson (1999) and Swanson and Leon (2001).

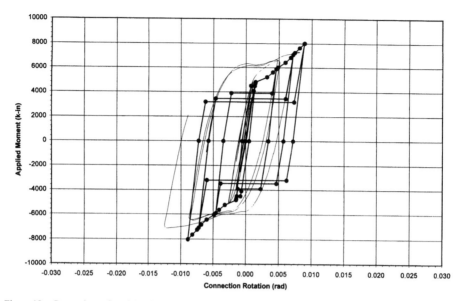

Figure 13 – Comparison of model and experimental moment-rotation curve for FS-08 (quadratic slip load decay).

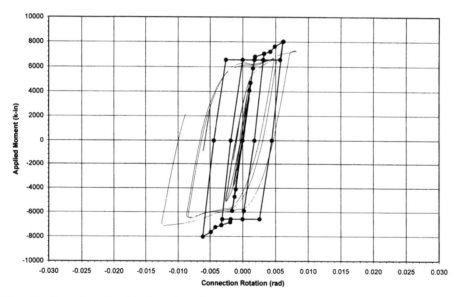

Figure 14 – Comparison of model and experimental moment-rotation curve for FS-08 (no slip load decay).

5 MODELING FOR CYCLIC LOADS

The monotonic spring models described above were extended to the cyclic case. The following changes had to be implemented to match the observed experimental behavior:

5.1 *Flange Spring*

In order to fully describe the cyclic behavior of the flange, a new parameter, the elastic unloading stiffness of the flange, was introduced. After sustaining inelastic deformation, the flange unloads at a stiffness that is lower than its initial elastic stiffness.

This reduction in stiffness is probably due to a change in the force transfer mechanism from the head of the bolt to the flange. For strength calculations and the monotonic stiffness model, it was assumed that the bolt force was transferred from the bolt to the flange as a point load acting at the inside edge of the bolt shank. After being loaded into the inelastic range, however, local bending effects in the bolt will tend to rotate the head of the bolt and will cause the point of application of the bolt force on the flange to shift towards the outside edge of the bolt shank. When the loading is reversed, the flange unloads with this new stiffness to zero load. Once in compression, the response follows a multi-linear compressive loading curve until the deformation reaches zero. After reaching a deformation of zero, an infinite or very large stiffness is used to represent the back of the T-stub flange bearing against the column flange. When unloading from compression, an infinite stiffness is used until the point on the load deformation curve is reached where the T-stub came into bearing with the column flange. Afterwards, the response follows a straight line to the point of maximum load on the virgin tensile curve and then follows the tensile monotonic curve to failure or to the next load reversal.

5.2 Stem Spring

Examination of experimental cyclic stem responses similar to that shown in Figure 7 shows that the unloading and reloading stiffnesses are approximately equal to the stem's initial stiffness. As a result, a simple cyclic model, very similar to basic traditional cyclic material models, was found to be sufficiently accurate to reproduce the main behavioral characteristics of the stem. In the model, two boundaries are established that represent yield lines. These yield lines have a slope equal to the secondary stiffness of the T-stem. The upper yield line passes through the point defined by the yield load and yield deformation that were defined for the monotonic case. The lower yield line lies a vertical distance of 2 times the yield load below the upper yield line. When the stem is loaded, the response travels along the elastic curve until it encounters one of the two yield lines. It then follows the yield line until failure occurs or until the load is reversed. If the load is reversed, the stem unloads along a line parallel to the initial elastic response. The scheme has the advantages of being simple and widely accepted for other applications. One shortcoming of the model, however, is that it does not account for a higher secondary stiffness in compression.

5.3 Slip and Bearing Spring

The slip behavior is characterized by hysteretic behavior that combines an increasing slip deformation and a decreasing slip load (Fig. 5). The slip deformation is a function of the amount of bearing deformation that has taken place and the slip load deterioration is due to a loss of pretension in the shear bolts. The loss of pretension can be attributed to two sources. First, shear and bearing stresses act to relax the tensile pretension in the shear bolts. Second, the thicknesses of the T-stem and beam flange decrease in areas of large strain around the bolt holes which leads to further relaxation of the bolt pretensions. Another source of the decreasing slip load that is not easily quantified is the changing frictional characteristics of the faying surfaces themselves. Figure 8 shows the main characteristics of the slip and bearing model. The cyclic slip/bearing rules can be summarized as follows:

- When loading in tension or compression to a new maximum load level, the response follows the virgin monotonic curve.
- When reloading to a tensile target force, the response maintains a stiffness of K_{br} while the load is in compression and a stiffness of K_{fi} while the load is in tension.
- When reloading to a compressive target force, the response maintains a stiffness of K_{br} while the load is in tension and a stiffness of K_{fi} while the load is in compression.
- The slip load decays according to one of three relationships that are based on the maximum achieved bearing deformation.

Figs. 9 and 10 show a comparison of the analytical model predictions and experimental data for the flange responses for T-stubs TA-03 and TC-12. T-stubs TA-03 and TC-12 were chosen for this comparison because the flanges of these two T-stubs demonstrated significant ductility. The analytical and experimental curves match relatively well in the tensile range for both specimens. The compressive curve, however, deviates from the experimental curve, especially after the flange has sustained significant inelastic deformations, as shown primarily for the case of TA-03. While loading in compression, the actual flange appears to encounter an appreciable stiffness at a positive deformation just before coming into contact with the column flange. This increasing stiffness is likely due to a shortening span length of the T-stub flange as more and more of it comes into contact with the column flange. The model does not currently capture this behavior correctly and the model is being improved to address this issue.

Figure 11 shows the comparison of the overall load-stem deformation curve for TA-09. The model appears to be accurate except that it does not pick up the material hardening; thus it underestimates the strength in the negative direction at large deformations.

Figure 12 shows a comparison of the model predictions with one of the full-scale specimens tested as part of this program. The model reproduces most of the behavioral characteristics well, but overestimates the slip. Thus comparison to total absorbed energy is not as good as expected. This is primarily due to the fact that the component tests, to which the model was calibrated, exhibited more slip and pre-load degradation than the full-scale tests.

Figures 13 and 14 show comparisons of two slip models. They indicate that simplified models such as the ones developed here can provide good but not perfect agreement to test data.

6 CONCLUSIONS

The cyclic model presented was shown to predict the strength and stiffness characteristics of partial and full strength T-stub connections with reasonable accuracy. The energy dissipating characteristics of the model were evaluated and were found to be extremely conservative and highly dependent on the slip behavior. Despite this shortcoming, the model represents a valuable tool for designers to use in analysis of buildings containing partially restrained connections and a foundation that can be used for future model refinements

ACKNOWLEDGEMENT

The experimental data used for this research were generated as part of a SAC research project (SAC Task 7.03) under the sponsorship of FEMA. The analytical work was funded by the Mid-America Earthquake (MAE) Center as part of the work under Task ST7.

REFERENCES

Kulak, G. L., Fisher, J. W., and Struik, J. H. A., 1987. Guide to Design Criteria for Bolted and Riveted Joints, 2nd Ed., John Wiley & Sons, New York.

Rex, C. O., and Easterling, W. S., 1996. "Behavior and Modeling of Single Bolt Lap Splice Connections," Report No. CE/VPI-ST 96/15, Virginia Polytechnic Institute and State University, Blacksburg, VA.

Swanson, J. A. and Leon, R. T., 2000. "Bolted Steel Connections: Tests on T-stub Components," J. Struc. Engrg., ASCE, Vol. 126, No. 1, pp. 50-56.

Swanson, J. A. 1999. "Characterization of the Strength, Stiffness and Ductility Behavior of T-stub Connections," Ph.D. Dissertation, Georgia Institute of Technology, Atlanta, GA.

Swanson, J. A. and Leon, R. T., 2001. "Stiffness Modeling of Bolted T-stub Connections," submitted for review to J. Struc. Engrg., ASCE.

Behaviour of Steel Structures in Seismic Areas, Mazzolani & Tremblay (eds) © 2000 Balkema, Rotterdam, ISBN 90 5809 130 9

Inelastic finite element studies of unreinforced welded beam-to-column moment connections

C. Mao, J. M. Ricles, L. W. Lu & J. W. Fisher
Department of Civil and Environmental Engineering, Lehigh University, Bethlehem, Pa., USA

ABSTRACT: This paper presents the results of a 3-D finite element study of welded beam-to-column moment connections in steel moment resisting frames. Computer models of connection subassemblies were developed using the general-purpose nonlinear finite element program ABAQUS. The analytical results have provided information related to basic performance and the effects that various connection parameters have on inelastic cyclic performance, thereby furthering the current understanding of welded moment connection behavior under seismic loading conditions and leading to improved design criteria. Based on the results of the study recommendations for the seismic design of connections are given related to weld access hole size and geometry, panel zone capacity, continuity plate requirements, and beam web-to-column attachment detail.

1 INTRODUCTION

Numerous welded beam-to-column moment connections in steel moment resisting frames (MRFs) failed during the 1994 Northridge Earthquake (Youssef et al. 1995). The failures raised many questions regarding the validity of current design and construction procedures for these connections. Since the earthquake several extensive analytical and experimental studies have been conducted at Lehigh University (Xue et al. 1996, Kaufmann et al. 1996, Lu et al. 1997). This paper presents the results of recent research that has focused on the seismic resistance of welded moment connections with improved details for MRFs. The improvements in the connection detail include the use of notch tough electrodes, properly contoured and sized weld access holes, beam web full penetration welds, and column continuity plates with adequate thickness.

Numerous weld access hole fractures have been reported in post-Northridge earthquake inspections and in laboratory tests, where conventional access hole configurations were employed in the connections. Previous finite element analysis (El-Tawil et al. 1998) has shown that small access holes result in less strain concentration around the holes. However, the reduced access hole tends to increase the size of the weld defects resulting from incomplete fusion at the root of the flange groove weld. For the beam lower flange, the critical location of major defects often occurs near the access hole, where fracture may initiate. A detailed analytical study was therefore undertaken to examine the effect of the size and

geometry of the access hole on the fracture potential of the material near the hole.

The typical shear tab of a pre-Northridge moment connection is welded to the column flange as well as bolted to the beam web. The shear tab is designed to resist the beam shear force, while the welded beam flanges resistant the beam bending moment. Previous experimental studies (Tsai et al. 1995, Lu et al. 1997) have shown that web supplemental fillet welds, which are provided along three edges of the shear tab, and full penetration groove welds between the beam web and column face enhance the strength, ductility and energy absorbing capacity of the connection. While the contribution of web welds are considered to be beneficial, it is not clear how much contribution they offer. Also, it is not clear what type of modification is sufficient to enhance the ductility of the connection. Additional experimental and theoretical studies are needed to develop a better understanding of the behavior and effect of bolted versus a welded shear tab on moment connection behavior.

Based on current seismic design provisions (AISC 1997) the use of continuity plates is recommended in all cases, unless tests have shown that other design features of a given connection are effective in reducing or redistributing flange stresses such that the connection performs satisfactory without them. Thin continuity plates will likely result in lower residual stresses, which will reduce the potential of weld cracking because of less heat input during welding. Some laboratory tests of connections show good performance without the use of any continuity plates. Thus, current continuity plate seismic design requirements need to be reevaluated.

The current studies at Lehigh University include a finite element analysis study of unreinforced welded beam-to-column moment connections. Figure 1 shows a typical detail of an exterior unreinforced welded beam-to-column connection (Specimen T1) that was tested by the authors and reported by Mao et al. (2000). Nonlinear finite element models were used to conduct a parametric study to investigate the effects of the following on inelastic response of the connection: (1) weld access hole size and geometry; (2) panel zone capacity; (3) beam web-to-column flange attachment detail; (4) column continuity plates. The analyses were conducted applying monotonic increasing static load and cyclic variable amplitude load to the models, respectively.

A companion paper (Mao et al. 2000) presents the experimental results of a complementary study of full-scale unreinforced welded beam-to-column connection subassemblies subjected to cyclic inelastic loading.

2 WELD ACCESS HOLE

The parametric study to investigate the effect of weld access hole size and geometry involved the modeling of the complete test specimen assembly of an exterior connection that was similar to that shown in Figure 1. Three-dimensional models of the test assembly were developed using the general-purpose nonlinear finite element analysis (FEA) program ABAQUS (1999). The three-dimensional finite element model is shown as Figure 2. The FEA results are sensitive to the type of elements and the mesh size and orientation used in the model. Mesh convergence studies were therefore performed. The results indicated that a eight-node brick element (element C3D8 in ABAQUS) with standard integration produced consistent results. A sub-modeling technique was applied to the model in the access hole region in order to obtain results that were sufficiently accurate to allow close examination of the ductile fracture potential of the various access hole configurations in the study. The sub-model analysis directly utilized the results of the global analysis as boundary conditions around the boundaries of the model. Geometric and material non-linearity were included in the analyses.

The global model of the assembly was loaded horizontally at the top of the column and the analysis continued until an plastic story drift of 3.0% rad. was reached. In the model the bottom of the column and end of the beam had pin and roller boundary conditions, respectively.

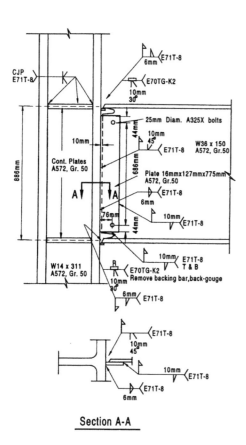

Section A-A

Figure 1. Details of Specimen T1.

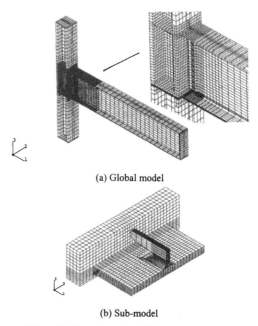

(a) Global model

(b) Sub-model

Figure 2. Three-dimensional finite element model.

The parametric study for the weld access hole involved evaluating the nine different weld access hole configurations shown in Figure 3. The configurations are identified as follows:
(1) Standard access hole
(2) No access hole (as a basis of comparison)
(3) 50 mm. long standard access hole
(4) 125 mm long standard access hole
(5) 75 mm long access hole, Type-I
(6) 75 mm long access hole, Type-II
(7) 75 mm long access hole, Type-III
(8) 50 mm long access hole, Type-I
(9) 50 mm long access hole, Type-II

The standard access hole is the minimum hole size for rolled sections permitted by the AISC-LRFD specification (1995). The diameter of the circular portion of the hole is 20 mm with a length equal to 1.5 times the thickness of the beam web. The size and geometry of the other configurations were developed from the standard hole with the intent to minimize plastic strain in order to reduce the potential for fracture of the beam flange near the hole region. The diameter of the holes in all the configurations was 20 mm, which is probably the smallest diameter that still permits proper welding of the beam flanges.

Figure 4 summarizes the maximum effective plastic strain (PEEQ) indices in the access hole region of all the nine configurations studied. The

PEEQ index is the effective plastic strain normalized by the yield strain. Note that the PEEQ index of the standard hole (configuration (1)) is twice that of configuration (6), which appears to be the best access hole configuration. This configuration is recommended for use in seismic-resistant structures and was used in fabricating the connection test specimen assemblies referred to previously. The testing of these specimens had no fracture occur in the access hole region of the beam flange, with plastic story drifts of up to 5.0% rad. being achieved (Mao et al. 2000).

3 BEAM WEB ATTACHMENT DETAILS

Analyses were conducted of exterior connection models with the web attachment detail shown in Figure 1, consisting of a full penetration groove welded beam web with supplemental shear tab welds. Analyses were also conducted on models having a bolted shear tab detail without the beam web welds. The latter was similar to that used in "pre-Northridge" moment connections in MRFs, with the exception that the weld access hole of both models was configuration (6). A horizontal force was applied at the top of the column in the model to reach the same plastic story drift in order to compare the results of the two models.

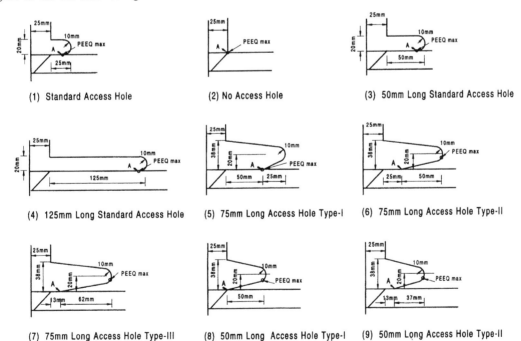

Figure 3. Various weld access hole configurations of parametric study.

713

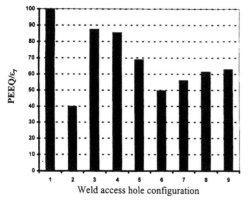

Figure 4. Maximum PEEQ index at the critical location within the access hole region.

A summary of the maximum values for the PEEQ index, pressure index (PI), triaxiality ratio (TR), and rupture index (RI) for the two models are given below in Table 1. The rupture index is based on the work of Hancock and MacKenzie (1976) on ductile fracture in steel under a multi-axial state of stress. The model with the welded beam web has smaller values for the indices compared to the bolted beam web model. The welded beam web detail is less susceptible to fracture by the fact that the maximum PEEQ index, PI, and RI are reduced by 25%, 9%, and 26%, respectively, compared to the bolted beam web model. It was determined that the supplement fillet welding of the shear tab to the beam web inhibited web local bucking and reduced the demand in the beam tension flange at the access hole region.

Based on the analysis results it is recommended that full penetration groove welds be used at the beam web-column interface with supplemental fillet welds around the edges of the shear tab.

Table 1. Summary of bolted web vs. welded web FEM analysis

Case	bolted web	welded web
PEEQ Index[1]	48.5	36.0
PEEQ Index[2]	63.2	46.5
Pressure Index[2]	0.722	0.654
Triaxiality Ratio[2]	0.469	0.471
Rupture Index[2]	0.212	0.157

1. at location of root of weld access hole
2. at location of maximum value
   PEEQ index = effective plastic strain/ yield strain
   Pressure index, PI = Hydrostatic stress/yield stress
   Triaxiality ratio, TR = Hydrostatic stress/von Mises stress
   Rupture Index, RI = PEEQ exp(1.5TR)

4 CONTROL OF PANEL ZONE DEFORMATION

In the exterior connection specimens tested by Mao et al. (2000) a majority of the total plastic deforma-

tion developed in the panel zone. During testing of Specimen T1 cracks were developed at the top and bottom edges of the fillet web connecting the shear tab to the column flange. The beam web groove weld at the edges of the shear tab also started to crack. The cracks grew in size and propagated vertically during the test. A ductile fracture initiated at an edge of the beam bottom flange in the heat-affected zone, which led to a complete separation of the beam flange from the column. The vertical crack in the beam web groove weld had extended to about 10% of the beam depth when this happened. The total plastic story drift achieved was 3.5% rad., 70% of which was due to panel zone shear deformation. A comparison of the results of this test with previous specimens tested by Lu et al. (1997), which had a relatively strong panel zone, would suggest that the cause of the early web weld fracture was due to the excessive panel zone deformation in Specimen T1.

To further investigate this phenomena, a detailed finite element study was made of the effect on panel-zone strength and stiffness on the magnitude and distribution of stress and strain and the ductile fracture potential of the beam web groove weld at the edge of the shear tab (see Figure 5(a)) and at critical locations (center and edges) of the beam bottom flange (see Figure 5(b)). The study used Specimen T1 as the reference analysis case, where cracking occurred at the locations identified in Figure 5. The discussion below pertains to the case when a 13 mm thick doubler plate with a yield stress of 345 MPa was added to the column web panel.

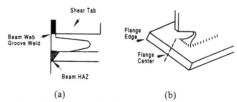

(a)   (b)

Figure 5. Fracture critical location of tension flange zone.

Comparisons of the PEEQ index, pressure index, and the rupture index in the weld at the edge of the shear tab are shown in Table 2. Table 2 shows that the demand at the web weld exceeds that near the region of the weld access hole having the geometry of configuration (6). However, the PEEQ and the rupture indices at the edge of the shear tab are decreased by approximately 50% when the panel zone is strengthened using two-13 mm-thick doubler plates. The reduction in these indices indicates a reduction in the likelihood of cracking developing in the web weld. The rupture indices at the two critical locations of the beam flange are also reduced in the connection with a stronger panel zone. The contribution of the panel zone deformation was reduced from 70% to 30% of the total plastic story drift of the connection subassembly.

Table 2. Stress and strain at the bottom edge of shear tab (Specimen T1).

Case	PEEQ Index	Pressure Index	Triaxiality Ratio	Rupture Index
no Doubler plates	55.5	2.658	1.304	0.654
13 mm Doubler Plates	25.0	2.339	1.287	0.285

5 CONTINUITY PLATES

To investigate the effects of continuity plates on connection behavior a finite element model of an interior connection subassembly was developed, and is shown in Figure 6. The exterior connection model (Figure 2) was also utilized in the continuity plate study. The finite element model for the interior connection was generated using the same brick element used in the exterior connection model. Due to computer capacity limitations, however, portions of the beams and columns away from the connection that remained elastic were modeled using 3-D beam-column elements.

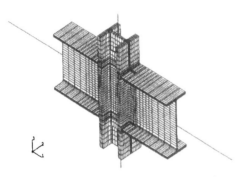

Figure 6. Finite element model of interior connection.

The bottom end of the column and ends of the beams away from the connection in the model had pin and roller boundary conditions, respectively. Cyclic load was horizontally applied to the top of column of the model, following the SAC load protocol (SAC 1997) until reaching a story drift of 5% rad. A nonlinear cyclic strain hardening constitutive relationship was employed in the model by combining the nonlinear kinematic and isotropic strain hardening relationships obtained from cyclic coupon test results of the steel.

Eight different cases of various continuity plate thickness were analyzed, as indicated in Table 3. In Table 3 $b_{f,bm}$ and $t_{f,col}$ are the beam flange width and column flange thickness, respectively. Of the eight cases, the first-five involved the analysis of an interior connection. A range of column flange thickness $t_{f,col}$ was considered, with the beam size equal to a

W36x150 for all cases, resulting in a beam flange width-to-column flange thickness $b_{f,bm}/t_{f,col}$ ratio of 4.2 to 6.8. The continuity plate thickness ranged from zero to one beam flange thickness $t_{f,bm}$ for the various cases as indicated in Table 3. A number of the analysis cases in Table 3 have been tested in the laboratory, as indicated by the presence of a test specimen label in Table 3.

Table 3. Analysis matrix of continuity plate thickness.

Case	Test Spec	Column size	Connection type	Thickness of cont. plates	$b_{f,bm}/t_{f,col}$
1	C1	W14x398	Interior	0	4.2
2	C2	W14x398	Interior	$t_{f,bm}$	4.2
3	C3	W27x258	Interior	0	6.8
4	-	W27x258	Interior	$0.5t_{f,bm}$	6.8
5	C4	W27x258	Interior	$t_{f,bm}$	6.8
6	-	W14x311	Exterior	0	5.3
7	-	W14x311	Exterior	$0.5t_{f,bm}$	5.3
8	T1	W14x311	Exterior	$t_{f,bm}$	5.3

* Beam size is W36x150

The analysis results indicate that the fractures that occurred in the test specimens were related to accumulated plastic strain causing low-cyclic fatigue failure. Shown in Figure 7 are the accumulated PEEQ across the beam flange at the root of the groove weld where the strain demand was largest (see Figure 5).

Figure 7(a) compares the results of cases for a stiff column flange (W14x398), where $b_{f,bm}/t_{f,col}$ = 4.2. It is apparent in Figure 7(a) that the maximum accumulated PEEQ is the same for both cases (cases 1 and 2 in Table 3) and that there is little difference between the results. The results for cases 3, 4, and 5 involving a flexible column flange (W27x258, $b_{f,bm}/t_{f,col}$ = 6.8), and an intermediate flexible column flange (W14x311, $b_{f,bm}/t_{f,col}$ = 5.3) for cases 6, 7, and 8 are shown in Figures 7(b) and 7(c), respectively. It is apparent in Figures 7(b) and 7(c) that the distribution of accumulated PEEQ across the beam flange width becomes more uniform when continuity plates are added to the model. For the W27x258 column the maximum accumulated PEEQ is also reduced whereas for the W14x311 column no change in the maximum value of accumulated PEEQ occurs when adding continuity plates. Furthermore, the distribution of accumulated PEEQ across the beam flange are similar for the cases when the continuity plate thickness is either $0.5t_{f,bm}$ (12 mm) or $1.0t_{f,bm}$ (24 mm).

The analysis results have a good agreement with the experimental results of Specimens C1, C2, C3 and C4 noted in Table 3. The test results of Specimens C1 and C2 showed no difference in local behavior at the beam-column interface, despite Specimen C1 not having continuity plates. Specimen C4 (W27x258 column) showed some improvement in behavior and connection ductility with the use of continuity plates, compared with Specimen C3 which did not have continuity plates and a W27x258 section for the column.

A plot of the analysis results for the maximum accumulated PEEQ at the center of the beam flange as a function of the $b_{f,bm}/t_{f,col}$ ratio is shown in Figure 8. Figure 8 indicates that continuity plates are not required when the $b_{f,bm}/t_{f,col}$ ratio is less than 5.2, for the maximum accumulated PEEQ is the same value for

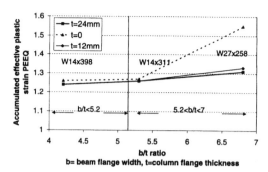

Figure 8. Accumulated PEEQ vs. $b_{f,bm}/t_{f,col}$ ratio.

all cases with and without continuity plates. However, for larger values of the $b_{f,bm}/t_{f,col}$ ratio the use of continuity plates reduces the maximum accumulated PEEQ. The use of a reduced continuity plate thickness of $0.5t_{f,bm} = 12$ mm is shown to be effective in reducing the maximum accumulated PEEQ when $b_{f,bm}/t_{f,col} = 6.8$.

Based on the nonlinear finite element analysis and experimental results, recommendations for continuity plates requirements are given under Table 4. In Table 4 t_c is the required continuity plate thickness. In addition to these requirements the criteria in the AISC LRFD specification (AISC 1994) should also be checked to determine continuity plate requirements for other loading conditions.

Table 4. Required thickness t_c of continuity plates.

$b_{f,bm}/t_{f,col}$	$t_c/t_{f,bm}$
less than 5.2	0
between 5.2 to 7.0	0.5
larger than 7.0	1.0

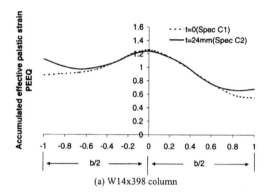

(a) W14x398 column

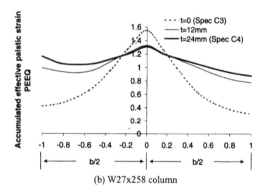

(b) W27x258 column

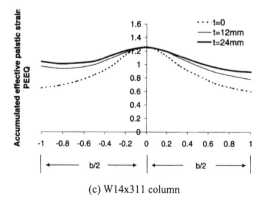

(c) W14x311 column

Figure 7. Accumulated PEEQ across the beam flange near the root of the groove weld at 5% rad. story drift.

6 RECOMMENDATIONS AND CONCLUSIONS

This paper identified four critical issues that have a strong effect on the ductility of welded moment connections and which should be carefully considered in design. The issues are: (1) size and geometry of weld access holes, (2) control of panel zone deformation, (3) beam web groove welds and supplemental web fillet welds, and (4) the thickness of continuity plates. These issues have been studied in various research programs conducted at Lehigh University following the 1994 Northridge earthquake.

The key conclusions and recommendations may be summarized as follows:

(1). In some connection assembly tests conducted in the past, a fracture, which initiated at the edges of weld access hole, was observed. An effort was made in the current study to investigate the influence of the size and geometry of the access hole on the potential of ductile fracture initiation near the holes. Nine different access hole configurations were included in an analytical study. The results indicate the importance of selecting a proper hole configuration. The particular configuration shown as configuration (6) in Figure 3 is recommended. This configuration was adopted in fabricating nine full-scale interior and exterior unreinforced welded beam-to-column test assemblies. No fracture near the access holes was observed during the testing of these specimens, which responded with good ductility.

(2). In some of the exterior connection tests initial fracture occurred in the vertical welds connecting the shear tab or beam web to the column flange. The test specimens in this series had a relatively weak column panel zone. Limiting the amount of panel zone deformation can control the problem of beam web weld fracture. In some applications, this would require strengthening of the panel zone with doubler plates. The connection rotation contributed by panel zone deformation should be designed not to exceed 50% of the total plastic rotation. This can be accomplished by basing the panel zone shear resistance on only the column web and ignoring the benefits of strain hardening effects from the column flanges.

(3). Based on the analysis results, full penetration groove welds at the beam web-column interface in conjunction with supplemental fillet welds around the edges of the shear tab reduce the accumulated effective plastic strain. It is recommended for special MRFs that this beam web attachment be used in order to develop adequate plastic deformation in the connections. The bolted shear tab details is recommended only for ordinary moment resistant frames.

(4).FEM analysis and test results conducted in the study indicate that continuity plates requirements can likely be relaxed for seismic design. Recommendations for continuity plate requirements were provided in Table 4. All specimens tested to date at Lehigh University without continuity plates have developed plastic story drift greater than 3.9% rad. These specimens had columns of different sizes and a beam flange width-to-column flange thickness ratio ranging from 4.2 to 6.8.

The results and conclusions presented are primarily based on experimental studies of nine full-exterior and interior connection assemblies, and a finite element parametric study of various connection designs. Further research is necessary to study the behavior of connections having different details, including beams with thicker flanges. Additional issues may emerge from this research.

7 ACKNOWLEDGEMENTS

The research described in this paper was supported by grants from the SAC Joint Venture, and the Department of Community and Economic Development of the Commonwealth of Pennsylvania through the Pennsylvania Infrastructure Technology Alliance. Dr. Eric J. Kaufmann of the ATLSS Center, a specialist in welding metallurgy, was most helpful in providing insights to some of the welding related issues.

8 REFERENCES

ABAQUS User's Manual (1999) Hibbit, Karlsson and Sorensen, Inc., Providence, Rhode Island.

American Institute of Steel Construction (1995) "Load and Resistance Factor Design," 2nd Edition, Chicago, Ill.

American Institute of Steel Construction (1997) "Seismic Provisions For Structural Steel Buildings," Chicago, Ill.

El-Tawil, Mikesell, T., Vidarsson, E. and Kunnath, S.K. (1998) "Strength And Ductility Of FR Welded-Bolted Connections." Report SAC/BD - 98/01, SAC Joint Venture.

Engelhardt, M. and Sabol, T.A. (1998) "Reinforcing Of Steel Moment Connections With Cover Plates: Benefits And Limitations." Engineering Structures, 20(4-6).

Hancock, J.W. and MacKenzie, A.C. (1976) "On The Mechanisms Of Ductile Failure In High-Strength Steels Subjected To Multi-Axial Stress States," Journal Mech. Phys. of Solids, 24.

Kaufmann, E.J., Xue, M., Lu, L.W. and Fisher, J.W. (1996) "Achieving Ductile Behavior Of Moment Connections." Modern Steel Construction, AISC.

Lu, L.W., Xue, M., Kaufmann, E.J. and Fisher, J.W. (1997) "Cracking, Repair And Ductility Enhancement Of Welded Moment Connections." Proceedings, NEHRP Conference and Workshop on Research on the Northridge, California Earthquake of January 17, 1994, Vol. III, Los Angeles.

Mao, C., Ricles, J.M., Lu, L.W., and Fisher, J.W. (2000) "Seismic Testing Of Welded Beam-To-Column Moment Connections," Proceedings, STESSA 2000, Montreal, Canada.

SAC Joint Venture (1997) "Protocol For Fabrication, Inspection, Testing, and Documentation of Beam-Column Connection Tests and Other Experimental Specimens," Report No. SAC/BD-97/02, Version 1.1, October.

Tsai, K.C., Wu, S., and Popov, E.P. (1995) "Experimental Performance Of Seismic Steel Beam-Column Moment Joints," Journal of Structural Engineering, Vol. 121, No. 6.

Xue, M., Kaufmann, E.J., Lu, L.W. and Fisher, J.W. (1996) "Achieving Ductile Behavior Of Moment Connections - Part II." Modern Steel Construction, AISC.

Youssef, N., Bonowitz, D. and Gross, J.L. (1995) "A Survey Of Steel Moment-Resisting Frame Buildings Affected By The 1994 Northridge Earthquake." NISTIR 5625, National Institute of Standard and Technology.

Behaviour of Steel Structures in Seismic Areas, Mazzolani & Tremblay (eds) © 2000 Balkema, Rotterdam, ISBN 90 5809 130 9

Seismic testing of welded beam-to-column moment connections

C. Mao, J. M. Ricles, L. W. Lu & J. W. Fisher
Department of Civil and Environmental Engineering, Lehigh University, Bethlehem, Pa., USA

ABSTRACT: This paper presents the results of an experimental study of the seismic performance of improved, unreinforced welded beam-to-column moment connections. The study involved the inelastic cyclic testing of nine full-scale connection specimens to evaluate the effects of weld access hole geometry, panel zone strength, continuity plates, and beam web attachment detail on connection cyclic performance. Tests were also performed on 15 beam tension flange-to-column flange *pull-plate* specimens to investigate the effects of notch toughness of the filler metal, use of grooved backing bars, and weld reinforcement. It is shown that the use of a notch-tough electrode is the single most important detail that improves connection ductility. Improved access hole details, in conjunction with notch-tough beam flange welds and a beam web groove weld detail is shown to enable a connection to achieve a cyclic plastic rotation more than 3% rads. Based on the results of the research improved guidelines are given for the seismic resistant design of welded connections for steel moment resisting frames.

1 INTRODUCTION

The widespread damage to welded moment connections in moment resisting frames (MRFs) during the 1994 Northridge Earthquake undermined the confidence in the ductility of steel special moment resisting frames. The failure of welded connections raised many questions regarding the validity of current design and construction procedures for these connections. As part of the SAC II Steel Project extensive analytical and experimental studies have been conducted at Lehigh University on unreinforced welded moment connections. The objectives of this study are: (1) to develop a more thorough understanding of the inelastic behavior of unreinforced welded connections and (2) develop improved guidelines for the seismic resistant design of unreinforced welded moment connections for new buildings. This paper presents the experimental results of this study. Nine full-scale connection specimens were tested, along with 15 tests conducted on beam tension flange-to-column flange specimens (called *pull-plate* specimens). All of the connection specimens were fabricated using the notch-tough E70TG-K2 filler metal, whereas the pull-plate specimens were fabricated using both this and the E70-T4 filler metal. The parameters studied in the connection specimen tests included: (1) access hole geometry; (2) panel zone strength; (3) continuity plate; and (4) beam web attachment detail. The pull-

plate specimen tests provided a simple, low-cost means to investigate material and welding variables associated with welded connection behavior. The parameters studied in pull-plate tests included: (1) notch toughness of weld filler metal; (2) grooved backing bars; and (3) beam flange weld reinforcement.

A companion paper (Mao et al, 2000) presents the analytical study consisting of finite element analysis of unreinforced welded beam-to-column connection subassemblies.

2 PULL-PLATE TESTS

A typical pull-plate specimen is shown in Figure 1. Each pull-plate specimen consists of a 234-mm long section of a W14 x 176 wide flange column shape with one flange removed. A 25-mm thick by 152-mm wide plate (called a *beam flange plate*) is groove welded to the column flange face to simulate the beam tension flange-to-column flange connection. A slotted steel pull plate, simulating a continuity plate, is welded to the web of the column section to permit the assemblage to be gripped in the test machine. A simulated coped web plate is tack welded to the beam plate to introduce welding access restrictions that are similar to the conditions present when field welding the bottom flange of a moment connection. The specimens were fabricated from steel with a nominal yield strength of 345 MPa.

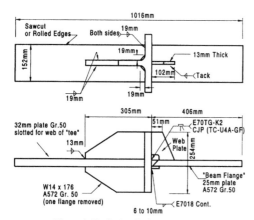

Figure 1. Typical pull-plate test specimen.

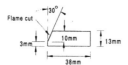

Figure 2. Grooved backing-bar.

E70T-4 and E70TG-K2 filler metal were used for the full penetration groove weld. Three specimens used the E70T-4 filler metal, a lower notch toughness filler metal, with the backing bar remaining in place along with a continuous E7018 fillet weld reinforcement added. Two specimens were fabricated using E70T-4 filler metal along with the grooved backing bar shown in Figure 2 that remained in place with no fillet reinforcement weld. The remaining ten specimens used the higher notch toughness E70TG-K2 filler metal. These ten specimens were tested in order to investigate the use of grooved backing bars, leaving the backing bar in place without a fillet reinforcement weld, and the presence of continuity plates on a joint welded with a notch-tough filler metal.

The specimens were tested in a computer-controlled 2670-kN capacity universal testing machine. The tests were conducted in a dynamic manner under displacement control in order to account for strain-rate effects. Each test specimen was loaded to failure by tension load applied in two ramp rate segments. A specimen was initially loaded at a displacement rate of 1 mm/sec to a load of 267 kN (corresponding to approximately a stress of 70 MPa in the beam flange plate) in order to seat the specimen. The displacement rate was then increased to 3.8 mm/sec to fail the specimen. The displacement rate in the vicinity of the weld joint corresponded to a strain rate of approximately 0.02 sec^{-1}. This strain rate resulted in loading through the elastic range with yielding occurring in about 1 second. In comparison, strain rates for static loading are typically of

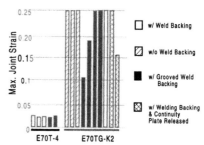

Figure 3. Summary and comparison of pull-plate specimen joint deformation.

the order of 0.001-0.0001 sec^{-1}. Each specimen was visually examined after testing for any evidence of cracking in the groove weld and to determine the origin and mode of fracture.

The ductility of the test specimens is. given in Figure 3. Specimens which fractured in the base metal of the beam flange plate had a ductile failure mode with over 20% elongation, whereas specimens which fractured in the groove weld had no ductility and were brittle. The pull-plate tests resulted in the following findings:

(1) Retrofitting full penetration groove welds of E70T-4 filler metal by the placement of a notch tough fillet reinforcement weld did not improve the performance, with the specimens failing in a brittle manner.

(2) The use of grooved backing bars did not improve the performance of the specimens with E70T-4 filler metal. The performance of the specimens with E70TG-K2 filler metal was affected by the positioning of the grooved backing bar. The intentional misalignment of the backing bar caused entrapment of mill scale and slag, creating a weld flaw and leading to a weld fracture and loss of ductility.

(3) Specimens using E70TG-K2 filler metal and a standard backing bar were ductile. Leaving the backing bar in place without a reinforcement fillet weld also produced a ductile response. However, the omission of the continuity plates effected the ductility of the joint.

(4) The test results support the newly proposed SAC recommendations (FEMA 1999) for filler metal used in critical joints to have a minimum CVN impact toughness of 27 Joules at -29 degrees Celsius and 54 Joules at 21 degrees Celsius in conjunction with good weld detailing, such as the removal of weld backing bars and control of weld flaws through inspection.

3 FULL-SCALE CONNECTION TESTS

The connection study consisted of testing both an exterior connection and an interior connection of an MRF. All specimens were fabricated in accordance with the SAC protocol (SAC, 1997). The specimens for the exterior connection consisted of a beam attached to a column, while the specimens for the interior connection were cruciform-shaped with a beam attached to each side of the column. All specimens were fabricated using notch-tough electrodes. The details for each specimen are described below. The test setup for the interior connection specimens is shown in Figure 4, where rigid links and cylindrical bearings were used to create roller and pin boundary conditions at the ends of the beams and column, respectively. The test setup for the exterior connection was similar to that shown in Figure 4, except that the specimen had only one beam. In each test the top of the column was subjected to the lateral displacement history prescribed by SAC (SAC 1997) to achieve targeted values of story drift. The displacement history consisted of a series of symmetric cycles with increasing amplitude. Each specimen was instrumented to measure member forces and displacements to enable separate determinations of the beam, column and panel zone rotations to be made.

3.1 Exterior Connection Specimens

Four exterior connection specimens were tested. The beam and column for each specimen were W36x150 and W14x311 sections, respectively, with a nominal yield strength of 345 MPa.

In the exterior connection tests the effects of the beam web attachment detail on cyclic performance and the reliability of the modified weld access hole shown in Figure 5 were investigated. All specimens used the modified weld access hole.

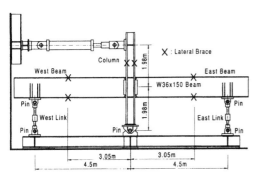

Figure 4. Test set-up.

Figure 5. Modified weld access hole.

The weld access hole was established by nonlinear finite element analysis and is described in a companion paper (Mao et al., 2000). All specimens had full penetration beam flange groove welds with E70TG-K2 filler metal. The beam top flange backing bar was left in place and a reinforcing fillet weld placed, whereas the beam bottom flange backing bar was removed, backgouged and a fillet weld was placed (see Figure 6). The shear tabs for all specimens were fillet welded to the column using E71T-8 filler metal, which is of high notch-toughness. The test matrix and web attachment detail for the four specimens are summarized in Table 1.

Table 1. Exterior connection test matrix and beam web attachment detail.

Specimen	Grooved welded web	Fillet weld shear tab	Bolted web
T1	Yes	Yes	-
T2	Yes	-	-
T3	-	Yes	-
T4	-	-	Yes

The beam web of Specimen T1 was groove welded with a E71T-8 filler metal directly to the column flange. A supplementary fillet weld was then placed around the edges of the shear tab. The weld details for Specimen T2 are the same as T1, except no supplementary fillet weld was placed around the edges of the shear tab. For Specimen T3, a supplementary fillet weld was placed around the edges of the shear tab with no groove weld placed between the beam web and column flange. The beam web of Specimen T4 was bolted to the shear tab with no supplemental fillet weld and web groove weld.

The applied lateral load vs. column top displacement hysteretic response for the exterior connection specimens is given in Figure 7. During the application of the 2% drift cycles to Specimen T1 cracks were observed at the top and bottom edges of the fillet weld connecting the shear tab to the column flange. The beam web groove weld at the edges of the shear tab also started to crack at the fusion line. The cracks in the web groove weld grew in size and propagated vertically during subsequent cycles of testing. The beam flanges near the column buckled locally and the panel zone showed extensive yielding during the 3% drift cycles. A ductile fracture started to develop at an edge of the beam bottom flange in the base metal (heat-affected zone) when the drift reached 5%, which eventually caused complete separation of the flange from the column. The crack in the groove weld at the bottom edge of the beam web had extended to about 10% of the weld length when this happened. The maximum plastic story drift $\theta_{p,max}$ achieved was 3.5% rad, about 70% of which was due to panel zone shear deformation.

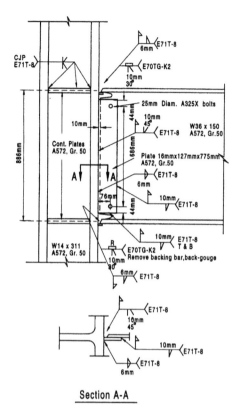

Section A-A

Figure 6. Typical detail of exterior connection specimen
(Specimen T1).

During the testing of Specimen T2 the beam top
flange fractured at the interface of the beam flange
groove weld and base metals, initiating at the edge
of the flange during the second cycle of 4% drift.
The top erection bolt and beam web groove weld
fractured during the first cycle of 5% drift. The $\theta_{p,max}$
achieved during the 4% drift cycles before fracture
occurred was 2.5% rad., of which 1.8% rad. was
plastic shear deformation in the panel zone.

The fillet weld of the shear tab of Specimen T3
fractured completely through the depth of the shear
tab at the column face during the first cycle of 3%
drift. The beam shear force was then resisted entirely
by the beam flanges. Large flexural deformations
were observed in the beam flanges near the HAZ
during the 3% drift cycles. The bottom flange fractured in the base metal during the second cycle of
3% drift. The $\theta_{p,max}$ achieved during the test was 2%
rad. The test results show that the fillet weld connecting the shear tab to the column is not strong
enough to resist the moment that is transferred from
the beam web to the shear tab by the supplemental
fillet welds.

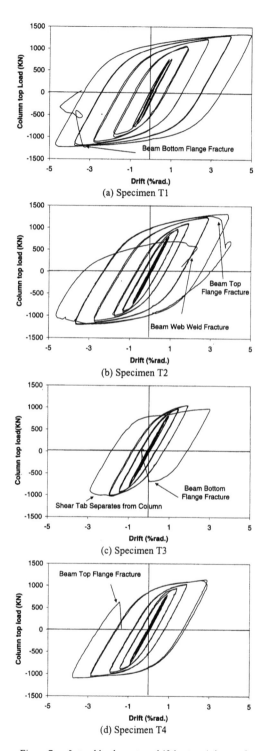

Figure 7.   Lateral load vs. story drift hysteresis loops of
exterior connection specimens.

During the test of Specimen T4 cracks appeared at the interface of the weld metal and the base metal of the beam flanges at the 2% drift cycles. These cracks propagated in subsequent cycles. Local buckling of the beam flanges occurred near the column face during the 3% drift cycles. Also, at the interface of the weld and base metals of the beam top flange a through-thickness crack initiated at the edge of the flange during the second cycle of 3% drift. The crack extended 75 mm across the beam flange. The beam top flange fractured in the base metal near the HAZ during the first cycle of 4% drift. The $\theta_{p,max}$ achieved during the 3% drift cycles before fracture occurred was 1.8% rad., which included 0.8% rad. of plastic shear deformation in the panel zone. The shear tab yielded locally near the bolt holes. No cracking was found in the fillet weld of the shear tab. Slip developed between the beam web and shear tab, which reached 11 mm during the 4% drift cycles. The test results indicate that this slip reduced the plastic deformation demand in the panel zone.

A relatively large portion of the total plastic deformation of the exterior connection specimens was due to panel zone deformation. The modified weld access hole performed well, with no cracking found in the region of the access hole. The test results indicate that the web attachment detail can significantly affect the ductility of the connection. Only Specimen T1 achieved a $\theta_{p,max}$ greater than the currently required AISC seismic design criteria (AISC 1997) of 3% rad. for moment connections in special MRFs.

3.2 Interior Connection Specimens

In the interior connection specimen tests the effects of continuity plates on connection performance were investigated. Five specimens were tested, with the test matrix given in Table 2. The beams for each specimen were W36x150 wide flange sections while the column was either a W14x398 or W27x258 steel section. All sections had a 345 MPa nominal yield strength. The column section, doubler plate and continuity plate thicknesses for each specimen are given in Table 2, where $t_{f,bm}$ designates beam flange thickness. The doubler plates were sized by using just the first term in Equation (9-1) of the *1997* ASIC LRFD Seismic Provisions (AISC 1997) to consider only the column web contribution to the panel zone shear resistance in order to control inelastic panel zone deformations. All of the interior connection specimens had full penetration beam flange groove welds using E70TG-K2 filler metal, a beam web groove weld, and supplemental fillet welds that were similar to the detail for Specimen T1. The modified access hole (Figure 5) was also used for all interior connection specimens. A schematic of Specimen C1 is shown in Figure 8. Specimen C5 had a 165 mm thick composite floor slab.

Table 2. Test matrix for interior connection specimens.

Specimen	Column Size	Doubler plate thick.	Continuity plate thick.	Concrete slab
C1	W14x398	2@19mm	-	-
C2	W14x398	2@19mm	$t_{f,bm}$	-
C3	W27x258	2@16mm	-	-
C4	W27x258	2@16mm	$t_{f,bm}$	-
C5	W14x398	2@19mm	-	165mm

Beam size is W36x150 A572 Gr.50

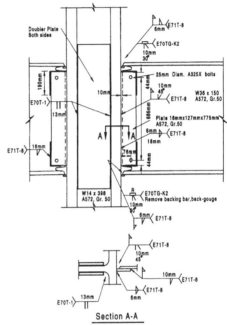

Figure 8. Typical detail of interior connection specimen (Specimen C1).

Specimens C1 was a specimen with stiff column flanges and no continuity plates. The beam flange width-to-column flange thickness ($b_{f,bm}/t_{f,col}$) ratio was equal to 4.2. During the test of Specimen C1 cracking initiated in the beam web groove weld on the top edge of the shear tab of both beams during the 2% drift cycles. These cracks propagated in subsequent cycles. Minor local buckling of the beam flanges occurred near the column face also during the 2% drift cycles, and the column flange of the panel zone began to yield. Severe local buckling occurred in the flanges and web of both beams during the 4% drift cycles. A crack initiated at the interface of the weld and base metals of the top flange of the west beam at the end of the 4% drift cycles, and propagated across 80% of the flange width by the end of the first cycle of 5% drift. After the first cycle of 5% drift was completed, the west beam reaction was released to continue the test with only the east beam. The crack in the web groove weld of the west beam had propagated to 45% of the length of the weld. During the second cycle of 5% drift the east beam top flange fractured in the HAZ of the base metal, propagating rapidly across the beam flange. The $\theta_{p,max}$ achieved was 3.9% rad. The panel zone developed a maximum plastic shear deformation of 0.7% rad.

Specimen C2 is similar to Specimen C1 except that Specimen C2 had continuity plates. In Specimen C2 the toe of the beam flange groove weld was burr ground to minimize the potential for micro-cracking. Also, weld tabs were used for the beam web vertical groove weld to maintain the quality of the weld at the edges of the beam web. The weld tabs were removed after placing the groove weld. Consequently, Specimen C2 had an improved performance compared to Specimen C1. Minor local buckling occurred in the beam flanges of Specimen C2 near the column face during the 2% drift cycles; also the column flange of the panel zone began to yield. Cracks initiated in the web groove welds on the top edge of the shear tab of both beams during the 3% drift cycles, but did not propagate in subsequent cycles. Severe local buckling occurred in the flanges and web of both beams during the 4% drift cycles. At the end of the 4% drift cycles a crack initiated at the interface of the weld and base metals at the edge of the top flange of both beams. The east beam bottom flange and west beam top flange fractured during the first half of the second cycle of 6% drift. The fracture in the east beam occurred in the base metal close to the HAZ, propagating from a crack that had initiated at the north edge of the weld-base metal interface of the beam bottom flange. The fracture in the west beam top flange occurred in the base metal at 152 mm from the column face where severe cyclic beam local flange buckling occurred. Figure 9 shows a photograph of Specimen C2 after completion of the test. Specimen C2 developed a $\theta_{p,max}$ of 5.0% rad. The panel zone developed a maximum plastic shear deformation of 0.7% rad.

Specimens C3 and C4 had a more flexible column flange ($b_{f,bm}/t_{f,col}$ = 6.8) without and with continuity plates, respectively. During the testing of

Figure 9. Specimen C2 after completion of test.

Specimen C3 cracks initiated in the beam web groove weld at the edges of the shear tab of both beams during the 2% drift cycles. These cracks did not propagate during subsequent cycles. Yielding occurred in the column flange within the panel zone and the vertical groove weld at the edge of the double plates also during the 2% drift cycles. By the end of the 3% drift cycles a crack initiated at the interface of the weld and base metals at the center of the bottom flange of both beams. During the 4% drift cycles severe local buckling occurred in the flanges and web of both beams. The west beam top flange fractured during the first cycle of 5.5% drift. The fracture occurred in the base metal close to the HAZ, propagating from a crack that had initiated at the south edge of the weld-base metal interface.

In Specimen C4, the column flange in the panel zone began to yield during the 2% drift cycles, with cracks initiating in the beam web groove welds at the edge of the shear tab of both beams. These cracks did not propagate in subsequent cycles. By the end of the 3% drift cycles cracks had initiated at the interface of the weld and base metals at the edge of both flanges of the west beam. During the 4% drift cycles severe local buckling occurred in the flanges and web of both beams. The top flange of both beams developed cracking at the end of the 6% drift cycles. These cracks occurred in the base metal at the location of severe local buckling of the top flanges.

The $\theta_{p,max}$ achieved in Specimens C3 and C4 was 4.1% rad., and 5.2% rad., respectively. In both of these specimens the panel zone developed a maximum plastic shear deformation of 0.1% rad.

The concrete slab in Specimen C5 cracked during the 0.75% drift cycle, and developed initial crushing at the column face at the 3% drift cycles. Also, the beam web groove welds on the bottom edge of the shear tab of the west beam started to crack during the 3% drift cycles, but did not propagate in subsequent cycles. Cracks initiated at the interface of the weld and base metals of the bottom flange of both beams, developing at the center of the beam flange width during the 4% drift cycles. Brittle fracture occurred in the top flange base metal of the east beam during the second cycle of 4% drift. The fracture occurred at 533 mm from the column face, where a shear stud was located on the beam flange and where severe local buckling had occurred. After the second cycle of 4% drift was completed, the east beam reaction was released to continue the test with only the west beam. The west beam top flange fractured in the base metal during the second cycle of 5% drift. This fracture occurred at 229 mm from the column face, where a shear stud was located on the beam flange and where severe local buckling had occurred. The composite slab increased the moment capacity of the beams by 11% compared to the companion bare steel specimen (Specimen C1), leading to large shear force and inelastic shear deformation in the panel zone.

The welding of the shear studs to the beam flange in the plastic hinge zone, where local buckling and cracks occurred, evidently led to brittle fracture of the beam flange by reducing the notch toughness of the base metal and thereby resistance to crack propagation.

The applied lateral load-story drift hysteretic response for the interior connection specimens is shown in Figure 10. The hysteresis loops of the interior connection specimens have a gradual deterioration in strength prior to fracture due to the severe local buckling that occurred in the beam flanges and web. This is in contrast to the response of the exterior connections that were dominated by panel zone deformation, where there was no deterioration in specimen strength prior to fracture (see Figure 7). The modified weld access hole performed well during the test. There was no cracking found in the access hole region. The practice of using weld tabs at the beam web improved the quality of the web groove weld and the performance of the connection. The cracking in the beam web groove weld was less severe than that of the test specimens that did not use the web weld tabs. Also, less cracking was found at the toe of the beam flange groove weld where the burr grinding was performed.

A summary of the $\theta_{p,max}$ and the contribution from the beam ($\theta_{p,bm}$) and the panel zone ($\theta_{p,pz}$) is given in Figure 11 and Table 3. Included in Table 3 is the $M_{pz}/\Sigma M_p$ ratio, where M_{pz} is the beam moment associated with panel zone shear capacity and ΣM_p the sum of the beam flexural capacity. M_{pz} is based on only the column web contribution. The results shown in Figure 10 and Table 3 indicate that for weak panel zone specimens (i.e., $M_{pz}/\Sigma M_p < 0.9$) that a majority of the plastic story drift occurs in the panel zone. Furthermore, by comparing $\theta_{p,max}$ of Specimen T1 with $\theta_{p,max}$ of the interior connection

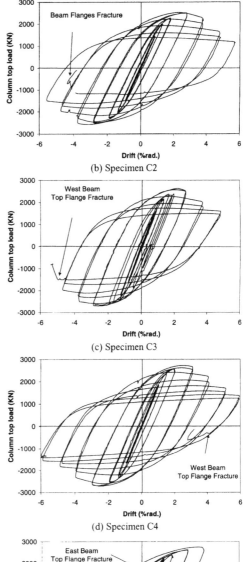

(b) Specimen C2

(c) Specimen C3

(d) Specimen C4

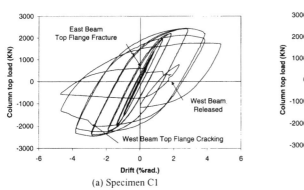

(a) Specimen C1

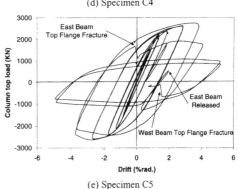

(e) Specimen C5

Figure 10. Lateral load vs. story drift hysteresis loops of interior connection specimens.

specimens, which had stronger panel zones ($M_{pz}/\Sigma M_p > 1.0$) and the same web groove weld detail as Specimen T1, it is apparent that the specimens with stronger panel zones performed better that the specimens with the weaker panel zones. Nonlinear finite element analysis studies which are presented in a companion paper (Mao et al, 2000) show that a moment connection with a strong panel zone can significantly reduce the plastic strain at the edge of the beam web groove weld.

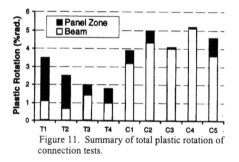

Figure 11. Summary of total plastic rotation of connection tests.

Table 3. Summary of specimen performance.

Specimen	Total Drift (%rad)	Plastic Rotation			M_{pz}
		$\theta_{p,max}$	$\theta_{p,bm}$ (%rad)	$\theta_{p,pz}$	$\overline{\Sigma M_p}$
T1	5.0	3.5	1.1	2.4	0.88
T2	5.0	2.5	0.7	1.8	0.88
T3	3.0	2.0	1.4	0.6	0.88
T4	3.0	1.8	1.0	0.8	0.88
C1	5.0	3.9	3.2	0.7	1.03
C2	6.0	5.0	4.3	0.7	1.03
C3	5.5	4.1	4.0	0.1	1.20
C4	6.0	5.2	5.1	0.1	1.20
C5	5.0	4.6	3.6	1.0	1.03

4 SUMMARY AND CONCLUSIONS

Based on the results of the experimental studies reported herein the following conclusions are noted:

(1). The pull-plate specimen test results indicate that a notch tough filler metal can significantly improve the ductility of the connection. Grooved backing bars can create defects in the weld due to entrapment of slag, which can lead to weld failure. The use of grooved backing bars is therefore not recommended.

(2). A modified weld access hole detail consisting of a flat region on the beam flange of one beam flange thickness improved the performance of welded unreinforced moment connections. No flange fracture initiated from the access hole region, which did occur in numerous other tests conducted by SAC where the specimens had the standard access hole and unreinforced beam flange.

(3). The web attachment detail can significantly affect the ductility of the connection. An attachment detail consisting of a groove welded beam web and supplemental fillet welds around the beam shear tab produced ductile specimen response. The total plastic story drift of specimens with this detail ranged from 3.5% to 5.2% rad. Weld tabs were found to significantly improve the quality of the web groove weld and thereby the connection performance.

(4). Test results to date indicate that continuity plates requirements can probably be relaxed for seismic design. The three specimens tested without continuity plates had a plastic story drift greater then 3.9% rad. These specimens had a $b_{f,bm}/t_{f,col}$ ratio of 4.2 (W36x150 beam + W14x398 column) and 6.8 (W36x150 beam + W27x258 column).

(5). Specimens with a stronger panel zone had better performance. Specimens with a weak panel zone, where $M_{pz}/\Sigma M_p = 0.88$, developed significant shear deformation in the panel zone causing a local prying of the beam flange welds as well as an increase in the plastic strain on the beam web welds. The test results indicate that better performance can be achieved if the panel zone strength is based on only the column web, with the contribution of the column flanges not considered.

(6). Shear studs were found to initiate fracture of the beam flange where local buckling occurs. Shear studs should, therefore, not be placed in the beam plastic hinge region where local beam flange buckling is expected.

5 ACKNOWLEDGEMENTS

This research described in this paper was supported by grants from the SAC Joint Venture, and the Department of Community and Economic Development of the Commonwealth of Pennsylvania through the Pennsylvania Infrastructure Technology Alliance.

6 REFERENCES

American Institute of Steel Construction (1997) "Seismic provisions for structural steel buildings," Chicago, Illinois.
FEMA (1999) "Recommendation Specifications for Moment-Resisting Steel Frames for Seismic Applications," 90% Draft, FEMA, Washington, D.C.
Mao, C., Ricles, J., Lu, L.W., and J. Fisher (2000) "Inelastic Finite Element Studies of Unreinforced Welded Beam-to-Column Moment Connections," STESSA 2000, Montreal, Canada
SAC Joint Venture (1997) "Protocol For Fabrication, Inspection, Testing, and Documentation of Beam-Column Connection Tests and Other Experimental Specimens", Report No. SAC/BD-97/02, Version 1.1.

Behaviour of Steel Structures in Seismic Areas, Mazzolani & Tremblay (eds) © 2000 Balkema, Rotterdam, ISBN 90 5809 130 9

Seismic design of sixteen-bolt extended stiffened moment end-plate connections

T.W. Mays, E.A. Sumner, R.H. Plaut & T.M. Murray
The Charles E. Via, Jr., Department of Civil and Environmental Engineering, Virginia Polytechnic Institute and State University, Blacksburg, USA

ABSTRACT: A 16-bolt extended stiffened (16ES) moment end-plate connection is considered, with special emphasis on conclusions from the SAC Steel Project and requirements for moment connections as stated in the *AISC Seismic Provisions for Structural Steel Buildings*. The purpose of the connection is to provide a moment end-plate configuration that can be designed to be stronger than the adjoining beam. Design procedures for the largest configurations currently available are unable to develop numerous practical beam sections. The 16ES connection allows for the design of an economical moment connection that will meet the requirements for ordinary moment frames as stated in the *AISC Seismic Provisions for Structural Steel Buildings*, while avoiding any type of experimental testing requirement. A simplified LRFD design procedure that considers the limit states of end-plate yielding and bolt rupture is presented, based on results obtained using yield line analysis and the finite element method, respectively.

1 INTRODUCTION

1.1 Background

With all the problems associated with welded moment connections uncovered after the Northridge earthquake, large bolted connections are becoming a much more attractive alternative for design in seismic regions. This paper presents some of the findings of an ongoing study to develop design procedures for large moment end-plate connections to be used in seismic areas. In particular, a large connection, shown in Figure 1, is considered with special emphasis on design requirements stated in the *AISC Seismic Provisions for Structural Steel Buildings* (1997). The connection is tagged 16ES moment end-plate connection to designate the 16-bolt extended stiffened configuration. Although the authors are currently developing much-needed design recommendations for the four-bolt wide moment end-plate configuration, results from the SAC Steel Project manifest the need for even larger connections.

The *AISC Seismic Provisions for Structural Steel Buildings* requires that for intermediate and special moment frames, the connection (comprising all limit states) be designed stronger than the adjoining beam. This is a result of conclusions drawn from Phase I of the SAC Steel Project. It can be quite easily shown, using all currently existing end-plate design procedures, that a connection strength greater than the strength of an adjoining deep beam (i.e., greater than 30 inches) is not possible in most cases when the

beam yield stress exceeds 50 ksi. Results from a four-bolt-wide parametric study lead to the same disappointing conclusion (Mays, 2000). In addition, even when the connection can be designed stronger than the beam, 1 1/2 in. diameter A490 bolts are often required. Such large bolts are simply not practical or economical to install. For ordinary moment frames, the connection can be designed weaker than the adjoining beam, as long as 0.01 radians of plastic rotation are shown to be obtainable. Nevertheless, test results from the SAC Steel Project show that this value is difficult to achieve without significant plastic behavior from the bolts (i.e., plastic elongation). Plasticity from such a brittle element is obviously questionable. A simplified design procedure is presented for the 16ES moment end-plate connection to effectively design thick end-plates to be part of ordinary moment frames. The procedure is then applied as a moment end-plate connection is designed for a W21x101 beam.

The ANSYS finite element package is used to model the beam-to-column connection. Since the problem is three-dimensional in nature, solid eight-node brick elements that include plasticity effects are used extensively to model the assembly. Contact elements are included between the end-plate and the column flange to represent the nonlinear behavior of this complex interaction problem. Bolt pretensioning effects are included and prying forces are tracked throughout the loading process.

1.2 Yield line analysis

Yield line analysis has proven to be a useful tool in determining the ultimate load carrying capacity of end-plates. The procedure is similar to the plastic analysis of beams where elastic deformations are considered negligible when compared to plastic deformations caused by the formation of plastic hinges. Yield lines are selected in any kinematically valid pattern to divide the end-plate into a series of rigid sections which form a mechanism. The lines are straight, and it is assumed that the moment along each line is constant and equal to the plastic moment capacity of the plate. The beam web centerline is given a virtual rotation about the bottom of the beam, and virtual work is then used to find the controlling mechanism. Since the connections are to be designed such that bolt yielding is not the controlling limit state, it is assumed that all of the internal work takes place in the end-plate itself. Therefore, the plate rotates with the beam web but has no deflection at the bolt hole locations. The plate deformations are consistent with the selected yield line mechanism. For a more complete description of this method, see, for example, Szilard (1974).

Borgsmiller (1995) summarizes the controlling yield line mechanisms for the most commonly used end-plate configurations. All the patterns were developed for gravity and wind loading. Such connections usually only have two bolts at the compression flange and it is assumed that the internal work done in this region is negligible. However, for seismic design, the connection should be able to resist a complete reversal of loading and the tension and compression flange bolt configurations are designed symmetrically. Nevertheless, Mays (2000) shows that the internal work done in the compression region is only about 10 percent of the total internal work, and neglecting this value is conservative and simplifies the analysis.

Lastly, yield line analysis predicts what is usually called "first yield moment" or simply "first yield". This moment is not the failure load of the connection. In fact, it can be well below the actual moment resulting in structural failure of the connection. However, in design, it is typically used as the design strength of the connection for the plate bending limit state. Theoretically, it is the moment capacity that would result in structural failure in the absence of any additional resistance. In reality, membrane stresses in the plate resist the additional applied moment at the connection and stretching of the plate takes place. Practically speaking, the first yield moment predicts the point at which plastic rotation due to inelastic plate behavior becomes significant.

2 PROCEDURE

2.1 End-plate bending

The controlling yield line pattern for the 16ES moment end-plate connection is shown in Figure 2. Here, it is assumed that internal work done in the proximity of the bottom flange is negligible. The letters A through J as shown in the figure correspond to the beam web centerline locations where the deflections from i = A through J represent the linear variation in deflection from the bottom of the beam

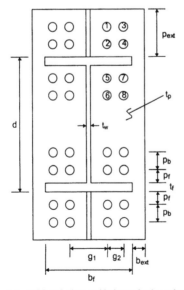

Figure 1. 16ES end-plate and bolt numbering scheme.

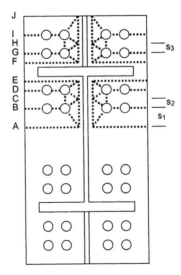

Figure 2. Controlling yield line pattern.

to the top of the end-plate. Using virtual work, and rotating the bottom flange a value of $\theta = 1/(d+p_{ext})$ so that the top of the end-plate deflects one unit, the strength of the plate can be obtained. Including the internal work done by all the yield lines of Figure 2, the end-plate bending strength is found as

$$M_n = 2M_{pl}[b_{ext}\{-1+\frac{d+p_{ext}}{p_{ext}-p_f-p_b}$$

$$+\frac{2(-p_f^2+s_1(2d-t_f)+p_f(d-p_b-t_f))}{p_f s_1}\}$$

$$+\frac{1}{2}(4g_1+\frac{1}{p_f(p_b-p_{ext}+p_f)s_1}(b_f(2dp_bp_f$$

$$-2p_b^2p_f-2dp_{ext}p_f+2p_bp_{ext}p_f$$

$$+2dp_f^2-4p_bp_f^2+2p_{ext}p_f^2-2p_f^3$$

$$+4dp_bs_1-4dp_{ext}s_1+3dp_fs_1-p_bp_fs_1 \tag{1}$$

$$-p_f^2s_1-2t_f(p_b-p_{ext}+p_f)(s_1+p_f)))$$

$$+\frac{8}{g_1}(dp_{ext}+dp_f+p_fp_{ext}-p_f^2+ds_1$$

$$-p_fs_1+s_2^2+s_3^2+dp_b+p_bp_{ext}$$

$$-2p_bp_f-p_bs_1-p_bs_2-p_bs_3-t_fp_b$$

$$-t_fp_f-s_1t_f)]$$

where $M_{pl}=F_{yp}t_p^2/4$ is the plastic moment capacity of the yield line per unit length, F_{yp} is the nominal yield stress of the end-plate, and the unknown lengths s_1, s_2, and s_3 are obtained by minimizing the internal work and are found to be

$$s_1 = \frac{1}{2}\sqrt{(2b_{ext}+b_f)(g_1)} \quad , \quad s_2 = s_3 = \frac{p_b}{2} \tag{2}$$

2.2 Bolt rupture

For moment end-plate connections, the design strength for the limit state of bolt rupture is quite difficult to determine. Prying forces must be included and, as shown in Mays (2000), they can change significantly as a function of the bolt geometry. Nevertheless, the problem is simplified here. The thick end-plate design minimizes prying forces, and the width of the beam flange only allows for certain geometric configurations regarding bolt layout. For a 16ES connection with a specified minimum p_f distance, the beam flange width, b_f, is the biggest deterrent to a good connection design. As the outside bolts (#4 and #7) approach the beam flange tip, their effectiveness in resisting the applied flange force is lost. Using fully tensioned high strength bolts and a minimum bolt pitch to beam flange distance of $p_f=d_b+1/2$ in., the authors have found that the following simplified relationship can be used to determine the number of bolts that are effective in resisting the design flange force:

$$n_{eff} = \begin{cases} 7 & \text{for} \quad 1/4" \le b_{edge} < 1/2" \\ 8 & \text{for} \quad 1/2" \le b_{edge} < 3/4" \\ 9 & \text{for} \quad 3/4" \le b_{edge} < 1.25" \\ 9.5 & \text{for} \quad b_{edge} \ge 1.25" \quad (d_b > 1") \\ 10.5 & \text{for} \quad b_{edge} \ge 1.25" \quad (d_b \le 1") \end{cases} \tag{3}$$

where $b_{edge}=[b_f-(g_1+2g_2)]/2$ and is defined as the distance from the centerline of the outer column of bolts to the beam flange tip. For the case when this distance is greater than 1.25 in., the number of effective bolts has been shown to be dependent on the bolt size. Apparently, in this range, smaller diameter bolts have better stress redistribution characteristics than do larger bolts. This must be related to the smaller bolt heads on the smaller diameter bolts. When b_{edge} is less than 1/4 in., the 16ES connection should not be used. SAC Steel Project tests of four-bolt wide specimens in this range have been shown to fail in a non-ductile manner. Also, the outside bolts are highly ineffective in resisting the applied flange force and the connection's response is more like that of an eight-bolt extended stiffened connection.

Given that 16 bolts are at the tension flange, the values 7 through 10.5 bolts seem like a relatively small number of effective bolts. However, it will soon be shown that the four bolts farthest from the intersection between the beam flange and the beam web or stiffener are not effective at all, regardless of the configuration. In other words, it is really 7 to 10.5 out of 12 bolts that are effective. The values in Equation 3 represent lower bounds for numerous 16ES connection geometries analyzed using the finite element method. These results are consistent with the superposition of four-bolt wide and eight-bolt extended stiffened results discussed in Mays (2000) and Murray and Kukreti (1988), respectively.

There are other design considerations. Bolts must be fully tensioned ASTM A325 or A490 high strength bolts that are less than 1 1/2 in. in diameter. End-plates with yield stresses between 36 and 60 ksi may be used. As a minimum, the weld designs should be in accord with the eight-bolt extended stiffened weld design procedure presented in the *AISC Manual of Steel Construction, Vol. II*. However, it is strongly recommended that for seismic applications, the beam-flange-to-end-plate weld and the stiffener-to-end-plate weld be made using full penetration groove welds. To provide an effective load path from the stiffener to the upper bolts (e.g., bolts #3 and #4), it is recommended that the stiffener extend a horizontal distance of $1.7p_{ext}$ along the beam flange. This is consistent with recommendations provided for the SAC Steel Project specimens.

729

2.3 *Sample application*

To show the simplicity and effectiveness of the design procedure discussed in the previous section, a connection will be designed to develop a W21x101 beam section to be part of an ordinary moment frame. This connection was chosen to represent a worst case lower bound design for a specific bolt spacing. For this purpose, the actual bolt tensile strength will be used in lieu of the design tensile strength. This will show the effectiveness of the design procedure. However, to account for strength variability in design using LRFD, an appropriate strength reduction factor must be used. Current SAC discussions suggest that the normal strength reduction factor (ϕ=0.75) for bolts in tension may be too conservative when combined with other load variability considerations. Also, stress redistribution between bolts increases the ability to more accurately predict the connection's strength.

Table 1 provides the details of the specimen considered here and shown in Figure 1. The values listed are based on minimum bolt spacing and tightening clearances. The symbols F_{yp}, F_{ys}, and F_{yb} represent the nominal yield stress of the end-plate, stiffener, and beam, respectively. Note that the tension flange (top flange) bolts are numbered in Figure 1. Individual bolts will be referred to by these numbers.

According to the *AISC Seismic Provisions for Structural Steel Buildings* (1997), to avoid any type of experimental testing requirement, the connections must be designed to resist $1.1R_yM_p$ or the maximum moment that can be delivered to the system, where R_y is the ratio of expected yield stress to nominal yield stress and M_p is the plastic moment capacity of the beam in question. This requirement is specified to ensure that beam hinging occurs during an earthquake.

Table 1. Connection details for W21x101 16ES connection.

Dimension	Value
d	21.36 in.
b_f	12.29 in.
t_f	0.80 in.
t_w	0.50 in.
g_1	5.00 in.
g_2	3.33 in.
b_{ext}	2.00 in.
p_{ext}	7.00 in.
p_f	1.75 in.
p_b	3.33 in.
t_p	1.00 in.
Bolt designation	1 ¼ in. A325
F_{yp}	36 ksi
F_{ys}	50 ksi
F_{yb}	50 ksi

For the W21x101 beam, the ultimate moment, M_u, is found as

$$M_u = 1.1R_yM_p$$
$$M_u = 1.1 \times 1.1 \times 1054 = 1276 \ k-ft \quad (4)$$

Next, the end-plate bending strength must be designed stronger than M_u. Using the data from Table 1 and Equation 1, a 1.00 in. thick end-plate results in a design strength of 1277 k-ft. For the purpose of illustrating the effectiveness of the method, the nominal strength was used here (i.e., use ϕ=0.90 for end-plate bending). Since 1277 k-ft is greater than 1276 k-ft, a 1.00 inch thick end-plate will be used. In the next section, a 7/8 in. end-plate will also be considered to show how the connection may be inadequate for developing the beam if the end-plate is not sufficiently thick. The end-plates will be tagged thick and thin plates based on their relative thickness.

The flange force, F_f, should now be determined:

$$F_f = \frac{M_u}{d-t_f} = \frac{1276 \times 12}{21.36-0.8} = 745 \ k \quad (5)$$

Equation 3 is then used to determine the number of effective bolts:

$$b_{edge} = [12.29-(5+2 \times 3.33)]/2 = 0.32 \ in. \quad (6)$$
$$n_{eff} = 7$$

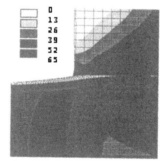

Figure 3. Hinging of W21x101 beam at maximum applied moment.

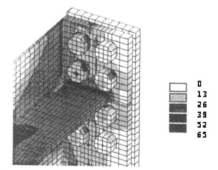

Figure 4. Von Mises stress (ksi) distribution at tension flange.

730

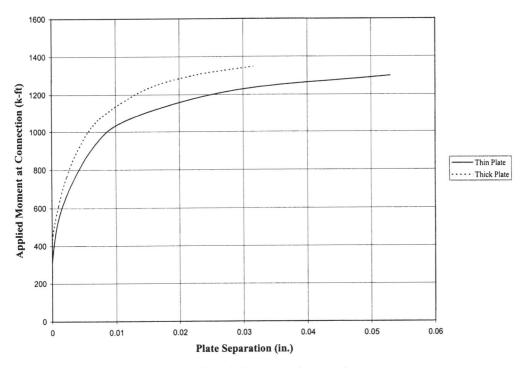

Figure 5. Applied moment vs. plate separation.

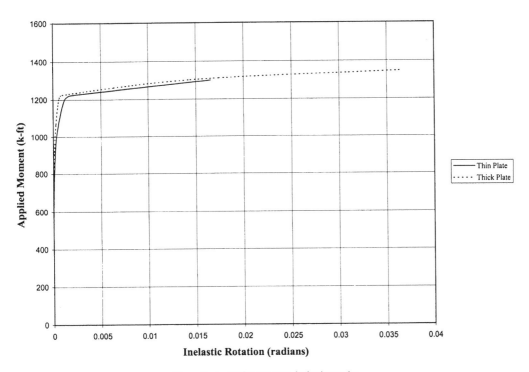

Figure 6. Applied moment vs. inelastic rotation.

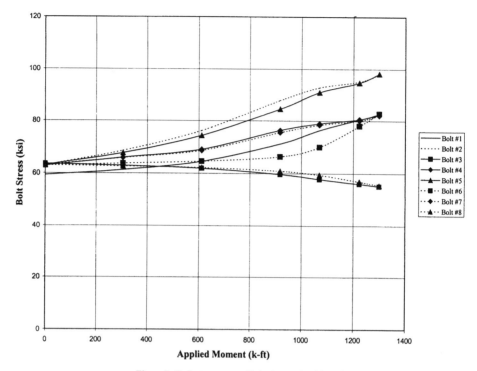

Figure 7. Bolt stress vs. applied moment for thin end-plate.

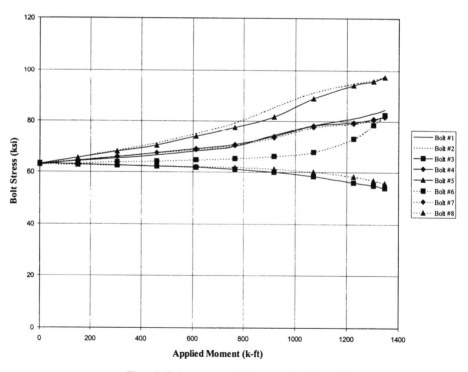

Figure 8. Bolt stress vs. applied moment for thick end-plate.

Since 0.32 in. is greater than 0.25 in., but less than 0.5 in., seven bolts are effective.

Dividing the flange force by seven results in 106.43 kips per bolt. From Table 8-15 of the *AISC Manual of Steel Construction* (1994), 1 1/4 in. A325 bolts are required. These bolts have actual tensile strengths of 82.8 / 0.75 = 110.4 kips per bolt (note that a strength reduction factor of 0.75 is used in this table). Sixteen 1 1/4 in. A325 bolts will be used. Again, it should be pointed out that the design bolt tensile strength of 82.8 kips should not be divided by the strength reduction factor in actual design. Once the proper strength reduction factor has been established by SAC, the actual bolt tensile strength should be multiplied by this factor and not $\phi = 0.75$.

3 FINITE ELEMENT RESULTS

Using the finite element method and loading the beam at the beam tip with increasing loads, it was found that the 16ES moment end-plate connection could successfully develop large beams. Figure 3 shows the formation of a plastic hinge in the W21x101 beam. Note that stresses in excess of the beam yield stress of 50 ksi occur just past the stiffener. Similar results have been obtained in the laboratory for extended stiffened connections of the SAC Steel Project.

Figure 4 shows the Von Mises stress distribution at the tension flange for the thin plate specimen's maximum applied moment. It is interesting to note the smooth load path that the beam flange and stiffener provide to the bolts. However, there is very little stress around bolt #3 and bolt #8, indicating that there is very little load distribution to these bolts. This is expected, as no source of stiffness (e.g.., beam flange, beam web, or stiffener) is in the proximity of these bolts. Figure 4 also shows the Von Mises stress distribution across the thin end-plate. Although not shown here, the stress distribution across the thick end-plate is quite similar at its maximum applied moment. The maximum applied moments were 1300 and 1350 k-ft for the thin and thick end-plate connections, respectively. Although both the thin and thick end-plate are able to develop the beam, the thin end-plate connection fails via bolt rupture prior to large inelastic deformations and at a lower maximum applied moment.

Figure 5 plots the applied moment vs. maximum plate separation for the W21x101 beam. As expected, for the same applied moment, the thin end-plate deflects more than the thick one. Figure 6

plots the applied moment vs. inelastic rotation for the W21x101 beam. It is apparent that the connection response is similar for the thick and thin plates. However, the thin plate solution diverges at around 0.015 radians of inelastic rotation, far less than the value 0.035 of its thick plate counterpart. Figures 7 and 8 plot the bolt stress in all the bolts vs. applied moment for the thin and thick end-plate, respectively. Since the end-plate thicknesses are so close, it is difficult to see much difference. However, the critical bolts reach their maximum values at an earlier applied moment for the thin end-plate. It is interesting that bolt #3 and bolt #8 actually lose load as a result of their location.

4 CONCLUSIONS

It has been shown that the 16ES moment end-plate connection can be used to develop large beams to be part of seismic lateral force resisting systems. The design procedure is simplified by using a thick end-plate and 16 bolts at each flange. As expected, a properly detailed connection allows beam hinging to provide the 0.01 radians of inelastic rotation as required for ordinary moment frames. In fact, for the thick end-plate case considered here, 0.03 radians of inelastic rotation, as required for special moment frames, is surpassed. The reason for the lack of any experimental testing requirement is readily apparent for this design approach. For ordinary moment frames, as long as the design engineer can ensure that the connection is stronger than the adjoining beam, beam hinging can provide enough ductility to the system. It has also been shown that bolts #3 and #8 are ineffective at resisting the flange force delivered to the connection. Hence, these bolts may be left out of the design without any loss in strength. Even as a source of redundancy, these bolts are questionable.

5 REFERENCES

American Institute of Steel Construction 1994. *LRFD Manual of Steel Construction*, Chicago.

American Institute of Steel Construction 1997. *Seismic Provisions for Structural Steel Buildings*, Chicago.

Borgsmiller, J. 1995. *Simplified Method for Design of Moment-End-Plate Connections*. M.S. Thesis, Virginia Polytechnic Institute and State University, Blacksburg, VA.

Mays, T. W. 2000. *Application of the Finite Element Method to the Seismic Design of Moment End-Plate Connections*. Ph.D. Dissertation, Virginia Polytechnic Institute and State University, Blacksburg, VA. (in progress)

Murray, T. M., and Kukreti, A. 1988. Design of eight-bolt stiffened moment end-plates. *Engineering Journal* (AISC), 25, 45-52.

Szilard, R. 1974. *Theory and Analysis of Plates*. Englewood Cliffs, NJ: Prentice-Hall.

Behaviour of Steel Structures in Seismic Areas, Mazzolani & Tremblay (eds) © 2000 Balkema, Rotterdam, ISBN 90 5809 130 9

'Heavy' moment end-plate connections subjected to seismic loading

Emmett A. Sumner, Timothy W. Mays & Thomas M. Murray
The Charles E. Via, Jr., Department of Civil and Environmental Engineering, Virginia Polytechnic and State University, Blacksburg, USA

ABSTRACT: The four bolt unstiffened extended and the eight bolt stiffened extended moment end-plate connections were tested as a part of the SAC Steel Project to determine their suitability for use in seismic force resisting moment frames. The connections were "heavy" connections between large hot-rolled members that included W24, W30 and W36 beam sections. It was determined that end-plate connections can be designed to be very ductile and to meet the inelastic rotation requirements of the AISC *Seismic Provisions for Structural Steel Buildings* (1997) for special moment frames. A validation study was conducted utilizing the finite element method and has been shown to successfully predict the behavior of the end-plate connection.

1 INTRODUCTION

Since the Northridge earthquake in 1994, a great deal of research has been conducted in the area of steel moment resisting frames. The first phase of the research focused on determining the cause of failure of a number of the fully welded beam-to-column connections in moment resisting frames. Results from the research indicated several problems with the fully welded connection. The second phase of research focused on connection configurations that could be used in lieu of the fully welded connection. The extended moment end-plate connection is one alternative. As a part of the SAC Steel Project, a series of full scale beam-to-column extended moment end-plate connection tests have been conduced at Virginia Polytechnic Institute and State University. The primary objective of the tests was to determine the suitability of the extended moment end-plate connections for use in seismic load resisting moment frames.

Moment end-plate connections consist of a plate that is shop-welded to the end of a beam that is then field bolted to the connecting member. The connections are primarily used to connect a beam to a column or to splice two beams together. The four bolt extended unstiffened and the eight bolt extended stiffened moment end-plate configurations will be discussed. The four bolt extended unstiffened end-plate connection consists of two rows of two bolts for each flange. One row is outside the flange on the extended part of the end-plate and the other is inside the flange. The eight bolt extended stiffened end-

plate connection consists of four rows of two bolts for each flange. Two rows are outside the flange on the extended part of the end-plate and the other two are inside the flange. The extended part of the end-plate is stiffened by a triangular stiffener centered over the web of the beam. Typical configurations for both connections are shown in Figure 1.

a) Four bolt unstiffened   b) Eight bolt stiffened
Figure 1. Extended end-plate connection configurations

Another result of the Northridge earthquake was the publication of the AISC *Seismic Provisions for Structural Steel Buildings* (AISC, 1997). The new provisions set forth additional requirements for welded and bolted connections that are subjected to seismic loading. These requirements include specific inelastic rotation requirements for the beam-to-column connections in moment resisting frames. The requirements are met by satisfying a series of qualifying criteria that are based on experimental testing of the connection. A specific requirement for bolted connections that are a part of the seismic force resisting system is that they be configured such that a ductile limit-state either in the connection or in the member controls the design. For end-plate connections, this means that the connection must be designed to either develop the nominal plastic moment capacity of the beam or have a ductile limit state such as end-plate yielding control the design. Brittle limit states such as bolt rupture must be avoided.

2 PROJECT DESCRIPTION

In this project, four-bolt extended unstiffened and eight-bolt extended stiffened end-plate configurations were tested under cyclic loading in accordance with the *Protocol for Fabrication, Inspection, Testing, and Documentation of Beam-column Connection Tests and Other Experimental Specimens*, (SAC, 1997). The connections were "heavy" connections between large hot-rolled shapes. Three beam and column combinations were used: W24x68 beam and W14x120 column, W30x99 beam and W14x193 column, and W36x150 beam and W14x257 column. The beam and column members were ASTM A572 Grade 50 steel and the end-plates were ASTM A36 steel. The connections utilized 1 ¼ inch diameter ASTM A325 and ASTM A490 bolts. For each test configuration, two different connection tests were performed. One test with the connection designed to develop 110 percent of the nominal plastic moment strength of the beam (strong plate connection). The other connection test was designed to develop 80 percent of the nominal plastic moment strength of the beam (weak plate connection). The test matrix for this series of tests is shown in Table 1.

Table 1. Test matrix

Specimen ID*	Beam	No. of Bolts (Material)	End-Plate (in)
	Column		
4E-1 1/4 -1 1/2-24	W24x68	8 (A490)	1 1/2
	W14x120		
4E-1 1/4-1 1/8-24	W24x68	8 (A325)	1 1/8
	W14x120		
8ES-1 1/4-1 3/4-30	W30x99	16 (A490)	1 3/4
	W14x193		
8ES-1 1/4-1-30	W30x99	16 (A325)	1
	W14x193		
8ES-1 1/4-2 1/2-36	W36x150	16 (A490)	2 1/2
	W14x257		
8ES-1 1/4-1 1/4-36	W36x150	16 (A325)	1 1/4
	W14x257		

* 4E designates a four bolt extended unstiffened
  8ES designates an eight bolt extended stiffened

3 DESIGN OF SPECIMENS

The design of the test connections was done utilizing existing design methods for connections subject to monotonic static loading. The design procedure for the four bolt extended unstiffened connections uses a combination of yield line analysis for determination of the end-plate thickness and the modified Kennedy method (Kennedy, 1981) for the calculation of bolt forces including the effects of prying action. The design of the eight bolt extended stiffened connections was done using the design procedure provided in the AISC Design Guide *Extended End-Plate Moment Connections* (Murray, 1990). The design guide procedure provides equations for the required plate thickness and the determination of bolt forces including the effects of prying action. The expected failure mode of the strong plate connections was local flange and web buckling of the beam and the expected failure mode of the weak plate connections was end-plate yielding followed by bolt rupture.

The columns were designed in accordance with the AISC *LRFD Manual of Steel Construction* (AISC, 1994) and the AISC *Seismic Provisions for Structural Steel Buildings* (AISC, 1997). The columns typically had continuity plates in line with both connecting beam flanges and a web doubler plate attached to one side of the web. The thickness of the continuity plates was approximately equal to the thickness of the connecting beam flanges.

4 FABRICATION

The test specimens were fabricated by a combination of university laboratory personnel and commercial fabricators. The end-plate to beam connection was made using full penetration groove welds for the flanges and fillet welds for the web. All welds were made in accordance with the AWS *Sturctural Welding Code* (AWS, 1998) using the Flux Cored Arc Welding (FCAW) process and E71T-1 welding electrodes. The flange welds were similar to AWS TC-U4b-GF utilizing a full depth 45-degree bevel and a minimal root opening, backed by a 3/8 in. fillet on the underside of the flanges. The root of the flange groove welds was backgouged after installation of the 3/8 fillet to remove any contaminants present from the fillet weld. As recommended by Meng (1996), weld access holes were not used. It is noted that this welding procedure results in an area of non-inspected flange groove weld above the web of the beam. The beam web to end-plate connection consisted of 5/16 in. fillet welds on both sides of the web. All welds were inspected in accordance with AWS specifications (AWS, 1998).

736

5 TESTING

The test setup used in the evaluation of the connections consisted of an exterior beam-column subassemblage with a single cantilever beam attached to the flange of a column. The tests were conduced in a horizontal position to allow use of the available reaction floor supports and for safety of testing. The typical test setup is shown in Figure 2. Lateral supports were provided for the beam at a spacing close enough to prevent lateral torsional buckling of the beam prior to development of its nominal plastic moment capacity. A roller was used to support the beam tip and to eliminate any bending moments caused by gravity forces perpendicular to the plane of loading. The load was applied to the free end of the beam using a double acting hydraulic ram. No axial load was applied to the column.

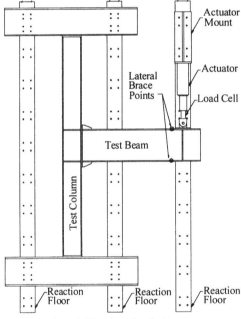

Figure 2. Plan view of typical test setup

Instrumentation of the test setup included a 200 kip tension-compression load cell, displacement transducers to measure the beam displacement at the point of loading, displacement transducers aligned with the column continuity plates to measure the rotation of the column at the connection, displacement transducers to measure the panel zone rotation, instrumented calipers to measure end-plate separation from the column flange, strain gages to measure the beam and column flange strains, and strain gage rosettes to measure the panel zone strains. Additional displacement transducers were used to measure any rigid body movements of the column ends due to shifting of the test frame. All of the connection bolts were instrumented with strain gages that were installed inside the unthreaded portion of the bolt. The bolts were calibrated prior to testing to establish the load-strain relationship for each bolt. This allowed for accurate tightening of the bolts during assembly of the connection and monitoring of the bolt strains during testing.

Prior to testing, the connection bolts were tightened to the minimum pretension specified in the AISC *LRFD Specification* (AISC, 1994). The specimen was then whitewashed to aid in the observation of yielding within the connection region. The instrumentation was then reset and the loading was applied. The specimen was loaded in a quasi-static or "slow cyclic" manner in accordance with the *SAC Loading Protocol*, (SAC, 1997). The loading protocol is based on the total interstory drift angle of the beam-column assembly. The interstory drift angles can be achieved by displacing either the beam tip or the column tip. In these tests, a hydraulic ram was used to displace the beam tip until the rotation angles specified by the loading protocol were achieved. Any rigid body rotation of the subassemblage due to shifting of the test frame was monitored and subtracted from the total rotation to ensure that the proper rotation angle was achieved. The prescribed interstory drift angles and number of cycles for each are shown in Figure 3. Data points were recorded at regular intervals throughout the duration of the test using a PC-based data acquisition system.

6 RESULTS

The six extended end-plate connections behaved as expected. The strong plate connections (110% of the beam capacity) resulted in failure of the beam with little or no distress observed within the connection region. The weak plate connections (80% of the beam capacity) resulted in failure of the connection with one exception, the W30x99 weak plate eight bolt extended stiffened connection (8ES-1.25-1-30) resulted in failure of the beam.

The strong plate tests exhibited a great deal of ductility and energy dissipation capacity. Beam flange local buckling, the primary failure mode, is a predictable limit state that provides a ductile failure mechanism. There was no yielding or other distress of the connection region observed during the tests, which indicates that the strong plate connections remained elastic throughout the duration of the test. The total interstory drift rotations for the strong plate tests ranged from 0.05 to 0.061 radians with inelastic rotations all greater than 0.04 radians. The inelastic interstory drift rotations exceed the AISC seismic connection prequalification requirements (AISC, 1997) of 0.03 radians interstory drift for special moment frames (the most stringent category). This would indicate that these connections should be adequate for use in special, intermediate and ordinary moment frames. Typical moment vs. total rotation and moment vs. inelastic rotation plots for the strong plate tests are shown in Figures 4 and 5 respectively.

Load Step #	Interstory Drift Angle, θ (rad)	Number of Loading Cycles
1	0.00375	6
2	0.005	6
3	0.0075	6
4	0.01	4
5	0.015	2
6	0.02	2
7	0.03	2
Continue with increments in θ of 0.01, and perform two cycles at each step		

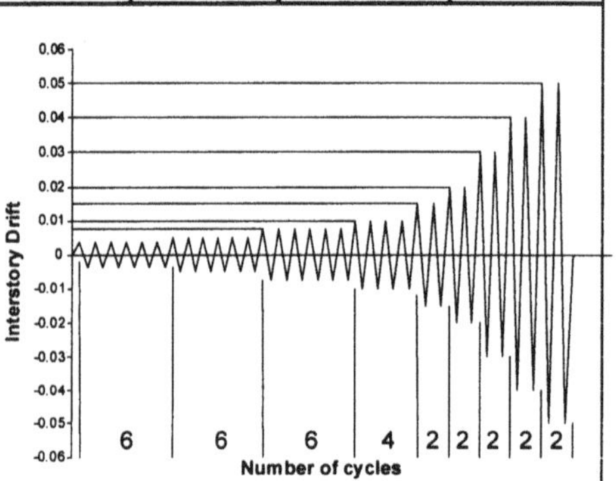

Figure 3. SAC loading protocol, (SAC, 1997).

As expected, the weak plate tests did not exhibit as much ductility as the strong plate tests. The behavior of the weak plate connections was controlled by yielding of the end-plates followed by bolt rupture. In some cases, yielding of the connection region was observed as early as the third load step (0.0075 radians). These controlling limit states are more variable than the beam failure and ultimately result in brittle failure mechanisms. The total interstory drift rotations for the weak plate tests ranged from 0.04 to 0.06 radians with inelastic rotations greater than 0.019 radians.

The behavior of the bolts was of particular interest during the test. The bolts in the strong plate tests gradually lost the majority of their pretension force as the load cycles increased in magnitude. The

maximum observed bolt strains for these connections were only slightly higher than the initial pretension strains. A typical bolt strain vs. moment plot is shown in Figure 6. The weak plate connection bolts were observed to have strains rising sharply above the initial pretension strains as the load steps increased in magnitude. The bolts would undergo yielding indicated by permanent set of the bolt strains. As the bolts approached failure, the bolt strains increased very sharply exceeding the capacity of the strain gages. There were no observed differences in ductility or behavior of the A325 and A490 bolts.

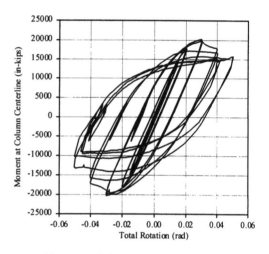

Figure 4. Typical Moment vs. Rotation Plot (8ES-1.25-1.75-30).

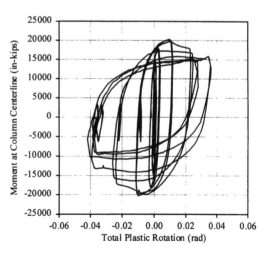

Figure 5. Typical Moment vs. Inelastic Rotation Plot (8ES-1.25-1.75-30)

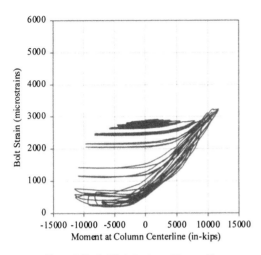

Figure 6. Typical Bolt Strain vs. Moment Plot (4E-1.25-1.5-24)

A summary of the test performance is shown Table 2. A photograph of a typical beam failure is shown in Figure 7. The local flange buckling and severe yielding of both flanges and the web is visible. A typical connection failure is shown in Figure 8. The severe yielding of the end-plate around the bolts and flange is clearly shown.

Table 2. Summary of Test Performance

Test ID	M_{max}/M_n *	θ_{Total} (rad)	$\theta_{P\,Max}$ (rad)
4E-1.25-1.5-24	1.05	0.061	0.044
4E-1.25-1.125-24	1.0	0.050	0.031
8ES-1.25-1.75-30	1.04	0.050	0.042
8ES-1.25-1-30	1.11	0.060	0.049
8ES-1.25-2.5-36	1.11	0.060	0.045
8ES-1.25-1.25-36	0.93	0.040	0.019

* M_{max} = Maximum applied moment at the face of column
$M_n = 1.1\,R_y\,F_y\,Z_x = 1.1(1.1)(50)\,Z_x$

7 PRELIMINARY VALIDATION

To examine the validity of the results obtained during the experimental program and to determine if the finite element method can be used to effectively predict the behavior of moment end-plate connections under cyclic loading, the ANSYS finite element program was used to model the connections. Since the problem is three-dimensional in nature, solid eight-node brick elements that include plasticity effects are used extensively to model the assembly. Contact elements are included between the end-plate and the column flange to represent the nonlinear behavior of this complex interaction problem. The effects of bolt pretension are included, and bolt prying forces are tracked throughout the loading process.

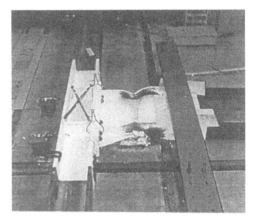

Figure 7. Typical beam failure (8ES-1.25-1.75-30).

Figure 8. Typical connection failure (4E-1.25-1.125-24).

Figure 9 plots bolt stress vs. applied moment for one bolt of the four-bolt extended unstiffened connection (4E-1 1/4-1 1/8-24). The experimental curve is taken as the backbone of the observed bolt behavior. The correlation between the experimental and finite element curves is readily apparent. Figure 10 shows the Von Mises stress distribution at the tension flange for the eight-bolt extended stiffened moment end-plate connection (8ES-1 1/4-1-30) at the maximum applied moment. High stresses can be seen in the end-plate, particularly below the bottom flange. Also, relatively high stresses are seen in the beam flange near the termination of the stiffener. This is in accord with testing which shows beam hinging to begin in this region.

The results from the validation study indicate that the finite element method can be used to effectively predict bolt forces and end-plate separation. In addition, the connection's ultimate limit strength can be predicted via divergence of the analysis.

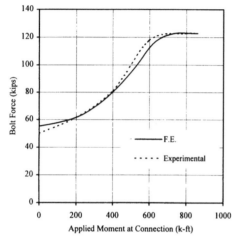

Figure 9. Bolt force vs. applied moment comparing the experimental and finite element method results

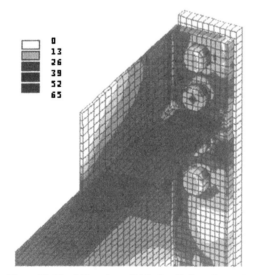

Figure 10. Von Mises stress distribution (ksi) at the tension flange for the eight-bolt extended stiffened moment end-plate connection (8ES-1 1/4-30)

8 SUMMARY

The results of the six extended moment end-plate connection tests indicated that the four bolt unstiffened and the eight bolt stiffened extended moment end-plate connections can be designed to successfully withstand seismic loading. The strong plate connections exhibited the most ductility while the weak plate connections typically failed in a brittle manner. The strong plate connections exceeded the inelastic rotation requirements for special moment frames specified in the AISC *Seismic Provisions for Sturctural Steel Buildings* (AISC, 1997). In addition, the finite element method has been shown to successfully predict the behavior of the end-plate connection.

9 ACKNOWLEDGMENTS

This research was funded by the SAC Joint Venture as a part of the SAC Steel Project. Appreciation is extended to FEI Limited and Cives Steel Company for their donation of labor for the fabrication of the test specimens. Appreciation is also extended to Nucor-Yamato Steel Company and Lincoln Arc Welding for the donation of test materials.

REFERENCES

AWS (1998), *"Structural Welding Code – Steel" AWS D1.1-98*, American Welding Society, Miami, 1998.

AISC (1994), *Load and Resistance Factor Design Manual of Steel Construction*, American Institute of Steel Construction, Chicago.

AISC (1997), *Seismic Provisions for Structural Steel Buildings*, American Institute of Steel Construction, Chicago.

Kennedy, N. A., Vinnakota, S., Sherbourne, A. (1981). "The Split-Tee Analogy in Bolted Splices and Beam-Column Connections", *Joints in Structural Steelwork*, John Wiley and Sons, New York, pp. 2.138-2.157.

Meng, R. (1996), *"Design of Moment End-Plate Connections for Seismic Loading"*. Doctoral Dissertation, Virginia Polytechnic Institute and State University, Blacksburg, Virginia.

Murray, T. M., (1990), *AISC Design Guide Extended End-Plate Moment Connections*, American Institute of Steel Construction, Chicago.

SAC (1997), *Protocol for Fabrication, Inspection, Testing and Documentation of Beam-Column Connection Tests and Other Experimental Specimens*, Report No. SAC/BD-97/02, SAC Joint Venture.

Behaviour of Steel Structures in Seismic Areas, Mazzolani & Tremblay (eds) © 2000 Balkema, Rotterdam, ISBN 90 5809 130 9

Seismic response of 3-D steel structure with Bi-directional columns

Hiroyuki Tagawa & Gregory A. MacRae
Department of Civil and Environmental Engineering, University of Washington, Seattle, USA

ABSTRACT: Three-dimensional non-linear dynamic time-history analyses of a three-story steel moment-resisting frame designed according to the Uniform Building Code for Los Angeles seismicity were carried out. Beams in both directions had moment connections to the hollow rectangular columns. Analyses were performed with design level ground motions applied at different angles to the structure. It was found that code drift limits, rather than code considerations for bi-directional horizontal shaking, governed the member sizes. The structure exceeded its yield drift in both directions simultaneously. Significant column yielding at levels above the base occurred because the columns were not required to be designed for the plastic flexural strength of the beams in both directions simultaneously. Orthogonal shaking increased drifts by 46%. Two dimensional analyses may therefore be non-conservative in estimating the behavior of this type of 3-D structure.

1 INTRODUCTION

Building structures in seismic zones in the US have traditionally been designed with a few seismic frames, to resist primarily lateral force, and with many gravity frames, to carry gravity forces only. As a result of the 1994 Northridge earthquake, fracture occurred at many welded beam-column connections in moment-resisting steel seismic frames. Large member sizes in these frames and the associated large welds are possible contributing factors to the fractures. Lack of redundancy and the possibility yield in gravity columns in frames designed according to the current codes are also of concern.

This paper describes the performance of a structure with one 3-D seismic frame in which the beams frame into the columns with moment-connections in two directions. This structure has smaller member and weld sizes as compared to traditional frames, increased redundancy, there are no gravity columns. However, bi-directional shaking effects are of concern.

2 FRAME DESIGN

The structure a redesign of a 3-story special steel moment frame structure designed by consultants for the SAC project for Los Angeles seismicity and described by MacRae & Mattheis (2000). The structure was redesigned (Tagawa, 2000) according to the

static lateral force procedure of the UBC code (ICBO, 1997) to have columns with moment connections to all beams.

The design considered the code level design forces considering the 30% rule and drift limits in the separate directions. A strength check is required for columns with the axial force ratio of greater than 0.3, according to the UBC (ICBO, 1997) code :

$$\frac{\sum Z_c \left(F_{yc} - \dfrac{P_u}{A} \right)}{\sum Z_b F_{yb}} \geq 1.0 \qquad (1)$$

Here, A is the gross cross-sectional area of the column, P_u is the axial force on the column, F_{yb} is the yield stress of the beam steel, F_{yc} is the yield stress of the column steel, Z_b is the beam plastic section modulus and Z_c is the column plastic section modulus. The summation signs indicate that the sum of the beams or columns framing into the joint in the plane being considered should be used.

It was found that the design level forces did not control the member sizes, even when the 30% rule was used. The frame drift limits controlled the member sizes. Equation 1 was not applied in the design since column axial force ratios were not greater than 0.27. However, the sizes selected did satisfy Equation 1 except at the roof. A plan and elevation are given in Figure 1. The 4 corner columns are hollow square sections which are 356mm x 356mm x 12.7mm thick. All other columns are 406mm x 406mm x 12.7mm hollow square sections. Beams are wide-flange members with second moment of area and plastic section modulus I=762x10^{-6}m^4 and Z=2901x10^{-6}m^3 respectively at Level 1 and 2. At the roof level, I=554x10^{-6}m^4 and Z=2671x10^{-6}m^3. Column and beam yield strengths are 345MPa and 248MPa respectively. All columns have fully restrained base connections at the bottom and beams have moment connections to the columns. All interstory heights were 3.692m. Loading for the building give floor seismic weights of 9995kN for floors 1 and 2 and 8632kN for the roof including the penthouse. The weight of the total building was therefore, 28,621kN.

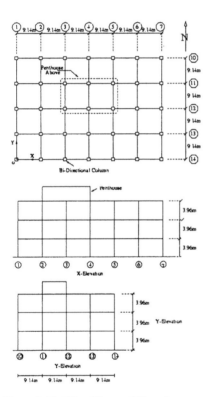

Figure 1. Building Plan and Elevations

3 FRAME MODELLING AND EARTHQUAKE RECORDS

Modeling of the frames in three dimensions was carried out using DRAIN-3DX (Powell et al., 1994). Center-line modeling was used at all beam-column connections. Stiff diagonal members were used to make a near-rigid floor diaphragm. The contribution of floor slabs to beam strength and stiffness was neglected. Beam elements were modeled to be elastic. Separate nodes were placed at the seismic beam ends. These beam end nodes were slaved to the column nodes in all degree-of-freedom (DOF) apart from the rotational DOF corresponding to beam strong axis bending. Rotational beam-end springs between these nodes had an elastic stiffness of approximately $1000x(EI/L)_{beam}$ to make them behave rigidly. The spring strength was set to the beam by giving the springs a rotational post-yield stiffness of $0.03x(6EI/L)_{beam}$. Beam fixed end loads were applied to the nodes at the ends of the seismic beams beside the member spring as nodal loads. Columns were modeled with the DRAIN-3DX column-fiber hinge model - element Type 08. Sections were split into 8 fiber elements in the hinge regions. This resulted in similar pushover behavior to that if 24 elements were used without the computational cost. The fiber strain hardening ratio was 0.03 and the plastic hinge length was 0.2477 of the column length (Tagawa, 2000). The Type 08 element capability to consider P-Δ effects was used throughout the analyses. Rayleigh damping (Clough & Penzien, 1993) of 2% was used based on the initial stiffness in the first mode and at a period of 0.2s. No damping was applied to the rotational springs or to the rigid floor struts. The damping ratio in the shortest period mode was 24%. Masses were lumped at the nodes at each level. Horizontal translational slaving was used in each floor on the column lines in both directions reducing the degrees of freedom from 978 to 804 without restraining torsion. The Newmark constant average acceleration integration method with a time step of 0.0005s was used to obtain convergence. Overshoot tolerances for the fiber hinge elements were initially set as 0.69MPa for columns and 0.069MPa for the beams. The periods of the structure are 0.794s and 0.791s in the principal directions of the frame.

A push-over analysis of the frame was conducted using the code defined lateral force distribution. It indicated that maximum base shear force, H_{max}, was approximately five times the design shear force, H_{des}. Since the design lateral force reduction factor, R, was 8.5, this implies that an equivalent system lateral force reduction factor was 8.5/5=1.7. A bilinear approximation made to the story lateral force-drift plot was used to define story "yield drift". One set of earthquake records was selected to analyze the structure from the SAC record suite (Somerville et al., 1997). Response spectra of the records are given in Figure 2. The record set LA05,06 consisted of design level records LA05 and LA06 for which the magnitude was determined based on an event with a probability of exceedance of 10% in 50 years in Los Angeles.

The angle at which the SN component of shaking was applied to the building is referred to as the attack angle. It is measured anti-clockwise from the x-axis.

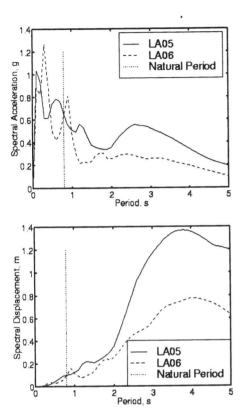

Figure 2. Response Spectra for LA05, 06

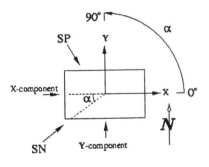

Figure 3. Earthquake Attack Angle

The SP component is applied in the direction 90° clockwise from the SN component as shown in Figure 3. Components of SN and SP shaking were applied to the frame separately in the x- and y- directions as well as combined in order to evaluate how well separate 2-D analyses in the x- and y- directions independently would estimate the combined response.

4 ELASTIC BEHAVIOR

Building first story drifts in the x and y directions during the record set LA05,06 applied at an attack angle of 45° are given in the upper digrams in Figure 4. Here the structure is forced to remain elastic. Drifts due to the components of shaking in the x- and y- directions separately occur predominantly in the direction of shaking. This is because the frame is almost symmetric. Also, the peak drift in the x- direction due to shaking in this direction is about 0.86% which is less than the story yield drift of 1.0%. For y-direction shaking, the drift is 1.44%, which is larger than the story yield drift.

The lower diagram shows the combined shaking due to both the x- and y- components of shaking being applied simultaneously. This plot contains lines based on the 30% rule (ICBO, 1997), on the SRSS method, and on the SAV (Sum-of-Absolute Values) method. A rigorous method to obtain these lines may be developed based on the Ray diagram (MacRae & Mattheis 2000). Tagawa (2000) has shown that the method is rather involved. However, for a near-symmetric structure where shaking in either the x- or y- directions causes drift predominantly in these directions respectively, the SAV method simply gives a rectangular envelope around the peak responses in the respective directions. The SRSS method gives a quarter of an ellipse in each quadrant. The 30% rule gives an oc-

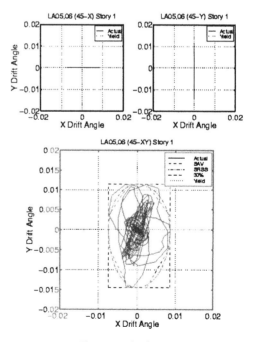

Figure 4. Elastic Response

tagonal envelope with corners when the distance between the zero drift and the SAV drift on the four sides is 30% of the SAV drift magnitude. It may be seen that the actual response was always within the SAV envelope showing that the SAV method was conservative for all attack angle. However, it was often greater than the SRSS and 30% rule predictions indicating some phase coherence of response.

5 INELASTIC BEHAVIOR

Building first story drifts in the x- and y- directions during the record set LA05,06 applied at an attack angle of 67.5° are given in Figure 5. In this case, the structure is permitted to yield. The story yield drift was not exceeded due to shaking in the individual directions. The y-drift increase of 46% from 0.96% to 1.40% due to simultaneous shaking in the x-direction. None of the estimation methods, which are based on elastic considerations, conservatively estimates the response. According to work conducted by MacRae & Mattheis (2000) the SAV is always a conservative estimate of the drifts in each direction due to combined shaking if torsional effects are small and only the beams yield. The fact that the methods were not conservative as a result of column biaxial flexural yielding is confirmed by the hinge pattern at the time of the peak drift in both directions. Column yielding oc-

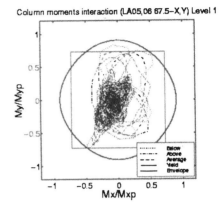

Figure 6. Moment Interaction

curred at the top and bottom of all internal columns in the first story of the frame as well as at the base of the external columns simultaneously during this design level shaking. Column moments above and below a first story internal joint as shown in Figure 6. The column moment yield envelope, "yield" shown is based on the plastic strength considering the axial force on the section and bi-axial bending for a box-section (Chen & Atsuta 1977). The "Envelope" lines represent the demands on the column from the beams if the beam plastic moments are distributed equally to columns above and below the joint. It may be seen that column yielding, may occur before beam yielding. Column yield occurred beneath the joint in this case. Design implications of this behavior are discussed elsewhere (Tagawa 2000).

6 CONCLUSIONS

A three-dimensional three-story near-symmetric steel framed structure designed for Los Angeles seismicity with moment connections in both directions to the hollow rectangular columns was subjected to design level earthquake motions. It was found that :

1) Code drift limits considering one-directional shaking governed the member sizes rather than any code provisions considering bi-directional horizontal shaking. Weak-column strong-beam behavior is possible with the code approach.
2) Design level earthquake shaking due to LA05,06 caused drifts as high as 1.95% but significant column yielding occurred.
3) Two-dimensional analyses would underestimate the drift by as much as 46% since orthogonal shaking increased the drift on one direction by this amount.

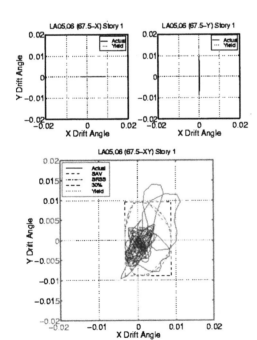

Figure 5. Inelastic Response

7 REFERENCES

Chen W. F. & Atsuta T. 1977. Theory of Beam Columns, *McGraw-Hill*, New York.

Clough R. & Penzien J. 1993. Dynamics of Structures, *2nd Edition, McGraw-Hill*.

ICBO. 1997.1997 Uniform Building Code, *International Conference of Building Officials, Whittier, CA*.

MacRae G. A. & Mattheis J. 2000. Three-Dimensional Steel Building Response to Near-Fault Motions, *ASCE Journal of Structural Engineering, January 2000, pp. 117-126*.

Powell G. H. & Campbell S. 1994. DRAIN-3DX Base Program Description and User Guide Version 1.10, *Department of Civil Engineering, University of California Berkeley, Report No. UCB/SEMM-94/07&08*.

Somerville, P., Smith N., Punyamurthula S. and Sun J. 1997. Development of Ground Motion Time Histories for Phase 2 of the FEMA/SAC Steel Project, *SAC Joint Venture Project Report No. SAC/BD-97/04*.

Tagawa H. 2000. Seismic Response of 3-D Steel Frames With Bi-Directional Columns, *Master Thesis, Department of Civil Engineering, University of Washington*.

Behaviour of Steel Structures in Seismic Areas, Mazzolani & Tremblay (eds) © 2000 Balkema, Rotterdam, ISBN 90 5809 130 9

Cyclic instability of steel moment connections with reduced beam sections

C.-M.Uang & C.-C.Fan
University of California, San Diego, La Jolla, USA

ABSTRACT: A statistical study was performed to evaluate the cyclic instability of steel beams with reduced beam sections (RBS). Based on test results of 55 full-scale specimens, regression analyses were performed to evaluate the relationships between response quantities (plastic rotation and strength degradation rate) and slenderness parameters for buckling. Both linear and nonlinear regressions show the response quantities are highly dependent on the slenderness ratio (h/t_w) of web local buckling, but not on lateral-torsional buckling. Relationships between the response quantities and the slenderness ratios of both web and flange local buckling modes were developed. For RBS beams of Grade 50 steel, a limiting value of $1155/\sqrt{F_y}$ is recommended for the h/t_w ratio to achieve a plastic rotation of 0.03 radian. The presence of a concrete slab was shown to increase both the strength and plastic rotation of the RBS beam under positive bending. However, slabs cannot be counted on to brace beams for negative bending.

1 INTRODUCTION

As the reduced beam section (RBS) moment connection is becoming popular after the Northridge, California earthquake in 1994, lateral bracing of RBS beams has become a concern. Some design engineers believe that additional lateral bracing should be provided near the RBS for the following reasons: (1) both AISC seismic and plastic design provisions require lateral bracing at the plastic hinge location, and (2) the torsional properties of the beam section are significantly reduced at the RBS location. Since additional bracing translates to higher construction cost, some argue against it because: (1) available test results show acceptable performance even though lateral bracing is not provided near the RBS location, (2) the lateral bracing requirement in the AISC Seismic Provisions (AISC 1997) may be too conservative, (3) the slab that exists in most practical applications is beneficial in controlling lateral-torsional buckling, and (4) the axial restraint in a building frame would limit the amount of beam buckling.

2 OBJECTIVE

The objective of this study was to evaluate the level of necessity for additional bracing near the RBS region. A database of RBS moment connections

(Engelhardt 1999, Bonowitz 1999) was established for this study. The database was limited to full-scale specimens tested after the 1994 Northridge earthquake. The effect of slenderness ratios on the plastic rotation and strength degradation rate was examined statistically. The effect of concrete slab on these response quantities was also examined.

3 DATABASE

The database consists of 55 full-scale RBS moment connection specimens (Uang & Fan 1999). RBS specimens with one or more of the following

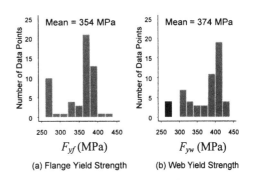

(a) Flange Yield Strength   (b) Web Yield Strength

Figure 1 Histograms of Measured Yield Strengths.

conditions were excluded: (1) reduced-scale specimens, (2) specimens with a plastic rotation less than 0.02 radian that experienced premature weld fracture, (3) specimens with an RBS at the bottom flange only, (4) RBS specimens with cover plates, and (5) specimens with a very weak panel zone such that beams did not experience buckling.

Measured yield strengths of the flange and the web (F_{yf} and F_{yw}) were based on the standard ASTM tensile coupon test results. Histograms of the flange and web yield strengths of the 55 specimens are shown in Figure 1.

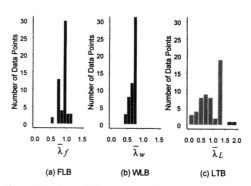

(a) FLB   (b) WLB   (c) LTB

Figure 2 Slenderness Histograms of Different Buckling Modes.

4 SLENDERNESS PARAMETERS

The definition of non-dimensional slenderness parameters (λ's) for flange local buckling (FLB), web local buckling (WLB), and lateral-torsional buckling (LTB) are summarized in Table 1. The limiting values (defined as λ_{ps} in this study) specified in the AISC Seismic Provisions (1997) are also listed, in which the measured F_y values were used to compute λ_{ps}. In this study, λ is further normalized by λ_{ps} for each buckling limit state; the normalized term is defined as $\bar{\lambda}$.

Table 1 Definition of Slenderness Parameters.

	FLB	WLB	LTB
λ	$\lambda_f = \dfrac{b_f}{2t_f}$	$\lambda_w = \dfrac{h}{t_w}$	$\lambda_L = \dfrac{L_b}{r_y}$
λ_{ps}	$\dfrac{137}{\sqrt{F_{yf}}}$	$\dfrac{1365}{\sqrt{F_{yw}}}$	$\dfrac{17235}{F_{yf}}$
$\bar{\lambda}$	$\bar{\lambda}_f = \dfrac{b_f}{2t_f}\dfrac{\sqrt{F_{yf}}}{137}$	$\bar{\lambda}_w = \dfrac{h}{t_w}\dfrac{\sqrt{F_{yw}}}{1365}$	$\bar{\lambda}_L = \dfrac{L_b}{r_y}\dfrac{F_{yf}}{17235}$

Note: Yield stress values are in MPa.

The distributions of the normalized slenderness parameters are presented in Figure 2. The database encompasses a wide range of slenderness ratios for both FLB and LTB. However, the database does not cover the upper range of the WLB slenderness ratio–a range shown later to be critical for RBS beams.

An alternate definition for the slenderness parameters that are based on the following reduced section properties was also used in this study: the reduced beam flange width (b_{fr}), and the associated radius of gyration (r_{yr}). The corresponding slenderness parameters are defined as λ_{fr} and λ_{Lr}, respectively; note that λ_w is unaffected. These alternate slenderness parameters are also normalized by the limiting values (λ_{ps}) and are denoted as $\bar{\lambda}_{fr}$ and $\bar{\lambda}_{Lr}$.

5 RESPONSE PARAMETERS

Plastic rotation and strength degradation rate were selected as response parameters for the statistical study. The definitions of these two response quantities are explained as follows.

Consider a sample hysteresis response plot shown in Figure 3. The ordinate is the resistance, R, which is generally expressed as the moment or the applied load. The abscissa is the total plastic rotation computed as Δ_p/L_c, where Δ_p is the plastic component of the beam tip deflection, and L_c is the distance from the measuring point of beam tip deflection to the column centerline. When the reported θ_p value from a referenced literature was

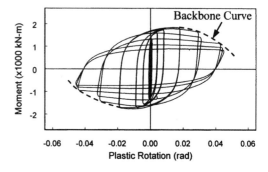

Figure 3 Construction of Backbone Curve.

748

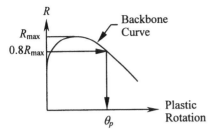

$$R_{max}$$
$$0.8R_{max}$$

Backbone Curve

$$\theta_p$$

Plastic Rotation

(a) Plastic Rotation

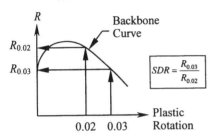

$$R_{0.02}$$
$$R_{0.03}$$

Backbone Curve

$$SDR = \frac{R_{0.03}}{R_{0.02}}$$

0.02   0.03

Plastic Rotation

(b) Strength Degradation Rate

Figure 4 Definitions of Response Parameters.

based on a span from the column face to the deflection measuring point, an adjustment of the reported θ_p values was made for consistency. A similar adjustment was also made if the plastic rotation was reported at an assumed plastic hinge location.

In Figure 3, a backbone curve (shown in dashed line) is constructed by connecting the peak response points of the first cycle for each of the deformation amplitudes. Once the backbone curve is established, the plastic rotation, θ_p, is defined as the total plastic rotation beyond which the connection strength starts to degrade below 80% of the peak strength (see Figure 4a). Note the definition of θ_p used in this study is different from those outlined in either Section 6.6.6.2 (Acceptance Criteria) of the SAC Interim Guidelines (1997) or Section 9 (Appendix S) of the AISC Seismic Provisions. Such deviation is necessary because the θ_p value determined from a criterion based on the nominal yield strength–an approach used by both SAC and AISC–was judged to be too "artificial" for this study.

Strength degradation due to buckling is measured by the strength degradation rate, SDR, defined as follows:

$$SDR = \frac{R_{0.03}}{R_{0.02}} \tag{1}$$

where $R_{0.02}$ and $R_{0.03}$ are the resistance at $\theta_p = 0.02$ and 0.03 radians, respectively (see Figure 4b). Response parameters for both positive and negative

bending directions were computed, and the average values were assigned to each beam.

6 REGRESSION ANALYSIS

6.1 Linear Models

Treating each buckling limit state as an independent phenomenon, linear regression between the normalized slenderness parameter (λ) and the plastic rotation, θ_p, was performed. In addition, an effective normalized slenderness parameter (λ_e) that considers the three basic buckling limit states as a combined buckling mode (Kemp 1996) was also used for the regression analysis:

$$\bar{\lambda}_e = \bar{\lambda}_f \bar{\lambda}_w \bar{\lambda}_L \tag{2}$$

Figure 5 shows the least-square fit of the plastic rotation for all three buckling limit states and the combined buckling mode. The slope is a measure of the dependence of θ_p with respect to each normalized slenderness parameter. Although significant scattering of the data exists, the results do show the strongest dependence between θ_p and the slenderness ratio for WLB. This comes as no surprise because WLB usually occurs first in a typical RBS test. On the contrary, LTB, which receives most of the attention on the use of RBS connections for seismic applications, shows a weak dependence on θ_p. Kemp (1996) showed that λ_e for the combined buckling mode of a prismatic wide-flange beam correlates well with the plastic rotation capacity under monotonic loading. However, Figure 5(d) shows that λ_e does not reduce the scattering of the RBS data.

6.2 Nonlinear Models

Laboratory observations generally indicate that buckling modes do interact with each other. In this section, regression of the following nonlinear model was performed to evaluate the relative contribution of key parameters to the response quantity of interest (θ_p in this case):

$$\theta_p = C \left(\frac{b_f}{2t_f} \right)^\alpha \left(\frac{h}{t_w} \right)^\beta \left(\frac{L_b}{r_y} \right)^\gamma \left(F_{yf} \right)^\delta \tag{3}$$

where exponents α, β, γ, δ, and the constant C remain to be determined from regression.

Note that only one yield strength F_{yf} is used in Equation 3 in order to simplify the model. The current mill practice is to report yield strength of the flange coupons. Therefore, F_{yf} is used in Equation 3. Taking the logarithm on both sides of the equation can linearize the model in Equation 3:

$$\log \theta_p = \log C + \alpha \log\left(\frac{b_f}{2t_f}\right) + \beta \log\left(\frac{h}{t_w}\right)$$

$$+ \gamma \log\left(\frac{L_b}{r_y}\right) + \delta \log F_{yf} \tag{4}$$

Least-square regression of the above equation results in the following expression with a standard deviation of 0.00748 radian:

$$\theta_p = 20.7 \left(\frac{b_f}{2t_f}\right)^{-0.04} \left(\frac{h}{t_w}\right)^{-0.64} \left(\frac{L_b}{r_y}\right)^{0.04} \left(F_{yf}\right)^{-0.67} \tag{5}$$

Several observations can be made from the regression result. First, the large exponent for WLB (−0.64) implies that θ_p is much more affected by WLB than by FLB or LTB. Second, the exponents for both FLB and LTB are close to zero. (A value of zero would indicate no dependence between a particular λ and θ_p). Although the exponent for LTB is small, the positive value goes against intuition. It is suspected that the positive exponent for LTB is due to the weak link between θ_p and L_b/r_y, the small size of the database, and the variability inherent in the database. Third, the exponent for F_{yf} (−0.67) is similar to the AISC format for local buckling

control, where limiting slenderness ratios for WLB and FLB (see Table 1) are inversely proportional to $\sqrt{F_y}$ (i.e., $F_y^{-0.5}$).

For design purposes, it is desirable to simplify Equation 5. The following simplifications were made: (1) the exponent for F_{yf} was assigned a value of −0.5, and (2) the exponent for LTB (0.04) was assumed to be zero. The regressed expression based on the simplified form is

$$\theta_p = 7.1 \left(\frac{b_f}{2t_f}\right)^{-0.09} \left(\frac{h}{t_w}\right)^{-0.57} \left(F_{yf}\right)^{-0.5} \tag{6}$$

Such an adjustment of the exponents decreased the standard deviation slightly (0.00747 versus 0.00748 radian). Since the exponent for WLB (−0.57) is close to −0.5, another simplification was made by assigning −0.5 to the exponent of h/t_w. The revised expression, with a standard deviation of 0.00739 radian, is

$$\theta_p = 5.8 \left(\frac{b_f}{2t_f}\right)^{-0.12} \left(\frac{h}{t_w}\right)^{-0.5} \left(F_{yf}\right)^{-0.5} \tag{7}$$

For a given plastic rotation, the above expression can be used to establish a limiting slenderness requirement for the RBS beam design. However,

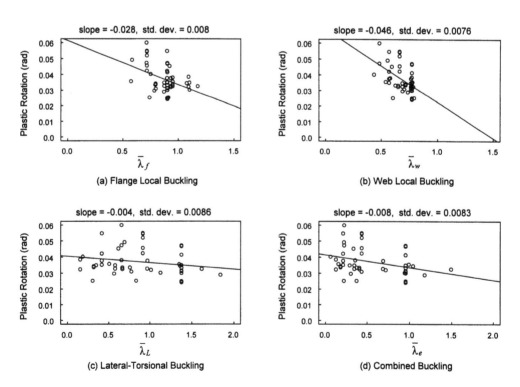

Figure 5 Linear Regression of Plastic Rotation.

since the definitions of θ_p used in this study and the AISC Seismic Provisions (θ_{pa}) are different, it is necessary to examine the relationship between these two quantities. Figure 6 shows the relationship between θ_p and θ_{pa} for a limited number of data points. Until further data become available, it is assumed that θ_p and θ_{pa} values are identical.

To demonstrate the use of Equation 7, consider a required plastic rotation of 0.03 radian:

$$0.03 = 5.8 \left(\frac{b_f}{2t_f} \right)^{-0.12} \left(\frac{h}{t_w} \right)^{-0.5} \left(F_{ye} \right)^{-0.5} \quad (8)$$

Substituting F_{ye} for $R_y F_y$ (AISC 1997), the above expression can be approximated and rewritten as:

$$\left(\frac{b_f}{2t_f} \right)^{1/8} \left(\frac{h}{t_w} \right)^{1/2} = \frac{202}{\sqrt{R_y F_y}} \quad (9)$$

Figure 7 shows the limiting surfaces for θ_p = 0.03 radian and three other θ_p values (Grade 50 steel with R_y = 1.1 is used). The curve for θ_p of 0.02 radian indicates that the plastic rotation can be achieved when the limiting slenderness ratios (θ_{ps}) as specified in the AISC Seismic Provisions for web and flange local buckling are satisfied. However, for θ_p = 0.03 radian, the curve shows that the h/t_w ratio for WLB needs to be reduced for RBS beam design. For this plastic rotation, a limiting h/t_w value of $1155/\sqrt{F_y}$ is recommended.

A regression analysis based on the reduced sectional properties was also performed:

$$\theta_p = 42.7 \left(\frac{b_{fr}}{2t_f} \right)^{-0.04} \left(\frac{h}{t_w} \right)^{-0.85} \left(\frac{L_b}{r_{yr}} \right)^{0.05} \left(F_{yf} \right)^{-0.66} \quad (10)$$

However, since the standard deviation for Equation 10 was similar to that for Equation 5 (0.00726 versus 0.00748), no attempt was made to simplify Equation 10.

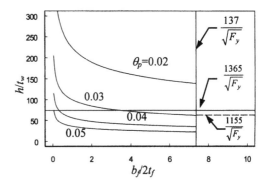

Figure 7 Limiting Slenderness Surfaces (F_y = 345 MPa).

A similar regression analysis was also performed on the strength degradation rate:

$$SDR = 6.7 \left(\frac{b_f}{2t_f} \right)^{0.09} \left(\frac{h}{t_w} \right)^{-0.21} \left(F_{yf} \right)^{-0.22} \quad (11)$$

In deriving the above expression, the L_b/r_y term was dropped out due to the weak link between SDR and L_b/r_y.

7 SLAB EFFECT

Test results of two nominally identical two-sided RBS moment connection specimens, one with and one without the concrete slab, were used to evaluate the bracing effect of the slab (Engelhardt and Venti 1999). Backbone curves for positive and negative bending directions were constructed separately. Each specimen is composed of two beams. For convenience, the backbone curves of both beams are shown in the same plot, in quadrants one and three.

7.1 Positive Bending

Figure 8 is the plot for positive bending. For comparison purposes, the moment axis is

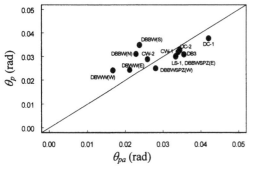

Figure 6 Relationship between θ_{pa} and θ_p.

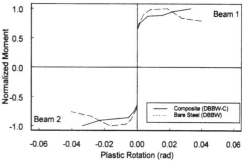

Figure 8 Comparison of backbone curves for positive bending.

normalized. The figure clearly shows the stabilizing effect that the slab provides to increase the positive plastic rotation capacity.

7.2 *Negative Bending*

In the development of lateral-torsional buckling criteria for steel beam design, it is generally assumed that the beam section is not distorted. With this assumption, Roeder and Assadi (1982) showed that, for elastic design, tension flange restraint as provided by a slab could significantly increase the buckling capacity of beams. However, the web of a composite beam can be distorted when the beam buckles laterally under negative bending. Dekker et al. (1995) showed that, under negative bending, the lateral-torsional buckling resistance of a steel beam is dependent on the extent to which the web can transmit the restraining action to the unstable bottom flange. Dekker et al. concluded that the distortional restraint provided by the slab could be quantified by using an effective length of 0.71 times the unbraced length of the compression flange.

Figure 9 shows a comparison of the backbone curves of two RBS specimens under negative bending. Contrary to the finding of Dekker et al., the normalized plots in Figure 9 show that the effect of slab only increased the plastic rotation capacity by a small amount. For seismic design, a slab cannot be counted on to provide lateral bracing for a beam under negative bending.

8 CONCLUSIONS

A statistical study on the cyclic instability of steel moment connections with reduced beam sections (RBS) was performed. A database of 55 full-scale RBS moment connection specimens that were tested cyclically was used for the regression analysis. Measured yield strengths as well as slenderness parameters for FLB, WLB, and LTB were considered as independent variables. These variables were regressed against two response parameters: plastic rotation and strength degradation rate. Both linear and nonlinear models were considered for regression. The following conclusions can be made. These conclusions are based on the member sizes considered (see Uang and Fan 1999), where the RBS geometry was similar to that recommended by Engelhardt (1999). Extrapolation of the conclusions to other cases (e.g., RBS connections with deep columns) should be made with care.

1. Treating each buckling mode as an independent limit state, the linear model showed that the slenderness ratio for WLB affects the plastic rotation and strength degradation rate more than the slenderness ratios for FLB and LTB. Interestingly, the slenderness ratio for LTB showed the least effect on response parameters, regardless of whether the unreduced or reduced sectional properties were used.

2. Considering that buckling modes of an RBS beam interact with each other, nonlinear regression was performed on a model (Equation 3) that considers the slenderness ratios of all three buckling modes. The regressed results showed again both the strong influence of the WLB and the weak influence of the LTB on response parameters.

3. Ignoring the effect of LTB, a simplified model (Equation 7) that relates the plastic rotation to the slenderness ratios of both FLB and WLB was established. Based on this relationship, limiting slenderness surfaces at different plastic rotation levels can be constructed (Figure 7). Expressions that are suitable for design can also be derived from Equation 7 (Equation 9 is one example). For Grade 50 steel (F_y = 345 Mpa), the results show that a plastic rotation of up to 0.02 radian can be achieved if the limiting slenderness ratios for local buckling, as specified in the AISC Seismic Provisions, are satisfied. However, the limiting h/t_w ratio for WLB needs to be further reduced for a plastic rotation of 0.03 radian. A limiting h/t_w ratio of $1155/\sqrt{F_y}$ is recommended for the seismic design of RBS beams. Although this study shows that LTB plays a minor role for the range of parameters investigated, the recommendations for local buckling control are based on the assumption that the slenderness ratio for LTB also satisfies the AISC Seismic Provisions ($17235/\sqrt{F_y}$).

4. The strength degradation ratio (*SDR*) as defined in this study can be evaluated by Equation 11.

5. The concrete slab would increase the plastic rotation of the RBS beam under positive bending

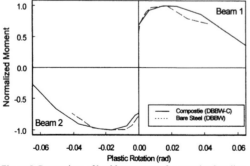

Figure 9 Comparison of backbone curves for negative bending.

(Figure 8). Nevertheless, such increase vanished under negative bending (Figure 9). For design purposes, the slab cannot be counted on to provide lateral bracing for RBS beams under negative bending.

ACKNOWLEDGEMENTS

This research was funded by SAC Joint Venture under Task 14 with Mr. J. Malley as the Project Director of Topical Investigations. Dr. C. Roeder, Team Leader of Connection Performance, initiated the study. Guidance provided by Dr. Roeder and Mr. Malley was greatly appreciated.

REFERENCE

AISC, Load and Resistance Factor Design Specification for Structural Steel Buildings (LRFD), American Institute of Steel Construction, Chicago, IL., 1993.

AISC, Seismic Provisions for Structural Steel Buildings, AISC, Chicago, IL., 1997.

Bonowitz, D., "The SAC Database of Beam-Column Connection Tests," SAC Joint Venture, Sacramento, CA, 1999.

Dekker, N. W., Kemp, A. R., and Trinchero, P., "Factors Influencing the Strength of Continuous Composite Beams in Negative Bending," *Journal of Constructional Steel Research*, Vol. 34, pp. 161-185, 1995.

Engelhardt, M. D., and Venti, M., "Brief Report of Steel Moment Connection Tests (Specimens DBBW and DBBW-C)," *Progress Report to SAC*, University of Texas at Austin, 1999a.

Engelhardt, M. D., and Venti, M., "Brief Report of Steel Moment Connection Tests (Specimens DBWP and DBWP-C)," *Progress Report to SAC*, University of Texas at Austin, 1999b.

Engelhardt, M., "Design of Reduced Beam Section Moment Connections," *Proceedings*, North American Steel Construction Conference, pp. 1-1 to 1-29, Toronto, Canada, 1999.

Kemp, A.R., "Inelastic Local and Lateral Buckling in Design Codes," *Journal of Structural Engineering*, Vol. 122, No. 4, pp. 374-382, ASCE, 1996.

Roeder, C. W. and Assadi, M., "Lateral Stability of I-Beams with Partial Support," *Journal of Structural Engineering*, Vol. 108, No. ST8, pp. 1768-1780, ASCE, 1982.

SAC, "Interim Guidelines: Evaluation, Repair, Modification and Design of Welded Steel Moment Frame Structures," *Report No. SAC-95-02*, SAC Joint Venture, CA, 1995.

Uang, C. M., and Fan, C. C., "Cyclic Instability of Steel Moment Connections with Reduced Beam Section," *Report No. SSRP-99/21*, University of California, San Diego, La Jolla, CA, 1999.

Behaviour of Steel Structures in Seismic Areas, Mazzolani & Tremblay (eds) © 2000 Balkema, Rotterdam, ISBN 90 5809 130 9

Effects of lateral bracing and system restraint on the behavior of RBS moment connections

Q.S.'Kent' Yu
Degenkolb Engineers, San Francisco, Calif., USA

C.-M.Uang
Department of Structural Engineering, University of California, San Diego, La Jolla, USA

ABSTRACT: An experimental and analytical study was conducted to evaluate the effects of lateral bracing and system restraint on the behavior of reduced beam section (RBS) moment connections in a moment frame. Two full-scale RBS moment connections, one with lateral bracing near the reduced beam region and one without, were cyclically tested to evaluate the lateral bracing effects. Both connections were able to deliver a plastic rotation of over 0.03 radian without welded joint fracture. Although adding lateral bracing near the RBS region did not significantly improve the cyclic response of the RBS connection, it did reduce the strength degradation rate and buckling amplitudes for a story drift ratio beyond 3%. To investigate the system axial restraining effects, non-linear finite element analysis of a one-and-a-half-bay beam-column subassembly representing a portion of a multistory frame was performed. The system axial restraint significantly improved the moment connection behavior by reducing the web local buckling amplitude. Minor strength degradation was observed for the axially restrained connections.

1 INTRODUCTION

Widespread damage of welded joints in steel moment frame structures during the 1994 Northridge earthquake led the engineers and researchers to develop alternative connection types. One such design–the reduced beam section (RBS) scheme which introduces a structural fuse in the steel beam by intentionally trimming some portion of the beam flanges, can significantly enhance the seismic behavior of the steel moment frame connection. Stable yielding of the beam and column panel zone can be developed by moving the beam plastic hinge region a short distance from the column face, thereby protecting the beam flange groove welds from brittle fracture.

The effectiveness of the RBS moment frame connections has been demonstrated through numerous large-scale experimental tests (Chen et al. 1996, Engelhardt et al. 1998, Plumier 1997, Zekioglu et al. 1997). As a result, RBS moment connections have been popular in the post-Northridge design era. But questions have arisen concerning the lateral bracing of RBS beams. Some design engineers believe that lateral bracing should be provided near the RBS region because (1) both AISC seismic and plastic de-

sign provisions require lateral bracing at the plastic hinge location, and (2) the torsional properties of the beam section are significantly reduced at the RBS location. Since adding lateral bracing near the RBS region translates to higher construction cost, some argue against it because: (1) available test results show acceptable performance even though such additional lateral bracing is not provided near the RBS location, (2) the lateral bracing requirement in the AISC Seismic Provisions (AISC 1997) may be too conservative, (3) the slab that exists in most practical applications is beneficial in controlling lateral-torsional buckling, and (4) the axial restraint in a building frame would limit the amount of beam buckling.

2 OBJECTIVES

The main objectives of this research study were (1) to evaluate experimentally the effects of lateral bracing near the RBS region on the cyclic response of steel beams, and (2) to investigate analytically the system axial restraining effects on the behavior of RBS moment connections. In the first part of the paper, the lateral bracing effect is assessed by comparing the performance of two full-scale specimens (Yu et al. 1999). In the second part of the paper, an analytical study of a one-and-a-half-bay beam-column subassembly representing a portion of a multistory moment frame is presented to evaluate the system axial restraining effects.

3 LATERAL BRACING EFFECTS

3.1 Specimen design and fabrication

Two nominally identical specimens, designated as LS-1 and LS-4, were designed, constructed, and tested using the setup shown in Figure 1. The design, which incorporated a reduced beam section, was based on a procedure recommended by Engelhardt (Engelhardt 1999) and the AISC Seismic Provisions (AISC 1997). The RBS moment connection details are illustrated in Figure 2.

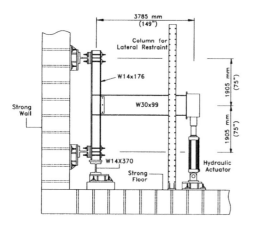

Figure 1. Test setup for lateral bracing effects

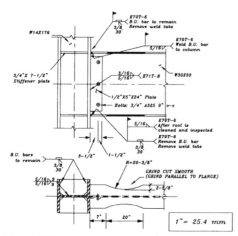

Figure 2. RBS moment connection details

The specimens were constructed by a commercial fabricator. To simulate the field condition, the beam was installed and the moment connection was welded with the column in an upright position. All welding was performed with the self-shielded flux-cored arc welding process, using electrodes that have a specified minimum Charpy V-Notch impact value of 27.1 Joule (20 ft-lbs) at –28.8°C (-20°F) (SAC 1996). A 1.8 mm (0.072 in) diameter E71T-8 (Lincoln NR-232) electrode was used for the complete joint penetration groove weld between the beam web and the column flange, and a 2.4 mm (3/32 in) diameter E70T-6 (Lincoln NR-305) electrode was used for the welds between the beam flanges and the column. The steel backing was removed from the beam bottom flange and the weld cleaned and covered with a fillet weld from below. The steel backing was left on the beam top flange, but a fillet weld was applied between it and the column flange.

A572 Grade 50 steel with special requirements (now called A992 Grade 50) per AISC Technical Bulletin (AISC 1997) was used for the test specimens. The columns of Specimens LS-1 and LS-4 were from different heats of steel. However, the beams of both specimens were from the same heat of steel. Tensile coupons were cut from the members and tested according to ASTM standard procedures, with the results shown in Table 1. The values from the certified mill test reports are also included in Table 1.

Table 1. Mechanical properties

Specimen	Member	Coupon	F_y (MPa)	F_u (MPa)	Elongation
Both	Beam	Flange	379	496	28%
	W30×99	Web	400	517	26%
		(Mill Cert.)	(386)	(510)	(26%)
LS-1	Column	Flange	386	510	31%
	W14×176	Web	372	503	28%
		(Mill Cert.)	(400)	(524)	(21%)
LS-4	Column	Flange	–	–	–
	W14×176	Web	–	–	–
		(Mill Cert.)	(441)	(579)	(26%)

Figure 3. Specimen LS-4 prior to testing

Both specimens were laterally braced near the end of the beam, but additional lateral bracing was provided to LS-4 at a location 152 mm (6 in) outside the RBS region. Specimen LS-4 with lateral bracing prior to testing is shown in Figure 3. Lateral bracing was designed for 6% of the strength of the compression flange of the RBS section. The bracing system was also designed to provide an equivalent axial stiffness of about 122.6 kN/mm (700 kips/in) at both top and bottom flange levels. Lateral bracing was designed to move up and down vertically with the beam.

3.2 Loading histories

Both specimens were tested with the standard SAC loading protocol in Figure 4 (Clark et al. 1997). The loading sequence was controlled by the story drift ratio, beginning with six cycles each of 0.375%, 0.5%, and 0.75% drift, sequentially. The next four cycles were at 1% drift, followed by two cycles of 1.5% drift. The sequence then completed two cycles each of successively increasing drift percentages (i.e., 2%, 3%, 4%, ...) until failure.

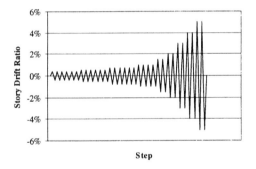

Figure 4. SAC standard loading history

3.3 Test results

Specimen LS-1 performed well, remaining elastic through the cycles of 0.75% story drift ratio. Panel zone yielding began at about 1% story drift ratio and minor web local buckling (WLB) was observed as early as 1.5% story drift ratio. Significant buckling began during the cycles of 3% story drift ratio, with lateral-torsional buckling (LTB) and WLB measured at 86 mm (3-1/4 in) and 30 mm (1-3/16 in), respectively. By the second cycle at 4% story drift ratio the LTB amplitude had increased to 114 mm (4-1/2 in), the WLB amplitude increased to 71 mm (2-13/16 in), and the strength at the column face degraded to just below 80% of the nominal plastic moment capacity of the unreduced section. The test was halted after the third cycle of 5% story drift ratio

due to strength degradation below 50% of the plastic moment capacity. Figure 5a is a plot of the load versus beam tip displacement relationship. At the end of the test, a beam shortening of 38 mm (1-1/2 in) and 76 mm (3 in) due to WLB and LTB was measured at beam top and bottom flanges, respectively.

Specimen LS-4 behaved similarly to LS-1 up to 3% story drift ratio. After the second cycle of 3% story drift ratio, LTB and WLB amplitudes were measured at 78 mm (3-1/16 in) and 29 mm (1-1/8 in), respectively. Beyond 3% drift, the connection strength began to degrade and the lateral bracing reduced the buckling amplitudes. At the end of 4% drift cycles, LTB and WLB amplitudes had increased to only 89 mm (3-1/2 in) and 60 mm (2-3/8 in), respectively, and the strength had not significantly degraded. The testing continued through 5% drift cycles and was stopped during the first cycle of 6% story drift because of a failure in the bracing system. Figure 5b is a plot of the load versus beam tip displacement.

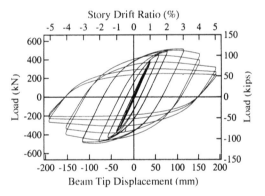

(a) Specimen LS-1

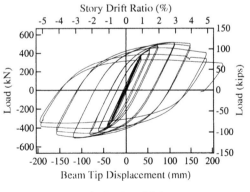

(b) Specimen LS-4

Figure 5. Load versus beam tip displacement relationships

3.4 *Response comparison*

Both specimens performed very well, and completed the cycles of 5% story drift ratio. Therefore, a direct comparison of the responses can be made for the lateral bracing effect.

3.4.1 *Load-displacement envelopes*

Figure 6 shows a comparison of the load-displacement envelopes of both specimens. It can be seen that the response envelopes of both specimens are similar up to about 3% story drift ratio, beyond which strength degradation took place due to significant WLB and LTB (see Figure 8). Note that lateral bracing did not increase the maximum strength of the beam (or the force demand in the beam flange groove welds). The beneficial effect of providing lateral bracing near the RBS region to reduce the rate of strength degradation became significant beyond 3% story drift ratio.

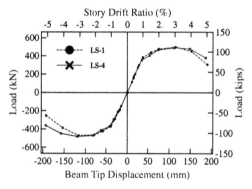

Figure 6. Response envelopes of specimens LS-1 and LS-4

3.4.2 *Plastic rotation*

The Acceptance Criteria in FEMA 273A (SAC 1996) was used to determine the plastic rotation capacity. LS-1 and LS-4 both reached a total plastic rotation over 0.03 radian. The plastic rotation in the panel zone was about 0.01 radian in both specimens. The total plastic rotation for Specimen LS-4 reached 0.04 radian because the capacity of the beam had not degraded below 80% of the beam nominal plastic moment during the last excursion to 6% story drift ratio. Had lateral bracing not failed, it is expected that the plastic rotation would have reached 0.05 radian. The plastic rotation for Specimen LS-1 was 0.03 radian because the beam capacity degraded slightly below 80% of the beam nominal capacity during the first cycle of 4% story drift ratio.

3.4.3 *Energy dissipation*

Figure 7a shows a comparison of the energy dissipated at each displacement amplitude up to 2 cycles of 5% story drift ratio. Up to 4% story drift ratio, the total energy dissipation was similar for both specimens. The effect of lateral bracing on energy dissipation became significant beyond 4% story drift ratio. Figure 7b presents the amount of energy dissipated at the end of 2 cycles of 5% story drift by each specimen, as well as the contributions of the beam and panel zone, separately. It can be seen that the majority of energy dissipation took place in the beam at the RBS region. For LS-1 the beam dissipated 80% of the total energy, while for LS-4 the beam contributed 71% of its total energy dissipation.

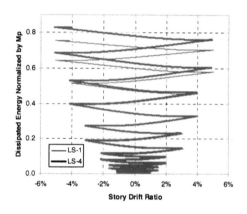

(a) Energy dissipation

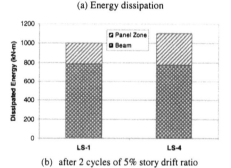

(b) after 2 cycles of 5% story drift ratio

Figure 7. Comparison of energy dissipation capacities

3.4.4 *Buckling amplitudes*

A comparison of the buckling amplitudes of each specimen at each level of story drift was performed (see Figure 8). Regardless of the presence of additional lateral bracing near the RBS, web local buckling occurred at 1.5% story drift ratio in both specimens, being followed by lateral-torsional buckling at 2% story drift ratio. Before 3% story drift ratio, the WLB amplitudes as well as LTB amplitudes were very similar. Beyond 3% story drift ratio, however, the extra lateral bracing near the RBS region was effective in reducing the amount of buckling distortions.

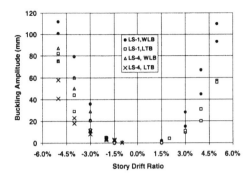

Figure 8. Comparison of buckling amplitudes

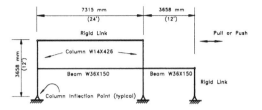

(a) One-and-a-half-bay beam-column assembly

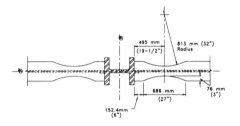

(b) RBS moment connection details

Figure 9. Structural assembly and connection details

4 SYSTEM RESTRAINT EFFECTS

4.1 General

A typical setup for the steel moment connection test is shown in Figure 1, where the beam is cantilevered from the column. At higher deformation levels, strength degradation would occur due to beam buckling. To accommodate the buckling, laboratory observations show that the beam is shortened. In an actual building frame, however, the beam is restrained axially at both ends by the columns. Therefore, the amount of buckling or strength degradation observed in the laboratory component test may be exaggerated. Since it is costly to conduct a system test, an analytical study was performed to evaluate the system restraint effect.

Figure 9a shows a portion of a multistory frame that was used for this purpose. The frame of one story height (3658 mm) is composed of one interior beam spanning between two columns. Unlike the interior beam which is restrained axially by the columns, only one-half length of an adjacent interior beam with an assumed inflection point at midspan is included in the model; this half-span beam can shorten freely due to the way that the assembly freebody is extracted from the multistory frame. To simulate the seismic effect, an equal amount of lateral displacement was imposed to the top end of both columns. Thus, the system restraint effect can be conveniently assessed by comparing the responses of the "interior" and "cantilever" beams.

4.2 Finite element modeling

The ABAQUS finite element analysis program (ABAQUS 1995) was used to model the assembly for large-deformation nonlinear analyses. Figure 10 shows the mesh of the finite element model. A quadrilateral 4-node thin shell elements (element type S4R5 in ABAQUS) with 5 degrees of freedom per node (i.e., three translational components and two in-plane rotations) was used in the modeling of both the beam and column webs and flanges. To capture the material non-linearity, each element was divided into five layers across the thickness of the element. The reduced integration scheme with one Gauss integration point in the center of each layer was used to formulate element properties. In the RBS region, a more refined mesh was used. The beam flanges and web were fully connected to the column flange, that is, the beam flange weld access holes and the beam web bolted connection were not modeled.

Both columns were supported at the bottom by a hinge, while the cantilever beam end was supported by a horizontal roller to allow for axial shortening. The beams were braced laterally at locations 2438 mm (8 ft) from the column face.

Elastic material properties [i.e., Young's modulus = 200 MPa (29,000 ksi), Poisson's ratio = 0.3] were assigned to the portions of the model that were not expected to yield. Nonlinear material properties (see Figure 11) idealized from coupon test results of a research project (Yu et al. 1997) were assigned to the elements in the regions expected to yield. The plasticity model was based on the Von Misses yielding criteria and the associated flow rule.

Both columns were loaded by simultaneously imposing an equal column tip displacement from zero to 152 mm (6 in). Because strength degradation due to buckling was expected to occur, the standard Newton method, which fails (i.e., diverges numeri-

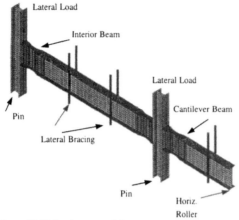

Figure 10. Finite element model

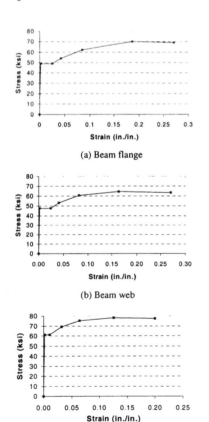

(a) Beam flange

(b) Beam web

(c) Column web

Figure 11. Assumed true stress vs. true strain relationships

cally) near the limit point of the force-displacement curve, was not adopted to solve numerically the non-linear equilibrium equations. Instead, an algorithm

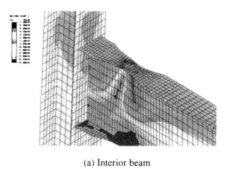

(a) Interior beam

(b) Cantilever beam

Figure 12. RBS beam deformation configurations (at 4% story drift ratio)

called the modified Riks method was implemented to avoid the divergence near the limit point.

4.3 Behavior comparison

Figure 12 shows the predicted deformation configurations of both beams at 4% story drift ratio. Buckling was less severe in the restrained interior beam. While the cantilever beam showed strength degradation, this was not the case for the interior beam.

The beam shear versus story drift ratio relationships of both beams are compared in Figure 13. Lacking axial restraint, the cantilever beam shows strength degradation because it is more prone to buckling.

Figure 14 shows the buckling amplitudes of both beams. At 4% story drift level, the WLB amplitude of the restrained beam is about half that of the cantilever beam.

The slab that exists in an actual building may also increase the axial restraining effect. As a parametric study, a separate analysis was performed by constraining equal lateral displacement at both ends of the interior beam. The results were very similar to the case without such constraint.

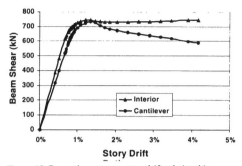

Figure 13. Beam shear versus story drift relationships

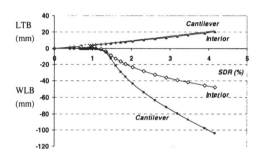

Figure 14. Comparison of LTB and WLB buckling amplitudes

5 SUMMARY AND CONCLUSIONS

An experimental study of two full-scale RBS moment connections with the SAC standard cyclic loading protocol was performed to investigate lateral bracing effects. A non-linear finite element analysis of a one-and-a-half-bay beam-column assembly representing a portion of a multistory frame was conducted to assess the system axial restraining effects on the behavior of RBS moment connections. The following conclusions can be made.

1. Both RBS moment connection specimens reached 0.03 radian plastic rotation needed for Special Moment-Resisting Frames (SMRFs) without suffering weld fracture.
2. Regardless of the presence of additional lateral bracing near the RBS region, WLB initiated at 1.5% story drift ratio, being followed by LTB at 2% story drift ratio. The additional lateral bracing became effective to reduce WLB and LTB amplitudes beyond 3% story drift ratio.
3. Adding lateral bracing near the RBS region did not increase the beam maximum strength (or the force demand in the welded joints) because strength degradation initiated before lateral bracing became effective. However, it would further increase the plastic rotation capacity beyond the minimum required plastic rotation because the

strength degradation occurred at a slower rate.
4. The energy dissipation capacities of two RBS test specimens were almost identical up to 4% story drift ratio. At higher displacement level, the presence of lateral bracing did increase the energy dissipation capacity of the RBS connection.
5. The system axial restraining effects to the beams could significantly reduce the WLB amplitude, and, therefore, dramatically reduced the strength degradation at higher displacement levels.
6. For an SMRF building where axial restraining effects typically exist, adding lateral bracing near the RBS region appears to be unwarranted. However, for the situation where the axial restraining effect does not exist or where unusually high story drift demand exists, adding lateral bracing can be considered as one option to improve the seismic performance of RBS moment connections.

6 ACKNOWLEDGEMENTS

This research was funded by SAC Joint Venture with Mr. James O. Malley as the Project Director of Topical Investigations. The authors would like to thank Nucor-Yamato Steel, AISC, PDM/Strocal, Asbury Steel, and TSI Inc. for their contributions to this project. Mr. Chad Gilton assisted in the experimental study.

REFERENCE

ABAQUS 1995. *User's manual, Vols. I and II.* Version 5.6, Hibbitt, Karlsson & Sorensen, Inc., Providence, RI.
AISC 1993. Load and resistance factor design specification for structural steel buildings (LRFD), American Institute of Steel Construction, Chicago, IL.
AISC 1997. Seismic Provisions for Structural Steel Buildings, American Institute of Steel Construction, Chicago, IL.
AISC 1997. Shape material (ASTM A572 Gr. 50 with special requirements). *Technical Bulletin No. 3*, American Institute of Steel Construction, Chicago, IL.
SAC 1996. Interim Guidelines Advisory No.1. *Report No. SAC-96-03 (Report No. FEMA-267A)*, SAC Joint Venture, Sacramento, CA.
Chen, S.-J., Yeh, C. H. & Chu, J. M. 1996. Ductile steel beam-to-column connections for seismic resistance. *Journal of Structural Engineering*, ASCE, 122(11): 1292-1299.
Clark, P., Frank, K., Krawinkler, H. & Shaw, R. 1997. Protocol for fabrication, inspection, and documentation of beam-column connection tests and other experimental specimens. *Report No. SAC/BD-97/02*, SAC Joint Venture, Sacramento, CA.
Engelhardt, M. D. 1999. Design of reduced beam section moment connections. *1999 North American Steel Construction Conference Proceedings*: 3-29, AISC, Chicago, IL.
Engelhardt, M. D., Winneberger, T., Zekany, A. J. & Potyraj, T. 1998. Experimental investigation of dogbone moment connections. *Engineering Journal*, AISC, 35(4): 128-139.

Plumier, A. 1997. The dogbone: back to the future. *Engineering Journal*, AISC, 34(2): 61-67.

Yu, Q.-S., Noel, S. & Uang, C.-M. 1997. Experimental studies on seismic rehabilitation of pre-Northridge steel moment connections: RBS and haunch approach. *Report No. SSRP-97/08*, University of California at San Diego, La Jolla, CA.

Yu, Q.-S., Gilton, C. S. & Uang, C.-M. 1999. Cyclic response of RBS moment onnections: loading sequence and lateral bracing effects. *Report No. SSRP 99-13*, University of California at San Diego, La Jolla, CA.

Zekioglu, A., Mozaffarian, H., Chang, K. L., Uang, C.-M. & Noel, S. 1997. Designing after northridge. *Modern Steel Construction*, AISC, 37(3): 36-42.

Behaviour of Steel Structures in Seismic Areas, Mazzolani & Tremblay (eds) © 2000 Balkema, Rotterdam, ISBN 90 5809 130 9

Authors index